McGraw-Hill
Dictionary of
Earth
Sciences

McGraw-Hill
Dictionary of
Earth
Sciences

Sybil P. Parker Editor in Chief

McGraw-Hill Book Company

New York   St. Louis   San Francisco

Auckland   Bogotá   Guatemala   Hamburg
Johannesburg   Lisbon   London   Madrid   Mexico
Montreal   New Delhi   Panama   Paris   San Juan
São Paulo   Singapore   Sydney   Tokyo   Toronto

1 2 3 4 5 6 7 8 9 0   DODO   8 9 1 0 9 8 7 6 5 4

ISBN 0-07-045252-0

Library of Congress Cataloging in Publication Data

McGraw-Hill dictionary of earth sciences.

   1. Earth sciences—Dictionaries.   I. Parker, Sybil P.
II. McGraw-Hill Book Company.   III. Title: Dictionary of
earth sciences.
QE5.M365   1984   550'.3'21   83-20362
ISBN 0-07-045252-0

Preface

The *McGraw-Hill Dictionary of Earth Sciences* concentrates on the vocabulary of those disciplines that constitute the earth sciences and related fields of science and engineering. With more than 15,000 terms defined, it serves as a major compendium of the specialized language that is essential to understanding the earth sciences.

The concept of this Dictionary is that earth sciences embrace many unique disciplines which are usually represented in specialized dictionaries and glossaries. However, specialists in one area are nonspecialists in others, and will therefore find a multidisciplinary approach invaluable for their professional needs. Similarly, engineers, students, teachers, librarians, writers, and general readers of scientific literature will appreciate the convenience of a single comprehensive reference.

Terms and definitions in the Dictionary represent 18 fields, including geology, oceanography, crystallography, petroleum engineering, mining engineering, geochemistry, hydrology, petrology, and mineralogy. Each definition is identified by the field in which it is primarily used. When the same definition is used in more than one field, it is identified by a more general field. For example, a term used in both petrology and mineralogy would be assigned to geology.

The terms selected for this Dictionary are, in the opinion of the editors, fundamental to understanding the earth sciences. Many definitions were drawn from the *McGraw-Hill Dictionary of Scientific and Technical Terms* (3d ed., 1984), and others were written especially for this work. Synonyms, acronyms, and abbreviations are given along with the definitions; they are also listed in the alphabetical sequence as cross-references to the defining term.

The *McGraw-Hill Dictionary of Earth Sciences* is a reference tool which the editors hope will facilitate the communication of ideas and information, and thus serve the needs of users with either professional or pedagogical interests in the earth sciences.

Sybil P. Parker
EDITOR IN CHIEF

Editorial Staff

Sybil P. Parker, *Editor in Chief*

Jonathan Weil, *Editor*
Betty Richman, *Editor*

Edward J. Fox, *Art director*

Ann D. Bonardi, *Art production supervisor*

Joe Faulk, *Editing manager*

Ann Jacobs, *Editing supervisor*
Patricia W. Albers, *Senior editing assistant*
Barbara Begg, *Editing assistant*

Consulting Editors

for the McGraw-Hill Dictionary of Scientific and Technical Terms

Dr. Charles B. Curtin—*Associate Professor of Biology, Creighton University.* BIOLOGY.
Robert L. Davidson—*Formerly, Editor in Chief, Business Books and Services, Professional and Reference Division, McGraw-Hill Book Company.* PETROLEUM ENGINEERING.
Alvin W. Knoerr—*Mining Engineer, U.S. Bureau of Mines, Washington, D.C.* MINING ENGINEERING.
Dr. Edward C. Monahan—*Statutory Lecturer in Physical Oceanography, University College, Galway, Ireland.* OCEANOGRAPHY.
Prof. Frederic Schwab—*Department of Geology, Washington and Lee University.* GEOLOGY; PHYSICAL GEOGRAPHY.
Dr. C. N. Touart—*Senior Scientist, Air Force Geophysics Laboratory.* GEOCHEMISTRY; GEOPHYSICS; METEOROLOGY.

How to Use the Dictionary

I. ALPHABETIZATION

The terms in the *McGraw-Hill Dictionary of Earth Sciences* are alphabetized on a letter-by-letter basis; word spacing, hyphen, comma, solidus, and apostrophe in a term are ignored in the sequencing. For example, an ordering of terms would be:

> **air-earth current**
> **air endway**
> **airglow**
> **air intake**
> **AKF diagram**

II. FORMAT

The basic format for a defining entry provides the term in boldface, the field in small capitals, and the single definition in lightface:

> **term** [FIELD] Definition.

A field may be followed by multiple definitions, each introduced by a boldface number:

> **term** [FIELD] **1.** Definition. **2.** Definition. **3.** Definition.

A term may have definitions in two or more fields:

> **term** [CLIMATOL] Definition. [GEOL] Definition.

A simple cross-reference entry appears as:

> **term** *See* another term.

A cross-reference may also appear in combination with definitions:

> **term** [CLIMATOL] Definition. [GEOL] *See* another term.

III. CROSS-REFERENCING

A cross-reference entry directs the user to the defining entry. For example, the user looking up "abyssal fan" finds:

> **abyssal fan** *See* submarine fan.

The user then turns to the "S" terms for the definition. Cross-references are also made from variant spellings, acronyms, abbreviations, and symbols.

> **abs** *See* absolute.
> **APR** *See* airborne profile recorder.
> **arenyte** *See* arenite.

IV. ALSO KNOW AS . . ., etc.

A definition may conclude with a mention of a synonym of the term, a variant spelling, an abbreviation for the term, or other such information, introduced by "Also known as . . .," "Also spelled . . .," "Abbreviated . . .," "Symbolized . . .," "Derived from . . . ." When a term has more than one definition, the positioning of any of these phrases conveys the extent of applicability. For example:

term [CLIMATOL] **1.** Definition. Also known as synonym. **2.** Definition. Symbolized T.

In the above arrangement, "Also known as . . ." applies only to the first definition: "Symbolized . . ." applies only to the second definition.

term [CLIMATOL] **1.** Definition. **2.** Definition.   [GEOL] Definition. Also known as synonym.

In the above arrangement "Also known as . . ." applies only to the second field.

term [CLIMATOL] Also known as synonym. **1.** Definition. **2.** Definition.   [GEOL] Definition.

In the above arrangement, "Also known as . . ." applies to both definitions in the first field.

term Also known as synonym. [CLIMATOL] **1.** Definition. **2.** Definition.   [GEOL] Definition.

In the above arrangement, "Also known as . . ." applies to all definitions in both fields.

Field Abbreviations

CLIMATOL	climatology
CRYSTAL	crystallography
ENG	engineering
GEOCHEM	geochemistry
GEOD	geodesy
GEOGR	geography
GEOL	geology
GEOPHYS	geophysics
HYD	hydrology
MAP	mapping
METEOROL	meteorology
MIN ENG	mining engineering
MINERAL	mineralogy
OCEANOGR	oceanography
PALEOBOT	paleobotany
PALEON	paleontology
PETR	petrology
PETRO ENG	petroleum engineering

Scope of Fields

climatology—That branch of meteorology concerned with the mean physical state of the atmosphere together with its statistical variations in both space and time as reflected in the weather behavior over a period of many years.

crystallography—The branch of science that deals with the geometric description of crystals, their internal arrangement, and their properties.

engineering—The science by which the properties of matter and the sources of power in nature are made useful to humans in structures, machines, and products.

geochemistry—The study of the chemical composition of the various phases of the earth and the physical and chemical processes which have produced the observed distribution of the elements and nuclides in these phases.

geodesy—A subdivision of geophysics which includes determinations of the size and shape of the earth, the earth's gravitational field, and the location of points fixed to the earth's crust in an earth-referred coordinate system.

geography—The science that deals with the description of land, sea, and air and the distribution of plant and animal life, including humans.

geology—The study or science of earth, its history, and its life as recorded in the rocks; includes the study of the geologic features of an area, such as the geometry of rock formations, weathering and erosion, and sedimentation.

geophysics—A branch of geology in which the principles and practices of physics are used to study the earth and its environment, that is, earth, air, and (by extension) space.

hydrology—The science that treats of the surface and groundwaters of the earth; their occurrence, circulation, and distribution; their chemical and physical properties; and their reaction with their environment.

mapping—The art and practice of making a drawing or other representation, usually on a flat surface, of the whole or part of an area (as the surface of the earth or some other planet), indicating relative position and size according to a specified scale or projection of selected features, as countries, cities, rock formations, or bodies of water.

meteorology—The science concerned primarily with the observation of the atmosphere and its phenomena, including temperature, density, winds, clouds, and precipitation.

mineralogy—The science concerning the study of natural inorganic substances called minerals, including origin, description, and classification.

mining engineering—A branch of engineering concerned with the location and evaluation of coal and mineral deposits, the survey of mining areas, the layout and equipment of mines, the supervision of mining operations, and the cleaning, sizing, and dressing of the product.

oceanography—The scientific study and exploration of the oceans and seas in all their aspects.

paleobotany—The study of fossil plants and vegetation of the geologic past.

paleontology—The study of life in the geologic past as recorded by fossil remains.

petroleum engineering—A branch of engineering concerned with the search for and extraction of oil, gas, and liquefiable hydrocarbons.

petrology—The branch of geology dealing with the origin, occurrence, structure, and history of rocks, especially igneous and metamorphic rocks.

aa channel [GEOL] A narrow, sinuous channel in which a lava river moves down and away from a central vent to feed an aa flow.

aa lava [GEOL] Type of lava with a rough, fragmental surface; consists of clinkers and scoria.

Aalenian [GEOL] Lowermost Middle or uppermost Lower Jurassic geologic time.

a axis [CRYSTAL] One of the crystallographic axes used as reference in crystal description, usually oriented horizontally, front to back. [GEOL] The direction of movement or transport in a tectonite.

abandon [ENG] To stop drilling and remove the drill rig from the site of a borehole before the intended depth or target is reached.

abandoned channel *See* oxbow.

abandoned mine *See* abandoned workings.

abandoned workings [MIN ENG] Deserted excavations, either caved or sealed, in which further mining is not intended, and opening workings which are not ventilated and inspected regularly. Also known as abandoned mine.

abandonment [MIN ENG] Failure to perform work, by conveyance, by absence, and by lapse of time, on a mining claim. [PETRO ENG] *See* abandonment contour.

abandonment contour [PETRO ENG] A graph of actual cumulative yield of an oil well compared with its estimated ultimate yield; useful in determining the most economic time to abandon an oil well. Also known as abandonment.

ABC system [GEOD] *See* airborne control system. [GEOPHYS] A procedure in seismic surveying to determine the effect of irregular weathering thickness.

ablation [GEOL] The wearing away of rocks, as by erosion or weathering.

ablation area [HYD] The section in a glacier or snowfield where ablation exceeds accumulation.

ablation cone [HYD] A debris-covered cone of ice, firn, or snow formed by differential ablation.

ablation factor [HYD] The rate at which a snow or ice surface wastes away.

ablation form [HYD] A feature on a snow or ice surface caused by melting or evaporation.

ablation moraine [GEOL] **1.** A layer of rock particles overlying ice in the ablation of a glacier. **2.** Drift deposited from a superglacial position through the melting of underlying stagnant ice.

ablatograph [ENG]   An instrument that records ablation by measuring the distance a snow or ice surface falls during the observation period.

ablykite [GEOL]   A clay mineral resembling halloysite, with the composition magnesium calcium potassium aluminosilicate.

abnormal anticlinorium [GEOL]   An anticlinorium with axial planes of subsidiary folds diverging upward.

abnormal fold [GEOL]   An anticlinorium in which there is an upward convergence of the axial surfaces of the subsidiary folds.

abnormal magnetic variation [GEOPHYS]   The anomalous value in magnetic compass readings made in some local areas containing unknown sources that deflect the compass needle from the magnetic meridian.

abnormal place [MIN ENG]   An area in a coal mine where geological conditions render mining uneconomical.

abnormal synclinorium [GEOL]   A synclinorium with axial planes of subsidiary folds converging downward.

a-b plane [GEOL]   The surface along which differential movement takes place.

Abraham's tree [METEOROL]   The popular name given to a form of cirrus radiatus clouds, consisting of an assemblage of long feathers and plumes of cirrus that seems to radiate from a single point on the horizon.

abrasion [GEOL]   Wearing away of sedimentary rock chiefly by currents of water laden with sand and other rock debris and by glaciers.

abrasion platform [GEOL]   An uplifted marine peneplain or plain, according to the smoothness of the surface produced by wave erosion, which is of large area.

abrasive [GEOL]   A small, hard, sharp-cornered rock fragment, used by natural agents in abrading rock material or land surfaces. Also known as abrasive ground.

abrasive drilling [MIN ENG]   A rotary drilling method in which drilling is effected by the abrasive action of the drill steel or drilling medium which rotates while being pressed against the rock.

abrasive ground *See* abrasive.

abrolhos [GEOL]   A mushroom-shaped barrier reef with the wider portion near the sea surface.

abrolhos squall [METEOROL]   A frontal rain or squall occurring near the Abrolhos Islands (off Brazil), generally from May through August.

abs *See* absolute.

absarokite [PETR]   An alkalic basalt of about equal portions of olivine, augite, labradorite, and sanidine with accessory biotite, apatite, and opaque oxides; leucite is occasionally present in small amounts.

absolute [METEOROL]   Referring to the highest or lowest recorded value of a meteorological element, whether at a single station or over an area, during a given period. Abbreviated abs.

absolute age [GEOL]   The geologic age of a fossil, or a geologic event or structure expressed in units of time, usually years. Also known as actual age.

absolute drought [METEOROL]   In Britain, a period of at least 15 consecutive days during which no measurable daily precipitation has fallen.

absolute geopotential topography *See* geopotential topography.

absolute instability [METEOROL] The state of a column of air in the atmosphere when it has a superadiabatic lapse rate of temperature, that is, greater than the dry-adiabatic lapse rate. Also known as autoconvective instability; mechanical instability.

absolute isohypse [METEOROL] A line that has the properties of both constant pressure and constant height above mean sea level.

absolute linear momentum *See* absolute momentum.

absolute magnetometer [ENG] An instrument used to measure the intensity of a magnetic field without reference to other magnetic instruments.

absolute momentum [METEOROL] The sum of the (vector) momentum of a particle relative to the earth and the (vector) momentum of the particle due to the earth's rotation. Also known as absolute linear momentum.

absolute roof [MIN ENG] The entire mass of strata overlying a subsurface point of reference.

absolute stability [METEOROL] The state of a column of air in the atmosphere when its lapse rate of temperature is less than the saturation-adiabatic lapse rate.

absolute time [GEOL] Geologic time measured in years, as determined by radioactive decay of elements.

absorption [HYD] The entrance of surface water into the lithosphere.

absorption hygrometer [ENG] An instrument with which the water vapor content of the atmosphere is measured by means of the absorption of vapor by a hygroscopic chemical.

abstraction [HYD] The draining of water from a stream by another having more rapid corroding action.

abstract theory [SCI TECH] A theory in which a system is described without specifying a structure.

Abukuma-type facies [PETR] A type of dynathermal regional metamorphism characterized by low pressure.

abyssal *See* plutonic.

abyssal-benthic [OCEANOGR] Pertaining to the bottom of the abyssal zone.

abyssal cave *See* submarine fan.

abyssal fan *See* submarine fan.

abyssal floor [GEOL] The ocean floor, or bottom of the abyssal zone.

abyssal gap [GEOL] A gap in a sill, ridge, or rise that lies between two abyssal plains.

abyssal hill [GEOL] A hill 2000 to 3000 feet (600 to 900 meters) high and a few miles wide within the deep ocean.

abyssal injection [GEOL] The process of driving magmas, originating at considerable depths, up through deep-seated contraction fissures in the earth's crust.

abyssal plain [GEOL] A flat, almost level area occupying the deepest parts of many of the ocean basins.

abyssal rock [GEOL] Plutonic, or deep-seated, igneous rocks.

abyssal theory [GEOL] A theory of the origin of ores involving the separation of ore silicates from the liquid stage during the cooling of the earth.

abyssal zone [OCEANOGR] The biogeographic realm of the great depths of the ocean beyond the limits of the continental shelf, generally below 1000 meters.

abyssolith [GEOL]   A molten mass of eruptive material passing up without a break from the zone of permanently molten rock within the earth.

abyssopelagic [OCEANOGR]   Pertaining to the open waters of the abyssal zone.

Acadian orogeny [GEOL]   The period of formation accompanied by igneous intrusion that took place during the Middle and Late Devonian in the Appalachian Mountains.

acanthite [MINERAL] Ag_2S A blackish to lead-gray silver sulfide mineral, crystallizing in the orthorhombic system.

acaustobiolith [PETR]   A noncombustible organic rock, or one formed by organic accumulation of minerals.

acaustophytolith [PETR]   An acaustobiolith resulting from plant activity, such as a pelagic ooze that contains diatoms.

accelerated erosion [GEOL]   The process of weathering at a rate greater than normal for the site, brought about by man, usually through reduction of the vegetation.

acceleration of free fall *See* acceleration of gravity.

acceleration of gravity [MECH]   The acceleration imparted to bodies by the attractive force of the earth; has an international standard value of 980.665 cm/s^2 but varies with latitude and elevation. Also known as acceleration of free fall; apparent gravity.

acceptability [ENG]   State or condition of meeting minimum standards for use, as applied to methods, equipment, or consumable products.

accessory [MECH ENG]   A part, subassembly, or assembly that contributes to the effectiveness of a piece of equipment without changing its basic function; may be used for testing, adjusting, calibrating, recording, or other purposes.

accessory cloud [METEOROL]   A cloud form that is dependent, for its formation and continuation, upon the existence of one of the major cloud genera; may be an appendage of the parent cloud or an immediately adjacent cloudy mass.

accessory ejecta [GEOL]   Pyroclastic material formed from solidified volcanic rocks that are from the same volcano as the ejecta.

accessory element *See* trace element.

accessory mineral [MINERAL]   A minor mineral in an igneous rock that does not affect its general character.

accident [HYD]   An interruption in a river that interferes with, or sometimes stops, the normal development of the river system.

accidental ejecta [GEOL]   Pyroclastic rock formed from preexisting nonvolcanic rocks or from volcanic rocks unrelated to the erupting volcano.

accidental error [SCI TECH]   In experimental observations, an error which does not always recur when an observation is repeated under the same conditions.

accidental inclusion *See* xenolith.

accident block [GEOL]   A solid chip of rock broken off from the subvolcanic basement and ejected from a volcano.

acclivity [GEOL]   A slope that is ascending from a reference point.

accommodation [MAP]   The limits or range within which a stereo-plotting instrument is capable of operating.

accordant [GEOL]   Referring to topographic features that have nearly the same elevation.

accordant drainage [HYD]   Drainage that has developed in a definite relationship to, and is consequent upon, the present geologic structure.

accordant fold [GEOL] One of several folds that are similarly oriented.

accordant summit level [GEOL] A hypothetical horizontal plane that can be drawn over a broad region connecting mountain summits of similar elevation.

accretion [GEOL] **1.** Gradual buildup of land on a shore due to wave action, tides, currents, airborne material, or alluvial deposits. **2.** The process whereby stones or other inorganic masses add to their bulk by adding particles to their surfaces. Also known as aggradation.

accretionary lava ball [GEOL] A rounded ball of lava that occurs on the surface of an aa lava flow.

accretionary limestone [PETR] A type of limestone formed by the slow accumulation of organic remains.

accretionary ridge [GEOL] A beach ridge located inland from the modern beach, indicating that the coast has been built seaward.

accretion topography [GEOL] Topographic features built by accumulation of sediment.

accretion vein [GEOL] A type of vein formed by the repeated filling of channels followed by their opening because of the development of fractures in the zone undergoing mineralization.

accretion zone [GEOL] Any beach area undergoing accretion.

accumulated discrepancy [ENG] The sum of the separate discrepancies which occur in the various steps of making a survey.

accumulated divergence [MAP] In making a map, the algebraic sum of the divergences for the sections of a line of levels, from the beginning of the line to any section end at which it is desired to compute the total divergence.

accumulation [HYD] The quantity of snow or other solid form of water added to a glacier or snowfield by alimentation. [MIN ENG] **1.** In coal mining, firedamp that collects in higher parts of mine workings and at the edge of wastes. **2.** Oil or gas in some form of trap.

accumulation area [HYD] The portion of a glacier above the firn line, where the accumulation exceeds ablation. Also known as firn field; zone of accumulation.

accumulation zone [GEOL] The area where the bulk of the snow contributing to an avalanche was originally deposited.

accuracy [SCI TECH] The extent to which the results of a calculation or the readings of an instrument approach the true values of the calculated or measured quantities, and are free from error.

accuracy checking [MAP] The procurement of presumptive evidence of a map's compliance with specified accuracy standards; indicates the relative (rather than the absolute) accuracy of map features.

accuracy testing [MAP] The procurement of confirmed evidence, on a sampling basis, of a map's compliance with specified accuracy standards; indicates both the relative and absolute accuracy of map features.

accurate contour [MAP] A contour line whose accuracy lies within one-half of the basic vertical interval. Also known as normal contour.

ACF diagram [PETR] A triangular diagram showing the chemical character of a metamorphic rock; the three components plotted are $A = Al_2O_3 + Fe_2O_3 - (Na_2O + K_2O)$, $C = CaO$, $F = FeO + MgO + MnO$.

a-c fracture [CRYSTAL] A type of tension fracture lying parallel to the a-c fabric plane and normal to plane b in a crystal.

a-c girdle [GEOL] A girdle of points in a petrofabric diagram that have a tread parallel with the plane of the a and c fabric axes.

Acheulean [ARCHEO] Lower Paleolithic archeological time, characterized by biface tools having cutting edges all around.

achondrite [GEOL] A stony meteorite that contains no chondrules.

achroite [MINERAL] A colorless variety of tourmalines found in Malagasy.

acicular ice [HYD] Fresh-water ice composed of many long crystals and layered hollow tubes of varying shape containing air bubbles. Also known as fibrous ice; satin ice.

acid clay [GEOL] A type of clay that gives off hydrogen ions when it dissolves in water.

acid dilution [PETRO ENG] Dilution of concentrated hydrochloric acid with water prior to oil-well acidizing.

acidic lava [GEOL] Extruded felsic igneous magma which is rich in silica (SiO_2 content exceeds 65%).

acidic rock [PETR] Igneous rock containing more than 66% SiO_2, silicic.

acidity coefficient [GEOCHEM] The ratio of the oxygen content of the bases in a rock to the oxygen content in the silica. Also known as oxygen ratio.

acidizing [PETRO ENG] Well-stimulation method to increase oil production by injecting hydrochloric acid into the oil-bearing formation; the acid dissolves rock to enlarge the porous passages through which the oil must flow.

acid jetting [PETRO ENG] The jetting, from a device lowered through oil-well tubing, of an acid spray onto bottom-hole rock to clean away mud and scale interfering with oil flow.

acid mine drainage [MIN ENG] Drainage from bituminous coal mines containing a large concentration of acidic sulfates, especially ferrous sulfate.

acid mine water [MIN ENG] Mine water with free sulfuric acid, due to the weathering of iron pyrites.

acid precipitation [METEOROL] Rain or snow with a pH of less than 5.6.

acid soil [GEOL] A soil with pH less than 7; results from presence of exchangeable hydrogen and aluminum ions.

acid spar [MINERAL] A grade of fluorspar containing over 98% CaF_2 and no more than 1% SiO_2; produced by flotation; used for the production of hydrofluoric acid.

acidulous water [HYD] Mineral water either with dissolved carbonic acid or dissolved sulfur compounds such as sulfates.

aclinal [GEOL] Without dip; horizontal.

aclinic [GEOPHYS] Referring to a situation where a freely suspended magnetic needle remains in a horizontal position.

acme [PALEON] The time of largest abundance or variety of a fossil taxon; the taxon may be either general or local.

acmite [MINERAL] $NaFeSi_2O_6$ A brown or green silicate mineral of the pyroxene group, often in long, pointed prismatic crystals; hardness is 6–6.5 on Mohs scale, and specific gravity is 3.50–3.55; found in igneous and metamorphic rocks.

acoubuoy [ENG] An acoustic listening device similar to a sonobuoy, used on land to form an electronic fence that will pick up sounds of enemy movements and transmit them to orbiting aircraft or land stations.

acoustic detection [ENG] Determination of the profile of a geologic formation, an ocean layer, or some object in the ocean by measuring the reflection of sound waves off the object.

acoustic ocean-current meter [ENG] An instrument that measures current flow in rivers and oceans by transmitting acoustic pulses in opposite directions parallel to the flow and measuring the difference in pulse travel times between transmitter-receiver pairs.

acoustic position reference system [ENG] An acoustic system used in offshore oil drilling to provide continuous information on ship position with respect to an ocean-floor acoustic beacon transmitting an ultrasonic signal to three hydrophones on the bottom of the drilling ship.

acoustic radar [ENG] Use of sound waves with radar techniques for remote probing of the lower atmosphere, up to heights of about 1500 meters, for measuring wind speed and direction, humidity, temperature inversions, and turbulence.

acoustic signature [ENG] In acoustic detection, the profile characteristic of a particular object or class of objects, such as a school of fish or a specific ocean-bottom formation.

acoustic spectrograph [ENG] A spectrograph used with sound waves of various frequencies to study the transmission and reflection properties of ocean thermal layers and marine life.

acoustic transponder [NAV] A device used in underwater navigation which, on being interrogated by coded acoustic signals, emits acoustic reply.

acoustic well logging [ENG] A ground exploration method that uses a high-energy sound source and a receiver, both underground.

acquisition [ENG] The process of pointing an antenna or a telescope so that it is properly oriented to allow gathering of tracking or telemetry data from a satellite or space probe.

acre-foot [HYD] The volume of water required to cover 1 acre to a depth of 1 foot, hence 43,560 cubic feet; a convenient unit for measuring irrigation water, runoff volume, and reservoir capacity.

acre foot per day [HYD] The United States unit of volume rate of water flow. Abbreviated acre ft/d.

acre-in. *See* acre-inch.

acre-inch [HYD] A unit of volume used in the United States for water flow, equal to 3630 cubic feet. Abbreviated acre-in.

acre-yield [GEOL] The average amount of oil, gas, or water taken from one acre of a reservoir.

acrobatholithic [GEOL] A stage in batholithic erosion where summits of cupolas and stocks are exposed without any exposure of the surface separating the barren interior of the batholith from the mineralized upper part.

acromorph [GEOL] A salt dome.

Acrotretacea [PALEON] A family of Cambrian and Ordovician inarticulate brachiopods of the suborder Acrotretidina.

actinogram [ENG] The record of heat from a source, such as the sun, as detected by a recording actinometer.

actinograph [ENG] A recording actinometer.

actinolite [MINERAL] $Ca_2(Mg,Fe)_5Si_8O_{22}(OH)_2$ A green, monoclinic rock-forming amphibole; a variety of asbestos occurring in needlelike crystals and in fibrous or columnar forms; specific gravity 3–3.2.

actinometer [ENG] Any instrument used to measure the intensity of radiant energy, particularly that of the sun.

actinometry [ASTROPHYS] The science of measurement of radiant energy, particularly that of the sun, in its thermal, chemical, and luminous aspects.

Actinostromariidae [PALEON] A sphaeractinoid family of extinct marine hydrozoans.

activated coal plough [MIN ENG] A type of power-operated cutting blade used for coal seams too hard to be sheared by a normal blade.

active cave See live cave.

active entry [MIN ENG] An entry in which coal is being mined from a portion or from connected sections.

active front [METEOROL] A front, or portion thereof, which produces appreciable cloudiness and, usually, precipitation.

active glacier [HYD] A glacier in which some of the ice is flowing.

active layer [GEOL] That part of the soil which is within the suprapermafrost layer and which usually freezes in winter and thaws in summer. Also known as frost zone.

active permafrost [GEOL] Permanently frozen ground (permafrost) which, after thawing by artificial or unusual natural means, reverts to permafrost under normal climatic conditions.

active sonar [ENG] A system consisting of one or more transducers to send and receive sound, equipment for the generation and detection of the electrical impulses to and from the transducer, and a display or recorder system for the observation of the received signals.

active volcano [GEOL] A volcano capable of venting lava, pyroclastic material, or gases.

active workings [MIN ENG] All places in a mine that are ventilated and inspected regularly.

activity ratio [GEOL] The ratio of plasticity index to percentage of clay-sized minerals in sediment.

actual age See absolute age.

actual elevation [METEOROL] The vertical distance above mean sea level of the ground at the meteorological station.

actual pressure [METEOROL] The atmospheric pressure at the level of the barometer (elevation of ivory point), as obtained from the observed reading after applying the necessary corrections for temperature, gravity, and instrumental errors.

actual relative movement See slip.

acute angle block [GEOL] A fault block in which the strike of strata on the down-dip side meets a diagonal fault at an acute angle.

acute bisectrix [MINERAL] A bisecting line of the acute angle of the optic axes of biaxial minerals.

adalert [GEOPHYS] An advance alert issued by a regional warning center to give prompt warning of a change in solar activity.

adamant [MINERAL] A general designaton for a hard mineral.

adamantine spar [MINERAL] A silky brown variety of corundum.

adamellite See quartz monzonite.

adamite   [MINERAL] $Zn_2(AsO_4)(OH)$ A colorless, white, or yellow mineral consisting of basic zinc arsenate, crystallizing in the orthorhombic system; hardness is 3.5 on Mohs scale, and specific gravity is 4.34–4.35.

adamsite   [MINERAL]   Greenish-black mica.   [ORG CHEM] $C_6H_4 \cdot NH \cdot C_6H_4 \cdot$ AsCl   A yellow crystalline arsenical; used in leather tanning and in warfare and riot control to produce skin and eye irritation, chest distress, and nausea; U.S. Army code is DM. Also known as diphenylaminechloroarsine; phenarsazine chloride.

ada mud   [ENG]   A conditioning material added to drilling mud to obtain satisfactory cores and samples of formations.

adding tape   [ENG]   A surveyor's tape that is calibrated from 0 to 100 by full feet (or meters) in one direction, and has 1 additional foot (or meter) beyond the zero end which is subdivided in tenths or hundredths.

addition solid solution   [CRYSTAL]   Random addition of atoms or ions in the interstices within a crystal structure.

adelite   [MINERAL] $CaMg(AsO_4)(OH,F)$ A colorless to gray, bluish-gray, yellowish-gray, yellow, or light green orthorhombic mineral consisting of a basic arsenate of calcium and magnesium; usually occurs in massive form.

ader wax   *See* ozocerite.

adfreezing   [HYD]   The process by which one object adheres to another by the binding action of ice; applied to permafrost studies.

adiabatic atmosphere   [METEOROL]   A model atmosphere characterized by a dry-adiabatic lapse rate throughout its vertical extent.

adiabatic chart   *See* Stuve chart.

adiabatic condensation pressure   *See* condensation pressure.

adiabatic equilibrium   [METEOROL]   A vertical distribution of temperature and pressure in an atmosphere in hydrostatic equilibrium such that an air parcel displaced adiabatically will continue to possess the same temperature and pressure as its surroundings, so that no restoring force acts on a parcel displaced vertically. Also known as convective equilibrium.

adiabatic equivalent temperature   *See* equivalent temperature.

adiabatic saturation pressure   *See* condensation pressure.

adiabatic system   [SCI TECH]   A body or system whose condition is altered without gaining heat from or losing heat to the surroundings.

adiagnostic   [PETR]   Pertaining to a rock texture in which identification of individual components is not possible macroscopically or microscopically; applied especially to igneous rock.

adinole   [GEOL]   An argillaceous sediment that has undergone albitization at the margin of a basic intrusion.

adipocerite   *See* hatchettite.

adipocire   *See* hatchettite.

adit   [MIN ENG]   A nearly horizontal tunnel for access, drainage, or ventilation of a mine. Also known as side drift.

adjacent sea   [GEOGR]   A sea connected with the oceans but semienclosed by land; examples are the Caribbean Sea and North Polar Sea.

adjoining sheets   [MAP]   Those maps that are contiguous to one or more sides and corners of a map series.

adjusted elevation [GEOD] **1.** The elevation resulting from the application of an adjustment correction to an orthometric elevation. **2.** The elevation resulting from the application of both an orthometric correction and an adjustment correction to a preliminary elevation.

adjusted position [MAP] An adjusted value of the coordinate position of a point.

adjusted stream [HYD] A stream which flows mostly parallel to the strike and as little as necessary in other courses.

adjusted value [SCI TECH] A value of a quantity derived from observed data by some orderly process which eliminates discrepancies arising from errors in those data.

adjustment [GEOD] **1.** The determination and application of corrections to orthometric differences of elevation or to orthometric elevations to make the elevation of all bench marks consistent and independent of the circuit closures. **2.** The placing of detail or control stations in their positions relative to other detail or control stations.

adlittoral [OCEANOGR] Of, pertaining to, or occurring in shallow waters adjacent to a shore.

administrative map [MAP] **1.** A map with graphically recorded information pertaining to administrative matters, such as supply and evacuation installations, medical facilities, service and maintenance areas, main supply roads, boundaries, and other details necessary to show the administrative situation in relation to the tactical situation. **2.** Any map on which are delineated political subdivisions and boundaries of a country or countries.

admiralty coal [MIN ENG] A high-quality, smokeless steam coal.

admixture [GEOL] One of the lesser or subordinate grades of sediment.

adobe [GEOL] Heavy-textured clay soil found in the southwestern United States and in Mexico.

adobe flats [GEOL] Broad flats that are floored with sandy clay and have been formed from sheet floods.

adolescence [GEOL] Stage in the cycle of erosion following youth and preceding maturity.

adolescent coast [GEOL] A type of shoreline characterized by low but nearly continuous sea cliffs.

adolescent river [HYD] A river with a graded bed and a well-cut channel that reaches base level at its mouth, its waterfalls and lakes of the youthful stage having been destroyed.

adolescent stream [HYD] A stream characterized by a well-cut, smoothly graded channel that may reach base level at its mouth.

adret [GEOL] A mountain slope oriented in such a way that it receives the maximum amount of light from the sun.

adularia [MINERAL] A weakly triclinic form of the mineral orthoclase occurring in transparent, colorless to milky-white pseudo-orthorhombic crystals.

adularization [GEOL] Replacement by or introduction of the mineral adularia.

ad valorem tax [PETRO ENG] Property tax for oil-producing properties, assessed at a flat rate for each net barrel of oil produced.

advance [GEOL] **1.** A continuing movement of a shoreline toward the sea. **2.** A net movement over a specified period of time of a shoreline toward the sea. [HYD] The forward movement of a glacier.

advanced dune [GEOL] A sand dune formed on the windward side of an attached dune but separate from it.

advanced gallery [MIN ENG] A small heading driven in advance of the main tunnel in tunnel excavation.

advance overburden [MIN ENG] Overburden in excess of the average overburden-to-ore ratio that must be removed in opencut mining.

advance stripping [MIN ENG] The removal of barren or sub-ore-grade earthy or rock materials to expose the minable grade of ore.

advance wave [MIN ENG] The air pressure wave preceding the flame in a coal dust explosion.

advancing [MIN ENG] Mining outward from the shaft toward the boundary.

advancing longwall [MIN ENG] Mining coal outward from the shaft pillar, with roadways maintained through the worked-out portion of the mine.

advection [METEOROL] The process of transport of an atmospheric property solely by the mass motion of the atmosphere. [OCEANOGR] The process of transport of water, or of an aqueous property, solely by the mass motion of the oceans, most typically via horizontal currents.

advection fog [METEOROL] A type of fog caused by the horizontal movement of moist air over a cold surface and the consequent cooling of that air to below its dew point.

advective hypothesis [METEOROL] The assumption that local temperature changes are the result only of horizontal or isobaric advection.

advective model [METEOROL] A mathematical or dynamic model of fluid flow which is characterized by the advective hypothesis.

advective thunderstorm [METEOROL] A thunderstorm resulting from static instability produced by advection of relatively colder air at high levels or relatively warmer air at low levels or by a combination of both conditions.

adventive cone [GEOL] A volcanic cone that is on the flank of and subsidiary to a larger volcano. Also known as lateral cone; parasitic cone.

adventive crater [GEOL] A crater opened on the flank of a large volcanic cone.

Aechminidae [PALEON] A family of extinct ostracods in the order Paleocopa in which the hollow central spine is larger than the valve.

Aeduellidae [PALEON] A family of Lower Permian palaeoniscoid fishes in the order Palaeonisciformes.

aegirine [MINERAL] $NaFe(SiO_3)_2$ A brown or green clinopyroxene occurring in alkali-rich igneous rocks. Also known as aegirite.

aegirite See aegirine.

Aegyptopithecus [PALEON] A primitive primate that is thought to represent the common ancestor of both the human and ape families.

aeolian See eolian.

Aepyornis [PALEON] A genus of extinct ratite birds representing the family Aepyornithidae.

Aepyornithidae [PALEON] The single family of the extinct avian order Aepyornithiformes.

Aepyornithiformes [PALEON]   The elephant birds, an extinct order of ratite birds in the superorder Neognathae.

aerial mapping [MAP]   The making of planimetric and contoured maps and charts on the basis of photographs of the ground surface from an aircraft, spacecraft, or rocket. Also known as aerocartography.

aerial photogrammetry [ENG]   Use of aerial photographs to make accurate measurements in surveying and mapmaking.

aerial photography [ENG]   The making of photographs of the ground surface from an aircraft, spacecraft, or rocket. Also known as aerophotography.

aerial survey [ENG]   A survey utilizing photographic, electronic, or other data obtained from an airborne station. Also known as aerosurvey; air survey.

aerocartography *See* aerial mapping.

AERO code [METEOROL]   An international code used to encode for transmission, in words five numerical digits long, synoptic weather observations of particular interest to aviation operations.

aerogeography [GEOGR]   The geographic study of earth features by means of aerial observations and aerial photography.

aerogeology [GEOL]   The geologic study of earth features by means of aerial observations and aerial photography.

aerograph [ENG]   Any self-recording instrument carried aloft by any means to obtain meteorological data.

aerography [METEOROL]   **1.** The study of the air or atmosphere. **2.** The practice of weather observation, map plotting, and maintaining records. **3.** *See* descriptive meteorology.

aerohydrous mineral [MINERAL]   A mineral containing water in small cavities.

aerological days [METEOROL]   Specified days on which additional upper-air observations are made; an outgrowth of the International Polar Year.

aerology [METEOROL]   **1.** Synonym for meteorology, according to official usage in the U.S. Navy until 1957. **2.** The study of the free atmosphere throughout its vertical extent, as distinguished from studies confined to the layer of the atmosphere near the earth's surface.

aeromagnetic surveying [GEOPHYS]   The mapping of the magnetic field of the earth through the use of electronic magnetometers suspended from aircraft.

aerometeorograph [ENG]   A self-recording instrument used on aircraft for the simultaneous recording of atmospheric pressure, temperature, and humidity.

aeronautical chart [MAP]   A basic map of countries or lands made primarily on Lambert conformal ionic projection and layer-tinted; air navigational data are overprinted.

aeronautical climatology [METEOROL]   The application of the data and techniques of climatology to aviation meteorological problems.

aeronautical meteorology [METEOROL]   The study of the effects of weather upon aviation.

aeronomy [GEOPHYS]   The study of the atmosphere of the earth or other bodies, particularly in relation to composition, properties, relative motion, and radiation from outer space or other bodies.

aerophotography *See* aerial photography.

aerosiderite [GEOL]   A meteorite composed principally of iron.

aerosurvey *See* aerial survey.

aerothermodynamic border [GEOPHYS] An altitude of about 100 miles (161 km), above which the atmosphere is so rarefied that the skin of an object moving through it at high speeds generates no significant heat.

Aetosauria [PALEON] A suborder of Triassic archosaurian quadrupedal reptiles in the order Thecodontia armored by rings of thick, bony plates.

affine deformation [GEOL] A type of deformation in which very thin layers slip against each other so that each moves equally with respect to its neighbors; generally does not result in folding.

affine strain [GEOPHYS] A strain in the earth that does not differ from place to place.

A frame [OCEANOGR] An A-shaped frame used for outboard suspension of oceanographic gear on a research vessel.

Africa [GEOGR] The second largest continent, with an area of 11,700,000 square miles (30,300,000 km²); bisected midway by the Equator, above and below which it shows symmetry of climate and vegetation zones.

afterbreak [MIN ENG] A phenomenon occurring during mine subsidence; material slides inward after the main break, assumed at right angles to the plane of the seam.

afterburst [MIN ENG] **1.** A tremor that sometimes follows a rock blast as the ground adjusts to the new stress distribution. **2.** A sudden collapse of rock in an underground mine subsequent to a rock burst.

afterdamp [MIN ENG] The mixture of gases which remains in a mine after a mine fire or an explosion of firedamp.

aftergases [MIN ENG] Gases produced by mine explosions or mine fires.

aftershock [GEOPHYS] A small earthquake following a larger earthquake and originating at or near the larger earthquake's epicenter.

Aftonian interglacial [GEOL] Post-Nebraska interglacial geologic time.

afwillite [MINERAL] $Ca_3Si_2O_4(OH)_6$ A colorless mineral consisting of a hydrous calcium silicate and occurring in monoclinic crystals; specific gravity is 2.6.

agalite [MINERAL] A mineral with the same composition as talc but with a less soapy feel; used as a filler in writing paper.

agalmatolite [GEOL] A soft, waxy, gray, green, yellow, or brown mineral or stone, such as pinite and steatite; used by the Chinese for carving images. Also known as figure stone; lardite; pagodite.

agaric mineral *See* rock milk.

Agassiz orogeny [GEOL] A phase of diastrophism confined to North America Cordillera occurring at the boundary between the Middle and Late Jurassic.

Agassiz trawl [OCEANOGR] A dredge consisting of a net attached to an iron frame with a hoop at each end that is used to collect organisms, particularly invertebrates, living on the ocean bottom.

Agassiz Valleys [GEOL] Undersea valleys in the Gulf of Mexico between Cuba and Key West.

agate [MINERAL] SiO_2 A fine-grained, fibrous variety of chalcedony with color banding or irregular clouding.

agate jasper [MINERAL] An impure variety of quartz consisting of jasper and agate. Also known as jaspagate.

agatized wood *See* silicified wood.

age [GEOL] **1.** Any one of the named epochs in the history of the earth marked by specific phases of physical conditions or organic evolution, such as the Age of Mammals. **2.** One of the smaller subdivisions of the epoch as geologic time, corresponding to the stage or the formation, such as the Lockport Age in the Niagara Epoch.

aged [GEOL] Of a ground configuration, having been reduced to base level.

age determination [GEOL] Identification of the geologic age of a biological or geological specimen by using the methods of dendrochronology or radiometric dating.

aged shore [GEOL] A shore long established at a constant level and adjusted to the waves and currents of the sea.

age of diurnal inequality [GEOPHYS] The time interval between the maximum semimonthly north or south declination of the moon and the time that the maximum effect of the declination upon the range of tide or speed of the tidal current occurs. Also known as age of diurnal tide; diurnal age.

age of diurnal tide *See* age of diurnal inequality.

Age of Fishes [GEOL] An informal designation of the Silurian and Devonian periods of geologic time.

Age of Mammals [GEOL] An informal designation of the Cenozoic era of geologic time.

Age of Man [GEOL] An informal designation of the Quaternary period of geologic time.

age of parallax inequality [GEOPHYS] The time interval between the perigee of the moon and the maximum effect of the parallax (distance of the moon) upon the range of tide or speed of tidal current. Also known as parallax age.

age of phase inequality [GEOPHYS] The time interval between the new or full moon and the maximum effect of these phases upon the range of tide or speed of tidal current. Also known as age of tide; phase age.

age of tide *See* age of phase inequality.

ageostrophic wind *See* geostrophic departure.

age ratio [GEOL] The ratio of the amount of daughter to parent isotope in a mineral being dated radiometrically.

agglomerate ice [HYD] Ice formed from a mixture of floating ice fragments.

agglomeration [METEOROL] The process in which particles grow by collision with and assimilation of cloud particles or other precipitation particles. Also known as coagulation.

agglomeration test [MIN ENG] A test of a button of coke whose results are used as a measure of the binding qualities of the coal.

agglutinate cone *See* spatter cone.

aggradation [GEOL] *See* accretion. [HYD] A process of shifting equilibrium of stream deposition, with upbuilding approximately at grade.

aggraded valley floor [GEOL] The surface of a flat deposit of alluvium which is thicker than the stream channel's depth and is formed where a stream has aggraded its valley.

aggraded valley plain *See* alluvial plain.

aggregate [GEOL] A collection of soil grains or particles gathered into a mass.

aggregate structure [GEOL] A mass composed of separate small crystals, scales, and grains that, under a microscope, extinguish at different intervals during the rotation of the stage.

aggressive magma [GEOL] A magma that forces itself into place.

aggressive water [HYD] Any of the waters which force their way into place.

Aglaspida [PALEON] An order of Cambrian and Ordovician merostome arthropods in the subclass Xiphosurida characterized by a phosphatic exoskeleton and vaguely trilobed body form.

agmatite [PETR] **1.** A migmatite that contains xenoliths. **2.** Fragmental plutonic rock with granitic cement.

agonic line [GEOPHYS] The imaginary line through all points on the earth's surface at which the magnetic declination is zero; that is, the locus of all points at which magnetic north and true north coincide.

agpaite [PETR] A group of igneous rocks containing feldspathoids; includes naujaite, lujavrite, and kakortokite.

agravic [GEOPHYS] Of or pertaining to a condition of no gravitation.

agricere [GEOL] A waxy or resinous organic coating on soil particles.

agricolite See eulytite.

agricultural geology [GEOL] A branch of geology that deals with the nature and distribution of soils, the occurrence of mineral fertilizers, and the behavior of underground water.

Agriochoeridae [PALEON] A family of extinct tylopod ruminants in the superfamily Merycoidodontoidea.

aguilarite [MINERAL] Ag_4SeS An iron-black mineral associated with argentite and silver in Mexico.

Agulhas Current [OCEANOGR] A fast current flowing in a southwestward direction along the southeastern coast of Africa.

ahlfeldite [MINERAL] $(Ni,Co)SeO_3 \cdot 2H_2O$ A triclinic mineral identified as green to yellow crystals with a reddish-brown coating, consisting of a hydrous selenite of nickel.

aiguille [GEOL] The needle-top of the summit of certain glaciated mountains, such as near Mont Blanc.

aikinite [MINERAL] $PbCuBiS_3$ A mineral crystallizing in the orthorhombic system and occurring massive and in gray needle-shaped crystals; hardness is 2 on Mohs scale, and specific gravity is 7.07. Also known as needle ore.

ailsyte [PETR] An alkalic microgranite containing a considerable amount of riebeckite. Also known as paisanite.

aiming circle [ENG] An instrument for measuring angles in azimuth and elevation in connection with artillery firing and general topographic work; equipped with fine and coarse azimuth micrometers and a magnetic needle.

aimless drainage [HYD] Drainage without a well-developed system, as in areas of glacial drift or karst topography.

air barrage [MIN ENG] An airtight wall dividing a ventilation gallery in a mine into two parts, so that air is led in through one part and out through the other.

air base [MAP] **1.** The line joining two air stations, or the length of this line. **2.** The distance, at the scale of the stereoscopic model, between adjacent perspective centers as reconstructed in the plotting instrument.

airblast [MIN ENG] A disturbance in underground workings accompanied by a strong rush of air.

airblasting [ENG] A blasting technique in which air at very high pressure is piped to a steel shell in a shot hole and discharged. Also known as air breaking.

airborne control system [GEOD] A survey system for fourth-order horizontal and vertical control surveys involving electromagnetic distance measurements and horizontal and vertical measurements from two or more known positions to a helicopter hovering over the unknown position. Also known as ABC system.

airborne magnetometer [ENG] An airborne instrument used to measure the magnetic field of the earth.

airborne profile [GEOD] Continuous terrain-profile data produced by an absolute altimeter in an aircraft which is making an altimeter-controlled flight along a prescribed course.

airborne profile recorder [ENG] An electronic instrument that emits a pulsed-type radar signal from an aircraft to measure vertical distances between the aircraft and the earth's surface. Abbreviated APR. Also known as terrain profile recorder (TPR).

air breaking *See* airblasting.

air casing [ENG] A metal casing surrounding a pipe or reservoir and having a space between to prevent heat transmission.

air composition [METEOROL] The kinds and amounts of the constituent substances of air, the amounts being expressed as percentages of the total volume or mass.

air course *See* airway.

aircraft ceiling [METEOROL] After United States weather observing practice, the ceiling classification applied when the reported ceiling value has been determined by a pilot while in flight within 1.5 nautical miles (2.8 kilometers) of any runway of the airport.

aircraft electrification [METEOROL] **1.** The accumulation of a net electric charge on the surface of an aircraft. **2.** The separation of electric charge into two concentrations of opposite sign on distinct portions of an aircraft surface.

aircraft icing [METEOROL] The accumulation of ice on the exposed surfaces of aircraft when flying through supercooled water drops (cloud or precipitation).

aircraft report *See* pilot report.

aircraft thermometry [METEOROL] The science of temperature measurement from aircraft.

aircraft weather reconnaissance [METEOROL] The making of detailed weather observations or investigations from aircraft in flight.

air crossing [MIN ENG] A mine passage in which two airways cross each other.

air current [GEOPHYS] *See* air-earth conduction current. [MIN ENG] The flow of air ventilating the workings of a mine. Also known as airflow.

air discharge [GEOPHYS] **1.** A form of lightning discharge, intermediate in character between a cloud discharge and a cloud-to-ground discharge, in which the multibranching lightning channel descending from a cloud base does not reach the ground, but succeeds only in neutralizing the space charge distributed in the subcloud layer. **2.** A type of diffuse electrical discharge occasionally reported as occurring in the region above an active thunderstorm.

air door [MIN ENG] A door placed in a mine roadway to prevent the passage of air.

air drainage [METEOROL] General term for gravity-induced, downslope flow of relatively cold air.

air drift [MIN ENG] A roadway, generally inclined, driven in stone for ventilation purposes.

air-earth conduction current [GEOPHYS] That part of the air-earth current contributed by the electrical conduction of the atmosphere itself; represented as a downward movement of positive space charge in storm-free regions all over the world. Also known as air current.

air-earth current [GEOPHYS] The transfer of electric charge from the positively charged atmosphere to the negatively charged earth; made up of the air-earth conduction current, a precipitation current, a convection current, and miscellaneous smaller contributions.

air endway [MIN ENG] A narrow roadway driven in a coal seam parallel and close to a winning headway for ventilation.

airfloat clay [MIN ENG] Fine particles of clay obtained by air separation from coarser particles following a grinding operation.

airfoil-vane fan [MIN ENG] A centrifugal-type mine fan; the vanes are curved backward from the direction of rotation.

air gap *See* wind gap.

airglow [GEOPHYS] The quasi-steady radiant emission from the upper atmosphere over middle and low latitudes, as distinguished from the sporadic emission of auroras which occur over high latitudes. Also known as light-of-the-night-sky; night-sky light; night-sky luminescence; permanent aurora.

air heave [GEOL] Deformation of plastic sediments on a tidal flat as a result of the growth of air pockets in them; the growth occurs by accretion of smaller air bubbles oozing through the sediment.

air hoar [HYD] A hoar growing on objects above the ground or snow.

airhole [MIN ENG] A small excavation or hole made to improve ventilation by communication with other workings or with the surface.

air intake [MIN ENG] A device for supplying a compressor with clean air at the lowest possible temperature.

air lift *See* air-lift pump.

air-lift pump [MECH ENG] A device composed of two pipes, one inside the other, used to extract water from a well; the lower end of the pipes is submerged, and air is delivered through the inner pipe to form a mixture of air and water which rises in the outer pipe above the water in the well; also used to move corrosive liquids, mill tailings, and sand. Also known as air lift.

airlight [METEOROL] In determinations of visual range, light from sun and sky which is scattered into the eyes of an observer by atmospheric suspensoids (and, to slight extent, by air molecules) lying in the observer's cone of vision.

air lock [MIN ENG] A casing atop an upcast mine shaft to minimize surface air leakage into the fan.

airman [MIN ENG] A man who constructs brattices. [ORD] An enlisted man in the military air force.

air mass [METEOROL] An extensive body of the atmosphere which approximates horizontal homogeneity in its weather characteristics, particularly with reference to temperature and moisture distribution.

air-mass analysis [METEOROL] In general, the theory and practice of synoptic surface-chart analysis by the so-called Norwegian methods, which involve the concepts of the polar front and of the broad-scale air masses which it separates.

air-mass climatology [CLIMATOL] The representation of the climate of a region by the frequency and characteristics of the air masses under which it lies; basically, a type of synoptic climatology.

air-mass precipitation [METEOROL]   Any precipitation that can be attributed only to moisture and temperature distribution within an air mass when that air mass is not, at that location, being influenced by a front or by orographic lifting.

air-mass shower [METEOROL]   A shower that is produced by local convection within an unstable air mass; the most common type of air-mass precipitation.

air-mass source region [METEOROL]   An extensive area of the earth's surface over which bodies of air frequently remain for a sufficient time to acquire characteristic temperature and moisture properties imparted by that surface.

air-measuring station [MIN ENG]   A place in a mine airway where the volume of air passing is measured periodically.

air mover [MIN ENG]   A portable compressed-air appliance, used as a blower or exhauster.

air parcel [METEOROL]   An imaginary body of air to which may be assigned any or all of the basic dynamic and thermodynamic properties of atmospheric air.

air pocket [METEOROL]   An expression used in the early days of aviation for a downdraft; such downdrafts were thought to be pockets in which there was insufficient air to support the plane.

air pressure [PHYS]   The force per unit area that the air exerts on any surface in contact with it, arising from the collisions of the air molecules with the surface.

air regulator [MIN ENG]   An adjustable door installed in permanent air stoppings to control ventilating current.

air sac *See* vesicle.

air sampling [ENG]   The collection and analysis of samples of air to measure the amounts of various pollutants or other substances in the air, or the air's radioactivity.

air shaft [MIN ENG]   A usually vertical earth bore or shaft to supply surface air to an underground facility such as a mine.

air shooting [GEOPHYS]   In seismic prospecting, the technique of applying a seismic pulse to the earth by detonating one or more charges in the air.

air shot [ENG]   A shot prepared by loading (charging) so that an air space is left in contact with the explosive for the purpose of lessening its shattering effect.

air sounding [METEOROL]   The act of measuring atmospheric phenomena or determining atmospheric conditions at altitude, especially by means of apparatus carried by balloons or rockets.

airspace [METEOROL]   **1.** Of or pertaining to both the earth's atmosphere and space. Also known as aerospace. **2.** The portion of the atmosphere above a particular land area, especially a nation or other political subdivision.

air station *See* camera station.

air survey *See* aerial survey.

air temperature [METEOROL]   **1.** The temperature of the atmosphere which represents the average kinetic energy of the molecular motion in a small region and is defined in terms of a standard or calibrated thermometer in thermal equilibrium with the air. **2.** The temperature that the air outside of the aircraft is assumed to have as indicated on a cockpit instrument.

air thermometer [ENG]   A device that measures the temperature of an enclosed space by means of variations in the pressure or volume of air contained in a bulb placed in the space.

air-track drill [MIN ENG]   A heavy drilling machine for quarry or opencast blasting, equipped with caterpillar tracks and operated by independent air motors.

air transport [MIN ENG] Movement from one place to another of the filling material in a mine through pneumatic pipelines. Also known as air transportation.

air transportation *See* air transport.

air turbulence [METEOROL] Highly irregular atmospheric motion characterized by rapid changes in wind speed and direction and by the presence, usually, of up and down currents.

air volcano [GEOL] An eruptive opening in the earth from which large volumes of gas emanate, in addition to mud and stones; a variety of mud volcano.

airwave [METEOROL] A wavelike oscillation in the pattern of wind flow aloft, usually with reference to the stronger portion of the westerly current.

airway [MIN ENG] A passage for air in a mine. Also known as air course.

airways code *See* U.S. airways code.

airways forecast *See* aviation weather forecast.

airways observation *See* aviation weather observation.

Airy isostasy [GEOPHYS] A theory of hydrostatic equilibrium of the earth's surface which contends that mountains are floating on a fluid lava of higher density, and that higher mountains have a greater mass and deeper roots.

Aistopoda [PALEON] An order of Upper Carboniferous amphibians in the subclass Lepospondyli characterized by reduced or absent limbs and an elongate, snakelike body.

Aitken dust counter [ENG] An instrument for determining the dust content of the atmosphere. Also known as Aitken nucleus counter.

Aitken nuclei [METEOROL] The microscopic particles in the atmosphere which serve as condensation nuclei for droplet growth during the rapid adiabatic expansion produced by an Aitken dust counter.

Aitken nucleus counter *See* Aitken dust counter.

Aitoff equal-area map projection [MAP] A Lambert equal-area azimuthal projection of a hemisphere converted into a map projection of the entire sphere by a manipulation suggested by Aitoff.

AJO breathing apparatus [MIN ENG] A breathing device consisting of a Siebe-Gorman mining gas mask with a small oxygen cylinder and a canister which neutralizes mining gases, such as carbon monoxide, sulfureted hydrogen, and nitrous fumes.

akaganeite [MINERAL] β-FeO(OH) A mineral found in meteorites and considered to be formed in flight or by alteration.

akenobeite [PETR] A form of aplite composed of orthoclase and oligoclase with quartz in the interstices.

akerite [PETR] A rock composed of quartz syenite containing soda microcline, oligoclase, and augite.

akermanite [MINERAL] $Ca_2MgSi_2O_7$ Anhydrous calcium-magnesium silicate found in igneous rocks; a melilite.

AKF diagram [PETR] A triangular diagram showing the chemical character of a metamorphic rock in which the three components plotted are A = Al_2O_3 + Fe_2O_3 + (CaO + Na_2O), K = K_2O, and F = FeO + MgO + MnO.

Akins' classifier [MIN ENG] A device for separating fine-size solids from coarser solids in a wet pulp; consists of an interrupted-flight screw conveyor, operating in an inclined trough.

akrochordite [MINERAL] $Mn_4Mg(AsO_4)_2(OH)_4\cdot4H_2O$ Mineral consisting of a hydrous basic manganese magnesium arsenate and occurring in reddish-brown rounded aggregates; hardness is 3 on Mohs scale, and specific gravity is 3.2.

aktological [GEOL] Nearshore shallow-water areas, conditions, sediments, or life.

aktology [GEOL] Study of nearshore and shallow-water areas, conditions, sediments, life, and environments.

alabandite [MINERAL] MnS A complex sulfide mineral that is a component of meteorites and usually occurs in iron-black massive or granular form. Also known as manganblende.

alabaster [MINERAL] **1.** $CaSO_4\cdot2H_2O$ A fine-grained, colorless gypsum. **2.** *See* onyx marble.

alamosite [MINERAL] $PbSiO_3$ A white or colorless monoclinic mineral consisting of lead silicate and occurring in radiating fibers; hardness is 4.5 on Mohs scale, and specific gravity is 6.5.

Alaska Current [OCEANOGR] A current that flows northwestward and westward along the coasts of Canada and Alaska to the Aleutian Islands.

alaskaite [MINERAL] A light lead-gray sulfide mineral consisting of a mixture of lead, silver, copper, and bismuth.

alaskite [PETR] A granitic rock composed mainly of quartz and alkali feldspar, with few dark mineral components.

albafite [MINERAL] Greenish to brownish bitumen which becomes white when exposed to air; contains up to 15% oxygen; fusible; insoluble in organic solvents; varies from soft to hard, porous to compact; atomic ratio H/C 1.75–2.25.

albanite [PETR] A melanocratic leucitite found near Rome, Italy.

albedo neutrons *See* albedo particles.

albedo particles [GEOPHYS] Neutrons or other particles, such as electrons or protons, which leave the earth's atmosphere, having been produced by nuclear interactions of energetic particles within the atmosphere. Also known as albedo neutrons.

Albers projection [MAP] An equal-area projection of the conical type, on which the meridians are straight lines that meet in a common point beyond the limits of the map, and the parallels are concentric circles whose center is at the point of intersection of the meridians.

Alberta low [METEOROL] A low centered on the eastern slope of the Canadian Rockies in the province of Alberta, Canada.

albertite [MINERAL] Jet-black, brittle natural hydrocarbon with conchoidal fracture, hardness of 1–2, and specific gravity of approximately 1.1. Also known as asphaltite coal.

Albian [GEOL] Uppermost Lower Cretaceous geologic time.

albic horizon [GEOL] A soil horizon from which clay and free iron oxides have been removed or in which the iron oxides have been segregated.

Albionian [GEOL] Lower Silurian geologic time.

albite [MINERAL] $NaAlSi_3O_8$ A colorless or milky-white variety of plagioclase of the feldspar group found in granite and various igneous and metamorphic rocks. Also known as sodaclase; sodium feldspar; white feldspar; white schorl.

albite-epidote-amphibolite facies [PETR] Rocks of metamorphic type formed under intermediate temperature and pressure conditions by regional metamorphism or in the outer contact metamorphic zone.

albite law [CRYSTAL] A rule specifying the orientation of alternating lamellae in multiple twin feldspar crystals; the twinning plane is brachypinacoid and is common in albite.

albitite [PETR] A porphyritic dike rock that is coarse-grained and composed almost wholly of albite; common accessory minerals are muscovite, garnet, apatite, quartz, and opaque oxides.

albitization [PETR] The formation of albite in a rock as a secondary mineral.

albitophyre [PETR] A porphyritic rock that contains albite phenocrysts in a ground-mass composed mostly of albite.

Alboll [GEOL] A suborder of the soil order Mollisol with distinct horizons, wet for some part of the year; occurs mostly on upland flats and in shallow depressions.

alboranite [PETR] Olivine-free hypersthene basalt.

alcove [GEOL] A large niche formed by a stream in a face of horizontal strata.

alcove lands [GEOL] Terrain where the mud rocks or sandy clays and shales that compose the hills (badlands) are interstratified by occasional harder beds; the slopes are terraced.

alee basin [GEOL] A basin formed in the deep sea by turbidity currents aggrading courses where the currents were deflected around a submarine ridge.

aleishtite [GEOL] A bluish or greenish mixture of dickite and other clay minerals.

aleurite [GEOL] A type of sedimentary deposit that is unconsolidated and has a texture intermediate between sand and clay.

Aleutian Current [OCEANOGR] A current setting southwestward along the southern coasts of the Aleutian Islands.

Aleutian low [METEOROL] The low-pressure center located near the Aleutian Islands on mean charts of sea-level pressure; represents one of the main centers of action in the atmospheric circulation of the Northern Hemisphere.

Alexandrian [GEOL] Lower Silurian geologic time.

alexandrite [MINERAL] A gem variety of chrysoberyl; emerald green in natural light but red in transmitted or artificial light.

Alfisol [GEOL] An order of soils with gray to brown surface horizons, a medium-to-high base supply, and horizons of clay accumulation.

algal [GEOL] Formed from or by algae.

algal biscuit [GEOL] A disk-shaped or spherical mass, up to 20 centimeters in diameter, made up of carbonate that is probably the result of precipitation by algae.

algal limestone [PETR] A type of limestone either formed from the remains of calcium-secreting algae or formed when algae bind together the fragments of other lime-secreting organisms.

algal pit [GEOL] An ablation depression that is small and contains algae.

algal reef [GEOL] An organic reef which has been formed largely of algal remains and in which algae are or were the main lime-secreting organisms.

algal ridge [GEOL] Elevated margin of a windward coral reef built by actively growing calcareous algae.

algal rim [GEOL] Low rim built by actively growing calcareous algae on the lagoonal side of a leeward reef or on the windward side of a patch reef in a lagoon.

algal structure [GEOL] A deposit, most frequently calcareous, with banding, irregular concentric structures, crusts, and pseudopisolites or pseudoconcretionary forms resulting from organic, colonial secretion and precipitation.

Algerian onyx *See* onyx marble.

alginite *See* algite.

algite [PETR] The petrological unit that constitutes algal material present in considerable amounts in algal or boghead coal. Also known as alginite.

algodonite [MINERAL] Cu_6As A steel gray to silver white mineral consisting of copper arsenide and occuring as minute hexagonal crystals or in massive and granular form.

Algoman orogeny [GEOL] Orogenic episode affecting Archean rocks of Canada about 2.4 billion years ago. Also known as Kenoran orogeny.

Algonkian [GEOL] Geologic time between the Archean and Paleozoic. Also known as Proterozoic.

alidade [ENG] **1.** An instrument for topographic surveying and mapping by the plane-table method. **2.** Any sighting device employed for angular measurement.

alignment [MAP] Representing of the correct direction, character, and relationships of a line or feature on a map. [MIN ENG] The act of laying out a tunnel or regulating by line; adjusting to a line.

alignment correction [ENG] A correction applied to the measured length of a line to allow for not holding the tape exactly in a vertical plane of the line.

alimentation [HYD] **1.** In glaciology, the combined processes which serve to increase the mass of a glacier or snowfield. **2.** *See* accumulation.

Aliva concrete sprayer [MIN ENG] A compressed-air machine for spraying concrete on the roof and sides of mine roadways.

alkali basalt [PETR] Basalt that contains olivine, augite, and calcic plagioclase.

alkalic Also known as alkali. [PETR] **1.** Of igneous rock, containing more than average alkali (K_2O and Na_2O) for that clan in which they are found. **2.** Of igneous rock, having feldspathoids or other minerals, such as acmite, so that the molecular ratio of alkali to silica is greater than 1:6. **3.** Of igneous rock, having a low alkali-lime index (51 or less).

alkali-calcic series [PETR] The series of igneous rocks with weight percentage of silica in the range 51–55, and weight percentages of CaO and $K_2O + Na_2O$ equal.

alkali emission [GEOPHYS] Light emission from free lithium, potassium, and especially sodium in the upper atmosphere.

alkali feldspar [MINERAL] A feldspar composed of potassium feldspar and sodium feldspar, such as orthoclase, microcline, albite, and anorthoclase; all are considered alkali-rich.

alkali flat [GEOL] A level lakelike plain formed by the evaporation of water in a depression and deposition of its fine sediment and dissolved minerals.

alkali lake [HYD] A lake with large quantities of dissolved sodium and potassium carbonates as well as sodium chloride.

alkali-lime index [PETR] The percentage by weight of silica in a sequence of igneous rocks on a variation diagram where the weight percentages of CaO and of K_2O and Na_2O are equal.

allactite [MINERAL] $Mn_7(AsO_4)_2(OH)_8$ Brownish-red mineral consisting of a basic manganese arsenate.

allalinite [PETR] An altered gabbro with original texture and euhedral pseudomorphs.

allanite [MINERAL] $(Ca,Ce,La,Y)_2(Al,Fe)_3Si_3O_{12}(OH)$ Monoclinic mineral distinguished from all other members of the epidote group of silicates by a relatively high content of rare earths. Also known as bucklandite; cerine; orthite; treanorite.

allcharite [MINERAL] A lead gray mineral, supposed to be a lead arsenic sulfide and known only crystallographically as orthorhombic crystals.

alleghanyite [MINERAL] $Mn_5(SiO_4)_2(OH)_2$ A pink mineral consisting of basic manganese silicate.

Alleghenian [GEOL] Lower Middle Pennsylvanian geologic time.

Alleghenian orogeny [GEOL] Pennsylvanian and Early Permian orogenic episode which deformed the rocks of the Appalachian Valley and the Ridge and Plateau provinces.

allemontite [MINERAL] AsSb Rhombohedric, gray or reddish, native antimony aresenide occurring in reniform masses. Also known as arsenical antimony.

allivalite [PETR] A form of gabbro composed of anorthite and olivine; accessories are augite, apatite, and opaque iron oxides.

allochem [GEOL] Sediment formed by chemical or biochemical precipitation within a depositional basin; includes intraclasts, oolites, fossils, and pellets.

allochemical metamorphism [PETR] Metamorphism accompanied by addition or removal of material so that the bulk chemical composition of the rock is changed.

allochetite [PETR] A porphyritic igneous rock composed of phenocrysts of labradorite, orthoclase, titanaugite, nepheline, magnetite, and apatite in a groundmass of augite, biotite, magnetite, hornblende, nepheline, and orthoclase.

allochromatic crystal [CRYSTAL] A crystal having photoconductive properties due to the presence of small particles within it.

allochthon [GEOL] A rock that was transported a great distance from its original deposition by some tectonic process, generally related to overthrusting, recumbent folding, or gravity sliding.

allochthonous [PETR] Of rocks whose primary constituents have not been formed in situ.

allochthonous coal [GEOL] A type of coal arising from accumulations of plant debris moved from their place of growth and deposited elsewhere.

allochthonous stream [HYD] A stream flowing in a channel that it did not form.

allogene [GEOL] A mineral or rock that has been moved to the site of deposition. Also known as allothigene; allothogene.

allogenic [GEOL] *See* allothogenic.

allomerism [CRYSTAL] A constancy in crystal form in spite of a variation in chemical composition.

allomorphism *See* paramorphism.

allomorphite [MINERAL] A mineral consisting of barite that is pseudomorphous after anhydrite.

allophane [GEOL] $Al_2O_3 \cdot SiO_2 \cdot nH_2O$ A clay mineral composed of hydrated aluminosilicate gel of variable composition; P_2O_5 may be present in appreciable quantity.

Allotheria [PALEON] A subclass of Mammalia that appeared in the Upper Jurassic and became extinct in the Cenozoic.

allothigene *See* allogene.

allothimorph [GEOL] A metamorphic rock constituent which retains its original crystal outlines in the new rock.

allothogene *See* allogene.

allothogenic [GEOL] Formed from preexisting rocks which have been transported from another location. Also known as allogenic.

allotrioblast *See* xenoblast.

allotriomorphic [MINERAL] Of minerals in igneous rock not bounded by their own crystal faces but having their outlines impressed on them by the adjacent minerals. Also known as anhedral; xenomorphic.

allotriomorphic-granular rock *See* xenomorphic-granular rock.

alluvial [GEOL] **1.** Of a placer, or its associated valuable mineral, formed by the action of running water. **2.** Pertaining to or consisting of alluvium, or deposited by running water.

alluvial cone [GEOL] An alluvial fan with steep slopes formed of loose material washed down the slopes of mountains by ephemeral streams and deposited as a conical mass of low slope at the mouth of a gorge. Also known as cone delta; cone of dejection; cone of detritus; debris cone; dry delta; hemicone; wash.

alluvial dam [GEOL] A sedimentary deposit which is built by an overloaded stream and dams its channel; especially characteristic of distributaries on alluvial fans.

alluvial fan [GEOL] A fan-shaped deposit formed by a stream either where it issues from a narrow moutain valley onto a plain or broad valley, or where a tributary stream joins a main stream.

alluvial flat [GEOL] A small alluvial plain having a slope of about 5 to 20 feet per mile and built of fine sandy clay or adobe deposited during flood.

alluvial mining [MIN ENG] The exploitation of alluvial deposits by dredging, hydraulicking, or drift mining.

alluvial ore deposit [GEOL] A deposit in which the valuable mineral particles have been transported and left by a stream.

alluvial plain [GEOL] A plain formed from the deposition of alluvium usually adjacent to a river that periodically overflows. Also known as aggraded valley plain; river plain; wash plain; waste plain.

alluvial slope [GEOL] A surface of alluvium which slopes down from mountainsides and merges with the plain or broad valley floor.

alluvial soil [GEOL] A soil deposit developed on floodplain and delta deposits.

alluvial terrace [GEOL] A terraced embankment of loose material adjacent to the sides of a river valley. Also known as built terrace; drift terrace; fill terrace; stream-built terrace; wave-built platform; wave-built terrace.

alluvial valley [GEOL] A valley filled with a stream deposit.

alluviation [GEOL] Deposition of sediment by a river.

alluvium [GEOL] The detrital materials eroded, transported, and deposited by streams; an important constituent of shelf deposits. Also known as alluvial deposit; alluvion.

almandine [MINERAL] $Fe_3Al_2(SiO_4)_3$ A variety of garnet, deep red to brownish red, found in igneous and metamorphic rocks in many parts of world; used as a gemstone and an abrasive. Also known as almandite.

almandite *See* almandine.

almerite *See* natroalunite.

alnoite [PETR]   A variety of biotite lamprophyres characterized by lepidomelane phenocrysts; it is feldspar-free but contains melitite, perovskite, olivine, and carbonate in the matrix.

aloisite [MINERAL]   A brown to violet mineral consisting of a hydrous subsilicate of calcium, iron, magnesium, and sodium, and occurring in amorphous masses.

alongshore current *See* littoral current.

alpenglow [METEOROL]   A reappearance of sunset colors on a mountain summit after the original mountain colors have faded into shadow; also, a similar phenomenon preceding the regular coloration at sunrise.

alphanumeric grid *See* atlas grid.

alpine glacier [HYD]   A glacier lying on or occupying a depression in mountainous terrain. Also known as mountain glacier.

Alpine orogeny [GEOL]   Jurassic through Tertiary orogeny which affected the Alpides.

alpine-type facies [PETR]   High-pressure, low-temperature (150–400°C) dynamothermal metamorphism characterized by the presence of the pumpellyite and glaucophane schist facies.

alpinotype tectonics [GEOL]   Tectonics of the alpine-type geosynclinal mountain belts characterized by deep-seated plastic folding, plutonism, and lateral thrusting.

alquifou [MINERAL]   A type of coarse-grained galena.

alsbachite [PETR]   A plutonic rock of sodic plagioclase, quartz, and subordinate orthoclase and accessory garnet, biotite, and muscovite; a variety of porphyritic granodiorite.

alstonite *See* bromlite.

alt *See* altitude.

alta [GEOL]   Black, shaly, highly sheared overburden accompanying quicksilver orebodies.

altaite [MINERAL]   PbTe   A tin-white lead-tellurium mineral occurring as isometric crystals with tin ores in central Asia.

altazimuth [ENG]   An instrument equipped with both horizontal and vertical graduated circles, for the simultaneous observation of horizontal and vertical directions or angles. Also known as astronomical theodolite; universal instrument.

alteration [PETR]   A change in a rock's mineral composition.

altiplanation [GEOL]   A phase of solifluction that may be seen as terracelike forms, flattened summits, and passes that are mainly accumulations of loose rock.

altiplanation surface [GEOL]   A flat area fronted by scarps a few to hundreds of feet in height; the area ranges from several square rods to hundreds of acres. Also known as altiplanation terrace.

altiplanation terrace *See* altiplanation surface.

altithermal [GEOPHYS]   Period of high temperature, particularly the postglacial thermal optimum.

Altithermal [GEOL]   A dry postglacial interval centered about 5500 years ago during which temperatures were warmer than at present.

altithermal soil   [GEOL]   Soil recording a period of rising or high temperature.

altitude   Abbreviated alt.   [ENG]   **1.** Height, measured as distance along the extended earth's radius above a given datum, such as average sea level.   **2.** Angular displacement above the horizon measured by an altitude curve.

altitude chamber   [ENG]   A chamber within which the air pressure, temperature, and so on can be adjusted to simulate conditions at different altitudes; used for experimentation and testing.

altitude-contour   See C factor.

altitude datum   [ENG]   The arbitrary level from which heights are reckoned.

altitude difference   [ENG]   The difference between computed and observed altitudes, or between precomputed and sextant altitudes. Also known as altitude intercept; intercept.

altitude intercept   See altitude difference.

altitude tints   See gradient tints.

alum   [MINERAL] $KAl(SO_4)_2 \cdot 12H_2O$ A colorless, white, astringent-tasting evaporite mineral.

alum coal   [GEOL]   Argillaceous brown coal rich in pyrite in which alum is formed on weathering.

aluminite   [MINERAL] $Al_2(SO_4)(OH)_4 \cdot 7H_2O$ Native monoclinic hydrous aluminum sulfate; used in tanning, papermaking, and water purification. Also known as websterite.

aluminum ore   [GEOL]   A natural material from which aluminum may be economically extracted.

alumite   See alunite.

alum rock   See alunite.

alum schist   See alum shale.

alum shale   [PETR]   A shale containing pyrite that is decomposed by weathering to form sulfuric acid, which acts on potash and alumina constituents to form alum. Also known as alum schist; alum slate.

alum slate   See alum shale.

alumstone   See alunite.

alunite   [MINERAL] $KAl_3(SO_4)_2(OH)_6$ A mineral composed of a basic potassium aluminum sulfate; it occurs as a hydrothermal-alteration product in feldspathic igneous rocks and is used in the manufacture of alum. Also known as alumite; alum rock; alumstone.

alunitization   [GEOL]   Introduction of or replacement by alunite.

alunogen   [MINERAL] $Al_2(SO_4)_3 \cdot 18H_2O$ A white mineral occurring as a fibrous incrustation of hydrated aluminum sulfate by volcanic action or decomposition of pyrite. Also known as feather alum; hair salt.

alurgite   [MINERAL]   A purple manganiferous variety of muscovite mica.

alyphite   [GEOL]   Bitumen that yields a high percentage of open-chain aliphatic hydrocarbons upon distillation.

amarantite   [MINERAL] $Fe(SO_4)(OH) \cdot 3H_2O$ An amaranth red to brownish- or orange-red triclinic mineral consisting of a hydrated basic sulfate of ferric iron.

amarillite   [MINERAL] $NaFe(SO_4)_2 \cdot 6H_2O$ A pale greenish-yellow mineral consisting of a hydrous sodium ferric sulfate.

amazonite   [MINERAL]   An apple-green, bright-green, or blue-green variety of microcline found in the United States and Soviet Union; sometimes used as a gemstone. Also known as amazon stone.

amazon stone *See* amazonite.

ambatoarinite   [MINERAL]   A mineral consisting of a carbonate of cerium metals and strontium.

amber   [MINERAL]   A transparent yellow, orange, or reddish-brown fossil resin derived from a coniferous tree; used for ornamental purposes; it is amorphous, has a specific gravity of 1.05–1.10, and a hardness of 2–2.5 on Mohs scale.

amberoid   [MINERAL]   A gem-quality mineral composed of small fragments of amber that have been reunited by heat or pressure.

ambient stress field   [GEOPHYS]   The distribution and numerical value of the stresses present in a rock environment prior to its disturbance by man. Also known as in-place stress field; primary stress field; residual stress field.

amblygonite   [MINERAL] $(Li,Na)AlPO_4(F,OH)$ A mineral occurring in white or greenish cleavable masses and found in the United States and Europe; important ore of lithium.

ambonite   [PETR]   Any of a group of hornblende-biotite andesites and dacites containing cordierite.

ambrite   [MINERAL]   A yellow-gray, semitransparent fossil resin resembling amber; found in large masses in New Zealand coal fields and regarded as a semiprecious stone.

ambrosine   [MINERAL]   A yellowish to clove-brown variety of amber rich in succinic acid; occurs as rounded masses in phosphate beds near Charleston, S.C.

Amebelodontinae   [PALEON]   A subfamily of extinct elephantoid proboscideans in the family Gomphotheriidae.

amemolite   [GEOL]   A stalactite with one or more changes in its axis of growth.

American boring system   [MIN ENG]   A rope system of percussive boring, with a derrick which enables the complete set of boring tools to be raised clear of the hole. Also known as American system.

American explosive   [MATER]   One of many explosives that have passed U.S. Bureau of Mines tests and are used under certain conditions.

American jade *See* californite.

American system *See* American boring system.

American system drill *See* churn drill.

amerikanite   [GEOL]   A natural glass found in South America and believed to be of volcanic origin.

amesite   [MINERAL] $(Mg,Fe)_4Al_4Si_2O_{10}(OH)_8$ Apple-green phyllosilicate mineral occurring in foliated hexagonal plates.

amethyst   [MINERAL]   The transparent purple to violet variety of the mineral quartz; used as a jeweler's stone.

amherstite   [PETR]   A syenodiorite containing andesine and antiperthite.

amianthus   [MINERAL]   A fine, silky variety of asbestos, such as chrysotile.

amictic lake   [HYD]   A lake that is perennially frozen.

Ammanian [GEOL]   Middle Upper Cretaceous geologic time.

ammonia dynamite [CHEM]   Dynamite with part of the nitroglycerin replaced by ammonium nitrate.

ammonia permissible [MATER]   A permissible explosive that is an ammonia dynamite.

ammonioborite [MINERAL] $(NH_4)_2B_{10}O_{16}\cdot5H_2O$ A white mineral consisting of a hydrous ammonium borite and occurring as aggregates of minute plates.

ammoniojarosite [MINERAL] $(NH_4)Fe_3(SO_4)_2(OH)_6$ Pale-yellow mineral consisting of basic ferric ammonium sulfate.

ammonite [MATER]   An explosive containing 70–95% ammonium nitrate.   [PALEON] A fossil shell of the cephalopod order Ammonoidea.

ammonoid [PALEON]   A cephalopod of the order Ammonoidea.

Ammonoidea [PALEON]   An order of extinct cephalopod mollusks in the subclass Tetrabranchia; important as index fossils.

amoeboid fold [GEOL]   A fold or structure, such as an anticline, having no prevailing trend or definite shape.

amoeboid glacier [HYD]   A glacier connected with its snowfield for a portion of the year only.

amorphous frost [HYD]   Hoar frost which possesses no apparent simple crystalline structure; opposite of crystalline frost.

amorphous mineral [MINERAL]   A mineral without definite crystalline structure.

amorphous peat [GEOL]   Peat composed of fine grains of organic matter; is plastic like wet, heavy soil, with all original plant structures destroyed by decomposition of cellulosic matter.

amorphous sky [METEOROL]   A sky characterized by an abundance of fractus clouds, usually accompanied by precipitation falling from a higher, overcast cloud layer.

amorphous snow [HYD]   A type of snow with irregular crystalline structure.

amosite [MINERAL]   A monoclinic amphibole form of asbestos having long fibers and a high iron content; used in insulation.

ampangabeite *See* samarskite.

ampelite [PETR]   A graphite schist containing silica, alumina, and sulfur; used as a refractory.

amphibole [MINERAL]   Any of a group of rock-forming, ferromagnesian silicate minerals commonly found in igneous and metamorphic rocks; includes hornblende, anthophyllite, tremolite, and actinolite (asbestos minerals).

amphibolite [PETR]   A crystalloblastic metamorphic rock composed mainly of amphibole and plagioclase; quartz may be present in small quantities.

amphibolite facies [PETR]   Rocks produced by medium- to high-grade regional metamorphism.

amphibolization [PETR]   Formation of amphibole in a rock as a secondary mineral.

Amphichelydia [PALEON]   A suborder of Triassic to Eocene anapsid reptiles in the order Chelonia; these turtles did not have a retractable neck.

Amphicyonidae [PALEON]   A family of extinct giant predatory carnivores placed in the infraorder Miacoidea by some authorities.

amphidromic [OCEANOGR]   Of or pertaining to progression of a tide wave or bulge around a point or center of little or no tide.

amphidromic point [MAP] On a chart of cotidal lines, a no-tide or nodal point from which the cotidal lines radiate.

amphidromic region [MAP] An area surrounding an amphidromic point in which the cotidal lines radiate from the no-tide point and progress through all hours of the tide cycle.

amphigene *See* leucite.

Amphilestidae [PALEON] A family of Jurassic triconodont mammals whose subclass is uncertain.

Amphimerycidae [PALEON] A family of late Eocene to early Oligocene tylopod ruminants in the superfamily Amphimerycoidea.

Amphimerycoidea [PALEON] A superfamily of extinct ruminant artiodactyls in the infraorder Tylopoda.

amphimorphic [GEOL] A rock or mineral formed by two geologic processes.

amphisapropel [GEOL] Cellulosic ooze containing coarse plant debris.

Amphissitidae [PALEON] A family of extinct ostracods in the suborder Beyrichicopina.

amphitheater [GEOGR] A valley or gulch having an oval or circular floor and formed by glacial action.

Amphitheriidae [PALEON] A family of Jurassic therian mammals in the infraclass Pantotheria.

amphoterite [GEOL] A stony meteorite containing bronzite and olivine with some oligoclase and nickel-rich iron.

amygdaloid [GEOL] Lava rock containing amygdules. Also known as amygdaloidal lava.

amygdaloidal lava *See* amygdaloid.

amygdule [GEOL] **1.** A mineral filling formed in vesicles (cavities) of lava flows; it may be chalcedony, opal, calcite, chlorite, or prehnite. **2.** An agate pebble.

Amynodontidae [PALEON] A family of extinct hippopotamuslike perissodactyl mammals in the superfamily Rhinoceratoidea.

anabatic wind [METEOROL] An upslope wind; usually applied only when the wind is blowing up a hill or mountain as the result of a local surface heating, and apart from the effects of the larger-scale circulation.

anabohitsite [PETR] A variety of olivine-pyroxenite containing hornblende and hypersthene and a high proportion (about 30%) of magnetite and ilmenite.

anabranch [HYD] A diverging branch of a stream or river that loses itself in sandy soil or rejoins the main flow downstream.

anaclinal [GEOL] Having a downward inclination opposite to that of a stratum.

anacoustic zone [GEOPHYS] The zone of silence in space, starting at about 100 miles (160 kilometers) altitude, where the distance between air molecules is greater than the wavelength of sound, and sound waves can no longer be propagated.

anaerobic sediment [GEOL] A highly organic sediment that is formed in the absence or near absence of oxygen in water and that is rich in hydrogen sulfide.

anafront [METEOROL] A front at which the warm air is ascending the frontal surface up to high altitudes.

analbite [MINERAL] A triclinic albite which is not stable and becomes monoclinic at about 700°C.

analcime [MINERAL] $NaAlSi_2O_6 \cdot H_2O$ A white or slightly colored isometric zeolite found in diabase and in alkali-rich basalts. Also known as analcite.

analcimite [PETR] An extrusive or hypabyssal rock that consists primarily of pyroxene and analcime.

analcimization [GEOL] The replacement in igneous rock of feldspars or feldspathoids by analcime.

analcite *See* analcime.

anallobaric center *See* pressure-rise center.

analog [METEOROL] A past large-scale synoptic weather pattern which resembles a given (usually current) situation in its essential characteristics.

analysis [METEOROL] A detailed study in synoptic meteorology of the state of the atmosphere based on actual observations, usually including a separation of the entity into its component patterns and involving the drawing of families of isopleths for various elements.

analytical aerotriangulation [ENG] Analytical phototriangulation, performed with aerial photographs.

analytical geomorphology *See* dynamic geomorphology.

analytical nadir-point triangulation [ENG] Radial triangulation performed by computational routines in which nadir points are utilized as radial centers.

analytical orientation [ENG] The computational steps required to determine tilt, direction of principal line, flight height, angular elements, and linear elements in preparing aerial photographs for rectification.

analytical photogrammetry [ENG] A method of photogrammetry in which solutions are obtained by mathematical methods.

analytical phototriangulation [ENG] A phototriangulation procedure in which the spatial solution is obtained by computational routines.

analytical radial triangulation [ENG] Radial triangulation performed by computational routines.

analytical three-point resection radial triangulation [ENG] A method of computing the coordinates of the ground principal points of overlapping aerial photographs by resecting on three horizontal control points appearing in the overlap area.

anamigmatism [GEOL] A process of high-temperature, high-pressure remelting of sediment to yield magma.

anamorphic zone [GEOL] The zone of rock flow, as indicated by reactions that may involve decarbonation, dehydration, and deoxidation; silicates are built up, and the formation of denser minerals and of compact crystalline structure takes place.

anamorphism [GEOL] A kind of metamorphism at considerable depths in the earth's crust and under great pressure, resulting in the formation of complex minerals from simple ones.

Anancinae [PALEON] A subfamily of extinct proboscidean placental mammals in the family Gomphotheriidae.

anapaite [MINERAL] $Ca_2Fe(PO_4)_2 \cdot 4H_2O$ A pale-green or greenish-white triclinic mineral consisting of a ferrous iron hydrous phosphate and occurring in crystals and massive forms; hardness is 3–4 on Mohs scale, and specific gravity is 3.81.

anapeirean *See* Pacific suite.

Anaplotheriidae [PALEON] A family of extinct tylopod ruminants in the superfamily Anaplotherioidea.

Anaplotherioidea [PALEON] A superfamily of extinct ruminant artiodactyls in the infraorder Tylopoda.

Anasca [PALEON] A suborder of extinct bryozoans in the order Cheilostomata.

anaseism [GEOL PHYS] Movement of the earth in a direction away from the focus of an earthquake.

Anaspida [PALEON] An order of extinct fresh- or brackish-water vertebrates in the class Agnatha.

anastatic water [HYD] That part of the subterranean water in the capillary fringe between the zone of aeration and the zone of saturation in the soil.

anatase [MINERAL] The brown, dark-blue, or black tetragonal crystalline form of titanium dioxide, TiO_2; used to make a white pigment. Also known as octahedrite.

anatexis [GEOL] A high-temperature process of metamorphosis by which plutonic rock in the lowest levels of the crust is melted and regenerated as a magma.

anathermal [GEOL] A period of time between the age of other strata or units of reference in which the temperature is increasing.

anauxite [MINERAL] $Al_2(SiO_7)(OH)_4$ A clay mineral that is a mixture of kaolinite and quartz. Also known as ionite.

anchieutectic [GEOL] A type of magma which is incapable of undergoing further notable main-stage differentiation because its mineral composition is practically in eutectic proportions.

anchimonomineralic [PETR] Of rock composed mostly of one kind of mineral.

anchor charge [ENG] A procedure that allows several charges to be preloaded in a seismic shot hole; the bottom charges are fired first, and the upper charges are held down by anchors.

anchored dune [GEOL] A sand dune stabilized by growth of vegetation.

anchor ice [HYD] Ice formed beneath the surface of water, as in a lake or stream, and attached to the bottom or to submerged objects. Also known as bottom ice; ground ice.

anchorite [PETR] A variety of diorite having nodules of mafic minerals and veins of felsic minerals.

anchor packer [PETRO ENG] A device used in oil wells to seal the annular space between the tubing and its surrounding casing to help control the oil-producing gas lift.

anchor stone [GEOL] A rock or pebble that has marine plants attached to it.

ancylite [MINERAL] $SrCe(CO_3)_2(OH)\cdot H_2O$ A mineral consisting of hydrous basic carbonate of cerium and strontium.

ancylopoda [PALEON] A suborder of extinct herbivorous mammals in the order Perissodactyla.

andalusite [MINERAL] Al_2SiO_5 A brown, yellow, green, red, or gray neosilicate mineral crystallizing in the orthorhombic system, usually found in metamorphic rocks.

Andean-type continental margin [GEOL] A continental margin, as along the Pacific coast of South America, where oceanic lithosphere descends beneath an adjacent continent producing andesitic continental margin volcanism.

Andept [GEOL] A suborder of the soil order Inceptisol, formed chiefly in volcanic ash or in regoliths with high components of ash.

andersonite [MINERAL] $Na_2Ca(UO_2)(CO_3)_3 \cdot 6H_2O$ Bright yellow-green secondary mineral consisting of a hydrous sodium calcium uranium carbonate.

andesine [MINERAL] A plagioclase feldspar with a composition ranging from $Ab_{70}An_{30}$ to $Ab_{50}An_{50}$, where $Ab = NaAlSi_3O_8$ and $An = CaAl_2Si_2O_8$; it is a primary constituent of intermediate igneous rocks, such as andesites.

andesite [PETR] Very finely crystalline extrusive rock of volcanic origin composed largely of plagioclase feldspar (oligoclase or andesine) with smaller amounts of dark-colored mineral (hornblende, biotite, or pyroxene), the extrusive equivalent of diorite.

andesite line [GEOL] The postulated geographic and petrographic boundary between the andesite-dacite-rhyolite rock association of the margin of the Pacific Ocean and the olivine-basalt-trachyte rock association of the Pacific Ocean basin.

andesitic glass [GEOL] A natural glass that is chemically equivalent to andesite.

Andes lightning [GEOPHYS] Electrical coronal discharges observable often as far as several hundred miles away, generally over any of the mountainous areas of the world when under disturbed electrical conditions. Also known as Andes glow.

andorite [MINERAL] $AgPbSb_3S_6$ A dark-gray or black orthorhombic mineral. Also known as sundtite.

andradite [MINERAL] The calcium-iron end member of the garnet group.

andrewsite [MINERAL] $(Cu,Fe^{2-})Fe_3^{3-}(PO_4)_3(OH)_2$ A bluish-green mineral consisting of a basic phosphate of iron and copper.

andrite [GEOL] A meteorite composed principally of augite with some olivine and troilite.

anemobiagraph [ENG] A recording pressure-tube anemometer in which the wind scale of the float manometer is linear through the use of springs; an example is the Dines anemometer.

anemoclast [GEOL] A clastic rock that was fragmented and rounded by wind.

anemoclastic [GEOL] Referring to rock that was broken by wind erosion and rounded by wind action.

anemoclinometer [ENG] A type of instrument which measures the inclination of the wind to the horizontal plane.

anemograph [ENG] **1.** An instrument which records wind velocities. **2.** A recording anemometer.

anemology [METEOROL] Scientific investigation of winds.

anemometer [ENG] A device which measures air speed.

anemometry [METEOROL] The study of measuring and recording the direction and speed (or force) of the wind, including its vertical component.

anemoscope [ENG] An instrument for indicating the direction of the wind.

anemovane [ENG] A combined contact anemometer and wind vane used in the Canadian Meteorological Service.

aneroid [ENG] **1.** Containing no liquid or using no liquid. **2.** *See* aneroid barometer.

aneroid barograph [ENG] An aneroid barometer arranged so that the deflection of the aneroid capsule actuates a pen which graphs a record on a rotating drum. Also known as aneroidograph; barograph; barometrograph.

aneroid barometer [ENG] A barometer which utilizes an aneroid capsule. Also known as aneroid.

aneroid capsule [ENG] A thin, disk-shaped box or capsule, usually metallic, partially evacuated and sealed, held extended by a spring, which expands and contracts with changes in atmospheric or gas pressure. Also known as bellows.

aneroid diaphragm [ENG] A thin plate, usually metal, covering the end of an aneroid capsule and moving axially as the ambient gas pressure increases or decreases.

aneroidograph *See* aneroid barograph.

angaralite [MINERAL] $Mg_2(Al,Fe)_{10}Si_6O_{29}$ A mineral of the chlorite group, occurring in thin black plates.

Angara Shield [GEOL] A shield area of crystalline rock in Siberia.

angle cut [MIN ENG] A drilling pattern in which drill holes converge, so that a core can be blasted out and an open or relieved cavity or free face be left for the following shots, which are timed to ensue with a fractional delay.

angle of attack [AERO ENG] The angle between a reference line fixed with respect to an airframe (usually the longitudinal axis) and the direction of movement of the body. [MIN ENG] The angle in a mine fan made by the direction of air approach and the chord of the aerofoil section.

angle of current [HYD] In stream gaging, the angular difference between 90° and the angle made by the current with a measuring section.

angle of depression [ENG] The angle in a vertical plane between the horizontal and a descending line. Also known as depression angle; descending vertical angle; minus angle.

angle of dip *See* dip.

angle of elevation [ENG] The angle in a vertical plane between the local horizontal and an ascending line, as from an observer to an object; used in astronomy, surveying, and so on. Also known as ascending vertical angle; elevation angle.

angle of shear [GEOL] The angle between the planes of maximum shear which is bisected by the axis of greatest compression.

angle of slide [MIN ENG] The slope, measured in degrees of deviation from the horizontal, on which loose or fragmented solid materials will start to slide.

angle set [MIN ENG] **1.** A timber set using an angle brace. **2.** One of a series of sets placed at angles to each other.

anglesite [MINERAL] $PbSO_4$ A mineral occurring in white or gray, tabular or prismatic orthorhombic crystals or compact masses. Also known as lead spar; lead vitriol.

Angoumian [GEOL] Upper middle Upper Cretaceous (Upper Turonian) geologic time.

angrite [GEOL] An achondrite stony meteorite composed principally of augite with a little olivine and troilite.

angular distortion [MAP] Distortion in a map projection because of nonconformality.

angular spreading [OCEANOGR] The lateral extension of ocean waves as they move out of the wave-generating area as swell.

angular-spreading factor [OCEANOGR] The ratio between the wave energy actually occurring at a point and that which would have been present if there were no angular spreading.

angular unconformity [GEOL] An unconformity in which the older strata dip at a different angle (usually steeper) than the younger strata.

angular wave number [METEOROL]   The number of waves of a given wavelength required to encircle the earth at the latitude of the disturbance. Also known as hemispheric wave number.

anhedral *See* allotriomorphic.

anhedron [PETR]   Rock that has the organized internal structure of a crystal without the external geometric form of a crystal.

anhydrite [MINERAL] $CaSO_4$ A mineral that represents gypsum without its water of crystallization, occurring commonly in white and grayish granular to compact masses; the hardness is 3–3.5 on Mohs scale, and specific gravity is 2.90–2.99. Also known as cube spar.

anhydrite evaporite [PETR] $CuSO_4$ A sedimentary rock composed chiefly of copper sulfate in compact granular form deposited by evaporation of water; resembles marble and differs from gypsum in lack of water of hydration and hardness.

anhydrock [PETR]   A sedimentary rock chiefly made of anhydrite.

Animikean [GEOL]   The middle subdivision of Proterozoic geologic time. Also known as Penokean; Upper Huronian.

animikite [GEOL]   An ore of silver, composed of a mixture of sulfides, arsenides, and antimonides, and containing nickel and lead; occurs in white or gray granular masses.

Anisean [GEOL]   Lower Middle Triassic geologic time.

anisodesmic [MINERAL]   Pertaining to crystals or compounds in which the ionic bonds are unequal in strength.

ankaramite [PETR]   A mafic olivine basalt primarily composed of pyroxene with smaller amounts of olivine and plagioclase and accessory biotite, apatite, and opaque oxides.

ankaratrite *See* olivine nephelinite.

ankerite [MINERAL] $Ca(Fe,Mg,Mn)(CO_3)_2$ A white, red, or gray iron-rich carbonate mineral associated with iron ores and found in thin veins in coal seams; specific gravity is 2.95–3.1. Also known as cleat spar.

Ankylosauria [PALEON]   A suborder of Cretaceous dinosaurs in the reptilian order Ornithischia characterized by short legs and flattened, heavily armored bodies.

annabergite [MINERAL] $(Ni,Co)_3(AsO_4)_2 \cdot 8H_2O$ A monoclinic mineral usually found as apple-green incrustations as an alteration product of nickel arsenides; it is isomorphous with erythrite. Also known as nickel bloom; nickel ocher.

annex point [MAP]   A point used to assist in the relative orientation of vertical and oblique photographs; selected in the overlap area between the vertical and its corresponding oblique photograph, about midway between the pass points.

annotated photograph [MAP]   A photograph on which planimetric, geologic, cultural, hydrographic, or vegetation information has been added to identify, classify, outline, clarify, or describe features that would not otherwise be apparent in examination of an unmarked photograph.

annual flood [HYD]   The highest flow at a point on a stream during any particular calendar year or water year.

annual inequality [OCEANOGR]   Seasonal variation in water level or tidal current speed, more or less periodic, due chiefly to meteorological causes.

annual labor *See* assessment drilling.

annual layer [GEOL] **1.** A sedimentary layer deposited, or presumed to have been deposited, during the course of a year; for example, a glacial varve. **2.** A dark layer in a stratified salt deposit containing disseminated anhydrite.

annual magnetic change *See* magnetic annual change.

annual magnetic variation *See* magnetic annual variation.

annual storage [HYD] The capacity of a reservoir that can handle a watershed's annual runoff but cannot carry over any portion of the water for longer than the year.

annular [MIN ENG] The space between the casing and wall of a hole or between a drill pipe and casing.

annular drainage pattern [HYD] A ringlike pattern subsequent in origin and associated with maturely dissected dome or basin structures.

anomalous magma [GEOL] Magma formed or obviously changed by assimilation.

anomaly [GEOL] A local deviation from the general geological properties of a region. [OCEANOGR] The difference between conditions actually observed at a serial station and those that would have existed had the water all been of a given arbitrary temperature and salinity.

anomaly finder [ENG] A computer-controlled data-plotting system used on ships to measure and record seismic, gravity, magnetic, and other geophysical data and water depth, time, course, and speed.

anomite [MINERAL] A variety of biotite different only in optical orientation.

Anomphalacea [PALEON] A superfamily of extinct gastropod mollusks in the order Aspidobranchia.

anorogenic [GEOL] Of a feature, forming during tectonic quiescence between orogenic periods, that is, lacking in tectonic disturbance.

anorogenic time [GEOL] Geologic time when no significant deformation of the crust occurred.

anorthic crystal *See* triclinic crystal.

anorthite [MINERAL] The white, grayish, or reddish calcium-rich end member of the plagioclase feldspar series; composition ranges from $Ab_{10}An_{90}$ to Ab_0An_{100}, where Ab = $NaAlSi_3O_8$ and An = $CaAl_2Si_2O_8$. Also known as calciclase; calcium feldspar.

anorthite-basalt [PETR] A rock composed of a basic variety of basalt with anorthite instead of labradorite.

anorthoclase [MINERAL] A triclinic alkali feldspar having a chemical composition ranging from $Or_{40}Ab_{60}$ to $Or_{10}Ab_{90}$ to about 20 mole % An, where Or = $KAlSi_3O_8$, Ab = $NaAlSi_3O_8$, and An = $CaAl_2Si_2O_8$. Also known as anorthose; soda microcline.

anorthose *See* anorthoclase.

anorthosite [PETR] A visibly crystalline plutonic rock composed almost entirely of plagioclase feldspar (andesine to anorthite) with minor amounts of pyroxene and olivine.

anorthositization [GEOL] A process of anorthosite formation by replacement or metasomatism.

Antarctica [GEOGR] A continent roughly centered on the South Pole and surrounded by an ocean consisting of the southern parts of the Atlantic, Pacific, and Indian oceans.

antarctic air [METEOROL] A type of air whose characteristics are developed in an antarctic region.

antarctic anticyclone [METEOROL]   The glacial anticyclone which has been said to overlie the continent of Antarctica; analogous to the Greenland anticyclone.

Antarctic Circle [GEOD]   The parallel of latitude approximately 66°32′ south of the Equator.

Antarctic Circumpolar Current [OCEANOGR]   The ocean current flowing from west to east through all the oceans around the Antarctic Continent. Also known as West Wind Drift.

Antarctic Convergence [OCEANOGR]   The oceanic polar front indicating the boundary between the subantarctic and subtropical waters. Also known as Southern Polar Front.

antarctic front [METEOROL]   The semipermanent, semicontinuous front between the antarctic air of the Antarctic continent and the polar air of the southern oceans; generally comparable to the arctic front of the Northern Hemisphere.

Antarctic Intermediate Water [OCEANOGR]   A water mass in the Southern Hemisphere, formed at the surface near the Antarctic Convergence between 45° and 55°S; it can be traced in the North Atlantic to about 25°N.

Antarctic Ocean [GEOGR]   A circumpolar ocean belt including those portions of the Atlantic, Pacific, and Indian oceans which reach the Antarctic continent and are bounded on the north by the Subtropical Convergence; not recognized as a separate ocean.

Antarctic Zone [GEOGR]   The region between the Antarctic Circle (66°32′S) and the South Pole.

antecedent moisture [HYD]   The quantity of moisture in a given soil sample at the start of the runoff cycle.

antecedent platform [GEOL]   A submarine platform 50 meters or more below sea level from which barrier reefs and atolls are postulated to grow toward the water's surface.

antecedent precipitation index [METEOROL]   A weighted summation of daily precipitation amounts; used as an index of soil moisture.

antecedent stream [HYD]   A stream that has retained its early course in spite of geologic changes since its course was assumed.

antecedent valley [GEOL]   A stream valley that existed before uplift, faulting, or folding occurred and which has maintained itself during and after these events.

anteconsequent stream [HYD]   A stream consequent to the form assumed by the earth's surface as the result of early movement of the earth but antecedent to later movement.

antediluvial [GEOL]   Formerly referred to time or deposits antedating Noah's flood.

antetheca [PALEON]   The last or exposed septum at any stage of fusulinid growth.

Anthocyathea [PALEON]   A class of extinct marine organisms in the phylum Archaeocyatha characterized by skeletal tissue in the central cavity.

anthodite [GEOL]   Gypsum or aragonite growing in clumps of long needle- or hairlike crystals on the roof or wall of a cave.

anthoinite [MINERAL] $Al_2W_2O_9 \cdot 3H_2O$ A white mineral consisting of a hydrous basic aluminum tungstate.

anthophyllite [MINERAL]   A clove-brown orthorhombic mineral of the amphibole group. A variety of asbestos occurring as lamellae, radiations, fibers, or massive in metamorphic rocks. Also known as bidalotite.

anthracite   [MINERAL]   A high-grade metamorphic coal having a semimetallic luster, high content of fixed carbon, and high density, and burning with a short blue flame and little smoke or odor. Also known as hard coal; Kilkenny coal; stone coal.

anthracite fines   [MIN ENG]   Small pieces of material from an anthracite coal preparation plant, usually below one-eighth-inch (3 millimeter) diameter.

Anthracosauria   [PALEON]   An order of Carboniferous and Permian labyrinthodont amphibians that includes the ancestors of living reptiles.

Anthracotheriidae   [PALEON]   A family of middle Eocene and early Pleistocene artiodactyl mammals in the superfamily Anthracotherioidea.

Anthracotherioidea   [PALEON]   A superfamily of extinct artiodactyl mammals in the suborder Paleodonta.

anthracoxene   [GEOL]   A brownish resin that occurs in brown coal; in ether it dissolves into an insoluble portion, anthrocoxenite, and a soluble portion, schlanite.

anthraxolite   [GEOL]   Anthracite-like asphaltic material occurring in veins in Precambrian slate of Sudbury District, Ontario.

anthraxylon   [GEOL]   The vitreous-appearing components of coal derived from the woody tissues of plants.

anthraxylous coal   [GEOL]   Coal in which the anthraxylon-to-attritus ratio is greater than 3:1.

anthropozoic   [GEOL]   Pertaining to that portion of geologic time that has elapsed since the appearance of humans.

anticenter   [GEOL]   The point on the surface of the earth that is diametrically opposite the epicenter of an earthquake. Also known as antiepicenter.

anticlinal   [GEOL]   Folded as in an anticline.

anticlinal axis   [GEOL]   The median line of a folded structure from which the strata dip on either side.

anticlinal bend   [GEOL]   An upwardly convex flexure of rock strata in which one limb dips gently toward the apex of the strata and the other dips steeply away from it.

anticlinal mountain   [GEOL]   Ridges formed by a convex flexure of the strata.

anticlinal theory   [GEOL]   A theory relating trapped underground oil accumulation to anticlinal structures.

anticlinal valley   [GEOL]   A valley that follows an anticlinal axis.

anticline   [GEOL]   A fold in which layered strata are inclined down and away from the axes.

anticlinorium   [GEOL]   A series of anticlines and synclines that form a general arch or anticline.

anticyclogenesis   [METEOROL]   The process which creates an anticyclone or intensifies an existing one.

anticyclolysis   [METEOROL]   Any weakening of anticyclonic circulation in the atmosphere.

anticyclone   [METEOROL]   High-pressure atmospheric closed circulation whose relative direction of rotation is clockwise in the Northern Hemisphere, counterclockwise in the Southern Hemisphere, and undefined at the Equator. Also known as high-pressure area.

anticyclonic [METEOROL]   Referring to a rotation about the local vertical that is clockwise in the Northern Hemisphere, counterclockwise in the Southern Hemisphere, undefined at the Equator.

anticyclonic shear [METEOROL]   Horizontal wind shear of such a nature that it tends to produce anticyclonic rotation of the individual air particles along the line of flow.

anticyclonic winds [METEOROL]   The winds associated with a high pressure area and constituting part of an anticyclone.

antidip stream [HYD]   A stream whose direction of flow is opposite to that of the general dip of the strata.

antidune [GEOL]   A temporary form of ripple on a stream bed analogous to a sand dune but migrating upcurrent.

antiepicenter *See* anticenter.

antiform [GEOL]   An anticlinelike structure whose stratigraphic sequence is not known.

antigorite [MINERAL] $Mg_3Si_2O_5(OH)_4$ Brownish-green variety of the mineral serpentine. Also known as baltimorite; picrolite.

Antilles Current [OCEANOGR]   A current formed by part of the North Equatorial Current that flows along the northern side of the Greater Antilles.

antimonite [MINERAL] Sb_2S_3 A lead-gray antimony sulfide mineral, the primary source of antimony; sometimes contains gold or silver; has a brilliant metallic luster, and occurs as prismatic orthorhombic crystals in massive forms. Also known as antimony glance; gray antimony; stibium; stibnite.

antimony blende *See* kermesite.

antimony glance *See* antimonite.

antipathetic crystals [MINERAL]   Crystals that are so far apart from each other in a crystallization sequence that they are not usually found in association.

antiperthite [GEOL]   Natural intergrowth of feldspars formed by separation of sodium feldspar (albite) and potassium feldspar (orthoclase) during slow cooling of molten mixtures; the potassium-rich phase is exsolved in a plagioclase host, exactly the inverse of perthite.

antipodes [GEOD]   Diametrically opposite points on the earth.

antistress mineral [MINERAL]   Minerals such as leucite, nepheline, alkalic feldspar, andalusite, and cordierite which cannot form or are unstable in an environment of high shearing stress, and hence are not found in highly deformed rocks.

antitrades [METEOROL]   A deep layer of westerly winds in the troposphere above the surface trade winds of the tropics.

antitriptic wind [METEOROL]   A wind for which the pressure force exactly balances the viscous force, in which the vertical transfers of momentum predominate.

antitwilight arch [METEOROL]   The pink or purplish band of about 3° vertical angular width which lies just above the antisolar point at twilight; it rises with the antisolar point at sunset and sets with the antisolar point at sunrise.

antlerite [MINERAL] $Cu_3SO_4(OH)_4$ Emerald- to blackish-green mineral occurring in aggregates of needlelike crystals; an ore of copper. Also known as vernadskite.

Antler orogeny [GEOL]   Late Devonian and Early Mississippian orogeny in Nevada, resulting in the structural emplacement of eugeosynclinal rocks over microgeosynclinal rocks.

anvil *See* incus.

anvil cloud [METEOROL] The popular name given to a cumulonimbus capillatus cloud, a thunderhead whose upper portion spreads in the form of an anvil with a fibrous or smooth aspect; it also refers to such an upper portion alone when it persists beyond the parent cloud.

Ao horizon [GEOL] That portion of the A horizon of a soil profile which is composed of pure humus.

Aoo horizon [GEOL] Uppermost portion of the A horizon of a soil profile which consists of undecomposed vegetable litter.

apachite [PETR] A phonolite consisting of enigmatite and hornblende in about the same quantity as the pyroxene, but of a later crystallization phase.

Apatemyidae [PALEON] A family of extinct rodentlike insectivorous mammals belonging to the Proteutheria.

apatite [MINERAL] A group of phosphate minerals that includes 10 mineral species and has the general formula $X_5(YO_4)_3Z$, where X is usually Ca^{2+} or Pb^{3+}, Y is P^{5+} or As^{5+}, and Z is F^-, Cl^-, or OH^-.

apex [GEOL] The part of a mineral vein nearest the surface of the earth.

aphaniphyric [PETR] Denoting a texture of porphyritic rocks with microaphanitic groundmasses. Also known as felsophyric.

aphanite [PETR] **1.** A general term applied to dense, homogeneous rocks whose constituents are too small to be distinguished by the unaided eye. **2.** A rock having aphanitic texture.

aphanitic [PETR] Referring to the texture of an igneous rock in which the crystalline components are not distinguishable by the unaided eye.

aphotic zone [OCEANOGR] The deeper part of the ocean where sunlight is absent.

Aphrosalpingoidea [PALEON] A group of middle Paleozoic invertebrates classified with the calcareous sponges.

aphrosiderite *See* ripiddite.

aphthitalite [MINERAL] $(K,Na)_3Na(SO_4)_2$ A white mineral crystallizing in the rhombohedral system and occurring massively or in crystals.

aphylactic map projection [MAP] A map projection which is neither conformal nor equal-area.

aphyric [PETR] Of the texture of fine-grained igneous rocks, showing two generations of the same mineral but without phenocrysts.

API unit [ENG] An arbitrary unit of the American Petroleum Institute for measuring natural radioactivity; used in certain well logging methods.

apjohnite [MINERAL] $MnAl_2(SO_4)_4 \cdot 22H_2O$ A white, rose-green, or yellow mineral containing water and occurring in crusts, fibrous masses, or efflorescences.

aplite [PETR] Fine-grained granitic dike rock made up of light-colored mineral constituents, mostly quartz and feldspar; used to manufacture glass and enamel.

apob [METEOROL] An observation of pressure, temperature, and relative humidity taken aloft by means of an aerometeorograph; a type of aircraft sounding.

apogean tidal currents [OCEANOGR] Tidal currents of decreased speed that occur in conjunction with apogean tides.

apogean tides [OCEANOGR] Tides of decreased range that occur when the moon nears apogee.

apophorometer [ENG] An apparatus used to identify minerals by sublimation.

apophyllite [MINERAL]   A hydrous calcium potassium silicate containing fluorine and occurring as a secondary mineral with zeolites in geodes and other igneous rocks; the composition is variable but approximates $KFCa_4(Si_2O_5)_4 \cdot 8H_2O$.   Also known as fish-eye stone.

Appalachia [GEOL]   Proposed borderland along the southeastern side of North America, seaward of the Appalachian geosyncline in Paleozoic time.

Appalachian orogeny [GEOL]   An obsolete term referring to Late Paleozoic diastrophism beginning perhaps in the Late Devonian and continuing until the end of the Permian; now replaced by Alleghenian orogeny.

apparent dip [GEOL]   Dip of a rock layer as it is exposed in any section not at a right angle to the strike.

apparent gravity *See* acceleration of gravity.

apparent movement of faults [GEOL]   The apparent motion observed to have occurred in any chance section across a fault.

apparent plunge [GEOL]   Inclination of a normal projection of lineation in the plane of a vertical cross section.

apparent polar wander path [GEOL]   A path traced on a map by connecting paleomagnetic poles of different ages in an ordered time sequence. Abbreviated APWP.

apparent precession *See* apparent wander.

apparent shoreline [MAP]   The outer edge of marine vegetation (marsh, mangrove, cypress) delineated on photogrammetric surveys where the actual shoreline is obscured.

apparent vertical [GEOPHYS]   The direction of the resultant of gravitational and all other accelerations. Also known as dynamic vertical.

apparent wander [GEOPHYS]   Apparent change in the direction of the axis of rotation of a spinning body, such as a gyroscope, due to rotation of the earth. Also known as apparent precession; wander.

apparent water table *See* perched water table.

appearance ratio *See* hyperstereoscopy.

appinite [PETR]   Hornblende-rich plutonic rock with high feldspar content.

apple coal [GEOL]   Easily mined soft coal that breaks into small pieces the size of apples.

Appleton layer *See* F_2 layer.

applied climatology [CLIMATOL]   The scientific analysis of climatic data in the light of a useful application for an operational purpose.

applied meteorology [METEOROL]   The application of current weather data, analyses, or forecasts to specific practical problems.

apposition beach [GEOL]   One of a series of parallel beaches formed on the seaward side of an older beach.

apposition fabric [PETR]   A primary orientation of the elements of a sedimentary rock that is developed or formed at time of deposition of the material; fabrics of most sedimentary rocks belong to this type. Also known as primary fabric.

approved flame safety lamp [MIN ENG]   A flame safety lamp which has been approved for use in gaseous coal mines.

approximate contour [MAP]   A contour substituted for a normal contour whenever there is a question as to its reliability; reliability usually is evaluated as exceeding one-half the contour interval.

APR *See* airborne profile recorder.

apron [GEOL] *See* outwash plain. [MIN ENG] A canvas-covered frame set at such an angle in the miner's rocker that the gravel and water passing over it are carried to the head of the machine.

Aptian [GEOL] Lower Cretaceous geologic time, between Barremian and Albian. Also known as Vectian.

APWP *See* apparent polar wander path.

Aqualf [GEOL] A suborder of the soil order Alfisol, seasonally wet and marked by gray or mottled colors; occurs in depressions or on rather wide flats in local landscapes.

aquamarine [MINERAL] A pale-blue or greenish-blue transparent gem variety of the mineral beryl.

Aquent [GEOL] A suborder of the soil order Entisol, bluish gray or greenish gray in color; under water until very recent times at the margins of oceans, lakes, or seas.

aqueo-igneous [GEOL] Pertaining to a rock or mineral crystallized from magma containing an influential presence of water.

aqueous lava [GEOL] Mud lava produced by the mixing of volcanic ash with condensing volcanic vapor or other water.

aqueous rock [PETR] A sedimentary rock deposited by or in water. Also known as hydrogenic rock.

Aquept [GEOL] A suborder of the soil order Inceptisol, which is wet or has been drained, and lacks silicate clay accumulation in the soil profiles; surface horizon varies in thickness.

aquiclude [GEOL] A porous formation that absorbs water slowly but will not transmit it fast enough to furnish an appreciable supply for a well or spring.

aquic soil [GEOL] Soil with almost no oxygen in its groundwater and a soil temperature greater than 5°C at 50 centimeters' depth.

aquifer [GEOL] A permeable body of rock capable of yielding quantities of groundwater to wells and springs. [HYD] A subsurface zone that yields economically important amounts of water to wells.

aquifuge [GEOL] An impermeable body of rock which contains no interconnected openings or interstices and therefore neither absorbs nor transmits water.

Aquitanian [GEOL] Lower lower Miocene or uppermost Oligocene geologic time.

aquitard [GEOL] A bed of low permeability adjacent to an aquifer; may serve as a storage unit for groundwater, although it does not yield water readily.

Aquod [GEOL] A suborder of the soil order Spodosol, with a black or dark brown horizon just below the surface horizon; seasonally wet, occupying depressional areas or wide flats from which water cannot escape easily.

Aquoll [GEOL] A suborder of the soil order Mollisol with thick surface horizons, formed under wet conditions and may be under water at times, but seasonally rather than continually wet.

Aquox [GEOL] A suborder of the soil order Oxisol, seasonally wet, found chiefly in shallow depressions; deeper soil profiles are dominantly gray, sometimes mottled and containing nodules or sheets of iron and aluminum oxides.

Aquult [GEOL] A suborder of the soil order Ultisol, seasonally wet, saturated with water for a significant part of the year unless drained; surface horizon of the soil profile is dark and varies in thickness, grading to gray in the deeper portions; it occurs in depressions or on wide upland flats from which water moves very slowly.

Araeoscelidia [PALEON]   A provisional order of extinct reptiles in the subclass Euryapsida.

Aragon spar *See* aragonite.

aragonite [MINERAL] $CaCO_3$ A white, yellowish, or gray orthorhombic mineral species of calcium carbonate but with a crystal structure different from those of vaterite and calcite, the other two polymorphs of the same composition. Also known as Aragon spar.

aramayoite [MINERAL] $Ag(Sb,Bi)S_2$ An iron-black mineral consisting of silver antimony bismuth sulfide.

aranilitic [PETR]   Resembling sandstone.

arapahite [PETR]   A dark-colored, porous, fine-grained basic basalt consisting of magnetite, bytownite, and augite.

Arbuckle orogeny [GEOL]   Mid-Pennsylvanian episode of diastrophism in the Wichita and Arbuckle Mountains of Oklahoma.

arc [GEOL]   A geologic or topographic feature that is repeated along a curved line on the surface of the earth.

arcanite [MINERAL] K_2SO_4 A colorless, vitreous orthorhombic sulfate mineral. Also known as glaserite.

Archaeoceti [PALEON]   The zeuglodonts, a suborder of aquatic Eocene mammals in the order Cetacea; the oldest known cetaceans.

Archaeocidaridae [PALEON]   A family of Carboniferous echinoderms in the order Cidaroida characterized by a flexible test and more than two columns of interambulacral plates.

Archaeocopida [PALEON]   An order of Cambrian crustaceans in the subclass Ostracoda characterized by only slight calcification of the carapace.

Archaeopteridales [PALEOBOT]   An order of Upper Devonian sporebearing plants in the class Polypodiopsida characterized by woody trunks and simple leaves.

Archaeopteris [PALEOBOT]   A genus of fossil plants in the order Archaeopteridales; used sometimes as an index fossil of the Upper Devonian.

Archaeopterygiformes [PALEON]   The single order of the extinct avian subclass Archaeornithes.

Archaeopteryx [PALEON]   The earliest known bird; a genus of fossil birds in the order Archaeopterygiformes characterized by flight feathers like those of modern birds.

Archaeornithes [PALEON]   A subclass of Upper Jurassic birds comprising the oldest fossil birds.

Archanthropinae [PALEON]   A subfamily of the Hominidae, set up by F. Weidenreich, which is no longer used.

arch blocks [MIN ENG]   Blocks applied to the wooden voussoirs used in framing a timber support for the roof when driving a tunnel.

Archean [GEOL]   A term, meaning ancient, which has been applied to the oldest rocks of the Precambrian; as more physical measurements of geologic time are made, the usage is changing; the term Early Precambrian is preferred.

arched iceberg [HYD]   An iceberg that has been eroded so that a large horizontal opening extends through the ice at the waterline.

archeological chemistry [ARCHEO]   The application of chemical theories and experimental procedures, especially of analytical chemistry, to the solution of problems in archeology.

Archeozoic [GEOL] **1.** The era during which, or during the latter part of which, the oldest system of rocks was made. **2.** The last of three subdivisions of Archean time, when the lowest forms of life probably existed; as more physical measurements of geologic time are made, the usage is changing; it is now considered part of the Early Precambrian.

archibenthic zone [OCEANOGR] The biogeographic realm of the ocean extending from a depth of about 200 meters to 800–1100 meters (665 feet to 2625–3610 feet).

archibole *See* positive element.

arching [GEOL] The folding of schists, gneisses, or sediments into anticlines. [MIN ENG] Curved support for roofs of openings in mines.

archipelagic apron [GEOL] A fan-shaped slope around an oceanic island differing from a deep-sea fan in having little, if any, sediment cover.

archipelago [GEOGR] **1.** A large group of islands. **2.** A sea that has a large group of islands within it.

arc measurement [GEOD] A survey method used to determine the size of the earth.

arc of parallel [GEOD] A part of an astronomic or a geodetic parallel.

arctic air [METEOROL] An air mass whose characteristics are developed mostly in winter over arctic surfaces of ice and snow.

arctic anticyclone *See* arctic high.

Arctic Circle [GEOD] The parallel of latitude 66°32′N (often taken as 66½°N).

arctic climate *See* polar climate.

arctic desert *See* polar desert.

arctic front [METEOROL] The semipermanent, semicontinuous front between the deep, cold arctic air and the shallower, basically less cold polar air of northern latitudes.

arctic haze [METEOROL] A condition of reduced horizontal and slant visibility (but unimpeded vertical visibility) encountered by aircraft in flight (up to more than 30,000 feet, or 9140 meters) over arctic regions.

arctic high [METEOROL] A weak high that appears on mean charts of sea-level pressure over the Arctic Basin during late spring, summer, and early autumn. Also known as arctic anticyclone; polar anticyclone; polar high.

arctic mist [METEOROL] A mist of ice crystals; a very light ice fog.

Arctic Ocean [GEOGR] The north polar ocean lying between North America, Greenland, and Asia.

arctic sea smoke [METEOROL] Steam fog; but often specifically applied to steam fog rising from small areas of open water within sea ice.

Arctic suite [PETR] A group of basic igneous rocks intermediate in composition between Atlantic and Pacific suites.

Arctic Zone [GEOGR] The area north of the Arctic Circle (66°32′N).

Arctocyonidae [PALEON] A family of extinct carnivorelike mammals in the order Condylarthra.

Arctolepiformes [PALEON] A group of the extinct joint-necked fishes belonging to the Arthrodira.

arc triangulation [ENG] A system of triangulation in which an arc of a great circle on the surface of the earth is followed in order to tie in two distant points.

arcuate delta [GEOL] A bowed or curved delta with the convex margin facing the body of water. Also known as fan-shaped delta.

arcuate fault [GEOL] A fault that is arched or bowed on any given transecting surface.

arcuation [GEOL] Production of an arc, as in rock flowage where movement proceeded in a fanlike manner.

arcus [METEOROL] A dense and horizontal roll-shaped accessory cloud, with more or less tattered edges, situated on the lower front part of the main cloud.

arcwall coal cutter [MIN ENG] A type of electric or compressed-air cutter for under- or overcutting a coal seam in narrow work.

Arcyzonidae [PALEON] A family of Devonian paleocopan ostracods in the superfamily Kirkbyacea characterized by valves with a large central pit.

ardealite [MINERAL] $Ca_2(HPO_4)(SO_4)\cdot4H_2O$ A white or light-yellow mineral consisting of a hydrous acid calcium phosphate-sulfate.

Ardennian orogeny [GEOL] A short-lived orogeny during the Ludlovian stage of the Silurian period of geologic time.

ardennite [MINERAL] $Mn_5Al_5(VO_4)(SiO_4)_5(OH)_2\cdot2H_2O$ A yellow to yellowish-brown mineral consisting of a hydrous silicate vanadate and arsenate of manganese and aluminum.

arduinite See mordenite.

area delimiting line [MAP] A line fixing the boundary of an area.

area forecast [METEOROL] A weather forecast for a specified geographic area; usually applied to a form of aviation weather forecast. Also known as regional forecast.

areal eruption [GEOL] Volcanic eruption resulting from collapse of the roof of a batholith; the volcanic rocks grade into parent plutonic rocks.

areal geology [GEOL] Distribution and form of rocks or geologic units of any relatively large area of the earth's surface.

areal pattern [PETRO ENG] Distribution pattern of oil-production wells and water- or gas-injection wells over a given oil reservoir.

areal sweep efficiency [PETRO ENG] Percentage of the total oil reservoir or pore volume which is within the area being swept of oil by a displacing fluid, as in a natural or artificial gas drive or water injection.

area triangulation [ENG] A system of triangulation designed to progress in every direction from a control point.

areic See arheic.

arenaceous [GEOL] Of sediment or sedimentary rocks that have been derived from sand or that contain sand. Also known as arenarious; psammitic; sabulous.

arenarious See arenaceous.

arendalite [MINERAL] A dark-green variety of epidote found in Arendal, Norway.

arenicolite [GEOL] A hole, groove, or other mark in a sedimentary rock, generally sandstone, interpreted as a burrow made by an arenicolous marine worm or a trail of a mollusk or crustacean.

Arenigian [GEOL] A European stage including Lower Ordovician geologic time (above Tremadocian, below Llanvirnian). Also known as Skiddavian.

arenite [PETR] Consolidated sand-texture sedimentary rock of any composition. Also known as arenyte; psammite.

Arent [GEOL] A suborder of the soil order Entisol consisting of soils formerly of other classifications that have been severely disturbed, with the sequence of horizons disrupted completely.

arenyte *See* arenite.

arête [GEOL] Narrow, jagged ridge produced by the merging of glacial cirques. Also known as arris; crib; serrate ridge.

ARFOR [METEOROL] A code word used internationally to indicate an area forecast; usually applied to an aviation weather forecast.

ARFOT [METEOROL] A code word used internationally to indicate an area forecast with units in the English system; usually applied to an aviation weather forecast.

arfvedsonite [MINERAL] A black monoclinic amphibole, containing sodium and silicon trioxide with occluded water and some calcium. Also known as soda hornblende.

argentine [MINERAL] A pearlescent variety of calcite.

argentite [MINERAL] Ag_2S A lustrous, lead-gray ore of silver; it is a monoclinic mineral and is dimorphous with acanthite. Also known as argyrite; silver glance; vitreous silver.

argentojarosite [MINERAL] $AgFe_3(SO_4)_2(OH)_6$ A yellow or brownish mineral consisting of basic silver ferric sulfate.

Argid [GEOL] A suborder of the soil order Aridisol, well drained, having a characteristically brown or red color and a silicate accumulation below the surface horizon; occupies the older land surfaces in deserts.

argillaceous [GEOL] Of rocks or sediments made of or largely composed of clay-size particles or clay minerals.

argillation [GEOL] Development of clay minerals by weathering of aluminum silicates.

argillic alteration [GEOL] A rock alteration in which certain minerals are converted to minerals of the clay group.

argilliferous [GEOL] Abounding in or producing clay.

argillite [PETR] A compact rock formed from siltstone, shale, or claystone but intermediate in degree of induration and structure between them and slate; argillite is more indurated than mudstone but lacks the fissility of shale.

Argovian [GEOL] Upper Jurassic (lower Lusitanian), a substage of geologic time in Great Britain.

argyrite *See* argentite.

argyrodite [MINERAL] Ag_8GeS_6 A steel-gray mineral, one of two germanium minerals and a source for germanium; crystallizes in the isometric system and is isomorphous with canfieldite.

arheic [GEOL] Referring to a region that has no surface streams because of low rainfall or lack of surface drainage. Also spelled areic.

arid climate [CLIMATOL] Any extremely dry climate.

arid erosion [GEOL] Erosion or wearing away of rock that occurs in arid regions, due largely to the wind.

Aridisol [GEOL] A soil suborder and a major soil of deserts characterized by an ochric epipedon and other pedogenic horizons.

aridity　[CLIMATOL]　The degree to which a climate lacks effective, life-promoting moisture.

aridity coefficient　[CLIMATOL]　A function of precipitation and temperature designed by W. Gorczynski to represent the relative lack of effective moisture (the aridity) of a place.

aridity index　[CLIMATOL]　An index of the degree of water deficiency below water need at any given station; a measure of aridity.

arid zone　*See* equatorial dry zone.

ariegite　[PETR]　A group of pyroxenites composed principally of clinopyroxene, orthopyroxene, and spinel.

Arikareean　[GEOL]　Lower Miocene geologic time.

Arizona ruby　[MINERAL]　A ruby-red pyrope garnet of igneous origin found in the southwestern United States.

arizonite　[MINERAL]　$Fe_2Ti_3O_9$　A steel-gray mineral containing iron and titanium and found in irregular masses in pegmatite.　[PETR]　A dike rock composed of mostly quartz, some orthoclase, and accessory mica and apatite.

Arkansas stone　[PETR]　A variety of novaculite quarried in Arkansas.

arkite　[PETR]　A feldspathoid-rich rock consisting largely of pseudoleucite and nepheline, subordinate melanite and pyroxene, and accessory orthoclase, apatite, and sphene.

arkose　[PETR]　A sedimentary rock composed of sand-size fragments that contain a high proportion of feldspar in addition to quartz and other detrital minerals.

arkose quartzite　*See* arkosite.

arkosic　[PETR]　Having wholly or partly the character of arkose.

arkosic bentonite　[PETR]　Bentonite derived from volcanic ash which contains 25–75% sandy impurities and whose detrital crystalline grains remain essentially unaltered. Also known as sandy bentonite.

arkosic limestone　[PETR]　An impure clastic limestone composed of a relatively high proportion of grains or crystals of feldspar.

arkosic sandstone　[PETR]　A sandstone in which much feldspar is present, ranging from unsorted products of granular disintegration of granite to partly sorted river-laid or even marine deposits.

arkosic wacke　*See* feldspathic graywacke.

arkosite　[PETR]　A quartzite with a high proportion of feldspar. Also known as arkose quartzite.

arksutite　*See* chiolite.

arm　[GEOL]　A ridge or spur that extends from a mountain.　[OCEANOGR]　A long, narrow inlet of water that extends inland.

armalcolite　[MINERAL]　A type of pseudobrookite occurring in samples of lunar rock.

armangite　[MINERAL]　$Mn_3(AsO_3)_2$　A black mineral crystallizing in the rhombohedral system and consisting of manganese arsenite.

armenite　[MINERAL]　$BaCa_2Al_6Si_8O_{28}\cdot2H_2O$　Mineral composed of a hydrous calcium barium aluminosilicate.

ARMET　[METEOROL]　An international code word used to indicate an area forecast with units in the metric system.

armored mud ball   [GEOL]   A large (1–50 centimeters in diameter) subspherical mass of silt or clay coated with coarse sand and fine gravel. Also known as pudding ball.

armored relict   [GEOL]   A metastable relict of a reaction series that is prevented from further reaction by the presence of a rim of reaction products.

Armorican orogeny   [GEOL]   Little-used term, now replaced by Hercynian or Variscan orogeny.

arnimite   [MINERAL] $Cu_5(SO_4)_2(OH)_6 \cdot 3H_2O$ Mineral consisting of a hydrous copper sulfate.

arquerite   [MINERAL]   A mineral consisting of a soft, malleable, silver-rich variety of amalgam, containing about 87% silver and 13% mercury.

arrastra   *See* arrastre.

arrastre   [MIN ENG]   A mill comprising a circular, rock-lined pit in which broken ore is pulverized by stones, attached to horizontal poles fastened in a central pillar, which stones are dragged around the pit. Also spelled arrastra.

arrested decay   [GEOL]   A stage in coal formation where biochemical action ceases.

arrhenite   [MINERAL]   A variety of fergusonite.

arris   [GEOL]   *See* arête.

arrival time   [GEOPHYS]   In seismological measurements, the time at which a given wave phase is detected by a seismic recorder.

arrojadite   [MINERAL] $Na_2(Fe,Mn)_5(PO_4)_4$ Dark-green mineral crystallizing in the monoclinic system, being isostructural with dickinsonite and occurring in masses.

arroyo   [GEOL]   Small, deep gully produced by flash flooding in arid and semiarid regions of the southwestern United State.

arsenical antimony   *See* allemontite.

arsenical nickel   *See* niccolite.

arsenic bloom   *See* arsenolite.

arseniopleite   [MINERAL]   A reddish-brown mineral consisting of a basic arsenate of manganese, calcium, iron, lead, and magnesium and occurring in cleavable masses.

arseniosiderite   [MINERAL] $Ca_3Fe_4(AsO_4)_4(OH)_4 \cdot 4H_2O$ A yellowish-brown mineral consisting of a basic iron calcium arsenate and occurring as concretions.

arsenobismite   [MINERAL] $Bi_2(AsO_4)(OH)_3$ A yellowish-green mineral consisting of a basic bismuth arsenate and occurring in aggregates.

arsenoclasite   [MINERAL] $Mn_5(AsO_4)_2(OH)_4$ A red mineral consisting of a basic manganese arsenate. Also spelled arsenoklasite.

arsenoklasite   *See* arsenoclasite.

arsenolamprite   [MINERAL]   FeAsS   A lead gray mineral consisting of nearly pure arsenic; occurs in masses with a fibrous foliated structure.

arsenolite   [MINERAL] As_2O_3 A mineral crystallizing in the isometric system and usually occurring as a white bloom or crust. Also known as arsenic bloom.

arsenopyrite   [MINERAL]   FeAsS   A white to steel-gray mineral crystallizing in the monoclinic system with pseudo-orthorhombic symmetry because of twinning; occurs in crystalline rock and is the principal ore of arsenic. Also known as mispickel.

arsoite   [PETR]   An olivine-bearing diopside trachyte.

arterite [PETR] **1.** A migmatite produced as a result of regional contact metamorphism during which residual magmas were injected into the host rock. **2.** Gneisses characterized by veins formed from the solution given off by deep-seated intrusions of molten granite. **3.** A veined gneiss in which the vein material was injected from a magma.

arteritic migmatite [GEOL] Injection gneiss supposedly produced by introduction of pegmatite, granite, or aplite into schist parallel to the foliation.

artesian aquifer [HYD] An aquifer that is bounded above and below by impermeable beds and that contains artesian water. Also known as confined aquifer.

artesian basin [HYD] A geologic structural feature or combination of such features in which water is confined under artesian pressure.

artesian discharge [HYD] Water discharge caused by artesian pressure from a well, spring, or aquifer.

artesian leakage [HYD] The slow percolation of water from artesian formations into the confining materials of a less permeable, but not strictly impermeable, character.

artesian spring [HYD] A spring whose water issues under artesian pressure, generally through some fissure or other opening in the confining bed that overlies the aquifer.

artesian water [HYD] Groundwater that is under sufficient pressure to rise above the level at which it encounters a well, but which does not necessarily rise to or above the surface of the ground.

artesian well [HYD] A well in which the water rises above the top of the water-bearing bed.

Arthrodira [PALEON] The joint-necked fishes, an Upper Silurian and Devonian order of the Placodermi.

articulite *See* itacolumite.

artifact [ARCHEO] Any manufactured object of common use that reflects the skills of humans in past cultures.

artificial lift [PETRO ENG] Any method of lifting oil out of underground reservoirs, usually by injecting gas into the rock or sand formation to force fluids from wells; an example is a gas lift.

artificial radiation belt [GEOPHYS] High-energy electrons trapped in the earth's magnetic field as a result of high-altitude nuclear explosions.

artificial ventilation [MIN ENG] The inducing of a flow of air through a mine or part of a mine by mechanical or other means.

artinite [MINERAL] $Mg_2CO_3(OH)_2 \cdot 3H_2O$ A snow-white mineral crystallizing in the orthorhombic system and occurring in crystals or fibrous aggregates.

Artinskian [GEOL] A European stage of geologic time including Lower Permian (above Sakmarian, below Kungurian).

Arundel method [MAP] A combination of graphical and analytical methods, based on radial triangulation, for point-by-point topographic mapping from aerial photographs.

arzrunite [MINERAL] A bluish-green mineral consisting of a basic copper sulfate with copper chloride and lead, and occurring as incrustations.

asar *See* esker.

asbestos [MINERAL] A general name for the useful, fibrous varieties of a number of rock-forming silicate minerals that are heat-resistant and chemically inert; two varieties exist: amphibole asbestos, the best grade of which approaches the composition $Ca_2Mg_5(OH)_2Si_8O_{22}$ (tremolite), and serpentine asbestos, usually chrysotile, $Mg_3Si_2(OH)_4O_5$.

asbolane *See* asbolite.

asbolite [MINERAL] A black, earthy mineral aggregate containing hydrated oxides of manganese and cobalt. Also known as asbolane; black cobalt; earthy cobalt.

ascending vertical angle *See* angle of elevation.

aschistic [GEOL] Pertaining to rocks of minor igneous intrusions that have not been differentiated into light and dark portions but that have essentially the same composition as the larger intrusions with which they are associated.

aseismic [GEOPHYS] Not subject to the occurrence or destructive effects of earthquakes.

aseismic ridge [GEOL] A portion of the continental crust consisting of a submarine ridge that is not part of the seismically active midocean ridge.

ashbed diabase [PETR] An igneous rock that has incorporated sandy material and bears a resemblance to conglomerate.

Ashby [GEOL] A North American stage of Middle Ordovician geologic time forming the upper subdivision of Chazyan, and lying above Marmor and below Porterfield.

ash cone [GEOL] A volcanic cone built primarily of unconsolidated ash and generally shaped somewhat like a saucer, with a rim in the form of a wide circle and a broad central depression often nearly at the same elevation as the surrounding country.

ash fall [GEOL] **1.** A fall of airborne volcanic ash from an eruption cloud; characteristic of Vulcanian eruptions. Also known as ash shower. **2.** Volcanic ash resulting from an ash fall and lying on the ground surface.

ash field [GEOL] A thick, extensive deposit of volcanic ash. Also known as ash plain.

ash flow [GEOL] **1.** An avalanche of volcanic ash, generally a highly heated mixture of volcanic gases and ash, traveling down the flanks of a volcano or along the surface of the ground. Also known as glowing avalanche; incandescent tuff flow. **2.** A deposit of volcanic ash and other debris resulting from such a flow and lying on the surface of the ground.

ash fusibility [GEOL] The gradual softening and melting of coal ash that takes place with increase in temperature as a result of the melting of the constituents and chemical reactions.

Ashgillian [GEOL] A European stage of geologic time in the Upper Orodovician (above Upper Caradocian, below Llandoverian of Silurian).

ash rock [GEOL] The material of arenaceous texture produced by volcanic explosions.

ash shower *See* ash fall.

ashstone [PETR] A rock composed of fine volcanic ash; particles are less than 0.06 millimeter in diameter.

ashtonite *See* mordenite.

ash viscosity [GEOL] The ratio of shearing stress to velocity gradient of molten ash; indicates the suitability of a coal ash for use in a slag-tap-type boiler furnace.

ashy grit [GEOL] **1.** Pyroclastic material of sand and smaller size. **2.** Mixture of ordinary sand and volcanic ash.

Asia [GEOGR] The largest continent, comprising the major portion of the broad east-west extent of the Northern Hemisphere land masses.

aso lava [GEOL] A type of indurated pyroclastic deposit produced during the explosive eruptions that formed the Aso Caldera of Kyushu, Japan.

asparagolite *See* asparagus stone.

asparagus stone [MINERAL]   A yellow-green variety of apatite occurring in crystals. Also known as asparagolite.

asphaltic sand [GEOL]   Deposits of sand grains cemented together with soft, natural asphalt.

asphaltite [GEOL]   Any of the dark-colored, solid, naturally occurring bitumens that are insoluble in water, but more or less completely soluble in carbon disulfide, benzol, and so on, with melting points between 250 and 600°F (121–316°C); examples are gilsonite and grahamite.

asphaltite coal *See* albertite.

asphalt rock [GEOL]   Natural rock asphalt or asphalt-containing rock, such as porous sandstones and dolomites. Also known as asphalt stone; bituminous rock; rock asphalt.

asphalt stone *See* asphalt rock.

Aspidorhynchidae [PALEON]   The single family of the Aspidorhynchiformes, an extinct order of holostean fishes.

Aspidorhynchiformes [PALEON]   A small, extinct order of specialized holostean fishes.

Aspinothoracida [PALEON]   The equivalent name for Brachythoraci.

aspirating screen [MIN ENG]   A vibrating screen from which light, liberated particles are removed by suction.

aspiration meteorograph [ENG]   An instrument for the continuous recording of two or more meteorological parameters, with the ventilation being provided by a suction fan.

aspiration psychrometer [ENG]   A psychrometer in which the ventilation is provided by a suction fan.

aspirator [MIN ENG]   A device made of wire gauze, of cloth, or of a fibrous mass held between pieces of meshed material and used to cover the mouth and nose to keep dusts from entering the lungs.

aspite [GEOL]   A cratered volcano with base wide in relation to height; for example, Mauna Loa.

assay plan [MIN ENG]   A mine map showing the assay, stope, width, and so forth of samples taken from positions marked.

assay pound [MIN ENG]   A weight which varies from time to time but is sometimes 0.5 gram, and is used by assayers to proportionately represent a pound.

assay ton [MIN ENG]   A unit of weight of ore equal to 29,167 milligrams; the number of milligrams of precious metal in this measure equals the number of troy ounces in a short ton.

assay value [MIN ENG]   The amount of gold or silver as shown by assay of any given sample and represented by ounces per ton of ore.

assay walls [MIN ENG]   The planes to which an ore body can be profitably mined, the limiting factor being the metal content of the country rock as determined from assays.

assemblage [ARCHEO]   All related cultural traits and artifacts associated with one archeological manifestation.

assemblage zone [PALEON]   A biostratigraphic unit defined and identified by a group of associated fossils rather than by a single index fossil.

assessment drilling [MIN ENG] Drilling to fulfill the requirement that a prescribed amount of work be done annually on an unpatented mining claim to retain title. Also known as annual labor.

assessment work [MIN ENG] Annual work at an unpatented mining claim in the public domain performed under law to maintain the claim title.

assimilation [GEOL] Incorporation of solid or fluid material that was originally in the rock wall into a magma.

Assmann psychrometer [ENG] A special form of the aspiration psychrometer in which the thermometric elements are well shielded from radiation.

associated corpuscular emission [GEOPHYS] The full complement of secondary charged particles associated with the passage of an x-ray or gamma-ray beam through air.

associated gas [PETRO ENG] Gaseous hydrocarbons occurring as a free-gas phase under original oil-reservoir conditions of temperature and pressure.

assumed plane coordinates [ENG] A local plane-coordinate system set up at the convenience of the surveyor.

assured mineral *See* developed reserves.

assyntite [PETR] A plutonic rock consisting largely of orthoclase and pyroxene, lesser amounts of sodalite and nepheline, and accessory biotite, sphene, apatite, and opaque oxides.

astatic galvanometer [ENG] A sensitive galvanometer designed to be independent of the earth's magnetic field.

astatized gravimeter [ENG] A gravimeter, sometimes referred to as unstable, where the force of gravity is maintained in an unstable equilibrium with the restoring force.

astel [MIN ENG] An overhead boarding or arching in a mine gallery.

asthenolith [GEOL] A body of magma locally melted anywhere at any time within any solid portion of the earth.

asthenosphere [GEOL] That portion of the upper mantle beneath the rigid lithosphere which is plastic enough for rock flowage to occur; extends from a depth of 50–100 kilometers to about 400 kilometers and is seismically equivalent to the low velocity zone.

Astian [GEOL] A European stage of geologic time: upper Pliocene, above Plaisancian, below the Pleistocene stage known as Villafranchian, Calabrian, or Günz.

astrakanite *See* bloedite.

Astrapotheria [PALEON] A relatively small order of large, extinct South American mammals in the infraclass Eutheria.

Astrapotheroidea [PALEON] A suborder of extinct mammals in the order Astrapotheria, ranging from early Eocene to late Miocene.

astrobleme [GEOL] A circular-shaped depression on the earth's surface produced by the impact of a cosmic body.

astrochanite *See* bloedite.

astrogeodetic [GEOD] Pertaining to direct measurements of the earth.

astrogeodetic datum orientation [GEOD] Adjustment of the ellipsoid of reference for a particular datum so that the sum of the squares of deflections of the vertical at selected points throughout the geodetic network is made as small as possible.

astrogeodetic deflection [GEOD] The angle at a point between the normal to the geoid and the normal to the ellipsoid of an astrogeodetically oriented datum. Also known as relative deflection.

astrogeodetic leveling [GEOD] A concept whereby the astrogeodetic deflections of the vertical are used to determine the separation of the ellipsoid and the geoid in studying the figure of the earth. Also known as astronomical leveling.

astrogeodetic undulations [GEOD] Separations between the geoid and astrogeodetic ellipsoid.

astrogravimetric leveling [GEOD] A concept whereby a gravimetric map is used for the interpolation of the astrogeodetic deflections of the vertical to determine the separation of the ellipsoid and the geoid in studying the figure of the earth.

astrogravimetric points [GEOD] Astronomical positions corrected for the deflection of the vertical by gravimetric methods.

astronomical azimuth [GEOD] The angle between the astronomical meridian plane of the observer and the plane containing the observed point and the true normal (vertical) of the observer, measured in the plane of the horizon, preferably clockwise from north.

astronomical equator [GEOD] An imaginary line on the surface of the earth connecting points having 0° astronomical latitude. Also known as terrestrial equator.

astronomical latitude [GEOD] Angular distance between the direction of gravity (plumb line) and the plane of the celestial equator; applies only to positions on the earth and is reckoned from the astronomical equator.

astronomical leveling *See* astrogeodetic leveling.

astronomical longitude [GEOD] The angle between the plane of the reference meridian and the plane of the local celestial meridian.

astronomical meridian [GEOD] A line on the surface of the earth connecting points having the same astronomical longitude. Also known as terrestrial meridian.

astronomical meridian plane [GEOD] A plane that contains the vertical of the observer and is parallel to the instantaneous rotation axis of the earth.

astronomical parallel [GEOD] A line connecting points having the same astronomical latitude.

astronomical position [GEOD] **1.** A point on the earth whose coordinates have been determined as a result of observation of celestial bodies. Also known as astronomical station. **2.** A point on the earth defined in terms of astronomical latitude and longitude.

astronomical refraction [GEOPHYS] The bending of a ray of celestial radiation as it passes through atmospheric layers of increasing density.

astronomical station *See* astronomical position.

astronomical surveying [GEOD] The celestial determination of latitude and longitude; separations are calculated by computing distances corresponding to measured angular displacements along the reference spheroid.

astronomical theodolite *See* altazimuth.

astronomical tide [OCEANOGR] An equilibrium tide due to attractions of the sun and moon. Also known as astronomic tide.

astronomical traverse [ENG] A survey traverse in which the geographic positions of the stations are obtained from astronomical observations, and lengths and azimuths of lines are obtained by computation.

astronomic tide *See* astronomical tide.

astrophyllite [MINERAL] $(K,Na)_3(Fe,Mn)_7Ti_2Si_8O_{24}(O,OH)_7$ A mineral composed of a basic silicate of potassium or sodium, iron or manganese, and titanium.

astrotectonic [GEOL]   Referring to terrestrial structures produced by cosmic bodies, for example, a deformation caused by a meteorite impact.

Asturian orogeny [GEOL]   Mid-Upper Carboniferous diastrophism.

asymmetrical bedding [GEOL]   An order in which lithologic types or facies follow one another in a circuitous arrangement so that, for example, the sequence of types 1–2–3–1–2–3–1–2–3 indicates asymmetry (while the sequence 1–2–3–2–1–2–3–2–1 indicates symmetrical bedding).

asymmetrical fold [GEOL]   A fold in which one limb dips more steeply than the other.

asymmetrical laccolith [GEOL]   A laccolith in which the beds dip at conspicuously different angles in different sectors.

asymmetrical ripple mark [GEOL]   The normal form of ripple mark, with short downstream slopes and comparatively long, gentle upstream slopes.

asymmetrical vein [GEOL]   A crustified vein of geologic material with unlike layers on each side.

asymptote of convergence *See* convergence line.

asymptotic cone of acceptance [GEOPHYS]   The solid angle in the celestial sphere from which particles have to come in order to contribute significantly to the counting rate of a given neutron monitor on the surface of the earth.

asymptotic direction of arrival [GEOPHYS]   The direction at infinity of a positively charged particle, with given rigidity, which impinges in a given direction at a given point on the surface of the earth, after passing through the geomagnetic field.

atacamite [MINERAL] $Cu_2Cl(OH)_3$ A native, green hydrous copper oxychloride crystallizing in the orthorhombic system.

atatschite [MINERAL] $Cu_2Cl(OH)_3$ A green mineral with orthorhombic crystal structure, formed by weathering of copper lodes under arid conditions.

ataxic [GEOL]   Pertaining to unstratified ore deposits.

ataxite [GEOL]   An iron meteorite that lacks the structure of either hexahedrite or octahedrite and contains more than 10% nickel.   [PETR]   A taxitic rock whose components are arranged in a breccialike manner, that is, there is no specific arrangement.

atectonic [GEOL]   Of an event that occurs when orogeny is not taking place.

atectonic pluton [GEOL]   A pluton that is emplaced when orogeny is not occurring.

atelestite [MINERAL] $Bi_8(AsO_4)_3O_5(OH)_5$ A yellow mineral consisting of basic bismuth arsenate and occurring in minute crystals; specific gravity is 6.82.

athrogenic [PETR]   Of or pertaining to pyroclastics.

Athyrididina [PALEON]   A suborder of fossil articulate brachiopods in the order Spiriferida characterized by laterally or, more rarely, ventrally directed spires.

Atlantic Ocean [GEOGR]   The large body of water separating the continents of North and South America from Europe and Africa and extending from the Arctic Ocean to the continent of Antarctica.

Atlantic series [PETR]   A great group of igneous rocks, based on tectonic setting, found in nonorogenic areas, often associated with block sinking and great crustal instability, and erupted along faults and fissures or through explosion vents. Also known as Atlantic suite.

Atlantic suite *See* Atlantic series.

Atlantic-type continental margin [GEOL] A continental margin typified by that of the Atlantic which is aseismic because oceanic and continental lithospheres are coupled.

atlantite [PETR] An olivine-bearing nepheline tephrite.

atlas [MAP] A collection of charts or maps kept loose or bound in a volume.

atlas grid [MAP] A reference system that permits the designation of the location of a point or an area on a map, photo, or other graphic in terms of numbers and letters. Also known as alphanumeric grid.

Atlas-Johnson tubing joint [PETRO ENG] A tapered, screw-on joint for connecting lengths of tubing for oil-well casing strings.

atmidometer *See* atmometer.

atmoclast [GEOL] A fragment of rock broken off in place by atmospheric weathering.

atmoclastic [PETR] Of a clastic rock, composed of atmoclasts that have been recemented without rearrangement.

atmogenic [GEOL] Of rocks, minerals, and other deposits derived directly from the atmosphere by condensation, wind action, or deposition from volcanic vapors; for example, snow.

atmolith [GEOL] A rock precipitated from the atmosphere, that is, an atmogenic rock.

atmometer [ENG] The general name for an instrument which measures the evaporation rate of water into the atmosphere. Also known as atmidometer; evaporation gage; evaporimeter.

atmophile element [METEOROL] **1.** Any of the most typical elements of the atmosphere (hydrogen, carbon, nitrogen, oxygen, iodine, mercury, and inert gases). **2.** Any of the elements which either occur in the uncombined state or, as volatile compounds, concentrate in the gaseous primordial atmosphere.

atmosphere [METEOROL] The gaseous envelope surrounding a planet or celestial body.

atmospheric absorption [GEOPHYS] The reduction in energy of microwaves by gases and moisture in the atmosphere.

atmospheric attenuation [GEOPHYS] A process in which the flux density of a parallel beam of energy decreases with increasing distance from the source as a result of absorption or scattering by the atmosphere.

atmospheric boundary layer *See* surface boundary layer.

atmospheric chemistry [METEOROL] The study of the production, transport, modification, and removal of atmospheric constituents in the troposphere and stratosphere.

atmospheric composition [METEOROL] The chemical abundance in the earth's atmosphere of its constituents, including nitrogen, oxygen, argon, carbon dioxide, water vapor, ozone, neon, helium, krypton, methane, hydrogen, and nitrous oxide.

atmospheric condensation [METEOROL] The transformation of water in the air from a vapor phase to dew, fog, or cloud.

atmospheric density [METEOROL] The ratio of the mass of a portion of the atmosphere to the volume it occupies.

atmospheric diffusion [METEOROL] The exchange of fluid parcels between regions in the atmosphere in the apparently random motions of a scale too small to be treated by equations of motion.

atmospheric disturbance [METEOROL]   Any agitation or disruption of the atmospheric steady state.

atmospheric duct [GEOPHYS]   A stratum of the troposphere within which the refractive index varies so as to confine within the limits of the stratum the propagation of an abnormally large proportion of any radiation of sufficiently high frequency, as in a mirage.

atmospheric electric field [GEOPHYS]   The atmosphere's electric field strength in volts per meter at any specified point in time and space; near the earth's surface, in fair-weather areas, a typical datum is about 100 and the field is directed vertically in such a way as to drive positive charges downward.

atmospheric electricity [GEOPHYS]   The electrical processes occurring in the lower atmosphere, including both the intense local electrification accompanying storms and the much weaker fair-weather electrical activity over the entire globe produced by the electrified storms continuously in progress.

atmospheric evaporation [HYD]   The exchange of water between the earth's oceans, lakes, rivers, ice, snow, and soil and the atmosphere.

atmospheric gas [METEOROL]   One of the constituents of air, which is a gaseous mixture primarily of nitrogen, oxygen, argon, carbon dioxide, water vapor, ozone, neon, helium, krypton, methane, hydrogen, and nitrous oxide.

atmospheric general circulation [METEOROL]   The statistical mean global flow pattern of the atmosphere.

atmospheric impurity [ENG]   An extraneous substance mixed as a contaminant with the air of the atmosphere.

atmospheric interference [GEOPHYS]   Electromagnetic radiation, caused by natural electrical disturbances in the atmosphere, which interferes with radio systems. Also known as atmospherics; sferics; strays.

atmospheric ionization [GEOPHYS]   The process by which neutral atmospheric molecules or atoms are rendered electrically charged chiefly by collisions with high-energy particles.

atmospheric layer *See* atmospheric shell.

atmospheric optics *See* meteorological optics.

atmospheric physics [GEOPHYS]   The study of the physical phenomena of the atmosphere.

atmospheric radiation [GEOPHYS]   Infrared radiation emitted by or being propagated through the atmosphere.

atmospheric refraction [GEOPHYS]   **1.** The angular difference between the apparent zenith distance of a celestial body and its true zenith distance, produced by refraction effects as the light from the body penetrates the atmosphere. **2.** Any refraction caused by the atmosphere's normal decrease in density with height.

atmospheric region *See* atmospheric shell.

atmospherics *See* atmospheric interference.

atmospheric scattering [GEOPHYS]   A change in the direction of propagation, frequency, or polarization of electromagnetic radiation caused by interaction with the atoms of the atmosphere.

atmospheric shell [METEOROL]   Any one of a number of strata or layers of the earth's atmosphere; temperature distribution is the most common criterion used for denoting the various shells. Also known as atmospheric layer; atmospheric region.

atmospheric sounding [METEOROL] A measurement of atmospheric conditions aloft, above the effective range of surface weather observations.

atmospheric structure [METEOROL] Atmospheric characteristics, including wind direction and velocity, altitude, air density, and velocity of sound.

atmospheric suspensoids [METEOROL] Moderately finely divided particles suspended in the atmosphere; dust is an example.

atmospheric tide [GEOPHYS] Periodic global motions of the earth's atmosphere, produced by gravitational action of the sun and moon; amplitudes are minute except in the upper atmosphere.

atmospheric turbulence [METEOROL] Apparently random fluctuations of the atmosphere that often constitute major deformations of its state of fluid flow.

Atokan [GEOL] A North American provincial series in lower Middle Pennsylvanian geologic time, above Morrowan, below Desmoinesian.

atoll [GEOGR] A ring-shaped coral reef that surrounds a lagoon without projecting land area and that is surrounded by open sea.

atoll moor [GEOL] A ring of peat bog completely surrounding a lake and itself surrounded completely by a ring of open water following the original lake shoreline.

atoll texture [GEOL] The surrounding of a ring of one mineral with another mineral, or minerals, within and without the ring. Also known as core texture.

atomic moisture meter [ENG] An instrument that measures the moisture content of coal instantaneously and continuously by bombarding it with neutrons and measuring the neutrons which bounce back to a detector tube after striking hydrogen atoms of water.

atopite [MINERAL] A yellow or brown variety of romeite that contains fluorine.

Atrypidina [PALEON] A suborder of fossil articulate brachiopods in the order Spiriferida.

attached dune [GEOL] A dune that has formed around a rock or other geological feature in the path of windblown sand.

attached groundwater [HYD] The portion of subsurface water adhering to pore walls in the soil.

attapulgite [MINERAL] $(Mg,Al)_2Si_4O_{10}(OH)\cdot4H_2O$ A clay mineral with a needlelike shape from Georgia and Florida; active ingredient in most fuller's earth, and used as a suspending agent, as an oil well drilling fluid, as a thickener in latex paint.

Atterberg scale [GEOL] A geometric and decimal grade scale for classification of particles in sediments based on the unit value of 2mm and involving a fixed ratio of 10 for each successive grade; subdivisions are geometric means of the limits of each grade.

Attican orogeny [GEOL] Late Miocene diastrophism.

attrital coal [GEOL] A bright coal composed of anthraxylon and of attritus in which the translucent cell-wall degradation matter or translucent humic matter predominates, with the ratio of anthraxylon to attritus being less than 1:3.

attrition [GEOL] The act of wearing and smoothing of rock surfaces by the flow of water charged with sand and gravel, by the passage of sand drifts, or by the movement of glaciers.

attritus [GEOL] **1.** Visible-to-ultramicroscopic particles of vegetable matter produced by microscopic and other organisms in vegetable deposits, particularly in swamps and bogs. **2.** The dull gray to nearly black, frequently striped portion of material that makes up the bulk of some coals and alternate bands of bright anthraxylon in well-banded coals.

aubrite [GEOL]   An enstatite achondrite (meteorite) consisting almost wholly of crystalline-granular enstatite (and clinoenstatite) poor in lime and practically free from ferrous oxide, with accessory oligoclase. Also known as bustite.

audio-modulated radiosonde [ENG]   A radiosonde with a carrier wave modulated by audio-frequency signals whose frequency is controlled by the sensing elements of the instrument.

auganite [PETR]   An olivine-free basalt (calcic plagioclase and augite are the essential mineral components) or an augite-bearing andesite.

augelite [MINERAL]   Natural, basic aluminum phosphate.

augen [PETR]   Large, lenticular eye-shaped mineral grain or mineral aggregate visible in some metamorphic rocks.

augen kohle *See* eye coal.

augen schist [PETR]   A mylonitic rock characterized by the presence of recrystallized minerals in schistose streaks and lenticles.

augen structure [PETR]   A structure found in some gneisses and granites in which certain of the constituents are squeezed into elliptic or lens-shaped forms and, especially if surrounded by parallel flakes of mica, resemble eyes.

auger boring [ENG]   **1.** The hole drilled by the use of auger equipment. **2.** *See* auger drilling.

auger drilling [ENG]   A method of drilling in which penetration is accomplished by the cutting or gouging action of chisel-type cutting edges forced into the substance by rotation of the auger bit. Also known as auger boring.

auget [ENG]   A priming tube, used in blasting. Also spelled augette.

augite [MINERAL] $(Ca,Mg,Fe)(Mg,Fe,Al)(Al,Si)_2O_6$ A general name for the monoclinic pyroxenes; occurs as dark green to black, short, stubby, prismatic crystals, often of octagonal outline.

augitite [PETR]   A volcanic rock consisting of abundant phenocrysts of augite in a glassy groundmass containing microlites of nepheline and plagioclase, with accessory biotite, apatite, and opaque oxides.

augitophyre [PETR]   A porphyritic rock in which the phenocrysts are augite and the groundmass is potash feldspar.

aulocogen [GEOL]   A major fault-bounded trough considered to be one part of a three-rayed fault system found on the domes above mantle hot spots; the other two rays open as proto-ocean basins.

Aulolepidae [PALEON]   A family of marine fossil teleostean fishes in the order Ctenothrissiformes.

Auloporidae [PALEON]   A family of Paleozoic corals in the order Tabulata.

aureole [GEOL]   A ring-shaped contact zone surrounding an igneous intrusion. Also known as contact aureole; contact zone; exomorphic zone; metamorphic aureole; metamorphic zone; thermal aureole.   [METEOROL]   A poorly developed corona in the atmosphere characterized by a bluish-white disk immediately around the luminous celestial body, as around the sun or moon in the fog.

aurichalcite [MINERAL] $(Zn,Cu)_5(CO_3)_2(OH)_6$ Pale-green or pale-blue mineral consisting of a basic copper zinc carbonate and occurring in crystalline incrustations. Also known as brass ore.

auriferous [GEOL] Of a substance, especially a mineral deposit, bearing gold.

aurora [GEOPHYS] The most intense of the several lights emitted by the earth's upper atmosphere, seen most often along the outer realms of the Arctic and Antarctic, where it is called the aurora borealis and aurora australis, respectively; excited by charged particles from space.

aurora australis [GEOPHYS] The aurora of southern latitudes. Also known as southern lights.

aurora borealis [GEOPHYS] The aurora of northern latitudes. Also known as northern lights.

auroral absorption event [GEOPHYS] A large increase in D-region electron density and associated radio-signal absorption, caused by electron-bombardment of the atmosphere during an aurora or a geomagnetic storm.

auroral caps [GEOPHYS] The regions surrounding the auroral poles, lying between the poles and the auroral zones.

auroral electrojet [GEOPHYS] An intense electric current in the magnetosphere, flowing along the auroral zones during a polar substorm.

auroral forms [GEOPHYS] Auroral display types, of which two are basic: ribbonlike bands and cloudlike surfaces.

auroral frequency [GEOPHYS] The percentage of nights on which an aurora is seen at a particular place, or on which one would be seen if clouds did not interfere.

auroral isochasm [GEOPHYS] A line connecting places of equal auroral frequency, averaged over a number of years.

auroral oval [GEOPHYS] An oval-shaped region centered on the earth's magnetic pole in which auroral emissions occur.

auroral poles [GEOPHYS] The points on the earth's surface on which the auroral isochasms are centered; coincide approximately with the magnetic-axis poles of the geomagnetic field.

auroral region [GEOPHYS] The region within 30° geomagnetic latitude of each auroral pole.

auroral storm [GEOPHYS] A rapid succession of auroral substorms, occurring in a short period, of the order of a day, during a geomagnetic storm.

auroral substorm [GEOPHYS] A characteristic sequence of auroral intensifications and movements occurring around midnight, in which a rapid poleward movement of auroral arcs produces a bulge in the auroral oval.

auroral zone [GEOPHYS] A roughly circular band around either geomagnetic pole within which there is a maximum of auroral activity; lies about 10–15° geomagnetic latitude from the geomagnetic poles.

auroral zone blackout [GEOPHYS] Communication fadeout in the auroral zone most often due to an increase of ionization in the lower atmosphere.

aurora polaris [GEOPHYS] A high-altitude aurora borealis or aurora australis.

aurosmiridium [MINERAL] A brittle, silver-white, isometric mineral consisting of a solid solution of gold and osmium in iridium.

auster *See* ostria.

austinite [MINERAL] $CaZnAsO_4(OH)$ A colorless or yellowish mineral crystallizing in the orthorhombic system; consists of a basic calcium zinc arsenate; hardness is 4.5 on Mohs scale, and specific gravity is 4.13.

austral [GEOD] Of or pertaining to south.

austral axis pole [GEOPHYS] The southern intersection of the geomagnetic axis with the earth's surface.

Australia [GEOGR] An island continent of 2,941,526 square miles (7,618,517 square kilometers), with low elevation and moderate relief, situated in the southern Pacific.

australite [GEOL] A tektite found in southern Australia, occurring as glass balls and spheroidal dumbbell forms of green and black, similar to obsidian and probably of cosmic origin.

Australopithecinae [PALEON] The near-men, a subfamily of the family Hominidae composed of the single genus *Australopithecus*.

Australopithecus [PALEON] A genus of near-men in the subfamily Australopithecinae representing a side branch of human evolution.

Austrian orogeny [GEOL] A short-lived orogeny during the end of the Early Cretaceous.

autallotriomorphic [PETR] Pertaining to an aplitic texture in which all mineral constituents crystallized simultaneously, preventing the development of euhedral crystals.

authalic latitude [MAP] A latitude based on a sphere having the same area as the spheroid, and such that areas between successive parallels of latitude are exactly equal to the corresponding areas on the spheroid. Also known as equal-area latitude.

authalic map projection *See* equal-area map projection.

authigene [MINERAL] A mineral which has not been transported but has been formed in place. Also known as authigenic mineral.

authigenic [GEOL] Of constituents that came into existence with or after the formation of the rock of which they constitute a part; for example, the primary and secondary minerals of igneous rocks.

authigenic mineral *See* authigene.

authigenic sediment [GEOL] Sediment occurring in the place where it was originally formed.

autobarotropy [METEOROL] The state of a fluid which is characterized by both barotropy and piezotropy when the coefficients of barotropy and piezotropy are equal.

autobrecciation [GEOL] The process whereby portions of the first consolidated crust of a lava flow are incorporated into the still-fluid portion.

autochthon [GEOL] A succession of rock beds that have been moved comparatively little from their original site of formation, although they may be folded and faulted extensively. [PALEON] A fossil occurring where the organism once lived.

autochthonous [GEOL] Formed or occurring in the place where found.

autochthonous coal [GEOL] Coal believed to have originated from accumulations of plant debris at the place where the plants grew. Also known as indigenous coal.

autochthonous stream [HYD] A stream flowing in its original channel.

autoclastic [GEOL] Of rock, fragmented in place by folding due to orogenic forces when the rock is not so heavily loaded as to render it plastic.

autoclastic schist [GEOL] Schist formed in place from massive rocks by crushing and squeezing.

autoconsequent falls [HYD] Waterfalls in streams carrying a heavy load of calcium carbonate in solution which develop at particular sites along the stream course where warming, evaporation, and other factors cause part of the solution load to be precipitated.

autoconsequent stream [HYD]   A stream in the process of building a fan or an alluvial plain, the course of which is guided by the slopes of the alluvium the stream itself has deposited.

autoconvection [METEOROL]   The phenomenon of the spontaneous initiation of convection in an atmospheric layer in which the lapse rate is equal to or greater than the autoconvective lapse rate.

autoconvection gradient *See* autoconvective lapse rate.

autoconvective instability *See* absolute instability.

autoconvective lapse rate [METEOROL]   The largely hypothetical environmental lapse rate of temperature in an atmosphere in which the density is constant with height (homogeneous atmosphere); 3°C per 100 meters in dry air. Also known as autoconvection gradient.

autogenetic drainage [HYD]   A self-established drainage system developed solely by headwater erosion.

autogenetic topography [GEOL]   Conformation of land due to the physical action of rain and streams.

autogeosyncline [GEOL]   A parageosyncline that subsides as an elliptical basin or trough nearly without associated highlands. Also known as intracratonic basin.

autoinjection *See* autointrusion.

autointrusion [GEOL]   A process wherein the residual liquid of a differentiating magma is drawn into rifts formed in the crystal mesh at a late stage by deformation of unspecified origin. Also known as autoinjection.

autolith [PETR]   **1.** A fragment of igneous rock enclosed in another igneous rock of later consolidation, each being regarded as a derivative from a common parent magma. **2.** A round, oval, or elongated accumulation of iron-magnesium minerals of uncertain origin in granitoid rock.

autolysis [GEOCHEM]   Return of a substance to solution, as of phosphate removed from seawater by plankton and returned when these organisms die and decay.

automatic casing hanger [PETRO ENG]   Unitized hanger-seal assembly latched at the lower end of an oil-well casing string to support the next smaller string and make a seal between the two strings.

automatic dam [MIN ENG]   In placer mining, a dam with a gate that automatically discharges the water when it reaches a certain height behind the dam.

automatic spider [MIN ENG]   A foot or hydraulically actuated drill-rod clamping device similar to a Wommer safety clamp.

automatic tank battery [PETRO ENG]   Interconnected system of storage tanks with automatic controls to direct incoming oil to empty tanks in a desired sequence.

automatic wagon control [MIN ENG]   A mechanism to keep the speed of wagons within certain designed limits; may consist of small hydraulic units fixed at intervals along the inside of the track.

automatic weather station [METEOROL]   A weather station at which the services of an observer are not required; usually equipped with telemetric apparatus.

autometamorphism [PETR]   Metamorphism of an igneous rock by the action of its own volatile fluids. Also known as autometasomatism.

autometasomatism *See* autometamorphism.

automolite [MINERAL]   A nearly black or dark green variety of gahnite.

automorphic [PETR] Of minerals in igneous rock bounded by their own crystal faces. Also known as euhedral; idiomorphic.

automorphosis [PETR] Metamorphosis of solidified igneous rock by solutions from its heated interior.

autophytograph [GEOL] An imprint on a rock surface made by chemical activity of a plant or plant part.

autopneumatolysis [GEOL] The occurrence of metamorphic changes at the pneumatolytic stage of a cooling magma when temperatures are approximately 400–600°C.

autostoper [MIN ENG] A stoper, or light compressed-air rock drill, mounted on an air-leg support which not only supports the drill but also exerts pressure on the drill bit.

autumn ice [OCEANOGR] Sea ice in early stage of formation; comparatively salty, and crystalline in appearance.

Autunian [GEOL] A European stage of geologic time: Lower Permian (above Stephanian of Carboniferous, below Saxonian).

autunite [MINERAL] $Ca(UO_2)_2(PO4)_2 \cdot 10H_2O$ A common fluorescent mineral that occurs as yellow tetragonal plates in uranium deposits; minor ore of uranium.

auxiliary fan [MIN ENG] A small fan installed underground for ventilating narrow coal drivages or hard headings which are not ventilated by the normal air current.

auxiliary fault [GEOL] A branch fault; a minor fault ending against a major one.

auxiliary mineral [MINERAL] A light-colored, relatively rare or unimportant mineral in an igneous rock; examples are apatite, muscovite, corundrum, fluorite, and topaz.

auxiliary plane [GEOL] A plane at right angles to the net slip on a fault plane as determined from analysis of seismic data for an earthquake.

auxiliary thermometer [ENG] A mercury-in-glass thermometer attached to the stem of a reversing thermometer and read at the same time as the reversing thermometer so that the correction to the reading of the latter, resulting from change in temperature since reversal, can be computed.

available moisture [HYD] Moisture in soil that is available for use by plants.

available relief [GEOL] The vertical distance after uplift between the altitude of the original surface and the level at which grade is first attained.

avalanche bedding [GEOL] Steeply inclined bedding formed by the rapid movement of sand down the slip face of a barchan or related form of dune.

avalanche blast [METEOROL] A very destructive avalanche wind.

avalanche boulder tongue [GEOL] A long, narrow formation of debris deposited by an avalanche.

avalanche breccia [GEOL] Breccia that has been formed by a rockfall.

avalanche chute [HYD] A trough cut by an avalanche through which debris such as snow, ice, rock, or soil passes.

avalanche talus [GEOL] An assemblage of rock fragments derived from avalanched snow and ice and carried from cliffs or steep rocky slopes.

avalanche track [HYD] The central corridor through which an avalanche has passed.

avalanche wind [METEOROL] The rush of air produced in front of an avalanche of dry snow or in front of a landslide.

aven [GEOL] *See* pothole. [MIN ENG] A vertical shaft leading upward from a cave passage, sometimes connecting with passages above.

aventurescence [MINERAL] Appearance of bright, vividly colored reflections in some translucent minerals caused by crystal components.

aventurine [MINERAL] **1.** A glass or mineral containing sparkling gold-colored particles, usually copper or chromic oxide. **2.** A shiny red or green translucent quartz having small, but microscopically visible, exsolved hematite or included mica particles.

average assay value [MIN ENG] The weighted result of assays obtained from a number of samples by multiplying the assay value of each sample by the width or thickness of the ore face over which it is taken and dividing the sum of these products by the total width of cross section sampled.

average discharge [HYD] A value used by the U.S. Geological Survey determined from the arithmetic average of water discharge of all years for which there are complete water records.

average discount factor *See* discount factor.

average igneous rock [PETR] A hypothetical rock whose composition is thought to be similar to the average chemical composition of the outermost 10-mile (16-kilometer) shell of the earth.

average-level anomaly [GEOPHYS] Gravity anomaly related to the average level of topography in a region. Also known as Putnam anomaly.

average limit of ice [OCEANOGR] The average extension of ice formation into the sea during a normal winter.

average velocity [GEOPHYS] For a seismic pulse, the ratio of the distance travelled along a ray path to the time required for the traversal. [HYD] The ratio of a stream discharge to the area of a cross section normal to the stream flow.

avezacite [PETR] A variety of plutonic rock with a composition intermediate between pyroxenite and hornblendite.

aviation weather forecast [METEOROL] A forecast of weather elements of particular interest to aviation, such as the ceiling, visibility, upper winds, icing, turbulence, and types of precipitation or storms. Also known as airways forecast.

aviation weather observation [METEOROL] An evaluation, according to set procedure, of those weather elements which are most important for aircraft operations. Also known as airways observation.

aviolite [PETR] A mica-cordierite-hornfels.

avogadrite [MINERAL] $(K,Cs)BF_4$ An orthorhombic fluoborate mineral occurring in small crystals on Vesuvian lava.

avulsion [HYD] A sudden change in the course of a stream by which a portion of land is cut off, as where a stream cuts across and forms an oxbow.

awaruite [MINERAL] Native nickel-iron alloy containing 57.7% nickel.

axial angle [CRYSTAL] **1.** The acute angle between the two optic axes of a biaxial crystal. Also known as optic angle; optic-axial angle. **2.** In air, the larger angle between the optic axes after refraction on leaving the crystal.

axial compression [GEOL] A compression applied parallel with the cylinder axis in experimental work with rock cylinders.

axial culmination [GEOL] Distortion of the fold axis upward in a form similar to an anticline.

axial dipole field [GEOPHYS] A postulated magnetic field for the earth, consisting of a dipolar field centered at the earth's center, with its axis coincident with the earth's rotational axis.

axial element [CRYSTAL] The lengths, length ratios, and angles which define a crystal's unit cell.

axial plane [CRYSTAL] **1.** A plane that includes two of the crystallographic axes. **2.** The plane of the optic axis of an optically biaxial crystal. [GEOL] A plane that intersects the crest or trough in such a manner that the limbs or sides of the fold are more or less symmetrically arranged with reference to it. Also known as axial surface.

axial plane cleavage [GEOL] Rock cleavage essentially parallel to the axial plane of a fold.

axial plane foliation [GEOL] Foliation developed in rocks parallel to the axial plane of a fold and perpendicular to the chief deformational pressure.

axial plane schistosity [GEOL] Schistosity developed parallel to the axial planes of folds.

axial plane separation [GEOL] The distance between axial planes of adjacent anticline and syncline.

axial ratio [CRYSTAL] The ratio obtained by comparing the length of a crystallographic axis with one of the lateral axes taken as unity.

axial stream [HYD] **1.** The chief stream of an intermontane valley, the course of which is along the deepest part of the valley and is parallel to its longer dimension. **2.** A stream whose course is along the axis of an anticlinal or a synclinal fold.

axial trace [GEOL] The intersection of the axial plane of a fold with the surface of the earth or any other specified surface; sometimes such a line is loosely and incorrectly called the axis.

axial trough [GEOL] Distortion of a fold axis downward into a form similar to a syncline.

axinite [MINERAL] $H_2(Ca,Fe,Mn)_4(BO)Al_2(SiO_4)_5$ Brown, blue, green, gray, or purplish gem mineral that commonly forms glassy triclinic crystals. Also known as glass schorl.

axinitization [GEOL] The replacement of rocks by axinite, as in the border zones of some granites.

axiolite [MINERAL] A variety of elongated spherulite in which there is an aggregation of minute acicular crystals arranged at right angles to a central axis.

axis [GEOL] **1.** A line where a folded bed has maximum curvature. **2.** The central portion of a mountain chain.

axis of homology [MAP] The intersection of the plane of the photograph with the horizontal plane of the map or the plane of reference of the ground. Also known as axis of perspective; map parallel; perspective axis.

axis of perspective *See* axis of homology.

azimuthal chart [MAP] A chart on an azimuthal projection. Also known as zenithal chart.

azimuthal equidistant chart [MAP] A chart on the azimuthal equidistant map projection.

azimuthal map projection [MAP] The transformation of a spherical representation of the earth's surface to a tangent or intersecting plane established perpendicular to a right line passing through the center of the spherical representation.

azimuthal orthomorphic projection *See* stereographic projection.

azimuth angle [ENG]   An angle in triangulation or in traverse through which the computation of azimuth is carried.

azimuth dial [ENG]   Any horizontal circle dial that reads azimuth.

azimuth equation [GEOD]   A condition equation which expresses the relationship between the fixed azimuths of two lines that are connected by triangulation or traverse.

azimuth line [ENG]   A radial line from the principal point, isocenter, or nadir point of a photograph, representing the direction to a similar point of an adjacent photograph in the same flight line; used extensively in radial triangulation.

azimuth transfer [ENG]   Connecting, with a straight line, the nadir points of two vertical photographs selected from overlapping flights.

azimuth traverse [ENG]   A survey traverse in which the direction of the measured course is determined by azimuth and verified by back azimuth.

Azoic [GEOL]   That portion of the earlier Precambrian time in which there is no trace of life.

azonal soil [GEOL]   Any group of soils without well-developed profile characteristics, owing to their youth, conditions of parent material, or relief that prevents development of normal soil-profile characteristics.

Azores high [METEOROL]   The semipermanent subtropical high over the North Atlantic Ocean, especially when it is located over the eastern part of the ocean; when in the western part of the Atlantic, it becomes the Bermuda high.

azulite [MINERAL]   A translucent pale-blue variety of smithsonite found in large masses in Arizona and Greece.

azure stone [MINERAL]   Generally, any blue mineral, such as lapis lazuli, lazulite, or azurite.

azurite [MINERAL] $Cu_3(CO_3)_2(OH)_2$ A blue monoclinic mineral consisting of a basic carbonate of copper; an ore of copper. Also known as blue copper ore; blue malachite; chessylite.

azurlite [MINERAL]   A type of chalcedony colored blue by the presence of chrysocolla; used for gemstones.

azurmalachite [MINERAL]   A mixture of azurite and malachite, usually occurring massive with concentric banding; used as an ornamental stone.

B

Babel quartz [MINERAL] A type of crystalline quartz named for its supposed resemblance to tiers on the Tower of Babel.

back balance [MIN ENG] **1.** A kind of self-acting incline in a mine. **2.** The means of maintaining tension on a rope transmission or haulage system, consisting of the tension carriage, attached weight, and supporting structure.

back bay [GEOGR] A small, shallow bay receiving the flow from coastal streams and connected to the sea through an opening between barrier islands.

back beach *See* backshore.

back-bent occlusion *See* bent-back occlusion.

backbone [GEOL] **1.** A ridge forming the principal axis of a mountain. **2.** The principal mountain ridge, range, or system of a region.

backcast stripping [MIN ENG] A stripping method using two draglines; one strips and casts the overburden, and the other recasts a portion of the overburden.

backdeep [GEOL] An epieugeosynclinal basin; a nonvolcanic postorogenic geosynclinal basin whose sediments are derived from an uplifted eugeosyncline.

back-door cold front [METEOROL] A front which leads a cold air mass toward the south and southwest along the Atlantic seaboard of the United States.

backflooding [HYD] A reversal of flow of water at the water table resulting from changes in precipitation.

backfolding [GEOL] Process in mountain forming in which the folds are overturned toward the interior of an orogenic belt. Also known as backward folding.

back furrow *See* esker.

backhand drainage [HYD] A drainage pattern in which the main stream and its tributaries flow in opposite directions.

back holes [MIN ENG] The holes which are shot last in mine shaft sinking.

backing [METEOROL] **1.** Internationally, a change in wind direction in a counterclockwise sense (for example, south to east) in either hemisphere of the earth. **2.** In United States usage, a change in wind direction in a counterclockwise sense in the Northern Hemisphere, clockwise in the Southern Hemisphere. [MIN ENG] **1.** Timbers across the top of a level, supported in notches cut in the rock. **2.** Rough masonry of a wall faced with finer work. **3.** Earth placed behind a retaining wall.

backing deals [MIN ENG] Boards, 1–4 inches (2.5–10 centimeters) thick, of sufficient length to bridge the space between timber or steel sets or between rings in skeleton tubbing.

backlands [GEOL] A section of a river floodplain lying behind a natural levee.

back lead [GEOL] A deposit of sand along the coastline above the high-water mark.

backlimb [GEOL] Of the two limbs of an asymmetrical anticline, the one that is more gently dipping.

back off [ENG] To withdraw the drill bit from a borehole.

back-pressure curve [PETRO ENG] A graph used to arrive at the capacity of a natural-gas well to deliver gas into a pipeline at a sustained rate; uses data from back-pressure testing.

back-pressure testing [PETRO ENG] Method of estimating open-flow capacity of natural-gas wells by relating a series of gas-flow rates and their corresponding stabilized pressures at the bottom of the well bore.

back-pressure valve [PETRO ENG] A check valve installed in a natural-gas well bore to shut off gas flow while replacing the blowout preventer (used during drilling) with a christmas tree piping arrangement, which controls gas flow out of the completed well.

back radiation *See* counterradiation.

back-reflection photography [CRYSTAL] A method of studying crystalline structure by x-ray diffraction in which the photographic film is placed between the source of x-rays and the crystal specimen.

back rush [OCEANOGR] Return of water seaward after the uprush of the waves.

backs [MIN ENG] Ore height available above a given working level.

back-set bed [GEOL] Cross bedding that dips in a direction against the flow of a depositing current.

backshore [GEOL] The upper shore zone that is beyond the advance of the usual waves and tides. Also known as back beach; backshore beach.

backshore beach *See* backshore.

backshore terrace *See* berm.

back shot [MIN ENG] A shot used for widening an entry, placed at some distance from the head of an entry.

backsight method [ENG] **1.** A plane-table traversing method in which the table orientation produces the alignment of the alidade on an established map line, the table being rotated until the line of sight is coincident with the corresponding ground line. **2.** Sighting two pieces of equipment directly at each other in order to orient and synchronize one with the other in azimuth and elevation.

back slope *See* dip slope.

back stoping *See* shrinkage stoping.

backswamp [GEOL] Swampy depressed area of a floodplain between the natural levees and the edge of the floodplain.

backswamp deposits [GEOL] Silt and clay deposited in thin layers in the flood basin behind a river levee.

backthrusting [GEOL] The thrusting in the direction of the interior of an orogenic belt, opposite the general structural trend.

backward folding *See* backfolding.

backwash [OCEANOGR] **1.** Water or waves thrown back by an obstruction such as a ship or breakwater. **2.** The seaward return of water after a rush of waves onto the beach foreshore.

backwash mark [GEOL] A crisscross ridge pattern in beach sand, caused by backwash.

backwash ripple mark [GEOL] Ripple marks that are broad and flat and parallel to the shoreline, with narrow, shallow troughs and crests about 30 centimeters apart; formed by backwash above the maximum wave retreat level.

backwasting [GEOL] A type of mass wasting which causes a slope to retreat without changing its declivity.

backwater [HYD] **1.** A series of connected lagoons, or a creek parallel to a coast, narrowly separated from the sea and connected to it by barred outlets. **2.** Accumulation of water resulting from and held back by an obstruction. **3.** Water reversed in its course by an obstruction.

backwater curve [HYD] The form of a stream surface, along a longitudinal profile, that is assumed above the point where depth is greater than normal due to a constriction or obstruction in the channel.

backweathering [GEOL] A type of weathering that contributes to the retreat of a slope.

back work [MIN ENG] Any kind of operation in a mine not immediately concerned with production or transport; literally, work behind the face, such as repairs to roads.

baculite [GEOL] A crystallite that looks like a dark rod.

baddeleyite [MINERAL] ZrO_2 A colorless, yellow, brown, or black monoclinic zirconium oxide mineral found in Brazil and Ceylon; used as heat- and corrosion-resistant linings for furnaces and muffles.

badlands [GEOGR] An erosive physiographic feature in semiarid regions characterized by sharp-edged, sinuous ridges separated by steep-sided, narrow, winding gullies.

baffling wind [METEOROL] A wind that is shifting so that nautical movement by sailing vessels is impeded.

bag powder [MATER] An explosive loaded in bags.

bahada *See* bajada.

bahamite [PETR] A consolidated limestone formed of sediment similar to a type currently found accumulating in the Bahamas.

bahiaite [PETR] Holocrystalline igneous rock formed mainly of hypersthene with subordinate hornblende and sometimes minor amounts of other minerals.

bahr [HYD] A body of water found in the Saharan region, frequently in the form of a deep natural spring.

bai [METEOROL] A yellow mist prevalent in China and Japan in spring and fall, when the loose surface of the interior of China is churned up by the wind, and clouds of sand rise to a great height and are carried eastward, where they collect moisture and fall as a yellow mist.

baikerinite [MINERAL] A tarry hydrocarbon which makes up about 30% of the naturally occurring substance of baikerite.

baikerite [MINERAL] A waxlike mineral from the vicinity of Lake Baikal, Siberia; apparently about 60% ozocerite with other tarry, waxy, and resinous hydrocarbons.

bailing [ENG] Removal of the cuttings from a well during cable-tool drilling, or of the liquid from a well, by means of a bailer.

bajada [GEOL]   An alluvial plain formed as a result of lateral growth of adjacent alluvial fans until they finally coalesce to form a continuous inclined deposit along a mountain front. Also spelled bahada.

bajada breccia [PETR]   An imperfectly stratified accumulation of coarse, angular rock fragments mixed with mud that formed in arid climates and results from a mudflow containing considerable water.

Bajocian [GEOL]   A European stage: the middle Middle or lower Middle Jurassic geologic time; above Toarcian, below Bathonian.

bakerite [MINERAL] $8CaO \cdot 5B_2O_3 \cdot 6SiO_2 \cdot 6H_2O$ White mineral, occurring in fine-grained, nodular masses, resembling marble and unglazed porcelain, and consisting of hydrous calcium borosilicate.

balance [HYD]   The change in mass of a glacier over a given period of time.   [MIN ENG]   The counterpoise or weight attached by cable to the drum of a winding engine to balance the weight of the cage and hoisting cable and thus assist the engine in lifting the load out of the shaft.

balance car [MIN ENG]   In quarrying, a car loaded with iron or stone and connected by means of a steel cable with a channeling machine operating on an inclined track; used to counteract the force of gravity and thus enable the channeling machine to operate with equal ease uphill and downhill.

balanced rock *See* perched block.

balance equation [METEOROL]   A diagnostic equation expressing a balance between the pressure field and the horizontal field of motion of the atmosphere.

balance shot [MIN ENG]   In coal mining, a shot for which the drill hole is parallel to the face of the coal that is to be broken by it.

balance year [HYD]   The period of time between the minimum mass of a glacier in one year to the minimum mass in the succeeding year.

balas ruby [MINERAL]   A pink or orange gem variety of spinel.

bald [GEOGR]   An elevated grassy, treeless area, as on the top of a mountain.   [MIN ENG]   **1.** Without framing.   **2.** A mine timber which has a flat end.

baldheaded anticline [GEOL]   An upfold with a crest that has been deeply eroded before later deposition.

Bali wind [METEOROL]   A strong east wind at the eastern end of Java.

ball [GEOL]   **1.** A low sand ridge, underwater by high tide, which extends generally parallel with the shoreline; usually separated by an intervening trough from the beach.   **2.** A spheroidal mass of sedimentary material.   **3.** Common name for a nodule, especially of ironstone.

ball-and-pillow structure [GEOL]   A primary sedimentary structure resembling balls and pillows and found in sandstones and some limestones.

ball-and-socket jointing *See* cup-and-ball joint.

ballas [MINERAL]   A spherical aggregate of small diamond crystals; used in diamond drill bits and other diamond tools.

ball coal [GEOL]   A variety of coal occurring in spheroidal masses.

ball ice [OCEANOGR]   Numerous floating spheres of sea ice having diameters of 1–2 in. (2.5–5 cm), generally occurring in belts similar to those of slush which forms at the same time.

ball ironstone [PETR]   A sedimentary rock that contains large, argillaceous ironstone nodules.

ball jasper [MINERAL] Either of two forms of jasper: one type occurs in spherical masses, and the other displays concentric banding of red and yellow.

ball lightning [GEOPHYS] A relatively rare form of lightning, consisting of a reddish, luminous ball, of the order of 1 foot (30 centimeters) in diameter, which may move rapidly along solid objects or remain floating in midair. Also known as globe lightning.

ballon [GEOL] A rounded, dome-shaped hill formed either by erosion or by uplift.

balloon ceiling [METEOROL] The ceiling classification applied when the ceiling height is determined by timing the ascent and disappearance of a ceiling balloon or pilot balloon in United States weather observing practice.

balloon drag [METEOROL] A small balloon, loaded with ballast and inflated so that it will explode at a predetermined altitude, which is attached to a larger balloon; frequently used to retard the ascent of a radiosonde during the early part of the flight so that more detailed measurements may be obtained.

ball sealers [PETRO ENG] Balls of rubber, plastic, or metal that are dropped down the well bore to aid the acidizing of impermeable zones of an oil reservoir; they wedge into and plug the bottomhole tubing perforations that are adjacent to the more permeable reservoir zones.

ballstone [GEOL] **1.** Large mass or concretion of fine, unstratified limestone resulting from growth of coral colonies. **2.** A nodule of rock, especially ironstone, in a stratified unit.

balm [GEOL] A concave cliff or precipice that forms a shelter together with overhanging rock.

Baltic Sea [GEOGR] An intracontinental, Mediterranean-type sea, connected with the North Sea and surrounded by Sweden, Denmark, Germany, Poland, the Baltic States, and Finland.

banakite [PETR] An alkalic basalt made up of plagioclase, sanidine, and biotite, with small quantities of analcime, augite, and olivine; quartz or leucite may be present.

banco [HYD] A meander or oxbow lake separated from a river by a change in its course.

band [GEOD] Any latitudinal strip, designated by accepted units of linear or angular measurement, which circumscribes the earth. [GEOL] A thin layer or stratum of rock that is noticeable because its color is different from the colors of adjacent layers.

bandaite [PETR] A dacite type of extrusive rock composed of hypersthene and labradorite.

banded [PETR] Pertaining to the appearance of rocks that have thin and nearly parallel bands of different textures, colors, and minerals.

banded agate [MINERAL] A variety of agate with parallel bands or stripes of varying widths and colors.

banded coal [GEOL] A variety of bituminous and subbituminous coal made up of a sequence of thin lenses of highly lustrous coalified wood or bark interspersed with layers of more or less striated bright or dull coal.

banded differentiate [PETR] A type of igneous rock made up of bands of different composition, frequently alternating between two varieties as in a layered intrusion.

banded gneiss [PETR] A type of gneiss made up of alternating layers that differ in composition and texture.

banded ore [GEOL] Ore made up of layered bands composed either of the same minerals that differ from band to band in color, textures, proportion, or of different minerals.

banded peat [GEOL] Peat formed of alternate layers of vegetable debris.

banded structure [METEOROL] The appearance of precipitation echoes in the form of long bands as presented on radar plan position indicator (PPI) scopes. [PETR] In igneous and metamorphic rocks, an outcrop feature that results from alternation of layers, stripes, flat lenses, or streaks that obviously differ in mineral composition or texture.

banded vein [GEOL] A vein composed of layers of different minerals that lie parallel to the walls. Also known as ribbon vein.

banding [HYD] In a glacier, the occurrence of alternate ice layers of different textures and appearance. [PETR] **1.** The series of layers occurring in a banded structure. **2.** In sedimentary rocks, the type of thin bedding of alternate layers owing to the deposit of various kinds of materials.

band lightning *See* ribbon lightning.

bandylite [MINERAL] $CuB_2O_4 \cdot CuCl_2 \cdot 4H_2O$ A tetragonal mineral that is deep blue with greenish lights and consists of a hydrated copper borate-chloride.

bank [GEOL] **1.** The edge of a waterway. **2.** The rising ground bordering a body of water. **3.** A steep slope or face, generally constituted of unconsolidated material. [MIN ENG] **1.** The top of the shaft. **2.** The surface around the mouth of a shaft. **3.** The whole, or sometimes only one side or one end, of a working place underground. **4.** To manipulate materials such as coal, gravel, or sand on a bank. **5.** A terrace-like bench in open-pit mining. [OCEANOGR] A raised portion of the sea floor, relatively flat-topped and at shallow depth, characteristically on the continental shelf or near an island.

bank caving [GEOL] The sliding of masses of sand, gravel, silt, or clay into a stream channel, caused by a turbulent current that undercuts the channel wall on the outside of the stream bed.

bank deposit [GEOL] Mounds, ridges, and terraces of sediment rising above and about the surrounding sea bottom.

banket [GEOL] A conglomerate containing valuable metal to be exploited.

bank-full stage [HYD] The flow stage of a river in which the stream completely fills its channel and the elevation of the water surface coincides with the bank margins.

bank height [MIN ENG] The vertical height of a bank as measured between its highest point or crest and its toe at the digging level or bench. Also known as bench height; digging height.

bank-inset reef [GEOL] A coral reef situated on island or continental shelves well inside the outer edges.

bank reef [GEOL] A reef which rises at a distance back from the outer margin of rimless shoals.

bank-run gravel [GEOL] A natural deposit of gravel or sand.

bank sand [GEOL] Deposits occurring in banks or pits and containing a low percentage of clay; used in core making.

bank slope [MIN ENG] The angle, measured in degrees of deviation from the horizontal, at which the earthy or rock material will stand in an excavated, terracelike cut in an open-pit mine or quarry. Also known as bench slope.

bank stability [GEOL] The degree of resistance to change in the slope and contour of a stream bank.

bank storage [HYD] Water absorbed in the permeable bed and banks of a lake, reservoir, or stream.

banner cloud   [METEOROL]   A cloud plume often observed to extend downwind from isolated mountain peaks, even on otherwise cloud-free days. Also known as cloud banner.

bar   [GEOL]   **1.** Any of the various submerged or partially submerged ridges, banks, or mounds of sand, gravel, or other unconsolidated sediment built up by waves or currents within stream channels, at estuary mouths, and along coasts. **2.** Any band of hard rock, for example, a vein or dike, that extends across a lode.   [MIN ENG]   *See* bar drill.

baraboo   [GEOL]   A monadnock buried by a series of strata and then reexposed by the partial erosion of these younger strata.

bararite   [MINERAL] $(NH_4)_2SiF_6$ A white, hexagonal mineral consisting of ammonium silicon fluoride; occurs in tabular, arborescent, and mammillary forms.

barat   [METEOROL]   A heavy northwest squall in Manado Bay on the north coast of the island of Celebes, prevalent from December to February.

barb   [METEOROL]   A means of representing wind speed in the plotting of a synoptic chart, being a short, straight line drawn obliquely toward lower pressure from the end of a wind-direction shaft. Also known as feather.

Barbados earth   [GEOL]   A deposit of fossil radiolarians.

bar beach   [GEOL]   A straight beach of offshore bars that are separated by shallow bodies of water from the mainland.

barbed drainage pattern   [HYD]   A drainage pattern in which tributaries join the main stream at sharp angles pointing upstream.

barbed tributary   [HYD]   A tributary that enters the main stream in an upstream direction instead of pointing downstream.

barber   [METEOROL]   A severe storm at sea during which spray and precipitation freeze onto the decks and rigging of ships.

barbertonite   [MINERAL] $Mg_6Cr_2(OH)_{16}CO_3 \cdot 4H_2O$ A lilac to rose pink, hexagonal mineral consisting of a hydrated carbonate-hydroxide of magnesium and chromium; occurs in massive form or in masses of fibers or plates.

barbierite   [MINERAL] $NaAlSi_3IO_8$ A hypothetical soda feldspar thought to be isomorphous with orthoclase.

barchan   [GEOL]   A crescent-shaped dune or drift of windblown sand or snow, the arms of which point downwind; formed by winds of almost constant direction and of moderate speeds. Also known as barchane; barkhan; crescentic dune.

barchane   *See* barchan.

bar drill   [MIN ENG]   A small diamond type or other type of rock drill mounted on a bar and used in an underground workplace. Also known as bar.

bar finger sand   [GEOL]   An elongated lenticular sand body that lies beneath a distributory in a birdfoot delta.

baric topography   *See* height pattern.

baric wind law   *See* Buys-Ballot's law.

barines   [METEOROL]   Westerly winds in eastern Venezuela.

baring   *See* overburden.

barite   [MINERAL] $BaSO_4$ A white, yellow, or colorless orthorhombic mineral occurring in tabular crystals, granules, or compact masses; specific gravity is 4.5; used in paints and drilling muds and as a source of barium chemicals; the principal ore of barium. Also known as baryte; barytine; cawk; heavy spar.

barite dollar [MINERAL]   Barite in the form of rounded disk-shaped masses; formed in a sandstone or sandy shale.

Barker method [CRYSTAL]   A method utilizing a number of convenient rules which allow two observers to choose the same reference system to describe the same noncubic crystal.

barkevikite [MINERAL]   A brown or black member of the amphibole mineral group; looks like basaltic hornblende but differs from it in its iron concentration.

barkhan *See* barchan.

bar lake [HYD]   A lake that has a sandbar across the outlet.

bar mining [MIN ENG]   The mining of river bars, usually between low and high waters, although the stream is sometimes deflected and the bar worked below water level.

barney [MIN ENG]   A small car or truck, attached to a rope or cable, used to push cars up a slope or an inclined plane. Also known as bullfrog; donkey; groundhog; larry; mule; ram; truck.

baroclinic disturbance [METEOROL]   Any migratory cyclone associated with strong baroclinity of the atmosphere, evidenced on synoptic charts by temperature gradients in the constant-pressure surfaces, vertical wind shear, tilt of pressure troughs with height, and concentration of solenoids in the frontal surface near the ground. Also known as baroclinic wave.

baroclinic field [METEOROL]   A distribution of atmospheric pressure and mass such that the specific volume, or density, of air is a function not solely of pressure.

baroclinic instability [METEOROL]   A hydrodynamic instability arising from the existence of a meridional temperature gradient (and hence of a thermal wind) in an atmosphere in quasi-geostrophic equilibrium and possessing static stability.

baroclinic model [METEOROL]   A concept of stratification in the atmosphere, involving surfaces of constant pressure intersecting surfaces of constant density.

baroclinic wave *See* baroclinic disturbance.

barogram [ENG]   The record of an aneroid barograph.

barograph *See* aneroid barograph.

barometer [ENG]   An absolute pressure gage specifically designed to measure atmospheric pressure.

barometer elevation [METEOROL]   The vertical distance above mean sea level of the ivory point (zero point) of a weather station's mercurial barometer; frequently the same as station elevation. Also known as elevation of ivory point.

barometric efficiency [HYD]   In a well, the ratio of water-level fluctuation to the change in atmospheric pressure that causes the fluctuation.

barometric hypsometry [ENG]   The determination of elevations by means of either mercurial or aneroid barometers.

barometric leveling [ENG]   The measurement of approximate elevation differences in surveying with the aid of a barometer; used especially for large areas.

barometric tendency *See* pressure tendency.

barometric tide [GEOPHYS]   A daily variation in atmospheric pressure due to the gravitational attraction of the sun and moon.

barometric wave [METEOROL]   Any wave in the atmospheric pressure field; the term is usually reserved for short-period variations not associated with cyclonic-scale motions or with atmospheric tides.

barometrograph *See* aneroid barograph.

barometry [ENG] The study of the measurement of atmospheric pressure, with particular reference to ascertaining and correcting the errors of the different types of barometer.

baroseismic storm [GEOPHYS] Microseisms that are caused by changes in atmospheric pressure.

barothermogram [ENG] The record made by a barothermograph.

barothermograph [ENG] An instrument which automatically records pressure and temperature.

barothermohygrogram [ENG] The record made by a barothermohygrograph.

barothermohygrograph [ENG] An instrument that produces graphs of atmospheric pressure, temperature, and humidity on a single sheet of paper.

barotropic disturbance [METEOROL] Also known as barotropic wave. **1.** A wave disturbance in a two-dimensional nondivergent flow; the driving mechanism lies in the variation of either vorticity of the basic current or the variation of the vorticity of the earth about the local vertical. **2.** An atmospheric wave of cyclonic scale in which troughs and ridges are approximately vertical.

barotropic field [METEOROL] A distribution of atmospheric pressures and mass such that the specific volume, or density, of air is a function solely of pressure.

barotropic model [METEOROL] Any of a number of model atmospheres in which some of the following conditions exist throughout the motion: coincidence of pressure and temperature surfaces, absence of vertical wind shear, absence of vertical motions, absence of horizontal velocity divergence, and conservation of the vertical component of absolute vorticity.

barotropic wave *See* barotropic disturbance.

bar plain [GEOL] A plain formed by a stream without a low-water channel or an alluvial cover.

barranca [GEOL] A hole or deep break made by heavy rain; a ravine.

barred basin *See* restricted basin.

barrel copper [MIN ENG] Copper in lumps small enough to be picked out of the mass of rock and put in the furnace without dressing.

Barremian [GEOL] Lower Cretaceous geologic age, between Hauterivian and Aptian.

barrens [GEOGR] An area that because of adverse environmental conditions is relatively devoid of vegetation compared with adjacent areas.

barrier bar [GEOL] Ridges whose crests are parallel to the shore and which are usually made up of water-worn gravel put down by currents in shallow water at some distance from the shore.

barrier basin [GEOL] A basin formed by natural damming, for example, by landslides or moraines.

barrier beach [GEOL] A single, long, narrow ridge of sand which rises slightly above the level of high tide and lies parallel to the shore, from which it is separated by a lagoon. Also known as offshore beach.

barrier chain [GEOL] A series of barrier spits, barrier islands, and barrier beaches extending along a coastline.

barrier flat [GEOL] An area which is relatively flat and frequently occupied by pools of water that separate the seaward edge of the barrier from a lagoon on the landward side.

barrier ice *See* shelf ice.

barrier island [GEOL] **1.** An island similar to an offshore bar but differing from it in having multiple ridges, areas of vegetation, and swampy terraces extending toward the lagoon. **2.** A detached portion of offshore bar between two inlets.

barrier lagoon [GEOGR] A shallow body of water that separates the shore and a barrier reef.

barrier lake [HYD] A small body of water that lies in a basin, retained there by a natural dam or barrier.

barrier reef [GEOL] A coral reef that runs parallel to the coast of an island or continent, from which it is separated by a lagoon.

barrier split [GEOL] A barrier of sand joined at one of its ends to the mainland.

barrier theory of cyclones [METEOROL] A theory of cyclone development, proposed by F.M. Exner, which states that a slow-moving mass of cold air in the path of rapidly eastward-moving warmer air will bring about the formation of low pressure on the lee side of the cold air; analogous to the formation of a dynamic trough on the lee side of an orographic barrier. Also known as drop theory.

barrier well [HYD] A recharge well used to build up a ridge of usable-quality water between wells which are used for water supply but which may be potential sources of contamination.

Barrovian metamorphism [GEOL] A regional metamorphism that can be zoned into facies that are metamorphic.

bar scale *See* graphic scale.

Barstovian [GEOL] Upper Miocene geologic time.

bar theory [GEOL] A theory that accounts for thick deposits of salt, gypsum, and other evaporites in terms of increased salinity of a solution in a lagoon caused by evaporation.

Bartonian [GEOL] A European stage: Eocene geologic time above Auversian, below Ludian. Also known as Marinesian.

Barychilinidae [PALEON] A family of Paleozoic crustaceans in the suborder Platycopa.

Barylambdidae [PALEON] A family of late Paleocene and early Eocene aquatic mammals in the order Pantodonta.

barysphere *See* centrosphere.

Barytheriidae [PALEON] A family of extinct proboscidean mammals in the suborder Barytherioidea.

Barytherioidea [PALEON] A suborder of extinct mammals of the order Proboscidea, in some systems of classification.

barytocalcite [MINERAL] $CaBa(CO_3)_2$ A colorless to white, grayish, greenish, or yellowish monoclinic mineral consisting of calcium and barium carbonate.

basal arkose [PETR] Partially reworked feldspathic residuum in the lower section of a sandstone that overlies granitic rock.

basal cleavage [CRYSTAL] Cleavage parallel to the base of the crystal structure or to the lattice plane which is normal to one of the lattice axes.

basal conglomerate [GEOL] A coarse gravelly sandstone or conglomerate forming the lowest member of a series of related strata which lie unconformably on older rocks; records the encroachment of the seabeach on dry land.

basal groundwater [HYD] A large body of groundwater that floats on and is in hydrodynamic equilibrium with sea water.

basal orientation [CRYSTAL] A crystal orientation in which the surface is parallel to the base of the lattice or to the lattice plane which is normal to one of the lattice axes.

basal plane [CRYSTAL] The plane perpendicular to the long, or c, axis in all crystals except those of the isometric system.

basal sapping [GEOL] The breaking away of rock fragments along the headwall of a cirque, due to frost action at the base of a crevice in the glacier.

basal sliding [HYD] Also known as basal slip. **1.** The sliding of a glacier on its bed. **2.** The velocity or rate of glacial sliding.

basal slip *See* basal sliding.

basalt [PETR] An aphanitic crystalline rock of volcanic origin, composed largely of plagioclase feldspar (labradorite or bytownite) and dark minerals such as pyroxene and olivine; the extrusive equivalent of gabbro.

basalt glass *See* tachylite.

basaltic dome *See* shield volcano.

basaltic hornblende [PETR] A black or brown variety of hornblende rich in ferric iron and occurring in basalts and other iron-rich basic igneous rocks. Also known as basaltine; lamprobolite; oxyhornblende.

basaltic lava [PETR] A volcanic fluid rock of basaltic composition.

basaltic magma [GEOL] Mobile rock material of basaltic composition.

basaltic rock [PETR] Igneous rock that is fine-grained and contains basalt, diabase, and dolerite; if andesite is included the rock is dark in color.

basaltic shell [GEOL] The lower crystal layer of basalt underlying the oceans and beneath the sialic layer of continents.

basaltiform [GEOL] Similar to basalt in form.

basaltine *See* basaltic hornblende.

basalt obsidian *See* tachylite.

basaluminite [MINERAL] $Al_4(SO_4)(OH)_{10} \cdot 5H_2O$ A white mineral consisting of hydrated basic aluminum sulfate; occurs in compact masses.

basal water table [HYD] The water table of basal groundwater.

basanite [PETR] A basaltic extrusive rock closely allied to chert, jasper, or flint. Also known as Lydian stone; lydite.

basculating fault *See* wrench fault.

base bullion [MET] Crude lead that has enough silver in it to make the extraction of silver worthwhile; gold may be present.

base-centered lattice [CRYSTAL] A space lattice in which each unit cell has lattice points at the centers of each of two opposite faces as well as at the vertices; in a monoclinic crystal, they are the faces normal to one of the lattice axes.

base conditions [PETRO ENG] Standard conditions of 14.65 psia pressure and 60°F (15.6°C) used to calculate the amount of gas contained in oil from a well (the gas-oil ratio).

base exchange [GEOCHEM] Replacement of certain ions by others in clay.

base flow [HYD]   The flow of water entering stream channels from groundwater sources in the drainage of large lakes.

base fracture [MIN ENG]   In quarrying, the broken condition of the base after a blast; it may be a good or bad base fracture.

base level [GEOL]   That critical plane of erosion and deposition represented by river level on continents and by wave or current base in the sea.

base-leveled plain [GEOL]   Any land surface changed almost to a plain by subaerial erosion. Also known as peneplain.

base-leveling epoch *See* gradation period.

base line [ENG]   A surveyed line, established with more than usual care, to which surveys are referred for coordination and correlation.

base map [MAP]   A map having essential outlines and onto which additional geographical or topographical data may be placed for comparison or correlation. Also known as mother map.

basement [GEOL]   **1.** A complex, usually of igneous and metamorphic rocks, that is overlain unconformably by sedimentary strata. **2.** A crustal layer beneath a sedimentary one and above the Mohorovičić discontinuity.

base of weathering [GEOL]   In seismic interpretation, the boundary between the low-velocity surface layer and an underlying layer of higher velocity.

base ore [MIN ENG]   Ore in which the gold is associated with sulfides, as contrasted with free-milling ores in which the sulfides have been removed by leaching.

base runoff [HYD]   Fair-weather runoff composed principally of effluent groundwater, but including some runoff delayed by passage through lakes or swamps.

base station [ENG]   The point from which a survey begins. [GEOD]   A geographic position whose absolute gravity value is known.

base surge [GEOL]   A ring-shaped cloud consisting of gas and suspended solids that moves radially outward at high velocity from the base of the vertical explosion column caused by a volcanic eruption or the formation of a crater.

bashing [MIN ENG]   **1.** The building of walls and nonporous stoppings for the complete isolation of a district of a mine in which a fire has occurred. **2.** The complete stowing of old mine workings or roadways after all equipment has been removed.

basic [PETR]   Of igneous rocks, having low silica content (generally less than 54%) and usually rich in iron, magnesium, or calcium.

basic border [PETR]   The region around the margin of an igneous intrusion that has a more basic composition than the rock mass.

basic front [GEOL]   An advancing zone of granitization enriched in calcium, magnesium, and iron.

basic hornfels [PETR]   A type of hornfels derived from a basic igneous rock.

basic plagioclase [MINERAL]   Plagioclase with a relatively low amount of silicic acid in the anhydrous form SiO_2.

basic rock [PETR]   An igneous rock with a relatively low silica content, and rich in iron, magnesium, or calcium.

basic schist [PETR]   A schistose rock that forms from the metamorphism of a basic igneous rock.

basic sediment and water [PETRO ENG]   Oil, water, and foreign matter that collects in the bottom of petroleum storage tanks. Abbreviated BS&W. Also known as bottoms; bottom settlings; sediment and water.

basification [GEOL] Development of a more basic rock, usually with more hornblende, biotite, and oligoclase, by contamination of a granitic magma in the assimilation of country rock.

basimesostasis [GEOL] A process of the partial or entire enclosure of plagioclase crystals in a diabase by augite.

basin [GEOL] **1.** A low-lying area, wholly or largely surrounded by higher land, that varies from a small, nearly enclosed valley to an extensive, mountain-rimmed depression. **2.** An entire area drained by a given stream and its tributaries. **3.** An area in which the rock strata are inclined downward from all sides toward the center. **4.** An area in which sediments accumulate. [OCEANOGR] Deep portion of sea surrounded by shallower regions.

basin accounting *See* hydrologic accounting.

basin-and-range structure [GEOL] Regional structure dominated by fault-block mountains separated by basins filled with sediment.

basin facies [GEOL] A stratigraphic facies made up of sediments which have been deposited beyond a land-bordering submarine shelf.

basin fold [GEOL] Synclinal and anticlinal folds in structural basins.

basining [GEOL] A settlement of earth in the form of basins due to the solution and transportation of underground deposits of salt and gypsum.

basin length [GEOL] Length in a straight line from the mouth of a stream to the farthest point on the drainage divide of its basin.

basin order [GEOL] A classification of basins according to stream drainage; for example, a first-order basin contains all of the drainage area of a first-order stream.

basin peat *See* local peat.

basin perimeter [MAP] The length of a map line that encloses the catchment area of a drainage basin.

basin range [GEOL] A mountain range characteristic of the Great Basin in the western United States and formed by a faulted and tilted block of strata.

basin relief [GEOL] The difference in elevation between a stream's mouth and the highest point in or around its drainage basin.

basin valley [GEOL] The filled-in depression of large intermountain areas; an example is Salt Lake Valley in Utah.

basiophitic [PETR] Pertaining to the texture of an ophitic rock in which the last-formed interstitial material is augite.

basite [PETR] An igneous rock with a basic composition.

bassanite [MINERAL] A white mineral consisting of hydrated calcium sulfate; a pseudomorph of gypsum.

basset [GEOL] The outcropping edge of a layer of rock exposed to the surface.

bassetite [MINERAL] A transparent, yellow, monoclinic mineral presumably consisting of a hydrated uranium phosphate containing divalent iron; occurs in groups of thin tablets.

bastion [GEOL] A mass of bedrock projecting far out into the main glacial valley from the mouth of a hanging glacial trough.

bastite [MINERAL] A hydrated magnesium silicate, a variety of serpentine occurring from the alteration of orthorhombic pyroxenes such as enstatite.

bastnaesite [MINERAL] (Ce,La)CO$_3$(F,OH) A greasy yellow to reddish-brown fluorocarbonate rare-earth metal mineral; source of rare earths, for example, cerium and lanthanum.

batea [MIN ENG] A conical-shaped wood unit (12.3 inches or 31 centimeters in diameter with about 150° apex angle) used to recover valuable metals from river channels and bars.

bathograd [GEOPHYS] A line on a geological map connecting all points that have been subject to equal metamorphic pressures.

batholite [GEOL] An older massive protrusion of magma that solidifies as coarse crystalline rock in the deep horizons of the earth's crust.

batholith [GEOL] A body of igneous rock, 40 square miles (100 square kilometers) or more in area, emplaced at great or intermediate depth in the earth's crust.

bathometer [ENG] A mechanism which measures depths in water.

Bathonian [GEOL] A European stage of geologic time: Middle Jurassic, below Callovian, above Bajocian. Also known as Bathian.

Bathornithidae [PALEON] A family of Oligocene birds in the order Gruiformes.

bathvillite [MINERAL] An oxygenated hydrocarbon mineral, found in Tortane Hill, Scotland, that is amorphous, fawn-brown, opaque, and quite friable.

bathyal zone [OCEANOGR] The biogeographic realm of the ocean depths between 100 and 1000 fathoms (180 and 1800 meters).

bathybenthic zone [OCEANOGR] The benthos of the bathyal zone.

bathyclinograph [ENG] A mechanism which measures vertical currents in the deep sea.

bathyconductograph [ENG] A device to measure the electrical conductivity of sea water at various depths from a moving ship.

bathydermal [GEOL] Referring to the deformation or gliding of the lower section of the sial.

bathygenesis [GEOL] Lowering of marine basins by tectonic movement.

bathygram [ENG] A graph recording the measurements of sonic sounding instruments.

bathylimnion [HYD] The deepest layer of the hypolimnion, characterized by uniform temperature at different depths.

bathymetric biofacies [GEOL] The lateral distribution and character of underwater sedimentary strata.

bathymetric chart [MAP] A topographic map of the floor of the ocean.

bathymetry [ENG] The science of measuring ocean depths in order to determine the sea floor topography.

bathyorographical [GEOD] Concerned with depths of oceans and heights of mountains.

bathypelagic zone [OCEANOGR] The biogeographic realm of the ocean lying between depths of 900 and 3700 meters.

bathyscaph [NAV ARCH] A free, crewed vehicle having a spherical cabin on the underside for exploring the deep ocean.

bathyseism [GEOPHYS] A deep-focus earthquake that is detected only by instruments.

bathysphere [NAV ARCH] A spherical chamber in which persons are lowered for observation and study of ocean depths.

bathythermogram [ENG] The record made by a bathythermograph.

bathythermograph [ENG] A device for obtaining a record of temperature against depth (actually, pressure) in the ocean from a ship underway. Abbreviated BT. Also known as bathythermosphere.

bathythermosphere *See* bathythermograph.

bathyvessel [NAV ARCH] A ship, such as a bathysphere or submarine, designed to operate far below the surface of the water.

batisite [MINERAL] $Na_2BaTi_2(Si_2O_7)_2$ A dark-brown mineral with orthorhombic structure.

batt [MIN ENG] A thin layer of coal occurring in the lower part of shale strata that lie above and close to a coal bed.

battery assay [MIN ENG] An assay of samples taken from ore as crushed in a stamp battery.

battery ore [GEOL] Manganese ore composed of a very pure crystalline form of manganese dioxide.

battery reefs *See* Kimberley reefs.

batture [GEOL] An elevation of the bed of a river under the surface of the water; sometimes used to signify the same elevation when it has risen above the surface.

batukite [PETR] A dark-colored extrusive rock principally composed of augite, with some olivine phenocrysts embedded in a groundmass of augite, magnetite, and leucite.

baumhauerite [MINERAL] $Pb_4As_6S_{13}$ A lead to steel gray, monoclinic mineral consisting of lead arsenic sulfide.

baum pot [MIN ENG] **1.** A concretion of calcareous material in the roof of a coal seam. **2.** A cavity remaining in the roof of a coal seam due to the falling of a cast of a fossil tree stump following the removal of coal.

bauxite [PETR] A whitish, grayish, brown, yellow, or reddish-brown rock composed of hydrous aluminum oxides and aluminum hydroxides and containing impurities such as free silica, silt, iron hydroxides, and clay minerals; the principal commercial source of aluminum.

bauxitization [GEOL] Bauxite development from either primary aluminum silicates or secondary clay minerals.

Baveno twin law [CRYSTAL] An uncommon twin law in feldspar, in which the twin plane and composition surface are (021); a Baveno twin usually consists of two individuals.

b axis [CRYSTAL] A crystallographic axis that is oriented horizontally, right to left. [PETR] A direction in the plane of movement that is at a right angle to the tectonic transport direction.

bay [GEOGR] **1.** A body of water, smaller than a gulf and larger than a cove in a recess in the shoreline. **2.** A narrow neck of water leading from the sea between two headlands.

bay bar *See* baymouth bar.

bay barrier [GEOL] A narrow shoal or small point of land projecting from the shore across the mouth of a bay and severing the bay's connection with the main body of water.

bay delta [GEOL]   A usually triangular alluvial deposit formed at the point where the mouth of a stream enters the head of a drowned valley.

bayerite [MINERAL]   Al(OH)$_3$ A polymorph of the mineral gibbsite.

bay head [GEOL]   A swampy region at the head of a bay.

bay head bar [GEOL]   A bar formed a short distance from the shore at the head of a bay.

bay head beach [GEOL]   A beach formed around a bay head by storm waves; layers of sediment cover the bay floor and bare rock benches front the headland cliffs.

bay head delta [GEOL]   A delta at the head of an estuary or a bay into which a river discharges because of the margin of the land's late partial submergence.

bay ice [OCEANOGR]   Sea ice that is young and flat but sufficiently thick to impede navigation.

bayldonite [MINERAL] $Cu_3(AsO_4)_2(OH)_2$ An apple green to yellowish-green monoclinic mineral consisting of a basic arsenate of copper and lead; occurs in minute mammillary concretions, in massive form, and as crusts.

bayleyite [MINERAL] $Mg_2(UO_2)(CO_3)_3 \cdot 18H_2O$ A sulfur yellow monoclinic mineral consisting of a hydrated carbonate of magnesium and uranium; occurs as minute, short-prismatic crystals.

baymouth bar [GEOL]   A bar extending entirely or partially across the mouth of a bay. Also known as bay bar.

bayou [HYD]   A small, sluggish secondary stream or lake that exists often in an abandoned channel or a river delta.

bay salt [GEOL]   A type of evaporite produced by shallow bays, lagoons, or ponds.

bayside beach [GEOL]   A beach formed at the side of a bay by materials eroded from nearby headlands and deposited by longshore currents.

bazzite [MINERAL] $Sc_2Be_3Si_6O_{18}$ An azure-blue mineral that crystallizes in the hexagonal system; the rare scandium analog of beryl.

B bit [MIN ENG]   A nonstandard core bit no longer in common use except in drilling deep boreholes to sample gold-bearing deposits in South Africa; the set outside and inside diameters are about 2$^1/_{16}$ and 1$^3/_8$ inches (52 and 35 millimeters), respectively.

b-c fracture [GEOL]   A tension fracture parallel with the fabric plane and normal to the *a* axis.

b-c plane [GEOL]   A plane that is perpendicular to the plane of movement and parallel to the *b* direction in that plane.

beach [GEOL]   The zone of unconsolidated material that extends landward from the low-water line to where there is marked change in material or physiographic form or to the line of permanent vegetation.

beach breccia [PETR]   A breccia of angular blocks formed by grinding of cliffs under conditions of rapid submergence; common on beaches with weak wave action.

beach concentrate [GEOL]   A concentration of heavy minerals in beach sand where they were originally present as accessory minerals.

beach crest [OCEANOGR]   A temporary ridge which marks the limit of normal wave activity on the shore.

beach cusp *See* cusp.

beach cycle [GEOL]   Periodic retreat and outbuilding of beaches resulting from waves and tides.

beach drift [GEOL] The material transported by drifting of beach.

beach face *See* foreshore.

beach firmness [GEOL] The ability of beach sand to resist pressure; it is a function of sand-particle size, trapped air, and the degree of packing and sorting of the sand.

beach gravel [GEOL] Gravels in which most of the particles cluster about one size.

beachline [GEOL] A shoreline comprising a number of well-developed beaches.

beach mining [MIN ENG] The mining of the heavy minerals, such as rutile, zircon, monazite, ilmenite, and sometimes gold, which occur in sand dunes, beaches, coastal plains, and deposits located inland from the shoreline.

beach plain [GEOL] Embankments of wave-deposited material added to a prograding shoreline.

beach platform *See* wave-cut bench.

beach pool [HYD] **1.** A small, temporary body of water between two beaches or beach ridges. **2.** A lagoon behind a beach ridge. **3.** A small body of water formed adjacent to a lake as the result of wave action.

beach profile [GEOL] Intersection of a beach's ground surface with a vertical plane perpendicular to the shoreline.

beach ridge [GEOL] A continuous mound of beach material behind the beach that was heaped up by waves or other action.

beachrock [PETR] A friable to well-cemented rock made of calcareous skeletal debris that is cemented together by calcium carbonate.

beach scarp [GEOL] A nearly vertical slope along the beach caused by wave erosion.

beaded esker [GEOL] An esker containing numerous bulges or swellings due to the formation of fans or deltas along its length.

beaded lake *See* paternoster lake.

beaker sampler [PETRO ENG] A small, cylindrical vessel with a tapered top used to collect crude oil samples; it is made of low-sparking metal or glass, the bottom is weighted, and there is a small stoppered opening at the top.

Beaman stadia arc [ENG] An attachment to an alidade consisting of a stadia arc on the outer edge of the visual vertical arc; enables the observer to determine the difference in elevation of the instrument and stadia rod without employing vertical angles.

beam-balanced pump [PETRO ENG] An oil well pumping unit having a center-pivoted beam with the sucker rod plunger (pump) at the front end and a counterweight on the rearward extension.

beam building [MIN ENG] A process of rock bolting in flat-lying deposits where the bolts are installed in bedded rock to bind the strata together to act as a single beam capable of supporting itself, thus stabilizing the overlying rock.

beam hanger [PETRO ENG] An attachment at the end of a walking beam above a well casing to lift the pump rods or sacked rods.

beam well [PETRO ENG] A well pumped by a walking beam.

bean ore [GEOL] A lenticular, pisolitic aggregate of limonite.

bear [MIN ENG] To underhole or undermine; to drive in at the top or side of a working.

Beaufort force [METEOROL] A number denoting the speed (or so-called strength) of the wind according to the Beaufort wind scale. Also known as Beaufort number.

Beaufort number *See* Beaufort force.

Beaufort wind scale [METEOROL]   A system of code numbers from 0 to 12 classifying wind speeds into groups from 0–1 mile per hour or 0–1.6 kilometers per hour (Beaufort 0) to those over 75 miles per hour or 121 kilometers per hour (Beaufort 12).

beaverite [MINERAL] $Pb(Cu,Fe,Al)_3(SO_4)_2(OH)_6$ A canary yellow, hexagonal mineral consisting of a basic sulfate of lead, copper, iron, and aluminum.

beaver meadow [ECOL]   Soft, moist ground resulting from the building of beaver dam.

bebedourite [PETR]   A type of pyroxenite that contains biotite along with the accessory minerals perovskite, apatite, and titanomagnetite.

Becke test [MINERAL]   A microscope test in which indices of refraction are compared for minerals; the Becke line appears to move toward the material of higher refractivity as the tube of the microscope is raised.

beckerite [MINERAL]   A brown variety of the fossil resin retinite having a very high oxygen content.

becquerelite [MINERAL] $CaU_6O_{19}\cdot11H_2O$ An orthorhombic mineral consisting of a hydrated oxide of uranium; occurs in tabular, elongated, striated, and massive form.

bed [GEOL]   **1.** The smallest division of a stratified rock series, marked by a well-defined divisional plane from its neighbors above and below. **2.** An ore deposit, parallel to the stratification, constituting a regular member of the series of formations; not an intrusion.   [HYD]   The bottom of a channel for the passage of water.

bedded [GEOL]   Pertaining to rocks exhibiting depositional layering or bedding formed from consolidated sediments.

bedded chert [PETR]   A chert of brittle, closed-jointed character comprising distinct, usually even-bedded layers separated by dark siliceous shale or by siderite; it is found in thick deposits over large areas.

bedded vein [GEOL]   A lode occupying the position of a bed that is parallel with the enclosing rock stratification.

bed detector [PETRO ENG]   Apparatus to detect and measure the extent of underground formations that are potential oil and gas reservoirs; methods include induction logs, gamma-ray logs, and sonic logs.

bedding [GEOL]   Condition where planes divide sedimentary rocks of the same or different lithology.

bedding cleavage [GEOL]   Cleavage parallel to the rock bedding.

bedding fault [GEOL]   A fault whose fault surface is parallel to the bedding plane of the constituent rocks. Also known as bedding-plane fault.

bedding fissility [GEOL]   Primary foliation parallel to the bedding of sedimentary rocks.

bedding joint [GEOL]   A joint parallel to the rock bedding.

bedding plane [GEOL]   Any of the division planes which separate the individual strata or beds in sedimentary or stratified rock.

bedding-plane fault *See* bedding fault.

bedding schistosity [GEOL]   Schistosity that is parallel to the rock bedding.

bedding thrust [GEOL]   A thrust fault parallel to bedding.

bedding void [GEOL]   A void formed between successive batches of lava that are discharged in a single short activity of a volcano, as well as between flows made a long time apart.

bede [MIN ENG] A miner's pick.

Bedford limestone *See* spergenite.

bediasite [GEOL] A black to brown tektite found in Texas.

bed load [GEOL] Particles of sand, gravel, or soil carried by the natural flow of a stream on or immediately above its bed.

Bedoulian [GEOL] Lower Cretaceous (lower Aptian) geologic time in Switzerland.

bedrock [GEOL] General term applied to the solid rock underlying soil or any other unconsolidated surficial cover.

beegerite [MINERAL] $Pb_6Bi_2S_9$ A light to dark gray mineral consisting of lead bismuth sulfide; usually occurs in granular to dense massive form.

beekite [MINERAL] **1.** A concretionary form of calcite or silica that occurs in small rings on the surface of a fossil shell which has weathered out of its matrix. **2.** White, opaque accretions of silica found on silicified fossils or along joint surfaces as a replacement of organic matter.

beerbachite [PETR] A hornfels with large poikiloblastic crystals of olivine.

beetle [MIN ENG] A powerful, cable-hauled propulsion unit, operated under remote control, for moving a train of wagons at the mine surface.

beetle stone *See* septarium.

before-breast [MIN ENG] The part of an orebody that lies ahead of the surface upon which mining operations are in progress.

beheaded stream [HYD] A water course whose upper portion, through erosion, has been cut off and captured by another water course.

beidellite [MINERAL] A clay mineral of the montmorillonite group in which Si^{4+} has been replaced by Al^{3+} and there is the virtual absence of Mg or Fe replacing Al.

belat [METEOROL] A strong land wind from the north or northwest which sometimes blows across the southeastern coast of Arabia and is accompanied by a hazy atmosphere due to sand blown from the interior desert.

Belemnoidea [PALEON] An order of extinct dibranchiate mollusks in the class Cephalopoda.

Belinuracea [PALEON] An extinct group of horseshoe crabs; arthropods belonging to the Limulida.

belite *See* larnite.

Bellerophontacea [PALEON] A superfamily of extinct gastropod mollusks in the order Aspidobranchia.

bell hole [MIN ENG] **1.** One of the holes or excavations made at the section joints of a pipeline for the purpose of repairs. **2.** A conical cavity in a coal mine roof caused by the falling of a large concretion.

bellingerite [MINERAL] $3Cu(IO_3)_2 \cdot 2H_2O$ A light green triclinic mineral consisting of hydrated copper iodate.

bell-metal ore *See* stannite.

bellows *See* aneroid capsule.

belonite [GEOL] A rod- or club-shaped microscopic embryonic crystal in a glassy rock.

below minimums [METEOROL] Below operational weather limits for aircraft.

belt [HYD]   A long area of pack ice, ranging from 1 kilometer to over 100 kilometers in width.

belted plain [GEOL]   A plain whose surface has been slowly worn down and sculptured into bands or belts of different levels.

belteroporic [GEOL]   Of crystals in rocks whose growth was determined by the direction of easiest growth.

belt of cementation *See* zone of cementation.

belt of no erosion [GEOL]   A zone adjacent to a drainage divide in which there is lack of erosion due to insufficient depth and velocity of flow, and to a degree of slope incapable of overcoming the initial resistance of the surface to sheet erosion.

belt of soil moisture *See* belt of soil water.

belt of soil water [GEOL]   The upper subdivision of the zone of aeration limited above by the land surface and below by the intermediate belt; this zone contains plant roots and water available for plant growth. Also known as belt of soil moisture; discrete film zone; soil-water belt; soil-water zone; zone of soil water.

bench [GEOL]   A terrace of level earth or rock that is raised and narrow and that breaks the continuity of a declivity.   [MIN ENG]   **1.** One of two or more divisions of a coal bench blasting seam, separated by slate and so forth or simply separated by the process of cutting the coal, one bench or layer being cut before the adjacent one. **2.** A long horizontal ledge of ore in an underground working place. **3.** A ledge in an open-pit mine from which excavation takes place at a constant level.

bench blasting [MIN ENG]   A mining system used either underground or in surface pits whereby a thick ore or waste zone is removed by blasting a series of successive horizontal layers called benches.

bench gravel [GEOL]   Gravel beds found on the sides of valleys above the present stream bottoms, representing parts of the bed of the stream when it was at a higher level.

bench height *See* bank height.

benching [MIN ENG]   A method of working small quarries or opencast pits in steps or benches.

bench lava [GEOL]   Semiconsolidated, crusted basaltic lava forming raised platforms and crags about the edges of lava lakes. Also known as bench magma.

bench magma *See* bench lava.

bench mark [ENG]   A relatively permanent natural or artificial object bearing a marked point whose elevation above or below an adopted datum, such as sea level, is known. Abbreviated BM.

bench placer [GEOL]   A placer in ancient stream deposits from 50 to 300 feet (15 to 90 meters) above present streams.

bench slope *See* bank slope.

bend [GEOL]   **1.** A curve (not yet developed into a meander) in a stream course, bed, or channel. **2.** The land that is partly enclosed by a bend or meander.

bending [HYD]   The upward or downward movement in sea ice as wind or tide creates lateral pressure.   [OCEANOGR]   The first stage in the formation of pressure ice caused by the action of current, wind, or tide or by air temperature changes.

Benguela Current [OCEANOGR]   A strong current flowing northward along the southwestern coast of Africa.

Benioff extensometer [ENG] A linear strainmeter for measuring the change in distance between two reference points separated by 20–30 meters or more; used to observe earth tides.

Benioff zone [GEOPHYS] A zone of earthquake hypocenters distributed on well-defined planes that dips from a shallow depth into the earth's mantle to depths as great as 700 kilometers.

benitoite [MINERAL] $BaTi(SiO_3)_3$ A blue to violet barium-titanium silicate mineral; at one time it was cut and sold as sapphire.

benjaminite [MINERAL] $Pb_2(Cu,Ag)_2Bi_4S_9$ A gray mineral occurring in granular massive form.

Bennettitales [PALEOBOT] An equivalent name for the Cycadeoidales.

Bennettitatae [PALEOBOT] A class of fossil gymnosperms in the order Cycadeoidales.

bent-back occlusion [METEOROL] An occluded front that has reversed its direction of motion as a result of the development of a new cyclone (usually near the point of occlusion) or, less frequently, as the result of the displacement of the old cyclone along the front. Also known as back-bent occlusion.

benthic [OCEANOGR] Of, pertaining to, or living on the bottom or at the greatest depths of a large body of water. Also known as benthonic.

benthogene [GEOL] Pertaining to sediments that are derived from benthic plants or animals, or that are chemically precipitated on the floor of the ocean.

benthograph [ENG] A spherical submersible container for photographic equipment used in deep-sea explorations.

benthonic See benthic.

benthos [OCEANOGR] The floor or deepest part of a sea or ocean.

bentonite [GEOL] A clay formed from volcanic ash decomposition and largely composed of montmorillonite and beidellite. Also known as taylorite.

bentu de soli [METEOROL] An east wind on the coast of Sardinia.

beraunite [MINERAL] $Fe^{2+}Fe^{3+}(PO_4)_3(OH)_5 \cdot 3H_2O$ A reddish-brown to blood red, monoclinic mineral consisting of hydrated basic phosphate of ferric and ferrous iron.

beresovite See phoenicochroite.

berg crystal See rock crystal.

Bergeron-Findeisen theory [METEOROL] The theoretical explanation that precipitation particles form within a mixed cloud (composed of both ice crystals and liquid water drops) because the equilibrium vapor pressure of water vapor with respect to ice is less than that with respect to liquid water at the same temperature. Also known as ice-crystal theory; Wedener-Bergeron process.

bergmehl See rock milk.

bergschrund [HYD] A type of crevice in a glacier; formed when ice and snow break away from a rock face.

bergy bit [HYD] A piece of floating ice less than 10 m in diameter and commonly less than 5 m above sea level in height.

bergy seltzer [HYD] A sound emitted by a melting glacier which is caused by release of air bubbles that had been retained under pressure by the ice.

Bering Sea [GEOGR]   A body of water north of the Pacific Ocean, bounded by Siberia, Alaska, and the Aleutian Islands.

berkeyite *See* lazulite.

berlinite [MINERAL] $Al(PO_4)$ A colorless to gray or pale rose, hexagonal mineral consisting of aluminum orthophosphate; occurs in massive form.

berm [GEOL]   **1.** A narrow terrace which originates from the interruption of an erosion cycle with rejuvenation of a stream in the mature stage of its development and renewed dissection.   **2.** A horizontal portion of a beach or backshore formed by deposit of material as a result of wave action. Also known as backshore terrace; coastal berm.

bermanite [MINERAL] $Mn^{2+}Mn_2^{3+}(PO_4)_2(OH)_2\cdot4H_2O$ A reddish-brown, orthorhombic mineral consisting of a hydrated basic phosphate of manganese; occurs in crystal aggregates and as lamellar masses.

berm crest [GEOL]   The seaward limit and usually the highest spot on a coastal berm. Also known as berm edge.

berm edge *See* berm crest.

Bermuda high [METEOROL]   The semipermanent subtropical high of the North Atlantic Ocean, especially when it is located in the western part of that ocean area.

bermudite [PETR]   A type of extrusive rock containing phenocrysts of biotite in a groundmass that is primarily composed of analcime, nepheline, and sanidine.

Berriasian [GEOL]   Part of or the underlying stage of the Valanginian at the base of the Cretaceous.

berthierite [MINERAL] $FeSb_2S_4$ A dark steel gray, orthorhombic mineral consisting of iron antimony sulfide.

berthonite *See* bournonite.

bertrandite [MINERAL] $Be_4Si_2O_7(OH)_2$ A colorless or pale-yellow mineral consisting of a beryllium silicate occurring in prismatic crystals; hardness is 6–7 on Mohs scale, and specific gravity is 2.59–2.60.

beryllonite [MINERAL] $NaBe(PO_4)$ A colorless or yellow mineral occurring in short, prismatic or tabular, monoclinic crystals with two good pinacoidal cleavages at right angles; hardness is 5.5–6 on Mohs scale, and specific gravity is 2.85.

berzelianite [MINERAL] Cu_2Se A silver-white mineral composed of copper selenide and found in igneous rock; specific gravity is 4.03.

Bessel ellipsoid of 1841 [GEOD]   The reference ellipsoid of which the semimajor axis is 6,377,397.2 meters, the semiminor axis is 6,356,079.0 meters, and the flattening or ellipticity equals 1/299.15. Also known as Bessel spheroid of 1841.

beta chalcocite *See* chalcocite.

betafite *See* ellsworthite.

beta plane [GEOPHYS]   The model, introduced by C.G. Rossby, of the spherical earth as a plane whose rate of rotation (corresponding to the Coriolis parameter) varies linearly with the north-south direction.

betrunked river [GEOL]   A river that is shorn of its lower course as a result of submergence of the land margin by the sea.

betwixt mountains *See* median mass.

beudantite [MINERAL] $PbFe_3(AsO_4)(SO_4)(OH)_6$ A black, dark green, or brown, hexagonal mineral consisting of a basic sulfate-arsenate of lead and ferric iron; occurs as rhombohedral crystals.

beveling [GEOL] Planing by erosion of the outcropping edges of strata.

beyerite [MINERAL] $(Ca,Pb)Bi_2(CO_3)_2O_2$ A bright yellow to lemon yellow, tetragonal mineral consisting of bismuth and calcium carbonate; occurs as thin plates and compact earthy masses.

Beyrichacea [PALEON] A superfamily of extinct ostracods in the suborder Beyrichicopina.

Beyrichicopina [PALEON] A suborder of extinct ostracods in the order Paleocopa.

Beyrichiidae [PALEON] A family of extinct ostracods in the superfamily Beyrichacea.

B girdle [PETR] A circular pattern in petrofabric diagrams that indicates a B axis.

B horizon [GEOL] The zone of accumulation in soil below the A horizon (zone of leaching). Also known as illuvial horizon; subsoil; zone of accumulation; zone of illuviation.

bianchite [MINERAL] $(Fe,Zn)SO_4 \cdot 6H_2O$ A white, monoclinic mineral consisting of iron and zinc sulfate hexahydrate; occurs in crusts of indistinct crystals.

biaxial crystal [CRYSTAL] A crystal of low symmetry in which the index ellipsoid has three unequal axes.

biaxial indicatrix [CRYSTAL] An ellipsoid whose three axes at right angles to each other are proportional to the refractive indices of a biaxial crystal.

bib [GEOL] A long tract of land which slopes into the sea.

bidalotite *See* anthophyllite.

bieberite [MINERAL] $CoSO_4 \cdot 7H_2O$ A rose red or flesh red, monoclinic mineral consisting of cobalt sulfate heptahydrate; occurs as crusts and stalactites.

bielenite [PETR] A peridotite that contains pyroxenes and olivine, as well as diallage, chromite, enstatite, and magnetite.

bifurcation ratio [HYD] The ratio of number of stream segments of one order to the number of the next higher order.

bight [GEOL] **1.** A long, gradual bend or recess in the coastline which forms a large, open receding bay. **2.** A bend in a river or mountain range. [OCEANOGR] An indentation in shelf ice, fast ice, or a floe.

big inch pipe [PETRO ENG] A pipeline 24 inches (61 centimeters) in diameter which carries oil or gas, usually for great distances.

bigwoodite [PETR] A medium-grained plutonic rock consisting of microcline, microcline-microperthite, sodic plagioclase, and hornblende, aegirine-augite, or biotite.

bilinite [MINERAL] $Fe^2Fe_2^3(SO_4)_4 \cdot 22H_2O$ A white to yellowish mineral consisting of a hydrated sulfate of divalent and trivalent iron; occurs in radial-fibrous aggregates.

Billingsellacea [PALEON] A group of extinct articulate brachiopods in the order Orthida.

billow cloud [METEOROL] Broad, nearly parallel lines of cloud oriented normal to the wind direction, with cloud bases near an inversion surface. Also known as undulatus.

binary granite [PETR] **1.** A granite made up of quartz and feldspar. **2.** A granite containing muscovite mica and biotite.

bindheimite [MINERAL] $Pb_2Sb_2O_6(O,OH)$ A hydrous lead antimonate mineral produced from natural oxidation of jamesonite; found in Nevada.

binding coal *See* caking coal.

bind-seize *See* freeze.

bing ore [GEOL]   The purest lead ore, with the largest crystals of galena.

biochemical deposit [GEOL]   A precipitated deposit formed directly or indirectly from vital activities of organisms, such as bacterial iron ore and limestone.

biochron [PALEON]   A fossil of relatively short range of time.

biochronology [GEOL]   The relative age dating of rock units based on their fossil content.

bioclastic rock [PETR]   Rock formed from material broken or arranged by animals, humans, or sometimes plants; a rock composed of broken calcareous remains of organisms.

biofacies [GEOL]   **1.** A rock unit differing in biologic aspect from laterally equivalent biotic groups. **2.** Lateral variation in the biologic aspect of a stratigraphic unit.

biofog [METEOROL]   A type of steam fog caused by contact between extremely cold air and the warm, moist air surrounding human or animal bodies or generated by human activity.

biogenic chert [PETR]   Chert derived from the tests of pelagic silica-secreting organisms, chiefly diatoms and radiolarians.

biogenic gas [GEOCHEM]   Hydrocarbons, ranging from methane (CH_4) to pentane (C_5H_{12}), in the early process of formation in rocks from the remains of living organisms.

biogenic mineral [MINERAL]   A mineral in sediments or sedimentary rock which represents the hard parts of dead organisms.

biogenic reef [GEOL]   A mass consisting of the hard parts of organisms, or of a biogenically constructed frame enclosing detrital particles, in a body of water; most biogenic reefs are made of corals or associated organisms.

biogenic sediment [GEOL]   A deposit resulting from the physiological activities of organisms.

biogeochemical cycle [GEOCHEM]   The chemical interactions that exist between the atmosphere, hydrosphere, lithosphere, and biosphere.

biogeochemical prospecting [GEOCHEM]   A prospecting technique for subsurface ore deposits based on interpretation of the growth of certain plants which reflect subsoil concentrations of some elements.

biogeochemistry [GEOCHEM]   A branch of geochemistry that is concerned with biologic materials and their relation to earth chemicals in an area.

bioherm [GEOL]   A circumscribed mass of rock exclusively or mainly constructed by marine sedimentary organisms such as corals, algae, and stromatoporoids. Also known as organic mound.

biohermal limestone [PETR]   Reefs or reeflike mounds of carbonate that accumulated much in the same fashion as modern reefs and atolls of the Pacific Ocean.

biohermite [PETR]   Limestone formed of debris from a bioherm.

biolite [GEOL]   A concretion formed of concentric layers through the action of living organisms. [PETR]   *See* biolith.

biolith [PETR]   A rock formed from or by organic material. Also known as biolite.

biolithite [PETR]   An inclusive category for all organic limestone.

biological oceanography [OCEANOGR]   The study of the flora and fauna of oceans in relation to the marine environment.

biological oil-spill control [ECOL] The use of cultures of microorganisms capable of living on oil as a means of degrading an oil slick biologically.

biologic weathering *See* organic weathering.

biomicrite [PETR] A limestone resembling biosparite except that the microcrystalline calcite matrix exceeds calcite cement.

biomicrosparite [PETR] **1.** Biomicrite in which the micrite groundmass has recrystallized to form microspar. **2.** Microsparite containing fossil fragments or fossils.

biomicrudite [PETR] Biomicrite with fossil fragments or fossils greater than 1 millimeter in diameter.

biopelite *See* black shale.

biopelmicrite [PETR] A limestone similar to biopelsparite but with a microcrystalline matrix that exceeds calcite cement.

biopelsparite [PETR] A limestone similar to biosparite but with the ratio of fossils and fossil fragments to pellets between 3:1 and 1:3.

biorhexistasy [GEOL] A theory of sediment production related to diversities in the vegetation on the land surface and characterized by long-term, stable, subtropical, deep-weathering conditions.

biosparite [PETR] A limestone made up of less than 25% oolites and less than 25% intraclasts, with the ratio by volume of fossils and fragments to pellets being more than 3:1 and the calcite cement content being greater than the microcrystalline calcite content.

biostratigraphic unit [GEOL] A stratum or body of strata that is defined and identified by one or more distinctive fossil species or genera, without regard to lithologic or other physical features or relations.

biostratigraphy [PALEON] A part of paleontology concerned with the study of the conditions and deposition order of sedimentary rocks.

biostromal limestone [GEOL] Biogenic carbonate accumulations that are laterally uniform in thickness, in contrast to the moundlike nature of bioherms.

biostrome [GEOL] A bedded structure or layer (bioclastic stratum) composed of calcite and dolomitized calcarenitic fossil fragments distributed over the sea bottom as fine lentils, independent of or in association with bioherms or other areas of organic growth.

biotite [MINERAL] A black, brown, or dark green, abundant and widely distributed species of rock-forming mineral of the mica group; its chemical composition is variable: $K_2[Fe(II),Mg]_{6-4}[Fe(III),Al,Ti]_{0-2}(Si_{6-5},Al_{2-3})O_{20-22}(OH,F)_{4-2}$. Also known as black mica; iron mica; magnesia mica; magnesium-iron mica.

biotite schist [PETR] A schist composed of biotite.

bioturbation [GEOL] The disruption of marine sedimentary structures by the activity of benthic organisms.

biozone [PALEON] The range of a single taxonomic entity in geologic time as reflected by its occurrence in fossiliferous rocks.

bipedal dinosaur [PALEON] A dinosaur having two long, stout hindlimbs for walking and two relatively short forelimbs.

Bird centrifuge [MIN ENG] A dewatering machine for fine coal or other fine materials such as potash minerals, clays, and cement rock. Also known as Bird coal filter.

Bird coal filter *See* Bird centrifuge.

bird-foot delta [GEOL]   A delta formed by the outgrowth of fingers or pairs of natural levees at the mouth of river distributaries; an example is the Mississippi delta.

bird-hipped dinosaur [PALEON]   Any member of the order Ornithischia, distinguished by the birdlike arrangement of their hipbones.

bischofite [MINERAL] $MgCl_2 \cdot 6H_2O$ A colorless to white, monoclinic mineral consisting of magnesium chloride hexahydrate.

biscuit cutter [MIN ENG]   A short (6–8 inches or 15–20 centimeters) core barrel that is sharpened at the bottom and forced into the rocks by the jars.

bisectrix [CRYSTAL]   A line that is the bisector of the angle between the optic axes of a biaxial crystal.

Bishop's ring [METEOROL]   A faint, broad, reddish-brown corona occasionally seen in dust clouds, especially those which result from violent volcanic eruptions.

bisilicate *See* metasilicate.

bismite [MINERAL] Bi_2O_3 A monoclinic mineral composed of bismuth trioxide; native bismuth ore, occurring as a yellow earth. Also known as bismuth ocher.

bismuth blende *See* eulytite.

bismuth glance *See* bismuthinite.

bismuthinite [MINERAL] Bi_2S_3 A mineral consisting of bismuth trisulfide, which has an orthorhombic structure and is usually found in fibrous or leafy masses that are lead gray with a yellowish tarnish and a metallic luster. Also known as bismuth glance.

bismuth ocher *See* bismite.

bismuth spar *See* bismutite.

bismutite [MINERAL] $(BiO)_2CO_3$ A dull-white, yellowish, or gray, earthy, amorphous mineral consisting of basic bismuth carbonate. Also known as bismuth spar.

bismutotantalite [MINERAL] $Bi(Ta,Nb)O_4$ A pitch black, orthorhombic mineral consisting of an oxide of bismuth and tantalum and occurring in crystals.

bisphenoid [CRYSTAL]   A form apparently consisting of two sphenoids placed together symmetrically.

bitter lake [HYD]   A lake rich in alkaline carbonates and sulfates.

bitumenite *See* torbanite.

bituminization *See* coalification.

bituminous coal [GEOL]   A dark brown to black coal that is high in carbonaceous matter and has 15–50% volatile matter. Also known as soft coal.

bituminous lignite [GEOL]   A brittle, lustrous bituminous coal. Also known as pitch coal.

bituminous rock *See* asphalt rock.

bituminous sand [GEOL]   Sand containing bituminous-like material, such as the tar sands at Athabasca, Canada, from which oil is extracted commercially.

bituminous sandstone [PETR]   A sandstone containing bituminous matter.

bituminous shale [PETR]   A shale containing bituminous material.

bituminous wood [GEOL]   A variety of brown coal having the fibrous structure of wood. Also known as board coal; wood coal; woody lignite; xyloid coal; xyloid lignite.

bixbyite [MINERAL] $(Mn,Fe)_2O_3$ A manganese-iron oxide mineral; black cubic crystals found in cavities in rhyolite. Also known as partridgeite; sitaparite.

black alkali [GEOL] A deposit of sodium carbonate that has formed on or near the surface in arid to semiarid areas.

black amber *See* jet coal.

blackband [GEOL] An earthy carbonate of iron that is present with coal beds.

black buran *See* karaburan.

black coal *See* natural coke.

black cobalt *See* asbolite.

blackdamp [MIN ENG] A nonexplosive mixture of carbon dioxide with other gases, especially with 85–90% nitrogen, which is heavier than air and cannot support flame or life. Also known as chokedamp.

black diamond *See* carbonado.

black durain [GEOL] A durain that has high hydrogen content and volatile matter, many microspores, and some vitrain fragments.

black frost [HYD] A dry freeze with respect to its effects upon vegetation, that is, the internal freezing of vegetation unaccompanied by the protective formation of hoarfrost. Also known as hard frost.

black granite *See* diorite.

black ice [HYD] A type of ice forming on lake or salt water; compact, and dark in appearance because of its transparency.

black lead *See* graphite.

black lignite [GEOL] A lignite with a fixed carbon content of 35–60% and a total carbon content of 73.6–76.2% that contains between 6300 and 8300 Btu per pound; higher in rank than brown lignite. Also known as lignite A.

black mica *See* biotite.

black mud [GEOL] A mud formed where there is poor circulation or weak tides, such as in lagoons, sounds, or bays; the color is due to iron sulfides and organic matter.

black ocher *See* wad.

black opal [MINERAL] A variety of gem-quality opal displaying internal reflections against a dark background.

black sand [GEOL] Heavy, dark, sandlike minerals found on beaches and in stream beds; usually magnetite and ilmenite and sometimes gold, platinum, and monazite are present.

Black Sea [GEOGR] A large inland sea, area 423,000 square kilometers (163,400 square miles), bounded on the north and east by the U.S.S.R., on the south and southwest by Turkey, and on the west by Bulgaria and Rumania.

black shale [PETR] Very thinly bedded shale rich in sulfides such as pyrite and organic material deposited under barred basin conditions so that there was an anaerobic accumulation. Also known as biopelite.

black silver *See* stephanite.

black snow [HYD] Snow that falls through a particulate-laden atmosphere.

black storm *See* karaburan.

black tellurium *See* nagyagite.

bladder [GEOL] *See* vesicle.

Blaine formation [GEOL] A Permian red bed formation containing red shale and gypsum beds of marine origin in Oklahoma, Texas, and Kansas.

blairmorite [PETR] A porphyritic extrusive rock consisting mainly of analcite phenocrysts in a groundmass of sanidine, analcite, and alkalic pyroxene, with accessory sphene, melanite, and nepheline.

blakeite [MINERAL] A deep reddish-brown to deep brown mineral consisting of anhydrous ferric tellurite; occurs in massive form, as microcrystalline crusts.

Blancan [GEOL] Upper Pliocene or lowermost Pleistocene geologic time.

blanket deposit [GEOL] A flat deposit of ore; its length and width are relatively great compared with its thickness.

blanket sand [GEOL] A relatively thin body of sand or sandstone covering a large area. Also known as sheet sand.

blaster [ENG] A device for detonating an explosive charge; usually consists of a machine by which an operator, by pressing downward or otherwise moving a handle of the device, may generate a powerful transient electric current which is transmitted to an electric blasting cap. Also known as blasting machine.

blasthole [ENG] **1.** A hole that takes a heavy charge of explosive. **2.** The hole through which water enters in the bottom of a pump stock.

blastic deformation [GEOL] Rock deformation involving recrystallization in which space lattices are destroyed or replaced.

blasting [ENG] **1.** Cleaning materials by a blast of air that blows small abrasive particles against the surface. **2.** The act of detonating an explosive. [GEOL] Abrasion caused by movement of fine particles against a stationary fragment.

blasting agent [MATER] A compound or mixture, such as ammonium nitrate or black powder, that detonates as a result of heat or shock; used in mining, blasting, pyrotechnics, and propellants.

blasting barrel [MIN ENG] A piece of iron pipe, usually about one-half inch in diameter, used to provide a smooth passageway through the stemming for the miner's squib; it is recovered after each blast and used until destroyed.

blasting cap [ENG] A copper shell closed at one end and containing a charge of detonating compound, which is ignited by electric current or the spark of a fuse; used for detonating high explosives.

blasting fuse [ENG] A core of gunpowder in the center of jute, yarn, and so on for igniting an explosive charge in a shothole.

blasting gelatin [MATER] A plastic dynamite that contains 5–10% nitrocellulose added to nitroglycerin; used principally in submarine work.

blasting machine *See* blaster.

blasting powder [MATER] A powder containing less nitrate, and in its place more charcoal than black powder; composition is 65–75% sodium or nitrate, potassium nitrate, 10–15% sulfur, and 15–20% charcoal.

blastogranitic rock [PETR] A metamorphic granitic rock which still has parts of the original granitic texture.

Blastoidea [PALEON] A class of extinct pelmatozoan echinoderms in the subphylum Crinozoa.

blastomylonite [PETR] Rock which has recrystallized after granulation.

blastopelitic [PETR] Descriptive of the structure of metamorphosed argillaceous rocks.

blastophitic [PETR]   A metamorphosed rock which once contained lath-shaped crystals partly or wholly enclosed in augite and in which part of the original texture remains.

blastoporphyritic [PETR]   Applied to the textures of metamorphic rocks that are derived from porphyritic rocks; the porphyritic character still remains as a relict feature.

blastopsammite [GEOL]   A relict fragment of sandstone that is contained in a metamorphosed conglomerate.

blastopsephitic [GEOL]   Descriptive of the structure of metamorphosed conglomerate or breccia.

blast roasting [MIN ENG]   The roasting of finely divided ores by means of a blast to maintain internal combustion within the charge. Also known as roast sintering.

bleach spot [GEOL]   A green or yellow area in red rocks formed by reduction of ferric oxide around an organic particle. Also known as deoxidation sphere.

blende *See* sphalerite.

blended unconformity [GEOL]   An unconformity that is not sharp because the original erosion surface was covered by a thick residual soil that graded downward into the underlying rock.

blind [GEOL]   Referring to a mineral deposit that does not include an outcrop at the surface.

blind apex [GEOL]   The upper end of a seam or vein of a mineral deposit.

blind coal *See* natural coke.

blind drainage *See* closed drainage.

blind drift [MIN ENG]   In a mine, a horizontal passage not yet connected with the other workings.

blinding [MIN ENG]   Interference with the functioning of a screen mesh by a matting of fine materials during screening. Also known as blocking; plugging.

blind island [GEOL]   A deposit of marl or organic matter found at shallow depth in a lake.

blind rollers [OCEANOGR]   Long, high swells which have increased in height, almost to the breaking point, as they pass over shoals or run in shoaling water. Also known as blind seas.

blind seas *See* blind rollers.

blind valley [GEOL]   A valley that has been made by a spring from an underground channel which emerged to form a surface stream, and that is enclosed at the head of the stream by steep walls.

blink [METEOROL]   A brightening of the base of a cloud layer, caused by the reflection of light from a snow- or ice-covered surface.

blister [GEOL]   A domelike protuberance caused by the buckling of the cooling crust of a molten lava before the flowing mass has stopped. [MIN ENG]   A protrusion, more or less circular in plan, extending downward into a coal seam.

blister hypothesis [GEOL]   A theory of the formation of compressional mountains by a process in which radiogenic heat expands and melts a portion of the earth's crust and subcrust, causing a domed regional uplift (blister) on a foundation of molten material that has no permanent strength.

blizzard [METEOROL] A severe weather condition characterized by low temperatures and by strong winds bearing a great amount of snow (mostly fine, dry snow picked up from the ground).

blob [METEOROL] In radar, oscilloscope evidence of a fairly small-scale temperature and moisture inhomogeneity produced by turbulence within the atmosphere.

block [MIN ENG] A division of a mine, usually bounded by workings but sometimes by survey lines or other arbitrary limits. [PETRO ENG] The subdivision of a sea area for the licensing of oil and gas exploration and production rights.

block caving [MIN ENG] A method of caving where a block, 150–250 feet (46–77 meters) on a side and several hundred feet high, is induced to cave in after it is undercut; the broken ore is drawn off at a bell-shaped draw point.

block clay See mélange.

block coal [MIN ENG] **1.** A bituminous coal that breaks into large lumps or cubical blocks. **2.** Coal that passes over 5-, 6-, and 8-inch (127, 152, and 203 millimeter) block screens; used in smelting iron.

block correction [MAP] A corrected reproduction of a small area of a nautical chart that is pasted to the chart for which it is issued.

blocked-out ore See developed reserves.

block faulting [GEOL] A type of faulting in which fault blocks are displaced at different orientations and elevations.

block field [GEOL] A thin blanket of rocks with no fine sizes visible, over solid or weathered bedrock or over colluvium, with no cliff or ledge rock above as a source; it is found on mountain slopes of less than 5°, usually above the tree line.

block glide [GEOL] A landslide in which the mass basically remains intact as it moves out and down.

blocking [MIN ENG] See blinding.

blocking and wedging [MIN ENG] A method of holding mine timber sets in place; blocks of wood are set on the caps directly over the post supports and have a grain of block parallel with the top of the cap; wedges are driven tightly between the blocks and the roof.

blocking anticyclone See blocking high.

blocking high [METEOROL] Any high (or anticyclone) that remains nearly stationary or moves slowly compared to the west-to-east motion upstream from its location, so that it effectively blocks the movement of migratory cyclones across its latitudes. Also known as blocking anticyclone.

blocking out [MIN ENG] In economic geology, the operation of exposing an orebody on three sides in order to develop it.

block lava [GEOL] Lava flows which occur as a tumultuous assemblage of angular blocks.

block mountain [GEOL] A mountain formed by the combined processes of uplifting, faulting, and tilting. Also known as fault-block mountain.

block movement [GEOL] The general collapse of a hanging wall.

block slope [GEOL] A thin blanket of rocks with no fine sizes visible, over solid or weathered bedrock or over colluvium, with no cliff or ledge rock above as a source; it is found on mountain slopes of more than 5° but less than 25°, usually above the tree line.

block system [MIN ENG] A system of pillars in which a series of entries, rooms, and crosscuts are driven to divide the coal into blocks of about equal size which are then extracted on retreat.

blocky iceberg [OCEANOGR] An iceberg with steep, precipitous sides and with either a horizontal or nearly horizontal upper surface.

blödite *See* bloedite.

bloedite [MINERAL] $MgSO_4 \cdot Na_2SO_4 \cdot 4H_2O$ A white or colorless monoclinic mineral consisting of magnessium sodium sulfate. Also spelled blödite. Also known as astrakanite; astrochanite.

blomstrandine *See* priorite.

blood rain [METEOROL] Rain of a reddish color caused by dust particles containing iron oxide that were picked up by the raindrops during descent.

bloodstone [MINERAL] **1.** A form of deep green chalcedony flecked with red jasper. Also known as heliotrope; oriental jasper. **2.** *See* hematite.

bloom [GEOL] *See* blossom. [MINERAL] *See* efflorescence.

blossom [GEOL] The oxidized or decomposed outcrop of a vein or coal bed. Also known as bloom.

blotter model [PETRO ENG] An analysis device in which the analogous movement of copper ammonium or zinc ammonium ions in blotting paper or gelatin indicates oil-well injection-fluid movement through porous underground reservoirs.

blowhole [GEOL] A longitudinal tunnel opening in a sea cliff, on the upland side away from shore; columns of sea spray are thrown up through the opening, usually during storms.

blow in [MET] To put a blast furnace into operation. [PETRO ENG] Of an oil well, to begin sending forth oil or gas.

blowing cave [GEOL] A cave with an alternating air movement. Also known as breathing cave.

blowing dust [METEOROL] Dust picked up locally from the surface of the earth and blown about in clouds or sheets.

blowing fan *See* forcing fan.

blowing sand [METEOROL] Sand picked up from the surface of the earth by the wind and blown about in clouds or sheets.

blowing snow [METEOROL] Snow lifted from the surface of the earth by the wind to a height of 6 feet (1.8 meters) or more (higher than drifting snow) and blown about in such quantities that horizontal visibility is restricted.

blowing spray [METEOROL] Spray lifted from the sea surface by the wind and blown about in such quantities that horizontal visibility is restricted.

blowout [GEOL] Any of the various trough-, saucer-, or cuplike hollows formed by wind erosion on a dune or other sand deposit. [HYD] A bubbling spring which bursts from the ground behind a river levee when water at flood stage is forced under the levee through pervious layers of sand or silt. Also known as sand boil. [PETRO ENG] A sudden, unplanned escape of oil or gas from a well during drilling.

blowout dune *See* parabolic dune.

blowout pond [HYD] A small, shallow, temporary body of water occupying a blowout.

blow well [HYD] A flowing artesian well.

blue amber [MINERAL] A type of osseous amber that has a blue tinge.

blue asbestos *See* crocidolite.

blue band [GEOL] **1.** A layer of bubble-free, dense ice found in a glacier. **2.** A bluish clay found as a thin, persistent bed near the base of No. 6 coal everywhere in the Illinois-Indiana coal basin.

blue cap [MIN ENG] The characteristic blue halo, or tip, of the flame of a safety lamp when firedamp is present in the air.

blue copper ore *See* azurite.

blue ground [GEOL] **1.** The decomposed peridotite or kimberlite that carries the diamonds in the South African mines. **2.** Strata of the coal measures, consisting principally of beds of hard clay or shale.

blue ice [HYD] Pure ice in the form of large, single crystals that is blue owing to the scattering of light by the ice molecules; the purer the ice, the deeper the blue.

blue iron earth *See* vivianite.

blue lead *See* galena.

blue magnetism [GEOPHYS] The magnetism displayed by the south-seeking end of a freely suspended magnet; this is the magnetism of the earth's north magnetic pole.

blue malachite *See* azurite.

blue metal [GEOL] The common fine-grained blue-gray mudstone which is part of many of the coal beds of England.

blue mud [GEOL] A combination of terrigenous and deep-sea sediments having a bluish gray color due to the presence of organic matter and finely divided iron sulfides.

blue ocher *See* vivianite.

blueschist facies [PETROL] High-pressure, low-temperature metamorphism associated with subduction zones which produces a broad mineral association including glaucophane, actinolite, jadeite, aegirine, lawsonite, and pumpellyite.

blue-sky scale *See* Linke scale.

blue spar *See* lazulite.

bluestone [MINERAL] *See* chalcanthite. [PETR] **1.** A sandstone that is highly argillaceous and of even texture and bedding. **2.** The commercial name for a feldspathic sandstone that is dark bluish gray; it is easily split into thin slabs and used as flagstone.

blue vitriol [MINERAL] *See* chalcanthite.

bluff [GEOGR] **1.** A steep, high bank. **2.** A broad-faced cliff.

BM *See* bench mark.

BM-AGA coal test [ENG] A laboratory test developed jointly by the United States Bureau of Mines and a committee of the American Gas Association which forecasts the quality of coke producible in commercial practice.

board coal *See* bituminous wood.

boathook bend [HYD] The portion of a tributary that displays a sharp curve as it joins the main stream in an upstream direction, following a barbed drainage pattern.

Bobasatranidae [PALEON] A family of extinct palaeonisciform fishes in the suborder Platysomoidei.

bobierrite [MINERAL] $Mg_3(PO_4)_2 \cdot 8H_2O$ A transparent, colorless or white, monoclinic mineral consisting of octahydrated magnesium phosphate.

bodenite [MINERAL] A metallic, steel-gray mineral consisting of cobalt, nickel, iron, arsenic, and bismuth; occurs in granular to fibrous masses.

bodily tide *See* earth tide.

body   [GEOGR]   A separate entity or mass of water, such as an ocean or a lake. [GEOL]   An ore body, or pocket of mineral deposit.

body-centered lattice   [CRYSTAL]   A space lattice in which the point at the intersection of the body diagonals is identical to the points at the corners of the unit cell.

body wave   [GEOPHYS]   A seismic wave that travels within the earth, as distinguished from one that travels along the surface.

boehmite   [MINERAL]   AlO(OH)   Gray, brown, or red orthorhombic mineral that is a major constituent of some bauxites.

boehm lamellae   [GEOL]   Lines or bands with dusty inclusions that are subparallel to the basal plane of quartz.

bog burst   [GEOL]   The bursting of a bog under pressure due to water retention caused by a marginal dam formed from growing vegetation.

bogen structure   [GEOL]   The structure of vitric tuffs composed largely of shards of glass.

bog flow   [GEOL]   A mudflow that issues from a bog burst.

boghead cannel shale   [GEOL]   A coaly shale that contains much waxy or fatty algae.

boghead coal   [GEOL]   Bituminous or subbituminous coal containing a large proportion of algal remains and volatile matter; similar to cannel coal in appearance and combustion.

bogie   [MIN ENG]   A small truck or trolley upon which a bucket is carried from the shaft to the spoil bank.

bog iron ore   [MINERAL]   A soft, spongy, porous deposit of impure hydrous iron oxides formed in bogs, marshes, swamps, peat mosses, and shallow lakes by precipitation from iron-bearing waters and by the oxidation action of algae, iron bacteria, or the atmosphere. Also known as lake ore; limnite; marsh ore; meadow ore; morass ore; swamp ore.

bog manganese *See* wad.

bog-mine ore *See* bog ore.

bog ore   [MINERAL]   A poorly stratified accumulation of earthy metallic mineral substances, consisting mainly of oxides, that are formed in bogs, marshes, swamps, and other low-lying moist places. Also known as bog-mine ore.

Bohemian ruby *See* rose quartz.

Bohemian topaz *See* citrine.

boiler plate   [GEOL]   A fairly smooth surface on a cliff, consisting of flush or overlapping slabs of rock, having little or no foothold.   [HYD]   Crusty, frozen surface of snow.

boiling spring   [HYD]   **1.** A spring which emits water at a high temperature or at boiling point. **2.** A spring located at the head of an interior valley and rising from the bottom of a residual clay basin. **3.** A rapidly flowing spring that develops strong vertical eddies.

bojite   [PETR]   **1.** A gabbro with primary hornblende substituting for augite. **2.** Hornblende diorite.

bole   [GEOL]   Any of various red, yellow, or brown earthy clays consisting chiefly of hydrous aluminum silicates. Also known as bolus; terra miraculosa.

boleite   [MINERAL]   A deep Prussian blue, tetragonal mineral consisting of a hydroxide-chloride of lead, copper, and silver.

boll-weevil hanger [PETRO ENG]   A screw-on connector used to connect and seal the lower end of a length of oil-well casing to the next smaller casing string.

bolson [GEOL]   In the southwestern United States, a basin or valley having no outlet.

bolson plain [GEOL]   A broad plain in the middle of a bolson or semibolson, made up of deep alluvial accumulations that have been washed into the basin from the mountains around it.

bolt *See* bolthole.

bolthole [MIN ENG]   A short, narrow opening made to connect the main working with the airhead or ventilating drift of a coal mine. Also known as bolt.

bolting [MIN ENG]   The use of vibrating sieves to separate particles of different sizes.

boltwoodite [MINERAL] $K_2(UO_2)_2(SiO_3)_2(OH)_2 \cdot 5H_2O$ Yellow mineral consisting of hydrous potassium uranyl silicate.

bolus *See* bole.

bolus alba *See* kaolin.

bomb [GEOL]   Any large (greater than 64 millimeters) pyroclast ejected while viscous.

bombiccite *See* hartite.

bomb sag [GEOL]   Depressed and deranged laminae mainly found in beds of fine-grained ash or tuff around an included volcanic bomb or block which fell on and became buried in the deposit.

Bone Age [ARCHEO]   A prehistoric period of human culture characterized by the use of implements made of bone and antler.

bone bed [GEOL]   Several thin strata or layers with many fragments of fossil bones, scales, teeth, and also organic remains.

bone breccia [PETR]   A rock formed from the accumulation of bones or bone fragments that have been cemented with calcium carbonate.

bone chert [PETR]   A weathered residual chert that appears chalky and porous with a white color but may be stained red or other colors.

bone coal [GEOL]   Argillaceous coal or carbonaceous shale that is found in coal seams.

bone phosphate [PETR]   A large bone bed containing enough calcium phosphate that it can be classified as phosphate rock.

boninite [PETR]   An andesitic rock that contains much glass and abundant phenocrysts of bronzite and less of olivine and augite.

Bonne projection [MAP]   A type of conical map projection; meridians are plotted as curves and the parallels are spaced along them at true distances.

Bononian [GEOL]   Upper Jurassic (lower Portlandian) geologic time.

bonus [PETRO ENG]   Payment by a lessee of an oil- or gas-production royalty to the landowner at a rate greater than the customary one-eighth of the value of the oil or gas withdrawn.

book *See* mica book.

book structure [GEOL]   A rock structure of numerous parallel sheets of slate alternating with quartz.

boomer [MIN ENG]   In placer mining, an automatic gate in a dam that holds the water until the reservoir is filled, then opens automatically and allows the escape of such a volume that the soil and upper gravel of the placer are washed away.

boomerang sediment corer [ENG] A device, designed for nighttime recovery of a sediment core, which automatically returns to the surface after taking the sample.

boorga *See* burga.

boost [ENG] To bring about a more potent explosion of the main charge of an explosive by using an additional charge to set it off.

booster stations [ENG] Booster pumps or compressors located at intervals along a liquid-products or gas pipeline to boost the pressure of the flowing fluid to keep it moving toward its destination.

boot [MIN ENG] **1.** A projecting portion of a reinforced concrete beam acting as a corbel to support the facing material, such as brick or stone. **2.** The lower end of a bucket elevator. [PETRO ENG] *See* surge column.

boothite [MINERAL] $CuSO_4 \cdot 7H_2O$ A blue, monoclinic mineral consisting of copper sulfate heptahydrate; usually occurs in massive or fibrous form.

bootleg [MIN ENG] A hole, shaped somewhat like the leg of a boot, caused by a blast that has failed to shatter the rock properly.

bora [METEOROL] A fall wind whose source is so cold that when the air reaches the lowlands or coast the dynamic warming is insufficient to raise the air temperature to the normal level for the region; hence it appears as a cold wind.

boracite [MINERAL] $Mg_3B_7O_{13}Cl$ A white, yellow, green, or blue orthorhombic borate mineral occurring in crystals which appear isometric in external form; it is strongly pyroelectric, has a hardness of 7 on Mohs scale, and a specific gravity of 2.9.

bora fog [METEOROL] A dense fog caused when the bora lifts a spray of small drops from the surface of the sea.

Boralf [GEOL] A suborder of the soil order Alfisol, dull brown or yellowish brown in color; occurs in cool or cold regions, chiefly at high latitudes or high altitudes.

borasca [MIN ENG] An unproductive area of a mine or orebody.

borate mineral [MINERAL] Any of the large and complex group of naturally occurring crystalline solids in which boron occurs in chemical combination with oxygen.

borax [MINERAL] $Na_2B_4O_7 \cdot 10H_2O$ A white, yellow, blue, green, or gray borate mineral that is an ore of boron and occurs as an efflorescence or in monoclinic crystals; when pure it is used as a cleaning agent, antiseptic, and flux. Also known as diborate; pyroborate; tincal.

border facies [GEOL] The marginal part of an igneous intrusion which is different in both texture and composition from the major portion of the intrusion.

borderland [GEOL] One of the crystalline, continental landmasses postulated to have existed on the exterior (oceanward) side of geosynclines.

borderland slope [GEOL] A declivity which indicates the inner margin of the borderland of a continent.

bore [OCEANOGR] **1.** A high, breaking wave of water, advancing rapidly up an estuary. Also known as eager; mascaret; tidal bore. **2.** A submarine sand ridge, in very shallow water, whose crest may rise to intertidal level.

borehole [ENG] A hole made by boring into the ground to study stratification, to obtain natural resources, or to release underground pressures.

borehole bit *See* noncoring bit.

borehole mining [PETRO ENG] Extraction of minerals as liquid or gas from the earth's crust by means of boreholes and suction pumps.

borehole survey [ENG]   Also known as drillhole survey. **1.** Determining the course of and the target point reached by a borehole, using an azimuth-and-dip recording apparatus small enough to be lowered into a borehole. **2.** The record of the information thereby obtained.

borickite [MINERAL] $CaFe_5(PO_4)_2(OH)_{11} \cdot 3H_2O$ A reddish-brown, isotropic mineral consisting of a hydrated basic phosphate of calcium and iron; occurs in compact reniform masses.

boring log *See* drill log.

bornite [MINERAL] Cu_5FeS_4 A primary mineral in many copper ore deposits; specific gravity 5.07; the metallic and brassy color of a fresh surface rapidly tarnishes upon exposure to air to an iridescent purple.

boroarsenate [MINERAL]   One of a group of borate minerals containing arsenic; cahnite is an example.

borolanite [PETR]   A hypabyssal rock that is essentially orthoclase and melanite with subordinate nepheline, biotite, and pyroxene.

Boroll [GEOL]   A suborder of the soil order Mollisol, characterized by a mean annual soil temperature of less than 8°C and by never being dry for 60 consecutive days during the 90-day period following the summer solstice.

boronatrocalcite *See* ulexite.

bort [MINERAL]   Imperfectly crystallized diamond material unsuitable for gems because of its shape, size, or color and because of flaws or inclusions; used for abrasive and cutting purposes.

bosporus [GEOGR]   A strait connecting two seas or a lake and a sea.

boss [GEOL]   A large, irregular mass of crystalline igneous rock that formed some distance below the surface but is now exposed by denudation.

bostonite [PETR]   A rock with coarse trachytic texture formed almost wholly of albite and microcline and with accessory pyroxene.

botallackite [MINERAL] $Cu_2(OH)_3Cl \cdot 3H_2O$ A pale bluish-green to green, orthorhombic mineral consisting of a basic copper chloride; occurs as crusts of crystals.

Bothriocidaroida [PALEON]   An order of extinct echinoderms in the subclass Perischoechinoidea in which the ambulacra consist of two columns of plates, the interambulacra of one column, and the madreporite is placed radially.

botryogen [MINERAL] $MgFe(SO_4)_2(OH) \cdot 7H_2O$ An orange-red, monoclinic mineral consisting of a hydrated basic sulfate of magnesium and trivalent iron.

botryoid [GEOL]   **1.** A mineral formation shaped like a bunch of grapes. **2.** Specifically, such a formation of calcium carbonate occurring in a cave. Also known as clusterite.

bottle spring [HYD]   A spring issuing fresh water from the bottom of a saline lake or pool.

bottle thermometer [ENG]   A thermoelectric thermometer used for measuring air temperature; the name is derived from the fact that the reference thermocouple is placed in an insulated bottle.

bottom [GEOL]   **1.** The bed of a body of running or still water. **2.** *See* root.

bottom break [MIN ENG]   The break or crack that separates a block of stone from a quarry floor.

bottomed hole [ENG]   A completed borehole, or a borehole in which drilling operations have been discontinued.

bottom flow [HYD] A density current that is denser than any section of the surrounding water and that flows along the bottom of the body of water. Also known as underflow.

bottom-hole cash [PETRO ENG] Cash which is contributed by mineral-rights lessees adjacent to a drilling lessee and which is payable when the well reaches a specified depth, regardless of whether or not the completed well is a producer.

bottom-hole packer [PETRO ENG] An anchored-in-place seal used to provide liquid-proof packing in the annular space between the outside of the oil-producing tubing and the inside of the drill casing.

bottom-hole pressure [PETRO ENG] Gas-drive pressure recorded at the bottom of an oil-well shaft; used to analyze oil-reservoir performance and evaluate the performance of downhole equipment.

bottom-hole samples [PETRO ENG] Fluid samples from gas-condensate-well reservoirs; used to study the state of the hydrocarbon system under reservoir conditions and to estimate total hydrocarbons in place.

bottom ice *See* anchor ice.

bottomland [GEOL] A lowland formed by alluvial deposit about a lake basin or a stream.

bottom load [GEOL] An obsolete term for the material rolled and pushed along the bottom of a stream.

bottom moraine *See* ground moraine.

bottom pillar [MIN ENG] A large block of solid coal left unworked around the shaft.

bottoms *See* basic sediment and water.

bottom sampler [ENG] Any instrument used to obtain a sample from the bottom of a body of water.

bottom sediment [PETRO ENG] A mixture of liquids and solids which form in the bottom of oil storage tanks.

bottomset beds [GEOL] Horizontal or gently inclined layers of finer material carried out and deposited on the bottom of a lake or sea in front of a delta.

bottom settlings *See* basic sediment and water.

bottom terrace [GEOL] A landform resulting from deposits by streams with moderate or small bottom loads of coarse sand and gravel, and characterized by a broad, gentle slope in the direction of flow and a steep escarpment facing downstream.

bottom water [HYD] Water lying beneath oil or gas in productive formations. [OCEANOGR] The water mass at the deepest part of a water column in the ocean.

boudin [GEOL] One of a series of sausage-shaped segments found in a boudinage.

boudinage [GEOL] A structure in which beds set in a softer matrix are divided by cross fractures into segments resembling pillows.

Bouguer correction *See* Bouguer reduction.

Bouguer gravity anomaly [GEOPHYS] A value that corrects the observed gravity for latitude and elevation variations, as in the free-air gravity anomaly, plus the mass of material above some datum (usually sea level) within the earth and topography.

Bouguer reduction [GEOL] A correction made in gravity work to take account of the station's altitude and the rock between the station and sea level. Also known as Bouguer correction.

Bouguer's halo [METEOROL] A faint, white circular arc of light of about 39° radius around the antisolar point. Also known as Ulloa's ring.

boulangerite [MINERAL] $Pb_5Sb_4S_{11}$ A bluish-lead-gray, monoclinic mineral consisting of lead antimony sulfide.

boulder [GEOL] A worn rock with a diameter exceeding 256 millimeters. Also spelled bowlder.

boulder barricade [GEOL] A group of large boulders that are visible along a coast between low and half tide.

boulder belt [GEOL] A long, narrow accumulation of boulders elongately transverse to the direction of glacier movement.

boulder buster [ENG] A heavy, pyramidical- or conical-point steel tool which may be attached to the bottom end of a string of drill rods and used to break, by impact, a boulder encountered in a borehole. Also known as boulder cracker.

boulder clay *See* till.

boulder cracker *See* boulder buster.

boulder pavement [GEOL] A surface of till with boulders; the till has been abraded to flatness by glacier movement.

boulder train [GEOL] Glacial boulders derived from one locality and arranged in a right-angled line or lines leading off in the direction in which the drift agency operated.

boule [CRYSTAL] A pure crystal, such as silicon, having the atomic structure of a single crystal, formed synthetically by rotating a small seed crystal while pulling it slowly out of molten material in a special furnace.

bounce cast [GEOL] A short ridge underneath a stratum fading out gradually in both directions.

boundary [GEOL] A line between areas occupied by rocks or formations of different type and age.

boundary line [MAP] A line of demarcation along which two areas meet. [PHYS CHEM] On a phase diagram, the line along which any two phase areas adjoin in a binary system, or the line along which any two liquidus surfaces intersect in a ternary system.

boundary map [MAP] A map constructed for the purpose of delineating a boundary line and adjacent territory.

boundary monument [ENG] A material object placed on or near a boundary line to preserve and identify the location of the boundary line on the ground.

boundary pillar [MIN ENG] A pillar left in mines between adjoining properties.

boundary survey [ENG] A survey made to establish or to reestablish a boundary line on the ground or to obtain data for constructing a map or plat showing a boundary line.

boundary wave [GEOPHYS] A seismic wave that propagates along a free surface or along an interface between definite layers.

bournonite [MINERAL] $PbCuSbS_3$ Steel-gray to black orthorhombic crystals; mined as an ore of copper, lead, and antimony. Also known as berthonite; cogwheel ore.

boussingaultite [MINERAL] $(NH_4)_2Mg(SO_4)_2 \cdot 6H_2O$ A colorless to yellowish-pink, monoclinic mineral consisting of a hydrated sulfate of ammonium and magnesium; usually occurs in massive form, as crusts or stalactites.

Bowen reaction series [MINERAL] A series of minerals wherein any early-formed phase will react with the melt later in the differentiation to yield a new mineral further in the series.

Bowie formula [GEOPHYS]   A correction used for calculation of the local gravity anomaly on earth.

bowk *See* hoppit.

bowlder *See* boulder.

bowlingite *See* saponite.

box canyon [GEOGR]   A canyon with steep rock sides and a zigzag course, that is usually closed upstream.

box fold [GEOL]   A fold in which the broad, flat top of an anticline or the broad, flat bottom of a syncline is bordered by steeply dipping limbs.

Box Hole [GEOL]   A meteorite crater in central Australia, 575 feet (244 meters) in diameter.

boxwork [GEOL]   Limonite and other minerals which formed at one time as blades or plates along cleavage or fracture planes, after which the intervening material dissolved, leaving the intersecting blades or plates as a network.

brace head [ENG]   A cross handle attached at the top of a column of drill rods by means of which the rods and attached bit are turned after each drop in chop-and-wash operations while sinking a borehole through overburden. Also known as brace key.

brace key *See* brace head.

brachyaxis [CRYSTAL]   The shorter lateral axis, usually the a axis, of an orthorhombic or triclinic crystal. Also known as brachydiagonal.

brachydiagonal *See* brachyaxis.

brachypinacoid [GEOL]   A pinacoid parallel to the vertical and the shorter lateral axis.

brachysyncline [GEOL]   A broad, short syncline.

Brachythoraci [PALEON]   An order of the extinct joint-necked fishes.

brackebuschite [MINERAL] $Pb_4MnFe(VO_4)_4 \cdot 2H_2O$ A dark brown to black, monoclinic mineral consisting of a hydrated vanadate of lead, manganese, and iron.

brackish [HYD]   **1.** Of water, having salinity values ranging from approximately 0.50 to 17.00 parts per thousand. **2.** Of water, having less salt than sea water, but undrinkable.

Bradford breaker [MIN ENG]   A machine which combines coal crushing and screening.

Bradfordian [GEOL]   Uppermost Devonian geologic time.

Bradford preferential separation process [MIN ENG]   A flotation process for the treatment of mixed sulfides in which certain mineral salts, such as thiosulfates, are added to the water used in the flotation cells.

bradleyite [MINERAL] $Na_3Mg(PO_4)(CO_3)$ A light gray mineral consisting of a phosphate-carbonate of sodium and magnesium; occurs as fine-grained masses.

Bradyodonti [PALEON]   An order of Paleozoic cartilaginous fishes (Chondrichthyes), presumably derived from primitive sharks.

braggite [MINERAL]   PtS A steel-gray platinum sulfide mineral with tetragonal crystals.

braided stream [HYD]   A stream flowing in several channels that divide and reunite.

brammalite [MINERAL]   A mica-type clay mineral that is different from illite because it has soda instead of potash; it is the sodium analog of illite. Also known as sodium illite.

branch [HYD]   A stream that flows into another stream, usually of larger size.

branching bay *See* estuary.

branch island   [GEOGR]   An island formed by the successive branching and rejoining of a stream.

branchite *See* hartite.

brandtite   [MINERAL] $Ca_2Mn(AsO_4)_2 \cdot 2H_2O$ A colorless to white, monoclinic mineral consisting of a hydrated arsenate of calcium and manganese.

brannerite   [MINERAL]   A complex, black, opaque titanite of uranium and other elements in which the weight of uranium exceeds that of titanium; monoclinic and possibly $(U,Ca,Fe,Y,Th)_3Ti_5O_6$.

brash ice   [HYD]   An accumulation of floating ice fragments, measuring 2 m or less in diameter, that represents wreckage of other ice forms.

brass   [GEOL]   A British term for sulfides of iron (pyrites) in coal. Also known as brasses.

brasses *See* brass.

brass ore *See* aurichalcite.

brattice   [MIN ENG]   A temporary board or cloth partition in any mine passage to confine the air and force it into the working places. Also spelled brattish; brettice; brettis.

brattice cloth   [MIN ENG]   Fire-resistant canvas or duck used to erect a brattice.

brattish *See* brattice.

braunite   [MINERAL] $3Mn_2O_3 \cdot MnSiO_3$ Brittle mineral that forms tetragonal crystals; commonly found as steel-gray or brown-black masses in the United States, Europe, and South America; it is an ore of manganese.

Bravais indices   [CRYSTAL]   A modification of the Miller indices; frequently used for hexagonal and trigonal crystalline systems; they refer to four axes: the c-axis and three others at 120° angles in the basal plane.

Bravais lattice   [CRYSTAL]   One of the 14 possible arrangements of lattice points in space such that the arrangement of points about any chosen point is identical with that about any other point.

brave west winds   [METEOROL]   A nautical term for the strong and rather persistent westerly winds over the oceans in temperate latitudes, between 40° and 50°S.

bravoite   [MINERAL] $(Ni,Fe)S_2$ A yellow sulfide ore of nickel containing iron.

Brazil Current   [OCEANOGR]   The warm ocean current that flows southward along the Brazilian coast below Natal; the western boundary current in the South Atlantic Ocean.

Brazilian emerald   [MINERAL]   A transparent green variety of tourmaline found in Brazil.

brazilianite   [MINERAL] $NaAl_3(PO_4)_2(OH)_4$ A chartreuse yellow to pale yellow, monoclinic mineral consisting of a basic phosphate of sodium and aluminum.

Brazilian ruby   [MINERAL]   A red mineral found only in Brazil that resembles a ruby but can be spinel, topaz, or tourmaline of the proper color.

Brazilian sapphire   [MINERAL]   A transparent blue variety of tourmaline found in Brazil.

breached anticline   [GEOL]   An anticline that has been more deeply eroded in the center. Also known as scalped anticline.

breached cone   [GEOL]   A cinder cone in which lava has broken through the sides and broken material has been carried away.

breadcrust [GEOL] A surficial structure resembling a crust of bread, as the concretions formed by evaporation of salt water.

breadcrust bomb [GEOL] A volcanic bomb with a cracked exterior.

break [GEOGR] A significant variation of topography, such as a deep valley. [GEOL] *See* knickpoint. [METEOROL] **1.** A sudden change in the weather; usually applied to the end of an extended period of unusually hot, cold, wet, or dry weather. **2.** A hole or gap in a layer of clouds. [MIN ENG] **1.** A plane of discontinuity in the coal seam such as a slip, fracture, or cleat; the surfaces are in contact or slightly separated. **2.** A fracture or crack in the roof beds as a result of mining operations.

breaker [MIN ENG] **1.** In anthracite mining, the structure in which the coal is broken, sized, and cleaned for market. Also known as coalbreaker. **2.** One of a row of drill holes above the mining holes in a tunnel face. [OCEANOGR] A wave breaking on a shore, over a reef, or other mass in a body of water.

breaker depth [OCEANOGR] The still-water depth determined at that point where a wave breaks. Also known as breaking depth.

breaker terrace [GEOL] A type of shore found in lakes in glacial drift; the terrace is formed from stones deposited by waves.

breaking depth *See* breaker depth.

breaking-drop theory [GEOPHYS] A theory of thunderstorm charge separation based upon the suggested occurrence of the Lenard effect in thunderclouds, that is, the separation of electric charge due to the breakup of water drops.

breaks in overcast [METEOROL] In United States weather observing practice, a condition wherein the cloud cover is more than 0.9 but less than 1.0.

breakthrough [MIN ENG] A passage cut through the pillar to allow the ventilating current to pass from one room to another; larger than a doghole. Also known as room crosscut.

breakthrough sweep efficiency [PETRO ENG] The completeness with which an oil-field waterflood sweeps through a reservoir area; related to the critical water saturation (point beyond which oil will not be pushed ahead of the water flow).

break thrust [GEOL] A thrust fault cutting across one limb of a fold.

breakup [HYD] The spring melting of snow, ice, and frozen ground; specifically, the destruction of the ice cover on rivers during the spring thaw.

breast [MIN ENG] **1.** In coal mines, a chamber driven in the seam from the gangway, for the extraction of coal. **2.** *See* face.

breathing cave *See* blowing cave.

breathing well [HYD] A water well that responds to changes in atmospheric pressure by alternately taking in and emitting a strong current of air.

breccia [PETR] A rock made up of very angular coarse fragments; may be sedimentary or may be formed by grinding or crushing along faults.

breccia dike [GEOL] A dike formed of breccia injected into the country rock.

breccia marble [PETR] Any marble containing angular fragments.

breccia pipe *See* pipe.

breeze [METEROL] **1.** A light, gentle, moderate, fresh wind. **2.** In the Beaufort scale, a wind speed ranging from 4 to 31 miles (6.4 to 49.6 kilometers) per hour.

breithauptite [MINERAL] NiSb A light copper red mineral consisting of nickel antimonide; commonly occurs in association with silver minerals.

Bretonian orogeny [GEOL]   Post-Devonian diastrophism that is found in Nova Scotia.

Bretonian strata [GEOL]   Upper Cambrian strata in Cape Breton, Nova Scotia.

brettice *See* brattice.

brettis *See* brattice.

breunnerite [MINERAL] $(Mg,Fe,Mn)CO_3$ A carbonate mineral consisting of an isomorphous system of the metallic components.

brewsterite [MINERAL] $Sr(Al_2Si_6O_{18})\cdot 5H_2O$ A member of the zeolite family of minerals; crystallizes in the monoclinic system and usually contains some calcium.

brickfielder [METEOROL]   A hot, dry, dusty north wind blowing from the interior across the southern coast of Australia.

bricking curb [MIN ENG]   A curb set in a circular shaft to support the brick walling.

brick walling [MIN ENG]   A permanent support for circular shafts by walling or casing.

bridal-veil fall [HYD]   A cataract of great height and small volume, with the water dissipated into a spray as it falls.

bridging [MIN ENG]   The obstruction of the receiving opening in a material-crushing device by two or more pieces wedged together, each of which could easily pass through.

Bridgman sampler [MIN ENG]   A mechanical device that automatically selects two samples as the ore passes through.

bridled-cup anemometer [ENG]   A combination cup anemometer and pressure-plate anemometer, consisting of an array of cups about a vertical axis of rotation, the free rotation of which is restricted by a spring arrangement; by adjustment of the force constant of the spring, an angular displacement can be obtained which is proportional to wind velocity.

bridled pressure plate [METEOROL]   An instrument for measuring air velocity in which the pressure on a plate exposed to the wind is balanced by the force of a spring, and the deflection of the plate is measured by an inductance-type transducer.

briefing *See* pilot briefing.

Briggs stretcher carriage [MIN ENG]   A stretcher used as an ambulance trolley in transporting casualties from underground workings.

bright band [METEOROL]   The enhanced echo of snow as it melts to rain, as displayed on a range-height indicator scope.

bright-banded coal *See* bright coal.

bright coal [GEOL]   A jet-black, pitchlike type of banded coal that is more compact than dull coal and breaks with a shell-shaped fracture; microscopic examination shows a consistency of more than 5% anthraxyllon and less than 20% opaque matter. Also known as bright-banded coal; brights.

brights *See* bright coal.

bright segment [GEOPHYS]   A faintly glowing band which appears above the horizon after sunset or before sunrise. Also known as crepuscular arch; twilight arch.

brimstone [MINERAL]   A common or commercial name for native sulfur.

brine [OCEANOGR]   Sea water containing a higher concentration of dissolved salt than that of the ordinary ocean.

brine cell [HYD]   A small tubular inclusion in sea ice, about 0.05 mm in diameter, containing liquid that is more saline than sea water. Also known as brine pocket.

brine pocket *See* brine cell.

brisa [METEOROL] **1.** A northeast wind which blows on the coast of South America or an east wind which blows on Puerto Rico during the trade wind season. **2.** The northeast monsoon in the Philippines. Also spelled briza.

brisa carabinera *See* carabine.

brisance index [ENG] The ratio of an explosive's power to shatter a weight of graded sand as compared to the weight of sand shattered by TNT.

brise carabinée *See* carabine.

brisote [METEOROL] The northeast trade wind when it is blowing stronger than usual on Cuba.

britannia cell [MIN ENG] In mineral processing, a pneumatic flotation cell 7 to 9 feet (2.1 to 2.7 meters) deep.

britholite [MINERAL] $(Na,Ce,Ca)_5(OH)[(P,Si)O_4]_3$ A rare-earth phosphate found in carbonatites in Kola Peninsula, Soviet Union.

brittle mica [MINERAL] Hydrous sodium, calcium, magnesium, and aluminum silicates; a group of more or less related minerals that resemble true micas but cleave to brittle flakes and contain calcium as the essential constituent.

brittle silver ore *See* stephanite.

briza *See* brisa.

brochanite *See* brochantite.

brochanthite *See* brochantite.

brochantite [MINERAL] $Cu_4(SO_4)(OH)_6$ A monoclinic copper mineral, emerald to dark green, commonly found with copper sulfide deposits; a minor copper ore. Also known as brochanite; brochanthite; warringtonite.

Brocken specter [METEOROL] The illusory appearance of a gigantic figure (actually, the observer's shadow projected on cloud surfaces), observed on the Brocken peak in the Hartz Mountains of Saxony, but visible from other mountaintops under suitable conditions.

broeboe [METEOROL] A strong, dry east wind in the southwestern part of the island of Celebes.

broken [METEOROL] Descriptive of a sky cover of from 0.6 to 0.9 (expressed to the nearest tenth).

broken belt [OCEANOGR] The transition zone between open water and consolidated ice.

broken ground *See* loose ground.

broken stone *See* crushed stone.

broken water [OCEANOGR] Water having a surface covered with ripples or eddies, and usually surrounded by calm water.

bromellite [MINERAL] BeO A white hexagonal mineral consisting of beryllium oxide; it is harder than zincite.

bromlite [MINERAL] $BaCa(CO_3)_2$ An orthorhombic mineral composed of a carbonate of barium and calcium. Also known as alstonite.

bromyrite [MINERAL] AgBr A secondary ore of silver that occurs in the oxidized zone of silver deposits; exists in crusts and coatings resembling a wax.

brontides [GEOPHYS] Low, rumbling, thunderlike sounds of short duration, most frequently heard in active seismic regions and believed to be of seismic origin.

Brontotheriidae [PALEON]   The single family of the extinct mammalian superfamily Brontotherioidea.

Brontotherioidea [PALEON]   The titanotheres, a superfamily of large, extinct perissodactyl mammals in the suborder Hippomorpha.

bronze mica *See* phlogopite.

bronzite [MINERAL] $(Mg,Fe)(SiO_3)$ An orthopyroxene mineral that forms metallic green orthorhombic crystals; a form of the enstatite-hypersthene series.

bronzitfels *See* bronzitite.

bronzitite [PETR]   A pyroxenite that is composed almost entirely of bronzite. Also known as bronzitfels.

Brookhill waffler [MIN ENG]   A coal cutter with the ordinary horizontal jib and also a shearing or mushroom jib.

brookite [MINERAL] TiO_2 A brown, reddish, or black orthorhombic mineral; it is trimorphous with rutile and anatase, has hardness of 5.5–6 on Mohs scale, and a specific gravity of 3.87–4.08. Also known as pyromelane.

brotocrystal [PETR]   A fragment of crystal from a previously consolidated rock that has been only partially assimilated in a later magma.

brown clay *See* red clay.

brown clay ironstone [GEOL]   Limonite in the form of concrete masses, often in concretionary nodules.

brown hematite *See* limonite.

brown iron ore *See* limonite.

brown lignite [GEOL]   A type of lignite with a fixed carbon content ranging from 30 to 55% and total carbon from 65 to 73.6%; contains 6300 Btu per pound (14.65 megajoules per kilogram). Also known as lignite B.

brown mica *See* phlogopite.

brown snow [METEOROL]   Snow intermixed with dust particles.

brown soil [GEOL]   Any of a zonal group of soils, with a brown surface horizon which grades into a lighter-colored soil and then into a layer of carbonate accumulation.

brown spar [GEOL]   Any light-colored crystalline carbonate that contains iron, such as ankerite or dolomite, and is therefore brown.

brownstone [PETR]   Ferruginous sandstone with its grains coated with iron oxide.

brubru [METEOROL]   A squall in Indonesia.

brucite [MINERAL] $Mg(OH)_2$ A hexagonal mineral; native magnesium hydroxide that appears gray and occurs in serpentines and impure limestones; hardness is 2.5 on Mohs scale, and specific gravity is 2.38–2.40.

Brückner cycle [CLIMATOL]   An alternation of relatively cool-damp and warm-dry periods, forming an apparent cycle of about 35 years.

brugnatellite [MINERAL] $Mg_6Fe(OH)_{13}CO_3 \cdot 4H_2O$ A flesh pink to yellowish- or brownish-white, hexagonal mineral consisting of a hydrated carbonate-hydroxide of magnesium and ferric iron; occurs in massive form.

bruma [METEOROL]   A haze that appears in the afternoons on the coast of Chile when sea air is transported inland.

Brunt-Douglas isallobaric wind *See* isallobaric wind.

brüscha [METEOROL]   A northwest wind in the Bergell Valley, Switzerland.

brushing shot [MIN ENG] A charge fired in the air of a mine to blow out obnoxious gases or to start an air current.

brushite [MINERAL] $CaHPO_4 \cdot 2H_2O$ A nearly colorless mineral that is a constituent of rock phosphates that crystallizes in slender or massive crystals.

Bruxellian [GEOL] Lower middle Eocene geologic time.

BS&W *See* basic sediments and water.

BT *See* bathythermograph.

B tectonite [PETR] Tectonite with a fabric dominated by linear elements indicating an axial direction rather than a slip surface.

bubble *See* bubble high.

bubble high [METEOROL] A small high, complete with anticyclonic circulation, of the order of 50 to 300 miles (80 to 480 kilometers) across, often induced by precipitation and vertical currents associated with thunderstorms. Also known as bubble.

bubble-point reservoir *See* dissolved-gas-drive reservoir.

bubble pulse [GEOPHYS] An extraneous effect during a seismic survey due to a bubble that is formed by a seismic charge, explosion, or spark fired in a body of water.

bubble train [GEOL] A string or strings of vesicles in lava, indicating the path of rising gas escaping a flow of lava.

bubble wall fragment [GEOL] A glassy volcanic shard revealing part of a vesicle surface which may be curved or flat.

bucaramangite [MINERAL] A pale yellow variety of retinite that looks like amber but is insoluble in alcohol.

buchite [PETR] A partially vitrified inclusion of sandstone in basalt.

buchonite [PETR] An extrusive rock formed of labradorite, titanaugite, and titaniferous hornblende, with nepheline and sodic sanidine and accessory biotite, apatite, and opaque oxides.

bucket drill [MIN ENG] An auger stem drill in which the drill bit is replaced by a bit incorporating a steel cylinder to confine the cutting.

bucket temperature [ENG] The surface temperature of ocean waters as measured by a bucket thermometer.

bucket thermometer [ENG] A thermometer mounted in a bucket and used to measure the temperature of water drawn into the bucket from the surface of the ocean.

bucking [MIN ENG] A hand process for crushing ore.

bucklandite *See* allanite.

buckle fold [GEOL] A double flexure of rock beds formed by compression acting in the plane of the folded beds.

buckwheat coal [GEOL] An anthracite coal that passes through 9/16-inch (14 millimeter) holes and over 5/16-inch (8 millimeter) holes in a screen.

buddle [MIN ENG] A device for concentrating ore that uses a circular arrangement from which the finely divided ore is delivered in water from a central point, the heavier particles sinking and the lighter particles overflowing.

Buddy [MIN ENG] A shortwall coal cutter designed for light duty such as stabling on longwall power-loaded faces and for subsidiary developments.

budget year [METEOROL] The 1-year period beginning with the start of the accumulation season at the firn line of a glacier or ice cap and extending through the following summer's ablation season.

buetschliite [MINERAL] $K_6Ca_2(CO_3)_5 \cdot 6H_2O$ A mineral that is probably hexagonal and consists of a hydrated carbonate of potassium and calcium.

bug dust [MIN ENG]   The fine coal or other material resulting from a boring or cutting of a drill, a mining machine, or even a pick.

buggy [MIN ENG]   A four-wheeled steel car used for hauling coal to and from chutes.

bughole *See* vug.

buhrstone [PETR]   A silicified fossiliferous limestone with abundant cavities previously occupied by fossil shells. Also known as millstone.

buildup curve [PETRO ENG]   Graph of bottom-hole pressure buildup versus shut-in time for a gas or oil well.

buildup pressure [PETRO ENG]   The increase in bottom-hole pressure up to an equilibrium value in a shut-in oil or gas well.

built terrace *See* alluvial terrace.

bulb glacier [HYD]   A glacier formed at the foot of a mountain and out into an open slope; the glacier ends spread out into an ice fan.

bulb of percussion [ARCHEO]   A cone-shaped bulge on a fractured flint surface that was made by a blow striking at an angle.

bulk flotation [MIN ENG]   The rising of a mineralized froth, of more than one mineral, in a single operation.

bulk mining [MIN ENG]   Mining in which large quantities of low-grade ore are taken without attempt to segregate the high-grade portions.

bulk plant [PETRO ENG]   A wholesale receiving and distributing facility for petroleum products; includes storage tanks, warehouses, railroad sidings, truck loading racks, and related elements. Also known as bulk terminal.

bulk terminal *See* bulk plant.

bull bit [MIN ENG]   A flat drill bit.

bullfrog *See* barney.

bulling bar [ENG]   A bar for ramming clay into cracks containing blasting charges which are about to be exploded.

bullion [MIN ENG]   A concretion found in some types of coal, composed of carbonate or silica stained by brown humic derivatives; often well-preserved plant structures form the nuclei.

bull's-eye squall [METEOROL]   A squall forming in fair weather, characteristic of the ocean off the coast of South Africa; it is named for the peculiar appearance of the small, isolated cloud marking the top of the invisible vortex of the storm.

bull shaker [MIN ENG]   A shaking chute where large coal from the dump is cleaned by hand.

bumps [MIN ENG]   Sudden, violent expulsion of coal from one or more pillars, accompanied by loud reports and earth tremors.

bunsenite [MINERAL] NiO A pistachio-green mineral consisting of nickel monoxide and occurring as octahedral crystals.

Bunter [GEOL]   Lower Triassic geologic time. Also known as Bundsandstein.

bunton [MIN ENG]   A steel or timber element in the lining of a rectangular shaft.

buoyancy pontoons [PETRO ENG]   Pontoons that buoy up offshore pipelines during the welding together of sections over bodies of water, after which the pontoons are removed and the pipeline is allowed to sink into position on the bottom.

buran [METEOROL] A violent northeast storm of south Russia and central Siberia, similar to the blizzard.

Burdigalian [GEOL] Upper lower Miocene geologic time.

burga [METEOROL] A storm of wind and sleet in Alaska. Also spelled boorga.

Burgers vector [CRYSTAL] A translation vector of a crystal lattice representing the displacement of the material to create a dislocation.

buried channel [GEOL] An old channel covered by surficial deposits; an example is a preglacial channel filled with glacial drift.

buried hill [GEOL] A hill of resistant older rock over which later sediments are deposited.

buried ice [HYD] A relatively distinct mass of ice embedded under the ground surface.

buried placer [GEOL] Old deposit of a placer which has been buried beneath lava flows or other strata.

buried river [GEOL] A river bed which has become buried beneath streams of alluvial drifts or basalt.

buried soil *See* paleosol.

burkeite [MINERAL] $Na_6(CO_3)(SO_4)_2$ A white to pale buff or gray mineral consisting of a carbonate-sulfate of sodium.

burl [GEOL] An oolith or nodule in fireclay, sometimes having a high alumina or iron oxide content.

burn cut *See* parallel cut.

burn-in *See* freeze.

burning line [PETRO ENG] A pipeline used to convey refinery fuel gas, as distinguished from gas intended for subsequent processing.

burn off [METEOROL] With reference to fog or low stratus cloud layers, to dissipate by daytime heating from the sun.

burnt shale [MIN ENG] Carbonaceous shale which has remained for a long period in a colliery tip and undergone spontaneous combustion and converted into a coppery slag material. Also known as oxidized shale.

burrow [MIN ENG] A refuse heap at a coal mine.

Busch lemniscate [METEOROL] The locus in the sky, or on a diagrammatic representation thereof, of all points at which the plane of polarization of diffuse sky radiation is inclined 45° to the vertical; a polarization isocline. Also known as neutral line.

bustite *See* aubrite.

butlerite [MINERAL] $Fe(SO_4)(OH) \cdot 2H_2O$ A deep orange, monoclinic mineral consisting of a hydrated basic ferric sulfate. Also known as parabutlerite.

butt [MIN ENG] Coal exposed at right angles to the face and, in contrast to the face, generally having a rough surface.

butt cable *See* hand cable.

butte [GEOGR] A detached hill or ridge which rises abruptly.

buttgenbachite [MINERAL] $Cu_{19}(NO_3)_2Cl_4(OH)_{32} \cdot 3H_2O$ An azure blue, hexagonal mineral consisting of a hydrated basic chloride-sulfate-nitrate of copper.

buttock [MIN ENG] A corner formed by two coal faces more or less at right angles, such as the end of a working face.

button balance [MIN ENG] A small, very delicate balance used for weighing assay buttons.

buttress [PALEON] A ridge on the inner surface of a pelecypod valve which acts as a support for part of the hinge.

buttress sands [GEOL] Sandstone bodies deposited above an unconformity; the upper portion rests upon the surface of the unconformity.

buttress-thread casing [PETRO ENG] A drill casing in which the ends of the sections are buttressed together and held in place with a short threaded outer sleeve; used where greater than normal clearance, strength, and leak resistance are needed.

buy-back crude [PETRO ENG] Oil in which the government of the production territory has a right to a share if it is a stockholder in the oil-producing company. Also known as participation crude.

Buys-Ballot's law [METEOROL] A law describing the relationship of the horizontal wind direction in the atmosphere to the pressure distribution: if one stands with one's back to the wind, the pressure to the left is lower than to the right in the Northern Hemisphere; in the Southern Hemisphere the relation is reversed. Also known as baric wind law.

byerite [GEOL] Bituminous coal that does not crack in fire and melts and enlarges upon heating.

byon [GEOL] Gem-bearing gravel, particularly that with brownish-yellow clay in which corundum, rubies, sapphires, and so forth occur.

bysmalith [GEOL] A body of igneous rock that is more or less vertical and cylindrical; it crosscuts adjacent sediments.

bytownite [MINERAL] A plagioclase feldspar with a composition ranging from $Ab_{30}An_{70}$ to $Ab_{10}An_{90}$, where $Ab = NaAlSi_3O_8$ and $An = CaAl_2Si_2O_8$; occurs in basic and ultrabasic igneous rock.

C

caballing [OCEANOGR] The mixing of two water masses of identical in situ densities but different in situ temperatures and salinities, such that the resulting mixture is denser than its components and therefore sinks.

cable *See* cable length.

cable length [OCEANOGR] A unit of distance, originally equal to the length of a ship's anchor cable, now variously considered to be 600 feet (183 meters), 608 feet (185.3 meters; one-tenth of a British nautical mile), or 720 feet or 120 fathoms (219.5 meters). Also known as cable.

cable system [MIN ENG] A drilling system involving a heavy string of tools suspended from a flexible cable.

cable-system drill *See* churn drill.

cable tools [MIN ENG] The bits and other bottom-hole tools and equipment used to drill boreholes by percussive action, using a cable, instead of rods, to connect the drilling bit with the machine on the surface.

cableway [MIN ENG] A cable system of material handling in which carriers are supported by a cable and not detached from the operating span.

cacimbo [METEOROL] Local name in Angola for the wet fogs and drizzles noted with onshore winds from the Benguela Current.

cacoxenite [MINERAL] $Fe_4(PO_4)_3(OH)_3 \cdot 12H_2O$ Yellow or brownish mineral consisting of a hydrous basic iron phosphate occurring in radiated tufts.

cadmium blende *See* greenockite.

cadmium ocher *See* greenockite.

cadwaladerite [MINERAL] $Al(OH)_2Cl \cdot 4H_2O$ A mineral consisting of a hydrous basic aluminum chloride.

Caenolestidae [PALEON] A family of extinct insectivorous mammals in the order Marsupialia.

cage guides [MIN ENG] Directive apparatus used to guide the cages in the mine shaft and to prevent their swinging or colliding with each other.

cage shoes [MIN ENG] Fittings attached on the side of a cage by bolts so that they engage the rigid guides in a shaft.

cahnite [MINERAL] $Ca_2B(OH)_4(AsO_4)$ A tetragonal borate mineral occurring in white, sphenoidal crystals.

Cainotheriidae [PALEON] The single family of the extinct artiodactyl superfamily Cainotherioidea.

Cainotherioidea [PALEON]   A superfamily of extinct, rabbit-sized tylopod ruminants in the mammalian order Artiodactyla.

cairngorm *See* smoky quartz.

caju rains [METEOROL]   In northeastern Brazil, light showers that occur during the month of October.

cake [MIN ENG]   **1.** Solidified drill sludge. **2.** That portion of a drilling mud adhering to the walls of a borehole. **3.** To form into a mass, as when ore sinters together in roasting, or coal cakes in coking.

caking coal [GEOL]   A type of coal which agglomerates and softens upon heating; after volatile material has been expelled at high temperature, a hard, gray cellular mass of coke remains. Also known as binding coal.

Calabrian [GEOL]   Lower Pleistocene geologic time.

calaite *See* turquoise.

calamine *See* hemimorphite; smithsonite.

Calamitales [PALEOBOT]   An extinct group of reedlike plants of the subphylum Sphenopsida characterized by horizontal rhizomes and tall, upright, grooved, articulated stems.

calaverite [MINERAL]   AuTe$_2$ A yellowish or tin-white, monoclinic mineral commonly containing gold telluride and minor amounts of silver.

calc-alkalic series [PETR]   Series of igneous rocks in which the weight percentage of silica is 55–61.

calcarenite [PETR]   A type of limestone or dolomite composed of coral or shell sand or of sand formed by erosion of older limestones, with particle size ranging from $\frac{1}{16}$ to 2 millimeters.

calcareous crust *See* caliche.

calcareous ooze [GEOL]   A fine-grained pelagic sediment containing undissolved sand- or silt-sized calcareous skeletal remains of small marine organisms mixed with amorphous clay-sized material.

calcareous schist [PETR]   A coarse-grained metamorphic rock derived from impure calcareous sediment.

calcareous sinter *See* tufa.

calcareous soil [GEOL]   A soil containing accumulations of calcium and magnesium carbonate.

calcareous tufa *See* tufa.

calcification [GEOCHEM]   Any process of soil formation in which the soil colloids are saturated to a high degree with exchangeable calcium, thus rendering them relatively immobile and nearly neutral in reaction.

calcilutite [PETR]   **1.** A dolomite or limestone formed of calcareous rock flour that is typically nonsiliceous. **2.** A rock of calcium carbonate formed of grains or crystals with average diameter less than $\frac{1}{16}$ millimeter.

calcimeter [ENG]   An instrument for estimating the amount of lime in soils.

calciocarnotite *See* tyuyamunite.

calcioferrite [MINERAL]   Ca$_2$Fe$_2$(PO$_4$)OH·7H$_2$O   A yellow or green mineral consisting of a hydrous basic calcium iron phosphate and occurring in nodular masses.

calciovolborthite [MINERAL]   CaCu(VO$_4$)(OH)   Green, yellow, or gray mineral consisting of a basic vanadate of calcium and copper. Also known as tangeite.

calcirudite [PETR]   Dolomite or limestone formed of worn or broken pieces of coral or shells or of limestone fragments coarser than sand; the interstices are filled with sand, calcite, or mud, the whole bound together with a calcareous cement.

calcite [MINERAL] $CaCO_3$ One of the commonest minerals, the principal constituent of limestone; hexagonal-rhombohedral crystal structure, dimorphous with aragonite. Also known as calcspar.

calcite bubble [GEOL]   A hollow sphere found in caves in which a gas bubble was trapped by calcite on the surface of a pond.

calcite compensation depth [GEOL]   The depth in the ocean (about 5000 meters) below which solution of calcium carbonate occurs at a faster rate than its deposition. Abbreviated CCD.

calcite dolomite [PETR]   A carbonate rock with a composition of 10–50% calcite and 90–50% dolomite.

calclacite [MINERAL] $CaCl_2Ca(C_2H_3O_2)_2\cdot10H_2O$ A white mineral consisting of a hydrated chloride-acetate of calcium; occurs as hairlike efflorescences.

Calclamnidae [PALEON]   A family of Paleozoic echinoderms of the order Dendrochirotida.

calclithite [PETR]   Limestone with 50% or more fragments of older limestone that was redeposited after being eroded from the land.

calcrete [GEOL]   A conglomerate of surficial gravel and sand cemented by calcium carbonate.

calc-silicate hornfels [PETR]   A metamorphic rock with a fine grain of calcium silicate minerals.

calc-silicate marble [PETR]   Marble having conspicuous calcium silicate or magnesium silicate minerals.

calcspar *See* calcite.

calcsparite *See* sparry calcite.

caldera [GEOL]   A more or less circular volcanic depression whose diameter is many times greater than that of the volcanic vent.

caldron [GEOL]   A small, steep-sided, circular depression on the ocean bottom.

Caledonian orogeny [GEOL]   Deformation of the crust of the earth by a series of diastrophic movements beginning perhaps in Early Ordovician and continuing through Silurian, extending from Great Britain through Scandinavia.

Caledonides [GEOL]   A mountain system formed in Late Silurian to Early Devonian time in Scotland, Ireland, and Scandinavia.

caledonite [MINERAL] $Cu_2Pb_5(SO_4)_3CO_3(OH)_6$ A mineral occurring as green, orthorhombic crystals composed of basic copper lead sulfate; found in copper-lead deposits.

calf *See* calved ice.

caliche [GEOL]   **1.** Conglomerate of gravel, rock, soil, or alluvium cemented with sodium salts in Chilean and Peruvian nitrate deposits; contains sodium nitrate, potassium nitrate, sodium iodate, sodium chloride, sodium sulfate, and sodium borate. **2.** A thin layer of clayey soil capping auriferous veins (Peruvian usage). **3.** Whitish clay in the selvage of veins (Chilean usage). **4.** A recently discovered mineral vein. **5.** A secondary accumulation of opaque, reddish brown to buff or white calcareous material occurring in layers on or near the surface of stony soils in arid and semiarid regions of the southwestern United States; called hardpan, calcareous duricrust, and kanker in different geographic regions. Also known as calcareous crust; croute calcaire; nari; sabach; tepetate.

California Current [OCEANOGR] The ocean current flowing southward along the western coast of the United States to northern Baja California.

California fog [METEOROL] Fog peculiar to the coast of California and its coastal valleys; off the coast, winds displace warm surface water, causing colder water to rise from beneath, resulting in the formation of fog; in the coastal valleys, fog is formed when moist air blown inland during the afternoon is cooled by radiation during the night.

California method [HYD] A form of frequency analysis which employs the return period, a parameter that measures the average time period between the occurrence of a quantity in hydrology and that of an equal or greater quantity, as the plotting position.

California sampler [MIN ENG] A drive sampler equipped with a piston that can be retracted mechanically to any desired point within the barrel of the sampler.

California-type dredge [MIN ENG] A single-lift dredge in which closely spaced buckets deliver to a trommel; oversize rocks are piled behind the dredge by a conveyor or stacker; the undersize are washed on gold-saving tables on the deck, and tailings discharge astern through sluices.

californite [MINERAL] $Ca_{10}Al_4(Mg,Fe)_2Si_9O_{34}(OH,F)_4$ A variety of vesuvianite that resembles jade; it is dark-, yellowish-, olive-, or grass-green and occurs in translucent to opaque compact or massive form. Also known as American jade.

calina [METEOROL] A haze prevalent in Spain during the summer, when the air becomes filled with dust swept up from the dry ground by strong winds.

caliper log [MIN ENG] A graphic record showing the diameter of a drilled hole at each depth; measurements are obtained by drawing a caliper upward through the hole and recording the diameter on quadrile paper.

Callao painter *See* painter.

callenia *See* stromatolite.

Callon's rule [MIN ENG] The rule stating that when a pillar is left in an inclined seam for support in a shaft or a structure on the surface, a greater width should be left on the rise side of the shaft or structure than on the dip side.

Callovian [GEOL] A stage in uppermost Middle or lowermost Upper Jurassic which marks a return to clayey sedimentation.

Callow flotation cell [MIN ENG] Nonmechanical apparatus for separation of floatable solid gangue from pulverized ore, with the mixture suspended in liquid and aerated by air bubbles coming up through a porous medium, so that the lighter gangue floats away from the heavier ore.

Callow process [MIN ENG] A flotation process in which agitation is provided by the forcing of air into the pulp through the canvas-covered bottom of the cell.

Callow screen [MIN ENG] A continuous belt system formed of fine screen wire that is used to separate fine solids from coarse ones.

calm [METEOROL] The absence of apparent motion of the air; in the Beaufort wind scale, smoke is observed to rise vertically, or the surface of the sea is smooth and mirrorlike; in U.S. weather observing practice, the wind has a speed under 1 mile per hour or 1 knot (1.6 kilometers per hour).

calm belt [METEOROL] A belt of latitude in which the winds are generally light and variable; the principal calm belts are the horse latitudes (the calms of Cancer and of Capricorn) and the doldrums.

calms of Cancer [METEOROL] One of the two light, variable winds and calms which occur in the centers of the subtropical high-pressure belts over the oceans; their usual position is about latitude 30°N, the horse latitudes.

calms of Capricorn [METEOROL] One of the two light, variable winds and calms which occur in the centers of the subtropical high-pressure belts over the oceans; their usual position is about latitude 30°S, the horse latitudes.

calomel [MINERAL] Hg_2Cl_2 A colorless, white, grayish, yellowish, or brown secondary, sectile, tetragonal mineral; used as a cathartic, insecticide, and fungicide. Also known as calomelene; calomelite; horn quicksilver; mercurial horn ore.

calomelene *See* calomel.

calomelite *See* calomel.

calved ice [OCEANOGR] A piece of ice floating in a body of water after breaking off from a mass of land ice or an iceberg. Also known as calf.

calving [GEOL] The breaking off of a mass of ice from its parent glacier, iceberg, or ice shelf. Also known as ice calving.

calyx drill [ENG] A rotary core drill with hardened steel shot for cutting rock. Also known as shot drill.

camanchaca *See* garúa.

camber [GEOL] **1.** A terminal, convex shoulder of the continental shelf. **2.** A structural feature that is caused by plastic clay beneath a bed flowing toward a valley so that the bed sags downward and seems to be draped over the sides of the valley.

Cambrian [GEOL] The lowest geologic system that contains abundant fossils of animals, and the first (earliest) geologic period of the Paleozoic era from 570 to 500 million years ago.

cameo mountain [GEOL] A mountain made of elevated horizontal strata deposited by two or more subparallel streams cutting deep channels that eventually come together.

camera station [MAP] In aerial photography, the point in space occupied by the camera lens at the moment of exposure. Also known as air station.

Camerata [PALEON] A subclass of extinct stalked echinoderms of the class Crinoidea.

Cammett table [MIN ENG] A side-moving table for concentrating ore.

Campanian [GEOL] European stage of Upper Cretaceous.

camptonite [PETR] A lamprophyre containing pyroxene, sodic hornblende, and olivine as dark constituents and labradorite as the light constituent; sodic orthoclase may be present.

Canadian Shield *See* Laurentian Shield.

canal [GEOGR] A long, narrow arm of the sea extending far inland, between islands, or between islands and the mainland.

Canary Current [OCEANOGR] The prevailing southward flow of water along the northwestern coast of Africa.

Canastotan [GEOL] Lower Upper Silurian geologic time.

cancrinite [MINERAL] $Na_3CaAl_3Si_3O_{12}CO_3(OH)_2$ A feldspathoid tectosilicate occurring in hexagonal crystals in nepheline syenites, usually in compact or disseminated masses.

candite *See* ceylonite.

Candlemas crack *See* Candlemas Eve winds.

Candlemas Eve winds [METEOROL]   Heavy winds often occurring in Great Britain in February or March (Candlemas is February 2). Also known as Candlemas crack.

canfieldite [MINERAL] Ag_8SnS_6 A black mineral of the argyrodite series consisting of silver thiostannate, with a specific gravity of 6.28; found in Germany and Bolivia.

can hoisting system [MIN ENG]   A hoisting method used in shallow lead and zinc mines in which cans are loaded below and hoisted to the surface where they are capsized and the load discharged; the operation is controlled at the top of the shaft.

cannel coal [GEOL]   A fine-textured, highly volatile bituminous coal distinguished by a greasy luster and blocky, conchoidal fracture; burns with a steady luminous flame. Also known as cannelite.

cannelite *See* cannel coal.

canneloid [GEOL]   **1.** Coal that resembles cannel coal. **2.** Coal intermediate between bituminous and cannel. **3.** Durain laminae in banded coal. **4.** Cannel coal of anthracite or semianthracite rank.

cannel shale [GEOL]   A black shale formed by the accumulation of an aquatic ooze rich in bituminous organic matter in association with inorganic materials such as silt and clay.

Canterbury northwester [METEOROL]   A strong northwest foehn descending the New Zealand Alps onto the Canterbury plains of South Island, New Zealand.

canyon [GEOGR]   A chasm, gorge, or ravine cut in the surface of the earth by running water; the sides are steep and form cliffs.

canyon bench [GEOL]   A steplike level of hard strata in the walls of deep valleys in regions of horizontal strata.

canyon fill [GEOL]   Loose, unconsolidated material which fills a canyon to a depth of 50 feet (15 meters) or more during periods between great floods.

canyon wind [METEOROL]   Also known as gorge wind. **1.** The mountain wind of a canyon; that is, the nighttime down-canyon flow of air caused by cooling at the canyon walls. **2.** Any wind modified by being forced to flow through a canyon or gorge; its speed may be increased as a jet-effect wind, and its direction is rigidly controlled.

cap [ENG]   A detonating or blasting cap.   [MIN ENG]   **1.** A piece of timber placed on top of a prop or post in a mine. **2.** The horizontal section of a set of timber that is used as a support in a mine roadway.

capacity correction [ENG]   The correction applied to a mercury barometer with a nonadjustable cistern in order to compensate for the change in the level of the cistern as the atmospheric pressure changes.

capacity of the wind [GEOL]   The total weight of airborne particles (soil and rock) of given size, shape, and specific gravity, which can be carried in 1 cubic mile (4.17 cubic kilometers) of wind blowing at a given speed.

cap cloud [METEOROL]   An approximately stationary cloud, or standing cloud, on or hovering above an isolated mountain peak; formed by the cooling and condensation of humid air forced up over the peak. Also known as cloud cap.

cap crimper [ENG]   A tool resembling a pliers that is used to press the open end of a blasting cap onto the safety fuse before placing the cap in the primer.

cape [GEOGR]   A prominent point of land jutting into a body of water. Also known as head; headland; mull; naze; ness; point; promontory.

cape doctor [METEOROL]   The strong southeast wind which blows on the South African coast.

Cape Horn Current [OCEANOGR] That part of the west wind drift flowing eastward in the immediate vicinity of Cape Horn, and then curving northeastward to continue as the Falkland Current.

Capell fan [MIN ENG] A centrifugal type of mine shaft fan.

capillarity correction [ENG] As applied to a mercury barometer, that part of the instrument correction which is required by the shape of the meniscus of the mercury.

capillary collector [ENG] An instrument for collecting liquid water from the atmosphere; the collecting head is fabricated of a porous material having a pore size of the order of 30 micrometers; the pressure difference across the water-air interface prevents air from entering the capillary system while allowing free flow of water.

capillary conductivity [GEOPHYS] The capability of an unsaturated soil or rock to transmit water or other liquids.

capillary control [PETRO ENG] Discarded theory of reservoir oil flow to a well hole through capillary pores; it attributed flow resistance to gas bubbles within the capillaries.

capillary ejecta *See* Pele's hair.

capillary equilibrium method [PETRO ENG] Test method to predict oil and gas flow through an oil reservoir core by dethrottling the flow to hold capillary flow in equilibrium between oil and gas within the reservoir.

capillary fringe [HYD] The lower subdivision of the zone of aeration that overlies the zone of saturation and in which the pressure of water in the interstices is lower than atmospheric pressure.

capillary migration [HYD] Movement of water produced by the force of molecular attraction between rock material and the water.

capillary pyrites *See* millerite.

capillary water [HYD] Soil water held by capillarity as a continuous film around soil particles and in interstices between particles above the phreactic line.

cap lamp [MIN ENG] The lamp a miner wears on his safety hat or cap for illumination.

capped column [HYD] A form of ice crystal consisting of a hexagonal column with plate or stellar crystals (so-called caps) at its ends and sometimes at intermediate positions; the caps are perpendicular to the column.

capped fuse [ENG] A length of safety fuse with the cap or detonator crimped on before it is taken to the place of use.

capped quartz [MINERAL] A type of quartz that contains thin layers of clay.

cappelenite [MINERAL] $(Ba,Ca,Na)(Y,La)_6B_6Si_{13}(O,OH)_{27}$ A greenish-brown hexagonal mineral consisting of a rare yttrium-barium borosilicate occurring in crystals.

cap piece [MIN ENG] A piece of wood fitted over a straight post or timber to provide more bearing surface.

capping [ENG] Preparation of a capped fuse. [GEOL] **1.** Consolidated barren rock overlying a mineral or ore deposit. **2.** *See* gossan. [MIN ENG] The attachment at the end of a winding rope. [PETRO ENG] **1.** The process of sealing or covering a borehole such as an oil or gas well. **2.** The material or device used to seal or cover a borehole.

cap rock [GEOL] **1.** An overlying, generally impervious layer or stratum of rock that overlies an oil- or gas-bearing rock. **2.** Barren vein matter, or a pinch in a vein, supposed to overlie ore. **3.** A hard layer of rock, usually sandstone, a short distance above a coal seam. **4.** An impervious body of anhydrite and gypsum in a salt dome.

Captorhinimorpha [PALEON]   An extinct subclass of primitive lizardlike reptiles in the order Cotylosauria.

capture [HYD]   The natural diversion of the headwaters of one stream into the channel of another stream having greater erosional activity and flowing at a lower level. Also known as piracy; river capture; river piracy; robbery; stream capture; stream piracy; stream robbery.

carabine [METEOROL]   In France and Spain, a sudden and violent wind. Also known as brisa carabinera; brise carabinée.

caracolite [MINERAL]   A rare, colorless mineral occurring as crystalline incrustations, and consisting of a sulfate and chloride of sodium and lead.

Caradocian [GEOL]   Lower Upper Ordovician geologic time.

carapace [GEOL]   The upper normal limb of a fold having an almost horizontal axial plane.

carbide lamp [MIN ENG]   A lamp that is charged with calcium carbide and water to form acetylene, which it burns.

carbide miner [MIN ENG]   An automated coal mining machine; the unit is a continuous miner controlled from outside the coal seam.

carbohumin *See* ulmin.

carbonaceous chondrite [GEOL]   A meteorite containing hydrated clay-type silica minerals in addition to organic substances such as aromatic and fatty acids, hydrocarbons, and porphyrins; large amounts of noble gases, particularly xenon; and almost no elemental nickel and iron.

carbonaceous meteorite [GEOL]   A meteorite yielding relatively large amounts of carbon when it is analyzed.

carbonaceous rock [PETR]   Rock with carbonaceous material included.

carbonaceous sandstone [PETR]   Sandstone rich in carbon.

carbonaceous shale [GEOL]   Shale rich in carbon.

carbonado [MINERAL]   A dark-colored, fine-grained diamond aggregate; valuable for toughness and absence of cleavage planes. Also known as black diamond; carbon diamond.

carbonate compensation depth [OCEANOGR]   The ocean depth at which the $CaCO_3$ content is reduced to 5%. Abbreviated CCD.

carbonate cycle [GEOCHEM]   The biogeochemical carbonate pathways, involving the conversion of carbonate to CO_2 and HCO_3, the solution and deposition of carbonate, and the metabolism and regeneration of it in biological systems.

carbonate mineral [MINERAL]   A mineral containing considerable amounts of carbonates.

carbonate reservoir [GEOL]   An underground oil or gas trap formed in reefs, clastic limestones, chemical limestones, or dolomite.

carbonate rock [PETR]   A rock composed principally of carbonates, especially if at least 50% by weight.

carbonate spring [HYD]   A type of spring containing dissolved carbon dioxide gas.

carbonatite [PETR]   **1.** Intrusive carbonate rock associated with alkaline igneous intrusive activity. **2.** A sedimentary rock composed of at least 80% calcium or magnesium.

carbon cycle [GEOCHEM] The cycle of carbon in the biosphere, in which plants convert carbon dioxide to organic compounds that are consumed by plants and animals, and the carbon is returned to the biosphere in the form of inorganic compounds by processes of respiration and decay.

carbon diamond *See* carbonado.

carbon dioxide indicator [MIN ENG] A detector of carbon dioxide in mines based on the gas's absorption by potassium hydroxide.

Carboniferous [GEOL] A division of late Paleozoic rocks and geologic time including the Mississippian and Pennsylvanian periods.

carbonification *See* coalification.

carbon isotope ratio [GEOL] Ratio of carbon-12 to either of the less common isotopes, carbon-13 or carbon-14, or the reciprocal of one of these ratios; if not specified, the ratio refers to $^{12}C/^{13}C$. Also known as carbon ratio.

carbonite *See* natural coke.

carbonization [GEOCHEM] **1.** In the coalification process, the accumulation of residual carbon by changes in organic material and their decomposition products. **2.** Deposition of a thin film of carbon by slow decay of organic matter underwater. **3.** A process of converting a carbonaceous material to carbon by removal of other components.

carbon-nitrogen-phosphorus ratio [OCEANOGR] The relatively constant relationship between the concentrations of carbon (C), nitrogen (N), and phosphorus (P) in plankton, and N and P in sea water, owing to removal of the elements by the organisms in the same proportions in which the elements occur and their return upon decomposition of the dead organisms.

carbon ratio [GEOL] **1.** The ratio of fixed carbon to fixed carbon plus volatile hydrocarbons in a coal. **2.** *See* carbon isotope ratio.

carbon ratio theory [GEOL] The theory that the gravity of oil in any area is inversely proportional to the carbon ratio of the coal.

carcenet [METEOROL] A very cold and violent gorge wind in the eastern Pyrenees (upper Aude valley).

cardinal point [GEOD] Any of the four principal directions: north, east, south, or west of a compass.

cardinal winds [METEOROL] Winds from the four cardinal points of the compass, that is, north, east, south, and west winds.

Caribbean Current [OCEANOGR] A water current flowing westward through the Caribbean Sea.

Caribbean Sea [GEOGR] One of the largest and deepest enclosed basins in the world, surrounded by Central and South America and the West Indian island chains.

Carinthian furnace [MET] A zinc distillation furnace with small, vertical retorts. [MIN ENG] A small reverberatory furnace with an inclined hearth, in which lead ore is treated by roasting and reaction, wood being the usual fuel.

Carlsbad law [CRYSTAL] A feldspar twin law in which the twinning axis is the c axis, the operation is rotation of 180°, and the contact surface is parallel to the side pinacoid.

Carlsbad turn [CRYSTAL] A twin crystal in the monoclinic system with the vertical axis as the turning axis.

carminite [MINERAL] $PbFe_2(AsO_4I)_2(OH)_2$ A carmine to tile-red mineral consisting of a basic arsenate of lead and iron.

carnallite [MINERAL] $KMgCl_3 \cdot 6H_2IO$ A milky-white or reddish mineral that crystallizes in the orthorhombic system and occurs in deliquescent masses; it is valuable as an ore of potassium.

carnegieite [MINERAL] $NaAlSiO_4$ An artificial mineral similar to feldspar; it is triclinic at low temperatures, isometric at elevated temperatures.

carnelian [MINERAL] A variety of chalcedony containing iron impurities which impart colors of varying shades of red to brown.

Carnian [GEOL] Lower Upper Triassic geologic time. Also spelled Karnian.

Carnosauria [PALEON] A group of large, predacious saurischian dinosaurs in the suborder Theropoda having short necks and large heads.

carnotite [MINERAL] $K(UO_2)_2(VO_4)_2 \cdot nH_2O$ A canary-yellow, fine-grained hydrous vanadate of potassium and uranium having monoclinic microcrystals; an ore of radium and uranium.

Carolina Bays [GEOGR] Shallow, marshy, often ovate depressions on the coastal plain of the mideastern and southeastern United States of unknown origin.

carpholite [MINERAL] $MnAl_2Si_2O_6(OH)_4$ A straw-yellow fibrous mineral consisting of a hydrous aluminum manganese silicate occurring in tufts; specific gravity is 2.93.

carphosiderite [MINERAL] A yellow mineral consisting of a basic hydrous iron sulfate occurring in masses and crusts.

car pincher [MIN ENG] A worker in a mine who uses a pinch bar to position cars under loading chutes.

Carpoidea [PALEON] Former designation for a class of extinct homalozoan echinoderms.

carpoids [PALEON] An assemblage of three classes of enigmatic, rare Paleozoic echinoderms formerly grouped together as the class Carpoidea.

Carrara marble [PETR] All marble quarried near Carrara, Italy, having a prevailing white to bluish color, or white with blue veins.

carry-over [HYD] The portion of the stream flow during any month or year derived from precipitation in previous months or years.

cartogram [MAP] A type of single-factor or topical map that is often diagrammatic to show traffic flow, movement of people or goods, or value by area, where areas of the political subdivisions are distorted so that their size is proportional to their monetary value.

cartographer [GRAPHICS] An individual who makes charts or maps.

cartographic satellite [AERO ENG] An applications satellite that is used to prepare maps of the earth's surface and of the culture on it.

cartography [GRAPHICS] The making of maps and charts for the purpose of visualizing spatial distributions over various areas of the earth.

carved-out payment [PETRO ENG] A proportionate royalty payment based on proceeds from oil or gas production from leased property that has been assigned (carved) out of total payments for the leased property.

caryinite [MINERAL] $(Ca,Pb,Na)_5(Mn,Mg)_4(AsO_4)_5$ A mineral consisting chiefly of a calcium manganese arsenate.

cascade [GEOL] A landform structure formed by gravity collapse, consisting of a bed that buckles into a series of folds as it slides down the flanks of an anticline. [HYD] A small waterfall or series of falls descending over rocks.

Cascadian orogeny [GEOL] Post-Tertiary deformation of the crust of the earth in western North America.

cascading glacier [HYD] A glacier broken by numerous crevasses because of passing over a steep irregular bed, giving the appearance of a cascading stream.

case [MIN ENG] A small fissure admitting water into the mine workings. [PETRO ENG] To line a borehole with steel tubing, such as casing or pipe.

case hardening [GEOL] Formation of a mineral coating on the surface of porous rock by evaporation of a mineral-bearing solution.

casing [PETRO ENG] A special steel tubing welded or screwed together and lowered into a borehole to prevent entry of loose rock, gas, or liquid into the borehole or to prevent loss of circulation liquid into porous, cavernous, or crevassed ground.

casing hanger *See* hanger.

casinghead [PETRO ENG] A fitting at the head of an oil or gas well that allows the pumping operation to take place, as well as the separation of oil and gas.

casinghead tank [PETRO ENG] Storage tank for natural gasoline or other liquids with vapor pressures between 4 and 40 pounds per square inch gage (28 and 280 kilopascals, gage); intermediate between a general-purpose tank and a compressed-gas tank.

casing joint [PETRO ENG] Joint or union that connects two lengths of pipe used to form an oil-well casing.

casing log [PETRO ENG] Recorded data of a down-hole inspection of an oil or gas well made to determine some characteristic of the formations penetrated by the drill hole; types of logs include resistivity, induction, radioactivity, geologic, temperature, and acoustic.

casing shoe [ENG] A ring with a cutting edge that is used on the bottom of a well casing.

casing spear [PETRO ENG] An instrument used for recovering casing which has accidentally fallen into the well.

casing tester [PETRO ENG] A closely fitting, rubber-flanged bucket or a similar tool let down in a well to determine the location of a leak in the casing.

Cassadagan [GEOL] Middle Upper Devonian geologic time, above Chemungian.

Casselian *See* Chattian.

Cassiar orogeny [GEOL] Orogenic episode in the Canadian Cordillera during late Paleozoic time.

cassidyite [MINERAL] $Ca_2(Ni,Mg)(PO_4)_2 \cdot 2H_2O$ A mineral which is found in meteorites.

cassiterite [MINERAL] SnO_2 A yellow, black, or brown mineral that crystallizes in the tetragonal system in prisms terminated by dipyramids; the most important ore of tin. Also known as tin stone.

cast [PALEON] A fossil reproduction of a natural object formed by infiltration of a mold of the object by waterborne minerals.

castellanus [METEOROL] A cloud species with at least a fraction of its upper part presenting some vertically developed cumuliform protuberances (some of which are more tall than wide) which give the cloud a crenellated or turreted appearance. Previously known as castellatus.

castellatus *See* castellanus.

castings [ENG] Any term or symbol designating a low-quality drill diamond. [GEOL] *See* fecal pellets.

castorite [MINERAL] A transparent variety of petalite occurring in crystals.

CAT *See* clear-air turbulence.

catachosis [GEOL] Fracturing or crushing of rock during metamorphism.

cataclasis [GEOL] Deformation of rock by fracture and rotation of aggregates or mineral grains.

cataclasite *See* cataclastic rock.

cataclastic metamorphism [PETR] Local metamorphism restricted to a region of faults and overthrusts involving purely mechanical forces resulting in cataclasis.

cataclastic rock [PETR] Rock containing angular fragments formed by cataclasis. Also known as cataclasite.

cataclastic structure *See* mortar structure.

catamorphism *See* katamorphism.

catapleiite [MINERAL] $(Na_2,Ca)ZrSi_3O_9 \cdot 2H_2O$ A yellow or yellowish-brown mineral crystallizing in the hexagonal system, consisting of a hydrous silicate of sodium, calcium, and zirconium, and occurring in thin tabular crystals; hardness is 6 on Mohs scale, and specific gravity is 2.8.

cataract [HYD] A waterfall of considerable volume with the vertical fall concentrated in one sheer drop.

catastrophism [GEOL] The theory that most features in the earth were produced by sudden, short-lived, worldwide events. [PALEON] The theory that the differences between fossils in successive stratigraphic horizons resulted from a general catastrophe followed by creation of the different organisms found in the next-younger beds.

catazone [GEOL] The deepest zone of rock metamorphism where high temperatures and pressures prevail.

catchment glacier *See* snowdrift glacier.

catch pit [MIN ENG] In mineral processing, a sump in a mill to which the floor slopes gently; spillage gravitates or is hosed to this area.

catoptrite [MINERAL] An iron black to jet black, monoclinic mineral consisting of a silicoantimonate of aluminum and divalent manganese. Also spelled katoptrite.

cat's paw [METEOROL] A puff of wind; a light breeze affecting a small area, as one that causes patches of ripples on the surface of water.

cauldron subsidence [GEOL] **1.** A structure formed by the lowering along a steep ring fracture of a more or less cylindrical block, usually 1 to 10 miles (1.6 to 16 kilometers) in diameter, into a magma chamber. **2.** The process of forming such a structure.

caustobiolith [GEOL] Combustible organic rock formed by direct accumulation of plant materials; includes coal peat.

cavaburd [METEOROL] Shetland Islands term for a thick fall of snow. Also spelled kavaburd.

cavaliers [METEOROL] The local term, in the vicinity of Montpelier, France, for the days near the end of March or the beginning of April when the mistral is usually strongest.

cave [GEOL] A natural, hollow chamber or series of chambers and galleries beneath the earth's surface, or in the side of a mountain or hill, with an opening to the surface. [MIN ENG] **1.** Fragmented rock materials, derived from the sidewalls of a borehole, that obstruct the hole or hinder drilling progress. Also known as cavings.

2. The partial or complete failure of borehole sidewalls or mine workings. Also known as cave-in.

cave breathing [GEOL] The back-and-forth flow of air in the constricted passages of caves, on a cycle of a few seconds to a few minutes.

cave breccia [GEOL] Sharp fragments of limestone debris deposited on the floor of a cave.

cave flower [GEOL] An elongated, curved formation of gypsum or epsomite that has been deposited on the wall of a cave.

cave formation *See* speleothem.

cave-in *See* cave.

cave-in lake [HYD] A lake which occupies a depression produced by the collapse of ground resulting from unequal thawing of permafrost.

Cavellinidae [PALEON] A family of Paleozoic ostracods within the suborder Platycopa.

cave pearl [GEOL] A small, smooth, rounded concretion of calcite or aragonite, formed by concentric precipitation about a nucleus and usually found in limestone caves.

caver [METEOROL] A gentle breeze in the Hebrides, west of Scotland. Also spelled kaver.

cavern [GEOL] An underground chamber or series of chambers of indefinite extent carved out by rock springs in limestone.

cavernous [GEOL] **1.** Having many caverns or cavities. **2.** Producing caverns. **3.** Of or pertaining to a cavern, that is, suggesting vastness.

cave system [GEOL] A network of caves that may be related hydrologically even though they are not connected physically.

caving [MIN ENG] A mining procedure, used when the surface is expendable, in which the ore body is undercut and allowed to fall, breaking into small pieces that are recovered by passages (drifts) driven for that purpose; sublevel caving, block caving, and top slicing are examples.

caving ground [MIN ENG] Rock formation that will not stand in the walls of an underground opening without support such as that offered by cementation, casing, or timber.

cavings *See* cave.

c axis [CRYSTAL] A vertically oriented crystal axis, usually the principal axis; the unique symmetry axis in tetragonal and hexagonal crystals. [GEOL] The reference axis perpendicular to the plane of movement of rock or mineral strata.

cay [GEOL] **1.** A flat coral island. **2.** A flat mound of sand built up on a reef slightly above high tide. **3.** A small, low coastal islet or emergent reef composed largely of sand or coral.

cay sandstone [GEOL] Firmly cemented or friable coral sand formed near the base of coral reef cays.

Caytoniales [PALEOBOT] An order of Mesozoic plants.

Cayugan [GEOL] Upper Silurian geologic time.

Cazenovian [GEOL] Lower Middle Devonian geologic time.

Cc *See* cirrocumulus cloud.

CCD *See* calcite compensation depth.

Cebochoeridae [PALEON]   A family of extinct palaeodont artiodactyls in the superfamily Entelodontoidae.

cebollite [MINERAL] $H_2Ca_4Al_2Si_3O_{16}$ A greenish to white mineral consisting of hydrous calcium aluminum silicate occurring in fibrous aggregates; hardness is 5 on Mohs scale, and specific gravity is 3.

cecilite [PETR]   A basaltic rock having few phenocrysts and consisting of at least 50% leucite with the addition of augite, melilite, nepheline, olivine, anorthite, magnetite, and apatite.

cedricite [MINERAL]   A variety of lamproite composed principally of diopside, leucite, and phlogopite and usually containing crystals of serpentine.

ceiling [METEOROL]   In the United States, the height ascribed to the lowest layer of clouds or of obscuring phenomena when it is reported as broken, overcast, or obscuration and not classified as thin or partial.

ceiling balloon [AERO ENG]   A small balloon used to determine the height of the cloud base; the height is computed from the ascent velocity of the balloon and the time required for its disappearance into the cloud.

ceiling classification [METEOROL]   As used in aviation weather observations, a description or an explanation of the manner in which the height of the ceiling is determined.

ceilometer [ENG]   An automatic-recording cloud-height indicator.

celadonite [MINERAL]   A soft, green variety of mica having high iron content and containing silicates of magnesium and potassium.

celestial geodesy [GEOD]   The branch of geodesy which utilizes observations of near celestial bodies and earth satellites in order to determine the size and shape of the earth.

celestine *See* celestite.

celestite [MINERAL] $SrSO_4$ A colorless or sky-blue mineral occurring in orthorhombic, tabular crystals and in compact forms; fracture is uneven and luster is vitreous; principal ore of strontium. Also known as celestine.

cell [MIN ENG]   A compartment in a flotation machine.

cellular [PETR]   Pertaining to igneous rock having a porous texture, usually with the cavities larger than pore size and smaller than caverns.

cellular convection [METEOROL]   An organized, convective, fluid motion characterized by the presence of distinct convection cells or convective units, usually with upward motion (away from the heat source) in the central portions of the cell, and sinking or downward flow in the cell's outer regions.

cellular soil *See* polygonal ground.

celsian [MINERAL] $BaAl_2Si_2O_8$ Colorless, monoclinic mineral consisting of barium feldspar.

cement [GEOL]   Any chemically precipitated material, such as carbonates, gypsum, and barite, occurring in the interstices of clastic rocks.

cementation [GEOL]   The precipitation of a binding material around minerals or grains in rocks.

cementation factor [PETRO ENG]   Mathematical expression in oil-reservoir analysis for the degree to which precipitated minerals have bound together the grains (for example, of sand).

cementation sinking [MIN ENG] A technique of shaft sinking through strata containing water by injecting liquid cement or chemicals into the ground.

cement gravel [GEOL] Gravel consolidated by clay, silica, calcite, or other binding material.

cement-lined casing [PETRO ENG] Steel-, oil-, or gas-well casing pipe with internal lining of special cement; used to withstand severe corrosive conditions.

cement log [PETRO ENG] Gamma-ray measurement and logging of the height and condition of cement surrounding down-hole oil-well casing.

cement rock [PETR] An argillaceous limestone containing lime, silica, and alumina in variable proportions and usually some magnesia; used in the manufacture of natural hydraulic cement.

Cenomanian [GEOL] Lower Upper Cretaceous geologic time.

cenote *See* pothole.

cenotypal rock [PETR] A fine-grained porphyritic igneous rock, such as that of Tertiary and Holocene age, which resembles fresh or nearly fresh extrusive rock.

Cenozoic [GEOL] The youngest of the eras, or major subdivisions of geologic time, extending from the end of the Mesozoic Era to the present, or Recent.

center counter [PETR] An instrument used in structural petrology to count the number of points on a point diagram.

center jump [METEOROL] The formation of a second low-pressure center within an already well-developed low-pressure center; the latter diminishes in magnitude as the center of activity shifts or appears to jump to the new center.

center of action [METEOROL] A semipermanent high or low atmospheric pressure system at the surface of the earth; fluctuations in the intensity, position, orientation, shape, or size of such a center are associated with widespread weather changes.

center of falls *See* pressure-fall center.

center-of-gravity map [PETR] A map displaying the relative, weighted mean position of a lithologic type expressed in terms of its distance from the top of a specific stratigraphic unit.

center of inversion [CRYSTAL] A point in a crystal lattice such that the lattice is left invariant by an inversion in the point.

center of rises *See* pressure-rise center.

central breaker [MIN ENG] A breaker where the coal from several mines in a district is prepared.

central pressure [METEOROL] At any given instant, the atmospheric pressure at the center of a high or low; the highest pressure in a high, the lowest pressure in a low.

central valley *See* rift valley.

central water [OCEANOGR] Upper water mass associated with the central region of oceanic gyre.

Centronellidina [PALEON] A suborder of extinct articulate brachiopods in the order Terebratulida.

centrosphere [GEOL] The central core of the earth. Also known as the barysphere.

Cephalaspida [PALEON] An equivalent name for the Osteostraci.

ceramicite [PETR] A porcelained pyrometamorphic rock composed of basic plagioclase and cordierite with a small amount of hypersthene and a groundmass of glass.

Ceramoporidae [PALEON]   A family of extinct, marine bryozoans in the order Cysto-porata.

cerargyrite [MINERAL]   AgCl   A colorless to pearl-gray mineral; crystallizes in the isometric system, but crystals, usually cubic, are rare; a secondary mineral that is an ore of silver. Also known as chlorargyrite; horn silver.

ceratite [PALEON]   A fossil ammonoid of the genus *Ceratites* distinguished by a type of suture in which the lobes are further divided into subordinate crenulations while the saddles are not divided and are smoothly rounded.

ceratitic [PALEON]   Pertaining to a ceratite.

Ceratodontidae [PALEON]   A family of Mesozoic lungfishes in the order Dipteri-formes.

Ceratopsia [PALEON]   The horned dinosaurs, a suborder of Upper Cretaceous rep-tiles in the order Ornithischia.

cerine *See* allanite.

cerite [MINERAL] $(Ca,Fe)Ce_3Si_3O_{12}\cdot H_2O$ A brown rare-earth hydrous silicate of cerium and other metals found in gneiss; hardness is 5.5 on Mohs scale, and specific gravity is 4.86.

cerolite [MINERAL]   A mixture of serpentine and stevensite occurring in yellow or greenish waxlike masses.

cers [METEOROL]   A term for the mistral in Catalonia, Narbonne, and parts of Provence (southern France and northeastern Spain).

cerussite [MINERAL] $PbCO_3$ A yellow or white member of the aragonite group occurring in orthorhombic crystals; produced by the action of carbon dioxide on lead ore.

cervantite [MINERAL] Sb_2O_4 A white or yellow secondary mineral crystallizing in the orthorhombic system and formed by oxidation of antimony sulfide.

cesarolite [MINERAL] $H_2PbMn_3O_8$ A steel-gray mineral consisting of a hydrous lead manganate occurring in spongy masses.

ceylonite [MINERAL]   A dark-green, brown, or black iron-bearing variety of spi-nel. Also known as candite; pleonaste; zeylanite.

C factor [MAP]   An empirical value in aerial photography which expresses the vertical measuring capability of a given stereoscopic system; the factor is the ratio of the flight height to the smallest contour interval accurately plottable; in planning for aerial photography the C factor is used to determine the flight height required for a specified contour interval, camera, and instrument system. Also known as altitude contour.

C figure *See* C index.

chabazite [MINERAL] $CaAl_2Si_4O_{12}\cdot 6H_2O$ A white to yellow or red member of the zeolite group occurring in glassy rhombohedral crystals; hardness is 4–5 on Mohs scale, and specific gravity is 2.08–2.16.

chadacryst *See* xenocryst.

Chaetetidae [PALEON]   A family of Paleozoic corals of the order Tabulata.

chain breat machine [MIN ENG]   A coal-cutting machine, so constructed that a series of cutting points attached to a circulating chain work their way for a certain distance under a seam.

chain coal cutter [MIN ENG]   A cutter which makes a groove in the coal by an endless chain moving around a flat plate called a jib.

chain gage [ENG] An instrument for gaging water-surface elevation, consisting of a tagged or indexed line, usually made of chain or tape.

chain lightning [GEOPHYS] A rare form of lightning in a long zigzag or apparently broken line.

chainwall [MIN ENG] A coal mining technique in which the mine roof is supported by coal pillars, between which the coal is mined away.

chairs *See* folding boards.

chalazoidite *See* mud ball.

chalcanthite [MINERAL] $CuSO_4 \cdot 5H_2O$ A blue to bluish-green mineral occurring in triclinic crystals or in massive fibrous veins or stalactites. Also known as bluestone; blue vitriol.

chalcedony [MINERAL] A cryptocrystalline variety of quartz; occurs as crusts with a rounded, mammillary, or botryoidal surface and as a major constituent of nodular and bedded cherts; varieties include carnelian and bloodstone.

chalcedonyx [MINERAL] A mineral consisting of onyx with alternating gray and white bands; valued as a semiprecious stone.

chalcoalumite [MINERAL] $CuAl_4(SO_4)(OH)_{12} \cdot 3H_2O$ A turquoise-green to pale-blue mineral consisting of a hydrous basic sulfate of copper and aluminum.

chalcocite [MINERAL] Cu_2S A fine-grained, massive mineral with a metallic luster which tarnishes to dull black on exposure; crystallizes in the orthorhombic system, the crystals being rare and small usually with hexagonal outline as a result of twinning; hardness is 2.5–3 on Mohs scale, and specific gravity is 5.5–5.8. Also known as beta chalcocite; chalcosine; copper glance; redruthite; vitreous copper.

chalcocyanite [MINERAL] $CuSO_4$ A white mineral consisting of copper sulfate. Also known as hydrocyanite.

chalcolite *See* torbernite.

chalcomenite [MINERAL] $CuSeO_3 \cdot 2H_2O$ A blue mineral consisting of copper selenite occurring in crystals.

chalcophanite [MINERAL] $(Zn,Mn,Fe)Mn_2O_5 \cdot nH_2O$ Black mineral with metallic luster consisting of hydrous manganese and zinc oxide.

chalcophile [GEOL] Having an affinity for sulfur and therefore massing in greatest concentration in the sulfide phase of a molten mass.

chalcophyllite [MINERAL] $Cu_{18}Al_2(AsO_4)_3(OH)_{27} \cdot 33H_2O$ A green mineral consisting of basic arsenate and sulfate of copper and aluminum occurring in tabular crystals or foliated masses. Also known as copper mica.

chalcopyrite [MINERAL] $CuFeS_2$ A major ore mineral of copper; crystallizes in the tetragonal crystal system, but crystals are generally small with diphenoidal faces resembling the tetrahedron; usually massive with a metallic luster and brass-yellow color; hardness is 3.5–4 on Mohs scale, and specific gravity is 4.1–4.3. Also known as copper pyrite; yellow pyrite.

chalcopyrrohite [MINERAL] $CuFe_4S_5$ A sulfide mineral occurring in meteorites.

chalcosiderite [MINERAL] $Cu(Fe,Al)_6(PO_4)_4(OH)_8 \cdot 4H_2O$ A green mineral, isomorphous with turquoise, consisting of a hydrous basic phosphate of copper, iron, and aluminum.

chalcosine *See* chalcocite.

chalcostibite [MINERAL] $CuSbS_2$ A lead-gray mineral consisting of antimony copper sulfide.

chalcotrichite [MINERAL]   A capillary variety of cuprite occurring in long needlelike crystals. Also known as hair copper; plush copper ore.

Chalicotheriidae [PALEON]   A family of extinct perissodactyl mammals in the superfamily Chalicotherioidea.

Chalicotherioidea [PALEON]   A superfamily of extinct, specialized perissodactyls having claws rather than hooves.

chalk [PETR]   A variety of limestone formed from pelagic organisms; it is very fine-grained, porous, and friable; white or very light-colored, it consists almost entirely of calcite.

chalkland [GEOL]   An area underlain by chalk deposits, characterized by rolling hills and plateaus, expansive pasturelands, and dry valleys.

challiho [METEOROL]   Strong southerly winds which blow for about 40 days in spring in some parts of India; sometimes the winds are violent, causing blinding dust storms.

chalmersite *See* cubanite.

chamber [MIN ENG]   **1.** The working place of a miner. **2.** A body of ore with definite boundaries apparently filling a preexisting cavern.

chambering [MIN ENG]   Increasing the size of a drill hole in a quarry by firing a succession of small charges, until the hole can take a proper explosive charge to bring down the face of the quarry.

chamosite [MINERAL]   A greenish-gray or black mineral consisting of silicate belonging to the chlorite group and having monoclinic crystals; found in many oolitic iron ores.

Champlainian [GEOL]   Middle Ordovician geologic time.

chance process [MIN ENG]   A method for separating clean coal from slate and other impurities in a mixture of sand and water.

Chandler motion *See* polar wandering.

Chandler period [GEOPHYS]   The period of the Chandler wobble.

Chandler wobble [GEOPHYS]   A movement in the earth's axis of rotation, the period of motion being about 14 months. Also known as Eulerian nutation.

chandui *See* chanduy.

chanduy [METEOROL]   A cool, descending wind at Guayaquil, Ecuador, which blows during the dry season (July to November). Also spelled chandui.

change chart [METEOROL]   A chart indicating the amount and direction of change of some meteorological element during a specified time interval; for example, a height-change chart or pressure-change chart. Also known as tendency chart.

change of tide [OCEANOGR]   A reversal of the direction of motion (rising or falling) of a tide, or in the set of a tidal current. Also known as turn of the tide.

channel [HYD]   The deeper portion of a waterway carrying the main current.

channel control [HYD]   A condition whereby the stage of a stream is controlled only by discharge and the general configuration of the stream channel, that is, the contours of its bed, banks, and floodplains.

channel fill [GEOL]   Accumulations of sand and detritus in a stream channel where the transporting capacity of the water is insufficient to remove the material as rapidly as it is delivered.

channel frequency *See* stream frequency.

channel geometry [HYD] The specification of the shape of a cross section of a river channel for a limited distance.

channel gradient ratio *See* stream gradient ratio.

channel-lag deposit [GEOL] Coarse residual material left as accumulations in the channel in the normal processes of the stream.

channel-mouth bar [GEOL] A bar formed where moving water enters a body of still water, due to decreased velocity.

channel net [HYD] Stream channel pattern within a drainage basin.

channel order *See* stream order.

channel recording [GEOPHYS] A system, chain, or cascade of interconnecting devices through which geophysical data are passed from source to a recording instrument.

channel roughness [GEOL] A measure of the resistivity offered by the material constituting stream channel margins to the flow of water.

channel sample *See* groove sample.

channel sand [GEOL] A sandstone or sand deposited in a stream bed or other channel eroded into the underlying bed.

channel segment *See* stream segment.

channel spring [HYD] A depression spring issuing from a stream bank where the channel has been cut below the water table.

channel width [GEOL] The distance across a stream or channel as measured from bank to bank near bankful stage.

chapeiro [GEOL] An isolated coral reef developing in small scattered patches, often rising like a tower and sometimes extending outward in a mushroom shape.

Chapman equation [GEOPHYS] A theoretical relation describing the distribution of electron density with height in the upper atmosphere. [STAT MECH] The relationship that the viscosity of a gas equals $(0.499)mv/[\sqrt{2}\,\pi\sigma^2\,(1 + C/T)]$, where m is the mass of a molecule, v its average speed, σ its collision diameter, C the Sutherland constant, and T the absolute temperature (Kelvin scale).

chapmanite [MINERAL] $Fe_2Sb(SiO_4)_2(OH)$ A mineral consisting of a silicate of iron and antimony.

Chapman region [GEOPHYS] A hypothetical region in the upper atmosphere in which the distribution of electron density with height can be described by Chapman's theoretical equation.

character [GEOPHYS] A distinctive aspect of a seismic event, for example, the waveform.

characteristic fossil [GEOL] A fossil species or genus that is associated with a specific stratigraphic unit or time unit.

Charmouthian [GEOL] Middle Lower Jurassic geologic time.

charnockite [PETR] Any of various faintly foliated, nearly massive varieties of quartzofeldspathic rocks containing hypersthene.

charnockite series [GEOL] A series of plutonic rocks compositionally similar to the granitic rock series but characterized by the presence of orthopyroxene.

chart [MAP]   **1.** A map, generally designed for navigation or other particular purposes, in which essential map information is combined with various other data critical to the intended use.   **2.** To prepare a chart or to engage in a charting operation.

chart convergence [MAP]   Convergence of the meridians as shown on a chart.

chart datum *See* datum plane.

charted depth [OCEANOGR]   The vertical distance from the tidal datum to the bottom.

chart projection [MAP]   A map projection used for a chart.

Charybdis *See* Galofaro.

chassignite [GEOL]   An achondritic stony meteorite composed chiefly of olivine (95%); resembles dunite.

chatoyant [MINERAL]   Of a mineral or gemstone, having a changeable luster or color marked by a band of light, resembling the eye of a cat in this respect.

chattermark [GEOL]   A scar on the surface of bedrock made by the abrasive action of drift carried at the base of a glacier.

Chattian [GEOL]   Upper Oligocene geologic time. Also known as Casselian.

Chautauquan [GEOL]   Upper Devonian geologic time, below Bradfordian.

Chazyan [GEOL]   Middle Ordovician geologic time.

check observation [METEOROL]   An aviation weather observation taken primarily for aviation radio broadcast purposes; usually abbreviated to include just those elements of a record observation that have an important affect on aircraft operations.

check screen *See* oversize control screen.

Cheiracanthidae [PALEON]   A family of extinct acanthodian fishes in the order Acanthodiformes.

chelogenic [GEOL]   Referring to a cycle of continental evolution.

chemical-cartridge respirator [MIN ENG]   An air purification device worn by miners that removes small quantities of toxic gases or vapors from the inspired air; the cartridge contains chemicals which operate by processes of oxidation, absorption, or chemical reaction.

chemical crystallography [CRYSTAL]   The geometric description, and study, of the internal arrangement of atoms in crystals formed from chemical compounds.

chemical demagnetization [GEOPHYS]   A technique of partial demagnetization involving treatment by acid or other reagents to selectively remove one magnetically ordered mineral while leaving the remanent magnetization of the others unaltered.

chemical denudation [GEOL]   Wasting of the land surface by water transport of soluble materials into the sea.

chemical exfoliation [PETR]   An exfoliation process caused by an increase in volume due to changes in the chemical composition of the rock material.

chemical gaging [HYD]   A stream-gaging technique in which velocity flow is determined by introducing a chemical of known saturation in a stream and then measuring the level of dilution.

chemical mining [MIN ENG]   The extraction of valuable components of an orebody, either in place or within the mining area, by chemical methods such as leaching.

chemical oceanography [OCEANOGR]   The study of the chemistry of ocean water, the dissolved and suspended substances in it, the level of acidity, and the geographic and temporal variation of its chemical characteristics.

chemical oxygen demand [HYD] The amount of oxygen required for maximum oxidation in a body of water.

chemical precipitates [GEOL] A sediment formed from precipitated materials as distinguished from detrital particles that have been transported and deposited.

chemical pump [PETRO ENG] Skid-mounted pumping unit used to feed chemicals into the power oil (used to operate bottom-hole pumps in oil wells) to reduce corrosion in the system and to assist in water removal when the power oil and well-produced oil reach the ground-level wash tank.

chemical remanent magnetization [GEOPHYS] Permanent magnetization of rocks acquired when a magnetic material, such as hematite, is grown at low temperature through the oxidation of some other iron mineral, such as magnetite or goethite; the growing mineral becomes magnetized in the direction of any field which is present. Abbreviated CRM.

chemical reservoir [GEOL] An underground oil or gas trap formed in limestones or dolomites deposited in quiescent geologic environments.

chemical rock [PETR] A sedimentary rock formed by chemical deposition of organic or inorganic materials.

chemical unconformity [GEOL] An unconformity that can be characterized by chemical analysis.

chemical weathering [GEOCHEM] A weathering process whereby rocks and minerals are transformed into new, fairly stable chemical combinations by such chemical reactions as hydrolysis, oxidation, ion exchange, and solution. Also known as decay; decomposition.

chemocline [HYD] The transition in a meromictic lake between the mixolimnion layer (at the top) and the monimolimnion layer (at the bottom).

chemogenic [PETR] Referring to rock materials that have been deposited entirely by chemical action.

chemosphere [METEOROL] The vaguely defined region of the upper atmosphere in which photochemical reactions take place; generally considered to include the stratosphere (or the top thereof) and the mesosphere, and sometimes the lower part of the thermosphere.

Chemungian [GEOL] Middle Upper Devonian geologic time, below Cassodagan.

chenevixite [MINERAL] $Cu_2Fe_2(AsO_4)_2(OH)_4 \cdot H_2O$ A dark-green to greenish-yellow mineral consisting of a hydrous copper iron arsenate occurring in masses.

chenier [GEOL] A continuous ridge of beach material built upon swampy deposits; often supports trees, such as pines or evergreen oaks.

chergui [METEOROL] An eastern or southeastern desert wind in Morocco (North Africa), especially in the north; it is persistent, very dry and dusty, hot in summer, cold in winter.

Chernozem [GEOL] One of the major groups of zonal soils, developed typically in temperate to cool, subhumid climate; the Chernozem soils in modern classification include Borolls, Ustolls, Udolls, and Xerolls. Also spelled Tchernozem.

cherry picker [MIN ENG] A small hoist used to facilitate car changing near the loader in a mine tunnel.

chert [PETR] A hard, dense, micro- or cryptocrystalline rock composed of chalcedony and microcrystalline quartz. Also known as hornstone; phthanite.

chertification [GEOL] A process of replacement by silica in limestone in the form of fine-grained quartz or chalcedony.

chessylite *See* azurite.

Chesterian [GEOL]   Upper Mississippian geologic time.

chestnut coal [GEOL]   Anthracite coal small enough to pass through a round mesh of 1⅝ inches (4.13 centimeters) but too large to pass through a round mesh of 1¾₁₆ inches (3.02 centimeters).

Chestnut soil [GEOL]   One of the major groups of zonal soils, developed typically in temperate to cool, subhumid to semiarid climate; the Chestnut soils in modern classification include Ustolls, Borolls, and Xerolls.

chevkinite [MINERAL] $(Fe,Ca)(Ce,La)_2(Si,Ti)_2O_8$ A mineral consisting of silicotitanate of iron, calcium, and rare-earth elements.

chevron cross-bedding [GEOL]   Rock layers that dip in different or opposite directions in alternating or superimposed beds, thus forming a chevron or herringbone pattern.

chevron dune [GEOL]   A V-shaped dune formed in a vegetated area by strong winds blowing in a constant direction.

chevron fold [GEOL]   An accordionlike fold with limbs of equal length.

Cheyes point counter [PETR]   An instrument for petrographic modal analysis in which minerals are identified at regularly spaced points and tabulated mechanically.

chiastolite [MINERAL]   A variety of andalusite whose crystals have a cross-shaped appearance in cross section due to the arrangement of carbonaceous impurities. Also known as macle.

chibli *See* ghibli.

chichili *See* chili.

Chideruan [GEOL]   Uppermost Permian geologic time.

childrenite [MINERAL] $(Fe,Mn)AlPO_4(OH)_2 \cdot H_2O$ A pale-yellowish to dark-brown orthorhombic mineral consisting of a hydrous basic iron aluminum phosphate occurring as translucent crystals; it is isomorphous with eosphorite; hardness is 4.5–5 on Mohs scale, and specific gravity is 3.18–3.24.

Chile niter *See* Chile saltpeter.

Chile saltpeter [MINERAL]   Also known as Chile niter. **1.** Soda niter found in large quantities in caliche in arid regions of northern Chile. **2.** Deposits of sodium nitrate.

chili [METEOROL]   A warm, dry, descending wind in Tunisia, resembling the sirocco; in southern Algeria it is called chichili.

chilled contact [PETR]   The finer-grained portion of an igneous rock found near its contact with older rock.

chill wind factor [METEOROL]   An arbitrary index, developed by the Canadian Army, to correlate the performance of equipment and personnel in an Arctic winter; it is equal to the sum of the wind speed in miles per hour and the negative of the Fahrenheit temperature; the term is not to be confused with wind chill.

chill zone [PETR]   The border or marginal zone of an igneous intrusion, which has finer grain than the interior of the rock mass, because of more rapid cooling.

Chilobolbinidae [PALEON]   A family of extinct ostracods in the superfamily Hollinacea showing dimorphism of the velar structure.

chimney *See* pipe; spouting horn.

chimney cloud [METEOROL]   A cumulus cloud in the tropics that has much greater vertical than horizontal extent.

chimney rock   [GEOL]   **1.** A chimney-shaped remnant of a rock cliff whose sides have been cut into and carried away by waves and the gravel beach.   **2.** A rock column rising above its surroundings.   [MATER]   A porous phosphate rock used principally in chimney construction.

chinook   [METEOROL]   The foehn on the eastern side of the Rocky Mountains.

chinook arch   [METEOROL]   A foehn cloud formation appearing as a bank of clouds over the Rocky Mountains, generally a flat layer of altostratus, heralding the approach of a chinook.

chiolite   [MINERAL] $Na_5Al_3F_{14}$ A snow white mineral resembling cryolite. Also known as arksutite.

chip sampling   [MIN ENG]   Taking small pieces of ore or coal from the width of an ore face exposure; may be done at random or along a line.

chiral twinning   *See* optical twinning.

Chirodidae   [PALEON]   A family of extinct chondrostean fishes in the suborder Platysomoidei.

Chirognathidae   [PALEON]   A family of conodonts in the suborder Neurodonti-formes.

Chitinozoa   [PALEON]   An extinct group of unicellular microfossils of the kingdom Protista.

chiviatite   [MINERAL] $Pb_2Bi_6S_{11}$ A lead-gray mineral consisting of a lead bismuth sulfide occurring in foliated masses.

chloanthite   [MINERAL] $NiAs_{2-3}$ A white or gray mineral with metallic luster forming crystals in the isometric system; it is isomorphous with nickel-skutterudite.

chloraluminite   [MINERAL] $AlCl_3 \cdot 6H_2O$ A mineral consisting of hydrous aluminum chloride.

chlorapatite   [MINERAL] $Ca_5(PO_4)_3Cl$ An apatite mineral containing chlorine.

chlorargyrite   *See* cerargyrite.

chlorastrolite   [MINERAL]   A mottled, green variety of pumpellyite occurring as grains or small nodules of a stellate structure in basic igneous rock in the Lake Superior region; used as a semiprecious stone.

chlorine log   *See* chlorinolog.

chlorinity   [OCEANOGR]   A measure of the chloride and other halogen content, by mass, of sea water.

chlorinolog   [PETRO ENG]   A record of the presence and concentration of chlorine in oil reservoirs, prepared as a method of locating salt-water strata. Also known as chlorine log.

chlorite schist   [PETR]   A metamorphic rock whose composition is dominated by members of the chlorite group.

chlorite-sericite schist   [PETR]   A low-grade, fine-grained variety of mica schist without biotite.

chloritoid   [MINERAL] $FeAl_4Si_2O_{10}(OH)_4$ A micaceous mineral related to the brittle mica group; has both monoclinic and triclinic modifications, a gray to green color, and weakly pleochroic crystals.

chloritoid schist   [PETR]   A variety of mica schist whose composition is dominated by chloritoid.

chlormanganokalite [MINERAL] K_4MnCl_6 A wine yellow to lemon or canary yellow, hexagonal mineral consisting of potassium and manganese chloride; occurs as rhombohedrons.

chlorocalcite [MINERAL] $KCaCl_3$ A white mineral consisting of a chloride of potassium and calcium. Also known as hydrophilite.

chloromagnesite [MINERAL] $MgCl_2$ A mineral consisting of anhydrous magnesium chloride, found on the volcano Vesuvius.

chloropal *See* nontronite.

chlorophoenicite [MINERAL] $(Mn,An)_5(AsO_4)(OH)_7$ Gray-green monoclinic mineral consisting of a basic arsenate of manganese and zinc occurring in crystals.

chlorosity [OCEANOGR] The chlorine and bromide content of one liter of sea water; equals the chlorinity of the sample times its density at 20°C.

chlorothionite [MINERAL] $K_2Cu(SO_4)Cl_2$ Bright-blue secondary mineral consisting of potassium copper sulfate chloride, found on the volcano Vesuvius.

chloroxiphite [MINERAL] $Pb_3CuCl_2(OH)_2O_2$ A dull-olive or pistachio-green mineral consisting of a basic chloride of lead and copper, found in the Mendip Hills of England.

choanate fish [PALEON] Any of the lobefins composing the subclass Crossopterygii.

chock [MIN ENG] A square pillar for supporting the roof in a mine, constructed of prop timber laid up in alternate cross layers, in log-cabin style, the center being filled with waste.

chocolate gale *See* chocolatero.

chocolatero [METEOROL] A moderate norther in the Gulf region of Mexico. Also known as chocolate gale.

Choeropotamidae [PALEON] A family of extinct palaeodont artiodactyls in the superfamily Entelodontoidae.

choke [PETRO ENG] A removable nipple inserted in a flow line to control oil or gas flow.

choke crushing [MIN ENG] A recrushing of fine ore.

chokedamp *See* blackdamp.

chondrite [GEOL] A stony meteorite containing chondrules.

chondritic coincidence [GEOPHYS] A near equality between the total heat that leaves the earth each second and the heat produced by the radioactive elements in a model of the earth that has the same chemical composition as chondritic meteorites.

chondrodite [MINERAL] $Mg_5(SiO_4)_2(F,OH)_2$ A monoclinic mineral of the humite group; has a resinous luster, is yellow-red in color, and occurs in contact-metamorphosed dolomites.

Chondrostei [PALEON] The most archaic infraclass of the subclass Actinopterygii, or rayfin fishes.

Chondrosteidae [PALEON] A family of extinct actinopterygian fishes in the order Acipenseriformes.

chondrule [GEOL] A spherically shaped body consisting chiefly of pyroxene or olivine minerals embedded in the matrix of certain stony meteorites.

Chonetidina [PALEON] A suborder of extinct articulate brachiopods in the order Strophomenida.

choppy sea [OCEANOGR] In popular usage, short, rough, irregular wave motion on a sea surface.

chorismite [PETR] A mixed rock whose fabric is macropolyschematic and which consists of petrologically dissimilar materials of varied origins.

Choristodera [PALEON] A suborder of extinct reptiles of the order Eosuchia composed of a single genus, *Champsosaurus*.

C horizon [GEOL] The portion of the parent material in soils which has been penetrated with roots.

chorography [MAP] All of the methods used to map a region or district.

Christmas tree [PETRO ENG] An assembly of valves, tees, crosses, and other fittings at the wellhead, used to control oil or gas production and to give access to the well tubing.

chromatic mineral [MINERAL] A mineral with color.

chrome iron ore *See* chromite.

chrome spinel *See* picotite.

chromite [MINERAL] $FeCr_2O_4$ A mineral of the spinel group; crystals and pure form are rare, and it usually is massive; the only important ore mineral of chromium. Also known as chrome iron ore.

chromocratic *See* melanocratic.

chron [GEOL] The time unit equivalent to the stratigraphic unit, subseries, and geologic name of a division of geologic time.

chronocline [PALEON] A cline shown by successive morphological changes in the members of a related group, such as a species, in successive fossiliferous strata.

chronolith *See* time-stratigraphic unit.

chronolithologic unit *See* time-stratigraphic unit.

chronostratic unit *See* time-stratigraphic unit.

chronostratigraphic unit *See* time-stratigraphic unit.

chrysoberyl [MINERAL] $BeAl_2O_4$ A pale green, yellow, or brown mineral that crystallizes in the orthorhombic system and is found most commonly in pegmatite dikes; used as a gem. Also known as chrysopal; gold beryl.

Chrysochloridae [PALEON] The golden moles, a family of extinct lipotyphlan mammals in the order Insectivora.

chrysocolla [MINERAL] $CuSiO_3 \cdot 2H_2O$ A silicate mineral ordinarily occurring in impure cryptocrystalline crusts and masses with conchoidal fracture; a minor ore of copper; luster is vitreous, and color is normally emerald green to greenish-blue.

chrysolite [MINERAL] **1.** A gem characterized by light-yellowish-green hues, especially the gem varieties of olivine, but also including beryl, topaz, and spinel. **2.** A variety of olivine having a magnesium to magnesium-iron ratio of 0.90–0.70.

chrysopal *See* chrysoberyl.

chrysoprase [MINERAL] An apple-green variety of chalcedony that contains nickel; used as a gem. Also known as green chalcedony.

chrysotile [MINERAL] $Mg_3Si_2O_5(OH)_4$ A fibrous form of serpentine that constitutes one type of asbestos.

chthonic [GEOL]   Referring to deep-sea clastic debris and sediment originating from preexisting rocks.

chubasco [METEOROL]   A severe thunderstorm with vivid lightning and violent squalls coming from the land on the west coast of Nicaragua and Costa Rica in Central America.

Chubb [GEOL]   A meteorite crater in Ungava, Quebec, Canada.

churada [METEOROL]   A severe rain squall in the Mariana Islands (western Pacific Ocean) during the northeast monsoon; these squalls occur from November to April or May, but especially from January through March.

churchite *See* weinschenkite.

churn drill [MECH ENG]   Portable drilling equipment, with drilling performed by a heavy string of tools tipped with a blunt-edge chisel bit suspended from a flexible cable, to which a reciprocating motion is imparted by its suspension from an oscillating beam or sheave, causing the bit to be raised and dropped. Also known as American system drill; cable-system drill.

churn hole *See* pothole.

chute [HYD]   A short channel across a narrow land area which bypasses a bend in a river; formed by the river's breaking through the land.

chute conveyor *See* jigging conveyor.

chute system [MIN ENG]   A method of mining by which ore is broken from the surface downward into chutes and is removed through passageways below. Also known as glory hole system; milling system.

Ci *See* cirrus cloud.

CI *See* temperature humidity index.

ciminite [PETR]   An extrusive rock consisting essentially of olivine with sanidine and pyroxene and basic plagioclase.

cimolite [MINERAL] $2Al_2O_3 \cdot 9SiO_3 \cdot 6H_2O$ A white, grayish, or reddish mineral consisting of hydrous aluminum silicate occurring in soft, claylike masses.

Cincinnatian [GEOL]   Upper Ordovician geologic time.

cinder [GEOL]   Fine-grained pyroclastic material ranging in diameter from 4 to 32 millimeters.

cinder coal *See* natural coke.

cinder cone [GEOL]   A conical elevation formed by the accumulation of volcanic debris around a vent.

C index [GEOPHYS]   A subjectively obtained daily index of geomagnetic activity, in which each day's record is evaluated on the basis of 0 for quiet, 1 for moderately disturbed, and 2 for very disturbed. Also known as C figure; magnetic character figure.

cinetheodolite [ENG]   A surveying theodolite in which 35-millimeter motion picture cameras with lenses of 60- to 240-inch (1.5 to 6.1 meter) focal length are substituted for the surveyor's eye and telescope; used for precise time-correlated observation of distant airplanes, missiles, and artificial satellites.

cinnabar [MINERAL]   HgS   A vermilion-red mineral that crystallizes in the hexagonal system, although crystals are rare, and commonly occurs in fine, granular, massive form; the only important ore of mercury. Also known as cinnabarite.

cinnabarite *See* cinnabar.

CIPW classification [PETR]   A designation for the Norm system of classifying igneous rocks; from the initial letters of the names of those who devised it: Cross, Iddings, Pirsson, and Washington.

circle haul [MIN ENG]   A haulage system in strip mining; empty units enter the mine over one lateral and leave, loaded, over the lateral nearest the tipple.

circle of equal altitude [GEOD]   A circle on the surface of the earth, on every point of which the altitude of a given celestial body is the same at a given instant; the pole of this circle is the geographical position of the body, and the great-circle distance from this pole to the circle is the zenith distance of the body.

circle of inertia *See* inertial circle.

circle of longitude [GEOD]   *See* parallel.

circular coal *See* eye coal.

circular shaft [MIN ENG]   A shaft excavated in a round shape.

circular vortex [METEOROL]   An atmospheric flow in parallel planes in which streamlines and other isopleths are concentric circles about a common axis; an atmospheric model of easterly and westerly winds is a circular vortex about the earth's polar axis.

circulating fluid [ENG]   A fluid pumped into a borehole through the drill stem, the flow of which cools the bit and transports the cuttings out of the borehole.

circulation [METEOROL]   For an air mass, in the line integral of the tangential component of the velocity field about a closed curve.   [OCEANOGR]   Water current flow occurring within a large area, usually a closed circular pattern.

circulation flux [METEOROL]   Flux due to mean atmospheric motion as opposed to eddy flux; the dominant flux in low latitudes.

circulation index [METEOROL]   A measure of the magnitude of one of several aspects of large-scale atmospheric circulation patterns; indices most frequently measured represent the strength of the zonal (east-west) or meridional (north-south) components of the wind, at the surface or at upper levels, usually averaged spatially and often averaged in time.

circulation pattern [METEOROL]   The general geometric configuration of atmospheric circulation usually applied, in synoptic meteorology, to the large-scale features of synoptic charts and mean charts.

circum-Pacific province *See* Pacific suite.

circumpolar [GEOGR]   Located around one of the polar regions of earth.

circumpolar westerlies *See* westerlies.

circumpolar whirl *See* polar vortex.

circumvallation [GEOL]   The process in which mountains are formed on a featureless plain by stream incision.

cirque [GEOL]   A steep elliptic to elongated enclave high on mountains in calcareous districts, usually forming the blunt end of a valley.

cirque lake [HYD]   A small body of water occupying a cirque.

cirriform [METEOROL]   Descriptive of clouds composed of small particles, mostly ice crystals, which are fairly widely dispersed, usually resulting in relative transparency and whiteness and often producing halo phenomena not observed with other cloud forms.

cirrocumulus cloud   [METEOROL]   A principal cloud type, appearing as a thin, white path of cloud without shadows, composed of very small elements in the form of grains, ripples, and so on. Abbreviated Cc.

cirrostratus cloud   [METEOROL]   A principal cloud type, appearing as a whitish veil, usually fibrous but sometimes smooth, which may totally cover the sky and often produces halo phenomena, either partial or complete. Abbreviated Cs.

cirrus cloud   [METEOROL]   A principal cloud type composed of detached cirriform elements in the form of white, delicate filaments, of white (or mostly white) patches, or narrow bands. Abbreviated Ci.

cistern barometer   [ENG]   A pressure-measuring device in which pressure is read by the liquid rise in a vertical, closed-top tube as a result of system pressure on a liquid reservoir (cistern) into which the bottom, open end of the tube is immersed.

citrine   [MINERAL]   An important variety of crystalline quartz, yellow to brown in color and transparent. Also known as Bohemian topaz; false topaz; quartz topaz; topaz quartz; yellow quartz.

cladodont   [PALEON]   Pertaining to sharks of the most primitive evolutionary level.

Cladoselachii   [PALEON]   An order of extinct elasmobranch fishes including the oldest and most primitive of sharks.

Claibornian   [GEOL]   Middle Eocene geologic time.

claim   *See* mining claim.

Clairaut's formula   [GEOD]   An approximate formula for gravity at the earth's surface, assuming that the earth is an ellipsoid; states that the gravity is equal to g_e [1 + (5/2 m' − f) sin^2 θ], where θ is the latitude, g_e is the gravity at the equator, m' is the ratio of centrifugal acceleration to gravity at the equatorial surface, and f is the earth's flattening, equal to $(a − b)/a$, where a is the semimajor axis and b is the semiminor axis.

clairite   *See* enargite.

clan   [PETR]   A category of igneous rocks defined in terms of similarities in mineralogical or chemical composition.

clarain   [GEOL]   A coal lithotype appearing as stratifications parallel to the bedding plane and usually having a silky luster and scattered or diffuse reflection. Also known as clarite.

Clarendonian   [GEOL]   Lower Pliocene or upper Miocene geologic time.

clarinite   [MINERAL]   A heterogeneous, generally translucent material making up the major micropetrological ingredient of clarain.

clarite   *See* clarain.

clarke   [GEOCHEM]   A unit of the average abundance of an element in the earth's crust, expressed as a percentage.

Clarkecarididae   [PALEON]   A family of extinct crustaceans within the order Anaspidacea.

Clarke ellipsoid of 1866   [GEOD]   The reference ellipsoid adopted by the U.S. Coast and Geodetic Survey of 1880 for charting North America.

clarkeite   [MINERAL] $(Na,Ca,Pb)_2U_2(O,OH)_7$ A dark reddish-brown or dark brown mineral consisting of a hydrous or hydrated uranium oxide.

clarodurain   [GEOL]   A transitional lithotype of coal composed of vitrinite and other macerals, principally micrinite and exinite.

clarofusain   [GEOL]   A transitional lithotype of coal composed of fusinite and vitrinite and other macerals.

clarovitrain [GEOL] A transitional lithotype of coal rock composed primarily of the maceral vitrinite, with lesser amounts of other macerals.

clast [GEOL] An individual grain, fragment, or constituent of detrital sediment or sedimentary rock produced by physical breakdown of a larger mass.

clastation *See* weathering.

clastic [GEOL] Rock or sediment composed of clasts which have been transported from their place of origin, as sandstone and shale.

clastic dike [GEOL] A tabular-shaped sedimentary dike composed of clastic material and transecting the bedding of a sedimentary formation; represents invasion by extraneous material along a crack of the containing formation.

clastic pipe [GEOL] A cylindrical body of clastic material, having an irregular columnar or pillar-like shape, standing approximately vertical through enclosing formations (usually in limestone), and measuring a few centimeters to 50 meters in diameter and 1 to 60 meters in height.

clastic ratio [GEOL] The ratio of the percentage of clastic rocks to that of nonclastic rocks in a geologic section. Also known as detrital ratio.

clastic reservoir [GEOL] An underground oil or gas trap formed in clastic limestone.

clastic sediment [GEOL] Deposits of clastic materials transported by mechanical agents. Also known as mechanical sediment.

clastic wedge [GEOL] The sediments of the exogeosyncline, derived from the tectonic landmasses of the adjoining orthogeosyncline.

clathrate [PETR] Pertaining to a condition, chiefly in leucite rock, in which clear leucite crystals are surrounded by tangential leucite crystals to give the rock an appearance of a net or a section of sponge.

claudetite [MINERAL] As_2O_3 A mineral containing arsenic that is dimorphous with arsenolite; crystallizes in the monoclinic system.

clausthalite [MINERAL] PbSe A mineral consisting of lead selenide and resembling galena; specific gravity is 7.6–8.8.

Clavatoraceae [PALEOBOT] A group of middle Mesozoic algae belonging to the Charophyta.

clay [GEOL] **1.** A natural, earthy, fine-grained material which develops plasticity when mixed with a limited amount of water; composed primarily of silica, alumina, and water, often with iron, alkalies, and alkaline earths. **2.** The fraction of an earthy material containing the smallest particles, that is, finer than 3 micrometers.

clay atmometer [ENG] An atmometer consisting of a porous porcelain container connected to a calibrated reservoir filled with distilled water; evaporation is determined by the depletion of water.

Clay Belt [GEOL] A lowland area bordering on the western and southern portions of Hudson and James bays in Canada, composed of clays and silts recently deposited in large glacial lakes during the withdrawal of the continental glaciers.

clay gall [GEOL] A dry, curled clay shaving derived from dried, cracked mud and embedded and flattened in a sand stratum.

clay ironstone [PETR] **1.** A clayey rock containing large quantities of iron oxide, usually limonite. **2.** A clayey-looking stone occurring among carboniferous and other rocks; contains 20–30% iron.

clayite [GEOL] A hydrous aluminum silicate thought to be the true clay substance in kaolin, and considered to be an amorphous (colloidal) material of the same chemical composition as kaolinite.

clay loam [GEOL] Soil containing 27–40% clay, 20–45% sand, and the remaining portion silt.

clay marl [GEOL] A chalky clay, whitish with a smooth texture.

clay mineral [MINERAL] One of a group of finely crystalline, hydrous silicates with a two- or three-layer crystal structure; the major components of clay materials; the most common minerals belong to the kaolinite, montmorillonite, attapulgite, and illite groups.

claypan [GEOL] A stratum of compact, stiff, relatively impervious noncemented clay; can be worked into a soft, plastic mass if immersed in water.

clay plug [GEOL] Sediment, with a great deal of organic muck, deposited in a cutoff river meander.

clay shale [GEOL] **1.** Shale composed wholly or chiefly of clayey material which becomes clay again on weathering. **2.** Consolidated sediment composed of up to 10% sand and having a silt to clay ratio of less than 1:2.

clay soil [GEOL] A fine-grained inorganic soil which forms hard lumps when dry and becomes sticky when wet.

claystone [GEOL] Indurated clay, consisting predominantly of fine material of which a major proportion is clay mineral.

clay vein [GEOL] A body of clay which is similar to an ore vein in form and fills a crevice in a coal seam. Also known as dirt slip.

cleanout auger *See* cleanout jet auger.

cleanout jet auger [ENG] An auger equipped with water-jet orifices designed to clean out collected material inside a driven pipe or casing before taking soil samples from strata below the bottom of the casing. Also known as cleanout auger.

cleanup [MIN ENG] **1.** The collecting of all the valuable product of a given period of operation in a stamp mill or in a hydraulic or placer mine. **2.** The valuable material resulting from a cleanup.

clear [METEOROL] **1.** After United States weather observing practice, the state of the sky when it is cloudless or when the sky cover is less than 0.1 (to the nearest tenth). **2.** To change from a stormy or cloudy weather condition to one of no precipitation and decreased cloudiness.

clear-air turbulence [METEOROL] A meteorological phenomenon occurring in the upper troposphere and lower stratosphere, in which high-speed aircraft are subject to violent updrafts and downdrafts. Abbreviated CAT.

clearance [MIN ENG] The space between the top or side of a car and the roof or wall. [PETRO ENG] The annular space between down-hole drill-string equipment, such as bits, core barrels, and casing, and the walls of the borehole with the down-hole equipment centered in the hole.

clear ice [HYD] Generally, a layer or mass of ice which is relatively transparent because of its homogeneous structure and small number and size of air pockets.

cleat [GEOL] Vertical breakage planes in coal. Also spelled cleet.

cleat spar *See* ankerite.

cleavage [GEOL] Splitting, or the tendency to split, along parallel, closely positioned planes in rock.

cleavage banding [GEOL] A compositional banding, usually formed from incompetent material such as argillaceous rocks, that is parallel to the cleavage rather than the bedding.

cleavage crystal [CRYSTAL] A crystal fragment bounded by cleavage faces giving it a regular form.

cleavage fracture [CRYSTAL] **1.** Manner of breaking a crystalline substance along the cleavage plane. **2.** The appearance of such a broken surface.

cleavage plane [CRYSTAL] A plane along which a crystalline substance may be split.

cleavelandite [MINERAL] A white, lamellar variety of albite that is almost pure $NaAlSi_3O_8$ and has a tabular habit, with individuals often showing mosaic developments and tending to occur in fan-shaped aggregates.

cleet See cleat.

cleft [GEOL] A sudden chasm, gash, breach, or other sharp opening, such as a craggy fissure, in a rock.

cliachite [MINERAL] A group of brownish, colloidal aluminum hydroxides that constitutes most bauxite.

cliff [GEOGR] A high, steep, perpendicular or overhanging face of a rock; a precipice.

cliff of displacement See fault scarp.

Cliftonian [GEOL] Middle Middle Silurian geologic time.

climagram See climatic diagram.

climagraph See climatic diagram.

climate [CLIMATOL] The long-term manifestations of weather.

climate control [CLIMATOL] Schemes for artificially altering or controlling the climate of a region.

climatic change [CLIMATOL] The long-term fluctuation in rainfall, temperature, and other aspects of the earth's climate.

climatic classification [CLIMATOL] The division of the earth's climates into a system of contiguous regions, each one of which is defined by relative homogeneity of the climate elements.

climatic controls [CLIMATOL] The relatively permanent factors which govern the general nature of the climate of a portion of the earth, including solar radiation, distribution of land and water masses, elevation and large-scale topography, and ocean currents.

climatic cycle [CLIMATOL] A long-period oscillation of climate which recurs with some regularity, but which is not strictly periodic. Also known as climatic oscillation.

climatic diagram [CLIMATOL] A graphic presentation of climatic data; generally limited to a plot of the simultaneous variations of two climatic elements, usually through an annual cycle. Also known as climagram; climagraph; climatograph; climogram; climograph.

climatic divide [CLIMATOL] A boundary between regions having different types of climate.

climatic factor [CLIMATOL] Climatic control, but regarded as including more local influences; thus city smoke and the extent of the builtup metropolitan area are climatic factors, but not climatic controls.

climatic forecast [CLIMATOL]   A forecast of the future climate of a region; that is, a forecast of general weather conditions to be expected over a period of years.

climatic optimum [CLIMATOL]   The period in history (about 5000–2500 B.C.) during which temperatures were warmer than they are at present in nearly all parts of the world.

climatic oscillation *See* climatic cycle.

climatic peat [GEOL]   Peat that is characteristic of a particular climatic zone.

climatic prediction [CLIMATOL]   The estimation of the chances that specified climatic conditions will occur in a certain time interval.

climatic province [CLIMATOL]   A region of the earth's surface characterized by an essentially homogeneous climate.

climatic snow line [METEOROL]   The altitude above which a flat surface (fully exposed to sun, wind, and precipitation) would experience a net accumulation of snow over an extended period of time; below this altitude, ablation would predominate.

climatic zone [CLIMATOL]   A belt of the earth's surface within which the climate is generally homogeneous in some respect; an elemental region of a simple climatic classification.

Climatiidae [PALEON]   A family of archaic tooth-bearing fishes in the suborder Climatioidei.

Climatiiformes [PALEON]   An order of extinct fishes in the class Acanthodii having two dorsal fins and large plates on the head and ventral shoulder.

Climatioidei [PALEON]   A suborder of extinct fishes in the order Climatiiformes.

climatochronology [GEOL]   The absolute age dating of recent geologic events by using the oxygen isotope ratios in ice, shells, and so on.

climatograph *See* climatic diagram.

climatography [CLIMATOL]   A quantitative description of climate, particularly with reference to the tables and charts which show the characteristic values of climatic elements at a station or over an area.

climatological forecast [METEOROL]   A weather forecast based upon the climate of a region instead of upon the dynamic implications of current weather, with consideration given to such synoptic weather features as cyclones and anticyclones, fronts, and the jet stream.

climatological station elevation [CLIMATOL]   The elevation above mean sea level chosen as the reference datum level for all climatological records of atmospheric pressure in a given locality.

climatological station pressure [CLIMATOL]   The atmospheric pressure computed for the level of the climatological station elevation, used to give all climatic records a common reference; it may or may not be the same as station pressure.

climatological substation [CLIMATOL]   A weather-observing station operated (by an unpaid volunteer) for the purpose of recording climatological observations.

climatology [METEOROL]   That branch of meteorology concerned with the mean physical state of the atmosphere together with its statistical variations in both space and time as reflected in the weather behavior over a period of many years.

climax avalanche [HYD]   The largest type of avalanche, containing a sizable quantity of old snow, and arising from conditions that developed over a year or more.

climbing dune [GEOL] A dune that develops on the windward side of mountains or hills.

climogram *See* climatic diagram.

climograph *See* climatic diagram.

clinker [GEOL] Burnt or vitrified stony material, as ejected by a volcano or formed in a furnace.

clinoamphibole [MINERAL] A group of amphiboles which crystallize in the monoclinic system.

clinoaxis [CRYSTAL] The inclined lateral axis that makes an oblique angle with the vertical axis in the monoclinic system. Also known as clinodiagonal.

clinochlore [MINERAL] $(Mg,Fe,Al)_3(Si,Al)_2O_5(OH)_4$ Green mineral of the chlorite group, occurring in monoclinic crystals, in folia or scales, or massive.

clinoclase [MINERAL] $Cu_3(AsO_4)(OH)_3$ A dark-green mineral consisting of basic copper arsenate occurring in translucent prismatic crystals or massive. Also known as clinoclasite.

clinoclasite *See* clinoclase.

clinodiagonal *See* clinoaxis.

clinoenstatite [MINERAL] $Mg_2(Si_2O_6)$ A monoclinic pyroxene consisting principally of magnesium silicate; occurs frequently in stony meteorites, but is rare in terrestrial environments.

clinoferrosilite [MINERAL] $Fe_2(Si_2O_6)$ A monoclinic pyroxene consisting of iron silicate.

clinoform [GEOL] A subaqueous landform, such as the continental slope of the ocean or the foreset bed of a delta.

clinograph [ENG] A type of directional surveying instrument that records photographically the direction and magnitude of deviations from the vertical of a borehole, well, or shaft; the information is obtained by the instrument in one trip into and out of the well.

clinographic projection [GRAPHICS] A method of representing objects, especially crystals, in which each point P of the object to be represented is projected onto the foot of a perpendicular from P to a plane which is located so that no plane surface of the object is represented by a line.

clinohedral class [CRYSTAL] A rare class of crystals in the monoclinic system having a plane of symmetry but no axis of symmetry. Also known as domatic class.

clinohedrite [MINERAL] $CaZnSiO_3(OH)_2$ Colorless, white, or purplish monoclinic mineral consisting of a calcium zinc silicate occurring in crystals; hardness is 5.5 on Mohs scale, and specific gravity is 3.33.

clinohumite [MINERAL] $Mg_9(SiO_4)_4(F,OH_2)$ A monoclinic mineral of the humite group.

clinometer [ENG] A hand-held surveying device for measuring vertical angles; consists of a sighting tube surmounted by a graduated vertical arc with an attached level bubble; used in meteorology to measure cloud height at night, in conjunction with a ceiling light, and in ordnance for boresighting.

clinopinacoid [CRYSTAL] A form of monoclinic crystal whose faces are parallel to the inclined and vertical axes.

clinopyroxene [MINERAL]   The general term for any of those pyroxenes that crystallize in the monoclinic system; on occasion, these pyroxenes have large amounts of calcium with or without aluminum and the alkalies. Also known as monopyroxene clinoaugite.

clinozoisite [MINERAL] $Ca_2Al_3(SiO_4)_3(OH)$ A grayish-white, pink, or green monoclinic mineral of the epidote group.

clint [GEOL]   A hard or flinty rock, such as a projecting rock or ledge.

Clintonian [GEOL]   Lower Middle Silurian geologic time.

clintonite [MINERAL] $Ca(Mg,Al)_3(Al,Si)O_{10}(OH)_2$ A reddish-brown, copper-red, or yellowish monoclinic mineral of the brittle mica group occurring in crystals or foliated masses. Also known as seybertite; xanthophyllite.

clog snow [HYD]   A skiing term for wet, sticky, new snow.

close [METEOROL]   Colloquially, descriptive of oppressively still, warm, moist air, frequently applied to indoor conditions.

closed basin [GEOL]   An enclosed area with no drainage outlet, so that water can escape by evaporation.

closed structure [GEOL]   A structure which, when represented on a topographic map, is surrounded by one or more closed contour lines.

closed-circuit grinder [MIN ENG]   A grinder connected to a size classifier (cyclone or screen) to return oversized particles to the grinding operation in closed-circuit pulverizing.

closed-circuit pulverizing [MIN ENG]   A process used in ore dressing in which the material discharged from the pulverizer is passed through an external classifier where the finished product is removed and oversized particles are returned to the pulverizer.

closed drainage [HYD]   Drainage in which the surface flow of water collects in sinks or lakes having no surface outlet. Also known as blind drainage.

closed fold [GEOL]   A fold whose limbs have been compressed until they are parallel, and whose structure contour lines form a closed loop. Also known as tight fold.

closed frame [MIN ENG]   A mine support frame that is completely closed; especially in inclined shafts, it is used to protect all sides from rock pressure.

closed high [METEOROL]   A high that may be completely encircled by an isobar or contour line.

closed low [METEOROL]   A low that may be completely encircled by an isobar or contour line, that is, an isobar or contour line of any value, not necessarily restricted to those arbitrarily chosen for the analysis of the chart.

closed rotative gas lift [PETRO ENG]   Oil-well control system in which high-pressure compressor gas is injected into a well to force oil fluids from the reservoir, with spent lift gas recompressed for reinjection.

close-joints cleavage See slip cleavage.

close-packed crystal [CRYSTAL]   A crystal structure in which the lattice points are centers of spheres of equal radius arranged so that the volume of the interstices between the spheres is minimal.

close work [MIN ENG]   Driving a tunnel or drifting between two coal seams.

closure [GEOL]   The vertical distance between the highest and lowest point on an anticline enclosed by contour lines.

cloud [METEOROL]   Suspensions of minute water droplets or ice crystals produced by the condensation of water vapor.

cloud absorption [GEOPHYS] The absorption of electromagnetic radiation by the waterdrops and water vapor within a cloud.

cloudage *See* cloud cover.

cloud band [METEOROL] A broad band of clouds, about 10 to 100 or more miles (16 to 161 kilometers) wide, and varying in length from a few tens of miles to hundreds of miles.

cloud bank [METEOROL] A fairly well-defined mass of cloud observed at a distance; covers an appreciable portion of the horizon sky, but does not extend overhead.

cloud bar [METEOROL] **1.** A heavy bank of clouds that appears on the horizon with the approach of an intense tropical cyclone (hurricane or typhoon); it is the outer edge of the central cloud mass of the storm. **2.** Any long, narrow, unbroken line of cloud, such as a crest cloud or an element of billow cloud.

cloud base [METEOROL] For a given cloud or cloud layer, that lowest level in the atmosphere at which the air contains a perceptible quantity of cloud particles.

cloudburst [METEOROL] In popular terminology, any sudden and heavy fall of rain, usually of the shower type, and with a fall rate equal to or greater than 100 millimeters (3.94 inches) per hour. Also known as rain gush; rain gust.

cloudburst flood [HYD] A flood of very brief duration that occurs during a cloudburst, usually in an arid or semiarid region.

cloud cap *See* cap cloud.

cloud classification [METEOROL] **1.** A scheme of distinguishing and grouping clouds according to their appearance and, where possible, to their process of formation. **2.** A scheme of classifying clouds according to their altitudes: high, middle, or low clouds. **3.** A scheme of classifying clouds according to their particulate composition: water clouds, ice-crystal clouds, or mixed clouds.

cloud cover [METEOROL] That portion of the sky cover which is attributed to clouds, usually measured in tenths of sky covered. Also known as cloudage; cloudiness.

cloud crest *See* crest cloud.

cloud deck [METEOROL] The upper surface of a cloud.

cloud discharge [GEOPHYS] A lightning discharge occurring between a positive charge center and a negative charge center, both of which lie in the same cloud. Also known as cloud flash; intracloud discharge.

cloud droplet [METEOROL] A particle of liquid water from a few micrometers to tens of micrometers in diameter, formed by condensation of atmospheric water vapor and suspended in the atmosphere with other drops to form a cloud.

cloud-drop sampler [ENG] An instrument for collecting cloud particles, consisting of a sampling plate or cylinder and a shutter, which is so arranged that the sampling surface is exposed to the cloud for a predetermined length of time; the sampling surface is covered with a material which either captures the cloud particles or leaves an impression characteristic of the impinging elements.

cloud echo [METEOROL] The radar target signal returned from clouds alone, as detected by cloud detection radars or other very-short-wavelength equipment.

clouded agate [MINERAL] A variety of transparent or light gray agate in which patches of darker gray resembling clouds appear.

cloud flash *See* cloud discharge.

cloud formation [METEOROL] **1.** The process by which various types of clouds are formed, generally involving adiabatic cooling of ascending moist air. **2.** A particular arrangement of clouds in the sky, or a striking development of a particular cloud.

cloud height [METEOROL] The absolute altitude of the base of a cloud.

cloud height indicator [ENG] General term for an instrument which measures the height of cloud bases.

cloudiness *See* cloud cover.

cloud layer [METEOROL] An array of clouds, not necessarily all of the same type, whose bases are at approximately the same level; may be either continuous or composed of detached elements.

cloud level [METEOROL] **1.** A layer in the atmosphere in which are found certain cloud genera; three levels are usually defined: high, middle, and low. **2.** At a particular time, the layer in the atmosphere bounded by the limits of the bases and tops of an existing cloud form.

cloud mirror *See* mirror nephoscope.

cloud modification [METEOROL] Any process by which the natural course of development of a cloud is altered by artificial means.

cloud particle [METEOROL] A particle of water, either a drop of liquid water or an ice crystal, comprising a cloud.

cloud-phase chart [METEOROL] A chart designed to indicate and distinguish supercooled water clouds from ice-crystal clouds.

cloud physics [METEOROL] The study of the physical and dynamical processes governing the structure and development of clouds and the release from them of snow, rain, and hail.

cloud seeding [METEOROL] Any technique carried out with the intent of adding to a cloud certain particles that will alter its natural development.

cloud shield [METEOROL] The principal cloud structure of a typical wave cyclone, that is, the cloud forms found on the cold-air side of the frontal system.

cloud street [METEOROL] A line of cumuliform clouds frequently one cumulus element wide, but ranging upward in width so that it is sometimes difficult to differentiate between streets and bands.

cloud symbol [METEOROL] One of a set of specified ideograms that represent the various cloud types of greatest significance or those most commonly observed, and entered on a weather map as part of a station model.

cloud system [METEOROL] An array of clouds and precipitation associated with a cyclonic-scale feature of atmospheric circulation, and displaying typical patterns and continuity. Also known as nephsystem.

cloud-to-cloud discharge [GEOPHYS] A lightning discharge occurring between a positive charge center of one cloud and a negative charge center of a second cloud. Also known as intercloud discharge.

cloud-to-ground discharge [GEOPHYS] A lightning discharge occurring between a charge center (usually negative) in the cloud and a center of opposite charge at the ground. Also known as ground discharge.

cloud top [METEOROL] The highest level in the atmosphere at which the air contains a perceptible quantity of cloud particles for a given cloud or cloud layer.

cloudy [METEOROL] The character of a day's weather when the average cloudiness, as determined from frequent observations, is more than 0.7 for the 24-hour period.

clough [GEOGR] A cleft in a hill; a ravine or narrow valley.

cluse [GEOL] A narrow, gorge, trench, or water gap with steep sides that cuts transversely through an otherwise continuous ridge.

clusterite *See* botryoid.

coagulation *See* agglomeration.

Coahuilan [GEOL] A North American provincial series in Lower Cretaceous geologic time, above the Upper Jurassic and below the Comanchean.

coal [GEOL] The natural, rocklike, brown to black derivative of forest-type plant material, usually accumulated in peat beds and progressively compressed and indurated until it is finally altered into graphite or graphite-like material.

coal auger [MIN ENG] A type of continuous miner which consists of a screw drill of large diameter and which cuts, transports, and loads the coal.

coal ball [GEOL] A subspherical mass containing mineral matter embedded with plant material, found in coal seams and overlying beds of the late Paleozoic.

coal bank [MIN ENG] A seam of coal that is exposed.

coal barrier [MIN ENG] A protective pillar composed of coal.

coal bed [GEOL] A seam or stratum of coal parallel to the rock stratification. Also known as coal rake; coal seam.

coal blasting [MIN ENG] Breaking coal with explosives.

coal breccia [GEOL] Angular fragments of coal within a coal bed.

coal clay *See* underclay.

coal cutter [MIN ENG] A power-operated machine which cuts out a thin strip of coal from the bottom of the seam; it draws itself by means of rope haulage along the coal face.

coal digger *See* faceman.

coal drill [MIN ENG] Usually, an electric drill of a compact, light design; however, also a light pneumatic drill.

coal dust [MIN ENG] A finely divided coal, sometimes defined as coal that will pass through 100-mesh screens (that is, 100 wires to the inch or 40 wires to the centimeter).

coalescence [METEOROL] In cloud physics, the merging of two or more water drops into a single larger drop.

coalescence efficiency [METEOROL] The fraction of all collisions which occur between waterdrops of a specified size and which result in actual merging of two drops into a single larger drop.

coalescence process [METEOROL] The growth of raindrops by the collision and coalescence of cloud drops or small precipitation particles.

coal face [MIN ENG] The mining face from which coal is extracted.

coalfield [MIN ENG] A region containing coal deposits.

coal getter *See* faceman.

coalification [GEOL] Formation of coal from plant material by the processes of diagenesis and metamorphism. Also known as bituminization; carbonification; incarbonization; incoalation.

Coal Measures [GEOL] The sequence of rocks typically containing coal of the Upper Carboniferous.

coal mining [MIN ENG] The technical and mechanical job of removing coal from the earth and preparing it for market.

coal paleobotany [PALEOBOT]   A branch of the paleobotanical sciences concerned with the origin, composition, mode of occurrence, and significance of fossil plant materials that occur in or are associated with coal seams.

coal pebbles [GEOL]   Rounded masses of coal occurring in sedimentary rock.

coal petrology [GEOL]   The science that deals with the origin, history, occurrence, structure, chemical composition, and classification of coal.

coal pipe [MIN ENG]   A cylinder-shaped part of a coal seam which extends into the overlying rock, and was produced from a tree stump that underwent rapid burial.

coal planer [MIN ENG]   A type of continuous coal mining machine for longwall mining; consists of a heavy steel plow with cutting knives, with power equipment to drag it back and forth across a coal face.

coal plough [MIN ENG]   A device with steel blades which shears off coal and pushes it onto the face conveyor.

coal rake *See* coal bed.

coal seam *See* coal bed.

coal-sensing probe [MIN ENG]   A nucleonic instrument to measure the thickness of coal left in the seam floor by means of a gamma-ray backscattering unit.

coal sizes [MIN ENG]   The sizes by which anthracite coal is marketed.

coal split *See* split.

coarse clay [GEOL]   Clay composed of particles with diameters ranging from 77 to 154 microinches ($1.96–3.91 \times 10^{-4}$ millimeter).

coarse fragment [GEOL]   Any particle of rock material with a diameter greater than 0.079 inch (2 millimeters).

coarse-grained [PETR]   *See* phaneritic.

coarse roll [MIN ENG]   A large roll for the preliminary crushing of large pieces of ore, rock, or coal.

coastal berm *See* berm.

coastal current [OCEANOGR]   An offshore current flowing generally parallel to the shoreline with a relatively uniform velocity.

coastal dune [GEOL]   A mobile mound of windblown material found along many sea and lake shores.

coastal ice *See* fast ice.

coastal landform [GEOGR]   The characteristic features and patterns of land in a coastal zone subject to marine and subaerial processes of erosion and deposition.

coastal plain [GEOGR]   Broad, low-level plain between a mountain range and a sea-shore.

coastal sediment [GEOL]   The mineral and organic deposits of deltas, lagoons, and bays, barrier islands and beaches, and the surf zone.

coast ice *See* fast ice.

coastline [GEOGR]   **1.** The line that forms the boundary between the shore and the coast. **2.** The line that forms the boundary between the water and the land.

coastlining [MAP]   The process of obtaining data from which the coastline can be drawn on a chart.

coast shelf *See* submerged coastal plain.

cob [MIN ENG]   To chip away waste material from an ore, using hand hammers.

cobalt bloom *See* erythrite.

cobalt glance *See* cobaltite.

cobaltite [MINERAL] CoAsS A silver-white mineral with a metallic luster that crystallizes in the isometric system, resembling crystals of pyrite; one of the chief ores of cobalt. Also known as cobalt glance; gray cobalt; white cobalt.

cobaltocalcite [MINERAL] A red, cobalt-bearing variety of calcite.

cobalt ocher *See* asbolite; erythrite.

cobaltomenite [MINERAL] $CoSeO_3 \cdot 2H_2O$ A mineral consisting of a hydrous cobalt selenium oxide.

cobalt pyrites *See* linnaeite.

cobber [MIN ENG] **1.** A device used to reject waste materials from ore concentrates. **2.** A person who breaks fibers from asbestos rocks or chips low-grade material from ore.

cobble beach *See* shingle beach.

Coblentzian [GEOL] Upper Lower Devonian geologic time.

coccolith ooze [GEOL] A fine-grained pelagic sediment containing undissolved sand- or silt-sized particles of coccoliths mixed with amorphous clay-sized material.

coccosphere [PALEOBOT] The fossilized remains of a member of Coccolithophorida.

Coccosteomorphi [PALEON] An aberrant lineage of the joint-necked fishes.

Cochliodontidae [PALEON] A family of extinct chondrichthian fishes in the order Bradyodonti.

cocinerite [MINERAL] Cu_4AgS A silver gray mineral consisting of copper and silver sulfide; occurs in massive form.

cockade ore [GEOL] A metallic ore in which successive crusts of minerals are deposited around rock fragments.

cockeyed bob [METEOROL] A thunder squall occurring during the summer, on the northwest coast of Australia.

cockpit karst *See* cone karst.

cocurrent line [OCEANOGR] A line through places having the same tidal current hour.

code-sending radiosonde [ENG] A radiosonde which transmits the indications of the meteorological sensing elements in the form of a code consisting of combinations of dots and dashes. Also known as code-type radiosonde; contracted code sonde.

code-type radiosonde *See* code-sending radiosonde.

Coelacanthidae [PALEON] A family of extinct lobefin fishes in the order Coelacanthiformes.

Coelolepida [PALEON] An order of extinct jawless vertebrates (Agnatha) distinguished by skin set with minute, close-fitting scales of dentine, similar to placoid scales of sharks.

Coelurosauria [PALEON] A group of small, lightly built saurischian dinosaurs in the suborder Theropoda having long necks and narrow, pointed skulls.

Coenopteridales [PALEOBOT] A heterogeneous group of fernlike fossil plants belonging to the Polypodiophyta.

coeruleolactite [MINERAL] $(Ca,Cu)Al_6(PO_4)_4(OH)_8 \cdot 4{-}5H_2O$ A milky-white to sky-blue mineral consisting of an aluminum phosphate.

coesite [MINERAL] A high-pressure polymorph of SiO_2 formed in nature only under unique physical conditions, requiring pressures of more than 20 kilobars (2×10^9 newtons per square meter); usually found in meteor impact craters.

coffinite [MINERAL] $USiO_4$ A black silicate important as a uranium ore; found in sandstone deposits and hydrothermal veins in New Mexico, Utah, and Wyoming.

cognate [GEOL] Pertaining to contemporaneous fractures in a system with regard to time of origin and deformational type.

cognate ejecta [GEOL] Essential or accessory pyroclasts derived from the magmatic materials of a current volcanic eruption.

cogwheel ore *See* bournonite.

cohenite [MINERAL] $(Fe,Ni,Co)_3C$ A tin-white, isometric mineral found in meteorites.

cohesionless [GEOL] Referring to a soil having low shear strength when dry, and low cohesion when wet. Also known as frictional; noncohesive.

cohesiveness [GEOL] Property of unconsolidated fine-grained sediments by which the particles stick together by surface forces.

cohesive soil [GEOL] A sticky soil, such as clay or silt; its shear strength equals about half its unconfined compressive strength.

coincidence rangefinder [OPTICS] An optical rangefinder in which one-eyed viewing through a single eyepiece provides the basis for manipulation of the rangefinder adjustment to cause two images of the target or parts of each, viewed over different paths, to match or coincide.

coke coal *See* natural coke.

cokeite *See* natural coke.

coking coal [GEOL] A very soft bituminous coal suitable for coking.

col [GEOL] A high, sharp-edged pass in a mountain ridge, usually produced by the headward erosion of opposing cirques.   [METEOROL] The point of intersection of a trough and a ridge in the pressure pattern of a weather map; it is the point of relatively lowest pressure between two highs and the point of relatively highest pressure between two lows. Also known as neutral point; saddle point.

colatitude [GEOD] Ninety degrees minus the latitude.

cold-air drop *See* cold pool.

cold-air outbreak *See* polar outbreak.

cold anticyclone *See* cold high.

cold-core cyclone *See* cold low.

cold-core high *See* cold high.

cold-core low *See* cold low.

cold dome [METEOROL] A cold air mass, considered as a three-dimensional entity.

cold drop *See* cold pool.

cold front [METEOROL] Any nonoccluded front, or portion thereof, that moves so that the colder air replaces the warmer air; the leading edge of a relatively cold air mass.

cold-front-like sea breeze [METEOROL] A sea breeze which forms over the ocean, moves slowly toward the land, and then moves inland quite suddenly. Also known as sea breeze of the second kind.

cold-front thunderstorm [METEOROL] A thunderstorm attending a cold front.

cold high [METEOROL]   At a given level in the atmosphere, any high that is generally characterized by colder air near its center than around its periphery. Also known as cold anticyclone; cold-core high.

cold low [METEOROL]   At a given level in the atmosphere, any low that is generally characterized by colder air near its center than around its periphery. Also known as cold-core cyclone; cold-core low.

cold pole [CLIMATOL]   The location which has the lowest mean annual temperature in its hemisphere.

cold pool [METEOROL]   A region of relatively cold air surrounded by warmer air; the term is usually applied to cold air of appreciable vertical extent that has been isolated in lower latitudes as part of the formation of a cutoff low. Also known as cold-air drop; cold drop.

cold tongue [METEOROL]   In synoptic meteorology, a pronounced equatorward extension or protrusion of cold air.

cold wall [OCEANOGR]   The line or surface along which two water masses of significantly different temperature are in contact.

cold-water sphere [OCEANOGR]   Those portions of the ocean water having a temperature below 8°C. Also known as oceanic stratosphere.

cold wave [METEOROL]   A rapid fall in temperature within 24 hours to a level requiring substantially increased protection to agriculture, industry, commerce, and social activities.

colemanite [MINERAL] $Ca_2B_6O_{11} \cdot 5H_2O$ A colorless or white hydrated borate mineral that crystallizes in the monoclinic system and occurs in massive crystals or as nodules in clay.

Coleodontidae [PALEON]   A family of conodonts in the suborder Neurodontiformes.

colk *See* pothole.

colla [METEOROL]   In the Philippines, a fresh or strong (less than 39–46 miles per hour, or 63–74 kilometers per hour) south to southwest wind, accompanied by heavy rain and severe squall. Also known as colla tempestade.

collada [METEOROL]   A strong wind (35–50 miles per hour, or 56–80 kilometers per hour) in the Gulf of California, blowing from the north or northwest in the upper part, and from the northeast in the lower part of the gulf.

collapse breccia [GEOL]   Angular rock fragments derived from the collapse of rock overlying a hollow space.

collapse caldera [GEOL]   A caldera formed primarily as a result of collapse due to withdrawal of magmatic support.

collapse sink [GEOL]   A sinkhole resulting from local collapse of a cavern that has been enlarged by solution and erosion.

collapse structure [GEOL]   A structure resulting from rock slides under the influence of gravity. Also known as gravity-collapse structure.

collar [MIN ENG]   The mouth of a mine shaft.

collared hole [ENG]   A started hole drilled sufficiently deep to confine the drill bit and prevent slippage of the bit from normal position.

collar locator log [PETRO ENG]   Down-hole nuclear-log measurement to locate drill-hole casing collars, usually for precise location of perforating points.

colla tempestade *See* colla.

collective [METEOROL] In aviation weather observations, a group of observations transmitted in prescribed order by stations on the same long-line teletypewriter circuit. Also known as sequence.

collenia [PALEOBOT] A convex, slightly arched, or turbinate stromatolite produced by late Precambrian blue-green algae of the genus *Collenia*.

colliery [MIN ENG] A whole coal mining plant; generally the term is used in connection with anthracite mining but sometimes to designate the mine, shops, and preparation plant of a bituminous operation.

collinite [GEOL] The maceral, of collain consistency, of jellified plant material precipitated from solution and hardened; a variety of euvitrinite.

collinsite [MINERAL] $Ca_2(Mg,Fe)(PO_4)_2$ A phosphate mineral occurring in concentric layers in phosphoric nodules; found in meteorites.

Collins miner [MIN ENG] A type of remote-controlled continuous miner for thin-seam extraction.

collision efficiency [METEOROL] The fraction of all water-drops which, initially moving on a collision course with respect to other drops, actually collide (make surface contact) with the other drops.

colloform [GEOL] Pertaining to the rounded, globular texture of mineral formed by colloidal precipitation.

colloidal instability [METEOROL] A property attributed to clouds, by which the particles of the cloud tend to aggregate into masses large enough to precipitate.

collophane [MINERAL] A massive, cryptocrystalline, carbonate-containing variety of apatite and a principal source of phosphates for fertilizers. Also known as collophanite.

collophanite *See* collophane.

colluvium [GEOL] Loose, incoherent deposits at the foot of a slope or cliff, brought there principally by gravity.

Collyritidae [PALEON] A family of extinct, small, ovoid, exocyclic Euechinoidea with fascioles or a plastron.

Colmol miner [MIN ENG] A continuous miner, in which the coal is completely augered by two banks of cutting arms fitted with picks; the arms rotate in opposite directions to assist in gathering up the cuttings for the central conveyor.

Coloradoan [GEOL] Middle Upper Cretaceous geologic time.

Colorado low [METEOROL] A low which makes its first appearance as a definite center in the vicinity of Colorado on the eastern slopes of the Rocky Mountains; analogous to the Alberta low.

columbite [MINERAL] $(Fe,Mn)(Cb,Ta)_2O_6$ An iron-black mineral with a submetallic luster that crystallizes in the orthorhombic system; the chief ore mineral of niobium (columbium); hardness is 6 on Mohs scale, and specific gravity is 5.4–6.5. Also known as dianite; greenlandite; niobite.

columnar coal [GEOL] Coal with a columnar fracture structure, usually caused by metamorphism by an igneous intrusion.

columnar jointing [GEOL] Parallel, prismatic columns that are formed as a result of contraction during cooling in basaltic flow and other extrusive and intrusive rocks. Also known as columnar structure; prismatic jointing; prismatic structure.

columnar resistance [GEOPHYS] The electrical resistance of a column of air 1 centimeter square, extending from the earth's surface to some specified altitude.

columnar section [GEOL] A vertical strip or scale drawing of the strip taken from a given area or locality showing the sequence of the rock units and their stratigraphic relationship, and indicating the thickness, lithology, age, classification, and fossil content of the rock units. Also known as section.

columnar structure [GEOL] *See* columnar jointing. [MINERAL] Mineral structure consisting of parallel columns of slender prismatic crystals. [PETR] A primary sedimentary structure consisting of columns arranged perpendicular to the bedding.

column pipe [MIN ENG] The large cast-iron (or wooden) pipe through which the water is conveyed from the mine pumps to the surface.

colusite [MINERAL] $Cu_3(As,Sn,V,Fe,Te)S_4$ A bronze-colored mineral consisting of a sulfide of copper and arsenic with vanadium, iron, and telluride substituting for arsenic; usually occurs in massive form.

comagmatic province *See* petrographic province.

Comanchean [GEOL] A North American provincial series in Lower and Upper Cretaceous geologic time, above Coahuilan and below Gulfian.

comber [OCEANOGR] A long, curling, deep-water ocean wave with a high, breaking crest that is propelled by a strong wind.

combination coefficient [GEOPHYS] A measure of the specific rate of disappearance of small ions in the atmosphere due to either union with neutral Aitken nuclei to form new large ions, or union with large ions of opposite sign to form neutral Aitken nuclei.

combination-drive reservoir [PETRO ENG] A type of reservoir in which hydrocarbons are swept (displaced) toward the drill hole by injection of water followed by liquefied-petroleum-gas or gas injection.

combined moisture [MIN ENG] Moisture in coal that cannot be removed by ordinary drying.

comb nephoscope [ENG] A direct-vision nephoscope constructed with a comb (a crosspiece containing equispaced vertical rods) attached to the end of a column 8–10 feet (2.4–3 meters) long and supported on a mounting that is free to rotate about its vertical axis; in use, the comb is turned so that the cloud appears to move parallel to the tips of the vertical rods.

combustion nucleus [METEOROL] A condensation nucleus formed as a result of industrial or natural combustion processes.

comendite [GEOL] A white, sodic rhyolite which contains alkalic amphibole or pyroxene.

comfort index *See* temperature humidity index.

Comleyan [GEOL] Lower Cambrian geologic time.

commercial mine [MIN ENG] A coal mine operated to supply purchasers in general, as contrasted with a captive mine.

commercial ore [MIN ENG] Mineralized material that is profitable at prevailing metal prices.

commission ore [MIN ENG] Uranium-bearing ore of 0.10% U_3O_8 or higher, for which the U.S. Atomic Energy Commission has an established price.

common establishment *See* high-water full and change.

common feldspar *See* orthoclase.

common mica *See* muscovite.

common pyrite *See* pyrite.

common salt *See* halite; sodium chloride.

compaction [GEOL]   Process by which soil and sediment mass loses pore space in response to the increasing weight of overlying material.

comparative rabal [ENG]   A rabal observation (that is, a radiosonde balloon tracked by theodolite) taken simultaneously with the usual rawin observation (tracking by radar or radio direction-finder), to provide a rough check on the alignment and operating accuracy of the electronic tracking equipment.

compartment [MIN ENG]   A section of a mine shaft separated by framed timbers and planking.

compass points [GEOD]   The 32 divisions of a compass at intervals of 11¼°, with each division further divided into quarter points. Also known as points of the compass.

compensation depth [OCEANOGR]   The depth at which the light intensity is sufficient to bring about a balance between the oxygen produced and that consumed by algae.

competence [GEOL]   The ability of the wind to transport solid particles either by rolling, suspension, or saltation (intermittent rolling and suspension); usually expressed in terms of the weight of a single particle.   [HYD]   The ability of a stream, flowing at a given velocity, to move the largest particles.

competent beds [GEOL]   Beds or strata capable of withstanding the pressures of folding without flowing or changing in original thickness.

complementary rocks [GEOL]   Rocks which are differentiated from the same magma, and whose average composition is the same as the parent magma.

complex [GEOL]   An assemblage of rocks that has been folded together, intricately mixed, involved, or otherwise complicated.   [MINERAL]   Composed of many ingredients.

complex climatology [CLIMATOL]   Analysis of the climate of a single space, or comparison of the climates of two or more places, by the relative frequencies of various weather types or groups of such types; a type is defined by the simultaneous occurrence within specified narrow limits of each of several weather elements.

complex dune [GEOL]   A dune of varying forms, often very large, and produced by variable, shifting winds and the merging of various dune types.

complex fold [GEOL]   A fold whose axial line is also folded.

complex low [METEOROL]   An area of low atmospheric pressure within which more than one low-pressure center is found.

complex mountain [GEOL]   A mountain comprising an assortment of intimately admixed structures and showing a great diversity of landforms.

complex stream [HYD]   A stream in a second or later cycle of erosion.

complex tombolo [GEOL]   A system resulting when several islands and the mainland are interconnected by a complex series of tombolos. Also known as tombolo cluster; tombolo series.

complex twin [CRYSTAL]   A twin crystal in feldspar that is formed by both normal and parallel twinning.

composite cone [GEOL]   A large volcanic cone constructed of lava and pyroclastic material in alternating layers.

composite dike [GEOL]   A dike consisting of several intrusions differing in chemical and mineralogical composition.

composite flash [GEOPHYS]   A lightning discharge which is made up of a series of distinct lightning strokes with all strokes following the same or nearly the same channel, and with successive strokes occurring at intervals of about 0.05 second. Also known as multiple discharge.

composite fold [GEOL]   A fold having smaller folds on its limbs.

composite gneiss [PETR]   A banded rock formed by intimate penetration of magma into country rocks.

composite grain [GEOL]   A sedimentary clast formed of two or more original particles.

composite map [MIN ENG]   A map in which several levels of a mine are shown on a single sheet.

composite sequence [GEOL]   An ideal sequence of cyclic sediments containing all the lithological types in their proper order.

composite sill [GEOL]   A sill consisting of several intrusions differing in chemical and mineralogical compositions.

composite topography [GEOL]   A combination of topographic features which have developed through two or more cycles of erosion.

composite unconformity [GEOL]   An unconformity representing more than one episode of nondeposition and possible erosion.

composite vein [GEOL]   A large fracture zone composed of parallel ore-filled fissures and converging diagonals, whose walls and intervening country rock have been replaced to a certain degree.

composite volcano *See* stratovolcano.

compositional maturity [GEOL]   Concept of a type of maturity in sedimentary rocks in which a sediment approaches the compositional end product to which formative processes drive it.

composition face *See* composition surface.

composition plane [CRYSTAL]   A planar composition surface in a crystal uniting two individuals of a contact twin.

composition surface [CRYSTAL]   The surface uniting individuals of a crystal twin; may or may not be planar. Also known as composition face.

compound alluvial fan [GEOL]   Structure formed by the lateral growth and merger of fans made by neighboring streams.

compound fault [GEOL]   A number of closely spaced, approximately parallel faults in series.

compound ripple marks [GEOL]   Complex ripple marks of great diversity which originate by simultaneous interference of wave oscillation with current action.

compound shaft [MIN ENG]   A shaft in which the upper stage is often a vertical shaft, while the lower stage, or stages, may be inclined and driven into the deposit.

compound twins [CRYSTAL]   Individuals of one mineral group united in accordance with two or more different twin laws.

compound valley glacier [HYD]   A glacier composed of several ice streams emanating from different tributary valleys.

compound volcano [GEOL]   **1.** A volcano consisting of a complex of two or more cones. **2.** A volcano with an associated volcanic dome.

compressed-air blasting [MIN ENG] A method for breaking down coal by compressed-air power.

compression [GEOD] *See* flattening.   [GEOL] A system of forces which tend to decrease the volume or shorten rocks.

compression plant [PETRO ENG] Gas-compression facility used to produce a high-pressure gas stream for injection into reservoir formations to increase oil yield; when the injected gas is that recovered from the well during oil production, the facility is called a gas-cycling plant.

concave bank [GEOGR] The outer bank of a concave curve in a stream.

concealed coalfield [GEOL] A coal deposit that does not have an outcrop.

concentrating table [MIN ENG] A device consisting of a riffled deck to which a reciprocating motion in a horizontal direction is imparted; the material to be separated is fed in a stream of water, the heavy particles collect between the riffles and are conveyed in the direction of the reciprocating motion, while the lighter particles are borne by the water over the riffles to be discharged laterally from the table.

concentration [HYD] The ratio of the area of a given sea ice cover to the total area of sea surface.   [MIN ENG] Separation and accumulation of economic minerals from gangue.

concentration time [HYD] The time required for water to travel from the most remote portion of a river basin to the basin outlet; it varies with the quantity of flow and channel conditions.

concentric faults [GEOL] Faults that are arranged concentrically.

concentric fold [GEOL] A fold in which the original thickness of the strata is unchanged during deformation. Also known as parallel fold.

concentric fractures [GEOL] A system of fractures concentrically arranged about a center.

concentric weathering *See* spheroidal weathering.

concession lease [MIN ENG] A lease form that conveys specified national or state permission to a lessee to explore for or produce minerals (such as oil, gas, or uranium) from specified properties.

conchoidal [GEOL] Having a smoothly curved surface; used especially to describe the fracture surface of a mineral or rock.

concordant body [GEOL] An intrusive igneous body whose contacts are parallel to the bedding of the country rock. Also known as concordant injection; concordant pluton.

concordant coastline [GEOL] A coastline parallel to the land structures which form the margin of an ocean basin.

concordant injection *See* concordant body.

concordant pluton *See* concordant body.

concordia [GEOCHEM] A time curve obtained by plotting the ratio $^{206}Pb/^{238}U$ versus the ratio $^{207}Pb/^{235}U$ of a closed U-Pb system; both ratios increase in value as uranium undergoes nuclear decay to lead with the passage of time.

concordia intercept [GEOCHEM] The points of intersection of the concordia curve and a straight line which is the plot of discordant U-Pb ages.

concretion [GEOL] A hard, compact mass of mineral matter in the pores of sedimentary or fragmental volcanic rock; represents a concentration of a minor constituent of the enclosing rock or of cementing material.

concretionary [GEOL] Tending to grow together, forming concretions.

concretioning [GEOL] The process of forming concretions.

concussion crack [HYD] A crack in sea ice, produced by the impact of one ice cake upon another.

concussion fracture [GEOL] Radiating system of fractures in a shock-metamorphosed rock.

concussion table [MIN ENG] An inclined table, agitated by a series of shocks and operating like a buddle. Also known as percussion table.

condensate field [GEOL] A petroleum field developed in predominantly gas-bearing reservoir rocks, but within which condensation of gas to oil commonly occurs with decreases in field pressure.

condensate well [PETRO ENG] A well that produces a natural gas highly saturated with condensable hydrocarbons heavier than methane and ethane.

condensation [METEOROL] The process by which water vapor becomes a liquid such as dew, fog, or cloud or a solid like snow; condensation in the atmosphere is brought about by either of two processes: cooling of air to its dew point, or addition of enough water vapor to bring the mixture to the point of saturation (that is, the relative humidity is raised to 100%).

condensation cloud [METEOROL] A mist or fog of minute water droplets that temporarily surrounds the fireball following an atomic detonation in a comparatively humid atmosphere.

condensation nucleus [METEOROL] A particle, either liquid or solid, upon which condensation of water vapor begins in the atmosphere.

condensation pressure [METEOROL] The pressure at which a parcel of moist air expanded dry adiabatically reaches saturation. Also called adiabatic condensation pressure; adiabatic saturation pressure.

condensation trail [METEOROL] A visible trail of condensed water vapor or ice particles left behind an aircraft, an airfoil, or such, in motion through the air. Also known as contrail; vapor trail.

condensing gas drive [PETRO ENG] Reservoir-oil displacement by gas where hydrocarbon components of the injected gas condense in the oil that it is displacing.

conditional instability [METEOROL] The state of a column of air in the atmosphere when its lapse rate of temperature is less than the dry adiabatic lapse rate but greater than the saturation adiabatic lapse rate.

conductive equilibrium *See* isothermal equilibrium.

conduit [GEOL] A completely water-filled subterranean passage always under hydrostatic pressure.

Condylarthra [PALEON] A mammalian order of extinct, primitive, hoofed herbivores with five-toed plantigrade to semidigitigrade feet.

cone [GEOL] A mountain, hill, or other landform having relatively steep slopes and a pointed top.

cone delta *See* alluvial cone.

cone dike *See* cone sheet.

cone-in-cone structure [GEOL] The structure of a concretion characterized by the development of a succession of cones one within another.

cone karst [GEOL] A type of karst typical of tropical regions, characterized by a pattern of steep, convex sides and slightly concave floors. Also known as cockpit karst; Kegel karst.

Conemaughian [GEOL] Upper Middle Pennsylvanian geologic time.

cone dejection *See* alluvial cone.

cone of depression [HYD] The depression in the water table around a well defining the area of influence of the well. Also known as cone of influence.

cone of detritus *See* alluvial cone.

cone of escape [GEOPHYS] A hypothetical cone in the exosphere, directed vertically upward, through which an atom or molecule would theoretically be able to pass to outer space without a collision.

cone of influence *See* cone of depression.

cone settler [MIN ENG] A conical vessel fed centrally with fine ore pulp, in which the apex discharge carries the larger-sized particles, and the peripheral top overflow carries the finer fraction of the solids.

cone sheet [GEOL] An accurate dike forming part of a concentric set that dips inward toward the center of the arc. Also known as cone dike.

Conewangoan [GEOL] Upper Upper Devonian geologic time.

confined aquifer *See* artesian aquifer.

confined groundwater [HYD] Groundwater which is under significantly greater pressure than atmospheric pressure and whose upper surface is underneath an impermeable bed or a bed of distinctly lower permeability than that of the material containing the water.

confining bed [GEOL] An impermeable bed adjacent to an aquifer.

confining pressure [GEOL] An equal, all-sided pressure, such as lithostatic pressure produced by overlying rocks in the crust of the earth.

confluence [HYD] **1.** A stream formed from the flowing together of two or more streams. **2.** The place where such streams join.

conformable [GEOL] **1.** Pertaining to the contact of an intrusive body when it is aligned with the internal structures of the intrusion. **2.** Referring to strata in which layers are formed above one another in an unbroken, parallel order.

conformal chart [MAP] A chart on a conformal map projection.

conformality [MAP] The retention of angular relationships at each point on a map projection.

conformal map projection [MAP] A map projection on which the shape of any small area of the surface mapped is preserved unchanged. Also known as orthomorphic map projection.

confused sea [OCEANOGR] A highly disturbed water surface without a single, well-defined direction of wave travel.

congelifraction [GEOL] The splitting or disintegration of rocks as the result of the freezing of the water contained. Also known as frost bursting; frost riving; frost shattering; frost splitting; frost weathering; frost wedging; gelifraction; gelivation.

congeliturbate [GEOL] Soil or unconsolidated earth which has been moved or disturbed by frost action.

congeliturbation [GEOL] The churning and stirring of soil as a result of repeated cycles of freezing and thawing; includes frost heaving and surface subsidence during thaws. Also known as cryoturbation; frost churning; frost stirring; geliturbation.

conglomerate [GEOL] Cemented, rounded fragments of water-worn rock or pebbles, bound by a siliceous or argillaceous substance.

conglomeratic mudstone *See* paraglomerate.

congressite [PETR] A light-colored, coarse-grained igneous rock, with nepheline as its main component.

congruent melting [GEOL] Melting of a solid substance to a liquid identical in composition.

Coniacian [GEOL] Lower Senonian geologic time.

conic chart [MAP] A chart on a conic projection.

conic chart with two standard parallels [MAP] A chart on the conic projection with two standard parallels. Also known as secant conic chart.

conichalcite [MINERAL] $CaCu(AsO_4)(OH)$ A grass green to yellowish-green or emerald green, orthorhombic mineral consisting of a basic arsenate of calcium and copper.

Coniconchia [PALEON] A class name proposed for certain extinct organisms thought to have been mollusks; distinguished by a calcareous univalve shell that is open at one end and by lack of a siphon.

conic projection [MAP] A map deformation pattern resulting from the transfer of the map to a tangent or intersecting cone.

conic projection with two standard parallels [MAP] A conic map projection in which the surface of a sphere or spheroid, such as the earth, is conceived as developed on a cone which intersects the sphere or spheroid along two standard parallels, the cone being spread out to form a plane; for example, the Lambert conformal projection. Also known as secant conic projection.

coning [PETRO ENG] The process in which water underneath an oil reservoir moves up into the oil column and enters the well.

conjugate [GEOL] **1.** Pertaining to fractures in which both sets of veins or joints show the same strike but opposite dip. **2.** Pertaining to any two sets of veins or joints lying perpendicular.

conjugate joint system [GEOL] Two joint sets with a symmetrical pattern arranged about another structural feature or an inferred stress axis.

connarite [MINERAL] A green mineral consisting of hydrous nickel silicate occurring as small crystals or grains.

connate [GEOL] In sedimentary processes, describing materials that developed simultaneously with adjacent materials.

connate water [HYD] Water entrapped in the interstices of igneous rocks when the rocks were formed; usually highly mineralized.

connellite [MINERAL] $Cu_{19}(SO_4)Cl_4(OH)_{32} \cdot 3H_2O$ A deep-blue striated copper mineral; crystals are in the hexagonal system. Also known as footeite.

Conoclypidae [PALEON] A family of Cretaceous and Eocene exocyclic Euechinoidea in the order Holectypoida having developed aboral petals, internal partitions, and a high test.

Conocyeminae [PALEON] A subfamily of Mesozoan parasites in the family Dicyemidae.

conodont [PALEON] A minute, toothlike microfossil, composed of translucent amber-brown, fibrous or lamellar calcium phosphate; taxonomic identity is controversial.

Conodontiformes [PALEON] A suborder of conodonts from the Ordovician to the Triassic having a lamellar internal structure.

Conodontophoridia   [PALEON]   The ordinal name for the conodonts.

conoplain *See* pediment.

Conrad discontinuity   [GEOPHYS]   A relatively abrupt discontinuity in the velocity of elastic waves in the earth, increasing from 6.1 to 6.4–6.7 kilometers per second; occurs at various depths and marks contact of granitic and basaltic layers.

Conrad machine   [MIN ENG]   Mechanized pit digger used in checking of alluvial boring; sections of tubing, 5 feet (152 centimeters) long and 2 feet (61 centimeters) in inside diameter, are worked into the ground while spoil is removed by means of a bucket or grab.

consanguineous   [GEOL]   Of a natural group of sediments or sedimentary rocks, having common or related origin.

consanguinity   [PETR]   The genetic relationship between igneous rocks occurring in a single petrographic province which are presumably derived from a common parent magma.

consequent   [GEOL]   Of, pertaining to, or characterizing movements of the earth resulting from the external transfer of material in the process of gradation.

consequent lake   [HYD]   A body of water found in a depression formed as an original inequality in any new land surface, such as a lake in a depression in glacial deposits.

consequent stream   [GEOL]   A stream whose course is determined by the slope of the land. Also known as superposed stream.

consequent valley   [GEOL]   **1.** A valley whose direction depends on corrugation. **2.** A valley formed by the widening of a trench cut by a consequent stream.

conservative concentrations   [OCEANOGR]   Concentrations such as of heat or salt in bodies of water that are altered locally, except at the boundaries, by diffusion and advection.

conservative elements   [OCEANOGR]   Inert or abundant elements which enter sea water in minute quantities as solids only, and are constant in terms of their relative abundance to each other.

consolidated ice   [OCEANOGR]   Ice which has been compacted into a solid mass by wind and ocean currents and covers an area of the ocean.

consolidation   [GEOL]   **1.** Processes by which loose, soft, or liquid earth becomes coherent and firm. **2.** Adjustment of a saturated soil in response to increased load; involves squeezing of water from the pores and a decrease in void ratio.

consolidation test   [MIN ENG]   A test in which the specimen is confined laterally in a ring and is compressed between porous plates which are saturated with water.

constancy *See* persistence.

constant-distance sphere   [ENG ACOUS]   The relative response of a sonar projector to variations in acoustic intensity, or intensity per unit band, over the surface of a sphere concentric with its center.

constant-height chart   [METEOROL]   A synoptic chart for any surface of constant geometric altitude above mean sea level (a constant-height surface), usually containing plotted data and analyses of the distribution of such variables as pressure, wind, temperature, and humidity at that altitude. Also known as constant-level chart; fixed-level chart; isohypsic chart.

constant-height surface   [METEOROL]   A surface of constant geometric or geopotential altitude measured with respect to mean sea level. Also known as constant-level surface; isohypsic surface.

constant-level chart *See* constant-height chart.

constant-level surface *See* constant-height surface.

constant of the cone [MAP] The chart convergence factor for a conic projection.

constant-pressure chart [METEOROL] The synoptic chart for any constant-pressure surface, usually containing plotted data and analyses of the distribution of height of the surface, wind temperature, humidity, and so on. Also known as isobaric chart; isobaric contour chart.

constant-pressure surface *See* isobaric surface.

Constellariidae [PALEON] A family of extinct, marine bryozoans in the order Cystoporata.

constituent number [OCEANOGR] One of the harmonic elements in a mathematical expression for the tide-producing force, and in corresponding formulas for the tide or tidal current.

constructive wave [OCEANOGR] A type of wave that helps create a beach by transporting material landward.

consumptive use [HYD] The total annual land water loss in an area, due to evaporation and plant use.

contact aureole *See* aureole.

contact breccia [PETR] Angular rock fragments resulting from shattering of wall rocks around laccolithic and other igneous masses.

contact log [PETRO ENG] Record of electrical-resistivity data pertaining to strata structures along the depth of a drill hole.

contact metamorphic rock [PETR] A rock formed by the processes of contact metamorphism.

contact metamorphism [PETR] Metamorphism that is genetically related to the intrusion or extrusion of magmas and takes place in rocks at or near their contact.

contact metasomatism [GEOL] One of the main local processes of thermal metamorphism that is related to intrusion of magmas; takes place in rocks or near their contact with a body of igneous rock.

contact mineral [MINERAL] A mineral formed by the processes of contact metamorphism.

contact twin [CRYSTAL] Twinned crystals whose members are symmetrically arranged about a twin plane.

contact vein [GEOL] **1.** A variety of fissure vein formed by deposition of minerals in a fault fissure at a rock contact. **2.** A replacement vein formed by mineralized solutions percolating along the more permeable surface areas of the contact.

contact zone *See* aureole.

contamination [GEOL] Alteration of a magma's chemical composition by the assimilation of country rocks.

contemporaneous [GEOL] **1.** Formed, existing, or originating at the same time. **2.** Of a rock, developing during formation of the enclosing rock.

continent [GEOGR] A protuberance of the earth's crustal shell, with an area of several million square miles and sufficient elevation so that much of it is above sea level.

continental accretion [GEOL] The theory that continents have grown by the addition of new continental material around an original nucleus, mainly through the processes of geosynclinal sedimentation and orogeny.

continental air [METEOROL] A type of air whose characteristics are developed over a large land area and which therefore has relatively low moisture content.

continental anticyclone *See* continental high.

continental borderland [GEOL] The area of the continental margin between the shoreline and the continental slope.

continental climate [CLIMATOL] Climate characteristic of the interior of a landmass of continental size, marked by large annual, daily, and day-to-day temperature ranges, low relative humidity, and a moderate or small irregular rainfall; annual extremes of temperature occur soon after the solstices.

continental crust [GEOL] The basement complex of rock, that is, metamorphosed sedimentary and volcanic rock with associated igneous rocks, mainly granitic, that underlies the continents and the continental shelves.

continental deposits [GEOL] Sedimentary deposits laid down within a general land area.

continental drift [GEOL] The concept of continent formation by the fragmentation and movement of land masses on the surface of the earth.

continental geosyncline [GEOL] A geosyncline filled with nonmarine sediments.

continental glacier [HYD] A sheet of ice covering a large tract of land, such as the ice caps of Greenland and the Antarctic.

continental growth [GEOL] The processes contributing to growth of continents at the expense of ocean basins.

continental heat flow [GEOPHYS] The amount of thermal energy escaping from the earth through the continental crust per unit area and unit time.

continental high [METEOROL] A general area of high atmospheric pressure which on mean charts of sea-level pressure is seen to overlie a continent during the winter. Also known as continental anticyclone.

continentality [CLIMATOL] The degree to which a point on the earth's surface is in all respects subject to the influence of a land mass.

continental margin [GEOL] Those provinces between the shoreline and the deep-sea bottom; generally consists of the continental borderland, shelf, slope, and rise.

continental mass [GEOGR] The continental land rising more or less abruptly from the ocean floor and also the shallow submerged areas surrounding this land.

continental nucleus [GEOL] A large area of basement rock consisting of basaltic and more mafic oceanic crust and periodotitic mantle from which it is postulated that continents have grown. Also known as cratogene.

continental plate [GEOL] Thick continental crust.

continental plateau *See* tableland.

continental platform *See* continental shelf.

continental polar air [METEOROL] Polar air having low surface temperature, low moisture content, and (especially in its source regions) great stability in the lower layers.

continental rise [GEOL] A transitional part of the continental margin; a gentle slope with a generally smooth surface, built up by the shedding of sediments from the continental block, and located between the continental slope and the abyssal plain.

continental shelf [GEOL] The zone around a continent, that part of the continental margin extending from the shoreline and the continental slope; composes with the continental slope the continental terrace. Also known as continental platform; shelf.

continental shield [GEOL] Large areas of Precambrian rocks exposed within the cratons of continents.

continental slope [GEOL] The part of the continental margin consisting of the declivity from the edge of the continental shelf extending down to the continental rise.

continental terrace [GEOL] The continental shelf and slope together.

continental tropical air [METEOROL] A type of tropical air produced over subtropical arid regions; it is hot and very dry.

continent formation [GEOL] A series of six or seven major episodes, resulting from the buildup of radioactive heat and then the melting or partial melting of the earth's interior; the molten rock melt rises to the surface, differentiating into less primitive lavas; the continent then nucleates, differentiates, and grows from oceanic crust and mantle.

continuity chart [METEOROL] A chart maintained for weather analysis and forecasting upon which are entered the positions of significant features (pressure centers, fronts, instability lines, through lines, ridge lines) of the regular synoptic charts at regular intervals in the past.

continuous cleavage [GEOL] A penetrative cleavage in rock material in which the planes of cleavage are continuous.

continuous coal cutter [MIN ENG] A coal mining machine that cuts the coal face without being withdrawn from the cut.

continuous gas lift [PETRO ENG] Oil production in which reservoir gas pressure (natural or injected) is sufficient to provide a continuous upward flow of oil through the well tubing.

continuous leader *See* dart leader.

continuous miner [MIN ENG] Machine designed to remove coal or other soft minerals from the face and to load it into cars or conveyors continuously, without the use of cutting machines, drills, or explosives.

continuous mining [MIN ENG] A type of mining in which the continuous miner cuts or rips coal or other soft minerals from the face and loads it in a continuous operation.

continuous permafrost zone [GEOL] Regional zone predominantly underlain by permanently frozen subsoil that is not interrupted by pockets of unfrozen ground.

continuous profiling [GEOL] A method of shooting in seismic exploration in which uniformly placed seismometer stations along a line are shot from holes spaced along the same line so that each hole records seismic ray paths geometrically identical with those from adjacent holes.

continuous reaction series [MINERAL] A branch of Bowen's reaction series comprising the plagioclase mineral group in which reaction of early-formed crystals with water takes place continuously, without abrupt changes in crystal structure.

continuous stream [HYD] An uninterrupted stream having no wet and dry reaches, as contrasted with an ephemeral stream.

continuous velocity log [ENG] A log that continuously records the velocity of acoustic waves or seismic waves over small intervals as the logging mechanism moves through a borehole.

contour *See* contour line.

contour-change line *See* height-change line.

contour current [OCEANOGR] An ocean current moving along isopycnic lines that are approximately parallel to the bathymetric contours.

contour line [MAP] A map line representing a contour, that is, connecting points of equal elevation above or below a datum plane, usually mean sea level. Also known as contour; isoheight; isohypse. [METEOROL] A line on a weather map connecting points of equal atmospheric pressure, temperature, or such.

contour map [MAP] A map displaying topographic or structural contour lines.

contour microclimate [CLIMATOL] That portion of the microclimate which is directly attributable to the small-scale variations of ground level.

contour plan [MIN ENG] A plan showing surface contours or calculated contours of coal seams to be developed.

contour value [MAP] A numerical value placed upon a contour line to denote its elevation relative to a given datum, usually mean sea level.

contracted code sonde *See* code-sending radiosonde.

contraction hypothesis [GEOL] Theory that shrinking of the earth is the cause of compression folding and thrusting.

contrail *See* condensation trail.

contrail-formation graph [METEOROL] A graph containing the parameters pressure, temperature, and relative humidity for critical values at which condensation trails (contrails) form; used as an aid in forecasting the formation of condensation trails.

contra solem [METEOROL] Characterizing air motion that is counterclockwise in the Northern Hemisphere and clockwise in the Southern Hemisphere; literally, against the sun.

contrasted differentiation [PETR] Natural sorting of magma into basic and acidic magmas, which are believed to react with each other to produce intermediate rock types.

contrastes [METEOROL] Winds a short distance apart blowing from opposite quadrants, frequent in the spring and fall in the western Mediterranean.

contributory *See* tributary.

control day [METEOROL] One of several days on which the weather is supposed (according to folklore) to provide the key for the weather of a subsequent period. Also known as key day.

control-tower visibility [METEOROL] The visibility observed from an airport control tower.

Conularida [PALEON] A small group of extinct invertebrates showing a narrow, four-sided, pyramidal-shaped test.

Conulidae [PALEON] A family of Cretaceous exocyclic Euechinoidea characterized by a flattened oral surface.

convection [METEOROL] Atmospheric motions that are predominantly vertical, resulting in vertical transport and mixing of atmospheric properties. [OCEANOGR] Movement and mixing of ocean water masses.

convectional stability *See* static stability.

convection cell [GEOPHYS] A concept in plate tectonics that accounts for the lateral or the upward and downward movement of subcrustal mantle material as due to heat variation in the earth. [METEOROL] An atmospheric unit in which organized convective fluid motion occurs.

convection current [GEOPHYS] Mass movement of subcrustal or mantle material as a result of temperature variations. [METEOROL] Any current of air involved in convection; usually, the upward-moving portion of a convection circulation, such as a thermal or the updraft in cumulus clouds. Also known as convective current.

convection theory of cyclones   [METEOROL]   A theory of cyclone development proposing that the upward convection of air (particularly of moist air) due to surface heating can be of sufficient magnitude and duration that the surface inflow of air will attain appreciable cyclonic rotation.

convective activity   [METEOROL]   Generally, manifestations of convection in the atmosphere, alluding particularly to the development of convective clouds and resulting weather phenomena, such as showers, thunderstorms, squalls, hail, and tornadoes.

convective cloud   [METEOROL]   A cloud which owes its vertical development, and possibly its origin, to convection.

convective-cloud-height diagram   [METEOROL]   A graph used as an aid in estimating the altitude of the base of convective clouds; since its basis is the same as that for the dew-point formula, only the surface temperature and dew point need be known to use the diagram.

convective condensation level   [METEOROL]   On a thermodynamic diagram, the point of intersection of a sounding curve (representing the vertical distribution of temperature in an atmospheric column) with the saturation mixing-ratio line corresponding to the average mixing ratio in the surface layer (that is, approximately the lowest 1500 feet, or 450 meters).

convective current   *See* convection current.

convective equilibrium   *See* adiabatic equilibrium.

convective instability   [METEOROL]   The state of an unsaturated layer or column of air in the atmosphere whose wet-bulb potential temperature (or equivalent potential temperature) decreases with elevation. Also known as potential instability.

convective overturn   *See* overturn.

convective precipitation   [METEOROL]   Precipitation from convective clouds, generally considered to be synonymous with showers.

convective region   [METEOROL]   An area particularly favorable for the formation of convection in the lower atmosphere, or one characterized by convective activity at a given time.

conventional mining   [MIN ENG]   The cycle which includes cutting the coal, drilling the shot holes, charging and shooting the holes, loading the broken coal, and installing roof support. Also known as cyclic mining.

convergence   [GEOL]   Diminution of the interval between geologic horizons. [HYD]   The line of demarcation between turbid river water and clear lake water. [METEOROL]   The increase in wind setup observed beyond that which would take place in an equivalent rectangular basin of uniform depth, due to changes in platform or depth.   [OCEANOGR]   An oceanic condition in which currents or water masses having different densities, temperatures, or salinities meet; results in the sinking of the colder or more saline water.

convergence constant   *See* convergence factor.

convergence factor   [MAP]   For a specific chart or map projection, the factor which, when multiplied by the difference of longitude between points on two meridians, will give the convergence of the meridians. Also known as convergence constant.

convergence line   [METEOROL]   Any horizontal line along which horizontal convergence of the airflow is occurring. Also known as asymptote of convergence.

convergence of meridians   [MAP]   **1.** The angular drawing together of the geographic meridians in passing from the equator to the poles. **2.** The relative difference of directions of meridians at specific points on the meridians; it is equal to the product of the difference of longitude and the convergence factor.

convergent precipitation [METEOROL] A synoptic type of precipitation caused by local updrafts of moist air.

convergent zone paths [OCEANOGR] The velocity structure of permanent deep sound channels that produces focusing regions at distant intervals from a shallow source.

convolute bedding [GEOL] The extremely contorted laminae usually confined to a single layer of sediment, resulting from subaqueous slumping.

convolution [GEOL] **1.** The development of convolute bedding. **2.** A structure that originates by a convolution process, for example, a small-scale but intricate fold.

cooling crack [GEOL] A joint formed as a result of the cooling of igneous rock.

cooling unit [GEOL] A flow, or rapid multiple flows, of volcanic material exhibiting characteristic welding and crystallization patterns.

cooperative observer [METEOROL] An unpaid observer who maintains a meteorological station for the U.S. National Weather Service.

cooperite [MINERAL] (Pt,Pd)S A steel-gray tetragonal mineral of metallic luster consisting of a sulfide of platinum, occurring in irregular grains in igneous rock.

coordinate conversion [MAP] Changing the map coordinate values from one system to those of another system.

coordinates [MAP] **1.** Linear or angular quantities which designate the position that a point occupies in a given reference frame or system. **2.** A general term to designate the particular kind of reference frame or system, such as plane rectangular coordinates or spherical coordinates.

coorongite [GEOL] A boghead coal in the peat stage.

Copenhagen water *See* normal water.

copiapite [MINERAL] **1.** $Fe_5(SO_4)_6(OH)_2 \cdot 20H_2O$ A yellow mineral occurring in granular or scalar aggregates. Also known as ihleite; knoxvillite; yellow copperas. **2.** A group of minerals containing hydrous iron sulfates.

coping [MIN ENG] **1.** Process of cutting and trimming the edges of stone slabs. **2.** Process of cutting a stone slab into two pieces.

Copodontidae [PALEON] An obscure family of Paleozoic fishes in the order Bradyodonti.

copper glance *See* chalcocite.

copperite [MINERAL] An important platinum mineral that is composed of platinum sulfide.

copper mica *See* chalcophyllite.

copper nickel *See* niccolite.

copper ore [GEOL] Rock containing copper minerals.

copper pyrite *See* chalcopyrite.

copper uranite *See* torbernite.

coprolite [GEOL] Petrified excrement.

coquimbite [MINERAL] $Fe_2(SO_4)_3 \cdot 9H_2O$ A white mineral that crystallizes in the hexagonal system; it is dimorphous with paracoquimbite.

coquina [PETR] A coarse-grained, porous, easily crumbled variety of limestone composed principally of mollusk shell and coral fragments cemented together as rock.

coquinoid [PETR] **1.** Of or pertaining to coquina. **2.** Lithified coquina. **3.** An autochthonous deposit of limestone made up of more or less whole mollusk shells.

coracite *See* uraninite.

coral head [GEOL] A small reef patch of coralline material. Also known as coral knoll.

coral knoll *See* coral head.

coral limestone [PETR] A limestone created from the calcareous skeletons of corals, often with fragments of other organisms included and often with calcium carbonate as a binding medium.

coral mud [GEOL] Fine-grade deposits of coral fragments formed around coral islands and coasts bordered by coral reefs.

coral pinnacle [GEOL] A sharply upward-projecting growth of coral rising from the floor of an atoll lagoon.

coral rag [PETR] A well-cemented, rubbly limestone consisting mainly of broken and rolled fragments from coral reefs.

coral reef [GEOL] A ridge or mass of limestone built up of detrital material deposited around a framework of skeletal remains of mollusks, colonial coral, and massive calcareous algae.

coral-reef lagoon [GEOGR] The central, shallow body of water of an atoll or the water separating a barrier reef from the shore.

coral-reef shoreline [GEOL] A shoreline formed by reefs which are composed of coral polyps. Also known as coral shoreline.

coral rock *See* reef limestone.

coral sand [GEOL] Coarse-grade deposits of coral fragments formed around coral islands and coasts bordered by coral reefs.

coral shoreline *See* coral-reef shoreline.

Cordaitaceae [PALEOBOT] A family of fossil plants belonging to the Cordaitales.

Cordaitales [PALEOBOT] An extensive natural grouping of forest trees of the late Paleozoic.

cordierite [MINERAL] $Mg_2(Al_4Si_5O_{18})$ A blue, orthorhombic magnesium aluminosilicate mineral frequently occurring in association with thermally metamorphosed rocks derived from argillaceous sediments.

cordillera [GEOGR] A mountain range or group of ranges, including valleys, plains, rivers, lakes, and so on, forming the main mountain axis of a continent.

cordilleran geosyncline [GEOL] The Devonian geosynclinal region of western North America.

cord of ore [MIN ENG] A unit of about 7 tons (6.35 metric tons), but measured by wagonloads and not by weight.

cordonazo [METEOROL] A southerly wind of hurricane force generated along the western coast of Mexico when a tropical cyclone passes offshore in a northerly direction.

Cordtex [MIN ENG] A trademark for a detonating fuse suitable for opencut and quarry mining; consists of an explosive core of pentaerythritol tetranitrate (PETN) contained within plastic covering.

cordylite [MINERAL] $(Ce,La)_2Ba(CO_3)_3F_2$ A colorless to wax-yellow mineral consisting of a carbonate and fluoride of cerium, lanthanum, and barium.

core [GEOL] **1.** Center of the earth, beginning at a depth of 2900 kilometers. Also known as earth core. **2.** A vertical, cylindrical boring of the earth from which composition and stratification may be determined; in oil or gas well exploration the presence of hydrocarbons or water are items of interest. [OCEANOGR] That region within a layer of ocean water where parameters such as temperature, salinity, or velocity reach extreme values.

core analysis [GEOL] The use of core samples taken from the borehole during drilling to give information on strata age, composition, and porosity, and the presence of hydrocarbons or water along the length of the borehole.

core barrel rod *See* guide rod.

core-catcher case *See* lifter case.

core-gripper case *See* lifter case.

core intersection [GEOL] **1.** The point in a borehole where an ore vein or body is encountered as shown by the core. **2.** The width or thickness of the ore body, as shown by the core. Also known as core interval.

core interval *See* core intersection.

core-lifter case *See* lifter case.

core logging [GEOL] The analysis of the strata through which a borehole passes by the taking of core samples at predetermined depth intervals as the well is drilled.

corer [ENG] An instrument used to obtain cylindrical samples of geological materials or ocean sediments.

core sample [GEOL] A sample of rock, soil, snow, or ice obtained by driving a hollow tube into the undisturbed medium and withdrawing it with its contained sample or core.

core-spring case *See* lifter case.

core texture *See* atoll texture.

coring [PETRO ENG] The use of a core barrel (hollow length of tubing) to take samples from the underground formation during the drilling operation; used for core analysis.

Coriolis parameter [GEOPHYS] Twice the component of the earth's angular velocity about the local vertical $2\Omega \sin\phi$, where Ω is the angular speed of the earth and ϕ is the latitude; the magnitude of the Coriolis force per unit mass on a horizontally moving fluid parcel is equal to the product of the Coriolis parameter and the speed of the parcel.

corneite [GEOL] A biotite-hornfels formed during deformation of shale by folding.

cornetite [MINERAL] $Cu_3(PO_4)(OH)_3$ A peacock-blue mineral consisting of basic copper phosphate.

Cornish rolls [MIN ENG] A geared pair of horizontal cylinders, one fixed in a frame and the other held by strong springs; used for grinding.

corn snow *See* spring snow.

cornstone [PETR] A calcareous concretion that is embedded in marl and grades into concretionary limestone.

cornwallite [MINERAL] $Cu_5(AsO_4)_2(OH)_4 \cdot H_2O$ A verdigris green to blackish-green mineral consisting of a hydrated basic arsenate of copper; occurs as small botryoidal crusts.

coromant cut [MIN ENG] A drill hole pattern; two overlapping holes about 2¼ inches (about 5.5 centimeters) in diameter are drilled in the tunnel center and left uncharged; they form a slot roughly 4 by 2 inches (10 by 5 centimeters) to which the easers can break.

coromell [METEOROL] A land breeze from the south at La Paz, Mexico, near the mouth of the Gulf of California, prevailing from November to May; it sets in at night and usually persists until 8 or 10 A.M.

corona [GEOL] A mineral zone that is usually radial about another mineral or at the area between two minerals. Also known as kelyphite. [METEOROL] A set of one or more prismatically colored rings of small radii, concentrically surrounding the disk of the sun, moon, or other luminary when veiled by a thin cloud; due to diffraction by numerous waterdrops.

coronadite [MINERAL] $Pb(Mn^{2+}, Mn^{4+})_8O_{16}$ A black mineral consisting of a lead and manganese oxide, occurring in massive form with fibrous structure; an important constituent of manganese ore.

corona method [GEOPHYS] A method of estimating drop sizes in clouds by utilizing measurements of the angular radii of the rings of a corona.

corrasion [GEOL] Mechanical wearing away of rock and soil by the action of solid materials moved along by wind, waves, running water, glaciers, or gravity. Also known as mechanical erosion.

corrasion valley [GEOL] An elongated hollow or trough formed by corrasive activity.

corrected altitude [METEOROL] The indicated altitude corrected for temperature deviation from the standard atmosphere. Also known as true altitude.

corrected establishment *See* mean high water lunitidal interval.

correcting wedge [MIN ENG] A deflection wedge used to deflect a crooked borehole back into its intended course.

correlation [GEOL] **1.** The determination of the equivalence or contemporaneity of geologic events in separated areas. **2.** As a step in seismic study, the selecting of corresponding phases, taken from two or more separated seismometer spreads, of seismic events seemingly developing at the same geologic formation boundary.

correlation shooting [ENG] A method of seismic shooting in which individual profiles are shot and correlated in order to obtain the relative structural positions of the horizons mapped.

correlative rights [PETRO ENG] Legal rights protecting property over a portion of a gas or oil reservoir from excessive or wasteful withdrawal of hydrocarbons by adjoining properties overlying the same reservoir.

corrosion [GEOCHEM] Chemical erosion by motionless or moving agents.

corrosion border *See* corrosion rim.

corrosion rim [MINERAL] A modification of the outlines of a porphyritic crystal due to the corrosive action of a magma on previously stable minerals. Also known as corrosion border.

corsite [PETR] A spheroidal variety of gabbro. Also known as miagite; napoleonite.

cortlandite [PETR] A peridotite consisting of large crystals of hornblende with poikilitically included crystals of olivine. Also known as hudsonite.

corundum [MINERAL] Al_2O_3 A hard mineral occurring in various colors and crystallizing in the hexagonal system; crystals are usually prismatic or in rounded barrel shapes; gem varieties are ruby and sapphire.

corvusite [MINERAL] $V_2O_4\text{–}6V_2O_5 \cdot n\,H_2O$ A blue-black to brown mineral consisting of a hydrous oxide of vanadium; occurs in massive form.

Coryphodontidae [PALEON] The single family of the Coryphodontoidea, an extinct superfamily of mammals.

Coryphodontoidea [PALEON] A superfamily of extinct mammals in the order Pantodonta.

cosalite [MINERAL] $Pb_2Bi_2S_5$ A lead-gray or steel-gray mineral consisting of lead, bismuth, and sulfur; specific gravity is 6.39–6.75.

cosmogenic isotope [GEOCHEM] A radioisotope produced through cosmic radiation.

cotidal chart [MAP] A chart of cotidal lines that show approximate locations of high water at hourly intervals as measured from a reference meridian, usually Greenwich.

cotidal line [MAP] A line on a chart passing through all points where high water occurs at the same time.

cotton ball *See* ulexite.

cotton-belt climate [CLIMATOL] A type of warm climate characterized by dry winters and rainy summers; that is, a monsoon climate, in contrast to a Mediterranean climate.

cotunnite [MINERAL] $PbCl_2$ An alteration product of galena; a soft, white to yellowish mineral that crystallizes in the orthorhombic crystal system.

Cotylosauria [PALEON] An order of primitive reptiles in the subclass Anapsida, including the stem reptiles, ancestors of all of the more advanced Reptilia.

coulee [GEOL] **1.** A thick, solidified sheet or stream of lava. **2.** A steep-sided valley or ravine, sometimes with a stream at the bottom.

counterradiation [GEOPHYS] The downward flux of atmospheric radiation passing through a given level surface, usually taken as the earth's surface. Also known as back radiation.

country rock [GEOL] **1.** Rock that surrounds and is penetrated by mineral veins. **2.** Rock that surrounds and is invaded by an igneous intrusion.

Couvinian [GEOL] Lower Middle Devonian geologic time.

covalent crystal [CRYSTAL] A crystal held together by covalent bonds. Also known as valence crystal.

cove [GEOGR] **1.** A small, narrow, sheltered bay, creek, or recess along a coast. **2.** A deep recess or hollow in a cliff or steep mountainside.

covellite [MINERAL] CuS An indigo-blue mineral of metallic luster that crystallizes in the hexagonal system; it is usually massive or occurs in disseminations through other copper minerals and represents an ore of copper. Also known as indigo copper.

cover [MIN ENG] The thickness of rock between the mine workings and the surface.

cover head [GEOL] A deep accumulation of debris, made of talus cones and alluvial fans, found on an elevated marine terrace.

cover hole [MIN ENG] One of a group of boreholes drilled in advance of mine workings to probe for and detect water-bearing fissures or structures.

covite [PETR] A rock of igneous origin composed of sodic orthoclase, hornblende, sodic pyroxene, nepheline, and accessory sphene, apatite, and opaque oxides.

cowshee *See* kaus.

coyote blasting [MIN ENG] A method of blasting in which large charges are fired in small adits or tunnels driven at the level of the floor, in the face of a quarry or the slope of an open-pit mine. Also known as coyote-hole blasting; gopher-hole blasting; heading blasting.

coyote-hole blasting *See* coyote blasting.

crab locomotive [MIN ENG] A type of trolley locomotive equipped with an electric motor, a drum, and haulage cable mounted on a small truck; used to haul mine cars from workings.

crachin [METEOROL] A period of light rain accompanied by low stratus clouds and poor visibility which frequently occurs in the China Sea between January and April.

cradle dump [MIN ENG] A tipple which dumps cars with a rocking motion.

crag [GEOL] A steep, rugged point or eminence of rock, as one projecting from the side of a mountain.

cranch [MIN ENG] A vein left unworked.

crandallite [MINERAL] $CaAl_3(PO_4)_2(OH)_5 \cdot H_2O$ A white to light-grayish mineral consisting of a hydrous phosphate of calcium and aluminum occurring in fine, fibrous masses.

crater [GEOL] **1.** A large, bowl-shaped topographic depression with steep sides. **2.** A rimmed structure at the summit of a volcanic cone; the floor is equal to the vent diameter.

crater cone [GEOL] A cone built around a volcanic vent by lava extruded from the vent.

crater cuts [MIN ENG] Cuts with one or more fully charged holes in which blasting is conducted toward the face of the tunnel.

crater lake [HYD] A fresh-water lake formed by the accumulation of rain and ground-water in a caldera or crater.

craton [GEOL] The large, relatively immobile portion of continents consisting of both shield and platform areas.

cream ice *See* sludge.

crednerite [MINERAL] $CuMn_2O_4$ A steel-gray to iron-black foliated mineral consisting of copper, manganese, and oxygen.

creedite [MINERAL] $Ca_3Al_2(SO_4)(F,OH)_{10} \cdot 2H_2O$ A white or colorless monoclinic mineral consisting of hydrous calcium aluminum fluoride with calcium sulfate, occurring in grains and radiating crystalline masses; hardness is 2 on Mohs scale, and specific gravity is 2.7.

creek [HYD] A natural stream of water smaller than a river but larger than a brook.

creep [GEOL] A slow, imperceptible downward movement of slope-forming rock or soil under sheer stress. [MIN ENG] *See* squeeze.

creeper [MIN ENG] An endless chain that catches mine car axles on projecting bars.

crenitic [GEOL] Relating to or resulting from the raising of subterranean minerals by the action of spring water.

crenulate shoreline [GEOGR] An irregular shoreline with small, rounded indentations and sharp headlands.

crenulation cleavage *See* slip cleavage.

Creodonta [PALEON] A group formerly recognized as a suborder of the order Carnivora.

crepuscular arch *See* bright segment.

crescent beach [GEOL] A crescent-shaped beach at the head of a bay or the mouth of a stream entering the bay, with the concave side facing the sea.

crescentic dune *See* barchan.

crescentic lake *See* oxbow lake.

crescentic mark [GEOL] A scour mark produced by a glacier with a characteristic crescent shape.

crescumulate texture [PETR] The texture of igneous rocks containing large, elongated crystals that have an orientation at right angles to the massed rock layering.

crestal injection *See* external gas injection.

crestal plane [GEOL] The plane formed by joining the crests of all beds of an anticline.

crest cloud [METEOROL] A type of standing cloud which forms along a mountain ridge, either on the ridge, or slightly above and leeward of it, and remains in the same position relative to the ridge. Also known as cloud crest.

crest length [OCEANOGR] The length of a wave measured along its crest. Also known as crest width.

crest line [GEOL] The line connecting the highest points on the same bed of an anticline in an infinite number of cross sections.

crest stage [HYD] The highest stage reached at a point along a stream culminating a rise by waters of that stream.

crest width *See* crest length.

Cretaceous [GEOL] The latest system of rocks or period of the Mesozoic Era, between 136 and 65 million years ago.

cretification [PETR] A process in which rock is converted into chalk.

crevasse [GEOL] An open, nearly vertical fissure in a glacier or other mass of land ice or the earth, especially after earthquakes.

crevasse deposit [GEOL] Kame deposited in a crevasse.

crevasse hoar [HYD] Ice crystals which form and grow in glacial crevasses and in other cavities where a large cooled space is formed and in which water vapor can accumulate under calm, still conditions.

criador [METEOROL] The rain-bringing west wind of northern Spain.

crib *See* arête.

crinkled bedding [GEOL] Minutely wrinkled bedding.

crinoidal limestone [PETR] A rock composed predominantly of crystalline joints of crinoids, with foraminiferans, corals, and mollusks.

cristobalite [MINERAL] SiO_2 A silicate mineral that is a high-temperature form of quartz; stable above 1470°C; crystallizes in the tetragonal system at low temperatures and the isometric system at high temperatures.

critical bottom slope [GEOL] The depth distribution in which depth d of an ocean increases with latitude ϕ according to an equation of the form $d = d_0 \sin \phi + \text{constant}$.

critical density [GEOL] The density of a saturated, granular material below which, under rapid deformation, the material weakens and above which it becomes stronger.

critical depth [HYD] The level of water in a channel at which flow is at its minimum energy in relation to the channel bottom.

critical elevation   [MAP]   That elevation which is the high point within the area of a chart.

critical flow prover   [PETRO ENG]   Device used to measure the velocity of gas flow during open-flow testing of gas wells.

critical frequency   [GEOPHYS]   The minimum frequency of a vertically directed radio wave which will penetrate a particular layer in the ionosphere; for example, all vertical radio waves with frequencies greater than the E-layer critical frequency will pass through the E layer. Also known as penetration frequency.

critical gas saturation *See* equilibrium gas saturation.

critical height   [GEOPHYS]   The maximum height at which a vertical or sloped soil bank will stand without support under specific conditions.

critical level of escape   [GEOPHYS]   **1.** That level, in the atmosphere, at which a particle moving rapidly upward will have a probability of $1/e$ (e is base of natural logarithm) of colliding with another particle on its way out of the atmosphere. **2.** The level at which the horizontal mean free path of an atmospheric particle equals the scale height of the atmosphere.

critical mineral   [MINERAL]   A mineral or a member of a mineral association that is stable only under the conditions of one metamorphic facies and will change if the facies changes.

critical slope   [HYD]   The channel slope or grade that is equivalent to the loss of head per foot resulting from flow at a depth that provides uniform flow at critical depth.

crivetz   [METEOROL]   A wind blowing from the northeast quadrant in Romania and southern Russia, especially a cold boralike wind from the north-northeast, characteristic of the climate of Romania.

CRM *See* chemical remanent magnetization.

crocidolite   [MINERAL]   A lavender-blue, indigo-blue, or leek-green asbestiform variety of riebeckite; occurs in fibrous, massive, and earthy forms. Also known as blue asbestos; krokidolite.

Crockett magnetic separator   [MIN ENG]   An assembly consisting of a continuous belt submerged in a tank through which ore pulp flows; magnetic solids adhere to the belt, which has a series of flat magnets attached to it, and the solids are dragged clear.

crocoisite *See* crocoite.

crocoite   [MINERAL] $PbCrO_4$ A yellow to orange or hyacinth-red secondary mineral occurring as monoclinic, prismatic crystals; it is also massive granular. Also known as crocoisite; red lead ore.

Croixian   [GEOL]   Upper Cambrian geologic time.

Cro-Magnon man   [PALEON]   **1.** A race of tall, erect Caucasoid men having large skulls; identified from skeletons found in southern France. **2.** A general term to describe all fossils resembling this race that belong to the upper Paleolithic (35,000–8000 B.C.) in Europe.

cromfordite *See* phosgenite.

Cromwell Current   [OCEANOGR]   An eastward-setting subsurface current that extends about 1½° north and south of the equator, and from about 150°E to 92°W.

cronstedtite   [MINERAL] $Fe_4^{2+}Fe_2^{3+}(Fe_2^{3+}Si_2)O_{10}(OH)_8$ A black to brownish-black mineral consisting of a hydrous iron silicate crystallizing in hexagonal prisms; specific gravity is 3.34–3.35.

crooked hole [PETRO ENG] A borehole drilled at an angle, often because of steeply dipping formations; not to be confused with holes deliberately deviated from the vertical to avoid obstacles or to tap otherwise unavailable reservoirs.

crookesite [MINERAL] $(Cu,Tl,Ag)_2Se$ An important selenium mineral occurring in lead-gray masses and having a metallic appearance.

crop coal [MIN ENG] Coal of inferior quality found near the surface.

crop out *See* outcrop.

cross-bedding [GEOL] The condition of having laminae lying transverse to the main stratification planes of the strata; occurs only in granular sediments. Also known as cross-lamination; cross-stratification.

crosscut [MIN ENG] **1.** A small passageway driven at right angles to the main entry of a mine to connect it with a parallel entry of air course. **2.** A passageway in a mine that cuts across the geological structure.

crosscutting relationships [GEOL] Relationships which may occur between two adjacent rock bodies, where the relative age may be determined by observing which rock "cuts" the other, for example, a granitic dike cutting across a sedimentary unit.

cross fault [GEOL] **1.** A fault whose strike is perpendicular to the general trend of the regional structure. **2.** A minor fault that intersects a major fault.

cross fold [GEOL] A secondary fold whose axis is perpendicular or oblique to the axis of another fold. Also known as subsequent fold; superimposed fold; transverse fold.

cross gateway *See* cross heading.

crosshead [MIN ENG] A runner or guide positioned just above a sinking bucket to restrict excessive swinging.

cross heading [MIN ENG] Mine passage driven for ventilation from the airway to the gangway, or from one breast through the pillar to the adjoining working. Also known as cross gateway; cross hole; headway.

cross hole *See* cross heading.

cross joint [GEOL] A fracture in igneous rock perpendicular to the lineation caused by flow magma. Also known as transverse joint.

cross-lamination *See* cross-bedding.

cross-level [ENG] To level at an angle perpendicular to the principal line of sight.

Crossopterygii [PALEON] A subclass of the class Osteichthyes comprising the extinct lobefins or choanate fishes and represented by one extant species; distinguished by two separate dorsal fins.

cross sea [OCEANOGR] A series of waves or swells crossing another wave system at an angle.

cross section [GEOL] **1.** A diagram or drawing that shows the downward projection of surficial geology along a vertical plane; for example, a portion of a stream bed drawn at right angles to the mean direction of the flow of the stream. **2.** An actual exposure or cut that reveals geological features. [MAP] A horizontal grid system laid out on the ground for determining contours, quantities of earthwork, and so on, by means of elevations of the grid points.

cross-stone *See* harmotome; staurolite.

cross-stratification *See* cross-bedding.

cross valley *See* transverse valley.

crosswind [METEOROL] A wind which has a component directed perpendicularly to the course (or heading) of an exposed, moving object.

croute calcaire *See* caliche.

crown [MIN ENG] A horizontal roof member of a timber up to 16 feet (4.9 meters) long and supported at each end by an upright.

crown block [PETRO ENG] A wooden or steel beam joined to the tops of derrick posts of an oil well to support pulleys.

crude oil [GEOL] A comparatively volatile liquid bitumen composed principally of hydrocarbon, with traces of sulfur, nitrogen, or oxygen compounds; can be removed from the earth in a liquid state.

crude ore [MIN ENG] The ore as it leaves the mine in an unconcentrated form.

crush [MIN ENG] **1.** A general settlement of the strata above a coal mine due to failure of pillars; generally accompanied by numerous local falls of roof in mine workings. **2.** To reduce ore or quartz by stamps, crushers, or rolls.

crush breccia [GEOL] A breccia formed in place by mechanical fragmentation of rock during movements of the earth's crust.

crush conglomerate [GEOL] Beds similar to a fault breccia, except that the fragments are rounded by attrition. Also known as tectonic conglomerate.

crushed stone [MIN ENG] Irregular fragments of rock crushed or ground to smaller sizes after quarrying. Also known as broken stone.

crush fold [GEOL] A fold of large dimensions that may involve considerable minor folding and faulting such as would produce a mountain chain or an oceanic deep.

crushing [MIN ENG] The quantity of ore pulverized or crushed at a single operation in processing.

crush zone [GEOL] A zone of fault breccia on fault gouge.

crust [GEOL] The outermost solid layer of the earth, mostly consisting of crystalline rock and extending no more than a few miles from the surface to the Mohorovičić discontinuity. Also known as earth crust. [HYD] A hard layer of snow lying on top of a soft layer.

crustal motion [GEOL] Movement of the earth's crust.

crustal plate [GEOL] One of the few major 100-kilometer-thick blocks into which the lithosphere can be divided according to plate tectonics.

crutter [MIN ENG] **1.** A worker who drills blasting holes and prepares the blasting charge. **2.** A worker who removes blasted rock.

cryoclinometer [ENG] An airborne instrument designed for measuring the horizontal expanse of a field of sea ice.

cryoconite [GEOL] A dark, powdery dust transported by wind and deposited on the surface of snow or ice; found, however, mainly in cryoconite holes. [MINERAL] A mixture composed of garnet, sillimanite, zircon, pyroxene, quartz, and various other minerals.

cryoconite hole [GEOL] A cylindrical dust well filled with cryoconite; absorbs solar radiation, causing melting of glacier ice around and below it.

cryogenic lake [HYD] A lake formed from the melt of a portion of permafrost.

cryogenic period [GEOL] A time period in geologic history during which large bodies of ice appeared at or near the poles and climate favored the formation of continental glaciers.

cryolaccolith *See* hydrolaccolith.

cryolite [MINERAL] Na_3AlF_6 A white or colorless mineral that crystallizes in the monoclinic system but has a pseudocubic aspect; found in masses of waxy luster;

hardness is 2.5 on Mohs scale, and specific gravity is 2.95–3.0; used chiefly as a flux in producing aluminum from bauxite and for making salts of sodium and aluminum and porcelaneous glass. Also known as Greenland spar; ice stone.

cryolithionite [MINERAL] $Na_3Li_3Al_2F_{12}$ A colorless mineral that crystallizes in the isometric system; found in the Ural Mountains.

cryolithology [HYD] The study of the development, physical characteristics, and structure of underground ice, especially in permafrost areas.

cryology [HYD] The study of ice and snow.

cryopedology [GEOL] A branch of geology that deals with the study of intensive frost action and permanently frozen ground.

cryoplanation [GEOL] Land erosion at high latitudes or elevations due to processes of intensive frost action.

cryostatic [GEOPHYS] Referring to frost-induced hydrostatic phenomena.

cryotectonic [GEOL] Referring to features and deposits characteristic of the borders of glaciers.

cryoturbation *See* congeliturbation.

cryptobatholithic [GEOL] Referring to a mineral deposit that occurs in the roof rocks of an unexposed batholith.

cryptoclastic [GEOL] Composed of extremely fine, almost submicroscopic, broken or fragmental particles.

cryptoclimate [ENG] The climate of a confined space, such as inside a house, barn, or greenhouse, or in an artificial or natural cave; a form of microclimate.

cryptoclimatology [CLIMATOL] The science of climates of confined spaces (cryptoclimates); basically, a form of microclimatology.

cryptocrystalline [GEOL] Having a crystalline structure but of such a fine grain that individual components are not visible with a magnifying lens.

cryptodepression [GEOL] A lake basin in which the bottom is below sea level.

cryptographic [PETR] Referring to a rock texture with such fine components that they are indistinguishable with a microscope.

cryptohalite [MINERAL] $(NH_4)_2SiF_6$ A colorless to white or gray, isometric mineral consisting of ammonium silicon fluoride; occurs in massive and arborescent forms.

cryptolite *See* monazite.

cryptomelane [MINERAL] $KMn_8O_{16} \cdot H_2O$ A usually massive mineral, common in manganese ores; contains an oxide of manganese and potassium and crystallizes in the monoclinic system.

cryptoolitic [PETR] Referring to a rock texture whose fine grains can be perceived only under a microscope.

cryptoperthite [MINERAL] A fine-grained, submicroscopic variety of perthite consisting of an intergrowth of potassic and sodic feldspar, detectable only by means of x-rays or with the aid of an electron microscope.

Cryptostomata [PALEON] An order of extinct bryozoans in the class Gymnolaemata.

cryptovolcanic [GEOL] A small, nearly circular area of highly disturbed strata in which there is no evidence of volcanic materials to confirm the origin as being volcanic.

cryptozoon [PALEOBOT] A hemispherical or cabbagelike reef-forming fossil algae, probably from the Cambrian and Ordovician.

crystal [CRYSTAL] A homogeneous solid made up of an element, chemical compound or isomorphous mixture throughout which the atoms or molecules are arranged in a regularly repeating pattern. [MINERAL] *See* rock crystal.

crystal axis [CRYSTAL] A reference axis used for the vectoral properties of a crystal.

crystal chemistry [CRYSTAL] The study of the crystalline structure and properties of a mineral or other solid.

crystal class [CRYSTAL] One of 32 categories of crystals according to the inversions, rotations about an axis, reflections, and combinations of these which leaves the crystal invariant. Also known as symmetry class.

crystal defect [CRYSTAL] Any departure from crystal symmetry caused by free surfaces, disorder, impurities, vacancies and interstitials, dislocations, lattice vibrations, and grain boundaries. Also known as lattice defect.

crystal face [CRYSTAL] One of the outward planar surfaces which define a crystal and reflect its internal structure. Also known as face.

crystal gliding [CRYSTAL] Slip along a crystal plane due to plastic deformation; often produces crystal twins. Also known as translation gliding.

crystal growth [CRYSTAL] The growth of a crystal, which involves diffusion of the molecules of the crystallizing substance to the surface of the crystal, diffusion of these molecules over the crystal surface to special sites on the surface, incorporation of molecules into the surface at these sites, and diffusion of heat away from the surface.

crystal habit [CRYSTAL] The size and shape of the crystals in a crystalline solid. Also known as habit.

crystal indices *See* Miller indices.

crystallaria [GEOL] In soil, a group of single crystals or arrangements of crystals of relatively pure fractions of the soil plasma that do not enclose the matrix of the soil material but instead form cohesive masses.

crystal lattice [CRYSTAL] A lattice from which the structure of a crystal may be obtained by associating with every lattice point an assembly of atoms identical in composition, arrangement, and orientation.

crystalline [CRYSTAL] Of, pertaining to, resembling, or composed of crystals.

crystalline frost [HYD] Hoarfrost that exhibits a relatively simple macroscopic crystalline structure.

crystalline-granular texture [PETR] A primary texture of an igneous rock due to crystallization from a fluid medium.

crystalline porosity [GEOL] Porosity in crystalline limestone and dolomite, making possible underground oil reservoirs.

crystalline rock [PETR] **1.** Rock made up of minerals in a clearly crystalline state. **2.** Igneous and metamorphic rock, as opposed to sedimentary rock.

crystallinity [CRYSTAL] The quality or state of being crystalline. [PETR] Degree of crystallization exhibited by igneous rock.

crystallite [GEOL] A small, rudimentary form of crystal which is of unknown mineralogic composition and which does not polarize light.

crystallization [CRYSTAL] The formation of crystalline substances from solutions or melts.

crystalloblast [MINERAL] A mineral crystal produced by metamorphic processes.

crystalloblastic series [GEOL] A series of metamorphic minerals ordered according to decreasing formation energy, so crystals of a listed mineral have a tendency to form idioblastic outlines at surfaces of contact with simultaneously developed crystals of all minerals in lower positions.

crystalloblastic strength *See* form energy.

crystalloblastic texture [GEOL] A crystalline texture resulting from metamorphic recrystallization under conditions of high viscosity and directed pressure.

crystallogram [CRYSTAL] A photograph of the x-ray diffraction pattern of a crystal.

crystallographic axis [CRYSTAL] One of three lines (sometimes four, in the case of a hexagonal crystal), passing through a common point, that are chosen to have definite relation to the symmetry properties of a crystal, and are used as a reference in describing crystal symmetry and structure.

crystallographic texture [MINERAL] A texture of replacement or exsolution deposits of minerals in which the form and distribution of the inclusions are dependent upon the crystallography of the host mineral.

crystal mold [GEOL] A cavity left by solution or sublimation of a crystal that was embedded in soft, fine-grained sediment.

crystal sandstone [GEOL] Siliceous sandstone in which deposited silica is precipitated upon the quartz grains in crystalline position.

crystal settling [GEOL] Sinking of crystals in magma from the liquid in which they formed, by the action of gravity.

crystal sorting [GEOPHYS] The separation of crystals from a magma or the separation of one crystal phase from another during the crystallization of the magma.

crystal structure [CRYSTAL] The arrangement of atoms or ions in a crystalline solid.

crystal symmetry [CRYSTAL] The existence of nontrivial operations, consisting of inversions, rotations around an axis, reflections, and combinations of these, which bring a crystal into a position indistinguishable from its original position.

crystal system [CRYSTAL] One of seven categories (cubic, hexagonal, tetragonal, trigonal, orthorhombic, monoclinic, and triclinic) into which a crystal may be classified according to the shape of the unit cell of its Bravais lattice, or according to the dominant symmetry elements of its crystal class.

crystal tube [GEOL] A formation of masses of crystals filling or partially filling tube-shaped openings in soil material.

crystal tuff [GEOL] Consolidated volcanic ash in which crystals and crystal fragments predominate.

crystal-vitric tuff [GEOL] Consolidated volcanic ash composed of 50–75% crystal fragments and 25–50% glass fragments.

crystal whiskers [CRYSTAL] Single crystals that have grown in a filamentary form.

crystosphene [HYD] A buried sheet or mass of ice, as in the tundra of northern America, formed by the freezing of rising and spreading springwater beneath alluvial deposits.

Cs *See* cirrostratus cloud.

CTD recorder *See* salinity-temperature-depth recorder.

Ctenothrissidae [PALEON] A family of extinct teleostean fishes in the order Ctenothrissiformes.

Ctenothrissiformes [PALEON] A small order of extinct teleostean fishes; important as a group on the evolutionary line leading from the soft-rayed to the spiny-rayed fishes.

cubanite [MINERAL] $CuFe_2S_3$ Bronze-yellow mineral that crystallizes in the orthorhombic system. Also known as chalmersite.

cube ore *See* pharmacosiderite.

cube spar *See* anhydrite.

cubic crystal [CRYSTAL]   A crystal whose lattice has a unit cell with perpendicular axes of equal length.

cubic packing [CRYSTAL]   The spacing pattern of uniform solid spheres in a clastic sediment or crystal lattice in which the unit cell is a cube.

cubic plane [CRYSTAL]   A plane that is at right angles to any one of the three crystallographic axes of the cubic system.

cubic system *See* isometric system.

cuesta [GEOGR]   A gently sloping plain which terminates in a steep slope on one side.

Cullender isochronal method [PETRO ENG]   Procedure for back-pressure testing to analyze gas wells that produce from low-permeability reservoirs with a consequent slow approach to stabilized producing conditions.

culm [MIN ENG]   Fine, refuse coal, screened and separated from larger pieces.

culmination [GEOL]   A high point on the axis of a fold.

cumberlandite [PETR]   A coarse-grained, ultramafic, ultrabasic rock composed principally of olivine crystals in a ground mass of magnetite and ilmenite with minor plagioclase.

cumbraite [PETR]   A variety of dacite or rhyodacite containing very calcic plagioclase and pyroxene in a glassy groundmass.

cumengite [MINERAL] $Pb_4Cu_4Cl_8(OH)_8 \cdot H_2O$ A deep-blue or light-indigo-blue tetragonal mineral consisting of a basic lead-copper chloride occurring in crystals.

cummingtonite [MINERAL] $(Fe,Mg)_7Si_8O_{22}(OH)_2$ A brownish mineral that crystallizes in the monoclinic system; usually occurs as lamellae or fibers in metamorphic rocks.

cum sole [GEOPHYS]   With the sun; hence anticyclonic or clockwise in the Northern Hemisphere.

cumulate [PETR]   Any igneous rock formed by the accumulation of crystals settling out of a magma.

cumulative gas [PETRO ENG]   Measurement of total gas produced from a reservoir, usually expressed in graphical relationship to total (cumulative) oil produced from the same reservoir.

cumuliform cloud [METEOROL]   A fundamental cloud type, showing vertical development in the form of rising mounds, domes, or towers.

cumulonimbus calvus cloud [METEOROL]   A species of cumulonimbus cloud evolving from cumulus congestus: the protuberances of the upper portion have begun to lose the cumuliform outline; they loom and usually flatten, then transform into a whitish mass with a more or less diffuse outline and vertical striation; cirriform cloud is not present, but the transformation into ice crystals often proceeds with great rapidity.

cumulonimbus capillatus cloud [METEOROL]   A species of cumulonimbus cloud characterized by the presence of distinct cirriform parts, frequently in the form of an anvil, a plume, or a vast and more or less disorderly mass of hair, and usually accompanied by a thunderstorm.

cumulonimbus cloud [METEOROL]   A principal cloud type, exceptionally dense and vertically developed, occurring either as isolated clouds or as a line or wall of clouds with separated upper portions.

cumulophyre [PETR]   An igneous rock with a cumulophyric texture.

cumulophyric [PETR]   Pertaining to an igneous rock containing irregular groups of phenocrysts.

cumulus cloud [METEOROL]   A principal cloud type in the form of individual, detached elements which are generally dense and possess sharp nonfibrous outlines; these elements develop vertically, appearing as rising mounds, domes, or towers, the upper parts of which often resemble a cauliflower.

cumulus congestus cloud [METEOROL]   A strongly sprouting cumulus species with generally sharp outline and sometimes a great vertical development, and with cauliflower or tower aspect.

cumulus humilis cloud [METEOROL]   A species of cumulus cloud characterized by small vertical development and a generally flattened appearance, vertical growth is usually restricted by the existence of a temperature inversion in the atmosphere, which in turn explains the unusually uniform height of the cloud. Also known as fair-weather cumulus.

cumulus mediocris cloud [METEOROL]   A cloud species unique to the species cumulus, of moderate vertical development, the upper protuberances or sproutings being not very marked; there may be a small cauliflower aspect; while this species does not give any precipitation, it frequently develops into cumulus congestus and cumulonimbus.

cup-and-ball joint [GEOL]   A dish-shaped transverse fracture which divides a basalt column into segments. Also known as ball-and-socket joint.

cup anemometer [ENG]   A rotation anemometer, usually consisting of three or four hemispherical or conical cups mounted with their diametral planes vertical and distributed symmetrically about the axis of rotation; the rate of rotation of the cups, which is a measure of the wind speed, is determined by a counter.

cup barometer [ENG]   A barometer in which one end of a graduated glass tube is immersed in a cup, both cup and tube containing mercury.

cup crystal [HYD]   A crystal of ice in the form of a hollow hexagonal cup; a common form of depth hoar.

cupola [GEOL]   An isolated, upward-projecting body of plutonic rock that lies near a larger body; both bodies are presumed to unite at depth.

cuprite [MINERAL] Cu_2O A red mineral that crystallizes in the isometric system and is found in crystals and fine-grained aggregates or is massive; a widespread supergene copper ore. Also known as octahedral copper ore; red copper ore; ruby copper ore.

cuprocopiapite [MINERAL] $CuFe_4(SO_4)_6(OH)_2 \cdot 20H_2O$ A sulfur yellow to orange-yellow, triclinic mineral consisting of a hydrated basic sulfate of copper and iron.

cuprodescloizite See mottramite.

cuprotungstite [MINERAL] $Cu_2(WO_4)(OH)_2$ A green mineral that forms compact masses; soluble in acids; the crystal system is not known.

cuprouranite See torbernite.

curb [MIN ENG]   A timber frame, circular or square, wedged in a shaft to make a foundation for walling or tubbing, or to support, with or without other timbering, the walls of the shaft.

curite [MINERAL] $Pb_2U_5O_{17} \cdot 4H_2O$ An orange-red radioactive mineral, occurring in acicular crystals, an alteration product of uraninite.

current-bedding [GEOL]   Cross-bedding resulting from water or air currents.

current chart [MAP]   A map of a water area depicting current speeds and directions by current roses, vectors, or other means.

current constants [OCEANOGR] Tidal current relations that remain practically constant for any particular locality.

current curve [OCEANOGR] In marine operations, a graphic representation of the flow of a current, consisting of a rectangular-coordinate graph on which speed is represented by the ordinates and time by the abscissas.

current cycle [OCEANOGR] A complete set of tidal current conditions, as those occurring during a tidal day, lunar month, or Metonic cycle.

current diagram [OCEANOGR] A graph showing the average speeds of flood and ebb currents throughout the current cycle for a considerable part of a tidal waterway.

current difference [OCEANOGR] In marine operations, the difference between the time of slack water or strength of current at a subordinate station and its reference station.

current drogue [ENG] A current-measuring assembly consisting of a weighted current cross, sail, or parachute, and an attached surface buoy.

current ellipse [OCEANOGR] In marine operations, a graphic representation of a rotary current, in which the speed and direction of the current at various hours of the current cycle are represented by radius vectors; a line connecting the ends of the radius vectors approximates an ellipse.

current hour [OCEANOGR] The average time interval between the moon's transit over the meridian of Greenwich and the time of the following strength of flood current modified by the times of slack water and strength of ebb.

current lineation *See* parting lineation.

current mark [GEOL] Any structure produced by the action of water current, either directly or indirectly, on a sedimentary surface.

current ripple [GEOL] A type of ripple mark having a long, gentle slope toward the direction from which the current flows, and a shorter, steeper slope on the lee side.

current rips [OCEANOGR] Small waves formed on the surface of water by the meeting of opposing ocean currents; vertical oscillation, rather than progressive waves, is characteristic of current rips.

current rose [MAP] A graphic presentation of ocean currents for specified areas, utilizing arrows at the cardinal and intercardinal compass points to show the direction toward which the prevailing current flows and the present frequency of set for a given period of time.

current tables [OCEANOGR] Tables listing predictions of the time and speeds of tidal currents at various places.

curtain [GEOL] **1.** A thin sheet of dripstone that hangs or projects from a cave wall. **2.** A rock formation connecting two neighboring bastions.

curvature correction [GEOD] The correction applied in some geodetic work to take account of the divergence of the surface of the earth (spheroid) from a plane.

curvilinear coordinates [MAP] Any linear coordinates which are not cartesian coordinates; frequently used curvilinear coordinates are polar coordinates and cylindrical coordinates.

cusp [GEOL] One of a series of low, crescent-shaped mounds of beach material separated by smoothly curved, shallow troughs spaced at more or less regular intervals along and generally perpendicular to the beach face. Also known as beach cusp. [GEOPHYS] A funnel-shaped region in the magnetosphere, extending from the front magnetopause to the polar ionosphere, which is filled with solar wind plasma.

cuspate bar [GEOL] A crescentic bar joining with the shore at each end.

cuspate ripple mark *See* linguoid ripple mark.

custard winds [METEOROL] Cold easterly winds on the northeastern coast of England.

cut [MIN ENG] **1.** To intersect a vein or working. **2.** To excavate coal. **3.** To shear one side of an entry or crosscut by digging out the coal from floor to roof with a pick.

cutan [GEOL] An alteration of the structure, texture, or fabric of a soil material along a natural surface within it, which is caused by a concentration of a specific soil constituent.

cut and fill [GEOL] **1.** Lateral corrosion of one side of a meander accompanied by deposition on the other. **2.** A sedimentary structure consisting of a small filled-in channel.

cutbank [GEOL] The concave bank of a winding stream that is maintained as a steep or even overhanging cliff by the action of water at its base.

cutinite [GEOL] A variety of exinite consisting of plant cuticles.

cutoff [GEOL] **1.** A new, relatively short channel formed when a stream cuts through the neck of an oxbow or horseshoe bend. **2.** A boundary, oriented normal to bedding planes, that marks the areal limit of a specific stratigraphic unit.   [MIN ENG] **1.** A quarryman's term for the direction along which the granite must be channeled, because it will not split. **2.** The number of feet a bit may be used in a particular type of rock (as specified by the drill foreman). **3.** Minimum percentage of mineral in an ore that can be mined profitably.

cutoff high [METEOROL] A warm high which has become displaced out of the basic westerly current, and lies to the north of this current.

cutoff lake *See* oxbow lake.

cutoff low [METEOROL] A cold low which has become displaced out of the basic westerly current, and lies to the south of this current.

cutout [GEOL] *See* horseback.

cut platform *See* wave-cut platform.

cut shot [MIN ENG] A shot designed to bring down coal which has been sheared or opened on one side.

cutter [MIN ENG] **1.** An operator of a coal-cutting or rock-cutting machine, or a worker engaged in underholing by pick or drill. **2.** A joint, usually a dip joint, running in the direction of working; usually in the plural.

cutting machine [MIN ENG] A power-driven apparatus used to undercut or shear the coal to help in its removal from the face.

cutting-off process [METEOROL] A sequence of events by which a warm high or cold low, originally within the westerlies, becomes displaced either poleward (cutoff high) or equatorward (cutoff low) out of the westerly current; this process is evident at very high levels in the atmosphere, and it frequently produces, or is part of the production of, a blocking situation.

cuttings [MIN ENG] Rock fragments broken from the penetrated rock during drilling operations.

Cuvieroninae [PALEON] A subfamily of extinct proboscidean mammals in the family Gomphotheriidae.

cyanite *See* kyanite.

cyanochroite [MINERAL] $K_2Cu(SO_4)_2 \cdot 6H_2O$ A blue mineral consisting of a hydrous sulfate of potassium and copper.

cyanotrichite [MINERAL] $Cu_4Al_2(SO_4)(OH)_{12}\cdot 2H_2O$ A bright-blue or sky-blue mineral consisting of a hydrous basic copper aluminum sulfate.

Cycadeoidaceae [PALEOBOT] A family of extinct plants in the order Cycadeoidales characterized by sparsely branched trunks and a terminal crown of leaves.

Cycadeoidales [PALEOBOT] An order of extinct plants that were abundant during the Triassic, Jurassic, and Cretaceous periods.

Cycadofilicales [PALEOBOT] The equivalent name for the extinct Pteridospermae.

cycle of erosion *See* geomorphic cycle.

cycle of sedimentation [GEOL] Also known as sedimentary cycle. **1.** A series of related processes and conditions appearing repeatedly in the same sequence in a sedimentary deposit. **2.** The sediments deposited from the beginning of one cycle to the beginning of a second cycle of the spread of the sea over a land area, consisting of the original land sediments, followed by those deposited by shallow water, then deep water, and then the reverse process of the receding water.

cycleology [GEOL] The determination and investigation of cycles in paleontologic and geologic phenomena.

cyclic mining *See* conventional mining.

cyclic salt [OCEANOGR] Salt removed from the sea as spray, blown inland, and returned to its source by land drainage.

cyclic sedimentation [GEOL] A process in which various types of sediment are deposited in repeated regular sequence.

cyclic twinning [CRYSTAL] Repeated twinning of three or more individuals in accordance with the same twinning law but without parallel twinning axes.

Cyclocystoidea [PALEON] A class of small, disk-shaped, extinct echinozoans in which the lower surface of the body probably consisted of a suction cup.

cyclogenesis [METEOROL] Any development or strengthening of cyclonic circulation in the atmosphere.

cyclolysis [METEOROL] The weakening or decay of cyclonic circulation in the atmosphere.

cyclone [METEOROL] A low-pressure region of the earth's atmosphere with roundish to elongated-oval ground plan, in-moving air currents, centrally upward air movement, and generally outward movement at various higher elevations in the troposphere.

cyclone family [METEOROL] A series of wave cyclones occurring in the interval between two successive major outbreaks of polar air, and traveling along the polar front, usually eastward and poleward.

cyclone wave [METEOROL] **1.** A disturbance in the lower troposphere, of wavelength 1000–2500 kilometers; cyclone waves are recognized on synoptic charts as migratory high- and low-pressure systems. **2.** A frontal wave at the crest of which there is a center of cyclonic circulation, that is, the frontal wave of a wave cyclone.

cyclonic [GEOPHYS] Having a sense of rotation about the local vertical that is the same as that of the earth's rotation: as viewed from above, counterclockwise in the Northern Hemisphere, clockwise in the Southern Hemisphere, undefined at the Equator.

cyclonic scale [METEOROL] The scale of the migratory high- and low-pressure systems (or cyclone waves) of the lower troposphere, with wavelengths of 1000–2500 kilometers. Also known as synoptic scale.

cyclonic shear [METEOROL] Horizontal wind shear of such a nature that it contributes to the cyclonic vorticity of the flow; that is, it tends to produce cyclonic rotation of the individual air particles along the line of flow.

cyclopean [PETR] *See* mosaic.

cyclosilicate [MINERAL] A silicate having the SiO_4 tetrahedra linked to form rings, with a silicon-oxygen ratio of 1:3, such as $Si_3O_9^{6-}$ or $Si_6O_{18}^{12-}$. Also known as ring silicate.

Cyclosteroidea [PALEON] A class of Middle Ordovician to Middle Devonian echinoderms in the subphylum Echinozoa.

cyclostrophic wind [METEOROL] The horizontal wind velocity for which the centripetal acceleration exactly balances the horizontal pressure force.

cyclothem [GEOL] A rock stratigraphic unit associated with unstable shelf of interior basin conditions, in which the sea has repeatedly covered the land.

cylindrical map projection [MAP] A map projection produced by projecting the geographic meridians and parallels onto a cylinder which is tangent to (or intersects) the surface of a sphere, and then developing the cylinder into a plane.

cylindrical structure [GEOL] A vertical sedimentary structure shaped like an irregular column or pillar.

cylindrite [MINERAL] $Pb_3Sn_4Sb_2S_{14}$ A blackish-gray mineral consisting of sulfur, lead, antimony, and tin, occurring in cylindrical forms that separate under pressure into distinct sheets or folia.

cymoid [GEOL] A vein of mineral deposit with a cross section forming a reverse curve.

cymrite [MINERAL] $Ba_2Al_5Si_5O_{19}(OH) \cdot 3H_2O$ Zeolite mineral consisting of a basic aluminosilicate of barium.

Cystoidea [PALEON] A class of extinct crinozoans characterized by an ovoid body that was either sessile or attached by a short aboral stem.

dachiardite [MINERAL] $(Na_2Ca)_2(Al_4Si_{20}O_{48})\cdot12H_2O$ A white to colorless mineral in the mordenite group of the zeolite family that crystallizes in the monoclinic system.

Dacian [GEOL] Lower upper Pliocene geologic time.

dacite [GEOL] Very fine crystalline or glassy rock of volcanic origin, composed chiefly of sodic plagioclase and free silica with subordinate dark-colored minerals.

dacite glass [GEOL] A natural glass formed by rapid cooling of dacite lava.

dactylitic [GEOL] Of a rock texture, characterized by fingerlike projections of a mineral that penetrate another mineral.

dadur [METEOROL] In India, a wind blowing down the Ganges Valley from the Siwalik hills at Hardwar.

daily forecast [METEOROL] A forecast for periods of from 12 to 48 hours in advance.

daily mean [METEOROL] The average value of a meteorological element over a period of 24 hours.

daily retardation [OCEANOGR] The amount of time by which corresponding tidal phases grow later day by day; averages approximately 50 minutes.

daily variation [GEOPHYS] Oscillation of the earth's magnetic field that occurs during a 1-day period.

Dakotan [GEOL] Lower Upper Cretaceous geologic time.

dalles [HYD] The rapids in a deep, narrow stream running between the walls of a canyon or ravine.

Dalmatian wettability [PETRO ENG] Theory of wettability that some in situ reservoir rocks are partly preferentially oil-wet and partly preferentially water-wet.

damkjernite [PETR] A melanocratic dike rock composed of biotite and pyroxene phenocrysts in a groundmass of pyroxene, biotite, and magnetite.

damp air [METEOROL] Air that has a high relative humidity.

damp haze [METEOROL] Small water droplets or very hygroscopic particles in the air, reducing the horizontal visibility somewhat, but to not less than 1¼ miles (2 kilometers); similar to a very thin fog, but the droplets or particles are more scattered than in light fog and presumably smaller.

damp sheet [MIN ENG] A curtain used in a mine to direct airflow, thus preventing gas accumulation.

danalite [MINERAL] $(Fe,Mn,Zn)_4Be_3(SiO_4)_3S$ A mineral consisting of a silicate and sulfide of iron and beryllium; it is isomorphous with helvite and genthelvite.

danburite　[MINERAL]　$CaB_2(SiO_4)_2$　An orange-yellow, yellowish-brown, grayish, or colorless transparent to translucent borosilicate mineral with a feldspar structure crystallizing in the orthorhombic system; it resembles topaz and is used as an ornamental stone.

dancing dervish *See* dust whirl.

dancing devil *See* dust whirl.

dangerous semicircle　[METEOROL]　The half of the circular area of a tropical cyclone having the strongest winds and heaviest seas, where a ship tends to be drawn into the path of the storm.

Danian　[GEOL]　Lowermost Paleocene or uppermost Cretaceous geologic time.

dannemorite　[MINERAL]　$(Fe,Mn,Mg)_7Si_8O_{22}(OH)_2$　A yellowish-brown to greenish-gray monoclinic mineral consisting of a columnar or fibrous amphibole.

daphnite　[MINERAL]　$(MgFe)_3(Fe,Al)_3(Si,Al)_4O_{10}(OH)_8$　A mineral of the chlorite group consisting of a basic aluminosilicate of magnesium, iron, and aluminum.

Daphoenidae　[PALEON]　A family of extinct carnivoran mammals in the superfamily Miacoidea.

darapskite　[MINERAL]　$Na_3(NO_3)(SO_4)·H_2O$　A naturally occurring hydrate mineral consisting of a hydrous nitrate and sulfate of sodium.

dark halo crater　[ASTRON]　A small lunar crater ringed by material having a lower albedo than that of the surrounding region.

dark-red silver ore *See* pyrargyrite.

dark-ruby silver *See* pyrargyrite.

dark segment　[METEOROL]　A bluish-gray band appearing along the horizon opposite the rising or setting sun and lying just below the antitwilight arch. Also known as earth's shadow.

Darling shower　[METEOROL]　A dust storm caused by cyclonic winds in the vicinity of the River Darling in Australia.

dart leader　[GEOPHYS]　The leader which, after the first stroke, initiates each succeeding stroke of a composite flash of lightning. Also known as continuous leader.

Darwin-Doodson system　[GEOPHYS]　A method for predicting tides by expressing them as sums of harmonic functions of time.

Darwin glass　[GEOL]　A highly siliceous, vesicular glass shaped in smooth blobs or twisted shreds, found in the Mt. Darwin range in western Tasmania. Also known as queenstownite.

dashkesanite　[MINERAL]　$(Na,K)Ca_2(Fe,Mg)_5(Si,Al)_8O_{22}Cl_2$　A monoclinic mineral of the amphibole group consisting of a chloroaluminosilicate of sodium, potassium, iron, and magnesium.

datolite　[MINERAL]　$CaBSiO_4(OH)$　A mineral nesosilicate crystallizing in the monoclinic system; luster is vitreous, and crystals are colorless or white with a greenish tinge.

datum　[GEOD]　The latitude and longitude of an initial point; the azimuth of a line from this point.　[GEOL]　The top or bottom of a bed of rock on which structure contours are drawn.

datum level *See* datum plane.

datum plane　[ENG]　A permanently established horizontal plane, surface, or level to which soundings, ground elevations, water surface elevations, and tidal data are referred. Also known as chart datum; datum level; reference level; reference plane.

datum point [MAP] Any reference point of known or assumed coordinates from which calculation or measurements may be taken.

daubree [GEOL] A unit of measure delineating the degree of wear of a sedimentary particle which is equivalent to the removal of 0.1 gram from a 100-gram sphere of quartz.

daubreeite [MINERAL] $FeCr_2S_4$ A mineral composed of a black chromium iron sulfide; occurs in some meteors.

Dauphine law [CRYSTAL] A twin law in which the twinned parts are related by a rotation of 180° around the c axis.

Davian [GEOL] A subdivision of the Upper Cretaceous in Europe; a limestone formation with abundant hydrocorals, bryozoans, and mollusks in Denmark; marine limestone and nonmarine rocks in southeastern France; and continental formations in the Davian of Spain and Portugal.

davidite [MINERAL] A black primary pegmatite uranium mineral of the general formula $A_6B_{15}(O,OH)_{36}$, where $A = Fe^{2+}$, rare earths, uranium, calcium, zirconium, and thorium, and $B =$ titanium, Fe^{3+}, vanadium, and chromium.

Davidson Current [OCEANOGR] A coastal countercurrent of the Pacific Ocean running north, inshore of the California Current, along the western coast of the United States (from northern California to Washington to at least latitude 48°N) during the winter months.

daviesite [MINERAL] An orthorhombic mineral consisting of a lead oxychloride, occurring in minute crystals.

Davis magnetic tester [MIN ENG] An instrument used to determine the magnetic contents of ores.

davisonite [MINERAL] $Ca_3Al(PO_4)_2(OH)_3 \cdot H_2O$ White mineral consisting of a hydrous basic phosphate of calcium and aluminum.

Davy lamp [MIN ENG] An early safety lamp with a mantle of wire gauze around the flame to dissipate the heat from the flame to below the ignition temperature of methane.

dawsonite [MINERAL] $NaAl(OH)_2CO_3$ A white, bladed mineral found in certain oil shales that contains large quantities of alumina; specific gravity is 2.40.

day drift [MIN ENG] A mine passageway that has one end at the surface.

4-D chart [METEOROL] A chart showing the field of D values (deviations of the actual altitudes along a constant-pressure surface from the standard atmosphere altitude of that surface) in terms of the three dimensions of space and one of time; it is a form of a four-dimensional display of pressure altitude; the space dimensions are represented by D-value contours, and the time dimension is provided by tau-value lines.

DDA value *See* depth-duration-area value.

dead [GEOL] In economic geology, referring to an area without economic value. [MIN ENG] An area of subsidence that has undergone complete settling and so is unlikely to move.

dead air [MIN ENG] Air in a mine when it is stagnant or contains carbonic acid.

dead-burn [MIN ENG] A calcination to produce a dense refractory substance; done at a higher temperature and for a longer time than for normal calcination.

dead cave [GEOL] A cave which no longer contains any moisture or exhibits any growth of moisture-related mineral deposits.

deadhead [MIN ENG] To begin a new cut without excavating the material from the preceding cut.

dead line [GEOL] The level above which a batholith is metalliferous and below which it is considered economically useless.

dead-roast [MIN ENG] **1.** A roasting process for driving off sulfur. Also known as sweet roast. **2.** Removing volatiles by roasting within a specified temperature range.

dead sea [HYD] A body of water that has undergone precipitation of its rock salt, gypsum, or other evaporites.

Dead Sea [GEOGR] A salt lake between Jordan and Israel.

dead water [OCEANOGR] The mass of eddying water associated with formation of internal waves near the keel of a ship; forms under a ship of low propulsive power when it negotiates water which has a thin layer of fresher water over a deeper layer of more saline water.

dead work [MIN ENG] Preparatory work which is for future operations and not directly productive.

deaister *See* doister.

debris [GEOL] Large fragments arising from disintegration of rocks and strata.

debris avalanche [GEOL] The sudden and rapid downward movement of incoherent mixtures of rock and soil on deep slopes.

debris cone [GEOL] **1.** A mound of ice or snow on a glacier covered with a thin layer of debris. **2.** *See* alluvial cone.

debris fall [GEOL] A relatively free downward or forward falling of unconsolidated or poorly consolidated earth or rocky debris from a cliff, cave, or arch.

debris flow [GEOL] A variety of rapid mass movement involving the downslope movement of high-density coarse clast-bearing mudflows, usually on alluvial fans.

debris glacier [HYD] A glacier formed from ice fragments that have fallen from a larger and taller glacier.

debris line *See* swash mark.

debris slide [GEOL] A type of landslide involving a rapid downward sliding and forward rolling of comparatively dry, unconsolidated earth and rocky debris.

debris slope *See* talus slope.

decay [GEOCHEM] *See* chemical weathering. [OCEANOGR] In ocean-wave studies, the loss of energy from wind-generated ocean waves after they have ceased to be acted on by the wind; this process is accompanied by an increase in length and a decrease in height of the wave. [PHYS] Gradual reduction in the magnitude of a quantity, as of current, magnetic flux, a stored charge, or phosphorescence.

decay area [OCEANOGR] The area into which ocean waves travel (as swell) after leaving the generating area.

decay distance [OCEANOGR] The distance through which ocean waves pass after leaving the generating area.

decay of waves [OCEANOGR] The decrease in height and increase in length of waves after they leave a generating area and pass through a calm, or region of lighter winds; accompanied by a loss of energy.

Deccan basalt [GEOL] Fine-grained, nonporphyritic, tholeiitic basaltic lava consisting essentially of labradorite, clinopyroxene, and iron ore; found in the Deccan region of southeastern India. Also known as Deccan trap.

Deccan trap *See* Deccan basalt.

decementation [GEOL] The removal of cement from a sedimentary rock by dissolving or leaching.

deck charge [MIN ENG] A charge that is separated into several smaller components and placed along a quarry borehole.

decking [MIN ENG] Changing tubs on a cage at both ends of a shaft.

deck loading [MIN ENG] The method of loading deck charges in a quarry borehole.

declination [GEOPHYS] The angle between the magnetic and geographical meridians, expressed in degrees and minutes east or west to indicate the direction of magnetic north from true north. Also known as magnetic declination; variation.

declination compass *See* declinometer.

declination variometer [ENG] An instrument that measures changes in the declination of the earth's magnetic field, consisting of a permanent bar magnet, usually about 1 centimeter long, suspended with a plane mirror from a fine quartz fiber 5–15 centimeters in length; a lens focuses to a point a beam of light reflected from the mirror to recording paper mounted on a rotating drum. Also known as D variometer.

declinometer [ENG] A magnetic instrument similar to a surveyor's compass, but arranged so that the line of sight can be rotated to conform with the needle or to any desired setting on the horizontal circle; used in determining magnetic declination. Also known as declination compass.

declivity [GEOL] **1.** A slope descending downward from a point of reference. **2.** A downward deviation from the horizontal.

décollement [GEOL] Folding or faulting of sedimentary beds by sliding over the underlying rock.

decomposition [GEOCHEM] *See* chemical weathering.

decrement [HYD] *See* groundwater discharge.

decrepitation [GEOPHYS] Breaking up of mineral substances when exposed to heat; usually accompanied by a crackling noise.

decussate structure [GEOL] A crisscross microstructure of certain minerals; most noticeable in rocks composed predominantly of minerals with a columnar habit.

dedolomitization [GEOL] Destruction of dolomite to form calcite and periclase, usually by contact metamorphism at low pressures.

dedusting [MIN ENG] Cleaning ore, using pneumatic means and screening, to remove dust and other fine impurities. Also known as aspirating.

deep [OCEANOGR] An area of great depth in the ocean, representing a depression in the ocean floor.

deep-casting [OCEANOGR] Sampling ocean water at great depths by lowering a number of self-sealing bottles, usually made of brass or bronze, on a cable.

deep coal [MIN ENG] Coal that is far enough beneath the land surface to necessitate extraction using underground mining techniques.

deep easterlies *See* equatorial easterlies.

deepening [METEOROL] A decrease in the central pressure of a pressure system on a constant-height chart, or an analogous decrease in height on a constant-pressure chart.

deep inland sea [GEOGR] A sea adjacent to but in restricted communication with the sea; depth exceeds 200 meters.

deep scattering layer [OCEANOGR] The stratified populations of organisms which scatter sound in most oceanic waters.

deep-sea basin [GEOL] A depression of the sea floor more or less equidimensional in form and of variable extent.

deep-sea channel [GEOL] A trough-shaped valley of low relief beyond the continental rise on the deep-sea floor. Also known as mid-ocean canyon.

deep-sea plain [GEOL] A broad, almost level area forming the predominant portion of the ocean floor.

deep-seated *See* plutonic.

deep-sea trench [GEOL] A long, narrow depression of the deep-sea floor having steep sides and containing the greatest ocean depths; formed by depression, to several kilometers' depth, of the high-velocity crustal layer and the mantle.

deep trades *See* equatorial easterlies.

deep water [OCEANOGR] An ocean area where depth of the water layer is greater than one-half the wave length.

deep-water wave [OCEANOGR] A surface wave whose length is less than twice the depth of the water.

Deerparkian [GEOL] A North American stage of geologic time in the Lower Devonian, above Helderbergian and below Onesquethawan.

deferment factor *See* discount factor.

defile [GEOL] A very long, narrow, steep-sided pass running through hills or mountains which often constitutes the entrance to a larger pass.

deflation [GEOL] The sweeping erosive action of the wind over the ground.

deflation basin [GEOL] A topographic depression formed by deflation.

deflation lake [HYD] A lake, usually occurring in an arid or semiarid region, which occupies a basin formed mainly by wind erosion.

deflection angle [GEOD] The angle at a point on the earth between the direction of a plumb line (the vertical) and the perpendicular (the normal) to the reference spheroid; this difference seldom exceeds 30 seconds of arc.

deformation fabric [GEOL] The space orientation of rock elements produced by external stress on the rock.

deformation lamella [GEOL] A type of slipband in the crystalline grains of a material (particularly quartz) produced by intracrystalline slip during tectonic deformation.

deglaciation [HYD] Exposure of an area from beneath a glacier or ice sheet as a result of shrinkage of the ice by melting.

degradation [GEOL] The wearing down of the land surface by processes of erosion and weathering. [HYD] **1.** Lowering of a steam bed. **2.** Shrinkage or disappearance of permafrost.

degraded illite [MINERAL] Illite with a depleted potassium content because of prolonged leaching. Also known as stripped illite.

degrading stream [HYD] A stream that is actively deepening its channel or valley and is able to carry more material than is being supplied.

degrees of frost [METEOROL] In England, the number of degrees Fahrenheit that the temperature falls below the freezing point; thus a day with a minimum temperature of 27°F may be designated as a day of five degrees of frost.

dehrnite [MINERAL] $(Ca,Na,K)_5(PO_4)_3(OH)$ A colorless to pale green, greenish-white, or gray, hexagonal mineral consisting of a basic phosphate of calcium, sodium, and potassium; occurs as botryoidal crusts and minute hexagonal prisms.

Deinotheriidae [PALEON] A family of extinct proboscidean mammals in the suborder Deinotherioidea; known only by the genus *Deinotherium*.

Deinotherioidea [PALEON] A monofamilial suborder of extinct mammals in the order Proboscidea.

Deister phase [GEOL] A subdivision of the late Ammerian phase of the Jurassic period between the Kimmeridgian and lower Portlandian.

delafossite [MINERAL] $CuFeO_2$ A mineral consisting of an oxide of copper and iron.

delay-action detonator See delay blasting cap.

delay blasting cap [ENG] A blasting cap which explodes at a definite time interval after the firing current has been passed by the exploder. Also known as delay-action detonator.

delayed development well See step-out well.

dell [GEOGR] A small, secluded valley or vale.

dellenite See rhyodacite.

Delmontian [GEOL] Upper Miocene or lower Pliocene geologic time.

delorenzite See tanteuxenite.

delta bar [GEOL] A bar formed by a tributary stream that is building a delta into the channel of the main stream.

delta geosyncline See exogeosyncline.

deltaic deposits [GEOL] Sedimentary deposits in a delta.

deltaite [MINERAL] A mixture of crandallite and hydroxylapatite.

delta kame [GEOL] A plateaulike, steep-sided hill of well-separated sand and gravel built up by a meltwater stream that flows into a proglacial lake.

delta moraine See ice-contact delta.

delta plain [GEOL] A plain formed by deposition of silt at the mouth of a stream or by overflow along the lower stream courses.

delta region [METEOROL] A region in the atmosphere characterized by difluence.

delta-T [GEOPHYS] With regard to seismic reflection and refraction stepout times, the observed or interpolated difference between two time values.

delta terrace [GEOL] A fan-shaped terrace remaining from the delta of a former stream.

Deltatheridia [PALEON] An order of mammals that includes the dominant carnivores of the early Cenozoic.

deltohedron [CRYSTAL] A polyhedron which has 12 quadrilateral faces, and is the form of a crystal belonging to the cubic system and having hemihedral symmetry. Also known as deltoid dodecahedron; tetragonal tristetrahedron.

deltoid dodecahedron See deltohedron.

delvauxite [MINERAL] A mineral, with the approximate formula $Fe_4(PO_4)_2(OH)_6 \cdot nH_2O$, consisting of a hydrous phosphate of iron.

demagnetization [MIN ENG] Deflocculation in dense-media process using ferrosilicon by passing the fluid through an alternating-current field.

demantoid [MINERAL] A lustrous, green variety of andradite; used as a gem.

demorphism See weathering.

dendrite [CRYSTAL] A crystal having a treelike structure.

dendritic drainage [HYD] Irregular stream branching, with tributaries joining the main stream at all angles.

dendritic glacier [HYD] A main glacier that has been joined by many tributary glaciers to form a pattern resembling a branching tree.

dendritic valleys [GEOL] Treelike extensions of the valleys in a region lying upon horizontally bedded rock.

dendrochronology [GEOL] The science of measuring time intervals and dating events and environmental changes by reading and dating growth layers of trees as demarcated by the annual rings.

dendroclimatology [GEOL] Reconstruction of past climates by comparative studies of annual variations in width of growth rings of trees and wood specimens.

dendrohydrology [HYD] The science of determining hydrologic occurrences by the comparison of tree ring thickness with streamflow or precipitation.

Dendroidea [PALEON] An order of extinct sessile, branched colonial animals in the class Graptolithina occurring among typical benthonic fauna.

denivellation [HYD] A change in water level, especially in a lake.

dense-media separator [MIN ENG] A device in which a heavy mineral is dispersed in water, causing heavier ores to sink and lighter ores to float.

density altitude [METEOROL] The altitude, in the standard atmosphere, at which a given density occurs.

density channel [METEOROL] A channel used to investigate a density current; for example, in experiments relating to the behavior of cold masses of air in the atmosphere and related frontal structures.

density correction [ENG] **1.** The part of the temperature correction of a mercury barometer which is necessitated by the variation of the density of mercury with temperature. **2.** The correction, applied to the indications of a pressure-tube anemometer or pressure-plate anemometer, which is necessitated by the variation of air density with temperature.

density current [METEOROL] Intrusion of a dense air mass beneath a lighter air mass; the usage applies to cold fronts. [OCEANOGR] *See* turbidity current.

density log [PETRO ENG] Radioactivity logging of reservoir structure densities down an oil-well bore by emission and detection of gamma rays.

density ratio [METEOROL] The ratio of the density of the air at a given altitude to the air density at the same altitude in a standard atmosphere.

denudation [GEOL] General wearing away of the land; laying bare of subjacent lands.

Denver cell [MIN ENG] A subaeration type of flotation cell, mechanized and self-aerating.

Denver jig [MIN ENG] A pulsion-suction diaphragm for separating sulfur from coal before flotation; hydraulic water is admitted through a rotary valve.

deoxidation sphere *See* bleach spot.

departure [METEOROL] The amount by which the value of a meteorological element differs from the normal value.

depegram [METEOROL] On a diagram having entropy and temperature as coordinates, a curve representing the distribution of the dew point as a function of pressure for a given sounding of the atmosphere.

dependable yield [HYD] The minimum amount of water for a given region which is available on demand and which may decrease an average of once every n years.

depéq [METEOROL] Strong winds over Loet Tawar (Sumatra, East Indies) during the southwest monsoon.

depergelation [HYD] The act or process of thawing permafrost.

Depertellidae [PALEON] A family of extinct perissodactyl mammals in the superfamily Tapiroidea.

depletion drive [PETRO ENG] Displacement mechanism (type of drive) to expel hydrocarbons from porous reservoir formations, that is, to remove more hydrocarbon from the reservoir; types of drives are gas or water (natural or injected) and injected LPG.

depletion-type reservoir [PETRO ENG] Oil reservoir which is initially in (and during depletion remains in) a state of equilibrium between the gas and liquid phases; includes single-phase gas, two-phase bubble-point, and retrograde-gas-condensate (or dewpoint) reservoirs.

depocenter [GEOL] A site of maximum deposition.

deposit [GEOL] Consolidated or unconsolidated material that has accumulated by a natural process or agent.

deposited snow [HYD] All snow lying on the ground or on other snow, firn, or ice, and subject to change by wind action.

depositional dip *See* primary dip.

depositional fabric [PETR] Arrangement of detrital particles settled from suspension or of crystals from a differentiating magma determined by the plane of the surface on which they come to rest.

depositional remanent magnetization [GEOPHYS] Remanent magnetization occurring in sedimentary rock following the depositional alignment of previously magnetized grains. Abbreviated DRM.

depositional sequence [GEOL] A major but informal assemblage of formations or groups and supergroups, bounded by regionally extensive unconformities at both their base and top and extending over broad areas of continental cratons.

depositional strike [GEOL] Sedimentary deposits that are continuous laterally on a gently sloping surface.

depression [GEOL] **1.** A hollow of any size on a plain surface having no natural outlet for surface drainage. **2.** A structurally low area in the crust of the earth. [METEOROL] An area of low pressure; usually applied to a certain stage in the development of a tropical cyclone, to migratory lows and troughs, and to upper-level lows and troughs that are only weakly developed. Also known as low.

depression angle *See* angle of depression.

depression spring [HYD] A type of gravity spring that flows onto the land surface because the surface slopes down to the water table.

depression storage [HYD] Water retained in puddles, ditches, and other depressions in the surface of the ground.

depth [OCEANOGR] The vertical distance from a specified sea level to the sea floor.

depth contour *See* isobath.

depth curve *See* isobath.

depth-duration-area value [METEOROL] The average depth of precipitation that has occurred within a specified time interval over an area of given size. Abbreviated DDA value.

depth hoar [HYD]   A layer of ice crystals formed between the ground and snow cover by sublimation. Also known as sugar snow.

depth ice [OCEANOGR]   Small ice particles formed below the surface of a sea agitated by wave activity.

depth marker [ENG]   A thin board or other lightweight substance used as a means of identifying the surface of snow or ice which has been covered by a more recent snowfall.

depth of compensation [GEOPHYS]   That depth at which density differences occurring in the earth's crust are compensated isostatically; calculated to be between 100 and 113–117 kilometers.   [HYD]   The depth in a body of water at which illuminance has diminished to the extent that oxygen production through photosynthesis and oxygen consumption through respiration by plants are equal; it is the lower boundary of the euphotic zone.

depth point [ENG]   In seismology, a position at which the depth of a mapped horizon has been determined.

depth sounder [ENG]   An instrument for mechanically measuring the depth of the sea beneath a ship.

depth zone of earth [GEOL]   A zone within the earth giving rise to different metamorphic assemblages.

deranged drainage pattern [HYD]   A distinctively disordered drainage pattern in a recently glaciated area where the former surface and preglacial drainage has been remodeled and effaced.

derbylite [MINERAL] $Fe_6Ti_6Sb_2O_{23}$ A black or brown orthorhombic mineral occurring in cinnabar-bearing gravels.

Derbyshire spar *See* fluorite.

derivative rock *See* sedimentary rock.

derived gust velocity [METEOROL]   The maximum velocity of a sharp-edged gust that would produce a given acceleration on a particular airplane flown in level flight at the design cruising speed of the aircraft at a given air density.

derrick barge [PETRO ENG]   A crane barge used in offshore drilling platform construction and suitable for work in rough seas.

descendant [GEOL]   A topographic feature cut from the mass underlying an older topographic structure now removed.

descending vertical angle *See* angle of depression.

descriptive climatology [CLIMATOL]   Climatology as presented by graphic and verbal description, without going into causes and theory.

descriptive meteorology [METEOROL]   A branch of meteorology which deals with the description of the atmosphere as a whole and its various phenomena, without going into theory. Also known as aerography.

desert [GEOGR]   **1.** A wide, open, comparatively barren tract of land with few forms of life and little rainfall. **2.** Any waste, uninhabited tract, such as the vast expanse of ice in Greenland.

desert armor [GEOL]   A desert pavement surfaced with stony fragments that protect underlying finer-grained material from wind erosion.

desert climate [CLIMATOL]   A climate type which is characterized by insufficient moisture to support appreciable plant life; that is, a climate of extreme aridity.

desert crust *See* desert pavement.

desert devil *See* dust whirl.

desert dome [GEOL] A convex rock surface with very regular and smooth slopes, resulting from desert erosion of a mountain mass.

desert mosaic [GEOL] A desert pavement made up of tightly interlocking and evenly set fragments resembling a mosaic.

desert pavement [GEOL] A mosaic of pebbles and large stones which accumulate as the finer dust and sand particles are blown away by the wind. Also known as desert crust.

desert peneplain *See* pediplain.

desert plain *See* pediplain.

desert polish [GEOL] A smooth, shining surface imparted to rocks and other hard substances by the action of windblown sand and dust of desert regions.

desert ripple [GEOL] One of a system of slightly arcuate ridges produced by the wind that are en echelon about 15 m apart with vegetated crests and caliche-lined troughs.

desert soil [GEOL] In early United States classification systems, a group of zonal soils that have a light-colored surface soil underlain by calcareous material and a hardpan.

desert varnish [GEOL] A brown or black stain or crust of manganese or iron oxide characterizing many exposed rock surfaces in the desert.

desert wind [METEOROL] A wind blowing off the desert, which is very dry and usually dusty, hot in summer but cold in winter, and with a large diurnal range of temperature.

desiccation [HYD] The permanent decrease or disappearance of water from a region, caused by a decrease of rainfall, a failure to maintain irrigation, or deforestation or overcropping. [SCI TECH] Thorough removal of water from a substance, often with the use of a desiccant.

desiccation breccia [GEOL] Fragments of a mud-cracked layer of sediment deposited with other sediments.

desiccation crack *See* mud crack.

desiccation polygon [GEOL] A small polygon, usually with three, four, or five sides, that has formed as the result of drying of moist sediment with a fine-grained, clay texture.

desilication [GEOCHEM] Removal of silica, as from rock or a magma.

desmine *See* stilbite.

Desmodonta [PALEON] An order of extinct bivalve, burrowing mollusks.

Des Moinesian [GEOL] Lower Middle Pennsylvanian geologic time.

Desmostylia [PALEON] An extinct order of large hippopotamuslike, amphibious, gravigrade, shellfish-eating mammals.

Desmostylidae [PALEON] A family of extinct mammals in the order Desmostylia.

destressing [MIN ENG] Relieving stress on the abutments of an excavation by drilling and blasting to loosen peak stress zones.

detached core [GEOL] The inner bed or beds of a fold that may become separated or pinched off from the main body of the strata due to extreme folding and compression.

detail log [ENG] An electric log plotted on a magnified scale so that minor flucuations in the formations penetrated by the borehole can be easily observed.

det drill *See* fusion-piercing drill.

detonating fuse [ENG]   A device consisting of a core of high explosive within a waterproof textile covering and set off by an electrical blasting cap fired from a distance by means of a fuse line; used in large, deep boreholes.

detonating relay [ENG]   A device used in conjunction with the detonating fuse to avoid short-delay blasting.

detonator [ENG]   A device, such as a blasting cap, employing a sensitive primary explosive to detonate a high-explosive charge.

detonator safety [ENG]   A fuse has detonator safety or is detonator safe when the functioning of the detonator cannot initiate subsequent explosive train components.

detonics [ENG]   The study of detonating and explosives performance.

detrainment [METEOROL]   The transfer of air from an organized air current to the surrounding atmosphere.

detrital minerals [MINERAL]   Grains of heavy minerals found in sediment, resulting from mechanical disintegration of the parent rock.

detrital ratio *See* clastic ratio.

detrital reservoir [GEOL]   A clastic or detrital-granular reservoir, classified by rock type and other factors such as sediments (quartzose-type, graywacke, or arkose sediments).

detrital sediment [GEOL]   Accumulations of the organic and inorganic fragmental products of the weathering and erosion of land transported to the place of deposition.

detritus [GEOL]   Any loose material removed directly from rocks and minerals by mechanical means, such as disintegration or abrasion.

deuteric [GEOL]   Of or pertaining to alterations in igneous rock during the later stages and as a direct result of consolidation of magma or lava. Also known as epimagmatic; paulopost.

DeVecchis process [MIN ENG]   A smelting process for pyrites in which the raw material is roasted, concentrated magnetically, and then reduced in a rotary kiln or electric furnace.

developed ore *See* developed reserves.

developed reserves [MIN ENG]   Ore that is exposed on three sides and for which tonnage yield and quality estimates have been made. Also known as assured mineral; blocked-out ore; developed ore; measured ore; ore in sight.

development [GEOL]   The progression of changes in fossil groups which have succeeded one another during deposition of the strata of the earth.   [METEOROL]   The process of intensification of an atmospheric disturbance, most commonly applied to cyclones and anticyclones.

development drift [MIN ENG]   A tunnel dug in a mine either from the surface or a point underground to get to coal or ore for exploitation or mining purposes.

development drilling [MIN ENG]   Drilling boreholes to locate, identify, and prove an ore body or coal seam.

development index [METEOROL]   An index used as an aid in forecasting cyclogenesis; the development index I is defined most frequently as the difference in divergence between two well-separated, tropospheric, constant-pressure surfaces. Also known as relative divergence.

development rock [MIN ENG]   Rock containing both barren and valuable rock, broken during development work.

development well [PETRO ENG] A well drilled to produce oil or gas from a proven productive area.

Devereaux agitator [MIN ENG] An agitator that utilizes an upthrust propeller to stir pulp; used in leach agitation of minerals.

deviation hole [PETRO ENG] Drilled hole with deviation from true vertical, usually limited by contract to 3–5°; not to be confused with a crooked hole, resulting from carelessness or a steeply dipping formation.

devillite [MINERAL] $Cu_4Ca(SO_4)_2(OH)_6 \cdot 3H_2O$ A dark-green mineral consisting of a hydrous basic sulfate of copper and calcium, occurring in six-sided platy crystals.

Devonian [GEOL] The fourth period of the Paleozoic Era.

dew [HYD] Water condensed onto grass and other objects near the ground, the temperatures of which have fallen below the dew point of the surface air because of radiational cooling during the night but are still above freezing.

dewatering [ENG] Removing or draining water from an enclosure or a structure, such as a riverbed, caisson, or mine shaft, by pumping or evaporation.

deweylite [MINERAL] A mixture of clinochrysolite and stevensite. Also known as gymnite.

dewindtite [MINERAL] $Pb(UO_2)_2(PO_4)_2 \cdot 3H_2O$ A canary-yellow secondary mineral consisting of a hydrous phosphate of lead and uranium.

de Witte relation [GEOPHYS] Graphical plot of the relation between electrical conductivity and distance over which the conductivity is measured through reservoir rock with clay minerals (the effect is similar to two parallel electrical circuits), the current passing through the conducting clay minerals and the water-filled pores.

dew point [CHEM] The temperature at which water vapor begins to condense. [METEOROL] The temperature at which air becomes saturated when cooled without addition of moisture or change of pressure; any further cooling causes condensation. Also known as dew-point temperature.

dew-point formula [METEOROL] A formula for the calculation of the approximate height of the lifting condensation level; employed to estimate the height of the base of convective clouds, under suitable atmospheric and topographic conditions.

dew-point reservoir [PETRO ENG] A hydrocarbon reservoir in which the temperature lies between the critical temperature and the cricondentherm (maximum temperature and pressure at which two phases can coexist) and in the one-phase region. Also known as retrograde gas-condensate reservoir.

dew-point spread [METEOROL] The difference in degrees between the air temperature and the dew point.

dew-point temperature *See* dew point.

dextral drag fold [GEOL] A drag fold in which the trace of a given surface bed is displaced to the right.

dextral fault [GEOL] A strike-slip fault in which an observer approaching the fault sees the opposite block as having moved to the right. Also known as right-lateral fault; right-lateral slip fault; right-slip fault.

dextral fold [GEOL] An asymmetric fold in which the long limb appears to be offset to the right to an observer looking along the long limb.

D horizon [GEOL] A soil horizon sometimes occurring below a B or C horizon, consisting of unweathered rock.

DI *See* temperature-humidity index.

diabantite [MINERAL] $(Mg,Fe^{2+},Al)_6(Si,Al)_4O_{10}(OH)_8$ Mineral of the chlorite group consisting of a basic silicate of magnesium, iron, and aluminum, occurring in cavities in basic igneous rock.

diabase [PETR] An intrusive rock consisting principally of labradorite and pyroxene.

diabase amphibolite [PETR] Amphibolite formed by dynamic metamorphism of diabase.

diabasic [PETR] Denoting igneous rock in which the interstices between the feldspar crystals are filled with discrete crystals or grains of pyroxene.

diaboleite [MINERAL] $Pb_2CuCl_2(OH)_4$ A sky-blue mineral consisting of a basic chloride of lead and copper.

diachronous [GEOL] Of a rock unit, varying in age in different areas or cutting across time planes or biostratigraphic zones. Also known as time-transgressive.

diaclinal [GEOL] Pertaining to a stream crossing a fold, perpendicular to the strike of the underlying strata it traverses.

Diacodectidae [PALEON] A family of extinct artiodactyl mammals in the suborder Palaeodonta.

diadochite [MINERAL] $Fe_2(PO_4)(SO_4)(OH)\cdot5H_2O$ A brown or yellowish mineral consisting of a basic hydrous ferric phosphate and sulfate.

diadochy [CRYSTAL] Replacement or ability to be replaced of one atom or ion by another in a crystal lattice.

diagenesis [GEOL] Chemical and physical changes occurring in sediments during and after their deposition but before consolidation.

diagnostic equation [METEOROL] Any equation governing a system which contains no time derivative and therefore specifies a balance of quantities in space at a moment of time; examples are a hydrostatic equation or a balance equation.

diagonal fault [GEOL] A fault whose strike is diagonal or oblique to the strike of the adjacent strata. Also known as oblique fault.

diagonal joint [GEOL] A joint having its strike oblique to the strike of the strata of the sedimentary rock, or to the cleavage plane of the metamorphic rock in which it occurs. Also known as oblique joint.

diallage [MINERAL] A green, brown, gray, or bronze-colored clinopyroxene characterized by prominent parting parallel to the front pinacoid *a* (100).

diamantine [MINERAL] Consisting of or resembling diamond.

diamictite [PETR] A calcareous, terrigenous sedimentary rock that is not sorted or poorly sorted and contains particles of many sizes. Also known as mixtite.

diamicton [PETR] A nonlithified diamictite. Also known as symmicton.

diamond [MINERAL] A colorless mineral composed entirely of carbon crystallized in the isometric system as octahedrons, dodecahedrons, and cubes; the hardest substance known; used as a gem and in cutting tools.

diamond coring [ENG] Obtaining core samples of rock by using a diamond drill.

diamond drill [DES ENG] A drilling machine with a hollow, diamond-set bit for boring rock and yielding continuous and columnar rock samples.

Diamond-Hinman radiosonde [ENG] A variable audio-modulated radiosonde used by United States weather services; the carrier signal from the radiosonde is modulated by audio signals determined by the electrical resistance of the humidity- and temperature-transducing elements and by fixed reference resistors; the modulating signals are transmitted in a fixed sequence at predetermined pressure levels by means of a baroswitch.

diamond structure [CRYSTAL] A crystal structure in which each atom is the center of a tetrahedron formed by its nearest neighbors.

diamond washer [MIN ENG] An apparatus for shaking and separating rock gravel containing diamonds, utilizing a vertical series of screens with 8-, 4-, 2-, and 1-millimeter mesh.

dianite *See* columbite.

Dianulitidae [PALEON] A family of extinct, marine bryozoans in the order Cystoporata.

diaphorite [MINERAL] $PB_2Ag_3Sb_3S_8$ A gray-black orthorhombic mineral consisting of sulfide of lead, silver, and antimony, occurring in crystals. Also known as ultrabasite.

diaphragm jig [MIN ENG] A jig having a flexible diaphragm to pulse water; used in gravity concentration of minerals.

diaphthoresis *See* retrograde metamorphism.

diaphthorite [PETR] Schistose rocks in which minerals have formed by retrograde metamorphism.

diapir [GEOL] A dome or anticlinal fold in which a mobile plastic core has ruptured the more brittle overlying rock. Also known as diapiric fold; piercement; piercement dome; piercing fold.

diapiric fold *See* diapir.

diaspore [MINERAL] AlO(OH) A mineral composed of some bauxites occurring in white, lamellar masses; crystallizes in the orthorhombic system.

diastem [GEOL] A temporal break between adjacent geologic strata that represents nondeposition or local erosion but not a change in the general regimen of deposition.

diastrophism [GEOL] **1.** The general process or combination of processes by which the earth's crust is deformed. **2.** The results of this deforming action.

diatomaceous earth [GEOL] A yellow, white, or light-gray, siliceous, porous deposit made of the opaline shells of diatoms; used as a filter aid, paint filler, adsorbent, abrasive, and thermal insulator. Also known as kieselguhr; tripolite.

diatomaceous ooze [GEOL] A pelagic, siliceous sediment composed of more than 30% diatom tests, up to 40% calcium carbonate, and up to 25% mineral grains.

diatomite [GEOL] Dense, chert-like, consolidated diatomaceous earth.

diatreme [GEOL] A circular volcanic vent produced by the explosive energy of gas-charged magmas.

Diatrymiformes [PALEON] An order of extinct large, flightless birds having massive legs, tiny wings, and large heads and beaks.

diborate *See* borax.

Dichobunidae [PALEON] A family of extinct artiodactyl mammals in the superfamily Dichobunoidea.

Dichobunoidea [PALEON] A superfamily of extinct artiodactyl mammals in the suborder Paleodonta composed of small- to medium-size forms with tri- to quadritubercular bunodont upper teeth.

Dickinsoniidae [PALEON]   A family that comprises extinct flat-bodied, mutisegmented coelomates; identified as ediacaran fauna.

dickinsonite [MINERAL] $H_2Na_6(Mn,Fe,Ca,Mg)_{14}(PO_4)_{12} \cdot H_2O$ A green mineral consisting of foliated hydrous acid phosphate, chiefly of manganese, iron, and sodium, and is isostructural with arrojadite; specific gravity is 3.34.

dickite [MINERAL] $Al_2Si_2O_5(OH)_4$ A mineral of the kaolin group found crystallized in clay in hydrothermal veins; it is polymorphous with kaolinite and nacrite.

Dictyonellidina [PALEON]   A suborder of extinct articulate brachiopods.

dictyonema bed [GEOL]   A thin shale bed rich in remains of graptolites of the genus *Dictyonema*.

Dictyospongiidae [PALEON]   A family of extinct sponges in the subclass Amphidiscophora having spicules resembling a one-ended amphidisc (paraclavule).

Didolodontidae [PALEON]   A family of extinct medium-sized herbivores in the order Condylarthra.

didymolite [MINERAL] $Ca_2Al_6Si_9O_{29}$ A dark-gray monoclinic mineral consisting of a calcium aluminum silicate, occurring in twinned crystals.

diesel squeeze [PETRO ENG]   A technique of forcing dry cement mixed with diesel oil through casing openings to repair water-bearing areas without affecting the oil-bearing areas.

dietrichite [MINERAL] $(Zn,Fe,Mn)Al_2(SO_4)_4 \cdot 22H_2O$ Mineral consisting of a hydrous sulfate of aluminum and one or more of the metals zinc, iron, and manganese.

dietzeite [MINERAL] $Ca_2(IO_3)_2(CrO_4)$ A dark-golden-yellow iodate mineral commonly in fibrous or columnar form as a component of caliche.

difference of latitude [GEOD]   The shorter arc of any meridian between the parallels of two places, expressed in angular measure.

difference of longitude [GEOD]   The smaller angle at the pole or the shorter arc of a parallel between the meridians of two places, expressed in angular measure.

difference of meridional parts *See* meridional difference.

differential analysis [METEOROL]   Synoptic analysis of change charts or of vertical differential charts (such as thickness charts) obtained by the graphical or numerical subtraction of the patterns of some meteorological variable at two times or two levels.

differential chart [METEOROL]   A chart showing the amount and direction of change of a meteorological quantity in time or space.

differential compaction [GEOL]   Compression in sediments, such as sand or limestone, as the weight of overburden causes reduction in pore space and forcing out of water.

differential entrapment [PETRO ENG]   The control of oil and gas migration and accumulation by selective trapping or gas flushing in interconnecting reservoirs.

differential erosion [GEOL]   Rapid erosion of one area of the earth's surface relative to another.

differential fault *See* scissor fault.

differential flotation [MIN ENG]   Separation of a complex ore into two or more mineral components and gangue by flotation.

differential infrared line-scan [ENG]   A remote-sensing system in which quantitative imagery is formed, that is, the gray scale of the images produced is not directly related to the incident flux striking the detector by a known energy-transfer function; temperature-calibration points on the terrain are required to establish a crude relationship between gray-scale increments.

differential leveling [ENG] A surveying process in which a horizontal line of sight of known elevation is intercepted by a graduated standard, or rod, held vertically on the point being checked.

differential melting [PETR] The partial melting of rock material caused by the different melting points of the various mineral components.

differential temperature survey [PETRO ENG] Well-temperature logging method that detects very small temperature anomalies; two thermometers, 6 feet (1.8 meters) apart, record the temperature gradient down the well bore, with small difference changes showing up anomalies.

differential weathering [GEOL] Irregular or uneven weathering caused by variations in rock composition or differences in intensity of weathering processes.

diffraction symmetry [CRYSTAL] Any symmetry in a crystal lattice which causes the systematic annihilation of certain beams in x-ray diffraction.

diffractometry [CRYSTAL] The science of determining crystal structures by studying the diffraction of beams of x-rays or other waves.

diffuse aurora [GEOPHYS] A widespread and relatively uniform type of aurora which is easily overlooked from the ground but is prominent in satellite pictures.

diffuse front [METEOROL] A front across which the characteristics of wind shift and temperature change are weakly defined.

diffusion [METEOROL] The exchange of fluid parcels (and hence the transport of conservative properties) between regions in space, in the apparently random motions of the parcels on a scale too small to be treated by the equations of motion; the diffusion of momentum (viscosity), vorticity, water vapor, heat (conduction), and gaseous components of the atmospheric mixture have been studied extensively.

diffusion diagram [METEOROL] A diagram for displaying the comparative properties of various diffusion processes, with coordinates of the mean free path or mixing length and mean molecular speed or diffusion velocity, for molecular or eddy diffusion, respectively; each point of the diagram determines diffusivity.

diffusion hygrometer [ENG] A hygrometer based upon the diffusion of water vapor through a porous membrane; essentially, it consists of a closed chamber having porous walls and containing a hygroscopic compound, whose absorption of water vapor causes a pressure drop within the chamber that is measured by a manometer.

diffusive equilibrium [METEOROL] The steady state resulting from the diffusion process, primarily of interest when external forces and sources and sinks exist within the field; in such a state the constituent gases of the atmosphere would be distributed independently of each other, the heavier decreasing more rapidly with height than the lighter; but the presence of turbulent mixing precludes establishment of complete diffusive equilibrium.

digger [MIN ENG] A person who digs in the ground; usually refers to a coal miner.

digging height *See* bank height.

digital log [ENG] A well log that has been sampled at various intervals and recorded on a magnetic tape for use in computerized plotting and interpretation.

dihexagonal [CRYSTAL] Of crystals, having a symmetrical form with 12 sides.

dihexagonal-dipyramidal [CRYSTAL] Characterized by the class of crystals in the hexagonal system in which any section perpendicular to the sixfold axis is dihexagonal.

dihexahedron [CRYSTAL] A type of crystal that has 12 faces, such as a double six-sided pyramid.

dike [GEOL] A tabular body of igneous rock that cuts across adjacent rocks or cuts massive rocks.

dike ridge [GEOL] Any small wall-like ridge created by differential erosion.

dike set [GEOL] A small group of dikes arranged linearly or parallel to each other.

dike swarm [GEOL] A large group of parallel, linear, or radially oriented dikes.

dilatancy [GEOL] Expansion of deformed masses of granular material, such as sand, due to rearrangement of the component grains.

dimensional orientation [PETR] In structural petrology, a tendency of elongate fabric elements to be oriented so that their longer axes are almost parallel.

dimictic lake [HYD] A lake which circulates twice a year.

dimmerfoehn [METEOROL] A rare form of foehn where, during a very strong upper wind from the south, a pressure difference of 12 millibars or more exists between the south and north sides of the Alps; a stormy foehn wind then overleaps the upper valleys in the northern slopes, reaches the ground in the lower parts of the valleys, and enters the foreground as a very strong wind; the foehn wall and the precipitation area extend beyond the crest across the almost calm surface area in the upper valleys.

dimorphite [MINERAL] As_4S_3 An orange-yellow mineral consisting of arsenic sulfide.

Dimylidae [PALEON] A family of extinct lipotyphlan mammals in the order Insectivora; a side branch in the ancestry of the hedgehogs.

Dinantian [GEOL] Lower Carboniferous geologic time. Also known as Avonian.

Dinarides [GEOGR] A mountain system, east of the Adriatic Sea, in Yugoslavia.

Dines anemometer [ENG] A pressure-tube anemometer in which the pressure head on a weather vane is kept facing into the wind, and the suction head, near the bearing which supports the vane, develops a suction independent of wind direction; the pressure difference between the heads is proportional to the square of the wind speed and is measured by a float manometer with a linear wind scale.

Dinocerata [PALEON] An extinct order of large, herbivorous mammals having semigraviportal limbs and hoofed, five-toed feet; often called uintatheres.

Dinornithiformes [PALEON] The moas, an order of extinct birds of New Zealand; all had strong legs with four-toed feet.

dinosaur [PALEON] The name, meaning terrible lizard, applied to the fossil bones of certain large, ancient bipedal and quadripedal reptiles placed in the orders Saurischia and Ornithischia.

dioctahedral [CRYSTAL] Pertaining to a crystal structure in which only two of the three available octahedrally coordinated positions are occupied.

diogenite [MINERAL] An achondritic stony meteorite composed essentially of iron-rich pyroxene minerals. Also known as rodite.

diopside [MINERAL] $CaMg(SiO_3)_2$ A white to green monoclinic pyroxene mineral which forms gray to white, short, stubby, prismatic, often equidimensional crystals. Also known as malacolite.

dioptase [MINERAL] $CuSiO_2(OH)_2$ A rare emerald-green mineral that forms hexagonal, hydrous crystals.

diorite [PETR]   A phaneritic plutonic rock with granular texture composed largely of plagioclase feldspar with smaller amounts of dark-colored minerals; used occasionally as ornamental and building stone. Also known as black granite.

dip [ENG]   The vertical angle between the sensible horizon and a line to the visible horizon at sea, due to the elevation of the observer and to the convexity of the earth's surface. Also known as dip of horizon.   [GEOL]   **1.** The angle that a stratum or fault plane makes with the horizontal. Also known as angle of dip; true dip. **2.** A pronounced depression in the land surface.

dip fault [GEOL]   A type of fault that strikes parallel with the dip of the strata involved.

diphead [MIN ENG]   A passage that follows the inclination of a coal seam.

dip inductor *See* earth inductor.

dip joint [GEOL]   A joint that strikes approximately at right angles to the cleavage or bedding of the constituent rock.

Diplacanthidae [PALEON]   A family of extinct acanthodian fishes in the suborder Diplacanthoidei.

Diplacanthoidei [PALEON]   A suborder of extinct acanthodian fishes in the order Climatiiformes.

Diplobathrida [PALEON]   An order of extinct, camerate crinoids having two circles of plates beneath the radials.

dip log [GEOL]   A log of the dips of formations traversed by boreholes.

diploid [CRYSTAL]   A crystal form in the isometric system having 24 similar quadrilateral faces arranged in pairs.

Diploporita [PALEON]   An extinct order of echinoderms in the class Cystoidea in which the thecal canals were associated in pairs.

dipmeter [ENG]   An instrument used to measure the direction and angle of dip of geologic formations.

dipmeter log [GEOL]   A dip log produced by reading of the direction and angle of formation dip as analyzed from impulses from a dipmeter consisting of three electrodes 120° apart in a plane perpendicular to the borehole.

dip of horizon *See* dip.

dip reversal *See* reversal of dip.

Diprotodontidae [PALEON]   A family of extinct marsupial mammals.

dip slip [GEOL]   The component of a fault parallel to the dip of the fault. Also known as normal displacement.

dip slope [GEOL]   A slope of the surface of the land determined by and conforming approximately to the dip of the underlying rocks. Also known as back slope; outface.

dip stream [HYD]   A consequent stream that flows in the direction of the dip of the strata it traverses.

dip-strike symbol [GEOL]   A geologic symbol used on maps to show the strike and dip of a planar feature.

dipyramid [CRYSTAL]   A crystal having the form of two pyramids that melt at a plane of symmetry.

dipyre *See* mizzonite.

direct cell [METEOROL]   A closed thermal circulation in a vertical plane in which the rising motion occurs at higher potential temperature than the sinking motion.

direct intake [HYD]   Direct recharge of an aquifer through the zone of saturation.

direction *See* trend.

directional log [PETRO ENG]   A record of the wellhole drift, from the vertical, and the direction of that drift.

directional structure [GEOL]   Any sedimentary structure having directional significance; examples are cross-bedding and ripple marks. Also known as vectorial structure.

directional well [PETRO ENG]   A well drilled at an angle up to 70° from the vertical to avoid obstacles over the reservoir, such as towns, beaches, or bodies of water.

direct-line drive [PETRO ENG]   Waterflood operation involving a network of wells in a direct (straight) line.

direct stratification *See* primary stratification.

direct tide [GEOPHYS]   A gravitational solar or lunar tide in the ocean or atmosphere which is in phase with the apparent motions of the attracting body, and consequently has its local maxima directly under the tide-producing body, and on the opposite side of the earth.

dirt band [GEOL]   A dark layer in a glacier representing a former surface, usually a summer surface, where silt and debris accumulated.

dirt bed [GEOL]   A buried soil containing partially decayed organic material; sometimes occurs in glacial drift.

dirt cone [HYD]   A mound of glacial ice or snow that is covered with a layer of silt that protects the material beneath from the ablation that has lowered the surrounding surface.

dirt slip *See* clay vein.

Disasteridae [PALEON]   A family of extinct burrowing, exocyclic Euechinoidea in the order Holasteroida comprising mainly small, ovoid forms without fascioles or a plastron.

Discoidiidae [PALEON]   A family of extinct conical or globular, exocyclic Euechinoidea in the order Holectypoida distinguished by the rudiments of internal skeletal partitions.

discomfort index *See* temperature-humidity index.

disconformity [GEOL]   Unconformity between parallel beds or strata.

discontinuity [GEOL]   **1.** An interruption in sedimentation. **2.** A surface that separates unrelated groups of rocks.   [GEOPHYS]   A boundary at which the velocity of seismic waves changes abruptly.

discontinuous amplifier [ELECTR]   Amplifier in which the input waveform is reproduced on some type of averaging basis.

discontinuous phase *See* disperse phase.

discontinuous precipitation [MET]   Precipitation principally at and away from the grain boundaries in a supersaturated solid solution; diffraction patterns show two lattice parameters, the solute in solution and the precipitate.

discontinuous reaction series [GEOL]   The branch of Bowen's reaction series that include olivine, pyroxene, amphibole, and biotite; each change in the series represents an abrupt change in phase.

discordance [GEOL]   An unconformity characterized by lack of parallelism between strata which touch without fusion.

discordant pluton [GEOL]   An intrusive igneous body that cuts across the bedding or foliation of the intruded formations.

discount factor [PETRO ENG]   The ratio of the present worth of one or a series of future payments to the total undiscounted amount of such future payments. Also known as average discount factor; deferment factor; present-worth factor.

discovery [MIN ENG]   Finding of a valuable mineral deposit.

discovery claim [MIN ENG]   The first claim for the finding of a mineral deposit.

discovery vein [MIN ENG]   The vein on which a mining claim is based.

discovery well [PETRO ENG]   A successful exploration well.

discrete-film zone *See* belt of soil water.

disequilibrium assemblage [MINERAL]   A grouping of minerals that are not in thermodynamic equilibrium.

disharmonic fold [GEOL]   A fold in which changes in form or magnitude occur with depth.

dishpan experiment [METEOROL]   A model experiment carried out by differential heating of fluid in a flat, rotating pan; it establishes similarity with the atmosphere and is used to reproduce many important features of the general circulation and, on a smaller scale, atmospheric motion.

disinclination [CRYSTAL]   A type of crystal imperfection in which one part of the crystal is rotated and therefore displaced relative to the rest of the crystal; observed in liquid crystals and protein coats of viruses.

disjunct endemism [PALEON]   A type of regionally restricted distribution of a fossil taxon in which two or more component parts are separated by a major physical barrier and hence not readily explicable in terms of present-day geography.

disk hardness gage [ENG]   An instrument designed to measure the hardness of snow in terms of its resistance to the horizontal pressure exerted by a disk attachment.

disk-wall packer [PETRO ENG]   A disklike seal between the outside of the well tubing and the inside of the well casing; used to prevent fluid movement from the pressure differential above and below the sealing point.

dislocation [CRYSTAL]   A defect occurring along certain lines in the crystal structure and present as a closed ring or a line anchored at its ends to other dislocations, grain boundaries, the surface, or other structural feature. Also known as line defect. [GEOL]   Relative movement of rock on opposite sides of a fault. Also known as displacement.

dislocation breccia *See* fault breccia.

dismembered drainage [HYD]   A complex drainage system which, through a series of geological changes, has been isolated into independent streams that empty into the sea by separate mouths.

dismembering [HYD]   The transformation of a tributary into an independent stream through a change of geologic conditions.

dismicrite [GEOL]   Fine-grained limestone of obscure origin, resembling micrite but containing sparry calcite bodies.

disorder [CRYSTAL]   Departures from regularity in the occupation of lattice sites in a crystal containing more than one element.

dispersal map [GEOL]   A stratigraphic map that indicates the inferred source area of clastic materials and their direction or distance of transportation.

dispersal pattern [GEOCHEM]   Distribution pattern of metals in soil, rock, water, or vegetation.

dispersed elements [GEOCHEM]   Elements which form few or no independent minerals but are present as minor ingredients in minerals of abundant elements.

dispersed gas injection [PETRO ENG]   Gas-injection pressure maintenance of an oil reservoir in which the injection wells are arranged geometrically to distribute the gas uniformly throughout the oil-productive portions of the reservoir.

dispersion [MINERAL]   In optical mineralogy, the constant optical values at different positions on the spectrum.

dispersion flow [GEOPHYS]   Flow of granular sediment in which fluidity is maintained by collisions among particles.

disphenoid [CRYSTAL]   **1.** A crystal form with four similar triangular faces combined in a wedge shape; can be tetragonal or orthorhombic. **2.** A crystal form with eight scalene triangles combined in pairs.

displaced ore body [GEOL]   An ore body which has been subjected to displacement or disruption after its initial deposition.

displacement [GEOL]   *See* dislocation.

displacement efficiency [PETRO ENG]   In a gas condensate reservoir, the proportion (by volume) of wet hydrocarbons swept out of pores during dry-gas cycling.

dissected topography [GEOGR]   Physical features marked by erosive cutting.

dissection [GEOL]   Destruction of the continuity of the land surface by erosive cutting of valleys or ravines into a relatively even surface.

dissepiment [PALEON]   One of the vertically positioned thin plates situated between the septa in extinct corals of the order Rugosa.

dissipation constant [GEOPHYS]   In atmospheric electricity, a measure of the rate at which a given electrically charged object loses its charge to the surrounding air.

dissolved gas *See* solution gas.

dissolved-gas drive *See* internal gas drive.

dissolved-gas-drive reservoir [PETRO ENG]   Oil reservoir in which the temperature of the liquid phase is below critical, and the liquid is driven from the reservoir by the expansion of dissolved gas. Also known as a bubble-point reservoir.

dissolved-gas reservoir *See* solution-gas reservoir.

dissolved load [HYD]   The amount of total stream load that is transported in solution.

Distacodidae [PALEON]   A family of conodonts in the suborder Conodontiformes characterized as simple curved cones with deeply excavated attachment scars.

disthene *See* kyanite.

distorted water [METEOROL]   A multimolecular layer of water, at the boundary between a mass of liquid water and the surrounding vapor, whose structure is not identical with that of bulk water.

distortional wave *See* S wave.

distributary [HYD]   An irregular branch flowing out from a main stream and not returning to it, as in a delta. Also known as distributary channel.

distributary channel *See* distributary.

distributed fault *See* fault zone.

distribution graph [HYD]   A statistically derived hydrograph for a storm of specified duration, graphically representing the percent of total direct runoff passing a point on a stream, as a function of time; usually presented as a histogram or table of percent runoff within each of successive short time intervals.

distributive fault *See* step fault.

district forecast [METEOROL]   In U.S. Weather Bureau usage, a general weather forecast for conditions over an established geographical "forecast district."

disturbance [GEOL]   Folding or faulting of rock or a stratum from its original position.   [METEOROL]   **1.** Any low or cyclone, but usually one that is relatively small in size and effect. **2.** An area where weather, wind, pressure, and so on show signs of the development of cyclonic circulation. **3.** Any deviation in flow or pressure that is associated with a disturbed state of the weather, such as cloudiness and precipitation. **4.** Any individual circulatory system within the primary circulation of the atmosphere.

diurnal [METEOROL]   Pertaining to meteorological actions which are completed within 24 hours and which recur every 24 hours.

diurnal arc [ASTRON]   That part of a celestial body's diurnal circle which lies above the horizon of the observer.

diurnal inequality [OCEANOGR]   The difference between the heights of the two high waters or the two low waters of a lunar day.

diurnal range *See* great diurnal range.

diurnal tide [OCEANOGR]   A tide in which there is only one high water and one low water each lunar day.

diurnal variation [GEOPHYS]   Daily variations of the earth's magnetic field at a given point on the surface, with both solar and lunar periods having their source in the horizontal movements of air in the ionosphere.

divergence [METEOROL]   The two-dimensional horizontal divergence of the velocity field.   [OCEANOGR]   A horizontal flow of water, in different directions, from a common center or zone.

diversion [HYD]   The process by which one stream causes changes in the drainage or course of another stream.

diverting agent [PETRO ENG]   A viscous gel or suspension of graded solids used during acidizing of an oil reservoir to temporarily block off the most permeable sections of the pay zone to force the acid into less permeable sections.

Divesian *See* Oxfordian.

divide [GEOGR]   A ridge or section of high ground between drainage systems.

divining [MIN ENG]   An unscientific method for searching for subsurface water or minerals by means of a divining rod. Also known as dowsing.

divining rod [MIN ENG]   An unscientific device in the form of a forked rod or tree branch that is supposed to dip when held over water or minerals, depending on the specialty of the operator, or dowser. Also known as dowsing rod; wiggle stick.

dixenite [MINERAL] $Mn_9(SiO_3)(AsO_3)(OH)_2$ A black hexagonal mineral consisting of a manganese arsenite and silicate, occurring in scales.

djalmaite *See* microlite.

djerfisherite [MINERAL] $K_3CuFe_{12}S_{14}$ A sulfide mineral which is found only in meteorites.

Djulfian [GEOL] Upper upper Permian geologic time.

D layer [GEOL] The lower mantle of the earth, between a depth of 1000 and 2900 km. [GEOPHYS] The lowest layer of ionized air above the earth, occurring in the D region only in the daytime hemisphere; reflects frequencies below about 50 kilohertz and partially absorbs higher-frequency waves.

dneprovskite *See* wood tin.

Dobson prop [MIN ENG] A hydraulic supporting post used in mine tunnel construction.

Dobson support system [MIN ENG] A self-advancing unit consisting of three Dobson props used to support longwall faces.

Docodonta [PALEON] A primitive order of Jurassic mammals of North America and England.

doctor [METEOROL] A cooling sea breeze in the tropics.

Dodge crusher [MIN ENG] A type of jaw crusher with the movable jaw hinged at the bottom, allowing a highly uniform product to be discharged.

Dodge pulverizer [MIN ENG] A hexagonal drum-shaped pulverizer that rotates on a horizontal axis and contains steel balls for reducing rock and ore.

dog days [CLIMATOL] The period of greatest heat in summer.

dogger [GEOL] Concretionary masses of calcareous sandstone or ironstone.

doghole [MIN ENG] A small opening in a mine.

doghole mine [MIN ENG] A small coal mine employing 15 or less miners.

dogleg [PETRO ENG] Bend or sudden direction change in a wellhole that can cause tubing wear and failure.

dogs *See* folding boards.

doister [METEOROL] In Scotland, a severe storm from the sea. Also known as deaister; dyster.

Dolan equation [PETRO ENG] Empirical equation for reservoir-permeability damage factor by the invasion of drilling mud or other foreign materials.

doldrums [METEOROL] A nautical term for the equatorial trough, with special reference to the light and variable nature of the winds. Also known as equatorial calms.

dolerophanite [MINERAL] $Cu_2(SO_4)O$ A brown, monoclinic mineral consisting of a basic copper sulfate, occurring in crystals.

Dolichothoraci [PALEON] A group of joint-necked fishes assigned to the Arctolepiformes in which the pectoral appendages are represented solely by large fixed spines.

dolimorphic [PETR] Referring to a type of igneous rock characteristically composed of released minerals.

dolocast [GEOL] The cast or impression of a dolomite crystal.

dolomite [MINERAL] $CaMg(CO_3)_2$ The carbonate mineral; white or colorless with hexagonal symmetry and a structure similar to that of calcite, but with alternate layers of calcium ions being completely replaced by magnesium.

dolomite rock *See* dolomitic limestone.

dolomitic limestone [PETR] A limestone whose carbonate fraction contains more than 50% dolomite. Also known as dolomite rock; dolostone.

dolomitization [GEOL] Conversion of limestone to dolomite rock by replacing a portion of the calcium carbonate with magnesium carbonate.

dolostone *See* dolomitic limestone.

domatic class *See* clinohedral class.

dome [CRYSTAL] An open crystal form consisting of two faces astride a symmetry plane. [GEOL] **1.** A circular or elliptical, almost symmetrical upfold or anticlinal type of structural deformation. **2.** A large igneous intrusion whose surface is convex upward.

Domerian [GEOL] Upper Charmouthian geologic time.

dome theory [MIN ENG] The theory that the movements of strata resulting from underground excavations are limited by a dome whose base is the area of excavation, and that the movements decrease in intensity as they extend upward from the center of the base.

domeykite [MINERAL] Cu_3As A tin-white or steel-gray mineral consisting of copper arsenide; specific gravity is 7.2–7.75.

Donau glaciation [GEOL] A Pleistocene glacial time unit in the Alps region in Europe.

donkey [MIN ENG]   *See* barney.

doodlebug [GEOL] Also known as douser. **1.** Any unscientific device or apparatus, such as a divining rod, used to locate subsurface water, minerals, gas, or oil. **2.** A scientific instrument used for locating minerals.   [MIN ENG] The treatment plant or washing unit of a dredge which is mounted on a pontoon and can be floated in an excavation dug by a dragline.

dopplerite [GEOL] A naturally occurring gel of humic acids found in peat bags or where an aqueous extract from a low-rank coal can collect.

doreite [GEOL] An extrusive lava containing equal parts of potassium and sodium.

Dorypteridae [PALEON] A family of Permian palaeonisciform fishes sometimes included in the suborder Platysomoidei.

Dosco miner [MIN ENG] A large crawler-tracked cutter-loader that is designed for longwall faces in seams over 4½ feet (1.4 meters) thick and takes a buttock 0.5 foot (0.15 meter) wide; rated at 200 horsepower (1.49 × 10⁵ watts); has seven cutter chains mounted on the cutterhead, with a capacity of over 400 tons per machine.

double drill column [MIN ENG] Two drill columns connected by a horizontal bar on which a drill machine can be mounted. Also known as double jack.

double ebb [OCEANOGR] An ebb current that has two maxima of velocity separated by a smaller ebb velocity.

double-entry method [MIN ENG] A mining arrangement involving twin entries in flat or gently dipping coal, so that rooms can be extended from both entryways.

doublehand drilling [ENG] A rock-drilling method performed by two men, one striking the rock with a long-handled sledge hammer while a second holds the drill and twists it between strokes. Also known as double jacking.

double headings [MIN ENG] A pair of coal headings driven parallel to each other and positioned side by side about 10–20 yards (9–18 meters) apart.

double jack *See* double drill column.

double jacking *See* doublehand drilling.

double-theodolite observation [ENG] A technique for making winds-aloft observations in which two theodolites located at either end of a base line follow the ascent of a pilot balloon; synchronous measurements of the elevation and azimuth angles of the balloon, taken at periodic intervals, permit computation of the wind vector as a function of height.

double tide [OCEANOGR] A high tide with two maxima of nearly identical height separated by a relatively small depression, or conversely, a low water with two minima separated by a relatively small elevation.

double valley [GEOL] A valley with a low divide on its floor, from which two streams flow in different directions.

double valves [PETRO ENG] Two valves in series used as subsurface traveling or standing valves in wells, the dual arrangement being more reliable than a single valve.

double-wedge cut [MIN ENG] A drill-hole pattern composed of a shallow wedge within a larger, outer wedge.

doubly plunging fold [GEOL] A fold that plunges in opposite directions, either away from or toward a central point.

douglasite [MINERAL] $K_2FeCl_4 \cdot 2H_2O$ Ore from Stassfurt, Germany; a member of the erythrosiderite group; orthorhombic, in the isomorphous series.

douse [MIN ENG] To locate and delineate subsurface resources such as water, oil, gas, or minerals.

douser *See* doodlebug.

down [GEOL] **1.** Hillock of sand thrown up along the coast by the sea or the wind. **2.** A flat eminence on the top of a hill or mountain.

downcast [MIN ENG] Intake shaft for air in a mine.

downdip [GEOL] Pertaining to a position parallel to or in the direction of the dip of a stratum or bed.

downhole drill [MIN ENG] A hammer or percussive drill in which a reciprocating pneumatic piston is located immediately behind the drill bit and can follow and enter the bit down the hole, for minimizing energy losses.

downhole equipment *See* drill fittings.

downrush [METEOROL] A term sometimes applied to the strong downward-flowing air current that marks the dissipating stages of a thunderstorm.

downthrow [GEOL] The side of a fault whose relative movement appears to have been downward.

downwarp [GEOL] A segment of the earth's crust that is broadly bent downward.

downwelling *See* sinking.

dowsing *See* divining.

dowsing rod *See* divining rod.

Dowtonian [GEOL] Uppermost Silurian or lowermost Devonian geologic time.

Dowty prop [MIN ENG] A self-contained hydraulic supporting post consisting of two telescoping tubes; the upper (inner) tube contains the oil, pump, yield valve, and other accessories.

Draeger breathing apparatus [MIN ENG] A long-service, self-contained oxygen-breathing apparatus with the oxygen feed governed by the lungs; allows the user to do hard work for up to 5 hours and normal work for 7 hours, and can sustain a resting individual for 18 hours.

drag [MIN ENG] Movement of the hanging wall with respect to the foot wall due to the weight of the arch block in an inclined slope.

drag cut [ENG] A drill hole pattern for breaking out rock, in which angled holes are drilled along a floor toward a parting, or on a free face and then broken by other holes drilled into them.

drag fold [GEOL] A minor fold formed in an incompetent bed by movement of a competent bed so as to subject it to couple; the axis is at right angles to the direction in which the beds slip.

drag mark [GEOL] Long, even mark usually having longitudinal striations produced by current drag of an object across a sedimentary surface.

dragonite [PETR] A rounded but angular quartz pebble with a dull exterior surface.

drag-stone mill [MIN ENG] A mill in which a heavy stone is dragged over ore to grind it.

drainage basin [HYD] An area in which surface runoff collects and from which it is carried by a drainage system, as a river and its tributaries. Also known as catchment area; drainage area; feeding ground; gathering ground; hydrographic basin.

drainage coefficient [HYD] The quantity of water, in terms of depth or other units, that is removed or drained from an area in 24 hours.

drainage density [HYD] Ratio of the total length of all channels in a drainage basin and the basin area.

drainage divide [GEOL] **1.** The border of a drainage basin. **2.** The boundary separating adjacent drainage basins.

drainage lake [HYD] An open lake in which water loss occurs through a surface outlet or by discharge of the effluent.

drainage pattern [HYD] The configuration of a natural or artificial drainage system; stream patterns reflect the topography and rock patterns of the area.

drainage ratio [HYD] The ratio of runoff to precipitation in a given area for a specific time period.

drainage system [HYD] A surface stream or a body of impounded surface water, together with all its tributaries, that drains a region.

drainage wind *See* gravity wind.

drakonite [PETR] A type of extrusive rock containing biotie, sanidine, or plagioclase phenocrysts or hornblende in a groundmass of feldspar, amphibole, or pyroxene.

draping [GEOL] Structural concordance of the strata overlying a limestone reef or other hard core to the surface of the reef or core.

draw [MIN ENG] **1.** To remove timber supports, allowing overhanging coal to fall down for collection. **2.** To allow ore to run down chutes from stopes, chambers, or ore bins. **3.** To collect broken coal in trucks. **4.** To hoist coal, rock, ore, or other materials to the surface. **5.** The horizontal distance to which creep extends on the surface beyond the stopes.

drawdown [HYD] The magnitude of the change in water surface level in a well, reservoir, or natural body of water resulting from the withdrawal of water. [PETRO ENG] The difference between the static and the flowing bottom-hole pressure.

drawhole [MIN ENG] The aperture in a battery through which coal or ore is drawn.

drawing timber [MIN ENG] The act of withdrawing timber and other supports from abandoned or worked-out mines.

draw works [PETRO ENG]   An oil-well drilling mechanism used to supply driving power and to lift heavy objects; consists of a countershaft and drum.

D region [GEOPHYS]   The region of ionosphere up to about 60 miles (97 kilometers) above the earth, below the E and F regions, in which the D layer forms.

dreikanter [GEOL]   A pebble shaped with three facets by sandblasting.

Drepanellacea [PALEON]   A monomorphic superfamily of extinct paleocopan ostracods in the suborder Beyrichicopina having a subquadrate carapace, many with a marginal rim.

Drepanellidae [PALEON]   A monomorphic family of extinct ostracods in the superfamily Drepanellacea.

Dresbachian [GEOL]   Lower Croixan geologic time.

dress [ELECTR]   The arrangement of connecting wires in a circuit to prevent undesirable coupling and feedback.   [MECH ENG]   **1.** To shape a tool.   **2.** To restore a tool to its original shape and sharpness.   [MIN ENG]   To sort, grind, clean, and concentrate ore.

drewite [GEOL]   Calcareous ooze composed of impalpable calcareous material.

dribbling [MIN ENG]   Fall of debris from the roof of an excavation, usually preceding a heavy fall or cave-in.

drift [ENG]   **1.** A gradual deviation from a set adjustment, such as frequency or balance current, or from a direction.   **2.** The deviation, or the angle of deviation, of a borehole from the vertical or from its intended course.   [GEOL]   **1.** Rock material picked up and transported by a glacier and deposited elsewhere.   **2.** Detrital material moved and deposited on a beach by waves and currents.   [OCEANOGR]   *See* drift current.

drift bolt [ENG]   **1.** A bolt used to force out other bolts or pins.   **2.** A metal rod used to secure timbers.

drift bottle [OCEANOGR]   A bottle which is released into the sea for studying currents; contains a card, identifying the date and place of release, to be returned by the finder with date and place of recovery. Also known as floater.

drift card [OCEANOGR]   A card, such as is used in a drift bottle, encased in a buoyant, waterproof envelope and released in the same manner as a drift bottle.

drift current [OCEANOGR]   **1.** A wide, slow-moving ocean current principally caused by winds. Also known as drift; wind drift; wind-driven current.   **2.** Current determined from the differences between dead reckoning and a navigational fix.

drift curve [GEOPHYS]   A graph plotting a series of gravity values read at the same location at different times.

drift dam [GEOL]   A dam formed by glacial drift in a stream valley.

drifter [MECH ENG]   A rock drill, similar to but usually larger than a jack hammer, mounted for drilling holes up to 4½ inches (11.4 centimeters) in diameter.   [MIN ENG]   **1.** A person who excavates mine drifts.   **2.** An air-driven rock drill used for excavating mine drifts and crosscuts.

drift glacier *See* snowdrift glacier.

drift ice [OCEANOGR]   Sea ice that has drifted from its place of formation.

drift ice foot *See* ramp.

drifting [MIN ENG]   Tunneling along the strike of a lode.

drifting snow [METEOROL]   Wind-driven snow raised from the surface of the earth to a height of less than 6 feet (1.8 meters).

drift mining [MIN ENG]   Working of shallow veins or beds through drifts or shafts from the surface.

drift sheet [GEOL]   A large, evenly spread body of glacial drift deposited during a single glaciation.

drift station [OCEANOGR]   **1.** A scientific station established on the ice of the Arctic Ocean, generally based on an ice flow. **2.** A set of observations made over a period of time from a drifting vessel.

drift terrace *See* alluvial terrace.

drill column [MIN ENG]   A steel pipe that can be wedged across an underground opening in a vertical or horizontal position to serve as a base on which to mount a diamond or rock drill.

drill cuttings [ENG]   Cuttings of rock and other subterranean materials brought to the surface during the drilling of wellholes.

drill doctor [MIN ENG]   **1.** A person who services drill bits, tools, and steels. **2.** A shop where the mechanic works.

drill extractor [ENG]   A tool for recovering broken drill pieces or a detached drill from a borehole.

drill fittings [ENG]   All equipment used in a borehole during drilling. Also known as downhole equipment.

drill floor [ENG]   A work area covered with planks around the collar of a borehole at the base of a drill tripod or derrick.

drill footage [ENG]   The lineal feet of borehole drilled.

drillhole pattern [ENG]   The number, position, angle, and depth of the shot holes forming the round in the face of a tunnel or sinking pit.

drill-in [MIN ENG]   The act or process of setting casting through overburden by using a drill machine.

drilling fluid *See* drilling mud.

drilling mud [MATER]   A suspension of finely divided heavy material, such as bentonite and barite, pumped through the drill pipe during rotary drilling to seal off porous zones and flush out chippings, and to lubricate and cool the bit. Also known as drilling fluid.

drilling platform [ENG]   The structural base upon which the drill rig and associated equipment is mounted during the drilling operation.

drilling time [ENG]   **1.** The time required in rotary drilling for the bit to penetrate a specified thickness (usually 1 foot) of rock. **2.** The actual time the drill is operating.

drilling time log [ENG]   Foot-by-foot record of how fast a formation is drilled.

drill log [ENG]   **1.** A record of the events and features of the formations penetrated during boring. Also known as boring log. **2.** A record of all occurrences during drilling that might help in a complete logging of the hole or in determining the cost of the drilling.

drill out [ENG]   **1.** To complete one or more boreholes. **2.** To penetrate or remove a borehole obstruction. **3.** To locate and delineate the area of a subsurface ore body or of petroleum by a series of boreholes.

drill-over [ENG]   The act or process of drilling around a casing lodged in a borehole.

drill pipe [MIN ENG] A pipe used for driving a revolving drill bit, used especially in drilling wells; consists of a casing within which tubing is run to conduct oil or gas to ground level; drilling mud flows in the annular space between casing and tubing during the drilling operation.

drill rod [ENG] The long rod that drives the drill bit in drilling boreholes.

drill runner [MIN ENG] A tunnel miner who operates rock drills.

drill stem *See* drill string.

drill-stem test [PETRO ENG] Bottom-hole pressure information obtained and used to determine formation productivity.

drill string [MECH ENG] The assemblage of drill rods, core barrel, and bit, or of drill rods, drill collars, and bit in a borehole, which is connected to and rotated by the drill collar of the borehole. Also known as drill stem.

drip [HYD] Condensed or otherwise collected moisture falling from leaves, twigs, and so forth. [PETRO ENG] A discharge mechanism installed at a low point in a gas transmission line to collect and remove liquid accumulations.

driphole [GEOL] A niche or small hole in clay or rock caused by the action of dripping water.

dripping drop atomization [HYD] A type of natural gravitational atomization process in which there is periodic emission of drops from the bottom side of a surface to which a liquid is fed continuously, as in dripping of water from leaves.

drive [MIN ENG] **1.** To excavate in a horizontal or inclined plane. **2.** A horizontal underground tunnel along or parallel to a lode, vein, or ore body.

driven snow [METEOROL] Snow which has been moved by wind and collected into snowdrifts.

drivepipe [ENG] A thick-walled casing pipe that is driven through overburden or into a deep drill hole to prevent caving.

drive rod [ENG] Hollow shaft in the swivel head of a diamond-drill machine through which energy is transmitted from the drill motor to the drill string. Also known as drive spindle.

drive sampling [ENG] The act or process of driving a tubular device into soft rock material for obtaining dry samples.

drive spindle *See* drive rod.

drizzle [METEOROL] Very small, numerous, and uniformly dispersed water drops that may appear to float while following air currents; unlike fog droplets, drizzle falls to the ground; it usually falls from low stratus clouds and is frequently accompanied by low visibility and fog.

drizzle drop [METEOROL] A drop of water of diameter 0.2 to 0.5 millimeter falling through the atmosphere; however, all water drops of diameter greater than 0.2 millimeter are frequently termed raindrops, as opposed to cloud drops.

DRM *See* depositional remanent magnetization.

drop [HYD] The difference in elevation of water surfaces between up- and downstream points, with reference to a constriction in the stream.

drop [MINERAL] A funnel-shaped downward intrusion of a sedimentary rock in the roof of a coal seam.

drop-bottom car [MIN ENG] A mine car designed so that flaps drop open in the bottom to allow the coal to fall out as the car passes over the dump; flaps close as the car leaves.

droplet [METEOROL]   A water droplet in the atmosphere; there is no defined size limit separating droplets from drops of water, but sometimes a maximum diameter of 0.2 millimeter is the limit for droplets.

drop log [MIN ENG]   A timber which can be dropped across a mine track by remote control to derail cars.

drop-size distribution [METEOROL]   The frequency distribution of drop sizes (diameters, volumes) that is characteristic of a given cloud or rainfall.

dropsonde [ENG]   A radiosonde dropped by parachute from a high-flying aircraft to measure weather conditions and report them back to the aircraft.

dropsonde dispenser [ENG]   A chamber from which dropsonde instruments are released from weather reconnaissance aircraft; used only for some models of equipment, ejection chambers being used for others.

dropsonde observation [METEOROL]   An evaluation of the significant radio signals received from a descending dropsonde, and usually presented in terms of height, temperature, and dew point at the mandatory and significant pressure levels; it is comparable to a radiosonde observation.

drop theory *See* barrier theory of cyclones.

drosometer [ENG]   An instrument used to measure the amount of dew deposited on a given surface.

drought [CLIMATOL]   A period of abnormally dry weather sufficiently prolonged so that the lack of water causes a serious hydrologic imbalance (such as crop damage, water supply shortage, and so on) in the affected area; in general, the term should be reserved for relatively extensive time periods and areas.

drowned atoll [GEOL]   An atoll which has not reached the water surface.

drowned coast [GEOL]   A shoreline transformed from a hilly land surface to an archipelago of small islands by inundation by the sea.

drowned river mouth *See* estuary.

drowned stream [HYD]   A stream that has been flooded over by the ocean. Also known as flooded stream.

drowned valley [GEOL]   A valley whose lower part has been inundated by the sea due to submergence of the land margin.

droxtal [HYD]   An ice particle measuring 10–20 micrometers in diameter, formed by direct freezing of supercooled water droplets at temperatures below $-30°C$.

drumlin [GEOL]   A hill of glacial drift or bedrock having a half-ellipsoidal streamline form like the inverted bowl of a spoon, with its long axis paralleling the direction of movement of the glacier that fashioned it.

drumlinoid [GEOL]   A rock drumlin that has not attained full drumlin form because of surface modification by moving ice.

drumloid [GEOL]   A glacial till ridge or oval hill with a shape similar to that of a drumlin.

drummy [MIN ENG]   Loose rock or coal, especially in a mine roof, that produces a hollow, weak sound when tapped with a bar.

drum separator [MIN ENG]   A cylindrical vessel which rotates slowly and separates run-of-mine coal into clean coal, middlings, and refuse, can be adjusted for different specific gravities.

druse [GEOL]   A small cavity in a rock or vein encrusted with aggregates of crystals of the same minerals which commonly constitute the enclosing rock.

drusy [GEOL] Of or pertaining to rocks containing numerous druses.

dry adiabat [METEOROL] A line of constant potential temperature on a thermodynamic diagram.

dry-adiabatic lapse rate [METEOROL] A special process lapse rate of temperature, defined as the rate of decrease of temperature with height of a parcel of dry air lifted adiabatically through an atmosphere in hydrostatic equilibrium. Also known as adiabatic lapse rate; adiabatic rate.

dry-adiabatic process [METEOROL] An adiabatic process in a system of dry air.

dry air [METEOROL] Air that contains no water vapor.

dry beach [GEOL] The part of a beach that is covered only by storm waves.

dry-bone ore *See* smithsonite.

dry-cell cap light [MIN ENG] A headlamp with a focusing lens lamp and a dry-cell battery unit clipped to the belt; to prevent explosion in a mine, the bulb is ejected automatically in case of its breakage.

dry climate [CLIMATOL] **1.** In W. Köppen's climatic classification, the major category which includes steppe climate and desert climate, defined strictly by the amount of annual precipitation as a function of seasonal distribution and of annual temperature. **2.** In C. W. Thornwaite's climatic classification, any climate type in which the seasonal water surplus does not counteract seasonal water deficiency, and having a moisture index of less than zero; included are the dry subhumid, semiarid, and arid climates.

dry delta *See* alluvial fan.

drydock iceberg *See* valley iceberg.

dry drilling [MIN ENG] Drilling in which chippings and cuttings are lifted out of a borehole by a current of air or gas.

dry fog [METEOROL] A fog that does not moisten exposed surfaces.

dry freeze [HYD] The freezing of the soil and terrestrial objects caused by a reduction of temperature when the adjacent air does not contain sufficient moisture for the formation of hoarfrost on exposed surfaces.

dry haze [METEOROL] Fine dust or salt particles in the air, too small to be individually apparent but in sufficient number to reduce horizontal visibility, and to give the atmosphere a characteristic hazy appearance.

dry hole [ENG] A hole driven without the use of water. [PETRO ENG] A well in which no oil or gas is found.

dry-hot-rock geothermal system [GEOL] A water-deficient hydrothermal reservoir dominated by the presence of rocks at depths in which large quantities of heat are stored.

dry mining [MIN ENG] Mining operation in which there is no moisture in the ventilating air.

dry ore [MIN ENG] An ore of gold or silver which requires added lead and fluxes for treatment.

dry permafrost [GEOL] A loose and crumbly permafrost which contains little or no ice.

dry placer [MIN ENG] A gold-bearing alluvial deposit found in arid regions; it cannot be mined due to lack of water.

dry quicksand [GEOL] A sand mass that cannot support heavy loads due to alternating layers of firmly packed sand and loose, soft sand.

dry sample   [MIN ENG]   A sample of ore obtained by dry drilling.

dry sand   [GEOL]   **1.** A formation underlying the production sand and containing oil which has escaped during well drilling. **2.** A nonproductive oil sand.

dry season   [CLIMATOL]   In certain types of climate, an annually recurring period of one or more months during which precipitation is at a minimum for the region.

dry spell   [CLIMATOL]   A period of abnormally dry weather, generally reserved for a less extensive, and therefore less severe, condition than a drought; in the United States, describes a period lasting not less than 2 weeks, during which no measurable precipitation was recorded.

drystone   [GEOL]   A stalagmite or stalactite formed by dropping water.

dry tabling   [MIN ENG]   A process similar to wet tabling, but without the water; used to separate two or more minerals based on specific gravity differences.

dry tongue   [METEOROL]   In synoptic meteorology, a pronounced protrusion of relatively dry air into a region of higher moisture content.

dry unit weight   [GEOL]   The weight of soil solids per unit of total volume of soil mass.

dry wash   [GEOL]   A wash, arroyo, or coulee whose bed lacks water.

dry washer   [MIN ENG]   A machine for extracting gold mined from dry placers.

DSM screens   [MIN ENG]   Dutch State Mines screens for dewatering very fine materials; the slurry is sluiced over a concave screen surface with slots that flare out radially to promote egress of particles.

Dst   [GEOPHYS]   The "storm-time" component of variation of the terrestrial magnetic field, that is, the component which correlates with the interval of time since the onset of a magnetic storm; used as an index of intensity of the ring current.

8D technique   [METEOROL]   A technique for using the radiosonde observation to determine the presence of liquid water-droplets in supercooled clouds in saturated or nearly saturated layers of air; for each reported level in the sounding, the negative value of eight times the dew-point spread ($-8D$) is plotted on the pseudoadiabatic chart (or equivalent chart); where the temperature sounding lies to the left of the $-8D$ curve, liquid droplet clouds are considered to be present, and icing is possible on aircraft flying in the cloud layer. Also known as frost-point technique.

dual completion well   [PETRO ENG]   Single well casing containing two production tubing strings, each in a different zone of the reservoir (one higher, one lower) and each separately controlled.

dual-seal tubing joint   [PETRO ENG]   Tubing connection joint with two sealing surfaces to assure a leak-free connection between sections.

dubiocrystalline   [PETR]   Referring to a rock texture whose crystallinity is either uncertain or determined only with great difficulty.

duck-billed dinosaur   [PALEON]   Any of several herbivorous, bipedal ornithopods having the front of the mouth widened to form a ducklike beak.

duct   [GEOPHYS]   The space between two air layers, or between an air layer and the earth's surface, in which microwave beams are trapped in ducting. Also known as radio duct; tropospheric duct.

ducting   [GEOPHYS]   An atmospheric condition in the troposphere in which temperature inversions cause microwave beams to refract up and down between two air layers, so that microwave signals travel 10 or more times farther than the normal line-of-sight limit. Also known as superrefraction; tropospheric ducting.

dufrenite [MINERAL] A blackish-green, fibrous ferric phosphate mineral; commonly massive or in nodules.

dufrenoysite [MINERAL] $Pb_2As_2S_5$ A lead gray to steel gray, monoclinic mineral consisting of lead arsenic sulfide.

duftite [MINERAL] $PbCu(AsO_4)(OH)$ Orthorhombic mineral that is composed of a basic arsenate of lead and copper.

dull coal [GEOL] A component of banded coal with a grayish color and dull appearance, consisting of small anthraxylon constituents in addition to cuticles and barklike constituents embedded in the attritus.

dumontite [MINERAL] $Pb_2(UO_2)_3(PO_4)_2(OH)_4 \cdot 3H_2O$ Yellow orthorhombic mineral consisting of a hydrated phosphate of uranium and lead, occurring in crystals.

dumortierite [MINERAL] $Al_8BSi_3O_{19}(OH)$ A pink, green, blue, or violet mineral that crystallizes in the orthorhombic system but commonly occurs in parallel or radiating fibrous aggregates; mined for the manufacture of high-grade porcelain.

dundasite [MINERAL] $PbAl_2(CO_3)_2(OH)_4 \cdot 2H_2O$ A white mineral consisting of a basic lead aluminum carbonate, occurring in spherical aggregates.

dune [GEOL] A mound or ridge of unconsolidated granular material, usually of sand size and of durable composition (such as quartz), capable of movement by transfer of individual grains entrained by a moving fluid.

dune complex [GEOGR] The totality of topographic forms, especially dunes, which comprise the moving landscape.

dunite [PETR] An ultrabasic rock consisting almost solely of a magnesium-rich olivine with some chromite and picotite; an important source of chromium.

dune movement [GEOL] In a stream, the downstream movement of sediment along the bed in a wave or dune form.

dune ridge [GEOL] A series of parallel dunes, developed along the shore of a retreating sea and stabilized by the growth of vegetation.

duplexite [MINERAL] $Ca_4BeAl_2Si_9O_{24}(OH)_2$ A white fibrous mineral consisting of hydrous beryllium calcium aluminosilicate. Also known as bavenite.

Dupont process [MIN ENG] A method for separation of minerals in which organic liquids of high specific gravity and low viscosity are used.

durain [GEOL] A hard, granular ingredient of banded coal which occurs in lenticels and shows a close, firm texture. Also known as durite.

durangite [MINERAL] $NaAlF(AsO_4)$ An orange-red, monoclinic mineral consisting of a fluoarsenate of sodium and aluminum; occurs in crystals.

Durargid [GEOL] A great soil group constituting a subdivision of the Argids, indicating those soils with a hardpan cemented by silica and called a duripan.

duricrust [GEOL] The case-hardened soil crust formed in semiarid climates by precipitation of salts; contains aluminous, ferruginous, siliceous, and calcareous material.

durinite [GEOL] The principal maceral of durain; a heterogeneous material, semiopaque in section (including all parts of plants); micrinite, exinite, cutinite, resinite, collinite, xylinite, suberinite, and fusinite may be present.

duripan [GEOL] A horizon in mineral soil characterized by cementation by silica.

durite *See* durain.

düsenwind [METEOROL] The mountain-gap wind of the Dardanelles; a strong east-northeast wind which blows out of the Dardanelles into the Aegean Sea, penetrating as far as the island of Lemnos, and caused by a ridge of high pressure over the Black Sea.

dussertite [MINERAL] $BaFe_3(AsO_4)_2(OH)_5$ A mineral consisting of a hydrous basic arsenate of barium and iron.

dust [GEOL] Dry solid matter of silt and clay size (less than $\frac{1}{16}$ millimeter).

dust and fume monitor [MIN ENG] An instrument designed to measure and record concentrations of dust, fume, and gas in mine environments over an extended period of time.

dust avalanche [GEOL] An avalanche of dry, loose snow.

dust bowl [CLIMATOL] A name given, early in 1935, to the region in the south-central United States afflicted by drought and dust storms, including parts of Colorado, Kansas, New Mexico, Texas, and Oklahoma, and resulting from a long period of deficient rainfall combined with loosening of the soil by destruction of the natural vegetation; dust bowl describes similar regions in other parts of the world.

dust devil [METEOROL] A small but vigorous whirlwind, usually of short duration, rendered visible by dust, sand, and debris picked up from the ground; diameters range from about 10–100 feet (3–30 meters), and average height is about 600 feet (180 meters).

dust-devil effect [GEOPHYS] In atmospheric electricity, rather sudden and short-lived change (positive or negative) of the vertical component of the atmospheric electric field that accompanies passage of a dust devil near an instrument sensitive to the vertical gradient.

dust horizon [METEOROL] The top of a dust layer which is confined by a low-level temperature inversion and has the appearance of the horizon when viewed from above, against the sky; the true horizon is usually obscured by the dust layer.

dust ring [GEOL] A feature occurring in a detrital sand grain which has undergone secondary enlargement, and seen in thin section as a ring of very small inclusions.

dust storm [METEOROL] A strong, turbulent wind carrying large clouds of dust.

dust veil [METEOROL] The stratospheric pall caused by ejection of dust as a result of volcanic eruptions.

dust well [HYD] A pit in an ice surface produced when small, dark particles on the ice are heated by sunshine and sink down into the ice.

dust whirl [METEOROL] A rapidly rotating column of air over a dry and dusty or sandy area, carrying dust, leaves, and other light material picked up from the ground; when well developed, it is known as a dust devil. Also known as dancing dervish; dancing devil; desert devil; sand auger; sand devil.

Dutchman's log [ENG] A buoyant object thrown overboard to determine the speed of a vessel; the time required for a known length of the vessel to pass the object is measured, and the speed can then be computed.

duty of water [HYD] The total volume of irrigation water required to mature a particular type of crop, including consumptive use, evaporation and seepage from ditches and canals, and the water eventually returned to streams by percolation and surface runoff.

D variometer *See* declination variometer.

dwey *See* dwigh.

dwigh [METEOROL] In Newfoundland, a sudden shower or snow storm. Also known as dwey; dwoy.

Dwight-Lloyd machine [MIN ENG] A continuous sintering machine in which the feed is moved on articulated plates pulled by chains in conveyor-belt fashion.

Dwight-Lloyd process [MIN ENG] Blast roasting, with air currents being drawn downward through the ore.

dwoy *See* dwigh.

Dwyka tillite [GEOL] A glacial Permian deposit that is widespread in South Africa.

dy [GEOL] A dark, gelatinous, fresh-water mud primarily of unhumified or peaty organic material.

dynamic capillary pressure [PETRO ENG] Capillary-pressure saturation curves of a core sample determined by the simultaneous steady-state flow of two fluids through the sample; capillarity pressures are determined by the difference in the pressures of the two fluids.

dynamic climatology [CLIMATOL] The climatology of atmospheric dynamics and thermodynamics, that is, a climatological approach to the study and explanation of atmospheric circulation.

dynamic forecasting *See* numerical forecasting.

dynamic geomorphology [GEOL] The quantitative analysis of steady-state, self-regulatory geomorphic processes. Also known as analytical geomorphology.

dynamic height [GEOPHYS] As measured from sea level, the distance above the geoid of points on the same equipotential surface expressed in linear units measured along a plumb line at a designated latitude (usually 45°).

dynamic-height anomaly [OCEANOGR] The excess of the actual geopotential difference, between two given isobaric surfaces, over the geopotential difference in a homogeneous water column of salinity 35 per mille and temperature 0°C. Also known as anomaly of geopotential difference.

dynamic metamorphism [GEOL] Metamorphism resulting exclusively or largely from rock deformation, principally faulting and folding. Also known as dynamometamorphism.

dynamic meteorology [METEOROL] The study of atmospheric motions as solutions of the fundamental equations of hydrodynamics or other systems of equations appropriate to special situations, as in the statistical theory of turbulence.

dynamic roughness [OCEANOGR] A quantity, designated z_o, dependent on the shape and distribution of the roughness elements of the sea surface, and used in calculations of wind at the surface. Also known as roughness length.

dynamic thickness [OCEANOGR] The vertical separation between two isobaric surfaces in the ocean.

dynamic topography [MAP] A topographic map indicating the dynamic depth of an isobaric surface.

dynamic trough [METEOROL] A pressure trough formed on the lee side of a mountain range across which the wind is blowing almost at right angles. Also known as lee trough.

dynamic vertical *See* apparent vertical.

dynamo effect [GEOPHYS] A process in the ionosphere in which winds and the resultant movement of ionization in the geomagnetic field give rise to induced current.

dynamofluidal [PETR] Referring to a dynamometamorphosed rock texture exhibiting a parallel arrangement in a single direction.

dynamo theory [GEOPHYS] The hypothesis which explains the regular daily variations in the earth's magnetic field in terms of electrical currents in the lower ionosphere, generated by tidal motions of the ionized air across the earth's magnetic field.

dynamothermal metamorphism   [GEOL]   Metamorphism that results from directed pressures, shearing stress, confining pressures, and a temperature range of about 400–800°C.

dysanalyte   [MINERAL]   A variety of the mineral perovskite in which Nb^{5+} substitutes for Ti^{5+}, and Na^+ for Ca^{2+} in the formula $Ca[TiO_3]$.

dyscrasite   [MINERAL] Ag_2Sb A gray mineral that forms rhombic crystals.

dyscrystalline   [PETR]   Referring to an igneous rock texture characterized by grains that are too small to be seen without a microscope.

dysodil   [GEOL]   A sapropelic coal of lignitic rank that burns with an unpleasant odor.

Dysodonta   [PALEON]   An order of extinct bivalve mollusks with a nearly toothless hinge and a ligament in grooves or pits.

dyster *See* doister.

dysyntribite   [MINERAL]   A hydrated sodium potassium aluminosilicate, considered to be a type of mica.

E

earlandite [MINERAL] $Ca_3(C_6H_5O_7)_2\cdot4H_2O$ A mineral consisting of a hydrous citrate of calcium; found in sediments in the Weddell Sea.

earth [GEOL] **1.** Solid component of the globe, distinct from air and water. **2.** Soil; loose material composed of disintegrated solid matter.

earth core *See* core.

earth crust *See* crust.

earth-current storm [GEOPHYS] Irregular fluctuations in an earth current in the earth's crust, often associated with electric field strengths as large as several volts per kilometer, and superimposed on the normal diurnal variation of the earth currents.

earth figure [GEOD] The shape of the earth.

earthflow [GEOL] A variety of mass movement involving the downslope slippage of soil and weathered rock in a series of subparallel sheets.

earth hummock [GEOL] A small, dome-shaped uplift of soil caused by the pressure of groundwater. Also known as earth mound.

earth inductor [ENG] A type of inclinometer that has a coil which rotates in the earth's field and in which a voltage is induced when the rotation axis does not coincide with the field direction; used to measure the dip angle of the earth's magnetic field. Also known as dip inductor; earth inductor compass; induction inclinometer.

earth inductor compass *See* earth inductor.

earth interior [GEOL] The portion of the earth beneath the crust.

earth-layer propagation [GEOPHYS] **1.** Propagation of electromagnetic waves through layers of the earth's atmosphere. **2.** Electromagnetic wave propagation through layers below the earth's surface.

earth movements [GEOPHYS] Movements of the earth, comprising revolution about the sun, rotation on the axis, precession of equinoxes, and motion of the surface of the earth relative to the core and mantle.

earth oscillations [GEOPHYS] Any rhythmic deformations of the earth as an elastic body; for example, the gravitational attraction of the moon and sun excite the oscillations known as earth tides.

earthquake [GEOPHYS] A series of suddenly generated elastic waves in the earth occurring in shallow depths to about 700 kilometers.

earthquake intensity [ENG] The measurement of the effects of an earthquake at a given location on humans and structures.

earthquake rent *See* reverse scarplet.

earthquake scarplet [GEOL] A low, nearly straight fault scarp or step which is formed in conjunction with an earthquake and is either the cause or result of that earthquake.

earthquake zone [GEOL] An area of the earth's crust in which movements, sometimes with associated volcanism, occur. Also known as seismic area.

earth radiation *See* terrestrial radiation.

earth shadow [METEOROL] Any shadow projecting into a hazy atmosphere from mountain peaks at times of sunrise or sunset.

earth's shadow *See* dark segment.

earth thermometer *See* soil thermometer.

earth tide [GEOPHYS] The periodic movement of the earth's crust caused by forces of the moon and sun. Also known as bodily tide.

earth wax *See* ozocerite.

earthy cobalt *See* asbolite.

earthy manganese *See* wad.

east [GEOD] The direction 90° to the right of north.

East Africa Coast Current [OCEANOGR] A current that is influenced by the monsoon drifts of the Indian Ocean, flowing southwestward along the Somalia coast in the Northern Hemisphere winter and northeastward in the Northern Hemisphere summer. Also known as Somali Current.

East Australia Current [OCEANOGR] The current which is formed by part of the South Equatorial Current and flows southward along the eastern coast of Australia.

Eastern Hemisphere [GEOGR] The half of the earth lying mostly to the east of the Atlantic Ocean, including Europe, Africa, and Asia.

East Greenland Current [OCEANOGR] A current setting south along the eastern coast of Greenland and carrying water of low salinity and low temperature.

eastonite [MINERAL] $K_2Mg_5AlSi_5Al_3O_{20}(OH_4)$ A mineral consisting of basic silicate of potassium, magnesium, and aluminum; it is an end member of the biotite system.

east point [GEOD] That intersection of the prime vertical with the horizon which lies to the right of the observer when facing north.

ebb-and-flow structure [GEOL] Rock strata with alternating horizontal and cross-bedded layers, believed to have been produced by ebb and flow of tides.

ebb current [OCEANOGR] The tidal current associated with the decrease in the height of a tide.

ebb tide [OCEANOGR] The portion of the tide cycle between high water and the following low water. Also known as falling tide.

ecdemite [MINERAL] $Pb_6As_2O_7Cl_4$ A greenish-yellow to yellow, tetragonal mineral consisting of an oxychloride of lead and arsenic; occurs as coatings of small tabular crystals and as coarsely foliated masses.

echelon faults [GEOL] Separate, parallel faults having steplike trends.

Echinocystitoida [PALEON] An order of extinct echinoderms in the subclass Perischoechinoidea.

echogram [ENG] The graphic presentation of echo soundings recorded as a continuous profile of the sea bottom.

echograph [ENG] An instrument used to record an echogram.

echo location *See* echo ranging.

echo ranging [ENG] Active sonar, in which underwater sound equipment generates bursts of ultrasonic sound and picks up echoes reflected from submarines, fish, and other objects within range, to determine both direction and distance to each target. Also known as echo location.

echo-ranging sonar [ENG] Active sonar, in which underwater sound equipment generates bursts of ultrasonic sound and picks up echoes reflected from submarines, fish, and other objects within range, to determine both direction and distance to each target.

echo sounder *See* sonic depth finder.

echo sounding [ENG] Determination of the depth of water by measuring the time interval between emission of a sonic or ultrasonic signal and the return of its echo from the sea bottom.

eckermannite [MINERAL] $Na_3(Mg,Li)_4(Al,Fe)Si_8O_{22}(OH,F)_2$ Mineral of the amphibole group containing magnesium, lithium, iron, and fluorine.

Eckert projection [MAP] One of six map projections of the earth's surface; in each projection the geographic poles are denoted by parallel straight lines one-half the length of the equator.

eclogite [PETR] A class of metamorphic rocks distinguished by their composition, consisting essentially of omphacite and pyrope with small amounts of diopside, enstatite, olivine, kyanite, rutile, and rarely, diamond.

eclogite facies [PETR] A type of facies composed of eclogite and formed by regional metamorphism at extremely high temperature and pressure.

ecnephias [METEOROL] A squall or thunderstorm in the Mediterranean.

economic geography [GEOGR] A branch of geography concerned with the relations of physical environment and economic conditions to the manufacture and distribution of commodities.

economic geology [GEOL] **1.** Application of geologic knowledge to materials usage and principles of engineering. **2.** The study of metallic ore deposits.

economic mineral [MINERAL] Mineral of commercial value.

ectinites [PETR] One of two major groups of metamorphic rocks comprising those formed with no accession or introduction of feldspathic material.

ectohumus [GEOL] An accumulation of organic matter on the soil surface with little or no mixing with mineral material. Also known as mor; raw humus.

Edaphosuria [PALEON] A suborder of extinct, lowland, terrestrial, herbivorous reptiles in the order Pelycosauria.

eddy correlation [METEOROL] A method of studying the effects of sea surface on the air above it by measuring simultaneous fluctuations of the horizontal and vertical components of the airflow from the mean.

eddy mill *See* pothole.

Edenian [GEOL] Lower Cincinnatian geologic stage in North America, above the Mohawkian and below Maysvillian.

edge dislocation [CRYSTAL] A dislocation which may be regarded as the result of inserting an extra plane of atoms, terminating along the line of the dislocation. Also known as Taylor-Orowan dislocation.

edge water [GEOL] Water that surrounds, borders, or lies below oil or gas in an oil-bearing rock formation.

edge wave [OCEANOGR] An ocean wave moving parallel to the coast, with crests normal to the coastline; maximum amplitude is at shore, with amplitude falling off exponentially farther from shore.

Ediacaran fauna [PALEON] The oldest known assemblage of fossil remains of soft-bodied marine animals; first discovered in the Ediacara Hills, Australia.

edingtonite [MINERAL] $BaAl_2Si_3O_{10} \cdot 4H_2O$ Gray zeolite mineral that forms rhombic crystals; sometimes contains large amounts of calcium.

Edrioasteroidea [PALEON] A class of extinct Echinozoa having ambulacral radial areas bordered by tube feet, and the mouth and anus located on the upper side of the theca.

eductor pump [MIN ENG] A pump which removes slurried material from a hydraulically disseminated subsurface ore matrix.

effective atmosphere *See* optically effective atmosphere.

effective decline rate [PETRO ENG] The drop in oil or gas production rate over a period of time; equal to unity (1 month or 1 year) divided by the production rate at the beginning of the period.

effective diameter [GEOL] The diameter of particles in a hypothetical rock or soil sample that would permit flow of water at the same rate as the sample of rock or soil under study.

effective gust velocity [METEOROL] The vertical component of the velocity of a sharp-edged gust that would produce a given acceleration on a particular airplane flown in level flight at the design cruising speed of the aircraft and at a given air density.

effective molecular weight [PETRO ENG] Empirical relationship of oil graphed against API gravity to give the effective (pseudoaverage) molecular weight of the oil for reservoir calculations.

effective pore volume [GEOL] The space occupied by pores in a soil or rock sample that is available for water circulation.

effective porosity [GEOL] A property of earth containing interconnecting interstices, expressed as a percent of bulk volume occupied by the interstices.

effective precipitable water [METEOROL] That part of the precipitable water which, in theory, can actually fall as precipitation.

effective precipitation [HYD] **1.** The part of precipitation that reaches stream channels as runoff. Also known as effective rainfall. **2.** In irrigation, the portion of the precipitation which remains in the soil and is available for consumptive use.

effective radiation *See* effective terrestrial radiation.

effective snowmelt [HYD] The part of snowmelt that reaches stream channels as runoff.

effective terrestrial radiation [GEOPHYS] The amount by which outgoing infrared terrestrial radiation of the earth's surface exceeds downcoming infrared counterradiation from the sky. Also known as effective radiation; nocturnal radiation.

efflorescence [MINERAL] A whitish powder, consisting of one or several minerals produced as an encrustation on the surface of a rock in an arid region. Also known as bloom.

effluent [HYD] **1.** Flowing outward or away from. **2.** Liquid which flows away from a containing space or a main waterway.

effluent stream [HYD] A stream that receives groundwater seepage.

effusive stage [GEOL] The second cooling stage for volcanic rocks.

eggstone *See* oolite.

eglestonite [MINERAL] Hg_4Cl_2O Rare mercuric oxide mineral; forms yellow-brown isometric crystals upon exposure to air.

Egnell's law [METEOROL] The rule stating that above any fixed place the velocity of straight or nearly straight winds in the upper half of the troposphere increases with height at roughly the same rate that the density of the air decreases.

eguëite [MINERAL] $CaFe_{14}(PO_4)_{10}(OH)_{14} \cdot 21H_2O$ A brownish-yellow mineral consisting of a hydrated basic phosphate of calcium and iron; occurs as small nodules.

Egyptian asphalt [GEOL] A glance pitch (bituminous mixture similar to asphalt) found in the Arabian Desert.

einkanter [GEOL] A stone shaped by windblown sand only upon one facet.

ejecta [GEOL] Material discharged by a volcano.

Ekman convergence [OCEANOGR] A zone of convergence of warm surface water caused by Ekman transport, creating a marked depression of the ocean's thermocline in the affected area.

Ekman current meter [ENG] A mechanical device for measuring ocean current velocity which incorporates a propeller and a magnetic compass and can be suspended from a moored ship.

Ekman layer [METEOROL] The layer of transition between the surface boundary layer of the atmosphere, where the shearing stress is constant, and the free atmosphere, which is treated as an ideal fluid in approximate geostrophic equilibrium. Also known as spiral layer.

Ekman spiral [METEOROL] A theoretical representation that a wind blowing steadily over an ocean of unlimited depth and extent and uniform viscosity would cause, in the Northern Hemisphere, the immediate surface water to drift at an angle of 45° to the right of the wind direction, and the water beneath to drift further to the right, and with slower and slower speeds, as one goes to greater depths.

Ekman transport [OCEANOGR] The movement of ocean water caused by wind blowing steadily over the surface; occurs at right angles to the wind direction.

Ekman water bottle [ENG] A cylindrical tube fitted with plates at both ends and used for deep-water samplings; when hit by a messenger it turns 180°, closing the plates and capturing the water sample.

elastic bitumen *See* elaterite.

elastic rebound theory [GEOL] A theory which attributes faulting to stresses (in the form of potential energy) which are being built up in the earth and which, at discrete intervals, are suddenly released as elastic energy; at the time of rupture the rocks on either side of the fault spring back to a position of little or no strain.

elaterite [GEOL] A light-brown to black asphaltic pyrobitumen that is moderately soft and elastic. Also known as elastic bitumen; mineral caoutchouc.

E layer [GEOPHYS] A layer of ionized air occurring at altitudes between 100 and 120 kilometers in the E region of the ionosphere, capable of bending radio waves back to earth. Also known as Heaviside layer; Kennelly-Heaviside layer.

elbow [GEOGR] A sharp change in direction of a coast line, channel, bank, or so on.

electrical blasting cap [ENG] A blasting cap ignited by electric current and not by a spark.

electrical log [ENG]   Recorded measurement of the conductivities and resistivities down the length of an uncased borehole; gives a complete record of the formations penetrated.

electrical logging [ENG]   The recording in uncased sections of a borehole of the conductivities and resistivities of the penetrated formations; used for geological correlation of the strata and evaluation of possibly productive horizons. Also known as electrical well logging.

electrical prospecting [ENG]   The use of downhole electrical logs to obtain subsurface information for geological analysis.

electrical storm [METEOROL]   A popular term for a thunderstorm.

electrical thickness [OCEANOGR]   The vertical measure between the surface of an ocean current and an isokinetic point having a value of about one-tenth the surface speed.

electrical well logging *See* electrical logging.

electric calamine *See* hemimorphite.

electric gathering locomotive *See* gathering motor.

electric surface-recording thermometer [PETRO ENG]   Device to measure temperatures during oil-well temperature surveying; has a thermocouple, resistance wire, or thermistor as the temperature-sensitive element.

electrification ice nucleus [METEOROL]   An ice nucleus that is formed by the fragmentation of dendritic crystals exposed to an electric field strength of several hundred volts per centimeter; it is a type of fragmentation nucleus.

electrodynamic drift [GEOPHYS]   Motion of charged particles in the upper atmosphere due to the combined effect of electric and magnetic fields; in the ionospheric F region and above, the drift velocity is perpendicular to both the electric and magnetic fields.

electroexplosive [ENG]   An initiator or a system in which an electric impulse initiates detonation or deflagration of an explosive.

electrofiltration [GEOL]   Counterprocess during electrical logging of well boreholes, in which mud filtrate forced through the mud cake produces an emf in the mud cake opposite a permeable bed, positive in the direction of filtrate flow.

electrogram [METEOROL]   A record, usually automatically produced, showing the time variations of the atmospheric electric field at a given point.

electrojet [GEOPHYS]   A stream of intense electric current moving in the upper atmosphere around the equator and in polar regions.

electrokinetograph [ENG]   An instrument used to measure ocean current velocities based on their electrical effects in the magnetic field of the earth.

electrolytic model [PETRO ENG]   Laboratory simulation of steady-state fluid flow through porous reservoir mediums; depends on the mobility of ions in absorbent mediums (gelatin or blotter), or through a liquid (potentiometric technique). Also known as gelatin model; oil-field model; potentiometric model.

electromagnetic logging [ENG]   A method of well logging in which a transmitting coil sets up an alternating electromagnetic field, and a receiver coil, placed in the drill hole above the transmitter coil, measures the secondary electromagnetic field induced by the resulting eddy currents within the formation. Also known as electromagnetic well logging.

electromagnetic prospecting *See* electromagnetic surveying.

electromagnetic surveying [ENG] Underground surveying carried out by generating electromagnetic waves at the surface of the earth; the waves penetrate the earth and induce currents in conducting ore bodies, thereby generating new waves that are detected by instruments at the surface or by a receiving coil lowered into a borehole. Also known as electromagnetic prospecting.

electromagnetic well logging *See* electromagnetic logging.

electrostatic coalescence [METEOROL] **1.** The coalescence of cloud drops induced by electrostatic attractions between drops of opposite charges. **2.** The coalescence of two cloud or rain drops induced by polarization effects resulting from an external electric field.

electrum [MINERAL] A naturally occurring alloy of gold with no less than 20% silver.

eleolite *See* nepheline.

elephant [METEOROL] *See* elephanta.

elephanta [METEOROL] A strong southeasterly wind on the Malabar coast of southwest India in September and October, at the end of the southwest monsoon, bringing thundersqualls and heavy rain. Also known as elephant; elephanter.

elephanter *See* elephanta.

elephant-hide pahoehoe [GEOL] A type of pahoehoe on whose surface are innumerable tummuli, broad swells, and pressure ridges which impart the appearance of elephant hide.

elerwind [METEOROL] A wind of Sun Valley north of Kufstein, in the Tyrol.

elevation angle [ENG] *See* angle of elevation.

elevation of ivory point *See* barometer elevation.

elevation tints *See* gradient tints.

ellestadite [MINERAL] A pale rose, hexagonal mineral consisting of an apatite-like calcium sulfate-silicate; occurs in granular massive form.

ellipsoidal lava *See* pillow lava.

elliptical projection [MAP] A map of the surface of the earth formed on an ellipse's interior.

ellsworthite [MINERAL] $(Ca,Na,U)_2(Nb,Ta)_2O_6(O,OH)$ A yellow, brown, greenish or black mineral of the pyrochlore group occurring in isometric crystals and consisting of an oxide of niobium, titanium, and uranium. Also known as betafite; hatchettolite.

El Niño [OCEANOGR] A warm current setting south along the coast of Peru; generally develops during February and March concurrently with a southerly shift in the tropical rain belt.

elpasolite [MINERAL] K_2NaAlF_6 Mineral composed of sodium potassium aluminum fluoride.

elpidite [MINERAL] $Na_2ZrSi_6O_{15}\cdot 3H_2O$ A white to brick-red mineral composed of hydrated sodium zirconium silicate.

Elsasser's radiation chart [METEOROL] A radiation chart developed by W. M. Elsasser for the graphical solution of the radiative transfer problems of importance in meteorology: given a radiosonde record of the vertical variation of temperature and water vapor content, one can find with this chart such quantities as the effective terrestrial radiation, net flux of infrared radiation at a cloud base or a cloud top, and radiative cooling rates.

eluvial [GEOL] Of, composed of, or relating to eluvium.

eluvial placer [GEOL]   A placer deposit that is concentrated in the vicinity of the decomposed outcrop of the source.

eluviation [HYD]   The process of transporting dissolved or suspended materials in the soil by lateral or downward water flow when rainfall exceeds evaporation.

eluvium [GEOL]   Disintegrated rock material formed and accumulated in situ or moved by the wind alone.

elvegust [METEOROL]   A cold descending squall in the upper parts of Norwegian fjords. Also known as sno.

embacle [HYD]   The piling up of ice in a stream after a refreeze, and the pile so formed.

embata [METEOROL]   A local onshore southwest wind caused by the reversal of the northeast trade winds in the lee of the Canary Islands.

embatholithic [GEOL]   Pertaining to ore deposits associated with a batholith where exposure of the batholith and country rock is about equal.

embayed [GEOGR]   Formed into a bay.   [NAV]   Pertaining to a vessel in a bay unable to put to sea or to put to sea safely because of wind, current, or sea.

embayed coastal plain [GEOL]   A coastal plain that has been partly sunk beneath the sea, thereby forming a bay.

embayed mountain [GEOL]   A mountain that has been depressed enough for sea water to enter the bordering valleys.

embayment [GEOGR]   Indentation in a shoreline forming a bay.   [GEOL]   **1.** Act or process of forming a bay.   **2.** A reentrant of sedimentary rock into a crystalline massif.

embolite [MINERAL] $Ag(Cl,Br)$ A yellow-green mineral resembling cerargyrite; composed of native silver chloride and silver bromide.

Embolomeri [PALEON]   An extinct side branch of slender-bodied, fish-eating aquatic anthracosaurs in which intercentra as well as centra form complete rings.

embouchure [GEOL]   **1.** The mouth of a river.   **2.** A river valley widened into a plain.

embrechites [PETR]   A type of migmatite in which structural features of crystalline shifts are preserved but often partially obliterated by metablastesis.

Embrithopoda [PALEON]   An order established for the unique Oligocene mammal *Arsinoitherium*, a herbivorous animal that resembled the modern rhinoceros.

emerald [MINERAL] $Al_2(Be_3Si_6O_{18})$ A brilliant-green to grass-green gem variety of beryl that crystallizes in the hexagonal system; green color is caused by varying amounts of chromium. Also known as smaragd.

emerged shoreline *See* shoreline of emergence.

emergence [GEOL]   **1.** Dry land which was part of the ocean floor.   **2.** The act or process of becoming an emergent land mass.

emery rock [PETR]   A rock that contains corundum and iron ores.

emissary sky [METEOROL]   A sky of cirrus clouds which are either isolated or in small, separated groups; so called because this formation often is one of the first indications of the approach of a cyclonic storm.

emmonsite [MINERAL] $FE_2Te_3O_9 \cdot 2H_2O$ Yellow-green mineral composed of a hydrous oxide of iron and tellurium.

emplacement [GEOL]   Intrusion of igneous rock or development of an ore body in older rocks.

emplectite [MINERAL] CuBiS₃ A grayish or white mineral that crystallizes in the orthorhombic system; occurs in masses.

empressite [MINERAL] AgTe An opaque, pale-bronze mineral whose crystal system is unknown.

emulsion texture [GEOL] An ore texture that displays minute inclusions of one mineral randomly distributed in another mineral.

Enaliornithidae [PALEON] A family of extinct birds assigned to the order Hesperornithiformes, having well-developed teeth found in grooves in the dentary and maxillary bones of the jaws.

enargite [MINERAL] A lustrous, grayish-black mineral which is found in orthorhombic crystals but is more commonly columnar, bladed, or massive; hardness is 3 on Mohs scale, specific gravity is 4.44; in some places enargite is a valuable copper ore. Also known as clairite; luzonite.

encrinal limestone [GEOL] A limestone consisting of more than 10% but less than 50% of fossil crinoidal fragments.

encrinite [PALEON] One of certain fossil crinoids, especially of the genus *Encrinus*.

endellite [MINERAL] Al₂Sl₂O₅(OH)₄·4H₂O Term used in the United States for a clay mineral, the more hydrous form of halloysite. Also known as hydrated halloysite; hydrohalloysite; hydrokaolin.

endlichite [MINERAL] A mineral similar to vanadinite, but with the vanadium replaced by arsenic.

end member [MINERAL] One of the two or more pure chemical compounds that enters into solid solution with other pure chemical compounds to make up a series of minerals of similar crystal structure (that is, an isomorphous, solid-solution series).

end moraine [GEOL] An accumulation of drift in the form of a ridge along the border of a valley glacier or ice sheet.

endobatholithic [GEOL] Pertaining to ore deposits along projecting portions of a batholith.

endocast *See* steinkern.

endogenetic *See* endogenic.

endogenic [GEOL] Of or pertaining to a geologic process, or its resulting feature such as a rock, that originated within the earth. Also known as endogenetic; endogenous.

endogenous [GEOL] *See* endogenic.

endometamorphism [GEOL] A phase of contact metamorphism involving changes in an igneous rock due to assimilation of portions of the rocks invaded by its magma.

endoreic [GEOL] Referring to a region which drains to an interior closed basin.

Endotheriidae [PALEON] A family of Cretaceous insectivores from China belonging to the Proteutheria.

Endothyracea [PALEON] A superfamily of extinct benthic marine foraminiferans in the suborder Fusulinina, having a granular or fibrous wall.

en echelon [GEOL] Referring to an overlapped or staggered arrangement of geologic features.

en echelon fault blocks [GEOL] A belt in which the individual fault blocks trend approximately 45° to the trend of the entire fault belt.

energy level [GEOL]   In an aqueous sedimentary environment, the kinetic energy that is supplied by the action of waves or current either at the interface of deposition or at a level several meters above.

energy transfer [METEOROL]   The transfer of energy of a given form among different scales of motion; for example, kinetic energy may be transferred between the zonal and meridional components of the wind, or between the mean and eddy components of the wind.

englacial [HYD]   Of or pertaining to the inside of a glacier.

englishite [MINERAL] $K_2Ca_4Al_8(PO_4)_8(OH)_{10} \cdot 9H_2O$ A white mineral composed of hydrous basic phosphate of potassium, calcium, and aluminum.

engysseismology [GEOPHYS]   Seismology dealing with earthquake records made close to the disturbance.

enigmatite [MINERAL] $Na_2Fe_5TiSi_6O_{20}$ A black amphibole mineral occurring in triclinic crystals; specific gravity is 3.14–3.80. Also spelled aenigmatite.

ensialic geosyncline [GEOL]   A geosyncline whose geosynclinal prism accumulates on a sialic crust and contains clastics.

ensimatic geosyncline [GEOL]   A geosyncline whose geosynclinal prism accumulates on a simatic crust and is composed largely of volcanic rock or sediments of volcanic debris.

enstatite [MINERAL] $MgOSiO_2$ A member of the pyroxene mineral group that crystallizes in the orthorhombic system; usually yellowish gray but becomes green when a little iron is present.

Enteletacea [PALEON]   A group of extinct articulate brachiopods in the order Orthida.

Entelodontidae [PALEON]   A family of extinct palaeodont artiodactyls in the superfamily Entelodontoidea.

Entelodontoidea [PALEON]   A superfamily of extinct piglike mammals in the suborder Palaeodonta having huge skulls and enlarged incisors.

Entisol [GEOL]   An order of soil having few or faint horizons.

Entomoconchacea [PALEON]   A superfamily of extinct marine ostracods in the suborder Myodocopa that are without a rostrum above the permanent aperture.

entrail pahoehoe [GEOL]   A type of pahoehoe having a surface that resembles an intertwined mass of entrails.

entrainment [HYD]   The pickup and movement of sediment as bed load or in suspension by current flow.   [METEOROL]   The mixing of environmental air into a preexisting organized air current so that the environmental air becomes part of the current.   [OCEANOGR]   The transfer of fluid by friction from one water mass to another, usually occurring between currents moving in respect to each other.

entrance region [METEOROL]   The region of confluence at the upwind extremity of a jet stream.

entrapment [GEOL]   The underground trapping of oil or gas reserves by folds, faults, domes, asphaltic seals, unconformities, and such.

entrenched meander [HYD]   A deepened meander of a river which is carried downward further below the valley surface in which the meander originally formed. Also known as inherited meander.

entrenched stream [HYD]   A stream that flows in a valley or narrow trench cut into a plain or relatively level upland. Also spelled intrenched stream.

environmental lapse rate [METEOROL] The rate of decrease of temperature with elevation in the atmosphere. Also known as atmospheric lapse rate.

environment of sedimentation [GEOL] A more or less destructive geomorphologic setting in which sediments are deposited as beach environment.

Eocambrian [GEOL] Pertaining to the thick sequences of strata conformably underlying Lower Cambrian fossils. Also known as Infracambrian.

Eocene [GEOL] The next to the oldest of the five major epochs of the Tertiary period (in the Cenozoic era).

Eocrinoidea [PALEON] A class of extinct echinoderms in the subphylum Crinozoa that had biserial brachioles like those of cystoids combined with a theca like that of crinoids.

Eogene *See* Paleogene.

Eohippus [PALEON] The earliest, primitive horse, included in the genus *Hyracotherium;* described as a small, four-toed species.

eolation [GEOL] Any action of wind on the land.

eolian [METEOROL] Pertaining to the action or the effect of the wind, as in eolian sounds or eolian deposits (of dust). Also spelled aeolian.

eolian anemometer [ENG] An anemometer which works on the principle that the pitch of the eolian tones made by air moving past an obstacle is a function of the speed of the air.

eolian dune [GEOL] A dune resulting from entrainment of grains by the flow of moving air.

eolian erosion [GEOL] Erosion due to the action of wind.

eolianite [GEOL] A sedimentary rock consisting of clastic material which has been deposited by wind.

eolian ripple mark [GEOL] A mark made in sand by the wind.

eolian sand [GEOL] Deposits of sand arranged by the wind.

eolian soil [GEOL] A type of soil ranging from sand dunes to loess deposits whose particles are predominantly of silt size.

Eomoropidae [PALEON] A family of extinct perissodactyl mammals in the superfamily Chalicotherioidea.

eosphorite [MINERAL] $(Mn,Fe)Al(PO_4)(OH)_2 \cdot H_2O$ A usually rose-pink mineral composed of hydrous aluminum manganese phosphate, found massive or in prismatic crystals.

Eosuchia [PALEON] The oldest, most primitive, and only extinct order of lepidosaurian reptiles.

eötvös [GEOPHYS] A unit of horizontal gradient of gravitational acceleration, equal to a change in gravitational acceleration of 10^{-9} galileo over a horizontal distance of 1 centimeter.

Eötvös torsion balance [ENG] An instrument which records the change in the acceleration of gravity over the horizontal distance between the ends of a beam; used to measure density variations of subsurface rocks.

epeiric sea *See* epicontinental sea.

epeirogeny [GEOL] Movements which affect large tracts of the earth's crust.

ephemeral stream [HYD] A stream channel which carries water only during and immediately after periods of rainfall or snowmelt.

epicenter [GEOL] A point on the surface of the earth which is directly above the seismic focus of an earthquake and where the earthquake vibrations reach first.

epiclastic [GEOL] Pertaining to the texture of mechanically deposited sediments consisting of detrital material from preexistent rocks.

epicontinental [GEOL] Located upon a continental plateau or platform.

epicontinental sea [OCEANOGR] That portion of the sea lying upon the continental shelf, and the portions which extend into the interior of the continent with similar shallow depths. Also known as epeiric sea; inland sea.

epidiorite [PETR] A dioritic rock formed by alteration of pyroxenic igneous rocks.

epidosite [PETR] A rare metamorphic rock composed of epidote and quartz.

epidote [MINERAL] A pistachio-green to blackish-green calcium aluminum sorosilicate mineral that crystallizes in the monoclinic system; the luster is vitreous, hardness is 6½ on Mohs scale, and specific gravity is 3.35–3.45.

epidote-amphibolite facies [PETR] Metamorphic rocks formed under pressures of 3000–7000 bars and temperatures of 250–450°C with conditions intermediate between those that formed greenschist and amphibolite, or with characteristics intermediate.

epidotization [GEOL] The introduction of epidote into, or the formation of epidote from, rocks.

epieugeosyncline [GEOL] Deep troughs formed by subsidence which have limited volcanic power and overlie a eugeosyncline.

epigene [GEOL] **1.** A geologic process originating at or near the earth's surface. **2.** A structure formed at or near the earth's surface.

epigenesis [GEOL] Alteration of the mineral content of rock due to outside influences.

epigenetic [GEOL] Produced or formed at or near the surface of the earth.

epigenite [MINERAL] $(Cu,Fe)_5AsS_6$ A steel gray, orthorhombic mineral consisting of copper and iron arsenic sulfide.

epilimnion [HYD] A fresh-water zone of relatively warm water in which mixing occurs as a result of wind action and convection currents.

epimagma [GEOL] A gas-free, vesicular to semisolid magmatic residue of pasty consistency formed by cooling and loss of gas from liquid lava in a lava lake.

epimagmatic *See* deuteric.

epipelagic [OCEANOGR] Of or pertaining to the portion of oceanic zone into which enough light penetrates to allow photosynthesis.

epipelagic zone [OCEANOGR] The region of an ocean extending from the surface to a depth of about 200 meters; light penetrates this zone, allowing photosynthesis.

episode [GEOL] A distinctive event or series of events in the geologic history of a region or feature.

epistilbite [MINERAL] $CaAl_2Si_6O_{16} \cdot 5H_2O$ A mineral of the zeolite family that contains calcium and aluminosilicate and crystallizes in the monoclinic system; occurs in white prismatic crystals or granular forms.

epithermal [GEOL] Pertaining to mineral veins and ore deposits formed from warm waters at shallow depth, at temperatures ranging from 50–200°C, and generally at some distance from the magnetic source.

epithermal deposit [GEOL] Ore deposit formed in and along openings in rocks by deposition at shallow depths from ascending hot solutions.

epizone [GEOL] **1.** The zone of metamorphism characterized by moderate temperature, low hydrostatic pressure, and powerful stress. **2.** The outer depth zone of metamorphic rocks.

epsomite [MINERAL] $MgSO_4 \cdot 7H_2O$ A mineral that occurs in clear, needlelike, orthorhombic crystals; commonly, it is massive or fibrous; luster varies from vitreous to milky, hardness is 2–2.5 on Mohs scale, and specific gravity is 1.68; it has a salty bitter taste and is soluble in water. Also known as epsom salt.

epsom salt *See* epsomite.

equal-area latitude *See* authalic latitude.

equal-area map projection [MAP] A map projection having a constant area scale; it is not conformal and is not used for navigation. Also known as authalic map projection; equivalent map projection.

equant [GEOL] Referring to a sedimentary particle whose length is less than 1.5 times its width. [PET] Referring to an igneous or sedimentary rock crystal in which the diameters are equal or almost equal in all directions.

equator [GEOD] The great circle around the earth, equally distant from the North and South poles, which divides the earth into the Northern and Southern hemispheres; the line from which latitudes are reckoned.

equatorial air [METEOROL] The air of the doldrums or the equatorial trough; distinguished somewhat vaguely from the tropical air of the trade-wind zones.

equatorial axis [GEOD] The diameter of the earth described between two points on the equator.

equatorial bulge [GEOD] The excess of the earth's equatorial diameter over the polar diameter.

equatorial calms *See* doldrums.

equatorial chart [MAP] A chart on an equatorial projection.

equatorial convergence zone *See* intertropical convergence zone.

Equatorial Countercurrent [OCEANOGR] An ocean current flowing eastward (counter to and between the westward-flowing North Equatorial Current and South Equatorial Current) through all the oceans.

Equatorial Current *See* North Equatorial Current; South Equatorial Current.

equatorial cylindrical orthomorphic chart *See* Mercator chart.

equatorial dry zone [CLIMATOL] An arid region existing in the equatorial trough; the most famous dry zone is situated a little south of the equator in the central Pacific. Also known as arid zone.

equatorial easterlies [METEOROL] The trade winds in the summer hemisphere when they are very deep, extending at least 8 to 10 kilometers in altitude, and generally not topped by upper westerlies; if upper westerlies are present, they are too weak and shallow to have an influence on the weather. Also known as deep easterlies; deep trades.

equatorial electrojet [GEOPHYS] A concentration of electric current in the atmosphere found in the magnetic equator.

equatorial front *See* intertropical front.

equatorial projection [MAP] A map projection centered on the equator.

equatorial radius [GEOD] The radius assigned to the great circle making up the terrestrial equator; approximately 6,378,099 ± 116 meters.

equatorial tide [OCEANOGR]   **1.** A lunar fortnightly tide. **2.** A tidal component with a period of 328 hours.

equatorial trough [METEOROL]   The quasicontinuous belt of low pressure lying between the subtropical high-pressure belts of the Northern and Southern hemispheres. Also known as meteorological equator.

Equatorial Undercurrent [OCEANOGR]   **1.** A subsurface current flowing from west to east in the Indian Ocean near the 150-meter depth at the equator during the time of the Northeast Monsoon. **2.** A permanent subsurface current in the equatorial region of the Atlantic and Pacific oceans.

equatorial vortex [METEOROL]   A closed cyclonic circulation with the equatorial trough.

equatorial wave [METEOROL]   A wavelike disturbance of the equatorial easterlies that extends across the equatorial trough.

equatorial westerlies [METEOROL]   The westerly winds occasionally found in the equatorial trough and separated from the mid-latitude westerlies by the broad belt of easterly trade winds.

equigeopotential surface *See* geopotential surface.

equiglacial line [GEOL]   A line on a map or chart showing a similarity in ice conditions at a specific time.

equigranular [PETR]   Pertaining to the texture of rocks whose essential minerals are all of the same order of size.

equilibrium gas saturation [PETRO ENG]   Condition of zero relative permeability of a nonwetting/wetting phase system in a reservoir; relation to the nonwetting phase (for example, oil) to the wetting phase (for example, water) when the nonwetting-phase saturation is so small that relatively few pores contain it. Also known as critical gas saturation.

equilibrium solar tide [GEOPHYS]   The form of the atmosphere which is determined solely by gravitational forces in the absence of any rotation of the earth relative to the sun.

equilibrium spheroid [GEOPHYS]   The shape that the earth would attain if it were entirely covered by a tideless ocean of constant depth.

equilibrium stage [ENG]   In hypsometric analysis of drainage basins, the stage in which a steady state is developed and maintained as relief slowly diminishes, corresponding to maturity and old age in the geomorphic cycle.

equilibrium theory [OCEANOGR]   A model of ocean waters which assumes the instant response of water bodies to the tide-producing forces produced by the moon and sun to form an equilibrium surface, and ignores the effects of friction, inertia, and irregular distribution of land masses.

equilibrium tide [OCEANOGR]   The hypothetical tide due to the tide-producing forces of celestial bodies, particularly the sun and moon.

equinoctial rains [METEOROL]   Rainy seasons which occur regularly at or shortly after the equinoxes in many places within a few degrees of the equator.

equinoctial storm [METEOROL]   In semipopular belief, a violent storm of wind and rain which is supposed, both in the United States and in Britain, to occur at or near the time of the equinox. Also known as line gale; line storm.

equinoctial tide [OCEANOGR]   A tide occurring near an equinox.

equiparte [METEOROL]   In Mexico, heavy cold rains during October to January, which last for several days. Also known as equipatos.

equipatos *See* equiparte.

equiphase zone [GEOPHYS] That region in space where the difference in phase of two radio signals is indistinguishable.

equipotential surface [GEOPHYS] A surface characterized by the potential being constant everywhere on it for the attractive forces concerned.

equipressure contour [PETRO ENG] Within a reservoir, a plot or map of the equal isopressure flow network; used to locate sites for water-injection wells for flood coverage of an areal reservoir pattern.

equity crude [PETRO ENG] Crude produced which belongs to an oil company that owns a concession jointly with a host government.

equivalent-barotropic model [METEOROL] A model atmosphere characterized by frictionless and adiabatic flow and by hydrostatic quasigeostrophic equilibrium, and in which the vertical shear of the horizontal wind is assumed to be proportional to the horizontal wind itself.

equivalent diameter *See* nominal diameter.

equivalent height *See* virtual height.

equivalent map projection *See* equal-area map projection.

equivalent potential temperature [METEOROL] The potential temperature corresponding to the adiabatic equivalent temperature.

equivalent resistance [ELEC] Concentrated or lumped resistance that would cause the same power loss as the actual small resistance values distributed throughout a circuit.

equivalent tail wind [NAV] A fictitious wind blowing along the track of an aircraft in the same direction as that of motion of the aircraft and of such speed that it would result in the same ground speed as that actually attained.

equivalent temperature [METEOROL] **1.** The temperature that an air parcel would have if all water vapor were condensed out at constant pressure, the latent heat released being used to heat the air. Also known as isobaric equivalent temperature. **2.** The temperature that an air parcel would have after undergoing the following theoretical process: dry-adiabatic expansion until saturated, pseudoadiabatic expansion until all moisture is precipitated out, and dry adiabatic compression to the initial pressure; this is the equivalent temperature as read from a thermodynamic chart and is always greater than the isobaric equivalent temperature. Also known as adiabatic equivalent temperature; pseudoequivalent temperature.

equivalent vapor volume [PETRO ENG] The volume occupied by a barrel of oil if all the oil were to become a vapor; expressed as cubic foot per barrel at 60°F (15.6°C).

equivoluminal wave *See* S wave.

era [GEOL] A division of geologic time of the highest order, comprising one or more periods.

eradiation *See* terrestrial radiation.

erathem [GEOL] The largest accepted time-stratigraphic unit.

Erian [GEOL] Middle Devonian geologic time; a North American provincial series.

Erian orogeny [GEOL] One of the orogenies during Phanerozoic geologic time, at the end of the Silurian; the last part of the Caledonian orogenic era. Also known as Hibernian orogeny.

erikite [MINERAL] A brown mineral consisting of a silicate and phosphate of cerium metals; occurs in orthorhombic crystals.

erinite [MINERAL] $Cu_5(OH)_4(AsO_4)_2$ Emerald-green mineral composed of basic copper arsenate.

erionite [MINERAL] A chabazite mineral of the zeolite family that contains calcium ions and crystallizes in the hexagonal system.

eroding stress [GEOL] The shear stress of overland flow which is available to dislodge soil material per unit area.

eroding velocity [GEOL] The minimum average velocity required for eroding homogeneous material of a given particle size.

erosion [GEOL] **1.** The loosening and transportation of rock debris at the earth's surface. **2.** The wearing away of the land, chiefly by rain and running.

erosional unconformity [GEOL] The surface that separates older, eroded rocks from younger, overlying sediments.

erosion cycle [GEOL] A postulated sequence of conditions through which a new landmass proceeds as it wears down, classically the concept of youth, maturity, and old age, as stated by W.M. Davis; an original landmass is uplifted above base level, cut by canyons, gradually converted into steep hills and wide valleys, and is finally reduced to a flat lowland at or near base level.

erosion integral [GEOL] An expression of the relative volume of a landmass erosionally depleted at a given contour; the inverse of the hypsometric integral.

erosion pavement [GEOL] A layer of pebbles and small rocks that prevents the soil underneath from eroding.

erosion platform *See* wave-cut platform.

erosion ridge [HYD] One of a group of ridges on the surface of snow; formed by the corrosive action of wind-blown snow.

erosion surface [GEOL] A land surface shaped by agents of erosion.

erratic [GEOL] A rock fragment that has been transported a great distance, generally by glacier ice or floating ice, and differs from the bedrock on which it rests.

ertor [METEOROL] The effective (radiational) temperature of the ozone layer (region).

eruption [GEOL] The ejection of solid, liquid, or gaseous material from a volcano.

eruption cloud [GEOL] A cloud formed by an erupting volcano, consisting of gases and solid particles or fragments of erupted material.

eruptive rock [PETR] **1.** Rock formed from a volcanic eruption. **2.** Igneous rock that reaches the earth's surface in a molten condition.

erythrine *See* erythrite.

erythrite [MINERAL] $Co_3(AsO_4)_2 \cdot 8H_2O$ A crimson, peach, or pink-red secondary oxidized cobalt mineral that occurs in monoclinic crystals, in globular and reniform masses, or in earthy forms. Also known as cobalt bloom; cobalt ocher; erythrine; peachblossom ore; red cobalt.

erythrosiderite [MINERAL] $K_2FeCl_5 \cdot H_2O$ Mineral composed of hydrous potassium iron chloride; occurs in lavas.

esboite [PETR] A form of diorite in which the dominant plagioclase is andesine or oligocase in orbicular form.

escar *See* esker.

escarpment [GEOL] A cliff or steep slope of some extent, generally separating two level or gently sloping areas, and produced by erosion or faulting. Also known as scarp.

eschar *See* esker.

eschwegeite *See* tanteuxenite.

eschynite [MINERAL] $(Ce,Ca,Fe,Th)(Ti,Cb)_2O_6$ A black mineral, occurring in prismatic crystals; a rare oxide of cesium, titanium, and other metals, which is isomorphous with priorite.

eskar *See* esker.

eskebornite [MINERAL] $CuFeSe_2$ The selenium analog of the mineral pyrrhotite $(Fe_{1-x}S)$.

esker [GEOL] A sinuous ridge of constructional form, consisting of stratified accumulations, glacial sand, and gravel. Also known as asar; back furrow; escar; eschar; eskar; osar; serpent kame.

eskolaite [MINERAL] Cr_2O_3 A mineral which is an isomorph of hematite.

espalier drainage *See* trellis drainage.

essential mineral [MINERAL] A mineral that is a necessary component in the classification and nomenclature of a rock, but that may not be present in large quantities.

essexite [PETR] A rock of igneous origin composed principally of plagioclase hornblende, biotite, and titanaugite.

establishment [OCEANOGR] The interval of time between the transit (upper or lower) of the moon and the next high water at a place.

estuarine deposit [GEOL] A sediment deposited at the heads and floors of estuaries.

estuarine environment [OCEANOGR] The physical conditions and influences of an estuary.

estuarine oceanography [OCEANOGR] The study of the chemical, physical, biological, and geological properties of estuaries.

estuary [GEOGR] A semienclosed coastal body of water which has a free connection with the open sea and within which sea water is measurably diluted with fresh water. Also known as branching bay; drowned river mouth; firth.

etching [GEOL] Generally, the development of a landform through erosion or chiseling.

etesian climate *See* Mediterranean climate.

etesians [METEOROL] The prevailing northerly winds in summer in the eastern Mediterranean, and especially the Aegean Sea; basically similar to the monsoon and equivalent to the maestro of the Adriatic Sea.

ethmolith [GEOL] A downward tapering, funnel-shaped, discordant intrusion of igneous rocks.

etindite [PETR] An extrusive rock having a composition that is intermediate between that of leucitite and nephelinite, with phenocrysts of augite in a dark, dense matrix.

ettringite [MINERAL] $Ca_6Al_2(SO_4)_3(OH)_{12} \cdot 26H_2O$ A mineral composed of hydrous basic calcium and aluminum sulfate.

eucairite [MINERAL] $CuAgSe$ A white, native selenide that crystallizes in the isometric crystal system.

euchlorin [MINERAL] $(K,Na)_8Cu_9(SO_4)_{10}(OH)_6$ An emerald-green mineral consisting of a basic sulfate of potassium, sodium, and copper; found in lava at Vesuvius.

euchroite [MINERAL] $Cu_2(AsO_4)(OH) \cdot 3H_2O$ An emerald green or leek green, orthorhombic mineral consisting of a hydrated basic copper arsenate.

euclase [MINERAL] $BeAlSiO_4(OH)$ A brittle, pale green, blue, yellow, or violet monoclinic mineral, occurring as prismatic crystals.

eucrite [MINERAL]   An olivine-bearing gabbro containing unusually calcic plagiocase; a meteorite component.

eucryptite [MINERAL] $LiAlSiO_4$ A colorless or white lithium aluminum silicate mineral, crystallizing in the hexagonal system; specific gravity is 2.67.

eudialite [MINERAL] $(Na,Ca,Fe)_6ZrSi_6O_{18}(OH,Cl)$ Hexagonal-crystalline silicate chloride mineral; color is red to brown.

eudidymite [MINERAL] $NaBeSi_3O_7(OH)$ A glassy white mineral composed of sodium beryllium silicate.

eugeogenous [PETR]   Referring to a rock that is very susceptible to weathering, thus producing a large amount of loose material.

eugeosyncline [GEOL]   The internal volcanic belt of an orthogeosyncline.

euhedral *See* automorphic.

Eulerian nutation *See* Chandler wobble.

Eulerian wind [METEOROL]   A wind motion only in response to the pressure force; the cyclostrophic wind is a special case of the Eulerian wind, which is limited in its meteorological applicability to those situations in which the Coriolis effect is negligible.

eulittoral [OCEANOGR]   A subdivision of the benthic division of the littoral zone of the marine environment, extending from high-tide level to about 60 meters, the lower limit for abundant growth of attached plants.

eulysite [PETR]   A granular pyroxene peridotite containing manganese-rich fayalite, garnet, and magnetite.

eulytine *See* eulytite.

eulytite [MINERAL] $Bi_4Si_3O_{12}$ A bismuth silicate mineral usually found as minute dark-brown or gray tetrahedral crystals; specific gravity is 6.11. Also known as agricolite; bismuth blende; eulytine.

Euomphalacea [PALEON]   A superfamily of extinct gastropod mollusks in the order Aspidobranchia characterized by shells with low spires, some approaching bivalve symmetry.

eupelagic *See* pelagic.

euphotic [OCEANOGR]   Of or constituting the upper levels of the marine environment down to the limits of effective light penetration for photosynthesis.

Euproopacea [PALEON]   A group of Paleozoic horseshoe crabs belonging to the Limulida.

Europe [GEOGR]   A great western peninsula of the Eurasian landmass, usually called a continent; its eastern limits are arbitrary and are conventionally drawn along the water divide of the Ural Mountains, the Ural River, the Caspian Sea, and the Caucasus watershed to the Black Sea.

Euryapsida [PALEON]   A subclass of fossil reptiles distinguished by an upper temporal opening on each side of the skull.

Eurychilinidae [PALEON]   A family of extinct dimorphic ostracods in the superfamily Hollinacea.

Eurymylidae [PALEON]   A family of extinct mammals presumed to be the ancestral stock of the order Lagomorpha.

Eurypterida [PALEON]   A group of extinct aquatic arthropods in the subphylum Chelicerata having elongate-lanceolate bodies encased in a chitinous exoskeleton.

eustacy [OCEANOGR]   Worldwide fluctuations of sea level due to changing capacity of the ocean basins or the volume of ocean water.

eutaxite [PETR]   A rock exhibiting eutaxitic structure.

eutaxitic [PETR]   Referring to the banded structure in some extrusive rocks resulting from alternation of layers of different texture, composition, or color.

eutectofelsite *See* eutectophyre.

eutectophyre [PETR]   A light-colored tufflike igneous rock exhibiting a network of interlocking quartz and orthoclase crystals. Also known as eutectofelsite.

Euthacanthidae [PALEON]   A family of extinct acanthodian fishes in the order Climatiiformes.

eutrophic [HYD]   Pertaining to a lake containing a high concentration of dissolved nutrients; often shallow, with periods of oxygen deficiency.

euxenite [MINERAL]   A brownish-black rare-earth mineral that crystallizes in the orthorhombic system, contains oxide of calcium, cerium, columbium, tantalum, titanium, and uranium, and has a metallic luster; hardness is 6.5 on Mohs scale, and specific gravity is 4.7–5.0.

euxinic [HYD]   Of or pertaining to an environment of restricted circulation and stagnant or anaerobic conditions.

evansite [MINERAL] $Al_3(PO_4)(OH)_6 \cdot 6H_2O$ A colorless to milky white mineral consisting of a hydrated basic aluminum phosphate; occurs in massive form and as stalactites.

evaporation capacity *See* evaporative power.

evaporation current [OCEANOGR]   An ocean current resulting from the accumulation of water through precipitation and river runoff at one point, and loss by evaporation at another point.

evaporation gage *See* atmometer.

evaporation power *See* evaporative power.

evaporative capacity *See* evaporative power.

evaporative power [METEOROL]   A measure of the degree to which the weather or climate of a region is favorable to the process of evaporation; it is usually considered to be the rate of evaporation, under existing atmospheric conditions, from a surface of water which is chemically pure and has the temperature of the lowest layer of the atmosphere. Also known as evaporation capacity; evaporation power; evaporative capacity; evaporativity; potential evaporation.

evaporativity *See* evaporative power.

evaporimeter *See* atmometer.

evaporite [GEOL]   Deposits of mineral salts from sea water or salt lakes due to evaporation of the water.

evapotranspiration [HYD]   Discharge of water from the earth's surface to the atmosphere by evaporation from lakes, streams, and soil surfaces and by transpiration from plants. Also known as fly-off; total evaporation; water loss.

evapotranspirometer [ENG]   An instrument which measures the rate of evapotranspiration; consists of a vegetation soil tank so designed that all water added to the tank and all water left after evapotranspiration can be measured.

event [GEOL]   An incident of probable tectonic significance, but whose full implications are unknown.

evjite [PETR] · A gabbro of hornblende in which the only light-colored mineral is labradorite or bytownite; hornblende must be primary, not uralitic.

evorsion [GEOL]   The process of pothole formation in riverbeds; plays an important role in denudation.

evorsion hollow *See* pothole.

excess argon [GEOCHEM]   Argon-40, not resulting from radioactive decay, that is trapped in rocks or minerals at the time of their crystallization or formation.

excessive precipitation [METEOROL]   Precipitation (generally in the form of rain) of an unusually high rate of fall; although often used qualitatively, several meteorological services have adopted quantitative limits.

excess pore pressure [GEOPHYS]   Transient pore pressure in an aquitard or aquiclude beyond the pressure that would exist at the point under consideration if steady-flow conditions had been attained throughout the bed.

exchange capacity [GEOL]   The ability of a soil material to participate in ion exchange; measured by the quantity of exchangeable ions in a given unit of the material.

exfoliation *See* sheeting.

exfoliation joint *See* sheeting structure.

exhalation [GEOPHYS]   The process by which radioactive gases escape from the surface layers of soil or loose rock, where they are formed by decay of radioactive salts.

exhaust trail [METEOROL]   A visible condensation trail (contrail) that forms when the water vapor of an aircraft exhaust is mixed with and saturates (or slightly supersaturates) the air in the wake of the aircraft.

exhumation [GEOL]   The uncovering by erosion of a preexisting surface or feature that had been buried by subsequent deposits.

exhumed *See* resurrected.

exinite [GEOL]   A hydrogen-rich maceral group consisting of spore exines, cuticular matter, resins, and waxes; includes sporinite, cutinite, alginite, and resinite. Also known as liptinite.

exit region [METEOROL]   The region of difluence at the downwind extremity of a jet stream.

exocline [GEOL]   An inverted anticline or syncline.

exogene effect [PETR]   The effect of an igneous mass upon the rock it invades.

exogenous inclusion *See* xenolith.

exogeosyncline [GEOL]   A parageosyncline that lies along the cratonal border and obtains its clastic sediments from erosion of the adjacent orthogeosynclinal belt outside the craton. Also known as deltageosyncline; foredeep; transverse basin.

exomorphic zone *See* aureole.

exomorphism [PETR]   A change in a rock mass caused by intrusion of external igneous material; in the usual sense, contact metamorphism.

exorheic [GEOL]   Referring to a basin or region characterized by external drainage.

exosphere [METEOROL]   An outermost region of the atmosphere, estimated at 500–1000 kilometers, where the density is so low that the mean free path of particles depends upon their direction with respect to the local vertical, being greatest for upward-traveling particles. Also known as region of escape.

exotic stream [HYD] A stream that crosses a desert as it flows to the sea, or any stream which derives most of its water from the drainage system of another region.

expanded foot [HYD] A broad lobe or fanlike mass of ice resulting from the lower part of a valley gacier extending onto an adjacent lowland at the foot of a mountain slope.

expansion fissures [GEOL] A system of fissures which radiate randomly and pass through feldspars and other minerals adjacent to olivine crystals that have been replaced by serpentine.

expansion joint [GEOL] *See* sheeting structure.

expansion system [PETRO ENG] Gas-liquid recovery system in which the refrigeration effect of rapidly depressurized well-stream effluent through a wellhead choke is used to obtain maximum removal of liquefiable hydrocarbons from the gas stream.

experimental petrology [PETR] A branch of petrology in which phenomena that occur during petrological processes are reproduced and studied in the laboratory.

exploding-bomb texture [MINERAL] A pattern of pyrite replacement by copper sulfides in mineral deposits, so that there remain scattered pyrite fragments surrounded by copper minerals.

exploitation [MIN ENG] The extraction from the earth and utilization of ore, gas, oil, and minerals found by exploration.

exploration [MIN ENG] The search for economic deposits of minerals, ore, gas, oil, or coal by geological surveys, geophysical prospecting, boreholes and trial pits, or surface or underground headings, drifts, or tunnels.

exploratory well [PETRO ENG] An oil well drilled for purposes of exploration for underlying petroleum.

explosion breccia [PETR] Breccia resulting from volcanic eruption or a phreatic explosion.

explosion crater [GEOL] A volcanic crater formed by explosion and commonly developed along rift zones on the flanks of large volcanoes.

explosion tuff [GEOL] A tuff whose constituent ash particles are in the place they fell after being ejected from a volcanic vent.

explosive [MATER] A substance, such as trinitrotoluene, or a mixture, such as gunpowder, that is characterized by chemical stability but may be made to undergo rapid chemical change without an outside source of oxygen, whereupon it produces a large quantity of energy generally accompanied by the evolution of hot gases.

explosive-actuated device [ENG] Any of various devices actuated by means of explosive; includes devices actuated either by high explosives or low explosives, whereas propellant-actuated devices include only the latter.

explosive echo ranging [ENG] Sonar in which a charge is exploded underwater to produce a shock wave that serves the same purpose as an ultrasonic pulse; the elapsed time for return of the reflected wave gives target range.

explosive index [GEOL] The percentage of pyroclastics in the material from a volcanic eruption.

exposure [METEOROL] The general surroundings of a site, with special reference to its openness to winds and sunshine.

exsolution [GEOL] A phenomenon during which molten rock solutions separate when cooled.

exsolution lamellae [GEOL] Layers of sedimentary rock that solidify from solution by either precipitation or secretion.

extended forecast [METEOROL] In general, a forecast of weather conditions for a period extending beyond 2 days from the day of issue. Also known as long-range forecast.

extended-range forecast *See* medium-range forecast.

extended stream [HYD] A stream lengthened by the extension of its downstream course; the course is through a newly emerged land such as a coastal plain.

extended succession [GEOL] A relatively thick, uninterrupted stratigraphic sequence in which deposits were rapidly accumulated.

extended valley [GEOL] **1.** A valley that is lengthened downstream by regression of the sea or by uplift of the coastal area. **2.** A valley eroded by or containing an extended stream.

extending flow [HYD] A flow pattern displayed by glaciers in which the velocity increases with distance downstream.

extensional fault *See* tension fault.

extension fracture [GEOL] A fracture that develops perpendicular to the direction of greatest stress and parallel to the direction of compression.

extension joints [GEOL] Fractures that form parallel to a compressive force.

extension ore *See* possible ore.

extensometer [ENG] **1.** A strainometer that measures the change in distance between two reference points separated 20–30 meters or more; used in studies of displacements due to seismic activities. **2.** An instrument designed to measure minute deformations of small objects subjected to stress.

external gas injection [PETRO ENG] Pressure-maintenance gas injection with wells located in the structurally higher positions of the reservoir, usually in the primary or secondary gas cap. Also known as crestal injection; gas-cap injection.

external upset casing [PETRO ENG] Special oil- or gas-well casing designed for extreme conditions requiring greater than usual strength and leak resistance. Also known as extreme line casing.

extinction [HYD] The drying up of a lake as water is lost or the basin is destroyed.

extraclast [GEOL] A calcareous sedimentary fragment resulting from erosion of an older rock outside its original area.

extraordinary wave [GEOPHYS] Magnetoionic wave component which, when viewed below the ionosphere in the direction of propagation, has clockwise or counterclockwise elliptical polarization respectively, accordingly as the earth's magnetic field has a positive or negative component in the same direction. Also known as X wave.

extratropical cyclone [METEOROL] Any cyclone-scale storm that is not a tropical cyclone. Also known as extratropical low; extratropical storm.

extratropical low *See* extratropical cyclone.

extratropical storm *See* extratropical cyclone.

extravasation [GEOL] The eruption of lava from a vent in the earth.

extreme [CLIMATOL] The highest, and in some cases the lowest, value of a climatic element observed during a given period or during a given month or season of that period; if this is the whole period for which observations are available, it is the absolute extreme.

extreme line casing *See* external upset casing.

extreme line tubing [PETRO ENG] Special oil- or gas-well tubing designed for extreme conditions requiring greater than usual strength and leak resistance.

extrusive rock *See* volcanic rock.

eye agate [MINERAL]   Agate with a dark center surrounded by a display of concentric bands, usually of various colors.

eye assay [MIN ENG]   An estimate of the valuable mineral content of a core or ore sample as based on visual inspection. Also known as eyeball assay.

eyeball assay *See* eye assay.

eye coal [GEOL]   Coal characterized by small, circular or elliptic structural disks that reflect light and are arranged in parallel planes either in or normal to the bedding. Also known as augen kohle; circular coal.

eye of the storm [METEOROL]   The center of a tropical cyclone, marked by relatively light winds, confused seas, rising temperature, lowered relative humidity, and often by clear skies.

eye of the wind [METEOROL]   The point or direction from which the wind is blowing.

F

Fabian system [MIN ENG]   The free-fall drilling system from which all other free-fall systems have originated.

fabric [GEOL]   The spatial orientation of the elements of a sedimentary rock. [PETR]   The sum of all the structural and textural features of a rock. Also known as petrofabric; rock fabric; structural fabric.

fabric analysis *See* structural petrology.

fabric diagram [PETR]   In structural petrology, a graphic representation of the data of fabric elements. Also known as petrofabric diagram.

fabric domain [PETR]   A three-dimensional area which is delimited by boundaries such as structural or compositional discontinuities and within which the rock fabric exhibits uniformity.

fabric element [PETR]   A surface or line of structural discontinuity in a rock fabric.

fabric-type dust collector [MIN ENG]   A collector which removes dust particles from ore by means of a filter made of fabric.

face [CRYSTAL]   *See* crystal face. [GEOL] **1.** The main surface of a landform. **2.** The original surface of a layer of rock.   [MIN ENG]   A surface on which mining operations are being performed. Also known as breast.

face area [MIN ENG]   The working area toward the interior of the last open crosscut in an entry or room.

face belt conveyor [MIN ENG]   A lightweight belt conveyor used at the working face in a mine.

face boss [MIN ENG]   A foreman in charge of operations at the working face in a bituminous coal mine.

face-centered cubic lattice [CRYSTAL]   A lattice whose unit cells are cubes, with lattice points at the center of each face of the cube, as well as at the vertices. Abbreviated fcc lattice.

face-centered orthorhombic lattice [CRYSTAL]   An orthorhombic lattice which has lattice points at the center of each face of a unit cell, as well as at the vertices.

face conveyor [MIN ENG]   Any type of mine conveyor used at and parallel to a working face.

face height [MIN ENG]   The vertical distance between the top and toe of a quarry or opencast face.

facellite *See* kaliophillite.

faceman [MIN ENG] A coal miner who performs the duties involved in drilling underground openings into which explosives are charged and set off, to extract coal, slate, and rock. Also known as coal digger; coal getter.

face mechanization [MIN ENG] The use of a cutter-loader on a longwall face.

face sampling [MIN ENG] Taking random samples of ore and rock from exposed faces of ore and waste.

face signal [MIN ENG] A wire stretched along the face and connected to a panel near the main gate to control the running of a face conveyor.

facet [GEOGR] Any part of an intersecting surface that forms a unit of geographic study; for example, a slope or flat.

faceted pebble [GEOL] A pebble with three or more faces naturally worn flat and meeting at sharp angles.

faceted spur [GEOL] A spur with an inverted-V face that was made by faulting or by the trimming or truncating action of streams, waves, or glaciers.

face timbering [MIN ENG] Positioning of safety posts at the working portion of a coal face to support the roof of the mine.

face worker [MIN ENG] A miner who works regularly at the face.

facies [GEOL] Any observable attribute or attributes of a rock or stratigraphic unit, such as overall appearance or composition, of one part of the rock or unit as contrasted with other parts of the same rock or unit.

facies map [GEOL] A stratigraphic map indicating distribution of sedimentary facies within a specific geologic unit.

facsimile chart [METEOROL] Any graphic form of weather information, usually a type of synoptic chart, which has been reproduced by facsimile equipment. Also known as fax chart; fax map.

Fagergren cell [MIN ENG] A froth-flotation cell in which a squirrel-cage rotor is driven concentrically in a vertical stator, so that air is drawn down the rotor shaft and dispersed into the pulp.

Fagersta cut [MIN ENG] A cut drilled with handheld equipment in two steps, first as a pilot hole and then as an enlargement of this hole.

fahlband [GEOL] A stratum containing metal sulfides; occurs in crystalline rock.

fahlore *See* tetrahedrite.

failed hole [ENG] A drill hole loaded with dynamite which did not explode. Also known as missed hole.

fair [METEOROL] Generally descriptive of pleasant weather conditions, with regard for location and time of year; it is subject to popular misinterpretation, for it is a purely subjective description; when this term is used in forecasts of the U.S. Weather Bureau, it is meant to imply no precipitation, less than 0.4 sky cover of low clouds, and no other extreme conditions of cloudiness or windiness.

fairchildite [MINERAL] $K_2Ca(CO_3)_2$ A mineral composed of potassium calcium carbonate; occurs in partly burned trees.

fairfieldite [MINERAL] $Ca_2Mn(PO_4)_2 \cdot 2H_2O$ A white or pale-yellow mineral composed of hydrous calcium manganese phosphate and occurring in foliated or fibrous form.

fair-weather cumulus *See* cumulus humilis cloud.

fairy stone *See* staurolite.

fake set *See* false set.

falaise [GEOL] An old low cliff, located on an emergent coast, that once again makes contact with the open sea.

Falkland Current [OCEANOGR] An ocean current flowing northward along the Argentine coast.

fall [MIN ENG] A mass of rock, coal, or ore which has fallen from the roof or side in any subterranean working or gallery.

fallback [GEOL] Fragments ejected from an impact or explosion crater, and then partly refilling the true crater almost immediately.

falling tide *See* ebb tide.

fall line [GEOL] **1.** The zone or boundary between resistant rocks of older land and weaker strata of plains. **2.** The line indicated by the edge over which a waterway suddenly descends, as in waterfalls.

falloff curve [PETRO ENG] Graphical representatation of bottom-hole pressure falloff for a shut-in well as the reservoir drainage area expands.

fall of ground [MIN ENG] The fall of rock from the roof into a mine opening.

fallout winds [METEOROL] Tropospheric winds that carry the radioactive fallout materials, observed by standard winds-aloft observation techniques.

fall streaks *See* virga.

Fallstreifen *See* virga.

fall wind [METEOROL] A strong, cold, downslope wind, differing from a foehn in that the initially cold air remains relatively cold despite adiabatic warming upon descent, and from the gravity wind in that it is a larger-scale phenomenon prerequiring an accumulation of cold air at high elevations.

false bedding [GEOL] An inclined bedding produced by currents.

false bottom [MIN ENG] A flat, hexagonal or cylindrical iron die upon which ore is crushed in a stamp mill.

false cirrus cloud [METEOROL] Cirrus composed of the debris of the upper frozen parts of a cumulonimbus cloud.

false cleavage [GEOL] **1.** A weak cleavage at an angle to the slaty cleavage. **2.** Spaced surfaces about a millimeter apart along which a rock splits.

false drumlin *See* rock drumlin.

false floor [GEOL] In a cave, a floorlike flowstone layer with open space underneath.

false form *See* pseudomorph.

false galena *See* sphalerite.

false ice foot [OCEANOGR] Ice that has formed along a beach terrace and has been attached to it just above the high-water mark; develops from water coming from melting snow above the terrace.

false lapis *See* lazulite.

false oolith *See* pseudooolith.

false set [MIN ENG] A light, temporary lagging set of timber supporting the side and roof lagging until the drive is advanced sufficiently to allow the heavy permanent set to be put, at which time the false set is taken out and used again in advance of the next permanent set. Also known as fake set.

false stull [MIN ENG] A stull so placed as to offer support or reinforcement for a stull, prop, or other timber.

false topaz *See* citrine.

false warm sector [METEOROL] The sector, in a horizontal plane, between the occluded front and a secondary cold front of an occluded cyclone.

famatinite [MINERAL] Cu_3SbS_4 A reddish-gray mineral composed of copper antimony sulfide.

fan [GEOL] A gently sloping, fan-shaped feature usually found near the lower termination of a canyon.

fan cut [ENG] A cut in which holes of equal or increasing length are drilled in a pattern on a horizontal plane or in a selected stratum to break out a considerable part of the plane or stratum before the rest of the round is fired.

fan drift [MIN ENG] The short tunnel connecting the upcast shaft with the exhaust fan.

fan-drift doors [MIN ENG] Isolation doors for each drift leading to each fan, when there are two fans at a mine.

fan drilling [ENG] **1.** Drilling boreholes in different vertical and horizontal directions from a single-drill setup. **2.** A radial pattern of drill holes from a setup.

fan fold [GEOL] A fold of strata in which both limbs are overturned, forming a syncline or anticline.

fanglomerate [GEOL] Coarse material in an alluvial fan, with the rock fragments being only slightly worn.

fan shaft [MIN ENG] The ventilating shaft to which a mine fan is connected.

fan-shaped delta *See* arcuate delta.

farinaceous [GEOL] Of a rock or sediment, having a texture that is mealy, soft, and friable; for example, a limestone or a pelagic ooze.

farmer's year [CLIMATOL] In Great Britain, the 12-month period starting with the Sunday nearest March 1.

farringtonite [MINERAL] $Mg_3(PO_4)_2$ Colorless, wax-white, or yellow phosphate mineral known only in meteorites.

fast-delay detonation [ENG] The firing of blasts by means of a blasting timer or millisecond delay caps.

fastest mile [METEOROL] Over a specified period (usually the 24-hour observational day), the fastest speed, in miles per hour, of any mile of wind, with its accompanying direction.

fast ice [HYD] Any type of sea, river, or lake ice attached to the shore (ice foot, ice shelf), beached (shore ice), stranded in shallow water, or frozen to the bottom of shallow waters (anchor ice). Also known as landfast ice. [OCEANOGR] Sea ice generally remaining in the position where originally formed and sometimes attaining a considerable thickness; it is attached to the shore or over shoals where it may be held in position by islands, grounded icebergs, or polar ice. Also known as coastal ice; coast ice.

fast ion *See* small ion.

fatal accident [MIN ENG] A coal mine accident in which less than five persons are killed and property damage is slight; excludes ignitions and mine fires.

fat clay [GEOL] A cohesive, highly plastic, and compressible clay that contains many minerals.

fathom [OCEANOGR] The common unit of depth in the ocean, equal to 6 feet (1.8288 meters).

fathom curve *See* isobath.

faujasite [MINERAL] $(Na_2,Ca)Al_2Si_4O_{12}\cdot6H_2O$ Zeolite mineral of the sodalite group, crystallizing in the cubic system.

fault [GEOL] A fracture in rock along which the adjacent rock surfaces are differentially displaced.

fault basin [GEOL] A region depressed in relation to surrounding regions and separated from them by faults.

fault block [GEOL] A rock mass that is bounded by faults; the faults may be elevated or depressed and not necessarily the same on all sides.

fault-block mountain *See* block mountain.

fault breccia [GEOL] The assembly of angular fragments found frequently along faults. Also known as dislocation breccia.

fault cliff *See* fault scarp.

fault escarpment *See* fault scarp.

faulting [GEOL] The fracturing and displacement processes which produce a fault.

fault ledge *See* fault scarp.

fault line [GEOL] Intersection of the fault surface with the surface of the earth or any other horizontal surface of reference. Also known as fault trace.

fault-line scarp [GEOL] A cliff produced when a soft rock erodes against hard rock at a fault.

fault plane [GEOL] A planar fault surface.

fault rock [GEOL] A rock often found along a fault plane and made up of fragments formed by the crushing and grinding which accompany a dislocation.

fault scarp [GEOL] A steep cliff formed by movement along one side of a fault. Also known as cliff of displacement; fault cliff; fault escarpment; fault ledge.

fault separation [GEOL] Apparent displacement of a fault measured on the basis of disrupted linear features.

fault strike [GEOL] The angular direction, with respect to north, of the intersection of the fault surface with a horizontal plane.

fault system [GEOL] Two or more fault sets which interconnect.

fault terrace [GEOL] A step on a slope, produced by displacement of two parallel faults.

fault throw [GEOL] The amount of vertical displacement of rocks due to faulting.

fault trace *See* fault line.

fault trap [GEOL] Oil or gas reservoir formed by a structural trap limited in one or more directions by subterranean geological faulting.

fault vein [GEOL] A mineral vein deposited in a fault fissure.

fault wall [GEOL] The mass of rock on a particular side of a fault.

fault zone [GEOL] A fault expressed as an area of numerous small fractures. Also known as distributed fault.

faunizone [GEOL] A bed characterized by fossils of a particular assemblage of fauna.

faunule [PALEON] The localized stratigraphic and geographic distribution of a particular taxon.

faustite [MINERAL] $(ZnCu)Al_6(PO_4)_4(OH)_8 \cdot 5H_2O$ An apple-green mineral that is the zinc analog of turquoise.

Faust jig [MIN ENG] A plunger-type jig, usually built with multiple compartments; distinguished by synchronized plungers on both sides of the screen plate, withdrawal of refuse through kettle valves in each compartment, and discharge of the hutch periodically by means of hand valves.

Favositidae [PALEON] A family of extinct Paleozoic corals in the order Tabulata.

fax chart *See* facsimile chart.

fax map *See* facsimile chart.

fayalite [MINERAL] Fe_2SiO_4 A brown to black mineral of the olivine group, consisting of iron silicate and found either massive or in crystals; specific gravity is 4.1.

fcc lattice *See* face-centered cubic lattice.

feasible ground [MIN ENG] Ground that can be easily worked and yet will stand without the support of timber or boards.

feather *See* barb.

feather joint [GEOL] One of a series of joints in a fault zone formed by shear and tension. Also known as pinnate joint.

feather ore *See* jamesonite.

fecal pellets [GEOL] Mainly the excreta of invertebrates occurring in marine deposits and as fossils in sedimentary rocks. Also known as castings.

feeder [GEOL] A small ore-bearing vein which merges with a larger one. [HYD] *See* tributary.

feeder beach [GEOL] A beach that has been artificially widened and nourishes down-drift beaches by natural littoral currents or forces.

feeder current [OCEANOGR] The current flowing parallel to the shore (inside the breakers) that joins other such currents to form the neck of the rip current.

feeder trough [MIN ENG] The trough connected to the conveyor pan line in a duckbill.

Feinc filter [MIN ENG] A vacuum-type drum filter in which a system of parallel strings is used to carry the filter cake away from the drum, instead of the usual filter cloth.

feldspar [MINERAL] A group of silicate minerals that make up about 60% of the outer 15 kilometers of the earth's crust; they are silicates of aluminum with the metals potassium, sodium, and calcium, and rarely, barium.

feldspathic graywacke [PETR] Sandstone containing less than 75% quartz and chert and 15–75% detrital clay matrix, and having feldspar grains in greater abundance than rock fragments. Also known as arkosic wacke; high-rank graywacke.

feldspathic sandstone [PETR] Sandstone rich in feldspar; intermediate in composition between arkosic sandstone and quartz sandstone, made up of 10–25% feldspar and less than 20% matrix material.

feldspathic shale [PETR] A typically well-laminated shale whose feldspar content is greater than 10% in the silt size with a finer matrix of kaolinitic clay minerals.

feldspathization [GEOL] Formation of feldspar in a rock usually as a result of metamorphism leading toward granitization.

feldspathoid [GEOL] Aluminosilicates of sodium, potassium, or calcium that are similar in composition to feldspars but contain less silica than the corresponding feldspar.

felsenmeer [GEOL] A flat or gently sloping veneer of angular rock fragments occurring on moderate mountain slopes above the timber line.

felsic [MINERAL] A light-colored mineral. [PETR] Of an igneous rock, having a mode containing light-colored minerals.

felsite [PETR] **1.** A light-colored, fine-grained igneous rock composed chiefly of quartz or feldspar. **2.** A rock characterized by felsitic texture.

felsöbányaite [MINERAL] $Al_4(SO_4)(OH)_{10} \cdot 5H_2O$ A yellow to white, probably orthorhombic mineral consisting of a hydrated basic sulfate of aluminum; occurs as aggregates of lamellar crystals.

felsophyric *See* aphaniphyric.

fen [GEOGR] Peat land covered by water, especially in the upper regions of old estuaries and around lakes, that can be drained only artificially.

fender [MIN ENG] A thin pillar of coal adjacent to the gob, left for protection while driving a lift through the mine pillar.

Fenestellidae [PALEON] A family of extinct fenestrated, cryptostomatous bryozoans which abounded during the Silurian.

fen peat *See* lowmoor peat.

fenster *See* window.

ferberite [MINERAL] $FeNO_4$ A black mineral of the wolframite solid-solution series occurring as monoclinic, prismatic crystals and having a submetallic luster; hardness is 4.5 on Mohs scale, and specific gravity is 7.5.

ferghanite [MINERAL] $U_3(VO_4)_2 \cdot 6H_2O$ Sulfur-yellow mineral composed of hydrated uranium vanadate, occurring in scales.

fergusite [PETR] A plutonic foidite containing a potassium feldspathoid (leucite) and 30–60% mafic minerals, such as olivine, apatite, and biotite, with accessory opaque oxides.

fergusonite [MINERAL] $Y_2O_3(Nb,Ta)_2O_5$ Brownish-black rare-earth mineral with a tetragonal crystal form; it is isomorphous with formanite.

fermorite [MINERAL] $(Ca,Sr)_5[(As,P)O_4]_3$ A white mineral composed of arsenate, phosphate, and fluoride of calcium and strontium, occurring in crystalline masses.

fernandinite [MINERAL] A dull green mineral composed of hydrous calcium vanadyl vanadate.

ferriamphibole [MINERAL] The ferric ion equivalent of the amphibole group of minerals.

ferricrete [GEOL] A conglomerate of surficial sand and gravel cemented by iron oxide originating from percolating solutions of iron salts.

ferrierite [MINERAL] $(Na,K)_2MgAl_3Si_{15}O_{36}(OH) \cdot 9H_2O$ A zeolite mineral crystallizing in the orthorhombic system.

ferriferous [GEOL] Of a sedimentary rock, iron-rich. [MINERAL] Of a mineral, iron-bearing.

ferrimolybdite [MINERAL] $Fe_2(MoO_4)_3 \cdot 8H_2O$ A colorless to canary yellow, probably orthorhombic mineral consisting of hydrated ferric molybdate; occurs in massive form, as crusts or aggregates.

ferrinatrite [MINERAL] $Na_3Fe(SO_4)_3 \cdot 3H_2O$ A greenish or white mineral composed of sodium ferric iron double sulfate; usually occurs in spherical forms.

ferrisicklerite [MINERAL] $(Li,Fe,Mn)(PO_4)$ Mineral composed of phosphate of lithium, ferric iron, and manganese, more iron being present than manganese; it is isomorphous with sicklerite.

ferrite [PETR] Grains or scales of unidentifiable, generally transparent amorphous iron oxide in the matrix of a porphyritic rock.

ferritremolite [MINERAL] The ferric ion equivalent of the monoclinic amphibole, tremolite.

ferritungstite [MINERAL] $Fe_2(WO_4)(OH)_4 \cdot 4H_2O$ A yellow ocher mineral composed of hydrous ferric tungstate, occurring as a powder.

ferroamphibole [MINERAL] The ferrous iron equivalent of the amphibole group of minerals.

ferroan dolomite [MINERAL] A species of ankerite having less than 20% of the manganese positions occupied by iron.

ferroaugite [MINERAL] A form of monoclinic pyroxene.

Ferrod [GEOL] A suborder of the soil order Spodosol that is well drained and contains an iron accumulation and little organic matter.

ferrodolomite [MINERAL] $CaFe(CO_3)_2$ A mineral composed of calcium iron carbonate, isomorphous with dolomite, and occurring in ankerite.

ferrogabbro [PETR] A gabbro rock in which the pyroxene and olivine constituents have an unusually high iron content.

ferrosilite [MINERAL] A mineral in the orthopyroxene group; the iron analog of enstatite; occurs in hypersthene, but is not found separately in nature.

ferrospinel *See* hercynite.

ferrotremolite [MINERAL] The ferrous iron equivalent of the monoclinic amphibole, tremolite.

ferruccite [MINERAL] $NaBF_4$ An orthorhombic boron mineral consisting of sodium fluoborate.

fersmanite [MINERAL] $(Na,Ca)_2(TI,Cb)Si(O,F)_6$ A brown mineral composed of a silicate fluoride of sodium, calcium, titanium, and columbium.

fersmite [MINERAL] $(Ca,Ce)(Cb,Ti)_2(O,F)_6$ A black mineral composed of an oxide and fluoride of calcium and columbium with cerium and titanium.

fervanite [MINERAL] $Fe_4V_4O_{16} \cdot 5H_2O$ Golden-brown mineral composed of a hydrated iron vanadate; although itself not radioactive, it occurs with radioactive minerals.

fetch [OCEANOGR] **1.** The distance traversed by waves without obstruction. **2.** An area of the sea surface over which seas are generated by a wind having a constant speed and direction. **3.** The length of the fetch area, measured in the direction of the wind in which the seas are generated. Also known as generating area.

fetch length [OCEANOGR] That horizontal distance over which a specific wind blows applied in forecasting waves.

fiamme [PETR] Dark, vitric lenses in piperno, averaging a few centimeters in length, perhaps formed by the collapse of fragments of pumice.

fiberizer [MIN ENG] A hammer mill which cracks open asbestos-bearing rock to yield a fibrous product.

fibratus [METEOROL] A cloud species characterized by a fine hairlike or striated composition, the filaments of which are usually distinctly separated from each other; the extremities of these filaments are always thin and never terminated by tufts or hooks. Also known as filosus.

Fibrist [GEOL] A suborder of the soil order Histosol, consisting mainly of recognizable plant residues or sphagnum moss and saturated with water for most of the year.

fibroblastic [PETR] Of a metamorphic rock, having a texture that is homeoblastic as a result of the development of minerals with a fibrous habit during recrystallization.

fibroferrite [MINERAL] $Fe(SO_4)(OH) \cdot 5H_2O$ A yellowish mineral composed of a hydrous basic ferric sulfate, occurring in fibrous form.

fibrolite *See* sillimanite.

fibrous ice *See* acicular ice.

FIDO [METEOROL] A system for artificially dissipating fog, in which gasoline or other fuel is burned at intervals along an airstrip to be cleared. Derived from fog investigation dispersal operations.

fiducial temperature [METEOROL] That temperature at which, in a specified latitude, the reading of a particular barometer requires no temperature or latitude correction.

fiducial time [GEOPHYS] A time on a seismograph record that has been corrected to correspond to a datum plane in space.

fiedlerite [MINERAL] $Pb_3(OH)_2Cl_4$ A colorless mineral composed of a hydroxychloride of lead, occurring as monoclinic crystals.

field [GEOL] An area characterized by a particular mineral resource, such as a gold field. [GEOPHYS] An area or space in which a specific geophysical effect occurs and is measureable, such as a gravity field or a magnetic field.

field changes [METEOROL] With regard to thunderstorm electricity, the rapid variations in the vertical component of the electric field strength at the earth's surface.

field focus [GEOPHYS] The total area or volume occupied by an earthquake source.

field moisture [HYD] Water in the ground that lies above the water table.

figure of the earth [GEOD] A precise geometric shape of the earth.

figure stone *See* agalmatolite.

filiform lapilli *See* Pele's hair.

fill *See* pack.

filled stopes [MIN ENG] Stopes filled with barren stone, low-grade ore, sand, or tailings (mill waste) after the ore has been extracted.

fillet lightning *See* ribbon lightning.

filling [METEOROL] An increase in the central pressure of a pressure system on a constant-height chart, or an analogous increase in height on a constant-pressure chart; the term is commonly applied to a low rather than to a high. [MIN ENG] Allowing a mine to fill with water.

fillowite [MINERAL] $H_2Na_6(Mn,Fe,Ca)_{14}(PO_4)_{12} \cdot H_2O$ A brown, yellow, or colorless mineral composed of a hydrous phosphate of manganese, iron, sodium, and other metals.

fill terrace *See* alluvial terrace.

film sizing table [MIN ENG] A table used in ore dressing for sorting fine material by means of a film of flowing water.

filosus *See* fibratus.

fine admixture [GEOL] The smaller size grades of a sediment of mixed size grades.

fine earth [GEOL] A soil which can be passed through a 2.0-millimeter sieve without grinding its primary particles.

fine gravel [GEOL] Gravel consisting of particles with a diameter range of 1 to 2 millimeters.

fine sand [GEOL] Sand grains between 0.25 and 0.125 millimeter in diameter.

finger [GEOL] The tendency for gas which is displacing liquid hydrocarbons in a heterogeneous reservoir rock system to move forward irregularly (in fingers), rather than on a uniform front. [PETRO ENG] A pair or set of bracketlike projections placed at a strategic point in a drill tripod or derrick to keep a number of lengths of drill rods or casing in place when they are standing in the tripod or derrick.

finger board [PETRO ENG] A board with projecting dowels or pipe fingers located in the upper part of the drill derrick or tripod to support stands of drill rod, drill pipe, or casing.

finger chute [MIN ENG] Steel rails hinged independently over an ore chute, to control rate of flow of rock.

finger coal *See* natural coke.

finger lake [HYD] A long, narrow lake occurring in a rock basin on a glacial-trough floor or contained by a morainal dam across the lower end of the valley.

finger raise [MIN ENG] Steeply sloping openings permitting caved ore to flow down raises through grizzlies to chutes on the haulage level.

finishing roll [MIN ENG] The last roll, or the one that does the finest crushing in ore dressing.

finite closed aquifer [HYD] The part of a subterranean reservoir containing water (aquifer) in which the aquifer is limited (finite), with no water flow across the exterior reservoir boundary.

finnemanite [MINERAL] $Pb_5Cl(AsO_3)_3$ A gray, olive-green, or black hexagonal mineral composed of arsenite and chloride of lead.

fiorite *See* siliceous sinter.

fire [ENG] To blast with gunpowder or other explosives. [MIN ENG] A warning that a shot is being fired.

fire boss [MIN ENG] An individual who examines a mine for gas and other dangers. Also known as mine examiner.

firebreak [MIN ENG] A strip across an area in which either no combustible material is employed or in which, if timber supports are used, sand is filled and packed tightly around them.

fireclay [GEOL] **1.** A clay that can resist high temperatures without becoming glassy. **2.** Soft, embedded, white or gray clay rich in hydrated aluminum silicates or silica and deficient in alkalies and iron.

firedamp [MIN ENG] **1.** A gas formed in mines by decomposition of coal or other carbonaceous matter; consists chiefly of methane and is combustible. **2.** An airtight stopping to isolate an underground fire and to prevent the inflow of fresh air and the outflow of foul air. Also known as fire wall.

firedamp alarm [MIN ENG] An instrument which gives a warning signal when the methane content in the mine atmosphere exceeds a known value.

firedamp detector [MIN ENG] A portable device to detect the presence and determine the percentage of firedamp in mine air.

firedamp drainage [MIN ENG] The collection of firedamp from coal strata, generally into pipes, with or without the use of suction. Also known as methane drainage.

firedamp drainage drill [MIN ENG] A heavy, compressed-air-operated, percussive, rotary or rotary-percussive drilling machine for putting up the boreholes in firedamp drainage.

firedamp explosion [MIN ENG] An explosion of a mixture of firedamp and air.

firedamp fringe [MIN ENG] The zone of contact between the coal gases and the ventilation air current at the face of the mine.

firedamp layer [MIN ENG] An accumulation of firedamp under the roof of a mine roadway where the ventilation is insufficient to dilute and remove the gas.

firedamp migration [MIN ENG] The movement of firedamp through the strata or coal of a mine.

firedamp pressure-chamber method [MIN ENG] A method of firedamp drainage in coal mines; pressure chambers built at the intake and the return of a worked-out area are used to trap firedamp, which is drawn off in pipes.

firedamp probe [MIN ENG] A flexible rubber tube connected to a rod, which can be thrust into roof cavities and breaks so that a sample of the air may be transferred to a methanometer and its firedamp content determined.

fire flooding [PETRO ENG] A method to improve secondary recovery in an oil reservoir; a combustion process is started in the reservoir at an injection well by continued introduction of gas containing oxygen or other material to support combustion, and the combustion wave is driven through the reservoir toward the production well.

fire opal [MINERAL] A translucent or transparent, orangy-yellow, brownish-orange, or red variety of opal that gives out fiery reflections in bright light and that may have a play of colors. Also known as pyrophane; sun opal.

firestone *See* flint.

fire weather [METEOROL] The state of the weather with respect to its effect upon the kindling and spreading of forest fires.

firing cable *See* shot-firing cable.

firing machine [ENG] An electric blasting machine.   [MECH ENG] A mechanical stoker used to feed coal to a boiler furnace.

firing mechanism [ENG] A mechanism for firing a primer; the primer may be for initiating the propelling charge, in which case the firing mechanism forms a part of the weapon; if the primer is for the purpose of initiating detonation of the main charge, the firing mechanism is a part of the ammunition item and performs the function of a fuse.

firn [HYD] Material transitional between snow and glacier ice; it is formed from snow after existing through one summer melt season and becomes glacier ice when its permeability to liquid water drops to zero. Also known as firn snow.

firn field [HYD] The accumulation area or upper region of a glacier where snow accumulates and firn is secreted. Also known as firn basin.

firn ice *See* iced firn.

firnification [HYD] The process of firn formation from snow and of transformation of firn into glacier ice.

firn limit *See* firn line.

firn line　[GEOL]　**1.** The regional snow line on a glacier. **2.** The line that divides the ablation area of a glacier from the accumulation area. Also known as firn limit.

firn snow *See* firn; old snow.

first arrival　[GEOPHYS]　As recorded by a seismograph, the first signal that is due to seismic-wave travel from a known source.

first bottom　[GEOL]　The floodplain of a river, below the first terrace.

first gust　[METEOROL]　The sharp increase in wind speed often associated with the early mature stage of a thunderstorm cell; it occurs with the passage of the discontinuity zone which is the boundary of the cold-air downdraft.

first-order climatological station　[METEOROL]　A meteorological station at which autographic records or hourly readings of atmospheric pressure, temperature, humidity, wind, sunshine, and precipitation are made, together with observations at fixed hours of the amount and form of clouds and notes on the weather.

first-order station　[METEOROL]　After U.S. National Weather Service practice, any meteorological station that is staffed in whole or in part by National Weather Service (Civil Service) personnel, regardless of the type or extent of work required of that station.

firth *See* estuary.

Fischer ellipsoid of 1960　[GEOD]　The reference ellipsoid of which the semimajor axis is 6,378,166.000 meters, the semiminor axis is 6,356,784.298 meters, and the flattening or ellipticity is 1/298.3. Also known as Fischer spheroid of 1960.

fischerite　[MINERAL]　A green mineral composed of a basic aluminum phosphate; may be identical to wavellite.

Fischer spheroid of 1960 *See* Fischer ellipsoid of 1960.

fish-eye stone *See* apophyllite.

fishhook dune　[GEOL]　A dune comprising a long, sinuous, sigmoidal ridge (the shaft) and a clearly defined crescent (the hook).

fishing　[ENG]　In drilling, the operation by which lost or damaged tools are secured and brought to the surface from the bottom of a well or drill hole.

fissile　[GEOL]　Capable of being split along the line of the grain or cleavage plane.

fission-track dating　[GEOL]　A method of dating geological specimens by counting the radiation-damage tracks produced by spontaneous fission of uranium impurities in minerals and glasses.

fissure　[GEOL]　**1.** A high, narrow cave passageway. **2.** An extensive crack in a rock.

fissure system　[GEOL]　A group of fissures equal in age and having more or less parallel strike and dip.

fissure vein　[GEOL]　A mineral deposit in a cleft or crack in the rock material of the earth's crust.

Fistuliporidae　[PALEON]　A diverse family of extinct marine bryozoans in the order Cystoporata.

fitness figure　[METEOROL]　In Great Britain, a measure of the "fitness" of the weather at an airport for the safe landing of aircraft; the figure F is computed on the basis of corrected values of visibility and cloud height; observed visibility is adjusted according to intensity of precipitation, and cloud height is corrected for height of nearby obstructions and cloud amount; further corrections are applied for the cross-runway component of the wind. Also known as fitness number.

fitness number *See* fitness figure.

five-and-ten system [METEOROL]   The most common system for representing wind speed, to the nearest 5 knots, in symbolic form on synoptic charts, consisting of drawing the appropriate number of half-barbs, barbs, and pennants from the end of the wind-direction shaft; in this system, a half-barb represents 5 knots, a barb 10 knots, a pennant 50 knots.

five-day forecast [METEOROL]   A forecast of the average weather conditions and large-scale synoptic features in a 5-day period; a type of extended forecast.

five-spot well pattern [PETRO ENG]   A symmetrical network pattern of five wells (one in center, four equally spaced in a square pattern) as used in water-injection pressure maintenance of reservoirs.

fixed-electrode method [ENG]   A geophysical surveying method used in a self-potential system of prospecting in which one electrode remains stationary while the other is grounded at progressively greater distances from it.

fixed-level chart *See* constant-height chart.

fixed-needle traverse [ENG]   In surveying, a traverse with a compass fitted with a sight line which can be moved above a graduated horizontal circle, so that the azimuth angle can be read, as with a theodolite.

fixed rent *See* minimum rent.

fixed screen [MIN ENG]   A stationary panel, commonly of wedge wire, used to remove a large proportion of water and fines from a suspension of coal in water.

fixed sonar [ENG]   Sonar in which the receiving transducer is not constantly rotated, in contrast to scanning sonar.

fizelyite [MINERAL]   A metallic, lead-gray mineral composed of a lead silver antimony sulfide, occurring as prisms.

fjord [GEOGR]   A narrow, deep inlet of the sea between high cliffs or steep slopes. Also spelled fiord.

fjord valley [GEOGR]   A deep, narrow channel occupied by the sea and extending inland about 50–100 miles (80–160 kilometers).

flaggy [GEOL]   **1.** Of bedding, consisting of strata 10–100 centimeters in thickness. **2.** Of rock, tending to split into layers of suitable thickness (1–5 centimeters) for use as flagstones.

flagstone [GEOL]   **1.** A hard, thin-bedded sandstone, firm shale, or other rock that splits easily along bedding planes or joints into flat slabs. **2.** A piece of flagstone used for making pavement or covering the side of a house.

flajolotite [MINERAL] $4FeSbO_4 \cdot 3H_2O$ A claylike, lemon-yellow mineral composed of a hydrous iron antimonate, occurring in nodular masses.

flaking [MIN ENG]   Breaking small chips from the face of a refractory, particularly chrome ore containing refractories.

flamboyant structure [GEOL]   The optical continuity of crystals or grains as disturbed by a structure that is divergent.

flame collector [ENG]   A device used in atmospheric electrical measurements for the removal of induction charge on apparatus; based upon the principle that products of combustion are ionized and will consequently conduct electricity from charged bodies.

flame structure [GEOL]   A sedimentary structure consisting of wave or flame-shaped plumes of mud that have been squeezed irregularly upward into an overlying layer. [PETR]   The presence of fiamme in a welded tuff.

Flanders storm   [METEOROL]   In England, a heavy fall of snow coming with the south wind.

Flandrian transgression   [OCEANOGR]   The rapid rise of the North Sea between 8000 and 3000 B.C. from about 180 feet (55 meters) below to about 20 feet (6 meters) below its present level.

flank   [GEOL]   *See* limb.

flank hole   [MIN ENG]   **1.** A hole bored in advance of a working place when approaching old workings.   **2.** A borehole driven from the side of an underground excavation, not parallel with the center line of the excavation, to detect water, gas, or other danger.

flare-type bucket   [MIN ENG]   A dragline bucket that has flared sides to allow heaped loading.

flaser   [GEOL]   Streaky layer of parallel, scaly aggregates that surrounds the lenticular bodies of granular material in flaser structure; caused by pressure and shearing during metamorphism.

flaser gabbro   [GEOL]   A cataclastic gabbro that contains augen of feldspar or quartz surrounded by flakes of mica or chlorite.

flaser structure   [GEOL]   **1.** A metamorphic structure in which small lenses and layers of granular material are surrounded by a matrix of sheared, crushed material, resembling a crude flow structure. Also known as pachoidal structure.   **2.** A primary sedimentary structure consisting of fine-sand or silt lenticles that are aligned and cross-bedded.

flash flood   [HYD]   A sudden local flood of short duration and great volume; usually caused by heavy rainfall in the immediate vicinity.

flash roast   [MIN ENG]   Rapid removal of sulfur from ore by having finely divided sulfide mineral fall through a heated oxidizing atmosphere. Also known as suspension roast.

flat   [GEOGR]   A level tract of land.   [GEOL]   *See* mud flat.   [MINERAL]   An inferior grade of rough diamonds.

flat-back stope   [MIN ENG]   An overhand stoping method in which the ore is broken in slices parallel with the levels.

flat cut   [MIN ENG]   A manner of placing the boreholes, for the first shot in a tunnel, in which they are started about 2 or 3 feet (60–90 centimeters) above the floor and pointed downward so that the bottom of the hole will be about level with the floor.

flat-lying   [GEOL]   Of mineral deposits and coal seams, having a relatively flat dip, up to 5°.

flattening   [GEOD]   The ratio of the difference between the equatorial and polar radii of the earth; the flattening of the earth is the ellipticity of the spheroid; the magnitude of the flattening is sometimes expressed as the numerical value of the reciprocal of the flattening. Also known as compression.   [MET]   Straightening of metal sheet by passing it through special rollers which flatten it without changing its thickness. Also known as roll flattening.

flaw   [METEOROL]   An English nautical term for a sudden gust or squall of wind.   [OCEANOGR]   **1.** The seaward edge of fast ice.   **2.** A shore lead just outside fast ice.

flaxseed ore   [GEOL]   Iron ore composed of disk-shaped oölites that have been partially flattened parallel to the bedding plane.

F layer   [GEOPHYS]   An ionized layer in the F region of the ionosphere which consists of the F_1 and F_2 layers in the day hemisphere, and the F_2 layer alone in the night hemisphere; it is capable of reflecting radio waves to earth at frequencies up to about 50 megahertz.

F₁ layer [GEOPHYS]   The ionosphere layer beneath the F_2 layer during the day, at a virtual height of 200–300 kilometers, being closest to earth around noon; characterized by a distinct maximum of free-electron density, except at high latitudes during winter, when the layer is not detectable.

F₂ layer [GEOPHYS]   The highest constantly observable ionosphere layer, characterized by a distinct maximum of free-electron density at a virtual height from about 225 kilometers in the polar winter to more than 400 kilometers in daytime near the magnetic equator. Also known as Appleton layer.

Flexibilia [PALEON]   A subclass of extinct stalked or creeping Crinoidea; characteristics include a flexible tegmen with open ambulacral grooves, uniserial arms, a cylindrical stem, and five conspicuous basals and radials.

flexible sandstone [GEOL]   A variety of itacolumite that consists of fine grains and occurs in thin layers.

flexible ventilation ducting [MIN ENG]   Flexible fabric tubes covered with rubber or polyvinyl chloride, used for auxiliary ventilation.

flexure [GEOL]   **1.** A broad, domed structure. **2.** A fold.

flight briefing *See* pilot briefing.

flight-weather briefing *See* pilot briefing.

flinkite [MINERAL] $Mn_3(AsO_4)(OH)_4$ Greenish-brown mineral composed of basic manganese arsenate, occurring in feathery forms.

flint [MINERAL]   A black or gray, massive, hard, somewhat impure variety of chalcedony, breaking with a conchoidal fracture. Also known as firestone.

flint clay [GEOL]   A hard, smooth, flintlike fireclay; when it is ground, it develops no plasticity, and it breaks with conchoidal fracture.

flist [METEOROL]   In Scotland, a keen blast or shower accompanied by a squall.

float [GEOL]   An isolated, displaced rock or ore fragment.

floatability [MIN ENG]   Response of a specific mineral to the flotation process.

float-and-sink analysis [MIN ENG]   Use of a series of heavy liquids diminishing (or increasing) in density by accurately controlled stages in order to divide a sample of crushed coal or other minerals or metals into fractions that are either equal-settling or equal-floating at each stage.

float barograph [ENG]   A type of siphon barograph in which the mechanically magnified motion of a float resting on the lower mercury surface is used to record atmospheric pressure on a rotating drum.

float coal [GEOL]   Small, irregularly shaped, isolated deposits of coal embedded in sandstone or in siltstone. Also known as raft.

floater *See* drift bottle.

floating ice [OCEANOGR]   Any form of ice floating in water, including grounded ice and drifting land ice.

floating sand [PETR]   An isolated grain of quartz sand that is not, or does not appear to be, in contact with neighboring sand grains scattered throughout the finer-grained matrix of a sedimentary rock.

float mineral [GEOL]   Small ore fragments carried from the ore bed by the action of water or by gravity; a float mineral often leads to discovery of mines.

float-type rain gage [ENG] A class of rain gage in which the level of the collected rainwater is measured by the position of a float resting on the surface of the water; frequently used as a recording rain gage by connecting the float through a linkage to a pen which records on a clock-driven chart.

floccus [METEOROL] A cloud species in which each element is a small tuft with a rounded top and a ragged bottom.

floe [OCEANOGR] A piece of floating sea ice other than fast ice or glacier ice; may consist of a single fragment or of many consolidated fragments, but is larger than an ice cake and smaller than an ice field. Also known as ice floe.

floeberg [OCEANOGR] A mass of hummocked ice formed by the piling up of many ice floes by lateral pressure; an extreme form of pressure ice; may be more than 50 feet (15 meters) high and resemble an iceberg.

floe till [GEOL] **1.** A glacial till resulting from the intact deposition of a grounded iceberg in a lake bordering an ice sheet. **2.** A lacustrine clay with boulders, stones, and other glacial matter dropped into it by melting icebergs. Also known as berg till.

flokite *See* mordenite.

flood [HYD] The condition that occurs when water overflows the natural or artificial confines of a stream or other body of water, or accumulates by drainage over low-lying areas. [OCEANOGR] The highest point of a tide.

flood basalt *See* plateau basalt.

flood basin [GEOL] **1.** The tract actually covered by water during the highest known flood in a specific land area. **2.** The broad, flat area situated between a sloping, low plain and a natural levee of a river.

flood coverage [PETRO ENG] The extent of subterranean coverage within an oil reservoir by the injection of pressure-maintenance (or water-drive) water.

flood current [OCEANOGR] The tidal current associated with the increase in the height of a tide.

flooded stream *See* drowned stream.

flood flow [HYD] Stream discharge during a flood.

flood fringe *See* pondage land.

flood icing *See* icing.

flooding [PETRO ENG] Technique of increasing recovery of oil (secondary recovery) from a reservoir by injection of water into the formation to drive the oil toward producing wellholes. Also known as waterflooding.

flooding ice *See* icing.

flood-out pattern [PETRO ENG] Pattern of subterranean water penetration and spread in an oil reservoir as a result of water injection.

floodplain [GEOL] The relatively smooth valley floors adjacent to and formed by alluviating rivers which are subject to overflow.

flood plane [HYD] The position that is reached by a stream's water surface during a specific flood.

flood pot test [PETRO ENG] Laboratory simulation of an oil reservoir to appraise the residual reservoir saturation after waterflooding.

flood routing [HYD] The process of computing the progressive time and shape of a flood wave at successive points along a river. Also known as storage routing; stream-flow routing.

flood stage [HYD] The stage, on a fixed river gage, at which overflow of the natural banks of the stream begins to cause damage in any portion of the reach for which the gage is used as an index.

flood tide [OCEANOGR] **1.** That period of tide between low water and the next high water. **2.** A tide at its highest point.

flood tuff *See* ignimbrite.

floor [GEOL] **1.** The rock underlying a stratified or nearly horizontal deposit, corresponding to the footwall of more steeply dipping deposits. **2.** A horizontal, flat ore body. [MIN ENG] Boards laid at the heading to receive blasted rocks and to facilitate ore loading.

floorboard [MIN ENG] A thick wooden plank constituting part of a drill platform or other work platform.

floor burst [MIN ENG] A type of outburst in longwall faces which is preceded by heavy weighting due to floor lift; gas evolved below the seam collects beneath an impervious layer of rock, and a gas blister forms beneath the face, giving the observed floor lift; later, the floor fractures and the firedamp escapes into the mine atmosphere.

floor cut [MIN ENG] A machine-made cut in the floor dirt just below the coal seam.

floor sill [MIN ENG] A large timber laid flat on the ground or in a level, shallow ditch, to which are fastened the drill-platform boards or planking, or which is used as the base for a full timber set.

flop gate [MIN ENG] An automatic gate used in placer mining when there is a shortage of water; the gate closes a reservoir until it is filled with water, then automatically opens and allows the water to flow into the sluices; when the reservoir is empty, the gate closes, and the operation is repeated.

florencite [MINERAL] $CeAl_3(PO_4)_2(OH)_6$ Pale-yellow mineral composed of basic phosphate of cerium and aluminum.

Florida Current [OCEANOGR] A fast current that sets through the Straits of Florida to a point north of Grand Bahama Island, where it joins the Antilles Current to form the Gulf Stream.

flotation [ENG] A process used to separate particulate solids by causing one group of particles to float; utilizes differences in surface chemical properties of the particles, some of which are entirely wetted by water, others are not; the process is primarily applied to treatment of minerals but can be applied to chemical and biological materials; in mining engineering it is referred to as froth flotation.

flotation cell [MIN ENG] The device in which froth flotation of ores is performed.

flowage [HYD] Flooding of water onto adjacent land.

flowage line [GEOL] A contour line at the edge of a body of water, such as a reservoir, representing a given water level.

flow banding [GEOL] An igneous rock structure resulting from flowing of magmas or lavas and characterized by alternation of mineralogically unlike layers.

flow bog [ECOL] A peat bog with a variable surface level depending on the rain and the tides.

flow breccia [GEOL] A breccia formed with the movement of lava flow while the flow is still in motion.

flow cast [PETR] One of a group of bedding plane structures formed in graywacke.

flow cleavage [GEOL]   Rock cleavage in which solid flow of rock accompanies re-crystallization. Also known as slaty cleavage.

flow earth *See* solifluction mantle.

flow fold [GEOL]   Folding in beds, composed of relatively plastic rock, that assume any shape impressed upon them by the more rigid surrounding rocks or by the general stress pattern of the deformed zone; there are no apparent surfaces of slip.

flowing-film concentration [MIN ENG]   A concentration based on the fact that liquid films in laminar flow possess a velocity which is not the same in all depths of the film; by this principle lighter particles of ore may be washed off while the heavier particles accumulate and are intermittently removed.

flowing pressure [PETRO ENG]   Pressure at the bottom of an oil-well bore (bottom-hole pressure) during normal oil production.

flowing-pressure gradient [PETRO ENG]   The slope of decreasing pressure plotted against distance measured for upward liquid flow in a continuous-flow gas-lift oil well.

flowing well [PETRO ENG]   Oil reservoir in which gas-drive pressure is sufficient to force oil flow up through and out of a wellhole.

flow layer [PETR]   A layer in an igneous rock differing in composition or texture from the adjacent layers.

flow line [HYD]   A contour of the water level around a body of water.   [PETR]   In an igneous rock, any internal structure produced by parallel orientation of crystals, mineral streaks, or inclusions.   [PETRO ENG]   A pipeline that takes oil from a single well or a series of wells to a gathering center.

flowmeter [ENG]   An instrument used to measure pressure, flow rate, and discharge rate of a liquid, vapor, or gas flowing in a pipe. Also known as fluid meter.

flow regime [HYD]   A range of streamflows involving similarity of bed forms, of resistance to flow, and of mode of sediment transport.

flow rock [PETR]   An igneous rock that had been liquid.

flow slide [GEOL]   A landslide of waterlogged material that lacks a well-defined slip surface.

flowstone [GEOL]   Deposits of calcium carbonate that accumulated against the walls of a cave where water flowed on the rock.

flow string [PETRO ENG]   Total length of oil- or gas-well tubing made up of a string of interconnected tubing sections.

flow structure [GEOL]   A primary sedimentary structure that develops from sub-aqueous slump or flow.

flow texture [PETR]   A pattern of an igneous rock that is formed when the stream or flow lines of a once-molten material have a subparallel arrangement of prismatic or tabular cyrstals or microlites. Also known as fluidal texture.

fluctuating current [ELEC]   Direct current that changes in value but not at a steady rate.

flowtill [GEOL]   Muddy sludge produced by meltwater around stagnant ice from a glacier.

flow velocity [GEOL]   In soil, a vector point function used to indicate rate and direction of movement of water through soil per unit of time, perpendicular to the direction of flow.

fluctuation [OCEANOGR]   **1.** Wavelike motion of water. **2.** The variations of water-level height from mean sea level that are not due to tide-producing forces.

fluellite [MINERAL] $AlF_3 \cdot H_2O$ A colorless or white mineral composed of aluminum fluoride, occurring in crystals.

fluidal texture *See* flow texture.

fluid coefficient [PETRO ENG] A measure of the flow resistance to the leaking off of reservoir fracturing fluids into the formation during the fracturing operation.

fluid geometry [GEOL] Fluid distribution in reservoir strata controlled by rock effective pore-size distribution, rock wettability characteristics in relation to the fluids present, method of producing saturation, and rock heterogeneity.

fluid inclusion [PETR] A tiny fluid-filled cavity in an igneous rock that forms by the entrapment of the liquid from which the rock crystallized.

fluid-loss test [PETRO ENG] Measure of fracturing fluid loss versus time (spurt loss) before the fluid-loss agent forms a nonpermeable layer in the reservoir pore matrix.

fluid meter *See* flowmeter.

fluid saturation [GEOL] Measure of the gross void space in a reservoir rock that is occupied by a fluid.

fluid viscosity ratio [PETRO ENG] Ratio of viscosity of a displacing gas to that of oil in a gas-drive reservoir; used in unit displacement efficiency calculations.

flume [GEOL] A ravine with a stream flowing through it.

flumed [MIN ENG] In hydraulic mining, pertaining to the transportation of solids by suspension or flotation in flowing water.

fluoborite [MINERAL] $Mg_3(BO_3)(F,OH)_3$ A colorless mineral composed of magnesium fluoborate; occurs in hexagonal prisms. Also known as nocerite.

fluocerite [MINERAL] $(Ce,La,Nd)F_3$ A reddish-yellow mineral composed of fluoride of cerium and related elements.

fluolite *See* pitchstone.

fluor *See* fluorite.

fluorapatite [MINERAL] **1.** $Ca_5(PO_4)_3F$ A mineral of the solid-solution series of the apatite group; common accessory mineral in igneous rocks. **2.** An apatite mineral in which the fluoride member dominates.

fluorite [MINERAL] CaF_2 A transparent to translucent, often blue or purple mineral, commonly found in crystalline cubes in veins and associated with lead, tin, and zinc ores; hardness is 4 on Mohs scale; the principal ore of fluorine. Also known as Derbyshire spar; fluor; fluorspar.

fluorocummingtonite [MINERAL] Cummingtonite with a high content of fluorine.

fluorspar *See* fluorite.

flurry [METEOROL] A brief shower of snow accompanied by a gust of wind, or a sudden, brief wind squall.

flushed-zone resistivity [PETRO ENG] Electrical resistivity of the reservoir area which surrounds a borehole to a distance of at least 3 inches (7.6 centimeters) and for which the original interstitial fluids have been flushed out by drilling-mud filtrate.

flushing period [HYD] The interval of time required for a quantity of water equal to the volume of a lake to pass through its outlet; computed by dividing lake volume by mean flow rate of the outlet.

flush-joint casing [PETRO ENG] Lengths of casing that when connected end to end form a smooth joint flush with the outer diameter of the remainder of the section length.

flute [GEOL] **1.** A natural groove running vertically down the face of a rock. **2.** A groove in a sedimentary structure formed by the scouring action of a turbulent, sediment-laden water current, and having a steep upcurrent end.

flute cast [GEOL] A raised, oblong, or subconical welt on the bottom surface of a siltstone or sandstone bed formed by the filling of a flute.

Fluvent [GEOL] A suborder of the soil order Entisol that is well drained, with marks of sedimentation still evident and no identifiable horizons; occurs in recently deposited alluvium along streams or in fans.

fluvial [HYD] **1.** Pertaining to or produced by the action of a stream or river. **2.** Existing, growing, or living in or near a river or stream.

fluvial cycle of erosion *See* normal cycle.

fluvial deposit [GEOL] A sedimentary deposit of material transported by, suspended in, or laid down by a river.

fluvial sand [GEOL] Sand laid down by a river or stream.

fluvial soil [GEOL] Soil laid down by a river or stream.

fluviatile [GEOL] Resulting from river action.

fluvioglacial [GEOL] Pertaining to or produced by the action of a glacier and meltwater flowing with the glacier.

fluviology [HYD] The science of rivers.

flying veins [GEOL] Mineral-deposit veins that overlap and intersect in a branchlike pattern.

fly-off *See* evapotranspiration.

flysch [GEOL] Deposits of dark, fine-grained, thinly bedded sandstone shales and of clay, thought to be deposited by turbidity currents and originally defined as rock formations on the northern and southern borders of the Alps.

foam *See* pumice.

foam crust [HYD] A snow surface feature that looks like small overlapping waves, like sea foam on a beach, occurring during the ablation of the snow surface and may further develop into a more pronounced wedge-shaped form, known as plowshares.

foam drilling [MIN ENG] A method of dust suppression in which thick foam is forced through the drill by means of compressed air, and the foam-and-dust mixture emerges from the mouth of the hole in the form of a thick sludge.

foam line [OCEANOGR] The front of a wave as it moves toward the shore, after the wave has broken.

foam mark [GEOL] A sedimentary structure comprising a pattern of faint ridges and hollows, formed where wind-generated sea foam passes over a surface of wet sand.

focus [GEOPHYS] The center of an earthquake and the origin of its elastic waves within the earth.

foehn [METEOROL] A warm, dry wind on the lee side of a mountain range, the warmth and dryness being due to adiabatic compression as the air descends the mountain slopes. Also spelled föhn.

foehn air [METEOROL] The warm, dry air associated with foehn winds.

foehn cloud [METEOROL] Any cloud form associated with a foehn, but usually signifying only those clouds of the lenticularis species formed in the lee wave parallel to the mountain ridge.

foehn cyclone [METEOROL] A cyclone formed (or at least enhanced) as a result of the foehn process on the lee side of a mountain range.

foehn island [METEOROL] An isolated area where the foehn has reached the ground, in contrast to the surrounding area where foehn air has not replaced colder surface air.

foehn nose [METEOROL] As seen on a synoptic surface chart, a typical deformation of the isobars in connection with a well-developed foehn situation; a ridge of high pressure is produced on the windward slopes of the mountain range, while a foehn trough forms on the lee side; the isobars "bulge" correspondingly, giving a noselike configuration.

foehn pause [METEOROL] **1.** A temporary cessation of the foehn at the ground, due to the formation or intrusion of a cold air layer which lifts the foehn above the valley floor. **2.** The boundary between foehn air and its surroundings.

foehn period [METEOROL] The duration of continuous foehn conditions at a given location.

foehn phase [METEOROL] One of three stages to describe the development of the foehn in the Alps: the preliminary phase, when cold air at the surface is separated from warm dry air aloft by a subsidence inversion; the anticyclonic phase, when the warm air reaches a station as the result of the cold air flowing out from the plain; and the stationary phase or cyclonic phase, when the foehn wall forms and the downslope wind becomes appreciable.

foehn storm [METEOROL] A type of destructive storm which frequently occurs in October in the Bavarian Alps.

foehn trough [METEOROL] The dynamic trough formed in connection with the foehn.

foehn wall [METEOROL] The steep leeward boundary of flat, cumuliform clouds formed on the peaks and upper windward sides of mountains during foehn conditions.

fog [METEOROL] Water droplets or, rarely, ice crystals suspended in the air in sufficient concentration to reduce visibility appreciably.

fogbank [METEOROL] A fairly well-defined mass of fog observed in the distance, most commonly at sea.

fog deposit [HYD] The deposit of an ice coating on exposed surfaces by a freezing fog.

fog dispersal [METEOROL] Artificial dissipation of a fog by means such as seeding or heating.

fog drip [HYD] Water dripping to the ground from trees or other objects which have collected the moisture from drifting fog; the dripping can be as heavy as light rain, as sometimes occurs among the redwood trees along the coast of northern California.

fog drop [METEOROL] An elementary particle of fog, physically the same as a cloud drop. Also known as fog droplet.

fog droplet *See* fog drop.

fog horizon [METEOROL] The top of a fog layer which is confined by a low-level temperature inversion so as to give the appearance of the horizon when viewed from above against the sky; the true horizon is usually obscured by the fog in such instances.

fog scale [METEOROL] A classification of fog intensity based on its effectiveness in decreasing horizontal visibility; such practice is not current in United States weather observing procedures.

fog wind [METEOROL]   Humid east wind which crosses the divide of the Andes east of Lake Titicaca and descends on the west in violent squalls; probably the same as puelche.

föhn *See* foehn.

foidite [PETR]   A plutonic or volcanic rock in which feldspathoids constitute 60–100% of the light-colored components.

fold [GEOL]   A bend in rock strata or other planar structure, usually produced by deformation; folds are recognized where layered rocks have been distorted into wave-like form.

fold belt *See* orogenic belt.

folding [GEOL]   Compression of planar structure in the formation of fold structures.

folding boards [MIN ENG]   A shifting frame on which the cage rests in a mine. Also known as chairs; dogs; keeps; keps.

fold system [GEOL]   A group of folds with common trends and characteristics.

folia [PETR]   Thin, leaflike layers that occur in rocks of gneissic or schistose type.

foliaceous [GEOL]   Having a leaflike or platelike structure composed of thin layers of minerals.

foliated ice [HYD]   Large masses of ice which grow in thermal contraction cracks in permafrost. Also known as ice wedge.

foliation [GEOL]   A laminated structure formed by segregation of different minerals into layers that are parallel to the schistosity.

Folist [GEOL]   A suborder of the soil order Histosol, consisting of wet forest litter resting on rock or rubble.

follower [ENG]   A drill used for making all but the first part of a hole, the first part being made with a drill of larger gage.

follower rail [MIN ENG]   The rail of a mine switch on the other side of the turnout corresponding to the lead rail.

following wind [METEOROL]   **1.** A wind blowing in the direction of ocean-wave advance. **2.** *See* tailwind.

fool's gold *See* pyrite.

footage [MIN ENG]   **1.** The number of feet of borehole drilled per unit of time, or that required to complete a specific project or contract. **2.** The payment of miners by the running foot of work.

footeite *See* connellite.

foot holes [MIN ENG]   Holes cut in the sides of shafts or winzes to enable miners to climb up or down.

footwall [GEOL]   The mass of rock that lies beneath a fault, an ore body, or a mine working. Also known as heading side; heading wall; lower plate.

footwall shaft *See* underlay shaft.

Foraky boring method [MIN ENG]   A percussive boring system; a closed-in derrick contains the crown pulley, over which a steel rope with the boring tools is passed from a drum; the drum moves the tools, which are vibrated by a walking beam.

Foraky freezing process [MIN ENG]   A method of shaft sinking through heavily watered sands by freezing the sands.

forbesite [MINERAL] $H(Ni,Co)AsO_4 \cdot 3\frac{1}{2}H_2O$ A grayish-white mineral composed of hydrous nickel cobalt arsenate; occurs in fibrocrystalline form.

forced auxiliary ventilation [MIN ENG] A system in which the duct delivers the intake air to the face.

forced-caving system [MIN ENG] A stoping system in which the ore is broken down by large blasts into the stopes that are kept partly full of broken ore.

force piece *See* foreset.

forcing fan [MIN ENG] A fan which forces the intake air into mine workings. Also known as blowing fan.

forecast [METEOROL] A statement of expected future meteorological occurrences.

forecast period [METEOROL] The time interval for which a forecast is made.

forecast-reversal test [METEOROL] A test used to evaluate the adequacy of a given method of forecast verification; the same verification method is applied, simultaneously, to a given forecast and to a fabricated forecast of opposite conditions; comparison of the verification scores gives an indication of the value of the verification system.

forecast verification [METEOROL] Any process for determining the accuracy of a weather forecast by comparing the predicted weather with the observed weather of the forecast period; used to test forecasting skills and methods.

foredeep [GEOL] **1.** A long, narrow depression that borders an orogenic belt, such as an island arc, on the convex side. **2.** *See* exogeosyncline.

fore drift [MIN ENG] That one of a pair of parallel headings which is kept a short distance in advance of the other.

foredune [GEOL] A coastal dune or ridge that is parallel to the shoreline of a large lake or ocean and is stabilized by vegetation.

foreign inclusion [PETR] A fragment of country rock that is enveloped in an igneous intrusion.

foreland [GEOGR] An extensive area of land jutting out into the sea.   [GEOL] **1.** A lowland area onto which piedmont glaciers have moved from adjacent mountains. **2.** A stable part of a continent bordering an orogenic or mobile belt.

foreland facies *See* shelf facies.

Forel scale [OCEANOGR] A scale of yellows, greens, and blues for recording the color of sea water as seen against the white background of a Secchi disk.

forepoling [MIN ENG] A timbering method for a very weak roof in which a bench of timbers is set and boards or long wedges are placed above the header; as the next bench of timbers is placed at the inbye end of the wedges, other like wedges are driven in under the first wedges and over the second header. Also known as spiling.

fore reef [GEOL] The distal or seaward surface of a reef, steeply inclined and littered with material broken off from the reef.

forerunner [OCEANOGR] Low, long-period ocean swell which commonly precedes the main swell from a distant storm, especially a tropical cyclone.

foreset [MIN ENG] **1.** To place a prop under the coal-face end of a bar. **2.** Timber set used for roof support at the working face. Also known as force piece.

foreset bed [GEOL] One of a series of inclined symmetrically arranged layers of a cross-bedding unit formed by deposition of sediments that rolled down a steep frontal slope of a delta or dune.

foreshaft sinking [MIN ENG]   The first 150 feet (46 meters) of shaft sinking from the surface; the plant and services for the main shaft are installed during this step.

foreshock [GEOPHYS]   A tremor which precedes a larger earthquake or main shock.

foreshore [GEOL]   The zone that lies between the ordinary high- and low-watermarks and is daily traversed by the rise and fall of the tide. Also known as beach face.

forest climate *See* humid climate.

forest wind [METEOROL]   A light breeze which blows from forests toward open country on calm clear nights.

forfeiture [MIN ENG]   Loss of a mining claim by operation of the law, without regard to the intention of the locator, whenever he fails to preserve his right by complying with the conditions imposed by law.

forked lightning [GEOPHYS]   A common form of lightning, in a cloud-to-ground discharge, which exhibits downward-directed branches from the main lightning channel.

formanite [MINERAL]   A mineral composed of an oxide of uranium, zirconium, thorium, calcium, tantalum, and niobium with some rare-earth metals.

formation [GEOL]   Any assemblage of rocks which have some common character and are mappable as a unit.

formation factor [GEOL]   A function of the porosity and internal geometry of a reservoir rock system, expressed as $F = \phi^{-m}$, where ϕ is the fractional porosity of the rock, and m is the cementation factor (pore-opening reduction).   [GEOCHEM]   The ratio between the conductivity of an electrolyte to that of a rock saturated with the same electrolyte. Also known as resistivity factor.

formation fracturing [PETRO ENG]   Method of applying hydraulic pressure to a reservoir formation to cause the rock to split open, that is to fracture; used to increase oil production.

formation resistivity [GEOPHYS]   Electrical resistivity of reservoir formations measured by electrical log sondes; used for clues to formation lithography and fluid content.

formation solubility [PETRO ENG]   Measure of formation rock solubility in oil-well acidizing solution (hydrochloric acid or hydrochloric-hydrofluoric acids).

formation tester [PETRO ENG]   Device for retrieval of samples of fluid from an oil-reservoir formation.

formation water [HYD]   Water present with petroleum or gas in reservoirs. Also known as oil-reservoir water.

form energy [GEOCHEM]   The potentiality of a mineral to develop its own crystal form against the resistance of the surrounding solid medium. Also known as crystalloblastic strength; power of crystallization.

Forrel cell [METEOROL]   A type of atmospheric circulation in which air moves away from the thermal equator at low latitude levels and in the opposite direction in higher latitudes.

Forrester machine [MIN ENG]   A pneumatic flotation cell in which pulp is aerated by low-pressure air, delivering a mineralized froth along the overflow, and tailings to the end weir.

forril farina *See* rock milk.

forsterite [MINERAL] Mg_2SiO_4 A whitish or yellowish, magnesium-rich variety of olivine. Also known as white olivine.

fortification agate *See* landscape agate.

fortnightly tide [OCEANOGR] A tide occurring at intervals of one-half the period of oscillation of the moon, approximately 2 weeks.

Forty Saints' storm [METEOROL] A southerly gale in Greece, occurring a little before the equinox in March.

forward scatter [GEOPHYS] The scattering of radiant energy into the hemisphere of space bounded by a plane normal to the direction of the incident radiation and lying on the side toward which the incident radiation was advancing.

foshagite [MINERAL] $Ca_5Si_3O_{10}(OH)_2 \cdot 2H_2O$ A white mineral composed of a basic hydrous calcium silicate.

fossil dune [GEOL] An ancient desert dune.

fossil fuel [GEOL] Any hydrocarbon deposit that may be used for fuel; examples are petroleum, coal, and natural gas.

fossil ice [HYD] **1.** Relatively old ground ice found in regions of permafrost. **2.** Underground ice in regions where present-day temperatures are not low enough to have formed it.

fossil permafrost *See* passive permafrost.

fossil reef [GEOL] An ancient reef.

fossil resin [GEOL] Any natural resin that is found in a geologic deposit as an exudate of long-buried plant life; examples are amber and retinite.

fossil soil *See* paleosol.

fossil wax *See* ozocerite.

Foster's formula [MIN ENG] The empirical formula $R = 3\sqrt{DT}$ for determining the radius R of a shaft pillar, where D = depth in feet and T = thickness of lode in feet.

foundation coefficient [GEOPHYS] A coefficient that indicates how many times stronger the effect of an earthquake is in a given rock that would have been true of an undisturbed crystalline rock under identical conditions.

founder [GEOL] To sink beneath the water because the land has been depressed or the sea level has been raised, usually in reference to large crustal masses, islands, or parts of continents.

fourchite [PETR] A monchiquite that lacks feldspar and olivine.

fourmarierite [MINERAL] An orange-red to brown mineral composed of a hydrous oxide of lead and uranium.

four-piece set [MIN ENG] Squared timber frame used in underground driving to give all-around support to weak ground.

four-way dip [GEOPHYS] In seismic prospecting, the dip as determined by geophones located at points in four directions from the shot point; three locations are essential, and the fourth provides a check.

fowan [METEOROL] A dry, scorching wind of Great Britain and the Isle of Man.

fowlerite [MINERAL] A zinc-bearing variety of rhodonite.

foyaite [PETR] A nepheline syenite composed chiefly of potassium feldspar.

fractional crystallization [PETR] Separation of a cooling magma into multiple minerals as the different minerals cool and congeal at progressively lower temperatures. Also known as crystallization differentiation; fractionation.

fractional gas-flow curve [PETRO ENG]   Graph of the fraction of free injected gas flowing through a reservoir formation versus the liquid saturation of the gas for various parameter values of oil viscosity; used to calculate displacement efficiency during gas injection.

fractional sampling [MIN ENG]   Mechanical selection of samples of uniformly graded material without segregation.

fractoconformity [GEOL]   The relation among conformable strata, where the older beds undergo faulting at the same time that the newer beds are being deposited.

fracture [GEOL]   A crack, joint, or fault in a rock due to mechanical failure by stress. Also known as rupture.   [MINERAL]   A break in a mineral other than along a cleavage plane.

fracture cleavage [GEOL]   Cleavage that occurs in deformed but only slightly metamorphosed rocks along closely spaced, parallel joints and fractures.

fractured formation [PETRO ENG]   Reservoir formation in which rock has been split by hydraulic pressure produced by injected fluids.

fracture dome [MIN ENG]   The zone of loose or semiloose rock which exists in the immediate hanging or footwall of a stope.

fracture plane inclination [GEOL]   Gradient or inclination of the plane of fracture formed in a reservoir formation.

fracture system [GEOL]   A stress-related group of contemporaneous fractures.

fracture zone [GEOL]   An elongate zone on the deep-sea floor that is of irregular topography and often separates regions of different depths; frequently crosses and displaces the midoceanic ridge by faulting.

fractus [METEOROL]   A cloud species in which the cloud elements are irregular but generally small in size, and which presents a ragged, shredded appearance, as if torn; these characteristics change ceaselessly and often rapidly.

fragipan [GEOL]   A natural subsurface layer of hard soil characterized by relatively slow permeability to water because of the layer's extreme density or compactness rather than high clay content or cementation.

fragmentation [MIN ENG]   The blasting of coal, ore, or rock into pieces small enough to load, handle, and transport without the need for hand-breaking or secondary blasting.

fragmentation nucleus [METEOROL]   A tiny ice particle broken from a large ice crystal, serving as an ice nucleus; that is, a growth center for a new ice crystal.

framboid [GEOL]   A microscopic aggregate of pyrite grains, often occurring in spheroidal clusters.

frame set [MIN ENG]   The arrangement of the legs and cap or crossbar so as to provide support for the roof of an underground passage. Also known as framing; set.

framework silicate See tectosilicate.

framing See frame set.

framing table [MIN ENG]   An inclined table on which ore slimes are separated by running water.

franckeite [MINERAL]   A dark-gray or black massive mineral composed of lead antimony tin sulfide.

francolite [MINERAL] $Ca_5(PO_4,CO_3)_3(F,OH)$ Colorless fluoride-bearing carbonate-apatite.

Franconian [GEOL] A North American stage of geologic time; the middle Upper Cambrian.

franklinite [MINERAL] $ZnFe_2O_4$ Black, slightly magnetic mineral member of the spinel group; usually possesses extensive substitution of divalent manganese and iron for the divalent zinc, and limited trivalent manganese for the trivalent iron.

Frank partial dislocation [CRYSTAL] A partial dislocation whose Burger's vector is not parallel to the fault plane, so that it can only diffuse and not glide, in contrast to a Schockley partial dislocation.

Frasch process [MIN ENG] A process to remove sulfur from sulfur beds; superheated water is forced under pressure into the sulfur bed, and the molten sulfur is thus forced to the surface.

Fraser's air-sand process [MIN ENG] A process in which dry, specific-gravity separation of coal from refuse is achieved by utilizing a flowing dense medium intermediate in density between coal and refuse.

frazil [HYD] Ice crystals which form in supercooled water that is too turbulent to permit coagulation of the crystals into sheet ice.

frazil ice [HYD] A spongy or slushy accumulation of frazil in a body of water. Also known as needle ice.

free air *See* free atmosphere.

free-air anomaly [GEOPHYS] A gravity anomaly calculated as the difference between the measured gravity and the theoretical gravity at sea level and a free-air coefficient determined by the elevation of the measuring station.

free-air gravity anomaly [GEOPHYS] A measure of the mass excesses and deficiencies within the earth; calculated as thedifference between the measured gravity and the theoretical gravityat sea level and a free-air coefficient determined by the elevation of the measuring station.

free-air temperature [METEOROL] Temperature of the atmosphere, obtained by a thermometer located so as to avoid as completely as practicable the effects of extraneous heating.

free atmosphere [GEOPHYS] That portion of the earth's atmosphere, above the planetary boundary layer, in which the effect of the earth's surface friction on the air motion is negligible and in which the air is usually treated (dynamically) as an ideal fluid. Also known as free air.

free convection *See* natural convection.

free crushing [MIN ENG] Crushing under conditions of speed and feed so that there is ample room for the fine ore to fall away from the coarser material and thereby escape further crushing.

freedom to mine [MIN ENG] The law by which anybody has the right to mine certain minerals when he has prospected for them and has filed a proper application for the right to mine them.

free end *See* free face.

free face [MIN ENG] The exposed surface of a mass of rock or of coal. Also known as free end.

free-fed [MIN ENG] In comminution, pertaining to rolls fed only enough ore to maintain a ribbon of material between them.

free foehn *See* high foehn.

free gas [PETRO ENG]   A hydrocarbon that exists in the gaseous phase at reservoir pressure and temperature and remains a gas when produced under normal conditions.   [PHYS]   Any gas at any pressure not in solution, or mechanically held in the liquid hydrocarbon phase.

free-gas saturation [PETRO ENG]   Proportion of oil-reservoir pore structure saturated by free (undissolved) gas.

Freeman-Nichols roaster [MIN ENG]   A unit in which pyrite flotation concentrates are flash-roasted.

free meander [HYD]   A stream meander that displaces itself very easily by lateral corrasion.

free milling [MIN ENG]   A process applied to ores which contain free gold or silver and can be reduced by crushing and amalgamation (by gravity or on blankets), without roasting or other chemical treatment.

free-milling ore [MIN ENG]   Ore containing gold which can be caught with mercury by a variety of gravity processes or on blankets.

free settling [MIN ENG]   In classification, the free fall of particles through fluid media.

free-traveling wave *See* progressive wave.

free-water content *See* water content.

free-water elevation *See* water table.

free-water surface *See* water table.

freeze [ENG]   **1.** To permit drilling tools, casing, drivepipe, or drill rods to become lodged in a borehole by reason of caving walls or impaction of sand, mud, or drill cuttings, to the extent that they cannot be pulled out. Also known as bind-seize. **2.** The act or process of drilling a borehole by utilizing a drill fluid chilled to -30 to $-40°F$ (-34 to $-40°C$) as a means of consolidating, by freezing, the borehole wall materials or core as the drill penetrates a water-saturated formation, such as sand or gravel.

freeze-out lake [HYD]   A quite shallow lake subject to being frozen over for extended periods.

freeze sinking [MIN ENG]   A method of shaft sinking in waterlogged strata by the use of cold brine circulating through a system of pipes until an ice wall is formed. Also known as freezing method.

freeze-up [HYD]   The formation of a continuous ice cover on a body of water.

freezing drizzle [METEOROL]   Drizzle that falls in liquid form but freezes upon impact with the ground to form a coating of glaze.

freezing level [METEOROL]   The lowest altitude in the atmosphere over a given location, at which the air temperature is $0°C$; the height of the $0°C$ constant-temperature surface.

freezing-level chart [METEOROL]   A synoptic chart showing the height of the $0°C$ constant-temperature surface by means of contour lines.

freezing nucleus [METEOROL]   Any particle which, when present within a mass of supercooled water, will initiate growth of an ice crystal about itself.

freezing precipitation [METEOROL]   Any form of liquid precipitation that freezes upon impact with the ground or exposed objects; that is, freezing rain or freezing drizzle.

freezing rain [METEOROL]   Rain that falls in liquid form but freezes upon impact to form a coating of glaze upon the ground and on exposed objects.

F region [GEOPHYS] The general region of the ionosphere in which the F_1 and F_2 layers tend to form.

freibergite [MINERAL] A steel-gray, silver-bearing variety of tetrahedrite.

freieslebenite [MINERAL] $Pb_3Ag_5Sb_5S_{12}$ A steel-gray to dark mineral composed of a sulfide of antimony, lead, and silver.

freirinite [MINERAL] $Na_3Cu_3(AsO_4)_2(OH)_3\cdot H_2O$ A lavender to turquoise-blue mineral composed of a basic hydrous arsenate of sodium and copper.

fremontite *See* natromontebrasite.

fresh [GEOL] Unweathered in reference to a rock or rock surface. [METEOROL] Pertaining to air which is stimulating and refreshing.

fresh breeze [METEOROL] In the Beaufort wind scale, a wind whose speed is 17 to 21 knots (19 to 24 miles per hour, or 31 to 39 kilometers per hour).

fresh-core technique [PETRO ENG] A method in which a core sample, fresh from the field, is subjected to waterflooding in the laboratory, and the resulting residual oil is determined; used to calculate total waterflood recovery of oil from a reservoir formation.

freshet [HYD] **1.** The annual spring rise of streams in cold climates as a result of melting snow. **2.** A flood resulting from either rain or melting snow; usually applied only to small streams and to floods of minor severity. **3.** A small fresh-water stream.

fresh gale [METEOROL] In the Beaufort wind scale, a wind whose speed is from 34 to 40 knots (39 to 46 miles per hour, or 63 to 74 kilometers per hour).

fresh ice *See* newly formed ice.

fresh water [HYD] Water containing no significant amounts of salts, such as in rivers and lakes.

friagem [METEOROL] A period of cold weather in the middle and upper parts of the Amazon Valley and in eastern Bolivia, occurring during the dry season in the Southern Hemisphere winter. Also known as vriajem.

frictional *See* cohesionless.

friction blocks [PETRO ENG] Thin blocks with cylindrical surfaces that drag on the inside of the well casing to prevent rotation of the packer (seal between the outside of tubing and inside of casing).

friction crack [GEOL] A short, crescent-shaped crack in glaciated rock produced by a localized increase in friction between rock and ice, oriented transverse to the direction of ice flow.

friction depth [OCEANOGR] The depth at which the velocity of wind-driven current becomes neglibible compared to the surface velocity; sometimes referred to as the depth of the Ekman layer.

friction feed [MIN ENG] Longitudinal movement or advance of a drill stem and bit accomplished by friction devices in a diamond-drill swivel head, as opposed to a system consisting entirely of meshing gears.

friction layer *See* surface boundary layer.

friction velocity [METEOROL] A reference wind velocity defined by the relation $u\ \sqrt{|\tau/\rho|}$, where τ is the Reynolds stress, ρ the density, and u the friction velocity.

friction yielding prop *See* mechanical yielding prop.

friedelite [MINERAL] $Mn_8Si_6O_{18}(OH,Cl)_4\cdot 3H_2O$ A rose-red mineral composed of manganese silicate with chlorine.

Friedel's law [CRYSTAL]   The law that x-ray or electron diffraction measurements cannot determine whether or not a crystal has a center of symmetry.

fringe region [METEOROL]   The upper portion of the exosphere, where the cone of escape equals or exceeds 180°; in this region the individual atoms have so little chance of collision that they essentially travel in free orbits, subject to the earth's gravitation, at speeds imparted by the last collision. Also known as spray region.

fringing reef [GEOL]   A coral reef attached directly to or bordering the shore of an island or continental landmass.

frog storm [METEOROL]   The first bad weather in spring after a warm period. Also known as whippoorwill storm.

frohbergite [MINERAL] $FeTe_2$ A mineral composed of iron telluride; it is isomorphous with marcasite.

frondelite [MINERAL] $MnFe_4(PO_4)_5(OH)_5$ A mineral composed of basic phosphate of manganese and iron; it is isomorphous with rockbridgeite.

front [METEOROL]   A sloping surface of discontinuity in the troposphere, separating air masses of different density or temperature.

front abutment pressure [GEOPHYS]   The release of energy in the superincumbent strata above the seam induced by the extraction of the seam.

frontal-advance performance [PETRO ENG]   The theory that, during the waterflood of a formation reservoir, displacement of oil causes a desaturation of the displaced fluids in accordance with relative-permeability relationships, and the displacement is linear.

frontal apron *See* outwash plain.

frontal contour [METEOROL]   The line of intersection of a front (frontal surface) with a specified surface in the atmosphere, usually a constant-pressure surface; with respect to only one surface, this line is usually called the front.

frontal cyclone [METEOROL]   Any cyclone associated with a front; often used synonymously with wave cyclone or with extratropical cyclone (as opposed to tropical cyclones, which are nonfrontal).

frontal drive [PETRO ENG]   In an oil reservoir with constant pressure (by gas injection or from a large gas-to-oil ratio), the driving of oil fluids into the wellbore by the free gas.

frontal fog [METEOROL]   Fog associated with frontal zones and frontal passages.

frontal inversion [METEOROL]   A temperature inversion in the atmosphere, encountered upon vertical ascent through a sloping front (or frontal zone).

frontal lifting [METEOROL]   The forced ascent of the warmer, less-dense air at and near a front, occurring whenever the relative velocities of the two air masses are such that they converge at the front.

frontal occlusion *See* occluded front.

frontal plain *See* outwash plain.

frontal precipitation [METEOROL]   Any precipitation attributable to the action of a front; used mainly to distinguish this type from air-mass precipitation and orographic precipitation.

frontal profile [METEOROL]   The outline of a front as seen on a vertical cross section oriented normal to the frontal surface.

frontal strip [METEOROL]   The presentation of a front, on a synoptic chart, as a frontal zone; that is, two lines, rather than a single line, are drawn to represent the boundaries of the zone; a rare usage.

frontal system [METEOROL]   A system of fronts as they appear on a synoptic chart.

frontal thunderstorm   [METEOROL]   A thunderstorm associated with a front; limited to thunderstorms resulting from the convection induced by frontal lifting.

frontal wave   [METEOROL]   A horizontal, wavelike deformation of a front in the lower levels, commonly associated with a maximum of cyclonic circulations in the adjacent flow; it may develop into a wave cyclone.

frontal zone   [METEOROL]   The three-dimensional zone or layer of large horizontal density gradient, bounded by frontal surfaces and surface front.

front of the fetch   [OCEANOGR]   In wave forecasting, that end of the generating area toward which the wind is blowing.

frontogenesis   [METEOROL]   **1.** The initial formation of a frontal zone or front.   **2.** The increase in the horizontal gradient of an air mass property, mainly density, and the formation of the accompanying features of the wind field that typify a front.

frontogenetic function   [METEOROL]   A kinematic measure of the tendency of the flow in an air mass to increase the horizontal gradient of a conservative property.

frontolysis   [METEOROL]   **1.** The dissipation of a front or frontal zone.   **2.** In general, a decrease in the horizontal gradient of an air mass property, principally density, and the dissipation of the accompanying features of the wind field.

front pinacoid   [CRYSTAL]   The {100} pinacoid in an orthorhombic, monoclinic, or triclinic crystal.   Also known as macropinacoid; orthopinacoid.

front slope   *See* scarp slope.

frost   [HYD]   A covering of ice in one of its several forms, produced by the sublimation of water vapor on objects colder than 32°F (0°C).

frost action   [GEOL]   **1.** The weathering process caused by cycles of freezing and thawing of water in surface pores, cracks, and other openings.   **2.** Alternate or repeated cycles of freezing and thawing of water contained in materials; the term is especially applied to disruptive effects of this action.

frost boil   [GEOL]   **1.** An accumulation of water and mud released from ground ice by accelerated spring thawing.   **2.** A low mound formed by local differential frost heaving at a location most favorable for the formation of segregated ice and accompanied by the absence of an insulating cover of vegetation.

frost bursting   *See* congelifraction.

frost churning   *See* congeliturbation.

frost climate   [CLIMATOL]   The coldest temperature province in C. W. Thornthwaite's climatic classification: the climate of the ice cap regions of the earth, that is, those regions perennially covered with snow and ice.

frost day   [METEOROL]   An observational day on which frost occurs.

frost feathers   *See* ice feathers.

frost flakes   *See* ice fog.

frost flowers   *See* ice flowers.

frost fog   *See* ice fog.

frost hazard   [METEOROL]   The risk of damage by frost, expressed as the probability or frequency of killing frost on different dates during the growing season, or as the distribution of dates of the last killing frost of spring or the first of autumn.

frost heaving   [GEOL]   The lifting and distortion of a surface due to internal action of frost resulting from subsurface ice formation; affects soil, rock, pavement, and other structures.

frostless zone [METEOROL] The warmest part of a slope above a valley floor, lying between the layer of cold air which forms over the valley floor on calm clear nights and the cold hill tops or plateaus; the air flowing down the slopes is warmed by mixing with the air above ground level, and to some extent also by adiabatic compression. Also known as green belt; verdant zone.

frost line [GEOL] **1.** The maximum depth of frozen ground during the winter. **2.** The lower limit of permafrost.

frost mound [GEOL] A hill and knoll associated with frozen ground in a permafrost region, containing a core of ice. Also known as soffosian knob; soil blister.

frost point [METEOROL] The temperature to which atmospheric moisture must be cooled to reach the point of saturation with respect to ice.

frost-point hygrometer [ENG] An instrument for measuring the frost point of the atmosphere; air under test is passed continuously across a polished surface whose temperature is adjusted so that a thin deposit of frost is formed which is in equilibrium with the air.

frost-point technique *See* 8D technique (in the D's).

frost riving *See* congelifraction.

frost shattering *See* congelifraction.

frost smoke [METEOROL] **1.** A rare type of fog formed in the same manner as a steam fog, but at colder temperatures so that it is composed of ice particles instead of water droplets. **2.** *See* steam fog.

frost splitting *See* congelifraction.

frost stirring *See* congeliturbation.

frost table [GEOL] An irregular surface in the ground which, at any given time, represents the penetration of thawing into seasonally frozen ground.

frost weathering *See* congelifraction.

frost wedging *See* congelifraction.

frost zone *See* seasonally frozen ground.

froth flotation [ENG] A process for recovery of particles of ore or other material, in which the particles adhere to bubbles and can be removed as part of the froth.

frothing collector [MIN ENG] An ore collector which in addition produces a stable foam.

frozen fog *See* ice fog.

frozen ground [GEOL] Soil having a temperature below freezing, generally containing water in the form of ice. Also known as gelisol; merzlota; taele; tjaele.

frozen precipitation [METEOROL] Any form of precipitation that reaches the ground in frozen form; that is, snow, snow pellets, snow grains, ice crystals, ice pellets, and hail.

Frue vanner [MIN ENG] A side-shake type of ore-dressing apparatus consisting of an inclined rubber belt on which the material is washed by a constant flow of water.

fuchsite [MINERAL] A bright-green variety of muscovite rich in chromium.

fucoid [GEOL] A tunnellike marking on a sedimentary structure identified as a trace fossil but not referred to a described genus.

fugitive air [MIN ENG] Air which moves through the ventilation fan but never reaches the mine workings.

fulchronograph [ENG] An instrument for recording lightning strokes, consisting of a rotating aluminum disk with several hundred steel fins on its rim; the fins are magnetized if they pass between two coils when these are carrying the surge current of a lightning stroke.

fulgurite [GEOL] A glassy, rootlike tube formed when a lightning stroke terminates in dry sandy soil; the intense heating of the current passing down into the soil along an irregular path fuses the sand.

fuller's earth [GEOL] A natural, fine-grained earthy material, such as a clay, with high adsorptive power; consists principally of hydrated aluminum silicates; used as an adsorbent in refining and decolorizing oils, as a catalyst, and as a bleaching agent.

full-face firing [MIN ENG] Drilling of small-diameter holes from top to bottom of the face.

full-seam mining [MIN ENG] A mining system in which the entire section is dislodged together and the coal is separated from the rock outside the mine by the cleaning plant.

full subsidence [MIN ENG] The greatest amount of subsidence occurring as a result of mine workings.

fully arisen sea *See* fully developed sea.

fully developed mine [MIN ENG] In coal mining, a mine where all development work has reached the boundaries and further extraction will be done on the retreat.

fully developed sea [OCEANOGR] The maximum ocean waves or sea state that can be produced by a given wind force blowing over sufficient fetch, regardless of duration. Also known as fully arisen sea.

fuloppite [MINERAL] $Pb_3Sb_8S_{15}$ A lead gray, monoclinic mineral consisting of lead antimony sulfide.

fumarole [GEOL] A hole, usually found in volcanic areas, from which vapors or gases escape.

fumulus [METEOROL] A very thin cloud veil at any level, so delicate that it may be almost invisible.

fundamental circle *See* primary great circle.

fundamental complex [GEOL] An agglomeration of metamorphic rocks underlying sedimentary or unmetamorphosed rocks; specifically, an agglomeration of Archean rocks supporting a geological column.

fundamental jelly *See* ulmin.

fundamental strength [GEOPHYS] Under specified conditions but regardless of time, that maximum stress that a geological structure can tolerate without creep.

fundamental substance *See* ulmin.

funnel cloud [METEOROL] The popular term for the tornado cloud, often shaped like a funnel with the small end nearest the ground.

furiani [METEOROL] A southwest wind that blows in the vicinity of the Po River, Italy, and is vehement and short-lived, followed by a gale from the south or southeast.

fusain [GEOL] The local lithotype strands or patches, characterized by silky luster, fibrous structure, friability, and black color. Also known as mineral charcoal.

fuse [ENG] Also spelled fuze. **1.** A device with explosive components designed to initiate a train of fire or detonation in an item of ammunition by an action such as hydrostatic pressure, electrical energy, chemical energy, impact, or a combination of these. **2.** A nonexplosive device designed to initiate an explosion in an item of ammunition by an action such as continuous or pulsating electromagnetic waves or acceleration.

fuse blasting cap [ENG]   A small copper cylinder closed at one end and charged with a fulminate.

fuse gage [ENG]   An instrument for slicing time fuses to length.

fusehead [ENG]   That part of an electric detonator consisting of twin metal conductors, bridged by fine resistance wire, and surrounded by a bead of igniting compound which burns when the firing current is passed through the bridge wire.

fuse lighter [ENG]   A device for facilitating the ignition of the powder core of a fuse.

fusinite [GEOL]   The micropetrological constituent of fusain which consists of carbonized woody tissue.

fusinization [GEOL]   The process of formation of fusain in coal.

fusion crust [GEOL]   A thin glasslike coating, often black and rarely more than 1 millimeter thick, that is formed by ablation on a meteorite surface.

fusion piercing [ENG]   A method of producing vertical blastholes by virtually burning holes in rock. Also known as piercing.

fusion-piercing drill [ENG]   A machine designed to use the fusion-piercing mode of producing holes in rock. Also known as det drill; jet-piercing drill; Linde drill.

Fusulinacea [PALEON]   A superfamily of large, marine extinct protozoans in the order Foraminiferida characterized by a chambered calcareous shell.

Fusulinidae [PALEON]   A family of extinct protozoans in the superfamily Fusulinacea.

Fusulinina [PALEON]   A suborder of extinct rhizopod protozoans in the order Foraminiferida having a monolamellar, microgranular calcite wall.

fuze *See* fuse.

G

gabbro [PETR] A group of dark-colored, intrusive igneous rocks with granular texture, composed largely of basic plagioclase and clinopyroxene.

gabion [ENG] A bottomless basket of wickerwork or strap iron filled with earth or stones; used in building fieldworks or as revetments in mining. Also known as pannier.

gad [MIN ENG] **1.** A heavy steel wedge, 6 or 8 inches (15 or 20 centimeters) long, with a narrow chisel point used in mining to cut samples, break out pieces of loose rock, and so on. **2.** A small iron punch with a wooden handle used to break up ore.

gadder [MIN ENG] A small car or platform with a drilling machine attached, to make a straight line of holes along its course in getting out dimension stone. Also known as gadding car; gadding machine.

gadding car *See* gadder.

gadding machine *See* gadder.

gadolinite [MINERAL] $Be_2FeY_2Si_2O_{10}$ A black, greenish-black, or brown rare-earth mineral; hardness is 6.5–7 on Mohs scale, and specific gravity is 4–4.5.

gagatite [GEOL] A coalified woody material similar to jet.

gagatization [GEOL] During coal formation, the impregnating of wood fragments with dissolved organic substances.

gageite [MINERAL] $(Mn,Mg,Zn)_8Si_3O_{14} \cdot 2H_2O$ (or $3H_2O$) A mineral composed of a hydrous silicate of manganese, magnesium, and zinc.

gage loss [MIN ENG] The diametrical reduction in the size of a bit or reaming shell caused by wear through use.

gahnite [MINERAL] $ZnAl_2O_4$ A usually dark-green, but sometimes yellow, gray, or black spinel mineral consisting of an oxide of zinc and aluminum. Also known as zinc spinel.

gaign [METEOROL] A cross-mountain wind that causes clouds to form on the crests of mountains in Italy.

galaxite [MINERAL] $MnAl_2O_4$ A black mineral of the spinel series composed of an oxide of manganese and aluminum.

gale [METEOROL] **1.** An unusually strong wind. **2.** In storm-warning terminology, a wind of 28–47 knots (52–87 kilometers per hour). **3.** In the Beaufort wind scale, a wind whose speed is 28–55 knots (52–102 kilometers per hour).

galena [MINERAL] PbS A bluish-gray to lead-gray mineral with brilliant metallic luster, specific gravity 7.5, and hardness 2.5 on Mohs scale; occurs in cubic or octahedral crystals, in masses, or in grains. Also known as blue lead; lead glance.

galenobismutite [MINERAL] $PbBi_2S_4$ A lead-gray or tin-white mineral consisting of bismuth sulfide; specific gravity is 6.9.

Galeritidae [PALEON]   A family of extinct exocyclic Euechinoidea in the order Holectypoida, characterized by large ambulacral plates with small, widely separated pore pairs.

galerna *See* galerne.

galerne [METEOROL]   A squally northwesterly wind that is cold, humid, and showery, occurring in the rear of a low-pressure area over the English Channel and off the Atlantic coast of France and northern Spain. Also known as galerna; galerno; giboulee.

galerno *See* galerne.

gale warning [METEOROL]   A storm warning for marine interests of impending winds from 28 to 47 knots (52–87 kilometers per hour), signaled by two triangular red pennants by day, and a white lantern over a red lantern by night.

gallego [METEOROL]   A cold, piercing, northerly wind in Spain and Portugal.

gallery [GEOL]   **1.** A horizontal, or nearly horizontal, underground passage, either natural or artificial.   **2.** A subsidiary passage in a cave at a higher level than the main passage.   [MIN ENG]   A level or drift.

gallery testing [MIN ENG]   A method of testing explosives; a test condition is achieved by firing light charges without any stemming, and heavier charges with only 1 inch (2.5 centimeters) of stemming.

Galloway sinking and walling stage *See* sinking and walling scaffold.

Galloway stage [MIN ENG]   A platform of several decks suspended near the shaft during the sinking operation.

gallows frame *See* headframe.

galmei *See* hemimorphite.

Galofaro [OCEANOGR]   A whirlpool in the Strait of Messina, between Sicily and Italy; formerly called Charybdis.

gamma-ray well logging [ENG]   Measurement of gamma-ray intensity versus depth down the wellbore; used to identify rock strata, their position, and their thicknesses.

Gampsonychidae [PALEON]   A family of extinct crustaceans in the order Palaeocaridacea.

gangue [GEOL]   The valueless rock or aggregates of minerals in an ore.

gangway [MIN ENG]   **1.** A principal underground haulage road.   **2.** A passageway into or out of an underground mine.

ganister [PETR]   A fine, hard quartzose sandstone; used to make refractory silica brick to line furnace reactors.

ganomalite [MINERAL] $(Ca_2)Pb_3Si_3O_{11}$ A colorless to gray silicate of lead with calcium crystallizing in the tetragonal system.

ganophyllite [MINERAL] $(Na,K)(Mn,Fe,Al)_5(Si,Al)_6O_{15}(OH)_5·2H_2O$ A brown, prismatic crystalline or foliated mineral composed of a hydrous silicate of manganese and aluminum.

gap [GEOGR]   Any sharp, deep notch in a mountain ridge or between hills.

garbin [METEOROL]   A sea breeze; in southwest France it refers to a southwesterly sea breeze which sets in about 9 A.M., reaches its maximum toward 2 P.M., and ceases about 5 P.M.

Garbutt rod   [PETRO ENG]   A device used to pull the standing valve out of a tubing-type oil-well sucker-rod pump.

Gardner crusher   [MIN ENG]   A swing-and-hammer crusher; the U-shaped hammers are thrown by a revolving shaft against the feed and a heavy anvil inside the housing.

garland [MIN ENG]   A channel fixed around a shaft in order to catch the water draining down the walls and conduct it to a lower level. Also known as water curb; water garland; water ring.

garnet [MINERAL]   A generic name for a group of mineral silicates that are isometric in crystallization and have the general formula $A_3B_2(SiO_4)_3$, where A is Fe^{2+}, Mn^{2+}, Mg, or Ca, and B is Al, Fe^{3+}, Cr^{3+}, or Ti^{3+}; used as a gemstone and abrasive.

garnierite [MINERAL] $(Ni,Mg)_3Si_2O_5(OH)_4$ An apple-green or pale-green, monoclinic serpentine; a gemstone and an ore of nickel. Also known as nepuite; noumeite.

garronite [MINERAL] $Na_2Ca_5Al_{12}Si_{20}O_{64} \cdot 27H_2O$ A zeolite mineral belonging to the phillipsite group; crystallizes in the tetragonal system.

garúa [METEOROL]   A dense fog or drizzle from low stratus clouds on the west coast of South America, creating a raw, cold atmosphere that may last for weeks in winter, and supplying a limited amount of moisture to the area. Also known as camanchaca.

gas alarm [MIN ENG]   A signal system which warns mine workers of dangerous concentration of firedamp.

gas anchor [PETRO ENG]   A downhole gas separator used to reduce gas-in-oil froth before the pump to increase pump efficiency.

gas and mist sampler [MIN ENG]   An instrument for automatic collection of one sample per hour of airborne contaminants such as sulfur dioxide or ammonia.

gas cap [GEOPHYS]   The gas immediately in front of a meteoroid as it travels through the atmosphere.   [PETRO ENG]   Gas occurring above liquid hydrocarbons in a reservoir under such trap conditions as the presence of water which prevents downward migration or the abutment of an impermeable formation against the reservoir.

gas-cap drive [PETRO ENG]   Driving liquid hydrocarbons through a porous reservoir and toward well holes by utilizing the pressure of gas overlying the liquid pool.

gas-cap expansion [PETRO ENG]   Process of reservoir-liquids displacement by the natural expansion of the reservoir gas cap to fill the voids vacated by recovered liquids.

gas-cap injection *See* external gas injection.

gas-cap reservoir [PETRO ENG]   Two-phase reservoir in which a free area of gas (a gas cap) is underlain by an oil or liquid phase.

gas-condensate reservoir [GEOL]   Hydrocarbon reservoir in which conditions of temperature and pressure have resulted in the condensation of the heavier hydrocarbon constituents from the reservoir gas.

gas-condensate well [PETRO ENG]   A well producing hydrocarbons from a gas-condensate reservoir.

gas coning [PETRO ENG]   The tendency of gas in a gas-drive reservoir to push oil downward in an inverse cone contour toward the casing perforations; at the extreme of coning, gas, not oil, will be produced from the well.

gas depletion drive *See* internal gas drive.

gas detector [MIN ENG]   A device which indicates the existence of firedamp or other combustible or noxious gas in a mine.

gas-drive reservoir [PETRO ENG]   An oil reservoir in which gas (either natural or reinjected) provides the driving force to sweep liquids through the formation and into the wellbore.

gas emission [MIN ENG]   The release of gas from the strata into the mine workings.

gaseous transfer [GEOL]   Separation, in a magma, of a gaseous phase which moves relative to the magma and releases dissolved substances, usually in the upper levels of the magma, when it enters an area of reduced pressure. Also known as volatile transfer.

gas explosion [MIN ENG]   An explosion of firedamp in a coal mine; coal dust apparently does not play a significant part.

gas field [PETRO ENG]   An area underlain with little or no interruption by one or more reservoirs of commercially valuable gas.

gas-filled porosity [GEOL]   A reservoir formation in which the pore space is filled by gas instead of liquid hydrocarbons.

gas ignition [MIN ENG]   The setting on fire of an accumulation of firedamp in a coal mine.

gas injection [PETRO ENG]   The injection of gas into a reservoir to maintain formation pressure and to drive liquid hydrocarbons toward the wellbores.

gas-injection well [PETRO ENG]   A well hole in a reservoir into which pressurized gas is injected to maintain formation pressure or to drive liquid hydrocarbons toward other well holes.

gas lift [PETRO ENG]   The injection of gas near the bottom of an oil well to aerate and lighten the column of oil to increase oil production from the well.

gas logging [PETRO ENG]   Hot-wire-detector or gas-chromatographic analysis and record of gas contained in the mud stream and cuttings for a well being drilled; a common way of detecting subsurface oil and gas shows.

gasman [MIN ENG]   An underground official who examines the mine for firedamp and has charge of its removal.

gas-oil ratio [PETRO ENG]   Approximation of oil-reservoir composition, expressed in cubic feet of gas per barrel of liquid at 14.7 psia and 60°F (15.6°C).

gas-oil separator [PETRO ENG]   An oil-field stock tank or series of tanks in which wellhead pressure is reduced so that the dissolved gas associated with reservoir oil is flashed off or separated as a separate phase. Also known as gas separator; oil-field separator; oil-gas separator; oil separator; separator.

gasoline locomotive [MIN ENG]   A mine locomotive which uses a gaseous fuel, is comparable to the steam locomotive in radius of travel, and has a speed of 3–12 miles per hour (5–19 kilometers per hour).

gaspeite [MINERAL] $NaCO_3$ An anhydrous normal carbonate mineral with calcite structure.

gas pocket [GEOL]   A gas-filled cavity in rocks, especially above an oil pocket.

gas-pressure maintenance [PETRO ENG]   The maintenance of oil-reservoir gas pressure, usually by gas injection, to increase hydrocarbon recovery and to improve reservoir production characteristics.

gas reservoir [GEOL]   An accumulation of natural gas found with or near accumulations of crude oil in the earth's crust.

gas rig [MIN ENG] A borehole drill, either rotary or churn type, driven by a combustion-type engine energized by a combustible liquid, such as gasoline, or a combustible gas, such as bottle gas.

gas separator *See* gas-oil separator.

gas-solubility factor [PETRO ENG] The number of standard cubic feet of gas liberated under specified gas-oil separator conditions that are in solution in one barrel of stock-tank oil at reservoir temperature and pressure.

gas spurt [GEOL] An accumulation of organic matter on certain strata caused by escaping gas.

gas stimulation [PETRO ENG] The detonation of a nuclear explosive in the strata of a natural-gas field to make the gas flow more freely.

gassy [MIN ENG] A coal mine rating by the U.S. Bureau of Mines, applicable when an ignition occurs or if a methane content exceeding 0.25% can be detected; work must be halted if the methane exceeds 1.5% in a return airway.

gas tracer [MIN ENG] Dust clouds, chemical smoke, or gaseous or radioactive tracers used to detect slowly moving air currents in a mine.

gas well [PETRO ENG] A well drilled for extraction of natural gas from a gas reservoir.

gas zone [GEOL] A rock formation containing gas under a pressure large enough to force the gas out if tapped from the surface.

gate conveyor [MIN ENG] A conveyor that carries coal from one source or face only, that is, from a single-unit or double-unit face.

gate interlock [MIN ENG] A system that prevents movement of shaft conveyances or transmission of action signals until all shaft gates are closed.

gate road bunker [MIN ENG] An appliance for coal storage from the face conveyors during peaks of production or during a stoppage of the outby transport.

gather [MIN ENG] **1.** To assemble loaded cars from several production points and deliver them to main haulage for transport to the surface or pit bottom. **2.** To drive a heading through disturbed or faulty ground so as to meet the seam of coal at a convenient level or point on the opposite side.

gathering area [PETRO ENG] The area, usually down the regional dip from a hydrocarbon trap, from which the oil or gas may have migrated updip into the trap.

gathering arm loader [MIN ENG] A machine for loading loose rock or coal; has a tractor-mounted chassis and carries a chain conveyor whose front end is built into a wedge-shaped blade.

gathering conveyor [MIN ENG] Any conveyor which gathers coal from other conveyors and delivers it either into mine cars or onto another conveyor.

gathering locomotive *See* gathering motor.

gathering mine locomotive *See* gathering motor.

gathering motor [MIN ENG] A lightweight type of electric locomotive used to haul loaded cars from the working places to the main haulage road and to replace them with empties. Also known as electric gathering locomotive; gathering locomotive; gathering mine locomotive.

gathering motorman [MIN ENG] In bituminous coal mining, one who operates a mine locomotive to haul loaded mine cars from working places to sidings.

gathering mule [MIN ENG] The mule used to collect the loaded cars from the separate working places and to return empties.

gathering pump [MIN ENG] A portable or semiportable pump that is required for removing water encountered while opening a new mine, for extending headings or entries in an operating mine, for pump rooms or rib sections lying in the dip, for collecting water from local pools, or for sinking a shaft.

gathering system [PETRO ENG] A pipeline system used to gather gas or oil production from a number of separate wells, bringing the combined production to a central storage, pipelining, or processing terminal.

gaufrage *See* plaiting.

gaussmeter [ENG] A magnetometer whose scale is graduated in gauss or kilogauss, and usually measures only the intensity, and not the direction, of the magnetic field.

gaylussite [MINERAL] $Na_2Ca(CO_3)_2 \cdot 5H_2O$ A translucent, yellowish-white hydrous carbonate mineral, with a vitreous luster, crystallizing in the monoclinic system; found in dry lakes.

geanticline [GEOL] A broad land uplift; refers to the land mass from which sediments in a geosyncline are derived.

gearksutite [MINERAL] $CaAl(OH)F_4 \cdot H_2O$ A clayey mineral composed of hydrous calcium aluminum fluoride, occurring with cryolite.

gebli *See* ghibli.

Geco sampler [MIN ENG] Straight-line cutter designed to traverse a falling stream of ore or pulp at regular intervals, so as to divert a representative sample to a holding vessel.

gedanite [MINERAL] A brittle, wine-yellow variety of amber containing little succinic acid; found on the shore of the Baltic Sea.

gedrite [MINERAL] An aluminous variety of the mineral anthophyllite.

geg [METEOROL] A desert dust whirl of China and Tibet.

gehlenite [MINERAL] Ca_2Al_2SiO A mineral of the melitite group that crystallizes in the tetragonal crystal system and is isomorphous with akermanite; a green, resinous material found with spinel.

geikielite [MINERAL] $MgTiO_3$ A bluish-black or brownish-black mineral that crystallizes in the rhombohedral system and occurs in the form of rolled pebbles; it is isomorphous with ilmenite.

geking [OCEANOGR] Obtaining measurements of ocean movements with a geomagnetic electrokinetograph (gek).

gelation model [PETRO ENG] Electrolytic analog of a reservoir; used to investigate the areal sweep movement of water from multiple injection wells; operates by movement of copper ammonium and zinc ammonium ions through the gelatin in a flat tray that simulates the reservoir.

gelifluction [GEOL] The slow, continuous downslope movement of rock debris and water-saturated soil that occurs above frozen ground, as in most polar regions and in many high mountain ranges.

gelifraction *See* congelifraction.

gelisol *See* frozen ground.

geliturbation *See* congeliturbation.

gelivation *See* congelifraction.

gel mineral *See* mineraloid.

Gelocidae [PALEON] A family of extinct pecoran ruminants in the superfamily Traguloidea.

gelose *See* ulmin.

gem [MINERAL] A natural or artificially produced mineral or other material that has sufficient beauty and durability for use as a personal adornment.

gemology [MINERAL] The science concerned with the identification, grading, evaluation, fashioning, and other aspects of gemstones.

gemstone [GEOL] A mineral or petrified organic matter suitable for use in jewelry.

Gemuendinoidei [PALEON] A suborder of extinct raylike placoderm fishes in the order Rhenanida.

gending [METEOROL] A local dry wind in the northern plains of Java that resembles the foehn, caused by a wind crossing the mountains near the south coast and pushing between the volcanoes.

generalized hydrostatic equation [GEOPHYS] The vertical component of the vector equation of motion in natural coordinates when the acceleration of gravity is replaced by the virtual gravity; for most purposes it is identical to the hydrostatic equation.

generalized transmission function [GEOPHYS] In atmospheric-radiation theory, a set of values, variable with wavelength, each one of which represents an average transmission coefficient for a small wavelength interval and for a specified optical path through the absorbing gas in question.

generating area *See* fetch.

genesis rocks [GEOL] Rocks that have retained their character from nearly 4.6×10^9 years ago, when planets were still occulting out of the cloud of dust and gas referred to as the solar nebula; examples are meteorites and asteroids.

Geniohyidae [PALEON] A family of extinct ungulate mammals in the order Hyracoidea; all members were medium to large-sized animals with long snouts.

Genoa cyclone [METEOROL] A cyclone, or low, which appears to have formed or developed in the vicinity of the Gulf of Genoa. Also known as Genoa low.

Genoa low *See* Genoa cyclone.

gentle breeze [METEOROL] In the Beaufort wind scale, a wind whose speed is from 7 to 10 knots (13–19 kilometers per hour).

gentnerite [MINERAL] $Cu_8Fe_3Cr_{11}S_{18}$ A sulfide mineral known only in meteorites.

geo [GEOGR] A narrow coastal inlet bordered by steep cliffs. Also spelled gio.

geobotanical prospecting [GEOL] The use of the distribution, appearance, and growth anomalies of plants in locating ore deposits.

geocentric vertical [GEOD] The direction of the radius vector drawn from the center of the earth through the location of the observer. Also known as geometric vertical.

geocerite [MINERAL] A white, waxy mineral composed of carbon, oxygen, and hydrogen, occurring in brown coal.

geochemical anomaly [GEOCHEM] Above-average concentration of a chemical element in a sample of rock, soil, vegetation, stream, or sediment; indicative of nearby mineral deposit.

geochemical balance [GEOCHEM] The study of the proportional distribution (and rates of migration) in the global fractionation of a specific element or compound; an example is the distribution of quartz in igneous rocks, its liberation through weathering processes, and its redistribution to sediments and to terrestrial waters and the oceans.

geochemical cycle [GEOCHEM]   During geologic changes, the sequence of stages in the migration of elements between the lithosphere, hydrosphere, and atmosphere.

geochemical evolution [GEOCHEM]   **1.** A change in any constituent of a rock beyond that amount present in the parent rock. **2.** A change in chemical composition of a major segment of the earth during geologic time, as the oceans.

geochemical prospecting [ENG]   The use of geochemical and biogeochemical principles and data in the search for economic deposits of minerals, petroleum, and natural gases.

geochemical well logging [ENG]   Well logging dependent on geochemical analysis of the data.

geochemistry [GEOL]   The study of the chemical composition of the various phases of the earth and the physical and chemical processes which have produced the observed distribution of the elements and nuclides in these phases.

geochronology [GEOL]   **1.** The dating of the events in the earth's history. **2.** A system of dating developed for the purposes of study of the earth's history.

geochronometry [GEOL]   The study of the absolute age of the rocks of the earth based on the radioactive decay of isotopes, such as ^{238}U, ^{235}U, ^{232}Th, ^{87}Rb, ^{40}K, and ^{14}C, present in minerals and rocks.

geocosmogony [GEOL]   The study of the origin of the earth.

geocronite [MINERAL] $Pb_5(Sb,As)_2S_3$ A mineral composed of lead-gray lead antimony arsenic sulfide.

geode [GEOL]   A roughly spheroidal, hollow body lined inside with inward-projecting, small crystals; found frequently in limestone beds but may occur in shale.

geodesy [GEOPHYS]   A subdivision of geophysics which includes determination of the size and shape of the earth, the earth's gravitational field, and the location of points fixed to the earth's crust in an earth-referred coordinate system.

geodetic astronomy [GEOD]   The branch of geodesy which utilizes astronomic observations to extract geodetic information.

geodetic coordinates [GEOD]   The quantities latitude, longitude, and elevation which define the position of a point on the surface of the earth with respect to the reference spheroid.

geodetic datum [GEOD]   A datum consisting of five quantities: the latitude and longitude of an initial point, the azimuth of a line from this point, and two constants necessary to define the terrestrial spheroid.

geodetic equator [GEOD]   The great circle midway between the poles of revolution of the earth, connecting points of 0° geodetic latitude.

geodetic gravimetry [GEOD]   Worldwide relative measurements of gravitational acceleration used in geodetic studies of the earth.

geodetic latitude [GEOD]   Angular distance between the plane of the equator and a normal to the spheroid; a geodetic latitude differs from the corresponding astronomical latitude by the amount of the meridional component of station error. Also known as geographic latitude; topographical latitude.

geodetic line [GEOD]   The shortest line between any two points on the surface of the spheroid.

geodetic longitude [GEOD]   The angle between the plane of the reference meridian and the plane through the polar axis and the normal to the spheroid; a geodetic longitude differs from the corresponding astronomical longitude by the amount of the prime-vertical component of station error divided by the cosine of the latitude. Also known as geographic longitude.

geodetic meridian [GEOD] A line on a spheroid connecting points of equal geodetic longitude. Also known as geographic meridian.

geodetic parallel [GEOD] A line connecting points of equal geodetic latitude. Also known as geographic parallel.

geodetic position [GEOD] **1.** A point on the earth, the coordinates of which have been determined by triangulation from an initial station, whose location has been established as a result of astronomical observations, the coordinates depending upon the reference spheroid used. **2.** A point on the earth, defined in terms of geodetic latitude and longitude.

geodetic satellite [AERO ENG] An artificial earth satellite used to obtain data for geodetic triangulation calculations.

geodetic survey [ENG] A survey in which the figure and size of the earth are considered; it is applicable for large areas and long lines and is used for the precise location of basic points suitable for controlling other surveys.

geodynamics [GEOPHYS] The science of the forces and processes of the earth's interior.

geoeconomy [GEOGR] The study of economic conditions that are influenced by geographic factors.

geoelectricity *See* terrestrial electricity.

geoflex *See* orocline.

geognosy [GEOL] The science dealing with the solid body of the earth as a whole, occurrences of minerals and rocks, and the origin of these and their relations.

geographical coordinates [GEOGR] Spherical coordinates, designating both astronomical and geodetic coordinates, defining a point on the surface of the earth, usually latitude and longitude. Also known as terrestrial coordinates.

geographical cycle *See* geomorphic cycle.

geographical position [GEOGR] Any position on the earth defined by means of its geographical coordinates, either astronomical or geodetic.

geographic latitude *See* geodetic latitude.

geographic longitude *See* geodetic longitude.

geographic parallel *See* geodetic parallel.

geographic position [GEOGR] The position of a point on the surface of the earth expressed in terms of geographical coordinates either geodetic or astronomical.

geographic range [GEOD] The extreme distance at which an object or light can be seen when limited only by the curvature of the earth and the heights of the object and the observer.

geographic vertical [GEOD] A line perpendicular to the surface of the geoid; it is the direction in which the force of gravity acts. [MAP] The direction of a line normal to the surface of the geoid. Also known as map vertical.

geography [SCI TECH] The study of all aspects of the earth's surface, comprising its natural and political divisions, the differentiation of areas, and, sometimes people in relationship to the environment.

geohydrology [HYD] The science dealing with underground water, often referred to as hydrogeology.

geoid [GEOD] The figure of the earth considered as a sea-level surface extended continuously over the entire earth's surface.

geoisotherm [GEOPHYS] The locus of points of equal temperature in the interior of the earth; a line in two dimensions or a surface in three dimensions. Also known as geotherm; isogeotherm.

geolith *See* rock-stratigraphic unit.

geologic age [GEOL] **1.** Any great time period in the earth's history marked by special phases of physical conditions or organic development. **2.** A formal geologic unit of time that corresponds to a stage. **3.** An informal geologic time unit that corresponds to any stratigraphic unit.

geological oceanography [GEOL] The study of the floors and margins of the oceans, including descriptions of topography, composition of bottom materials, interaction of sediments and rocks with air and sea water, the effects of movements in the mantle on the sea floor, and action of wave energy in the submarine crust of the earth. Also known as marine geology; submarine geology.

geological survey [GEOL] **1.** An organization making geological surveys and studies. **2.** A systematic geologic mapping of a terrain.

geological transportation [GEOL] Shifting of material by the action of moving water, ice, or air.

geologic climate *See* paleoclimate.

geologic column [GEOL] **1.** The vertical sequence of strata of various ages found in an area or region. **2.** The geologic time scale as represented by rocks.

geologic erosion *See* normal erosion.

geologic log [GEOL] A graphic presentation of the lithologic or stratigraphic units or both traversed by a borehole; used in petroleum and mining engineering as well as geological surveys.

geologic map [GEOL] A representation of the geologic surface or subsurface features by means of signs and symbols and with an indicated means of orientation; includes nature and distribution of rock units, and the occurrence of structural features, mineral deposits, and fossil localities.

geologic noise [GEOPHYS] Disturbances in observed data caused by random inhomogeneities in surface and near-surface material.

geologic province [GEOL] An area in which geologic history has been the same.

geologic section [GEOL] Any succession of rock units found at the surface or below ground in an area. Also known as section.

geologic structure [GEOL] The total structural features in an area.

geologic thermometer *See* geothermometer.

geologic thermometry *See* geothermometry.

geologic time [GEOL] The period of time covered by historical geology, from the end of the formation of the earth as a separate planet to the beginning of written history.

geologic time scale [GEOL] The relative age of various geologic periods and the absolute time intervals.

geologist [GEOL] An individual who specializes in the geological sciences.

geolograph [ENG] A device that records the penetration rate of a bit during the drilling of a well.

geology [SCI TECH] The study or science of the earth, its history, and its life as recorded in the rocks; includes the study of geologic features of an area, such as the geometry of rock formations, weathering and erosion, and sedimentation.

geomagnetic coordinates [GEOPHYS] A system of spherical coordinates based on the best fit of a centered dipole to the actual magnetic field of the earth.

geomagnetic cutoff [GEOPHYS] The minimum energy of a cosmic-ray particle able to reach the top of the atmosphere at a particular geomagnetic latitude.

geomagnetic dipole [GEOPHYS] The magnetic dipole caused by the earth's magnetic field.

geomagnetic electrokinetograph [ENG] An instrument that can be suspended from the side of a ship to measure the direction and speed of ocean currents while the ship is under way by measuring the voltage induced in the moving conductive seawater by the magnetic field of the earth.

geomagnetic equator [GEOPHYS] That terrestrial great circle which is 90° from the geomagnetic poles.

geomagnetic field [GEOPHYS] The earth's magnetic field.

geomagnetic field reversal [GEOPHYS] Reversed magnetization in sedimentary and igneous rock, that is, polarized opposite to the mean geomagnetic field.

geomagnetic latitude [GEOPHYS] The magnetic latitude that a location would have if the field of the earth were to be replaced by a dipole field closely approximating it.

geomagnetic longitude [GEOPHYS] Longitude that is determined around the geomagnetic axis instead of around the rotation axis of the earth.

geomagnetic meridian [GEOPHYS] A semicircle connecting the geomagnetic poles.

geomagnetic pole [GEOPHYS] Either of two antipodal points marking the intersection of the earth's surface with the extended axis of a powerful bar magnet assumed to be located at the center of the earth and having a field approximating the actual magnetic field of the earth.

geomagnetic reversal [GEOPHYS] Reversed magnetization of the earth's magnetic dipole.

geomagnetic secular variation *See* secular variation.

geomagnetic storm [GEOPHYS] A large disturbance of the earth's magnetic field.

geomagnetic variation [GEOPHYS] Temporal changes in the geomagnetic field, both long-term (secular) and short-term (transient).

geomagnetism [GEOPHYS] **1.** The magnetism of the earth. Also known as terrestrial magnetism. **2.** The branch of science that deals with the earth's magnetism.

geometrical dip [GEOD] The vertical angle, at the eye of an observer, between the horizontal and a straight line tangent to the surface of the earth.

geometrical horizon [GEOD] The intersection of the celestial sphere and an infinite number of straight lines radiating from the eye of the observer and tangent to the earth's surface.

geometric vertical *See* geocentric vertical.

geomorphic cycle [GEOL] The cycle of change in the surface configuration of the earth. Also known as cycle of erosion; geographical cycle.

geomorphology [GEOL] The study of the origin of secondary topographic features which are carved by erosion in the primary elements and built up of the erosional debris.

geopetal [PETR]　Pertaining to the top-to-bottom relations in rocks at the time of formation.

geopetal fabric [PETR]　The internal structure of a rock indicating the original orientation of the top-to-bottom strata.

geophysical engineering [ENG]　A branch of engineering that applies scientific methods for locating mineral deposits.

geophysical prospecting [ENG]　Application of quantitative concepts and principles of physics and mathematics in geologic explorations to discover the character of and mineral resources in underground rocks in the upper portions of the earth's crust.

geophysicist [GEOPHYS]　An individual who specializes in geophysics.

geophysics [GEOL]　The physics of the earth and its environment, that is, earth, air, and (by extension) space.

geopotential [PHYS]　The potential energy of a unit mass relative to sea level, numerically equal to the work that would be done in lifting the unit mass from sea level to the height at which the mass is located, against the force of gravity.

geopotential height [GEOPHYS]　The height of a given point in the atmosphere in units proportional to the potential energy of unit mass (geopotential) at this height, relative to sea level.

geopotential number [GEOPHYS]　The numerical value C that is assigned to a given geopotential surface when expressed in geopotential units (1 gpu = 1 meter × 1 kilogal).

geopotential surface [GEOPHYS]　A surface of constant geopotential, that is, a surface along which a parcel of air could move without undergoing any changes in its potential energy. Also known as equigeopotential surface; level surface.

geopotential thickness [GEOPHYS]　The difference in the geopotential height of two constant-pressure surfaces in the atmosphere, proportional to the appropriately defined mean air temperature between the two surfaces.

geopotential topography [GEOPHYS]　The topography of any surface as represented by lines of equal geopotential; these lines are the contours of intersection between the actual surface and the level surfaces (which everywhere are normal to the direction of the force of gravity), and are spaced at equal intervals of dynamic height. Also known as absolute geopotential topography.

geopotential unit [GEOPHYS]　A unit of gravitational potential used in describing the earth's gravitational field; it is equal to the difference in gravitational potential of two points separated by a distance of 1 meter when the gravitational field has a strength of 10 meters per second squared and is directed along the line joining the points. Abbreviated gpu.

geopressurized geothermal system [GEOL]　A geothermal system dominated by the presence of hot fluids under high pressure (brine plus methane) and having higher-than-normal temperatures because of their low thermal conductivity, the presence of interbedded shale layers, or the existence of local, exothermic chemical reactions.

Georges Banks [GEOL]　An elevation beneath the sea east of Cape Cod, Mass.

georgiadesite [MINERAL]　$Pb_3(AsO_4)Cl_3$　A white or brownish-yellow mineral composed of lead chloroarsenate, occurring in orthorhombic crystals.

georgiaite [GEOL]　Any of a group of North American tektites, 134 million years of age, found in Georgia.

geosere [GEOL]　A series of ecological climax communities following each other in geologic time and changing in response to changing climate and physical conditions.

geosphere [GEOL] **1.** The solid mass of earth, as distinct from the atmosphere and hydrosphere. **2.** The lithosphere, hydrosphere, and atmosphere combined.

geostatic pressure *See* ground pressure.

geostrophic [GEOPHYS] Pertaining to deflecting force resulting from the earth's rotation.

geostrophic approximation [GEOPHYS] The assumption that the geostrophic current can represent the actual horizontal current. Also known as geostrophic assumption.

geostrophic current [GEOPHYS] A current defined by assuming the existence of an exact balance between the horizontal pressure gradient force and the Coriolis force.

geostrophic departure [METEOROL] A vector representing the difference between the real wind and the geostrophic wind. Also known as ageostrophic wind; geostrophic deviation.

geostrophic deviation *See* geostrophic departure.

geostrophic distance [METEOROL] The distance (in degrees latitude) along a constant-pressure surface over which the change in height (in feet) is equal to the geostrophic wind speed (in knots).

geostrophic equation [GEOPHYS] An equation, used to compute geostrophic current speed, which represents a balance between the horizontal pressure gradient force and the Coriolis force.

geostrophic equilibrium [GEOPHYS] A state of motion of a nonviscous fluid in which the horizontal Coriolis force exactly balances the horizontal pressure force at all points of the field so described.

geostrophic flow [GEOPHYS] A form of gradient flow where the Coriolis force exactly balances the horizontal pressure force.

geostrophic flux [METEOROL] The transport of an atmospheric property by means of the geostrophic wind.

geostrophic vorticity [METEOROL] The vorticity of the geostrophic wind.

geostrophic wind [METEOROL] That horizontal wind velocity for which the Coriolis acceleration exactly balances the horizontal pressure force.

geostrophic-wind level [METEOROL] The lowest level at which the wind becomes geostrophic in the theory of the Ekman spiral. Also known as gradient-wind level.

geostrophic-wind scale [METEOROL] A graphical device used for the determination of the speed of the geostrophic wind from the isobar or contour-line spacing on a synoptic chart; it is a nomogram representing solutions of the geostrophic-wind equation.

geosynclinal couple *See* orthogeosyncline.

geosynclinal cycle *See* tectonic cycle.

geosynclinal facies [GEOL] A sedimentary facies characterized by great thickness, predominantly argillaceous character, and paucity of carbonate rocks.

geosyncline [GEOL] A part of the crust of the earth that sank deeply through time.

geotechnology [ENG] Application of the methods of engineering and science to exploitation of natural resources.

geotectogene *See* tectogene.

geotectonic cycle *See* orogenic cycle.

geotectonics *See* tectonics.

geotherm *See* geoisotherm.

geothermal [GEOPHYS]   Pertaining to heat within the earth.

geothermal gradient [GEOPHYS]   The change in temperature with depth of the earth.

geothermal prospecting [ENG]   Exploration for sources of geothermal energy.

geothermal system [GEOL]   Any regionally localized geological setting where naturally occurring portions of the earth's internal heat flow are transported close enough to the earth's surface by circulating steam or hot water to be readily harnessed for use; examples are the Geysers Region of northern California and the hot brine fields in the Imperial Valley of southern California.

geothermal well logging [ENG]   Measurement of the change in temperature of the earth by means of well logging.

geothermometer [ENG]   A thermometer constructed to measure temperatures in boreholes or deep-sea deposits.   [GEOL]   A mineral that yields information about the temperature range within which it was formed. Also known as geologic thermometer.

geothermometry [GEOL]   Measurement of the temperatures at which geologic processes occur or occurred. Also known as geologic thermometry.

gerhardtite [MINERAL] $Cu_2(NO_3)(OH)_3$ An emerald-green mineral composed of basic copper nitrate.

germanite [MINERAL] $Cu_3(Ge,Ga,Fe)(S,As)_4$ Reddish-gray mineral occurring in massive form; an important source of germanium.

German tubbing [MIN ENG]   A form of tubbing, with internal flanges and bolts, for lining circular shafts sunk through heavily watered strata.

germination *See* grain growth.

gersdorffite [MINERAL]   NiAsS A silver-white to steel-gray mineral, crystallizing in the isometric system; resembles cobaltite and may contain some iron and cobalt. Also known as nickel glance.

gestalt [METEOROL]   A complex of weather elements occuring in a familiar form, and though not necessarily referring to basic hydrodynamical or thermodynamical quantities, may persist for an appreciable length of time and is often considered to be an entity in itself.

geyser [HYD]   A natural spring or fountain which discharges a column of water or steam into the air at more or less regular intervals.

geyserite *See* siliceous sinter.

gharbi [METEOROL]   A fresh westerly wind of oceanic origin in Morocco.

gharra [METEOROL]   Hard squalls from the northeast in Libya and Africa that are sudden and frequent, and are accompanied by heavy rain and thunder.

ghibli [METEOROL]   A hot, dust-bearing, desert wind in North Africa, similar to the foehn. Also known as chibli; gebli; gibleh; gibli; kibli.

ghost [PETR]   The discernible outline of the shape of a former crystal or of another rock structure that has been partly obliterated and has as its boundaries inclusions, bubbles or other foreign matter. Also known as phantom.

ghost crystal *See* phantom crystal.

giant *See* hydraulic monitor.

giant granite *See* pegmatite.

gib [MIN ENG]   **1.** A temporary support at the face to prevent coal from falling before the cut is complete, either by hand or by machine. **2.** A prop put in the holing of a seam while being undercut. **3.** A piece of metal often used in the same hole with a wedge-shaped key for holding pieces together.

gibbsite [MINERAL]   Al(OH)₃ A white or tinted mineral, crystallizing in the monoclinic system; a principal constituent of bauxite; Also known as hydrargillite.

gibleh *See* ghibli.

gibli *See* ghibli.

giboulee *See* galerne.

Gibraltar stone *See* onyx marble.

Giesler coal test [ENG]   A plastometric method for estimating the coking properties of coals.

gillespite [MINERAL]   BaFeSi₄O₁₀ A micalike mineral composed of barium and iron silicate.

gilsonite [MINERAL]   A variety of asphalt; it has black color, brilliant luster, brown streaks, and conchoidal fracture.

ginging [MIN ENG]   **1.** Lining a shaft with masonry or brick. **2.** The brick or masonry of a shaft lining.

ginorite [MINERAL]   Ca₂B₁₄O₂₃·8H₂O A white monoclinic mineral composed of hydrous borate of calcium.

gin pit [MIN ENG]   A shallow mine, the hoisting from which is done by a gin.

gio *See* geo.

giobertite *See* magnesite.

gipsy winch [MIN ENG]   A small winch that may be attached to a post and operated by a rotary motion or the reciprocating action of a handle having a pair of pawls and a ratchet.

girdle [PETR]   With reference to a fabric diagram or equal-area projection net, a belt showing concentration of points which is approximately coincident with a great circle of the net and which represents orientation of the fabric elements.

girt [ENG]   A brace member running horizontally between the legs of a drill tripod or derrick.   [MIN ENG]   In square-set timbering, a horizontal brace running parallel to the drift.

gismondite [MINERAL]   CaAl₂Si₂O₈·4H₂O A light-colored mineral composed of hydrous calcium aluminum silicate, occurring in pyramidal crystals.

glacial [GEOL]   Pertaining to an interval of geologic time which was marked by an equatorward advance of ice during an ice age; the opposite of interglacial; these intervals are variously called glacial periods, glacial epochs, glacial stages, and so on.   [HYD]   Pertaining to ice, especially in great masses such as sheets of land ice or glaciers.

glacial abrasion [GEOL]   Alteration of portions of the earth's surface as a result of glacial flow.

glacial accretion [GEOL]   Deposition of material as a result of glacial flow.

glacial advance [GEOL]   **1.** Increase in the thickness and area of a glacier. **2.** A time period equal to that increase.

glacial anticyclone [METEOROL] A type of semipermanent anticyclone which overlies the ice caps of Greenland and Antarctica. Also known as glacial high.

glacial boulder [GEOL] A boulder moved to a point distant from its original site by a glacier.

glacial deposit [GEOL] Material carried to a point beyond its original location by a glacier.

glacial drift [GEOL] All rock material in transport by glacial ice, and all deposits predominantly of glacial origin made in the sea or in bodies of glacial meltwater, including rocks rafted by icebergs.

glacial epoch [GEOL] **1.** Any of the geologic epochs characterized by an ice age; thus, the Pleistocene epoch may be termed a glacial epoch. **2.** Generally, an interval of geologic time which was marked by a major equatorward advance of ice; the term has been applied to an entire ice age or (rarely) to the individual glacial stages which make up an ice age.

glacial erosion [GEOL] Movement of soil or rock from one point to another by the action of the moving ice of a glacier. Also known as ice erosion.

glacial flour *See* rock flour.

glacial flow *See* glacier flow.

glacial geology [GEOL] The study of land features resulting from glaciation.

glacial high *See* glacial anticyclone.

glacial ice [HYD] Ice that is flowing or that exhibits evidence of having flowed.

glacial lake GEOL] A lake that exists because of the effects of the glacial period.

glacial lobe [HYD] A tonguelike projection from a continental glacier's main mass.

glacial maximum [GEOL] The time or position of the greatest extent of any glaciation; most frequently applied to the greatest equatorward advance of Pleistocene glaciation.

glacial mill *See* moulin.

glacial outwash *See* outwash.

glacial period [GEOL] **1.** Any of the geologic periods which embraced an ice age; for example, the Quaternary period may be called a glacial period. **2.** Generally, an interval of geologic time which was marked by a major equatorward advance of ice.

glacial plucking *See* plucking.

glacial retreat [GEOL] A condition occurring when backward melting at the front of a glacier takes place at a rate exceeding forward motion.

glacial scour [GEOL] Erosion resulting from glacial action, whereby the surface material is removed and the rock fragments carried by the glacier abrade, scratch, and polish the bedrock. Also known as scouring.

glacial striae [GEOL] Scratches, commonly parallel, on smooth rock surfaces due to glacial abrasion.

glacial till *See* till.

glacial varve *See* varve.

glaciated terrain [GEOL] A region that once bore great masses of glacial ice; a distinguishing feature is marks of glaciation.

glaciation [GEOL] Alteration of any part of the earth's surface by passage of a glacier, chiefly by glacial erosion or deposition. [METEOROL] The transformation of cloud particles from waterdrops to ice crystals, as in the upper portion of a cumulonimbus cloud.

glaciation limit   [GEOPHYS]   For a given locality, the lowest altitude at which glaciers can develop.

glacier   [HYD]   A mass of land ice, formed by the further recrystallization of firn, flowing slowly (at present or in the past) from an accumulation area to an area of ablation.

glacieret *See* snowdrift ice.

glacier flow   [HYD]   The motion that exists within a glacier's body. Also known as glacial flow.

glacier front   [HYD]   The leading edge of a glacier.

glacier ice   [HYD]   Any ice that is or was once a part of a glacier, consolidated from firn by further melting and refreezing and by static pressure; for example, an iceberg.

glacier mill *See* moulin.

glacier pothole *See* moulin.

glacier table   [GEOL]   A stone block supported by an ice pedestal above the surface of a glacier.

glacier well *See* moulin.

glacier wind   [METEOROL]   A shallow gravity wind along the icy surface of a glacier, caused by the temperature difference between the air in contact with the glacier and free air at the same altitude.

glaciofluvial   [GEOL]   Pertaining to streams fed by melting glaciers, or to the deposits and landforms produced by such streams.

glaciolacustrine   [GEOL]   Pertaining to lakes fed by melting glaciers, or to the deposits forming therein.

glaciology   [GEOL]   The study of existing or modern glaciers in their entirety.

glacon   [OCEANOGR]   A piece of sea ice which is smaller than a medium-sized floe.

gladite   [MINERAL] $PbCuBi_5S_9$ A lead gray mineral consisting of lead and copper bismuth sulfide; occurs as prismatic crystals.

glaebule   [GEOL]   A three-dimensional unit, usually prolate to equant in shape, within the matrix of a soil material, recognizable by its greater concentration of some constituent, by its difference in fabric as compared with the enclosing soil material, or by its distinct boundary with the enclosing soil material.

glance pitch   [GEOL]   A variety of asphaltite having brilliant conchoidal fracture, and resembling gilsonite but having higher specific gravity and percentage of fixed carbon.

glare ice   [HYD]   Ice with a smooth, shiny surface.

glaserite *See* arcanite.

glass   [METEOROL]   In nautical terminology, a contraction for "weather glass" (a mercury barometer).

glass porphyry *See* virtophyre.

glass schorl *See* axinite.

glassy feldspar *See* sanidine.

glauberite   [MINERAL] $Na_2Ca(SO_4)_2$ A brittle, gray-yellow monoclinic mineral having vitreous luster and saline taste.

glaucocerinite [MINERAL]   A mineral composed of a hydrous basic sulfate of copper, zinc, and aluminum.

glaucochroite [MINERAL] $CaMnSiO_4$ A bluish-green mineral that is related to monticellite, is composed of calcium manganese silicate, and occurs in prismatic crystals.

glaucodot [MINERAL] $(Co,Fe)AsS$ A grayish-white, metallic-looking mineral composed of cobalt iron sulfarsenide, occurring in orthorhombic crystals.

glauconite [MINERAL] $K_{15}(Fe,Mg,Al)_{4-6}(Si,Al)_8O_{20}(OH)_4$ A type of clay mineral; it is dioctohedral and occurs in flakes and as pigmentary material.

glauconitic sandstone [PETR]   A quartz sandstone or an arkosic sandstone that has many glauconite grains.

glaucophane [MINERAL] $Na_2Mg_3Al_2Si_8$ A blue to black monoclinic sodium amphibole; blue to black coloration with marked pleochroism.

glaucophane schist [PETR]   Metamorphic schist that contains glaucophane.

glave *See* glaves.

glaves [METEOROL]   A foehnlike wind of the Faroe Islands. Also known as glave; glavis.

glavis *See* glaves.

glaze [HYD]   A coating of ice, generally clear and smooth but usually containing some air pockets, formed on exposed objects by the freezing of a film of supercooled water deposited by rain, drizzle, or fog, or possibly condensed from supercooled water vapor. Also known as glaze ice; glazed frost; verglas.

glazed frost *See* glaze.

glaze ice *See* glaze.

glessite [GEOL]   Fossil resin similar to amber.

gley [GEOL]   A sticky subsurface layer of clay in some waterlogged soils.

glide plane [CRYSTAL]   A lattice plane in a crystal on which translation or twin gliding occurs. Also known as slip plane.

glime [HYD]   An ice coating with a consistency intermediate between glaze and rime.

glimmer ice [HYD]   Ice newly formed within the cracks or holes of old ice, or on the puddles on old ice.

global radiation [GEOPHYS]   The total of direct solar radiation and diffuse sky radiation received by a unit horizontal surface.

global sea [OCEANOGR]   All the seawaters of the Earth considered as a single ocean, constantly intermixing.

globe [MAP]   A sphere on the surface of which is a map of the world.

globe lightning *See* ball lightning.

globigerina ooze [GEOL]   A pelagic sediment consisting of than 30% calcium carbonate in the form of foraminiferal tests of which *Globigerina* is the dominant genus.

globular *See* spherulitic.

globular projection [MAP]   A projection, in perspective, of a hemisphere upon a plane parallel to the base of the hemisphere.

globulite [GEOL]   A small, isotropic, globular of spherulelike crystallite; usually dark in color and found in glassy extrusive rocks.

glockerite [MINERAL] A brown, ocher yellow, black, or dull green mineral consisting of a hydrated basic sulfate of ferric iron; occurs in stalactitic, encrusting, or earthy forms.

gloom [METEOROL] The condition existing when daylight is very much reduced by dense cloud or smoke accumulation above the surface, the surface visibility not being materially reduced.

glory hole [MIN ENG] An opening formed by the removal of soft or broken ore through an underground passage.

glory hole system *See* chute system.

glossopterid flora [PALEOBOT] Permian and Triassic fossil ferns of the genus *Glossopteris*.

gloup [GEOL] An opening in the roof of a sea cave.

glowing avalanche *See* ash flow.

glowing cloud *See* nuée ardente.

Glyphocyphidae [PALEON] A family of extinct echinoderms in the order Temnopleuroida comprising small forms with a sculptured test, perforate crenulate tubercles, and diademoid ambulacral plates.

Glyptocrinina [PALEON] A suborder of extinct crinoids in the order Monobathrida.

gmelinite [MINERAL] $(Na_2Ca)Al_2Si_4O_{12} \cdot 6H_2O$ Zeolite mineral that is colorless or lightly colored and crystallizes in the hexagonal system.

Gnathobelodontinae [PALEON] A subfamily of extinct elephantoid proboscideans containing the shovel-jawed forms of the family Gomphotheriidae.

Gnathodontidae [PALEON] A family of extinct conodonts having platforms with large, cup-shaped attachment scars.

gneiss [PETR] A variety of rocks with a banded or coarsely foliated structure formed by regional metamorphism.

gneissic granodiorites [PETR] Granodiorite rocks with gneissic characteristics.

gnomonic chart [MAP] A chart on the gnomonic projection where great circles project as straight lines. Also known as great-circle chart.

gnomonic projection [CRYSTAL] A projection for displaying the poles of a crystal in which the poles are projected radially from the center of a reference sphere onto a plane tangent to the sphere. [MAP] A projection on a plane tangent to the surface of a sphere having the point of projection at the center of the sphere.

goaf [MIN ENG] **1.** That part of a mine from which the coal has been worked away and the space more or less filled up. **2.** The refuse or waste left in the mine. Also known as gob.

gob *See* goaf.

gobi [GEOL] Sedimentary deposits in a synclinal basin.

Gobiatheriinae [PALEON] A subfamily of extinct herbivorous mammals in the family Uintatheriidae known from one late Eocene genus; characterized by extreme reduction of anterior dentition and by lack of horns.

gob stink [MIN ENG] **1.** The odor from the burning coal given off by an underground fire. **2.** The odor given off by the spontaneous heating of coal, not necessarily in the gob. Also known as stink.

goethite [MINERAL] FeO(OH) A yellow, red, or dark-brown mineral crystallizing in the orthorhombic system, although it is usually found in radiating fibrous aggregates; a common constituent of natural rust or limonite. Also known as xanthosiderite.

gold beryl *See* chrysoberyl.

goldschmidtine *See* stephanite.

goldschmidtite *See* sylvanite.

Goldschmidt's mineralogical phase rule [GEOL] The rule that the probability of finding a system with degrees of freedom less than two is small under natural rock-forming conditions.

golfada [METEOROL] A heavy gale of the Mediterranean.

Gomphotheriidae [PALEON] A family of extinct proboscidean mammals in the suborder Elephantoidea consisting of species with shoveling or digging specializations of the lower tusks.

Gomphotheriinae [PALEON] A subfamily of extinct elephantoid proboscideans in the family Gomphotheriidae containing species with long jaws and bunomastodont teeth.

gon *See* grade.

Gondwanaland [GEOL] The ancient continent that is supposed to have fragmented and drifted apart to form eventually the present continents.

goniometer [ENG] **1.** An instrument used to measure the angles between crystal faces. **2.** An instrument which uses x-ray diffraction to measure the angular positions of the axes of a crystal.

gonnardite [MINERAL] $Na_2CaAl_4Si_6O_{20}\cdot7H_2O$ Zeolite mineral occurring in fibrous, radiating spherules; specific gravity is 2.3.

Goodman loader [MIN ENG] **1.** An electrohydraulic power shovel designed for loading coal where the seams are 6 feet (1.8 meters) or more in thickness. **2.** A loader designed for loading coal from thin seams; has a telescoping fan-shaped apron that extends from the entry of the room to the working face.

goongarrite [MINERAL] $Pb_4Bi_2S_7$ A mineral composed of a sulfide of lead and bismuth.

gooseberry stone *See* grossularite.

gopher hole [ENG] Horizontal T-shaped opening made in rock in preparation for blasting. Also known as coyote hole. [MIN ENG] An irregular pitting hole made during prospecting.

gopher-hole blasting *See* coyote blasting.

gophering [MIN ENG] A method of breaking up a sandy medium-hard overburden where usual blastholes tend to cave in, by firing an explosive charge in each of a series of shallow holes; debris is cleared, and holes are made deeper for further charges, until they are deep enough to take enough explosives to break up the deposit.

gorceixite [MINERAL] $BaAl_3(PO_4)_2(OH)_5\cdot H_2O$ A brown mineral composed of a hydrous basic phosphate of barium and aluminum.

gordonite [MINERAL] $MgAl_2(PO_4)_2(OH)_2\cdot8H_2O$ A colorless mineral composed of a hydrous basic phosphate of magnesium and aluminum.

gorge [GEOGR] A narrow passage between mountains or the walls of a canyon, especially one with steep, rocky walls. [OCEANOGR] A collection of solid matter obstructing a channel or a river, as an ice gorge.

gorge wind *See* canyon wind.

goslarite   [MINERAL] $ZnSO_4 \cdot 7H_2O$ A white mineral composed of hydrous zinc sulfate.

gosling blast   [METEOROL]   A sudden squall of rain or sleet in England. Also known as gosling storm.

gossan   [GEOL]   A rusty, ferruginous deposit filling the upper regions of mineral veins and overlying a sulfide deposit; formed by oxidation of pyrites. Also known as capping; gozzan; iron hat.

Gotlandian   [GEOL]   A geologic time period recognized in Europe to include the Ordovician; it appears before the Devonian.

gouge   [GEOL]   Soft, pulverized mixture of rock and mineral material found along shear (fault) zones and produced by the differential movement across the plane of slippage.   [MIN ENG]   A layer of soft material along the wall of a vein which favors miners by enabling them, after gouging it out with a pick, to attack the solid vein from the side.

gowk storm   [METEOROL]   In England, a storm or gale occurring at about the end of April or the beginning of May.

goyazite   [MINERAL] $SrAl_3(PO_4)_2(OH)_5 \cdot H_2O$ A granular, yellowish-white mineral composed of a hydrous strontium aluminum phosphate.

gozzan   *See* gossan.

gpu   *See* geopotential unit.

grab   [ENG]   An instrument for extricating broken boring tools from a borehole.

graben   [GEOL]   A block of the earth's crust, generally with a length much greater than its width, that has dropped relative to the blocks on either side.

grab sample   [MIN ENG]   A random mode of sampling; the samples may be taken from the pile broken in the process of mining, or from a truck or car of ore or coal.

gradation period   [GEOL]   The time during which the base level of the sea remains in one position. Also known as baseleveling epoch.

grade   [ENG]   The degree of strength of a high explosive.   [GEOL]   The slope of the bed of a stream, or of a surface over which water flows, upon which the current can just transport its load without either eroding or depositing.   [MIN ENG]   **1.** A classification of ore according to recoverable amount of a valuable metal. **2.** To sort and classify diamonds.

graded   [GEOL]   Brought to or established at grade.

graded bedding   [GEOL]   A stratification in which each stratum displays a gradation in the size of grains from coarse below to fine above.

graded stream   [HYD]   A stream in which, over a period of years, slope is adjusted to yield the velocity required for transportation of the load supplied from the drainage basin.

grade of coal   [MIN ENG]   A classification of coal based on the amount and nature of the ash and sulfur content.

grade scale   [GEOL]   A continuous scale of particle sizes divided into a series of size classes.

gradient   [GEOL]   The rate of descent or ascent (steepness of slope) of any topographic feature, such as streams or hillsides.

gradient current   [OCEANOGR]   A current defined by assuming that the horizontal pressure gradient in the sea is balanced by the sum of the Coriolis and bottom frictional forces; at some distance from the bottom the effect of friction becomes negligible, and above this the gradient and geostrophic currents are equivalent. Also known as slope current.

gradienter [ENG] An attachment placed on a surveyor's transit to measure angle of inclination in terms of the tangent of the angle.

gradient flow [METEOROL] Horizontal frictionless flow in which isobars and streamlines coincide, or equivalently, in which the tangential acceleration is everywhere zero; the balance of normal forces (pressure force, Coriolis force, centrifugal force) is then given by the gradient wind equation.

gradient of equal traction [MIN ENG] The gradient at which the tractive force necessary to pull an empty tram inby (slightly uphill) is equal to that force required to pull a loaded tram outby.

gradient tints [MAP] A series of color tints used on maps or charts to indicate relative heights or depths. Also known as altitude tints; elevation tints; hypsometric tints.

gradient wind [METEOROL] A wind for which Coriolis acceleration and the centripetal acceleration exactly balance the horizontal pressure force.

gradient-wind level *See* geostrophic-wind level.

graftonite [MINERAL] $(Fe,Mn,Ca)_3(PO_4)_2$ A salmon-pink mineral, crystallizing in the monoclinic system, and found as laminated intergrowths of triphylite; hardness is 5 on Mohs scale, and specific gravity is 3.7.

grahamite [GEOL] *See* mesosiderite. [MINERAL] A solid, jetblack hydrocarbon that occurs in veinlike masses; soluble in carbon disulfide and chloroform.

grain [GEOL] The particles or discrete crystals that make up a sediment or rocks. [HYD] The particles which make up settled snow, firn, and glacier ice.

grain growth [PETR] Enlargement of some individual crystals in a monomineralic rock, producing a coarser texture. Also known as germination.

grain size [GEOL] Average size of mineral particles composing a rock or sediment. [MET] Average size of grains in a metal expressed as average diameter, or grains per unit area or volume.

gramenite *See* nontronite.

Granby car [MIN ENG] An automatically dumped car for hand loading or power-shovel loading; a wheel attached to its side engages an inclined track at the dumping point.

Grand Banks [GEOGR] Banks off southeastern Newfoundland, important for cod fishing.

grandite [MINERAL] A garnet that is intermediate in chemical composition between grossular and androdite.

granite [PETR] A visibly crystalline plutonic rock with granular texture; composed of quartz and alkali feldspar with subordinate plagioclase and biotite and hornblende.

granite-gneiss [PETR] A banded metamorphic rock derived from igneous or sedimentary rocks mineralogically equivalent to granite.

granite pegmatite *See* pegmatite.

granite porphyry *See* quartz porphyry.

granite series [GEOL] A sequence of products that evolved continuously during crustal fusion, earlier products tending to be deep-seated, syntectonic, and granodioritic, and later products tending to be shallower, late-syntectonic, or postsyntectonic, and more potassic.

granite wash [GEOL]   Material eroded from granites and redeposited, forming a rock with the same major mineral constituents as the original rock.

granitic batholith [GEOL]   A granitic shield mass intruded as the fusion of older formations.

granitic layer *See* sial.

granitic magma [PETR]   A coarse-grained igneous rock.

granitization [PETR]   A process whereby various types of rock may be converted to granite or closely related material.

granoblastic fabric [PETR]   The texture of metamorphic rocks composed of equi-dimensional elements formed during recrystallization.

granodiorite [PETR]   A visibly crystalline plutonic rock composed chiefly of sodic plagioclase, alkali feldspar, quartz, and subordinate dark-colored minerals.

granogabbro [PETR]   Plutonic rock composed of quartz, basic plagioclase, potash-feldspar, and at least one ferromagnesian mineral; intermediate between a granite and a gabbro, and in a strict sense, a granodiorite with more that 50% boric plagioclase.

granophyre [PETR]   A quartz porphyry or fine-grained porphyritic granite.

granular ice [HYD]   Ice composed of many tiny, opaque, white or milky pellets or grains frozen together and presenting a rough surface; this is the type of ice deposited as rime and compacted as névé.

granularity [PETR]   The feature of rock texture relating to the size of the constituent grains or crystals.

granular snow *See* snow grains.

granule [GEOL]   A somewhat rounded rock fragment ranging in diameter from 2 to 4 millimeters; larger than a coarse sand grain and smaller than a pebble.

granulite [PETR]   **1.** Granite that contains muscovite. **2.** A relatively coarse, granuloblastic rock formed at the high temperatures and pressures of the granulite facies.

granulite facies [PETR]   A group of gneissic rocks characterized by a granoblastic fabric and formed by regional dynamothermal metamorphism at temperatures above 650°C and pressures of 3000–12,000 bars.

granulometry [PETR]   Measurement of grain sizes of sedimentary rock.

grapestone [GEOL]   A cluster of sand-size grains, such as calcareous pellets, held together by incipient cementation shortly after deposition; the outer surface is lumpy, resembling a bunch of grapes.

grapevine drainage *See* trellis drainage.

graphic granite [PETR]   A distinct type of pegmatite in which quartz and orthoclase crystals grew together along a parallel axis. Also known as Hebraic granite; runite.

graphic intergrowth [PETR]   An intergrowth of crystals, commonly feldspar and quartz, that produces a type of poikilitic texture in which the larger crystals have a fairly regular geometric outline and orientation, resembling cuneiform writing.

graphic scale [MAP]   A graduated line that indicates the length of miles or kilometers as they appear on a map; the line has the advantage of remaining true after the map is enlarged or reduced in reproduction. Also known as bar scale.

graphic tellurium *See* sylvanite.

graphic texture [GEOL]   A pattern of rocks that is similar to cuneiform characters.

graphite [MINERAL] A mineral consisting of a low-pressure allotropic form of carbon; it is soft, black, and lustrous and has a greasy feeling; it occurs naturally in hexagonal crystals or massive or can be synthesized from petroleum coke; hardness is 1–2 on Mohs scale, and specific gravity is 2.09–2.23; used in pencils, crucibles, lubricants, paints, and polishes. Also known as black lead; plumbago.

graptolite shale [GEOL] Shale containing an abundance of extinct colonial marine organisms known as graptolites.

Graptolithina [PALEON] A class of extinct colonial animals believed to be related to the class Pterobranchia of the Hemichordata.

Graptoloidea [PALEON] An order of extinct animals in the class Graptolithina including branched, planktonic forms described from black shales.

Graptozoa [PALEON] The equivalent name for Graptolithina.

grassland climate *See* subhumid climate.

grass minimum [METEOROL] The minimum temperature shown by a minimum thermometer exposed in an open situation with its bulb on the level of the tops of the grass blades of short turf.

grass-roots deposit [MIN ENG] A deposit that is discovered in surface croppings, is easily exploited, and can pay for its own development while in progress.

grass-roots mining [MIN ENG] Also known as mining on a shoestring. **1.** Inadequately financed mining operation, with catch-as-catch-can practices. **2.** Mining from surface down to bedrock.

grass temperature [METEOROL] The temperature registered by a thermometer with its bulb at the level of the tops of the blades of grass in short turf.

graticule [MAP] A network of lines representing the earth's parallels of latitude and meridians of longitude on a map, chart, or plotting sheet.

gratonite [MINERAL] $Pb_9As_4S_{15}$ A mineral composed of lead arsenic sulfide, occurring in rhombohedral crystals.

graupel *See* snow pellets.

gravel [GEOL] A loose or unconsolidated deposit of rounded pebbles, cobbles, or boulders.

gravel bank [GEOL] A natural mound or exposed face of gravel, particularly such a place from which gravel is dug.

gravel mine [MIN ENG] A mine extracting gold from sand or gravel. Also known as placer mine.

gravimeter [ENG] A highly sensitive weighing device used for relative measurement of the force of gravity by detecting small weight differences of a constant mass at different points on the earth. Also known as gravity meter.

gravimetric geodesy [GEOD] The science that utilizes measurements and characteristics of the earth's gravity field, as well as theories regarding this field, to deduce the shape of the earth and, in combination with arc measurements, the earth's size.

gravitational convection *See* thermal convection.

gravitational settling [GEOL] A movement of sediment resulting from gravitational forces.

gravitational sliding [GEOL] Extensive sliding of strata down a slope of an uplifted area. Also known as sliding.

gravitational tide [OCEANOGR] An atmospheric tide due to gravitational attraction of the sun and moon.

gravity anomaly   [GEOPHYS]   The difference between the observed gravity and the theoretical or predicted gravity.

gravity bar   [MIN ENG]   A 5-foot (1.5-meter) length of heavy half-round rod forming the link between the wedge-oriented coupling and the drill-rod swivel coupling on an assembled Thompson retrievable borehole-deflecting wedge.

gravity classification   [MIN ENG]   The grading of ores, or the separation of waste from coal, by the differences in specific gravities of the substances.

gravity concentration   [ENG]   The separation of liquid-liquid dispersions based on settling out of the dense phase by gravity.

gravity drainage   [HYD]   Withdrawal of water from strata as a result of gravitational forces.

gravity drainage reservoir   [GEOL]   A reservoir in which production is significantly affected by gas, oil, and water separating under the influence of gravity while production takes place.

gravity erosion   *See* mass erosion.

gravity fault   *See* normal fault.

gravity flow   [HYD]   A form of glacier movement in which the flow of the ice results from the downslope gravitational component in an ice mass resting on a sloping floor.

gravity-flow gathering system   [PETRO ENG]   The use of gravity (downhill flow) through pipelines to transport and collect liquid at a central location; used for gathering of waste water from waterflooding operations for treatment prior to reuse or disposal.

gravity haulage   [MIN ENG]   A type of haulage system in which the set of full cars is lowered at the end of a rope, and gravity force pulls up the empty cars, the rope being passed around a sheave at the top of the incline. Also known as self-acting incline.

gravity incline   [MIN ENG]   An opening made in the direction of, and along the same gradient as, the dip of the deposit.

gravity map   [GEOPHYS]   A map of gravitational variations in an area displaying gravitational highs and lows.

gravity meter   [ENG]   **1.** U-tube-manometer type of device for direct reading of solution specific gravities in semimicro quantities. **2.** An electrical device for measuring variations in gravitation through different geologic formations; used in mineral exploration. **3.** *See* gravimeter.

gravity prospecting   [ENG]   Identifying and mapping the distribution of rock masses of different specific gravity by means of a gravity meter.

gravity separation   [ENG]   Separation of immiscible phases (gas-solid, liquid-solid, liquid-liquid, solid-solid) by allowing the denser phase to settle out under the influence of gravity; used in ore dressing and various industrial chemical processes.

gravity spring   [HYD]   A spring that issues under the influence of gravity, not internal pressure.

gravity stamp   [MIN ENG]   Unit in a stamp battery which directs a heavy falling weight onto a die on which rock is crushed.

gravity station   [ENG]   The site of installation of gravimeters.

gravity survey   [ENG]   The measurement of the differences in gravity force at two or more points.

gravity tide   [GEOPHYS]   Cyclic motion of the earth's surface caused by interaction of gravitational forces of the moon, sun, and earth.

gravity wind [METEOROL]   A wind (or component thereof) directed down the slope of an incline and caused by greater air density near the slope than at the same levels some distance horizontally from the slope. Also known as drainage wind; katabatic wind.

gray antimony *See* antimonite; jamesonite.

gray cobalt *See* cobaltite.

gray copper ore *See* tetrahedrite.

gray filter *See* neutral-density filter.

gray hematite *See* specularite.

Grayloc tubing joint [PETRO ENG]   Special wellbore-tubing joint that has greater leak resistance and strength than standard API tubing joints.

gray manganese ore *See* manganite.

graywacke [PETR]   An argillaceous sandstone characterized by an abundance of unstable mineral and rock fragments and a fine-grained clay matrix binding the larger, sand-size detrital fragments.

greased-deck concentration [MIN ENG]   A separation process based on selective adhesion of certain grains (diamonds) to quasi-solid grease.

grease ice [HYD]   A kind of slush with a greasy appearance, formed from the congelation of ice crystals in the early stages of freezing. Also known as ice fat; lard ice.

grease table [MIN ENG]   An apparatus for concentrating minerals, such as diamonds, which adhere to grease; usually a shaking table coated with grease or wax over which an aqueous pulp is flowed.

greasy quartz *See* milky quartz.

Great Basin high [METEOROL]   A high-pressure system centered over the Great Basin of the western United States; it is a frequent feature of the surface chart in the winter season.

great circle [GEOD]   A circle, or near circle, described on the earth's surface by a plane passing through the center of the earth.

great-circle chart *See* gnomonic chart.

great-circle direction [GEOD]   Horizontal direction of a great circle, expressed as angular distance from a reference direction.

great-circle distance [GEOD]   The length of the shorter arc of the great circle joining two points.

great diurnal range [OCEANOGR]   The difference in height between mean higher high water and mean lower low water. Also known as diurnal range.

greater ebb [OCEANOGR]   The stronger of two ebb currents occurring during a tidal day.

greater flood [OCEANOGR]   The stronger of two flood currents occurring during a tidal day.

Great Ice Age [GEOL]   The Pleistocene epoch.

great soil group [GEOL]   A group of soils having common internal soil characteristics; a subdivision of a soil order.

great tropic range [OCEANOGR]   The difference in height between tropic higher high water and tropic lower low water.

greco [METEOROL]   An Italian name for the northeast wind.

Greenburg-Smith impinger [MIN ENG] A dust-sampling apparatus based on the principle of impingement of the dust-carrying air at high velocity upon a wetted glass surface; also involves bubbling the air through a liquid medium.

greenhouse [BOT] Glass-enclosed, climate-controlled structure in which young or out-of-season plants are cultivated and protected.

green chalcedony *See* chrysoprase.

greenhouse effect [METEOROL] The effect of the earth's atmosphere in trapping heat from the sun; the atmosphere acts like a greenhouse.

Greenland anticyclone [METEOROL] The glacial anticyclone which is supposed to overlie Greenland; analogous to the Antarctic anticyclone.

greenlandite *See* columbite.

Greenland spar *See* cryolite.

green lead ore *See* pyromorphite.

green mud [GEOL] **1.** A fine-grained, greenish terrigenous mud or oceanic ooze found near the edge of a continental shelf at depths of 300–7500 feet (90–2300 meters). **2.** A deep-sea terrigenous deposit characterized by the presence of a considerable proportion of glauconite and calcium carbonate.

greenockite [MINERAL] CdS A green or orange mineral that crystallizes in the hexagonal system; occurs as an earthy encrustation and is dimorphous with hawleyite. Also known as cadmium blende; cadmium ocher; xanthochroite.

green roof [MIN ENG] A mine roof which has not broken down or which shows no sign of taking weight.

greensand [GEOL] A greenish sand consisting principally of grains of glauconite and found between the low-water mark and the inner mud line. [PETR] Sandstone composed of greensand with little or no cement.

greenschist [PETR] A schistose metamorphic rock with abundant chlorite, epidote, or actinolite present, giving it a green color.

greenschist facies [PETR] Any schistose rock containing an abundance of green minerals and produced under conditions of low to intermediate temperatures (300–500°C) and low to moderate hydrostatic pressures (3000–8000 bars).

green sky [METEOROL] A greenish tinge to part of the sky, supposed by seamen to herald wind or rain, or in some cases, a tropical cyclone.

green snow [HYD] A snow surface that has attained a greenish tint as a result of the growth within it of certain microscopic algae.

greenstone [MINERAL] *See* nephrite. [PETR] Any altered basic igneous rock which is green due to the presence of chlorite, hornblende, or epidote.

greenstone belts [GEOL] Oceanic and island arclike sequences that are similar to, and run to the south and north of, the Swaziland System.

greenstone schist [PETR] Greenstone with a foliated structure.

Greenwell formula [MIN ENG] A formula used for calculating the thickness of tubing and involving the required thickness of tubing in feet, the vertical depth in feet, the diameter of the shaft in feet, and an allowance for possible flaws or corrosion.

Greenwich meridian [GEOD] The meridian passing through Greenwich, England, and serving as the reference for Greenwich time; it also serves as the origin of measurement of longitude.

gregale [METEOROL] The Maltese and best-known variant of a term for a strong northeast wind in the central and western Mediterranean and adjacent European land areas; it occurs either with high pressure over central Europe or the Balkans and low pressure over Libya, when it may continue for up to 5 days, or with the passage of a depression to the south or southeast, when it lasts only 1 or 2 days; it is most frequent in winter.

greisen [PETR] A pneumatolytically altered granite consisting of mainly quartz and a light-green mica.

grenatite *See* leucite; staurolite.

Grenville orogeny [GEOL] A Precambrian mountain-forming epoch.

grid [MAP] A system of uniformly spaced perpendicular lines and horizontal lines running north and south, and east and west on a map, chart, or aerial photograph; used in locating points. [MIN ENG] Imaginary line used to divide the surface of an area when following a checkerboard placement of boreholes.

grid nephoscope [ENG] A nephoscope constructed of a grid work of bars mounted horizontally on the end of a vertical column and rotating freely about the vertical axis; the observer rotates the grid and adjusts the position until some feature of the cloud appears to move along the major axis of the grid; the azimuth angle at which the grid is set is taken as the direction of the cloud motion.

griffithite [MINERAL] A micalike mineral containing magnesium, iron, calcium, and aluminosilicate.

grinding [MECH ENG] **1.** Reducing a material to relatively small particles. **2.** Removing material from a workpiece with a grinding wheel. [MIN ENG] The act or process of continuing to drill after the bit or core barrel is blocked, thereby crushing and destroying any core that might have been produced.

griphite [MINERAL] $(Na,Al,Ca,Fe)_6Mn_4(PO_4)_5(OH)_4$ Mineral composed of a basic phosphate of sodium, calcium, iron, aluminum, and manganese.

griquaite [PETR] A hypabyssal rock that contains garnet and diopside, and sometimes olivine or phlogopite, and is found in kimberlite pipes and dikes.

grit [GEOL] **1.** A hard, sharp granule, as of sand. **2.** A coarse sand. [PETR] A sandstone composed of angular grains of different sizes.

grizzly chute [MIN ENG] A chute equipped with grizzlies which separate fine from coarse material as it passes through the chute.

grizzly worker [MIN ENG] In metal mining, a laborer who works underground at a grizzly.

Groeberiidae [PALEON] A family of extinct rodentlike marsupials.

groove [GEOL] Glaciated marks of large size on rock.

groove casts [GEOL] Rounded or sharp, crested, rectilinear ridges that are a few millimeters high and a few centimeters long; found on the undersurfaces of sandstone layers lying on mudstone.

groove face [MIN ENG] The portion of a surface of a member that is included in a groove.

groove sample [MIN ENG] A sample of coal or ore obtained by cutting appropriate grooves along or across the road exposures. Also known as channel sample.

gross-austausch [METEOROL] The exchange of air mass properties and the associated momentum and energy transports produced on a worldwide scale by the migratory large-scale disturbances of middle latitudes.

gross recoverable value   [MIN ENG]   The part of the total metal recovered from an ore multiplied by the price.

grossular *See* grossularite.

grossularite   [MINERAL] $Ca_3Al_2(SiO_4)_3$ The colorless or green, yellow, brown, or red end member of the garnet group, often occurring in contact-metamorphosis impure limestones. Also known as gooseberry stone; grossular.

gross unit value   [MIN ENG]   The weight of metal per long or short ton as determined by assay or analysis, multiplied by the market price of the metal.

grothite *See* sphene.

ground   [GEOL]   **1.** Any rock or rock material. **2.** A mineralized deposit. **3.** Rock in which a mineral deposit occurs.

ground coal   [MIN ENG]   The bottom of a coal seam.

ground discharge *See* cloud-to-ground discharge.

grounded ice *See* stranded ice.

ground fog   [METEOROL]   A fog that hides less than 0.6 of the sky and does not extend to the base of any clouds that may lie above it.

ground frost   [METEOROL]   In British usage, a freezing condition injurious to vegetation, which is considered to have occurred when a minimum thermometer exposed to the sky at a point just above a grass surface records a temperature (grass temperature) of 30.4°F (-0.9°C) or below.

groundhog *See* barney.

ground ice   [HYD]   **1.** A body of clear ice in frozen ground, most commonly found in more or less permanently frozen ground (permafrost), and may be of sufficient age to be termed fossil ice. Also known as stone ice; subsoil ice; subterranean ice; underground ice. **2.** *See* anchor ice.

ground ice mound   [GEOL]   A frost mound containing bodies of ice. Also known as ice mound.

ground inversion *See* surface inversion.

ground layer *See* surface boundary layer.

ground magnetic survey   [ENG]   A determination of the magnetic field at the surface of the earth by means of ground-based instruments.

groundman *See* mucker.

groundmass *See* matrix.

ground moraine   [GEOL]   Rock material carried and deposited in the base of a glacier. Also known as bottom moraine; subglacial moraine.

ground noise   [GEOPHYS]   In seismic exploration, disturbance of the ground due to some cause other than the shot.

ground pressure   [GEOPHYS]   The pressure to which a rock formation is subjected by the weight of the superimposed rock and rock material or by diastrophic forces created by movements in the rocks forming the earth's crust. Also known as geostatic pressure; lithostatic pressure; rock pressure.

ground sluice   [MIN ENG]   A channel through which gold-bearing earth is passed in placer mining.

ground streamer [METEOROL] An upward advancing column of high-ion density which rises from a point on the surface of the earth toward which a stepped leader descends at the start of a lightning discharge.

ground swell [OCEANOGR] Swell as it passes through shallow water, characterized by a marked increase in height in water shallower than one-tenth wavelength.

ground-to-cloud discharge [GEOPHYS] A lightning discharge in which the original streamer processes start upward from an object located on the ground.

ground visibility [METEOROL] In aviation terminology, the horizontal visibility observed at the ground, that is, surface visibility or control-tower visibility.

groundwater [HYD] All subsurface water, especially that part that is in the zone of saturation.

groundwater decrement *See* groundwater discharge.

groundwater depletion curve [HYD] A recession curve of streamflow, so adjusted that the slope of the curve represents the runoff (depletion rate) of the groundwater; it is formed by the observed hydrograph during prolonged periods of no precipitation. Also known as groundwater recession.

groundwater discharge [HYD] **1.** Water released from the zone of saturation. **2.** Release of such water. Also known as decrement; groundwater decrement; phreatic-water discharge.

groundwater flow [HYD] That portion of the precipitation that has been absorbed by the ground and has become part of the groundwater.

groundwater increment *See* recharge.

groundwater level [HYD] **1.** The level below which the rocks and subsoil are full of water. **2.** *See* water table.

groundwater recession *See* groundwater depletion curve.

groundwater recharge *See* recharge.

groundwater replenishment *See* recharge.

groundwater surface *See* water table.

groundwater table *See* water table.

group number [OCEANOGR] The first two numbers in the argument number in A. T. Doodson's scheme for predicting tides.

groutite [MINERAL] $HMnO_2$ A mineral of the diaspore group, composed of manganese, hydrogen, and oxygen; it is polymorphous with manganite.

growl [GEOPHYS] Noise heard when strata are subjected to great pressure.

growler [OCEANOGR] A small piece of floating sea ice, usually a fragment of an iceberg or floeberg; it floats low in the water, and its surface often is heavily pitted; it often appears greenish in color.

growth fabric [PETR] Orientation of fabric elements independent of the influences of stress and resultant movement.

grubstake [MIN ENG] In the United States, supplies or money furnished to a mining prospector for a share in his discoveries.

grubstake contract [MIN ENG] An agreement between two or more persons to locate mines upon the public domain by their joint aid, effort, labor, or expense, with each to acquire by virtue of the act of location such an interest in the mine as agreed upon in the contract.

gruenlingite [MINERAL] Bi_4TeS_3 A mineral composed of sulfide and telluride of bismuth.

grunerite [MINERAL] $(Mg,Fe)_7Si_8O_{22}(OH)_2$ Variety of amphibole; forms monoclinic crystals.

grus *See* gruss.

gruss [GEOL] A loose accumulation of fragmental products formed from the weathering of granite. Also spelled grus.

Guadalupian [GEOL] A North American provincial series in the Lower and Upper Permian, above the Leonardian and below the Ochoan.

guanajuatite [MINERAL] Bi_2Se_3 Bluish-gray mineral composed of bismuth selenide, occurring in crystals or masses.

guard [MIN ENG] A support in front of a roll train to guide the bar into the groove.

guard-electrode system [PETRO ENG] System of extra electrodes used during electrical logging of reservoir formations to confine the surveying current from the measuring electrode to a generally horizontal path.

guard magnet [MIN ENG] A magnet employed in a crushing system to remove or arrest tramp iron ahead of the machinery.

guard screen *See* oversize control screen.

guba [METEOROL] In New Guinea, a rain squall on the sea.

gudmundite [MINERAL] FeSbS A silver-white to steel-gray orthorhombic mineral composed of a sulfide and antimonide of iron.

guest element *See* trace element.

Guiana Current [OCEANOGR] A current flowing northwestward along the northeastern coast of South America.

guide bracket [MIN ENG] A steel bracket fixed to a bunton to secure rigid guides in a shaft.

guide coupling [MIN ENG] A short coupling with a projecting reamer guide or pup to which is attached a reaming bit, which it couples to a reaming barrel.

guide frame [MIN ENG] A frame held rigidly in place by roof jacks or timbers, with provisions for attaching a shaker conveyor pan line to the movable portion of the frame; prevents jumping or side movement of the pan line.

guide ring [MIN ENG] A longitudinally grooved, annular ring made almost full borehole size, which is fitted to an extension coupling between the core barrel and the first drill rod.

guide rod [MIN ENG] A heavy drill rod coupled to and having the same diameter as a core barrel on which it is used; gives additional rigidity to the core barrel and helps to prevent deflection of the borehole. Also known as core barrel rod; oversize rod.

guides [MIN ENG] **1.** Steel, wood, or steel-wire rope conductors in a mine shaft to guide the movement of the cages. **2.** Timber, rope, or metal tracks in a hoisting shaft, which are engaged by shoes on the cage or skip so as to steady it in transit. **3.** The holes in a crossbeam through which the stems of the stamps in a stamp mill rise and fall.

guildite [MINERAL] $(Cu,Fe)_3(Fe,Al)_4(SO_4)_7(OH)_4 \cdot 15H_3O$ A dark-brown mineral composed of a basic hydrated sulfate of copper, iron, and aluminum.

Guinea Current [OCEANOGR] A current flowing eastward along the southern coast of northwestern Africa into the Gulf of Guinea.

guitermanite [MINERAL] $Pb_{10}Ar_6S_{19}$ A bluish-gray mineral composed of lead, arsenic, and sulfur, occurring in compact masses.

gulch [GEOGR] A gulley, sometimes occupied by a torrential stream.

gulch claim [MIN ENG] A claim laid upon and along the bed of an unnavigable stream winding through a canyon, with precipitous, nonmineral, and uncultivable bands wherein have accumulated placer deposits.

gulf [GEOGR] **1.** An abyss or chasm. **2.** A large extension of the sea partially enclosed by land.

Gulfian [GEOL] A North American provincial series in Upper Cretaceous geologic time, above Comanchean and below the Paleocene of the Tertiary.

Gulf Stream [OCEANOGR] A relatively warm, well-defined, swift, relatively narrow, northward-flowing ocean current which originates north of Grand Bahama Island where the Florida Current and the Antilles Current meet, and which eventually becomes the eastward-flowing North Atlantic Current.

Gulf Stream Countercurrent [OCEANOGR] **1.** A surface current opposite to the Gulf Stream, one current component on the Sargasso Sea side and the other component much weaker, on the inshore side. **2.** A predicted, but as yet unobserved, large current deep under the Gulf Stream but opposite to it.

Gulf Stream eddy [OCEANOGR] A cutoff meander of the Gulf Stream.

Gulf Stream front [OCEANOGR] The pronounced horizontal temperature gradient that defines a cross section of the Gulf Stream.

Gulf Stream meander [OCEANOGR] One of the changeable, winding bends in the Gulf Stream; such bends intensify as the Gulf Stream merges into North Atlantic Drift and break up into detached eddies at times, at about 40°N.

Gulf Stream system [OCEANOGR] The Florida Current, Gulf Stream, and North Atlantic Current, collectively.

gully [GEOGR] A narrow ravine.

gully erosion [GEOL] Erosion of soil by running water.

gully-squall [METEOROL] A nautical term for a violent squall of wind from mountain ravines on the Pacific side of Central America.

gumbo [GEOL] A soil that forms a sticky mud when wet.

gumboti [GEOL] Deoxidized, leached clay that contains siliceous stones.

gummite [MINERAL] Any of various yellow, orange, red, or brown secondary minerals containing hydrous oxides of uranium, thorium, and lead. Also known as uranium ocher.

Günz [GEOL] A European stage of geologic time, in the Pleistocene (above Astian of Pliocene, below Mindel); it is the first stage of glaciation of the Pleistocene in the Alps.

Günz-Mindel [GEOL] The first interglacial stage of the Pleistocene in the Alps, between Günz and Mindel glacial stages.

gusset [MIN ENG] A V-shaped cut in the face of a heading.

gust [METEOROL] A sudden, brief increase in the speed of the wind; it is of a more transient character than a squall and is followed by a lull or slackening in the wind speed.

gustiness [METEOROL] A quality of airflow characterized by gusts.

gustiness components [METEOROL] **1.** The ratios, to the mean wind speed, of the average magnitudes of the component fluctuations of the wind along three mutually perpendicular axes. **2.** The ratios of the root-mean-squares of the eddy velocities to the mean wind speed. Also known as intensity of turbulence.

gustiness factor [METEOROL] A measure of the intensity of wind gusts; it is the ratio of the total range of wind speeds between gusts and the intermediate periods of lighter wind to the mean wind speed, averaged over both gusts and lulls.

guti weather [METEOROL] In Rhodesia, a dense stratocumulus overcast, frequently with drizzle, occurring mainly in early summer, and associated with easterly winds that invade the interior, bringing in cool and stable maritime air when an anticyclone moves eastward south of Africa.

gutter [BUILD] A trough along the edge of the eaves of a building to carry off rainwater.   [CIV ENG] A shallow trench provided beside a canal, bordering a highway, or elsewhere, for surface drainage.   [MET] A groove along the periphery of a die impression to allow for excess flash during forging.   [MIN ENG] A drainage trench cut along the side of a mine shaft to conduct the water back into a lodge or sump.

guttering [ENG] A process of quarrying stone in which channels, several inches wide, are cut by hand tools, and the stone block is detached from the bed by pinch bars. [MIN ENG] The process of cutting gutters in a mine shaft.

guttra [METEOROL] In Iran, sudden squalls in May.

guxen [METEOROL] A cold wind of the Alps in Switzerland.

guyot [GEOL] A seamount, usually deeper than 100 fathoms (180 meters), having a smooth platform top. Also known as tablemount.

guzzle [METEOROL] In the Shetland Islands, an angry blast of wind, dry and parching.

Gymnarthridae [PALEON] A family of extinct lepospondylous amphibians that have a skull with only a single bone representing the tabular and temporal elements of the primitive skull roof.

gymnite *See* deweylite.

Gymnocodiaceae [PALEOBOT] A family of fossil red algae.

gypsite [GEOL] A variety of gypsum consisting of dirt and sand; found as an efflorescent deposit in arid regions, overlying gypsum. Also known as gypsum earth.

gypsum [MINERAL] $CaSO_4 \cdot 2H_2O$ A mineral, the commonest sulfate mineral; crystals are monoclinic, clear, white to gray, yellowish, or brownish in color, with well-developed cleavages; luster is subvitreous to pearly, hardness is 2 on Mohs scale, and specific gravity is 2.3; it is calcined at 190–200°C to produce plaster of paris.

gypsum earth *See* gypsite.

Gyracanthididae [PALEON] A family of extinct acanthodian fishes in the suborder Diplacanthoidei.

gyre [OCEANOGR] A closed circulatory system, but larger than a whirlpool or eddy.

gyrogonite [PALEOBOT] A minute, ovoid body that is the residue of the calcareous encrustation about the female sex organs of a fossil stonewort.

gyttja [GEOL] A fresh-water anaerobic mud containing an abundance of organic matter; capable of supporting aerobic life.

Haanel depth rule [GEOPHYS] A rule for estimating the depth of a magnetic body, provided the body may be considered magnetically equivalent to a single pole; the depth of the pole is then equal to the horizontal distance from the point of maximum vertical magnetic intensity to the points where the intensity is one-third of the maximum value.

haar [METEOROL] A wet sea fog or very fine drizzle which drifts in from the sea in coastal districts of eastern Scotland and northeastern England; it occurs most frequently in summer.

Haase system [MIN ENG] Shaft sinking in loose ground or quicksand by piles in the form of iron tubes connected by webs; downward movement is facilitated by water forced down the tubes to wash away loose material beneath their points.

habit [CRYSTAL] *See* crystal habit.

habit plane [CRYSTAL] The crystallographic plane or system of planes along which certain phenomena such as twinning occur.

haboob [METEOROL] A strong wind and sandstorm or duststorm in the northern and central Sudan, especially around Khartum, where the average number is about 24 haboobs a year.

hachure [GEOL] A short line at some angle to contours, used to denote slopes of the ground in geologic maps.

hackmanite [MINERAL] A mineral of the sodalite family containing a small amount of sulfur; fluoresces orange or red in ultraviolet light.

hadal [OCEANOGR] Pertaining to the environment of the ocean trenches, over 6.5 kilometers in depth.

hade [GEOL] **1.** The angle of inclination of a fault as measured from the vertical. **2.** The inclination angle of a vein or lode.

Hadley cell [METEOROL] A direct, thermally driven, and zonally symmetric circulation first proposed by George Hadley as an explanation for the trade winds; it consists of the equatorward movement of the trade winds between about latitude 30° and the equator in each hemisphere, with rising wind components near the equator, poleward flow aloft, and finally descending components at about latitude 30° again.

hadrosaur [PALEON] A duck-billed dinosaur.

Hadsel mill [MIN ENG] An early autogenous grinding mill in which comminution was caused by the fall of ore on ore that was rotating in a large-diameter horizontal cylinder.

haff [GEOGR] A freshwater lagoon separated from the sea by a sandbar.

haidingerite [MINERAL] $HCaAsO_4 \cdot H_2O$ A white mineral composed of hydrous calcium arsenate.

hail [METEOROL] Precipitation in the form of balls or irregular lumps of ice, always produced by convective clouds, nearly always cumulonimbus.

hail stage [METEOROL] The thermodynamic process of freezing of suspended water drops in adiabatically rising air with temperature below the freezing point, under the assumption that release of latent heat of fusion maintains constant temperature until all water is frozen.

hailstone [METEOROL] A single unit of hail, ranging in size from that of a pea to that of a grapefruit, or from less than ¼ inch (6 millimeters) to more than 5 inches (13 centimeters) diameter; may be spheroidal, conical, or generally irregular in shape.

hair copper *See* chalcotrichite.

hair pyrites *See* millerite.

hair salt *See* alunogen.

hairstone [GEOL] Quartz embedded with hairlike crystals of rutile, actinolite, or other mineral.

halcyon days [METEOROL] A period of fine weather.

half-arc angle [METEOROL] The elevation angle of that point which a given observer regards as the bisector of the arc from his zenith to his horizon; a measure of the apparent degree of flattening of the dome of the sky.

half-course [MIN ENG] The drift or opening driven at an angle of about 45° to the strike and in the plane of the seam.

half-header [MIN ENG] A large cap piece; used by sawing a header in two and placing, generally, two timbers under the half header on one side of the haulage, with the end extending over the haulage.

half set [MIN ENG] In mine timbering, one leg piece and a collar.

half tide [OCEANOGR] The condition when the tide is at the level between any given high tide and the following or preceding low tide. Also known as mean tide.

half-tide level [OCEANOGR] The level midway between mean high water and mean low water.

Halimond tube [MIN ENG] A miniature pneumatic flotation cell used for examination of small ore samples under closely controllable conditions.

halite [MINERAL] NaCl Native salt; an evaporite mineral occurring as isometric crystals or in massive, granular, or compact form. Also known as common salt; rock salt.

Halitheriinae [PALEON] A subfamily of extinct sirenian mammals in the family Dugongidae.

Hallett table [MIN ENG] A Wilfley-type concentrating table having the tops of the riffles in the same plane as the cleaning planes, and riffles inclined toward the waste water side.

Hallinger shield [MIN ENG] A tunneling shield valuable for working in very soft ground; incorporates a mechanical excavator and does not entail the use of timbering to protect the miners.

halloysite [MINERAL] $Al_2Si_2O_5(OH)_4 \cdot 2H_2O$ Porcelainlike clay mineral whose composition is like that of kaolinite but contains more water and is structurally distinct; varieties are known as metahalloysites.

halo [GEOL] A ring or cresent surrounding an area of opposite sign; it is a diffusion of a high concentration of the sought mineral into surrounding ground or rock; it is encountered in mineral prospecting and in magnetic and geochemical surveys. [METEOROL] Any one of a large class of atmospheric optical phenomena which appear as colored or whitish rings and arcs about the sun or moon when seen through an ice crystal cloud or in a sky filled with falling ice crystals.

halocline [OCEANOGR] A well-defined vertical gradient of salinity in the oceans and seas.

halogen mineral [MINERAL] Any of the naturally occurring compounds containing a halogen as the sole or principal anionic constituent.

halomorphic [GEOCHEM] Describing an intrazonal soil whose characteristics have been greatly affected by neutral or alkali salts or both.

halotrichite [MINERAL] **1.** $FeAl_2(SO_4)_4 \cdot 22H_2O$ A mineral composed of hydrous sulfate of iron and aluminum. Also known as butter rock; feather alum; iron alum; mountain butter. **2.** Any sulfate mineral resembling halotrichite in structure and habit.

Halysitidae [PALEON] A family of extinct Paleozoic corals of the order Tabulata.

hambergite [MINERAL] Be_2BO_3OH A grayish-white or colorless mineral composed of beryllium borate and occurring as prismatic crystals; hardness is 7.5 on Mohs scale, and specific gravity is 2.35.

hammarite [MINERAL] $Pb_2Cu_2Bi_4S_9$ A monoclinic mineral whose color is a steel gray with red tone; consists of lead and copper bismuth sulfide.

hammer pick [MIN ENG] A pneumatic hand-held machine used to break up the harder rocks in a mine; consists of a pick which is driven by a hammer set in a cylinder which receives compressed air.

hammock *See* hummock.

hancockite [MINERAL] A complex silicate mineral containing lead, calcium, strontium, and other minerals; it is isomorphous with epidote.

Hancock jig [MIN ENG] A moving-screen jig used to treat lead-zinc ores; the material is jigged in a tank of water, and the heavy layer settles through slots.

hand cable [MIN ENG] A flexible cable for electrical connection between a mining machine and a truck carrying a reel of portable cable. Also known as butt cable; head cable.

hand electric lamp [MIN ENG] In mining, a portable, battery operated hand lamp with a tungsten-filament light source that forms a self-contained unit.

hand hammer drill [ENG] A hand-held rock drill.

hand jig [MIN ENG] A moving-screen jig operated by hand and used to treat small batches of ore; the jig box is attached to a rocking beam and moved in a tank of water.

hand level [ENG] A hand-held surveyor's level, basically a telescope with a bubble tube attached so that the position of the bubble can be seen when looking through the telescope.

hand loader [MIN ENG] A miner who uses a shovel, rather than a machine to load coal.

handpicking [MIN ENG] Manual removal of a selected fraction of coarse run-of-mine ore, after washing and screening away waste.

hand sampling [MIN ENG] Using manual methods to detach and reduce in size representative samples of ore; one of the major methods in sampling small batches of ore, others being grab sampling, trench or channel sampling, fractional selection, coning and quartering, and pipe sampling.

hanger [GEOL]   *See* hanging wall.   [PETRO ENG]   **1.** A device to seat in the bowl of a lowermost casing head to suspend the next-smaller casing string and form a seal between the two. Also known as casing hanger. **2.** A device to provide a seal between the tubing and the tubing head. Also known as tubing hanger.

hanging [GEOL]   *See* hanging wall.

hanging bolt [MIN ENG]   A bolt used to suspend wall plates in shaft construction.

hanging coal [MIN ENG]   A portion of the coal seam which, by undercutting, has had its natural support removed.

hanging glacier [HYD]   A glacier lying above a cliff or steep mountainside; as the glacier advances, calving can cause ice avalanches.

hanging sets [MIN ENG]   Timbers from which cribs are suspended in working through soft strata.

hanging sheave [MIN ENG]   The grooved wheel or pulley which is suspended from the drill tripod clevis or from the roof or side of a haulage road, and over which the hoist line runs to minimize friction.

hanging side *See* hanging wall.

hanging valley [GEOL]   A valley whose floor is higher than the level of the shore or other valley to which it leads.

hanging wall [GEOL]   The rock mass above a fault plane, vein, lode, ore body, or other structure. Also known as hanger; hanging; hanging side.

hang-up [ENG]   A virtual leak resulting from the release of entrapped tracer gas from a leak detector vacuum system.   [MIN ENG]   Blockage of the movement of ore by rock in an underground chute.

hanksite [MINERAL] $Na_{22}K(SO_4)_9(CO_3)_2Cl$ A white or yellow mineral crystallizing in the hexagonal system; found in California.

hannayite [MINERAL] $Mg_3(NH_4)_2H_2(PO_4)_4 \cdot 8H_2O$ Mineral composed of hydrous acid ammonium magnesium phosphate; occurs as yellow crystals in guano.

Haplolepidae [PALEON]   A family of Carboniferous chondrostean fishes in the suborder Palaeoniscoidei having a reduced number of fin rays and a vertical jaw suspension.

haplopore [PALEON]   Any randomly distributed pore on the surface of fossil cystoid echinoderms.

harbor [GEOGR]   Any body of water of sufficient depth for ships to enter and find shelter from storms or other natural phenomena. Also known as port.

harbor reach [GEOGR]   The stretch of a river or estuary which leads directly to the harbor.

hard bottom [MIN ENG]   A condition encountered in some opencut mines wherein the rock occasionally does not break down to grade because of an extra-hard streak of ground or because insufficient powder is used.

hardcap [MIN ENG]   In mining, the upper foot or two (0.3–0.6 meter) of a bauxite deposit; usually used as a roof during the mining operations.

hard coal *See* anthracite.

hard-coal plough [MIN ENG]   A plough-type cutter-loader consisting of rigid or swiveling kerfing bits which precut the coal in hard-coal seams.

hard freeze [HYD]   A freeze in which seasonal vegetation is destroyed, the ground surface is frozen solid underfoot, and heavy ice is formed on small water surfaces such as puddles and water containers.

hard frost *See* black frost.

hard ground [MIN ENG] Ground that is difficult to work.

hardness number [ENG] A number representing the relative hardness of a mineral, metal, or other material as determined by any of more than 30 different hardness tests.

hardpan *See* caliche.

hard rime [HYD] Opaque, granular masses of rime deposited chiefly on vertical surfaces by a dense super-cooled fog; it is more compact and amorphous than soft rime, and may build out into the wind as glazed cones or feathers.

hard rock [GEOL] Rock which needs drilling and blasting for removal.

hard-rock driller [MIN ENG] A worker who operates a drill in a mine where the rocks are generally igneous or metamorphosed and considered hard, such as rocks in which coal and salt are generally found.

hard-rock mine [MIN ENG] A mine located in hard rock, especially a mine difficult to drill, blast, and square up.

hard-rock miner [MIN ENG] A worker competent to mine in hard rock, usually an expert miner.

hard-rock tunnel boring [MIN ENG] A tunneling method utilizing a mole to cut out 7-feet-diameter (2.1-meter-diameter) drifts in hard rock at an average rate of 5 feet (1.5 meters) per hour.

Hardwick conveyor loader head [MIN ENG] A dust collector for belt conveyors used at the loading station; a scraper chain runs at the bottom of a coal hopper and collects underbelt fines.

Harlechian [GEOL] A European stage of geologic time: Lower Cambrian.

harlequin opal [MINERAL] Opal with small, close-set angular (mosaiclike) patches of play of color of similar size.

harmatan *See* harmattan.

harmattan [METEOROL] A dry, dust-bearing wind from the northeast or east which blows in West Africa especially from late November until mid-March; it originates in the Sahara as a desert wind and extends southward to about 5°N in January and 18°N in July. Also spelled harmatan; harmetan; hermitan.

harmetan *See* harmattan.

harmless-depth theory [MIN ENG] Formerly, hypothesis that there was a certain depth below which mining could be carried on without risk of damage to the surface.

harmonic decline [PETRO ENG] One of three types of decline in oil or gas production rate (the others are constant-percentage and hyperbolic), in which the nominal decline in production rate per unit of time expressed as a fraction of the production rate is proportional to the production rate itself.

harmonic folding [GEOL] Folding in the earth's surface, with no sharp changes with depth in the form of the folds.

harmonic tide plane *See* Indian spring low water.

harmotome [MINERAL] $(K,Ba)(Al,Si)_2(Si_6O_{16})\cdot6H_2O$ A zeolite mineral with ion-exchange properties that forms cruciform twin crystals. Also known as cross-stone.

harstigite [MINERAL] $Be_2Ca_3Si_3O_{11}$ A mineral composed of silicate of beryllium and calcium.

hartite [GEOL] A white, crystalline, fossil resin that is found in lignites. Also known as bombiccite; branchite; hofmannite; josen.

harzburgite [PETR] A peridotite consisting principally of olivine and orthopyroxene.

Harz jig [MIN ENG]   A device used to separate coal and foreign matter which gives pulsion intermittently with suction.

Hasenclever turntable [MIN ENG]   A turntable that is made to rotate by the friction between the positively driven pulley, the car, and the table; used as an alternative to the shunt-back or the traverser for changing the direction of mine cars or tubs, either on the surface or underground.

haster [METEOROL]   In England, a violent rain storm.

hastingsite [MINERAL] $NaCa_2(Fe,Mg)_5Al_2Si_6O_{22}(OH)_2$ A mineral of the amphibole group crystallizing in the monoclinic system and composed chiefly of sodium, calcium, and iron, but usually with some potassium and magnesium.

hatchettine *See* hatchettite.

hatchettite [MINERAL] $C_{38}H_{78}$ A yellow-white mineral paraffin wax, melting at 55–65°C in the natural state and 79°C in the pure state; occurs in masses in ironstone nodules or in cavities in limestone.   Also known as adipocerite; adipocire; hatchettine; mineral tallow; mountain tallow; naphthine.

hatchettolite *See* ellsworthite.

hatchite [MINERAL]   A lead-gray mineral composed of sulfide of lead and arsenic; occurs in triclinic crystals.

haud [METEOROL]   In Scotland, a squall.

hauerite [MINERAL] MnS_2 A reddish-brown or brownish-black mineral composed of native manganese sulfide; occurs massive or in octahedral or pyritohedral crystals.

haughtonite [PETR]   A black variety of biotite that is rich in iron.

haulage [MIN ENG]   The movement, in cars or otherwise, of men, supplies, ore, and waste, underground and on the surface.

haulage conveyor [MIN ENG]   A conveyor used to transport material between the gathering conveyor and the outside.

haulage curve [MIN ENG]   A bend in a haulage road in any direction.

haulage drum [MIN ENG]   A cylinder on which steel haulage rope is coiled.

haulage level [MIN ENG]   An underground level, either along and inside the ore body or closely parallel to it, and usually in the footwall, in which mineral is loaded into trams and moved out to the hoisting shaft.

haulage stage [MIN ENG]   A mine roadway along which a load is moved by one form of haulage without coupling or uncoupling of cars.

haulageway [MIN ENG]   The gangway, entry, or tunnel through which loaded or empty mine cars are hauled by animal or mechanical power.

haul-cycle time [MIN ENG]   The time required for the scraper to haul a load to the dumping area and to return to its position in the loading area.

haul road [MIN ENG]   A road built to carry heavily loaded trucks at a good speed; the grade is limited and usually kept to less than 17% of climb.

hausmannite [MINERAL] Mn_3O_4 Brownish-black, opaque mineral composed of manganese tetroxide.

Hauterivian [GEOL]   A European stage of geologic time, in the Lower Cretaceous, above Valanginian and below Barremian.

Haüy law [CRYSTAL]   The law that for a given crystal there is a set of ratios such that the ratios of the intercepts of any crystal plane on the crystal axes are rational fractions of these ratios.

haüyne [MINERAL] $(Na,Ca)_{4-8}(Al_6Si_6O_{24})(SO_4,S)_{1-2}$ An isometric silicate mineral of the sodalite group occurring as grains embedded in various igneous rocks; hardness is 5.5–6 on Mohs scale, and specific gravity is 2.4–2.5. Also known as haüynite.

havgul See havgull.

havgula See havgull.

havgull [METEOROL] Cold, damp wind blowing from the sea during summer in Scotland and Norway. Also known as havgul; havgula.

haycock [HYD] An isolated ice cone rising above land ice or shelf ice because of the pressure or ice movement.

haze [METEOROL] Fine dust or salt particles dispersed through a portion of the atmosphere; the particles are so small that they cannot be felt, or individually seen with the naked eye, but they diminish horizontal visibility and give the atmosphere a characteristic opalescent appearance that subdues all colors. [OPTICS] The degree of cloudiness in a solution, cured plastic material, or coating material.

haze factor [METEOROL] The ratio of the luminance of a mist or fog through which an object is viewed to the luminance of the object.

haze horizon [METEOROL] The top of a haze layer which is confined by a low-level temperature inversion and has the appearance of the horizon when viewed from above against the sky.

haze layer [METEOROL] A layer of haze in the atmosphere, usually bounded at the top by a temperature inversion and frequently extending downward to the ground.

haze level See haze line.

haze line [METEOROL] The boundary surface in the atmosphere between a haze layer and the relatively clean, transparent air above the top of a haze layer. Also known as haze level.

hazel sandstone [GEOL] An arkosic, iron-bearing redbed sandstone from the Precambrian found in western Texas.

head [GEOGR] See headland.

headblock [MIN ENG] **1.** A stop at the head of a slope or shaft to keep cars from going down the shaft or slope. **2.** A cap piece.

headboard [MIN ENG] **1.** A wooden wedge placed against the hanging wall, and against which one end of the stull is jammed. **2.** A board in the roof of a heading, contacting the earth above and supported by a headtree on each side.

head cable See hand cable.

header [MIN ENG] **1.** An entry-boring machine that bores the entire section of the entry in one operation. **2.** A rock that heads off or delays progress. **3.** A blasthole at or above the head.

headframe [MIN ENG] **1.** The frame at the top of a shaft, on which is mounted the hoisting pulley. Also known as gallows; gallows frame; headstock; hoist frame. **2.** The shaft frame, sheaves, hoisting arrangements, dumping gear, and connected works at the top of a shaft. Also known as headgear.

headgear See headframe.

headhouse [MIN ENG] **1.** A timber framing located at the top of a shaft and receiving the shaft guides that carry the cage or elevator. **2.** A structure that houses the headframe.

heading-and-bench mining [MIN ENG]   A stoping method used in thicker ore where it is customary first to take out a slice or heading 7 or 8 feet (2.1 or 2.4 meters) high directly under the top of the ore, and then to bench or stope down the ore between the bottom of the heading and the bottom of the ore or floor of the level.

heading blasting *See* coyote blasting.

heading–overhand bench method [MIN ENG]   A tunneling method in which the heading is the lower part of the section and is driven at least a round or two in advance of the upper part (bench), which is taken out by overhand excavating. Also known as inverted heading and bench method.

heading side *See* footwall.

heading wall *See* footwall.

headland [GEOGR]   **1.** A high, steep-faced promontory extending into the sea. Also known as head; mull. **2.** High ground surrounding a body of water.

headline [MIN ENG]   In dredging, the line which is anchored ahead of the dredge pond and holds the dredge up to its digging front.

headman [MIN ENG]   **1.** A person who brings coal to the tramway from the workings. **2.** One who engages or disengages grips on mine cars at the top of a haulage slope.

head mast [MIN ENG]   The tower carrying the working lines of a cable excavator.

headroom [MIN ENG]   **1.** Distance between the drill platform and the bottom of the sheave wheel. **2.** Height between the floor and the roof in a mine opening.

headrope [MIN ENG]   In rope haulage, that rope used to pull the loaded transportation device toward the discharge point.

heads [MIN ENG]   **1.** Material removed from the ore in the treatment plant and containing the valuable metallic constituents. **2.** The feed to a concentrating system in ore dressing.

head sheave [MIN ENG]   Pulley in the headgear of a winding shaft over which the hoisting rope runs.

headtree [MIN ENG]   The horizontal timber placed at each side of a rectangular heading to support the headboard.

head value [MIN ENG]   Assay value of the feed to a concentrating system.

headwall [GEOL]   The steep cliff at the back of a cirque.

headward erosion [GEOL]   Erosion caused by water flowing at the head of a valley. Also known as head erosion; headwater erosion.

headwaters [HYD]   The source and upstream waters of a stream.

headway [MIN ENG]   *See* cross heading.

heaped capacity [MIN ENG]   In scraper loading, the volume of heaped material that a scraper will hold.

heap roasting [MIN ENG]   A process in which ore with a high sulfur content is roasted by the combustion of the sulfur.

heap sampling [MIN ENG]   Method of ore sampling in which the material is shoveled into a conical heap which is then flattened with a spade and shoveled into four equal heaps, two of which are retained, crushed, mixed, and formed into another, smaller cone; the process is repeated until the required small sample is produced.

hearth roasting [MIN ENG]   A process in which ore or concentrate enters at the top of a multiple hearth roaster and drops from hearth until it is discharged at the bottom.

heat balance [GEOPHYS] The equilibrium which exists on the average between the radiation received by the earth and atmosphere from the sun and that emitted by the earth and atmosphere.

heat budget [GEOPHYS] Amount of heat needed to raise a lake's water from the winter temperature to the maximum summer temperature.

heat equator [METEOROL] **1.** The line which circumscribes the earth and connects all points of highest mean annual temperature for their longitudes. **2.** The parallel of latitude of 10°N, which has the highest mean temperature of any latitude. Also known as thermal equator.

heating degree-day [METEOROL] A form of degree-day used as an indication of fuel consumption; in United States usage, one heating degree-day is given for each degree that the daily mean temperature departs below the base of 65°F (where the Celsius scale is used, the base is usually 19°C).

heat lightning [GEOPHYS] Nontechnically, the luminosity observed from ordinary lightning too far away for its thunder to be heard.

heat low *See* thermal low.

heat storage [OCEANOGR] The tendency of the ocean to act as a heat reservoir; results in smaller daily and annual variations in temperature over the sea.

heat thunderstorm [METEOROL] In popular terminology, a thunderstorm of the air mass type which develops near the end of a hot, humid summer day.

heave [GEOL] **1.** The horizontal component of the slip, measured at right angles to the strike of the fault. **2.** A predominantly upward movement of the surface of the soil due to expansion or displacement. [MIN ENG] A rising of the floor of a mine caused by its being too soft to resist the weight on the pillars. [OCEANOGR] The motion imparted to a floating body by wave action.

Heaviside layer *See* E layer.

heavy crude [PETRO ENG] Crude oil having a high proportion of viscous, high-molecular-weight hydrocarbons, and often having a high sulfur content.

heavy ground [MIN ENG] Dangerous hanging wall requiring vigilance against possible rock fall.

heavy ion *See* large ion.

heavy-liquid separation [MIN ENG] A laboratory technique for separating ore particles by allowing them to settle through, or float above, a fluid of intermediate density.

heavy-media separation [MIN ENG] A series of processes for the concentration of ore developed at one time, but now used in coal cleaning; uses suspensions of magnetic materials such as magnetite.

heavy mineral [MINERAL] A mineral with a density above 2.9, which is the density of bromoform, the liquid used to separate the heavy from the light minerals.

heavy-mineral prospecting [MIN ENG] Locating the source of an economic mineral by determining the relative amounts of the mineral in stream sediments and tracing the drainage upstream.

heazlewoodite [MINERAL] Ni_3S_2 A meteorite mineral consisting of a sulfide of nickel.

Hebraic granite *See* graphic granite.

hecatolite *See* moonstone.

hectorite [MINERAL] $(Mg,Li)_3Si_4O_{10}(OH)_2$ A trioctohedral clay mineral of the montmorillonite group composed of a hydrous silicate of magnesium and lithium.

hedenbergite [MINERAL] $CaFeSi_2O_6$ A black mineral consisting of calcium-iron pyroxene and occurring at the contacts of limestone with granitic masses.

hedleyite [MINERAL] A mineral composed of an alloy of bismuth and tellurium.

hedreocraton [GEOL] A craton that influenced later continental development.

hedyphane [MINERAL] $(Ca,Pb)_5Cl(AsO_4)_3$ Yellowish-white mineral composed of lead and calcium arsenate and chloride; occurs in monoclinic crystals.

heel of a shot [ENG] **1.** In blasting, the front or face of a shot farthest from the charge. **2.** The distance between the mouth of the drill hole and the corner of the nearest free face. **3.** That portion of a drill hole which is filled with the tamping. [MIN ENG] That portion of the coal to be fractured which is outside the powder.

height-change chart [METEOROL] A chart indicating the change in height of a constant-pressure surface over a specified previous time interval; comparable to a pressure-change chart.

height-change line [METEOROL] A line of equal change in height of a constant-pressure surface over a specified previous interval of time; the lines drawn on a height-change chart. Also known as contour-change line; isallohypse.

height of tide [OCEANOGR] Vertical distance from the chart datum to the level of the water at any time; it is positive if the water level is higher than the chart datum.

height pattern [METEOROL] The general geometric characteristics of the distribution of height of a constant-pressure surface as shown by contour lines on a constant-pressure chart. Also known as baric topography; isobaric topography; pressure topography.

Helderbergian [GEOL] A North American stage of geologic time, in the lower Lower Devonian.

held in common [MIN ENG] Pertaining to a claim whereof there is more than one owner.

helical steel support [MIN ENG] A continuous, screw-shaped steel joist lining used for staple shafts.

Helicoplacoidea [PALEON] A class of free-living, spindle- or pear-shaped, plated echinozoans known only from the Lower Cambrian of California.

heliolite See sunstone.

Heliolitidae [PALEON] A family of extinct corals in the order Tabulata.

heliophyllite [MINERAL] $Pb_6As_2O_7Cl_4$ A yellow to greenish-yellow, orthorhombic mineral consisting of an oxychloride of lead and arsenic; occurs in massive and tabular form and as crystals.

heliosphere [GEOPHYS] The region in the ionosphere where helium ions are predominant (sometimes there may be no region in which helium ions dominate).

heliotrope [ENG] An instrument that reflects the sun's rays over long distances; used in geodetic surveys. [MINERAL] See bloodstone.

heliotropic wind [METEOROL] A subtle, diurnal component of the wind velocity leading to a diurnal shift of the wind or turning of the wind with the sun, produced by the east-to-west progression of daytime surface heating.

hellandite [MINERAL] Mineral composed of silicate of metals in the cerium group with aluminum, iron, manganese, and calcium.

Helmert's formula [GEOPHYS]   A formula for the acceleration due to gravity in terms of the latitude and the altitude above sea level.

helm wind [METEOROL]   A strong, cold northeasterly wind blowing down into the Eden valley from the western slope of the Crossfell Range in northern England.

Helodontidae [PALEON]   A family of extinct ratfishes conditionally placed in the order Bradyodonti.

helper grade [MIN ENG]   A grade on which helper engines are required to assist road locomotives. Also known as pusher grade.

helper set [MIN ENG]   A set of timbers to reinforce the normal set of timbers in a mine.

helvine *See* helvite.

helvite [MINERAL] $(Mn,Fe, Zn)_4Be_3(SiO_4)_3S$ A silicate mineral isomorphous with danalite and genthelvite. Also known as helvine.

hemafibrite [MINERAL] $Mn_3(AsO_4)(OH)_3 \cdot H_2O$ A brownish to garnet-red mineral composed of basic manganese arsenate.

hematite [MINERAL] Fe_2O_3 An iron mineral crystallizing in the rhombohedral system; the most important ore of iron, it is dimorphous with maghemite, occurs in black metallic-looking crystals, in reniform masses or fibrous aggregates, or in reddish earthy forms. Also known as bloodstone; red hematite; red iron ore; red ocher; rhombohedral iron ore.

hematolite [MINERAL] $(Mn,Mg)_4Al(AsO_4)(OH)_8$ A brownish-red mineral composed of aluminum manganese arsenate; occurs in rhombohedral crystals.

hematophanite [MINERAL] $Pb_5Fe_4O_{10}(Cl,OH)_2$ A mineral composed of oxychloride lead and iron.

Hemicidaridae [PALEON]   A family of extinct Echinacea in the order Hemicidaroida distinguished by a stirodont lantern, and ambulacra abruptly widened at the ambitus.

Hemicidaroida [PALEON]   An order of extinct echinoderms in the superorder Echinacea characterized by one very large tubercle on each interambulacral plate.

hemicone *See* alluvial cone.

hemicrystalline *See* hypocrystalline.

hemihedral symmetry [CRYSTAL]   The possession by a crystal of only half of the elements of symmetry which are possible in the crystal system to which it belongs.

hemiholohedral [CRYSTAL]   Of hemihedral form but with half the octants having the full number of planes.

hemimorphic crystal [CRYSTAL]   A crystal with no transverse plane of symmetry and no center of symmetry; composed of forms belonging to only one end of the axis of symmetry.

hemimorphite [MINERAL] $Zn_4Si_2O_7(OH)_2 \cdot H_2O$ A white, colorless, pale-green, blue, or yellow mineral having an orthorhombic crystal structure; an ore of zinc. Also known as calamine; electric calamine; galmei.

hemipelagic region [OCEANOGR]   The region of the ocean extending from the edge of a shelf to the pelagic environment; roughly corresponds to the bathyal zone, in which the bottom is 200 to 1000 meters below the surface.

hemipelagic sediment [GEOL]   Deposits containing terrestrial material and the remains of pelagic organisms, found in the ocean depths.

hemiprism [CRYSTAL]   A pinacoid that cuts two crystallographic axes.

hemisphere [GEOGR]   A half of the earth divided into north and south sections by the equator, or into an east section containing Europe, Asia, and Africa, and a west section containing the Americas.

hemispherical pyrheliometer [ENG]   An instrument for measuring the total solar energy from the sun and sky striking a horizontal surface, in which a thermopile measures the temperature difference between white and black portions of a thermally insulated horizontal target within a partially evacuated transparent sphere or hemisphere.

hemispheric wave number *See* angular wave number.

Hemist [GEOL]   A suborder of the soil order Histosol, consisting of partially decayed plant residues and saturated with water most of the time.

hermitan *See* harmattan.

hemitropic [CRYSTAL]   Pertaining to a twinned structure in which, if one part were rotated 180°, the two parts would be parallel.

Hemizonida [PALEON]   A Paleozoic order of echinoderms of the subclass Asteroidea having an ambulacral groove that is well defined by adambulacral ossicles, but with restricted or undeveloped marginal plates.

Hepplewhite-Gray lamp [MIN ENG]   A lamp which drew its air from the top down through four tubular pillars into the base, where the air fed the flame through a gauze ring; the outlet was through a metal chimney closed by a gauze disk.

hercularc lining [MIN ENG]   A German method of lining mine roadways subjected to heavy pressures; a closed circular arch of wedge-shaped precast concrete blocks made in two sizes are erected so that alternate blocks offer their wedge action in opposite direction, the larger blocks toward the center of the roadway and the smaller outward.

Hercules stone *See* lodestone.

Hercynian geosyncline [GEOL]   A principal area of geosynclinal sediment accumulation in Devonian time; found in south-central and southern Europe and northern Africa.

Hercynian orogeny *See* Variscan orogeny.

hercynite [MINERAL] $(Fe, Mg)Al_2O_4$ A black mineral of the spinel group; crystallizes in the isometric system. Also known as ferrospinel; iron spinel.

herderite [MINERAL] $CaBe(PO_4)(F,OH)$ A colorless to pale-yellow or greenish-white mineral consisting of phosphate and fluoride of calcium and beryllium; hardness is 7.5–8 on Mohs scale, and specific gravity is 3.92.

Hermann-Mauguin symbols [CRYSTAL]   Symbols representing the 32 symmetry classes, consisting of series of numbers giving the multiplicity of symmetry axes in descending order, with other symbols indicating inversion axes and mirror planes.

herringbone stoping [MIN ENG]   Method used in flattish Rand stope panels 500–1000 feet (150–300 meters) long for breaking and moving the ore; the stope is divided into 20-feet (6-meter) panels, and a different gang works each panel; a tramming system delivers the cut rock to a central scraper system.

herringbone timbering [MIN ENG]   A method of timber support in a roadway with a weak roof and strong sides, using neither arms nor side uprights; the crossbar is notched into the sides and supported at its center by a bar under it and parallel with the roadway; the bar is supported by struts notched into the sides at about half height.

hervidero *See* mud volcano.

Hesperornithidae [PALEON]   A family of extinct North American birds in the order Hesperornithiformes.

Hesperornithiformes [PALEON]   An order of ancient extinct birds; individuals were large, flightless, aquatic diving birds with the shoulder girdle and wings much reduced and the legs specialized for strong swimming.

hessite [MINERAL] Ag_2Te A lead-gray sectile mineral crystallizing in the isometric system; usually massive and often auriferous.

hetaerolite [MINERAL] $ZnMn_2O_4$ A black mineral consisting of zinc-manganese oxide found with chalcophanite.

Heteractinida [PALEON]   A group of Paleozoic sponges with calcareous spicules; probably related to the Calcarea.

heteroblastic [PETR]   Pertaining to rocks in which the essential constituents are of two distinct orders of magnitude of size.

heterochronism [GEOL]   That phenomenon by which two analogous geologic deposits are of different age though their processes of formation were essentially identical.

Heterocorallia [PALEON]   An extinct small, monofamilial order of fossil corals with elongate skeletons; found in calcareous shales and in limestones.

heterodesmic [CRYSTAL]   Pertaining to those atoms bonded in more than one way in crystals.

heterogeneous reservoir [GEOL]   Formation with two or more noncommunicating sand members, each possibly with different specific- and relative-permeability characteristics.

heterogenite [MINERAL] $CoO(OH)$ A black cobalt mineral, sometimes with some copper and iron, found in mammillary masses. Also known as stainierite.

heteromorphite [MINERAL] $Pb_7Sb_8S_{19}$ An iron black, monoclinic mineral consisting of lead antimony sulfide.

Heterophyllidae [PALEON]   The single family of the extinct coelenterate order Heterocorallia.

heterosite [MINERAL]   A mineral composed of phosphate of iron and manganese; it is isomorphous with purpurite.

heterosphere [METEOROL]   The upper portion of a two-part division of the atmosphere (the lower portion is the homosphere) according to the general homogeneity of atmospheric composition; characterized by variation in composition, and in mean molecular weight of constituent gases; starts at 80–100 kilometers above the earth and therefore closely coincides with the ionosphere and the thermosphere.

Heterostraci [PALEON]   An extinct group of ostracoderms, or armored, jawless vertebrates; armor consisted of bone lacking cavities for bone cells.

heulandite [MINERAL] $CaAl_2Si_6O_{16}\cdot5H_2O$ A zeolite mineral that crystallizes in the monoclinic system; often occurs as foliated masses or in crystal form in cavities of decomposed basic igneous rocks.

hewettite [MINERAL] $CaV_6O_{16}\cdot9H_2O$ A deep-red mineral composed of hydrated calcium vanadate; found in silky orthorhombic crystal aggregates in Colorado, Utah, and Peru.

hexagonal close-packed structure [CRYSTAL]   Close-packed crystal structure characterized by the regular alternation of two layers; the atoms in each layer lie at the vertices of a series of equilateral triangles, and the atoms in one layer lie directly above the centers of the triangles in neighboring layers. Abbreviated hcp structure.

hexagonal column [METEOROL]   One of the many forms in which ice crystals are found in the atmosphere; this crystal habit is characterized by hexagonal cross-section in a plane perpendicular to the long direction (principal axis, optic axis, or c-axis) of the columns; it differs from that found in hexagonal platelets only in that environmental conditions have favored growth along the principal axis rather than perpendicular to that axis.

hexagonal lattice [CRYSTAL] A Bravais lattice whose unit cells are right prisms with hexagonal bases and whose lattice points are located at the vertices of the unit cell and at the centers of the bases.

hexagonal platelet [METEOROL] A small ice crystal of the hexagonal tabular form; the distance across the crystal from one side of the hexagon to the opposite side may be as large as about 1 millimeter, and the thickness perpendicular to this dimension is of the order of one-tenth as great; this crystal form is usually formed at temperatures of -10 to $-20°C$ by sublimation; at higher temperatures the apices of the hexagon grow out and develop dendritic forms.

hexagonal system [CRYSTAL] A crystal system that has three equal axes intersecting at 120° and lying in one plane; a fourth, unequal axis is perpendicular to the other three.

hexahedrite [GEOL] An iron meteorite composed of single crystals or aggregates of kamacite, usually containing 4–6% nickel in the metal phase.

hexahydrite [MINERAL] $MgSO_4·6H_2O$ A white or greenish-white monoclinic mineral composed of hydrous magnesium sulfate.

hexoctahedron [CRYSTAL] A cubic crystal form that has 48 equal triangular faces, each of which cuts the three crystallographic axes at different distances.

hextetrahedron [CRYSTAL] A 24-faced form of crystal in the tetrahedral group of the isometric system.

hibernal [METEOROL] Of or pertaining to winter.

Hibernian orogeny *See* Erian orogeny.

hiddenite [MINERAL] A transparent green or yellowish-green spodumene mineral containing chromium and valued as a gem.

hiemal climate [CLIMATOL] Climate pertaining to winter.

hieratite [MINERAL] K_2SiF_6 A grayish mineral composed of potassium fluosilicate; occurs as deposits in volcanic holes.

hieroglyph [GEOL] Any sedimentary mark or structure occurring on a bedding plane.

high [METEOROL] An area of high pressure, referring to a maximum of atmospheric pressure in two dimensions (closed isobars) in the synoptic surface chart, or a maximum of height (closed contours) in the constant-pressure chart; since a high is, on the synoptic chart, always associated with anticyclonic circulation, the term is used interchangeably with anticyclone.

high aloft *See* upper-level anticyclone.

high-altitude station [METEOROL] A weather observing station at a sufficiently high elevation to be nonrepresentative of conditions near sea level; 6500 feet (about 2000 meters) has been given as a reasonable lower limit.

high-angle fault [GEOL] A fault with a dip greater than 45°.

high clouds [METEOROL] Types of clouds whose mean lower level is above 20,000 feet (6100 meters); principal clouds in this group are cirrus, cirrocumulus, and cirrostratus.

higher high water [OCEANOGR] The higher of two high tides occurring during a tidal day.

higher low water   [OCEANOGR]   The higher of two low tides occurring during a tidal day.

high foehn   [METEOROL]   The occurrence of warm, dry air above the level of the general surface, accompanied by clear skies, resembling foehn conditions; it is due to subsiding air in an anticyclone, above a cold surface layer; in such circumstances the mountain peaks may be warmer than the lowlands. Also known as free foehn.

high fog   [METEOROL]   The frequent fog on the slopes of the coastal mountains of California, especially applied when the fog overtops the range and extends as stratus over the leeward valleys.

high-grade   [MIN ENG]   To steal or pilfer ore or gold from a mine or miner.

high index   [METEOROL]   A relatively high value of the zonal index which, in middle latitudes, indicates a relatively strong westerly component of wind flow and the characteristic weather features attending such motion; a synoptic circulation pattern of this type is commonly called a high-index situation.

highland   [GEOGR]   **1.** Any relatively large area of elevated or mountainous land standing prominently above adjacent low areas. **2.** The higher land of a region.   [GEOL]   **1.** A lofty headland, cliff, or other high platform. **2.** A dissected mountain region composed of old folded rocks.

highland climate   *See* mountain climate.

highland glacier   [HYD]   A semicontinuous ice cap or glacier that covers the highest or central portion of a mountainous area and partly reflects irregularities of the land surface lying beneath it. Also known as highland ice.

highland ice   *See* highland glacier.

high-level anticyclone   *See* upper-level anticyclone.

high-level cyclone   *See* upper-level cyclone.

high-level thunderstorm   [METEOROL]   Generally, a thunderstorm based at a comparatively high altitude in the atmosphere, roughly 8000 feet (2400 meters) or higher.

high-level trough   *See* upper-level trough.

high plain   [GEOGR]   A large area of level land situated above sea level.

high-pressure area   *See* anticyclone.

high-pressure gas injection   [PETRO ENG]   Oil reservoir pressure maintenance by injection of gas at pressures higher than those used in conventional equilibrium gas drives.

high-pressure separator   [PETRO ENG]   A horizontal vessel through which a low-temperature, high-pressure gas stream is fed, and in which free liquids separate out from the gas.

high-pressure well   [PETRO ENG]   A well with a shut-in wellhead pressure of more than 2000 psia (1.4×10^7 newtons per square meter, absolute).

high quartz   [MINERAL]   Quartz that was formed at high temperatures.

high-rank coal   [GEOL]   Coal consisting of less than 4% moisture when air-dried, or more than 84% carbon.

high-rank graywacke   *See* feldspathic graywacke.

high tide   [OCEANOGR]   The maximum height reached by a rising tide. Also known as high water.

high-volatile bituminous coal [GEOL]   A bituminous coal composed of more than 31% volatile matter.

highwall [MIN ENG]   The unexcavated face of exposed overburden and coal or ore in an opencast mine or the face or bank of the uphill side of a contour strip-mine excavation.

high water *See* high tide.

high-water full and change [GEOPHYS]   The average interval of time between the transit (upper or lower) of the full or new moon and the next high water at a place. Also known as common establishment; vulgar establishment.

high-water inequality [OCEANOGR]   The difference between the heights of the two high tides during a tidal day.

high-water line [OCEANOGR]   The intersection of the plane of mean high water with the shore.

high-water lunitidal interval [GEOPHYS]   The interval of time between the transit (upper or lower) of the moon and the next high water at a place.

high-water platform *See* wave-cut bench.

high-water quadrature [OCEANOGR]   The average high-water interval when the moon is at quadrature.

high-water springs *See* mean high-water springs.

high-water stand [OCEANOGR]   The condition at high tide when there is no change in the height of the water.

hilgardite [MINERAL] $Ca_6(B_6O_{11})_3Cl_4 \cdot H_2O$ Colorless mineral composed of hydrous borate and chloride of calcium; occurs as monoclinic domatic crystals.

hill [GEOGR]   A land surface feature characterized by strong relief; it is a prominence smaller than a mountain.

hill creep [GEOL]   Slow gravity movement of rock and soil waste down a steep hillside. Also known as hillside creep.

hillebrandite [MINERAL] $Ca_2SiO_3(OH)_2$ A white mineral composed of hydrous calcium silicate; occurs in masses.

hillock [GEOL]   A small, low hill.

hillside creep *See* hill creep.

hillside quarry [MIN ENG]   A quarry cut along a hillside.

Hilt's law [GEOL]   The law that in a small area the deeper coals are of higher rank than those above them.

hindered settling [MIN ENG]   Settling of particles in a thick suspension in water through which their fall is hindered by rising water.

hindered-settling ratio [MIN ENG]   The ratio of the specific gravity of a mineral to that of the suspension of ore raised to a power between one-half and unity.

hinged bar [MIN ENG]   A steel extension bar placed in contact with the mine roof perpendicular to the longwall face and supported by yielding steel props. Also known as link bar.

hinge fault [GEOL]   A fault whose movement is an angular or rotational one on a side of an axis that is normal to the fault plane.

hinge line [GEOL]   **1.** The line separating the region in which a beach has been thrust upward from that in which it is horizontal. **2.** A line in the plane of a hinge fault separating the part of a fault along which thrust or reverse movement occurred from that having normal movement.

hinsdalite [MINERAL] $(Pb,Sr)Al_3(PO_4)(SO_4)(OH)_6$ A dark-gray or greenish rhombohedral mineral composed of basic lead and strontium aluminum sulfate and phosphate; occurs in coarse crystals and masses.

hinterland [GEOL] **1.** The region behind the coastal district. **2.** The terrain on the back of a folded mountain chain. **3.** The moving block which forces geosynclinal sediments toward the foreland.

Hirschback method [MIN ENG] A method for draining firedamp from coal seams by means of superjacent entries located 80–138 feet (24–42 meters) above the seams and supplemented by boreholes drilled at right angles to the entry walls. Also known as superjacent roadway system.

hisingerite [MINERAL] $Fe_2^{3-}Si_2O_5(OH)_4 \cdot 2H_2O$ A black, amorphous mineral composed of hydrous ferric silicate; an iron ore.

historical climate [CLIMATOL] A climate of the historical period (the past 7000 years).

historical geology [GEOL] A branch of geology concerned with the systematic study of bedded rocks and their relations in time and the study of fossils and their locations in a sequence of bedded rocks.

Histosol [GEOL] An order of wet soils consisting mostly of organic matter, popularly called peats and mucks.

hitch [GEOL] **1.** A fault of strata common in coal measures, accompanied by displacement. **2.** A minor dislocation of a vein or stratum not exceeding in extent the thickness of the vein or stratum. [MIN ENG] **1.** A step cut in the rock face to hold timber support in an underground working. **2.** A hole cut in side rock solid enough to hold the cap of a set of timbers, permitting the leg to be dispensed with.

hitcher [MIN ENG] **1.** The worker who runs trams into or out of the cages, gives the signals, and attends at the shaft when workers are riding in the cage. **2.** A worker at the bottom of a haulage slope or plane who engages the clips or grips by means of which mine cars are attached to a hoisting cable or chain. Also known as hitcher-on.

hitcher-on *See* hitcher.

hitch timbering [MIN ENG] Installing timbers in hitches either cut or drilled in the rock.

hjelmite [MINERAL] A black mineral containing yttrium, iron, manganese, uranium, calcium, columbium, tantalum, tin, and tungsten oxide; often occurs with crystal structure disrupted by radiation.

hoar crystal [HYD] An individual ice crystal in a deposit of hoarfrost; always grows by sublimation.

hoarfrost [HYD] A deposit of interlocking ice crystals formed by direct sublimation on objects. Also known as white frost.

hodgkinsonite [MINERAL] $MnZnSiO_5 \cdot H_2O$ A pink to reddish-brown mineral composed of hydrous zinc manganese silicate; occurs as crystals.

hoegbomite [MINERAL] $Mg(Al,Fe,Ti)_4O_7$ A black mineral composed of an oxide of magnesium, aluminum, iron, and titanium. Also spelled högbomite.

hoernesite [MINERAL] $Mg_3As_2O_8 \cdot H_2O$ A white, monoclinic mineral composed of hydrous magnesium arsenate; occurs as gypsumlike crystals.

hoe scraper [MIN ENG] A scraper-loader consisting of a box-sided hoe pulled by cables and used in mining to gather and transport severed rock.

hofmannite *See* hartite.

hogback [GEOL]   Alternate ridges and ravines in certain areas of mountains, caused by erosive action of mountain torrents.

högbomite *See* hoegbomite.

hohmannite [MINERAL] $Fe_2(SO_4)_2(OH)_2 \cdot 7H_2O$ A chestnut brown to burnt orange and amaranth red, triclinic mineral consisting of a hydrated basic sulfate of iron.

hoisting compartment [MIN ENG]   The section of a shaft used for hoisting the mined mineral to the surface.

hoisting cycle [MIN ENG]   The periods of acceleration, uniform speed, retardation, and rest; the deeper the shaft, the greater is the ratio of the time of full-speed hoisting to the whole cycle.

hoistman [ENG]   One who operates steam or electric hoisting machinery to lower and raise cages, skips, or instruments into a mine or an oil or gas well. Also known as hoist operator; winch operator.

hoist operator *See* hoistman.

holddown [PETRO ENG]   A device to anchor an oil well rod pump in its position.

holdenite [MINERAL]   A red, orthorhombic mineral composed of basic manganese zinc arsenate with a small amount of calcium, magnesium, and iron.

Holectypidae [PALEON]   A family of extinct exocyclic Euechinoidea in the order Holectypoida; individuals are hemispherical.

hole director [MIN ENG]   A steel framework used in underground tunneling to set the angle at which holes for a blasting round are to be drilled.

hole layout [MIN ENG]   In quarrying, an arrangement of vertical and horizontal holes.

hole-through [MIN ENG]   The meeting of two approaching tunnel heads.

holing [MIN ENG]   **1.** The working of a lower part of a bed of coal to bring down the upper mass. **2.** The final act of connecting two workings underground. **3.** The meeting of two mine roadways driven to intersect. Also known as thirling.

hollandite [MINERAL] $Ba(Mn^{2+},Mn^{4+})_8O_{16}$ A silvery-gray to black mineral composed of manganate of barium and manganese; occurs as crystals.

Hollinacea [PALEON]   A dimorphic superfamily of extinct ostracods in the suborder Beyrichicopina including forms with sulci, lobation, and some form of velar structure.

Hollinidae [PALEON]   An extinct family of ostracods in the superfamily Hollinacea distinguished by having a bulbous third lobe on the valve.

hollow drill [DES ENG]   A drill rod or stem having an axial hole for the passage of water or compressed air to remove cuttings from a drill hole. Also known as hollow rod; hollow stem.

hollow-plunger pump [MIN ENG]   A pump for mining and quarrying in gritty and muddy water.

hollow reamer [ENG]   A tool or bit used to correct the curvature in a crooked borehole.

hollow rod *See* hollow drill.

hollow-rod churn drill [MECH ENG]   A churn drill with hollow rods instead of steel wire rope.

hollow-rod drilling [ENG]   A modification of wash boring in which a check valve is introduced at the bit so that the churning action may be also used to pump the cuttings up the drill rods.

hollow shafting [MECH ENG]   Shafting made from hollowed-out rods or hollow tubing to minimize weight, allow internal support, or permit other shafting to operate through the interior.

hollow stem *See* hollow drill.

Holman counterbalanced drill rig [MIN ENG] A drill rig consisting of a rail-track carriage with a counterbalanced boom 10 feet (3 meters) long.

Holman dust extractor [MIN ENG] A dust-trapping system in which the dust and chippings from percussive drilling operations are drawn back through the hollow drill rod and along a hose to a metal container with filter elements.

Holman stamp [MIN ENG] A crushing stamp raised by a crank and accelerated in its fall by compressed air.

Holme mud sampler [ENG] A scooplike device which can be lowered by cable to the ocean floor to collect sediment samples.

holmquisite [MINERAL] $(Na,K,Ca)Li(Mg,Fe)_3Al_2Si_8O_{22}(OH)_2$ A bluish-black, orthorhombic mineral composed of alkali and silicate of iron, magnesium, lithium, and aluminum.

holoaxial [CRYSTAL] Having all possible axes of symmetry.

holoclastic [PETR] Being or belonging to ordinary (sedimentary) clastic rock.

holocrystalline [PETR] Pertaining to igneous rocks that are entirely crystallized minerals, without glass.

holohedral [CRYSTAL] Pertaining to a crystal structure having the highest symmetry in each crystal class. Also known as holosymmetric; holosystemic.

holohedron [CRYSTAL] A crystal form of the holohedral class, having all the faces needed for complete symmetry.

holohyaline [PETR] Pertaining to an entirely glassy rock.

holomictic lake [HYD] A lake whose water circulates completely from top to bottom.

Holoptychidae [PALEON] A family of extinct lobefin fishes in the order Osteolepiformes.

holosymmetric *See* holohedral.

holosystemic *See* holohedral.

Holuridae [PALEON] A group of extinct chondrostean fishes in the suborder Palaeoniscoidei distinguished in having lepidotrichia of all fins articulated but not bifurcated, fins without fulcra, and the tail not cleft.

Homacodontidae [PALEON] A family of extinct palaeodont mammals in the superfamily Dichobunoidea.

homeoblastic [PETR] Of a metamorphic crystalloblastic texture, having constituent minerals of approximately the same size.

homilite [MINERAL] $(Ca_2(Fe,Mg)B_2Si_2O_{10}$ A black or blackish brown mineral composed of iron calcium borosilicate.

homobront *See* isobront.

homocline [GEOL] Any rock unit in which the strata exhibit the same dip.

homodesmic [CRYSTAL] Of a crystal, having atoms bonded in a single way.

homogeneous atmosphere [METEOROL] A hypothetical atmosphere in which the density is constant with height.

Homoistela [PALEON] A class of extinct echinoderms in the subphylum Homalozoa.

homometric pair [CRYSTAL] A pair of crystal structures whose x-ray diffraction patterns are identical.

homopause [GEOPHYS] The level of transition between the homosphere and the heterosphere; it lies about 80 to 90 kilometers above the earth.

homosphere [METEOROL]   The lower portion of a two-part division of the atmosphere (the upper portion is the heterosphere) according to the general homogeneity of atmospheric composition; the region in which there is no gross change in atmospheric composition, that is, all of the atmosphere from the earth's surface to about 80–100 kilometers.

honeycomb coral [PALEON]   The common name for members of the extinct order Tabulata; has prismatic sections arranged like the cells of a honeycomb.

Honigmann process [MIN ENG]   A method of shaft sinking through water-bearing sand; the shaft is bored in stages, increasing in size from the pilot hole, about 4 feet in diameter, to the final size.

hook [GEOGR]   The end of a spit of land that is turned toward shore. Also known as recurved spit.

hooked spit *See* hook.

hooker [MIN ENG]   A worker who detaches empty downcoming buckets and hookloads buckets or cans onto the hoisting rope.

hook tender [MIN ENG]   In bituminous coal mining, a worker who attaches the hook of a hoisting cable to the link of a trip of cars to be hauled up or lowered down an incline. Also known as rope cutter.

hook-wall packer [PETRO ENG]   Fluid-proof seal between the outside of oil well tubing and the inside of the casing; hooks hold it in place.

Hooper jig [MIN ENG]   A pneumatic jig, used when water is scarce or the ore must be kept dry, to concentrate values from sands.

hopeite [MINERAL] $Zn_3(PO_4)_2 \cdot 4H_2O$ A gray, orthorhombic mineral composed of hydrous phosphate of zinc; specific gravity is 2.76–2.85; dimorphous with parahopeite.

hoppit [MIN ENG]   A large bucket, usually up to about 80 cubic feet (2.3 cubic meters), used in shaft sinking for hoisting men, rock, materials, and tools. Also known as bowk; kibble; sinking bucket.

horadiam drilling [MIN ENG]   The drilling of a number of horizontal boreholes radiating outward from a common center. Also known as horizontal-ring drilling.

horizon [GEOL]   **1.** The surface separating two beds. **2.** One of the layers, each of which is a few inches to a foot thick, that make up a soil.

horizon distance [GEOD]   The distance, at any given azimuth, to the point on the earth's surface constituting the horizon for some specified observer.

horizon mining [MIN ENG]   A system of mining suitable for inclined, and perhaps faulted, coal seams; main stone headings are driven, at predetermined levels, from the winding shaft to intersect the seams to be developed.

horizontal borer [MIN ENG]   A machine that makes holes 2–6 inches (5–15 centimeters) in diameter, used for drilling at opencut coal mines.

horizontal crosscut *See* horizontal drive.

horizontal displacement *See* strike slip.

horizontal drive [MIN ENG]   An opening with a small inclination directed toward the shaft to drain the water and facilitate hauling of full cars to the shaft. Also known as horizontal crosscut.

horizontal fold *See* nonplunging fold.

horizontal intensity [GEOPHYS]   The strength of the horizontal component of the earth's magnetic field.

horizontal intensity variometer [ENG] Essentially a declination variometer with a larger, stiffer fiber than in the standard model; there is enough torsion in the fiber to cause the magnet to turn 90° out of the magnetic meridian; the magnet is aligned with the magnetic prime vertical to within 0.5° so it does not respond appreciably to changes in declination. Also known as H variometer.

horizontal magnetometer [ENG] A measuring instrument for ascertaining changes in the horizontal component of the magnetic field intensity.

horizontal pressure force [GEOPHYS] The horizontal pressure gradient per unit mass, $-\alpha\nabla_H\,p$, where α is the specific volume, p the pressure, and ∇_H the horizontal component of the del operator; this force acts normal to the horizontal isobars toward lower pressure; it is one of the three important forces appearing in the horizontal equations of motion, the others being the Coriolis force and friction.

horizontal-ring drilling *See* horadiam drilling.

horizontal separation *See* strike slip.

horizontal separator [PETRO ENG] Horizontal tank used to separate free oil well gas from liquid hydrocarbons.

horn [GEOL] A topographically high, sharp, pyramid-shaped mountain peak produced by the headward erosion of mountain glaciers; the Matterhorn is the classic example.

hornblende [MINERAL] A general name given to the monoclinic calcium amphiboles that form an extensive solid-solution series between the various metals in the generalized formula $(Ca,Na)_2(Mg,Fe,Al)_5(Al,Si)_8O_{22}(OH,F)_2$.

hornblendite [PETR] A plutonic rock consisting mainly of hornblende.

horned dinosaur [PALEON] Common name for extinct reptiles of the suborder Ceratopsia.

horned-toad dinosaur [PALEON] The common name for extinct reptiles composing the suborder Ankylosauria.

hornfels [PETR] A common name for a class of metamorphic rocks produced by contact metamorphism and characterized by equidimensional grains without preferred orientation.

hornfels facies [PETR] Rock formed at depths in the earth's crust not exceeding 10 kilometers at temperatures of 250–800°C; includes albite-epidote hornfels facies, pyroxene-hornfels facies, and hornblende-hornfels facies.

horn lead *See* phosgenite.

horn quicksilver *See* calomel.

horn silver *See* cerargyrite.

horn spoon [MIN ENG] A troughlike section cut from a cow horn and scraped thin; used for washing auriferous gravel and pulp when exacting tests are to be performed.

hornstone *See* chert.

horse [GEOL] A large rock caught along a fault.   [MIN ENG] *See* horseback.

horseback [GEOL] A low and sharp ridge of sand, gravel, or rock. [MIN ENG] **1.** Shale or sandstone occurring in a channel that was cut by flowing water in a coal seam. Also known as cutout; horse; roll; swell; symon fault; washout. **2.** To move or raise a heavy piece of machinery or timber by using a pinch bar as a lever. Also known as pinch.

horsehead [MIN ENG] Timbers or steel joists used to support planks in tunneling through loose ground.

horse latitudes [METEOROL] The belt of latitudes over the oceans at approximately 30–35°N and S where winds are predominantly calm or very light and weather is hot and dry.

horseshoe bend See oxbow.

horseshoe lake See oxbow lake.

horsetail ore [GEOL] Ore occurring in fractures which diverge from a larger fracture.

horsfordite [MINERAL] Cu_5Sb A silver-white mineral composed of copper-antimony alloy.

horst [GEOL] **1.** A block of the earth's crust uplifted along faults relative to the rocks on either side. **2.** A mass of the earth's crust limited by faults and standing in relief. **3.** One of the older mountain masses limiting the Alps on the west and north. **4.** A knobby ledge of limestone beneath a thin soil mantle.

Horton number [HYD] A dimensionless number that is formed by the product of runoff intensity and erosion proportionality factor; expresses the relative intensity of erosion on the slopes of a drainage basin.

hortonolite [MINERAL] $(Fe,Mg,Mn)_2SiO_4$ A dark mineral composed of silicate of iron, magnesium, and manganese; a member of the olivine series.

Hoskold formula [MIN ENG] A two-rate valuation formula formerly used to determine present value of mining properties or shares, with redemption of invested capital.

host rock [GEOL] Rock which serves as a host for other rocks or for mineral deposits.

hot acid [PETRO ENG] The use of hot hydrochloric acid (200–300°F; 93–149°C) for oil well acidizing where wellbore scale is slow-dissolving and hard to remove.

hot belt [CLIMATOL] The belt around the earth within which the annual mean temperature exceeds 20°C.

hot wind [METEOROL] General term for winds characterized by intense heat and low relative humidity, such as summertime desert winds or an extreme foehn.

hourly observation See record observation.

howardite [GEOL] An achondritic stony meteorite composed chiefly of calcic plagioclase and orthopyroxene.

howlite [MINERAL] $Ca_2Bi_5SiO_9(OH)_5$ A white mineral occurring in nodular or earthy form.

Hubbard Glacier [GEOL] A valley glacier which reaches tidewater from a source area of Mount St. Elias of Alaska and the Yukon.

hudsonite See cortlandite.

huebnerite [MINERAL] $MnWO_4$ A brownish-red to black manganese member of the wolframite series, occurring in short, monoclinic, prismatic crystals; isomorphous with ferberite.

hühnerkobelite [MINERAL] $(Na,Ca)(Fe,Mn)_2(PO_4)_2$ A mineral composed of phosphate of sodium, calcium, iron, and manganese; it is isomorphous with varulite.

hulsite [MINERAL] $(Fe^{2+},Mg)_2(Fe^{3+},Sn)(BO_3)O_2$ A black mineral composed of iron calcium magnesium tin borate.

Humble gage [PETRO ENG] Device to measure oil well bottom-hole pressure; a piston acts through a stuffing box against a helical spring in tension.

Humble relation [PETRO ENG] Equation used by oil companies to estimate porosity of the oil-bearing formation from measurements made with a contact resistivity device, such as a microlog.

Humboldt Current *See* Peru Current.

Humboldt Glacier [HYD] The largest Arctic iceberg, at latitude 79°, with a seaward front extending 65 miles (105 kilometers).

humboldtine [MINERAL] $FeC_2O_4 \cdot 2H_2O$ A mineral composed of hydrous ferrous oxalate. Also known as humboldtite; oxalite.

Humboldt jig [MIN ENG] An ore jig with a movable screen.

humic-cannel coal *See* pseudocannel coal.

humic coal [GEOL] A coal whose attritus is composed mainly of transparent humic degradation material.

humid climate [CLIMATOL] A climate whose typical vegetation is forest. Also known as forest climate.

humidity [METEOROL] Atmospheric water vapor content, expressed in any of several measures, such as relative humidity.

humidity coefficient [METEOROL] A measure of the precipitation effectiveness of a region; it recognizes the exponential relationship of temperature versus plant growth and is expressed as humidity coefficient $= P/(1.07)^t$, where P is the precipitation in centimeters, and t is the mean temperature in degrees Celsius for the period in question; the denominator approximately doubles with each 10°C rise in temperature.

humidity index [CLIMATOL] An index of the degree of water surplus over water need at any given station; it is calculated as humidity index $= 100s/n$, where s (the water surplus) is the sum of the monthly differences between precipitation and potential evapotranspiration for those months when the normal precipitation exceeds the latter, and where n (the water need) is the sum of monthly potential evapotranspiration for those months of surplus.

humidity mixing ratio [METEOROL] The amount of water vapor mixed with one unit mass of dry air, usually expressed as grams of water vapor per kilogram of air.

humidity province [CLIMATOL] A region in which the precipitation effectiveness of its climate produces a definite type of biological consequence, in particular the climatic climax formations of vegetation (rain forest, tundra, and the like).

humification [GEOL] Formation of humus.

humin *See* ulmin.

humite [MINERAL] **1.** A humic coal mineral. **2.** A series of magnesium neosilicate minerals closely related in crystal structure and chemical composition.

hummer screen [MIN ENG] An electrically vibrated ore screen for sizing moderately small material.

hummock [GEOL] A rounded or conical knoll, mound, hillock, or other small elevation, generally of equidimensional shape and not ridgelike. Also known as hammock. [HYD] A mound, hillock, or pile of broken floating ice, either fresh or weathered, that has been forced upward by pressure, as in an ice field or ice floe.

hummocked ice [OCEANOGR] Pressure ice, characterized by haphazardly arranged mounds or hillocks; it has less definite form, and show the effects of greater pressure, than either rafted ice or tented ice, but in fact may develop from either of those.

hummocky [GEOL] Any topographic surface characterized by rounded or conical mounds.

Humod [GEOL] A suborder of the soil order Spodosol containing aluminum-rich humus and no iron.

humodurite *See* translucent attritus.

humogelite *See* ulmin.

Humox [GEOL] A suborder of the soil order Oxisol that is high in organic matter, well drained but moist all or nearly all year, and restricted to relatively cool climates and high altitudes for Oxisols.

Humphrey's spiral [MIN ENG] An ore concentrator consisting of a stationary spiral trough through which ore pulp gravitates; heavy particles stay on the inside and lighter ones climb to the outside.

Humult [GEOL] A suborder of the soil order Ultisol having a moderately thick surface horizon and being well drained; formed under rather high rainfall distributed evenly over the year; common in southeastern Brazil.

humus [GEOL] The amorphous, ordinarily dark-colored, colloidal matter in soil; a complex of the fractions of organic matter of plant, animal, and microbial origin that are most resistant to decomposition.

Hunt continuous filter [MIN ENG] A continuous-vacuum filter consisting of a horizontally revolving, annular filter bed on which pulp is washed and then vacuum-dried.

Huntington-Heberlein process [MIN ENG] A sink-float process employing a galena medium, which is recovered by froth flotation.

huntite [MINERAL] $CaMg_3(CO_3)_4$ A white mineral consisting of calcium magnesium carbonate.

hurdle sheet [MIN ENG] A brattice-cloth screen across a roadway below a roof cavity or at the ripping lip to divert air current upward, thus diluting and removing firedamp.

hureaulite [MINERAL] $Mn_5H_2(PO_4)_4 \cdot 4H_2O$ A monoclinic mineral of varying colors consisting of a hydrated acid phosphate of manganese.

Huronian [GEOL] The lower system of the restricted Proterozoic.

hurricane [METEOROL] A tropical cyclone of great intensity; any wind reaching a speed of more than 73 miles per hour (117 kilometers per hour) is said to have hurricane force.

hurricane air stemmer [MIN ENG] A mechanical device for rapidly tamping shotholes; consists of a funnel connected by a T piece to the charge tube, with a connection to a compressed-air column; sand is put in the funnel and injected into the shothole as the charge tube is withdrawn.

hurricane band *See* spiral band.

hurricane beacon [ENG] An air-launched balloon designed to be released in the eye of a tropical cyclone, to float within the eye at predetermined levels, and to transmit radio signals.

hurricane-force wind [METEOROL] In the Beaufort wind scale, a wind whose speed is 64 knots (117 kilometers per hour) or higher.

hurricane monitoring buoy [METEOROL] A free-floating automatic weather station designed as an expendable instrument in connection with hurricane and typhoon monitoring and forecasting services.

hurricane radar band *See* spiral band.

hurricane surge *See* hurricane wave.

hurricane tide *See* hurricane wave.

hurricane tracking [ENG]   Recording of the movement of individual hurricanes by means of airplane sightings and satellite photography.

hurricane warning [METEOROL]   A warning of impending winds of hurricane force; for maritime interests, the stormwarning signals for this condition are two square red flags with black centers by day, and a white lantern between two red lanterns by night.

hurricane watch [METEOROL]   An announcement for a specific area that hurricane conditions pose a threat; residents are cautioned to take stock of their preparedness needs but, otherwise, are advised to continue normal activities.

hurricane wave [OCEANOGR]   As experienced on islands and along a shore, a sudden rise in the level of the sea associated with a hurricane. Also known as hurricane surge; hurricane tide.

hurricane wind [METEOROL]   In general, the severe wind of an intense tropical cyclone (hurricane or typhoon); the term has no further technical connotation, but is easily confused with the strictly defined hurricane-force wind. Also known as typhoon wind.

Hurst formula [PETRO ENG]   Relationship used in reservoir material-balance analysis; interrelates field pressure and production data at a number of different times.

Hurst method [PETRO ENG]   A calculation method for the bottom-hole static pressure of a well; uses graphical extrapolation of pressure buildup over a short period of time.

hutch [MIN ENG]   The bottom compartment of a jig used in ore dressing.

hutchinsonite [MINERAL] $(Pb,Tl)_2(Cu,Ag)As_5S_{10}$ Red mineral composed of sulfide of lead, copper, and arsenic, with varying amounts of thallium and silver, occurring in small orthorhombic crystals.

hutch product [MIN ENG]   Fine, heavy materials that pass through the screen of a jig and collect in the hutch.

huttonite [MINERAL] $ThSiO_4$ A colorless to pale-green monoclinic mineral composed of silicate of thorium; it is dimorphous with thorite.

Huwood loader [MIN ENG]   A machine consisting of a number of horizontal rotating flight bars working near the floor of the seam which push prepared coal up a ramp onto a low, bottom-loaded coveyor belt.

H variometer *See* horizontal intensity variometer.

hyacinth *See* zircon.

Hyaenodontidae [PALEON]   A family of extinct carnivorous mammals in the order Deltatheridia.

hyaline [GEOL]   Transparent and resembling glass.

hyalinocrystalline [PETR]   Of porphyritic rock texture, having the phenocrysts lying in a glassy ground mass.

hyalite [MINERAL]   A colorless, clear or translucent variety of opal occurring as globular concretions or botryoidal crusts in cavities or cracks of rocks. Also known as Müller's glass; water opal.

hyalobasalt *See* tachylite.

hyaloclastite [GEOL]   A tufflike deposit formed by the flowing of basalt under water and ice and its consequent fragmentation. Also known as aquagene tuff.

hyaloophitic [PETR]   Of the texture of igneous rocks, being composed principally of a glassy ground mass with little interstitial texture.

hyalophane [MINERAL] $BaAl_2Si_2O_8$ A colorless feldspar mineral crystallizing in the monoclinic system; isomorphous with adularia. Also known as baryta feldspar.

hyalopsite *See* obsidian.

Hyalospongia [PALEON] A class of extinct glass sponges, equivalent to the living Hexactinellida, having siliceous spicules made of opaline silica.

hyalotekite [MINERAL] $(Pb,Ca,Ba)_4BSi_6O_{17}(OH,F)$ A white gray mineral composed of borosilicate and fluoride of lead, barium, and calcium, occurring in crystalline masses.

Hybodontoidea [PALEON] An ancient suborder of extinct fossil sharks in the order Selachii.

hybrid [PETR] Pertaining to a rock formed by the assimilation of two magmas.

hydatogenesis [GEOL] Crystallization and deposition of minerals from aqueous solutions.

hydrargillite *See* gibbsite.

hydrated halloysite *See* endellite.

hydrated silica *See* silicic acid.

hydraulic blasting [MIN ENG] Fracturing coal by means of a hydraulic cartridge.

hydraulic bottom-hole pump [PETRO ENG] Liquid power–operated oil production pump; the liquid is oil, piped under pressure to the bottom of the well to operate the engine that drives the pump.

hydraulic cartridge [MIN ENG] A device used in mining to split coal or rock and having 8–12 small hydraulic rams in the sides of a steel cylinder.

hydraulic chock [MIN ENG] A steel face-support structure consisting of one to four hydraulic legs mounted in a steel frame with a large head and base plate.

hydraulic circulating system [MIN ENG] A method used to drill a borehole wherein water or a mud-laden liquid is circulated through the drill string.

hydraulic current [OCEANOGR] A current in a channel, due to a difference in the water level at the two ends.

hydraulic discharge [HYD] The direct discharge of groundwater from the zone of saturation upon the land or into a body of surface water.

hydraulic excavation *See* hydraulicking.

hydraulic extraction *See* hydraulicking.

hydraulic filling [MIN ENG] The use of water to wash waste material into stopes in order to prevent failure of rock walls and subsidence.

hydraulic flume transport [MIN ENG] The transport of coal, pulp, or minerals in water flowing in semicircular or rectangular channels.

hydraulic fracturing [PETRO ENG] A method in which sand-water mixtures are forced into underground wells under pressure; the pressure splits the petroleum-bearing sandstone, thereby allowing the oil to move toward the wells more freely.

hydraulic giant *See* hydraulic monitor.

hydraulicking [MIN ENG] Excavating alluvial or other mineral deposits by means of high-pressure water jets. Also known as hydraulic excavation; hydraulic extraction; hydroextraction.

hydraulic loading [MIN ENG] The flushing of coal or other material broken down by jets of water along the mine floor and into flumes.

hydraulic locomotive [MIN ENG] A diesel locomotive in which traction wheels are driven by hydraulic motors powered by a hydraulic system on the unit; used in mine haulage.

hydraulic mine [MIN ENG] A placer mine worked by means of a water stream directed against a bank.

hydraulic monitor [MIN ENG] A device for directing a high-pressure jet of water in hydraulicking; essentially, a swivel-mounted, counterweighted nozzle attached to a tripod or other type of stand and so designed that one worker can easily control and direct the vertical and lateral movements of the nozzle. Also known as giant; hydraulic giant; monitor.

hydraulic packer holddown [PETRO ENG] A pressure-actuated anchor located below a production packer (the seal between tubing and casing) to prevent well pressure from forcing the packer upward in the casing.

hydraulic profile [HYD] A vertical section of an aquifer's potentiometric surface.

hydraulic prop [MIN ENG] A supporting device consisting of two telescoping steel cylinders extended by hydraulic pressure provided by a built-in hand pump.

hydraulic ratio [GEOL] The weight of a heavy mineral multiplied by 100 and divided by the weight of a hydraulically equivalent light mineral.

hydrobasaluminite [MINERAL] $Al_4(SO_4)(OH)_{10} \cdot 36H_2O$ Mineral composed of a hydrous sulfate and hydroxide of aluminum.

hydrobiotite [MINERAL] A light-green, trioctahedral clay mineral of mixed layers of biotite and vermiculite.

hydroboracite [MINERAL] $CaMgB_6O_{11} \cdot 6H_2O$ A white mineral composed of hydrous calcium magnesium borate, occurring in fibrous and foliated masses.

hydrocalumite [MINERAL] $Ca_2Al(OH)_7 \cdot 3H_2O$ A colorless to light-green mineral composed of a hydrous hydroxide of calcium and aluminum.

hydrocarbon-mud log [PETRO ENG] Record of oil, gas, or cuttings released into mud during rock drilling; used to detect the presence of hydrocarbon-bearing strata.

hydrocarbon pore volume [PETRO ENG] The pore volume in a reservoir formation available to hydrocarbon intrusion.

hydrocarbon stabilization [PETRO ENG] The stepwise pressure reduction of a well stream to allow the release of dissolved gases until the liquid is stable at storage-tank conditions.

hydrocerussite [MINERAL] $Pb_3(OH)_2(CO_3)_2$ A colorless mineral composed of basic lead carbonate, occurring as crystals in thin hexagonal plates.

hydrocyanite See chalcocyanite.

hydroextraction See hydraulic king.

hydrogarnets [MINERAL] A group of minerals which has the general formula $A_3B_2(SiO_4)_{3-x}(OH)_{4x}$; isomorphous with certain garnets.

hydrogenic rock See aqueous rock.

hydrogeochemistry [GEOCHEM] The study of the chemical characteristics of ground and surface waters as related to areal and regional geology.

hydrogeology [HYD] The science dealing with the occurrence of surface and ground water, its utilization, and its functions in modifying the earth, primarily by erosion and deposition.

hydrograph [HYD] A graphical representation of stage, flow, velocity, or other characteristics of water at a given point as a function of time.

hydrographic chart [MAP]   A map designed from data obtained by hydrographic surveys for purposes of navigation.

hydrographic cruise [OCEANOGR]   Exploration of a body of water for hydrographic surveys.

hydrographic sextant [ENG]   A surveying sextant similar to those used for celestial navigation but smaller and lighter, constructed so that the maximum angle that can be read is slightly greater than that on the navigating sextant; usually the angles can be read only to the nearest minute by means of a vernier; it is fitted with a telescope with a large object glass and field of view. Also known as sounding sextant; surveying sextant.

hydrographic survey [OCEANOGR]   Survey of a water area with particular reference to tidal currents, submarine relief, and any adjacent land.

hydrographic table [OCEANOGR]   Tabular arrangement of data relating sea-water density to salinity, temperature, and pressure.

hydrography [GEOGR]   Science which deals with the measurement and description of the physical features of the oceans, lakes, rivers, and their adjoining coastal areas, with particular reference to their control and utilization.

hydrohalite [MINERAL] $Na_2Cl \cdot 2H_2O$ A mineral composed of hydrated sodium chloride, formed only from salty water cooled below $0°C$.

hydrohalloysite *See* endellite.

hydrohetaerolite [MINERAL] $Zn_2Mn_4O_8 \cdot H_2O$ A dark brown to brownish-black mineral consisting of a hydrated oxide of zinc and manganese; occurs in massive form.

hydrokaolin *See* endellite.

hydrolaccolith [GEOL]   A frost mound, 0.1–6 meters in height, having a core of ice and resembling a laccolith in section. Also known as cryolaccolith.

hydrolith [PETR]   **1.** A chemically precipitated aqueous rock, such as rock salt. **2.** A rock that is free of organic material.

hydrologic accounting [HYD]   A systematic summary of the terms (inflow, outflow, and storage) of the storage equation as applied to the computation of soil-moisture changes, groundwater changes, and so forth; an evaluation of the hydrologic balance of an area. Also known as basin accounting; water budget.

hydrologic cycle [HYD]   The complete cycle through which water passes, from the oceans, through the atmosphere, to the land, and back to the ocean. Also known as water cycle.

hydrologist [HYD]   An individual who specializes in hydrology.

hydrology [GEOPHYS]   The science that treats the occurrence, circulation, distribution, and properties of the waters of the earth, and their reaction with the environment.

hydrolyzate [GEOL]   A sediment characterized by elements such as aluminum, potassium, or sodium which are readily hydrolyzed.

hydromagnesite [MINERAL] $Mg_4(OH)_2(CO_3)_3 \cdot 3H_2O$ White, earthy mineral crystallizing in the monoclinic system and found in small crystals, amorphous masses, or chalky crusts.

hydrometamorphism [GEOL]   Alteration of rocks by material carried in solution by water without the influence of high temperature or pressure.

hydrometeor [HYD]   **1.** Any product of condensation or sublimation of atmospheric water vapor, whether formed in the free atmosphere or at the earth's surface. **2.** Any water particles blown by the wind from the earth's surface.

hydrometeorology   [METEOROL]   That part of meteorology of direct concern to hydrologic problems, particularly to flood control, hydroelectric power, irrigation, and similar fields of engineering and water resources.

hydromica   [GEOL]   Any of several varieties of muscovite, especially illite, which are less elastic than mica, have a pearly luster, and sometimes contain less potash and more water than muscovite. Also known as hydrous mica.

hydrophilite   *See* chlorocalcite.

hydrosphere   [HYD]   The water portion of the earth as distinguished from the solid part (lithosphere) and from the gaseous outer envelope (atmosphere).

hydrostatic approximation   [METEOROL]   The assumption that the atmosphere is in hydrostatic equilibrium.

hydrostatic assumption   [GEOPHYS]   **1.** The assumption that the pressure of seawater increases by 1 atmosphere (101,325 newtons per square meter) over approximately 10 meters of depth, the exact value depending on the water density and the local acceleration of gravity.   **2.** Specifically, the assumption that fluid is not undergoing vertical accelerations, hence the vertical component of the passive gradient force per unit mass is equal to g, the local acceleration due to gravity.

hydrostatic stability   *See* static stability.

hydrotalcite   [MINERAL] $Mg_6Al_2(OH)_{16}(CO_3)\cdot4H_2O$ Pearly-white mineral composed of hydrous aluminum and magnesium hydroxide and carbonate.

hydrothermal   [GEOL]   Of or pertaining to heated water, to its action, or to the products of such action.

hydrothermal alteration   [GEOL]   Rock or mineral phase changes that are caused by the interaction of hydrothermal liquids and wall rock.

hydrothermal deposit   [GEOL]   A mineral deposit precipitated from a hot, aqueous solution.

hydrothermal solution   [GEOL]   Hot, residual watery fluids derived from magmas during the later stages of their crystallization and commonly containing large amounts of dissolved metals which are deposited as ore veins in fissures along which the solutions often move.

hydrothermal synthesis   [GEOL]   Mineral synthesis in the presence of heated water.

hydrotroilite   [MINERAL] $FeS\cdot NH_2O$ A black, finely divided colloidal material reported in many muds and clays; thought to be formed by bacteria on bottoms of marine basins.

hydrotungstite   [MINERAL] $H_2WO_4\cdot H_2O$ A mineral composed of hydrous tungstic acid.

hydrous   [MINERAL]   Indicating a definite proportion of combined water.

hydrous mica   *See* hydromica.

Hydrox   [MIN ENG]   An explosive device, used in some English coal mines, consisting of a steel tube that has a thin steel shearing disk and is filled with a charge of ammonium chloride and sodium nitrate; when the charge is ignited, the powder is gasified and the disk is sheared.

hydroxylapatite   [MINERAL] $Ca_5(PO_4)_3OH$ A rare form of the apatite group that crystallizes in the hexagonal system.

hydroxylherderite   [MINERAL] $CaBe(PO_4)(OH)$ A monoclinic mineral composed of a phosphate and hydroxide of calcium and beryllium; isomorphous with herderite.

hydrozincite [MINERAL] $Zn_5(OH)_5(CO_3)_2$ A white, grayish, or yellowish mineral composed of basic zinc carbonate, occurring as masses or crusts.

Hyeniales [PALEOBOT]   An order of Devonian plants characterized by small, dichotomously forked leaves borne in whorls.

Hyeniopsida [PALEOBOT]   An extinct class of the division Equisetophyta.

hyetal coefficient *See* pluviometric coefficient.

hyetal equator [CLIMATOL]   A line (or transition zone) which encircles the earth (north of the geographical equator) and lies between two belts that typify the annual time distribution of rainfall in the lower latitudes of each hemisphere; a form of meteorological equator.

hyetal region [CLIMATOL]   A region in which the amount and seasonal variation of rainfall are of a given type.

hyetograph [CLIMATOL]   A map or chart displaying temporal or areal distribution of precipitation.

hyetography [CLIMATOL]   The study of the annual variation and geographic distribution of precipitation.

hyetology [METEOROL]   The science which treats of the origin, structure, and various other features of all the forms of precipitation.

hygrogram [ENG]   The record made by a hygrograph.

hygrograph [ENG]   A recording hygrometer.

hygrokinematics [METEOROL]   The descriptive study of the motion of water substances in the atmosphere.

hygrology [METEOROL]   The study which deals with the water vapor content (humidity) of the atmosphere.

hygrometer [ENG]   An instrument for giving a direct indication of the amount of moisture in the air or other gas, the indication usually being in terms of relative humidity as a percentage which the moisture present bears to the maximum amount of moisture that could be present at the location temperature without condensation taking place.

hygrometry [ENG]   The study which treats of the measurement of the humidity of the atmosphere and other gases.

hygroscopic coefficient [HYD]   The percentage of water that a soil will absorb and hold in equilibrium in a saturated atmosphere.

hygroscopic water [HYD]   The component of soil water that is held adsorbed on the surface of soil particles and is not available to vegetation.

hygrothermograph [ENG]   An instrument for recording temperature and humidity on a single chart.

Hyopssodontidae [PALEON]   A family of extinct mammalian herbivores in the order Condylarthra.

hypabyssal rock [PETR]   Those igneous rocks that rose from great depths as magmas but solidified as minor intrusions before reaching the surface.

hypautomorphic *See* hypidiomorphic.

hyperacoustic zone [GEOPHYS]   The region in the upper atmosphere, between 100 and 160 kilometers, where the distance between the rarefied air molecules roughly equals the wavelength of sound, so that sound is transmitted with less volume than at lower levels.

hyperbolic decline   [PETRO ENG]   One of three types of decline in oil or gas production rate (the others are constant-percentage and harmonic decline).

hypergene *See* supergene.

hyperpycnal inflow   [HYD]   A denser inflow that occurs when a sediment-laden fluid flows down the side of a basin and along the bottom as a turbidity current.

hypersaline   [GEOL]   Geologic material with high salinity.

hyperstereoscopy   [MAP]   Stereoscopic viewing of a map in which the scale (usually vertical) along the line of sight is exaggerated in comparison with the scale perpendicular to the line of sight. Also known as appearance ratio.

hypersthene   [MINERAL] $(Mg,Fe)SiO_3$ A grayish, greenish, black, or dark-brown rock-forming mineral of the orthopyroxene group, with bronzelike luster on the cleavage surface.

hypersthenfels *See* norite.

Hypertragulidae   [PALEON]   A family of extinct chevrotainlike pecoran ruminants in the superfamily Traguloidea.

hypidiomorphic   [PETR]   Of the texture of igneous rocks, having the crystals bounded partly by the crystal faces characteristic of the mineral species. Also known as hypautomorphic; subidiomorphic.

hypocrystalline   [PETR]   Pertaining to the texture of igneous rock characterized by crystalline components in an amorphous groundmass. Also known as hemicrystalline; hypohyaline; merocrystalline; miocrystalline; semicrystalline.

hypogene   [GEOL]   **1.** Of minerals or ores, formed by ascending waters. **2.** Of geologic processes, originating within or below the crust of the earth.

hypohyaline *See* hypocrystalline.

hypolimnion   [HYD]   The lower level of water in a stratified lake, characterized by a uniform temperature that is generally cooler than that of other strata in the lake.

hypomagma   [GEOL]   Relatively immobile, viscous lava that forms at depth beneath a shield volcano, is undersaturated with gases, and activates volcanic activity.

hypopycnal inflow   [HYD]   Flowing water of lower density than the body of water into which it flows.

hypothermal   [GEOL]   Referring to the high-temperature (300–500°C) environment of hypothermal deposits.

hypothermal deposit   [MINERAL]   Mineral deposit formed at great depths and high (300–500°C) temperatures.

hypsithermal interval   [PALEON]   A warm postglacial interval when the average annual temperatures were warmer than those of the present.

hypsographic map   [MAP]   A chart showing topographic relief in reference to a given datum, usually sea level.

hypsography   [GEOGR]   The science of measuring or describing elevations of the earth's surface with reference to a given datum, usually sea level.

hypsometer   [ENG]   An instrument for measuring atmospheric pressure to ascertain elevations by determining the boiling point of liquids.

hypsometric formula   [GEOPHYS]   A formula, based on the hydrostatic equation, for either determining the geopotential difference or thickness between any two pressure levels, or for reducing the pressure observed at a given level to that at some other level.

hypsometric map [MAP] In topographic surveying, a map giving elevations by contours, or sometimes by means of shading, tinting, or batching.

hypsometric tints *See* gradient tints.

hypsometry [ENG] The measuring of elevation with reference to sea level.

Hyracodontidae [PALEON] The running rhinoceroses, an extinct family of perissodactyl mammals in the superfamily Rhinoceratoidea.

Hystrichospherida [PALEON] A group of protistan microfossils.

I

IAC *See* international analysis code.

ianthinite [MINERAL] $2UO_2 \cdot 7H_2O$ A violet mineral composed of hydrous uranium dioxide, occurring as orthorhombic crystals.

Ibe wind [METEOROL] A local strong wind which blows through the Dzungarian Gate (western China), a gap in the mountain ridge separating the depression of Lakes Balkash and Ala Kul from that of Lake Ebi Nor; the wind resembles the foehn and brings a sudden rise of temperature, in winter from about -15 to about $30°F$ (-20 to $-1°C$).

ice accretion [HYD] The process by which a layer of ice builds up on solid objects which are exposed to freezing precipitation or to supercooled fog or cloud droplets.

ice-accretion indicator [ENG] An instrument used to detect the occurrence of freezing precipitation, usually consisting of a strip of sheet aluminum about 1½ inches (4 centimeters) wide, and is exposed horizontally, face up, in the free air a few meters above the ground.

ice age [GEOL] **1.** A major interval of geologic time during which extensive ice sheets (continental glaciers) formed over many parts of the world. **2.** *See* Pleistocene.

ice apron [HYD] **1.** The snow and ice attached to the walls of a cirque. **2.** The ice that is flowing from an ice sheet over the edge of a plateau. **3.** A piedmont glacier's lobe. **4.** Ice that adheres to a wall of a valley below a hanging glacier.

ice band [HYD] A layer of ice in firn or snow.

ice barrier [HYD] The periphery of the Antarctic ice sheet; or used generally for any ice dam.

ice bay [OCEANOGR] A baylike recess in the edge of a large ice floe or ice shelf. Also known as ice bight.

ice belt [OCEANOGR] A band of fragments of sea ice in otherwise open water. Also known as ice strip.

iceberg [OCEANOGR] A large mass of detached land ice floating in the sea or stranded in shallow water.

ice blink [METEOROL] A relatively bright, usually yellowish-white glare on the underside of a cloud layer, produced by light reflected from an ice-covered surface such as pack ice; used in polar regions with reference to the sky map; ice blink is not as bright as snow blink, but much brighter than water sky or land sky.

ice boundary [HYD] At any given time, the demarcation between fast ice and pack ice or between areas of pack ice having different ice concentrations.

ice bridge   [OCEANOGR]   Surface river ice of sufficient thickness to impede or prevent navigation.

ice cake   [HYD]   A single, usually relatively flat piece of ice of any size in a body of water.

ice calving *See* calving.

ice canopy *See* pack ice.

ice cap   [HYD]   **1.** A perennial cover of ice and snow in the shape of a dome or plate on the summit area of a mountain through which the mountain peaks emerge.   **2.** A perennial cover of ice and snow on a flat land mass such as an Arctic island.

ice-cap climate *See* perpetual frost climate.

ice cascade *See* icefall.

ice cave   [GEOL]   A cave that is cool enough to hold ice through all or most of the warm season.   [HYD]   A cave in ice such as a glacier formed by a stream of melted water.

ice clearing *See* polyn'ya.

ice concentration   [OCEANOGR]   In sea ice reporting, the ratio of the areal extent of ice present and the total areal extent of ice and water. Also known as ice cover.

ice-contact delta   [GEOL]   A delta formed by a stream flowing between a valley slope and the margin of glacial ice. Also known as delta moraine; morainal delta.

ice cover *See* ice concentration.

ice crust   [HYD]   A type of snow crust; a layer of ice, thicker than a film crust, upon a snow surface, formed by the freezing of meltwater or rainwater which has flowed onto it.

ice-crystal cloud   [METEOROL]   A cloud consisting entirely of ice crystals, such as cirrus (in this sense distinguished from water clouds and mixed clouds), and having a diffuse and fibrous appearance quite different from that typical of water droplet clouds.

ice-crystal fog *See* ice fog.

ice-crystal haze   [METEOROL]   A type of very light ice fog composed only of ice crystals and at times observable to altitudes as great as 20,000 feet (6100 meters), and usually associated with precipitation of ice crystals.

ice-crystal theory *See* Bergeron-Findeisen theory.

ice day   [CLIMATOL]   A day on which the maximum air temperature in a thermometer shelter does not rise above 32°F (0°C), and ice on the surface of water does not thaw.

ice desert   [CLIMATOL]   Any polar area permanently covered by ice and snow, with no vegetation other than occasional red snow or green snow.

iced firn   [HYD]   A mixture of glacier ice and firn; firn permeated with meltwater and then refrozen. Also known as firn ice.

ice erosion   [GEOL]   **1.** Erosion due to freezing of water in rock fractures.   **2.** *See* glacial erosion.

icefall   [HYD]   That portion of a glacier where a sudden steepening of descent causes a chaotic breaking up of the ice. Also known as ice cascade.

ice fat *See* grease ice.

ice feathers   [HYD]   A type of hoarfrost formed on the windward side of terrestrial objects and on aircraft flying from cold to warm air layers. Also known as frost feathers.

ice field [HYD] A mass of land ice resting on a mountain region and covering all but the highest peaks. [OCEANOGR] A flat sheet of sea ice that is more than 5 miles (8 kilometers) across.

ice floe *See* floe.

ice flowers [HYD] **1.** Formations of ice crystals on the surface of a quiet, slowly freezing body of water. **2.** Delicate tufts of hoarfrost that occasionally form in great abundance on an ice or snow surface. Also known as frost flowers. **3.** Frost crystals resembling a flower, formed on salt nuclei on the surface of sea ice as a result of rapid freezing of sea water. Also known as salt flowers.

ice fog [METEOROL] A type of fog composed of suspended particles of ice, partly ice crystals 20–100 micrometers in diameter but chiefly, especially when dense, droxtals 12–20 micrometers in diameter; occurs at very low temperatures and usually in clear, calm weather in high latitudes. Also known as frost flakes; frost fog; frozen fog; ice-crystal fog; pogonip; rime fog.

ice foot [OCEANOGR] Sea ice firmly frozen to a polar coast at the high-tide line and unaffected by tide; this fast ice is formed by the freezing of seawater during ebb tide, and of spray, and it is separated from the floating sea ice by a tide crack.

ice-free [HYD] **1.** Referring to a harbor, river, estuary, and so on, when there is not sufficient ice present to interfere with navigation. **2.** Descriptive of a water surface completely free of ice.

ice fringe [HYD] An ice deposit on plant surfaces, not of hoarfrost from atmospheric water vapor, but of moisture exuded from the stems of plants and appearing as frosted fringes or ribbons. Also known as ice ribbon. [OCEANOGR] A belt of sea ice extending a short distance from the shore.

ice gland [HYD] A column of ice in the granular snow at the top of a glacier.

ice gruel [HYD] A type of slush formed by the irregular freezing together of ice crystals.

ice island [OCEANOGR] A large tabular fragment of shelf ice found in the Arctic Ocean and having an irregular surface, thickness of 15–50 meters, and an area between a few thousand square meters and 500 square kilometers or more.

ice island iceberg [OCEANOGR] An iceberg having a conical or dome-shaped summit, often mistaken by mariners for ice-covered islands.

ice jam [HYD] **1.** An accumulation of broken river ice caught in a narrow channel, frequently producing local floods during a spring breakup. **2.** Fields of lake or sea ice thawed loose from the shores in early spring, and blown against the shore, sometimes exerting great pressure.

ice-laid drift *See* till.

Iceland agate *See* obsidian.

Iceland crystal *See* Iceland spar.

Icelandic low [METEOROL] **1.** The low-pressure center located near Iceland (mainly between Iceland and southern Greenland) on mean charts of sea-level pressure. **2.** On a synoptic chart, any low centered near Iceland.

Iceland spar [MINERAL] A pure, transparent form of calcite found particularly in Iceland; easily cleaved to form rhombohedral crystals that are doubly refracting. Also known as Iceland crystal.

ice layer [HYD] An ice crust covered with new snow; when exposed at a glacier front or in crevasses, the ice layers viewed in cross section are termed ice bands.

ice mantle *See* ice sheet.

ice mound *See* ground ice mound.

ice nucleus [METEOROL]   Any particle which may act as a nucleus in formation of ice crystals in the atmosphere.

ice pack *See* pack ice.

ice pellets [METEOROL]   A type of precipitation consisting of transparent or translucent pellets of ice 5 millimeters or less in diameter; may be spherical, irregular, or (rarely) conical in shape.

ice period [CLIMATOL]   The interval between the first appearance and the final dissipation of ice during any year in a given locale.

ice pillar [HYD]   A column of glacial ice covered with stones or debris which tend to protect the ice from melting.

ice pole [GEOGR]   The approximate center of the most consolidated portion of the arctic pack ice, near 83 or 84°N and 160°W. Also known as pole of inaccessibility.

ice push [GEOL]   Lateral pressure that is caused by expansion of shoreward-moving ice on a lake or a bay of the sea and that follows a rise in temperature. Also known as ice shove; ice thrust.

icequake [HYD]   The crash or concussion that accompanies breakup of ice masses, often due to contraction of the ice because of the extreme cold.

icerafting [GEOL]   The transporting of rock and other minerals, of all sizes, on or within icebergs, ice floes, river drift, or other forms of floating ice.

ice ribbon *See* ice fringe.

ice rind [HYD]   A thin but hard layer of sea ice, river ice, or lake ice, which is either a new encrustation upon old ice or a single layer of ice usually found in bays and fiords, where fresh water freezes on top of slightly colder sea water.

ice run [HYD]   The initial stage in the spring or summer breakup of river ice, being an exceedingly rapid process, seldom taking more than 1 day.

ice sheet [HYD]   A thick glacier, more than 50,000 square kilometers in area, forming a cover of ice and snow that is continuous over a land surface and moving outward in all directions. Also known as ice mantle.

ice shelf [OCEANOGR]   A thick sheet of ice with a fairly level or undulating surface, formed along a polar coast and in shallow bays and inlets, fastened to the shore along one side but mostly afloat and nourished by annual accumulation of snow and by the seaward extension of land glaciers.

ice shove *See* ice push.

ice spar *See* sanidine.

ice stone *See* cryolite.

ice storm [METEOROL]   A storm characterized by a fall of freezing precipitation, forming a glaze on terrestrial objects that creates many hazards. Also known as silver storm.

ice stream [HYD]   A current of ice flowing in an ice sheet or ice cap; usually moves toward an ocean or to an ice shelf.

ice strip *See* ice belt.

ice thrust *See* ice push.

ice tongue [HYD]   Any narrow extension of a glacier or ice shelf, such as a projection floating in the sea or an outlet glacier of an ice cap.

ice wedge *See* foliated ice.

ichnofossil *See* trace fossil.

ichor [GEOL] A fluid rich in mineralizers.

Ichthyodectidae [PALEON] A family of Cretaceous marine osteoglossiform fishes.

Ichthyopterygia [PALEON] A subclass of extinct Mesozoic reptiles composed of predatory fish-finned and sea-swimming forms with short necks and a porpoiselike body.

Ichthyornis [PALEON] The type genus of Ichthyornithidae.

Ichthyornithes [PALEON] A superorder of fossil birds of the order Ichthyornithiformes according to some systems of classification.

Ichthyornithidae [PALEON] A family of extinct birds in the order Ichthyornithiformes.

Ichthyornithiformes [PALEON] An order of ancient fossil birds including strong flying species from the Upper Cretaceous that possessed all skeletal characteristics of modern birds.

Ichthyosauria [PALEON] The only order of the reptilian subclass Ichthyopterygia, comprising the extinct predacious fish-lizards; all were adapted to a sea life in having tail flukes, paddles, and dorsal fins.

Ichthyostegalia [PALEON] An extinct Devonian order of labyrinthodont amphibians, the oldest known representatives of the class.

icicle [HYD] Ice shaped like a narrow cone, hanging point downward from a roof, fence, or other sheltered or heated source from which water flows and freezes in below-freezing air.

icing [HYD] **1.** Any deposit or coating of ice on an object, caused by the impingement and freezing of liquid (usually supercooled) hydrometeors. **2.** A mass or sheet of ice formed on the ground surface during the winter by successive freezing of sheets of water that may seep from the ground, from a river, or from a spring. Also known as flood icing; flooding ice.

icing level [METEOROL] The lowest level in the atmosphere at which an aircraft in flight does, or could, encounter aircraft icing conditions over a given locality.

ICL *See* lifting condensation level.

icositetrahedron *See* trapezohedron.

Ictidosauria [PALEON] An extinct order of mammallike reptiles in the subclass Synapsida including small carnivorous and herbivorous terrestrial forms.

iddingsite [MINERAL] A reddish-brown mixture of silicates, forming patches in basic igneous rocks.

ideal productivity index [PETRO ENG] Theoretical straight-line relationship between oil production from a reservoir and the resultant pressure drop within that reservoir.

idioblast [GEOL] A mineral constituent of a metamorphic rock formed by recrystallization which is bounded by its own crystal faces.

idiochromatic [MINERAL] Having characteristic color, usually applied to minerals.

idiomorphic *See* automorphic.

idocrase *See* vesuvianite.

idrialite [MINERAL] A mineral composed of crystalline hydrocarbon, $C_{22}H_{14}$.

IFR terminal minimums [METEOROL] The operational weather limits concerned with minimum conditions of ceiling and visibility at an airport under which aircraft may legally approach and land under instrument flight rules; these minimums fre quently are in the form of a sliding scale, and also vary with aircraft type, pilot experience, and from airport to airport.

IFR weather *See* instrument weather.

igneous [PETR]   Pertaining to rocks which have congealed from a molten mass.

igneous complex [PETR]   An assemblage of igneous rocks that are intimately associated and roughly contemporaneous.

igneous facies [PETR]   A part of an igneous rock differing in structure, texture, or composition from the main mass.

igneous meteor [GEOPHYS]   A visible electric discharge in the atmosphere; lightning is the most common and important type, but types of corona discharge are also included.

igneous mineral [MINERAL]   Mineral material forming igneous rock.

igneous petrology [PETR]   The study of igneous rocks, their occurrence, composition, and origin.

igneous province *See* petrographic province.

ignimbrite [PETR]   A silicic volcanic rock that forms thick, compact, lavalike sheets over a wide area of New Zealand. Also known as flood tuff.

igniter [ENG]   **1.** A device for igniting a fuel mixture. **2.** A charge, as of black powder, to facilitate ignition of a propelling or bursting charge.

igniter cord [ENG]   A cord which passes an intense flame along its length at a uniform rate to light safety fuses in succession.

IGY *See* International Geophysical Year.

ijolite [PETR]   A plutonic rock of nepheline and 30–60% mafic materials, generally sodic pyroxene, with accessory apatite, sphene, calcite, and titaniferous garnet.

ilesite [MINERAL] $(Mn,Zn,Fe)SO_4 \cdot 4H_2O$ A green mineral composed of hydrous manganese zinc iron sulfate.

Ilgner flywheel [MIN ENG]   A flywheel in the Ward-Leonard control of winding engines in mine hoists.

Ilgner system [MIN ENG]   A variation of the Ward-Leonard speed control system; the Ilgner flywheel, carried on the motor generator shaft, smooths out peak loads.

Illinoian [GEOL]   The third glaciation of the Pleistocene in North America, between the Yarmouth and Sangamon interglacial stages.

illite [MINERAL]   A group of gray, green, or yellowish-brown micalike clay minerals found in argillaceous sediments; intermediate in composition and structure between montmorillonite and muscovite.

illumination climate [METEOROL]   Also known as light climate. **1.** The worldwide distribution of natural light from the sun and sky (direct solar radiation plus diffuse sky radiation) as received on a horizontal surface. **2.** The character of total illumination at any given place.

illuvial [GEOL]   Pertaining to a region or material characterized by the accumulation of soil by the illuviation of another zone or material.

illuvial horizon *See* B horizon.

illuviation [GEOL]   The deposition of colloids, soluble salts, and small mineral particles in an underlying layer of soil.

illuvium [GEOL]   Material leached by chemical or other processes from one soil horizon and deposited in another.

ilmenite [MINERAL] $FeTiO_3$ An iron-black, opaque, rhombohedral mineral that is the principal ore of titanium. Also known as mohsite; titanic iron ore.

ilsemannite [MINERAL]   A black, blue-black, or blue mineral composed of hydrous molybdenum oxide or perhaps sulfate, occurring in earthy massive form.

imbricate structure [GEOL]   **1.** A sedimentary structure characterized by shingling of pebbles all inclined in the same direction with the upper edge of each leaning downstream or toward the sea. Also known as shingle structure. **2.** Tabular masses that overlap one another and are inclined in the same direction. Also known as schuppen structure; shingle-block structure.

imbrication [GEOL]   Formation of an imbricate structure. Also known as shingling.

imerinite [MINERAL] $Na_2(Mg,Fe)_6Si_8O_{22}(O,OH)_2$ A colorless to blue mineral composed of a basic silicate of sodium, iron, and magnesium, occurring as acicular crystals.

immature soil [GEOL]   A soil in which erosion has exceeded the rate at which the soil develops downward.

impactite [GEOL]   Glassy fused rock or meteor fragments resulting from heat of impact of a meteor on the earth.

impactor [MIN ENG]   A rotary hammermill which crushes ore by impacting it against crushing plates or elements.

impression [GEOL]   A form left on a soft soil surface by plant parts; the soil hardens and usually the imprint is a concave feature.

imprint *See* overprint.

impsonite [GEOL]   A black, asphaltic pyrobitumen with a high fixed-carbon content derived from the metamorphosis of petroleum.

inactive front [METEOROL]   A front, or portion thereof, that produces very little cloudiness and no precipitation, as opposed to an active front. Also known as passive front.

Inadunata [PALEON]   An extinct subclass of stalked Paleozoic Crinozoa characterized by branched or simple arms that were free and in no way incorporated into the calyx.

inby [MIN ENG]   Away from the shaft or mine entrance and therefore toward the working face.

incandescent tuff flow *See* ash flow.

incarbonization *See* coalification.

Inceptisol [GEOL]   A soil order in which materials other than carbonates or amorphous silica are altered or removed from the pedogenic horizon.

incised meander [GEOL]   A deep, tortuous valley cut by a meandering stream that was rejuvenated.

inclination [GEOL]   The angle at which a geological body or surface deviates from the horizontal or vertical; often used synonymously with dip.

inclination of the wind [METEOROL]   The angle between the direction of the wind and the isobars.

inclined bedding [GEOL]   A type of bedding in which the strata dip in the direction of current flow.

inclined contact [GEOL]   A contact plane of gas or oil with water underlying, in which the plane slopes or is inclined.

inclined drilling [ENG]   The drilling of blastholes at an angle with the vertical.

inclined skip hoist [MIN ENG] A skip hoist that operates on steeply inclined rails placed on a mine pit slope or wall.

incline man [MIN ENG] In anthracite coal mining, bituminous coal mining, and metal mining, a laborer who controls the movement of cars on a self-acting incline.

incline shaft [MIN ENG] A shaft which has been dug at an angle to the vertical to follow the depth of the lode.

inclinometer [ENG] **1.** An instrument for measuring the angle between the earth's magnetic field vector and the horizontal plane. **2.** An apparatus used to ascertain the direction of the magnetic field of the earth with reference to the plane of the horizon.

inclusion [CRYSTAL] **1.** A crystal or fragment of a crystal found in another crystal. **2.** A small cavity filled with gas or liquid in a crystal.   [PETR] A fragment of older rock enclosed in an igneous rock.

incoalation *See* coalification.

incoherent [GEOL] Pertaining to a rock or deposit that is loose or unconsolidated or that is unable to hold together firmly or solidly.

incompetent bed [GEOL] A bed not combining sufficient firmness and flexibility to transmit a thrust and to lift a load by bending.

incompetent rock [ENG] Soft or fragmented rock in which an opening, such as a borehole or an underground working place, cannot be maintained unless artificially supported by casing, cementing, or timbering.

incongruous [GEOL] Of a drag fold, having an axis and axial surface not parallel to the axis and axial surface of the main fold to which it is related.

increment [HYD]   *See* recharge.

incumbent [GEOL] Lying above, said of a stratum that is superimposed or overlies another stratum.

incus [METEOROL] A supplementary cloud feature peculiar to cumulonimbus capillatus; the spreading of the upper portion of cumulonimbus when this part takes the form of an anvil with a fibrous or smooth aspect. Also known as anvil; thunderhead.

indefinite ceiling [METEOROL] After United States weather observing practice, the ceiling classification applied when the reported ceiling value represents the vertical visibility upward into surface-based, atmospheric phenomena (except precipitation), such as fog, blowing snow, and all of the lithometeors. Formerly known as ragged ceiling.

inderborite [MINERAL] $CaMgB_6O_{11} \cdot 11H_2O$ A monoclinic mineral composed of hydrous calcium and magnesium borate.

inderite [MINERAL] $Mg_2B_6O_{11} \cdot 15H_2O$ A hydrated borate mineral.

index cycle [METEOROL] A roughly cyclic variation in the zonal index.

index fossil [PALEON] The ancient remains and traces of an organism that lived during a particular geologic time and that geologically date the containing rocks.

index mineral [PETR] A mineral whose first appearance in passing from low to higher grades of metamorphism indicates the outer limit of a zone.

index of aridity [CLIMATOL] A measure of the precipitation effectiveness or aridity of a region, given by the following relationship: index of aridity = $P/(T + 10)$, where P is the annual precipitation in centimeters, and T the annual mean temperature in degrees Celsius.

index plane [GEOL] A surface used as a reference point in determining geological structure.

indialite [MINERAL] $Mg_2Al_4Si_5O_{18}$ A hexagonal cordierite mineral; it is isotypic with beryl.

indianaite [MINERAL] A white porcelainlike clay mineral; a variety of halloysite found in Indiana.

Indiana limestone *See* spergenite.

Indian Ocean [GEOGR] The smallest and geologically the most youthful of the three oceans, whose surface area is 75,900,000 square kilometers; it is bounded on the north by India, Pakistan, and Iran; on the east by the Malay Peninsula; on the south by Antarctica; and on the west by the Arabian peninsula and Africa.

Indian spring low water [OCEANOGR] An arbitrary tidal datum approximating the level of the mean of the lower low waters at spring time, first used in waters surrounding India. Also known as harmonic tide plane; Indian tide plane.

Indian summer [CLIMATOL] A period, in mid- or late autumn, of abnormally warm weather, generally clear skies, sunny but hazy days, and cool nights; in New England, at least one killing frost and preferably a substantial period of normally cool weather must precede this warm spell in order for it to be considered a true Indian summer; it does not occur every year, and in some years there may be two or three Indian summers; the term is most often heard in the northeastern United States, but its usage extends throughout English-speaking countries.

Indian tide plane *See* Indian spring low water.

indicated air temperature [METEOROL] The uncorrected reading from the free air temperature gage. Also known as outside air temperature.

indicated ore [MIN ENG] A known mineral deposit for which quantitative estimates are made partly from inference and partly from specific sampling. Also known as probable ore.

indicolite [MINERAL] An indigo-blue variety of tourmaline that is used as a gemstone. Also known as indigolite.

indigenous coal *See* autochthonous coal.

indigo copper *See* covellite.

indirect cell [METEOROL] A closed circulation in a vertical plane in which the rising motion occurs at lower potential temperature than the descending motion, thus forming an energy sink.

indirect stratification *See* secondary stratification.

induced magnetization [GEOPHYS] That component of a rock's magnetization which is proportional to, and has the same direction as, the ambient magnetic field.

induction inclinometer *See* earth inductor.

induction method [GEOPHYS] In studies of the radioactivity of the atmosphere, a technique for estimating the concentration of the radioactive gases by exposing a negatively charged wire to the air and then using an ionization chamber to count the activity of the radioactive deposit formed on the wire.

induration [GEOL] **1.** The hardening of a rock material by the application of heat or pressure or by the introduction of a cementing material. **2.** A hardened mass formed by such processes. **3.** The hardening of a soil horizon by chemical action to form a hardpan.

industrial climatology [CLIMATOL] A type of applied climatology which studies the effect of climate and weather on industry's operations; the goal is to provide industry with a sound statistical basis for all administrative and operational decisions which involve a weather factor.

Industrial diamond [MINERAL]   Diamond that is too hard or too radial-grained to be used for jewel cutting.

industrial geography [GEOGR]   A branch of geography that deals with location, raw materials, products, and distribution, as influenced by geography.

industrial jewel [MINERAL]   A hard stone, such as ruby or sapphire, used for bearings and impulse pins in instruments and for recording needles.

industrial meteorology [METEOROL]   The application of meteorological information and techniques to industrial problems.

inertia currents [OCEANOGR]   Currents resulting after the cessation of wind in a generating area or after the water movement has left the generating area; circular currents with a period of one-half pendulum day.

inertial circle [METEOROL]   A loop in the path of an air parcel in inertial flow, which is approximately circular if the latitudinal displacement is small. Also known as circle of inertia.   [OCEANOGR]   The circle described by inertial motion in a body of ocean water and having a radius $R = C/f$, where C is the particle velocity in a given direction and f is the Coriolis parameter.

inertial flow [GEOPHYS]   Frictionless flow in a geopotential surface in which there is no pressure gradient; the centrifugal and Coriolis accelerations must therefore be equal and opposite, and the constant inertial wind speed V_i is given by $V_i = fR$, where f is the Coriolis parameter and R the radius of curvature of the path.

inertial theory [OCEANOGR]   The theory concerning motion of an ocean current under influences of inertia and Coriolis force, causing it to take a circular path.

inertia period [OCEANOGR]   The time required for a given particle to complete an inertia circle.

inertinite [GEOL]   A carbon-rich maceral group, which includes micrinite, sclerotinite, fusinite, and semifusinite.

inesite [MINERAL] $Ca_2Mn_7Si_{10}O_{28}(OH)_2 \cdot 5H_2O$ A pale-red mineral composed of hydrous manganese calcium silicate, occurring in small prismatic crystals or massive.

infancy [GEOL]   The initial (youthful) or very early stage of the cycle of erosion characterized by smooth, nearly level erosional surfaces dissected by narrow stream gorges, numerous depressions filled by marshy lakes and ponds, and shallow streams. Also known as topographic infancy.

inferred ore [MIN ENG]   An ore whose estimate of tonnage and grade is based largely on knowledge of the deposit's geological character and to a lesser degree on samples and other data.

infiltration [GEOL]   Deposition of mineral matter among the pores or grains of a rock by permeation of water carrying the matter in solution.   [HYD]   Movement of water through the soil surface into the ground.

infiltration capacity [HYD]   The maximum rate at which water enters the soil or other porous material in a given condition.

infiltration vein [GEOL]   Vein deposited in rock by percolating water.

infinite aquifer [HYD]   The portion of a formation that contains water, and for which the exterior boundary is at an effectively infinite distance from the oil reservoir.

infinite reservoir [PETRO ENG]   In reservoir unsteady-state liquid-diffusion calculations, a reservoir in which the outer boundary is considered to be effectively at an infinite distance from the inner boundary at the well of an aquifer.

inflatable packer [PETRO ENG] A packer (downhole pressure seal between tubing and casing) set and held in place by an element that is inflated with hydraulic pressure.

influence function [PETRO ENG] Mathematical statement of the influence on pressure and production of each oil reservoir pool in a multipool aquifer.

influent stream [HYD] A stream that contributes water to the zone of saturation of groundwater and develops bank storage. Also known as losing stream.

influx *See* mouth.

Infracambrian *See* Eocambrian.

infralateral tangent arcs [METEOROL] Two oblique, colored arcs, convex toward the sun and tangent to the halo of 46° at points below the altitude of the sun, produced by refraction (90° effective prism angle) in hexagonal columnar ice crystals whose principal axes are horizontal but randomly directed in azimuth; if the sun's elevation exceeds about 68°, the arcs cannot appear.

infrared window [GEOPHYS] A frequency region in the infrared where there is good transmission of electromagnetic radiation through the atmosphere.

infusorial earth [GEOL] Formerly, and incorrectly, a soft rock or an earthy substance composed of siliceous remains of diatoms.

Ingersoll-Rand jumbo columns [MIN ENG] Columns held in place by the pressure of air-operated pistons against the roof; drills are attached to movable arms mounted on the columns.

ingrown meander [GEOL] A meander of a stream with an undercut bank on one side and a gentle slope on the other.

inherent bursts [MIN ENG] Rock bursts that occur in development.

initial condition [METEOROL] A prescription of the state of a dynamical system at some specified time; for all subsequent times the equations of motion and boundary conditions determine the state of the system; the appropriate synoptic weather charts, for example, constitute a (discrete) set of initial conditions for a forecast; in many contexts, initial conditions are considered as boundary conditions in the dimension of time.

initial detention *See* surface storage.

initial dip *See* primary dip.

initial saturation [PETRO ENG] A reservoir's initial relative content (saturation) of water, oil, and gas.

injected [PETR] Pertaining to intrusive igneous rock or other mobile rock that has erupted through rock walls to neighboring older rocks.

injected gas [PETRO ENG] Gas that has been pumped into an oil-producing reservoir to provide a gas-drive for increased oil production.

injected hole [MIN ENG] A borehole into which a cement slurry or grout has been forced by high-pressure pumps and allowed to harden.

injection [GEOL] Also known as intrusion; sedimentary injection. **1.** A process by which sedimentary material is forced under abnormal pressure into a preexisting rock or deposit. **2.** A structure formed by an injection process. [MIN ENG] The introduction under pressure of a liquid or plastic material into cracks, cavities, or pores in a rock formation.

injection fluid [PETRO ENG]   Gas or water, depending on the nature of the reservoir and its fluid content, for injection into the formation to increase hydrocarbon production.

injection-fluid front [PETRO ENG]   The moving interfacial contact between an injected fluid (gas or water) and the natural fluid content of the reservoir formation.

injection gas-fluid ratio [PETRO ENG]   The ratio of gas injected into a reservoir formation to the fluid hydrocarbons produced by the resultant gas lift.

injection gneiss [PETR]   A composite rock with banding entirely or partly caused by layer-by-layer injection of granitic magma into rock layers.

injection pressure [PETRO ENG]   Pressure of fluid injected into oil formations for waterflood (water) or pressure maintenance (gas).

injection well [PETRO ENG]   In secondary recovery of petroleum, a well in which a fluid such as gas or water is injected to provide supplemental energy to drive the oil remaining in the reservoir to the vicinity of production wells.

injection-well plugging [PETRO ENG]   Plugging of the sand face of an injection well because of lubricant or corrosion-product carryover from surface lines or well equipment.

injectivity index [PETRO ENG]   The number of barrels per day of gross liquid pumped into an injection well per psi (pound per square inch) pressure differential between the mean injection pressure and the mean formation pressure.

injectivity test [PETRO ENG]   A test series of reservoir water injection rates at different pressures to predict the performance of an injection well.

inland [GEOGR]   Interior land, not bordered by the sea.

inland ice [HYD]   Ice composing the inner portion of a continental glacier or large ice sheet; applied particularly to Greenland ice.

inland sea *See* epicontinental sea.

inland water [GEOGR]   **1.** A lake, river, or other body of water wholly within the boundaries of a state. **2.** An interior body of water not bordered by the sea.

inlet [GEOGR]   **1.** A short, narrow waterway connecting a bay or lagoon with the sea. **2.** A recess or bay in the shore of a body of water. **3.** A waterway flowing into a larger body of water.

inlier [GEOL]   A circular or elliptical area of older rocks surrounded by strata that are younger.

inner barrel *See* inner tube.

inner harbor [GEOGR]   The part of a harbor more remote from the sea, as contrasted with the outer harbor; this expression is normally used only in a harbor that is clearly divided into parts, by a narrow passageway or artificial structure; the inner harbor generally has additional protection and is often the principal berthing area.

inner mantle *See* lower mantle.

inner tube [MIN ENG]   The inside tube which acts as the core container of a double-tube core barrel; used to obtain core samples for analysis of an ore formation. Also known as inner barrel.

inner-tube extension *See* lifter case.

inorganic chert [PETR]   Chert derived from siliceous colloids precipitated from silica-saturated waters.

inosilicate [GEOL]   A class or structural type of silicate in which the SiO_4 tetrahedrons are linked together by the sharing of oxygens to form linear chains of indefinite length.

in-place stress field *See* ambient stress field.

inselberg [GEOL] A large, steep-sided residual hill, knob, or mountain, generally rocky and bare, rising abruptly from an extensive, nearly level lowland erosion surface in arid or semiarid regions. Also known as island mountain.

insequent stream [HYD] A stream that has developed on the present surface, but not consequent upon it, and seemingly not controlled or adjusted by the rock structure and surface features.

insert pump *See* rod pump.

inshore [GEOGR] **1.** Located near the shore. **2.** Indicating a shoreward position.

inshore current [OCEANOGR] The horizontal movement of water inside the surf zone, including longshore and rip currents.

inshore zone [GEOL] The zone of variable width extending from the shoreline at low tide through the breaker zone.

in situ combustion [PETRO ENG] A method of driving high-viscosity, low-gravity ore otherwise unrecoverable from a formation by setting fire to the oil sand and thereby heating the oil in the horizon to increase its mobility by reducing its viscosity.

insoluble residue [GEOL] Material remaining after a geological specimen is dissolved in hydrochloric or acetic acid.

inspector [MIN ENG] One employed to make examinations of and to report upon mines and surface plants relative to compliance with mining laws, rules and regulations, and safety methods.

instability line [METEOROL] Any nonfrontal line or band of convective activity in the atmosphere; this is the general term and includes the developing, mature, and dissipating stages; however, when the mature stage consists of a line of active thunderstorms, it is properly termed a squall line; therefore, in practice, instability line often refers only to the less active phases.

instantaneous detonator [ENG] A type of detonator that does not have a delay period between the passage of the electric current through the detonator and its explosion.

instantaneous fuse [ENG] A fuse with an ignition rate of several thousand feet per minute; an example is PETN.

instrument weather [METEOROL] Route or terminal weather conditions of sufficiently low visibility to require the operation of aircraft under instrument flight rules (IFR). Also known as IFR weather.

intake [HYD] *See* recharge.

integral-joint casing [PETRO ENG] Oil well casing lengths on whose ends the connection joints are integrally formed.

integrated drainage [HYD] Drainage resulting after folding and faulting of a surface under arid conditions; the streams by working headward have joined basins across intervening mountains or ridges.

integrated train [MIN ENG] A long string of cars, permanently coupled together, that shuttles endlessly between one mine and one generating plant, not even stopping to load and unload, since rotary couplers permit each car to be flipped over and dumped as the train moves slowly across a trestle.

intensifier [PETRO ENG] Hydrofluoric acid added to hydrochloric acid for oil well acidizing; the fluoride destroys silica films that are insoluble by hydrochloric acid.

interbedded [GEOL] Having beds lying between other beds with different characteristics.

intercalation [GEOL]   A layer located between layers of different character.

intercardinal point [GEOD]   Any of the four directions midway between the cardinal points, that is, northeast, southeast, southwest, and northwest. Also known as quadrantal point.

intercept [CRYSTAL]   One of the distances that are cut off a crystal's reference axis by planes.   [MAP]   *See* altitude difference.

interception [HYD]   **1.** The process by which precipitation is caught and retained on vegetation or structures and subsequently evaporated without reaching the ground. **2.** That part of the precipitation intercepted by vegetation.   [METEOROL]   **1.** The loss of sunshine, a part of which may be intercepted by hills, trees, or tall buildings. **2.** The depletion of part of the solar spectrum by atmospheric gases and suspensoids; this commonly refers to the absorption of ultraviolet radiation by ozone and dust.

interceptometer [ENG]   A rain gage which is placed under trees or in foliage to determine the rainfall in that location; by comparing this catch with that from a rain gage set in the open, the amount of rainfall which has been intercepted by foliage is found.

intercloud discharge *See* cloud-to-cloud discharge.

intercommunication [PETRO ENG]   Flow interconnection between the reservoir areas being drained by adjacent wells.

intercontinental sea [GEOGR]   A large body of salt water extending between two continents.

interface [GEOPHYS]   *See* seismic discontinuity.

interfacial angle [CRYSTAL]   The angle between two crystal faces.

interference ripple mark [GEOL]   A pattern resulting from two sets of symmetrical ripples formed by waves crossing at right angles.

interference test [PETRO ENG]   Test of pressure interrelationships (interference) between wells serving the same formation.

interflow [HYD]   The water, derived from precipitation, that infiltrates the soil surface and then moves laterally through the upper layers of soil above the water table until it reaches a stream channel or returns to the surface at some point downslope from its point of infiltration.

intergelisol *See* pereletok.

interglacial [GEOL]   Pertaining to or formed during a period of geologic time between two successive glacial epochs or between two glacial stages.

intergrowth [MINERAL]   A state of interlocking of different mineral crystals because of simultaneous cyrstallization.

interlobate moraine *See* intermediate moraine.

intermediate haulage *See* relay haulage.

intermediate haulage conveyor [MIN ENG]   A type of conveyor, usually 500 to 3000 feet (150 to 900 meters) in length, that transports material between the gathering conveyor and the main haulage conveyor.

intermediate ion [METEOROL]   An atmospheric ion of size and mobility intermediate between the small ion and the large ion.

intermediate layer *See* sima.

intermediate moraine [GEOL]   A type of lateral moraine formed at the junction of two adjacent glacial lobes. Also known as interlobate moraine.

intermittent current [OCEANOGR] A unidirectional current interrupted at intervals.

intermittent gas lift [PETRO ENG] A gas-drive oil reservoir that is valved and timed for intermittent activity.

intermittent spring [HYD] A spring that ceases flow after a long dry spell but flows again after heavy rains.

intermittent stream [HYD] A stream which carries water a considerable portion of the time, but which ceases to flow occasionally or seasonally because bed seepage and evapotranspiration exceed the available water supply.

intermontane [GEOL] Located between or surrounded by mountains.

intermontane glacier [GEOL] A glacier that is formed by the confluence of several valley glaciers and occupies a trough between separate ranges of mountains.

intermontane trough [GEOL] **1.** A subsiding area in an island arc of the ocean, lying between the stable elements of a region. **2.** A basinlike area between mountains.

internal cast *See* steinkern.

internal drift current [OCEANOGR] Motion in an underlying layer of water caused by shearing stresses and friction created by current in a top layer that has different density.

internal gas drive [PETRO ENG] A primary oil recovery process in which oil is displaced from the reservoir by the expansion of the gas originally dissolved in the liquid. Also known as dissolved-gas drive; gas depletion drive; solution gas drive.

international analysis code [METEOROL] An internationally recognized code for communicating details of synoptic chart analyses. Abbreviated IAC.

international ellipsoid of reference [GEOD] The reference ellipsoid, based upon the Hayford spheroid, the semimajor axis of which is $6,378,388$ meters; the flattening or ellipticity equals $1/297$; by computation the semiminor axis is $6,356,911.946$ meters. Also known as international spheroid.

International Geophysical Year [GEOPHYS] An internationally accepted period, extending from July 1957 through December 1958, for concentrated and coordinated geophysical exploration, primarily of the solar and terrestrial atmospheres. Abbreviated IGY.

international gravity formula [GEOD] A formula for the acceleration of gravity at the earth's surface, stating that acceleration of gravity is equal to $9.780318[1 + 5.3024 \times 10^{-3} \sin^2 \phi - 5.8 \times 10^{-6} \sin^2 2\phi]$ m/s^2, where ϕ is the latitude.

International Ice Patrol [OCEANOGR] An organization established in 1914 to protect shipping by providing iceberg warnings.

international index numbers [METEOROL] A system of designating meteorological observing stations by number, established and administered by the World Meteorological Organization; under this scheme, specified areas of the world are divided into blocks, each bearing a two-number designator; stations within each block have an additional unique three-number designator, the numbers generally increasing from east to west and from south to north.

International Polar Year [METEOROL] The years 1882 and 1932, during which participating nations undertook increased observations of geophysical phenomena in polar (mostly arctic) regions; the observations were largely meteorological, but included such as auroral and magnetic studies.

International Quiet Sun Year [GEOPHYS] An international cooperative effort, similar to the International Geophysical Year and extending through 1964 and 1965, to study the sun and its terrestrial and planetary effects during the minimum of the 11-year cycle of solar activity. Abbreviated IQSY. Also known as the International Year of the Quiet Sun.

international spheroid *See* international ellipsoid of reference.

international synoptic code [METEOROL] A synoptic code approved by the World Meteorological Organization in which the observable meteorological elements are encoded and transmitted in words of five numerical digits length.

internides [GEOL] The internal part of an orogenic belt, farthest away from the craton, commonly the site of a eugeosyncline during its early phases, and subjected later to plastic folding and plutonism. Also known as primary arc.

interpenetration twin [CRYSTAL] Two or more individual crystals so twinned that they appear to have grown through one another. Also known as penetration twin.

interpluvial [GEOL] Pertaining to an episode or period of geologic time that was dryer than the pluvial period occurring before or after it.

intersertal [PETR] Referring to the texture of a porphyritic igneous rock in which the groundmass forms a small proportion of the rock, filling the interstices between unoriented feldspar laths.

interstadial [GEOL] Pertaining to a period during a glacial stage in which the ice retreated temporarily.

interstitial [CRYSTAL] A crystal defect in which an atom occupies a position between the regular lattice positions of a crystal.

interstitial atom [CRYSTAL] A displaced atom which is forced into a nonequilibrium site within a crystal lattice.

interstitial water [HYD] Subsurface water contained in pore spaces between the grains of rock and sediments.

interstitial water saturation [HYD] The water content of a subterranean reservoir formation.

intertidal zone [OCEANOGR] The part of the littoral zone above low-tide mark.

intertropical convergence zone [METEOROL] The axis, or a portion thereof, of the broad trade-wind current of the tropics; this axis is the dividing line between the southeast trades and the northeast trades (of the Southern and Northern hemispheres, respectively). Also known as equatorial convergence zone; meteorological equator.

intertropical front [METEOROL] The interface or transition zone occurring within the equatorial trough between the Northern and Southern hemispheres. Also known as equatorial front; tropical front.

interurban [GEOGR] Connecting or extending between urban areas.

in-the-seam mining [MIN ENG] The usual method of mining characterized by the driving of development shafts into the coal seam.

into the solid [MIN ENG] Of a shot, going into the coal beyond the point to which the coal can be broken by the blast. Also known as on the solid.

intraclast [GEOL] A fragment of limestone formed by erosion within a basin of deposition and redeposited there to form a new sediment.

intracloud discharge *See* cloud discharge.

intraformational breccia [PETR] A rock resulting from cracking and desiccation-shrinking of a mud after withdrawal of water followed by almost contemporaneous sedimentation.

intraformational conglomerate [GEOL] **1.** A conglomerate in which clasts and the matrix are contemporaneous in origin. **2.** A conglomerate formed in the midst of a geologic formation.

intrastratal solution [GEOCHEM] A chemical attrition of the constituents of a rock after deposition.

intratelluric [GEOL] **1.** Pertaining to a phenocryst that is formed earlier than its matrix. **2.** Pertaining to a period in which igneous rocks crystallized prior to their eruption. **3.** Located, formed, or originating at great depths within the earth.

intrazonal soil [GEOL] A group of soils with well-developed characteristics that reflect the dominant influence of some local factor of relief, parent material, or age over the usual effect of vegetation and climate.

introductory column [MIN ENG] The highest and first column that is inserted in casing a borehole.

intrusion [GEOL] **1.** The process of emplacement of magma in preexisting rock. Also known as injection; invasion; irruption. **2.** A large-scale sedimentary injection. Also known as sedimentary intrusion. **3.** Any rock mass formed by an intrusive process.

intrusive [PETR] Pertaining to material forced while still in a fluid state into cracks or between layers of rock.

inundation [HYD] Flooding, by the rise and spread of water, of a land surface that is not normally submerged.

invaded zone [PETRO ENG] Transitional downhole area between the area invaded completely by drilling mud and uncontaminated bulk of the reservoir.

invasion [GEOL] **1.** The movement of one material into a porous reservoir area that has been occupied by another material. **2.** *See* intrusion; transgression.

invasion efficiency [PETRO ENG] Completeness of invasion of a reservoir formation by a fluid.

inverna [METEOROL] A southeast wind of Lake Maggiore, Italy.

inverse cylindrical orthomorphic chart *See* transverse Mercator chart.

inverse cylindrical orthomorphic projection *See* transverse Mercator projection.

inverse Mercator chart *See* transverse Mercator chart.

inverse Mercator projection *See* transverse Mercator projection.

inverse parallel *See* transverse parallel.

inverse rhumb line *See* transverse rhumb line.

inversion [CRYSTAL] A change from one crystal polymorph to another. Also known as transformation. [GEOL] **1.** Development of inverted relief through which anticlines are transformed into valleys and synclines are changed into mountains. **2.** The occupancy by a lava flow of a ravine or valley that occurred in the side of a volcano. [METEOROL] A departure from the usual decrease or increase with altitude of the value of an atmospheric property, most commonly temperature.

inversion axis *See* rotation-inversion axis.

inversion layer [METEOROL] The atmosphere layer through which an inversion occurs.

inverted *See* overturned.

inverted heading and bench method *See* heading–overhand bench method.

inverted plunge [GEOL]   A plunge of a fold whose inclination has been carried past the vertical, so that the plunge is less than 90° in the direction opposite from the original attitude; younger rocks plunge beneath the older rocks.

inyoite [MINERAL] $Ca_2B_6O_{11} \cdot 13H_2O$ A colorless, monoclinic mineral consisting of a hydrous calcium borate; hardness is 2 on Mohs scale, and specific gravity is 2.

iodargyrite [MINERAL]   AgI   A yellowish or greenish hexagonal mineral composed of native silver iodide, usually occurring in thin plates. Also known as iodyrite.

iodobromite [MINERAL]   Ag(Br,Cl,I)   An isometric mineral composed of chloride, iodide, and bromide of silver; it is isomorphous with cerargyrite and bromyrite.

iodyrite *See* iodargyrite.

ion cloud [GEOPHYS]   An inhomogeneity or patch of unusually great ion density in one of the regular regions of the ionosphere; such patches occur quite often in the E region.

ion column [GEOPHYS]   The trail of ionized gases in the trajectory of a meteoroid entering the upper atmosphere; a part of the composite phenomenon known as a meteor. Also known as meteor trail.

ionic crystal [CRYSTAL]   A crystal in which the lattice-site occupants are charged ions held together primarily by their electrostatic interaction.

ionic lattice [CRYSTAL]   The lattice of an ionic crystal.

ionic ratios [OCEANOGR]   The ratios by weight of major constituents of seawater to the chloride ion content; for example, $SO_4/Cl = 0.1396$, $Ca/Cl = 0.02150$, $Mg/Cl = 0.06694$.

ionite *See* anauxite.

ionized layers [GEOPHYS]   Layers of increased ionization within the ionosphere produced by cosmic radiation; responsible for absorption and reflection of radio waves and important in connection with communications and tracking of satellites and other space vehicles.

ionosonde [ENG]   A radar system for determining the vertical height at which the ionosphere reflects signals back to earth at various frequencies; a pulsed vertical beam is swept periodically through a frequency range from 0.5 to 20 megahertz, and the variation of echo return time with frequency is photographically recorded.

ionosphere [GEOPHYS]   That part of the earth's upper atmosphere which is sufficiently ionized by solar ultraviolet radiation so that the concentration of free electrons affects the propagation of radio waves; its base is at about 70 or 80 kilometers and it extends to an indefinite height.

ionospheric disturbance [GEOPHYS]   A temporal variation in electron concentration in the ionosphere that is caused by solar activity and that makes the heights of the ionosphere layers go beyond the normal limits for a location, date, and time of day.

ionospheric D scatter meteor burst [GEOPHYS]   Phenomenon affecting ionospheric scatter communications resulting from the penetration of meteors through the D region of the ionospheric layer.

ionospheric storm [GEOPHYS]   A turbulence in the F region of the ionosphere, usually due to a sudden burst of radiation from the sun; it is accompanied by a decrease in the density of ionization and an increase in the virtual height of the region.

Iowan glaciation [GEOL]   The earliest substage of the Wisconsin glacial stage; occurred more than 30,000 years ago.

ipsonite [GEOL]   The final stage of weathered asphalt; a black, infusible substance, only slightly soluble in carbon disulfide, containing 50–80% fixed carbon and very little oxygen.

IQSY *See* International Quiet Sun Year.

iridescent clouds [METEOROL] Ice-crystal clouds which exhibit brilliant spots or borders of colors, usually red and green, observed up to about 30° from the sun.

irisation [METEOROL] The coloration exhibited by iridescent clouds and at times along the borders of lenticular clouds.

Irish Sea [GEOGR] A marginal sea of the Atlantic Ocean between Ireland and England, approximately 53°N latitude and 5°W longitude.

IRM *See* isothermal remanent magnetization.

Irminger Current [OCEANOGR] An ocean current that is one of the terminal branches of the Gulf Stream system, flowing west off the southern coast of Iceland.

iron and steel sheet piling [MIN ENG] A technique that uses iron and steel piling instead of wood to drive a shaft through loose, wet ground near the surface.

iron formation [GEOL] Sedimentary, low-grade iron ore bodies consisting mainly of chert or fine-grained quartz and ferric oxide segregated in bands or sheets irregularly mingled.

iron glance *See* specularite.

iron mica *See* lepidomelane.

iron ore [GEOL] Rocks or deposits containing compounds from which iron can be extracted.

iron pyrite *See* pyrite.

ironshot [MINERAL] Pertaining to a mineral with streaks or spots of iron or iron ore.

iron spinel *See* hercynite.

ironstone [PETR] An iron-rich sedimentary rock, either deposited directly as a ferruginous sediment or resulting from chemical replacement.

iron-stony meteorite *See* stony-iron meteorite.

iron winds [METEOROL] Northeasterly winds of Central America, prevalent during February and March, and blowing steadily for several days at a time.

irreducible saturation [PETRO ENG] In a permeable reservoir, that condition in which the nonwetting phase saturation is so large that the wetting phase can be reduced no more.

irregular crystal [METEOROL] A snow particle, sometimes covered by a coating of rime, composed of small crystals randomly grown together; generally, component crystals are so small that the crystalline form of the particle can be seen only through a magnifying glass or microscope.

irregular iceberg *See* pinnacled iceberg.

irrespirable atmosphere [MIN ENG] Atmosphere in a coal mine requiring workers to wear breathing apparatus because of poisonous gas or insufficient oxygen as a result of an explosion from firedamp or coal dust, or mine fires.

irrotational strain [GEOL] Strain in which the orientation of the axes of strain does not change. Also known as nonrotational strain.

irruption *See* intrusion.

Irvingtonian [GEOL] A stage of geologic time in southern California, in the lower Pleistocene, below the Rancholabrean.

isabnormal [METEOROL] A line on a chart linking points having the same difference from normal (usually temperature) or indicating same difference between actual values and calculated values at different parallels.

isallobar [METEOROL] A line of equal change in atmospheric pressure during a specified time interval; an isopleth of pressure tendency; a common form is drawn for the three-hourly local pressure tendencies on a synoptic surface chart.

isallobaric [METEOROL] Of equal or constant pressure change; this may refer either to the distribution of equal pressure tendency in space or to the constancy of pressure tendency with time.

isallobaric high *See* pressure-rise center.

isallobaric low *See* pressure-fall center.

isallobaric maximum *See* pressure-rise center.

isallobaric minimum *See* pressure-fall center.

isallobaric wind [METEOROL] The wind velocity whose Coriolis force exactly balances a locally accelerating geostrophic wind. Also known as Brunt-Douglas isallobaric wind.

isallohypse *See* height-change line.

isallohypsic wind [METEOROL] An isallobaric wind, using the height tendency in a constant-pressure surface instead of the pressure tendency in a constant-height surface.

isallotherm [METEOROL] A line connecting points of equal change in temperature within a given time period.

isanabat [METEOROL] A line drawn through points of equal vertical component of wind velocity; positive values indicate upward motion, negative values indicate downward motion.

isanakatabar [METEOROL] A line on a chart of equal atmospheric-pressure range during a specified time interval.

isanomal [METEOROL] A line on a chart linking points with the same anomalies of temperature, pressure, and so forth.

isarithm *See* isopleth.

isaurore *See* isochasm.

Ischnacanthidae [PALEON] The single family of the acanthodian order Ischnacanthiformes.

Ischnacanthiformes [PALEON] A monofamilial order of extinct fishes of the order Acanthodii; members were slender, lightly armored predators with sharp teeth, deeply inserted fin spines, and two dorsal fins.

Isectolophidae [PALEON] A family of extinct ceratomorph mammals in the superfamily Tapiroidea.

isentropic chart [METEOROL] A constant-entropy chart; a synoptic chart presenting the distribution of meteorological elements in the atmosphere on a surface of constant potential temperature (equivalent to an isentropic surface); it usually contains the plotted data and analysis of such elements as pressure (or height), wind, temperature, and moisture at that surface.

isentropic condensation level *See* lifting condensation level.

isentropic map [GEOL] A map indicating constant entropy function for facies.

isentropic mixing [METEOROL] Any atmospheric mixing process which occurs within an isentropic surface; the fact that many atmospheric motions are reversible adiabatic processes renders this type of mixing important, and exchange coefficients have been computed therefor.

isentropic surface [METEOROL] A surface in space in which potential temperature is everywhere equal.

isentropic thickness chart [METEOROL] A thickness chart of an atmospheric layer bounded by two selected isentropic surfaces (surfaces of constant potential temperature); the thickness of such a layer is directly proportional to the static instability of that layer; hence, these charts have been called instability charts. Also known as thick-thin chart.

isentropic weight chart [METEOROL] A chart of atmospheric pressure difference between two selected isentropic surfaces (surfaces of constant potential temperature); the greater the pressure difference the greater the weight of the air column separating the two surfaces.

ishikawaite [MINERAL] A black, orthorhombic mineral consisting essentially of uranium, iron, rare earth, and columbium oxide.

ishkyldite [MINERAL] $Mg_{15}Si_{11}O_{27}(OH)_{20}$ A mineral composed of a basic silicate of magnesium.

island [GEOGR] A tract of land smaller than a continent and surrounded by water; normally in an ocean, sea, lake, or stream.

island arc [GEOGR] A group of islands usually with a curving archlike pattern, generally convex toward the open ocean, having a deep trench or trough on the convex side and usually enclosing a deep basin on the concave side.

island mountain *See* inselberg.

isobar [METEOROL] A line drawn through all points of equal atmospheric pressure along a given reference surface, such as a constant-height surface (notably mean sea level on surface charts), an isentropic surface, or the vertical plan of a synoptic cross section.

isobaric chart *See* constant-pressure chart.

isobaric contour chart *See* constant-pressure chart.

isobaric divergence [METEOROL] The horizontal divergence in a constant-pressure surface; expressed in a system of coordinates with pressure as an independent variable.

isobaric equivalent temperature *See* equivalent temperature.

isobaric map [METEOROL] A map depicting points in the atmosphere of equal barometric pressure.

isobaric surface [METEOROL] A surface on which the pressure is uniform. Also known as constant-pressure surface.

isobaric topography *See* height pattern.

isobaric vorticity [METEOROL] Relative vorticity in a constant-pressure surface, that is, expressed in a system of coordinates with pressure as an independent variable.

isobath [OCEANOGR] A contour line connecting points of equal water depths on a chart. Also known as depth contour; depth curve; fathom curve.

isobathytherm [OCEANOGR] A line or surface showing the depth in oceans or lakes at which points have the same temperatures.

isobront [METEOROL] A line drawn through geographical points at which a given phase of thunderstorm activity occurred simultaneously. Also known as homobront.

isocarb [GEOCHEM] A line on a map that connects points of equal content of fixed carbon in coal.

isoceraunic [METEOROL] Indicating or having equal frequency or intensity of thunderstorm activity. Also spelled isokeraunic.

isoceraunic line [METEOROL]   A line drawn through geographical points at which some phenomenon connected with thunderstorms has the same frequency or intensity; used for lines of equal frequency of lightning discharges.

isochasm [GEOPHYS]   A line connecting points on the earth's surface at which the aurora is observed with equal frequency. Also known as isaurore.

isochemical metamorphism [PETR]   Theoretically, a metamorphism involving no great change in its chemical composition. Also known as treptomorphism.

isochemical series [PETR]   A series of rocks with identical chemical compositions.

isochronal test [PETRO ENG]   Short-time back-pressure test for low-permeability reservoirs that otherwise require excessively long times for pressure stabilization when wells are shut in.

isoclasite [MINERAL] $Ca_2(PO_4)(OH) \cdot 2H_2O$ A white mineral composed of a basic hydrous calcium phosphate; occurring in small crystals or columnar forms.

isoclinal chart [GEOPHYS]   A chart showing isoclinic lines. Also known as isoclinic chart.

isocline [GEOL]   A fold of strata so tightly compressed that parts on each side dip in the same direction.

isoclinic chart *See* isoclinal chart.

isoclinic line [GEOPHYS]   A line connecting points on the earth's surface which have the same magnetic dip. Also known as isoclinal.

isodee [METEOROL]   A line on a chart linking points of equal difference between pressure altitude and absolute altitude above sea level.

isodrosotherm [METEOROL]   An isogram of dew-point temperature.

isodynamic line [GEOPHYS]   One of the lines on a map of a magnetic field that connect points having equal strengths of the earth's field.

isofacies map [GEOL]   A stratigraphic map showing the distribution of one or more facies within a particular stratigraphic unit.

isofronts-preiso code [METEOROL]   A code in which data on isobars and fronts at sea level (or earth's surface) are encoded and transmitted; a modified form of the international analysis code.

isogeotherm *See* geoisotherm.

isogonic line [GEOPHYS]   **1.** Any of the lines on a chart or map showing the same direction of the wind vector. **2.** Any of the lines on a chart or map connecting points of equal magnetic variation.

isogor [PETRO ENG]   Constant gas-oil ratio.

isogor map [PETRO ENG]   Oil reservoir contour-line map that shows constant gas-oil ratios.

isograd [GEOL]   A line on a map joining those rocks comprising the same metamorphic grade.

isogradient [METEOROL]   A line connecting points having the same horizontal gradient of atmospheric pressure, temperature, and so on.

isogram *See* isopleth.

isohaline [OCEANOGR]   **1.** Of equal or constant salinity. **2.** A line on a chart connecting all points of equal salinity.

isoheight *See* contour line.

isohel [METEOROL] A line drawn through geographical points having the same duration of sunshine (or other function of solar radiation) during any specified time period.

isohume [GEOL] A line of a map or chart connecting points of equal moisture content in a coal bed. [METEOROL] A line drawn through points of equal humidity on a given surface; an isopleth of humidity; the humidity measures used may be the relative humidity or the actual moisture content (specific humidity or mixing ratio).

isohyet [METEOROL] A line drawn through geographic points recording equal amounts of precipitation for a specified period or for a particular storm.

isohypse *See* contour line.

isohypsic chart *See* constant-height chart.

isohypsic surface *See* constant-height surface.

isokeraunic *See* isoceraunic.

isokinetic *See* isotach.

isokinetic sampling [ENG] Any technique for collecting airborne particulate matter in which the collector is so designed that the airstream entering it has a velocity equal to that of the air passing around and outside the collector.

isolith [GEOL] A line on a contour-type map that denotes the aggregate thickness of a single lithology in a stratigraphic succession composed of one or more lithologies.

isolith map [GEOL] A contour-line map depicting the thickness of an exclusive lithology.

isomagnetic [GEOPHYS] Of or pertaining to lines connecting points of equality in some magnetic element.

isometric system [CRYSTAL] The crystal system in which the forms are referred to three equal, mutually perpendicular axes. Also known as cubic system.

isomorph *See* isomorphic mineral.

isomorphic mineral [MINERAL] Any two or more crystalline mineral compounds having different chemical composition but identical structure, such as the garnet series or the feldspar group Also known as isomorph.

isoneph [METEOROL] A line drawn through all points on a map having the same amount of cloudiness.

isopach map [GEOL] Map of the areal extent and thickness variation of a stratigraphic unit; used in geological exploration for oil and for underground structural analysis.

isopachous line [GEOL] One of the lines drawn on a map to indicate equal thickness.

isopag [HYD] A line on a chart linking points where ice occurs for the same number of days per year.

isopectic [CLIMATOL] A line on a map connecting points at which ice begins to form at the same time of winter.

isoperimetric line [MAP] A line on a map projection that indicates no variation from exact scale.

isoperm [PETRO ENG] One of the lines of equal (constant) permeability plotted on a reservoir map.

isopiestic line [HYD] A line indicating on a map the piezometric surface of an aquifer.

isopleth [METEOROL] **1.** A line of equal or constant value of a given quantity with respect to either space or time. Also known as isogram. **2.** More specifically, a line drawn through points on a graph at which a given quantity has the same numerical value (or occurs with the same frequency) as a function of the two coordinate variables. Also known as isarithm.

isopluvial [METEOROL]   A line on a map drawn through geographical points having the same pluvial index.

isopor [GEOPHYS]   An imaginary line connecting points on the earth's surface having the same annual change in a magnetic element.

isopotential level *See* potentiometric surface.

isopotential map [PETRO ENG]   A contour-line map to show the initial or calculated daily rate of oil well production in a multiwell field.

isopycnic [METEOROL]   A line on a chart connecting all points of equal or constant density.

isopycnic level [METEOROL]   Specifically, a level surface in the atmosphere, at about 8 kilometers altitude, where the air density is approximately constant in space and time; this level corresponds to the maximum upper-tropospheric interdiurnal pressure variation.

isoreactiongrad [GEOL]   An isograd based on a specific mineralogical reaction.

isoseismal [GEOPHYS]   Pertaining to points having equal intensity of earthquake shock, or to a line on a map of the earth's surface connecting such points.

isoshear [METEOROL]   A line on a chart of equal magnitude of vertical wind shear.

isostasy [GEOPHYS]   A theory of the condition of approximate equilibrium in the outer part of the earth, such that the gravitational effect of masses extending above the surface of the geoid in continental areas is approximately counterbalanced by a deficiency of density in the material beneath those masses, while deficiency of density in ocean waters is counterbalanced by an excess in density of the material under the oceans.

isostatic adjustment *See* isostatic compensation.

isostatic anomaly [GEOPHYS]   A gravity anomaly based on a generalized hypothesis that the gravitational effect of masses above sea level is approximately compensated by a density deficiency of the subsurface materials.

isostatic compensation [GEOL]   The process in which lateral transport at the surface of the earth by erosion or deposition is compensated by lateral movements in a subcrustal layer. Also known as isostatic adjustment; isostatic correction.

isostatic correction *See* isostatic compensation.

isostructural [CRYSTAL]   Pertaining to crystalline materials that have corresponding atomic positions, and have a considerable tendency for ionic substitution.

isosulfur map [PETRO ENG]   A contour-line map to show the percentage of sulfur in underground crude oil.

isotach [METEOROL]   A line in a given surface connecting points with equal wind speed. Also known as isokinetic; isovel.

isotach chart [METEOROL]   A synoptic chart showing the distribution of wind by means of isotachs.

isothere [CLIMATOL]   A line on a map connecting points having the same mean summer temperature.

isotherm [GEOPHYS]   A line on a chart connecting all points of equal or constant temperature.

isothermal atmosphere [METEOROL] An atmosphere in hydrostatic equilibrium, in which the temperature is constant with height and the pressure decreases exponentially upward. Also known as exponential atmosphere.

isothermal chart [GEOPHYS] A map showing the distribution of air temperature (or sometimes sea-surface or soil temperature) over a portion of the earth or at some level in the atmosphere; places of equal temperature are connected by lines called isotherms.

isothermal equilibrium [METEOROL] The state of an atmosphere at rest, uninfluenced by any external agency, in which the conduction of heat from one part to another has produced, after a sufficient length of time, a uniform temperature throughout its entire mass. Also known as conductive equilibrium.

isothermal layer [METEOROL] The approximately isothermal region of the atmosphere immediately above the tropopause.

isothermal remanent magnetization [GEOPHYS] Remanent magnetization that is produced solely by applying a magnetic field, with no change of temperature. Abbreviated IRM.

isothermobath [OCEANOGR] A line connecting points having the same temperature in a diagram of a vertical section of the ocean.

isotherm ribbon [METEOROL] A zone of crowded isotherms on a synoptic upper-level chart; the temperature gradient is many times greater than normally encountered in the atmosphere.

isotimic [METEOROL] Pertaining to a quantity which has equal value in space at a particular time.

isotimic line [METEOROL] On a given reference surface in space, a line connecting points of equal value of some quantity; most of the lines drawn in the analysis of synoptic charts are isotimic lines.

isotimic surface [METEOROL] A surface in space on which the value of a given quantity is everywhere equal; isotimic surfaces are the common reference surfaces for synoptic charts, principally constant-pressure surfaces and constant-height surfaces.

isotropic fabric [PETR] A random orientation in space of the elements that compose a rock.

isotypic [CRYSTAL] Pertaining to a crystalline substance whose chemical formula is analogous to, and whose structure is like, that of another specified compound.

isovel *See* isotach.

isthmus [GEOGR] A narrow strip of land having water on both sides and connecting two large land masses.

itacolumite [PETR] A fine-grained, thin-bedded sandstone or a schistose quartzite that contains mica, chlorite, and talc and that exhibits flexibility when split into slabs. Also known as articulite.

J

jacinth *See* zircon.

jack [MINERAL] *See* sphalerite.

jackbit [DES ENG] A drilling bit used to provide the cutting end in rock drilling; the bit is detachable and either screws on or is taper-fitted to a length of drill steel. Also known as ripbit.

jacket [PETRO ENG] The support structure of a steel offshore production platform; it is fixed to the seabed by piling, and the superstructure is mounted on it.

jack line [PETRO ENG] A steel cable or rod that connects the arms of the central pumping engine to the two or more wells that are being pumped.

jack post [MIN ENG] Timber used where a coal seam is separated by a rock band and one bench is loaded out before the other.

jack timber [MIN ENG] A timber such as a rafter that is shorter than others with which it is used.

Jacobshavn Glacier [HYD] A glacier on the west coast of Greenland at latitude 68°N; it is the most productive glacier in the Northern Hemisphere, calving about 1400 icebergs yearly.

jacobsite [MINERAL] $MnFe_2O_4$ A black magnetic mineral composed of an oxide of manganese and iron; a member of the magnetite series.

jacupirangite [PETR] An ultramafic plutonic rock that is part of the ijolite series, composed chiefly of titanaugite and magnetite with a smaller amount of nepheline.

jade [MINERAL] A hard, compact, dark-green or greenish-white gemstone composed of either jadeite or nephrite. Also known as jadestone.

jadeite [MINERAL] $NaAl(SiO_3)_2$ A clinopyroxene mineral occurring as green, fibrous monoclinic crystals; the most valuable variety of jade.

jadestone *See* jade.

James concentrator [MIN ENG] A concentration table whose deck is divided into two sections; one section contains riffles for the coarse material, and the other section is smooth to allow settling of the fine particles which will not settle on a riffled surface.

jamesonite [MINERAL] $Pb_4FeSb_6S_{14}$ A lead-gray to gray-black mineral that crystallizes in the orthorhombic system, occurs in acicular crystals with fibrous or featherlike forms, and has a metallic luster. Also known as feather ore; gray antimony.

January thaw [CLIMATOL] A period of mild weather popularly supposed to recur each year in late January in New England and other parts of the northeastern United States.

jar coupling *See* jars.

jarlite [MINERAL] $NaSr_3Al_3F_{16}$ A colorless to brownish mineral composed of aluminofluoride of sodium and strontium.

jarosite [MINERAL] $KFe_3(SO_4)_2(OH)_6$ An ocher-yellow or brown alunite mineral having rhombohedral crystal structure. Also known as utahite.

jars [PETRO ENG] A series of links in the drill string to connect drill cables to the drill bit; sets up the uneven motion on the upstroke that helps free the string of tools. Also known as jar coupling.

jaspagate *See* agate jasper.

jasper [PETR] A dense, opaque to slightly translucent cryptocrystalline quartz containing iron oxide impurities; characteristically red. Also known as jasperite; jasperoid; jaspis.

jasperite *See* jasper.

jasperoid *See* jasper.

jaspilite [PETR] A compact siliceous rock resembling jasper and containing iron oxides in bands.

jaspis *See* jasper.

jaspoid *See* tachylite.

jauch *See* jauk.

jauk [METEOROL] A local name for the foehn in the Klagenfurt basin of Austria; it may come from the south, but is developed as a north foehn. Also spelled jauch.

jaw [GEOL] The side of a narrow passage such as a gorge.

jeffersonite [MINERAL] $Ca(Mn,Zn,Fe)Si_2O_6$ A dark-green or greenish-black mineral composed of pyroxene.

Jeffrey crusher [MIN ENG] A crusher to break soft minerals, such as limestone. Also known as whizzer mill.

Jeffrey diaphragm jig [MIN ENG] A plunger-type jig with the plunger beneath the screen.

Jeffrey molveyor [MIN ENG] A string of short conveyors on driven wheels connected together to run alongside a heading or room conveyor; used to keep a continuous miner in operation at all times.

Jeffrey single-roll crusher [MIN ENG] A simple type of crusher for coal, with a drum to which are bolted toothed segments designed to grip the coal, thus forcing it down into the crushing opening.

Jeffrey swing-hammer crusher [MIN ENG] A crusher with swing arms on a revolving shaft for crushing coal, ore, or other material against the iron casing of the crusher; a screen at the bottom allows sufficiently fine pieces to pass through.

Jeffrey-Traylor vibrating feeder [MIN ENG] A feed chute vibrated electromagnetically in a direction oblique to its surface; rate of movement of rock depends on amplitude and frequency of vibration.

Jeffrey-Traylor vibrating screen [MIN ENG] A vibrating screen whose action results from an oscillating armature and a stationary coil.

jelly *See* ulmin.

jeremejevite [MINERAL] $AlBO_3$ A colorless or yellowish mineral composed of aluminum borate that occurs in hexagonal crystals.

jet coal [GEOL] A hard, lustrous, pure black variety of lignite, occurring in isolated masses in bituminous shale; thought to be derived from waterlogged driftwood. Also known as black amber.

jet-effect wind [METEOROL] A wind which is increased in speed through the channeling of air by some mountainous configuration, such as a narrow mountain pass or canyon.

jet-flame drill [MIN ENG] A mining drill that utilizes a high-velocity flame to spall out a hole.

jet hole [ENG] A borehole drilled by use of a directed, forceful stream of fluid or air.

jet-piercing drill *See* fusion-piercing drill.

jet stream [METEOROL] A relatively narrow, fast-moving wind current flanked by more slowly moving currents; observed principally in the zone of prevailing westerlies above the lower troposphere, and in most cases reaching maximum intensity with regard to speed and concentration near the troposphere.

jetting tool [PETRO ENG] Downhole device that jets a high-pressure, sand-laden fluid stream to clean out wellbore holes, to disintegrate perforating pipe, and to perform other operations.

Jevons effect [METEOROL] The effect upon the measurement of rainfall caused by the presence of the rain gage; in 1861 W.S. Jevons pointed out that the rain gage causes a disturbance in airflow past it, and this carries part of the rain past the gage which would normally be captured.

jezekite *See* morinite.

J function [GEOPHYS] A dimensionless mathematical relationship to correlate capillary pressure data of similar geologic formations.

jib end [MIN ENG] The delivery end in conveyor systems in which a jib is fitted to deliver the load in advance of and remote from the drive.

jig [MIN ENG] A vibrating device in which coal is cleaned and ore is concentrated in water.

jigger *See* jigging conveyor.

jigger boss [MIN ENG] A first-line supervisor in some western United States mines.

jigging [MIN ENG] A gravity method which separates mineral from gangue particles by utilizing an effective difference in settling rate through a periodically dilated bed.

jigging conveyor [MIN ENG] A series of steel troughs suspended from the roof of the stope or laid on rollers on its floor, and given reciprocating motion mechanically, to move mineral. Also known as chute conveyor; jigger; pan conveyor.

jig washer [MIN ENG] A coal or mineral washer for relatively coarse material; the broken ore is placed on a screen and pulsed vertically with water; the heavy portion passes through the screen and the light portion goes over the sides.

joaquinite [MINERAL] $NaBa_2Ce_2Fe(Ti,Nb)_2Si_8O_{26}(OH,F)_2$ A honey-yellow mineral composed of sodium iron titanium silicate, occurring in orthorhombic crystals.

jochwinde [METEOROL] The mountain-gap wind of the Tauern Pass in the Alps.

joggle [MIN ENG] A truss or set of timbers joined for taking pressure at right angles.

johannite [MINERAL] $Cu(UO_2)_2(SO_4)_2 \cdot 6H_2O$ An emerald green to apple green, triclinic mineral consisting of a hydrated basic copper and uranium sulfate.

johannsenite [MINERAL] $CaMnSi_2O_6$ A clove-brown, grayish, or greenish clinopyroxene mineral composed of a silicate of calcium and manganese; a member of the pyroxene group.

Johnson concentrator [MIN ENG] A device used to separate heavy particles such as metallic gold from auriferous pulp; composed of a shell in the shape of a cylinder that is lined with rubber grooves parallel to the inclined axis.

johnstrupite [MINERAL] A mineral composed of a complex silicate of cerium and other metals, approximately $(Ca,Na)_3(Ce,Ti,Zr)(SiO_4)_2F$; occurs in prismatic crystals.

joint [GEOL] A fracture that traverses a rock and does not show any discernible displacement of one side of the fracture relative to the other.

joint drag *See* kink band.

joint set [GEOL] A group of parallel joints in a geologic formation.

joint system [GEOL] Two or more joint sets.

joint vein [GEOL] A small vein in a joint.

Jolly balance [ENG] A spring balance used to measure specific gravity of mineral specimens by weighing a specimen when in the air and when immersed in a liquid of known density.

Jones riffle [MIN ENG] An apparatus used to reduce the size of a sample to a desired weight; consists of a hopper which passes samples to a series of open-bottom pockets, each of which divides the sample into two equal parts, and the next pass of each part gives a quarter of the original sample, and so on until the desired sample is obtained.

Jones splitter [MIN ENG] A device used to reduce the volume of a sample, consisting of a belled, rectangular container, the bottom of which is fitted with a series of narrow slots or alternating chutes designed to cast material in equal quantities to opposite sides of the device.

Jones sucker rod [PETRO ENG] Connecting rod between the subsurface pump and the lifting or pumping device on the surface; serves to lift oil out of the cased hole.

joran *See* juran.

jordanite [MINERAL] $(Pb,Tl)_{13}As_7S_{23}$ A lead-gray mineral composed of lead arsenic sulfide, occurring as monoclinic crystals.

Jordan sunshine recorder [ENG] A sunshine recorder in which the time scale is supplied by the motion of the sun; it consists of two opaque metal semicylinders mounted with their curved surfaces facing each other; each of the semicylinders has a short narrow slit in its flat side; sunlight entering one of the slits falls on light-sensitive paper (blueprint paper) which lines the curved side of the semicylinder.

joseite [MINERAL] $Bi_3Te(Si,S)$ A mineral composed of telluride of bismuth containing sulfur and selenium.

josen *See* hartite.

josephinite [MINERAL] A mineral consisting of an alloy of iron and nickel; occurs naturally in stream gravel.

Joy double-ended miner [MIN ENG] A cutter-loader for continuous mining on a longwall face.

Joy extensible conveyor [MIN ENG] A type of belt conveyor consisting of a head section and a tail section, each mounted on crawler tracks and independently driven; used between a loader or continuous miner and the main transport.

Joy extensible steel band [MIN ENG] A hydraulically driven system linking a continuous miner to the main transport; the steel band is coiled on the drivehead.

Joy loader [MIN ENG] A loading machine which uses mechanical arms to collect coal or ore onto an apron that is pushed onto the broken material.

Joy longwall loading machine [MIN ENG] A modified Joy loader comprising a hydraulically elevated loading head fitted with mechanical gathering arms.

Joy microdyne [MIN ENG] A dust collector that wets and traps dust pulled into it and releases the dust as a slurry to be removed by pumps; used at the return end of tunnels or hard headings.

Joy miner [MIN ENG] A continuous miner weighing about 15 tons (13,600 kilograms) and made up of a turntable, a ripper bar, and a discharge boom conveyor; used mainly in coal headings and in extraction of coal pillars.

Joy-Sullivan hydrodrill rig [MIN ENG] A drill rig set on a jib or boom which can be moved to and locked in any position by hydraulic power controlled from the drill carriage.

Joy transloader [MIN ENG] A rubber-tired, self-propelled machine for loading, transporting, and dumping.

Joy walking miner [MIN ENG] A continuous miner designed to make thin seams; a walking mechanism is used instead of caterpillar tracks for moving the miner.

julienite [MINERAL] $Na_2Co(SCN)_4 \cdot 8H_2O$ A blue, tetragonal mineral consisting of a hydrated sodium cobalt thiocyanate.

jumping a claim [MIN ENG] **1.** Taking possession of a mining claim which has been abandoned. **2.** Taking possession of a mining claim that is liable to forfeiture because the requirements of the law are unfulfilled. **3.** Taking possession of a mining claim by stealth, fraud, or force.

junction streamer [GEOPHYS] The streamer process by which negative charge centers at successively higher altitudes in a thundercloud are believed to be "tapped" for discharge by lightning.

junk wind [METEOROL] A south or southeast monsoon wind, favorable for the sailing of junks; the wind is known in Thailand, China, and Japan.

junta [METEOROL] A wind blowing through Andes Mountain passes, sometimes reaching hurricane force.

juran [METEOROL] A wind blowing from the Jura Mountains in Switzerland from the northwest toward Lake Geneva; it is a cold and snowy wind and may be very turbulent, especially in spring. Also spelled joran.

Jurassic [GEOL] Also known as Jura. **1.** The second period of the Mesozoic era of geologic time. **2.** The corresponding system of rocks.

jurupaite [MINERAL] $(Ca,Mg)_2(Si_2O_5)(OH)_2$ A mineral composed of hydrous calcium magnesium silicate.

jury rig [MIN ENG] Any makeshift or temporary device, rig, or piece of equipment.

juvenile water *See* magmatic water.

juvite [PETR] A light-colored nepheline syenite in which the feldspar is exclusively or predominantly orthoclase and in which the potassium oxide content is higher than the sodium oxide.

K-A age [GEOL] The radioactive age of a rock determined from the ratio of potassium-40 (^{40}K) to argon-40 (^{40}A) present in the rock.

kaavie [METEOROL] In Scotland, a heavy fall of snow.

kachchan [METEOROL] A hot, dry west or southwest wind of foehn type in the lee of the Sri Lanka (Ceylon) hills during the southwest monsoon in June and July; it is well developed at Batticaloa on the east coast, where it is strong enough to overcome the sea breeze and bring maximum temperatures of nearly 100°F (38°C).

kainite [MINERAL] $MgSO_4 \cdot KCl \cdot 3H_2O$ A white, gray, pink, or black monoclinic mineral, occurring in irregular granular masses; used as a fertilizer and as a source of potassium and magnesium compounds.

kainosite [MINERAL] $Ca_2(Ce,Y)_2(SiO_4)_3CO_3 \cdot H_2O$ A yellowish-brown mineral composed of a hydrous silicate and carbonate of calcium, cerium, and yttrium.

kal Baisakhi [METEOROL] A short-lived dusty squall at the onset of the southwest monsoon (April-June) in Bengal.

kalema [OCEANOGR] A very heavy surf breaking on the Guinea coast of Africa during the winter.

kaliborite [MINERAL] $HKMg_2B_{12}O_{21} \cdot 9H_2O$ A colorless to white mineral composed of a hydrous borate of potassium and magnesium. Also known as paternoite.

kalicinite [MINERAL] $KHCO_3$ A colorless to white or yellowish, monoclinic mineral consisting of potassium bicarbonate; occurs in crystalline aggregates.

kalinite [MINERAL] $KAl(SO_4)_2 \cdot 11H_2O$ A birefringent mineral of the alum group composed of a hydrous sulfate of potassium and aluminum, occurring in fibrous form. Also known as potash alum.

kaliophilite [MINERAL] $KAlSiO_4$ A rare hexagonal tectosilicate mineral found in volcanic rocks; high in potassium and low in silica, it is dimorphous with kalsilite. Also known as facellite; phacellite.

kalkowskite [MINERAL] $Fe_2Ti_3O_9$ A rare, brownish or black mineral composed of an oxide of iron and titanium, usually with small amounts of rare-earth elements, niobium, and tantalum.

kalsilite [MINERAL] $KAlSiO_4$ A rare mineral from volcanic rocks in southwestern Uganda; the crystal system is hexagonal; kalsilite is dimorphous with kaliophilite and sometimes contains sodium.

kalunite [MINERAL] The naturally occurring form of alum.

kamacite [MINERAL] A mineral composed of a nickel-iron alloy and comprising with taenite the bulk of most iron meteorites.

kame [GEOL] A low, long, steep-sided mound of glacial drift, commonly stratified sand and gravel, deposited as an alluvial fan or delta at the terminal margin of a melting glacier.

kame terrace [GEOL] A terracelike ridge deposited along the margins of glaciers by meltwater streams flowing adjacent to the valley walls.

kansite *See* mackinawite.

kaolin [MINERAL] Any of a group of clay minerals, including kaolinite, nacrite, dickite, and anauxite, with a two-layer crystal in which silicon-oxygen and aluminum-hydroxyl sheets alternate; approximate composition is $Al_2O_3 \cdot 2SiO_2 \cdot 2H_2O$. [PETR] A soft, nonplastic white rock composed principally of kaolin minerals. Also known as bolus alba; white clay.

kaolinite [MINERAL] $Al_2Si_2O_5(OH)_4$ The principal mineral of the kaolin group of clay minerals; a white, gray, or yellowish high-alumina mineral consisting of sheets of tetrahedrally coordinated silicon linked by an oxygen shared with octahedrally coordinated aluminum.

kaolinization [GEOL] The forming of kaolin by the weathering of aluminum silicate minerals or other clay minerals.

karaburan [METEOROL] A violent northeast wind of Central Asia occurring during spring and summer; it carries clouds of dust (which darken the sky) instead of snow. Also known as black buran; black storm.

karajol [METEOROL] On the Bulgarian coast, a west wind which usually follows rain and persists 1–3 days.

karema [METEOROL] A violent east wind on Lake Tanganyika in Africa.

karif [METEOROL] A strong southwest wind on the southern shore of the Gulf of Aden, especially at Berbera, Somaliland, during the southwest monsoon.

karoo *See* karroo.

karroo [GEOGR] A dry, broad, level, elevated area found especially in southern Africa, often rising to considerable elevations in terrace formations; does not support vegetation in the dry season but supports grass during the wet season. Also spelled karoo.

Karroo System [GEOL] Glaciated strata formed in Permian times in southern Africa.

karst [GEOL] A topography formed over limestone, dolomite, or gypsum and characterized by sinkholes, caves, and underground drainage.

karstbora [METEOROL] The bora of the Yugoslavian coast.

karst plain [GEOL] A plain on which karst features are developed.

kasolite [MINERAL] $Pb(UO_2)SiO_4 \cdot H_2O$ Yellow-ocher mineral composed of a hydrous lead uranium silicate, occurring in monoclinic crystals.

katabaric *See* katallobaric.

katabatic wind *See* gravity wind.

katafront [METEOROL] A front (usually a cold front) at which warm air descends the frontal surface (except, presumably, in the lowest layers).

katallobaric [METEOROL] Of or pertaining to a decrease in atmospheric pressure. Also known as katabaric.

katallobaric center *See* pressure-fall center.

katazone [GEOL] The lowest depth zone of metamorphism, exhibiting high temperatures (500–700°C), generally strong hydrostatic pressure, and low or no shearing stress.

katoptrite *See* catoptrite.

kaus [METEOROL]   A moderate to gale-force southeasterly wind in the Persian Gulf, accompanied by gloomy weather, rain, and squalls; it is most frequent between December and April. Also known as cowshee; sharki.

kavaburd *See* cavaburd.

kaver *See* caver.

kay *See* key.

Kazanian [GEOL]   A European stage of geologic time: Upper Permian (above Kungurian, below Tatarian).

K bentonite *See* potassium bentonite.

keeps *See* folding boards.

Keewatin [GEOL]   A division of the Archeozoic rocks of the Canadian Shield.

Kegel karst *See* cone karst.

kehoeite [MINERAL]   An amorphous mineral composed of a basic hydrous calcium aluminum zinc phosphate, occurring massive.

Keilor skull [PALEON]   An Australian fossil type specimen of *Homo sapiens* from the Pleistocene.

kelly [PETRO ENG]   A pipe attached to the top of a drill string and turned during drilling; transmits twisting torque from the rotary machinery to the drill string and ultimately to the bit.

kelsher [METEOROL]   In England, a heavy fall of rain.

Kelvin wave [OCEANOGR]   A type of wave progression in relatively confined water bodies where because of Coriolis force the wave is higher to the right of direction of advance (in the Northern Hemisphere).

kelyphite *See* corona.

kelyphytic border *See* kelyphytic rime.

kelyphytic rime [PETR]   A peripheral zone of pyroxene or amphibole developed around olivine in some igneous rocks. Also known as kelyphytic border.

kempite [MINERAL] $Mn_2(OH)_3Cl$ An emerald-green orthorhombic mineral composed of a basic manganese oxychloride, occurring in small crystals.

Kennelly-Heaviside layer *See* E layer.

Kenoran orogeny *See* Algoman orogeny.

kentrolite [MINERAL] $Pb_2Mn_2Si_2O_9$ A dark reddish-brown mineral composed of a lead manganese silicate.

kep interlock [MIN ENG]   A system designed to prevent the lowering of a shaft conveyance before all keps are fully withdrawn, and to indicate the position of the keps.

keps *See* folding boards.

kerabitumen *See* kerogen.

keratophyre [PETR]   Any dike rock or salic lava that is characterized by the presence of albite or albite oligoclase, chlorite, epidote, and calcite.

kerf [MIN ENG]   A narrow, deep cut made in the face of coal to facilitate mining.

kermesite [MINERAL] Sb_2S_2O A cherry-red mineral occurring as tufts of capillary crystals, and formed from an alteration of stibnite. Also known as antimony blende; purple blende; pyrostibite; red antimony.

kernel ice [HYD]   In aircraft icing, an extreme form of rime ice, that is, very irregular, opaque, and of low density; it forms at temperatures of $-15°C$ and lower.

kernite [MINERAL] $Na_2B_4O_7 \cdot 4H_2O$ A colorless to white hydrous borate mineral crystallizing in the monoclinic system and having vitreous luster; an important source of boron. Also known as rasorite.

kerogen [GEOL]   The complex, fossilized organic material present in sedimentary rocks, especially in shales; converted to petroleum products by distillation. Also known as kerabitumen; petrologen.

kerogen shale *See* oil shale.

kerosine shale *See* torbanite.

kersantite [PETR]   Dark dike rocks consisting mostly of biotite, plagioclase, and augite.

kettle [GEOL]   **1.** A bowl-shaped depression with steep sides in glacial drift deposits that is formed by the melting of glacier ice left behind by the retreating glacier and buried in the drift. Also known as kettle basin; kettle hole. **2.** *See* pothole.

kettle basin *See* kettle.

kettle hole *See* kettle.

Keuper [GEOL]   A European stage of geologic time, especially in Germany; Upper Triassic.

Kew barometer [ENG]   A type of cistern barometer; no adjustment is made for the variation of the level of mercury in the cistern as pressure changes occur; rather, a uniformly contracting scale is used to determine the effective height of the mercury column.

Keweenawan [GEOL]   The younger of two Precambrian time systems that constitute the Proterozoic period in Michigan and Wisconsin.

key [GEOL]   A cay, especially one of the islets off the south of Florida. Also spelled kay.

key bed [GEOL]   Also known as index bed; key horizon; marker bed. **1.** A stratum or body of strata that has distinctive characteristics so that it can be easily identified. **2.** A bed whose top or bottom is employed as a datum in the drawing of structure contour maps.

key day *See* control day.

khamsin [METEOROL]   A dry, dusty, and generally hot desert wind in Egypt and over the Red Sea; it is generally southerly or southeasterly, occurring in front of depressions moving eastward across North Africa or the southeastern Mediterranean.

khibinite *See* mosandrite.

kibble *See* hoppit.

kibli *See* ghibli.

kidney ore [MINERAL]   A form of hematite found in compact masses, concretions, or nodules that are kidney-shaped.

kidney stone *See* nephrite.

kieselguhr *See* diatomaceous earth.

kieserite [MINERAL] $MgSO_4 \cdot H_2O$ A white mineral that crystallizes in the monoclinic system, is composed of hydrous magnesium sulfate, and occurs in saline residues.

Kikuchi lines   [CRYSTAL]   A pattern consisting of pairs of white and dark parallel lines, obtained when an electron beam is scattered (diffracted) by a crystalline solid; the pattern gives information on the structure of the crystal.

Kilkenny coal *See* anthracite.

Kimberley reefs   [GEOL]   Gold-bearing reefs in southern Africa that lie above the Main reef and Bird reef groups. Also known as battery reefs.

kimberlite   [PETR]   A form of mica peridolite that is formed mainly of phenocrysts, olivine, phlogopite, and subordinate melilite with minor amounts of pyroxene, apatite, perovskite, and opaque oxides.

Kimmeridgian   [GEOL]   A European stage of geologic time; middle Upper Jurassic, above Oxfordian, below Portlandian.

kimzeyite   [MINERAL] $Ca_3(Zr,Ti)_2(Al,Si)_3O_{12}$ A mineral of the garnet group.

Kind-Chaudron process   [MIN ENG]   A technique used to sink a large-diameter deep shaft; a pilot shaft of smaller diameter is first dug, then enlarged until the full diameter is reached; a lining with a moss box at the bottom is forced into place when water is found.

Kinderhookian   [GEOL]   Lower Mississippian geologic time, above the Chautauquan of Devonian, below Osagian.

kink band   [GEOL]   A deformation band in a single crystal or in foliated rocks in which the orientation is changed due to slipping on several parallel slip planes. Also known as joint drag; knick band; knick zone.

kinzigite   [PETR]   A coarse-grained metamorphic rock that is formed principally of garnet and biotite, with K feldspar, quartz, mica, cordierite, and sillimanite.

Kirkbyacea   [PALEON]   A monomorphic superfamily of extinct ostracods in the suborder Beyrichicopina, all of which are reticulate.

Kirkbyidae   [PALEON]   A family of extinct ostracods in the superfamily Kirkbyacea in which the pit is reduced and lies below the middle of the valve.

kirovite   [MINERAL] $(Fe,Mg)SO_4 \cdot 7H_2O$ A mineral composed of a hydrous sulfate of iron and magnesium; it is isomorphous with malanterite and pisanite.

Kiruna method   [MIN ENG]   A borehole-inclination survey method whereby the electrolytic deposition of copper from a solution is used to make a mark on the inside of a metal container.

kite observation   [METEOROL]   An atmospheric sounding by means of instruments carried aloft by a kite.

klaprothite   [MINERAL] $Cu_6Bi_4S_9$ A gray mineral composed of copper bismuth sulfide.

klebelsbergite   [MINERAL]   A mineral composed of basic antimony sulfate, occurring between crystals of stibnite.

kleinite   [MINERAL]   A yellow to orange mineral composed of a basic oxide, sulfate, and chloride of mercury and ammonium.

Klinkenberg correction   [PETRO ENG]   Mathematical conversion of laboratory air-permeability measurements (made on formation material) into equivalent liquid-permeability values.

klint   [GEOL]   An exhumed coral reef or bioherm that is more resistant to the processes of erosion than the rocks that enclose it so that the core remains in relief as hills and ridges.

klintite [GEOL] The dense, hard dolomite composing a klint; gives to the core a strength and resistance to erosion.

klippe [GEOL] A block of rock that is separated from underlying rocks by a fault that usually has a gentle dip.

klockmannite [MINERAL] CuSe A slate gray mineral consisting of copper selenide; occurs in granular aggregates.

Kloedenellacea [PALEON] A dimorphic superfamily of extinct ostracods in the suborder Kloedenellocopina having the posterior part of one dimorph longer and more inflated than the other dimorph.

Kloedenellocopina [PALEON] A suborder of extinct ostracods in the order Paleocopa characterized by a relatively straight dorsal border with a gently curved or nearly straight ventral border.

kloof wind [METEOROL] A cold southwest wind of Simons Bay, South Africa.

knebelite [MINERAL] $(Fe,Mn)_2SiO_4$ A mineral composed of an iron manganese silicate.

knick *See* knickpoint.

knick band *See* kink band.

knickpoint [GEOL] A point of sharp change of slope, especially in the longitudinal profile of a stream or of its valley. Also known as break; knick; nick; nickpoint; rejuvenation head; rock step.

knick zone *See* kink band.

knik wind [METEOROL] Local name for a strong southeast wind in the vicinity of Palmer in the Matanuska Valley of Alaska; it blows most frequently in the winter, although it may occur at any time of year.

knoll [GEOL] A mound rising less than 1000 meters from the sea floor. Also known as sea knoll.

knopite [MINERAL] A cerium-bearing variety of perovskite.

Knox and Oxborne furnace [MIN ENG] A continuously working shaft furnace for roasting quicksilver ores, having the fireplace built in the masonry at one side; the fuel is wood.

knuckle [MIN ENG] The place on an incline where there is a sudden change in grade.

knuckle man [MIN ENG] A worker who connects mine cars to and disconnects them from cables and also couples cars into trains.

Knudsen reversing water bottle [ENG] A type of frameless reversing bottle for collecting water samples; carries reversing thermometers.

Knudsen's tables [OCEANOGR] Hydrographical tables published by Martin Knudsen in 1901 to facilitate the computation of results of seawater chlorinity titrations and hydrometer temperature readings, and their conversion to salinity and density.

kobellite [MINERAL] $Pb_2(Bi,Sb)_2S_5$ A blackish-gray mineral composed of antimony bismuth lead sulfide.

Koch freezing process [MIN ENG] A process used to sink a shaft through a formation such as clay that will not sustain a shaft; magnesium chloride cooled to about $-30°C$ is circulated through pipes sunk in the ground until the ground is frozen.

koechlinite [MINERAL] Bi_2MoO_6 A greenish-yellow orthorhombic mineral composed of a bismuth molybdate.

Koehler lamp [MIN ENG] A naptha-burning flame safety lamp for use in gaseous mines.

koembang [METEOROL] A dry foehnlike wind from the southeast or south in Cheribon and Tegal in Java, caused by the east monsoon which develops a jet effect in passing through the gaps in the mountain ranges and descends on the leeward side.

koenenite [MINERAL] $Mg_5Al_2(OH)_{12}Cl_4$ A very soft mineral composed of a basic magnesium aluminum chloride.

Koepe hoist *See* Koepe winder.

Koepe shear [MIN ENG] A wheel used in place of a winding drum in the Koepe winder; made up of a cast steel hub with steel arms and a welded rim.

Koepe winder [MIN ENG] A hoisting system in which the winding drum is replaced by large wheels or sheaves over which passes an endless rope. Also known as Koepe hoist.

Koepe winder brake [MIN ENG] A device that works directly on the Koepe shear to slow or stop the hoist; can be applied by the engineman's brake lever or by safety devices.

koettigite [MINERAL] $Zn_3(AsO_4)_2 \cdot 8H_2O$ A carmine mineral composed of a hydrated zinc arsenate.

koktaite [MINERAL] $(NH_4)_2Ca(SO_4)_2 \cdot H_2O$ A mineral composed of a hydrous calcium ammonium sulfate.

kolbeckite [MINERAL] A blue to gray mineral composed of a hydrous beryllium aluminum calcium silicate and phosphate. Also known as sterrettite.

Kolmogoroff inertial subrange [OCEANOGR] The middle portion of the turbulence spectrum, between the low-wave-number (long-wave) part and the high-wave-number (short-wave) part.

kona [METEOROL] A stormy, rain-bringing wind from the southwest or south-southwest in Hawaii; it blows about five times a year on the southwest slopes, which are in the lee of the prevailing northeast trade winds.

kona cyclone [METEOROL] A slow-moving extensive cyclone which forms in subtropical latitudes during the winter season. Also known as kona storm.

kona storm *See* kona cyclone.

kongsbergite [MINERAL] A silver-rich variety of a native amalgam composed of silver (95%) and mercury (5%).

konimeter [ENG] An air-sampling device used to measure dust as in a cement mill or a mine; a measured volume of air drawn through a jet impacts on a glycerin-jelly-coated glass surface; the particles are counted with a microscope.

koninckite [MINERAL] $FePO_4 \cdot 3H_2O$ A yellow mineral composed of a hydrous ferric phosphate.

Köppen-Supan line [METEOROL] The isotherm connecting places which have a mean temperature of 10°C (50°F) for the warmest month of the year.

koppite [MINERAL] Mineral composed of a form of pyrochlore containing cerium, iron, and potassium.

Korfmann arch saver [MIN ENG] A machine that uses a controlled hydraulic system to withdraw steel arches.

Korfmann power loader [MIN ENG] A cutter-loader that is able to cart and load in both directions; its components are four drilling heads and one cutter chain surrounding them.

kornelite [MINERAL] $Fe_2(SO_4)_3 \cdot 7H_2O$ A colorless to brown mineral composed of hydrous ferric sulfate.

kornerupine [MINERAL] $(Mg,Fe,Al)_{20}(Si,B)_9O_{43}$ A colorless, yellow, brown, or sea-green mineral composed of magnesium iron borosilicate.

kossava [METEOROL]   A cold, very squally wind descending from the east or southeast in the region of the Danube "Iron Gate" through the Carpathians, continuing westward over Belgrade, then spreading northward to the Rumanian and Hungarian borderlands and southward as far as Nish.

kotoite [MINERAL] $Mg_3(BO_3)_2$ An orthorhombic borate mineral; it is isostructural with jimboite.

Kozeny's equation [PETRO ENG]   Mathematical relationship of flow network permeability to capillary pore dimensions; used for reservoir calculations.

Krakatao winds [METEOROL]   A layer of easterly winds over the tropics at an altitude of about 18 to 24 kilometers, which tops the mid-tropospheric westerlies (the antitrades), is at least 6 kilometers deep, and is based at about 2 kilometers above the tropopause.

Krassowski ellipsoid of 1938 [GEOD]   The reference ellipsoid of which the semimajor axis is 6,378,245 meters and the flattening or ellipticity equals 1/298.3.

krausite [MINERAL] $KFe(SO_4)_2 \cdot H_2O$ Yellowish-green mineral that is composed of hydrous potassium iron sulfate.

kremastic water *See* vadose water.

kremersite [MINERAL] $[NH_4,K]_2FeCl_5 \cdot H_2O$ A red mineral composed of hydrous potassium ammonium iron chloride, occurring in octahedral crystals.

krennerite [MINERAL] $AuTe_2$ A silver-white to pale-yellow mineral composed of gold telluride and often containing silver. Also known as white tellurium.

kribergite [MINERAL] $Al_5(PO_4)_3(SO_4)(OH)_4 \cdot 2H_2O$ White, chalklike mineral composed of hydrous basic aluminum sulfate and phosphate.

krohnkite [MINERAL] $Na_2Cu(SO_4)_2 \cdot 2H_2O$ An azure-blue monoclinic mineral composed of hydrous copper sodium sulfate, occurring in massive form.

krokidolite *See* crocidolite.

Krupp ball mill [MIN ENG]   An ore pulverizer in which the grinding is done by chilled iron or steel balls of various sizes moving against each other and the die ring, composed of five perforated spiral plates, each of which overlaps the next; material is discharged through a cylindrical screen.

kryolithionite [MINERAL] $Na_3Li_3(AlF_6)_2$ Variety of spodumene found in Greenland; has a crystal structure resembling that of garnet.

Kuehneosauridae [PALEON]   The gliding lizards, a family of Upper Triassic reptiles in the order Squamata including the earliest known aerial vertebrates.

kukersite [GEOL]   An organic sediment rich in remains of the alga *Gloeoxapsamorpha prisca*; found in the Ordovician of Estonia.

Kungurian [GEOL]   A European stage of geologic time; Middle Permian, above Artinskian, below Kazanian.

kunzite [MINERAL]   A pinkish gem variety of spodumene.

kurnakovite [MINERAL] $Mg_2B_6O_{11} \cdot 13H_2O$ A white mineral composed of hydrous magnesium borate.

Kuroshio Countercurrent [OCEANOGR]   A component of the Kuroshio system flowing south and southwest between latitudes 155° and 160°E about 70 kilometers from the coast of Japan on the right-hand side of the Kuroshio Current.

Kuroshio Current [OCEANOGR] A fast ocean current (2–4 knots) flowing northeast-ward from Taiwan to the Ryukyu Islands and close to the coast of Japan to about 150°E. Also known as Japan Current.

Kuroshio extension [OCEANOGR] A general term for the warm, eastward-transi-tional flow that connects the Kuroshio and the North Pacific currents.

Kuroshio system [OCEANOGR] A system of ocean currents which includes part of the North Equatorial Current, the Tsushima Current, the Kuroshio Current, and the Kuroshio extension.

kutnahorite [MINERAL] $Ca(Mn,Mg,Fe)(CO_3)_2$ A rare carbonate of calcium and manganese, found with some magnesium and iron substituting for manganese; forms rhombohedral crystals and is isomorphous with dolomite.

Kutorginida [PALEON] An order of extinct brachiopod mollusks that is unplaced tax-onomically.

kyanite [MINERAL] Al_2SiO_5 A blue or light-green neosilicate mineral; crystallizes in the triclinic system, and luster is vitreous to pearly; occurs in long, thin bladed crystals and crystalline aggregates. Also known as cyanite; disthene; sappare.

kyrohydratic point [OCEANOGR] The temperature at which a particular salt crystal-lizes in brine which is trapped by frozen seawater, the eutectic temperature of that salt.

L

labbé [METEOROL] An infrequent, moderate to strong southwest wind that occurs only in March in Provence (southeastern France), bringing mild, humid, and very cloudy or rainy weather, while on the coast it raises a rough sea.

labite [MINERAL] $MgSi_3O_6(OH)_2 \cdot H_2O$ A mineral composed of hydrous basic silicate of magnesium.

La Bour centrifugal pump [MIN ENG] A self-priming centrifugal pump with a trap which ensures suffcient water for the pump to function, and a separator to remove the entrained air in the water.

Labrador Current [OCEANOGR] A current that flows southward from Baffin Bay, through the Davis Strait, and southwestward along the Labrador and Newfoundland coasts.

labradorite [MINERAL] A gray, blue, green, or brown plagioclase feldspar with composition ranging from $Ab_{50}An_{50}$ to $Ab_{30}An_{70}$, where $Ab = NaAlSi_3O_8$ and $An = CaAl_2Si_2O_8$; in the course of formation when the natural material cools, the feldspar sometimes exhibits a variously colored luster. Also known as Labrador spar.

Labrador spar *See* labradorite.

Labyrinthodontia [PALEON] A subclass of fossil amphibians descended from crossopterygian fishes, ancestral to reptiles, and antecedent to at least part of other amphibian types.

laccolith [GEOL] A body of igneous rock intruding into sedimentary rocks so that the overlying strata have been notably lifted by the force of intrusion.

lacroixite [MINERAL] A pale yellowish-green mineral composed of basic phosphate of aluminum, calcium, manganese, and sodium (often with fluorine), occurring as crystals.

LACT *See* lease automatic custody transfer.

lacunaris *See* lacunosus.

lacunosus [METEOROL] A cloud variety characterized more by the appearance of the spaces between the cloud elements than by the elements themselves, the gaps being generally rounded, often with fringed edges, and the overall appearance being that of a honeycomb or net; it is the negative of clouds composed of separate rounded elements. Formerly known as lacunaris.

lacustrine [GEOL] Belonging to or produced by lakes.

lacustrine sediments [GEOL] Sediments that are deposited in lakes.

lacustrine soil [GEOL] Soil that is uniform in texture but variable in chemical composition and that has been formed by deposits in lakes which have become extinct.

ladder road *See* ladderway.

ladderway [MIN ENG]   Also known as ladder road. **1.** Mine shaft between two main levels, equipped with ladders. **2.** The particular shaft, or compartment of a shaft, containing ladders. Also known as manway.

Ladinian [GEOL]   A European stage of geologic time: upper Middle Triassic (above Anisian, below Carnian).

Lafond's Tables [METEOROL]   A set of tables and associated information for correcting reversing thermometers and computing dynamic height anomalies, compiled by E. C. Lafond and published by the U.S. Navy Hydrographic Office.

lag deposit [GEOL]   Residual accumulation of coarse, unconsolidated rock and mineral debris left behind by the winnowing of finer material.

lag fault [GEOL]   A minor low-angle thrust fault occurring within an overthrust; it develops when one part of the mass is thrust farther than an adjacent higher or lower part.

lag gravel [GEOL]   Residual accumulations of particles that are coarser than the material that has blown away.

lagoon [GEOGR]   **1.** A shallow sound, pond, or lake generally near but separated from or communicating with the open sea. **2.** A shallow fresh-water pond or lake generally near or communicating with a larger body of fresh water.

Lagrangian current measurement [OCEANOGR]   Observation of the speed direction of an ocean current by means of a device, such as a parachute drogue, which follows the water movement.

lahar [GEOL]   **1.** A mudflow or landslide of pyroclastic material occurring on the flank of a volcano. **2.** The deposit of mud or land so formed.

lake [HYD]   An inland body of water, small to moderately large, with its surface water exposed to the atmosphere.

lake breeze [METEOROL]   A wind, similar in origin to the sea breeze but generally weaker, blowing from the surface of a large lake onto the shores during the afternoon; it is caused by the difference in surface temperature of land and water, as in the land and sea breeze system.

lake effect [METEOROL]   Generally, the effect of any lake in modifying the weather about its shore and for some distance downwind; in the United States, this term is applied specifically to the region about the Great Lakes.

lake effect storm [METEOROL]   A severe snowstorm over a lake caused by the interaction between the warmer water and unstable air above it.

lake peat [GEOL]   A sedimentary peat formed near lakes.

lake plain [GEOL]   One of the surfaces of the earth that represent former lake bottoms; these featureless surfaces are formed by deposition of sediments carried into the lake by streams.

lamb-blasts *See* lambing storm.

Lambert conformal chart [MAP]   A chart on the Lambert conformal projection.

Lambert conformal projection [MAP]   A conformal conic projection with two standard parallels, or a conformal conic map projection in which the surface of a sphere or spheroid, such as the earth, is conceived as developed on a cone which intersects the sphere or spheroid at two standard parallels; the cone is then spread out to form a plane which is the map.

lambing storm [METEOROL]   A slight fall of snow in the spring in England. Also known as lamb-blasts; lamb-showers; lamb-storm.

lamb-showers *See* lambing storm.

lamb-storm *See* lambing storm.

lamellar crystal   [CRYSTAL]   A polycrystalline substance whose grains are in the form of thin sheets.

lamina   [GEOL]   A thin, clearly differentiated layer of sedimentary rock or sediment, usually less than 1 centimeter thick.

lampadite   [MINERAL]   A mineral composed chiefly of hydrous manganese oxide with as much as 18% copper oxide and often cobalt oxide.

lamp-charging rack   [MIN ENG]   Mine-lamp-charging racks which allow miners to store lamp units for recharging after daily use.

lamping   [MIN ENG]   Use of a portable ultraviolet lamp to reveal fluorescent minerals in prospecting.

lampman   [MIN ENG]   A person responsible for maintaining and servicing miners' lamps.

lamprobolite *See* basaltic hornblende.

lamp room   [MIN ENG]   A room or building at the surface of a mine for charging, servicing, and issuing all cap, hand, and flame safety lamps. Also known as lamp cabin; lamp station.

lamprophyllite   [MINERAL] $Na_2SrTiSi_2O_8$ A mineral composed of titanium strontium sodium silicate.

lamprophyre   [PETR]   Any of a group of igneous rocks characterized by a porphyritic texture in which abundant, large crystals of dark-colored minerals appear set in a not visibly crystalline matrix.

Lanarkian   [GEOL]   A European stage of geologic time forming part of the lower Upper Carboniferous, above Lancastrianand below Yorkian, equivalent to lowermost Westphalian.

lanarkite   [MINERAL] Pb_2OSO_4 A white, greenish, or gray monoclinic mineral consisting of basic lead sulfate, with specific gravity of 6.92; formed by action of heat and air on galena.

Lancastrian   [GEOL]   A European stage of geologic time forming part of the lower Upper Carboniferous, above Viséan and below Lanarkian.

land   [GEOGR]   The portion of the earth's surface that stands above sea level.

land and sea breeze   [METEOROL]   The complete cycle of diurnal local winds occurring on seacoasts due to differences in surface temperature of land and sea; the land breeze component of the system blows from land to sea, and the sea breeze blows from sea to land.

landblink   [METEOROL]   A yellowish glow observed over snow-covered land in the polar regions.

land breeze   [METEOROL]   A coastal breeze blowing from land to sea, caused by the temperature difference when the sea surface is warmer than the adjacent land; therefore, the land breeze usually blows by night and alternates with a sea breeze which blows in the opposite direction by day.

land bridge   [GEOGR]   A strip of land linking two landmasses, often subject to temporary submergence, but permitting intermittent migration of organisms.

Landenian   [GEOL]   A European stage of geologic time: upper Paleocene (above Montian, below Ypresian of Eocene).

lander   [MIN ENG]   **1.** A worker at one of the levels of a mine shaft to unload rock and to load drilling and blasting supplies to be lowered. **2.** In the quarry industry, one who guides, steadies, and loads trucks or railroad cars with the blocks of stone hoisted from the quarry floor. Also known as top hooker. **3.** In metal mining, one who cleans skips by directing a blast of compressed air into them through a hose, records number of loaded skips hoisted to surface, and loads railroad cars with ore from bins. **4.** In coal mining, one who works with shaft sinking crew at the top of the shaft or at a level immediately above shaft bottom, dumping rock into mine cars from a bucket in which it is raised. Also known as bucket dumper; landing tender; top lander. **5.** The worker who receives the loaded bucket or tub at the mouth of the shaft. Also known as banksman.

landesite [MINERAL] A brown mineral consisting of a hydrated phosphate of iron and manganese.

landfast ice *See* fast ice.

landform [GEOGR] All the physical, recognizable, naturally formed features of land, having a characteristic shape; includes major forms such as a plain, mountain, or plateau, and minor forms such as a hill, valley, or alluvial fan.

landform map *See* physiographic diagram.

land hemisphere [GEOGR] The half of the globe, with its pole located at 47.25°N 2.5°W, in which most of the earth's land area is concentrated.

land ice [HYD] Any part of the earth's seasonal or perennial ice cover which has formed over land as the result, principally, of the freezing of precipitation.

landing [MIN ENG] **1.** Level stage in a shaft at which cages are loaded and discharged. **2.** The top or bottom of a slope, shaft, or inclined plane.

landlocked [GEOGR] Pertaining to a harbor which is surrounded or almost completely surrounded by land.

land pebble *See* land pebble phosphate.

land pebble phosphate [GEOL] A pebble phosphate in a clay or sand bed below the ground surface; a small amount of uranium is often present and is recovered as a byproduct; used as a source of phosphate fertilizer. Also known as land pebble; land rock; matrix rock.

land rock *See* land pebble phosphate.

landscape [GEOGR] The distinct association of landforms that can be seen in a single view.

landscape agate [MINERAL] A type of chalcedony that is translucent and contains inclusions which give it an appearance reminiscent of familiar natural scenes. Also known as fortification agate.

land sky [METEOROL] The relatively dark appearance of the underside of a cloud layer when it is over land that is not snow-covered, used largely in polar regions with reference to the sky map; it is brighter than water sky, but much darker than iceblink or snowblink.

landslide [GEOL] The perceptible downward sliding or falling of a relatively dry mass of earth, rock, or combination of the two under the influence of gravity. Also known as landslip.

landslide track [GEOL] The exposed path apparent in the rock or earth after a landslide has occurred.

landslip *See* landslide.

langbanite [MINERAL] An ironblack hexagonal mineral composed of silicate and oxide of manganese, iron, and antimony, occurring in prismatic crystals.

langbeinite [MINERAL] $K_2Mg_2(SO_4)_3$ Colorless, yellowish, reddish, or greenish hexagonal mineral with vitreous luster, found in salt deposits; used in the fertilizer industry as a source of potassium sulfate.

Langevin ion *See* large ion.

langite [MINERAL] A blue to green mineral composed of basic hydrous copper sulfate.

lansan [METEOROL] A strong southeast trade wind of the New Hebrides and East Indies.

lansfordite [MINERAL] $MgCO_3 \cdot 5H_2O$ A mineral composed of hydrous basic carbonate of magnesium when extracted from the earth, changing to nesquehovite after exposure to the air.

lanthanite [MINERAL] $(La,Ce)_2(CO_3)_3 \cdot 8H_2O$ A colorless, white, pink, or yellow mineral composed of hydrous lanthanum carbonate, occurring in crystals or in earthy form.

lapilli [GEOL] Pyroclasts that range from 1 to 64 millimeters in diameter.

lapilli-tuff [GEOL] A pyroclastic deposit that is indurated and consists of lapilli in a fine tuff matrix.

lapis lazuli [PETR] An azure-blue, violet-blue, or greenish-blue, translucent to opaque crystalline rock used as a semiprecious stone; composed chiefly of lazurite and calcite with some haüyne, sodalite, and other minerals. Also known as lazuli.

lapse line [METEOROL] A curve showing the variation of temperature with height in the free air.

lapse rate [METEOROL] **1.** The rate of decrease of temperature in the atmosphere with height. **2.** Sometimes, the rate of change of any meteorological element with height.

Laramic orogeny *See* Laramidian orogeny.

Laramide orogeny *See* Laramidian orogeny.

Laramide revolution *See* Laramidian orogeny.

Laramidian orogeny [GEOL] An orogenic era typically developed in the eastern Rocky Mountains; phases extended from Late Cretaceous until the end of the Paleocene. Also known as Laramic orogeny; Laramide orogeny; Laramide revolution.

larderillite [MINERAL] $(NH_4)B_5O_8 \cdot 2H_2O$ A white mineral composed of hydrous ammonium borate, occurring as a crystalline powder.

lard ice *See* grease ice.

lardite *See* agalmatolite.

large ion [METEOROL] An ion created by a small ion attaching to an Aitken nucleus; it is characterized by relatively large mass and low mobility. Also known as heavy ion; Langevin ion; slow ion.

large nuclei [OCEANOGR] Particles of concentrated seawater or crystalline salt in the marine atmosphere having radii larger than 10^{-5} centimeter.

large scale [MAP] A scale of sufficient size to permit the plotting of much detail with exactness. [METEOROL] A scale such that the curvature of the earth may not be considered negligible; this scale is applicable to the high tropospheric long-wave patterns, with four or five waves around the hemisphere in the middle latitudes.

large-scale convection [METEOROL] Organized vertical motion on a larger scale than atmospheric free convection associated with cumulus clouds; the patterns of vertical motion in hurricanes or in migratory cyclones are examples of such convection.

larnite [MINERAL] β-Ca₂SiO₄ A gray mineral that is a metastable monoclinic phase of calcium orthosilicate, stable from 520 to 670°C. Also known as belite.

larry [MIN ENG] **1.** A car with a hopper bottom and adjustable chutes for feeding coke ovens. Also known as lorry. **2.** *See* barney.

larsenite [MINERAL] PbZnSiO₄ A colorless or white mineral composed of lead zinc silicate, occurring in orthorhombic crystals.

Larsen's pile [MIN ENG] A collection of hollow cylinders that increases resistance against bending and crumpling; useful for sinking a shaft in sand or gravel.

Larsen's spiles [MIN ENG] Steel sheet made in various forms to resist bending; used in place of wooden spiles in timbering a weak roof.

larvikite [PETR] An alkali syenite consisting of cryptoperthite or anorthoclase in rhombic crystals; used as an ornamental building material.

Lasater's bubble-point pressure correction [PETRO ENG] Relation of the gas-oil ratio in a high-pressure oil reservoir to the bubble-point-pressure factor.

laser anemometer [ENG] An anemometer in which the wind being measured passes through two perpendicular laser beams, and the resulting change in velocity of one or both beams is measured.

laser ceilometer [ENG] A ceilometer in which the time taken by a light pulse from a ground laser to travel straight up to a cloud ceiling and be reflected to a receiving photomultiplier is measured and converted into a cathode-ray display that indicates cloud-base height.

laser earthquake alarm [ENG] An early-warning system proposed for earthquakes, involving the use of two lasers with beams at right angles, positioned across a known geologic fault for continuous monitoring of distance across the fault.

laser seismometer [ENG] A laser interferometer system that detects seismic strains in the earth by measuring changes in distance between two granite piers located at opposite ends of an evacuated pipe through which a helium-neon or other laser beam makes a round trip; movements as small as 80 nanometers (one-eighth the wavelength of the 632.8-nanometer helium-neon laser radiation) can be detected.

laser transit [ENG] A transit in which a laser is mounted over the sighting telescope to project a clearly visible narrow beam onto a small target at the survey site.

lashing [MIN ENG] Planks nailed inside of frames or sets in a shaft to keep them in place. Also known as listing.

lashing chain [MIN ENG] A short chain to attach tubs to an overrope in endless rope haulage by wrapping the chain around the rope.

latent instability [METEOROL] The state of that portion of a conditionally unstable air column lying above the level of free convection; latent instability is released only if an initial impulse on a parcel gives it sufficient kinetic energy to carry it through the layer below the level of free convection, within which the environment is warmer than the parcel.

lateral [MIN ENG] **1.** In horizon mining, a hard heading branching off a horizon along the strike of the seams. **2.** A horizontal mine working.

lateral accretion [GEOL] The digging away of material at the outer bank of a meandering stream and the simultaneous building up to the water level by deposition of material brought there by pushing and rolling along the stream bottom.

lateral cone *See* adventive cone.

lateral erosion [GEOL] The action of a stream in undermining a bank on one side of its channel so that material falls into the stream and disintegrates; simultaneously, the stream shifts toward the bank that is being undercut.

lateral moraine [GEOL] Drift material, usually thin, that was deposited by a glacier in a valley after the glacier melted.

lateral planation [GEOL] Reduction in land in interstream areas in a plane parallel to the stream profile; the reduction is caused by lateral movement of the stream against its banks.

lateral secretion [GEOL] A supposed phenomenon whereby a lode's or vein's mineral content is derived from the adjacent wall rock.

laterite [GEOL] Weathered material composed principally of the oxides of iron, aluminum, titanium, and manganese; laterite ranges from soft, earthy, porous soil to hard, dense rock.

lateritic soil [GEOL] **1.** Soil containing laterite. **2.** Any reddish soil developed from weathering. Also known as latosol.

laterization [GEOL] Those conditions of weathering that lead to removal of silica and alkalies, resulting in a soil or rock with high concentrations of iron and aluminum oxides (laterite).

laterlog [ENG] A downhole resistivity measurement method wherein electric current is forced to flow radially through the formation in a sheet of predetermined thickness; used to measure the resistivity in hard-rock reservoirs as a method of determining subterranean structural features.

lath crib *See* lath frame.

lath door-set *See* lath frame.

lath frame [MIN ENG] A weak construction of laths, surrounding a main crib, the space between being for the insertion of piles. Also known as lath crib; lath door-set.

latite [PETR] A not visibly crystalline rock of volcanic origin composed chiefly of sodic plagioclase and alkali feldspar with subordinate quantities of dark-colored minerals in a finely crystalline to glassy groundmass.

latitude [GEOD] Angular distance from a primary great circle or plane, as on the celestial sphere or the earth.

latitude effect [GEOPHYS] The variation of a quantity with latitude; applied particularly to the increase in cosmic-ray intensity with increasing magnetic latitude.

latitude variation [GEOPHYS] A periodic change in the latitude of any position on the earth's surface, caused by the polar variation.

latosol *See* lateritic soil.

latrappite [MINERAL] $(Ca,Na)(Nb,Ti,Fe)O_3$ A variety of the mineral perovskite.

latrine cleaner [MIN ENG] A laborer who brings toilet cars in a mine to the surface on a cage and flushes the contents into a sewer. Also called sanitary nipper.

lattice [CRYSTAL] A regular periodic arrangement of points in three-dimensional space; it consists of all those points P for which the vector from a given fixed point to P has the form $n_1 \mathbf{a} + n_2 \mathbf{b} + n_3 \mathbf{c}$, where n_1, n_2, and n_3 are integers, and $\mathbf{a}$, $\mathbf{b}$, and $\mathbf{c}$ are fixed, linearly independent vectors. Also known as periodic lattice; space lattice.

lattice constant [CRYSTAL] A parameter defining the unit cell of a crystal lattice, that is, the length of one of the edges of the cell or an angle between edges. Also known as lattice parameter.

lattice defect *See* crystal defect.

lattice drainage pattern *See* rectangular drainage pattern.

lattice parameter *See* lattice constant.

Lattorfian *See* Tongrian.

laubmannite [MINERAL] $Fe_3Fe_6(PO_4)_4(OH)_2$ Mineral composed of basic ferrous iron phosphate and ferric iron phosphate.

Laue camera [CRYSTAL]   The apparatus used in the Laue method; the x-ray beam usually enters through a hole in the x-ray film, which records beams bent through an angle of nearly 180° by the crystal; less commonly, the film is placed beyond the crystal.

Laue condition [CRYSTAL]   **1.** The condition for a vector to lie in a Laue plane: its scalar product with a specified vector in the reciprocal lattice must be one-half of the scalar product of the latter vector with itself. **2.** *See* Laue equations.

Laue equations [CRYSTAL]   Three equations which must be satisfied for an x-ray beam of specified wavelength to be diffracted through a specified angle by a crystal; they state that the scaler products of each of the crystallographic axial vectors with the difference between unit vectors in the directions of the incident and scattered beams, are integral multiples of the wavelength. Also known as Laue condition.

Laue method [CRYSTAL]   A method of studying crystalline structures by x-ray diffraction, in which a finely collimated beam of polychromatic x-rays falls on a single crystal whose orientation can be set as desired, and diffracted beams are recorded on a photographic film.

Laue pattern [CRYSTAL]   The characteristic photographic record obtained in the Laue method.

Laue plane [CRYSTAL]   A plane which is the perpendicular bisector of a vector in the reciprocal lattice; such planes form the boundaries of Brillouin zones.

Laue theory [CRYSTAL]   A theory of diffraction of x-rays by crystals, based on the Laue equations.

Laugiidae [PALEON]   A family of Mesozoic fishes in the order Coelacanthiformes.

laumontite [MINERAL] $CaAl_2Si_4O_{12} \cdot 4H_2O$ A white zeolite mineral crystallizing in the monoclinic system; loses water on exposure to air, eventually becoming opaque and crumbling. Also known as laumonite; lomonite; lomontite.

launder screen [MIN ENG]   A screen used in a launder for the sizing and dewatering of small sizes of anthracite.

launder separation process [MIN ENG]   A hydraulic process for separating heavy gravity product from the lighter product that flows above it; a stream of fluid carries material down a channel that has draws to separate the heavy material from the light material.

launder washer [MIN ENG]   A type of coal washer in which the coal is separated from the refuse by stratification due to hindered settling while being carried in aqueous suspension through the launder.

lauoho o pele *See* Pele's hair.

Laurasia [GEOL]   A continent theorized to have existed in the Northern Hemisphere; supposedly it broke up to form the present northern continents about the end of the Pennsylvania period.

Laurentian Plateau *See* Laurentian Shield.

Laurentian Shield [GEOL]   A Precambrian plateau extending over half of Canada from Labrador southwest along Hudson Bay and northwest to the Arctic Ocean. Also known as Canadian Shield; Laurentian Plateau.

Laurentide ice sheet [HYD]   A major recurring glacier that at its maximum completely covered North America east of the Rockies from the Arctic Ocean to a line passing through the vicinity of New York, Cincinnati, St. Louis, Kansas City, and the Dakotas.

laurionite [MINERAL] Pb(OH)Cl A colorless mineral composed of basic lead chloride, occurring in prismatic crystals; it is dimorphous with paralaurionite.

laurite [MINERAL] RuS_2 A black mineral composed of ruthenium sulfide (often with osmium), occurring as small crystals or grains.

lausenite [MINERAL] $Fe_2(SO_4)_3 \cdot 6H_2O$ A white, monoclinic mineral consisting of hydrated ferric sulfate; occurs in lumpy aggregates of fibers.

lautarite [MINERAL] $Ca(IO_3)_2$ A monoclinic mineral composed of calcium iodate that occurs in prismatic crystals.

lautite [MINERAL] CuAsS A mineral composed of copper sulfide and copper arsenide.

lava [GEOL] **1.** Molten extrusive material that reaches the earth's surface through volcanic vents and fissures. **2.** The rock mass formed by consolidation of molten rock issuing from volcanic vents and fissures, consisting chiefly of magnesium silicate; used for insulators.

lava blisters [GEOL] Small, steep-sided swellings that are hollow and raised on the surfaces of some basaltic lava flows; formed by gas bubbles pushing up the lava's viscous surface.

lava cone [GEOL] A volcanic cone that was formed of lava flows.

lava dome *See* shield volcano.

lava field [GEOL] A wide area of lava flow; it is commonly several square kilometers in area and forms along the base of a large compound volcano or on the flanks of shield volcanoes.

lava flow [GEOL] **1.** A lateral, surficial stream of molten lava issuing from a volcanic cone or from a fissure. **2.** The solidified mass of rock formed when a lava stream congeals.

lava fountain [GEOL] A jetlike eruption of lava that issues vertically from a volcanic vent or fissure. Also known as fire fountain.

lava lake [GEOL] A lake of lava that is molten and fluid; usually contained within a summit volcanic crater or in a pit crater on the flanks of a shield volcano.

lava plateau [GEOL] An elevated tableland or flat-topped highland that is several hundreds to several thousands of square kilometers in area; underlain by a thick succession of lava flows.

lava tube [GEOL] A long, tubular opening under the crust of solidified lava.

lavenite [MINERAL] $(Na,Ca)_3Zr(Si_2O_7)(O,OH,F)_2$ A mineral composed of complex silicate, occurring in prismatic crystals.

law of constant angles [CRYSTAL] The law that the angles between the faces of a crystal remain constant as the crystal grows.

law of rational intercepts *See* Miller law.

law of storms [METEOROL] Historically, the general statement of the manner in which the winds of a cyclone rotate about the cyclone's center, and the way that the entire disturbance moves over the earth's surface.

law of superposition [GEOL] The law that strata underlying other strata must be the older if there has been neither overthrust nor inversion.

lawrencite [MINERAL] $(Fe,Ni)Cl_2$ A brown or green mineral composed of ferrous chloride and found as an abundant accessory mineral in iron meteorites.

lawsonite [MINERAL] $CaAl_2(Si_2O_7)(OH)_2 \cdot H_2O$ A colorless or grayish-blue mineral crystallizing in the orthorhombic system; found in gneisses and schists.

lay-by [MIN ENG] Siding in single-track underground tramming road.

layer [GEOL] A tabular body of rock, ice, sediment, or soil lying parallel to the supporting surface and distinctly limited above and below.   [GEOPHYS] One of several strata of ionized air, some of which exist only during the daytime, occurring at altitudes between 50 and 400 kilometers; the layers reflect radio waves at certain frequencies and partially absorb others.

layer depth [OCEANOGR] **1.** The thickness of the mixed layer in an ocean. **2.** The depth to the top of the thermocline.

layering of firedamp [MIN ENG] The formation of a layer of firedamp at the roof of a mine working and above the ventilating air current.

layer lattice *See* layer structure.

layer loading [MIN ENG] A procedure whereby the coal is placed in the railroad cars in horizontal layers.

layer of no motion [OCEANOGR] A layer, assumed to be at rest, at some depth in the ocean.

layer silicate *See* phyllosilicate.

layer structure [CRYSTAL] A crystalline structure found in substances such as graphites and clays, in which the atoms are largely concentrated in a set of parallel planes, with the regions between the planes comparatively vacant. Also known as layer lattice.

lazuli *See* lapis lazuli.

lazulite [MINERAL] $(Mg,Fe)Al_2(OH)_2(PO_4)_2$ A violet-blue or azure-blue mineral with vitreous luster; composed of basic aluminum phosphate and occurring in small masses or monoclinic crystals; hardness is 5–6 on Mohs scale, and specific gravity is 3.06–3.12. Also known as berkeyite; blue spar; false lapis.

lazurite [MINERAL] $(Na,Ca)_8(Al,Si)O_{24}(S,SO_4)$ A blue or violet-blue feldspathoid mineral crystallizing in the isometric system; the chief mineral constituent of lapis lazuli.

LCL *See* lifting condensation level.

leaching [GEOCHEM] The separation or dissolving out of soluble constituents from a rock or ore body by percolation of water. [MIN ENG] Dissolving soluble minerals or metals out of the ore, as by the use of percolating solutions such as cyanide or chlorine solutions, acids, or water. Also known as lixiviation.

leach material [MIN ENG] Material sufficiently mineralized to be economically recoverable by leaching.

leach pile [MIN ENG] Mineralized materials stacked so as to permit wanted minerals to be effectively and selectively dissolved by leaching.

lead [GEOL] A small, narrow passage in acave.

leader [GEOPHYS] The streamer which initiates the first phase of each stroke of a lightning discharge; it is a channel of very high ion density which propagates through the air by the continual establishment of an electron avalanche ahead of its tip. Also known as leader streamer.

leader streamer *See* leader.

leader stroke [GEOPHYS] The entire set of events associated with the propagation of any leader between cloud and ground in a lightning discharge.

lead glance *See* galena.

leadhillite [MINERAL] $Pb_4(SO_4)(CO_3)_2(OH)_2$ A yellowish or greenish- or grayish-white monoclinic mineral consisting of basic sulfate and carbonate of lead; dimorphous with susanite.

lead-in-air indicator   [MIN ENG]   An instrument that utilizes reagents to measure the concentration of lead in the air.

leading heading   [MIN ENG]   **1.** The heading of a pair of parallel headings that is a short distance in front of the other; used to drain a mining area.   **2.** A heading dug into the solid coal before the advance of the general face.

leading stone   *See* lodestone.

lead marcasite   *See* sphalerite.

lead ocher   *See* massicot.

lead spar   *See* anglesite.

lead vitriol   *See* anglesite.

leaf mold   [GEOL]   A soil layer or compost consisting principally of decayed vegetable matter.

leaf peat   *See* paper peat.

leakage   [MIN ENG]   An unintentional diversion of ventilation air from its designed path.

leakage halo   [GEOCHEM]   The dispersion of elements along channels and paths followed by mineralizing solutions leading into and away from the central focus of mineralization.

leakage intake system   [MIN ENG]   A ventilation circuit with two adjacent intake roadways leading to the coal face.

leapfrog system   [MIN ENG]   A system used in mining coal on a longwall face; self-advancing supports are used, with alternate supports advancing as each web of coal is removed.

lease automatic custody transfer   [PETRO ENG]   Automatic unattended system to receive and record oil produced from a drilling lease, then to transfer the contents to a pipeline. Abbreviated LACT.

lease-distribution system   [PETRO ENG]   Any of the electrical distribution systems serving oil-field pump motors and other electrical equipment.

lease tank   [PETRO ENG]   An oil-field storage tank that stores oil flowing from designated wells.

lechatelierite   [MINERAL]   A natural silica glass, occurring in fulgurites and impact craters and formed by the melting of quartz sand at high temperatures generated by lightning or by the impact of a meteorite.

lecontite   [MINERAL] $Na(NH_4,K)SO_4·2H_2O$ Colorless mineral composed of a hydrous sodium potassium ammonium sulfate; found in bat guano.

ledge   [GEOL]   **1.** A narrow, shelflike ridge or projection of rock, usually horizontal and much greater in length than in height, occurring in a rock wall or on a cliff face.   **2.** An underwater ridge of rocks, especially a ridge close to the shore or connecting with and fringing the shore.

Ledian   [GEOL]   Lower upper Eocene geologic time. Also known as Auversian.

lee dune   [GEOL]   A dune formed to the leeward of a source of sand or of an obstacle.

Lee-Norse miner   [MIN ENG]   A continuous miner for driving headings in medium or thick coal seams.

lee tide   *See* leeward tidal current.

lee trough   *See* dynamic trough.

leeward tidal current [OCEANOGR] A tidal current setting in the same direction as that in which the wind is blowing. Also known as lee tide; leeward tide.

leeward tide *See* leeward tidal current.

left bank [GEOGR] The bank of a stream or river on the left of an observer when he is facing in the direction of flow, or downstream.

left-handed [CRYSTAL] Having a crystal structure with a mirror-image relationship to a right-handed structure.

left lateral fault [GEOL] A fault in which movement is such that an observer walking toward the fault along an index plane (a bed, vein, or dike) would turn to the left to find the other part of the displaced index plane. Also known as sinistral fault.

leg [GEOPHYS] A single cycle of more or less periodic motion in a wave train on a seismogram. [MIN ENG] **1.** In mine timbering, a prop or upright member of a set or frame. **2.** A stone that has to be wedged out from beneath a larger one.

legrandite [MINERAL] $Zn_{14}(OH)(AsO_4)_9 \cdot 12H_2O$ A yellow to nearly colorless mineral composed of basic hydrous zinc arsenate.

leg wire [ENG] One of the two wires forming a part of an electric blasting cap or squib.

Lehigh jig [MIN ENG] A plunger-type jig in which check valves open on the upstroke of the plunger, makeup water is introduced with the feed, the screen plate is at two levels, and the bottom of the discharge end is hinged; used to wash anthracite.

lehite [MINERAL] $(Na,K)_2Ca_5Al_8(PO_4)_8(OH)_{12} \cdot 6H_2O$ White mineral composed of hydrous basic calcium aluminum phosphate.

Lehmann process [MIN ENG] A process for treating coal by disintegration and separation of the petrographic constituents.

leifite [MINERAL] $Na_2AlSi_4O_{10}F$ A colorless mineral composed of fluoride and silicate of sodium and aluminum.

leightonite [MINERAL] $K_2Ca_2Cu(SO_4)_4 \cdot 2H_2O$ A pale-blue mineral composed of hydrous sulfate of copper, calcium, and potassium.

lengenbachite [MINERAL] $Pb_6(Ag,Cu)_2As_4S_{13}$ A steel gray mineral consisting of lead, silver, and copper arsenic sulfide.

length of record [CLIMATOL] The period during which observations have been maintained at a meteorological station, and which serves as the frame of reference for climatic data at that station.

length of shot [ENG] The depth of the shothole, in which powder is placed, or the size of the block of coal or rock to be loosened by a single blast, measured parallel with the hole. [MIN ENG] In open-pit mining, the distance from the first drill hole to the last drill hole along the bank.

lens [GEOL] A geologic deposit that is thick in the middle and converges toward the edges, resembling a convex lens.

lenticle [GEOL] A bed or rock stratum or body that is lens-shaped.

lenticular cloud *See* lenticularis.

lenticularis [METEOROL] A cloud species, the elements of which have the form of more or less isolated, generally smooth lenses; the outlines are sharp. Also known as lenticular cloud.

lentil [GEOL] **1.** A rock body that is lens-shaped and enclosed in a stratum of different material. **2.** A rock stratigraphic unit that is a subdivision of a formation and has limited geographic extent; it thins out in all directions.

Leonardian [GEOL] A North American provincial series: Lower Permian (ab ve Wolfcampian, below Guadalupian).

Leon firedamp tester [MIN ENG] A device, based on a form of Wheatstone bridge, that is used to detect firedamp.

leonite [MINERAL] $K_2Mg(SO_4)_2 \cdot 4H_2O$ A colorless, white, or yellowish mineral composed of hydrous magnesium potassium sulfate, occurring in monoclinic crystals.

leopoldite *See* sylvite.

Leperditicopida [PALEON] An order of extinct ostracods characterized by very thick, straight-backed valves which show unique muscle scars and other markings.

Leperditillacea [PALEON] A superfamily of extinct paleocopan ostracods in the suborder Kloedenellocopina including the unisulcate, nondimorphic forms.

lepidoblastic [PETR] Of the texture of a metamorphic rock, having a fabric of minerals characterized as flaky or scaly, such as mica.

lepidocrocite [MINERAL] α-FeO(OH) A ruby- or blood-red mineral crystallizing in the orthorhombic system; it is associated with limonite in iron ores and is a component of meteorites.

Lepidodendrales [PALEOBOT] The giant club mosses, an order of extinct lycopods (Lycopodiopsida) consisting primarily of arborescent forms characterized by dichotomous branching, small amounts of secondary vascular tissue, and heterospory.

lepidolite [MINERAL] $K(Li,Al)_3(Si,Al)_4O_{10}(F,OH)_2$ A rose-colored mineral of the mica group crystallizing in the monoclinic system. Also known as lithionite; lithium mica.

lepidomelane [MINERAL] A black variety of biotite that is characterized by the presence of large amounts of ferric iron. Also known as iron mica.

lepisphere [PETR] A microspherical aggregate of platy, blade-shaped crystals of opal-CT.

Lepospondyli [PALEON] A subclass of extinct amphibians including all forms in which the vertebral centra are formed by ossification directly around the notochord.

Leptictidae [PALEON] A family of extinct North American insectivoran mammals belonging to the Proteutheria which ranged from the Cretaceous to middle Oligocene.

leptite [PETR] A quartz-feldspathic metamorphic rock that is fine-grained with little or no foliation; formed by regional metamorphism of the highest grade.

Leptochoeridae [PALEON] An extinct family of palaeodont artiodactyl mammals in the superfamily Dichobunoidea.

leptogeosyncline [GEOL] A deep oceanic trough that has not been filled with sedimentation and is associated with volcanism.

Leptolepidae [PALEON] An extinct family of fishes in the order Leptolepiformes representing the first teleosts as defined on the basis of the advanced structure of the caudal skeleton.

Leptolepiformes [PALEON] An extinct order of small, ray-finned teleost fishes characterized by a relatively strong, ossified axial skeleton, thin cycloid scales, and a preopercle with an elongated dorsal portion.

lesser ebb [OCEANOGR] The weaker of two ebb currents occurring during a tidal day.

lesser flood [OCEANOGR] The weaker of two flood currents occurring during a tidal day.

leste [METEOROL]   Spanish nautical term for east wind, specifically the hot, dry, dusty easterly or southeasterly wind which blows from the Atlantic coast of Morocco out to Madeira and the Canary Islands; it is a form of sirocco, occurring in front of depressions advancing eastward.

letovicite [MINERAL] $(NH_3)_3H(SO_4)_2$ A mineral composed of acid ammonium sulfate.

leucite [MINERAL] $KAlSi_2O_6$ A white or gray rock-forming mineral belonging to the feldspathoid group; at ordinary temperatures the mineral exists as aggregates of trapezohedral crystals with glassy fracture; hardness is 5.5–6.0 on Mohs scale, and specific gravity is 2.45–2.50. Also known as amphigene; grenatite; vesuvian; Vesuvian garnet; white garnet.

leucite phonolite [PETR]   An extrusive rock composed of alkali feldspar, mafic minerals, and leucite.

leucitite [PETR]   A fine-grained or porphyritic extrusive rock or hypabyssal igneous rock composed mostly of pyroxene and leucite.

leucitohedron *See* trapezohedron.

leucochalcite *See* olivenite.

leucocratic [PETR]   Light-colored as applied to igneous rock containing 0–50% dark-colored minerals.

leucophanite [MINERAL] $(Na,Ca)_2BeSi_2(O,F,OH)_7$ Greenish mineral composed of beryllium sodium calcium silicate containing fluorine and occurring in glassy, tabular crystals.

leucophosphite [MINERAL] $K_2Fe_4(PO_4)_4(OH)_2 \cdot 9H_2O$ White mineral composed of hydrous basic phosphate of potassium and iron.

leucopyrite *See* loellingite.

leucosphenite [MINERAL] $Na_4BaTi_2Si_{10}O_{27}$ A white mineral composed of sodium barium silicotitanate and occurring as wedge-shaped crystals.

leucoxene [MINERAL]   A mineral composed of rutile with some anatase or sphene; occurs in igneous rocks, usually as an alteration product of ilmenite.

leuneburgite [MINERAL] $Mg_3B_2(PO_4)_2(OH)_6 \cdot 5H_2O$ A colorless mineral consisting of a hydrous basic phosphate of magnesium and boron.

levante [METEOROL]   The Spanish and most widely used term for an east or northeast wind occurring along the coast and inland from southern France to the Straits of Gibraltar; it is moderate or fresh, mild, very humid, and rainy, and occurs with a depression over the western Mediterranean Sea.

levantera [METEOROL]   A persistent east wind of the Adriatic, usually accompanied by cloudy weather.

Levantine Basin [OCEANOGR]   A basin in the Mediterranean Ocean between Asia Minor and Egypt.

leveche [METEOROL]   A warm wind in Spain, either a foehn or a hot southerly wind in advance of a low-pressure area moving from the Sahara Desert.

levee [GEOL]   **1.** An embankment bordering one or both sides of a sea channel or the low-gradient seaward part of a canyon or valley. **2.** A low ridge sometimes deposited by a stream on its sides.

level [MIN ENG]   **1.** Mine workings that are at the same elevation. **2.** A gutter for the water to run in.

level crosscut [MIN ENG]   A horizontal crosscut.

level fold *See* nonplunging fold.

leveling [MIN ENG] Measurement of rises and falls, heights, and contour lines.

level of free convection [METEOROL] The level at which a parcel of air lifted dry and adiabatically until saturated, and lifted saturated and adiabatically thereafter, would first become warmer than its surroundings in a conditionally unstable atmosphere. Abbreviated LFC.

level of nondivergence [METEOROL] A level in the atmosphere throughout which the horizontal velocity divergence is zero; although in some meteorological situation there may be several such surfaces (not necessarily level), the level of nondivergence usually considered is that mid-tropospheric surface which separates the major regions of horizontal convergence and divergence associated with the typical vertical structure of the migratory cyclonic-scale weather systems.

level of saturation *See* water table.

level surface [GEOPHYS] *See* geopotential surface.

levitation [MIN ENG] In froth flotation, raising of particles in a froth to the surface of the pulp, to facilitate separation of selected minerals in the froth.

levyine *See* levynite.

levyite *See* levynite.

levyne *See* levynite.

levynite [MINERAL] $NaCa_3Al_7Si_{11}O_{36} \cdot 15H_2O$ A white or light-colored mineral of the zeolite group, composed of hydrous silicate of aluminum, sodium, and calcium, and occurring in rhombohedral crystals. Also known as levyine; levyite; levyne.

lewis hole [MIN ENG] A series of two or more holes drilled as closely together as possible, but then connected by knocking out the thin partition between them, thus forming one wide hole having its greatest diameter in a plane with the desired rift.

lewisite [MINERAL] $(Ca,Fe,Na)_2$ A titanian romeite mineral. [ORG CHEM] $C_2H_2AsCl_3$ An oily liquid, colorless to brown or violet; forms a toxic gas, used in World War I.

lewistonite [MINERAL] $(Ca,K,Na)_5(PO_4)_3(OH)$ White mineral composed of basic calcium potassium sodium phosphate.

lherzolite [PETR] Peridotite composed principally of olivine with orthopyroxene and clinopyroxene.

Liassic [GEOL] The Lower Jurassic period of geologic time. Also known as Lias.

libeccio [METEOROL] Italian name for a southwest wind, used in northern Corsica for the west or southwest wind which blows throughout the year, especially in winter when it is often stormy.

libethenite [MINERAL] $Cu_2(O_4)OH$ An olive-green mineral composed of basic copper sulfate, occurring as small prismatic crystals or in masses.

liebigite [MINERAL] $Ca_2U(CO_3)_4 \cdot 10H_2O$ An apple- or yellow-green mineral composed of hydrous uranium calcium carbonate; occurs as a coating or concretion in rock.

Liesegang banding [GEOL] Colored or compositional rings or bands in a fluid-saturated rock due to rhythmic precipitation. Also known as Liesegang rings.

Liesegang rings *See* Liesegang banding.

life of mine [MIN ENG] The time in which, through the employment of the available capital, the ore reserves—or such reasonable extension of the ore reserves as conservative geological analysis may justify—will be extracted.

lift [MIN ENG] **1.** The vertical height traveled by a cage in a shaft. **2.** The distance between the first level and the surface or between any two levels. **3.** Any of the various gangways from which coal is raised at a slope colliery.

lifter [MIN ENG] A shothole drilled near the floor when tunneling, and fired subsequent to the cut and relief holes.

lifter case [MIN ENG] The sleeve or tubular part attached to the lower end of the inner tube of M-design core barrels and some other types of core barrels, in which is fitted a core lifter. Also known as core-catcher case; core-gripper case; core-lifter case; core-spring case; inner-tube extension; ring-lifter case; spring-lifter case.

lifting condensation level [METEOROL] The level at which a parcel of moist air lifted dry adiabatically would become saturated. Abbreviated LCL. Also known as isentropic condensation level (ICL).

lifting dog [ENG] **1.** A component part of the overshot assembly that grasps and lifts the inner tube or a wire-line core barrel. **2.** A clawlike hook for grasping cylindrical objects, such as drill rods or casing, while raising and lowering them.

lifting guard [MIN ENG] Fencing placed around the mouth of a shaft and lifted out of the way by the ascending cage.

light crude [PETRO ENG] Crude oil having a high proportion of low-viscosity, low-molecular-weight hydrocarbons.

light freeze [METEOROL] The condition when the surface temperature of the air drops to below the freezing point of water for a short time period, so that only the tenderest plants and vines are adversely affected.

light frost [HYD] A thin and more or less patchy deposit of hoarfrost on surface objects and vegetation.

light ion *See* small ion.

light mineral [MINERAL] **1.** A rock with minerals that have a specific gravity lower than a standard, usually 2.85. **2.** A light-colored mineral.

lightning [GEOPHYS] The large spark produced by an abrupt discontinuous discharge of electricity through the air, resulting most often from the creation and separation of electric charge in cumulonimbus clouds.

lightning channel [GEOPHYS] The irregular path through the air along which a lightning discharge occurs.

lightning discharge [GEOPHYS] The series of electrical processes by which charge is transferred within the atmosphere along a channel of high ion density between electric charge centers of opposite sign.

lightning flash [GEOPHYS] In atmospheric electricity, the total observed luminous phenomenon accompanying a lightning discharge.

lightning stroke [GEOPHYS] Any one of a series of repeated discharges comprising a single lightning discharge (or lightning flash); specifically, in the case of the cloud-to-ground discharge, a leader plus its subsequent return streamer.

light-of-the-night-sky *See* airglow.

light pillar *See* sun pillar.

light-red silver ore *See* proustite.

light-ruby silver *See* proustite.

lignite [GEOL] Coal of relatively recent origin, intermediate between peat and bituminous coal; often contains patterns from the wood from which it formed. Also known as brown coal.

lignite A *See* black lignite.

lignite B *See* brown lignite.

lillianite [MINERAL] $Pb_3Bi_2S_6$ A steel-gray mineral composed of lead bismuth sulfide.

lily-pad ice *See* pancake ice.

limb [GEOL] One of the two sections of an anticline or syncline on either side of the axis. Also known as flank.

limburgite [PETR] A dark, glass-rich igneous rock with abundant large crystals of olivine and pyroxene and with little or no feldspar.

lime-pan playa [GEOL] A playa with a smooth, hard surface composed of calcium carbonate.

limestone [PETR] **1.** A sedimentary rock composed dominantly (more than 95%) of calcium carbonate, principally in the form of calcite; examples include chalk and travertine. **2.** Any rock containing 80% or more of calcium carbonate or magnesium carbonate.

limestone log [ENG] A log that employs an electrical resistivity element in the form of four symmetrically arranged current electrodes to give accurate readings in borehole surveying of hard formations.

limestone pebble conglomerate [GEOL] A well-sorted conglomerate composed of limestone pebbles resulting from special conditions involving rapid mechanical erosion and short transport distances.

limit of the atmosphere [GEOPHYS] The level at which the atmospheric density becomes the same as the density of interplanetary space, which is usually taken to be about one particle per cubic centimeter.

limnimeter [ENG] A type of tide gage for measuring lake level variations.

limnograph [ENG] A recording made on a limnimeter.

limonite [MINERAL] A group of brown or yellowish-brown, amorphous, naturally occurring ferric oxides of variable composition; commonly formed secondary material by oxidation of iron-bearing minerals; a minor ore of iron. Also known as brown hematite; brown iron ore.

linarite [MINERAL] $PbCu(SO_4)(OH)_2$ A deep-blue mineral composed of basic lead copper sulfate and occurring as monoclinic crystals.

lindackerite [MINERAL] $Cu_6Ni_3(AsO_4)_4(SO_4)(OH)_4$ A light-green or apple-green mineral composed of hydrous basic sulfate and arsenate of nickel and copper; occurs in tabular crystals or massive.

Linde drill *See* fusion-piercing drill.

lindgrenite [MINERAL] $Cu_3(MoO_4)_2(OH)_2$ A green mineral composed of basic copper molybdate.

lindstromite [MINERAL] $PbCuBi_3S_6$ A lead-gray to tin-white mineral composed of bismuth copper lead sulfide.

lineage structure [CRYSTAL] An imperfection structure characterizing a crystal, parts of which have slight differences in orientation.

lineament [GEOL] A straight or gently curved, lengthy topographic feature expressed as depressions or lines of depressions. Also known as linear.

linear *See* lineament.

linear cleavage [GEOL]   The property of metamorphic rocks of breaking into long planar fragments.

linear flow structure *See* platy flow structure.

lineation [GEOL]   Any linear structure on or within a rock; examples are ripple marks and flow lines.

line blow [METEOROL]   A strong wind on the equator side of an anticyclone, probably so called because there is little shifting of wind direction during the blow, as contrasted with the marked shifting which occurs with a cyclonic windstorm.

line brattice [MIN ENG]   A partition in an opening to divide it into intake and return airways.

line clinometer [ENG]   A clinometer designed to be inserted between rods at any point in a string of drill rods.

line defect *See* dislocation.

line drilling [MIN ENG]   The combined methods of drilling and broaching for the primary cut in quarrying; deep, closely spaced holes are drilled in a straight line by means of a reciprocating drill, and webs between holes are removed by a drill or a flat broaching tool.

line gale *See* equinoctial storm.

line map *See* planimetric map.

line of strike *See* strike.

liner [MIN ENG]   **1.** A foot piece for uprights in timber sets. **2.** Timber supports erected to reinforce existing sets which are beginning to collapse due to heavy strata pressure. **3.** A bar put up between two other bars to assist in carrying the roof. **4.** Replaceable facings inside a grinding mill.

line squall [METEOROL]   A squall that occurs along a squall line.

line storm *See* equinoctial storm.

line timbers [MIN ENG]   Timbers placed along the sides of the track of a working place in rows according to a predetermined plan.

line up [MIN ENG]   **1.** A command signifying that the drill runner wants the hoisting cable attached to the drill stem, threaded through the sheave wheel, or wound on the hoist drum. **2.** To reposition a drill so that the drill stem is centered over the parallel to a newly collared drill hole.

linguloid ripple mark *See* linguoid ripple mark.

linguoid current ripple *See* linguoid ripple mark.

linguoid ripple mark [GEOL]   An aqueous current ripple mark with tonguelike projections which are formed by action of a current of water and which point into the current. Also known as cuspate ripple mark; linguloid ripple mark; linguoid current ripple.

lining sight [MIN ENG]   An instrument consisting of a plate with a slot in the middle, and the means of suspending it; used with a plumbline for directing the courses of underground drifts or headings.

link bar *See* hinged bar.

Linke scale [METEOROL]   A type of cyanometer; used to measure the blueness of the sky; it is simply a set of eight cards of different standardized shades of blue, numbered (evenly) 2 to 16; the odd numbers are used by the observer if the sky color lies between any of the given shades. Also known as blue-sky scale.

linnaeite [MINERAL] $(Co,Ni)_3S_4$ A steel-gray mineral with a coppery-red tarnish, occurring in isometric crystals; an ore of cobalt. Also known as cobalt pyrites; linneite.

linneite *See* linnaeite.

Lipalian [GEOL] A hypothetical geologic period that supposedly antedated the Cambrian.

Lipostraca [PALEON] An order of the subclass Branchiopoda erected to include the single fossil species *Lepidocaris rhyniensis*.

lipper [OCEANOGR] **1.** Slight roughness or ruffling occurring on a water surface. **2.** Light spray originating from small waves.

liptinite *See* exinite.

liquid blocking [PETRO ENG] The blocking or plugging of the sand around an injection-well borehole, usually caused by lubricant carryover from compressors.

liquid-dominated hydrothermal reservoir [GEOL] Any geothermal system mainly producing superheated water (often termed brines); hot springs, fumaroles, and geysers are the surface expressions of hydrothermal reservoirs; an example is the hot-brine region in the Imperial Valley–Salton Sea area of southern California.

liquid-filled porosity [GEOL] The condition in porous rock or sand formations in which pore spaces contain fresh or salt water, liquid petroleum, pressure-liquefied butane or propane, or tar.

liquid-in-glass thermometer [ENG] A thermometer in which the thermally sensitive element is a liquid contained in a graduated glass envelope; the indication of such a thermometer depends upon the difference between the coefficients of thermal expansion of the liquid and the glass; mercury and alcohol are liquids commonly used in meteorological thermometers.

liquid-in-metal thermometer [ENG] A thermometer in which the thermally sensitive element is a liquid contained in a metal envelope, frequently in the form of a Bourdon tube.

liquid limit [GEOL] The moisture content boundary that exists between the plastic and semiliquid states of a sediment.

liquid-water content *See* water content.

liroconite [MINERAL] $Cu_2Al(AsO_4)(OH)_4 \cdot 4H_2O$ A light-blue or yellowish-green mineral composed of basic hydrous aluminum copper arsenate, occurring in monoclinic crystals.

liskeardite [MINERAL] $(Al,Fe)_3(AsO_4)(OH)_6 \cdot 5H_2O$ A soft, white mineral composed of basic hydrous aluminum iron arsenate.

listing *See* lashing.

litharenite [PETR] A sandstone containing less than 75% quartz and metamorphic quartzite and more than 25% fine-grained volcanic, metamorphic and sedimentary rock fragments, including chert.

lithian muscovite [MINERAL] A form of the mineral lepidolite containing 3–4% lithium oxide and having a modified two-layer monoclinic muscovite structure.

lithic [PETR] Pertaining to stone.

lithic graywacke [PETR] A low-grade graywacke, that is, containing an abundance of unstable materials, especially a sandstone containing less than 75% quartz and chert, 15–75% detrital clay matrix, and more rock fragments than feldspar grains.

lithiclast *See* lithoclast.

lithic sandstone [PETR] A sandstone that contains more rock fragments than feldspar grains.

lithic tuff [GEOL] **1.** A tuff that is mostly crystalline rock fragments. **2.** An indurated volcanic ash deposit whose fragments are composed of previously formed rocks that first solidified in the volcanic vent and were then blown out.

lithifaction *See* lithification.

lithification [GEOL] **1.** Conversion of a newly deposited sediment into an indurated rock. Also known as lithifaction. **2.** Compositional change of coal to bituminous shale or other rock.

lithionite *See* lepidolite.

lithiophilite [MINERAL] $Li(Mn,Fe)PO_4$ A salmon-pink or clove-brown mineral crystallizing in the orthorhombic system; isomorphous with triphylite.

lithiophorite [MINERAL] $(Al,Li)MnO_2(OH)_2$ Mineral composed of basic manganese aluminum lithium oxide.

Lithistida [PALEON] An order of fossil sponges in the class Demospongia having a reticulate skeleton composed of irregular and knobby siliceous spicules.

lithium mica *See* lepidolite.

lithoclase [GEOL] A naturally produced rock fracture.

lithoclast [GEOL] A carbonate rock fragment, mechanically formed and deposited, usually measuring more than 2 millimeters in diameter, and derived from an older, lithified limestone or dolomite within, adjacent to, or outside the depositional site. Also spelled lithiclast.

lithofacies [GEOL] A subdivision of a specified stratigraphic unit distinguished on the basis of lithologic features.

lithofacies map [GEOL] The facies map of an area based on lithologic characters; shows areal variation in all aspects of the lithology of a stratigraphic unit.

lithogenesis [PETR] The branch of science dealing with the formation of rocks, especially the formation of sedimentary rocks.

lithogeochemical survey [GEOCHEM] A geochemical survey that involves the sampling of rocks.

lithographic limestone [GEOL] A dense, compact, fine-grained crystalline limestone having a pale creamy-yellow or grayish color. Also known as lithographic stone; litho stone.

lithographic stone *See* lithographic limestone.

lithographic texture [GEOL] The texture of certain calcareous sedimentary rocks characterized by grain size of less than 1/256 millimeter and having a smooth appearance.

lithologic map [GEOL] A type of geologic map showing the rock types of a particular area.

lithologic unit *See* rock-stratigraphic unit.

lithology [GEOL] The description of the physical character of a rock as determined by eye or with a low-power magnifier, and based on color, structures, mineralogic components, and grain size.

lithometeor [METEOROL] The general term for dry atmospheric suspensoids, including dust, haze, smoke, and sand.

lithomorphic [GEOL]   Referring to a soil whose characteristics are derived from events or conditions of a former period.

lithophile [GEOCHEM]   **1.** Pertaining to elements that have become concentrated in the silicate phase of meteorites or the slag crust of the earth. **2.** Pertaining to elements that have a greater free energy of oxidation per gram of oxygen than iron. Also known as oxyphile.

lithophysa [GEOL]   A large spherulitic hollow or bubble in glassy basalts and certain rhyolites. Also known as stone bubble.

lithosiderite *See* stony-iron meteorite.

lithosol [GEOL]   A group of shallow soils lacking well-defined horizons and composed of imperfectly weathered fragments of rock.

lithospar [MINERAL]   A naturally occurring mixture composed of spodumene and feldspar.

lithosphere [GEOL]   **1.** The rigid outer crust of rock on the earth about 50 miles (80 kilometers) thick, above the asthenosphere. Also known as oxysphere. **2.** Since the development of plate tectonics theory, a term referring to the rigid, upper 100 kilometers of the crust and upper mantle, above the asthenosphere.

lithostatic pressure *See* ground pressure.

litho stone *See* lithographic limestone.

lithostratic unit *See* rock-stratigraphic unit.

lithostratigraphic unit *See* rock-stratigraphic unit.

lithotope [GEOL]   **1.** The environment under which a sediment is deposited. **2.** An area of uniform sedimentation.

lithotype [GEOL]   A macroscopic band in humic coals, analyzed on the basis of physical characteristics rather than botanical origin.

Litopterna [PALEON]   An order of hoofed, herbivorous mammals confined to the Cenozoic of South America; characterized by a skull without expansion of the temporal or squamosal sinuses, a postorbital bar, primitive dentition, and feet that were three-toed or reduced to a single digit.

lit-par-lit [GEOL]   Pertaining to the penetration of bedded, schistose, or other foliate rocks by innumerable narrow sheets and tongues of granitic rock.

little brother [METEOROL]   A subsidiary tropical cyclone that sometimes follows a more severe disturbance.

little giant [MIN ENG]   A jointed iron nozzle used in hydraulic mining.

Little Ice Age [GEOL]   A period of expansion of mountain glaciers, marked by climatic deterioration, that began about 5500 years ago and extended to as late as A.D. 1550–1850 in some regions, as the Alps, Norway, Iceland, and Alaska.

littoral current [OCEANOGR]   A current, caused by wave action, that sets parallel to the shore; usually in the nearshore region within the breaker zone. Also known as alongshore current; longshore current.

littoral drift [GEOL]   Materials moved by waves and currents of the littoral zone. Also known as longshore drift.

littoral sediments [GEOL]   Deposits of littoral drift.

littoral transport [GEOL]   The movement of littoral drift.

littoral zone [ECOL]   Of or pertaining to the biogeographic zone between the high- and low-water marks.

Littorinacea [PALEON]   An extinct superfamily of gastropod mollusks in the order Prosobranchia.

live cave [GEOL]   A cave characterized by moisture and growth of mineral deposits associated with moisture. Also known as active cave.

livingstonite [MINERAL] $HgSb_4S_7$ A lead-gray mineral with red streak and metallic luster; a source of mercury.

lixiviation *See* leaching.

lizard-hipped dinosaur [PALEON]   The name applied to members of the Saurichia because of the comparatively unspecialized three-pronged pelvis.

Llandellian [GEOL]   Upper Middle Ordovician geologic time.

Llandoverian [GEOL]   Lower Silurian geologic time.

Llanvirnian [GEOL]   Lower Middle Ordovician geologic time.

llebetjado [METEOROL]   In northeastern Spain, a hot, squally wind descending from the Pyrenees and lasting for a few hours.

llerzolite [PETR]   A form of periodotite containing olivine and monoclinic and orthorhombic pyroxenes.

load [MIN ENG]   Unit of weight of ore used in the South African diamond mines; equal to 1600 pounds (725 kilograms); the equivalent of about 16 cubic feet (0.453 cubic meter) of broken ore.

load cast [GEOL]   An irregularity at the base of an overlying stratum, usually sandstone, that projects into an underlying stratum, usually shale or clay.

load controller [MIN ENG]   A device to control the load and prevent spillage on a gathering conveyor receiving coal or mineral from several loading points or subsidiary conveyors; it is a simplified weightometer.

loading density [ENG]   The number of pounds of explosive per foot length of drill hole.

loading pan [MIN ENG]   A box or scoop into which broken rock is shoveled in a sinking shaft while the hoppit is traveling in the shaft.

load metamorphism *See* static metamorphism.

loadstone *See* lodestone.

loam [GEOL]   Soil mixture of sand, silt, clay, and humus.

loaming [GEOCHEM]   A prospecting method in which samples of surficial material are tested for traces of a particular metal; presence of the metal presumably indicates a near-surface orebody.

lobate rill mark [GEOL]   A flute cast formed by current action.

lobe [HYD]   A curved projection on the margin of a continental ice sheet.

local attraction *See* local magnetic disturbance.

local base level *See* temporary base level.

local change [OCEANOGR]   The time rate of change of a scalar quantity (such as temperature, salinity, pressure, or oxygen content) in a fixed locality.

local extra observation [METEOROL]   An aviation weather observation taken at specified intervals, usually every 15 minutes, when there are impending aircraft operations and when weather conditions are below certain operational weather limits; the observation includes ceiling, sky condition, visibility, atmospheric phenomena, and pertinent remarks.

local forecast [METEOROL] Generally, any weather forecast of conditions over a relatively limited area, such as a city or airport.

local inflow [HYD] The water that enters a stream between two stream-gaging stations.

local magnetic disturbance [GEOPHYS] An anomaly of the magnetic field of the earth, extending over a relatively small area, due to local magnetic influences. Also known as local attraction.

local peat [GEOL] Peat formed by groundwater. Also known as basin peat.

local relief [GEOL] The vertical difference in elevation between the highest and lowest points of a land surface within a specified horizontal distance or in a limited area. Also known as relative relief.

local storm [METEOROL] A storm of mesometeorological scale; thus, thunderstorms, squalls, and tornadoes are often put in this category.

local winds [METEOROL] Winds which, over a small area, differ from those which would be appropriate to the general pressure distribution, or which possess some other peculiarity.

locate [MIN ENG] To mark out the boundaries of a mining claim and establish the right of possession.

location damages [MIN ENG] Compensation by an operator to the surface owner for injury to the surface or to growing crops, resulting from the drilling of a well.

location notice [MIN ENG] A written sign placed prominently on a claim, showing the locator's name and describing the claim's extent and boundaries.

location plan [MIN ENG] A scale map of the projected mine development indicating, among other things, proposed shafts and works in relation to existing surface features.

location work [MIN ENG] Labor required by law to be done on mining claims within 60 days of location, in order to establish ownership.

locomotive gradient [MIN ENG] The gradient set by law for a locomotive haulage; maximum is 1 in 15, but the limit for practical purposes is 1 in 25.

locomotive haulage [MIN ENG] The use of locomotive-hauled mine cars to carry coal ore, workers, and materials in a mine.

lode [GEOL] A fissure in consolidated rock filled with mineral; usually applied to metalliferous deposits.

lode claim [MIN ENG] That portion of a vein or lode, and of the adjoining surface, which has been acquired by a compliance with the law, both Federal and state.

lodestone [MINERAL] The naturally occurring magnetic iron oxide, or magnetite, possessing polarity, and attracting iron objects to itself. Also known as Hercules stone; leading stone; loadstone.

lodgement [GEOL] An accumulation of glacial deposits that remains firmly in place.

lodos [METEOROL] A southerly wind on the Black Sea coast of Bulgaria.

lodranite [GEOL] A stony iron meteorite composed of bronzite and olivine within a fine network of nickel-iron.

loellingite [MINERAL] $FeAs_2$ A silver-white to steel-gray mineral composed of iron arsenide with some cobalt, nickel, antimony, and sulfur; isomorphous with arsenopyrite; a source of arsenic. Also known as leucopyrite; löllingite.

loess [GEOL]   An essentially unconsolidated, unstratified calcareous silt; commonly it is homogeneous, permeable, and buff to gray in color, and contains calcareous concretions and fossils.

loess kindchen [GEOL]   An irregular or spheroidal nodule of calcium carbonate that is found in loess.

loeweite [MINERAL] $Na_4Mg_2(SO_4)_4 \cdot 5H_2O$ A white to pale-yellow mineral composed of hydrous sulfate of sodium and magnesium.

Lofco car feeder [MIN ENG]   A carrying chain running between the rails that controls mine cars at loading points, for marshaling trains and for loading cars into cages and tipplers.

log [ENG]   The record of, or the act or process of recording, events or the type and characteristics of the rock penetrated in drilling a borehole as evidenced by the cuttings, core recovered, or information obtained from electronic devices.

Logan slabbing machine [MIN ENG]   A machine that has three cutting chains; two are horizontal—one at the base of the coal seam, the other at a distance from the floor; the third is mounted vertically and shears off the coal at the back of the cut; a short conveyor transfers the coal to the face conveyor.

logarithmic velocity profile [METEOROL]   The theoretical variation of the mean wind speed with height in the surface boundary layer under certain assumptions.

logging [ENG]   Continuous recording versus depth of some characteristic datum of the formations penetrated by a drill hole; for example, resistivity, spontaneous potential, conductivity, fluid content, radioactivity, or density.

löllingite *See* loellingite.

lolly ice [OCEANOGR]   Saltwater frazil, a heavy concentration of which is called sludge.

lombarde [METEOROL]   An easterly wind (from Lombardy) that predominates along the French-Italian frontier, and comes from the High Alps; in winter it is violent and forms snow drifts in the mountain valleys; in the plains it is gentle and very dry.

Lomonosov ridge [GEOL]   An undersea ridge which subdivides the Arctic Basin, extending from Ellesmere Land to the New Siberian Islands.

long hole [MIN ENG]   An underground borehole and blasthole exceeding 10 feet (3 meters) in depth or requiring the use of two or more lengths of drill steel or rods coupled together to attain the desired depth.

long-hole drill [MIN ENG]   A rotary- or a percussive-type drill used to drill long holes.

long-hole jetting [MIN ENG]   A hydraulic mining system consisting essentially of drilling a hole down the pitch of the vein, replacing the drilling head with a jet cutting head, and then retracting the drill column with the jets in operation to remove the coal.

longitude [GEOD]   Angular distance, along the Equator, between the meridian passing through a position and, usually, the meridian of Greenwich.

longitudinal dune [GEOL]   A type of linear dune ridge that extends parallel to the direction of the dominant dune-building winds.

longitudinal fault [GEOL]   A fault parallel to the trend of the surrounding structure.

longitudinal stream *See* subsequent stream.

long-period tide [OCEANOGR]   A tide or tidal current constituent with a period which is independent of the rotation of the earth but which depends upon the orbital movement of the moon or of the earth.

long-pillar work   [MIN ENG]   A coal-winning technique used in underground mining in which large pillars of coal are left as the face is advanced; finally all the pillars are removed together.

long-range forecast   [METEOROL]   A weather forecast covering periodsfrom 48 hours to a week in advance (medium-range forecast) and ranging to even longer forecasts over periods of month, a season, and so on.

longshore bar   [GEOL]   A ridge of sand, gravel, or mud built on the seashore by waves and currents, generally parallel to the shore and submerged by high tides. Also known as offshore bar.

longshore current *See* littoral current.

longshore trough   [GEOL]   A long, wide, shallow depression of the sea floor parallel to the shore.

long tom   [MIN ENG]   A trough, longer than a rocker, for washing gold-bearing earth.

longwall coal cutter   [MIN ENG]   Compact machine driven by electricity or compressed air which cuts into the coal on relatively long faces, with its jib at right angles to its body.

longwall peak stoping   [MIN ENG]   An underland stoping method in which the rapid advance of the face goes on, and by working the faces at a 60° angle to the strike, the peak travels down the dip at twice the face advance rate.

longwall pillar working   [MIN ENG]   A technique used to extract coal pillars left behind in long-pillar working.

longwall retreating   [MIN ENG]   A longwall working system in which all the roadways are in the solid coal seam and the waste areas are left behind; developing headings are driven close to the boundary or limit, and the coal is taken out by the longwall retreating toward the shaft.

longwall system   [MIN ENG]   A method of mining in which the faces are advanced from the shaft toward the boundary, and the roof is allowed to cave in behind the miners as work progresses.

long wave   [METEOROL]   With regard to atmospheric circulation, a wave in the major belt of westerlies which is characterized by large length (thousands of kilometers) and significant amplitude; the wavelength is typically longer than that of the rapidly moving individual cyclonic and anticyclonic disturbances of the lower troposphere. Also known as major wave; planetary wave.

lonsdaleite   [MINERAL]   A mineral composed of a form of carbon; found in meteorites.

loo   [METEOROL]   A hot wind from the west in India.

loom   [METEOROL]   The glow of light below the horizon produced by greater-than-normal refraction in the lower atmosphere; it occurs when the air density decreases more rapidly with height than in the normal atmosphere.

looped pipeline   [PETRO ENG]   A pipeline that is paralleled (looped) by a second pipeline, both of which serve the same liquid or gas source and destination.

loop lake *See* oxbow lake.

loop rating   [HYD]   A rating curve that has higher values of discharge for a certain stage when the river is rising than it does when the river is falling; thus, the curve (stage versus discharge) describes a loop with each rise and fall of the river.

loop-type pit bottom   [MIN ENG]   An arrangement at the pit bottom in which loaded cars are fed to the cage on one side only, and the empties are returned by a loop roadway to the same place.

loopway [MIN ENG] A double-track loop in a main single-track haulage plane at which mine cars may pass.

loose ground [MIN ENG] Broken, fragmented, or loosely cemented bedrock material that tends to slough from sidewalls into a borehole and must be supported, as with timber sets. Also known as broken ground.

loparite [MINERAL] $(Ce,Na,Ca)_2(Ti,Nb)_2O_6$ A brown to black mineral; a variety of perovskite containing alkalies and cerium.

lopezite [MINERAL] $K_2Cr_2O_7$ An orange-red mineral composed of potassium dichromate.

Lophialetidae [PALEON] A family of extinct perissodactyl mammals in the superfamily Tapiroidea.

Lophiodontidae [PALEON] An extinct family of perissodactyl mammals in the superfamily Tapiroidea.

lopolith [GEOL] A large, floored intrusive body that is sunken centrally into the shape of a basin due to sagging of the underlying country rock.

lorandite [MINERAL] $TlAsS_2$ A cochineal- to carmine-red or dark lead-gray mineral composed of thallium sulfarsenide, occurring in monoclinic form.

loranskite [MINERAL] $(Y,Ce,Ca,Zr)TaO_4$ A black mineral composed of an oxide of yttrium, cerium, calcium, tantalum, and zirconium.

lorettoite [MINERAL] $Pb_7O_6Cl_2$ A honey-yellow to brownish-yellow mineral composed of lead oxychloride.

lorry *See* larry.

lose water *See* lost circulation.

loseyite [MINERAL] $(Mn,Zn)_7(CO_3)_2(OH)_{10}$ A bluish-white or brownish, monoclinic mineral consisting of a basic carbonate of manganese and zinc.

losing stream *See* influent stream.

lost circulation [PETRO ENG] A condition that occurs when the drilling fluid escapes into crevices or porous sidewalls of a borehole and does not return to the collar of the drill hole. Also called lose returns; lose water; lost returns; lost water.

lost mountain [GEOL] An isolated mountain in a desert apparently having no connection with the nearest main mass of mountains.

lost returns *See* lost circulation.

lost stream [HYD] **1.** A surface stream that flows into an underground channel and does not reappear in the same or any adjacent drainage basin. **2.** A dried-up stream in an arid environment.

lost water *See* lost circulation.

lotrite *See* pumpellyite.

loughlinite [MINERAL] $Na_2Mg_3Si_6O_{16}\cdot 8H_2O$ A pearly-white mineral that resembles asbestos, consisting of a hydrous silicate of sodium and magnesium.

lovchorrite *See* mosandrite.

Love wave [GEOPHYS] A horizontal dispersive surface wave, multireflected between internal boundaries of an elastic body, applied chiefly in the study of seismic waves in the earth's crust.

lovozerite [MINERAL] $(Na,K)_2(Mn,Ca)ZrSi_6O_{16}\cdot 3H_2O$ Mineral composed of hydrous silicate of sodium, potassium, manganese, calcium, and zirconium.

low *See* depression.

low aloft *See* upper-level cyclone.

low-angle fault [GEOL] A fault that dips at an angle less than 45°.

low-angle thrust *See* overthrust.

low clouds [METEOROL] Types of clouds, the mean level of which is between the surface and 6500 feet (1980 meters); the principal clouds in this group are stratocumulus, stratus, and nimbostratus.

low-energy environment [GEOL] An aqueous sedimentary environment in which there is standing water and a general lack of wave or current action, permitting accumulation of very fine-grained sediment.

lower atmosphere [METEOROL] That part of the atmosphere in which most weather phenomena occur (that is, the troposphere and lower stratosphere); in other contexts, the term implies the lower troposphere.

Lower Cambrian [GEOL] The earliest epoch of the Cambrian period of geologic time, ending about 540,000,000 years ago.

Lower Cretaceous [GEOL] The earliest epoch of the Cretaceous period of geologic time, extending from about 140- to 120,000,000 years ago.

Lower Devonian [GEOL] The earliest epoch of the Devonian period of geologic time, extending from about 400- to 385,000,000 years ago.

lower high water [OCEANOGR] The lower of two high tides occurring during a tidal day.

Lower Jurassic [GEOL] The earliest epoch of the Jurassic period of geologic time, extending from about 185- to 170,000,000 years ago.

lower low water [OCEANOGR] The lower of two low tides occurring during a tidal day.

lower mantle [GEOL] The portion of the mantle below a depth of about 1000 kilometer. Also known as inner mantle; mesosphere; pallasite shell.

Lower Mississippian [GEOL] The earliest epoch of the Mississippian period of geologic time, beginning about 350,000,000 years ago.

Lower Ordovician [GEOL] The earliest epoch of the Ordovician period of geologic time, extending from about 490- to 460,000,000 years ago.

Lower Pennsylvanian [GEOL] The earliest epoch of the Pennsylvanian period of geologic time, beginning about 310,000,000 years ago.

Lower Permian [GEOL] The earliest epoch of the Permian period of geologic time, extending from about 275- to 260,000,000 years ago.

lower plate *See* footwall.

Lower Silurian [GEOL] The earliest epoch of the Silurian period of geologic time, beginning about 420,000,000 years ago.

Lower Triassic [GEOL] The earliest epoch of the Triassic period of geologic time, extending from about 230- to 215,000,000 years ago.

low index [METEOROL] A relatively low value of the zonal index which, in middle latitudes, indicates a relatively weak westerly component of wind flow (usually implying stronger north-south motion), and the characteristic weather attending such motion; a circulation pattern of this type is commonly called a low-index situation.

low-moor bog [GEOL] A bog that is at or slightly below the ground water table.

low-moor peat [GEOL]   Peat found in low-moor bogs or swamps and containing little or no sphagnum. Also known as fen peat.

low-pressure well [PETRO ENG]   An oil or gas well with a shut-in wellhead pressure of less than 2000 psia (1.38 × 10⁷ newtons per square meter).

low quartz [MINERAL]   Quartz that has been formed below 573°C; the tetrahedral crystal structure is less symmetrically arranged than a quartz formed at a higher temperature.

low-rank graywacke [PETR]   A graywacke that is nonfeldspathic.

low-rank metamorphism [GEOL]   A metamorphic process that occurs under conditions of low to moderate pressure and temperature.

low tide *See* low water.

low-tide terrace [GEOL]   A flat area of a beach adjacent to the low-water line.

low-velocity layer [GEOPHYS]   A layer in the solid earth in which seismic wave velocity is lower than the layers immediately below or above.

low-volatile coal [GEOL]   A coal that is nonagglomerating, has 78% to less than 86% fixed carbon, and 14% to less than 22% volatile matter.

low water [OCEANOGR]   The lowest limit of the surface water level reached by the lowering tide. Also known as low tide.

low-water inequality [OCEANOGR]   The difference between the heights of two successive low tides.

low-water interval *See* low-water lunitidal interval.

low-water lunitidal interval [GEOPHYS]   For a specific location, the interval of time between the transit (upper or lower) of the moon and the next low water. Also known as low-water interval.

low-water neaps *See* mean low-water neaps.

low-water springs *See* mean low-water springs.

loxodrome *See* rhumb line.

Loxonematacea [PALEON]   Extinct superfamily of gastropod mollusks in the order Prosobranchia.

LPF process [MIN ENG]   Recovery of metals from tailings by a sequence of leaching, precipitation, and flotation.

Ludian [GEOL]   A European stage of uppermost Eocene geologic time, above the Bartonian and below the Tongrian of the Oligocene.

ludlamite [MINERAL] $(Fe,Mg,Mn)_3(PO_4)_2 \cdot 4H_2O$ A green mineral crystallizing in the monoclinic system and occurring in small, transparent crystals.

Ludlovian [GEOL]   A European stage of geologic time; Upper Silurian, below Gedinnian of Devonian, above Wenlockian.

ludwigite [MINERAL] $(Mg,Fe)_2FeBO_5$ Blackish-green mineral that crystallizes in the monoclinic system and occurs in fibrous masses; isomorphous with ronsenite.

lueneburgite [MINERAL] $Mg_3B_2(OH)_6(PO_4)_2 \cdot 6H_2O$ A colorless mineral composed of hydrous basic phosphate of magnesium and boron.

lueshite [MINERAL] $NaNbO_3$ An orthorhombic mineral having perovskite-type structure; it is dimorphous with natroniobite.

luganot [METEOROL]   A strong south or south-southeast wind of Lake Garda, Italy.

Luisian [GEOL] A North American stage of geologic time: Miocene (above Relizian, below Mohnian).

luminous cloud *See* sheet lightning.

luminous meteor [METEOROL] According to United States weather observing practice, any one of a number of atmospheric phenomena which appear as luminous patterns in the sky, including halos, coronas, rainbows, aurorae, and their many variations, but excluding lightning (an igneous meteor or electrometeor).

lump coal [MIN ENG] Bituminous coal that passes through a 6-inch (15-centimeter) round mesh in initial screening.

lunar atmospheric tide [METEOROL] An atmospheric tide due to the gravitational attraction of the moon; the only detectable components are the 12-lunar-hour or semidiurnal component, as in the oceanic tides, and two others of very nearly the same period; the amplitude of this atmospheric tide is so small that it is detected only by careful statistical analysis of a long record.

lunar inequality [GEOPHYS] A minute fluctuation of a magnetic needle from its mean position, caused by the moon.

lunar tide [OCEANOGR] The portion of a tide produced by forces of the moon.

lunate bar [GEOL] A crescent-shaped bar of sand that is frequently found off the entrance to a harbor.

lunette [GEOL] A broad, low crescentic mound of windblown fine silt and clay.

lunisolar tides [OCEANOGR] Harmonic tidal constituents derived partly from the development of both the lunar and the solar tide, and partly from the lunisolar synodic fortnightly constituent.

lunitidal interval [OCEANOGR] The period between the moon's upper or lower transit over a specified meridian and a specified phase of the tidal current following the transit.

luster mottlings [GEOL] The spotted, shimmering appearance of certain rocks caused by reflection of light from cleavage faces of crystals that contain small inclusions of other minerals.

lutaceous [GEOL] Claylike.

lutecite [GEOL] A fibrous, chalcedonylike quartz with optical anomalies that have led to its being considered a distinct species.

lutite [GEOL] A consolidated rock or sediment formed principally of clay or clay-sized particles.

luzonite *See* enargite.

L wave [GEOPHYS] A phase designation for an earthquake wave that is a surface wave, without respect to type.

Lydian stone *See* basanite.

lydite *See* basanite.

Lyginopteridaceae [PALEOBOT] An extinct family of the Lyginopteridales including monostelic pteridosperms having one or two vascular traces entering the base of the petiole.

Lyginopteridales [PALEOBOT] An order of the Pteridospermae.

Lyginopteridatae [PALEOBOT] The equivalent name for Pteridospermae.

lysimeter [ENG]   An instrument for measuring the water percolating through soils and determining the materials dissolved by the water.

lysocline [OCEANOGR]   The ocean depth at which calcium carbonate content begins decreasing.

maar [GEOL] A volcanic crater that was created by violent explosion but not accompanied by igneous extrusion; frequently, it is filled by a small circular lake.

macaluba *See* mud volcano.

macedonite [MINERAL] $PbTiO_3$ A mineral composed of an oxide of lead and titanium. [PETR] A basaltic rock that contains orthoclase, sodic plagioclase, biotite, olivine, and rare pyriboles.

maceral [GEOL] The microscopic organic constituents found in coal.

macgovernite [MINERAL] $Mn_5(AsO_3)SiO_3(OH)_2$ A mineral composed of basic manganese arsenite and silicate. Also spelled mcgovernite.

machine cut [MIN ENG] A groove or slot made horizontally or vertically in a coal seam by a coal cutter, as a step to shot firing.

mackayite [MINERAL] $FeTe_2O_5(OH)$ A green mineral composed of basic iron tellurite.

mackerel sky [METEOROL] A sky with considerable cirrocumulus or small-element altocumulous clouds, resembling the scales on a mackerel.

mackinawite [MINERAL] $(Fe,Ni)S$ A tetragonal mineral occurring as a corrosion product in iron pipes. Also known as kansite.

Macky effect [METEOROL] The reduction of the effective dielectric strength of air when waterdrops are present.

Mac-Lane system [MIN ENG] A means of conveying dirt to the top of a heap; an inclined rail track goes from the loading station at the bottom of the heap to an extending frame at the top; the rope that hauls the tubs with dirt up the rail track passes around a return sheave in the extending frame; a tub is tipped over at the top by a gear and the dirt discharged.

macle [CRYSTAL] A twinned crystal. [MINERAL] **1.** A dark or discolored spot in a mineral specimen. **2.** *See* chiastolite.

Macraucheniidae [PALEON] A family of extinct herbivorous mammals in the order Litopterna; members were proportioned much as camels are, and eventually lost the vertebral arterial canal of the cervical vertebrae.

macroclastic [PETR] Rock that is composed of fragments that are visible without magnification.

macroclimate [CLIMATOL] The climate of a large geographic region.

macrocrystalline [PETR] **1.** Pertaining to the texture of holocrystalline rock in which the constituents are visible without magnification. **2.** Pertaining to the texture of a rock with grains or crystals greater than 0.75 millimeter in diameter in recrystallized sediment.

macrodome [CRYSTAL] Dome of a crystal in which planes are parallel to the longer lateral axis.

macrofacies [GEOL] A collection of sedimentary facies that are related genetically.

macrofossil [PALEON] A fossil large enough to be observed with the naked eye.

macrometeorology [METEOROL] The study of the largest-scale aspects of the atmosphere, such as the general circulation, and weather types.

macropore [GEOL] A pore in soil of a large enough size so that water is not held in it by capillary attraction.

macroscopic anisotropy [ENG] Phenomenon in electrical downhole logging wherein electric current flows more easily along sedimentary strata beds than perpendicular to them.

maculose [GEOL] Of a group of contact-metamorphosed rocks or their structures, having spotted or knotted character.

maelstrom [OCEANOGR] A powerful and often destructive water current caused by the combined effects of high, wind-generated waves and a strong, opposing tidal current.

Maestrichtian [GEOL] A European stage of geologic time: Upper Cretaceous (above Menevian, below Fastiniogian).

maestro [METEOROL] A northwesterly wind with fine weather which blows, especially in summer, in the Adriatic, most frequently on the western shore; it is also found on the coasts of Corsica and Sardinia.

mafic mineral [MINERAL] **1.** A mineral that is composed predominantly of the ferromagnesian rock-forming silicates. **2.** In general, any dark mineral.

maghemite [MINERAL] γ-Fe_2O_3 A mineral form of iron oxide that is strongly magnetic and a member of the magnetite series.

magma [GEOL] The molten rock material from which igneous rocks are formed.

magma chamber [GEOL] A larger reservoir in the crust of the earth that is occupied by a body of magma.

magma geothermal system [GEOL] A geothermal system in which the dominant source of heat is a large reservoir of igneous magma within an intrusive chamber or lava pool; an example is the Yellowstone Park area of Wyoming.

magma province *See* petrographic province.

magmatic differentiation [PETR] **1.** The process by which the different types of igneous rocks are derived from a single parent magma. **2.** The process by which ores are formed by solidification from magma. Also known as magmatic segregation.

magmatic rock [PETR] A rock derived from magma.

magmatic segregation *See* magmatic differentiation.

magmatic stoping [GEOL] A process of igneous intrusion in which magma gradually works its way upward by breaking off and engulfing blocks of the country rock.

magmatic water [HYD] Water derived from or existing in molten igneous rock or magma. Also known as juvenile water.

magmatism [PETR] The formation of igneous rock from magma.

magmosphere *See* pyrosphere.

magnafacies [GEOL] A major, continuous belt of deposits that is homogeneous in lithologic and paleontologic characteristics and that extends obliquely across time planes or through several time-stratigraphic units.

magnesia mica *See* biotite.

magnesian calcite [MINERAL] (Ca,Mg)CO₃ A variety of calcite consisting of randomly substituted magnesium carbonate in a disordered calcite lattice. Also known as magnesium calcite.

magnesian limestone [PETR] Limestone with at least 90% calcite, a maximum of 10% dolomite, an approximate magnesium oxide equivalent of 1.1–2.1%, and an approximate magnesium carbonate equivalent of 2.3–4.4%.

magnesian marble [PETR] A type of magnesian limestone that has been metamorphosed; contains some dolomite. Also known as dolomitic marble.

magnesiochromite [MINERAL] $MgCr_2O_4$ A mineral of the spinel group composed of magnesium chromium oxide; it is isomorphous with chromite. Also known as magnochromite.

magnesiocopiapite [MINERAL] $MgFe_4(SO_4)_6(OH)_2·20H_2O$ A mineral of the copiapite group composed of hydrous basic magnesium and iron sulfate; it is isomorphous with copiapite and cuprocopiapite.

magnesioferrite [MINERAL] $(Mg,Fe)Fe_2O_4$ Black, strongly magnetic mineral of the magnetite series in the spinel group. Also known as magnoferrite.

magnesite [MINERAL] $MgCO_3$ The mineral form of magnesium carbonate, usually massive and white, with hexagonal symmetry; specific gravity is 3, and hardness is 4 on Mohs scale. Also known as giobertite.

magnesium calcite *See* magnesian calcite.

magnesium-iron mica *See* biotite.

magnet grate [MIN ENG] A series of magnetized bars used to trap and remove tramp iron from a flow of pulverized or granulated dry solids passing over the grate; used to protect crushing or grinding equipment.

magnetic annual change [GEOPHYS] The amount of secular change in the earth's magnetic field which occurs in 1 year. Also known as annual magnetic change.

magnetic annual variation [GEOPHYS] The small, systematic temporal variation in the earth's magnetic field which occurs after the trend for secular change has been removed from the average monthly values. Also known as annual magnetic variation.

magnetic bay [GEOPHYS] A small magnetic disturbance whose magnetograph resembles an indentation of a coastline; on earth, magnetic bays occur mainly in the polar regions and have a duration of a few hours.

magnetic character figure *See* C index.

magnetic daily variation *See* magnetic diurnal variation.

magnetic declination *See* declination.

magnetic direction [GEOD] Horizontal direction expressed as angular distance from magnetic north.

magnetic diurnal variation [GEOPHYS] Oscillations of the earth's magnetic field which have a periodicity of about a day and which depend to a close approximation only on local time and geographic latitude. Also known as magnetic daily variation.

magnetic element [GEOPHYS] Magnetic declination, dip, or intensity at any location on the surface of the earth.

magnetic iron ore *See* magnetite.

magnetic latitude [GEOPHYS] Angular distance north or south of the magnetic equator.

magnetic local anomaly [GEOPHYS] A localized departure of the geomagnetic field from its average over the surrounding area.

magnetic meridian [GEOPHYS] A line which is at any point in the direction of horizontal magnetic force of the earth; a compass needle without deviation lies in the magnetic meridian.

magnetic north [GEOPHYS] At any point on the earth's surface, the horizontal direction of the earth's magnetic lines of force (direction of a magnetic meridian) toward the north magnetic pole; a particular direction indicated by the needle of a magnetic compass.

magnetic observatory [GEOPHYS] A geophysical measuring station employing some form of magnetometer to measure the intensity of the earth's magnetic field.

magnetic prime vertical [GEOPHYS] The vertical circle through the magnetic east and west points of the horizon.

magnetic profile [GEOPHYS] A profile of a geologic structure showing magnetic anomalies.

magnetic prospecting [ENG] Carrying out airborne or ground surveys of variations in the earth's magnetic field, using a magnetometer or other equipment, to locate magnetic deposits of iron, nickel, or titanium, or nonmagnetic deposits which either contain magnetic gangue minerals or are associated with magnetic structures.

magnetic reversal [GEOPHYS] A reversal of the polarity of the earth's magnetic field that has occurred at irregular intervals on the order of 1,000,000 years.

magnetic secular change [GEOPHYS] The gradual variation in the value of a magnetic element which occurs over a period of years.

magnetic separator [ENG] A machine for separating magnetic from less magnetic or nonmagnetic materials by using strong magnetic fields; used for example, in tramp iron removal, or concentration and purification.

magnetic station [GEOPHYS] A facility equipped with instruments for measuring local variations in the earth's magnetic field.

magnetic storm [GEOPHYS] A worldwide disturbance of the earth's magnetic field; frequently characterized by a sudden onset, in which the magnetic field undergoes marked changes in the course of an hour or less, followed by a very gradual return to normalcy, which may take several days.

magnetic stratigraphy *See* paleomagnetic stratigraphy.

magnetic survey [GEOPHYS] **1.** Magnetometer map of variations in the earth's total magnetic field; used in petroleum exploration to determine basement-rock depths and geologic anomalies. **2.** Measurement of a component of the geomagnetic field at different locations.

magnetic temporal variation [GEOPHYS] Any change in the earth's magnetic field which is a function of time.

magnetic variation [GEOPHYS] Small changes in the earth's magnetic field in time and space.

magnetic wind direction [METEOROL] The direction, with respect to magnetic north, from which the wind is blowing; distinguished from true wind direction.

magnetite [MINERAL] An opaque iron-black and streak-black isometric mineral and member of the spinel structure type, usually occurring in octahedrals or in granular to massive form; hardness is 6 on Mohs scale, and specific gravity is 5.20. Also known as magnetic iron ore; octahedral iron ore.

magneto anemometer [ENG] A cup anemometer with its shaft mechanically coupled to a magnet; both the frequency and amplitude of the voltage generated are proportional to the wind speed, and may be indicated or recorded by suitable electrical instruments.

magnetoionic duct [GEOPHYS] Duct along the geomagnetic lines of force which exhibits waveguide characteristics for radio-wave propagation between conjugate points on the earth's surface.

magnetoionic theory [GEOPHYS] The theory of the combined effect of the earth's magnetic field and atmospheric ionization on the propagation of electromagnetic waves.

magnetoionic wave component [GEOPHYS] Either of the two elliptically polarized wave components into which a linearly polarized wave incident on the ionosphere is separated because of the earth's magnetic field.

magnetometer [ENG] An instrument for measuring the magnitude and sometimes also the direction of a magnetic field, such as the earth's magnetic field.

magnetopause [GEOPHYS] A boundary that marks the transition from the earth's magnetosphere to the interplanetary medium.

magnetoplumbite [MINERAL] $(PbMn)_2Fe_6O_{11}$ Black mineral consisting of a ferric oxide of plumbite and manganese, and occurring in acute metallic hexagonal crystals.

magnetosheath [GEOPHYS] The relatively thin region between the earth's magnetopause and the shock front in the solar wind.

magnetosphere [GEOPHYS] The region of the earth in which the geomagnetic field plays a dominant part in controlling the physical processes that take place; it is usually considered to begin at an altitude of about 100 kilometers and to extend outward to a distant boundary that marks the beginning of interplanetary space.

magnetospheric plasma [GEOPHYS] A low-energy plasma with particle energies less than a few electron volts that permeates the entire region of the earth's magnetosphere.

magnetospheric ring current [GEOPHYS] A belt of charged particles moving around the earth, in the magnetosphere, whose perturbations are associated with ionospheric storms.

magnetospheric substorm [GEOPHYS] A disturbance of particles and magnetic fields in the magnetosphere; occurs intermittently, lasts 1 to 3 hours, and is accompanied by various phenomena sensible from the earth's surface, such as intense auroral displays and magnetic disturbances, particularly in the nightside polar regions.

magnetotail [GEOPHYS] The portion of the magnetosphere extending from earth in the direction away from the sun for a variable distance of the order of 1000 earth radii.

magnitude [GEOPHYS] A measure of the amount of energy released by an earthquake.

magnochromite *See* magnesiochromite.

magnoferrite *See* magnesioferrite.

magnophorite [MINERAL] $NaKCaMg_5Si_8O_{23}OH$ A monoclinic mineral composed of a basic silicate of sodium, potassium, calcium, and magnesium; member of the amphibole group.

main-and-tail haulage [MIN ENG]   A single-track haulage system that is operated by a haulage engine with two drums, each with a separate rope.

main crosscut [MIN ENG]   A crosscut that traverses the entire mining field and penetrates all deposits.

main entry [MIN ENG]   The principal entry or set of entries driven through the coalbed from which cross entries, room entries, or rooms are turned.

main fans [MIN ENG]   Fans that produce the general ventilating current of the mine, being of large capacity and permanently installed.

main haulage [MIN ENG]   The section of the haulage system which moves the coal from the secondary or intermediate haulage system to the shaft or mine opening.

main haulage conveyor [MIN ENG]   A conveyor used to transport material in the main haulage section of the mine, between the intermediate haulage conveyor and a car-loading point or the outside.

main-line locomotive [MIN ENG]   A large, high-powered locomotive which hauls trains of cars over the main haulage system.

main return [MIN ENG]   The main return airway of a mine.

main rope *See* pull rope.

main-rope haulage system [MIN ENG]   A system of haulage for hauling loaded trains of tubs or cars up, or lowering them down, a comparatively steep gradient which is not steep enough, in the latter case, for a self-acting incline.

main stream [HYD]   The principal or largest stream of a given area or drainage system. Also known as master stream; trunk stream.

main stroke *See* return streamer.

main thermocline [OCEANOGR]   A thermocline that is deep enough in the ocean to be unaffected by seasonal temperature changes in the atmosphere. Also known as permanent thermocline.

Majac mill [MIN ENG]   A mill for dry-grinding mica by means of fluid energy; consists of a chamber which contains two horizontal, directly opposing jets and into which mica is fed continuously from a screw conveyor.

major fold [GEOL]   A large-scale fold with which minor folds are usually associated.

major trough [METEOROL]   A long-wave trough in the large-scale pressure pattern of the upper troposphere.

major wave *See* long wave.

malachite [MINERAL] $Cu_2(OH)_2(CO_3)$ A bright-green monoclinic mineral consisting of a basic carbonate of copper and usually occurring in massive forms or in bundles of radiating fibers; specific gravity is 4.05, and hardness is 3.5–4 on Mohs scale.

malacolite *See* diopside.

malchite [PETR]   A fine-grained lamprophyre with small, rare phenocrysts or hornblende, labradorite, and sometimes biotite embedded in a matrix of hornblende, andesine, and some quartz.

maldonite [MINERAL] Au_2Bi A pinkish silver-white mineral consisting of gold and bismuth; occurs in massive granular form.

malignite [PETR]   A mafic nepheline syenite which has more than 5% nepheline and roughly equal amounts of pyroxene and potassium feldspar.

malladrite [MINERAL] Na_2SiF_6 A hexagonal mineral composed of sodium fluosilicate, occurring as small crystals in volcanic holes in Vesuvius.

mallardite [MINERAL] $MnSO_4 \cdot 7H_2O$ A pale-rose, monoclinic mineral composed of hydrous manganese sulfate.

malloseismic [GEOPHYS] Referring to an area that is likely to experience destructive earthquakes several times in a century.

Malm [GEOL] The Upper Jurassic geologic series, above Dogger and below Cretaceous.

malysite [MINERAL] $FeCl_3$ A halogen mineral deposited by sublimation; found most commonly at Mount Vesuvius, Italy.

mamelon [GEOL] A small, rounded volcano that develops over a vent by the slow extrusion of silicic lava.

mammillary [MINERAL] Of or pertaining to an aggregate of crystals in the form of a rounded mass.

mammoth [PALEON] Any of various large Pleistocene elephants having long, up-curved tusks and a heavy coat of hair.

Mammutinae [PALEON] A subfamily of extinct proboscidean mammals in the family Mastodontidae.

manandonite [MINERAL] $Li_4Al_{14}B_4Si_6O_29(OH)_{24}$ A white mineral composed of basic borosilicate of lithium and aluminum.

manasseite [MINERAL] $MgAl_2(OH)_{16}(CO_3) \cdot 4H_2O$ A hexagonal mineral composed of basic hydrous magnesium and aluminum carbonate; it is dimorphous with hydrotalcite.

man cage [MIN ENG] A special cage for raising and lowering workers in a mine shaft.

man car [MIN ENG] A kind of car for transporting miners up and down the steeply inclined shafts of some mines.

mandatory layer [METEOROL] A layer of the atmosphere between two consecutive (or any two) specified mandatory levels.

mandatory level [METEOROL] One of several constant-pressure levels in the atmosphere for which a complete evaluation of data from upper-air observations is required. Also known as mandatory surface.

mandatory surface *See* mandatory level.

mandrel hanger [PETRO ENG] A device used to provide a liquid- or gas-tight seal (blowout preventer) between oil-well tubing and the tubing head.

manganese epidote *See* piemontite.

manganese nodule [GEOL] Small, irregular black to brown concretions consisting chiefly of manganese salts and manganese oxide minerals; formed in oceans as a result of pelagic sedimentation or precipitation.

manganite [MINERAL] $MnO(OH)$ A brilliant steel-gray or black polymorphous mineral; crystallizes in the orthorhombic system. Also known as gray manganese ore.

manganolangbeinite [MINERAL] $K_2Mn_2(SO_4)_3$ A rose-red, isometric mineral composed of potassium manganese sulfate; occurs in lava on Vesuvius.

manganosite [MINERAL] MnO An emerald-green isometric mineral occurring in small octahedrons that blacken on exposure; hardness is 5–6 on Mohs scale, and specific gravity is 5.18.

mankato stone [PETR] A variety of limestone containing more than 49% calcium carbonate, with about 4.5% alumina and some silica.

mansfieldite [MINERAL] $Al(AsO_4)\cdot 2H_2O$ A white to pale-gray orthorhombic mineral composed of hydrous aluminum arsenate; it is isomorphous with scorodite.

mantle [GEOL] The intermediate shell zone of the earth below the crust and above the core (to a depth of 3480 kilometers).

mantled gneiss dome [GEOL] A dome in metamorphic terrains that has a remobilized core of gneiss surrounded by a concordant sheath of the basal part of the overlying metamorphic sequence.

mantle rock *See* regolith.

manto [GEOL] A sedimentary or igneous ore body occurring in flat-lying depositional layers.

man trip [MIN ENG] A trip made by mine cars and locomotives to take workers, rather than coal, to and from the working places.

manual casing hanger [PETRO ENG] A device in the bowl of the lowermost (or intermediate) casing head to suspend the next smaller casing string and to provide a seal between the suspended casing and the casing-head bowl.

manway *See* ladderway.

map [GRAPHICS] A representation, usually on a plane surface, of all or part of the surface of the earth, celestial sphere, or other area; shows relative size and position, according to a given projection, of the physical features represented and such other information as may be applicable to the purpose intended.

map chart [MAP] A representation of a land-sea area, using the characteristics of a map to represent the land area and the characteristics of a chart to represent the sea area, with such special characteristics as to make the map chart most useful in military operations, particularly amphibious operations.

map parallel *See* axis of homology.

map plotting [METEOROL] The process of transcribing weather information onto maps, diagrams, and so on; it usually refers specifically to decoding synoptic reports and entering those data in conventional station-model form on synoptic charts. Also known as map spotting.

map projection *See* projection.

map reading [MAP] Interpretation of the symbols, lines, abbreviations, and terms appearing on maps.

map scale [MAP] The ratio between a distance on a map and the corresponding distance on the earth, often represented as 1:80,000 (natural scale) or 30 miles (48.27 kilometers) to an inch.

map spotting *See* map plotting.

map symbol [MAP] A character, letter, or similar graphic representation used on a map to indicate some object, characteristic, and so on.

map vertical *See* geographic vertical.

marble [PETR] **1.** Metamorphic rock composed of recrystallized calcite or dolomite. **2.** Commercially, any limestone or dolomite taking polish.

marble shot [PETRO ENG] An explosive shot in open-hole well completions in which glass marbles are packed around the explosive in the wellbore; the marbles become projectiles that help break up the formation.

marcasite [MINERAL] FeS_2 A pale bronze-yellow to nearly white mineral, crystallizing in the orthorhombic system; hardness is 6–6.5 on Mohs scale, and specific gravity is 4.89.

march [METEOROL] The variation of any meteorological element throughout a specific unit of time, such as a day, month, or year; as the daily march of temperature, the complete cycle of temperature during 24 hours.

Marconaflo slurry transport [MIN ENG] A system which recovers solids from ore or tailings piles and mixes them with water to produce a transportable slurry in pipelines.

Marcy mill [MIN ENG] A ball mill with a vertical grate diaphragm placed near the discharge end; screens for sizing the material are located between the diaphragm and the end of the tube.

marekanite [GEOL] Rounded to subangular obsidian bodies that occur in masses of perlite.

mare's tail *See* precipitation trajectory.

margarite [GEOL] A string of beadlike globulites; commonly found in glassy igneous rocks. [MINERAL] $CaAl_2(Al_2Si_2)O_{10}(OH)_2$ A pink, reddish, or yellow, brittle mica mineral.

margarosanite [MINERAL] $PbCa_2(SiO_3)_3$ A colorless or snow-white triclinic mineral composed of lead calcium silicate, occurring in lamellar masses.

margin [GEOGR] The boundary around a body of water.

marginal escarpment [GEOL] A seaward slope of a marginal plateau with a gradient of 1:10 or more.

marginal fissure [GEOL] A magma-filled fracture bordering an igneous intrusion.

marginal moraine *See* terminal moraine.

marginal plain *See* outwash plain.

marginal plateau [GEOL] A relatively flat shelf adjacent to a continent and similar topographically to, but deeper than, a continental shelf.

marginal salt pan [GEOL] A natural, coastal salt pan.

marginal sea [GEOGR] A semiclosed sea adjacent to a continent and connected with the ocean at the water surface.

marginal thrust [GEOL] One of a series of faults bordering an igneous intrusion and crossing both the intrusion and the wall rock. Also known as marginal upthrust.

marginal upthrust *See* marginal thrust.

Margules equation *See* Witte-Margules equation.

marialite [MINERAL] $3NaAlSi_3O_8 \cdot NaCl$ A scapolite mineral that is isomorphous with meronite.

Marietta miner [MIN ENG] Trade name for a continuous miner mounted on caterpillar treads; has two cutter arms and two cutter chains at the working face, and a conveyor system to carry the broken coal to be loaded onto cars.

marigram [OCEANOGR] A graphic record of the rising and falling movements of the tide expressed as a curve.

marigraph [ENG] A self-registering gage that records the heights of the tides.

marine [OCEANOGR] Pertaining to the sea.

marine abrasion [GEOL] Erosion of the ocean floor by sediment moved by ocean waves. Also known as wave erosion.

marine cave *See* sea cave.

marine climate [CLIMATOL]   A regional climate which is under the predominant influence of the sea, that is, a climate characterized by oceanity; the antithesis of a continental climate. Also known as maritime climate; oceanic climate.

marine-cut terrace [GEOL]   A terrace or platform cut by wave erosion of marine origin. Also known as wave-cut terrace.

marine forecast [METEOROL]   A forecast, for a specified oceanic or coastal area, of weather elements of particular interest to maritime transportation, including wind, visibility, the general state of the weather, and storm warnings.

marine geology *See* geological oceanography.

marine meteorology [METEOROL]   That part of meteorology which deals mainly with the study of oceanic areas, including island and coastal regions; in particular, it serves the practical needs of surface and air navigation over the oceans.

marine salina [GEOGR]   A body of salt water found along an arid coast and separated from the sea by a sand or gravel barrier.

Marinesian *See* Bartonian.

marine snow [OCEANOGR]   Concentrated living and dead organic material and inorganic debris suspended in the sea at density boundaries such as the thermocline.

marine swamp [GEOGR]   An area of low, salty, or brackish water found along the shore and characterized by abundant grasses, mangrove trees, and similar vegetation. Also known as paralic swamp.

marine terrace [GEOL]   A seacoast terrace formed by the merging of a wave-built terrace and a wave-cut platform. Also known as sea terrace; shore terrace.

marine transgression *See* transgression.

marine weather observation [METEOROL]   The weather as observed from a ship at sea, usually taken in accordance with procedures specified by the World Meteorological Organization.

maritime air [METEOROL]   A type of air whose characteristics are developed over an extensive water surface and which, therefore, has the basic maritime quality of high moisture content in at least its lower levels.

maritime climate *See* marine climate.

maritime polar air [METEOROL]   Polar air initially possessing similar properties to those of continental polar air, but in passing over warmer water it becomes unstable with a higher moisture content.

maritime tropical air [METEOROL]   The principal type of tropical air, produced over the tropical and subtropical seas; it is very warm and humid, and is frequently carried poleward on the western flanks of the subtropical highs.

marker bed [GEOL]   **1.** A stratified unit with distinctive characteristics making it an easily recognized geologic horizon. **2.** A rock layer which accounts for a characteristic portion of a seismic refraction time-distance curve. **3.** *See* key bed.

marl [GEOL]   A deposit of crumbling earthy material composed principally of clay with magnesium and calcium carbonate; used as a fertilizer for lime-deficient soils.

marlite *See* marlstone.

marlstone [PETR]   **1.** A consolidated rock that has about the same composition as marl; considered to be an earthy or impure argillaceous limestone. Also known as marlite. **2.** A hard ferruginous rock of the Middle Lias in England.

marly [GEOL]   Pertaining to, containing, or resembling marl.

marmatite [MINERAL] A dark-brown to black mineral composed of iron-bearing sphalerite. Also known as christophite.

marmolite [MINERAL] A pale-green serpentine mineral, occurring in thin laminations; a variety of chrysotile.

Marmor [GEOL] A North American stage of Middle Ordovician geologic time forming the lower subdivision of Chazyan, above Whiterock and below Ashby.

marrite [MINERAL] $PbAgAsS_3$ A monoclinic mineral, occurring as small crystals in Valais, Switzerland.

Marsden chart [METEOROL] A system for showing the distribution of meteorological data on a chart, especially over the oceans; using a Mercator map projection, the world between 80°N and 70°S latitudes is divided into Marsden "squares," each of 10° latitude by 10° longitude and systematically numbered to indicate position; each square may be divided into quarter squares, or into 100 one-degree subsquares numbered from 00 to 99 to give the position to the nearest degree.

marsh gas [GEOCHEM] Combustible gas, consisting chiefly of methane, produced as a result of decay of vegetation in stagnant water.

marshite [MINERAL] CuI A reddish, oil-brown isometric mineral composed of cuprous iodide and occurring as crystals; hardness is 2.5 on Mohs scale, and specific gravity is 5.6.

martite [MINERAL] Hematite occurring in iron-black octahedral crystals pseudomorphous after magnetite.

Marvin sunshine recorder [ENG] A sunshine recorder in which the time scale is supplied by a chronograph, and consisting of two bulbs (one of which is blackened) that communicate through a glass tube of small diameter, which is partially filled with mercury and contains two electrical contacts; when the instrument is exposed to sunshine, the air in the blackened bulb is warmed more than that in the clear bulb; the warmed air expands and forces the mercury through the connecting tube to a point where the electrical contacts are shorted by the mercury; this completes the electrical circuit to the pen on the chronograph.

mascagnite [MINERAL] $(NH_4)_2SO_4$ A yellowish-gray mineral found in guano, near burning coal beds, or as lava incrustation; specific gravity is 1.77; hardness is 2–2.5 on Mohs scale.

mascon [GEOL] A large, high-density mass concentration below a ringed mare on the surface of the moon.

mass attraction vertical [GEOPHYS] The vertical which is a function only of the distribution of mass and is unaffected by forces resulting from the motions of the earth.

mass erosion [GEOL] Any process which causes soil and rock materials to fall and then move downslope by direct application of gravitational body stresses. Also known as gravity erosion.

massicot [MINERAL] PbO A yellow, orthorhombic mineral consisting of lead monoxide; found in the western and southern United States. Also known as lead ocher.

massif [GEOL] A massive block of rock within an erogenic belt, generally more rigid than the surrounding rocks, and commonly composed of crystalline basement or younger plutons.

massive [GEOL] Of a mineral deposit, having a large concentration of ore in one place. [MINERAL] Of a mineral, lacking an internal structure. [PALEON] Of corallum, composed of closely packed corallites. [PETR] **1.** Of a competent rock, being homogeneous, isotropic, and elastically perfect. **2.** Of a metamorphic rock, having constituents which do not show parallel orientation and are not arranged in layers. **3.** Of igneous rocks, being homogeneous over wide areas and lacking layering, foliation, cleavage, or similar features.

mass movement [GEOL]   Movement of a portion of the land surface as a unit.

mass wasting [GEOL]   Dislodgement and downslope transport of loose rock and soil material under the direct influence of gravitational body stresses.

master stream *See* main stream.

mastodon [PALEON]   A member of the Mastodontidae, especially the genus *Mammut*.

Mastodontidae [PALEON]   An extinct family of elephantoid proboscideans that had low-crowned teeth with simple ridges and without cement.

mat [MIN ENG]   An accumulation of broken mine timbers, rock, earth, and other debris coincident with the caving system of mining.

Matanuska wind [METEOROL]   A strong, gusty, northeast wind which occasionally occurs during the winter in the vicinity of Palmer, Alaska.

matched terrace *See* paired terrace.

mathematical climate [CLIMATOL]   An elementary generalization of the earth's climatic pattern, based entirely on the annual cycle of the sun's inclination; this early climatic classification recognized three basic latitudinal zones (the summerless, intermediate, and winterless), which are now known as the Frigid, Temperate, and Torrid Zones, and which are bounded by the Arctic and Antarctic Circles and the Tropics of Cancer and Capricorn.

mathematical forecasting *See* numerical forecasting.

mathematical geography [GEOGR]   The branch of geography that deals with the features and processes of the earth, and their representations on maps and charts.

matildite [MINERAL] $AgBiS_2$ An iron black to gray, orthorhombic mineral consisting of silver bismuth sulfide; occurrence is massive or granular.

matinal [METEOROL]   The morning winds, that is, an east wind.

matlockite [MINERAL] $PbFCl$ A mineral consisting of lead chloride and fluoride.

mat packs [MIN ENG]   Timbers laid side by side into a solid mass 2 by 2½ feet (61 by 76 centimeters) square and 4–6 inches (10–15 centimeters) thick, kept together by wires through holes drilled through the edges of the mass, carried into a mine and built up to make effective roof supports.

matrix [PETR]   The continuous, fine-grained material in which large grains of a sediment or sedimentary rock are embedded. Also known as groundmass.

matrix porosity [GEOL]   Core-sample porosity determined from a small sample of the core, in contrast to total porosity, where the whole core is used.

matrix rock *See* land pebble phosphate.

matrix velocity [GEOPHYS]   The velocity of sound through a formation's rock matrix during an acoustic-velocity log.

mature [GEOL]   **1.** Pertaining to a topography or region, and to its landforms, having undergone maximum development and accentuation of form. **2.** Pertaining to the third stage of textural maturity of a clastic sediment.

matureland [GEOL]   The land surface characteristic of the stage of maturity in the erosion cycle.

mature soil *See* zonal soil.

maturity [GEOL] **1.** The second stage of the erosion cycle in the topographic development of a landscape or region characterized by numerous and closely spaced mature streams, reduction of level surfaces to slopes, large well-defined drainage systems, and the absence of swamps or lakes on the uplands. Also known as topographic maturity. **2.** A stage in the development of a shore or coast that begins with the attainment of a profile of equilibrium. **3.** The extent to which the texture and composition of a clastic sediment approach the ultimate end product. **4.** The stage of stream development at which maximum vigor and efficiency has been reached.

maturity index [GEOL] A measure of the progress of a clastic sediment in the direction of chemical or mineralogic stability; for example, a high ratio of quartz + cherts/feldspar + rock fragments indicates a highly mature sediment.

maucherite [MINERAL] $Ni_{11}As_8$ A reddish silver-white mineral composed of nickel arsenide.

maximum ebb [OCEANOGR] The greatest speed of an ebb current.

maximum flood [OCEANOGR] The greatest speed of a flood current.

maximum producible oil index [PETRO ENG] An approximation of the maximum amount of oil per bulk formation volume that is producible with water drive.

maximum subsidence [GEOL] The maximum amount of subsidence in a basin.

maximum-wind and shear chart [METEOROL] A synoptic chart on which are plotted the altitudes of the maximum wind speed, the maximum wind velocity (wind direction optional), plus the velocity of the wind at mandatory levels both above and below the level of maximum wind. Also known as max-wind and shear chart.

maximum-wind level [METEOROL] The height at which the maximum wind speed occurs, determined in a winds-aloft observation. Also known as max-wind level.

maximum-wind topography [METEOROL] The topography of the surface of maximum wind speed. Also known as max-wind topography.

maximum zonal westerlies [METEOROL] The average west-to-east component of wind over the continuous 20° belt of latitude in which this average is a maximum; it is usually found, in the winter season, in the vicinity of 40–60° north latitude.

max-wind and shear chart *See* maximum-wind and shear chart.

max-wind level *See* maximum-wind level.

McCaa breathing device [MIN ENG] A self-contained, compressed-oxygen breathing apparatus designed by the U.S. Bureau of Mines for mine rescue work; it is carried on the back in a protective aluminum cover.

McGinty [MIN ENG] Three sheaves over which a rope is passed so as to take a course somewhat like that of the letter M.

mcgovernite *See* macgovernite.

McLuckie gas detector [MIN ENG] A portable nonautomatic means of analyzing air that can be used underground or aboveground if the air sample is brought out of the mine in small rubber bladders.

McNally-Carpenter centrifuge [MIN ENG] A machine that removes water from fine coal; it has a cone-shaped rotating vertical element, into the top of which the wet feed is put by gravity; a distributing disk forces the material onto the screen or basket lining the cone; as the material approaches the cone bottom, the drying action increases until the dry coal is discharged from the bottom.

McNally-Norton jig [MIN ENG] A device used to clean raw coal by carrying it to a wash box and using water whose level rises and falls as a result of air pulsations; the incoming coal is suspended by the pulsating water while the heavier refuse sinks to the bottom; the coal then spills over into the next wash box where the process is repeated, and the clean coal is discharged by the dewatering screens.

McNally-Vissac dryer [MIN ENG]   A coal dryer that operates on the convection from the heavy forced draft from a coal-fired furnace; it consists of an inclined reciprocating screen over which coal moves; moisture is removed by the hot air from the furnace passing down through the bed of coal.

mean chart [METEOROL]   Any chart on which isopleths of the mean value of a given meteorological element are drawn. Also known as mean map.

mean depth [HYD]   Average water depth in a stream channel or conduit computed by dividing the cross-sectional area by the surface width.

meander [HYD]   A sharp, sinuous loop or curve in a stream, usually part of a series. [OCEANOGR]   A deviation of the flow pattern of a current.

meander bar *See* point bar.

meander belt [GEOL]   The zone along the floor of a valley across which a meandering stream periodically shifts its channel.

meander core [GEOL]   A hill encircled by a stream meander. Also known as rock island.

meandering stream [HYD]   A stream having a pattern of successive meanders. Also known as snaking stream.

meander niche [GEOL]   A conical or crescentic opening in the wall of a cave formed by downward and lateral stream erosion.

meander plain [GEOL]   A plain built by the meandering process, or a plain of lateral accretion.

meander scar [GEOL]   A crescentic, concave mark on the face of a bluff or valley wall formed by a meandering stream.

meander spur [GEOL]   An undercut projection of high land that extends into the concave part of, and is enclosed by, a meander.

mean diurnal high-water inequality [OCEANOGR]   Half the average difference between the heights of the two high waters of each tidal day over a 19-year period; it is obtained by substracting the mean of all high waters from the mean of the higher high waters.

mean diurnal low-water inequality [OCEANOGR]   Half the average difference between the heights of the two low waters of each tidal day over a 19-year period; it is obtained by subtracting the mean of all lower low waters from the mean of the low waters.

mean higher high water [OCEANOGR]   The average height of higher high waters at a place over a 19-year period.

mean high water [OCEANOGR]   The average height of all high waters recorded at a given place over a 19-year period.

mean high-water lunitidal interval [OCEANOGR]   The average interval of time between the transit (upper or lower) of the moon and the next high water at a place. Also known as corrected establishment.

mean high-water neaps [OCEANOGR]   The average height of the high waters of neap tides. Also known as neap high water.

mean high-water springs [OCEANOGR]   The average height of the high waters of spring tides. Also known as high-water springs; spring high water.

mean latitude [GEOD] Half the arithmetical sum of the latitudes of two places on the same side of the equator; mean latitude is labeled N or S to indicate whether it is north or south of the equator.

mean lower low water [OCEANOGR] The average height of the lower low waters at a place over a 19-year period.

mean lower low-water springs [OCEANOGR] The average height of lower low-water springs at a place.

mean low water [OCEANOGR] The average height of all low waters recorded at a given place over a 19-year period.

mean low-water lunitidal interval [OCEANOGR] The average interval of time between the transit (upper or lower) of the moon and the next low water at a place.

mean low-water neaps [OCEANOGR] The average height of the low water at neap tides. Also known as low-water neaps; neap low water.

mean low-water springs [OCEANOGR] The average height of the low waters of spring tides; this level is used as a tidal datum in some areas. Also known as low-water springs; spring low water.

mean map *See* mean chart.

mean neap range *See* neap range.

mean neap rise [OCEANOGR] The height of mean high-water neaps above the chart datum.

mean range [OCEANOGR] The difference in the height between mean high water and mean low water.

mean rise interval [OCEANOGR] The average interval of time between the transit (upper or lower) of the moon and the middle of the period of rise of the tide at a place; it may be either local or Greenwich, depending on the transit to which it is referred, but the local interval is assumed unless otherwise specified.

mean rise of tide [OCEANOGR] The height of mean high water above the chart datum.

mean river level [HYD] The average height of the surface of a river at any point for all stages of the tide over a 19-year period.

mean sea level [OCEANOGR] The average sea surface level for all stages of the tide over a 19-year period, usually determined from hourly height readings from a fixed reference level.

mean spring range *See* spring range.

mean spring rise [OCEANOGR] The height of mean high-water springs above the chart datum.

mean temperature [METEOROL] The average temperature of the air as indicated by a properly exposed thermometer during a given time period, usually a day, month, or year.

mean tide *See* half tide.

mean tide level [OCEANOGR] The tide level halfway between mean high water and mean low water.

mean water level [OCEANOGR] The average surface level of a body of water.

measured drilling depth [ENG] The apparent depth of a borehole as measured along its longitudinal axis.

measured ore *See* developed reserves.

measuring chute [MIN ENG] An ore bin or coal bin installed adjacent to the shaft bottom in skip winding and having a capacity equal to that of the skip used; ensures rapid, correct loading of skips without spillage. Also known as measuring pocket.

measuring day [MIN ENG] The day when work is measured and recorded for assessing the wages.

measuring pocket *See* measuring chute.

mechanical erosion *See* corrasion.

mechanical filter [PETRO ENG] Granule-packed steel shell used to filter suspended floc or undissolved solids out of treated waterflood water; granules can be graded sand and gravel, anthracite coal, graphitic ore, or aluminum-oxide plates with granular filter medium.

mechanical flotation cell [MIN ENG] A device that separates minerals from ore water pulp; it consists of a cell in which the pulp is kept mixed and moving by an impeller at the bottom of the cell; the impeller pulls air down the standpipe and disperses it as bubbles through the pulp; the floatable minerals concentrate in the froth above, and the pulp is removed by a scraper.

mechanical instability *See* absolute instability.

mechanical mucking [ENG] Loading of dirt or stone in tunnels or mines by machines.

mechanical oil valve [PETRO ENG] A float-operated liquid-level control valve used to control liquid flow out of oil-lease gas-oil separator tank systems.

mechanical sediment *See* clastic sediment.

mechanical yielding prop [MIN ENG] A steel prop in which yield is controlled by friction between two sliding surfaces or telescopic tubes. Also known as friction yielding prop.

medial moraine [GEOL] **1.** An elongate moraine carried in or upon the middle of a glacier and parallel to its sides. **2.** A moraine formed by glacial abrasion of a rocky protuberance near the middle of a glacier.

median mass [GEOL] A less disturbed structural block in the middle of an orogenic belt, bordered on both sides by orogenic structure, thrust away from it. Also known as betwixt mountains; Zwischengebirge.

median particle diameter [GEOL] The middlemost particle diameter of a rock or sediment, larger than 50% of the diameter in the distribution and smaller than the other 50%.

medina quartzite [MINERAL] A variety of quartz containing 97.8% silica; melting point is about 1700°C.

mediterranean *See* mesogeosyncline.

Mediterranean climate [CLIMATOL] A type of climate characterized by hot, dry, sunny summers and a winter rainy season; basically, this is the opposite of a monsoon climate. Also known as etesian climate.

mediterranean sea [GEOGR] A deep epicontinental sea that is connected with the ocean by a narrow channel.

Mediterranean Sea [GEOGR] A sea that lies between Europe, Asia Minor, and Africa and is completely landlocked except for the Strait of Gibraltar, the Bosporus, and the Suez Canal; total water area is 2,501,000 square kilometers.

medium-range forecast [METEOROL] A forecast of weather conditions for a period of 48 hours to a week in advance. Also known as extended-range forecast.

medium-volatile bituminous coal [GEOL] Bituminous coal consisting of 23–31% volatile matter.

Medullosaceae [PALEOBOT] A family of seed ferns; these extinct plants all have large spirally arranged petioles with numerous vascular bundles.

meerschaum *See* sepiolite.

megacryst [PETR] Any crystal or grain in an igneous or metamorphic rock that is significantly larger than the surrounding matrix.

megacyclothem [GEOL] A cycle of or combination of related cyclothems.

megaripple [GEOL] A large sand wave.

megatectonics [GEOL] The tectonics of the very large structural features of the earth.

meionite [MINERAL] $3CaAl_2Si_2O_8 \cdot CaCO_3$ A scapolite mineral composed of calcium aluminosilicate and calcium carbonate; it is isomorphous with marialite.

Melanesia [GEOGR] A group of islands in the Pacific Ocean northeast of Australia.

mélange [GEOL] A heterogeneous medley or mixture of rock materials; specifically, a mappable body of deformed rocks consisting of a pervasively sheared, fine-grained, commonly pelitic matrix, thoroughly mixed with angular and poorly sorted inclusions of native and exotic tectonic fragments, blocks, or slabs, of diverse origins and geologic ages, that may be as much as several kilometers in length. Also known as block clay.

melanic *See* melanocratic.

melanocerite [MINERAL] $(Ca,Ce,Y)_8(BO_3)(SiO_4)_4(F,OH)_4$ A brown or black rhombohedral mineral composed of complex silicate, borate, fluoride, tantalate, or other anion of cerium, yttrium, calcium, and other metals; occurs as crystals.

melanocratic [GEOL] Dark-colored, referring to igneous rock containing at least 50–60% mafic minerals. Also known as chromocratic; melanic.

melanophlogite [MINERAL] A mineral composed chiefly of silicon dioxide and containing some carbon and sulfur.

melanostibian [MINERAL] $Mn(Sb,Fe)O_3$ A black mineral consisting of iron and manganese antimonite; occurs as foliated masses and as striated crystals.

melanotekite [MINERAL] $Pb_2Fe_2Si_2O_9$ A black or dark-gray mineral composed of lead iron silicate.

melanovanadite [MINERAL] $Ca_2V_{10}O_{25}$ A black mineral composed of a complex oxide of calcium and vanadium.

melanterite [MINERAL] $FeSO_4 \cdot 7H_2O$ A green mineral occurring mainly in fibrous or concretionary masses, or in short, monoclinic, prismatic crystals; hardness is 2 on Mohs scale, and specific gravity is 1.90.

melaphyre [PETR] Altered basalt, especially of Carboniferous and Permian age.

melilite [MINERAL] A sorosilicate mineral group having the complex composition $[(Na, Ca)_2(Mg, Al)(Si, Al)_2O_7]$ and crystallizing in the tetragonal system; luster is vitreous to resinous, and color is white, yellow, greenish, reddish, or brown; hardness is 5 on Mohs scale, and specific gravity varies from 2.95 to 3.04.

melilitite [PETR] An extrusive rock that is generally olivine-free and composed of more than 90% mafic mineral such as melilite and augite, with minor amounts of feldspathoids and sometimes plagioclase.

meliphane *See* meliphanite.

meliphanite [MINERAL] $(Ca,Na)_2Be(Si,Al)_2(O,OH,F)_7$ A yellow, red, or black mineral composed of sodium calcium beryllium fluosilicate. Also known as meliphane.

mellite [MINERAL] $Al_2[C_6(COO)_6]\cdot18H_2O$ A honey-colored mineral with resinous luster composed of the hydrous aluminum salt of mellitic acid, occurring as nodules in brown coal; it is in part a product of vegetable decomposition.

melonite [MINERAL] $NiTe_2$ A reddish-white mineral composed of nickel telluride.

melting level [METEOROL]   The altitude at which ice crystals and snowflakes melt as they descend through the atmosphere.

meltwater [HYD]   Water derived from melting ice or snow, especially glacier ice.

member [GEOL]   A rock stratigraphic unit of subordinate rank comprising a specially developed part of a varied formation.

mendip [GEOL]   **1.** A buried hill that is exposed as an inlier.   **2.** A coastal-plain hill that was originally an offshore island.

mendipite [MINERAL] $Pb_3Cl_2O_2$ A white orthorhombic mineral consisting of an oxide and chloride of lead.

mendozite [MINERAL] $NaAl(SO_4)_2\cdot11H_2O$ A monoclinic mineral of the alum group composed of hydrous sodium aluminum sulfate.

meneghinite [MINERAL] $CuPb_{13}Sb_7S_{24}$ A blackish lead gray mineral consisting of lead antimony sulfide.

Meniscotheriidae [PALEON]   A family of extinct mammals of the order Condylarthra possessing selenodont teeth and molarized premolars.

Meramecian [GEOL]   A North American provincial series of geologic time: Upper Mississippian (above Osagian, below Chesterian).

meraspis [PALEON]   Advanced larva of a trilobite; stage in which the pygidium begins to form.

Mercalli scale [GEOPHYS]   A 12-point scale for classifying the magnitude of an earthquake.

mercallite [MINERAL] $KHSO_4$ A colorless or sky blue, orthorhombic mineral consisting of potassium acid sulfate; occurs as stalactites composed of minute crystals.

Mercator chart [MAP]   A chart on the Mercator projection, commonly used for marine navigation. Also known as equatorial cylindrical orthomorphic chart.

Mercator projection [MAP]   A conformal cylindrical map projection in which the surface of a sphere or spheroid, such as the earth, is conceived as developed on a cylinder tangent along the Equator; meridians appear as equally spaced vertical lines, and parallels as horizontal lines drawn farther apart as the latitude increases, such that the correct relationship between latitude and longitude scales at any point is maintained.

mercurial horn ore *See* calomel.

mercury barometer [ENG]   An instrument which determines atmospheric pressure by measuring the height of a column of mercury which the atmosphere will support; the mercury is in a glass tube closed at one end and placed, open end down, in a well of mercury. Also known as Torricellian barometer.

Merian's formula [OCEANOGR]   A formula for the period of a seiche, $T = (1/n)(2L/\sqrt{gd})$, where n is the number of nodes, L is the horizontal dimension of the basin measured in the direction of wave motion, g is the acceleration of gravity, and d is the depth of the water.

meridional [GEOL]   Pertaining to longitudinal movements or directions, that is, northerly or southerly.

meridional cell [GEOPHYS] A very large-scale convection circulation in the atmosphere or ocean which takes place in a meridional plane, with northward and southward currents in opposite branches of the cell, and upward and downward motion in the equatorward and poleward ends of the cell.

meridional circulation [METEOROL] An atmospheric circulation in a vertical plane oriented along a meridian; it consists, therefore, of the vertical and the meridional (north or south) components of motion only. [OCEANOGR] The exchange of water masses between northern and southern oceanic regions.

meridional difference [MAP] The difference between the meridional parts of any two given parallels of latitude; this difference is found by subtraction if the two parallels are on the same side of the equator, and by addition if on opposite sides. Also known as difference of meridional parts.

meridional flow [METEOROL] A type of atmospheric circulation pattern in which the meridional (north and south) component of motion is unusually pronounced; the accompanying zonal component is usually weaker than normal. [OCEANOGR] Current moving along a meridian.

meridional front [METEOROL] A front in the South Pacific separating successive migratory subtropical anticyclones; such fronts are essentially in the form of great arcs with meridians of longitudes as chords; they have the character of cold fronts.

meridional index [METEOROL] A measure of the component of air motion along meridians, averaged, without regard to sign, around a given latitude circle.

meridional parts [MAP] The length of the arc of a meridian between the equator and a given parallel on a mercator chart, expressed in units of 1 minute of longitude at the equator.

meridional wind [METEOROL] The wind or wind component along the local meridian, as distinguished from the zonal wind.

merismite [PETR] A type of chorismite in which penetration of the diverse units is irregular.

merocrystalline *See* hypocrystalline.

merohedral [CRYSTAL] Of a crystal class in a system, having a general form with only one-half, one-fourth, or one-eighth the number of equivalent faces of the corresponding form in the holohedral class of the same system. Also known as merosymmetric.

meromictic [HYD] Of or pertaining to a lake whose water is permanently stratified and therefore does not circulate completely throughout the basin at any time during the year.

merosymmetric *See* merohedral.

merrihueite [MINERAL] $(K,Na)_2(Fe,Mg)_5Si_{12}O_{30}$ A silicate mineral found only in meteorites.

merrillite [MINERAL] $Ca_3(PO_4)_2$ Colorless phosphate mineral found only in meteorites.

merwinite [MINERAL] $Ca_3MgSi_2O_8$ A rare colorless or pale-green neosilicate mineral crystallizing in the monoclinic system; occurs in granular aggregates showing polysynthetic twinning; hardness is 6 on Mohs scale, and specific gravity is 3.15.

Merycoidodontidae [PALEON] A family of extinct tylopod ruminants in the superfamily Merycoidodontoidea.

Merycoidodontoidea [PALEON] A superfamily of extinct ruminant mammals in the infraorder Tylopoda which were exceptionally successful in North America.

merzlota *See* frozen ground.

mesa [GEOGR]   A broad, isolated, flat-topped hill bounded by a steep cliff or slope on at least one side; represents an erosion remnant.

mesa-butte [GEOGR]   A butte formed as the result of erosion and reduction of a mesa.

Mesacanthidae [PALEON]   An extinct family of primitive acanthodian fishes in the order Acanthodiformes distinguished by a pair of small intermediate spines, large scales, superficially placed fin spines, and a short branchial region.

mesa plain [GEOGR]   A flat-topped summit of a hilly mountain.

mesh [MIN ENG]   **1.** A closed path traversed through the network in ventilation surveys. **2.** The size of diamonds as determined by sieves.

mesh-texture *See* reticulate.

mesobenthos [OCEANOGR]   Of or pertaining to the sea bottom at depths of 180–900 meters (100–500 fathoms).

mesoclimate [CLIMATOL]   **1.** The climate of small areas of the earth's surface which may not be representative of the general climate of the district. **2.** A climate characterized by moderate temperatures, that is, in the range 20–30°C. Also known as mesothermal climate.

mesoclimatology [CLIMATOL]   The study of mesoclimates.

mesocratic [PETR]   Of igneous rock, being intermediate in color between leucocratic and melanocratic due to equal amounts of light and dark constituents.

mesocrystalline [PETR]   Of a crystalline rock, containing crystals whose diameters are intermediate between microcrystalline and macrocrystalline rock.

mesocyclone [METEOROL]   A cyclonic circulation interior to a convective storm.

mesogeosyncline [GEOL]   A geosyncline between two continents. Also known as mediterranean.

Mesohippus [PALEON]   An early ancestor of the modern horse; occurred during the Oligocene.

mesolite [MINERAL] $Na_2Ca_2Al_6Si_9O_{30} \cdot 8H_2O$ Zeolite mineral composed of hydrous sodium calcium aluminosilicate, usually found in white or colorless tufts of acicular crystals; used as cation exchangers or molecular sieves.

mesometeorology [METEOROL]   That portion of the science of meteorology concerned with the study of atmospheric phenomena on a scale larger than that of micrometeorology, but smaller than the cyclonic scale.

Mesonychidae [PALEON]   A family of extinct mammals of the order Condylarthra.

mesopause [METEOROL]   The top of the mesosphere; corresponds to the level of minimum temperature at 80 to 95 kilometers.

mesopeak [METEOROL]   The temperature maximum at about 50 kilometers in the mesosphere.

mesopore [PALEON]   A tube paralleling the autopore or chamber in fossil bryozoans.

Mesosauria [PALEON]   An order of extinct aquatic reptiles known from a single genus, *Mesosaurus*, characterized by a long snout, numerous slender teeth, small forelimbs, and webbed hindfeet.

mesoscale eddies *See* mode eddies.

mesosiderite [GEOL]   A stony-iron meteorite containing about equal amounts of silicates and nickel-iron, with considerable troilite. Also known as grahamite.

mesosphere [GEOL] *See* lower mantle. [METEOROL] The atmospheric shell between about 45–55 kilometers and 80–95 kilometers, extending from the top of the stratosphere to the mesopause; characterized by a temperature that generally decreases with altitude.

mesostasis [GEOL] The last-formed interstitial material, either glassy or aphanitic, of an igneous rock.

Mesosuchia [PALEON] A suborder of extinct crocodiles of the Late Jurassic and Early Cretaceous.

mesothermal [MINERAL] Of a hydrothermal mineral deposit, formed at great depth at temperatures of 200–300°C.

mesothermal climate *See* mesoclimate.

mesotil [GEOL] A semiplastic or semifriable derivative of chemically weathered till; forms beneath a partially drained area.

Mesozoic [GEOL] A geologic era from the end of the Paleozoic to the beginning of the Cenozoic; commonly referred to as the Age of Reptiles.

mesozone [PETR] The intermediate depth zone of metamorphism in metamorphic rock characterized by moderate temperatures (300–500°C), hydrostatic pressure, and shearing stress.

metaanthracite [GEOL] Anthracite coal containing at least 98% fixed carbon.

metabentonite [GEOL] Altered bentonite, formed by compaction or metamorphism; it swells very little and lacks the usual high colloidal properties of bentonite.

metacinnabar [MINERAL] HgS A black isometric mineral that represents an ore of mercury. Also known as metacinnabarite.

metacinnabarite *See* metacinnabar.

Metacopina [PALEON] An extinct suborder of ostracods in the order Podocopida.

metacryst [PETR] A large crystal, such as garnet, formed in metamorphic rock by recrystallization. Also known as metacrystal.

metacrystal *See* metacryst.

metahalloysite [GEOL] A term used in Europe for the less hydrous form of halloysite. Also known as halloysite in the United States.

metaharmosis *See* metharmosis.

metahewettite [MINERAL] $CaV_6O_{16} \cdot 9H_2O$ A deep red, probably orthorhombic mineral consisting of hydrated calcium vanadate; occurs as pulverulent masses.

metahohmannite [MINERAL] $Fe_2(SO_4)_2(OH)_2 \cdot 3H_2O$ An orange mineral consisting of a hydrated basic iron sulfate; occurs as pulverulent masses.

metalimnion *See* thermocline.

metalliferous [MINERAL] Pertaining to mineral deposits from which metals can be extracted.

metallized slurry blasting [ENG] The breaking of rocks by using slurried explosive medium containing a powdered metal, such as powdered aluminum.

metallogenic province [GEOL] A region characterized by a particular mineral assemblage, or by one or more specific types of mineralization. Also known as metallographic province.

metallographic province *See* metallogenic province.

metal mining [MIN ENG]   The industry that supplies the various metals and associated products.

metamict [MINERAL]   Of a radioactive mineral, exhibiting lattice disruption due to radiation damage while the original external morphology is retained.

metamorphic aureole *See* aureole.

metamorphic breccia [PETR]   Breccia formed by metamorphism.

metamorphic differentiation [PETR]   Processes by which different mineral assemblages develop in some sequence from an initially uniform parent rock.

metamorphic facies [PETR]   All rocks of any composition that have reached chemical equilibrium with respect to certain ranges of pressure and temperature during metamorphism, characterized by the stability of specific index minerals. Also known as densofacies.

metamorphic facies series [PETR]   A group of metamorphic facies characteristic of an individual area, represented in a pressure-temperature diagram by a curve or group of curves illustrating the range of the different types of metamorphism and metamorphic facies.

metamorphic overprint *See* overprint.

metamorphic rock [PETR]   A rock formed from preexisting solid rocks by mineralogical, structural, and chemical changes, in response to extreme changes in temperature, pressure, and shearing stress.

metamorphic rock reservoir [GEOL]   Uncommon type of formation for oil reservoir; developed when secondary porosity results from fracturing or weathering.

metamorphic zone *See* aureole.

metamorphism [PETR]   The mineralogical and structural changes of solid rock in response to environmental conditions at depth in the earth's crust.

metaquartzite [MINERAL]   A quartzite formed by metamorphic recrystallization.

metaripple [GEOL]   An asymmetrical sand ripple.

metarossite [MINERAL] $CaV_2O_6 \cdot 2H_2O$ A light yellow mineral consisting of hydrated calcium vanadate; occurs as masses and veinlets.

metasediment [GEOL]   A sediment or sedimentary rock which shows evidence of metamorphism.   [PETR]   Metamorphic rock formed from sedimentary rock.

metasideronatrite [MINERAL] $Na_4Fe_2(SO_4)_4(OH)_2 \cdot 3H_2O$ A yellow mineral composed of basic hydrous iron sodium sulfate.

metasilicate [MINERAL]   A salt of the hypothetical metasilicic acid H_2SiO_3. Also known as bisilicate.

metasomatic [PETR]   Pertaining to the process or the result of metasomatism.

metasomatism [PETR]   A variety of metamorphism in which one mineral or a mineral assemblage is replaced by another of different composition without melting.

metatorbernite [MINERAL] $Cu(UO_2)_2(PO_4)_2 \cdot 8H_2O$ A green secondary mineral composed of hydrous copper uranium phosphate; similar to torbernite, but with less water content.

metavariscite [MINERAL] $AlPO_6 \cdot 2H_2O$ A green monoclinic mineral composed of hydrous aluminum phosphate; it is isomorphous with phosphosiderite.

metavauxite [MINERAL] $FeAl_2(PO_4)_2(OH)_2 \cdot 8H_2O$ A colorless mineral composed of hydrous basic phosphate of iron and aluminum; similar to vauxite, but with more water.

metavoltine [MINERAL] A yellowish-brown or orange-brown to greenish-brown, hexagonal mineral consisting of a hydrated basic sulfate of iron and potassium; occurs in tabular form or as aggregates.

metazeunerite [MINERAL] $Cu(UO_2)_2(AsO_4)_2 \cdot 8H_2O$ A grass to emerald green, tetragonal mineral consisting of a hydrated arsenate of copper and uranium; occurs in tabular form.

meteoric water [HYD] Groundwater which originates in the atmosphere and reaches the zone of saturation by infiltration and percolation.

meteorite [GEOL] Any meteoroid that has fallen to the earth's surface.

meteorite crater [GEOL] An impact crater on the surface of the earth or of a celestial body caused by a meteorite; a characteristic feature on the earth is the upturned rim, which formed as the rocks rebounded following the impact.

meteorogram [ENG] A record obtained from a meteorograph. [METEOROL] A chart in which meteorological variables are plotted against time.

meteorograph [ENG] An instrument that measures and records meteorological data such as air pressure, temperature, and humidity.

meteorological [METEOROL] Of or pertaining to meteorology or weather.

meteorological balloon [ENG] A balloon, usually of high-quality neoprene, polyethylene, or Mylar, used to lift radiosondes to high altitudes.

meteorological chart [METEOROL] A weather map showing the spatial distribution, at an instant of time, of atmospheric highs and lows, rain clouds, and other phenomena.

meteorological data [METEOROL] Facts pertaining to the atmosphere, especially wind, temperature, and air density.

meteorological equator *See* equatorial trough.

meteorological instrumentation [ENG] Apparatus and equipment used to obtain quantitative information about the weather.

meteorological minima [METEOROL] Minimum values of meteorological elements prescribed for specific types of flight operation.

meteorological optics [OPTICS] A branch of atmospheric physics or physical meteorology in which optical phenomena occurring in the atmosphere are described and explained. Also known as atmospheric optics.

meteorological radar [ENG] Radar which is used to study the scattering of radar waves by various types of atmospheric phenomena, for making weather observations and forecasts.

meteorological range [METEOROL] An empirically consistant measure of the visual range of a target; a concept developed to eliminate from consideration the threshold contrast and adaptation luminance, both of which vary from observer to observer. Also known as standard visibility; standard visual range.

meteorological rocket [ENG] Small rocket system used to extend observation of atmospheric character above feasible limits for balloon-borne observing and telemetering instruments. Also known as rocketsonde.

meteorological satellite [ENG] Earth-orbiting spacecraft carrying a variety of instruments for measuring visible and invisible radiations from the earth and its atmosphere.

meteorological solenoid [METEOROL] A hypothetical tube formed in space by the intersection of a set of surfaces of constant pressure and a set of surfaces of constant specific volume of air. Also known as solenoid.

meteorological tide [OCEANOGR] A change in water level caused by local meteorological conditions, in contrast to an astronomical tide, caused by the attractions of the sun and moon.

meteorology [SCI TECH] The science concerned with the atmosphere and its phenomena; the meteorologist observes the atmosphere's temperature, density, winds, clouds, precipitation, and other characteristics and aims to account for its observed structure and evolution (weather, in part) in terms of external influence and the basic laws of physics.

meteor trail *See* ion column.

metering installation [PETRO ENG] Oil-production receiving system that includes with the tank battery a metering separator, metering treater, or other type of meter used in conjunction with test separators or emulsion treaters.

metering separator [PETRO ENG] Oil-field process vessel that performs the dual functions of gas-oil separation and liquids metering.

methane drainage *See* firedamp drainage.

methane indicator [MIN ENG] A portable analytical instrument that can determine the methane content in the mine air at the place where the sample is taken; air is brought into the instrument through an aspirator bulb and passed through a cartridge filter to remove moisture.

methane monitoring system [MIN ENG] A system that samples methane content in mine air continuously and feeds this information into an electrical device that cuts off power in each mining machine when the methane content rises above a predetermined level.

metharmosis [GEOL] Changes that occur in a buried sediment after uplift or consolidation but before the onset of weathering. Also spelled metaharmosis.

method of images [PETRO ENG] Method of calculating the interference between reservoirs by assuming a mirror image of one reservoir on the far side of a geologic fault.

Mexican onyx *See* onyx marble.

meyerhofferite [MINERAL] $Ca_2B_6O_{11}\cdot7H_2O$ A colorless, hydrated borate mineral that crystallizes in the triclinic system.

Miacidae [PALEON] The single, extinct family of the carnivoran superfamily Miacoidea.

Miacoidea [PALEON] A monofamilial superfamily of extinct carnivoran mammals; a stem group thought to represent the progenitors of the earliest member of modern carnivoran families.

miagite *See* corsite.

miargyrite [MINERAL] $AgSbS_2$ An iron-black to steel-gray mineral that crystallizes in the monoclinic system.

miarolithite [PETR] A chorismite type of igneous rock having miarolitic cavities or vestiges thereof.

miarolitic [PETR] Of igneous rock, characterized by small irregular cavities into which well-formed crystals of the rock-forming mineral protrude.

mica [MINERAL] A group of phyllosilicate minerals (with sheetlike structures) of general formula $(K,Na,Ca)(Mg,Fe,Li,Al)_{2-3}(Al,Si)_4O_{10}(OH,F)_2$ characterized by low hardness (2–2½) and perfect basal cleavage.

mica book [MINERAL] A crystal of mica, usually large and irregular, whose cleavage plates resemble the leaves of a book. Also known as book.

micaceous [GEOL] Consisting of, containing, pertaining to, or resembling mica.

micaceous arkose [PETR] A sandstone containing 25–90% feldspars and feldspathic crystalline rock fragments, 10–50% micas and micaceous metamorphic rock fragments, and 0–65% quartz, chert, and metamorphic quartzite.

mica schist [PETR] A schist which is composed essentially of mica and quartz and whose characteristic foliation is mainly due to the parallel orientation of the mica flakes.

micellar flooding [PETRO ENG] A two-step enhanced oil recovery process in which a surfactant slug is injected into the well followed by a larger slug of water containing a high-molecular-weight polymer which pushes the chemicals through the field and improves mobility and sweep efficiency. Also known as microemulsion flooding; surfactant flooding.

Michigan cut [MIN ENG] A technique used to break off ore at a heading; a large hole or series of small holes at the center of the heading are drilled parallel to the tunnel direction but not charged with explosive; other holes are drilled in the heading and charged so that upon detonation they break out toward the uncharged holes.

Michigan tripod [MIN ENG] A support for a drilling outfit; consists of three debarked pine or fir timber poles about 25 feet (7.6 meters) long whose butt ends are about 12 inches (30 centimeters) in diameter; a sheave suspended from a clevis at the top of the tripod is aligned over the hoisting drum and the borehole; there is a minimum of 22 feet (6.7 meters) of headroom above the drill floor.

micrinite [PETR] An opaque granular variety of inertinite of medium hardness showing no plant-cell structure.

micrite [PETR] A semiopaque crystalline limestone matrix that consists of chemically precipitated calcite mud, whose crystals are generally 1–4 micrometers in diameter.

micritic limestone [PETR] A limestone consisting of more than 90% micrite or less than 10% allochems.

microbarm [GEOPHYS] That portion of the record of a microbarograph between any two or a specified small number of the successive crossings of the average pressure level in the same direction; analogous to microseism.

microbarogram [ENG] The record or trace made by a microbarograph.

microbarograph [METEOROL] A type of aneroid barograph designed to record atmospheric pressure variations of very small magnitude.

microbreccia [GEOL] A poorly sorted sandstone containing large, angular sand particles in a fine silty or clayey matrix.

microcaliper log [PETRO ENG] A detailed and accurate record of drill-hole diameter; used to detect caved sections and to verify the presence of mud cake.

microclimate [CLIMATOL] The local, rather uniform climate of a specific place or habitat, compared with the climate of the entire area of which it is a part.

microclimatology [CLIMATOL] The study of a microclimate, including the study of profiles of temperature, moisture and wind in the lowest stratum of air, the effect of the vegetation and of shelterbelts, and the modifying effect of towns and buildings.

microcline [MINERAL] $KAlSi_3O_8$ A triclinic potassium-rich feldspar, usually containing minor amounts of sodium; may be clear, white, pale-yellow, brick-red, or green, and is generally characterized by crosshatch twinning.

microcoquina [PETR] A clastic limestone composed wholly or partially of cemented sand-size particles of shell detritus.

microcrystalline [CRYSTAL] Composed of or containing crystals that are visible only under the microscope.

microemulsion flooding *See* micellar flooding.

microfacies [PETR] The composition, features, or appearance of a rock or mineral in thin section under the microscope.

microfossil [PALEON] A small fossil which is studied and identified by means of the microscope.

microlaterolog [PETRO ENG] Modification of the downhole microlog in which extra electrodes focus electric current into a trumpet-shaped area; gives greater resistivity-measurement resolution than does the microlog.

microlayer [OCEANOGR] The thin zone beneath the surface of the ocean or any free water surface within which physical processes are modified by proximity to the air-water boundary.

microlite [CRYSTAL] A microscopic crystal which polarizes light. Also known as microlith. [MINERAL] $(Na,Ca)_2(Ta, Nb)_2O_6(O,OH,F)$ A pale-yellow, reddish, brown, or black isometric mineral composed of sodium calcium tantalum oxide with a small amount of fluorine; it is isomorphous with pyrochlore. Also known as djalmaite.

microlith [CRYSTAL] *See* microlite.

microlithology [PETR] Microscopic study of the characteristics of rocks.

microlitic [PETR] Of the texture of a porphyritic igneous rock, having a groundmass composed of an aggregate of microlites in a generally glassy base.

microlog [PETRO ENG] A drill-hole resistivity log recorded with electrodes mounted at short distances from each other in the face of a rubber-padded microresistivity sonde.

micrometeorology [METEOROL] That portion of the science of meteorology that deals with the observation and explanation of the smallest-scale physical and dynamic occurrences within the atmosphere; studies are confined to the surface boundary layer of the atmosphere, that is, from the earth's surface to an altitude where the effects of the immediate underlying surface upon air motion and composition become negligible.

micropaleontology [PALEON] A branch of paleontology that deals with the study of microfossils.

micropegmatite [PETR] Microcrystalline graphic granite.

microperthite [MINERAL] Perthite in which the lamellae are visible only under the microscope.

microphyric [PETR] Of the texture of an igneous rock, containing microscopic phenocrysts (longest dimension 0.2 millimeter). Also known as microporphyritic.

micropoikilitic [PETR] Of the texture of an igneous rock, having poikilitic character visible only under the microscope.

micropore [GEOL] A pore small enough to hold water against the pull of gravity and to retard water flow.

microporphyritic *See* microphyric.

micropulsation [GEOPHYS] A short-period geomagnetic variation in the range of about 0.2–600 seconds, typically exhibiting an oscillatory waveform.

microrelief [GEOGR] Irregularities of the land surface causing variations in elevation amounting to no more than a few feet.

microresistivity survey [PETRO ENG] General term for downhole resistivity surveys of oil-bearing formations; includes microlog and microlaterolog surveys.

Microsauria [PALEON] An order of Carboniferous and early Permian lepospondylous amphibians.

microscopic anisotropy [PETRO ENG] Phenomenon in electrical downhole logging wherein electric current flows most easily along the water-filled interstices, usually parallel to sedimentary bed strata.

microseism [GEOPHYS] A weak, continuous, oscillatory motion in the earth having a period of 1–9 seconds and caused by a variety of agents, especially atmospheric agents; not related to an earthquake.

microseismic instrument [MIN ENG] An instrument for the study of roof strata and supports; it is inserted in holes, drilled at selected points, for listening to subaudible vibrations that precede rock failure.

microspherulitic [PETR] Of the texture of an igneous rock, having spherulitic character visible only under the microscope.

microstylolite [PETR] A stylolite in which the surface relief is less than 1 millimeter.

microtectonics *See* structural petrology.

microthermal climate [CLIMATOL] A temperature province in both of C.W. Thornthwaite's climatic classifications, generally described as a "cool" or "cold winter" climate.

Microtragulidae [PALEON] A group of saltatorial caenolistoid marsupials that appeared late in the Cenozoic and paralleled the small kangaroos of Australia.

microvitrain [GEOL] A coal lithotype; fine vitrainlike lenses or laminae in clarain.

mid-Atlantic ridge [GEOL] The mid-oceanic ridge in the Atlantic.

Middle Cambrian [GEOL] The geologic epoch between Upper and Lower Cambrian, beginning approximately 540,000,000 years ago.

middle clouds [METEOROL] Types of clouds the mean level of which is between 6500 and 20,000 feet (1980 and 6100 meters); the principal clouds in this group are altocumulus and altostratus.

Middle Cretaceous [GEOL] The geologic epoch between the Upper and Lower Cretaceous, beginning approximately 120,000,000 years ago.

Middle Devonian [GEOL] The geologic epoch between the Upper and Lower Devonian, beginning approximately 385,000,000 years ago.

Middle Jurassic [GEOL] The geologic epoch between the Upper and Lower Jurassic, beginning approximately 170,000,000 years ago.

middle latitude Also known as mid-latitude. [GEOGR] A point of latitude that is midway on a north-and-south line between two parallels. [NAV] The latitude at which the arc length of the parallel separating the meridians passing through two specific points is exactly equal to the departure in proceeding from one point to the other by middle-latitude sailing.

middle-latitude westerlies *See* westerlies.

Middle Mississippian [GEOL] The geologic epoch between the Upper and Lower Mississippian.

Middle Ordovician [GEOL] The geologic epoch between the Upper and Lower Ordovician, beginning approximately 460,000,000 years ago.

Middle Pennsylvanian [GEOL] The geologic epoch between the Upper and Lower Pennsylvanian.

Middle Permian [GEOL] The geologic epoch between the Upper and Lower Permian, beginning approximately 260,000,000 years ago.

Middle Silurian [GEOL] The geologic epoch between the Upper and Lower Silurian.

Middle Triassic [GEOL] The geologic epoch between the Upper and Lower Triassic, beginning approximately 215,000,000 years ago.

middling [MIN ENG] An ore product intermediate in mineral content between a concentrate and a tailing.

mid-extreme tide [OCEANOGR] A level midway between the extreme high water and extreme low water occurring at a place.

midfan [GEOL] The portion of an alluvial fan between the fanhead and the outer, lower margins.

midget impinger [MIN ENG] A dust-sampling impinger requiring only a 12-inch (30-centimeter) head of water for its operation.

mid-latitude westerlies *See* westerlies.

mid-ocean canyon *See* deep-sea channel.

mid-oceanic ridge [GEOL] A continuous, median, seismic mountain range on the floor of the ocean, extending through the North and South Atlantic oceans, the Indian Ocean, and the South Pacific Ocean; the topography is rugged, elevation is 1–3 kilometers (km), width is about 1500 km, and length is over 84,000 km. Also known as mid-ocean ridge; mid-ocean rise; oceanic ridge.

mid-ocean ridge *See* mid-oceanic ridge.

mid-ocean rift *See* rift valley.

mid-ocean rise *See* mid-oceanic ridge.

miersite [MINERAL] (Cu,Ag)I A canary yellow, isometric mineral consisting of copper and silver iodide.

migma [GEOL] A mixture of solid rock materials and rock melt with mobility or potential mobility.

migmatite [PETR] A mixed rock exhibiting crystalline textures in which a truly metamorphic component is streaked and mixed with obviously once-molten material of a more or less granitic character.

migmatization [PETR] Formation of migmatite; involves either injection or in-place melting.

migration [GEOL] **1.** Movement of a topographic feature from one place to another, especially movement of a dune by wind action. **2.** Movement of liquid or gaseous hydrocarbons from their source into reservoir rocks. [HYD] Slow, downstream movement of a system of meanders.

migratory [METEOROL] Commonly applied to pressure systems embedded in the westerlies and, therefore, moving in a general west-to-east direction.

milarite [MINERAL] $K_2Ca_4Be_4Al_2Si_{24}O_{62} \cdot H_2O$ A colorless to greenish, glassy, hexagonal mineral composed of a hydrous silicate of potassium, calcium, beryllium, and aluminum, occurring in crystals.

milky quartz [MINERAL] An opaque, milk-white variety of crystalline quartz, often with a greasy luster; milkiness is due to the presence of air-filled cavities. Also known as greasy quartz.

milky weather *See* whiteout.

mill [MIN ENG] **1.** An excavation made in the country rock, by a crosscut from the workings on a vein, to obtain waste for filling; it is left without timber so that the roof can fall in and furnish the required rock. **2.** A passage connecting a stope or upper level with a lower level intended to be filled with broken ore that can then be drawn out at the bottom as desired for further transportation.

Miller indices [CRYSTAL] Three integers identifying a type of crystal plane; the intercepts of a plane on the three crystallographic axes are expressed as fractions of the crystal parameters; the reciprocals of these fractions, reduced to integral proportions, are the Miller indices. Also known as crystal indices.

millerite [MINERAL] NiS A brass to bronze-yellow mineral that crystallizes in the hexagonal system and usually contains trace amounts of cobalt, copper, and iron; hardness is 3–3.5 on Mohs scale, and specific gravity is 5.5; it generally occurs in fine crystals, chiefly as nodules in clay ironstone. Also known as capillary pyrites; hair pyrites; nickel pyrites.

Miller law [CRYSTAL] If the edges formed by the intersections of three faces of a crystal are taken as the three reference axes, then the three quantities formed by dividing the intercept of a fourth face with one of these axes by the intercept of a fifth face with the same axis are proportional to small whole numbers, rarely exceeding 6. Also known as law of rational intercepts.

mill-head ore *See* run-of-mill.

milling [MIN ENG] A combination of open-cut and underground mining, wherein the ore is mined in open cut and handled underground.

milling system *See* chute system.

milling width [MIN ENG] Width of lode designated for treatment in the mill, as calculated with regard to daily tonnage.

millisecond delay cap [ENG] A delay cap with an extremely short (20–500 thousandths of a second) interval between passing of current and explosion. Also known as short-delay detonator.

millisite [MINERAL] $(Na,K)CaAl_6(PO_4)_4(OH)_9 \cdot 3H_2O$ White mineral composed of a basic hydrous phosphate of sodium, potassium, calcium, and aluminum.

mill ore [MIN ENG] An ore that must be given some preliminary treatment before a marketable grade or a grade suitable for further treatment can be obtained.

mill run [MIN ENG] **1.** A given quantity of ore tested for its quality by actual milling. **2.** The yield of such a test.

Mills-Crowe process [MIN ENG] Method of regeneration of cyanide liquor from the gold leaching process; the barren solution is acidified, liberating hydrocyanic acid which is separated and reabsorbed in alkaline solution.

millsite [MIN ENG] A plot of ground suitable for the erection of a mill, or reduction works, to be used in connection with mining operations.

millstone *See* buhrstone.

Mima mound [GEOGR] A term used in the northwestern United States for any of the numerous low, circular or oval domes composed of loose silt and soil material; probably built by pocket gophers.

mimetene *See* mimetite.

mimetesite *See* mimetite.

mimetic [CRYSTAL] Pertaining to a crystal that is twinned or malformed but whose crystal symmetry appears to be of a higher grade than it actually is. [PETR] Of a tectonite, having a deformation fabric, formed by mimetic crystallization, that reflects and is influenced by preexisting anisotropic structure.

mimetic crystallization [PETR]   Recrystallization or neomineralization in metamorphism which reproduces preexistent structures.

mimetite [MINERAL] $Pb_5(AsO_4)_3Cl$ A yellow to yellowish-brown mineral of the apatite group, commonly containing calcium or phosphate; a minor ore of lead. Also known as mimetene; mimetesite.

minable [MIN ENG]   Material that can be mined under present-day mining technology and economics.

minasragrite [MINERAL] $(VO)_2H_2(SO_4)_3 \cdot 15H_2O$ A blue, monoclinic mineral consisting of hydrated acid vanadyl sulfate; occurs in efflorescences and as aggregates or masses.

Mindel glaciation [GEOL]   The second glacial stage of the Pleistocene in the Alps.

Mindel-Riss interglacial [GEOL]   The second interglacial stage of the Pleistocene in the Alps; follows the Mindel glaciation.

mine [MIN ENG]   An opening or excavation in the earth for extracting minerals.

mine captain [MIN ENG]   **1.** The director of work in a mine, with or without superior officials or subordinates. **2.** *See* mine superintendent.

mine characteristic [MIN ENG]   The relation between pressure p and volume Q in the ventilation of a mine of resistance R, expressed as $p = RQ^2$.

mine characteristic curve [MIN ENG]   A graph derived by plotting the static or total mine head, or both, against the quantity; used to solve problems in mine ventilation.

mine development [MIN ENG]   The operations involved in preparing a mine for ore extraction, including tunneling, sinking, crosscutting, drifting, and raising.

mine dust [MIN ENG]   Dust from drilling, blasting, or handling rock.

mined volume [MIN ENG]   A statistic used in mine subsidence, computed by multiplying the mined area by the mean thickness of the bed or of that part of the bed which has been dug out.

mine examiner *See* fire boss.

mine fan signal system [MIN ENG]   A system which indicates by electric light or electric audible signal, or both, the slowing down or stopping of a mine ventilating fan.

mine fire truck [MIN ENG]   A low-slung railcar designed to fight fires in mines; has a water supply and a pump to supply high pressure to the fire hoses.

mine foreman [MIN ENG]   The worker charged with the general supervision of the underground workings of a mine and the persons employed therein.

mine hoist [MIN ENG]   A device for raising and lowering ore, rock, or coal in a mine, and men and supplies.

mine inspector [MIN ENG]   Generally, the state mine inspector, as contrasted to the Federal mine inspector; inspects mines to find fire and dust hazards and inspects the safety of working areas, electric circuits, and mine equipment.

mine jeep [MIN ENG]   An electrically driven car for underground transportation of officials, inspectors, rescue workers, and repair, maintenance, and surveying crews.

mine locomotive [MIN ENG]   A low, heavy haulage engine designed for underground operation; usually propelled by electricity, gasoline, or compressed air.

mine props [MIN ENG]   Sections of wood used for holding up pieces of rock in the roof of mines.

mine radio telephone system [MIN ENG] A communication system between the dispatcher and locomotive operators; radio impulses travel along the trolley wire and down the trolley pole to the radio telephone.

mineragraphy *See* ore microscopy.

mineral [GEOL] A naturally occurring substance with a characteristic chemical composition expressed by a chemical formula; may occur as individual crystals or may be disseminated in some other mineral or rock; most mineralogists include the requirements of inorganic origin and internal crystalline structure.

mineral caoutchoric *See* elaterite.

mineral charcoal *See* fusain.

mineral deposit [GEOL] A mass of naturally occurring mineral material, usually of economic value.

mineral economics [MIN ENG] Study and application of the processes used in management and finance connected with the discovery, exploitation, and marketing of minerals.

mineral engineering *See* mining engineering.

mineral facies [PETR] Rocks of any origin whose components have been formed within certain temperature-pressure limits characterized by the stability of certain index minerals.

mineralization [GEOL] **1.** The process of fossilization whereby inorganic materials replace the organic constituents of an organism. **2.** The introduction of minerals into a rock, resulting in a mineral deposit.

mineralize [GEOL] To convert to, or impregnate with, mineral material; applied to processes of ore vein deposition and of fossilization.

mineralizer [GEOL] A gas or fluid dissolved in a magma that aids in the concentration and crystallization of ore minerals.

mineral land [MIN ENG] Land which is worth more for mining than for agriculture or other use.

mineral lease *See* mining lease.

mineral monument [MIN ENG] A permanent monument established in a mining district to provide for an accurate description of mining claims and their location.

mineralogenetic epoch [GEOL] A geologic time period during which mineral deposits formed.

mineralogenetic province [GEOL] Geographic region where conditions were favorable for the concentration of useful minerals.

mineralogical phase rule [MINERAL] Any of several variations of the Gibbs phase rule, taking into account the number of degrees of freedom consumed by the fixing of physical-chemical variables in the natural environment; it assumes that temperature and pressure are fixed externally and that consequently the number of phases (minerals) in a system (rock) will not usually exceed the number of components.

mineralogist [MINERAL] A person who studies the occurrence, description, mode of formation, and uses of minerals.

mineralography *See* ore microscopy.

mineraloid [MINERAL] A naturally occurring, inorganic material that is amorphous and is therefore not considered to be a mineral. Also known as gel mineral.

mineral processing [MIN ENG] Procedures, such as dry and wet crushing and grinding of ore or other products containing minerals, to raise the concentration of the substance being mined.

mineral resources [GEOL] Valuable mineral deposits of an area that are presently recoverable and may be so in the future; includes known ore bodies and potential ore.

mineral right *See* mining right.

mineral sequence *See* paragenesis.

mineral soil [GEOL] Soil composed of mineral or rock derivatives with little organic matter.

mineral spring [HYD] A spring whose water has a definite taste due to a high mineral content.

mineral suite [MINERAL] **1.** A group of associated minerals in one deposit. **2.** A representative group of minerals from a certain locality. **3.** A group of specimens showing variations, as in color or form, in a single mineral species.

mineral tallow *See* hatchettite.

mineral water [HYD] Water containing naturally or artificially supplied minerals or gases.

mineral wax *See* ozocerite.

mine rescue apparatus [MIN ENG] Certain types of apparatus worn by workers and permitting them to perform in noxious or irrespirable atmospheres, such as during mine fires or following mine explosions.

mine rescue crew [MIN ENG] A crew consisting usually of five to eight persons who are thoroughly trained in the use of mine rescue apparatus, which they wear in rescue or recovery work in a mine following an explosion or during a fire.

mine rescue lamp [MIN ENG] A particular type of electric safety hand lamp used in rescue operations; equipped with a lens for concentrating or diffusing the light.

mine resistance [MIN ENG] Resistance by a mine to the passage of an air current.

miner's friend [MIN ENG] **1.** The Davy safety lamp. **2.** A steam engine once used to pump water from underground.

miner's hammer [MIN ENG] A hammer used to break ore.

miner's hand lamp [MIN ENG] A self-contained mine lamp with handle for carrying.

miner's helmet [MIN ENG] A hat designed for miners to provide head protection and to hold the cap lamp.

miner's horn [MIN ENG] A metal spoon or horn to collect ore particles in gold washing.

miner's inch [MIN ENG] The quantity of water that will escape from an aperture 1-inch (2.54-centimeter) square through a 2-inch-thick (5.08-centimeter-thick) plank, with a steady flow of water standing 6 inches (15.24 centimeters) above the top of the escape aperture, the quantity so discharged amounting to 2274 cubic feet (64.39 cubic meters) in 24 hours.

miner's lamp [MIN ENG] Any one of a variety of lamps used by a miner to furnish light, such as oil lamps, carbide lamps, flame safety lamps, and cap lamps.

miner's right [MIN ENG] An annual permit from the government to occupy and work mineral land.

miners' rules [MIN ENG] Rules and regulations proclaimed by the miners of any district, relating to the location of, recording of, and the work necessary to hold possession of a mining claim.

miner's self-rescuer [MIN ENG]   A pocket gas mask effective against carbon monoxide; air passes through a cannister containing fused calcium chloride before entering the mouth.

mine run [MIN ENG]   The unscreened output of a mine. Also known as run-of-mine.

mine sample [MIN ENG]   Coal or mineral extracted at underground exposures for analysis.

mine signal system [MIN ENG]   Signal lights installed at individual track switches immediately indicating to the motorman whether or not he can safely proceed.

mine skips [MIN ENG]   Skips used to bring mined ore to the surface of a shaft.

mine superintendent [MIN ENG]   A mine manager or group manager. Also known as mine captain.

mine surveyor [MIN ENG]   The official who periodically surveys the mine workings and prepares plans for the manager.

mine track devices [MIN ENG]   Track devices to provide maximum safety for haulage trains in mines.

mine tractor [MIN ENG]   A trackless, self-propelled vehicle used to transport equipment and supplies.

minette [PETR]   A syenitic variety of lamprophyre composed principally of biotite phenocrysts in a matrix of orthoclase and biotite.

mine valuation [MIN ENG]   Properly weighing the financial considerations to place a present value on mineral reserves.

mine ventilating fan [MIN ENG]   A motor-driven disk, propeller, or wheel for blowing or exhausting air to provide ventilation of a mine; large units are used for stationary systems, while small portable types provide fresh air in inaccessible locations, such as dead ends.

mine ventilation system [MIN ENG]   A combination of connecting airways with air pressure sources and governing devices that are instrumental in making and controlling airflow.

mine water [MIN ENG]   Water pumped from mines.

minimum duration [OCEANOGR]   The time that is required for steady-state wave conditions to develop for a specific wind velocity over a specific fetch length.

minimum ebb [OCEANOGR]   The least speed of a current that runs continuously ebb.

minimum flood [OCEANOGR]   The least speed of a current that runs continuously flood.

minimum rent [MIN ENG]   The right to work coal acquired by a mine owner by the payment of an annual rent and a royalty to the landowner (the coal or mineral owner). Also known as fixed rent.

mining [MIN ENG]   The technique and business of mineral discovery and exploitation.

mining camp [MIN ENG]   A term loosely applied to any mining town.

mining claim [MIN ENG]   That portion of the public mineral lands which a miner, for mining purposes, takes and holds in accordance with mining laws. Also known as claim.

mining engineer [MIN ENG]   One qualified by education, training, and experience in mining engineering.

mining engineering [ENG]   Engineering concerned with the discovery, development, and exploitation of coal, ores, and minerals, as well as the cleaning, sizing, and dressing of the product. Also known as mineral engineering.

mining geology [MIN ENG]   The study of the structure and occurrence of mineral deposits and the geologic aspects of mine planning.

mining ground [MIN ENG]   Land from which a mineral substance is extracted by the process of mining.

mining hazard [MIN ENG]   Any of the dangers unique to the winning and working of minerals and coal.

mining lease [MIN ENG]   A contract to work a mine and extract mineral or other deposits from it under specified conditions. Also known as mineral lease.

mining machine truck [MIN ENG]   A truck used to transport shortwall mining machines.

mining partnership [MIN ENG]   The arrangement whereby two or more persons acquire a mining claim and actually engage in working it.

mining property [MIN ENG]   Property valued for its mining possibilities.

mining retreating [MIN ENG]   A process of mining by which the ore or coal is untouched until after all the gangways and such are driven, at which time the work of extraction begins at the boundary and progresses toward the shaft.

mining right [MIN ENG]   A right to enter upon and occupy a specific piece of ground for the purpose of working it, either by underground excavations or open workings, to obtain the mineral ores which may be deposited therein. Also known as mineral right.

mining shield [MIN ENG]   A canopy or cover for the protection of workers and machines at the face of a mechanized coal heading.

mining title [MIN ENG]   A claim, exclusive prospecting license, right, concession, or lease.

mining town [MIN ENG]   A town that has arisen next to a mine or mines.

mining width [MIN ENG]   The minimum width needed to extract ore regardless of the actual width of ore-bearing rock.

minium [MINERAL] Pb_3O_4 A scarlet or orange-red mineral consisting of an oxide of lead; found in Wisconsin and the western United States. Also known as red lead.

minor trough [METEOROL]   A pressure trough of smaller scale than a long-wave trough; it ordinarily moves rapidly and is associated with a migratory cyclonic disturbance in the lower troposphere.

minuano [METEOROL]   A cold southwesterly wind of southern Brazil, occurring during the Southern Hemisphere winter (June to September).

minus angle *See* angle of depression.

minus-cement porosity [GEOL]   The porosity that would characterize a sedimentary material if it contained no chemical cement.

minyulite [MINERAL] $KAl_2(PO_4)_2(OH,F)\cdot4H_2O$ A white mineral composed of hydrous basic potassium aluminum phosphate.

Miocene [GEOL]   A geologic epoch of the Tertiary period, extending from the end of the Oligocene to the beginning of the Pliocene.

miocrystalline *See* hypocrystalline.

miogeosyncline [GEOL]   The nonvolcanic portion of an orthogeosyncline, located adjacent to the craton.

Miosireninae [PALEON]   A subfamily of extinct sirenian mammals in the family Dugongidae.

mirabilite [MINERAL] $Na_2SO_4 \cdot 10H_2O$ A yellow or white monoclinic mineral consisting of hydrous sodium sulfate, occurring as a deposit from saline lakes, playas, and springs, and as an efflorescence; the pure crystals are known as Glauber's salt.

mire [GEOL] Wet spongy earth, as of a marsh, swamp, or bog.

mirror glance *See* wehrlite.

mirror nephoscope [ENG] A nephoscope in which the motion of a cloud is observed by its reflection in a mirror. Also known as cloud mirror; reflecting nephoscope.

mirror stone *See* muscovite.

mirror plane of symmetry *See* plane of mirror symmetry.

miscible-phase displacement [PETRO ENG] Method of increasing reservoir oil recovery by displacement with an oil-miscible driving fluid, such as gas or liquefied petroleum gas.

miscible-slug process [PETRO ENG] A miscible-phase displacement in which reservoir oil recovery is increased by displacement with liquefied petroleum gas as the driving fluid.

misenite [MINERAL] $K_8H_6(SO_4)_7$ A white mineral composed of native acid potassium sulfate.

miser [PETRO ENG] A well-boring bit that is tubular with a valve at the bottom, and has a screw for forcing the earth upward. Also spelled mizer.

misfit stream [HYD] A stream whose meanders are either too large or too small to have eroded the valley in which it flows.

mispickel *See* arsenopyrite.

missed hole *See* failed hole.

missed round [ENG] A round in which all or part of the explosive has failed to detonate.

Mississippian [GEOL] A large division of late Paleozoic geologic time, after the Devonian and before the Pennsylvanian, named for a succession of highly fossiliferous marine strata consisting largely of limestones found along the Mississippi River between southeastern Iowa and southern Illinois; approximately equivalent to the European Lower Carboniferous.

Missourian [GEOL] A North American provincial series of geologic time: lower Upper Pennsylvanian (above Desmoinesian, below Virgilian).

mist droplet [METEOROL] A particle of mist, intermediate between a haze droplet and a fog drop.

mist projector [MIN ENG] An appliance to form a mist spray to allay dust and fume during blasting operations in a tunnel.

mistral [METEOROL] A north wind which blows down the Rhone Valley south of Valence, France, and into the Gulf of Lions. Strong, squally, cold, and dry, it is the combined result of the basic circulation, a fall wind, and jet-effect wind.

mitscherlichite [MINERAL] $K_2CuCl_4 \cdot 2H_2O$ A greenish-blue, tetragonal mineral consisting of potassium copper chloride dihydrate.

mix crystal *See* mixed crystal.

mixed cloud [METEOROL] A cloud containing both water drops and ice crystals, hence a cloud whose composition is intermediate between that of a water cloud and that of an ice-crystal cloud.

mixed crystal [CRYSTAL] A crystal whose lattice sites are occupied at random by different ions or molecules of two different compounds. Also known as mix crystal.

mixed current [OCEANOGR] A type of tidal current characterized by a conspicuous difference in speed between the two flood currents or two ebb currents usually occurring each tidal day.

mixed-flow fan [MIN ENG] A mine fan in which the flow is both radial and axial.

mixed layer [OCEANOGR] The layer of water which is mixed through wave action or thermohaline convection. Also known as surface water.

mixed-layer mineral [MINERAL] A mineral having an interstratified structure consisting of alternating layers of two different clays or of a clay and some other mineral.

mixed nucleus [METEOROL] A condensation nucleus of intermediate efficacy which, as a result of particle coagulation, contains both soluble hygroscopic matter and insoluble but wettable matter.

mixed ore [GEOL] Any ore with both oxidized and unoxidized minerals.

mixed tide [OCEANOGR] A tide in which the presence of a diurnal wave is conspicuous by a large inequality in the heights of either the two high tides or the two low tides usually occurring each tidal day.

mixing ratio [METEOROL] In a system of moist air, the dimensionless ratio of the mass of water vapor to the mass of dry air; for many purposes, the mixing ratio may be approximated by the specific humidity.

mixite [MINERAL] $Cu_{11}Bi(AsO_4)_5(OH)_{10} \cdot 6H_2O$ A green to whitish mineral composed of a hydrous basic arsenate of copper and bismuth.

Mixodectidae [PALEON] A family of extinct insectivores assigned to the Proteutheria; a superficially rodentlike group confined to the Paleocene of North America.

mixolimnion [HYD] The upper layer of a meromictic lake, characterized by low density and free circulation; this layer is mixed by the wind.

mixtite *See* diamictite.

mizer *See* miser.

mizzonite [MINERAL] A mineral of the scapolite group, composed of 54 to 57% silica. Also known as dipyre.

moat [GEOL] **1.** A ringlike depression around the base of a seamount. **2.** A valleylike depression around the inner side of a volcanic cone, between the rim and the lava dome. [HYD] **1.** A glacial channel in the form of a deep, wide trench. **2.** *See* oxbow lake.

mobile belt [GEOL] A long, relatively narrow crustal region of tectonic acitivity.

mobile drill [MIN ENG] A drill unit mounted on wheels or crawl-type tracks to facilitate moving.

mobile filling [MIN ENG] Filling which is supplemented only from above, and which sinks, filling the mined-out rooms.

mobility ratio [PETRO ENG] The ratio of the mobility of the driving fluid (such as water) to that of the driven fluid (such as gas) in a petroleum reservoir.

mobilization [GEOL] Any process by which solid rock becomes sufficiently soft and plastic to permit it to flow or to permit geochemical migration of the mobile components.

mock fog [METEOROL] A simulation of true fog by atmospheric refraction.

mock lead *See* sphalerite.

mock ore *See* sphalerite.

mode [PETR] The mineral composition of a rock, usually expressed as percentages of total weight or volume.

mode eddies [OCEANOGR] Densely packed, irregularly oval high- and low-pressure centers roughly 400 kilometers in diameter in which current intensities are typically tenfold greater than the local means. Also known as mesoscale eddies.

model atmosphere [METEOROL] Any theoretical representation of the atmosphere, particularly of vertical temperature distribution.

moderate breeze [METEOROL] In the Beaufort wind scale, a wind whose speed is from 11 to 16 knots (13 to 18 miles per hour or 20 to 30 kilometers per hour).

moderate gale [METEOROL] In the Beaufort wind scale, a wind whose speed is from 28 to 33 knots (32 to 38 miles per hour or 52 to 61 kilometers per hour).

modified index of refraction [METEOROL] An atmospheric index of refraction mathematically modified so that when its gradient is applied to energy propagation over a hypothetical flat earth, it is substantially equivalent to propagation over the true curved earth with the actual index of refraction. Also known as modified refractive index; refractive modulus.

modified Lambert conformal chart [MAP] A chart on the modified Lambert conformal projection. Also known as Ney's chart.

modified Lambert conformal projection [MAP] A modification of the Lambert conformal projection for use in polar regions, one of the standard parallels being at latitude 89°59′58″ and the other at latitude 71° or 74°, and the parallels being expanded slightly to form complete concentric circles. Also known as Ney's projection.

modified refractive index *See* modified index of refraction.

Moeritheriidae [PALEON] The single family of the extinct order Moeritherioidea.

Moeritherioidea [PALEON] A suborder of extinct sirenian mammals considered as primitive proboscideans by some authorities and as a sirenian offshoot by others.

mofette [GEOL] A small opening emitting carbon dioxide in an area of late-stage volcanic activity.

mohavite *See* tincalconite.

Mohawkian [GEOL] A North American stage of Middle Ordovician geologic time, above Chazyan and below Edenian.

Mohnian [GEOL] A North American stage of geologic time: Miocene (above Luisian, below Delmontian).

Moho *See* Mohorovičić discontinuity.

Mohole drilling [GEOL] Drilling aimed at penetration of the earth's crust, through the Mohorovičić discontinuity, to sample the mantle.

Mohorovičić discontinuity [GEOPHYS] A seismic discontinuity that separates the earth's crust from the subjacent mantle, inferred from travel time curves indicating that seismic waves undergo a sudden increase in velocity. Also known as Moho.

mohsite *See* ilmenite.

Mohs scale [MINERAL] An empirical scale consisting of 10 minerals with reference to which the hardness of all other minerals is measured; it includes, from softest (designated 1) to hardest (10): talc, gypsum, calcite, fluorite, apatite, orthoclase, quartz, topaz, corundum, and diamond.

moil [MIN ENG] A long steel wedge with a rounded point used for breaking up rocks in a mine.

moissanite [MINERAL] SiC A carbide mineral found in meteorites; identical with artificial carborundum.

moist adiabat *See* saturation adiabat.

moist-adiabatic lapse rate *See* saturation-adiabatic lapse rate.

moist air [METEOROL] **1.** In atmospheric thermodynamics, air that is a mixture of dry air and any amount of water vapor. **2.** Generally, air with a high relative humidity.

moist climate [CLIMATOL] In C.W. Thornthwaite's climatic classification, any type of climate in which the seasonal water surplus counteracts seasonal water deficiency; thus it has a moisture index greater than zero.

moisture [CLIMATOL] The quantity of precipitation or the precipitation effectiveness. [METEOROL] The water vapor content of the atmosphere, or the total water substance (gaseous, liquid, and solid) present in a given volume of air.

moisture adjustment [METEOROL] The adjustment of observed precipitation in a storm by the ratio of the estimated probable maximum precipitable water over the basin under study to the actual precipitable water calculated for the particular storm.

moisture factor [METEOROL] One of the simplest measures of precipitation effectiveness: moisture factor = P/T, where P is precipitation in centimeters and T is temperature degrees centigrade for the period in question.

moisture inversion [METEOROL] An increase with height of the moisture content of the air; specifically, the layer through which this increase occurs, or the altitude at which the increase begins.

molasse [GEOL] A paralic sedimentary facies consisting mainly of shale, subgraywacke sandstone, and conglomerate; it is more clastic and less rhythmic than the preceding flysch and is generally postorogenic.

mold [PALEON] An impression made in rock or earth material by an inner or outer surface of a fossil shell or other organic structure; a complete mold would be the hollow space.

moldauite *See* moldavite.

moldavite [GEOL] A translucent, olive- to brownish-green or pale-green tektite from western Czechoslovakia, characterized by surface sculpturing due to solution etching. Also known as moldauite; pseudochrysolite; vitavite. [MINERAL] A variety of ozocerite from Moldavia.

mole mining [MIN ENG] A method of working coal seams about 30 inches (75 centimeters) thick; a small continuous-miner type of machine is used, which is remote-controlled from the roadway, without associated supports.

Mollisol [GEOL] An order of soils having dark or very dark, friable, thick A horizons high in humus and bases such as calcium and magnesium; most have lighter-colored or browner B horizons that are less friable and about as thick as the A horizons; all but a few have paler C horizons, many of which are calcareous.

molybdenite [MINERAL] MoS_2 A metallic, lead-gray mineral that crystallizes in the hexagonal system and is commonly found in scales or foliated masses; hardness is 1.5 on Mohs scale, and specific gravity is 4.7; it is chief ore of molybdenum.

molybdic ocher *See* molybdite.

molybdine *See* molybdite.

molybdite [MINERAL] MoO_3 A mineral, much of which is actually ferrimolybdite. Also known as molybdic ocher; molybdine.

molybdophyllite [MINERAL] $(Pb,Mg)_2SiO_4 \cdot H_2O$ A colorless, white, or pale-green mineral composed of a silicate of lead and magnesium.

molysite [MINERAL] $FeCl_3$ A brownish-red or yellow mineral composed of native ferric chloride, occurring in lava at Vesuvius.

Momertz-Lentz system [MIN ENG] Placement of two winding engines alongside the top of the mine shaft using the shaft collar as a common foundation; results in practically vertical ropes and less rope oscillation.

monadnock [GEOL] A remnant hill of resistant rock rising abruptly from the level of a peneplain; commonly represents an outcrop of rock that has withstood erosion. Also known as torso mountain.

monalbite [MINERAL] A modification of albite with monoclinic symmetry that is stable under equilibrium conditions at temperatures (about 1000°C) near the melting point.

monazite [MINERAL] A yellow or brown rare-earth phosphate monoclinic mineral with appreciable substitution of thorium for rare-earths and silicon for phosphorus; the principal ore of the rare earths and of thorium. Also known as cryptolite.

monchiquite [PETR] A lamprophyre composed of olivine, pyroxene, and usually mica or amphibole phenocrysts embedded in a glass or analcime groundmass.

monetite [MINERAL] $CaHPO_4$ A yellowish-white mineral consisting of an acid calcium hydrogen phosphate, occurring in crystals.

monimolimnion [HYD] The dense bottom stratum of a meromictic lake; it is stagnant and does not mix with the water above.

monimolite [MINERAL] $(Pb,Ca)_3Sb_2O_8$ Yellowish to brownish or greenish mineral composed of lead calcium antimony oxide; it may contain ferrous iron.

monitor *See* hydraulic monitor.

monkey [MIN ENG] **1.** An appliance for mechanically gripping or releasing the rope in rope haulage. **2.** An airway in an anthracite mine.

monkey drift [MIN ENG] A small drift driven in for prospecting purposes, or a crosscut driven to an airway above the gangway.

monkey heading [MIN ENG] A narrow, low passage in the coal, providing refuge for miners while coal is blasted.

monkey ladder [MIN ENG] A ladder of saplings; the widely separated steps rest in the coal.

monkey winch [MIN ENG] A device for exerting a strong pull; consists of a framework containing a hand-operated drum, around which a steel rope 50 feet (15 meters) long is wound.

Monobothrida [PALEON] An extinct order of monocyclic camerate crinoids.

monocline [GEOL] A stratigraphic unit that dips from the horizontal in one direction only, not as part of an anticline or syncline.

monoclinic system [CRYSTAL] One of the six crystal systems characterized by a single, two-fold symmetry axis or a single symmetry plane.

Monocyathea [PALEON] A class of extinct parazoans in the phylum Archaeocyatha containing single-walled forms.

monogeosyncline [GEOL] A primary geosyncline that is long, narrow, and deeply subsided; composed of the sediments of shallow water and situated along the inner margin of the borderlands.

monomineralic [PETR] Of a rock, composed entirely or principally of a single mineral.

Mono pump [MIN ENG]   A pump designed and manufactured of special materials to meet mining conditions; consists of a rubber stator shaped like a double internal helix and a single helical rotor which travels in the stator with a slightly eccentric motion.

monopyroxene clinoaugite *See* clinopyroxene.

monotrophic [CRYSTAL]   Of crystal pairs, having one of the pair always metastable with respect to the other.

monsoon [METEOROL]   A large-scale wind system which predominates or strongly influences the climate of large regions, and in which the direction of the wind flow reverses from winter to summer; an example is the wind system over the Asian continent.

monsoon climate [CLIMATOL]   The type of climate which is found in regions subject to monsoons.

monsoon current [OCEANOGR]   A seasonal wind-driven current occurring in the northern part of the Indian Ocean.

monsoon fog [METEOROL]   An advection type of fog occurring along a coast where monsoon winds are blowing, when the air has a high specific humidity and there is a large difference in the temperature of adjacent land and sea.

monsoon low [METEOROL]   A seasonal low found over a continent in the summer and over the adjacent sea in the winter.

montanite [MINERAL] $Bi_2O_3 \cdot TeO_3 \cdot 2H_2O$ A yellowish mineral consisting of a hydrated tellurate of bismuth; occurs in soft and earthy to compact form.

montebrasite [MINERAL] $LiAlPO_4(OH)$ A mineral composed of basic lithium aluminum phosphate; it is isomorphous with amblygonite and natromontebrasite.

montgomeryite [MINERAL] $Ca_2Al_2(PO_4)_3(OH) \cdot 7H_2O$ A green to colorless mineral composed of hydrous basic calcium aluminum phosphate.

Montian [GEOL]   A European stage of geologic time: Paleocene (above Danian, below Thanetian).

monticellite [MINERAL] $CaMgSiO_4$ A colorless or gray mineral of the olivine structure type; isomorphous with kirsch steinite.

montmorillonite [MINERAL]   **1.** A group name for all clay minerals with an expanding structure, except vermiculite. **2.** The high-alumina end member of the montmorillonite group; it is grayish, pale red, or blue and has some replacement of aluminum ion by magnesium ion. **3.** Any mineral of the montmorillonite group.

montroydite [MINERAL] HgO Natural mercury oxide mineral from Texas.

monzonite [PETR]   A phaneritic (visibly crystalline) plutonic rock composed chiefly of sodic plagioclase and alkali feldspar, with subordinate amounts of dark-colored minerals, intermediate between syenite and dorite.

moon pillar [METEOROL]   A halo consisting of a vertical shaft of light through the moon.

moonstone [MINERAL]   An alkali feldspar or cryptoperthite that is semitransparent to translucent and exhibits a bluish to milky-white, pearly, or opaline luster; used as a gemstone if flawless. Also known as hecatolite.

moor coal [GEOL]   A friable lignite or brown coal.

mooreite [MINERAL] $(Mg,Zn,Mn)_8(SO_4)_4(OH)_{14} \cdot 4H_2O$ A glassy white mineral composed of hydrous basic magnesium zinc manganese sulfate.

mor *See* ectohumus.

morainal apron *See* outwash plain.

morainal delta *See* ice-contact delta.

morainal lake [HYD] A glacial lake filling a depression resulting from irregular deposition of drift in a terminal or ground moraine of a continental glacier.

morainal plain *See* outwash plain.

moraine [GEOL] An accumulation of glacial drift deposited chiefly by direct glacial action and possessing initial constructional form independent of the floor beneath it.

moraine bar [GEOL] A terminal moraine serving as a bar, rising out of deep water at some distance from the shore.

moraine kame [GEOL] One of a group of kames characterized by the same topography, constitution, and position as a terminal moraine.

moraine plateau [GEOL] A relatively flat area within a hummocky moraine, generally at the same elevation as, or a little higher than, the summits of surrounding knobs.

moravite [MINERAL] $Fe_2(N,Fe)_4Si_7O_{20}(OH)_4$ A black mineral of the chlorite group, composed of basic iron aluminum silicate, occurring as fine scales.

mordenite [MINERAL] $(Ca,Na_2,K_2)_4Al_8Si_{40}O_{96}\cdot28H_2O$ A zeolite mineral crystallizing in the orthorhombic system and found in minute crystals or fibrous concretions. Also known as arduinite; ashtonite; flokite; ptilolite.

morencite *See* nontronite.

morenosite [MINERAL] $NiSO_4\cdot7H_2O$ An apple-green or light-green mineral composed of hydrous nickel sulfate, occurring in crystals or fibrous crusts. Also known as nickel vitriol.

morganite *See* vorobyevite.

morinite [MINERAL] $Na_2Ca_3Al_3H(PO_4)_4F_6\cdot8H_2O$ A mineral composed of hydrous acid phosphate of sodium, calcium, and aluminum. Also known as jezekite.

Morisette expansion reamer [MIN ENG] A reaming device with three tapered lugs or cutters designed so that drilling pressure necessary to penetrate rock with a noncoring pilot bit forces the diamond-faced cutters of the reamer to expand outward, thereby enlarging the pilot hole sufficiently to allow the casing to follow the reamer as drilling progresses.

morphogenetic region [GEOL] A region in which, under certain climatic conditions, the predominant geomorphic processes will contribute regional characteristics to the landscape that contrast with those of other regions formed under different climatic conditions.

morphographic map *See* physiographic diagram.

morphotropism [CRYSTAL] Similarity of structure, axial ratios, and angles between faces of one or more zones in crystalline substances whose formulas can be derived one from another by substitution.

mortar structure [PETR] A cataclastic structure produced by dynamic metamorphism of crystalline rocks and characterized by a mica-free aggregate of finely crushed grains of quartz and feldspar filling the interstices between or forming borders on the edges of larger, rounded relicts. Also known as cataclastic structure; murbruk structure; porphyroclastic structure.

mortlake *See* oxbow lake.

morvan [GEOL] The area where two peneplains intersect. Also known as skiou.

mosaic [PETR] **1.** Pertaining to a granoblastic texture in a rock formed by dynamic metamorphism in which the boundaries between individual grains are straight or slightly curved. Also known as cyclopean. **2.** Pertaining to a texture in a crystalline sedimentary rock in which contacts at grain boundaries are more or less regular.

mosaic structure [CRYSTAL]   In crystals, a substructure in which neighboring regions are oriented slightly differently.

mosandrite [MINERAL]   A reddish-brown or yellowish-brown mineral composed of a silicate of sodium, calcium, titanium, zirconium, and cerium. Also known as khibinite; lovchorrite; rinkite; rinkolite.

mosasaur [PALEON]   Any reptile of the genus *Mosasaurus*; large, aquatic, fish-eating lizards from the Cretaceous which are related to the monitors but had paddle-shaped limbs.

moschellandsbergite [MINERAL] Ag_2Hg_3 A silver-white mineral consisting of a silver and mercury compound; occurs in dodecahedral crystals and in massive and granular forms.

moscovite *See* muscovite.

mosesite [MINERAL] $Hg_2N(SO_4,MoO_4)\cdot H_2O$ Mineral composed of a hydrous nitride of mercury and various anions.

moss agate [MINERAL]   A milky or almost transparent chalcedony containing dark inclusions in a dendritic pattern.

mossite [MINERAL] $Fe(Nb,Ta)_2O_6$ A mineral composed of an iron tantalum oxide; it is isomorphous with tapiolite.

mother lode [GEOL]   A main unit of mineralized matter that may not have economic value but to which workable veins are related.

mother map *See* base map.

mother-of-emerald *See* prase.

mother rock *See* source rock.

motive column [MIN ENG]   The ventilating pressure in a mine in units of feet of air column; the height of a column of air whose density is the same as the air in the downcast shaft, which exerts a pressure equal to the ventilating pressure.

mottled [GEOL]   Referring to a soil irregularly marked with spots of different colors. [PETR]   Of a sedimentary rock, marked with spots of various colors.

mottramite [MINERAL] $(Cu,Zn)Pb(VO_4)(OH)$ A mineral composed of a basic lead copper zinc vanadate; it is isomorphous with descloizite. Also known as cuprodescloizite; psittacinite.

moulin [HYD]   A shaft or hole in the ice of a glacier which is roughly cylindrical and nearly vertical, formed by swirling meltwater pouring down from the surface. Also known as glacial mill; glacier mill; glacier pothole; glacier well; pothole.

mound [GEOL]   **1.** A low, isolated, rounded natural hill, usually of earth. Also known as tuft. **2.** A structure built by fossil colonial organisms.

mountain [GEOGR]   A feature of the earth's surface that rises high above the base and has generally steep slopes and a relatively small summit area.

mountain and valley winds [METEOROL]   A system of diurnal winds along the axis of a valley, blowing uphill and upvalley by day, and downhill and downvalley by night; they prevail mostly in calm, clear weather.

mountain breeze [METEOROL]   A breeze that blows down a mountain slope due to the gravitational flow of cooled air.

mountain chain *See* mountain system.

mountain climate [CLIMATOL] Very generally, the climate of relatively high elevations; mountain climates are distinguished by the departure of their characteristics from those of surrounding lowlands, and the one common basis for this distinction is that of atmospheric rarefaction; aside from this, great variety is introduced by differences in latitude, elevation, and exposure to the sun; thus, there exists no single, clearly defined, mountain climate. Also known as highland climate.

mountain cork [MINERAL] **1.** A white or gray variety of asbestos composed of thick, interwoven fibers and having a corklike weight and texture. Also known as rock cork. **2.** A fibrous clay mineral, such as sepiolite.

mountain crystal *See* rock crystal.

mountain-gap wind [METEOROL] A local wind blowing through a gap between mountains.

mountain glacier *See* alpine glacier.

mountain mahogany *See* obsidian.

mountain pediment [GEOL] A plain of combined erosion and transportation at the base of and surrounding a desert mountain range; at a distance it has the appearance of a broad triangular mass.

mountain range [GEOGR] A succession of mountains or narrowly spaced mountain ridges closely related in position, direction, and geologic features.

mountain slope [GEOGR] The inclined surface that forms a mountainside.

mountain soap *See* saponite.

mountain system [GEOGR] A group of mountain ranges tied together by common geological features. Also known as mountain chain.

mountain tallow *See* hatchettite.

mountain wood [GEOL] **1.** A compact, fibrous, gray to brown variety of asbestos that looks like dry wood. Also known as rock wood. **2.** A fibrous clay mineral such as sepiolite.

Mount Rose snow sampler [ENG] A particular pattern of snow sampler having an internal diameter of 1.485 inches (3.7727 centimeters), so that each inch of water in the sample weighs 1 ounce (28.349 grams).

mouth [GEOGR] **1.** The place where one body of water discharges into another. Also known as influx. **2.** The entrance or exit of a geomorphic feature, such as of a cave or valley. [MIN ENG] **1.** The end of a shaft, adit, drift, entry, or tunnel emerging at the surface. **2.** The collar of a borehole.

Mozambique Current [OCEANOGR] The portion of the South Equatorial Current that turns and flows along the coast of Africa in the Mozambique Channel, forming one of the western boundary currents in the Indian Ocean.

muck [GEOL] Dark, finely divided, well-decomposed, organic matter intermixed with a high percentage of mineral matter, usually silt, forming a surface deposit in some poorly drained areas. [MIN ENG] *See* waste rock.

mucker [MIN ENG] A worker who loads broken mineral into trams or pushes them from stope chute to shaft. Also known as groundman.

mucking [ENG] Clearing and loading broken rock and other excavated materials, as in tunnels or mines.

mud [GEOL] An unindurated mixture of clay and silt with water; it is slimy with a consistency varying from that of a semifluid to that of a soft and plastic sediment. [PETR] The silt plus clay portion of a sedimentary rock.

mud ball [GEOL]   A rounded mass of mud or mudstone up to 20 centimeters in diameter in a sedimentary rock. Also known as chalazoidite; tuff ball.

mud blasting [ENG]   The detonation of sticks of explosive stuck on the side of a boulder with a mud covering, so that little of the explosive energy is used in breaking the boulder.

mud cake [ENG]   A caked layer of clay adhering to the walls of a well or borehole, formed where the water in the drilling mud filtered into a porous formation during rotary drilling.

mud-cake resistivity [PETRO ENG]   Resistivity of drilling mud cake pressed from a sample of the mud; important in mud log interpretation.

mudcap [ENG]   A quantity of wet mud, wet earth, or sand used to cover a charge of dynamite or other high explosive fired in contact with the surface of a rock in mud blasting.

mud conditioning [PETRO ENG]   In a well drilling operation, the treatment and control of drilling mud to ensure proper gel strength, viscosity, density and so on.

mud cone [GEOL]   A cone of sulfurous mud built around the opening of a mud volcano or mud geyser, with slopes as steep as 40° and diameters ranging upward to several hundred yards. Also known as puff cone.

mud crack [GEOL]   An irregular fracture formed by shrinkage of clay, silt, or mud under the drying effects of atmospheric conditions at the surface. Also known as desiccation crack; sun crack.

mud crack polygon See mud polygon.

mud drilling [MIN ENG]   Drilling operations in which a mud-laden circulation fluid is used. Also known as mud flush drilling.

mud flat [GEOL]   A relatively level, sandy or muddy coastal strip along a shore or around an island; may be alternately covered and uncovered by the tide or may be covered by shallow water. Also known as flat.

mudflow [GEOL]   A flowing mass of fine-grained earth material having a high degree of fluidity during movement.

mud flush drilling See mud drilling.

mud flush test [MIN ENG]   A test carried out at the boring site to determine whether the mud solution is of the correct viscosity and density.

mud log [PETRO ENG]   A continuous record of changes in oil or gas contents of circulating drilling mud while drilling a well.

mudlump [GEOL]   A diapiric sedimentary structure consisting of clay or silt and forming an island in deltaic areas; produced by the loading action of rapidly deposited delta front sands upon lighter-weight prodelta clays.

mud pit See slushpit.

mud polygon [GEOL]   A nonsorted polygon whose center lacks vegetation but whose peripheral fissures contain peat and plants. Also known as mud crack polygon.

mud pot [GEOL]   A type of hot spring which contains boiling mud, usually sulfurous and often multicolored; commonly associated with geysers and other hot springs found in volcanic areas. Also known as sulfur-mud pool.

mud–removal acid [PETRO ENG]   Mixture of hydrochloric and hydrofluoric acids with inhibitors, surfactants, and demulsifiers; used to dissolve drilling-mud clays away from the drillhole face.

mudslide [GEOL] A relatively slow-moving mudflow in which movement occurs essentially by sliding upon a discrete-boundary shear surface.

mudstone [GEOL] An indurated equivalent of mud in the form of a blocky or massive, fine-grained sedimentary rock containing approximately equal proportions of silt and clay; lacks the fine lamination or fissility of shale.

mud volcano [GEOL] A conical accumulation of variable admixtures of sand and rock fragments, the whole resulting from eruption of wet mud and impelled upward by fluid or gas pressure. Also known as hervidero; macaluba.

mugearite [PETR] A dark-colored, fine-grained igneous rock in which the chief feldspar is oligoclase, plus orthoclase and olivine with some apatite and opaque oxides; originates by differentiation and volcanic crystallization of the primary magma.

muggy [METEOROL] Colloquially descriptive of warm and especially humid weather.

mule *See* barney.

mule skinner [MIN ENG] A mule driver.

mull [GEOGR] *See* headland. [GEOL] Granular forest humus that is incorporated with mineral matter.

mullion [GEOL] A columnar structure occurring in folded sedimentary and metamorphic rocks in which rock columns appear to intersect.

mullite [MINERAL] $Al_6Si_2O_{13}$ An orthorhombic mineral consisting of an aluminum silicate that is resistant to corrosion and heat; used as a refractory. Also known as porcelainite.

mullock *See* waste rock.

multicompletion well *See* multiple-completion well.

multicycle [GEOL] Pertaining to a landscape or landform produced by more than one cycle of erosion.

multideck cage [MIN ENG] A cage with two or more compartments or platforms to hold the mine cars.

multideck screen [MIN ENG] A screen with two or more superimposed screening surfaces mounted within a common frame.

multideck sinking platform [MIN ENG] A sinking platform of several decks so that various shaft-sinking operations may be performed simultaneously.

multideck table [MIN ENG] A type of shaking table that is double-decked; while each deck is fed and discharged independently, one mechanism vibrates both.

multifuse igniter [ENG] A black powder cartridge that allows several fuses to be fired at the same time by lighting a single fuse.

multiple-completion packer [PETRO ENG] A device used to provide a seal between the outside surfaces of the two or more parallel tubing strings in a multiple-completion well, and the inside surface of the common wellbore casing.

multiple-completion well [PETRO ENG] Oil well in which there is production from more than one oil-bearing zone (different depths) with parallel tubing strings within a single wellbore casing string. Also known as multicompletion well.

multiple-current hypothesis [OCEANOGR] The proposal that the Gulf Stream, instead of being composed of a single tortuous current, actually consists of many quasipermanent currents, countercurrents, and eddies.

multiple discharge *See* composite flash.

multiple-entry system   [MIN ENG]   A system of access or development openings generally in bituminous coal mines involving more than one pair of parallel entries, one for haulage and fresh-air intake and the other for return air.

multiple fault   *See* step fault.

multiple firing   [ENG]   Electrically firing with delay blasting caps in a number of holes at one time.

multiple hydraulic pump   [PETRO ENG]   Oil-well pump arrangement by which a single pump can be used in alternating operation to lift oil from two producing zones within a single multiple completion well.

multiple openings   [MIN ENG]   Any series of underground openings separated by rib pillars or connected at frequent intervals to form a system of rooms and pillars.

multiple-parallel-tubing string   [PETRO ENG]   Two or more parallel and closely packed oil-well tubing strings used in multiple-completion wells. Also known as multiple-tubing string.

multiple reflection   [GEOPHYS]   A seismic wave which has more than one reflection. Also known as repeated reflection; secondary reflection.

multiple-row blasting   [ENG]   The drilling, charging, and firing of rows of vertical boreholes.

multiple series   [ENG]   A method of wiring a large group of blasting charges by connecting small groups in series and connecting these series in parallel. Also known as parallel series.

multiple shooting   [ENG]   The firing of an entire face at one time by means of connecting shot holes in a single series and shooting all holes at the same instant.

multiple-shot survey   [PETRO ENG]   The determining and recording of drill-hole direction.

multiple-stage separator   [PETRO ENG]   Oilwell gas-oil separator in which wellhead pressure is reduced in several stages, with the flashing off of gas at each pressure reduction.

multiple tropopause   [GEOPHYS]   A frequent condition in which the tropopause appears not as a continuous single "surface" of discontinuity between the troposphere and the stratosphere, but as a series of quasi-horizontal "leaves," which are partly overlapping in steplike arrangement.

multiple-tubing string   *See* multiple-parallel-tubing string.

multiple zone   [PETRO ENG]   Two or more discrete oil or gas reservoirs in the same geographic area, but at different depths.

Multituberculata   [PALEON]   The single order of the nominally mammalian suborder Allotheria; multituberculates had enlarged incisors, the coracoid bones were fused to the scapula, and the lower jaw consisted of the dentary bone alone.

multiwell gas-lift system   [PETRO ENG]   An installation to allow gas-lift production from the various tubing strings involved in a multiple completion well.

mundic   *See* pyrite.

murbruk structure   *See* mortar structure.

Murchisoniacea   [PALEON]   An extinct superfamily of gastropod mollusks in the order Prosobranchia.

muromontite   [MINERAL] $Be_2FeY_2(SiO_4)_3$ A mineral composed of yttrium iron beryllium silicate.

Muschelkalk [GEOL] A European stage of geologic time equivalent to the Middle Triassic, above Bunter and below Keuper.

muscovite [MINERAL] $KAl_2(AlSi_3)O_{10}(OH)_2$ One of the mica group of minerals, occurring in some granites and abundant in pegmatites; it is colorless, whitish, or pale brown, and the crystals are tabular sheets with prominent base and hexagonal or rhomboid outline; hardness is 2–2.5 on Mohs scale, and specific gravity is 2.7–3.1. Also known as common mica; mirror stone; moscovite; Muscovy glass; potash mica; white mica.

Muscovy glass *See* muscovite.

Muskat equation [PETRO ENG] Equation used to calculate oil reservoir permeability from the pressure buildup curve recorded when a producing well is shut in after a flow test.

mustard-seed coal [GEOL] Anthracite that will pass through circular holes 3/64 inch (1.2 millimeters) in diameter in a screen.

muthmannite [MINERAL] $(Ag,Au)Te$ A bright brass yellow mineral consisting of silver-gold telluride; occurs as tabular crystals.

mylonite [PETR] A hard, coherent, often glassy-looking rock that has suffered extreme mechanical deformation and granulation but has remained chemically unaltered; appearance is flinty, banded, or streaked, but the nature of the parent rock is easily recognized.

mylonite gneiss [PETR] A metamorphic rock intermediate in character between mylonite and schist.

mylonitic structure [PETR] A structure characteristic of mylonites, produced by extreme microbrecciation and shearing which gives the appearance of a flow structure.

mylonitization [GEOL] Rock deformation produced by intense microbrecciation without appreciable chemical alteration of granulated materials.

myrmekite [PETR] Intergrowth of plagioclase feldspar and vermicular quartz in an igneous rock.

myrmekitic [PETR] **1.** Pertaining to the texture of an igneous rock marked by intergrowths of feldspar and vermicular quartz. **2.** Having characteristic properties of myrmekite.

nacrite [MINERAL] $Al_2Si_2O_5(OH)_4$ A crystallized clay mineral of the kaolinite group; structurally distinct in being the most closely stacked in the *c*-axis direction.

nadorite [MINERAL] $PbSbO_2Cl$ A smoky brown or brownish-yellow to yellow, orthorhombic mineral consisting of an oxychloride of lead and antimony.

naegite [MINERAL] A variety of zircon containing thorium and uranium in addition to zirconium and silicon.

nagatelite [MINERAL] Black mineral composed of phosphosilicate of an aluminum, rare-earth elements, calcium, and iron; occurs in tabular masses.

nagyagite [MINERAL] $Pb_5Au(Te,Sb)_4S_{5-8}$ A lead-gray mineral consisting of a sulfide of lead, gold, tellurium, and antimony. Also known as black tellurium; tellurium glance.

nahcolite [MINERAL] $NaHCO_3$ A white, monoclinic mineral consisting of natural sodium bicarbonate.

naif [MINERAL] Referring to a gemstone that has a true or natural luster when uncut. Also spelled naife.

naife *See* naif.

nailhead spar [MINERAL] A variety of calcite in the form of crystals which are a combination of hexagonal prisms with flat rhombohedrons.

nailhead striation [GEOL] A glacial striation which has a well-defined, blunt head and then tapers in the direction of ice movement to an indefinite end, thus resembling a nail in shape.

naked karst [GEOL] Karst with well-exposed topographic features due to its formation in a region lacking soil cover.

naked-light mine [MIN ENG] A coal mine that is nongassy, where naked lights can be used by miners.

nakhlite [GEOL] An achondritic stony meteorite composed of an aggregate of diopside and olivine.

Namurian [GEOL] A European stage of geologic time; divided into a lower stage (Lower Carboniferous or Upper Mississippian) and an upper stage (Upper Carboniferous or Lower Pennsylvanian).

Nansen bottle [ENG] A bottlelike water-sampling device with valves at both ends that is lowered into the water by wire; at the desired depth it is activated by a messenger which strikes the reversing mechanism and inverts the bottle, closing the valves and trapping the water sample inside. Also known as Petterson-Nansen water bottle; reversing water bottle.

Nansen cast [OCEANOGR] A series of Nansen-bottle water samples and associated temperature observations resulting from one release of a messenger.

naphthine *See* hatchettite.

napoleonite *See* corsite.

Napoleonville [GEOL] A North American (Gulf Coast) stage of geologic time; a subdivision of the Miocene, above Anahuac and below Duck Lake.

nappe [GEOL] A sheetlike, allochthonous rock unit formed by thrust faulting or recumbent folding or both. [HYD] A sheet of water flowing over a dam.

narbonnais [METEOROL] A wind coming from Narbonne; a north wind in the Roussillon region of southern France resembling the tramontana; if associated with an influx of arctic air, it may be very stormy with heavy falls of rain or snow.

narcotine *See* noscapine.

nari *See* caliche.

Narizian [GEOL] A North American stage of geologic time; a subdivision of the upper Eocene, above Ulatisian and below Fresnian.

narrow [GEOGR] A constricted section of a mountain pass, valley, or cave, or a gap or narrow passage between mountains.

narrows [GEOGR] A navigable narrow part of a bay, strait, or river.

narsarsukite [MINERAL] $Na_2(Ti,Fe)Si_4(O,F)$ Mineral composed of sodium titanium iron fluoride and silicate.

n'aschi [METEOROL] A northeast wind which occurs in winter on the Iranian coast of the Persian Gulf, especially near the entrance to the Gulf, and also on the Makran (West Pakistan) coast; it is probably associated with an outflow from the central Asiatic anticyclone which extends over the high land of Iran.

nasonite [MINERAL] $Ca_4Pb_6Si_6O_{21}Cl_2$ A white mineral composed of silicate and chloride of calcium and lead and occurring in granular masses.

nasturan *See* pitchblende.

Nathansohn's theory [OCEANOGR] The theory that nutrient salts in the lighted surface layers of the ocean are consumed by plants, accumulate in the deep ocean through sinking of dead plant and animal bodies, and eventually return to the euphotic layer through diffusion and vertical circulation of the water.

native asphalt [GEOL] Exudations or seepages of asphalt occurring in nature in a liquid or semiliquid state. Also known as natural asphalt.

native coal *See* natural coke.

native element [GEOL] Any of 20 elements, such as copper, gold, and silver, which occur naturally uncombined in a nongaseous state; there are three groups—metals, semimetals, and nonmetals.

native metal [GEOCHEM] A metallic native element; includes silver, gold, copper, iron, mercury, iridium, lead, palladium, and platinum.

native paraffin *See* ozocerite.

native uranium [GEOCHEM] Uranium as found in nature; a mixture of the fertile uranium-238 isotope (99.3%), the fissionable uranium-235 isotope (0.7%), and a minute percentage of other uranium isotopes. Also known as natural uranium; normal uranium.

natric horizon [GEOL] A soil horizon that has the properties of an argillic horizon, but also displays a blocky, columnar or prismatic structure and has a subhorizon with an exchangeable-sodium saturation of over 15%.

natroalunite [MINERAL] $NaAl_3(SO_4)_2(OH)_6$ Mineral composed of basic sodium aluminum sulfate. Also known as almerite.

natrochalcite [MINERAL] $NaCu_2(SO_4)(OH)\cdot H_2O$ An emerald-green mineral composed of hydrous basic sulfate of sodium and copper.

natrolite [MINERAL] $Na_2Al_2Si_3O_{10}\cdot 2H_2O$ A zeolite mineral composed of hydrous silicate of sodium and aluminum; usually occurs in slender acicular or prismatic crystals.

natromontebrasite [MINERAL] $(Na,Li)Al(PO_4)(OH,F)$ Mineral composed of hydrous basic phosphate of sodium, lithium, and aluminum; it is isomorphous with montebrasite and amblygonite. Also known as fremontite.

natron [MINERAL] $Na_2CO_3\cdot 10H_{20}$ A white, yellow, or gray mineral that crystallizes in the monoclinic system, is soluble in water, and generally occurs in solution or in saline residues.

natronborocalcite *See* ulexite.

natron lake *See* soda lake.

natrophilite [MINERAL] $NaMn(PO_4)$ A mineral composed of sodium manganese phosphate.

natural arch [GEOL] **1.** A landform resembling a natural bridge but formed by means other than erosion. **2.** *See* natural bridge.

natural asphalt *See* native asphalt.

natural bitumen [GEOL] Native mineral pitch, tar, or asphalt.

natural bridge [GEOL] An archlike rock formation spanning a ravine or valley and formed by erosion. Also known as natural arch.

natural coke [GEOL] Coal that has been naturally carbonized by contact with an igneous intrusion or by natural combustion. Also known as black coal; blind coal; carbonite; cinder coal; coke coal; cokeite; finger coal; native coal.

natural glass [GEOL] An amorphous, vitreous inorganic material that has solidified from magma too quickly to crystallize.

natural harbor [GEOGR] A harbor where the configuration of the coast provides the necessary protection.

natural levee [GEOL] An elongate embankment compounded of sand and silt and deposited along both banks of a river channel during times of flood.

natural load [HYD] The quantity of sediment carried by a stable stream.

natural radio-frequency interference [GEOPHYS] Natural terrestrial phenomena of an electromagnetic nature, or natural electromagnetic disturbances originating outside the atmosphere, which interfere with radio communications.

natural remanent magnetization [GEOPHYS] The magnetization of rock which exists in the absence of a magnetic field and has been acquired from the influence of the earth's magnetic field at the time of their formation or, in certain cases, at later times. Abbreviated NRM.

natural splitting [MIN ENG] In mine ventilation, a flow of air dividing among the branches, of its own accord and without regulation, in inverse relation to the resistance of each airway.

natural uranium *See* native uranium.

natural ventilation [MIN ENG] The weak and varying ventilation in a mine caused by the difference in air density between shafts.

naujaite [PETR]   A hypidiomorphic-granular sodalite-rich nepheline syenite containing microcline and small amounts of albite, analcime, acmite, and sodium amphiboles, and exhibiting a characteristic coarse, poikilitic texture.

naumannite [MINERAL] Ag_2Se An iron-black mineral that crystallizes in the isometric system; consists of silver selenide, and occurs massive or in crystals; specific gravity is 8.

Navajo sandstone [GEOL]   A fossil dune formation of Jurassic age found in the Colorado Plateau of the United States.

naval meteorology [METEOROL]   The branch of meteorology which studies the interaction between the ocean and the overlying air mass, and which is concerned with atmospheric phenomena over the oceans, the effect of the ocean surface on these phenomena, and the influence of such phenomena on shallow and deep seawater.

Navarroan [PALEON]   A North American (Gulf Coast) stage of Upper Cretaceous geologic time, above the Tayloran and below the Midwayan of the Tertiary.

navigable semicircle [METEOROL]   That half of a cyclonic storm area in which the rotary and progressive motions of the storm tend to counteract each other, and the winds are in such a direction as to blow a vessel away from the storm track.

navite [MINERAL]   A porphyritic basalt containing phenocrysts of altered olivine, augite, and basic plagioclase in a groundmass of labradorite and augite.

Ne *See* neon.

neap high water *See* mean high-water neaps.

neap low water *See* mean low-water neaps.

neap range [OCEANOGR]   The mean semidiurnal range of tide when neap tides are occurring; the mean difference in height between neap high water and neap low water. Also known as mean neap range.

neap rise [OCEANOGR]   The height of neap high water above the chart datum.

neaps *See* neap tide.

neap tidal currents [OCEANOGR]   Tidal currents of decreased speed occurring at the time of neap tides.

neap tide [OCEANOGR]   Tide of decreased range occurring about every 2 weeks when the moon is in quadrature, that is, during its first and last quarter. Also known as neaps.

near earthquake [GEOPHYS]   An earthquake with the epicenter located about 1000–1200 kilometers from the detector.

nearest neighbors [CRYSTAL]   Any pair of atoms in a crystal lattice which are as close to each other, or closer to each other, than any other pair.

nearshore [OCEANOGR]   An indefinite zone that extends from the shoreline to somewhat beyond the breaker zone.

nearshore circulation [OCEANOGR]   Ocean circulation consisting of both the nearshore currents and the coastal currents.

nearshore current system [OCEANOGR]   The current system produced mostly by wave action in and near the breaker zone; it has four components: The shoreward mass transport of water, longshore currents, seaward return flow (including rip currents), and longshore movement of the expanded heads or rip currents.

Nebraskan drift [GEOL]   Rock material transported during the Nebraskan glaciation; it is buried below the Kansan drift in Iowa.

Nebraskan glaciation [GEOL] The first glacial stage of the Pleistocene epoch in North America, beginning about 1,000,000 years ago, and preceding the Aftonian interglacial stage.

nebulite [PETR] A chorismite in which one of the textural elements occurs in nebulitic lenticular masses.

nebulitic [PETR] **1.** Having indistinct boundaries between textural elements. **2.** Of or pertaining to a nebulite.

nebulosus [METEOROL] A cloud species with the appearance of a nebulous veil, showing no distinct details; found principally in the genera cirrostratus and stratus.

neck [GEOGR] A narrow strip of land, especially one connecting two larger areas. [GEOL] *See* pipe. [OCEANOGR] The narrow band of water forming the part of a rip current where feeder currents converge and flow swiftly through the incoming breakers and out to the head.

neck cutoff [GEOGR] A high-angle meander cutoff formed where a stream breaks through or across a narrow meander neck, as where downstream migration of one meander has been slowed and the next meander upstream has overtaken it.

Necrolestidae [PALEON] An extinct family of insectivorous marsupials.

necronite [MINERAL] A blue, pearly variety of orthoclase that gives off a foul odor when hammered.

Nectridea [PALEON] An order of extinct lepospondylous amphibians characterized by vertebrae in which large fan-shaped hemal arches grow directly downward from the middle of each caudal centrum.

needle ice *See* frazil ice.

needle iron ore *See* needle ironstone.

needle ironstone [MINERAL] A variety of goethite occurring in fibrous aggregates of acicular crystals. Also known as needle iron ore.

needle ore [MINERAL] **1.** Iron ore of very high metallic luster, found in small quantities, which may be separated into long, slender filaments resembling needles. **2.** *See* aikinite.

negative area [GEOG] An area that is almost uncultivable or uninhabitable. [GEOL] *See* negative element.

negative center [GEOPHYS] The central area of negative potential in an observed anomaly exhibiting spontaneous polarization.

negative element [GEOL] A large structural feature or part of the earth's crust, characterized through a long geologic time period by frequent and conspicuous downward movement (subsidence) or by extensive erosion, or by an uplift that is considerably less rapid or less frequent than that of adjacent positive elements. Also known as negative area.

negative elongation [CRYSTAL] In a section of an anisotropic crystal, a sign of elongation that is parallel to the faster of the two plane-polarized rays.

negative landform [GEOL] **1.** A relatively depressed or low-lying landform, such as a valley, basin, or plain. **2.** A volcanic structure formed by a lack of material, such as a caldera.

negative movement [GEOL] **1.** A downward movement of the earth's crust in relation to an adjacent part such as that produced by subsidence. **2.** A relative lowering of the sea level with respect to the land, which can be produced either by an upward movement of the crust or by a retreat of the sea.

negative rain [METEOROL]   Rain which exhibits a net negative electric charge.

negative shoreline *See* shoreline of emergence.

negative strip *See* Vening Meinesz zone.

neighbor [CRYSTAL]   One of a pair of atoms or ions in a crystal which are close enough to each other for their interaction to be of significance in the physical problem being studied.

nelsonite [PETR]   A group of hypabyssal rocks composed mainly of ilmenite and apatite.

nemalite [MINERAL]   A fibrous brucite that contains ferrous oxide.

nematoblastic [PETR]   Pertaining to a metamorphic rock with a homeoblastic texture due to development during recrystallization of slender prismatic crystals.

Nematophytales [PALEOBOT]   A group of fossil plants from the Silurian and Devonian periods that bear some resemblance to the brown seaweeds (Phaeophyta).

nemere [METEOROL]   In Hungary, a stormy, cold fall wind.

Neoanthropinae [PALEON]   A subfamily of the Hominidae in some systems of classification, set up to include *Homo sapiens* and direct ancestors of *H. sapiens*.

Neocathartidae [PALEON]   An extinct family of vulturelike diurnal birds of prey (Falconiformes) from the Upper Eocene.

Neocomian [GEOL]   A European stage of Lower Cretaceous geologic time; includes Berriasian, Valanginian, Hauterivian, and Barremian.

neocryst [GEOL]   A single crystal of a secondary mineral occurring in an evaporite.

neoformation *See* neogenesis.

Neogene [GEOL]   An interval of geologic time incorporating the Miocene and Pliocene of the Tertiary period; the Upper Tertiary.

neogenesis [GEOL]   The formation of new minerals through such processes as metamorphism or diagenesis. Also known as neoformation.

neoglaciation [GEOL]   The removal of glacier ice growth in certain mountain areas during the Little Ice Age, following its shrinkage or disappearance during the Altithermal interval.

neolensic texture [GEOL]   The secondary, nonporphyritic, roughly laminated texture of evaporites.

neomagma [GEOL]   Magma formed by partial or total remelting of previously existing rocks subject to conditions of plutonic metamorphism.

neomineralization [GEOCHEM]   Chemical reaction within a rock in which the mineral components are converted into new mineral species.

neoporphyrocrystic texture [GEOL]   A texture of an evaporite characterized by large neocrysts embedded in a finer-grained matrix.

neosilicate [MINERAL]   A structural type of silicate mineral characterized by linkage of isolated SiO_4 tetrahedra by ionic bonding only; an example is olivine.

neosome [GEOL]   A geometric element of a composite rock or mineral deposit, appearing to be of younger age than the main rock mass.

neotectonic map [GEOL]   A map depicting neotectonic structures.

neotectonics [GEOL]   The study of the most recent structures and structural history of the earth's crust, after the Miocene.

neotocite [MINERAL] A mineral composed of a hydrous silicate of manganese and iron with uncertain formula.

neovolcanic [PETR] Pertaining to extrusive rocks of Tertiary age or younger.

nephanalysis [METEOROL] The analysis of a synoptic chart in terms of the types and amount of clouds and precipitation; cloud systems are identified both as entities and in relation to the pressure pattern, fronts, and other aspects.

nephcurve [METEOROL] In nephanalysis, a line bounding a significant portion of a cloud system, for example, a clear-sky line, precipitation line, cloud-type line, or ceiling-height line.

nepheline [MINERAL] $(Na,K)AlSiO_4$ A mineral of the feldspathoid group crystallizing in the hexagonal system and occurring as glassy or coarse crystals or colorless grains or green to brown masses of greasy luster in alkalic igneous rocks; hardness is 5.5–6 on Mohs scale. Also known as eleolite; nephelite.

nepheline basalt *See* olivine nephelinite.

nepheline monzonite [PETR] A nepheline syenite in which sodic plagioclase exceeds the quantity of alkali feldspar.

nepheline phonolite [PETR] The fine-grained equivalent of nepheline syenite.

nepheline syenite [PETR] A phaneritic plutonic rock with granular texture, composed largely of alkali feldspar, nepheline, and dark-colored materials.

nephelinite [PETR] A dark-colored, aphanitic rock of volcanic origin, composed essentially of nepheline and pyroxene; texture is usually porphyritic with large crystals of augite and nepheline in a very-fine-grained matrix.

nephelite *See* nepheline.

nepheloid zone [OCEANOGR] In the North Atlantic Ocean, the water layer near the bottom of the continental rise and slope containing suspended sediment of the clay fraction and organic matter.

nepheloscope [ENG] An instrument for the production of clouds in the laboratory by condensation or expansion of moist air.

nephlinolith [PETR] An intrusive igneous rock solely nepheline.

nephology [METEOROL] The study of clouds.

nephometer [ENG] A general term for instruments designed to measure the amount of cloudiness; an early type consists of a convex hemispherical mirror mapped into six parts; the amount of cloud coverage on the mirror is noted by the observer.

nephoscope [ENG] An instrument for determining the direction of cloud motion.

nephrite [MINERAL] An exceptionally tough, compact, fine-grained, greenish or bluish amphibole constituting the less valuable type of jade; formerly worn as a remedy for kidney diseases. Also known as greenstone; kidney stone.

nephsystem *See* cloud system.

neptunian dike [GEOL] A dike formed by the depositing of sediment, generally sand, in a fissure or hollow in the ocean.

neptunianism *See* neptunism.

neptunian theory *See* neptunism.

neptunic rock [GEOL] **1.** A rock that is formed in the sea. **2.** *See* sedimentary rock.

neptunism [GEOL] The obsolete theory that all rocks of the earth's crust were deposited from or crystallized out of water. Also known as neptunianism; neptunian theory.

neptunite [MINERAL] $(Na,K)_2(Fe,Mn)TiSi_4O_{12}$ Black mineral composed of silicate of sodium, potassium, iron, manganese, and titanium.

nepuite *See* garnierite.

neritic [OCEANOGR] Of or pertaining to the region of shallow water adjoining the seacoast and extending from low-tide mark to a depth of about 200 meters.

nesophitic texture [PETR] The ophitic texture of an igneous rock containing pyroxene interstitial to plagiocase in isolated areas.

Nesophontidae [PALEON] An extinct family of large, shrewlike lipotyphlans from the Cenozoic found in the West Indies.

nesosilicate [MINERAL] A mineral (such as olivine) composed of independent silicon-oxygen tetrahedra bonded by ionic bonds, without sharing of oxygens.

nesquehonite [MINERAL] $MgCo_3 \cdot 3H_2O$ A colorless to white, orthorhombic mineral consisting of hydrated magnesium carbonate.

nest [GEOL] A concentration of some relatively conspicuous element of a geologic feature, such as pebbles or inclusions, within a sand layer or igneous rock.

nested [GEOL] **1.** Referring to volcanic cones, craters, or calderas that occur one within another. **2.** Referring to two or more calderas that intersect, having been formed at different times or by different explosions.

net [GEOL] **1.** In structural petrology, coordinate network of meridians and parallels, projected from a sphere at intervals of 2°; used to plot points whose spherical coordinates are known and to study the distribution and orientation of planes and points. Also known as projection net; stereographic net. **2.** A form of horizontal patterned ground whose mesh is intermediate between a circle and a polygon.

net balance [HYD] The mass difference of a glacier from the point of minimum mass in one year to the point in the folowing year. Also known as net budget.

net budget *See* net balance.

net slip [GEOL] On a fault, the distance between two formerly adjacent points on either side of the fault; defines direction and relative amount of displacement. Also known as total slip.

Neumann's principle [CRYSTAL] The principle that the symmetry elements of the point group of a crystal are included among the symmetry elements of any property of the crystal.

Neurodontiformes [PALEON] A suborder of Conodontophoridia having a lamellar internal structure.

neutral estuary [GEOGR] An estuary in which neither fresh-water inflow nor evaporation dominates.

neutral line *See* Busch lemniscate.

neutral point *See* col.

neutral pressure *See* neutral stress.

neutral shoreline [GEOL] A shoreline whose essential features are independent of either the submergence of a former land surface or the emergence of a former underwater surface.

neutral stress [HYD] The stress transmitted through the interstitial fluid of a soil or rock mass. Also known as neutral pressure; pore pressure; pore-water pressure.

neutron-gamma well logging [ENG] Neutron well logging in which the varying intensity of gamma rays produced artificially by neutron bombardment is recorded.

neutron logging *See* neutron well logging.

neutron well logging [ENG] Study of formation-fluid-content properties down a wellhole by neutron bombardment and detection of resultant radiation (neutrons or gamma rays). Also known as neutron logging.

neutrosphere [METEOROL] The atmospheric shell from the earth's surface upward, in which the atmospheric constituents are for the most part un-ionized, that is, electrically neutral; the region of transition between the neutrosphere and the ionosphere is somewhere between 70 and 90 kilometers, depending on latitude and season.

nevada [METEOROL] A cold wind descending from a mountain glacier or snowfield, for example, in the higher valleys of Ecuador.

Nevadan orogeny [GEOL] Orogenic episode during Jurassic and Early Cretaceous geologic time in the western part of the North American Cordillera. Also known as Nevadian orogeny; Nevadic orogeny.

Nevadian orogeny *See* Nevadan orogeny.

Nevadic orogeny *See* Nevadan orogeny.

névé [GEOGR] A geographic area of perennial snow.   [HYD] An accumulation of compacted, granular snow in transition from soft snow to ice; it contains much air; the upper portions of most glaciers and ice shelves are usually composed of névé.

newberyite [MINERAL] $MgH(PO_4)\cdot3H_2O$ A white, orthorhombic member of the brushite mineral group; it is isostructural with gypsum.

new global tectonics [GEOL] Comprehensive theory relating the formation of mountain belts, island arcs, and ocean trenches to the relative movement of regionally extensive lithospheric plates which are delineated by the major seismic belts of the earth.

newly formed ice [HYD] Ice in the first stage of formation and development. Also known as fresh ice.

New Red Sandstone [GEOL] The red sandstone facies of the Permian and Triassic systems exposed in the British Isles.

new snow [METEOROL] **1.** Fallen snow whose original crystalline structure has been retained and is therefore recognizable. **2.** Snow which has fallen in a single day.

Newson's boring method [MIN ENG] A method of boring small shafts, up to 5½ feet (1.7 meters) in diameter using the principle of chilled-shot drilling on a large scale.

Ney's chart *See* modified Lambert conformal chart.

Ney's projection *See* modified lambert conformal projection.

ngavite [GEOL] A chondritic stony meteorite composed of bronzite and olivine in a friable, breccialike mass of chondrules.

Niagaran [GEOL] A North American provincial geologic series, in the Middle Silurian.

niccolite [MINERAL] NiAs A pale-copper-red, hexagonal mineral with metallic luster; an important ore of nickel; hardness is 5–5.5 on Mohs scale. Also known as arsenical nickel; copper nickel; nickeline.

niche [GEOL] A shallow cave or reentrant produced by weathering and erosion near the base of a rock face or cliff, or beneath a waterfall.

niche glacier [HYD] A common type of small mountain glacier, occupying a funnel-shaped hollow or irregular recess in a mountain slope.

nick *See* knickpoint.

nickel-antimony glance *See* ullmannite.

nickel bloom *See* annabergite.

nickel glance *See* gersdorffite.

nickeline *See* niccolite.

nickel ocher *See* annabergite.

nickel pyrites *See* millerite.

nickel vitriol *See* morenosite.

nickpoint *See* knickpoint.

nicolo [MINERAL] A variety of onyx with black or brown base and a bluish-white or faint-bluish top layer.

nieve penitente [GEOL] A jagged pinnacle or spike of snow or firn, up to several meters in height. Also known as penitent.

nife [PETR] The rock material composed of nickel and iron and found in the core of the earth. Also known as nifel.

nifel *See* nife.

niggliite [MINERAL] PtSn or PtTe A silver-white mineral consisting of a platinum telluride compound.

nightglow [GEOPHYS] A subdivision of airglow in which energy comes from reactions of atomic oxygen between 70 and 100 kilometers, and from ionic recombination around 300 kilometers.

night-sky light *See* airglow.

night-sky luminescence *See* airglow.

night wind [METEOROL] Dry squalls which occur at night in southwest Africa and the Congo; the term is loosely applied to other diurnal local winds such as mountain wind, land breeze, and midnight wind.

nigrine [MINERAL] A black variety of rutile containing some iron, in addition to titanium.

niklesite [PETR] A pyroxenite containing the three pyroxenes: diopside, enstatite, and diallage.

nilas [HYD] A thin elastic crust of gray-colored ice formed on a calm sea, having a matte surface, and easily bent by waves and thrust into a pattern of interlocking fingers.

niligongite [PETR] A plutonic foidite intermediate in composition between fergusite and ijolite, containing approximately equal amounts of nepheline and leucite and 30–60% mafic minerals.

nimbostratus [METEOROL] A principal cloud type, or cloud genus, gray-colored and often dark, rendered diffuse by more or less continuously falling rain, snow, or sleet of the ordinary varieties, and not accompanied by lightning, thunder, or hail; in most cases the precipitation reaches the ground.

niningerite [MINERAL] (Mg,Fe,Mn)S A mineral found only in meteorites.

niobite *See* columbite.

nip [GEOL] **1.** A small, low cliff or break in slope produced by wavelets at the high-water mark. **2.** The point on the bank of a meander lake where erosion takes place due to crowding of the stream current toward the lake. **3.** Thinning of a coal seam, particularly if caused by tectonic movements. Also known as want. [MIN ENG] *See* squeeze.

Nissen stamp [MIN ENG] **1.** Machine used in crushing rock to sand sizes. **2.** An individual stamp worked in its own circular mortar box.

nitrate mineral [MINERAL] Any of several generally rare minerals characterized by a fundamental ionic structure of NO_3^-; examples are soder niter, niter, and nitrocalcite.

nitratine *See* soda niter.

nitromagnesite [MINERAL] $Mg(NO_3)_2 \cdot 6H_2O$ Mineral consisting of magnesium nitrate, occurring as an efflorescence in limestone caverns.

nival gradient [GEOL] The angle between a nival surface and the horizon.

nival surface [GEOL] The hypothetical planar surface containing all of the different snowlines of the same geologic time period.

nivation [GEOL] Rock or soil erosion beneath a snowbank or snow patch, due mainly to frost action but also involving chemical weathering, solifluction, and meltwater transport of weathering products. Also known as snow patch erosion.

nivation cirque *See* nivation hollow.

nivation glacier [HYD] A small, newly formed glacier; represents the initial stage of glaciation. Also known as snowbank glacier.

nivation hollow [GEOL] A small, shallow depression formed, and occupied during part of the year, by a snow patch or snowbank that, through nivation, is thought to initiate glaciation. Also known as nivation cirque; snow niche.

nivation ridge *See* winter-talus ridge.

niveal [GEOL] Pertaining to features and effects resulting from the action of snow and ice.

nivenite [MINERAL] UO_2 A velvet-black member of the uranite group; contains rare-earth metals cerium and yttrium; a source of uranium.

niveoglacial [GEOL] Pertaining to the combined action of snow and ice.

niveolian [GEOL] Pertaining to simultaneous accumulation and intermixing of snow and airborne sand at the side of a gentle slope.

nivo-karst [GEOL] A karstlike topography characteristic of periglacial areas that results from differential chemical weathering beneath snowbanks.

no-bottom sounding [ENG] A sounding in the ocean in which the bottom is not reached.

nocerite *See* fluoborite.

noctilucent cloud [METEOROL] A cloud of unknown composition which occurs at great heights and high altitudes; photometric measurements have located such clouds between 75 and 90 kilometers; they resemble thin cirrus, but usually with a bluish or silverish color, although sometimes orange to red, standing out against a dark night sky.

nocturnal radiation *See* effective terrestrial radiation.

no-cut rounds [MIN ENG] Set of holes drilled straight into the face for blasting underground.

nodal points [OCEANOGR] The no-tide points occurring in amphidromic regions.

nodal zone [OCEANOGR] An area at which the primary direction of the littoral transport changes.

node [GEOL]   That point along a fault at which the direction of apparent displacement changes.

nodular chert [GEOL]   Chert occurring as nodular or concretionary segregations (chert nodules).

nodule   [GEOL]   A small, hard mass or lump of a mineral or mineral aggregate characterized by a contrasting composition from and a greater hardness than the surrounding sediment or rock matrix in which it is embedded.

Noeggerathiales [PALEOBOT]   A poorly defined group of fossil plants whose geologic range extends from Upper Carboniferous to Triassic.

nog   [MIN ENG]   **1.** Roof support for stopes, formed of rectangular piles of logs squared at the ends and filled with waste rock. **2.** A wood block wedged tightly into the cut in a coal seam after the coal cutter has passed; it forms a temporary support.

nominal diameter   [GEOL]   The diameter computed for a hypothetical sphere which would have the same volume as the calculated volume for a specific sedimentary particle. Also known as equivalent diameter.

nonassociated-gas reservoir [PETRO ENG]   Formation in which gaseous hydrocarbons exist as a free phase in a reservoir that is not commercially productive of crude oil.

nonbanded coal [GEOL]   Coal without lustrous bands, composed mainly of clarain or durain without nitrain.

noncaking coal [GEOL]   Hard or dull coal that does not cake when heated. Also known as free-burning coal.

Noncalcic Brown soil [GEOL]   A great soil group having a slightly acidic, light-pink or reddish-brown A horizon and a light-brown or dull-red B horizon, and developed under a mixture of grass and forest vegetation in a subhumid climate. Also known as Shantung soil.

noncapillary porosity [GEOL]   The property of a volume of large interstices in a rock or soil that do not hold water by capillarity.

nonclastic [GEOL]   Of the texture of a sediment or sedimentary rock, formed chemically or organically and showing no evidence of a derivation from preexisting rock or mechanical deposition. Also known as nonmechanical.

noncohesive *See* cohesionless.

nonconformity [GEOL]   A type of unconformity in which rocks below the surface of unconformity are either igneous or metamorphic.

nonconservative element [OCEANOGR]   An element in sea water which is so uncommon that a large proportion of its total composition enters and leaves the particulate phase.

noncontributing area [HYD]   An area with closed drainage.

noncoring bit [ENG]   A general type of bit made in many shapes which does not produce a core and with which all the rock cut in a borehole is ejected as sludge; used mostly for blasthole drilling and in the unmineralized zones in a borehole where a core sample is not wanted. Also known as borehole bit; plug bit.

noncyclic terrace [GEOL]   One of several terraces representing former valley floors formed during periods when continued valley deepening accompanied lateral erosion.

nondepositional unconformity *See* paraconformity.

nondivergent flow [OCEANOGR]   Fluid flow in which the divergence of the ocean current field is zero.

nonesite [PETR] A porphyritic basalt composed of enstatite, labradorite, and augite phenocrysts in a groundmass of plagiocase and augite.

nonfrontal squall line *See* prefrontal squall line.

nonmechanical *See* nonclastic.

nonpenetrative [GEOL] Of a type of deformation, affecting only part of a rock, such as kink bands.

nonplunging fold [GEOL] A fold with a horizontal axial surface. Also known as horizontal fold; level fold.

nonrecording rain gage [ENG] A rain gage which indicates but does not record the amount of precipitation.

nonrotational strain *See* irrotational strain.

nonsegregated reservoir [PETRO ENG] Solution-gas-drive oil reservoir in which the gas does not separate from the oil as a function of height or upward movement.

nonselective mining [MIN ENG] Mining methods permitting low cost, generally by using a cheap stoping method combined with large-scale operations; can be used in deposits where the individual stringers, bands, or lenses of high-grade ore are numerous and so irregular in occurrence and separated by such thin lenses of waste that a selective method cannot be employed.

nonsorted polygon [GEOL] A form of patterned ground which has a dominantly polygonal mesh and an unsorted appearance due to the absence of border stones, and whose borders are generally marked by wedge-shaped fissures narrowing downward.

nonsymmetrical aquifer [PETRO ENG] In an oil reservoir formation, a water-containing part that is irregular, being neither radial nor linear.

nontectonite [PETR] Any rock whose fabric shows no influence of movement of adjacent grains; for example, a rock formed by mechanical settling.

nontidal current [OCEANOGR] Any current due to causes other than tidal, as a permanent ocean current.

nontronite [MINERAL] $Na(Al,Fe,Si)O_{10}(OH)_2$ An iron-rich clay mineral of the montmorillonite group that represents the end member in which the replacement of aluminum by ferric ion is essentially complete. Also known as chloropal; gramenite; morencite; pinguite.

nonwetting phase [PETRO ENG] The oil phase contained in a reservoir pore structure when the reservoir fluid is two-phase (oil and water) or three-phase (oil, water, and gas).

nonwetting sand [GEOL] Sand that resists infiltration of water; consists of angular particles of various sizes and occurs as a tightly packed lens.

norbergite [MINERAL] $Mg_3SiO_4(F,OH)_2$ A yellow or pink orthorhombic mineral composed of magnesium silicate with fluoride and hydroxyl; it is a member of the humite group.

nordenskioldine [MINERAL] $CaSn(BO_3)_2$ A colorless or sulfur-, lemon-, or wine-yellow, hexagonal mineral consisting of a borate of calcium and tin; occurs in tabular form and as lenslike crystals.

Nordenskjöld line [CLIMATOL] The line connecting all places at which the mean temperature of the warmest month is equal (in degrees Celsius) to $9 - 0.1k$, where k is the mean temperature of the coldest month (in degrees Fahrenheit it becomes $51.4 - 0.1k$).

nordmarkite [PETR]   A quartz-bearing alkalic syenite that has microperthite as its main component with smaller amounts of oligocase, quartz, and biotite, and is characterized by granitic or trachytoid texture.

nordsjoite [PETR]   A juvite that contains abundant nepheline and orthoclase but no microperthite.

nor'easter *See* northeaster.

Norian [GEOL]   A European stage of Upper Triassic geologic time that lies above the Carnian and below the Rhaetian.

norilskite [MINERAL]   A mineral consisting of platinum with high contents of iron and nickel.

norite [PETR]   A coarse-grained plutonic rock composed principally of basic plagioclase with orthopyroxene (hypersthene) as the dominant mafic material. Also known as hypersthenfels.

norm [PETR]   The theoretical mineral composition of a rock expressed in terms of standard mineral molecules as determined by means of chemical analyses.

normal [METEOROL]   The average value of a meteorological element over any fixed period of years that is recognized as standard for the country and element concerned.

normal aeration [GEOL]   The complete renewal of soil air to a depth of 20 centimeters about once each hour.

normal anticlinorium [GEOL]   An anticlinorium in which axial surfaces of the subsidiary folds converge downward.

normal chart [METEOROL]   Any chart that shows the distribution of the official normal values of a meteorological element. Also known as normal map.

normal consolidation [GEOL]   Consolidation of a sedimentary material in equilibrium with overburden pressure.

normal contour *See* accurate contour.

normal cycle [GEOL]   A cycle of erosion whereby a region is reduced to base level by running water, especially by the action of rivers. Also known as fluvial cycle of erosion.

normal dip *See* regional dip.

normal dispersion [GEOPHYS]   The dispersion of seismic waves in which the recorded wave period increases with time.

normal displacement *See* dip slip.

normal erosion [GEOL]   Erosion effected by prevailing agencies of the natural environment, including running water, rain, wind, waves, and organic weathering. Also known as geologic erosion.

normal fault [GEOL]   A fault, usually of 45–90°, in which the hanging wall appears to have shifted downward in relation to the footwall. Also known as gravity fault; normal slip fault; slump fault.

normal fold *See* symmetrical fold.

normal horizontal separation *See* offset.

normal hydrostatic pressure [HYD]   In porous strata or in a well, the pressure at a given point that is approximately equal to the weight of a column of water extending from the surface to that point.

normal-incidence pyrheliometer [ENG] An instrument that measures the energy in the solar beam; it usually measures the radiation that strikes a target at the end of a tube equipped with a shutter and baffles to collimate the beam.

normal map *See* normal chart.

normal-plate anemometer [ENG] A type of pressure-plate anemometer in which the plate, restrained by a stiff spring, is held perpendicular to the wind; the wind-activated motion of the plate is measured electrically; the natural frequency of this system can be made high enough so that resonance magnification does not occur.

normal polarity [GEOPHYS] Natural remanent magnetism nearly identical to the present ambient field.

normal pressure *See* standard pressure.

normal pressure surface [HYD] A potentiometric surface that coincides with the upper surface of the zone of saturation.

normal ripple mark [GEOL] An aqueous current ripple mark consisting of a simple asymmetrical ridge that may have various ground plans.

normal sandstone [PETR] A sandstone composed almost exclusively of quartz and subordinate amounts of other minerals.

normal slip fault *See* normal fault.

normal soil [GEOL] A soil having a profile that is more or less in equilibrium with the environment.

normal synclinorium [GEOL] A synclinorium in which the axial surfaces of the subidiary folds converge upward.

normal water [OCEANOGR] Water whose chlorinity lies between 19.30 and 19.50 parts per thousand and has been determined to within ±0.001 per thousand. Also known as Copenhagen water; standard seawater.

norte [METEOROL] **1.** The winter north wind in Spain. **2.** A strong, cold north-easterly wind which blows in Mexico and on the shores of the Gulf of Mexico, and results from an outbreak of cold air from the north; actually, the Mexican extension of a norther.

north [GEOD] The direction of the north terrestrial pole; the primary reference direction on the earth; the direction indicated by 000° in any system other than relative.

North America [GEOGR] The northern of the two continents of the New World or Western Hemisphere, extending from narrow parts in the tropics to progressively broadened portions in middle latitudes and Arctic polar margins.

North American anticyclone *See* North American high.

North American high [METEOROL] The relatively weak general area of high pressure which, as shown on mean charts of sea-level pressure, covers most of North America during winter. Also known as North American anticyclone.

North Atlantic Current [OCEANOGR] A wide, slow-moving continuation of the Gulf Stream originating in the region east of the Grand Banks of Newfoundland.

North Cape Current [OCEANOGR] A warm current flowing northeastward and east-ward around northern Norway, and curving into the Barents Sea.

Northeast Drift Current [OCEANOGR] A North Atlantic Ocean current flowing northeastward toward the Norwegian Sea, gradually widening and, south of Iceland, branching and continuing as the Irminger Current and the Norwegian Current; it is the northern branch of the North Atlantic Current.

northeaster [METEOROL]   A northeast wind, particularly a strong wind or gale. Also spelled nor'easter.

northeast storm [METEOROL]   A cyclonic storm of the east coast of North America, so called because the winds over the coastal area are from the northeast; they may occur at any time of year but are most frequent and most violent between September and April.

northeast trades [METEOROL]   The trade winds of the Northern Hemisphere.

North Equatorial Current [OCEANOGR]   Westward ocean currents driven by the northeast trade winds blowing over tropical oceans of the Northern Hemisphere. Also known as Equatorial Current.

norther [METEOROL]   A northerly wind.

Northern Hemisphere [GEOGR]   The half of the earth north of the Equator.

north foehn [METEOROL]   A foehn condition sustained by wind flow across the Alps from north to south.

north frigid zone [GEOGR]   That part of the earth north of the Arctic Circle.

north geographic pole *See* North Pole.

north geomagnetic pole *See* north pole.

north magnetic pole *See* north pole.

North Pacific Current [OCEANOGR]   The warm branch of the Kuroshio Extension flowing eastward across the Pacific Ocean.

north pole [GEOPHYS]   The geomagnetic pole in the Northern Hemisphere, at approximately latitude 78.5°N, longitude 69°W. Also known as north geomagnetic pole; north magnetic pole.

North Pole [GEOGR]   The geographic pole located at latitude 90°N in the Northern Hemisphere of the earth; it is the northernmost point of the earth, and the northern extremity of the earth's axis of rotation. Also known as north geographic pole.

northupite [MINERAL] $Na_3MgCl(CO_3)_2$ A white, yellow, gray, or colorless isometric mineral composed of magnesium sodium carbonate; occurs in octahedral crystals.

northwester [METEOROL]   A northwest wind. Also spelled nor'wester.

Norway Current [OCEANOGR]   A continuation of the North Atlantic Current, which flows northward along the coast of Norway. Also known as Norwegian Current.

Norwegian Current *See* Norway Current.

nor'wester *See* northwester.

nose [GEOL]   **1.** A plunging anticline that is short and without closure. **2.** A projecting and generally overhanging buttress of rock. **3.** The projecting end of a hill, spur, ridge, or mountain. **4.** The central forward part of a parabolic dune.

nosean *See* noselite.

noseanite [PETR]   A feldspar-and-olivine-free basalt that contains abundant noselite.

noseanolith [PETR]   An extrusive rock composed almost entirely of noselite.

noselite [MINERAL] $Na_4Al_3Si_3O_{12} \cdot SO_4$ A gray, blue or brown mineral of the sodalite group; similar to haüynite; hardness is 5.5 on Mohs scale. Also known as nosean.

noselitite [PETR]   An extrusive rock composed chiefly of noselite and a pyroxene or amphibole, or both.

nose-out   [GEOL]   A nose-shaped stratum, as seen in outcrop.

notch   [GEOGR]   A narrow passage between mountains or through a ridge, hill, or mountain.

Nothosauria   [PALEON]   A suborder of chiefly marine Triassic reptiles in the order Sauropterygia.

Notiomastodontinae   [PALEON]   A subfamily of extinct elephantoid proboscidean mammals in the family Gomphotheriidae.

Notioprogonia   [PALEON]   A suborder of extinct mammals comprising a diversified archaic stock of Notoungulata.

Notoryctidae   [PALEON]   An extinct family of Australian insectivorous mammals in the order Marsupialia.

Notoungulata   [PALEON]   An extinct order of hoofed herbivorous mammals, characterized by a skull with an expanded temporal region, primitive dentition, and primitive feet with five toes, the weight borne mainly by the third digit.

noumeite *See* garnierite.

nourishment   [GEOL]   The replenishment of a beach, either naturally (such as by littoral transport) or artificially (such as by deposition of dredged materials).   [HYD] *See* accumulation.

novaculite   [GEOL]   A siliceous sedimentary rock that is dense, hard, even-textured, light-colored, and characterized by dominance of microcrystalline quartz over chalcedony. Also known as razor stone.

novaculitic chert   [GEOL]   A gray chert that fragments into slightly rough, splintery pieces.

NRM *See* natural remanent magnetization.

NRM wind scale   [METEOROL]   A wind scale adapted by the United States Forest Service for use in the forested areas of the Northern Rocky Mountains (NRM); it is an adaptation of the Beaufort wind scale; the difference between these two scales lies in the specification of the visual effects of the wind; the force numbers and the corresponding wind speeds are the same in both.

nuclear magnetometer   [ENG]   Any magnetometer which is based on the interaction of a magnetic field with nuclear magnetic moments, such as the proton magnetometer. Also known as nuclear resonance magnetometer.

nuclear resonance magnetometer *See* nuclear magnetometer.

nuclear twin-probe gage *See* profiling snow gage.

nucleus   [HYD]   A particle of any nature upon which, or a locus at which, molecules of water or ice accumulate as a result of a phase change to a more condensed state.

nuée ardente   [GEOL]   A turbulent, rapidly flowing, and sometimes incandescent gaseous cloud erupted from a volcano and containing ash and other pyroclastics in its lower part. Also known as glowing cloud; Pelean cloud.

nugget   [GEOL]   A small mass of metal found free in nature.

numerical forecasting   [METEOROL]   The forecasting of the behavior of atmospheric disturbances by the numerical solution of the governing fundamental equations of hydrodynamics, subject to observed initial conditions. Also known as dynamic forecasting; mathematical forecasting; numerical weather prediction; physical forecasting.

numerical weather prediction *See* numerical forecasting.

nunakol   [GEOL]   A nunatak rounded by glacial erosion. Also known as rognon.

nunatak   [GEOL]   An isolated hill, knob, ridge, or peak of bedrock projecting prominently above the surface of a glacier and completely surrounded by glacial ice.

O

oasis [GEOGR] An isolated fertile area, usually limited in extent and surrounded by desert, and marked by vegetation and a water supply.

oberwind [METEOROL] A night wind from mountains or the upper ends of lakes; a wind of Salzkammergut in Austria.

oblique chart [MAP] A chart on an oblique projection.

oblique cylindrical orthomorphic projection *See* oblique Mercator projection.

oblique equator [MAP] A great circle, the plane of which is perpendicular to the axis of an oblique projection; an oblique equator serves as the origin for measurement of oblique latitude; on an oblique Mercator projection, the oblique equator is the tangent great circle.

oblique fault *See* diagonal fault.

oblique graticule [MAP] A fictitious graticule based upon an oblique projection.

oblique joint *See* diagonal joint.

oblique Mercator projection [MAP] A conformal cylindrical map projection in which points on the surface of a sphere or spheroid, such as the earth, are conceived as developed by Mercator principles on a cylinder tangent along an oblique great circle. Also known as oblique cylindrical orthomorphic projection.

oblique meridian [MAP] A great circle perpendicular to an oblique equator; the reference oblique meridian is called the prime oblique meridian.

oblique parallel [MAP] A circle or line parallel to an oblique equator, connecting all points of equal oblique latitude.

oblique pole [MAP] One of the two points 90° from an oblique equator.

oblique projection [MAP] A map projection with its axis at an oblique angle to the plane of the equator.

oblique rhumb line [MAP] **1.** A line making the same oblique angle with all fictitious meridians of an oblique Mercator projection; oblique parallels and meridians may be considered special cases of the oblique rhumb line. **2.** Any rhumb line, real or fictitious, making an oblique angle with its meridians; in this sense the expression is used to distinguish such rhumb lines from parallels and meridians, real or fictitious, which may be included in the expression "rhumb line."

oblique slip fault [GEOL] A fault which has slippage along both the strike and dip of the fault plane.

Obolellida [PALEON] A small order of Early and Middle Cambrian inarticulate brachiopods, distinguished by a shell of calcium carbonate.

obscuration [METEOROL]   In United States weather observing practice, the designation for the sky cover when the sky is completely hidden by surface-based obscuring phenomena, such as fog. Also known as obscured sky cover.

obscured sky cover *See* obscuration.

obscuring phenomenon [METEOROL]   In United States weather observing practice, any atmospheric phenomenon (not including clouds) which restricts the vertical visibility or slant visibility, that is, which obscures a portion of the sky from the point of observation.

obsequent [GEOL]   Of a stream, valley, or drainage system, being in a direction opposite to that of the original consequent drainage.

obsequent fault-line scarp [GEOL]   A fault-line scarp which faces in the direction opposite to that of the original fault scarp or in which the structurally upthrown block is topographically lower than the downthrown block.

observational day [GEOPHYS]   Any 24-hour period selected as the basis for climatological or hydrological observations.

obsidian [GEOL]   A jet-black volcanic glass, usually of rhyolitic composition, formed by rapid cooling of viscous lava; generally forms the upper parts of lava flows. Also known as hyalopsite; Iceland agate; mountain mahogany.

obsidianite *See* tektite.

obstructed stream [GEOL]   A stream whose valley has been blocked, as by a landslide, glacial moraine, sand dune, or lava flow; frequently consists of a series of ponds or small lakes.

obstruction moraine [GEOL]   A moraine formed where the movement of ice is obstructed, as by a ridge of bedrock.

obstruction to vision [METEOROL]   In United States weather observing practice, one of a class of atmospheric phenomena, other than the weather class of phenomena, which may reduce horizontal visibility at the earth's surface; examples are fog, smoke, and blowing snow.

occluded cyclone [METEOROL]   Any cyclone (or low) within which there has developed an occluded front.

occluded front [METEOROL]   A composite of two fronts, formed as a cold front overtakes a warm front or quasi-stationary front. Also known as frontal occlusion; occlusion.

occluded gases [MIN ENG]   Gases entering the mine atmosphere from feeders and blowers, and also from blasting operations.

occlusion [METEOROL]   *See* occluded front.

occult mineral [MINERAL]   A mineral component of rock which cannot be seen through a microscope, but whose presence can be detected by chemical analyses.

ocean [GEOGR]   A major primary subdivision of the intercommunicating body of salt water occupying the depressions of the earth's surface; bounded by continents and imaginary lines. Also known as sea.

ocean basin [GEOL]   The great depression occupied by the ocean on the surface of the lithosphere.

ocean circulation [OCEANOGR]   **1.** Water current flow in a closed circular pattern within an ocean. **2.** Large-scale horizontal water motion within an ocean.

ocean current [OCEANOGR]   A net transport of ocean water along a definable path.

ocean engineering [ENG] The branch of engineering concerned with the development of new equipment and the improvement of techniques to permit humans to operate beneath the ocean surface in order to exploit relevant resources.

ocean floor [GEOL] The near-horizontal surface of the ocean basin.

ocean floor spreading *See* sea floor spreading.

oceanic anticyclone *See* subtropical high.

oceanic basalt [PETR] Rocks of the oceanic island volcanoes.

oceanic climate *See* marine climate.

oceanic crust [GEOL] A thick mass of igneous rock which lies under the ocean floor.

oceanic delta [GEOL] A delta built into a tide-influenced sea and characterized by a concave delta plain.

oceanic heat flow [GEOPHYS] The amount of thermal energy escaping from the earth through the ocean floor per unit area and unit time.

oceanic high *See* subtropical high.

oceanic island [GEOL] Any island which rises from the deep-sea floor rather than from shallow continental shelves.

oceanicity [CLIMATOL] The degree to which a point on the earth's surface is in all respects subject to the influence of the sea; it is the opposite of continentality; oceanicity usually refers to climate and its effects; one measure for this characteristic is the ratio of the frequencies of maritime to continental types of air mass. Also known as oceanity.

oceanic ridge *See* mid-oceanic ridge.

oceanic rise [GEOL] A long, broad elevation of the bottom of the ocean.

oceanic stratosphere *See* cold-water sphere.

oceanic zone [OCEANOGR] The biogeographic area of the open sea.

oceanite [PETR] A picritic basalt in which olivine is a great deal more abundant than plagioclase.

oceanity *See* oceanicity.

oceanization [GEOL] Process by which continental crust (sial) is converted into oceanic crust (sima).

oceanographic dredge [ENG] An apparatus used aboard ship to bring up large samples of deposits and sediments from the ocean bottom.

oceanographic equator [OCEANOGR] **1.** The region of maximum temperature of the ocean surface. **2.** The region in which the temperature of the ocean surface is greater than 28°C.

oceanographic model [OCEANOGR] A theoretical representation of the marine environment which relates physical, chemical, geological, biological, and other oceanographic properties.

oceanographic platform [ENG] A man-made structure with a flat horizontal surface higher than the water, on which oceanographic equipment is suspended or installed.

oceanographic station [OCEANOGR] A geographic location at which oceanographic observations are taken from a stationary ship.

oceanographic survey [OCEANOGR] A study of oceanographic conditions with reference to physical, chemical, biological, geological, and other properties of the ocean.

oceanography [GEOPHYS] The scientific study and exploration of the oceans and seas in all their aspects. Also known as oceanology.

oceanology *See* oceanography.

ocean weather station [METEOROL] As defined by the World Meteorological Organization, a specific maritime location occupied by a ship equipped and staffed to observe weather and sea conditions and report the observations by international exchange.

ocellar [PETR] Of the texture of an igneous rock, having crystalline aggregates of phenocrysts arranged radially or tangentially around larger euhedral crystals or which form rounded branching forms.

ocellus [PETR] A phenocryst in an ocellar rock.

ocher [MINERAL] A yellow, brown, or red earthy iron oxide, or any similar earthy, pulverulent metallic oxides used as pigments.

Ochoan [GEOL] A North American provincial series that is the uppermost in the Permian, lying above the Guadalupian and below the lower Triassic.

Ochrept [GEOL] A suborder of the soil order Inceptisol with horizon below the surface, lacking clay, sesquioxides, or humus; widely distributed, occurring from the margins of the tundra region through the temperate zone, but not in the tropics.

Ocoee [GEOL] A provincial series of the Precambrian that is found in Virginia, Tennessee, North Carolina, and Georgia.

octahedral borax *See* magnetite.

octahedral cleavage [CRYSTAL] Crystal cleavage in the four planes parallel to the face of the octahedron.

octahedral coordination [MINERAL] An atomic structure where six cations surround every anion, and vice versa.

octahedral copper ore *See* cuprite.

octahedral iron ore *See* magnetite.

octahedral plane [CRYSTAL] The plane in a cubic lattice having three numerically equal Miller indices.

octahedrite [GEOL] The most common iron meteorite, containing 6–18% nickel in the metal phase and having intimate intergrowths lying parallel to the octahedral planes. [MINERAL] *See* anatase.

octaphyllite [MINERAL] **1.** A group of mica minerals that contain eight cations per ten oxygen atoms and two hydroxyl ions. **2.** Any mineral of this group, such as biotite.

Odontognathae [PALEON] An extinct superorder of the avian subclass Neornithes, including all large, flightless aquatic forms and other members of the single order Hesperornithiformes.

Oehman's survey instrument [ENG] A drill-hole surveying apparatus that makes a photographic record of the compass and clinometer readings.

Oepikellacea [PALEON] A dimorphic superfamily of extinct ostracods in the order Paleocopa, distinguished by convex valves and the absence of any trace of a major sulcus in the external configuration.

Oetling freezing method [MIN ENG] A method of shaft sinking by freezing the wet ground in sections as the sinking proceeds.

off-highway truck [MIN ENG] A truck of such size, weight, or dimensions that is cannot be used on public highways.

officials' inspection lamp [MIN ENG] A portable combined electric lamp and battery, fitted with a reflector to provide directional illumination.

offlap [GEOL] The successive lateral contraction extent of strata (in an upward sequence) due to their deposition in a shrinking sea or on the margin of a rising landmass. Also known as regressive overlap.

off-line [ENG] **1.** A condition existing when the drive rod of the drill swivel head is not centered and parallel with the borehole being drilled. **2.** A borehole that has deviated from its intended course. **3.** A condition existing wherein any linear excavation (shaft, drift, borehole) deviates from a previously determined or intended survey line or course.

off-reef [GEOL] Pertaining to the seaward margin or zone of a reef.

off-reef facies [GEOL] Facies of the inclined strata made up of reef detritus deposited along the seaward margin of a reef.

offset [GEOL] **1.** The movement of an upcurrent part of a shore to a more seaward position than a downcurrent part. **2.** A spur from a mountain range. **3.** A level terrace on the side of a hill. **4.** The horizontal displacement component in a fault, measured parallel to the stroke of the fault. Also known as normal horizontal separation. [MAP] During construction of a map projection, the small distance added to the length of meridians on either side of the central meridian in order to determine the chart's latitude.   [MIN ENG] **1.** A short drift or crosscut driven from a main gangway or level. **2.** The horizontal distance between the outcrops of a dislocated bed.

offset deposit [GEOL] A mineral deposit, especially of sulfides, formed partly by magmatic segregation and partly by hydrothermal solution, near the source rock.

offset drilling [PETRO ENG] The drilling of a well on property under which oil is being drained away by a well on adjacent property to make up for the loss of oil from the first property.

offset ridge [GEOL] A ridge consisting of resistant sedimentary rock that has been made discontinuous as a result of faulting.

offset stream [GEOL] A stream displaced laterally or vertically by faulting.

offset well [PETRO ENG] An oil well drilled near the boundary of a property and opposite to a producing or completed well on an adjoining property, for the purpose of preventing the drainage of oil or gas to the earlier productive well.

offshore [GEOL] The comparatively flat zone of variable width extending from the outer margin of the shoreface to the edge of the continental shelf.

offshore bar *See* longshore bar.

offshore beach *See* barrier beach.

offshore current [OCEANOGR] **1.** A prevailing nontidal current usually setting parallel to the shore outside the surf zone. **2.** Any current flowing away from shore.

offshore drilling [PETRO ENG] The drilling of oil or gas wells into water-covered locations, usually on submerged continental shelves.

offshore gas [PETRO ENG] Nautral gas produced from reservoirs under the offshore continental shelves.

offshore oil [PETRO ENG] Oil produced from reservoirs under the offshore continental shelves.

offshore slope [GEOL] The frontal slope below the outer edge of an offshore terrace.

offshore survey [PETRO ENG] Seismic geophysical survey procedures conducted over water-covered continental-shelf areas in the search for possible oil reservoirs.

offshore terrace   [GEOL]   A wave-built terrce in the offshore zone composed of gravel and coarse sand.

offshore water   [OCEANOGR]   Water adjacent to land in which the physical properties are slightly influenced by continental conditions.

offshore wind   [METEOROL]   Wind blowing from the land toward the sea.

off-the-road equipment   [MIN ENG]   Tires and earthmoving equipment designed for off-highway duty in surface mines and quarries.

off-the-road hauling   [MIN ENG]   Hauling off the public highways, and generally on the mining site or excavation site.

ogive   [GEOL]   One of a periodically repeated series of dark, curved structures occurring down a glacier that resemble a pointed arch.

Ohio sampler   [MIN ENG]   A single tube or pipe with a thread on top, and the bottom beveled and hardened for driving into the ground to obtain a soil sample.

oikocryst   [PETR]   One of the enclosing crystals in a poikilitic fabric.

oil   *See* petroleum.

oil accumulation   *See* oil pool.

oil-base mud   [PETRO ENG]   Drilling mud made with oil as the solvent carrier for the solids content.

oil derrick   [PETRO ENG]   Tower structure used during oil well drilling to aid in raising and lowering of drill and piping strings.

oil field   [PETRO ENG]   The surface boundaries of an area from which petroleum is obtained; may correspond to an oil pool or may be circumscribed by political or legal limits.

oil field brine   [HYD]   Connate waters, usually containing a high concentration of calcium and sodium salts and found during deep rock penetration by the drill.

oil field emulsion   [PETRO ENG]   A crude oil that reaches the surface as an oil-water emulsion.

oil field model   [PETRO ENG]   Laboratory simulation of steady-state fluid flow through porous reservoir media by electrical (Ohm's-law system), electronic (graphite-impregnated cloth), or electrolytic (gelatin, blotter, potentiometric-liquid) models.

oil field separator   *See* gas-oil separator.

oil-gas separator   *See* gas-oil separator.

oil gas tar   [MATER]   A type of tar produced during the oil gas process by cracking the oil vapors at high temperatures.

oil in place   [PETRO ENG]   The total volume of oil estimated to be present in an oil reservoir.

oil isoperms   [PETRO ENG]   Reservoir-map plotting areas of equal oil permeability.

oil pool   [GEOL]   An accumulation of petroleum locally confined by subsurface geologic features. Also known as oil accumulation; oil reservoir.

oil reclaiming   [ENG]   **1.** A process in which oil is passed distilled from cloves; thickens and darkens with time; boils at reuse, in the same manner that crank case oil is cleaned by an engine filter. **2.** A method in which solids are removed from oil by treatment in settling tanks.

oil reservoir   *See* oil pool.

oil-reservoir water   *See* formation water.

oil rock [GEOL] A rock stratum containing oil.

oil sand [GEOL] An unconsolidated, porous sand formation or sandstone containing or impregnated with petroleum or hydrocarbons.

oil saturation [PETRO ENG] Measurement of the degree of saturation of reservoir pore structure by reservoir oil.

oil seep [GEOL] The emergence of liquid petroleum at the land surface as a result of the slow migration from its buried source through minute pores or fissure networks. Also known as petroleum seep.

oil separator *See* gas-oil separator.

oil shale [GEOL] A finely layered brown or black shale that contains kerogen and from which liquid or gaseous hydrocarbons can be distilled. Also known as kerogen shale.

oil trap [GEOL] An accumulation of petroleum which, by a combination of physical conditions, is prevented from escaping laterally or vertically. Also known as trap.

Oiluvium *See* Pleistocene.

oil-water contact *See* oil-water surface.

oil-water interface *See* oil-water surface.

oil-water surface [GEOL] The datum of a two-dimensional oil-water interface. Also known as oil-water contact; oil-water interface.

oil well [PETRO ENG] A hole drilled (usually vertically) into an oil reservoir for the purpose of recovering the oil trapped in porous formations.

oil-well drive *See* reservoir drive mechanism.

oil-well pump [PETRO ENG] Device for artificial or secondary (non-gas-lift) oil production; about 85% of the production is by ground-level sucker-rod pumps, the remainder by downhole hydraulic lift pumps.

oil window [GEOL] In a basin containing hydrocarbons, an oil-prone zone that lies between two gas-prone zones.

okaite [PETR] An ultramafic igneous rock composed chiefly of melilite and haüyne, with accessory biotite, perovskite, apatite, calcite, and opaque oxides.

okenite [MINERAL] $CaSi_2O_4(OH)_2 \cdot H_2O$ A whitish mineral consisting of calcium silicate and occurring in fibrous masses.

old age [GEOL] The last stage of the erosion cycle in the development of the topography of a region in which erosion has reduced the surface almost to base level and the land forms are marked by simplicity of form and subdued relief. Also known as topographic old age.

old-from-birth peneplain [GEOL] A peneplain that was formed over a long period of time during an uplift of such extreme slowness that vertical corrasion was outpaced by valley-side grading and by general downwearing of the interstream uplands, resulting in a landscape which was old from the beginning, or which lacked any features characteristic of youth or maturity.

Oldhaminidina [PALEON] A suborder of extinct articulate brachiopods in the order Strophomenida distinguished by a highly lobate brachial valve seated within an irregular convex pedicle valve.

oldhamite [MINERAL] CaS A pale-brown mineral known only from meteorites; unstable under earth conditions; member of the galena group with face-centered isometric structure.

Oldham-Wheat lamp [MIN ENG] A cap lamp designed for full self-service.

old ice [OCEANOGR] Floating sea ice that is more than 2 years old.

old lake [GEOL] **1.** A lake in an advanced stage of filling by sediments. **2.** An eutrophic or dystrophic lake. **3.** A lake whose shoreline exhibits an advanced stage of development.

oldland [GEOL] **1.** An extensive area (such as the Canadian Shield) of ancient crystalline rocks which were reduced to low relief by long, continuous erosion from which the materials of later sedimentary rocks were derived. **2.** A region of older land, projected above sea level behind a coastal plain, that supplied the material of which the coastal-plain strata were formed.

old mountain [GEOL] A mountain that was formed before the beginning of the Tertiary Period.

Old Red Sandstone [GEOL] A Devonian formation in Great Britain and northwestern Europe, of nonmarine, predominantly red sedimentary rocks, consisting principally of sandstone, conglomerates, and shales.

old snow [HYD] Deposited snow in which the original crystalline forms are no longer recognizable, such as firn or spring snow. Also known as firn snow.

old stream [GEOL] A stream developed during the stage of old age.

old wives' summer [METEOROL] A period of calm, clear weather, with cold nights and misty mornings but fine warm days, which sets in over central Europe toward the end of September; comparable to Indian summer.

old workings [MIN ENG] Mines which have been abandoned, allowed to collapse, and sometimes sealed off.

Olenellidae [PALEON] A family of extinct arthropods in the class Trilobita.

Oligocene [GEOL] The third of the five major worldwide divisions (epochs) of the Tertiary period (Cenozoic era), extending from the end of the Eocene to the beginning of the Miocene.

oligoclase [MINERAL] A plagioclase feldspar mineral with a composition ranging from $Ab_{90}An_{10}$ to $Ab_{70}An_{30}$, where $Ab = NaAlSi_3O_8$ and $An = CaAl_2O_8$.

oligoclasite [PETR] A granular plutonic rock composed almost entirely of oligoclase. Also known as oligosite.

oligomictic [HYD] Pertaining to a lake that circulates only at rare, irregular intervals during abnormal cold spells. [PETR] Of a clastic sedimentary rock, composed of a single rock type.

oligonite [MINERAL] A variety of siderite containing up to 40% manganese carbonate.

oligopelic [GEOL] Pertaining to a lake bottom deposit which contains very little clay.

oligophyre [PETR] A light-colored diorite containing oligoclase phenocrysts in a groundmass of the same minerals.

Oligopygidae [PALEON] An extinct family of exocyclic Euchinoidia in the order Holectypoida which were small ovoid forms of the early Tertiary.

oligosiderite [GEOL] A meteorite containing only a small amount of metallic iron.

oligosite *See* oligoclasite.

oligotrophic [HYD] Of a lake, lacking plant nutrients and usually containing plentiful amounts of dissolved oxygen without marked stratification.

olistoglyph [GEOL] A hieroglyph produced by sliding or interlaminar gliding.

olistolith [GEOL] An exotic block or other rock mass that has been transported by submarine gravity sliding or slumping and is included in the binder of an olistostrome.

olistostrome [GEOL]   A sedimentary deposit composed of a chaotic mass of heterogeneous material that is intimately mixed; accumulated in the form of a semifluid body by submarine gravity sliding or slumping of unconsolidated sediments.

oliveiraite [MINERAL] $Zr_3Ti_2O_{10}\cdot2H_2O$ An isotropic mineral consisting of an oxide of titanium and zirconium.

olivenite [MINERAL] $Cu_2(AsO_4)(OH)$ An olive-green, dull-brown, gray, or yellow mineral crystallizing in the orthorhombic system and consisting of a basic arsenate of copper.   Also known as leucochalcite; wood copper.

Oliver filter [MIN ENG]   A continuous-type filter made in the form of a cylindrical drum with filter cloth stretched over the convex surface of the drum.

olivine [MINERAL] $(Mg,Fe_2)SiO_4$ A neosilicate group of olive-green magnesium-iron silicate minerals crystallizing in the orthorhombic system and having a vitreous luster; hardness is 6½–7 on Mohs scale; specific gravity is 3.27–3.37.

olivine basalt [PETR]   Any of a group of olivine-bearing basalts.

olivine diabase [PETR]   An igneous rock composed principally of olivine and formed from tholeiitic magmas by differentiation in thick sills.

olivine nephelinite [PETR]   An extrusive igneous rock differing in composition from nephelinite only by the presence of olivine.   Also known as ankaratrite; nepheline basalt.

olivinoid [MINERAL]   An olivinelike substance found in meteorites.

ollenite [PETR]   A type of hornblende schist characterized by abundant epidote, sphene, and rutile.

ombrogenous [GEOL]   Pertaining to a peat deposit whose moisture content depends upon rainfall.

ombrometer *See* rain gage.

ombroscope [ENG]   An instrument consisting of a heated, water-sensitive surface which indicates by mechanical or electrical techniques the occurrence of precipitation; the output of the instrument may be arranged to trip an alarm or to record on a time chart.

ombrotiphic [HYD]   Pertaining to a short-lived pond whose water is derived from rainfall.

omission [GEOL]   The elimination or nonexposure of certain stratigraphic beds at the surface of any specified section because of disruption and displacement of the beds by faulting.

omission solid solution [CRYSTAL]   A crystal with certain atomic sites incompletely filled.

omnimeter [ENG]   A theodolite with a microscope that can be used to observe vertical angular movement of the telescope.

omphacite [MINERAL]   A grassy- to pale-green, granular or foliated, high-temperature aluminous clinopyroxene mineral with a vitreous luster that commonly occurs in the rock eclogite; a variety of augite.

oncolite [GEOL]   A small, variously shaped (often spheroidal), concentrically laminated, calcareous sedimentary structure, resembling an oolith, and formed by accretion of successive layered masses of gelatinous sheaths of blue-green algae.

one-dimensional lattice [CRYSTAL]   A simplified model of a crystal lattice consisting of particles lying along a straight line at either equal or periodically repeating distances.

onegite [PETR]   A pale amethyst-colored sagenitic quartz penetrated by needles of goethite.

one-piece set [MIN ENG]   A single stick of timber used as a post, stull, or prop.

Onesquethawan [GEOL]   A North American stage in the Lower and Middle Devonian, lying above the Deerparkian and below the Cazenovian.

one-year ice [OCEANOGR]   Sea ice formed the previous season, not yet 1 year old.

onion-skin weathering [GEOL]   A type of spheroidal weathering in which the successive shells of decayed rock so produced resemble the layers of an onion.

onkilonite [PETR]   A nepheline-leucite basalt that also contains olivine, augite, and perovskite, but no feldspar.

onlap [GEOL]   A type of overlap characterized by regular and progressive pinching out of the strata toward the margins of a depositional basin; each unit transgresses and extends beyond the point of reference of the underlying unit. Also known as transgressive overlap.

on-line analyzer [MIN ENG]   An instrument which monitors the content of materials at various stages in flotation or other mineral-processing flow sheets.

onofrite [MINERAL]   A variety of metacinnabar that contains some selenium, in addition to mercury and sulfur.

onsetter [MIN ENG]   The worker in charge of hoisting coal, ore, men, or materials in a mine shaft.

onshore [GEOGR]   Pertaining to, in the direction toward, or located on the shore. Also known as shoreside.

onshore wind [METEOROL]   Wind blowing from the sea toward the land.

on the run [MIN ENG]   Manner of working a seam of coal when there is sufficient inclination to cause the coal, as worked toward the rise, to fall by gravity to the gangways for loading into cars.

on the solid [MIN ENG]   **1.** Pertaining to the practice of blasting heavy charges of explosives, in lieu of undercutting or channeling. **2.** *See* into the solid.

Onychodontidae [PALEON]   A family of Lower Devonian lobefin fishes in the order Osteolepiformes.

onyx [MINERAL]   **1.** Banded chalcedonic quartz, in which the bands are straight and parallel; natural colors are usually red or brown with white, although black is occasionally encountered. **2.** *See* onyx marble.

onyx agate [MINERAL]   A banded agate with straight, parallel, alternating bands of white and different tones of gray.

onyx marble [MINERAL]   A hard, compact, dense, generally translucent variety of calcite resembling true onyx and usually banded. Also known as alabaster; Algerian onyx; Gibraltar stone; Mexican onyx; onyx; oriental alabaster.

onyx opal [MINERAL]   Common opal with straight, parallel markings.

oolicastic porosity [PETR]   The porosity produced in an oolitic rock by removal of the ooids and formation of oolicasts.

oolite [PETR]   A sedimentary rock, usually a limestone, composed principally of cemented ooliths. Also known as eggstone; roestone.

oolith [PETR]   A small (0.25–2.0 millimeters), rounded accretionary body in a sedimentary rock; generally formed of calcium carbonate by inorganic precipitation or by replacement; ooliths generally exhibit concentric or radial internal structure.

oolitic chert [PETR] Chert composed chiefly of ooliths.

oolitic limestone [PETR] An even-textured limestone made up almost entirely of calcareous ooliths with essentially no matrix.

oomicrite [PETR] A limestone which contains at least 25% ooliths and no more than 25% intraclasts and in which the carbonate-mud matrix (micrite) is more abundant than the sparry-calcite cements.

oomicrudite [PETR] An oomicrite containing ooliths that are more than 1 millimeter in diameter.

oophasmic [PETR] Pertaining to a dolomite or recrystallized limestone which contains vague but unmistakable traces of oolitic texture.

oospararenite [PETR] An oosparite containing sand-sized (medium sand or coarse sand) ooliths.

oosparite [PETR] A limestone which contains at least 25% ooliths and no more than 25% intraclasts and in which the sparry-calcite cement is more abundant than the carbonate-mud matrix.

oosparrudite [PETR] An oosparite containing ooliths that are more than 1 millimeter in diameter.

oovoid [PETR] A void in the center of an incompletely replaced oolith.

ooze [GEOL] **1.** A soft, muddy piece of ground, such as a bog, usually resulting from the flow of a spring or brook. **2.** A marine pelagic sediment composed of at least 30% skeletal remains of pelagic organisms, the rest being clay minerals. **3.** Soft mud or slime, typically covering the bottom of a lake or river.

opacite [PETR] Swarms of opaque, microscopic grains in rocks, particularly in the groundmass of an igneous rock.

opacus [METEOROL] A variety of cloud (sheet, layer, or patch), the greater part of which is sufficiently dense to obscure the sun; found in the genera altocumulus, altostratus, stratocumulus, and stratus; cumulus and cumulonimbus clouds are inherently opaque.

opal [MINERAL] A natural hydrated form of silica; it is amorphous, usually occurs in botryoidal or stalactic masses, has a hardness of 5–6 on Mohs scale, and specific gravity is 1.9–2.2.

opal agate [PETR] A variety of banded opal that displays different shades of color, is agatelike in structure, and consists of alternating layers of opal and chalcedony.

opal-CT [PETR] A poorly ordered crystalline form of silica thought to be the intermediate phase in quartz chert formation.

opalized wood *See* silicified wood.

opaque attritus [GEOL] Attritus that does not contain large quantities of transparent humic degradation matter.

opaque sky cover [METEOROL] In United States weather observing practice, the amount (in tenths) of sky cover that completely hides all that might be above it; opposed to transparent sky cover.

opdalite [PETR] A type of granodiorite containing hypersthene and biotite.

open bay [GEOGR] An indentation between two capes or headlands, so broad and open that waves coming directly into it are nearly as high near its center as on adjacent parts of the open sea.

open-cast mining *See* open-pit mining.

open coast [GEOGR] A coast that is not sheltered from the sea.

opencut [[MIN ENG] **1.** To drive headings out, or to commence working in the coal after sinking the shafts. **2.** To commence longwall working. **3.** To increase the size of a shaft when it intersects a drift.

opencut mining *See* open-pit mining.

open-end method [MIN ENG] A technique of mining pillars in which no stump is left.

open fault [GEOL] A fault, or section of a fault, whose two walls have become separated along the fault surface.

open fire [MIN ENG] A fire at a roadway or at the coal face in a mine.

open-flow potential [PETRO ENG] Gas flow rate in thousands of cubic feet of gas per 24 hours that would be produced by a well if the only pressure against the face of the producing formation wellbore were atmospheric pressure.

open-flow test [PETRO ENG] The flowing of wells wide open to the atmosphere with simultaneous measurement of gas-flow rate and pressure drop; used to analyze the potential of a reservoir to deliver gas to the wellbore.

open-flow well [PETRO ENG] Gas well flowing open to the atmosphere.

open fold [GEOL] A fold having only moderately compressed limbs.

open form [CRYSTAL] A crystal form in which the crystal faces do not entirely enclose a space.

open harbor [GEOGR] An unsheltered harbor exposed to the sea.

open ice [OCEANOGR] On navigable waters, ice that has broken apart sufficiently to permit passage of vessels.

opening [GEOGR] A break in a coastline or a passage between shoals, and so forth. [MIN ENG] **1.** A widening of a crevice, in consequence of a softening or decomposition of the adjacent rock, so as to leave a vacant space. **2.** A short heading driven between two or more parallel headings or levels for ventilation. **3.** An area in a coal mine between pillars, or between pillars and ribs.   [OCEANOGR] Any break in sea ice which reveals the water.

open lake [HYD] **1.** A lake that has a stream flowing out of it. **2.** A lake whose water is free of ice or emergent vegetation.

open pack ice [OCEANOGR] Floes of sea ice that are seldom in contact with each other, generally covering between four-tenths and six-tenths of the sea surface.

open-pit mining [MIN ENG] Extracting metal ores and minerals that lie near the surface by removing the overlying material and breaking and loading the ore. Also known as open-cast mining; opencut mining.

open-pit quarry [MIN ENG] A quarry in which the opening is the full size of the excavation.

open rock [GEOL] Any stratum sufficiently open or porous to contain a significant amount of water or to convey it along its bed.

open sand [GEOL] A formation of sandstone that has porosity and permeability sufficient to provide good storage for oil.

open sea [GEOGR] **1.** That part of the ocean not enclosed by headlands, not within narrow straits, and so on. **2.** That part of the ocean outside the territorial jurisdiction of any country.

open stope [MIN ENG] Underground working place that is unsupported, or supported by timbers or pillars of rock.

open-stope method [MIN ENG] Stoping in which no regular artificial method of support is employed, although occasional props or cribs may be used to hold local patches of insecure ground.

open structure [GEOL] A structure which, when represented on a map by contour lines, is not surrounded by closed contours.

open system [HYD] A condition of freezing of the ground in which additional groundwater is available either through free percolation or through capillary movement.

open timbering [MIN ENG] A method of supporting the ground in a mine shaft or tunnel; supports are several feet apart, with the ground between them secured by struts.

open traverse [ENG] A surveying traverse in which the last leg, because of error, does not terminate at the origin of the first leg.

open water [ECOL] Lake water that is free from emergent vegetation, artificial obstruction, or tangled masses of underwater vegetation at very shallow depths. [HYD] Lake water that does not freeze during the winter. [OCEANOGR] Water less than one-tenth covered with floating ice.

open workings [MIN ENG] Surface workings, for example, a quarry or open-cast mine.

operational weather limits [METEOROL] The limiting values of ceiling, visibility, and wind, or runway visual range, established as safety minima for aircraft landings and takeoffs.

ophicalcite [PETR] A recrystallized limestone composed of calcite and serpentine, formed by dedolomitization of a siliceous dolomite.

Ophiocistioidea [PALEON] A small class of extinct Echinozoa in which the domed aboral surface of the test was roofed by polygonal plates and carried an anal pyramid.

ophiolite [PETR] A group of mafic and ultramafic igneous rocks, including spilite, basalt, gabbro, peridotite, and their metamorphic alternation products such as serpentine.

ophiolitic eclogite [PETR] Any of the eclogites which are products of early orogenic volcanism and which by later metamorphism transformed into rocks of the high-pressure facies series.

ophite [PETR] A diabase in which the ophitic structure is retained even though the pyroxene is altered to uralite.

ophitic [PETR] Of the holocrystalline, hypidiomorphic-granular texture of an igneous rock, exhibiting lath-shaped plagioclase crystals partly or wholly included within pyroxene crystals.

opisometer [ENG] An instrument for measuring the length of curved lines, such as those on a map; a wheel on the instrument is traced over the line.

opoka [PETR] A porous, flinty and calcareous sedimentary rock, with conchoidal or irregular fracture, consisting of fine-grained opaline silica (up to 90%), and hardened by the presence of silica of organic origin.

opposing wind [OCEANOGR] In wave forecasting, a wind blowing in the direction opposite to that in which the waves are progressing.

opposite tide [OCEANOGR] The high tide at a corresponding place on the opposite side of the earth accompanying a direct tide.

opthalmite [PETR] Chorismite characterized by augen or other lenticular aggregates of minerals.

optical air mass   [GEOPHYS]   A measure of the length of the path through the atmosphere to sea level traversed by light rays from a celestial body, expressed as a multiple of the path length for a light source at the zenith.

optical calcite   [MINERAL]   The type of calcite used to make Nicol prisms.

optical depth   *See* optical thickness.

optical horizon   [GEOD]   Locus of points at which a straight line from the given point becomes tangential to the earth's surface.

optically effective atmosphere   [GEOPHYS]   That portion of the atmosphere lying below the altitude (50–60 kilometers) from which scattered light at twilight still reaches the observer with sufficient intensity to be discerned. Also known as effective atmosphere.

optical oceanography   [OCEANOGR]   That aspect of physical oceanography which deals with the optical properties of sea water and natural light in sea water.

optical rangefinder   [ENG]   An optical instrument for measuring distance, usually from its position to a target point, by measuring the angle between rays of light from the target, which enter the rangefinder through the windows spaced apart, the distance between the windows being termed the baselength of the rangefinder; the two types are coincidence and stereoscopic.

optical square   [ENG]   A surveyor's hand instrument used for laying of right angles; employs two mirrors at a 45° angle.

optical thickness   [METEOROL]   **1.** In calculations of the transfer of radiant energy, the mass of a given absorbing or emitting material lying in a vertical column of unit cross-sectional area and extending between two specified levels. Also known as optical depth. **2.** Subjectively, the degree to which a cloud prevents light from passing through it; depends upon the physical constitution (crystals, drops, droplets), the form, the concentration of particles, and the vertical extent of the cloud.

optical twinning   [CRYSTAL]   Growing together of two crystals which are the same except that the structure of one is the mirror image of the structure of the other. Also known as chiral twinning.

optic angle   *See* axial angle.

optic-axial angle   *See* axial angle.

ora   [METEOROL]   A regular valley wind at Lake Garda in Italy.

orange sapphire   [MINERAL]   An orange variety of gem corundum (sapphire). Also known as padparadsha.

orangite   [MINERAL]   A bright orange-yellow variety of thorite.

oranite   [PETR]   A lamellar intergrowth of a potassium feldspar and a plagiocase near anorthite.

orbicular   [PETR]   Of the structure of a rock, containing large quantities of orbicules.

orbicule   [GEOL]   A nearly spherical body, up to 2 or more centimeters in diameter, in which the components are arranged in concentric layers.

orbit   [OCEANOGR]   The path of a water particle affected by wave motion, being almost circular in deep-water waves and almost elliptical in shallow-water waves.

orbital current   [OCEANOGR]   The flow of water that accompanies the orbital movement of the water particles in a wave.

orbite   [PETR]   An igneous rock containing large phenocrysts of hornblende, or plagioclase and hornblende, in a groundmass having the composition of malachite.

ordanchite [PETR]   An extrusive rock containing phenocrysts of sodic plagioclase, haüyne, hornblende, augite, and some olivine.

ordinary chert [PETR]   A generally homogeneous smooth chert with an even fracture surface, approaching opacity, and having slight granularity or crystallinity.

ordinary cut [MIN ENG]   The arrangement of drill holes in a mine in which the drill holes are symmetrical with respect to the vertical center-line of the section, extend horizontally, and make a large angle with the working face.

ordinary tides [OCEANOGR]   Waves with periods of 12 to 24 hours.

ordinary-wave component [GEOPHYS]   One of the two components into which an electromagnetic wave entering the ionosphere is divided under the influence of the earth's magnetic field; it has characteristics more nearly like those expected in the absence of a magnetic field. Also known as O-wave component.

ordosite [PETR]   A dark-colored acmite syenite.

Ordovician [GEOL]   The second period of the Paleozoic era, above the Cambrian and below the Silurian, from approximately 500 million to 440 million years ago.

ore [GEOL]   **1.** The naturally occurring material from which economically valuable minerals can be extracted. **2.** Specifically, a natural mineral compound of the elements, of which one element at least is a metal. **3.** More loosely, all metalliferous rock, though it contains the metal in a free state. **4.** Occasionally, a compound of nonmetallic substances, as sulfur ore.

ore bed [GEOL]   An economic aggregation of minerals occurring between or in rocks of sedimentary origin.

ore bin [MIN ENG]   A receptacle for ore awaiting treatment or shipment.

ore block [MIN ENG]   A vein of ore which is bound above, below, and at one or both ends; it is ready for excavation.

ore blocked out [MIN ENG]   Ore exposed on three sides within a reasonable distance of each other.

orebody [GEOL]   Generally, a solid and fairly continuous mass of ore, which may include low-grade ore and waste as well as pay ore, but is individualized by form or character from adjoining country rock.

ore bridge [MIN ENG]   A gantry crane used to load and unload stockpiles of ore.

ore car [MIN ENG]   A mine car for carrying ore or waste rock.

ore chimney *See* pipe.

ore chute [MIN ENG]   An inclined passage for the transfer of ore to a lower level.

ore cluster [GEOL]   A group of interconnected ore bodies.

ore control [GEOL]   A geologic feature that has influenced the ore deposition.

ore crusher [MIN ENG]   A machine for breaking up masses of ore, usually previous to passing through other size-reduction equipment.

ore deposit [GEOL]   Rocks containing minerals of economic value in such amount that they can be profitably exploited.

ore developed [MIN ENG]   Ore exposed on four sides in blocks variously prescribed.

ore district [GEOL]   A combination of several ore deposits into one common whole or system.

ore dressing [MIN ENG]   The cleaning of ore by the removal of certain valueless portions, as by jigging, cobbing, or vanning.

ore expectant [MIN ENG]   The whole or any part of the ore below the lowest level or beyond the range of vision.

ore faces [MIN ENG]   Those ore bodies that are exposed on one side, or show only one face.

ore grader [MIN ENG]   In metal mining, a person who directs the storage of iron ores in bins at shipping docks so that the various grades in each bin will contain approximate percentages of iron.

ore in sight [MIN ENG]   **1.** Ore exposed on at least three sides within reasonable distance of each other. **2.** Ore which may be reasonably assumed to exist, though not actually blocked out. **3.** *See* developed reserves.

ore intersection [MIN ENG]   **1.** The point at which a borehole, crosscut, or other underground opening encounters an ore vein or deposit. **2.** The thickness of the ore-bearing deposit so traversed.

ore-lead age [GEOL]   Measurement of the age of the earth by comparing the relative progress of the two radioactive decay schemes $^{235}U–^{207}Pb$ and $^{238}U–^{206}Pb$.

ore microscopy [MINERAL]   The use of a reflecting microscope to study polished sections of ore minerals. Also known as mineragraphy; mineralography.

orendite [PETR]   A porphyritic extrusive rock containing phlogopite phenocrysts in a nepheline-free reddish-gray groundmass of leucite, sanidine, phlogopite, amphibole, and diopside.

oreodont [PALEON]   Any member of the family Merycoidodontidae.

ore of sedimentation *See* placer.

ore pass [MIN ENG]   A vertical or inclined passage for the downward transfer of ore.

ore pipe *See* pipe.

ore pocket [MIN ENG]   **1.** Excavation near the hoisting shaft into which ore from stopes is moved, preliminary to hoisting. **2.** An unusual concentration of ore in the lode.

ore reduction [MIN ENG]   The size reduction of solids by crushing and grinding in mineral processing plants.

ore reserve [MIN ENG]   The total tonnage and average value of proved ore, plus the total tonnage and value (assumed) of the probable ore.

ore roll [GEOL]   Uranium or vanadium ore bodies within sedimentary rock, especially sandstone, that are discordant, forming S-shaped or C-shaped cross sections.

ore sampling [MIN ENG]   The process in which a portion of ore is selected so that its composition will represent the average composition of the entire bulk of ore.

ore shoot [GEOL]   **1.** A large, generally vertical, pipelike ore body that is economically valuable. Also known as shoot. **2.** A large and usually rich aggregation of mineral in a vein.

organic geochemistry [GEOCHEM]   That branch of geochemistry concerned with naturally occurring carbonaceous and biologically derived substances which are of geological interest.

organic mound *See* bioherm.

organic reef [GEOL]   A sedimentary rock structure of significant dimensions erected by, and composed almost exclusively of the remains of, corals, algae, bryozoans, sponges, and other sedentary or colonial organisms.

organic rock [PETR]   A sedimentary rock composed principally of the remains of plants and animals.

organic soil   [GEOL]   Any soil or soil horizon consisting chiefly of, or containing at least 30% of, organic matter; examples are peat soils and muck soils.

organic texture   [GEOL]   A sedimentary texture resulting from the activity of organisms such as the secretion of skeletal material.

organic weathering   [GEOL]   Biologic processes and changes that contribute to the breakdown of rocks. Also known as biologic weathering.

organogenic   [GEOL]   Property of a rock or sediment derived from organic substances.

organolite   [GEOL]   Any rock consisting mainly of organic material.

oriental alabaster   *See* onyx marble.

oriental amethyst   [MINERAL]   A violet to purple variety of sapphire.

oriental jasper   *See* bloodstone.

oriental topaz   [MINERAL]   A yellow variety of corundum, used as a gem.

orientation   [CRYSTAL]   The directions of the axes of a crystal lattice relative to the surfaces of the crystal, to applied fields, or to some other planes or directions of interest.

orientation diagram   [GEOL]   Any point or contour diagram used in structural petrology.

oriented   [GEOL]   Pertaining to a specimen that is so marked as to show its exact, original position in space.

orifice well tester   [PETRO ENG]   Velocity-type meter used to measure gas flow quantity from a gas well; static pressure differences before and after a sharp-edged orifice are converted to flow values.

original dip   *See* primary dip.

original interstice   [PETR]   An interstice that formed contemporaneously with the enclosing rock. Also known as primary interstice.

original valley   [GEOL]   A valley formed by hypogene action or by epigene action other than that of running water.

origofacies   [GEOL]   Facies of the primary sedimentary environment.

Ornithischia   [PALEON]   An order of extinct terrestrial reptiles, popularly known as dinosaurs; distinguished by a four-pronged pelvis, and a median, toothless predentary bone at the front of the lower jaw.

Ornithopoda   [PALEON]   A suborder of extinct reptiles in the order Ornithischia including all bipedal forms in the order.

orocline   [GEOL]   An orogenic belt with a change in horizontal direction, either a horizontal curvature or a sharp bend. Also known as geoflex.

oroclinotath   [GEOL]   An orogenic belt that has been subjected to both substantial bending and substantial stretching.

orocratic   [GEOL]   Pertaining to a period of time in which there is much diastrophism.

orogen   *See* orogenic belt.

orogene   *See* orogenic belt.

orogenesis   *See* orogeny.

orogenic belt   [GEOL]   A linear region that has undergone folding or other deformation during the orogenic cycle. Also known as fold belt; orogen; orogene.

orogenic cycle [GEOL]   A time interval during which a mobile belt evolved into an orogenic belt, passing through preorogenic, orogenic, and postorogenic stages. Also known as geotectonic cycle.

orogenic sediment [GEOL]   Any sediment that is produced as the result of an orogeny or that is directly attributable to the orogenic region in which it is later found.

orogenic unconformity [GEOL]   An angular unconformity produced locally in a region affected by mountain-building movements.

orogeny [GEOL]   The process or processes of mountain formation, especially the intense deformation of rocks by folding and faulting which, in many mountainous regions, has been accompanied by metamorphism, invasion of molten rock, and volcanic eruption; in modern usage, orogeny produces the internal structure of mountains, and epeirogeny produces the mountainous topography. Also known as orogenesis; tectogenesis.

orogeosyncline [GEOL]   A geosyncline that later became an area of orogeny.

orograph [ENG]   A machine that records both distance and elevations as it is pushed across land surfaces; used in maing topographic maps.

orographic [GEOL]   Pertaining to mountains, especially in regard to their location and distribution.

orographic cloud [METEOROL]   A cloud whose form and extent is determined by the disturbing effects of orography upon the passing flow of air; because these clouds are linked with the form of the terrestrial relief, they generally move very slowly, if at all, although the winds at the same level may be very strong.

orographic lifting [METEOROL]   The lifting of an air current caused by its passage up and over surface elevations.

orographic occlusion [METEOROL]   An occluded front in which the occlusion process has been hastened by the retardation of the warm front along the windward slopes of a mountain range.

orographic precipitation [METEOROL]   Precipitation which results from the lifting of moist air over an orographic barrier such as a mountain range; strictly, the amount so designated should not include that part of the precipitation which would be expected from the dynamics of the associated weather disturbance, if the disturbance were over flat terrain.

orography [GEOGR]   The branch of geography dealing with mountains.

orohydrography [HYD]   A branch of hydrography dealing with the relations of mountains to drainage.

orometer [ENG]   A barometer with a scale that indicates elevation above sea level.

orometry [GEOL]   The measurement of mountains.

orotath [GEOL]   An orogenic belt that has been stretched substantially in the direction of its length.

orotvite [PETR]   A syenite composed of hornblende, biotite, plagioclase, nepheline, and cancrinite, with accessory sphene, ilmenite, and apatite.

orpiment [MINERAL] As_2S_3 A lemon-yellow mineral, crystallizing in the monoclinic system, and generally occurring in foliated or columnar masses; luster is resinous and pearly on the cleavage surface, hardness is 1.5–2 on Mohs scale, and specific gravity is 3.49. Also known as yellow arsenic.

Orthacea [PALEON]   An extinct group of articulate brachiopods in the suborder Orthidina in which the delthyrium is open.

Orthent   [GEOL]   A suborder of the soil order Entisol, well drained and of medium or fine texture, usually lacking evidence of horizonation and being shallow to bedrock; occurs mostly on strong slopes.

Orthid   [GEOL]   A suborder of the soil order Aridisol, mostly well drained, gray or brownish gray with little change from top to bottom of the soil profile; occupies younger, but not the youngest, land surfaces in deserts.

Orthida   [PALEON]   An order of extinct articulate brachiopods which includes the oldest known representatives of the class.

Orthidina   [PALEON]   The principal suborder of the extinct Orthida, including those articulate brachiopods characterized by biconvex, finely ribbed shells with a straight hinge line and well-developed interareas on both valves.

orthite   [MINERAL]   Allanite in the form of slender prismatic or acicular crystals.

orthoandesite   [PETR]   An andesite containing orthopyroxene. Also known as sanukitoid.

orthoantigorite   [MINERAL]   A six-layer orthorhombic form of antigorite.

orthoaxis   [CRYSTAL]   The diagonal or lateral axis perpendicular to the vertical axis in the monoclinic system.

orthobituminous coal   [GEOL]   Bituminous coal that contains 87–89% carbon, analyzed on a dry, ash-free basis.

orthochem   [GEOCHEM]   A precipitate formed within a depositional basin or within the sediment itself by direct chemical action.

orthochlorite   [MINERAL]   Any distinctly crystalline form of chlorite, such as clinochlore or penninite.

orthochronology   [GEOL]   Geochronology based on a standard succession of biostratigraphically significant faunas or floras, or based on irreversible evolutionary processes.

orthochrysotile   [MINERAL]   An orthorhombic form of chrysotile.

orthoclase   [MINERAL] $KAlSi_3O_8$ A colorless, white, cream-yellow, flesh-reddish, or gray potassium feldspar that usually contains some sodium feldspar, either as albite or analbite or in some intermediate state; it is or appears to be monoclinic. Also known as common feldspar; orthose; pegmatolite.

orthoclasite   [PETR]   An orthoclase-bearing porphyritic extrusive rock, such as granite or syenite.

orthoconglomerate   [GEOL]   A conglomerate with an intact gravel framework held together by mineral cement and deposited by ordinary water currents.

orthocumulate   [PETR]   A cumulate composed chiefly of one or more cumulus minerals plus the crystallization products of the intercumulus liquid.

Orthod   [GEOL]   A suborder of the soil order Spodosol accumulations of humus, aluminum, and iron, widespread in Canada and the Soviet Union.

orthodolomite   [PETR]   **1.** A primary dolomite, or one formed by sedimentation. **2.** A dolomite rock so well cemented that the particles interlock.

orthoferrosilite   [MINERAL]   An orthopyroxene consisting of the orthorhombic silicate $FeSiO_3$.

orthogeosyncline   [GEOL]   A linear geosynclinal belt lying between continental and oceanic cratons, and having internal volcanic belts (eugeosynclinal) and external nonvolcanic belts (miogeosynclinal). Also known as geosynclinal couple; primary geosyncline.

orthogneiss [GEOL]   Gneiss originating from igneous rock.

orthogonal crystal [CRYSTAL]   A crystal whose axes are mutually perpendicular.

orthographic projection [CRYSTAL]   A projection for displaying the poles of a crystal in which the poles are projected from a reference sphere onto an equatorial plane by dropping perpendiculars from the poles to the plane.   [MAP]   A perspective azimuthal projection of one hemisphere produced by straight parallel lines from any point desired from an infinite distance; it is true to scale at the center only.

orthohydrous coal [GEOL]   Coal that contains 5–6% hydrogen, analyzed on a dry, ash-free basis.

ortholignitous coal [GEOL]   Coal that contains 75–80% carbon, analyzed on a dry, ash-free basis.

orthomagmatic stage [GEOL]   The principal stage in the crystallization of silicates from a typical magma; up to 90% of the magma may crystallize during this stage.   Also known as orthotectic stage.

orthometric correction [ENG]   A systematic correction that must be applied to a measured difference in elevation since level surfaces at varying elevations are not absolutely parallel.

orthometric height [ENG]   The distance above sea level measured along a plumb line.

orthomimic feldspars [MINERAL]   A group of feldspars that by repeated twinning simulate a higher degree of symmetry with rectangular cleavages.

orthomorphic [MAP]   Preserving the correct shape.

orthomorphic chart [MAP]   A chart on which very small shapes are correctly represented.

orthophyric [PETR]   Of the texture of the matrix of certain igneous rocks, having feldspar crystals with quadratic or short and stumpy rectangular cross sections.

Orthopsidae [PALEON]   A family of extinct echinoderms in the order Hemicidaroida distinguished by a camarodont lantern.

orthopyroxene [MINERAL]   A series of pyroxene minerals crystallizing in the orthorhombic system; members include enstatite, bronzite, hypersthene, ferrohypersthene, eulite, and orthoferrosilite.

orthoquartzite [PETR]   A clastic sedimentary rock composed almost entirely of detrital quartz grains; a quartzite of sedimentary origin. Also known as orthoquartzitic sandstone; sedimentary quartzite.

orthoquartzitic conglomerate [GEOL]   A lithologically homogeneous, light-colored orthoconglomerate composed of quartzose residues that is commonly interbedded with pure quartz sandstone. Also known as quartz-pebble conglomerate.

orthoquartzitic sandstone *See* orthoquartzite.

orthorhombic lattice [CRYSTAL]   A crystal lattice in which the three axes of a unit cell are mutually perpendicular, and no two have the same length.   Also known as rhombic lattice.

orthorhombic pyroxene [MINERAL]   A member of the mineral series enstatite ($Mg_2[Si_2O_6]$) to orthoferrosilite ($Fe_2[Si_2O_6]$), crystallizing in the orthorhombic system, space group *Pbca*.

orthorhombic system [CRYSTAL] A crystal system characterized by three axes of symmetry that are mutually perpendicular and of unequal length. Also known as rhombic system.

orthoschist [PETR] A schist derived from igneous rocks.

orthose *See* orthoclase.

orthosite [PETR] A light-colored coarse-grained igneous rock composed almost entirely of orthoclase.

orthosparite [PETR] A sparite cement developed by physicochemical precipitation in open voids.

orthostratigraphy [GEOL] Standard stratigraphy based on fossils which identify recognized biostratigraphic zones.

orthosymmetric crystal [CRYSTAL] A crystal that has orthorhombic symmetry.

orthotectic stage *See* orthomagmatic stage.

orthotill [GEOL] A till formed by immediate release of material from transported ice, such as by ablation and melting.

Orthox [GEOL] A suborder of the soil order Oxisol that is moderate to low in organic matter, well drained, and moist all or nearly all year; believed to be extensive at low altitudes in the heart of the humid tropics.

orvietite [PETR] An extrusive rock composed of approximately equal amounts of plagioclase and sanidine, along with leucite, augite, minor biotite and olivine, and accessory apatite and opaque oxides.

Osagean [GEOL] A provincial series of geologic time in North America; Lower Mississippian (above Kinderhookian, below Meramecian).

osar *See* esker.

osazone [BIOCHEM] Any of the compounds that contain two phenylhydrazine residues and are produced by a reaction between a reducing sugar and phenylhydrazine.

osbornite [MINERAL] TiN A mineral found only in meteorites.

oscillation ripple *See* oscillation ripple mark.

oscillation ripple mark [GEOL] A symmetric ripple mark having a sharp, narrow, and relatively straight crest between broadly rounded troughs, formed by the motion of water agitated by oscillatory waves on a sandy base at a depth shallower than wave base. Also known as oscillation ripple; oscillatory ripple mark; wave ripple mark.

oscillatory ripple mark *See* oscillation ripple mark.

oscillatory twinning [CRYSTAL] Repeated, parallel twinning.

Osos wind [METEOROL] In California, a strong northwest wind blowing from the Loa Osos valley to the San Luis valley.

osseous amber [MINERAL] Opaque or cloudy amber containing numerous minute bubbles.

ossipite [PETR] A coarse-grained variety of troctolite containing labradorite, olivine, magnetite, and a small amount of diallage.

Osteolepidae [PALEON] A family of extinct fishes in the order Osteolepiformes.

Osteolepiformes [PALEON] A primitive order of fusiform lobefin fishes, subclass Crossopterygii, generally characterized by rhombic bony scales, two dorsal fins placed well back on the body, and a well-ossified head covered with large dermal plating bones.

osteolith [PALEON] A fossil bone.

Osteostraci [PALEON] An order of extinct jawless vertebrates; they were mostly small, with the head and part of the body encased in a solid armor of bone, and the posterior part of the body and the tail covered with thick scales.

ostracoderm [PALEON] Any of various extinct jawless vertebrates covered with an external skeleton of bone which together with the Cyclostomata make up the class Agnatha.

ostria [METEOROL] A warm southerly wind on the Bulgarian coast; it is considered a precursor of bad weather. Also known as auster.

osumilite [MINERAL] $(K,Na)(Mg,Fe^{2+})_2(Al,Fe^{3+})_3(Si,Al)_{12}O_{30} \cdot H_2O$ A mineral that crystallizes in the hexagonal system and is commonly mistaken for cordierite.

otavite [MINERAL] $CdCO_3$ A mineral that crystallizes in the hexagonal system and is isostructural with calcite.

ottrelite [MINERAL] A gray to black variety of chloritoid containing manganese.

ouachitite [PETR] A biotite monchiquite with no olivine and a glassy or analcime groundmass.

ouari [METEOROL] A south wind of Somaliland, Africa; it is similar to the khamsin.

ouenite [PETR] A fine-grained igneous rock resembling eucrite and containing green augite, anorthite, and smaller amounts of hypersthene and olivine.

outage [PETRO ENG] The difference between the full or rated capacity of a barrel, tank, or tank car as compared to actual content.

outage method [PETRO ENG] Deduction of the liquid content of a tank by measurement of the distance from the top of the tank to the surface of the liquid; in contrast to the innage method.

outburst [MIN ENG] The sudden issue of gases, chiefly methane (sometimes accompanied by coal dust), from the working face of a coal mine.

outby [MIN ENG] Toward the mine entrance or shaft and therefore away from the working face.

outcrop [GEOL] Exposed stratum or body of ore at the surface of the earth. Also known as cropout.

outcrop map [GEOL] A type of geologic map that shows the distribution and shape of actual outcrops, leaving those areas without outcrops blank.

outcrop water [HYD] Rain and surface water which seeps downward through outcrops of porous and fissured rock, fault planes, old shafts, or surface drifts.

outer atmosphere [METEOROL] Very generally, the atmosphere at a great distance from the earth's surface; possibly best usage of the term is as an approximate synonym for exosphere.

outer bar [GEOL] A bar formed at the mouth of an ebb channel of an estuary.

outer beach [GEOL] That part of a beach that is ordinarily dry, reached only by the waves generated by violent storms.

outer harbor [GEOGR] The part of a harbor toward the sea, through which a vessel enters the inner harbor.

outer mantle *See* upper mantle.

outface *See* dip slope.

outflow cave [GEOL] A cave from which a stream issues, or is known to have issued.

outlet glacier [HYD]   A stream of ice from an ice cap to the sea.

outlet head [HYD]   The place where water leaves a lake and enters an effluent.

outlier [GEOL]   A group of rocks separated from the main mass and surrounded by outcrops of older rocks.

outline map [MAP]   A map that presents minimal geographic information, usually only coastlines, principal streams, major civil boundaries, and large cities, leaving as much space as possible for the reception of particular additional data.

outside air temperature *See* indicated air temperature.

outwash [GEOL]   **1.** Sand and gravel transported away from a glacier by streams of meltwater and either deposited as a floodplain along a preexisting valley bottom or broadcast over a preexisting plain in a form similar to an alluvial fan. Also known as glacial outwash; outwash drift; overwash. **2.** Soil material washed down a hillside by rainwater and deposited on more gently sloping land.

outwash apron *See* outwash plain.

outwash cone [GEOL]   A cone-shaped deposit consisting chiefly of sand and gravel found at the edge of shrinking glaciers and ice sheets.

outwash drift *See* outwash.

outwash fan [GEOL]   A fan-shaped accumulation of outwash deposited by meltwater streams in front of the terminal moraine of a glacier.

outwash plain [GEOL]   A broad, outspread flat or gently sloping alluvial deposit of outwash in front of or beyond the terminal moraine of a glacier. Also known as apron; frontal apron; frontal plain; marginal plain; morainal apron; morainal plain; outwash apron; overwash plain; sandur; wash plain.

outwash-plain shoreline [GEOL]   A prograding shoreline formed where the outwash plain in front of a glacier is built into a lake or sea.

outwash terrace [GEOL]   A dissected and incised valley train or benchlike deposit extending along a valley downstream from an outwash plain or terminal moraine.

outwash train *See* valley train.

ouvarovite *See* uvarovite.

oven [GEOL]   **1.** A rounded, saclike, chemically weathered pit or hollow in a rock (especially a granitic rock), having an arched roof and resembling an oven. **2.** *See* spouting horn.

overall drilling time [MIN ENG]   The total time for rock drilling including time for setting up, withdrawing, and moving drills, time for mechanical delays, and the time for the actual drilling.

overarching weight [MIN ENG]   The pressure of the rocks over the active mine workings.

overbank deposit [GEOL]   Fine-grained sediment (silt and clay) deposited from suspension on a floodplain by floodwaters that cannot be contained within the stream channel.

overburden [GEOL]   **1.** Rock material overlying a mineral deposit or coal seam. Also known as baring; top. **2.** Material of any nature, consolidated or unconsolidated, that overlies a deposit of useful materials, ores, or coal, especially those deposits that are mined from the surface by open cuts. **3.** Loose soil, sand, or gravel that lies above the bedrock.   [MIN ENG]   To charge in a furnace too much ore and flux in proportion to the amount of fuel.

overburdened stream *See* overloaded stream.

overcast [METEOROL] **1.** Pertaining to a sky cover of 1.0 (95% or more) when at least a portion of this amount is attributable to clouds or obscuring phenomena aloft, that is, when the total sky cover is not due entirely to surface-based obscuring phenomena. **2.** Cloud layer that covers most or all of the sky; generally, a widespread layer of clouds such as that which is considered typical of a warm front. [MIN ENG] **1.** An enclosed airway to permit one air current to pass over another without interruption. **2.** To move overburden removed from coal mined from surface mines to an area from which the coal has been mined.

overconsolidation [GEOL] Consolidation of sedimentary material exceeding that which is normal for the existing overburden.

overcut [MIN ENG] A machine cut made along the top or near the top of a coal seam; sometimes used in a thick seam or a seam with sticky coal.

overcutting machine [MIN ENG] A coal-cutting machine designed to make the cut at a desired place in the coal seam some distance above the floor.

overdeepening [GEOL] The erosive process by which a glacier deepens and widens an inherited preglacial valley to below the level of the subglacial surface.

overfalls [OCEANOGR] Short, breaking waves occurring when a strong current passes over a shoal or other submarine obstruction or meets a contrary current or wind.

overfeed [MIN ENG] To attempt to make a diamond- or rock-drill bit penetrate rock at a rate in excess of that at which the optimum economical performance of the bit is attained, needlessly damaging the bit and shortening its life.

overflow channel [GEOL] A channel or notch cut by the overflow water of a lake, especially the channel draining meltwater from a glacially dammed lake.

overflow ice [HYD] Ice formed during high spring tides by water rising through cracks in the surface ice and then freezing.

overflow spring [HYD] A type of contact spring that develops where a permeable deposit dips beneath an impermeable mantle.

overflow stream [HYD] **1.** A stream containing water that has overflowed the banks of a river or stream. Also known as spill stream. **2.** An effluent from a lake, carrying water to a stream, sea, or another lake.

overfold [GEOL] A fold that is overturned.

overgrinding [MIN ENG] Grinding an ore to a smaller particle size than that necessary to free the desired mineral from other materials.

overgrowth [CRYSTAL] A crystal growth in optical and crystallographic continuity around another crystal of different composition. [MINERAL] A mineral deposited on and growing in oriented, crystallographic directions on the surface of another mineral.

overhand cut and fill [MIN ENG] A method of mining ore in which material removed from the roof of a drive drops through chutes to a lower drive, from which the material is removed.

overhand stope [MIN ENG] A stope in which the ore above the point of entry is attacked, so that severed ore tends to gravitate toward discharge chutes, and the stope is self-draining.

overhand stoping [MIN ENG] A method of mining in which the ore is blasted from a series of ascending stepped benches; both horizontal and vertical holes may be employed.

overhead cableway [MIN ENG] A type of equipment for the removal of soil or rock, consisting of a strong overhead cable which is usually attached to towers at either end, and on which a car or traveler may run back and forth; from this car a pan or bucket may be lowered to the surface, then raised and locked to the car and transported to any position on the cable where it is desired to dump.

overite [MINERAL] $Ca_3Al_8(PO_4)_8(OH)_6 \cdot 15H_2O$ A mineral composed of hydrous basic calcium aluminum phosphate.

overland flow [HYD] Water flowing over the ground surface toward a channel; upon reaching the channel, it is called surface runoff. Also known as surface flow.

overlap [GEOL] **1.** Movement of an upcurrent part of a shore to a position extending seaward beyond a downcurrent part. **2.** Extension of strata over or beyond older underlying rocks. **3.** The horizontal component of separation measured parallel to the strike of a fault.

overlap fault [GEOL] A fault structure in which the displaced strata are doubled back upon themselves.

overload [GEOL] The amount of sediment that exceeds the ability of a stream to transport it and is thereby deposited.

overloaded stream [HYD] A stream so heavily loaded with sediment that its velocity is lessened and it is forced to deposit part of its load. Also known as overburdened stream.

overloader [MIN ENG] A loading machine which digs with a bucket, raises the bucket, and swings it in a wide horizontal arc to the dumping point.

overprint [GEOCHEM] A complete or partial disturbance of an isolated radioactive system by thermal, igneous, or tectonic activities, which results in loss or gain of radioactive or radiogenic isotopes and, hence, a change in the radiometric age that will be given the disturbed system. [GEOL] The development or superposition of metamorphic structures on original structures. Also known as imprint; metamorphic overprint; superprint.

overrunning [METEOROL] A condition existing when an air mass is in motion aloft above another air mass of greater density at the surface; this term usually is applied in the case of warm air ascending the surface of a warm front or quasi-stationary front.

oversaturated *See* silicic.

overseeding [METEOROL] Cloud seeding in which an excess of nucleating material is released; as the term is normally used, the excess is relative to that amount of nucleating material which would, theoretically, maximize the precipitation received at the ground.

oversize control screen [MIN ENG] A screen used to prevent the entry into a machine of coarse particles which might interfere with its operation. Also known as check screen; guard screen.

oversize rod *See* guide rod.

oversteepening [GEOL] The process by which an eroding alpine glacier steepens the sides of an inherited preglacial valley.

overstep [GEOL] **1.** An overlap characterized by the regular truncation of older units of a complete sedimentary sequence by one or more later units of the sequence. **2.** A stratum deposited on the upturned edges of underlying strata.

overstressed area [MIN ENG] In strata control, an area where the force is concentrated on pillars.

overthrust [GEOL] **1.** A thrust fault that has a low dip or a net slip that is large. Also known as low-angle thrust; overthrust fault. **2.** A thrust fault with the active element being the hanging wall.

overthrust black *See* overthrust nappe.

overthrust fault *See* overthrust.

overthrust nappe [GEOL] The body of rock making up the hanging wall of a large-scale overthrust. Also known as overthrust block; overthrust sheet; overthrust slice.

overthrust sheet *See* overthrust nappe.

overthrust slice *See* overthrust nappe.

overtide [OCEANOGR] A harmonic tidal component which has a speed that is an exact multiple of the speed of one development of the tide-producing force.

overtub system [MIN ENG] An endless-rope system in which the rope runs over the tubs or cars in the center of the rails.

overturn [HYD] Renewal of bottom water in lakes and ponds in regions where winter temperatures are cold; in the fall, cooled surface waters become denser and sink, until the whole body of water is at 4°C; in the spring, the surface is warmed back to 4°C, and the lake is homothermous. Also known as convective overturn.

overturned [GEOL] Of a fold or the side of a fold, tilted beyond the perpendicular. Also known as inverted; reversed.

overwash *See* outwash.

overwash mark [GEOL] A narrow, tonguelike ridge of sand formed by overwash on the landward side of a berm.

overwash plain *See* outwash plain.

overwash pool [OCEANOGR] A tidal pool between a berm and a beach scarp which water enters only at high tide.

O-wave component *See* ordinary-wave component.

owyheeite [MINERAL] $Ag_2Pb_5Sb_6S_{15}$ A steel gray to silver white mineral that is found in metallic fibrous masses and acicular crystals. Also known as silver jamesonite.

oxammite [MINERAL] $(NH_4)_2C_2O_4 \cdot H_2O$ A yellowish-white, orthorhombic mineral consisting of ammonium oxalate monohydrate; occurs as lamellar masses.

oxbow [HYD] **1.** A closely looping, U-shaped stream meander whose curvature is so extreme that only a neck of land remains between the two parts of the stream. Also known as horseshoe bend. **2.** *See* oxbow lake. [GEOL] The abandoned, horseshoe-shaped channel of a former stream meander after the stream formed a neck cutoff. Also known as abandoned channel.

oxbow lake [HYD] The crescent-shaped body of water located alongside a stream in an abandoned oxbow after a neck cutoff is formed and the ends of the original bends are silted up. Also known as crescentic lake; cutoff lake; horseshoe lake; loop lake; moat; mortlake; oxbow.

Oxfordian [GEOL] A European stage of geologic time, in the Upper Jurassic (above Callovian, below Kimmeridgean). Also known as Divesian.

oxidate [GEOL] A sediment made up of iron and manganese oxides and hydroxides crystallized from aqueous solution.

oxide mineral [MINERAL] A naturally occurring material in oxide form such as silicon dioxide, SiO_2, magnetite, Fe_3O_4, or lime, CaO.

oxidized shale *See* burnt shale.

oxidized zone [GEOL] A region of mineral deposits which has been altered by oxidizing surface waters.

Oxisol [GEOL] A soil order characterized by residual accumulations of inactive clays, free oxides, kaolin, and quartz; mostly tropical.

oxoferrite [GEOL] A variety of naturally occurring iron with some ferrous oxide (FeO) in solid solution.

Oxyaenidae [PALEON] An extinct family of mammals in the order Deltatheridea; members were short-faced carnivores with powerful jaws.

oxybasiophitic rock [PETR] An ophitic rock that can be either oxyophitic or basiophitic.

oxybiotite [MINERAL] Phenocrystic biotite with increased amounts of Fe(III).

oxygen deficit [GEOCHEM] The difference between the actual amount of dissolved oxygen in lake or sea water and the saturation concentration at the temperature of the water mass sampled.

oxygen distribution [OCEANOGR] The concentration of dissolved oxygen in ocean water as a function of depth, ranging from as much as 5 milliliters of oxygen per liter at the surface to a fraction of that value at great depths.

oxygen isotope fractionation [GEOCHEM] The temperature-dependent variation of the oxygen-18/oxygen-16 isotope ratio in the carbonate shells of marine organisms, used as a measure of the water temperature at the time of deposition.

oxygen minimum layer [HYD] A subsurface layer in which the content of dissolved oxygen is very low (or absent), lower than in the layers above and below.

oxygen ratio *See* acidity coefficient.

oxyheeite [MINERAL] $Pb_5Ag_2Sb_6S_{15}$ A light steel gray to silver white mineral consisting of lead and silver antimony sulfide; occurs as acicular needles or in massive form.

oxyhornblende *See* basaltic hornblende.

oxymesostasis [GEOL] The mesostasis (quartz, orthoclase, or micropegmatite) of an oxyophitic rock.

oxyophitic rock [PETR] An ophitic rock whose mesostasis is composed of quartz or orthoclase or both.

oxyphile *See* lithophile.

Oyashio [OCEANOGR] A cold current flowing from the Bering Sea southwest along the coast of Kamchatka, past the Kuril Islands, continuing close to the northeast coast of Japan, and reaching nearly 35°N.

Ozawainellidae [PALEON] A family of extinct protozoans in the superfamily Fusulinacea.

ozocerite [GEOL] A natural, brown to jet black paraffin wax occurring in irregular veins; consists principally of hydrocarbons, is soluble in water, and has a variable melting point. Also known as ader wax; earth wax; fossil wax; mineral wax; native paraffin; ozokerite.

ozokerite *See* ozocerite.

ozone cloud [METEOROL] A limited region in which the total ozone content of the ozonosphere is greater than normal.

ozone layer *See* ozonosphere.

ozonosphere   [METEOROL]   The general stratum of the upper atmosphere in which there is an appreciable ozone concentration and in which ozone plays an important part in the radiative balance of the atmosphere; lies roughly between 10 and 50 kilometers, with maximum ozone concentration at about 20 to 25 kilometers. Also known as ozone layer.

paar [GEOL] A depression produced by the moving apart of crustal blocks rather than by subsidence within a crustal block.

pachnolite [MINERAL] NaCaAlF$_6$·H$_2$O Colorless to white mineral composed of hydrous sodium calcium aluminum fluoride, occurring in monoclinic crystals.

pachycephalosaur [PALEON] A bone-headed dinosaur, composing the family Pachycephalosauridae.

Pachycephalosauridae [PALEON] A family of ornithischian dinosaurs characterized by a skull with a solid rounded mass of bone 10 centimeters thick above the minute brain cavity.

Pacific anticyclone *See* Pacific high.

Pacific Equatorial Countercurrent [OCEANOGR] The Equatorial Countercurrent flowing east across the Pacific Ocean between 3° and 10°N.

Pacific high [METEOROL] The nearly permanent subtropical high of the North Pacific Ocean, centered, in the mean, at 30–40°N and 140–150°W. Also known as Pacific anticyclone.

Pacific North Equatorial Current [OCEANOGR] The North Equatorial Current which flows westward between 10° and 20°N in the Pacific Ocean.

Pacific Ocean [GEOGR] The largest division of the hydrosphere, having an area of 165,000,000 square kilometers and covering 46% of the surface of the total extent of the oceans and seas; it is bounded by Asia and Australia on the west and North and South America on the east.

Pacific South Equatorial Current [OCEANOGR] The South Equatorial Current flowing westward between 3°N and 10°S in the Pacific Ocean.

Pacific suite [PETR] A large group of igneous rocks characterized by calcic and calc-alkalic rocks, especially in the region of the circum-Pacific orogenic belt. Also known as anapeirean; circum-Pacific province.

Pacific-type continental margin [GEOL] A continental margin typified by that of the western Pacific where oceanic lithosphere descends beneath an adjacent continent and produces an intervening island arc system.

pack [MIN ENG] **1.** A pillar built in the waste area or roadside within a mine to support the mine roof; constructed from loose stones and dirt. **2.** Waste rock or timber used to support the roof or underground workings or used to fill excavations. Also known as fill. [OCEANOGR] *See* pack ice.

pack builder [MIN ENG] **1.** One who builds packs or pack walls. **2.** In anthracite and bituminous coal mining, one who fills worked-out rooms, from which coal has been mined, with rock, slate, or other waste to prevent caving of walls and roofs, or who builds rough walls and columns of loose stone, heavy boards, timber, or coal along haulageways and passageways and in rooms where coal is being mined to prevent caving of roof or walls during mining operations. Also known as packer; pillar man; timber packer; waller.

packer *See* pack builder; production packer.

packer fluid [PETRO ENG] Fluid inserted in the annulus between the tubing and casing above a packer in order to reduce pressure differentials between the formation and the inside of the casing and across the packer.

pack ice [OCEANOGR] Any area of sea ice, except fast ice, composed of a hetero- geneous mixture of ice of varying ages and sizes, and formed by the packing together of pieces of floating ice. Also known as ice canopy; ice pack; pack.

packing [CRYSTAL] The arrangement of atoms or ions in a crystal lattice. [GEOL] The arrangement of solid particles in a sediment or in sedimentary rock.

packing density [GEOL] A measure of the extent to which the grains of a sedimentary rock occupy the gross volume of the rock in contrast to spaces between the grains, equal to the cumulative grain-intercept length along a traverse in a thin section.

packing index [CRYSTAL] The volume of ion divided by the volume of the unit cell in a crystal.

packing proximity [GEOL] In a sedimentary rock, an estimate of the number of grains that are in contact with adjacent grains; equal to the total percentage of grain-to-grain contacts along a traverse measured on a thin section.

packing radius [CRYSTAL] One-half the smallest approach distance of atoms or ions.

packsand [PETR] A very fine-grained sandstone that is so loosely consolidated by a slight calcareous cement that it can be readily cut by a spade.

packstone [GEOL] A carbonate rock related to reefs and including closely packed fragments of organic skeletal material.   [PETR] A sedimentary carbonate rock whose granular material is arranged in a self-supporting framework, yet also contains some matrix of calcareous mud.

pack wall [MIN ENG] A wall of dry stone built along the side of a roadway, or in the waste area, of a coal or metal mine to help support the roof and to retain the packing material and prevent its spreading into the roadway.

padparadsha *See* orange sapphire.

paesa [METEOROL] A violent north-northeast wind of Lake Garda in Italy.

paesano [METEOROL] A northerly night breeze, blowing down from the mountains, of Lake Garda in Italy.

pagodite *See* agalmatolite.

paha [GEOL] A low, elongated, rounded glacial ridge or hill which consists mainly of drift, rock, or windblown sand, silt, or clay but is capped with a thick cover of loess.

pahoehoe [GEOL] A type of lava flow whose surface is glassy, smooth, and undulat- ing; the lava is basaltic, glassy, and porous. Also known as ropy lava.

paigeite [MINERAL] $(Fe,Mg)FeBO_5$ A black mineral composed of iron magnesium borate, occurring as fibrous aggregates.

painter [METEOROL] A fog frequently experienced on the coast of Peru; the brownish deposit which it often leaves upon exposed surfaces is sometimes called Peruvian paint. Also known as Callao painter.

paint pot [GEOL] A mud pot containing multicolored mud.

paired terrace [GEOL] One of two stream terraces that face each other at the same elevation from opposite sides of the stream valley and that represent the remnants of the same floodplain or valley floor. Also known as matched terrace.

paisanite *See* ailsyte.

Palaeacanthaspidoidei [PALEON] A suborder of extinct, placoderm fishes in the order Rhenanida; members were primitive, arthrodire-like species.

Palaechinoida [PALEON] An extinct order of echinoderms in the subclass Perischoechinoidea with a rigid test in which the ambulacra bevel over the adjoining interambulacra.

Palaeoconcha [PALEON] An extinct order of simple, smooth-hinged bivalve mollusks.

Palaeocopida [PALEON] An extinct order of crustaceans in the subclass Ostracoda characterized by a straight hinge and by the anterior location for greatest height of the valve.

Palaeoisopus [PALEON] A singular, monospecific, extinct arthropod genus related to the pycnogonida, but distinguished by flattened anterior appendages.

Palaeomastodontinae [PALEON] An extinct subfamily of elaphantoid proboscidean mammals in the family Mastodontidae.

Palaeomerycidae [PALEON] An extinct family of pecoran ruminants in the superfamily Cervoidea.

Palaeonisciformes [PALEON] A large extinct order of chondrostean fishes including the earliest known and most primitive ray-finned forms.

Palaeoniscoidei [PALEON] A suborder of extinct fusiform fishes in the order Palaeonisciformes with a heavily ossified exoskeleton and thick rhombic scales on the body surface.

Palaeopantopoda [PALEON] A monogeneric order of extinct marine arthropods in the subphylum Pycnogonida.

Palaeoryctidae [PALEON] A family of extinct insectivorous mammals in the order Deltatheridia.

Palaeospondyloidea [PALEON] An ordinal name assigned to the single, tiny fish *Palaeospondylus*, known only from Middle Devonian shales in Cairthness, Scotland.

Palaeotheriidae [PALEON] An extinct family of perissodactylous mammals in the superfamily Equoidea.

palagonite [GEOL] A brown to yellow altered basaltic glass found as interstitial material or amygdules in pillow lavas.

palagonite tuff [PETR] A pyroclastic rock composed of angular fragments of palagonite.

palasite [GEOL] The most abundant of the intermediate types of meteorites, consisting of olivine enclosed in a nickel-iron matrix.

Palatinian orogeny *See* Pfälzian orogeny.

paleic surface [GEOL] A smooth, preglacial erosion surface.

paleoaktology [GEOL] Study of ancient seashore and shallow-water environments.

paleoautochthon [GEOL] The original autochthon or basement of a tectonic region.

paleobathymetry [GEOL] Topography of the ancient ocean floor.

paleobiochemistry [PALEON] The study of chemical processes used by organisms that lived in the geologic past.

paleobioclimatology [PALEON]　The study of climatological events affecting living organisms for millennia or longer.

paleobotanic province [GEOL]　A large region defined by similar fossil floras.

paleobotany [PALEON]　The branch of paleontology concerned with the study of ancient and fossil plants and vegetation of the geologic past.

paleoceanography [OCEANOGR]　The study of the history of the circulation, chemistry, biogeography, fertility, and sedimentation of the oceans.

Paleocene [GEOL]　A major worldwide division (epoch) of geologic time of the Tertiary period; extends from the end of the Cretaceous period to the Eocene epoch.

paleochannel [GEOL]　A remnant of a stream channel cut in older rock and filled by the sediments of younger overlying rock.

Paleocharaceae [PALEOBOT]　An extinct group of fossil plants belonging to the Charophyta distinguished by sinistrally spiraled gyrogonites.

paleoclimate [GEOL]　The climate of a given period of geologic time. Also known as geologic climate.

paleoclimatic sequence [GEOL]　The sequence of climatic changes in geologic time; it shows a succession of oscillations between warm periods and ice ages, but superimposed on this are numerous shorter oscillations.

paleoclimatology [GEOL]　The study of climates in the geologic past, involving the interpretation of glacial deposits, fossils, and paleogeographic, isotopic, and sedimentologic data.

Paleocopa [PALEON]　An order of extinct ostracods distinguished by a long, straight hinge.

paleocrystic ice [HYD]　Sea ice generally considered to be at least 10 years old, especially well-weathered polar ice.

paleocurrent [GEOL]　Ancient fluid current flow whose orientation can be inferred by primary sedimentary structures and textures.

paleodepth [PALEON]　The water level at which an ancient organism or group of organisms flourished.

paleodrainage pattern [GEOL]　A drainage pattern representing the distribution of a valley stream as it existed at a given moment of geologic time.

paleoecology [PALEON]　The ecology of the geologic past.

paleoenvironment [GEOL]　An environment in the geologic past.

paleoequator [GEOL]　The position of the earth's equator in the geologic past as defined for a specific geologic period and based on geologic evidence.

paleofluminology [GEOL]　The study of ancient stream systems.

Paleogene [GEOL]　A geologic time interval comprising the Oligocene, Eocene, and Paleocene of the lower Tertiary period. Also known as Eogene.

paleogeographic event *See* palevent.

paleogeographic stage *See* palstage.

paleogeography [GEOL]　The geography of the geologic past; concerns all physical aspects of an area that can be determined from the study of the rocks.

paleogeologic map [GEOL]　An areal map of the geology of an ancient surface immediately below a buried unconformity, showing the geology as it appeared at some time in the geologic past at the time the surface of unconformity was completed and before the overlapping strata were deposited.

paleogeology [GEOL] The geology of the past, applied particularly to the interpretation of the rocks at a surface of unconformity.

paleogeomorphology [GEOL] A branch of geomorphology concerned with the recognition of ancient erosion surfaces and the study of ancient topographies and topographic features that are now concealed beneath the surface and have been removed by erosion. Also known as paleophysiography.

paleoherpetology [PALEON] The study of fossil reptiles.

paleohydrology [GEOL] The study of ancient hydrologic features preserved in rock.

paleoichnology [PALEON] The study of trace fossils in the fossil state. Also spelled palichnology.

paleoisotherm [GEOL] The locus of points of equal temperature for some former period of geologic time.

paleokarst [GEOL] A rock or area at has been karsified and subsequently buried under sediments.

paleolatitude [GEOL] The latitude of a specific area on the earth's surface in the gologic past.

paleolimnology [GEOL] **1.** The study of the past conditions and processes of ancient lakes. **2.** The study of the sediments and history of existing lakes.

paleolithologic map [GEOL] A paleogeologic map indicating lithologic variations at a buried horizon or within a restricted zone at a specific time in the geologic past.

paleomagnetics [GEOPHYS] The study of the direction and intensity of the earth's magnetic field throughout geologic time.

paleomagnetic stratigraphy [GEOPHYS] The use of natural remanent magnetization in the identification of stratigraphic units. Also known as magnetic stratigraphy.

paleometeoritics [GEOL] The study of variation of extraterrestrial debris as a function of time over extended parts of the geologic record, especially in deep-sea sediments and possibly in sedimentary rocks, and, for more recent periods, in ice.

paleomorphology [PALEON] The study of the form and structure of fossil remains in order to describe the original anatomy of an organism.

paleomycology [PALEOBOT] The study of fossil fungi.

Paleonthropinae [PALEON] A former subfamily of fossil man in the family Hominidae; set up to include the Neanderthalers together with Rhodesian man.

paleopalynology [PALEON] A field of palynology concerned with fossils of microorganisms and of dissociated microscopic parts of megaorganisms.

Paleoparadoxidae [PALEON] A family of extinct hippopotamuslike animals in the order Desmostylia.

paleopedology [GEOL] The study of soils of past geologic ages, including determination of their ages.

paleophysiography *See* paleogeomorphology.

paleoplain [GEOL] An ancient degradational plain that is buried beneath later deposits.

paleopole [GEOL] A pole of the earth, either magnetic or geographic, in past geologic time.

paleosalinity [GEOL] The salinity of a body of water in the geologic past, as evaluated on the basis of chemical analyses of sediment or formation water.

paleoslope [GEOL]   The direction of initial dip of a former land surface, such as an ancient continental slope.

paleosol [GEOL]   A soil horizon that formed on the surface during the geologic past, that is, an ancient soil. Also known as buried soil; fossil soil.

paleosome [GEOL]   A geometric element of a composite rock or mineral deposit which appears to be older than an associated younger rock element.

paleostructure [GEOL]   The geologic structure of a region or sequence of rocks in the geologic past.

paleotectonic map [GEOL]   Regional map that shows the structural patterns that existed during a particular period of geologic time, for example, the Lower Cretaceous in western Canada.

paleotemperature [GEOL]   **1.** The temperature at which a geologic process took place in ancient past. **2.** The mean climatic temperature at a given time or place in the geologic past.

paleothermal [GEOL]   Pertaining to warm climates of the geologic past.

paleothermometry [GEOL]   Measurement or estimation of temperatures.

paleotopography [GEOL]   The topography of a given area in the geologic past.

paleovolcanic rock [GEOL]   An extrusive rock of pre-Tertiary age.

Paleozoic [GEOL]   The era of geologic time from the end of the Precambrian (600 million years before present) until the beginning of the Mesozoic era (225 million years before present).

paleozoology [PALEON]   The branch of paleontology concerned with the study of ancient animals as recorded by fossil remains.

palette [GEOL]   A broad sheet of calcite representing a solutional remnant in a cave. Also known as shield.

palevent [GEOL]   A relatively sudden and short-lived paleogeographic happening, such as the short, static existence of a particular depositional environment, or a rapid geographic change separating two palstages. Also known as paleogeographic event.

palimpsest [GEOL]   **1.** Referring to a kind of drainage in which a modern, anomalous drainage pattern is superimposed upon an older one, clearly indicating different topographic and possibly structural conditions at the time of development. **2.** In sedimentology, autochthonous sediment deposits which exhibit some of the attributes of the source sediment.   [PETR]   Of a metamorphic rock, having remnants of the original structure or texture preserved.

palingenesis [PETR]   In-place formation of new magma by the melting of preexisting rock material.

palingenetic *See* resurrected.

palinspastic map [GEOL]   A paleogeographic or paleotectonic map showing restoration of the features to their original geographic positions, before thrusting or folding of the crustal rocks.

Palisade disturbance [GEOL]   Appalachian orogenic episode occurring during Triassic time which produced a series of faultlike basins.

palisades [GEOL]   A series of sharp cliffs.

palladium amalgam *See* potarite.

pallasite [GEOL]   **1.** A stony-iron meteorite composed essentially of large single glassy crystals of olivine embedded in a network of nickel-iron. **2.** An ultramafic rock, whether of meteoric or terrestrial origin, which contains more than 60% iron in the meteoric type, or more iron oxides than silica in the terrestrial type.

pallasite shell *See* lower mantle.

palmierite [MINERAL] $(K,Na)_2Pb(SO_4)_2$ A white hexagonal mineral composed of potassium sodium lead sulfate.

palouser [METEOROL] A dust storm of northwestern Labrador.

palstage [GEOL] A period of time when paleogeographic conditions were changing gradually and progressively, with relation to such factors as sea level, surface relief, or distance from shore. Also known as paleogeographic stage.

palygorskite [MINERAL] **1.** A chain-structure type of clay mineral. **2.** A group of lightweight, tough, fibrous clay minerals showing extensive substitution of aluminum for magnesium.

palynology [PALEON] The study of spores, pollen, microorganisms, and microscopic fragments of megaorganisms that occur in sediments.

palynostratigraphy [PALEON] The stratigraphic application of palynologic methods.

pampero [METEOROL] A wind of gale force blowing from the southwest across the pampas of Argentina and Uruguay, often accompanied by squalls, thundershowers, and a sudden drop of temperature; it is comparable to the norther of the plains of the United States.

pan [GEOL] **1.** A shallow, natural depression or basin containing a body of standing water. **2.** A hard, cementlike layer, crust, or horizon of soil within or just beneath the surface; may be compacted, indurated, or very high in clay content. [MIN ENG] **1.** A shallow, circular, concave steel or porcelain dish in which drillers or samplers wash the drill sludge to gravity-separate the particles of heavy, dense minerals from the lighter rock powder as a quick visual means of ascertaining if the rocks traversed by the borehole contain minerals of value. **2.** The act or process of performing the above operation. [OCEANOGR] *See* pancake ice.

panabase *See* tetrahedrite.

pan-amalgamation process [MIN ENG] A process for extracting gold or silver from their ores; the ore is crushed and mixed with salt, copper sulfate, and mercury, and the gold or silver amalgamize with the mercury.

Pan-American jig [MIN ENG] Mineral jig developed to treat alluvial sands; the jig cell is pulsated vertically on a flexible diaphragm seated above the stationary hutch.

panas oetara [METEOROL] A strong, warm, dry north wind in February in Indonesia.

panautomorphic rock *See* panidiomorphic rock.

pancake [MIN ENG] A concrete disk employed in stope support. [OCEANOGR] *See* pancake ice.

pancake auger [DES ENG] An auger having one spiral web, 12 to 15 inches (30 to 38 centimeters) in diameter, attached to the bottom end of a slender central shaft; used as removable deadman to which a drill rig or guy line is anchored.

pancake ice [OCEANOGR] One or more small, newly formed pieces of sea ice, generally circular with slightly raised edges and about 1 to 10 feet (0.3 to 3 meters) across. Also known as lily-pad ice; pan; pancake; pan ice; plate ice.

pan coefficient [METEOROL] The ratio of the amount of evaporation from a large body of water to that measured in an evaporation pan.

pan conveyor *See* jigging conveyor.

pandermite *See* priceite.

panel [MIN ENG] **1.** A system of coal extraction in which the ground is laid off in separate districts or panels, pillars of extra-large size being left between. **2.** A large rectangular block or pillar of coal.

panethite [MINERAL] A phosphate mineral known only in meteorites; contains sodium, potassium, magnesium, calcium, iron, and manganese.

panfan *See* pediplain.

Pangea [GEOL] Postulated former supercontinent supposedly composed of all the continental crust of the earth, and later fragmented by drift into Laurasia and Gondwana.

panidiomorphic rock [GEOL] An igneous rock that is completely or predominantly idiomorphic. Also known as panautomorphic rock.

pannier *See* gabion.

Pannonian [GEOL] A European stage of geologic time comprising the lower Pliocene.

pannus [METEOROL] Numerous cloud shreds below the main cloud; may constitute a layer separated from the main part of the cloud or attached to it.

pan out [MIN ENG] To give a result, especially as compared with expectations; for example, in mining, the gravel may be said to pan out.

panplain [GEOL] A broad, level plain formed by coalescence of several adjacent flood plains. Also spelled panplane.

panplanation [GEOL] The action or process of formation or development of a panplain.

panplane *See* panplain.

pantellerite [PETR] A green to black extrusive rock characterized by acmite-augite or diopside, anorthoclase, and cossyrite phenocrysts in an acmite or feldspar matrix that is either pumiceous, partly glassy, fine-grained holocrystalline trachytic, or microlitic.

Panthalassa [GEOL] The hypothetical proto-ocean surrounding Pangea; supposed by some geologistis to have combined all the oceans or areas of oceanic crust of the earth at an early time in the geologic past.

Pantodonta [PALEON] An extinct order of mammals which included the first large land animals of the Tertiary.

Pantolambdidae [PALEON] A family of middle to late Paleocene mammals of North America in the superfamily Pantolambdoidea.

Pantolambdodontidae [PALEON] A family of late Eocene mammals of Asia in the superfamily Pantolambdoidea.

Pantolambdoidea [PALEON] A superfamily of extinct mammals in the order Pantodonta.

Pantolestidae [PALEON] An extinct family of large aquatic insectivores referred to the Proteutheria.

pantometer [ENG] An instrument that measures all the angles necessary for determining distances and elevations.

Pantotheria [PALEON] An infraclass of carnivorous and insectivorous Jurassic mammals; early members retained many reptilian features of the jaws.

pan-type car [MIN ENG] A vehicle for removing material from quarries; it is doorless, is reversible in direction, and can be dumped from either side.

Panzer-Forderer snaking conveyor [MIN ENG]   An armored conveyor that is moved forward behind the coal plough by means of a traveling wedge pulled along by the plough or by means of jacks or compressed-air-operated rams attached at intervals to the conveyor structure.

papagayo [METEOROL]   A violent, northeasterly fall wind on the Pacific coast of Nicaragua and Guatemala; it consists of the cold air mass of a norte which has overridden the mountains of Central America and, being a descending wind, it brings fine, clear weather.

paper peat [GEOL]   A thinly laminated peat. Also known as leaf peat.

paper shale [GEOL]   A shale that easily separates on weathering into very thin, tough, uniform, somewhat flexible layers or laminae suggesting sheets of paper.

paper spar [GEOL]   A crystallized variety of calcite occurring in thin lamellae or paperlike plates.

Pappotheriidae [PALEON]   A family of primitive, tenreclike Cretaceous insectivores assigned to the Proteutheria.

papule [GEOL]   A prolate to equant, somewhat rounded glaebule composed dominantly of clay minerals with a continuous or lamellar fabric and having sharp external boundaries.

parabituminous coal [GEOL]   Bituminous coal that contains 84–87% carbon, analyzed on a dry, ash-free basis.

parabolic dune [GEOL]   A long, scoop-shaped sand dune having a ground plan approximating the form of a parabola, with the horns pointing windward (upwind). Also known as blowout dune.

parabutlerite See butlerite.

parachronology [GEOL]   **1.** Practical dating and correlation of stratigraphic units. **2.** Geochronology based on fossils that supplement, or are used instead of, biostratigraphically significant fossils.

parachute [MIN ENG]   A kind of safety catch for mine shaft cages.

parachute weather buoy [ENG]   A general-purpose automatic weather station which can be air-dropped; it is 10 feet (3 meters) long and 22 inches (56 centimeters) in diameter, and is designed to operate for 2 months on a 6-hourly schedule, transmitting station identification, wind speed, wind direction, barometric pressure, air temperature, and sea-water temperature.

paraclinal [GEOL]   Referring to a stream or valley that is oriented in a direction parallel to the fold axes of a region.

paraconformity [GEOL]   A type of unconformity in which strata are parallel; there is little apparent erosion and the unconformity surface resembles a simple bedding plane. Also known as nondepositional unconformity; pseudoconformity.

paraconglomerate [GEOL]   A conglomerate that is not a product of normal aqueous flow but deposited by such modes of mass transport as subaqueous turbidity currents and glacier ice, that is characterized by a disrupted gravel framework, that is often unstratified, and that is notable for a content of matrix greater than gravel-sized fragments.

paracoquimbite [MINERAL] $Fe_2(SO_4)_3 \cdot 9H_2O$ A pale-violet rhombohedral mineral composed of hydrous ferric iron sulfate; it is dimorphous with coquimbite.

Paracrinoidea [PALEON]   A class of extinct Crinozoa characterized by the numerous, irregularly arranged plates, uniserial armlike appendages, and no clear distinction between adoral and aboral surfaces.

paradelta [GEOL]   The landward or upper part of a delta, or that part undergoing degradation.

paraffin coal [GEOL]   A type of light-colored bituminous coal from which oil and paraffin are produced.

paraffin dirt [GEOL]   A clay soil with rubbery or curdlike texture in the upper several inches of a soil profile near gas seeps; probably formed through biodegradation of natural gas.

paragenesis [MINERAL]   **1.** The association and order of crystallization of minerals in a rock or vein. **2.** The effect of one mineral on the development of another. Also known as mineral sequence; paragenetic sequence.

paragenetic mineralogy [MINERAL]   The study of mineral paragenesis, usually accompanying the analysis of the general geologic structures within and around the ore body.

paragenetic sequence See paragenesis.

parageosyncline [GEOL]   An epeirogenic geosynclinal basin located within a craton or stable area.

paraglomerate [GEOL]   A conglomerate which contains more matrix than gravel-sized fragments and was deposited by subaqueous turbidity flows and glacier ice rather than normal aqueous flow. Also known as conglomeratic mudstone.

paragneiss [GEOL]   A gneiss showing a sedimentary parentage.

paragonite [MINERAL] $NaAl_2(AlSi_3)O_{10}(OH)_2$ A yellowish or greenish mica species that contains sodium and usually occurs in metamorphic rock. Also known as soda mica.

parahilgardite [MINERAL] $Ca_8(B_6O_{11})_3Cl\cdot4H_2O$ A triclinic mineral composed of hydrous borate and chloride of calcium; it is dimorphous with hilgardite.

parahopeite [MINERAL] $Zn_3(PO_4)_2\cdot4H_2O$ A colorless mineral composed of hydrous phosphate of zinc, occurring in tabular triclinic crystals; it is dimorphous with hopeite.

paralaurionite [MINERAL] $PbCl(OH)$ A white mineral composed of basic lead chloride; it is dimorphous with laurionite.

paraliageosyncline [GEOL]   A geosyncline developing along a present-day continental margin, such as the Gulf Coast geosyncline.

paralic [GEOL]   Pertaining to deposits laid down on the landward side of a coast.

paralic coal basin [GEOL]   Coal deposits formed along the margin of the sea.

paralic swamp See marine swamp.

paralimnion [HYD]   The littoral part of a lake, extending from the margin to the deepest limit of rooted vegetation.

parallax age See age of parallax inequality.

parallax inequality [OCEANOGR]   The variation in the range of tide or in the speed of tidal currents due to the continual change in the distance of the moon from the earth.

parallel [GEOD]   A circle on the surface of the earth, parallel to the plane of the equator and connecting all points of equal latitude. Also known as circle of longitude; parallel of latitude.

parallel cut [ENG]   A group of parallel holes, not all charged with explosive, to create the initial cavity to which the loaded holes break in blasting a development round. Also known as burn cut.

parallel drainage pattern [HYD] A drainage pattern characterized by regularly spaced streams flowing parallel to one another over a large area.

parallel entry [MIN ENG] An intake airway parallel to the haulageway.

parallel firing [ENG] A method of connecting together a number of detonators which are to be fired electrically in one blast.

parallel fold *See* concentric fold.

parallel growth *See* parallel intergrowth.

parallel intergrowth [CRYSTAL] Intergrowth of two or more crystals in such a way that one or more axes in each crystal are approximately parallel. Also known as parallel growth.

parallelkanter [GEOL] A faceted pebble whose faces or edges are parallel.

parallel of latitude *See* parallel.

parallel ripple mark [GEOL] A ripple mark characterized by a relatively straight crest and an asymmetric profile.

parallel roads [GEOL] A series of horizontal beaches or wave-cut terraces occurring parallel to each other at different levels on each side of a glacial valley.

parallel series *See* multiple series.

parallel wire method [MIN ENG] An electrical prospecting method employing equipotential lines or curves in searching for ore bodies.

parallochthon [GEOL] Rocks that were brought from intermediate distances and deposited near an allochthonous mass during transit.

paramelaconite [MINERAL] A black tetragonal mineral composed of cupric and cuprous oxides, occurring in pyramidal crystals.

parameter [CRYSTAL] Any of the axial lengths or interaxial angles that define a unit cell.

parametric hydrology [HYD] That branch of hydrology dealing with the development and anlysis of relationships among the physical parameters involved in hydrologic events and the use of these relationships to generate, or synthesize, hydrologic events.

paramorph [MINERAL] A mineral exhibiting paramorphism.

paramorphism [MINERAL] The property of a mineral whose internal structure has changed without change in composition or external form. Formerly known as allomorphism.

Paranyrocidae [PALEON] An extinct family of birds in the order Anseriformes, restricted to the Miocene of South Dakota.

Paraparchitacea [PALEON] A superfamily of extinct ostracods in the suborder Kloedenellocopina including nonsulcate, nondimorphic forms.

paraquartzite [GEOL] A quartzite derived chiefly by contact metamorphism.

pararammelsbergite [MINERAL] $NiAs_2$ A tin white, orthorhombic or pseudoorthorhombic mineral consisting of nickel diarsenide; occurrence is usually in massive form.

pararipple [GEOL] A large, symmetric ripple whose surface slopes gently and which shows no assortment of grains.

paraschist [PETR] A schist derived from sedimentary rocks.

Paraseminotidae [PALEON]   A family of Lower Triassic fishes in the order Palaeonisciformes.

parasitic cone *See* adventive cone.

parastratigraphy [GEOL]   **1.** Supplemental stratigraphy based on fossils other than those governing the prevalent orthostratigraphy. **2.** Stratigraphy based on operational units.

Parasuchia [PALEON]   The equivalent name for Phytosauria.

paratacamite [MINERAL] $Cu_2(OH)_3Cl$ Rhombohedral mineral composed of basic copper chloride; it is dimorphous with tacamite.

Parathuramminacea [PALEON]   An extinct superfamily of foraminiferans in the suborder Fusulinina, with a test having a globular or tubular chamber and a simple, undifferentiated wall.

paratill [GEOL]   A till formed by ice-rafting in a marine or lacustrine environment; includes deposits from ice floes and icebergs.

para-time-rock unit [GEOL]   A working time-stratigraphic unit that is biostratigraphic and rock-stratigraphic in character and therefore is intrinsically transgressive with respect to time. Also known as para-time-stratigraphic unit.

para-time-stratigraphic unit *See* para-time-rock unit.

parautochthonous [GEOL]   Pertaining to a mobilized part of an autochthonous granite moved higher in the crust or into a tectonic area of lower pressure, and characterized by variable and diffuse contacts with country rocks.   [PETR]   Pertaining to a rock that is intermediate in tectonic character between autochthonous and allochthonous.

paravauxite [MINERAL] $FeAl_2(PO_4)_2(OH)_2 \cdot 8H_2O$ A colorless mineral composed of hydrous basic iron aluminum phosphate; contains more water than vauxite.

parawollastonite [MINERAL] $CaSiO_3$ A monoclinic mineral composed of silicate of calcium; it is dimorphous with wollastonite.

parcel method [METEOROL]   A method of testing for instability in which a displacement is made from a steady state under the assumption that only the parcel or parcels displaced are affected, the environment remaining unchanged.

parchettite [PETR]   An extrusive rock similar in composition to leucite tephrite but containing more leucite and some orthoclase.

Pareiasauridae [PALEON]   A family of large, heavy-boned terrestrial reptiles of the late Permian, assigned to the order Cotylosauria.

parental magma [GEOL]   The naturally occurring mobile rock material from which a particular igneous rock solidified or from which another magma was derived.

parent material [GEOL]   The unconsolidated mineral or organic material from which the true soil develops.

parent rock [GEOL]   **1.** The rock mass from which parent material is derived. **2.** *See* source rock.

paragsite [MINERAL]   A green or blue-green variety of hornblende containing sodium and found in contact-metamorphosized rocks.

parisite [MINERAL] $(Ce,La)_2Ca(CO_3)_3F_2$ A brownish-yellow secondary mineral composed of a carbonate and a fluoride of calcium, cerium, and lanthanum.

parkerite [MINERAL] $Ni_3(Bi,Pb)_2S_2$ A bright-bronze mineral composed of nickel bismuth lead sulfide.

parogenetic [GEOL] Formed previously to the enclosing rock; especially said of a concretion formed in a different (older) rock from its present (younger) host.

paroxysmal eruption *See* Vulcanian eruption.

parsettensite [MINERAL] $Mn_5Si_6O_{13}(OH)_8$ Copper-red mineral composed of hydrous silicate of manganese.

partial dislocation [CRYSTAL] The line at the edge of an extended dislocation where a slip through a fraction of a lattice constant has occurred.

partial-duration series [GEOPHYS] A series composed of all events during the period of record which exceed some set criterion; for example, all floods above a selected base, or all daily rainfalls greater than a specified amount.

partial obscuration [METEOROL] In United States weather observing practice, the designation for sky cover when part (0.1 to 0.9) of the sky is completely hidden by surface-based obscuring phenomena.

partial pluton [GEOL] That part of a composite intrusion representing a single intrusive episode.

partial potential temperature [METEOROL] The temperature that the dry-air component of an air parcel would attain if its actual partial pressure were changed to 1000 millibars (10^5 newtons per square meter).

partial pressure maintenance [PETRO ENG] The partial replacement of produced gas in an oil reservoir by gas injection to maintain a portion of the initial reservoir pressure.

partial tide [OCEANOGR] One of the harmonic components composing the tide at any point. Also known as tidal component; tidal constituent.

participation crude *See* buy-back crude.

particle diameter [GEOL] The diameter of a sedimentary particle considered as a sphere.

particle size [GEOL] The general dimensions of the particles or mineral grains in a rock or sediment based on the premise that the particles are spheres; commonly measured by sieving, by calculating setting velocities, or by determining areas of microscopic images.

particle-size analysis [GEOL] A determination of the distribution of particles in a series of size classes of a soil, sediment, or rock. Also known as size analysis; size-frequency analysis.

particle velocity [OCEANOGR] In ocean wave studies, the instantaneous velocity of a water particle undergoing orbital motion.

parting [GEOL] **1.** A bed or bank of waste material dividing mineral veins or beds. **2.** A soft, thin sedimentary layer following a surface of separation between thicker strata of different lithology. **3.** A surface along which a hard rock can be readily separated or is naturally divided into layers. [MINERAL] Fracturing a mineral along planes weakened by deformation or twinning.

parting cast [GEOL] A sand-filled tension crack produced by creep along the sea floor.

parting lineation [GEOL] A small-scale primary sedimentary structure made up of a series of parallel ridges and grooves formed parallel to the current. Also known as current lineation.

parting-plane lineation [GEOL] A parting lineation on a laminated surface, consisting of subparallel, linear, shallow grooves and ridges of low relief, generally less than 1 millimeter.

parting-step lineation [GEOL] A parting lineation characterized by subparallel, step-like ridges where the parting surface cuts across several adjacent laminae.

partiversal [GEOL] Pertaining to formations that dip in different directions roughly as far as a semicircle.

partly cloudy [METEOROL] **1.** The character of a day's weather when the average cloudiness, as determined from frequent observations, has been from 0.1 to 0.5 for the 24-hour period. **2.** In popular usage, the state of the weather when clouds are conspicuously present, but do not completely dull the day or the sky at any moment.

partridgeite *See* bixbyite.

parvafacies [GEOL] A body of rock constituting the part of any magnafacies that occurs between designated time-stratigraphic planes or key beds traced across the magnafacies.

pascoite [MINERAL] $Ca_2V_6O_{17} \cdot 11H_2O$ A dark-red-orange to yellow-orange mineral composed of hydrous vanadate of calcium.

pass [GEOGR] **1.** A natural break, depression, or other low place providing a passage through high terrain, such as a mountain range. **2.** A navigable channel leading to a harbor or river. **3.** A narrow opening through a barrier reef, atoll, or sand bar. [MIN ENG] **1.** A mine opening through which coal or ore is delivered from a higher to a lower level. **2.** A passage left in old workings for workers to travel as they move from one level to another. **3.** A treatment of the whole ore sample in a sample divider. **4.** A passage of an excavation or grading machine. **5.** In surface mining, a complete excavator cycle in removing overburden.

passage [GEOGR] A navigable channel, especially one through reefs or islands.

passage bed [GEOL] A stratum marking a transition from rocks of one geological system to those of another.

passing point [MIN ENG] The point at which two vehicles, such as coal cars or mine elevators, pass each other while going in opposite directions.

passive fold [GEOL] A fold in which the mechanism of folding, either flow or slip, crosses the boundaries of the strata at random.

passive front *See* inactive front.

passive glacier [HYD] A glacier with sluggish movement, generally occurring in a continental environment at a high latitude, where accumulation and ablation are both small.

passive permafrost [GEOL] Permafrost that will not refreeze under present climatic conditions after being disturbed or destroyed. Also known as fossil permafrost.

pastplain [GEOL] A region which has been uplifted and dissected so that it is no longer a true plain.

patch reef [GEOL] **1.** A small, irregular organic reef with a flat top forming a part of a reef complex. **2.** A small, thick, isolated lens of limestone or dolomite surrounded by rocks of different facies. **3.** *See* reef patch.

Patellacea [PALEON] An extinct superfamily of gastropod mollusks in the order Aspidobranchia which developed a cap-shaped shell and were specialized for clinging to rock.

patented claim [MIN ENG] A mining claim to which a patent has been secured from the government by compliance with the laws relating to such claims.

Paterinida [PALEON] A small extinct order of inarticulated brachiopods, characterized by a thin shell of calcium phosphate and convex valves.

paternoite *See* kaliborite.

paternoster lake [HYD]   One of a linear chain or series of small circular lakes occupying rock basins, usually at different levels, in a glacial valley, separated by morainal dams or riegels, but connected by streams, rapids, or waterfalls to resemble a rosary or string of beads. Also known as beaded lake; rock-basin lake; step lake.

patina [GEOL]   A thin, colored film produced on a rock surface by weathering.

patronite [MINERAL]   A black vanadium sulfide mineral; mined as a vanadium ore in Minasragra, Peru.

patterned ground [GEOL]   Any of several well-defined, generally symmetrical forms, such as circles, polygons, and steps, that are characteristic of surficial material subject to intensive frost action.

patterned sedimentation [GEOL]   Sedimentation characterized by a systematic sequence of beds.

pattern flood [PETRO ENG]   Waterflood of a petroleum reservoir in which there is an areal pattern of injection wells located so as to sweep oil from the area toward the bores of producing wells.

pattern shooting [ENG]   In seismic prospecting, firing of explosive charges arranged in geometric pattern.

paulingite [MINERAL]   An isometric zeolite mineral consisting of an aluminosilicate of potassium, calcium, and sodium.

paulopost *See* deuteric.

pavement [GEOL]   A bare rock surface that suggests a paved surface in smoothness, hardness, horizontality, surface extent, or close packing of its units.

pavonite [MINERAL] $AgBi_3S_5$ A mineral composed of silver bismuth sulfide.

pawdite [PETR]   A dark-colored, fine-grained, granular hypabyssal rock composed of magnetite, titanite, biotite, hornblende, calcic plagioclase, and traces of quartz.

pay dirt [MIN ENG]   Profitable mineral-rich earth or ore.

pay ore [MIN ENG]   Ore which can be mined, concentrated, or smelted at current cost of exploitation profitably at ruling market value of products.

pay streak [MIN ENG]   A layer of oil, ore, or other mineral that can be mined profitably.

pay zone [PETRO ENG]   The reservoir rock in which oil and gas are found in exploitable quantities.

peachblossom ore *See* erythrite.

pea coal [GEOL]   A size of anthracite that will pass through a 13/16-inch (20.6-millimeter) round mesh but not through a 9/16-inch (14.3-millimeter) round mesh.

peacock copper *See* peacock ore.

peacock ore [MINERAL]   A copper mineral, such as bornite, having an iridescent tarnished surface upon exposure to air. Also known as peacock copper.

peak [GEOL]   **1.** The more or less conical or pointed top of a hill or mountain. **2.** An individual mountain or hill taken as a whole, especially when isolated or having a pointed, conspicuous summit.   [METEOROL]   The point of intersection of the cold and warm fronts of a mature extra-tropical cyclone.

peak gust [METEOROL] After United States weather observing practice, the highest instantaneous wind speed recorded at a station during a specified period, usually the 24-hour observational day; therefore, a peak gust need not be a true gust of wind.

peakless pumping [MIN ENG] Spreading the pumping load over the entire day in a mine.

peak plain [GEOL] A high-level plain formed by a series of summits of approximately the same elevation, often explained as an uplifted and fully dissected peneplain. Also known as summit plain.

peak zone [PALEON] An informal biostratigraphic zone consisting of a body of strata characterized by the exceptional abundance of one or more taxa or representing the maximum development of some taxon.

pea ore [MINERAL] A variety of pisolitic limonite or bean ore occurring in small, rounded grains or masses about the size of a pea.

pearceite [MINERAL] $Ag_{16}As_2S_{11}$ A black mineral composed of sulfide of arsenic and silver.

pearlite See perlite.

pearl sinter See siliceous sinter.

pearl spar [MINERAL] A crystalline carbonate having a pearly luster; an example is ankerite.

pearlstone See perlite.

pea-soup fog [METEOROL] Any particularly dense fog.

peat [GEOL] A dark-brown or black residuum produced by the partial decomposition and disintegration of mosses, sedges, trees, and other plants that grow in marshes and other wet places.

peat bed See peat bog.

peat bog [GEOL] A bog in which peat has formed under conditions of acidity. Also known as peat bed; peat moor.

peat breccia [GEOL] Peat that has been broken up and then redeposited by water. Also known as peat slime.

peat formation [GEOCHEM] Decomposition of vegetation in stagnant water with small amounts of oxygen, under conditions intermediate between those of putrefaction and those of moldering.

peat moor See peat bog.

peat-sapropel [GEOL] A product of the degradation of organic matter that is transitional between peat and sapropel. Also known as sapropel-peat.

peat slime See peat breccia.

peat soil [GEOL] Soil containing a large amount of peat; it is rich in humus and gives an acid reaction.

pebble [GEOL] A clast, larger than a granule and smaller than a cobble having a diameter in the range of 4–64 millimeters. Also known as pebblestone. [MINERAL] See rock crystal.

pebble armor [GEOL] A desert armor made up of rounded pebbles.

pebble bed [GEOL] Any pebble conglomerate, especially one in which the pebbles weather conspicuously and fall loose. Also known as popple rock.

pebble coal [GEOL] Coal that is transitional between peat and brown coal.

pebble conglomerate [PETR] A consolidated rock consisting mainly of pebbles.

pebble dent [GEOL] A depression formed by a pebble on an unconsolidated sedimentary surface, represented by a downward curvature of laminae beneath the pebble.

pebble dike [GEOL] **1.** A clastic dike composed largely of pebbles. **2.** A tabular body containing sedimentary fragments in an igneous matrix.

pebble peat [GEOL] Peat that is formed in a semiarid climate by the accumulation of moss and algae, no more that 1/4 inch (6 millimeters) in thickness, under the surface pebbles of well-drained soils.

pebble phosphate [GEOL] A secondary phosphorite of either residual or transported origin, consisting of pebbles or concretions of phosphatic material.

pebblestone *See* pebble.

pebbly mudstone [GEOL] A delicately laminated till-like conglomeratic mudstone.

pebbly sand [GEOL] An unconsolidated sedimentary deposit containing at least 75% sand and up to a maximum of 25% pebbles.

pebbly sandstone [GEOL] A sandstone that contains 10–20% pebbles.

pectolite [MINERAL] $NaCa_2Si_3O_8(OH)$ A colorless, white, or gray inosilicate, crystallizing in the monoclinic system and having a vitreous to silky luster; hardness is 5 on Mohs scale, and specific gravity is 2.75.

ped [GEOL] A naturally formed unit of soil structure.

pedalfer [GEOL] A soil in which there is an accumulation of sesquioxides; it is characteristic of a humid region.

pedality [GEOL] The physical nature of a soil as expressed by the features of its constituent peds.

pedestal boulder [GEOL] A rock mass supported on a rock pedestal. Also known as pedestal rock.

pedestal rock *See* pedestal boulder.

pediment [GEOL] A piedmont slope formed from a combination of processes which are mainly erosional; the surface is chiefly bare rock but may have a covering veneer of alluvium or gravel. Also known as conoplain; piedmont interstream flat.

pedimentation [GEOL] The actions or processes by which pediments are formed.

pediment gap [GEOL] A broad opening formed by the enlargement of a pediment pass.

pediment pass [GEOL] A flat, narrow tongue that extends from a pediment on one side of a mountain to join a pediment on the other side.

pediocratic [GEOL] Pertaining to a period of time in which there is little diastrophism.

pedion [CRYSTAL] A crystal form with only one face; member of the asymmetric class of the triclinic system.

pediplain [GEOL] A rock-cut erosion surface formed in a desert by the coalescence of two or more pediments. Also known as desert peneplain; desert plain; panfan.

pediplanation [GEOL] The actions or processes by which pediplanes are formed.

pediplane [GEOL] Any planate erosion surface formed in the piedmont area of a desert, either bare or covered with a veneer of alluvium.

pedocal [GEOL]   A soil containing a concentration of carbonates, usually calcium carbonate; it is characteristic of arid or semiarid regions.

pedode [GEOL]   A spheroidal, discrete glaebule with a hollow interior, often with a drusy lining of crystals like that of a geode.

pedogenesis *See* soil genesis.

pedogenics [GEOL]   The study of the origin and development of soil.

pedogeochemical survey [GEOCHEM]   A geochemical prospecting survey in which the materials sampled are soil and till.

pedogeography [GEOL]   The study of the geographic distribution of soils.

pedography [GEOL]   The systematic description of soils; an aspect of soil science.

pedolith [GEOL]   A surface formation that has undergone one or more pedogenic processes.

pedologic age [GEOL]   The relative maturity of a soil profile.

pedologic unit [GEOL]   A soil considered without regard to its stratigraphic relations.

pedology *See* soil science.

pedon [GEOL]   The smallest unit or volume of soil that represents or exemplifies all the horizons of a soil profile; it is usually a horizontal, more or less hexagonal area of about 1 square meter, but may be larger.

pedorelic [GEOL]   Referring to a soil feature that is derived from a preexisting soil horizon.

pedosphere [GEOL]   That shell or layer of the planet earth in which soil-forming processes occur.

pedotubule [GEOL]   A soil feature consisting of skeleton grain (sometimes plus plasma) and having a tubular external form (either single tubes or branching systems of tubes) characterized by relatively sharp boundaries and by a relatively uniform cross-sectional size and shape (circular or elliptical).

peel-off time [ENG]   In seismic prospecting, the time correction to be applied to observed data to adjust them to a depressed reference datum.

peel thrust [GEOL]   A sedimentary sheet peeled off a sedimentary sequence, essentially along a bedding plane.

peesweep storm [METEOROL]   An early-spring storm in Scotland and England.

pegmatite [PETR]   Any extremely coarse-grained, igneous rock with interlocking crystals; pegmatites are relatively small, are relatively light colored, and range widely in composition, but most are of granitic composition; they are principal sources for feldspar, mica, gemstones, and rare elements. Also known as giant granite; granite pegmatite.

pegmatitic stage [GEOL]   A stage in the normal sequence of crystallization of magma containing volatiles when the residual fluid is sufficiently enriched in volatile materials to permit the formation of coarse-grained rocks, that is, pegmatites.

pegmatitization [GEOL]   Formation of or replacement by a pegmatite.

pegmatoid [PETR]   An igneous rock that has the coarse-grained texture of a pegmatite but that lacks graphic intergrowths or typically granitic composition.

pegmatolite *See* orthoclase.

peg model [GEOL]   Three-dimensional model used to illustrate and study stratigraphic and structural conditions of subsurface geology; consists of a flat platform onto which vertical pegs of varying heights are mounted to represent the contours of various strata.

pegostylite [GEOL] A speleothem that is formed by ascending waters.

peiroglyph [GEOL] A cross-cutting sedimentary structure, such as a sandstone dike.

Peking man [PALEON] *Sinanthropus pekinensis.* An extinct human type; the braincase was thick, with a massive basal and occipital torus structure and heavy browridges.

pekovskite [MINERAL] $CaTiO_3$ Lustrous, yellowish to gray-black, isometric or orthorhombic crystals with a hardness of 5.5; decomposes in hot sulfuric acid.

pelagic [GEOL] Pertaining to regions of a lake at depths of 10–20 meters or more, characterized by deposits of mud or ooze and by the absence of vegetation. Also known as eupelagic. [OCEANOGR] Pertaining to water of the open portion of an ocean, above the abyssal zone and beyond the outer limits of the littoral zone.

pelagic limestone [GEOL] A fine-textured limestone formed in relatively deep water by the concentration of calcareous tests of pelagic Foraminifera.

pelagochthonous [GEOL] Referring to coal derived from a submerged forest or from driftwood.

pelagosite [GEOL] A superficial calcareous crust a few millimeters thick, generally white, gray, or brownish, with a pearly luster, formed in the intertidal zone by ocean spray and evaporation, and composed of calcium carbonate accompanied by contents of magnesium carbonate, strontium carbonate, calcium sulfate, and silica that are higher than those found in normal limy sediments.

peldon [PETR] A very hard, smooth, compact sandstone with conchoidal fracture, occurring in coal measures.

Pelean cloud *See* nuée ardente.

pelelith [GEOL] Vesicular or pumiceous lava in the throat of a volcano.

Pele's hair [GEOL] A spun volcanic glass formed naturally by blowing out during quiet fountaining of fluid lava. Also known as capillary ejecta; filiform lapilli; lauoho o pele.

Pele's tears [GEOL] Volcanic glass in the form of small, solidified drops which precede pendants of Pele's hair.

pelinite [GEOL] A hydrous aluminum silicate throught to be the true clay substance in clays other than the kaolins, and considered to be an amorphous (colloidal) and plastic material of varying composition but of generally higher silica content than that in clayite, and also with appreciable alkalies or alkaline earths.

pelite [GEOL] A sediment or sedimentary rock, such as mudstone, composed of fine, clay- or mud-size particles. Also spelled pelyte.

pelitic [GEOL] Pertaining to, characteristic of, or derived from pelite.

pelitic hornfels [PETR] A fine-grained metamorphic rock derived from pelite.

pelitic schist [PETR] A foliated crystalline metamorphic rock derived from pelite.

pellet [GEOL] A fine-grained, sand-size, spherical to elliptical aggregate of clay-sized calcareous material, devoid of internal structure, and contained in the body of a well-sorted carbonate rock.

pelleting [ENG] Method of accelerating solidification of cast explosive charges by blending precast pellets of the explosives into the molten charge.

pelletization [MIN ENG] Forming aggregates of about 1/2 inch (13 millimeter) diameter from finely divided ore or coal.

pellicular water [HYD]   Films of groundwater adhering to particles or cavities above the water table.

pell-mell structure [GEOL]   A sedimentary structure characterized by absence of bedding in a coarse deposit of waterworn material; it may occur where deposition is too rapid for sorting or where slumping has destroyed the layered arrangement.

pellodite *See* pelodite.

pelmicrite [GEOL]   A limestone containing less than 25% each of intraclasts and ooliths, having a volume ratio of pellets to fossils greater than 3 to 1, and with the micrite matrix more abundant than the sparry-calcite cement.

pelodite [GEOL]   A lithified glacial rock flour which is composed of glacial pebbles in a silt or clay matrix and which was formed by redeposition of the fine fraction of a till. Also spelled pellodite.

pelogloea [GEOL]   Marine detrital slime from settled plankton.

pelphyte [GEOL]   A lake-bottom deposit consisting mainly of fine, nonfibrous plant remains.

pelsparite [PETR]   A limestone containing less than 25% each of intraclasts and ooliths, having a volume ratio of pellets to fossils greater than 3 to 1, and with the sparry-calcite cement more abundant than the micrite matrix.

Pelycosauria [PALEON]   An extinct order of primitive, mammallike reptiles of the subclass Synapsida, characterized by a temporal fossa that lies low on the side of the skull.

pelyte *See* pelite.

pencatite [PETR]   A recrystallized limestone containing periclase (or brucite) and calcite in approximately equal molecular proportions.

pencil cave [ENG]   A driller's term for hard, closely jointed shale that caves into a well in pencil-shaped fragments.

pencil cleavage [GEOL]   Cleavage in which fracture produces long slender pieces of rock.

pencil ganister [PETR]   A variety of ganister characterized by fine carbonaceous streaks or markings.

pencil gneiss [GEOL]   A gneiss that splits into thin, rodlike quartz-feldspar crystal aggregates.

pencil ore [GEOL]   Hard, fibrous masses of hematite that can be broken up into splinters.

pencil stone *See* pyrophyllite.

pendant *See* roof pendant.

pendant cloud *See* tuba.

pendent terrace [GEOL]   A connecting ribbon of sand that joins an isolated point of rock with a neighboring coast.

pendular ring [PETRO ENG]   Distribution of two nonmiscible liquids in a porous system; the pendular ring is a state of reservoir saturation in which the wetting phase is not continuous and the nonwetting phase is in contact with some of the solid surface.

pendular water [HYD]   Capillary water ringing the contact points of adjacent rock or soil particles in the zone of aeration.

pendulum anemometer [ENG]   A pressure-plate anemometer consisting of a plate which is free to swing about a horizontal axis in its own plane above its center of gravity;

the angular deflection of the plate is a function of the wind speed; this instrument is not used for station measurements because of the false reading which results when the frequency of the wind gusts and the natural frequency of the swinging plate coincide.

pendulum seismograph [ENG] A seismograph that measures the relative motion between the ground and a loosely coupled inertial mass; in some instruments, optical magnification is used whereas others exploit electromagnetic transducers, photocells, galvanometers, and electronic amplifiers to achieve higher magnification.

penecontemporaneous [GEOL] Of a geologic process or the structure or mineral that is formed by the process, occurring immediately following deposition but before consolidation of the enclosing rock.

peneplain See base-leveled plain.

peneplanation [GEOL] The actions or processes by which peneplains are formed.

penetration frequency See critical frequency.

penetration funnel [GEOL] An impact crater, generally funnel-shaped, formed by a small meteorite striking the earth at a relatively low velocity and containing nearly all the impacting mass within it.

penetration of fractures [PETRO ENG] The depth to which artificially produced fractures penetrate the fractured reservoir formation.

penetration test [ENG] A test to determine the relative values of density of noncohesive sand or silt at the bottom of boreholes.

penetration twin See interpenetration twin.

penetrative [GEOL] Referring to a texture of deformation that is uniformly distributed in a rock, without notable discontinuities; for example, slaty cleavage.

penfieldite [MINERAL] $Pb_2(OH)Cl_3$ A white hexagonal mineral composed of basic chloride of lead, occurring in hexagonal prisms.

penikkavaarite [PETR] An intrusive rock composed chiefly of augite, barkevikite, and green hornblende in a feldspathic groundmass.

peninsula [GEOGR] A body of land extending into water from the mainland, sometimes almost entirely separated from the mainland except for an isthmus.

penitent See nieve penitente.

penitent ice [HYD] A jagged spike or pillar of compacted firn caused by differential melting and evaporation; necessary for this formation are air temperature near freezing, dew point much below freezing, and strong insolation.

penitent snow [HYD] A jagged spike or pillar of compacted snow caused by differential melting and evaporation.

pennant [METEOROL] A means of representing wind speed in the plotting of a synoptic chart; it is a triangular flag, drawn pointing toward lower pressure from a wind-direction shaft.

pennantite [MINERAL] $Mn_9Al_6Si_5O_{20}(OH)_{16}$ Orange mineral composed of basic manganese aluminum silicate; member of the chlorite group; it is isomorphous with thuringite.

penninite [MINERAL] $(Mg,Fe,Al)_6(Si,Al)_4O_{11}(OH)_8$ An emerald-green, olive-green, pale-green, or bluish mineral of the chlorite group crystallizing in the monoclinic system, with a hardness of 2–2.5 on Mohs scale, and specific gravity of 2.6–2.85.

Pennsylvanian [GEOL] A division of late Paleozoic geologic time, extending from 320 to 280 million years ago, varyingly considered to rank as an independent period or as an epoch of the Carboniferous period; named for outcrops of coal-bearing bases embedded in the soft substratum of the sea.

Penokean *See* Animikean.

penroseite [MINERAL] (Ni,Co,Cu)Se$_2$ A lead gray, isometric mineral consisting of a selenide of nickel, copper, and cobalt; occurs in reniform masses.

pentad [CLIMATOL] A period of 5 consecutive days, often preferred to the week for climatological purposes since it is an exact factor of the 365-day year.

pentagonal dodecahedron *See* pyritohedron.

pentahydrite [MINERAL] MgSO$_4$·5H$_2$O A triclinic mineral composed of hydrous magnesium sulfate; it is isostructural with chalcanthite.

Pentamerida [PALEON] An extinct order of articulate brachiopods.

Pentameridina [PALEON] A suborder of extinct brachiopods in the order Pentamerida; dental plates associated with the brachiophores were well developed, and their bases enclosed the dorsal adductor muscle field.

pentice [MIN ENG] **1.** A rock pillar left, or a heavy timber bulkhead placed, in the bottom of a deep shaft of two or more compartments; the shaft is then further sunk through the pentice. **2.** In shaft sinking, a solid rock pillar left in the bottom of the shaft for overhead protection of miners while the shaft is being extended by sinking.

pentlandite [MINERAL] (Fe,Ni)$_9$S$_8$ A yellowish-bronze mineral having a metallic luster and crystallizing in the isometric system; hardness is 3.5–4 on Mohs scale, and specific gravity is 4.6–5.0; the major ore of nickel.

Penutian [GEOL] A North American stage of geologic time: lower Eocene (above Bulitian, below Ulatasian).

peperite [GEOL] A breccialike material in marine sedimentary rock, considered to be either a mixture of lava with sediment, or shallow intrusions of magma into wet sediment.

peralkaline [PETR] Of igneous rock, having a molecular proportion of aluminum lower than that of sodium oxide and potassium oxide combined.

peraluminous [PETR] Of igneous rock, having a molecular proportion of aluminum oxide greater than that of sodium oxide and potassium oxide combined.

perbituminous [GEOL] Referring to bituminous coal containing more than 5.8% hydrogen, analyzed on a dry, ash-free basis.

percentage depletion [PETRO ENG] Oil- or gas-reservoir depletion allowance calculated on the basis of unit sales and initial depletable leasehold cost.

percentage extraction [MIN ENG] The proportion of a coal seam or other ores which is removed from a mine.

percentage log [ENG] A sample log in which the percentage of each type of rock (except obvious cavings) present in each sample of cuttings is estimated and plotted.

percentage map [PETRO ENG] Contoured map in which the percentage of one component (such as sand) is compared with the total unit (such as sand plus shale).

percentage subsidence [MIN ENG] The measured amount of subsidence expressed as a percentage of the thickness of coal extracted.

percentage support [MIN ENG] The percentage of the total wall area which will actually be covered by supports.

perched aquifer [HYD] An aquifer that contains perched water.

perched block [GEOL] A large, detached rock fragment presumed to have been transported and deposited by a glacier, and perched in a conspicuous and precarious position on the side of a hill. Also known as balanced rock; perched boulder; perched rock.

perched boulder *See* perched block.

perched groundwater *See* perched water.

perched lake [HYD] A perennial lake whose surface level lies at a considerably higher elevation than those of other bodies of water, including aquifers, directly or closely associated with the lake.

perched rock *See* perched block.

perched spring [HYD] A spring that arises from a body of perched water.

perched stream [HYD] A stream whose surface level is above that of the water table and that is separated from underlying groundwater by an impermeable bed in the zone of aeration.

perched water [HYD] Groundwater that is unconfined and separated from an underlying main body of groundwater by an unsaturated zone. Also known as perched groundwater.

perched water table [HYD] The water table or upper surface of a body of perched water. Also known as apparent water table.

perching bed [GEOL] A body of rock, generally stratiform, that suports a body of perched water.

percolation [HYD] Gravity flow of groundwater through the pore spaces in rock or soil. [MIN ENG] Gentle movement of a solvent through an ore bed in order to extract a mineral.

percolation leaching [MIN ENG] The selective removal of a mineral by causing a suitable solvent to seep into and through a mass or pile of material containing the desired soluble mineral.

percolation zone [HYD] The area on a glacier or ice sheet where a limited amount of surface melting occurs, but the meltwater refreezes in the same snow layer and the whole thickness of the snow layer is not completely soaked or brought up to the melting temperature.

percussion figure [CRYSTAL] Radiating lines on a crystal section produced by a sharp blow.

percussion mark [GEOL] A small, crescent-shaped scar produced on a hard, dense pebble by a blow.

percussion side-wall sampling [PETRO ENG] A method of side-wall core sampling from wellbores in softer reservoir formations.

percussion table *See* concussion table.

percylite [MINERAL] $PbCuCl_2(OH)_2$ Mineral made up of a basic chloride of copper and lead and occurring as cubic blue crystals, with a hardness of 2.5.

pereletok [GEOL] A frozen layer of ground, at the base of the active layer, which may persist for one or several years. Also known as intergelisol.

perennial lake [HYD] A lake that retains water in its basin throughout the year and is not usually subject to extreme water-level fluctuations.

perennial spring [HYD] A spring that flows continuously, as opposed to an intermittent spring or a periodic spring.

perennial stream [HYD] A stream which contains water at all times except during extreme drought.

perezone [GEOL] A zone in which sediments accumulate along coastal lowlands; includes lagoons and brackish-water bays.

perfect crystal [CRYSTAL] A crystal without lattice defects; it is an unattained ideal or standard.

perfect prognostic [METEOROL] The observed pressure pattern at the verifying time of a forecast of some element other than pressure; used in objective forecast studies in which a forecast of the element is based on a simultaneous relation between this element and the pressure pattern plus a forecast of the pressure pattern at some future time.

perfemic rock [GEOL] An igneous rock in which the ratio of salic minerals to femic is less than 1:7.

perforated crust [HYD] A type of snow crust containing pits and hollows produced by ablation.

perforating [PETRO ENG] Special oil-well downhole procedure to make holes in tubing walls and surrounding cement; used to allow formation oil or gas to enter the wellbore tubing, or to allow water to be forced out into the formation to cause fracturing (hydraulic fracturing).

perforation deposit [GEOL] An isolated kame consisting of material that accumulated in a vertical shaft which pierced a glacier and affored no outlet for water at the bottom.

perforations [PETRO ENG] Downwell holes made in well tubing, usually by shot-and-explosive or shaped-charge techniques; used for oil or gas production from desired horizons, or for injection of acidizing or fracturing fluids into the formation at predetermined depths.

pergelation [HYD] The act or process of forming permafrost.

pergelic [GEOL] Referring to a soil temperature regime in which the mean annual temperature is less than 0°C and there is permafrost.

pergelisol table *See* permafrost table.

perhumid climate [CLIMATOL] As defined by C. W. Thornthwaite in his climatic classification, a type of climate which has humidity index values of +100 and above; this is his wettest type of climate (designated A), and compares closely to the "wet climate" which heads his 1931 grouping of humidity provinces.

perhydrous coal [GEOL] Coal that contains more than 6% hydrogen, analyzed on a dry, ash-free basis.

perhydrous maceral [GEOL] A maceral with high hydrogen content, such as exinite or resinite.

periblinite [GEOL] A variety of provitrinite consisting of cortical tissue.

periclase [MINERAL] MgO Native magnesia; a mineral occurring in granular forms or isometric crystals, with hardness of 6 on Mohs scale, and specific gravity of 3.67–3.90. Also known as periclasite.

periclasite *See* periclase.

periclinal [GEOL] Referring to strata and structures that dip radially outward from, or inward toward, a center, forming a dome or a basin.

pericline [GEOL] A fold characterized by central orientation of the dip of the beds. [MINERAL] A variety of albite elongated, and often twinned, along the *b*-axis.

pericline ripple mark [GEOL] A ripple mark arranged in an orthogonal pattern either parallel to or transverse to the current direction and having a wavelength up to 80 centimeters and amplitude up to 30 centimeters.

pericline twin law [CRYSTAL]   A parallel twin law in triclinic feldspars, in which the *b*-axis is the twinning axis and the composition surface is a rhombic section.

peridot [MINERAL]   **1.** A gem variety of olivine that is transparent to translucent and pale-, clear-, or yellowish-green in color. **2.** A variety of tourmaline approaching olivine in color.

peridotite [PETR]   A dark-colored, ultrabasic phaneritic igneous rock composed largely of olivine, with smaller amounts of pyroxene or hornblende.

peridotite shell *See* upper mantle.

perigean range [OCEANOGR]   The average range of tide at the time of perigean tides, when the moon is near perigee; the perigean range is greater than the mean range.

perigean tidal currents [OCEANOGR]   Tidal currents of increased speed occurring at the time of perigean tides.

perigean tide [OCEANOGR]   Tide of increased range occurring when the moon is near perigee.

perigenic [GEOL]   Referring to a rock constituent or mineral formed at the same time as the rock it is part of, but not formed at the specific location it now occupies in the rock.

periglacial [GEOL]   Of or pertaining to the outer perimeter of a glacier, particularly to the fringe areas immediately surrounding the great continental glaciers of the geologic ice ages, with respect to environment, topography, areas, processes, and conditions influenced by the low temperature of the ice.

periglacial climate [CLIMATOL]   The climate which is characteristic of the regions immediately bordering the outer perimeter of an ice cap or continental glacier; the principal climatic feature is the high frequency of very cold and dry winds off the ice area; it is also thought that these regions offer ideal conditions for the maintenance of a belt of intense cyclonic activity.

perimagmatic [GEOL]   Referring to a hydrothermal mineral deposit located near its magmatic source.

perimeter blasting [MIN ENG]   A method of blasting in tunnels, drifts, and raises, designed to minimize overbreak and leave clean-cut solid walls; the outside holes are loaded with very light continuous explosive charges and fired simultaneously, so that they shear from one hole to the other.

perimeter of airway [MIN ENG]   In mine ventilation, the linear distance in feet of the airway perimeter rubbing surface at right angles to the direction of the airstream.

period [GEOL]   A unit of geologic time constituting a subdivision of an era; the fundamental unit of the standard geologic time scale.

periodic current [OCEANOGR]   Current produced by the tidal influence of moon and sun or by any other oscillatory forcing function.

periodic lattice *See* lattice.

periodic spring [HYD]   A spring that ebbs and flows periodically, apparently due to natural siphon action.

peripediment [GEOL]   The segment of a pediplane extending across the younger rocks or alluvium of a basin which is always beyond but adjacent to the segment developed on the older upland rocks.

peripheral stream [HYD]   A stream that flows parallel to the edge of a glacier, usually just beyond the moraine.

Periptychidae [PALEON] A family of extinct herbivorous mammals in the order Condylarthra distinguished by specialized, fluted teeth.

peristerite [MINERAL] A gem variety of albite (An_2–An_{24}) that resembles moonstone and has a blue or bluish-white luster characterized by sharp internal reflections of blue, green, and yellow.

perlite [GEOL] A rhyolitic glass with abundant spherical or convolute cracks that cause it to break into small pearllike masses or pebbles, usually less than a centimeter across; it is commonly gray or green with a pearly luster and has the composition of rhyolite. Also known as pearlite; pearlstone.

perlitic [PETR] **1.** Of the texture of a glassy igneous rock, exhibiting small spheruloids formed from cracks due to contraction during cooling. **2.** Pertaining to or characteristic of perlite.

perlucidus [METEOROL] A cloud variety, usually of the species stratiformis, in which distinct spaces between its elements permit the sun, moon, blue sky, or higher clouds to be seen.

perm [PETRO ENG] A unit indicating the degree of permeability of a porous reservoir structure; the unit is expressed as bbl day^{-1} ft^{-2} psi^{-1} ft cp or ft^3 day^{-1} ft^{-2} psi^{-1} ft cp.

permafrost [GEOL] Perennially frozen ground, occurring wherever the temperature remains below 0°C for several years, whether the ground is actually consolidated by ice or not and regardless of the nature of the rock and soil particles of which the earth is composed.

permafrost drilling [ENG] Boreholes drilled in subsoil and rocks in which the contained water is permanently frozen.

permafrost island [GEOL] A small, shallow, isolated patch of permafrost surrounded by unfrozen ground.

permafrost line [GEOL] A line on a map representing the border of the arctic permafrost.

permafrost table [GEOL] The upper limit of permafrost. Also known as pergelisol table.

permanent aurora *See* airglow.

permanent-completion packer [PETRO ENG] A packer able to withstand large pressure differentials to allow for its permanent installation in a producing well.

permanent current [OCEANOGR] A current which continues with relatively little periodic or seasonal change.

permanent extinction [GEOL] The extinction of a lake by destruction of the lake basin, such as due to deposition of sediments, erosion of the basin rim, filling with vegetation, or catastrophic events.

permanent ice foot [HYD] An ice foot that does not melt completely in summer.

permanent monument [MIN ENG] A monument of a lasting character for marking a mining claim; it may be a mountain, hill, or ridge.

permanent pump [MIN ENG] A pump on which the mine depends for the final disposal of its drainage.

permanent thermocline *See* main thermocline.

permanent water [HYD] A source of water that remains constant throughout the year.

permeability [GEOL] The capacity of a porous rock, soil, or sediment for transmitting a fluid without damage to the structure of the medium. Also known as perviousness.

permeability-block method [PETRO ENG] Calculation method for oil recovery from water-drive oil fields in which there are variable-permeability distributions.

permeability profile [PETRO ENG] A graphical plot of porous reservoir permeability versus distance down the wellbore.

permeability trap [GEOL] An oil trap formed by lateral variation within a reservoir bed which seals the contained hydrocarbons through a change of permeability.

permeable bed [GEOL] A porous reservoir formation through which hydrocarbon fluids (oil or gas) or water (waterflood or interstitial) can flow.

permeation gneiss [PETR] A gneiss formed as a result of, or modified by, the passage of geochemically mobile materials through or into solid rock.

Permian [GEOL] The last period of geologic time in the Paleozoic era, from 280 to 225 million years ago.

permineralization [GEOL] A fossilization process whereby additional minerals are deposited in the pore spaces of originally hard animal parts.

permissible [MIN ENG] Said of equipment completely assembled and conforming in every respect with the design formally approved by the U.S. Bureau of Mines for use in gassy and dusty mines.

permissible lamp [MIN ENG] A lamp that meets the standards of the U.S. Bureau of Mines.

permissible machine [MIN ENG] A machine, such as a drill, mining machine, loading machine, conveyor, or locomotive, that meets the standards of the U.S. Bureau of Mines.

permissive [GEOL] Referring to a magmatic intrusion, and to the magma itself, whose emplacement is in spaces created by forces other than its own, such as orogenic forces. Also known as suctive.

Permo-Carboniferous [GEOL] **1.** The Permian and Carboniferous periods considered as one unit. **2.** The Permian and Pennsylvanian periods considered as a single unit. **3.** The rock unit, or the period of geologic time, transitional between the Upper Pennsylvanian and the Lower Permian periods.

perm-plug method [PETRO ENG] Laboratory method of measuring the permeability of reservoir core samples (or plugs) by the measurement of airflow through the sample at several flow rates.

perovskite [MINERAL] $Ca[TiO_3]$ A natural, yellow, brownish-yellow, reddish, brown, or black mineral and a structure type which includes no less than 150 synthetic compounds; the crystal structure is ideally cubic, it occurs as rounded cubes modified by the octahedral and dodecahedral forms, luster is subadamantine to submetallic, hardness is 5.5 on Mohs scale, and specific gravity is 4.0.

perpendicular slip [GEOL] The component of a fault slip measured at right angles to the trace of the fault on any intersecting surface.

perpendicular slope [GEOL] A very steep slope or precipitous face, as on a mountain.

perpendicular throw [GEOL] The distance between two points which were formerly adjacent in a faulted bed, vein, or other surface, measured at right angles to the surface.

perpetual frost climate [CLIMATOL] The climate of the ice cap regions of the world; thus, it requires temperatures sufficiently cold so that the annual accumulation of snow and ice is never exceeded by ablation. Also known as ice-cap climate.

Perret phase [GEOL] That stage of a volcanic eruption characterized by the emission of much high-energy gas that may significantly enlarge the volcanic conduit.

perry [METEOROL] In England, a sudden, heavy fall of rain; a squall, sometimes referred to as "half a gale."

perryite [MINERAL] $(Ni,Fe)_5(Si,P)_2$ A mineral found only in meteorites.

persalic rock [GEOL] An igneous rock in which the ratio of salic to femic minerals is greater than 7:1.

persilicic See silicic.

persistence [METEOROL] With respect to the long-term nature of the wind at a given location, the ratio of the magnitude of the mean wind vector to the average speed of the wind without regard to direction. Also known as constancy; steadiness.

persistence forecast [METEOROL] A forecast that the future weather condition will be the same as the present condition; often used as a standard of comparison in measuring the degree of skill of forecasts prepared by other methods.

perspective axis See axis of homology.

perspective chart [MAP] A chart on a perspective projection.

Pers sunshine recorder [ENG] A type of sunshine recorder in which the time scale is supplied by the motion of the sun.

perthite [GEOL] A parallel to subparallel intergrowth of potassium and sodium feldspar; the potassium-rich phase is usually the host from which the sodium-rich phase evolves.

perthitic [GEOL] Of a texture produced by perthite, exhibiting sodium feldspar as small strings, blebs, films, or irregular veinlets in a host of potassium feldspar.

perthitoid [GEOL] Referring to perthitic texture produced by minerals other than the feldspars.

perthosite [PETR] A light-colored syenite composed almost entirely of perthite, with less than 3% mafic minerals.

Peru Current [OCEANOGR] The cold ocean current flowing north along the coasts of Chile and Peru. Also known as Humboldt Current.

Peru saltpeter See soda niter.

perviousness See permeability.

Petalichthyida [PALEON] A small order of extinct dorsoventrally flattened fishes belonging to the class Placodermi; the external armor is in two shields of large plates.

petalite [MINERAL] $LiAlSi_4O_{10}$ A white, gray, or colorless monoclinic mineral composed of silicate of lithium and aluminum, occurring in foliated masses or as crystals.

Petalodontidae [PALEON] A family of extinct cartilaginous fishes in the order Bradyodonti distinguished by teeth with deep roots and flattened diamond-shaped crowns.

peter out [ENG] To fail gradually in size, quantity, or quality; for example, a mine may be said to have petered out.

Petersen grab [ENG] A bottom sampler consisting of two hinged semicylindrical buckets held apart by a cocking device which is released when the grab hits the ocean floor.

petra [GEOL] The rock materials produced in specific sedimentary organic environments.

petrifaction [GEOL] A fossilization process whereby inorganic matter dissolved in water replaces the original organic materials, converting them to a stony substance.

petrified wood See silicified wood.

petroblastesis [GEOL] Formation of rocks chiefly as the result of crystallization of diffusing ions.

petrocalcic [GEOL] Pertaining to a soil horizon that is characterized by an induration of calcium carbonate, sometimes with magnesium carbonate.

petrochemistry [GEOCHEM] An aspect of geochemistry that deals with the study of the chemical composition of rocks.

petrofabric *See* fabric.

petrofabric analysis *See* structural petrology.

petrofabric diagram *See* fabric diagram.

petrofabrics *See* structural petrology.

petrofacies *See* petrographic facies.

petrogenesis [PETR] That branch of petrology dealing with the origin of rocks, particularly igneous rocks. Also known as petrogeny.

petrogenic grid [PETR] A diagram whose coordinates are parameters of the rock-forming environment on which equilibrium curves are plotted indicating the limits of the stability fields of specific minerals and mineral assemblages.

petrogeny *See* petrogenesis.

petrogeometry *See* structural petrology.

petrographer [GEOL] An individual who does petrography.

petrographic facies [GEOL] Facies distinguished principally by composition and appearance. Also known as petrofacies.

petrographic period [GEOL] The extension in time of a rock association.

petrographic province [GEOL] A broad area in which similar igneous rocks are formed during the same period of igneous activity. Also known as comagmatic region; igneous province; magma province.

petrography [GEOL] The branch of geology that deals with the description and systematic classification of rocks, especially by means of microscopic examination.

petroleum [GEOL] A naturally occurring complex liquid hydrocarbon which after distillation yields combustible fuels, petrochemicals, and lubricants; can be gaseous (natural gas), liquid (crude oil, crude petroleum), solid (asphalt, tar, bitumen), or a combination of states.

petroleum engineer [PETRO ENG] An engineer whose primary objective is to find and produce oil or gas from petroleum reserves.

petroleum engineering [ENG] The application of almost all types of engineering to the drilling for and production of oil, gas, and liquefiable hydrocarbons.

petroleum geology [GEOL] The branch of economic geology dealing with the origin, occurrence, movement, accumulation, and exploration of hydrocarbon fuels.

petroleum secondary engineering [PETRO ENG] The process of removing oil from its native reservoirs by the use of supplemental energies after the natural energies causing oil production have been depleted.

petroleum seep *See* oil seep.

petroleum trap [GEOL] Stable underground formation (geological or physical) of such nature as to trap and hold liquid or gaseous hydrocarbons; usually consists of sand or porous rock surrounded by impervious rock or clay formations.

petroliferous [GEOL] Containing petroleum.

petrologen *See* kerogen.

petrologist [GEOL]   An individual who studies petrology.

petrology [GEOL]   The branch of geology concerned with the origin, occurrence, structure, and history of rocks, principally igneous and metamorphic rock.

petromict [GEOL]   Of a sediment, composed of metastable rock fragments.

petromorph [GEOL]   A speleothem or cave formation that is exposed to the surface by erosion of the limestone in which the cave was formed.

petromorphology *See* structural petrology.

petrophysics [GEOL]   Study of the physical properties of reservoir rocks.

petrotectonics [GEOL]   Extension of the field of structural petrology to include analysis of the movements that produced the rock's fabric. Also known as tectonic analysis.

Petterson-Nansen water bottle *See* Nansen bottle.

petzite [MINERAL] Ag_3AuTe_2 A steel-gray to iron-black mineral consisting of a silver gold telluride; hardness is 2.5–3 on Moh's scale, and specific gravity is 8.7–9.0.

peuroseite [MINERAL] $(Ni,Cu,Pb)Se_2$ A gray mineral composed of nickel copper lead selenide, occurring in columnar masses.

Pfälzian orogeny [GEOL]   A short-lived orogeny that occurred at the end of the Permian Period. Also known as Palatinian orogeny.

phacellite *See* kaliophilite.

phacolite [MINERAL]   A variety of chabazite, characterized by colorless lenticular crystals.

phacolith [GEOL]   A minor, concordant, lens-shaped, and usually granitic intrusion into folded sedimentary strata.

phanerite [PETR]   An igneous rock having phaneritic texture.

phaneritic [PETR]   Of the texture of an igneous rock, being visibly crystalline. Also known as coarse-grained; phanerocrystalline; phenocrystalline.

phanerocryst *See* phenocryst.

phanerocrystalline *See* phaneritic.

Phanerorhynchidae [PALEON]   A family of extinct chondrostean fishes in the order Palaeonisciformes having vertical jaw suspension.

Phanerozoic [GEOL]   The part of geologic time for which there is abundant evidence of life, especially higher forms, in the corresponding rock, essentially post-Precambrian.

phantom [GEOL]   A bed or member that is absent from a specific stratigraphic section but is usually present in a characteristic position in a sequence of similar geologic age.

phantom bottom [OCEANOGR]   A false bottom indicated by an echo sounder, some distance above the actual bottom; such an indication, quite common in the deeper parts of the ocean, is due to large quantities of small organisms.

phantom crystal [CRYSTAL]   A crystal containing an earlier stage of crystallization outlined by dust, minute inclusions, or bubbles. Also known as ghost crystal.

phantom horizon [GEOL]   In seismic reflection prospecting, a line constructed so that it is parallel to the nearest actual dip segment at all points along a profile.

pharmacolite [MINERAL] $CaH(AsO_4)\cdot 2H_2O$ A white to grayish monoclinic mineral composed of hydrous acid arsenate of calcium, occurring in fibrous form.

pharmacosiderite [MINERAL] $Fe_3(AsO_4)_2(OH)_3 \cdot 5H_2O$ Green or yellowish-green mineral composed of a hydrous basic iron arsenate and commonly found in cubic crystals. Also known as cube ore.

phase age *See* age of phase inequality.

phase behavior [PETRO ENG] The equilibrium relationships between water, liquid hydrocarbons, and dissolved or free gas, either in reservoirs or as liquids and gases are separated above ground in gas-oil separator systems.

phase inequality [OCEANOGR] Variations in the tide or tidal currents associated with changes in the phase of the moon.

phase lag [OCEANOGR] Angular retardation of the maximum of a constituent of the observed tide behind the corresponding maximum of the same constituent of the hypothetical equilibrium tide. Also known as tidal epoch.

phenacite *See* phenakite.

Phenacodontidae [PALEON] An extinct family of large herbivorous mammals in the order Condylarthra.

phenakite [MINERAL] Be_2SiO_4 A colorless, white, wine-yellow, pink, blue, or brown glassy mineral that crystallizes in the rhombohedral system; used as a minor gemstone. Also spelled phenacite.

phengite [MINERAL] A variety of muscovite with a high silica content.

phenicochroite *See* phoenicochroite.

phenoclastic rock [PETR] A nonuniformly sized clastic rock containing phenoclasts.

phenoclasts [PETR] The larger, conspicuous fragments in a sediment or sedimentary rock, such as cobbles in a conglomerate.

phenocryst [PETR] A large, conspicuous crystal in a porphyritic rock. Also known as phanerocryst.

phenocrystalline *See* phaneritic.

phenology [CLIMATOL] The science which treats of periodic biological phenomena with relation to climate, especially seasonal changes; from a climatologic viewpoint, these phenomena serve as bases for the interpretation of local seasons and the climatic zones, and are considered to integrate the effects of a number of bioclimatic factors.

phenoplast [PETR] A large rock fragment in a rudaceous rock that was plastic at the time of its incorporation in the matrix.

phi grade scale [GEOL] A logarithmic transformation of the Wentworth grade scale in which the diameter value of the particle is replaced by the negative logarithm to the base 2 of the particle diameter (in millimeters).

philipstadite [MINERAL] $Ca_2(Fe,Mg)_5(Si,Al)_8O_{22}(OH)_2$ Monoclinic mineral composed of basic silicate of calcium, iron, magnesium, and aluminum; member of the amphibole group.

phillipsite [MINERAL] $(K_2,Na_2CA)Al_2Si_4O_{12} \cdot H_2O$ A white or reddish zeolite mineral crystallizing in the orthorhombic system; occurs in complex fibrous crystals, which make up a large part of the red-clay sediments in the Pacific Ocean.

phlebite [PETR] Roughly banded or veined metamorphite or migmatite.

phlogopite [MINERAL] $K_2[Mg,Fe(II)]_6(Si_6,Al_2)O_{20}(OH)_4$ A yellow-brown to copper mineral of the mica group occurring in disseminated flakes, foliated masses, or large crystals; hardness is 2.5–3.0 on Mohs scale, and specific gravity is 2.8–3.0. Also known as bronze mica; brown mica.

phoenicite *See* phoenicochroite.

phoenicochroite [MINERAL] Pb_2CrO_5 A red mineral composed of basic chromate of lead, occurring in crystals and masses. Also known as beresovite; phenicochroite; phoenicite.

Pholidophoridae [PALEON]   A generalized family of extinct fishes belonging to the Pholidophoriformes.

Pholidophoriformes [PALEON]   An extinct actinopterygian group composed of mostly small fusiform marine and fresh-water fishes of an advanced holostean level.

phonolite [PETR]   A light-colored, aphanitic rock of volcanic origin, composed largely of alkali feldspar, feldspathoids, and smaller amounts of mafic minerals.

phorogenesis [GEOL]   The shifting or slipping of the earth's crust relative to the mantle.

phosgenite [MINERAL] $Pb_2Cl_2(CO_3)$ A white, yellow, or grayish mineral that crystallizes in the tetragonal system, has adamantine luster, hardness of 3 on Mohs scale, and specific gravity of 6–6.3. Also known as cromfordite; horn lead.

phosphate [MINERAL]   A mineral compound characterized by a tetrahedral ionic group of phosphate and oxygen, PO_4^{3-}.

phosphate recovery process [MIN ENG]   A process developed by the U.S. Bureau of Mines for recovering phosphate from low-grade phosphorus-bearing shales.

phosphate rock *See* phosphorite.

phosphatic nodule [GEOL]   A dark, usually black, earthy mass or pebble of variable size and shape, having a hard shiny surface and occurring in marine strata.

phosphatite [PETR]   A sedimentary rock composed of the mineral apatite in its various forms.

phosphatization [GEOCHEM]   Conversion to a phosphate or phosphates; for example, the diagenetic replacement of limestone, mudstone, or shale by phosphate-bearing solutions, producing phosphates of calcium, aluminum, or iron.

phosphoferrite [MINERAL] $(Fe,Mn)_3(PO_4)_2 \cdot 3H_2O$ A white or greenish orthorhombic mineral composed of hydrous phosphate of ferrous iron manganese phosphate; exhibits micalike cleavage.

phosphophyllite [MINERAL] $Zn_2(FeMn)(PO_4)_2 \cdot 4H_2O$ Colorless to pale-blue mineral composed of hydrous zinc ferrous iron manganese phosphate; exhibits micalike cleavage.

phosphorite [PETR]   A sedimentary rock composed chiefly of phosphate minerals. Also known as phosphate rock; rock phosphate.

phosphorization [GEOCHEM]   Impregnation or combination with phosphorus or a compound of phosphorus; for example, the diagenetic process of phosphatization.

phosphorroesslerite [MINERAL] $MgH(PO_4) \cdot 7H_2O$ A yellowish, monoclinic mineral consisting of a hydrated acid magnesium phosphate.

phosphorus-nitrogen ratio [OCEANOGR]   The proportion, by weight, of phosphorus to nitrogen in seawater or in plankton; the ratio is approximately 7:1.

phosphosiderite [MINERAL] $FePO_4 \cdot 2H_2O$ A pinkish-red mineral crystallizing in the monoclinic system, dimorphous with strengite and isomorphous with metavariscite.

phosphuranylite [MINERAL] $(UO_2)(PO_4)_2 \cdot 6H_2O$ A yellow secondary mineral composed of hydrous uranyl phosphate, occurring in powder form; it is phosphorescent when exposed to radium emanations.

photochemical smog [METEOROL]   Chemical pollutants in the atmosphere resulting from chemical reactions involving hydrocarbons and nitrogen oxides in the presence of sunlight.

photoclinometer [ENG]   A directional surveying instrument which records photographically the direction and magnitude of well deviations from the vertical.

photoclinometry [GEOL]   A technique for ascertaining slope information from an image brightness distribution, used especially for studying the amount of slope to a lunar crater wall or ridge by measuring the density of its shadow.

photogeologic anomaly [GEOL]   Any systematic deviation of a photogeologic factor from the expectable norm in a given area.

photogeologic map [GEOL]   A compilation of interpretations of a series of aerial photographs, including annotations of geologic features.

photogeology [GEOL]   The geologic interpretation of landforms by means of aerial photographs.

photogeomorphology [GEOL]   The study of landforms by means of aerial photographs.

photographic barograph [ENG]   A mercury barometer arranged so that the position of the upper or lower meniscus may be measured photographically.

photographic surveying [ENG]   Photographing of plumb bobs, clinometers, or magnetic needles in borehole surveying to provide an accurate permanent record.

photon curve [PETRO ENG]   A graphical plot of depth versus gamma radiation (photon) scatter during the radioactive logging of a well bore; used to detect differences in density at various reservoir depths.

phreatic [GEOL]   Of a volcanic explosion of material such as steam or mud, not being incandescent.

phreatic cycle [HYD]   The period of time during which the water table rises and then falls.

phreatic gas [GEOL]   A gas formed by the contact of atmospheric or surface water with ascending magma.

phreatic surface See water table.

phreatic water [HYD]   Groundwater in the zone of saturation.

phreatic-water discharge See groundwater discharge.

phreatic zone See zone of saturation.

phreatomagmatic [GEOL]   Pertaining to a volcanic explosion that extrudes both magmatic gases and steam; it is caused by the contact of the magma with groundwater or with ocean water.

phthanite See chert.

phyllarenite [PETR]   A litharenite composed chiefly of foliated, phyllosilicate-rich, metamorphic rock fragments, such as of slate, phyllite, and schist.

phyllite [PETR]   A metamorphic rock intermediate in grade between slate and schist, and derived from argillaceous sediments; has a silky sheen on the cleavage surface.

Phyllolepida [PALEON]   A monogeneric order of placoderms from the late Upper Devonian in which the armor is broad and low with a characteristic ornament of concentric and transverse ridges on the component plates.

phyllomorphic stage [GEOL] The most advanced geochemical stage of diagenesis, characterized by authigenic development of micas, feldspars, and chlorites at the expense of clays.

phyllonite [PETR] A metamorphic rock occupying an intermediate position between phyllite and mylonite.

phyllonitization [GEOL] The processes of mylonitization and recrystallization which together produce a phyllonite.

phyllosilicate [MINERAL] A structural type of silicate mineral in which flat sheets are formed by the sharing of three of the four oxygen atoms in each tetrahedron with neighboring tetrahedrons. Also known as layer silicate; sheet mineral; sheet silicate.

physical climate [CLIMATOL] The actual climate of a place, as distinguished from a hypothetical climate, such as the solar climate or mathematical climate.

physical climatology [CLIMATOL] The major branch of climatology, which deals with the explanation of climate, rather than with presentation of it (climatography).

physical exfoliation [GEOL] A type of exfoliation caused by physical forces, such as by the freezing of water that penetrated fine cracks in the rock or by the removal of overburden concealing deeply buried rocks.

physical forecasting *See* numerical forecasting.

physical geography [GEOGR] The branch of geography which deals with the description, analysis, classification, and genetic interpretation of the natural features and phenomena of the earth's surface.

physical geology [GEOL] That branch of geology concerned with understanding the composition of the earth and the physical changes occurring in it, based on the study of rocks, minerals, and sediments, their structures and formations, and their processes of origin and alteration.

physical meteorology [METEOROL] That branch of meteorology which deals with optical, electrical, acoustical, and thermodynamic phenomena of the atmosphere, its chemical composition, the laws of radiation, and the explanation of clouds and precipitation.

physical oceanography [OCEANOGR] The study of the physical aspects of the ocean, the movements of the sea, and the variability of these factors in relationship to the atmosphere and the ocean bottom.

physical residue [GEOL] A residue which results from physical, as opposed to chemical, weathering processes.

physical stratigraphy [GEOL] Stratigraphy based on the physical aspects of rocks, especially the sedimentologic aspects.

physical time [GEOL] Geologic time as measured by some physical process, such as radioactive decay of elements.

physiographic diagram [GEOL] A small-scale map showing landforms by the systematic application of a standardized set of simplified pictorial symbols that represent the appearances such forms would have if viewed obliquely from the air at an angle of about 45°. Also known as landform map; morphographic map.

physiographic feature [GEOL] A prominent or conspicuous physiographic form or noticeable part thereof.

physiographic form [GEOL] A landform considered with regard to its origin, cause, or history.

physiographic province [GEOL] A region having a pattern of relief features or landforms that differs significantly from that of adjacent regions.

phytoclimatology [CLIMATOL] The study of the microclimate in the air space occupied by plant communities, on the surfaces of the plants themselves and, in some cases, in the air spaces within the plants.

phytocollite [GEOL] A black, gelatinous, nitrogenous humic body occurring beneath or within peat deposits.

phytolith [PALEON] A fossilized part of a living plant that secreted mineral matter.

phytophoric rock [GEOL] A rock that consists of plant remains.

Phytosauria [PALEON] A suborder of Late Triassic long-snouted aquatic thecodonts resembling crocodiles but with posteriorly located external nostrils, absence of a secondary palate, and a different structure of the pelvic and pectoral girdles.

Piacention *See* Plaisancian.

pibal *See* pilot-balloon observation.

Picatinny test [ENG] An impact test used in the United States for evaluating the sensitivity of high explosives; a small sample of the explosive is placed in a depression in a steel die cup and capped by a thin brass cover, a cylindrical steel plug is placed in the center of the cover, and a 2-kilogram weight is dropped from varying heights on the plug; the reported sensitivity figure is the minimum height, in inches, at which at least 1 firing results from 10 trials.

pick-a-back conveyor [MIN ENG] A short conveyor that advances with a loader or continuous miner at the face of a mine and loads coal on the main haulage system.

picker [MIN ENG] **1.** An employee who picks or discards slate and other foreign matter from the coal in an anthracite breaker or at a picking table. **2.** A mechanical arrangement for removing slate from coal.

pickeringite [MINERAL] $MgAl_2(SO_4)_4 \cdot 22H_2O$ A white or faintly colored mineral composed of hydrous sulfate of magnesium and aluminum, occurring in fibrous masses.

picking [MIN ENG] **1.** Removal of waste material from an ore. **2.** Extraction of the lightest-grade ore from a mine. **3.** Emission of particles from the roof of a mine on the verge of collapse.

picking conveyor [MIN ENG] A continuous belt or apron conveyor used to carry a relatively thin bed of material past pickers who hand-sort or pick the material being conveyed.

picking table [MIN ENG] A flat or slightly inclined platform on which the coal or ore is run to be picked free from slate or gangue.

pick miner [MIN ENG] **1.** In anthracite and bituminous coal mining, one who uses hand tools to extract coal in underground working places. **2.** One who cuts out a channel under the bottom of the working face of coal with a pick.

picotite [MINERAL] A dark-brown variety of hercynite that contains chromium and is commonly found in dunites. Also known as chrome spinel.

picrite [PETR] A medium- to fine-grained igneous rock composed chiefly of olivine, with smaller amounts of pyroxene, hornblende, and plagioclase felspar.

picromerite [MINERAL] $K_2Mg(SO_4)_2 \cdot 6H_2O$ A white mineral composed of hydrous sulfate of magnesium and potassium, occurring as crystalline encrustations.

picropharmacolite [MINERAL] $(Ca,Mg)_3(AsO_4)_2 \cdot 6H_2O$ Mineral composed of hydrous calcium magnesium arsenate.

piecemeal stoping [GEOL] Magmatic stoping in which only isolated blocks of roof rock are assimilated.

piedmont [GEOL]   Lying or formed at the base of a mountain or mountain range, as a piedmont terrace or a piedmont pediment.

piedmont angle [GEOL]   The sharp break of slope between a hill and a plain, such as the angle at the junction of a mountain front and the pediment at its base.

piedmont bench *See* piedmont step.

piedmont benchland [GEOL]   One of several successions or systems of piedmont steps. Also known as piedmont stairway; piedmont treppe.

piedmont bulb [HYD]   The lobe or fan of ice formed when a glacier spreads out on a plain at the lower end of a valley.

piedmont flat *See* piedmont step.

piedmont glacier [HYD]   A thick, continuous ice sheet formed at the base of a mountain range by the spreading out and coalescing of valley glaciers from higher mountain elevations.

piedmont gravel [GEOL]   Coarse gravel derived from high ground by mountain torrents and spread out on relatively flat ground where the velocity of the water is decreased.

piedmont ice [HYD]   An ice sheet formed by the joining of two or more glaciers on a comparatively level plain at the base of the mountains down which the glaciers descended; it may be partly afloat.

piedmont interstream flat *See* pediment.

piedmontite *See* piemontite.

piedmont lake [HYD]   An oblong lake occupying a partly overdeepened basin excavated from rock by a piedmont glacier, or dammed by a glacial moraine.

piedmont plateau [GEOL]   A plateau lying between the mountains and the plains or the ocean.

piedmont scarp [GEOL]   A small, low cliff formed in alluvium on a piedmont slope at the foot of a steep mountain range; due to dislocation of the surface, especially by faulting. Also known as scarplet.

piedmont stairway *See* piedmont benchland.

piedmont step [GEOL]   A terracelike or benchlike piedmont feature that slopes outward or downvalley. Also known as piedmont bench; piedmont flat.

piedmont treppe *See* piedmont benchland.

piemontite [MINERAL] $Ca_2(Al,Mn^{3-},Fe)_3Si_3O_{12}(OH)$ Reddish-brown epidote mineral that contains manganese. Also known as manganese epidote; piedmontite.

pienaarite [PETR]   A sphene-rich malignite in which the feldspar is anorthoclase.

piercement *See* diapir.

piercement dome *See* diapir.

piercing *See* fusion piercing.

piercing fold *See* diapir.

piezocrystallization [GEOL]   Crystallization of a magma under pressure, such as pressure associated with orogeny.

piezogene [GEOL]   Pertaining to the formation of minerals primarily under the influence of pressure.

piezometric surface *See* potentiometric surface.

pigeonite [MINERAL] $(Mg,Fe^{2+},Ca)(MgFe^{2+})Si_{20}$ Clinopyroxene mineral species intermediate in composition between clinoenstatite and diopside, found in basic igneous rocks.

pigment minerals [MINERAL] Those minerals economically valuable as coloring agents.

pike [GEOL] A mountain or hill having a peaked summit.

pilandite [PETR] A hypabyssal rock containing abundant anorthoclase phenocrysts in a groundmass of the same mineral.

pillar [MIN ENG] An area of coal or ore left to support the overlying strata or hanging wall in a mine.

pillar-and-breast system [MIN ENG] A system of coal mining in which the working places are rectangular rooms usually five or ten times as long as they are broad, opened on the upper side of the gangway.

pillar-and-room system [MIN ENG] A system of mining whereby solid blocks of coal are left on either side of working places to support the roof until first-mining has been completed, when the pillar coal is then recovered.

pillar-and-stall system [MIN ENG] A system of working coal and other minerals where the first stage of excavation is accomplished with the roof sustained by coal or ore.

pillar burst [MIN ENG] A failure of a pillar, by crushing.

pillar drive [MIN ENG] A wide irregular drift or entry, in firm dry ground, in which the roof is supported by pillars of the natural earth, or by artificial pillars of stone, no timber being used.

pillar extraction [MIN ENG] Removal of the pillars of coal left over from mining by the pillar-and-stall method. Also known as pillar mining.

pillaring [MIN ENG] The process of extracting pillars. Also called pillar robbing; pulling pillars; robbing pillars.

pillar line [MIN ENG] Air currents which have definitely coursed through an inaccessible abandoned panel or area or which have ventilated a pillar line or a pillar area, regardless of the methane content, or absence of methane, in such air.

pillar man *See* pack builder.

pillar mining *See* pillar extraction.

pillar split [MIN ENG] An opening or crosscut driven through a pillar in the course of extraction of ore.

pillow lava [GEOL] Any lava characterized by pillow structure and presumed to have formed in a subaqueous environment. Also known as ellipsoidal lava.

pillow structure [GEOL] A primary sedimentary structure that resembles a pillow in size and shape. Also known as mammillary structure. [PETR] A pillow-shaped structure visible in some extrusive lavas attributed to the congealment of lava under water.

pilmer [METEOROL] In England, a heavy shower of rain.

pilotaxitic [GEOL] Pertaining to the texture of the groundmass of holocrystalline igneous rock in which lath-shaped microlites (usually of plagioclase) are arranged in a glass-free felty mesh, often aligned along the flow lines.

pilot-balloon observation [METEOROL] A method of winds-aloft observation, that is, the determination of wind speeds and directions in the atmosphere above a station;

involves reading the elevation and azimuth angles of a theodolite while visually tracking a pilot balloon. Also known as pibal.

pilot briefing [METEOROL] Oral comment on the observed and forecast weather conditions along a route, given by a forecaster to the pilot, navigator, or other air crew member prior to takeoff. Also known as briefing; flight briefing; flight-weather briefing.

pilot flood [PETRO ENG] A test waterflood operation (water-injection well) designed to evaluate procedures and to give advance information prior to instituting an extensive, multipoint waterflood.

pilot production [PETRO ENG] Limited or test production of oil or gas from a field to determine reservoir and product characteristics before commencing full-scale recovery operations.

pilot report [METEOROL] A report of in-flight weather by an aircraft pilot or crew member; a complete pilot report includes the following information in this order: location or extent of reported weather phenomena, time of observation, description of phenomena, altitude of phenomena, type of aircraft (only with reports of turbulence or icing). Also known as aircraft report; pirep.

pilot streamer [GEOPHYS] A relatively slow-moving, nonluminous lightning streamer, the existence of which has been postulated to help account for the observed mode of advance of a stepped leader as it initiates a lightning discharge.

pilot tunnel [ENG] A small tunnel or shaft excavated in advance of the main drivage in mining and tunnel building to gain information about the ground, create a free face, and thus simplify the blasting operations.

pimple mound [GEOL] A low, flattened, roughly circular or elliptical dome consisting of sandy loam that is entirely distinct from the surrounding soil; peculiar to the Gulf coast of eastern Texas and southwestern Louisiana.

pimple plain [GEOL] A plain distinguished by the presence of numerous, conspicuous pimple mounds.

pinacoid [CRYSTAL] An open crystal form that comprises two parallel faces.

pinacoidal class [CRYSTAL] That crystal class in the triclinic system having only a center of symmetry.

pinacoidal cleavage [CRYSTAL] A type of crystal cleavage that is parallel to one of the crystal's pinacoidal surfaces.

pinakiolite [MINERAL] $Mg_3Mn_3B_2O_{10}$ A black mineral composed of borate of magnesium and manganese; it is polymorphous with orthopinakiolite.

pinch [GEOL] Thinning of a rock layer, as where a vein narrows. [MIN ENG] *See* horseback; squeeze.

pinch-and-swell structure [GEOL] A structural condition common in pegmatites and veins of quartz in metamorphosed rocks; the vein is pinched at frequent intervals, leaving expanded parts between.

piner [METEOROL] In England, a rather strong breeze from the north or northeast.

pinfire opal [MINERAL] Opal in which the patches (small pinpoints) of play of color are very small and close together and usually less regularly spaced than the color patches in harlequin opal.

pingo [HYD] A frost mound resembling a volcano, being a relatively large and conical mound of soil-covered ice, elevated by hydrostatic pressure of water within or below the permafrost of arctic regions.

pingo ice [HYD]   Clear or relatively clear ice that occurs in permafrost; originates from groundwater under pressure.

pingo remnant [GEOL]   A rimmed depression formed by the rupturing of a pingo summit which results in the exposure of the ice core to melting followed by partial or total collapse. Also known as pseudokettle.

pinguite *See* nontronite.

pinhole chert [PETR]   Chert containing weathered pebbles pierced by minute holes or pores.

pinite [MINERAL]   A compact gray, green, or brown mica, chiefly muscovite derived from other minerals such as cordierite.

pinnacle [GEOL]   **1.** A sharp-pointed rock rising from the bottom, which may extend above the surface of the water, and may be a hazard to surface navigation; due to the sheer rise from the sea floor, no warning is given by sounding. **2.** Any high tower or spire-shaped pillar of rock, alone or cresting a summit.

pinnacled iceberg [OCEANOGR]   An iceberg weathered in such manner as to produce spires or pinnacles. Also known as irregular iceberg; pyramidal iceberg.

pinnate drainage [HYD]   A dendritic drainage pattern in which the main stream receives many closely spaced, subparallel tributaries that join it at acute angles, resembling in plan a feather.

pinnate joint *See* feather joint.

pinnoite [MINERAL] $Mg(BO_2)_2 \cdot 3H_2O$ A yellow mineral composed of hydrous borate of magnesium, occurring in nodular masses.

pinolite [PETR]   A metamorphic rock containing magnesite (breunnerite) as crystals and as granular aggregates in a schistose matrix (phyllite or talc schist).

pintadoite [MINERAL] $Ca_2V_2O_7 \cdot 9H_2O$ A green mineral consisting of a hydrated calcium vanadate; occurs as an efflorescence.

pin timbering [MIN ENG]   A method of mine roof support in which bolts are driven up into strong material, thus supporting lower weak layers.

pioneer tunnel [MIN ENG]   A small tunnel parallel to but ahead of a main tunnel and used to make crosscuts to the path that the main tunnel will follow.

piotine *See* saponite.

pipe [GEOL]   **1.** A vertical, cylindrical ore body. Also known as chimney; neck; ore chimney; ore pipe; stock. **2.** A tubular cavity of varying depth in calcareous rocks, often filled with sand and gravel. **3.** A vertical conduit through the crust of the earth below a volcano, through which magmatic materials have passed. Also known as breccia pipe.

pipe amygdule [GEOL]   An elongate amygdule occurring toward the base of a lava flow, probably formed by the generation of gases or vapor from the underlying material.

pipe clay [GEOL]   A mass of fine clay, usually lens-shaped, which forms the surface of bedrock, and upon which the gravel of old river beds often rests.

piperno [PETR]   A welded tuff characterized by flame structure.

pipernoid texture [GEOL]   The eutaxitic texture of certain extrusive igneous rocks in which dark patches and stringer occur in a light-colored groundmass.

pipe rock [PETR]   A marine sandstone containing abundant scolites.

pipestone [PETR]   A pink or mottled argillaceous stone; carved by the Indians into tobacco pipes.

pipe vesicle [GEOL]   A slender vertical cavity a few centimeters or tens of centimeters in length extending upward from the base of a lava flow.

piracy *See* capture.

pirate valley [GEOL]   A valley that appropriated the waters of another valley.

pirep *See* pilot report.

pirssonite [MINERAL] $Na_2Ca(CO_3)_2 \cdot 2H_2O$ A colorless or white orthorhombic mineral composed of hydrous carbonate of sodium and calcium.

pisanite [MINERAL] $(Fe,Cu)SO_4 \cdot 7H_2O$ A blue mineral composed of hydrous sulfate of copper and iron; it is isomorphous with kirovite and melanterite.

pisolite [PETR]   A sedimentary rock composed principally of pisoliths.

pisolith [GEOL]   Small, more or less spherical particles found in limestones and dolomites, having a diameter of 2–10 millimeters and often formed of calcium carbonate.

pisolitic [PETR]   Pertaining to pisolite or to the characteristic texture of such a rock.

pisolitic tuff [GEOL]   Of a tuff, composed of accretionary lapilli or pisolites.

pisoparite [PERT]   A limestone which contains at least 25% pisoliths and no more than 25% intraclasts, and in which the sparry-calcite cement is more abundant than the carbonate-mud matrix (micrite).

pistacite [GEOL]   A pistachio green variety of epidote, rich inferric iron.

pit [MIN ENG]   **1.** A coal mine; the term is not commonly used by the coal industry, except in reference to surface mining where the workings may be known as a strip pit. **2.** Any quarry, mine, or excavation area worked by the open-cut method to obtain material of value.

pitch *See* plunge.

pitchblende [MINERAL]   A massive, brown to black, and fine-grained, amorphous, or microcrystalline variety of uraninite which has a pitchy to dull luster and contains small quantities of uranium. Also known as nasturan; pitch ore.

pitch coal *See* bituminous lignite.

pitch mining [MIN ENG]   Mining coal beds with steep slopes.

pitch opal [MINERAL]   A yellowish to brownish inferior quality of common opal displaying a luster resembling that of pitch.

pitch ore *See* pitchblende.

pitch peat [GEOL]   Peat that resembles asphalt.

pitchstone [GEOL]   A type of volcanic glass distinguished by a waxy, dull, resinous, pitchy luster. Also known as fluolite.

pit limits [MIN ENG]   The vertical and lateral extent to which the mining of a mineral deposit by open pitting may be carried economically.

pitot-tube anemometer [ENG]   A pressure-tube anemometer consisting of a pitot tube mounted on the windward end of a wind vane and a suitable manometer to measure the developed pressure, and calibrated in units of wind speed.

pit-run gravel [GEOL]   A natural deposit of a mixture of gravel, sand, and foreign materials.

pit sampling [MIN ENG]   Using small untimbered pits to gain access to shallow alluvial deposits or ore dumps for purpose of testing or valuation.

pit slope [MIN ENG] The angle at which the wall of an open pit or cut stands as measured along an imaginary plane extended along the crests of the berms or from the slope crest to its toe.

pitted outwash plain [GEOL] An outwash plain characterized by numerous depressions such as kettles, shallow pits, and potholes.

pitted pebble [GEOL] A pebble having marked concavities not related to the texture of the rock in which it appears or to differential weathering.

pitticite [MINERAL] A mineral of varying color composed of a hydrous sulfate-arsenate of iron.

pitting [MIN ENG] The act of digging or sinking a pit.

Pityaceae [PALEOBOT] A family of fossil plants in the order Cordaitales known only as petrifactions of branches and wood.

pivotability [GEOL] A measure of roundness of sedimentary particles, expressed by the ease with which a particle can be dislodged from a surface or by the tendency of a particle to start rolling on a slope.

pivotal fault *See* rotary fault.

placanticline [GEOL] A gentle, anticlinal-like uplift of the continental platform, usually asymmetric and without a typical outline.

placer [GEOL] A mineral deposit at or near the surface of the earth, formed by mechanical concentration of mineral particles from weathered debris. Also known as ore of sedimentation.

placer claim [MIN ENG] A mining claim located upon gravel or ground whose mineral contents are extracted by the use of water, as by sluicing, or hydraulicking.

placer dredge [MIN ENG] A dredge for mining metals from placer deposits; it consists of a chain of closely connected buckets passing over an idler tumbler and an upper or driving tumbler, mounted on a structural-steel ladder which carries a series of rollers.

placer location [MIN ENG] Location of a tract of land for the sake of loose mineral-bearing or other valuable deposits on or near its surface, rather than within lodes or veins in rock in place.

placer mining [MIN ENG] **1.** The extraction and concentration of heavy metals from placers. **2.** Mining of gold by washing the sand, gravel, or talus.

placic horizon [GEOL] A black to dark red soil horizon that is usually cemented with iron and is not very permeable.

Placodermi [PALEON] A large and varied class of Paleozoic fishes characterized by a complex bony armor covering the head and the front portion of the trunk.

Placodontia [PALEON] A small order of Triassic marine reptiles of the subclass Euryapsida characterized by flat-crowned teeth in both the upper and lower jaws and on the palate.

Plaggept [GEOL] A suborder of the soil order Inceptisol, with very thick surface horizons of mixed mineral and organic materials resulting from manure or human wastes added over long periods of time.

plagiaplite [PETR] An aplite composed chiefly of plagioclase (oligoclase to andesine), possibly green hornblende, and accessory quartz, biotite, and muscovite.

Plagiaulacida [PALEON] A primitive, monofamilial suborder of multituberculate mammals distinguished by their dentition (dental formula I 3/0 C 0/0 Pm 5/4 M 2/2), having cutting premolars and two rows of cusps on the upper molars.

Plagiaulacidae [PALEON]   The single family of the extinct mammalian suborder Plagiaulacida.

plagioclase [MINERAL]   **1.** A type of triclinic feldspars having the general formula $(Na,Ca)Al(Si,Al)Si_2O_8$; they are common rock-forming minerals. **2.** A series in the plagioclase group which can be divided into a number of varieties based on the relative proportion of the solid solution end members, albite and anorthite (An): albite (An 0–10) oligoclase (An 10–30), andesine (An 30–50), labradorite (An 50–70), bytownite (An 70–90), and anorthite (An 90–100). Also known as sodium-calcium feldspar.

plagiohedral [CRYSTAL]   Pertaining to obliquely arranged spiral faces; in particular, to a member of a group in the isometric system with 13 axes but no center or planes.

plagionite [MINERAL] $Pb_5Sb_8S_{17}$ A lead-gray mineral with metallic appearance, composed of sulfide of lead and antimony.

Plagiosauria [PALEON]   An aberrant Triassic group of labyrinthodont amphibians.

plain [GEOGR]   An extensive, broad tract of level or rolling, almost treeless land with a shrubby vegetation, usually at a low elevation.   [GEOL]   A flat, gently sloping region of the sea floor, Also known as submarine plain.

plain of denudation [GEOL]   A surface that has been reduced to, or just above, sea level by the agents of erosion (usually considered to be of subaerial origin).

plain of lateral planation [GEOL]   An extensive, smooth, apronlike surface developed at the base of a mountain or escarpment by the widening of valleys and the coalescence of floodplains as a result of lateral planation.

plain of marine denudation [GEOL]   A plane or nearly plane surface worn down by the gradual encroachment of ocean waves upon the land, or a plane or nearly plane imaginary surface representing such a plain after uplift and partial subaerial erosion. Also known as plain of submarine denudation.

plain of marine erosion [GEOL]   A largely theoretical platform representing a plane surface of unlimited width, produced below sea level by the cutting away altogether of the land by marine processes acting over a very long period of stillstand.

plain of submarine denudation See plain of marine denudation.

plains-type fold [GEOL]   An anticlinal or domelike structure of the continental platform which has no typical outline and for which there is no corresponding synclinal structure.

plain tract [GEOL]   The lower part of a stream, characterized by a low gradient and a wide floodplain.

Plaisancian [GEOL]   A European stage of geologic time: lower Pliocene (above Pontian of Miocene, below Astian). Also known as Piacention; Plaisanzian.

Plaisanzian See Plaisancian.

plaiting [GEOL]   A texture in some schists that results from the intersection of relict bedding planes with well-developed cleavage planes. Also known as gaufrage.

planar cross-bedding [GEOL]   Cross-bedding characterized by planar surfaces of erosion in the lower bounding surface.

planar flow structure See platy flow structure.

planation [GEOL]   Erosion resulting in flat surfaces, caused by meandering streams, waves, ocean currents, wind, or glaciers.

planation stream piracy [HYD]   Capture effected by the lateral planation of a stream invading and diverting the upper part of a smaller stream.

plane atmospheric wave   [METEOROL]   An atmospheric wave represented in two-dimensional rectangular cartesian coordinates, in contrast to a wave considered on the spherical earth.

plane bed   [GEOL]   A sedimentary bed without elevations or depressions larger than the maximum size of the bed material.

plane defect   [CRYSTAL]   A type of crystal defect that occurs along the boundary plane of two regions of crystal, or between two grains.

plane dendrite   *See* plane-dendritic crystal.

plane-dendritic crystal   [CRYSTAL]   An ice crystal exhibiting an elaborately branched (dendritic) structure of hexagonal symmetry, with its much larger dimension lying perpendicular to the principal (*c*-axis) of the crystal.   Also known as plane dendrite; stellar crystal.

plane fault   [GEOL]   A fault whose surface is planar rather than curved.

plane jet   [HYD]   A stream flow pattern characteristic of hyperpycnal inflow, in which the inflowing water spreads as a parabola whose width is about three times the square root of the distance downstream from the mouth.

plane of contemporaneity   [GEOL]   The horizontal or nearly horizontal line between stratigraphic units (primarily formations) as seen in section.

plane of mirror symmetry   [CRYSTAL]   In certain crystals, a symmetry element whereby reflection of the crystal through a certain plane leaves the crystal unchanged. Also known as mirror plane of symmetry; plane of symmetry; reflection plane; symmetry plane.

plane of saturation   *See* water table.

plane of symmetry   *See* plane of mirror symmetry.

planerite   [MINERAL]   **1.** A variety of coeruleolactite containing copper. **2.** A variety of turquoise containing calcium.

plane surveying   [ENG]   Measurement of areas on the assumption that the earth is flat.

plane table   [ENG]   A surveying instrument consisting of a drawing board mounted on a tripod and fitted with a compass and a straight-edge ruler; used to graphically plot survey lines directly from field observations.

plane-table method   [MIN ENG]   A method of measuring areas of mine roadways; a drawing board is set up on a tripod in the plane of the mine section to be measured; the distance from a central point on the board to the perimeter of the roadway is measured with a tape along various offsets; the distance measured is scaled on the drawing board along the proper offset line.

planetary boundary layer   [METEOROL]   That layer of the atmosphere from the earth's surface to the geostrophic wind level, including, therefore, the surface boundary layer and the Ekman layer; above this layer lies the free atmosphere.

planetary geology   [GEOL]   A science that applies geologic principles and techniques to the study of planets and their natural satellites. Also known as planetary geoscience.

planetary geoscience   *See* planetary geology.

planetary vorticity effect   [GEOPHYS]   The effect of the variation of the earth's vorticity with latitude in altering the relative vorticity of a flow with a meridional component; a fluid with a free surface in a rotating cylinder exhibits a corresponding effect, owing to the shrinking or stretching of radially displaced columns.

planetary wave *See* long wave; Rossby wave.

planetary wind [METEOROL] Any wind system of the earth's atmosphere which owes its existence and direction to solar radiation and to the rotation of the earth.

planform [GEOGR] The shape of a body of water according to the still-water line.

planimetric map [MAP] A map indicating only the horizontal positions of features, without regard to elevation, in contrast with a topographic map, which indicates both horizontal and vertical positions. Also known as line map.

planisphere [MAP] A representation, on a plane, of the celestial sphere, especially one on a polar projection, with means provided for making certain measurements such as altitude and azimuth.

plankton net [ENG] A net for collecting plankton.

planoclastic rock [PETR] An even-grained or uniformly sized clastic rock.

planoconformity [GEOL] The relation between conformable strata that are approximately uniform in thickness and sensibly parallel throughout.

Planosol [GEOL] An intrazonal, hydromorphic soil having a clay pan or hardpan covered with a leached surface layer; developed in a humid to subhumid climate.

plash [HYD] A shallow, standing, usually short-lived pool or small pond resulting from a flood, heavy rain, or melting snow.

plasma mantle [GEOPHYS] A thick layer of plasma just inside the magnetopause characterized by a tailward bulk flow with a speed of 100 to 200 kilometers per second and by a gradual decrease of density, temperature, and speed as the depth inside the magnetosphere increases.

plasmapause [GEOPHYS] The sharp outer boundary of the plasmasphere, at which the plasma density decreases by a factor of 100 or more.

plasma sheet [GEOPHYS] A region of relatively hot plasma outside the plasmasphere, which reaches, during quiet times, from an altitude of about 50,000 kilometers to at least past the moon's orbit in a long tail extending away from the sun; composed of particles with typical thermal energies of 2 to 4 kiloelectronvolts.

plasmasphere [GEOPHYS] A region of relatively dense, cold plasma surrounding the earth and extending out to altitudes of approximately 2 to 6 earth radii, composed predominantly of electrons and protons, with thermal energies not exceeding several electron volts.

plaster conglomerate [GEOL] A conglomerate composed entirely of boulder derived from, and forming a wedgelike mass on the flank of, a partially exhumed monadnock.

plastering-on [GEOL] The addition of material to a ground moraine by the melting of ice at the base of a glacier.

plaster shooting [ENG] A surface blasting method used when no rock drill is necessary or one is not available; consists of placing a charge of gelignite, primed with safety fuse and detonator, in close contact with the rock or boulder and covering it completely with stiff damp clay.

plastic equilibrium [GEOL] State of stress within a soil mass or a portion thereof that has been deformed to such an extent that its ultimate shearing resistance is mobilized.

plasticity index [GEOL] The percent difference between moisture content of soil at the liquid and plastic limits.

plasticlast [GEOL] An intraclast consisting of calcareous mud that has been torn up while still soft.

plastic limit [GEOL] The water content of a sediment, such as a soil, at the point of transition between the plastic and semisolid states.

plat [MAP] A plan that shows land ownership, boundaries, and subdivisions together with data for description and identification of various parts.

plate [GEOL] **1.** A smooth, thin, flat fragment of rock, such as a flagstone. **2.** A large rigid, but mobile, block involved in plate tectonics; thickness ranges from 50 to 250 kilometers and includes both crust and a portion of the upper mantle.

plate anemometer *See* pressure-plate anemometer.

plateau [GEOGR] An extensive, flat-surfaced upland region, usually more than 150–300 meters in elevation and considerably elevated above the adjacent country and limited by an abrupt descent on at least one side. [GEOL] A broad, comparatively flat and poorly defined elevation of the sea floor, commonly over 200 meters in elevation.

plateau basalt [GEOL] One or a succession of high-temperature basaltic lava flows from fissure eruptions which accumulate to form a plateau. Also known as flood basalt.

plateau eruptions [GEOL] Successive lava flows from fissures that spread in sheets over a large area.

plateau glacier [HYD] A highland glacier that overlies a generally flat mountain tract; usually overflows its edges in hanging glaciers.

plateau gravel [GEOL] A sheet, spread, or patch of surficial gravel, often compacted, occupying a flat area on a hilltop, plateau, or other high region at a height above that normally occupied by a stream-terrace gravel.

plateau level [PETRO ENG] The peak production level reached by an oil field.

plateau mountain [GEOL] A pseudomountain produced by the dissection of a plateau.

plateau plain [GEOL] An extensive plain surmounted by a sublevel summit area and bordered by escarpments.

plate crystal [HYD] An ice crystal exhibiting typical hexagonal (rarely triangular) symmetry and having comparatively little thickness parallel to its principal axis (c-axis); as such crystals fall through the clouds in which they form, they may encounter conditions causing them to develop dendritic extensions, that is, to become plane-dendritic crystals.

plate ice *See* pancake ice.

plate tectonics [GEOL] Global tectonics based on a model of the earth characterized by a small number (10–25) of semirigid plates which float on some viscous underlayer in the mantle; each plate moves more or less independently and grinds against the others, concentrating most deformation, volcanism, and seismic activity along the periphery. Also known as raft tectonics.

platform [GEOL] **1.** Any level or almost level surface; a small plateau. **2.** A continental area covered by relatively flat or gently tilted, mainly sedimentary strata which overlay a basement of rocks consolidated during earlier deformations; platforms and shields together constitute cratons. [MIN ENG] A wooden floor on the side of a gangway at the bottom of an inclined seam, to which the coal runs by gravity, and from which it is shoveled into mine cars.

platform beach [GEOL] A looped bar or ridge of sand and gravel formed on a wave-cut platform.

platform facies *See* shelf facies.

platform reef [GEOL]   An organic reef, generally small but more extensive than a patch reef, with a flat upper surface.

platiniridium [MINERAL]   A silver-white cubic mineral composed of platinum, iridium, and related metals, occurring in grains.

platinite *See* platynite.

platte [GEOL]   A resistant knob of rock in a glacial valley or rising in the midst of an existing glacier, often causing a glacier to split near its snout.

plattnerite [MINERAL] PbO_2 An iron-black mineral consisting of lead dioxide, occurring in masses with submetallic luster.

platy [GEOL]   **1.** Referring to a sedimentary particle whose length is more than three times its thickness. **2.** Referring to a sandstone or limestone that splits into laminae having thicknesses in the range of 2 to 10 millimeters.

Platybelondoninae [PALEON]   A subfamily of extinct elephantoid mammals in the family Gomphotheriidae consisting of species with digging specializations of the lower tusks.

Platyceratacea [PALEON]   A specialized superfamily of extinct gastropod mollusks which adapted to a coprophagous life on crinoid calices.

platy flow structure [PETR]   Structure of an igneous rock characterized by tabular sheets which suggest stratification, and formation by contraction during cooling. Also known as linear flow structure; planar flow structure.

platynite [MINERAL] $PbBi_2(Se,S)_3$ An iron-black mineral composed of selenide and sulfide of lead and bismuth; occurs in thin metallic plates resembling graphite. Also spelled platinite.

Platysomidae [PALEON]   A family of extinct palaeonisciform fishes in the suborder Platysomoidei; typically, the body is laterally compressed and rhombic-shaped, with long dorsal and anal fins.

Platysomoidei [PALEON]   A suborder of extinct deep-bodied marine and fresh-water fishes in the order Palaeonisciformes.

playa [GEOL]   **1.** A low, essentially flat part of a basin or other undrained area in an arid region. **2.** A small, generally sandy land area at the mouth of a stream or along the shore of a bay. **3.** A flat, alluvial coastland, as distinguished from a beach.

playa lake [HYD]   A shallow temporary sheet of water covering a playa in the wet season.

Playfair's law [GEOL]   The law that each stream cuts its own valley, the valley being proportional in size to its stream, and the stream junctions in the valley are accordant in level.

Pleistocene [GEOL]   An epoch of geologic time of the Quaternary period, following the Tertiary and before the Holocene. Also known as Ice Age; Oiluvium.

pleomorphism *See* polymorphism.

pleonaste *See* ceylonite.

pleonastite [PETR]   An igneous rock similar in structure to diabase and composed of ceylonite, hercynite, and clinochlore surrounding corundum crystals.

plerotic water [HYD]   That part of subsurface water that forms the zone of saturation, including underground streams.

Plesiocidaroida [PALEON]   An extinct order of echinoderms assigned to the Euechinoidea.

Plesiosauria [PALEON]   A group of extinct reptiles in the order Sauropterygia constituting a highly specialized offshoot of the nothosaurs.

plessite [MINERAL]   A meteorite mineral consisting of an intimate, fine-grained intergrowth of kamacite and taenite.

Pleuracanthodii [PALEON]   An order of Paleozoic sharklike fishes distinguished by two-pronged teeth, a long spine projecting from the posterior braincase, and direct backward extension of the tail.

Pleuromeiaceae [PALEOBOT]   A family of plants in the order Pleuromiales, but often included in the Isoetales due to a phylogenetic link.

Pleuromeiales [PALEOBOT]   An order of Early Triassic lycopods consisting of the genus *Pleuromeia*; the upright branched stem had grasslike leaves and a single terminal strobilus.

Pleurotomariacea [PALEON]   An extinct superfamily of gastropod mollusks in the order Aspidobranchia.

plication [GEOL]   Intense, small-scale folding.

Pliensbachian [GEOL]   A European stage of geologic time: Lower Jurassic (above Sinemurian, below Toarcian).

Plinian eruption *See* Vulcanian eruption.

plinthite [GEOL]   In a soil, a material consisting of a mixture of clay and quartz with other diluents, that is rich in sesquioxides, poor in humus, and highly weathered.

Pliocene [GEOL]   A worldwide epoch of geologic time of the Tertiary period, extending from the end of the Miocene to the beginning of the Pleistocene.

Pliohyracinae [PALEON]   An extinct subfamily of ungulate mammals in the family Procaviidae.

pliothermic [GEOL]   Pertaining to a period in geologic history characterized by more than average climatic warmth.

plough [MIN ENG]   **1.** A continuous mining machine in which cutting blades, moved over the face being worked, bite into the coal as they are pulled along and discharge it on an accompanying conveyor. **2.** A V-shaped scraper that presses against the return belt of a conveyor, removing coal and debris from it.

plough wind *See* plow wind.

plowshare [HYD]   A wedge-shaped feature developed on a snow surface by further ablation of foam crust.

plow sole [GEOL]   A pressure pan representing a layer of soil compacted by repeated plowing to the same depth.

plow wind [METEOROL]   A term used in the midwestern United States to describe strong, straight-line winds associated with squall lines and thunderstorms; resulting damage is usually confined to narrow zones like that caused by tornadoes; however, the winds are all in one direction. Also spelled plough wind.

plucking [GEOL]   A process of glacial erosion which involves the penetration of ice or rock wedges into subglacial niches, crevices, and joints in the bedrock; as the glacier moves, it plucks off pieces of jointed rock and incorporates them. Also known as glacial plucking; quarrying.

pluck side [GEOL]   The downstream, or lee, side of a roche moutonnée, roughened and steepened by glacial plucking.

plug [GEOL]   **1.** A vertical pipelike magmatic body representing the conduit to a former volcanic vent. **2.** A crater filling of lava, the surrounding material of which has been removed by erosion. **3.** A mass of clay, sand, or other sediment filling the part of a stream channel abandoned by the formation of a cutoff.   [MIN ENG]   A watertight seal in a shaft formed by removing the lining and inserting a concrete dam, or by placing a plug of clay over ordinary debris used to fill the shaft up to the location of the plug.

plug-and-feather method   [MIN ENG]   A method of breaking large quarry stones into smaller blocks; a row of holes is drilled in the stone along a line where the break is desired; a pair of feathers (semicircular cross-section rods) is inserted in each hole; a plug (steel wedge) is inserted between each feather pair; the plugs are hammered in succession until the stone fractures.

plug back   [PETRO ENG]   To place cement or a mechanical plug in a well bottom for the purpose of excluding bottom water, sidetracking, or producing from a formation already drilled through.

plug bit   *See* noncoring bit.

plug dome   [GEOL]   A volcanic dome characterized by an upheaved, consolidated conduit filling.

plugging   [MIN ENG]   *See* blinding.   [PETRO ENG]   The act or process of stopping the flow of water, oil, or gas in strata penetrated by a borehole or well so that fluid from one stratum will not escape into another or to the surface; especially the sealing up of a well that is dry and is to be abandoned.

plugging agent   [PETRO ENG]   A chemical used to plug or block off selected permeable zones of a reservoir formation; used during formation acidizing to direct the acid to the tighter (less permeable) zones; examples are viscous gels, suspensions of graded solids, and finely ground vegetable material.

plugging-back   [PETRO ENG]   The act or process of cementing off a lower section of casing, or of blocking fluids below from rising in the casing to a higher section being tested.

plughole   [MIN ENG]   **1.** A passageway left open while an old portion of a mine is sealed off, to help maintain normal ventilation; it is sealed when the work is finished. **2.** A hole for an explosive charge or for a bolt.

plug reef   [GEOL]   A small, triangular reef that grows with its apex pointing seaward through openings between linear shelf-edge reefs.

plum   [GEOL]   A clast embedded in a matrix of a different kind, especially a pebble in conglomerate.

plumasite   [PETR]   A coarsely xenomorphic-granular hypabyssal rock of variable composition, but chiefly of corundum crystals enclosed in oligoclase grains.

plumbago   *See* graphite.

plumb line   [GEOPHYS]   A continuous curve to which the direction of gravity is everywhere tangential.

plumboferrite   [MINERAL] $PbFe_4O_7$ A dark hexagonal mineral composed of lead iron oxide.

plumbogummite   [MINERAL]   **1.** $PbAl_3(PO_4)_2(OH)_5 \cdot H_2O$ A mineral composed of hydrous basic lead aluminum phosphate. **2.** A group of isostructural minerals, that includes gorceixite, goyazite, crandallite, deltaite, florencite, and dussertite, as well as plumbogummite.

plumbojarosite   [MINERAL] $PbFe_6(SO_4)_4(OH)_{12}$ A mineral composed of basic lead iron sulfate; it is isostructural with jarosite.

plumboniobite   [MINERAL]   A dark brown to black mineral of complex composition, consisting of a niobate of yttrium, uranium, lead, iron, and rare earths.

plumose mica [MINERAL] A feathery variety of muscovite mica.

plumose ore *See* plumosite.

plumosite [MINERAL] An antimony-sulfide mineral having a feathery form, for example, jamesonite and boulangerite. Also known as plumose ore.

plunge [ENG] **1.** To set the horizontal cross hair of a theodolite in the direction of a grade when establishing a grade between two points of known level. **2.** *See* transit. [GEOL] The inclination of a geologic structure, especially a fold axis, measured by its departure from the horizontal. Also known as pitch; rake.

plunge basin [GEOL] A deep, large hollow or cavity scoured in the bed of a stream at the foot of a waterfall or cataract by the force and eddying effect of the falling water.

plunge point [OCEANOGR] The point at which a plunging wave curls over and falls as it moves toward the shore.

plunge pool [HYD] **1.** The water in a plunge basin. **2.** A deep, circular lake occupying a plunge basin after the waterfall has ceased to exist or the stream has been diverted. Also known as waterfall lake. **3.** A small, deep plunge basin.

plunger jig washer [MIN ENG] A machine for washing ore, coal, or stones in which water is forced alternately up or down by a plunger.

plunger lift [PETRO ENG] A method of lifting oil by using compressed gas to drive a free piston from the lower end of the tubing string to the surface.

plunger overtravel [PETRO ENG] Excessive upward or downward movement in a reciprocating-plunger-type sucker-rod oil-well pump.

plunging breaker [OCEANOGR] A breaking wave whose crest curls over and collapses suddenly. Also known as spilling breaker; surging breaker.

plunging cliff [GEOL] A sea cliff bordering directly on deep water, having a base that lies well below water level.

plush copper ore *See* chalcotrichite.

plutology [GEOL] The study of the interior of the earth.

pluton [GEOL] **1.** An igneous intrusion. **2.** A body of rock formed by metasomatic replacement.

plutonian *See* plutonic.

plutonic [GEOL] Pertaining to rocks formed at a great depth. Also known as abyssal; deep-seated; plutonian.

plutonic breccia [GEOL] Breccia consisting of older annular rock fragments enclosed in younger plutonic rock.

plutonic metamorphism [GEOL] Deep-seated regional metamorphism at high temperatures and pressures, often accompanied by strong deformation.

plutonic rock [GEOL] A rock formed at considerable depth by crystallization of magma or by chemical alteration.

plutonic water [HYD] Juvenile water in, or derived from, magma at a considerable depth, probably several kilometers.

plutonism [GEOL] **1.** Pertaining to the processes associated with pluton formation. **2.** The theory that the earth formed by solidification of a molten mass.

pluvial [GEOL] Of a geologic process or feature, effected by rain action. [METEOROL] Pertaining to rain, or more broadly, to precipitation, particularly to an abundant amount thereof.

pluvial lake [GEOL]   A lake formed during a period of exceptionally heavy rainfall; specifically, a Pleistocene lake formed during a period of glacial advance and now either extinct or only a remnant.

pluviofluvial [GEOL]   Pertaining to the combined action of rainwater and streams.

pluviograph *See* recording rain gage.

pluviometer *See* rain gage.

pluviometric coefficient [METEOROL]   For any month at a given station, the ratio of the monthly normal precipitation to one-twelfth of the annual normal precipitation. Also known as hyetal coefficient.

pneumatic filling [MIN ENG]   A filling method using compressed air to blow filling material into the mined-out stope.

pneumatic injection [MIN ENG]   A method for fighting underground coal fires, developed by the U.S. Bureau of Mines; this air-blowing technique involves the injection of incombustible mineral, like rock wool or dry sand, through 6-inch (15 centimeter) boreholes drilled from the surface to intersect underground passageways in the mines.

pneumatic lighting [MIN ENG]   Lighting of underground chambers by a compressed-air turbomotor driving a small dynamo.

pneumatic method [MIN ENG]   A method of flotation in which gas is introduced near the bottom of the flotation vessel.

pneumatic stowing [MIN ENG]   A method of filling used mine cavities with crushed rock; which is forced by compressed air into the cavity.

pneumatic tank switcher [PETRO ENG]   Pneumatic actuated valving system for oil-field tanks to shut off crude-oil flow to a filled tank and then to direct incoming crude flow to the next available empty tank.

pneumatogenic [GEOL]   Referring to a rock or mineral deposit formed by a gaseous agent.

pneumatolysis [GEOL]   Rock alteration or mineral crystallization effected by gaseous emanations from solidifying magma.

pneumatolytic [GEOL]   Formed by gaseous agents.

pneumatolytic metamorphism [PETR]   Contact metamorphism by the chemical action of magmatic gases.

pneumatolytic stage [GEOL]   The stage in the cooling of a magma in which the solid and gaseous phases are in equilibrium.

pneumotectic [GEOL]   Referring to processes and products of magmatic consolidation affected to some degree by gaseous constituents of the magma.

pocket [GEOL]   **1.** A localized enrichment. **2.** An enclosed or sheltered place along a coast, such as a reentrant between rocky, cliffed headlands or a bight on a lee shore. [MIN ENG]   A receptacle from which coal, ore, or waste is loaded into wagons or cars.

pocket beach [GEOL]   A small, narrow beach formed in a pocket, commonly crescentic in plan, with the concave edge toward the sea, and displaying well-sorted sands.

pocket valley [GEOL]   A valley whose head is enclosed by steep walls at the base of which underground water emerges as a spring.

pod [GEOL]   An orebody of elongate, lenticular shape. Also known as podiform orebody.

podiform orebody *See* pod.

Podzol [GEOL] A soil group characterized by mats of organic matter in the surface layer and thin horizons of organic minerals overlying gray, leached horizons and dark-brown illuvial horizons; found in coal forests to temperate coniferous or mixed forests.

podzolization [GEOL] The process by which a soil becomes more acid because of the depletion of bases, and develops surface layers that have been leached of clay.

Poetsch process [MIN ENG] Shaft sinking in which brine at subzero temperature is circulated through boreholes to freeze running water through which a shaft or tunnel is to be driven, during development of a waterlogged mine.

pogonip *See* ice fog.

poikilitic [PETR] Of the texture of an igneous rock, having small crystals of one mineral randomly scattered without common orientation in larger crystals of another mineral.

poikiloblast [GEOL] A large crystal (xenoblast) formed by recrystallization during metamorphism and containing numerous inclusions of small idioblasts.

poikiloblastic [PETR] Of a metamorphic texture, simulating the poikilitic texture of igneous rocks in having small idioblasts of one constituent lying within larger xenoblasts. Also known as sieve texture.

poikilocrystallic *See* poikilotopic.

poikilotope [GEOL] A large crystal enclosing smaller crystals of another mineral in a sedimentary rock showing poikilotopic fabric.

poikilotopic [GEOL] Referring to the fabric of a crystalline sedimentary rock in which the contituent crystals are of more than one size and in which larger crystals enclose smaller crystals of another mineral. Also known as poikilocrystallic.

point [GEOGR] A tapering piece of land projecting into a body of water; it is generally less prominent than a cape.

point bar [GEOL] One of a series of low, arcuate sand and gravel ridges formed on the inside of a growing meander by the gradual addition of accretions. Also known as meander bar.

point defect [CRYSTAL] A departure from crystal symmetry which affects only one, or, in some cases, two lattice sites.

point diagram [PETR] A fabric diagram in which a point represents the preferred orientation of each individual fabric element. Also known as scatter diagram.

point group [CRYSTAL] A group consisting of the symmetry elements of an object having a single fixed point; 32 such groups are possible.

point rainfall [METEOROL] The rainfall during a given time interval (or often one storm) measured in a rain gage, or an estimate of the amount which might have been measured at a given point.

point sample [GEOL] A sample of the sediment contained at a single point in a body of water.

points of the compass *See* compass points.

poised stream [HYD] A stream that is neither eroding nor depositing sediment.

polacke [METEOROL] A cold, dry, northeasterly katabatic wind in Bohemia descending from the Sudeten Mountains (from the direction of Poland).

polar air [METEOROL] A type of air whose characteristics are developed over high latitudes; there are two types: continental polar air and maritime polar air.

polar anticyclone *See* arctic high; subpolar high.

polar automatic weather station [METEOROL]   An automatic weather station which measures meteorological elements and transmits them by radio; the station is designed to function primarily in frigid or polar climates in order to fill the need for weather reports from inaccessible regions where manned stations are not practicable; since the equipment is designed to operate on ice or slush, the main structure is in the form of a sled with external pontoons for added stability.

polar axis [CRYSTAL]   An axis of crystal symmetry which does not have a plane of symmetry perpendicular to it.

polar-cap ice *See* polar ice.

polar chart [MAP]   **1.** A chart of polar areas. **2.** A chart on a polar projection; the projections most used for polar charts are the gnomonic, stereographic, azimuthal equidistant, transverse Mercator, and modified Lambert conformal.

polar circle [GEOD]   A parallel of latitude whose distance from the pole is equal to the obliquity of the ecliptic (approximately 23° 27′).

polar climate [CLIMATOL]   The climate of a geographical polar region, most commonly taken to be a climate which is too cold to support the growth of trees. Also known as arctic climate; snow climate.

polar continental air [METEOROL]   Air of an air mass that originates over land or frozen ocean areas in the polar regions; characterized by low temperature, stability, low specific humidity, and shallow vertical extent.

polar convergence [OCEANOGR]   The line of convergence of polar and subpolar water masses in the ocean.

polar cyclone *See* polar vortex.

polar desert [GEOGR]   A high-latitude desert where the existing moisture is frozen in ice sheets and is thus unavailable for plant growth. Also known as arctic desert.

polar easterlies [METEOROL]   The rather shallow and diffuse body of easterly winds located poleward of the subpolar low-pressure belt; in the mean in the Northern Hemisphere, these easterlies exist to an appreciable extent only north of the Aleutian low and Icelandic low.

polar-easterlies index [METEOROL]   A measure of the strength of the easterly wind between the latitudes of 55° and 70°N; the index is computed from the average sea-level pressure difference between these latitudes and is expressed as the east to west component of geostrophic wind in meters and tenths of meters per second.

polar electrojet [GEOPHYS]   An intense current that flows in a relatively narrow band of the auroral zone ionosphere during disturbances of the magnetosphere.

polar firn [HYD]   Firn formed at low temperatures with no melting or liquid water present. Also known as dry firn.

polar front [METEOROL]   The semipermanent, semicontinuous front separating air masses of tropical and polar origin; this is the major front in terms of air mass contrast and susceptibility to cyclonic disturbance.

polar-front theory [METEOROL]   A theory whereby a polar front, separating air masses of polar and tropical origin, gives rise to cyclonic disturbances which intensify and travel along the front, passing through various phases of a characteristic life history.

polar glacier [HYD]   A glacier whose temperature is below freezing throughout its mass, and on which there is no melting during any season.

polar high *See* arctic high; subpolar high.

polar ice [OCEANOGR]   Sea ice that is more than 1 year old; the thickest form of sea ice. Also known as polar-cap ice.

polarity epoch   [GEOPHYS]   A period of time during which the earth's magnetic field was predominantly of a single polarity.

polarity event   [GEOPHYS]   A period of no more than about 100,000 years when the earth's magnetic polarity was opposite to the predominant polarity of that polarity epoch.

polarization isocline   [METEOROL]   A locus of all points at which the inclination to the vertical of the plane of polarization of the diffuse sky radiation has the same value.

polar lake   [HYD]   A lake whose surface temperature never exceeds 4°C.

polar low   *See* polar vortex.

polar maritime air   [METEOROL]   Air of an air mass that originates in the polar regions and is then modified by passing over a relatively warm ocean surface; characterized by moderately low temperature, moderately high surface specific humidity, and a considerable degree of vertical instability.

polar meteorology   [METEOROL]   The application of meteorological principles to a study of atmospheric conditions in the earth's high latitudes or polar-cap regions, northern and southern.

polar migration   *See* polar wandering.

polar outbreak   [METEOROL]   The movement of a cold air mass from its source region; almost invariably applied to a vigorous equatorward thrust of cold polar air, a rapid equatorward movement of the polar front. Also known as cold-air outbreak.

polar projection   [MAP]   A map projection centered on a pole.

polar regions   [GEOGR]   The regions near the geographic poles; no definite limit for these regions is recognized.

polar trough   [METEOROL]   In tropical meteorology, a wave trough in the circumpolar westerlies having sufficient amplitude to reach the tropics in the upper air; at the surface it is reflected as a trough in the tropical easterlies, but at moderate elevations it is characterized by westerly winds.

polar variation   [GEOPHYS]   A small movement of the Earth's axis of rotation relative to the geoid, the resultant of the Chandler wobble and other smaller movements.

polar vortex   [METEOROL]   The large-scale cyclonic circulation in the middle and upper troposphere centered generally in the polar regions; specifically, the vortex has two centers in the mean, one near Baffin Island and another over northeastern Siberia; the associated cyclonic wind system comprises the westerlies of middle latitudes. Also known as circumpolar whirl; polar cyclone; polar low.

polar wandering   [GEOL]   Migration during geologic time of the earth's poles of rotation and magnetic poles. Also known as Chandler motion; polar migration.

polar westerlies   *See* westerlies.

pole   [CRYSTAL]   A direction perpendicular to one of the faces of a crystal.

pole of inaccessibility   *See* ice pole.

polestar recorder   [ENG]   An instrument used to determine approximately the amount of cloudiness during the dark hours; consists of a fixed long-focus camera positioned so that Polaris is permanently within its field of view; the apparent motion of the star appears as a circular arc on the photograph and is interrupted as clouds come between the star and the camera.

pole tide   [OCEANOGR]   An ocean tide, theoretically 6 millimeters in amplitude, caused by the Chandler wobble of the earth; has a period of 428 days.

poling [MIN ENG]   The act or process of temporarily protecting the face of a level, drift, or cut by driving poles or planks along the sides of the yet unbroken ground.

poling back [MIN ENG]   Carrying out excavation behind timbering already in place.

polished-joint hanger [PETRO ENG]   A type of tubing hanger that is slipped over or assembled around the top tubing joint in an oil well tubing string.

pollenite [PETR]   An igneous rock similar in composition to tautirite but containing olivine and having a glassy groundmass.

pollucite [MINERAL] $(Cs,Na)_2Al_2Si_4O_{12} \cdot H_2O$ A colorless, transparent zeolite mineral composed of hydrous silicate of cesium, sodium, and aluminum, occurring massive or in cubes; used as a gemstone. Also known as pollux.

pollux *See* pollucite.

polyargyrite [MINERAL] $Ag_{24}Sb_2S_{15}$ A gray to black mineral composed of antimony silver sulfide.

polybasite [MINERAL] $(Ag,Cu)_{16}Sb_2S_{11}$ An iron-black to steel-gray metallic-looking mineral; an ore of silver.

polyclinal fold [GEOL]   One of a group of adjacent folds whose axial surfaces are oriented randomly, but which have similar surface axes.

polyconic chart [MAP]   A chart on the polyconic projection.

polyconic projection [MAP]   A conic map projection in which the surface of a sphere or spheroid, such as the earth, is conceived as developed on a series of tangent cones, which are then spread out to form a plane; a separate cone is used for each small zone.

polycrase [MINERAL] $(Y,Ca,Ce,U,Th)(Ti,Cb,Ta)_2O_6$ Black mineral composed of titanate, columbate, and tantalate of yttrium-group metals; it is isomorphous with euxenite and occurs in granite pegmatites.

Polydolopidae [PALEON]   A Cenozoic family of rodentlike marsupial mammals.

polydymite [MINERAL] Ni_3S_4 A mineral of the linnaeite group consisting of nickel sulfide.

polygene [GEOL]   Describing an igneous rock composed of two or more minerals. Also known as polymere.

polygenetic [GEOL]   **1.** Resulting from more than one process of formation or derived from more than one source, or originating or developing at various places and times. **2.** Consisting of more than one type of material, or having a heterogeneous composition. Also known as polygenic.

polygenic *See* polygenetic.

polygeosyncline [GEOL]   A geosynclinal-geoanticlinal belt that lies along the continental margin and receives sediments from a borderland on its oceanic side.

Polygnathidae [PALEON]   A family of Middle Silurian to Cretaceous conodonts in the suborder Conodontiformes, having platforms with small pitlike attachment scars.

polygonal ground [GEOL]   A ground surface consisting of polygonal arrangements of rock, soil, and vegetation formed on a level or gently sloping surface by frost action. Also known as cellular soil.

polygonal karst [GEOL]   A karst pattern that is characteristic of tropical types such as cone karsts, with the surface completely divided into a polygonal network.

polygonal method [MIN ENG]   A method of estimating ore reserves in which it is assumed that each drill hole has an area of influence extending halfway to the neighboring drill holes.

polyhalite [MINERAL] $K_2MgCa_2(SO_4)_4\cdot2H_2O$ A sulfate mineral usually found in fibrous brick-red masses due to iron.

polylithionite [MINERAL] $KLi_2AISi_4O_{10}(F,OH)_2$ A mineral of the mica group, related to lepidolite.

polymere *See* polygene.

polymetamorphic diaphthoresis [GEOL] Retrograde changes during a second phase of metamorphism that is clearly separated from a previous, higher-grade metamorphic period.

polymetamorphism [GEOL] Polyphase or multiple metamorphism whereby two or more successive metamorphic events have left their imprint upon the same rocks.

polymictic [HYD] Pertaining to or characteristic of a lake having no stabile thermal stratification. [PETR] Of a clastic sedimentary rock, being made up of many rock types or of more than one mineral species.

polymignite *See* polymignyte.

polymignyte [MINERAL] $(Ca,Fe,Y,Zr,Th)(Nb,Ti,Ta)O_4$ A black mineral composed of niobate, titanate, and tantalate of cerium-group metals, with calcium and iron. Also spelled polymignite.

polymorph [CRYSTAL] One of the crystal forms of a substance displaying polymorphism. Also known as polymorphic modification.

polymorphic modification *See* polymorph.

polymorphism [CRYSTAL] The property of a chemical substance crystallizing into two or more forms having different structures, such as diamond and graphite. Also known as pleomorphism.

polyn'ya [OCEANOGR] A Russian term for a water area, other than a lead, lane, or crack, which is surrounded by sea ice; the term "window" is sometimes used for a similar open area in river ice. Also known as ice clearing.

polyschematic [GEOL] Referring to a mineral deposit having more than one textural element.

polyschematic chondrule [GEOL] A chondrule consisting of several crystals.

polysynthetic twinning [CRYSTAL] Repeated twinning that involves three or more individual crystals according to the same twin law and on parallel twin planes.

polytropic atmosphere [METEOROL] A model atmosphere in hydrostatic equilibrium with a constant nonzero lapse rate.

polytypism [CRYSTAL] Thr property of a mineral to crystallize in more than one form, due to more than one possible mode of atomic packing.

polzenite [PETR] **1.** A group of lamprophyres characterized by the presence of olivine and melilite. **2.** Any rock in the group.

pond [GEOGR] A small natural body of standing fresh water filling a surface depression, usually smaller than a lake.

pondage [HYD] Water held in a reservoir for short periods to regulate natural flow, usually for hydroelectric power.

pondage land [GEOL] Land on which water is stored as dead water during flooding, and which does not contribute to the downstream passage of flow. Also known as flood fringe.

ponded stream [HYD] A stream in which a pond forms due to an interruption of the normal streamflow.

ponente [METEOROL] A west wind on the French Mediterranean coast, the northern Roussillon region, and Corsica.

pontic [GEOL] Pertaining to sediments or facies deposited in comparatively deep and motionless water, such as an association of black shales and dark limestones deposited in a stagnant basin.

pony set [MIN ENG] A small timber set or frame incorporated in the main sets of a haulage level to accommodate an ore chute or other equipment from above or below.

ponzite [PETR] A feldspathoid-free trachyte containing augite and pyroxenes which may be rimmed with acmite or acmite-augite.

pool [GEOL] Underground accumulation of petroleum. [HYD] A small deep body of water, often fed by a spring. [MIN ENG] **1.** To wedge for splitting in quarrying or mining. **2.** To undermine or undercut.

pool spring [HYD] A spring fed from deep pools, probably related to faults.

pool stage [HYD] As used along the Ohio and upper Mississippi Rivers of the United States, a low-water condition with the navigation dams up so that the river is a series of shallow pools; when this condition exists, the river is said to be "in pool"; river depth is regulated by the dams so as to be adequate for navigation.

pop [MIN ENG] A drill hole blasted to reduce larger pieces of rock or to trim a working face. Also known as pop hole; pop shot.

pop hole *See* pop.

poppet [MIN ENG] A pulley frame or the headgear over a mine shaft.

popping [MIN ENG] Exploding a stick of dynamite on a boulder so as to break it for easy removal from a quarry or opencast mine.

popple rock *See* pebble bed.

pop shot *See* pop.

porcelainite *See* mullite.

porcelain jasper [GEOL] A hard, naturally baked, impure clay (or porcellanite) which because of its red color had long been considered a variety of jasper.

porcelaneous [GEOL] Resembling unglazed porcelain.

porcelaneous chert [PETR] A hard, opaque to subtranslucent smooth chert, having a smooth fracture surface and a typically white appearance resembling chinaware or glazed porcelain.

porcellanite [PETR] A hard, dense siliceous rock, such as impure chert or indurated clay or shale.

pore [GEOL] An opening or channelway in rock or soil.

pore compressibility [GEOL] The fractional change in reservoir-rock pore volume with a unit change in pressure upon that rock.

pore ice [HYD] Ice which fills or partially fills pore spaces in permafrost; forms by freezing soil water in place, with no addition of water.

pore pressure *See* neutral stress.

pore-size distribution [GEOL] Variations in pore sizes in reservoir formations; each type of rock has its own typical pore size and related permeability.

pore space [GEOL] The pores in a rock or soil considered collectively. Also known as pore volume.

pore volume *See* pore space.

pore-water pressure *See* neutral stress.

poriaz [METEOROL]   Violent northeast winds on the Black Sea near the Bosporus in the Soviet Union.

Porlezzina [METEOROL]   An east wind on Lake Lugano (Italy and Switzerland), blowing from the Gulf of Porlezza.

porosimeter [ENG]   Laboratory compressed-gas device used for measurement of the porosity of reservoir rocks.

porosity feet [PETRO ENG]   Reservoir porosity fraction multiplied by net pay in feet, where porosity fraction is the portion of the reservoir that is porous, and net pay is the depth and areal extent of the hydrocarbons-containing reservoir.

porosity trap *See* stratigraphic trap.

porous reservoir model [PETRO ENG]   Scaled laboratory model of porous reservoir used for the study of reservoir areal waterflood efficiencies.

Poroxylaceae [PALEOBOT]   A monogeneric family of extinct plants included in the Cordaitales.

porpezite [MINERAL]   A mineral consisting of a native alloy of palladium (5–10%) and gold. Also known as palladium gold.

porphrite *See* porphyry.

porphyritic [PETR]   Pertaining to or resembling porphyry.

porphyroaphanitic [PETR]   Referring to the texture of a porphyritic igneous rock (especially an extrusive rock), having large macroscopic phenocrysts in an aphanitic groundmass.

porphyroblast [PETR]   A relatively large crystal formed in a metamorphic rock.

porphyroblastic [PETR]   Pertaining to the texture of recrystallized metamorphic rock having large idioblasts of minerals possessing high form energy in a finer-grained crystalloblastic matrix.

porphyrocrystallic *See* porphyrotopic.

porphyroclastic structure *See* mortar structure.

porphyrogranulitic [PETR]   Referring to ophitic texture characterized by large phenocrysts of feldspar and augite or olivine in a groundmass of smaller lath-shaped feldspar crystals and irregular augite grains; a combination of porphyritic and intergranular textures.

porphyroid [PETR]   **1.** A blastoporphyritic, or sometimes porphyroblastic, metamorphic rock of igneous origin. **2.** A feldspathic metasedimentary rock having the appearance of a porphyry.

porphyroskelic [GEOL]   Pertaining to an arrangement in a soil fabric whereby the plasma occurs as a dense matrix in which skeleton grains are set like phenocrysts in a porphyritic rock.

porphyrotope [GEOL]   A large crystal enclosed in a finer-grained matrix in a sedimentary rock showing porphyrotopic fabric.

porphyrotopic [GEOL]   Referring to the fabric of a crystalline sedimentary rock in which the constituent crystals are of more than one size and larger crystals are enclosed in a finer-grained matrix. Also known as porphyrocrystallic.

porphyry [PETR]   An igneous rock in which large phenocrysts are enclosed in a very-fine-grained to aphanitic matrix. Formerly known as porphrite.

porphyry copper [GEOL]   A copper deposit in which the copper-bearing minerals occur in disseminated grains or in veinlets through a large volume of rock.

port *See* harbor.

portal [MIN ENG]   **1.** An entrance to a mine. **2.** The rock face at which a tunnel is started.

Porterfield [GEOL]   A North American geologic stage of the Middle Ordovician, forming the lower division of the Mohawkian and lying above Ashby and below Wilderness.

Portlandian [GEOL]   A European geologic stage of the Upper Jurassic, above Kimmeridgian, below Berriasian of Cretaceous.

portlandite [MINERAL] $Ca(OH)_2$ A colorless, hexagonal mineral consisting of calcium hydroxide; occurs as minute plates.

position blocks [MIN ENG]   Unproved mining claims that are in a position to contain a lode if the lode continues in the direction in which it has been proved in other claims.

positive area *See* positive element.

positive axis [METEOROL]   In tropical synoptic analysis, a locus of maximum streamline curvature in an easterly wave; used primarily in the analysis of waves that span the equatorial trough (equatorial waves); a positive axis corresponds to a trough line in the Northern Hemisphere and a ridge line in the Southern Hemisphere.

positive derail [MIN ENG]   A device installed in or on a mine track to derail runaway cars or trips.

positive element [GEOGR]   A large structural feature of the earth's crust characterized by long-term upward movement (uplift, emergence) or subsidence less rapid than that of adjacent negative elements.   Also known as archibole; positive area.

positive estuary [HYD]   An estuary in which there is a measurable dilution of seawater by land drainage.

positive movement [GEOL]   **1.** Uplift or emergence of the earth's crust relative to an adjacent area of the crust. **2.** A relative rise in sea level with respect to land level.

positive ore [MIN ENG]   Ore exposed on four sides in blocks of a size variously prescribed.

positive shoreline *See* shoreline of submergence.

possible ore [MIN ENG]   A class of ore whose existence is a reasonable possibility, based upon geologic-mineralogic relationships and the extent of ore bodies already developed.   Also known as extension ore.

possible reserves [PETRO ENG]   Primary petroleum reserves that may exist, but available data do not confirm their presence.

post [MIN ENG]   **1.** A mine timber, or any upright timber, more commonly the uprights which support the roof crosspieces. **2.** The support fastened between the roof and floor of a coal seam, used with certain types of mining machines or augers. **3.** A pillar of coal or ore.

postglacial [GEOL]   Referring to the interval of geologic time since the total disappearance of continental glaciers in middle latitudes or from a particular area.

posthumous structure [GEOL]   Folds, faults, and other structural features in covering strata which revive or mimic the structure of older underlying rocks that are generally more deformed.

postmagmatic [GEOL]   Pertaining to geologic reactions or events occurring after the bulk of the magma has crystallized.

postmineral [GEOL] In economic geology, describing a structural or other feature formed after mineralization.

postobsequent stream [HYD] A strike stream developed after the obsequent stream into which it flows.

postorogenic [GEOL] Of a geologic process or event, occurring after a period of orogeny.

pot *See* pothole.

potamogenic rock [PETR] A sedimentary rock formed by precipitation from river water.

potarite [MINERAL] PdHg A silver-white isometric mineral composed of palladium and mercury alloy. Also known as palladium amalgam.

potash alum *See* kalinite.

potash bentonite *See* potassium bentonite.

potash lake [HYD] An alkali lake whose waters contain a high content of dissolved potassium salts.

potash mica *See* muscovite.

potash regulations [MIN ENG] Rules governing the prospecting and exploitation of land containing potash.

potassic [PETR] Referring to a rock which contains a significant amount of potassium.

potassium-argon dating [GEOL] Dating of archeological, geological, or organic specimens by measuring the amount of argon accumulated in the matrix rock through decay of radioactive potassium.

potassium bentonite [GEOL] A clay of the illite group that contains potassium and is formed by alteration of volcanic ash. Also known as K bentonite; potash bentonite.

potassium feldspar [MINERAL] Any alkali feldspar (orthoclase, microcline, sonidine, adularia) containing the molecule $KAlSi_3O_8$. Incorrectly known as K feldspar; potash feldspar.

potato stone [GEOL] A potato-shaped geode, especially one consisting of hard silicified limestone with an internal lining of quartz crystals.

potential evaporation *See* evaporative power.

potential evaportranspiration [HYD] Generally, the amount of moisture which, if available, would be removed from a given land area by evapotranspiration; expressed in units of water depth.

potential index of refraction [METEOROL] An atmospheric index of refraction so formulated that it would have no height variation in an adiabatic atmosphere. Also known as potential refractive index.

potential instability *See* convective instability.

potential ore [MIN ENG] **1.** As yet undiscovered mineral deposits. **2.** A known mineral deposit for which recovery is not yet economically feasible.

potential refractive index *See* potential index of refraction.

potentiometric model study [PETRO ENG] Analogic electrical-resistance model (electrolyte in a contoured container) of an underground reservoir based on Darcy's law, that is, the steady-state flow of liquids through porous media is analogous to the flow of current through an electrical conductor; used to predict conditions in gas-condensate oil reservoirs.

potentiometric surface   [HYD]   An imaginary surface that represents the static head of groundwater and is defined by the level to which water will rise. Also known as isopotential level; piezometric surface; pressure surface.

pothole   [GEOL]   **1.** A shaftlike cave opening upward to the surface. **2.** Any bowl-shaped, cylindrical, or circular hole formed by the grinding action of a stone in the rocky bed of a river or stream. Also known as churn hole; colk; eddy mill; evorsion hollow; kettle; pot. **3.** A vertical, or nearly vertical shaft in limestone. Also known as aven; cenote. **4.** A small depression with steep sides in a coastal marsh; contains water at or below low-tide level. Also known as rotten spot.   [HYD]   *See* moulin.

potrero   [GEOL]   An elongate, islandlike beach ridge, surrounded by mud flats and separated from the coast by a lagoon and barrier island, and made up of a series of accretionary dune ridges.

Poulter seismic method   [GEOPHYS]   A type of air shooting in which the explosive is set on poles above the ground.

powder box   [MIN ENG]   A wooden box used by miners to store explosive powder and blasting caps.

powder diffraction camera   [CRYSTAL]   A metal cylinder having a window through which an x-ray beam of known wavelength is sent by an x-ray tube to strike a finely ground powder sample mounted in the center of the cylinder; crystal planes in this powder sample diffract the x-ray beam at different angles to expose a photographic film that lines the inside of the cylinder; used to study crystal structure. Also known as x-ray powder diffractometer.

powderman   [MIN ENG]   A man in charge of explosives in an operation of any nature requiring their use.

powder mine   [MIN ENG]   An excavation filled with powder for the purpose of blasting rocks.

powder snow   [HYD]   A cover of dry snow that has not been compacted in any way.

powder train   [ENG]   **1.** Train, usually of compressed black powder, used to obtain time action in older fuse types. **2.** Train of explosives laid out for destruction by burning.

powellite   [MINERAL] $Ca(WMo)O_4$ A commercially important tungsten mineral, crystallizing in the tetragonal system; isomorphous with scheelite ($CaWO_4$).

power equation   [MIN ENG]   The relationship indicating that the natural ventilating power plus the power required to force air through a mine is equal to the power used in lifting water out of the mine plus the power lost in the kinetic energy of air leaving the mine plus the power converted to heat in overcoming friction.

power grizzly   [MIN ENG]   Power-operated machine for removing dirt and fine particles from ore before it is crushed.

power-law profile   [METEOROL]   A formula for the variation of wind with height in the surface boundary layer.

power loader   [MIN ENG]   Power-operated machine for loading ore, coal, or other material into a car, conveyor, or other collector.

power of crystallization   *See* form energy.

power-shovel mining   [MIN ENG]   A technique utilizing power shovels to mine ores by mining or stripping and taking away overburden.

pozzolan   [GEOL]   A finely ground burnt clay or shale resembling volcanic dust, found near Pozzuoli, Italy; used in cement because it hardens under water.

praecipitatio   [METEOROL]   Precipitation falling from a cloud and apparently reaching the earth's surface; this supplementary cloud feature is mostly encountered in altostratus, nimbostratus, stratocumulus, stratus, cumulus, and cumulonimbus.

prairie [GEOGR] An extensive level-to-rolling treeless tract of land in the temperate latitudes of central North America, characterized by deep, fertile soil and a cover of coarse grass and herbaceous plants.

prairie climate *See* subhumid climate.

prairie soil [GEOL] A group of zonal soils which has a surface horizon that is dark or grayish brown which grades through brown soil into lighter-colored parent material; it is 0.6–1.5 m thick and develops under tall grass in a temperate and humid climate.

prase [MINERAL] **1.** A translucent and dull leek-green or light-grayish yellow-green variety of chalcedony. **2.** Crystalline quartz containing a multitude of green hairlike crystals of actinolite. Also known as mother-of-emerald.

prasinite [PETR] A greenschist in which the proportions of the hornblende-chlorite-epidote assemblage are more or less equal.

prealpine facies [GEOL] A geosynclinal facies characteristic of neritic areas, displaying thick limestone deposits and coarse terrigenous material, and resembling epicontinental platform sediments.

Precambrian [GEOL] All geologic time prior to the beginning of the Paleozoic era (before 600,000,000 years ago); equivalent to about 90% of all geologic time.

precious stone [MINERAL] **1.** Any genuine gemstone. **2.** A gemstone of high commercial value because of its beauty, rarity, durability, and hardness; examples are diamond, ruby, sapphire, and emerald.

precipice [GEOL] A very steeply inclined, vertical, or overhanging wall or surface of rock.

precipitable water [METEOROL] The total atmospheric water vapor contained in a vertical column of unit cross-sectional area extending between any two specified levels, commonly expressed in terms of the height to which that water substance would stand if completely condensed and collected in a vessel of the same unit cross section. Also known as precipitable water vapor.

precipitable water vapor *See* precipitable water.

precipitation [METEOROL] **1.** Any or all of the forms of water particles, whether liquid or solid, that fall from the atmosphere and reach the ground. **2.** The amount, usually expressed in inches of liquid water depth, of the water substance that has fallen at a given point over a specified period of time.

precipitation area [METEOROL] **1.** On a synoptic surface chart, an area over which precipitation is falling. **2.** In radar meteorology, the region from which a precipitation echo is received.

precipitation ceiling [METEOROL] After United States weather observing practice, a ceiling classification applied when the ceiling value is the vertical visibility upward into precipitation; this is necessary when precipitation obscures the cloud base and prevents a determination of its height.

precipitation cell [METEOROL] In radar meteorology, an element of a precipitation area over which the precipitation is more or less continuous.

precipitation current [METEOROL] The downward transport of charge, from cloud region to earth, that occurs in a fall of electrically charged rain or other hydrometeors.

precipitation echo [METEOROL] A type of radar echo returned by precipitation.

precipitation effectiveness *See* precipitation-evaporation ratio.

precipitation electricity [GEOPHYS] **1.** That branch of the study of atmospheric electricity concerned with the electric charges carried by precipitation particles and with the manner in which these charges are acquired. **2.** The electric charge borne by precipitation particles.

precipitation-evaporation ratio [CLIMATOL]   For a given locality and month, an empirical expression devised for the purpose of classifying climates numerically on the basis of precipitation and evaporation. Abbreviated P-E ratio. Also known as precipitation effectiveness.

precipitation excess [HYD]   The volume of water from precipitation that is available for direct runoff.

precipitation facies [GEOL]   Facies characteristics that provide evidence of depositional conditions; revealed mainly by sedimentary structures (such as cross-bedding and ripple marks) and by primary constituents (especially fossils).

precipitation gage [ENG]   Any device that measures the amount of precipitation; principally, a rain gage or snow gage.

precipitation-generating element [METEOROL]   In radar meteorology, a relatively small volume of supercooled cloud droplets in which ice crystals form and grow much more rapidly than in a lower, larger cloud mass.

precipitation intensity [METEOROL]   The rate of precipitation, usually expressed in inches per hour.

precipitation inversion [METEOROL]   As found in some mountain areas, a decrease of precipitation with increasing elevation of ground above sea level. Also known as rainfall inversion.

precipitation physics [METEOROL]   The study of the formation and precipitation of liquid and solid hydrometeors from clouds; a branch of cloud physics and of physical meteorology.

precipitation station [METEOROL]   A station at which only precipitation observations are made.

precipitation trails *See* virga.

precipitation trajectory [METEOROL]   In radar meteorology, a characteristic echo observed on range-height indicator scopes and time-height sections which represents the height-range pattern of snow falling from isolated precipitation-generating elements of a few miles in diameter. Also known as mare's tail.

precision depth recorder [ENG]   A machine that plots sonar depth soundings on electrosensitive paper; can plot variations in depth over a range of 400 fathoms (730 meters) on a paper 18.85 inches (47.9 centimeters) wide. Abbreviated PDR. Also known as precision graphic recorder (PGR).

precision graphic recorder *See* precision depth recorder.

pre-cold-frontal squall line *See* prefrontal squall line.

preconsolidation pressure [GEOL]   The greatest effective stress exerted on a soil; result of this pressure from overlying materials is compaction. Also known as prestress.

predazzite [PETR]   A recrystallized limestone that resembles pencatite, but contains less brucite than calcite.

prediction [METEOROL]   **1.** The act of making a weather forecast. **2.** The forecast itself.

predozzite [PETR]   Limestone rich in periclase and brucite.

preferred orientation [PETR]   The nonrandom orientation of planar or linear fabric elements in structural petrology.

prefrontal squall line [METEOROL]   A squall line or instability line located in the warm sector of a wave cyclone, about 50 to 300 miles (80 to 480 kilometers) in advance of the cold front, usually oriented roughly parallel to the cold front, and moving in about the same manner as the cold front. Also called nonfrontal squall line; pre-cold-frontal squall line.

preglacial   [GEOL]   **1.** Pertaining to the geologic time immediately preceding the Pleistocene epoch. **2.** Of material, underlying glacial deposits.

prehnite   [MINERAL] $Ca_2Al_2Si_3O_{10}(OH)_2$ A light-green to white mineral sorosilicate crystallizing in the orthorhombic system and generally found in reniform and stalactitic aggregates with crystalline surface; it has a vitreous luster, hardness is 6–6.5 on Mohs scale, and specific gravity is 2.8–2.9.

preliminary waves   [GEOPHYS]   The body of waves of an earthquake, including both P and S waves.

premineral   [GEOL]   In economic geology, describing a structural or other feature extant before mineralization.

preorogenic   [GEOL]   The initial phase of an orogenic cycle during which geosynclines form.

preparatory work   [MIN ENG]   Various excavations within a deposit so that actual mining can begin; includes inclines, drives between levels, crosscuts, and chutes.

present value   [MIN ENG]   The sum of money which, if expended on a mine for purchase, development, and equipment, would produce over the life of the mine a return of the original investment plus a commensurate profit.

present-worth factor *See* discount factor.

presque isle   [GEOGR]   A promontory or peninsula extending into a lake, nearly or almost forming an island, its head or end section connected with the shore by a sag or low gap only slightly above water level, or by a strip of lake bottom exposed as a land surface by a drop in lake level.

pressolved   [GEOL]   Referring to a sedimentary bed or rock in which the grains have undergone pressure solution.

pressolution *See* pressure solution.

pressolved   [GEOL]   Referring to a sedimentary bed or rock in which the grains have undergone pressure solution.

pressure altitude   [METEOROL]   The height above sea level at which the existing atmospheric pressure would be duplicated in the standard atmosphere; atmospheric pressure expressed as height according to a standard scale.

pressure-altitude variation   [METEOROL]   The pressure difference, in feet or meters, between mean sea level and the standard datum plane.

pressure arch   [HYD]   A wavelike prominence, formed by pressure, on the surface of a glacier.

pressure block   [MIN ENG]   The pressure on pillars, walls, and other supports in a mine caused by removal of surrounding formations from masses of rocks or by natural geological formations.

pressure bomb   [PETRO ENG]   Pipe-and-valve device used to capture downhole pressurized gas samples from oil wells; used to measure downhole pressure.

pressure breccia *See* tectonic breccia.

pressure bump   [MIN ENG]   Sudden failure of a coal pillar overloaded by the weight of the rock above it.

pressure burst   [MIN ENG]   A rockburst produced under stresses exceeding the elastic strength of the rock.

pressure center [METEOROL] **1.** On a synoptic chart (or on a mean chart of atmospheric pressure), a point of local minimum or maximum pressure; the center of a low or high. **2.** A center of cyclonic or anticyclonic circulation.

pressure chamber [MIN ENG] An enclosed space that seals off a part of a mine and in which the air pressure can be raised or lowered.

pressure-change chart [METEOROL] A chart indicating the change in atmospheric pressure of a constant-height surface over some specified interval of time. Also known as pressure-tendency chart.

pressure contour [METEOROL] A line connecting points of equal height of a given barometric pressure; the intersection of a constant pressure surface by a plane parallel to mean sea level.

pressure decline [PETRO ENG] The loss or decline in reservoir pressure resulting from pressure drawdown during the production of gas or oil. Also known as pressure depletion.

pressure depletion *See* pressure decline.

pressure depth [OCEANOGR] The depth at which an ocean sample was taken, as inferred from the difference in readings on protected and unprotected thermometers on the sampler; the higher reading is on the unprotected thermometer due to the effect of pressure on the mercury column at the sampling depth.

pressure distribution [PETRO ENG] The relative pressures (pressure gradients) between various portions of a producing reservoir zone; lowest pressures are nearest the producing wellbores.

pressure drawdown [PETRO ENG] The drop in reservoir pressure related to the withdrawal of gas from a producing well; for low-permeability formations, pressures near the wellbores can be much lower than in the main part of the reservoir; leads to pressure decline in the reservoir and ultimate pressure depletion.

pressure-fall center [METEOROL] A point of maximum decrease in atmospheric pressure over a specified interval of time; on synoptic charts, a point of greatest negative pressure tendency. Also known as center of falls; isallobaric low; isallobaric minimum; katallobaric center.

pressure fan [MIN ENG] A fan that forces fresh air into a mine as distinguished from one that exhausts air from the mine.

pressure field [OCEANOGR] A representation of a pressure gradient as isobar contours, parallel to which ocean currents flow.

pressure force *See* pressure-gradient force.

pressure fringe *See* pressure shadow.

pressure-gradient force [METEOROL] The force due to differences of pressure within the atmosphere; it usually refers only to the horizontal component of the force. Also known as pressure force.

pressure ice [OCEANOGR] Ice, especially sea ice, which has been deformed or altered by the lateral stresses of any combination of wind, water currents, tides, waves, and surf; may include ice pressed against the shore, or one piece of ice upon another.

pressure ice foot [HYD] An ice foot formed along a shore by the freezing together of stranded pressure ice.

pressure interface [PETRO ENG] The interrelation of several individual reservoir pressures whose productions are supported by water influx from a common aquifer; pressure depletion from withdrawal of oil from one reservoir will affect the position of the common aquifer and thus affect the pressures and gas-oil or water-oil contacts in the other reservoirs.

pressure jump [METEOROL] A steady-state propagation of a sudden finite change of inversion height, in analogy to the shock wave in a compressible fluid or to a hydraulic jump; the prefrontal squall line has been interpreted as a pressure jump, with the cold front providing the initial pistonlike impetus.

pressure jump line [METEOROL] A fast-moving line of sudden rise in atmospheric pressure, followed by a higher pressure level than that which preceded the jump; under suitable moisture conditions, sudden instability of the atmosphere conducive to the formation of thunderstorms can result.

pressure maintenance [PETRO ENG] The maintenance of gas pressure in a reservoir by an active water drive, water injection, gas injection, or a combination of the foregoing.

pressure pan [GEOL] An induced soil pan having a higher bulk density and a lower total porosity than the soil directly above or below it, produced as a result of pressure applied by normal tillage operation or by other artificial means.

pressure pattern [METEOROL] The general geometric characteristics of atmospheric pressure distribution as revealed by isobars on a constant-height chart; usually applied to cyclonic-scale features of a surface chart.

pressure penitente [GEOL] A nieve penitente composed of brilliantly white ice shaped into a slender ridge by lateral pressure of converging morainal streams and by melting of the adjacent debris-covered ice.

pressure pillow [ENG] A mechanical-hydraulic snow gage consisting of a circular rubber or metal pillow filled with a slolution of antifreeze and water, and containing either a pressure transducer or a riser pipe to record increase in pressure of the snow.

pressure-plate anemometer [ENG] An anemometer which measures wind speed in terms of the drag which the wind exerts on a solid body; may be classified according to the means by which the wind drag is measured. Also known as plate anemometer.

pressure plateau [GEOL] An uplifted area of a thick lava flow, measuring up to 3 or 4 meters, the uplift of which is due to the intrusion of new lava from below that does not reach the surface.

pressure radius [PETRO ENG] The effective radius of increased reservoir pressure surrounding a water-injection well.

pressure release [GEOPHYS] The outward-expanding force of pressure which is released within rock masses by unloading, as by erosion of superincumbent rocks, or by removel of glacial ice.

pressure-release jointing [GEOL] Exfoliation that occurs in once deeply buried rock that erosion has brought nearer the surface, thus releasing its confining pressure.

pressure ridge [GEOL] **1.** A seismic feature resulting from transverse pressure and shortening of the land surface. **2.** An elongate upward movement of the congealing crust of a lava flow. **3.** A ridge of glacier ice. [OCEANOGR] A ridge or wall of hummocks where one ice floe has been pressed against another.

pressure ring [MIN ENG] A ring about a large excavated area, evidenced by distortion of the openings near the main excavation.

pressure-rise center [METEOROL] A point of maximum increase in atmospheric pressure over a specified interval of time; on synoptic charts, a point of maximum positive pressure tendency. Also known as anallobaric center; center of rises; isallobaric high; isallobaric maximum.

pressure shadow [PETR]   In structural petrology, an area adjoining a prophyroblast, characterized by a growth fabric rather than a deformation fabric, as seen in a section perpendicular to the *b* axis of the fabric.   Also known as pressure fringe; strain shadow.

pressure solution [PETR]   In a sedimentary rock, solution occurring preferentially at the grain boundary surfaces,   Also known as pressolution.

pressure surface *See* potentiometric surfaces.

pressure survey [MIN ENG]   A study to determine the pressure distribution or pressure losses along consecutive lengths or sections of a ventilation circuit.   [PETRO ENG]   The measurement of static bottomhole pressures in an oil field with producing wells shut in for a time interval sufficient for reservoir pressure buildup to stabilize.

pressure system [METEOROL]   An individual cyclonic-scale feature of atmospheric circulation, commonly used to denote either a high or a low, less frequently a ridge or a trough.

pressure tendency [METEOROL]   The character and amount of atmospheric pressure change for a 3-hour or other specified period ending at the time of observation.   Also known as barometric tendency.

pressure-tendency chart *See* pressure-change chart.

pressure topography *See* height pattern.

pressure traverse [PETRO ENG]   Measurement of reservoir pressures at progressive depths.

pressure tube [HYD]   A deep, slender, cylindrical hole formed in a glacier by the sinking of an isolated stone that has absorbed more solar radiation than the surrounding ice.

pressure-tube anemometer [ENG]   An anemometer which derives wind speed from measurements of the dynamic wind pressures; wind blowing into a tube develops a pressure greater than the static pressure, while wind blowing across a tube develops a pressure less than the static; this pressure difference, which is proportional to the square of the wind speed, is measured by a suitable manometer.

pressure wall [HYD]   A snow escarpment at the side of an avalanche.

pressure wave [METEOROL]   A wave or periodicity which exists in the variation of atmospheric pressure on any time scale, usually excluding normal diurnal or seasonal trends.

pressurized stoppings [MIN ENG]   Stoppings which are erected in the intake and return roadways of a district to isolate an open fire or spontaneous heating and in which the pressures on both sides of each stopping are made equal by the use of auxiliary fans.

prester [METEOROL]   A whirlwind or waterspout accompanied by lightning in the Mediterranean Sea and Greece.

prestress *See* preconsolidation pressure.

presuppression [GEOPHYS]   In seismic prospecting, the suppression of the early events on a seismic record, for control of noise and reflections on that portion of the record.

prevailing current [OCEANOGR]   The ocean current most frequently observed during a given period, as a month, season, or year.

prevailing visibility [METEOROL]   In United States weather observing practice, the greatest horizontal visibility equaled or surpassed throughout half of the horizon circle; in the case of rapidly varying conditions, it is the average of the prevailing visibility while the observation is being taken.

prevailing westerlies [METEOROL] The prevailing westerly winds on the poleward sides of the subtropical high-pressure belts.

prevailing wind *See* prevailing wind direction.

prevailing wind direction [METEOROL] The wind direction most frequently observed during a given period; the periods most often used are the observational day, month, season, and year. Also known as prevailing wind.

previtrain [GEOL] The woody lenses in lignite that are equivalent to vitrain in coal of higher rank.

Priabonian [GEOL] A European stage of geologic time in the upper Eocene, believed to consist of Auversian and Bartonian.

priceite [MINERAL] $Ca_4B_{10}O_{19} \cdot 7H_2O$ A snow-white earthy mineral composed of hydrous calcium borate, occurring as a massive. Also known as pandermite.

prill [MIN ENG] **1.** The best ore after cobbing. **2.** A circular particle about the size of buckshot. **3.** Compressed and sized explosives such as ammonium nitrate.

primary [GEOL] **1.** A young shoreline whose features are produced chiefly by nonmarine agencies. **2.** Of a mineral deposit, unaffected by supergene enrichment.

primary arc [GEOL] **1.** A curved segment of elongated mountain zones that are the areas of the earth's major and most recent tectonic activity. **2.** *See* internides.

primary basalt [PETR] A presumed original magma from which all other rock types are supposedly obtained by various processes.

primary circle *See* primary great circle.

primary circulation [METEOROL] The prevailing fundamental atmospheric circulation on a planetary scale which must exist in response to radiation differences with latitude, to the rotation of the earth, and to the particular distribution of land and oceans, and which is required from the viewpoint of conservation of energy.

primary clay *See* residual clay.

primary cyclone [METEOROL] Any cyclone (or low), especially a frontal cyclone, within whose circulation one or more secondary cyclones have developed. Also known as primary low.

primary dip [GEOL] The slight dip assumed by a bedded deposit at its moment of deposition. Also known as depositional dip; initial dip; original dip.

primary dolomite [PETR] A dense, finely textured (particle diameters less than 0.01 millimeter) dolomite rock made up of crystals formed in place by direct chemical or biochemical precipitation from sea water or lake water, recognized as a well-stratified (thinly laminated) and wholly unfossiliferous dolomite associated with other primary sediments and commonly interbedded with anhydrite, clay, and micritic limestone.

primary drilling [ENG] The process of drilling holes in a solid rock ledge in preparation for a blast by means of which the rock is thrown down.

primary fabric *See* apposition fabric.

primary flowage [PETR] Movement within an igneous rock that is still partly fluid.

primary front [METEOROL] The principal, and usually original, front in any frontal system in which secondary fronts are found.

primary geosyncline *See* orthogeosyncline.

primary gneiss [PETR] A rock that exhibits planar or linear structures characteristic of metamorphic rocks but lacks observable granulation or recrystallization and is therefore considered to be of igneous origin.

primary gneissic banding [PETR]   A kind of banding developed in certain igneous (plutonic) rocks of heterogeneous composition, produced by the admixture of two magmas only partly miscible or by magma intimately admixed with country rock into which it has been injected along planes of bedding or foliation.

primary great circle [GEOD]   A great circle used as the origin of measurement of a coordinate; particularly, such a circle 90° from the poles of a system of spherical coordinates, as the equator. Also known as fundamental circle; primary circle.

primary haulage [MIN ENG]   A short haul in which there is no secondary- or main-line haulage. Also known as face haulage.

primary high explosive [MATER]   An explosive which is extremely sensitive to heat and shock and is normally used to initiate a secondary high explosive; examples are mercury fulminate, lead azide, lead styphnate, and tetracene.

primary interstice *See* original interstice.

primary low *See* primary cyclone.

primary magma [GEOL]   A magma that originates below the earth's crust.

primary mineral [MINERAL]   A mineral that is formed at the same time as the rock in which it is contained, and that retains its original form and composition.

primary orogeny [GEOL]   Orogeny that is characteristic of the internides and that involves deformation, regional metamorphism, and granitization.

primary porosity [GEOL]   Natural porosity in petroleum reservoir sands or rocks.

primary reserve [PETRO ENG]   Petroleum reserve recoverable commercially at current prices and costs by conventional methods and equipment as a result of the natural energy inherent in the reservoir.

primary rocks [PETR]   Rocks whose constituents are newly formed particles that have never been constituents of previously formed rocks and that are not the products of alteration or replacement, such as limestones formed by precipitation from solution.

primary sedimentary structure [GEOL]   A sedimentary structure produced during deposition, such as ripple marks and graded bedding.

primary stratification [GEOL]   Stratification which develops when sediments are first deposited. Also known as direct stratification.

primary stratigraphic trap [GEOL]   A stratigraphic trap formed by the deposition of clastic materials (such as shoestring sands, lenses, sand patches, bars, or cocinas) or through chemical deposition (such as organic reefs or biostromes).

primary stress field *See* ambient stress field.

primary structure [GEOL]   A structure, in an igneous rock, that formed at the same time as the rock, but before its final consolidation.

primary tectonite [PETR]   A tectonite with depositional fabric.

primary wave [GEOPHYS]   The first seismic wave that reaches a station from an earthquake.

prime meridian [GEOD]   The meridian of longitude 0°, used as the origin for measurement of longitude; the meridian of Greenwich, England, is almost universally used for this purpose.

primer [ENG]   In general, a small, sensitive initial explosive train component which on being actuated initiates functioning of the explosive train, and will not reliably initiate high explosive charge; classified according to the method of initiation, for example, percussion primer, electric primer, or friction primer.

primer cup [ENG]   A small metal cup, into which the primer mixture is loaded.

primer-detonator [ENG] A unit, in a metal housing, in which are assembled a primer, a detonator, and when indicated, an intervening delay charge.

primer leak [ENG] Defect in a cartridge which allows partial escape of the hot propelling gases in a primer, caused by faulty construction or an excessive charge.

primer mixture [MATER] An explosive mixture containing a sensitive explosive and other ingredients, used in a primer.

primeval [GEOL] Describing an entity that has persisted from the earliest ages of the earth without appreciable change.

prime vertical circle [GEOD] The vertical circle through the east and west points of the horizon.

priming of the tides [OCEANOGR] The acceleration in the times of occurrence of high and low tides when the sun's tidal effect comes before that of the moon.

Primitiopsacea [PALEON] A small dimorphic superfamily of extinct ostracods in the suborder Beyrichicopina; the velum of the male was narrow and uniform, but that of the female was greatly expanded posteriorly.

primitive cell [CRYSTAL] A parallelepiped whose edges are defined by the primitive translations of a crystal lattice; it is a unit cell of minimum volume.

primitive translation [CRYSTAL] For a space lattice, one of three translations which can be repeatedly applied to generate any translation which leaves the lattice unchanged.

primitive water [HYD] Water that has been imprisoned in the earth's interior, in either molecular or dissociated form, since the formation of the earth.

principal axis [CRYSTAL] The longest axis in a crystal.

principal vertical circle [GEOD] The vertical circle through the north and south points of the horizon, coinciding with the celestial meridian.

principle of uniformity *See* uniformitarianism.

Prioniodidae [PALEON] A family of conodonts in the suborder Conodontiformes having denticulated bars with a large denticle at one end.

Prioniodinidae [PALEON] A family of conodonts in the suborder Conodontiformes characterized by denticulated bars or blades with a large denticle in the middle third of the specimen.

priorite [MINERAL] $(Y,Ca,Th)(Ti,Nb)_2O_6$ A mineral composed of titanoniobate of rare-earth metals; it is isomorphous with eschynite. Also known as blomstrandine.

prism [CRYSTAL] A crystal which has three, four, six, eight, or twelve faces, with the face intersection edges parallel, and which is open only at the two ends of the axis parallel to the intersection edges. [GEOL] A long, narrow, wedge-shaped sedimentary body with a width-thickness ratio greater than 5 to 1 but less than 50 to 1.

prismatic cleavage [CRYSTAL] A type of crystal cleavage that occurs parallel to the faces of a prism.

prismatic jointing *See* columnar jointing.

prismatic structure *See* columnar jointing.

prism crack [GEOL] A mud crack that develops in regular or irregular polygonal patterns on the surface of drying mud puddles and that breaks the sediment into prisms standing normal to bedding.

prism level [ENG] A surveyor's level with prisms that allow the levelman to view the level bubble without moving his eye from the telescope.

private stream [HYD]   Any stream which diverts part or all of the drainage of another stream.

Proanura [PALEON]   Triassic forerunners of the Anura.

probability forecast [METEOROL]   A forecast of the probability of occurrence of one or more of a mutually exclusive set of weather contingencies, as distinguished from a series of categorical statements.

probable maximum precipitation [METEOROL]   The theoretically greatest depth of precipitation for a given duration that is physically possible over a particular drainage area at a certain time of year; in practice, this is derived over flat terrain by storm transposition and moisture adjustment to observed storm patterns.

probable ore [MIN ENG]   **1.** A mineral deposit adjacent to a developed ore but not yet proved by development. **2.** *See* indicated ore.

probable reserves [PETRO ENG]   Primary petroleum reserves based on limited evidence, but not proved by a commercial oil-production rate.

probe [ENG]   A small tube containing the sensing element of electronic equipment, which can be lowered into a borehole to obtain measurements and data.

probertite [MINERAL] $NaCaB_5O_9 \cdot 5H_2O$ A colorless mineral crystallizing in the monoclinic system, consisting of hydrous sodium calcium borate.

problematicum [GEOL]   A marking, object, structure, or other feature in a rock whose nature is doubtful or obscure.

process lapse rate [METEOROL]   The rate of decrease of the temperature of an air parcel as it is lifted, expressed as $-dT/dz$, where z is the altitude, or occasionally dT/dp, where p is pressure; the concept may be applied to other atmospheric variables, such as the process lapse rate of density.

prod cast [GEOL]   The cast of a prod mark. Also known as impact cast.

prodelta [GEOL]   The part of a delta lying beyond the delta front, and sloping gently down to the basin floor of the delta; it is entirely below the water level.

prodelta clay [GEOL]   Fine sand, silt, and clay transported by the river and deposited on the floor of a sea or lake beyond the main body of a delta.

Prodinoceratinae [PALEON]   A subfamily of extinct herbivorous mammals in the family Untatheriidae; animals possessed a carnivorelike body of moderate size.

prod mark [GEOL]   A short tool mark oriented parallel to the current and gradually deepening downcurrent. Also known as impact mark.

producing gas-oil ratio [PETRO ENG]   The ratio of gas to oil (GOR, or gas-oil ratio) from a producing well; an increase in GOR is a danger signal in the efficient control of reservoir performance.

producing horizon [PETRO ENG]   A reservoir bed within the stratigraphic series of an oil province from which gas or liquid hydrocarbons can be obtained by drilling a well.

producing reserves [PETRO ENG]   Developed (proved) petroleum reserves to be produced by existing wells in that portion of a reservoir subjected to full-scale secondary-recovery operations.

Productinida [PALEON]   A suborder of extinct articulate brachiopods in the order Strophomenida characterized by the development of spines.

production-decline curve [PETRO ENG]   A graphical means to estimate the ultimate recovery (oil or gas) from a reservoir; cumulative production is plotted against time, the curve being extrapolated to an end point (that is, ultimate recovery).

production packer [PETRO ENG] A downhole tool used to assist in the efficient production of oil and gas from a well having one or more productive horizons; the function is to provide a seal between the outside of the tubing and the inside of the casing to prevent movement of fluids past that point. Also known as packer.

productivity [PETRO ENG] Measure of an oil well's ability to produce liquid or gaseous hydrocarbons; categories include relative, specific, ultimate, and fractured-well productivity.

productivity index [PETRO ENG] The number of barrels of oil produced per day per decline in well bottom-hole pressure in pounds per square inch.

productivity ratio [PETRO ENG] **1.** The amount of damage or improvement to reservoir formation permeability adjacent to the borehole (due to invasion or reduction of drilling mud present, drilling-fluid filtrate water, swollen clay particles, or salt or wax deposition). **2.** The ratio of permeability calculated from the productivity index to the permeability calculated from reservoir buildup pressure.

productivity test [PETRO ENG] Graphical relation of bottomhole static pressure (calculated or measured) versus producing pressure for various gas flow rates; used to predict future oil well behavior.

profile [GEOL] **1.** The outline formed by the intersection of the plane of a vertical section and the ground surface. Also known as topographic profile. **2.** Data recorded by a single line of receivers from one shot point in seismic prospecting. [GEOPHYS] A graphic representation of the variation of one property, such as gravity, usually as ordinate, with respect to another property, usually linear, such as distance. [HYD] A vertical section of a potentiometric surface, such as a water table. [PETR] In structural petrology, a cross section of a homoaxial structure.

profile line [GEOL] The top line of a profile section, representing the intersection of a vertical plane with the surface of the ground.

profile of equilibrium [GEOL] **1.** The slope of the floor of a sea, ocean, or lake, taken in a vertical plane, when deposition of sediment is balanced by erosion. **2.** The longitudinal profile of a graded stream. Also known as equilibrium profile; graded profile.

profile section [GEOL] A diagram or drawing that shows along a given line the configuration or slope of the surface of the ground as it would appear if intersected by a vertical plane.

profiling snow gage [HYD] A type of radioactive gage for measuring the water equivalent and density/depth distribution of a snowpack, consisting of a radioactive source and a radioactivity detector which move up and down in two adjacent vertical pipes surrounded by snow. Also known as nuclear twin-probe gage.

profit in sight [MIN ENG] Probable gross profit from a mine's ore reserves, as distinct from the ground that is still to be blocked out.

Proganosauria [PALEON] The equivalent name for Mesosauria.

proglacial [GEOL] Of streams, deposits, and other features, being immediately in front of or just beyond the outer limits of a glacier or ice sheet, and formed by or derived from glacier ice.

proglyph [GEOL] A hieroglyph that consists of a cast, especially a groove cast.

prognostic chart [METEOROL] A chart showing, principally, the expected pressure pattern (or height pattern) of a given synoptic chart at a specified future time; usually, positions of fronts are also included, and the forecast values of other meteorological elements may be superimposed.

prognostic equation   [METEOROL]   Any equation governing a system which contains a time derivative of a quantity and therefore can be used to determine the value of that quantity at a later time when the other terms in the equation are known (for example, the vorticity equation).

progradation   [GEOL]   Seaward buildup of a beach, delta, or fan by nearshore deposition of sediments transported by a river, by accumulation of material thrown up by waves, or by material moved by longshore drifting.

prograde metamorphism   [GEOL]   Metamorphic changes in response to a higher pressure or temperature than that to which the rock last adjusted itself.

prograding shoreline   [GEOL]   A shoreline that is being built seaward by accumulation or deposition.

progressive metamorphism   [GEOL]   Systematic change in metamorphic grade from lower to higher in any metamorphic terrain.

progressive sand wave   [GEOL]   A sand wave characterized by downcurrent migration.

progressive sorting   [GEOL]   Sorting of sedimentary particles in the downcurrent direction, resulting in a systematic downcurrent decrease in the mean grain size of the sediment.

progressive wave   [METEOROL]   A wave or wavelike disturbance which moves relative to the earth's surface.

Progymnospermopsida   [PALEON]   A class of plants intermediate between ferns and gymnosperms; comprises the Denovian genus *Archaeopteris*.

projection   [MAP]   A system for presenting on a plane surface the spherical surface of the earth or the celestial sphere; some of these systems are conic, cylindrical, gnomonic, Mercator, orthographic, and stereographic. Also known as map projection.

projection net   *See* net.

Prolacertiformes   [PALEON]   A suborder of extinct terrestrial reptiles in the order Eosuchia distinguished by reduction of the lower temporal arcade.

prolapsed bedding   [GEOL]   Bedding characterized by a series of flat folds with near-horizontal axial planes contained entirely within a bed having undisturbed boundaries.

proluvium   [GEOL]   A complex, friable, deltaic sediment accumulated at the foot of a slope as a result of an occasional torrential washing of fragmental material.

promontory   [GEOL]   **1.** A high, prominent projection or point of land, or cliff of rock, jutting out boldly into a body of water beyond the coastline. **2.** A cape, either low-lying or of considerable height, with a bold termination. **3.** A bluff or prominent hill overlooking or projecting into a lowland.

prong reef   [GEOL]   A wall reef that has developed irregular buttresses normal to its axis in both leeward and (to a smaller degree) seaward directions.

prop   [MIN ENG]   Underground supporting post set across the lode, seam, bed, or other opening.

propaedeutic stratigraphy   *See* prostratigraphy.

propagated blast   [ENG]   A blast of a number of unprimed charges of explosives plus one hole primed, generally for the purpose of ditching, where each charge is detonated by the explosion of the adjacent one, the shock being transmitted through the wet soil.

propagation forecasting   [METEOROL]   Forecasting in which the known or predicted vertical distribution of the index of refraction over an area is used to forecast the propagation performance of radars or any microwave radio equipment operating in that area.

prop-crib timbering [MIN ENG] Shaft timbering with cribs kept apart at the proper distance by means of props.

prop-free [MIN ENG] A face with no posts between the coal and the conveyor used to remove it in longwall mining of a coal seam.

prop-free front [MIN ENG] In coal mining, longwall working in which support to the roof is given by roof beams cantilevered from behind the working face.

propylite [PETR] A modified andesite, altered by hydrothermal processes, resembling a greenstone and consisting of calcite, epidote, serpentine, quartz, pyrite, and iron ore.

propylization [PETR] A hydrothermal process by which propylite is formed from andesite by the introduction of or replacement by an assemblage of minerals.

Prorastominae [PALEON] A subfamily of extinct dugongs (Dugongidae) which occur in the Eocene of Jamaica.

Prosauropoda [PALEON] A division of the extinct reptilian suborder Sauropodomorpha; they possessed blunt teeth, long forelimbs, and extremely large claws on the first finger of the forefoot.

prosopite [MINERAL] $CaAl_2(F,OH)_8$ A colorless mineral composed of basic calcium aluminum fluoride.

prospect [MIN ENG] **1.** To search for minerals or oil by looking for surface indications, by drilling boreholes, or both. **2.** A plot of ground believed to be mineralized enough to be of economic importance.

prospecting seismology [PETRO ENG] The application of seismology to the exploration for natural resources, especially gas and oil.

prospector [MIN ENG] A person engaged in exploring for valuable minerals, or in testing supposed discoveries of the same.

prospect pit [MIN ENG] A pit excavated for the purpose of prospecting mineral-bearing ground.

prospect shaft [MIN ENG] A shaft constructed for the purpose of excavating mineral-bearing ground.

prostratigraphy [GEOL] Preliminary stratigraphy, including lithologic and paleontologic studies without consideration of the time factor. Also known as propaedeutic stratigraphy; protostratigraphy.

protactinium-ionium age method [GEOL] A method of calculating the ages of deep-sea sediments formed during the last 150,000 years from measurements of the ratio of protactinium-231 to ionium (thorium-230), based on the gradual change with time of this ratio because of the difference in half-lives.

protalus rampart [GEOL] An arcuate ridge consisting of boulders and other coarse debris marking the downslope edge of an existing or melted snowbank.

protectite [PETR] A rock formed by the crystallization of a primary magma.

proterobase [PETR] A diabase in which the mafic mineral is primary hornblende.

Proterosuchia [PALEON] A suborder of moderate-sized thecodont reptiles with lightly built triangular skulls, downturned snouts, and palatal teeth.

Proterotheriidae [PALEON] A group of extinct herbivorous mammals in the order Litopterna which displayed an evolutionary convergence with the horses in their dentition and in reduction of the lateral digits of their feet.

Proterozoic *See* Algonkian.

Protoceratidae [PALEON]   An extinct family of pecoran ruminants in the superfamily Traguloidea.

protoclastic [PETR]   Of igneous rocks, characterized by granulation and deformation of the earlier-formed minerals due to differential flow of the magma before solidification.

protodolomite [MINERAL]   A crystalline calcium-magnesium carbonate with a disordered lattice in which the metallic ions occur in the same crystallographic layers instead of in alternate layers as in the dolomite mineral.

Protodonata [PALEON]   An extinct order of huge dragonflylike insects found in Permian rocks.

protoenstatite [MINERAL]   An artificial, unstable, altered form of $MgSiO_3$ produced by thermal decomposition of talc; convertible to enstatite by grinding or heating to a high temperature.

Protoeumalacostraca [PALEON]   The stem group of the crustacean series Eumalacostraca.

protogenic [PETR]   Referring to an older crystalline rock believed to have been formed by igneous activity.

protointraclast [GEOL]   A limestone component that resulted from a premature attempt at resedimentation while it was still in an unconsolidated and viscous or plastic state, and that never existed as a free clastic entity.

protolith [PETR]   The original, unmetamorphosed rock from which a given metamorphic rock is formed.

protomylonite [PETR]   A mylonitic rock that develops from contact-metamorphosed rock; granulation and flowage are caused by overthrusts following the contact surfaces between the intrusion and the country rock.

Protopteridales [PALEOBOT]   An extinct order of ferns, class Polypodiatae.

protoquartzite [PETR]   A well-sorted sandstone that is intermediate in composition between subgraywacke and orthoquartzite, consisting of 75–95% quartz and chert, with less than 15% detrital clay matrix and 5–25% unstable materials in which there is a greater abundance of rock fragments than feldspar grains. Also known as quartzose subgraywacke.

protore [MIN ENG]   **1.** A primary mineral deposit which, through enrichment, can be modified to form an economic ore. **2.** A deposit which could become economically workable if technological change occurred or prices were increased.

Protosireninae [PALEON]   An extinct superfamily of sirenian mammals in the family Dugongidae found in the middle Eocene of Egypt.

protostratigraphy *See* prostratigraphy.

Protosuchia [PALEON]   A suborder of extinct crocodilians from the Late Triassic and Early Jurassic.

proustite [MINERAL] Ag_3AsS_3 A cochineal-red mineral that crystallizes in the rhombohedral system, consists of silver arsenic sulfide, is isomorphous with pyrargyrite, and occurs massively and in crystals. Also known as light-red silver ore; light-ruby silver.

proved ore [MIN ENG]   Ore in which there is practically no risk of failure of continuity.

proved reserves [PETRO ENG]   Reserves (primary or secondary) that have been proved by production at commercial flow rates.

provenance [GEOL] The location, topography, and composition of the source area for any sedimentary rock. Also known as source area; sourceland.

province [OCEANOGR] A region composed of a group of similar bathymetric features whose characteristics are markedly in contrast to surrounding areas.

provincial series [GEOL] A time-stratigraphic series recognized only in a particular region and involving a major division of time within a period.

provitrinite [GEOL] A variety of vitrinite characteristic of provitrain and including the varieties periblinite, suberinite, and xylinite.

provitrain [GEOL] Vitrain in which some plant structure can be discerned by microscope. Also known as telain.

provitrinite [GEOL] A variety of vitrinite characteristic of provitrain and including the varieties periblinite, suberinite, and xylinite.

proximal [GEOL] Of a sedimentary deposit, composed of coarse clastics and formed near the source.

proximate admixture [GEOL] An admixture (in a sediment of several size grades) whose particles are most similar in size to those of the dominant or maximum grade.

Psamment [GEOL] A suborder of the soil order Entisol, characterized by a sandy texture and a thin surface horizon grading into thicker horizons below.

psammite *See* arenite.

psammitic *See* arenaceous.

Psammodontidae [PALEON] A family of extinct cartilaginous fishes in the order Bradyodonti in which the upper and lower dentitions consisted of a few large quadrilateral plates arranged in two rows meeting in the midline.

psephicity [GEOL] A coefficient of roundability of a pebble- or sand-size mineral fragment, expressed as the ratio of specific gravity to hardness (as measured in air) or the quotient of specific gravity minus one divided by hardness (as measured in water).

psephite [GEOL] A sediment or sedimentary rock composed of fragments that are coarser than sand and which are set in a qualitatively and quantitatively varying matrix; equivalent to a rudite or, generally, a conglomerate.

psephyte [GEOL] A lake-bottom deposit consisting mainly of coarse, fibrous plant remains.

pseudoadiabat [METEOROL] On a thermodynamic diagram, a line representing a pseudoadiabatic expansion of an air parcel; in practice, approximate computations are employed, and the resulting lines represent, ambiguously, pseudoadiabats and saturation adiabats.

pseudoadiabatic chart *See* Stuve chart.

pseudoadiabatic expansion [GEOPHYS] A saturation-adiabatic process in which the condensed water substance is removed from the system, and which therefore is best treated by the thermodynamics of open systems; meteorologically, this process corresponds to rising air from which the moisture is precipitating.

pseudoallochem [GEOL] An object resembling an allochem but produced in place within a calcareous sediment by a secondary process such as recrystallization.

Pseudoborniales [PALEOBOT] An order of fossil plants found in Middle and Upper Devonian rocks.

pseudobreccia [PETR] Limestone that is partially and irregularly dolomitized and is characterized by a mottled, breccialike appearance. Also known as recrystallization breccia.

pseudobrookite [MINERAL] Fe_2TiO_5 A brown or black mineral consisting of iron titanium oxide and occurring in orthorhombic crystals; specific gravity is 4.4–4.98.

pseudocannel coal [GEOL] Cannel coal that contains much humic matter. Also known as humic-cannel coal.

pseudochrysolite *See* moldavite.

pseudocol [GEOL] A landform that is represented by a constriction of the valley of a stream diverted by glacial ponding, and that is formed by the cutting through of a cover of drift and subsequent exposure of a former col.

pseudo cold front *See* pseudo front.

pseudoconcretion [GEOL] A subspherical, secondary sedimentary structure resembling a true concretion but not formed by orderly precipitation of mineral matter in the pores of a sediment.

pseudoconformity *See* paraconformity.

pseudoconglomerate [GEOL] A rock that resembles, or may easily be mistaken for, a true or normal (sedimentary) conglomerate.

pseudocotunnite [MINERAL] K_2PbCl_4 A yellow or yellowish-green, orthorhombic mineral consisting of a potassium lead chloride.

pseudo cross-bedding [GEOL] **1.** An inclined bedding produced by deposition in response to ripple-mark migration and characterized by foreset beds that appear to dip into the current. **2.** A structure resembling cross-bedding, caused by distortion-free slumping and sliding of a semiconsolidated mass of sediments such as sandy shales.

pseudodiffusion [GEOL] Mixing of thin superpositioned layers of slowly accumulated marine sediments by the action of water motion or subsurface organisms.

pseudoequivalent temperature *See* equivalent temperature.

pseudofault [GEOL] A faultlike feature resulting from weathering along joint, shrinkage, or bedding planes.

pseudofibrous peat [GEOL] Peat that is fibrous in texture but is plastic and incoherent.

pseudofjord [GEOGR] A nonglaciated fjordlike valley.

pseudo front [METEOROL] A small-scale front, formed in association with organized severe convective activity, between a mass of rain-cooled air from the thunderstorm clouds and the warm surrounding air. Also known as pseudo cold front.

pseudogalena *See* sphalerite.

pseudogley [GEOL] A densely packed, silty soil that is alternately waterlogged and rapidly dried out.

pseudogradational bedding [GEOL] A structure in metamorphosed sedimentary rock in which the original textural graduation (coarse at the base, finer at the top) appears to be reversed, due to the formation of porphyroblasts in the finer-grained part of the rock.

pseudokarst [GEOL] A topography that resembles karst but that is not formed by the dissolution of limestone, usually a rough-surfaced lava field in which ceilings of lava tubes have collapsed.

pseudokettle *See* pingo remnant.

pseudoleucite [MINERAL] A pseudomorph after leucite consisting of a mixture of nepheline, orthoclase, and analcime.

pseudoliquid density [PETRO ENG] For reservoir studies, the calculated value of a pseudodensity for reservoir liquid at atmospheric conditions, followed by application of suitable correction factors to obtain an approximate value for actual liquid density in the reservoir.

pseudomalachite [MINERAL] $Cu_5(PO_4)_2(OH)_4\cdot H_2O$ An emerald green to dark green and blackish-green, monoclinic mineral consisting of a hydrated basic copper phosphate. Also known as tagilite.

pseudomicroseism [GEOPHYS] A microseism due to instrumental effects.

pseudomorph [MINERAL] An altered mineral whose crystal form has the outward appearance of another mineral species. Also known as false form.

pseudomountain [GEOL] A mountain formed by differential erosion, in contrast to one produced by uplift.

pseudonodule [GEOL] A primary sedimentary structure consisting of a ball-like mass of sandstone enclosed in shale or mudstone, characterized by a rounded base with upturned or inrolled edges, and resulting from the settling of sand into underlying clay or mud which welled up between isolated sand masses. Also known as sand roll.

pseudo oolith [GEOL] A spherical or roundish pellet or particle (generally less than 1 millimeter in diameter) in a sedimentary rock, externally resembling an oolith in size or shape but of secondary origin and amorphous or crypto or microcrystalline, and lacking the radial or concentric internal structure of an oolith. Also known as false oolith.

pseudophite [MINERAL] A general name for compact, massive chlorites resembling serpentines, in part clinochlore and in part penninite.

pseudoporphyritic [PETR] Pertaining to a rock that is not a true porphyry, but resembles one because of rapid growth of some of the crystals.

pseudoporphyroblastic [GEOL] Pertaining to a structure that resembles porphyroblastic texture but that is due to processes other than growth, such as to differential granulation.

pseudo ripple mark [GEOL] A bedding-plane feature that resembles a ripple mark but is formed by lateral pressure caused by slumping or by local, small-scale tectonic deformation.

pseudorutile [MINERAL] $Fe_2Ti_3O_9$ A mineral that is an oxidation product of ilmenite and is common in beach sands.

pseudoslickenside [GEOL] In a tectonite, a surface resembling slickenside, developed by rotation without any indication of the direction of movement

pseudosparite [GEOL] A limestone consisting of relatively large, clear calcite crystals that have developed by recrystallization or by grain growth.

pseudospharolith [MINERAL] A spherulite consisting of two minerals, one with parallel and one with inclined extinction, growing from the same center.

pseudostatic SP *See* pseudostatic spontaneous potential.

pseudostatic spontaneous potential [PETRO ENG] Theoretical maximum spontaneous potential current that can be measured in a downhole, mud-column log in shaly sand. Abbreviated pseudostatic SP; PSP.

pseudo steady-state pressure distribution [PETRO ENG] The condition when the declining pressure distribution within a reservoir system closed at one boundary is declining at a uniform rate everywhere (or, there is constant pressure gradient within the reservoir).

pseudostratification *See* sheeting structure.

Pseudosuchia [PALEON] A suborder of extinct reptiles of the order Thecodontia comprising bipedal, unarmored or feebly armored forms which resemble dinosaurs in many skull features but retain a primitive pelvis.

pseudosymmetry [CRYSTAL]   Apparent symmetry of a crystal, resembling that of another system; generally due to twinning.

pseudotachylite [PETR]   A black rock that resembles tachylite; carries fragmental enclosures and shows evidence of having been at high temperature.

pseudotill [GEOL]   A nonglacial deposit resembling a glacial till.

pseudotillite [GEOL]   A nonglacial tillitelike rock, such as a pebbly mudstone, formed on land by flow of nonglacial mud or deposited by a subaqueous turbidity flow.

pseudounconformity [GEOL]   A stratigraphic relationship which appears unconformable but is characterized by superabundance or excess accumulation of sediment, such as that due to submarine slumping penecontemporaneous with sedimentation off the sides of a rising anticline or dome.

pseudovitrinite [GEOL]   A maceral of coal that is superficially similar to vitrinite but is higher in reflectance from polished surfaces in oil immersion, and has slitted structure, remnant cellular structures, uncommon fracture patterns, higher relief, and paucity or absence of pyrite inclusions.

pseudovitrinoid [GEOL]   Pseudovitrinite occurring in bituminous coal.

psilomelane [MINERAL] $BaMn_9O_{16}(OH)_4$ A massive, hard, black, botryoidal manganese oxide mineral mixture with a specific gravity ranging from 3.7 to 4.7.

Psilophytales [PALEOBOT]   A group formerly recognized as an order of fossil plants.

Psilophytineae [PALEON]   The equivalent name for Rhyniopsida.

psittacinite *See* mottramite.

PSP *See* pseudostatic spontaneous potential.

Pteridospermae [PALEOBOT]   Seed ferns, a class of the Cycadicae comprising extinct plants characterized by naked seeds borne on large fernlike fronds.

Pteridospermophyta [PALEOBOT]   The equivalent name for Pteridospermae.

pterodactyl [PALEON]   The common name for members of the extinct reptilian order Pterosauria.

Pterodactyloidea [PALEON]   A suborder of Late Jurassic and Cretaceous reptiles in the order Pterosauria distinguished by lacking tails and having increased functional wing length due to elongation of the metacarpels.

pteropod ooze [GEOL]   A pelagic sediment containing at least 45% calcium carbonate in the form of tests of marine animals, particularly pteropods.

Pterosauria [PALEON]   An extinct order of flying reptiles of the Mesozoic era belonging to the subclass Archosauria; the wing resembled that of a bat, and a large heeled sternum supported strong wing muscles.

Ptilodontoidea [PALEON]   A suborder of extinct mammals in the order Multituberculata.

ptilolite *See* mordenite.

Ptyctodontida [PALEON]   An order of Middle and Upper Devonian fishes of the class Placodermi in which both the head and trunk shields are present, and the joint between them is a well-differentiated and variable structure.

ptygma [GEOL]   Pegmatitic material with migmatite or gneiss, resembling disharmonic folds. Also known as ptygmatic fold.

ptygmatic fold *See* ptygma.

pucherite [MINERAL] $BiVO_4$ A reddish-brown orthorhombic mineral composed of bismuth vanadate, occurring as small crystals.

pudding ball *See* armored mud ball.

puddingstone [GEOL] In Great Britain, a conglomerate consisting of rounded pebbles whose colors are in marked contrast with the matrix, giving a section of the rock the appearance of a raisin pudding.

puddle [ENG] To apply water in order to settle loose dirt.

puelche [METEOROL] An east wind which has crossed the Andes; the Andean foehn of the South American west coast.

puff cone *See* mud cone.

puffer [MIN ENG] A small stationary engine used in coal mines for hoisting material.

puff of wind [METEOROL] A slight local breeze which causes a patch of ripples on the surface of the sea.

puglianite [PETR] A coarse-grained igneous rock composed of euhedral augite, leucite, anorthite, sanidine, hornblende, and biotite.

pull-apart [GEOL] A precompaction sedimentary structure having the appearance of boudinage and consisting of beds that have been stretched and pulled apart into relatively short slabs.

pull rope [MIN ENG] **1.** The rope that pulls a journey of loaded cars on a haulage plane. **2.** The rope that pulls the loaded scoop or bucket in a scraper loader layout. Also known as main rope.

pull tube [PETRO ENG] Tube used in rod-type, traveling-barrel oil well pumps to connect the pump plunger with the seating anchor.

pulps [MINERAL] A fine, mealy, opaline silica, much like sand.

pulse-time-modulated radiosonde [ENG] A radiosonde which transmits the indications of the meteorological sensing elements in the form of pulses spaced in time; the meteorological data are evaluated from the intervals between the pulses. Also known as time-interval radiosonde.

pulverite [PETR] A sedimentary rock composed of silt- or clay-sized aggregates of nonclastic origin, simulating in texture a lutite of clastic origin.

pumice [GEOL] A rock froth, formed by the extreme puffing up of liquid lava by expanding gases liberated from solution in the lava prior to and during solidification. Also known as foam; pumice stone; pumicite; volcanic foam.

pumiceous [GEOL] Pertaining to the texture of a pyroclastic rock, such as pumice, characterized by numerous small cavities presenting a spongy, frothy appearance.

pumice fall [GEOL] Pumice falling from a volcano eruption cloud.

pumice stone *See* pumice.

pumicite *See* pumice.

pumilith [GEOL] A lithified deposit of volcanic ash.

pumpellyite [MINERAL] $Ca_2Al_3Si_3O_{12}(OH)$ A greenish epidotelike mineral that is probably related to clinozoisite. Also known as lotrite; zonochlorite.

pumpellyite-prehnite-quartz facies [PETR] A variety of low-temperature, moderate-pressure metamorphism.

Pumpelly's rule [GEOL] The generalization that the axial surfaces of minor folds of an area are in accord with those of the major fold structures.

pumping pressure [PETRO ENG] Pressure required to inject (pump) water, gas, or acid into a pressurized petroleum reservoir.

pumping well [PETRO ENG]   A producing oil well in which liquid products are recovered from the reservoir by means of a pump, rather than by gas lift.

Purbeckian [GEOL]   A stage of geologic time in Great Britain: uppermost Jurassic (above Bononian, below Cretaceous).

pure coal *See* vitrain.

purga [METEOROL]   A severe storm similar to the blizzard and buran, which rages in the tundra regions of northern Siberia in winter.

purl [HYD]   A swirling or eddying stream or rill, moving swiftly around obstructions.

purple blende *See* kermesite.

purple light [GEOPHYS]   The faint purple glow observed on clear days over a large region of the western sky after sunset and over the eastern sky before sunrise.

purpurite [MINERAL] $(Mn,Fe)PO_4$ A dark-red or purple mineral composed of ferric-manganic phosphate; it is isomorphous with heterosite.

pusher grade *See* helper grade.

pusher tractor [MIN ENG]   A bulldozer exerting pressure on the rear of a scraper-loader while the loader is digging and loading unconsolidated ground during opencast mining.

push moraine [GEOL]   A broad, smooth, arc-shaped morainal ridge consisting of material mechanically pushed or shoved along by an advancing glacier.   Also known as push-ridge moraine; shoved moraine; thrust moraine; upsetted moraine.

push-ridge moraine *See* push moraine.

Pustulosa [PALEON]   An extinct suborder of echinoderms in the order Phanerozonida found in the Paleozoic.

Putnam anomaly *See* average-level anomaly.

puy [GEOL]   A small, remnant volcanic cone.

pycnite [MINERAL]   A variety of topaz occurring in massive columnar aggregations.

pycnocline [OCEANOGR]   A region in the ocean where water density increases relatively rapidly with respect to depth.

Pycnodontiformes [PALEON]   An extinct order of specialized fishes characterized by a laterally compressed, disk-shaped body, long dorsal and anal fins, and an externally symmetrical tail.

Pygasteridae [PALEON]   The single family of the extinct order Pygasteroida.

Pygasteroida [PALEON]   An order of extinct echinoderms in the superorder Diadematacea having four genital pores, noncrenulate tubercles, and simple ambulacral plates.

pyralmandite [MINERAL]   A garnet intermediate in chemical composition between pyrope and almandine.

pyramid [CRYSTAL]   An open crystal having three, four, six, eight, or twelve nonparallel faces that meet at a point.

pyramidal cleavage [CRYSTAL]   A type of crystal cleavage that occurs parallel to the faces of a pyramid.

pyramidal iceberg *See* pinnacled iceberg.

pyranometer [ENG]   An instrument used to measure the combined intensity of incoming direct solar radiation and diffuse sky radiation; compares heating produced by the radiation on blackened metal strips with that produced by an electric current. Also known as solarimeter.

pyrargyrite [MINERAL] Ag_3SbS_3 A deep ruby-red to black mineral, crystallizing in the hexagonal system, occurring in massive form and in disseminated grains, and having an adamantine luster; hardness is 2.5 on Mohs scale, and specific gravity is 5.85; an important silver ore. Also known as dark-red silver ore; dark ruby silver.

Pyrenean orogeny [GEOL] A short-lived orogeny that occurred during the late Eocene, between the Bartonian and Ludian stages.

pyrgeometer [ENG] An instrument for measuring radiation from the surface of the earth into space.

pyribole [PETR] A rock which contains either pyroxene or amphibole or both.

pyrite [MINERAL] FeS_2 A hard, brittle, brass-yellow mineral with metallic luster, crystallizing in the isometric system; hardness is 6–6.5 on Mohs scale, and specific gravity is 5.02. Also known as common pyrite; fool's gold; iron pyrite; mundic.

pyrite roasting [MIN ENG] Thermal processing of iron pyrite (FeS_2, iron disulfide) in the presence of air to produce iron oxide sinter (used in steel mills) and elemental sulfur.

pyritization [GEOL] A common process of hydrothermal alteration involving introduction of or replacement by pyrite.

pyritobitumen [GEOL] Any of various dark-colored, relatively hard, nonvolatile hydrocarbon substances often associated with mineral matter, which decompose upon heating to yield bitumens. Also known as pyrobitumen.

pyritohedron [CRYSTAL] A dodecahedral crystal with 12 irregular pentagonal faces; it is characteristic of pyrite. Also known as pentagonal dodecahedron; pyritoid; regular dodecahedron.

pyritoid *See* pyritohedron.

pyroaurite [MINERAL] $Mg_6Fe_2(OH)_{16}CO_3 \cdot 4H_2O$ A goldlike or brownish rhombohedral mineral composed of hydrous basic magnesium iron carbonate.

pyrobelonite [MINERAL] $PbMn(VO_4)(OH)$ A fire-red to deep brilliant-red mineral composed of basic vanadate of manganese and lead, occurring as crystal needles.

pyrobiolite [PETR] An organic rock containing organic remains that have been altered by volcanic action.

pyrobitumen *See* pyritobitumen.

pyroborate *See* borax.

pyrochlore [MINERAL] $(Na,Ca)_2(Nb,Ta)_2O_6(OH,F)$ Pale-yellow, reddish, brown, or black mineral, crystallizing in the isometric system, and occurring in pegmatites derived from alkalic igneous rocks. Also known as pyrrhite.

pyrochroite [MINERAL] $Mn(OH)_2$ A hexagonal mineral composed of naturally occurring manganese hydroxide; it is white when fresh, but darkens upon exposure.

pyroclast [GEOL] An individual pyroclastic fragment or clast.

pyroclastic flow [GEOL] Ash flow not involving high-temperature conditions.

pyroclastic ground surge [GEOL] The relatively thin mantle of rock found around a volcanic vent; the thickness is not uniform, the internal stratification is not parallel to the top and bottom of the layer, and the extent is a few kilometers from the source.

pyroclastic rock [PETR] A rock that is composed of fragmented volcanic products ejected from volcanoes in explosive events.

pyrogenesis [GEOL] The intrusion and extrusion of magma and its derivatives.

pyrogenetic mineral [MINERAL] An anhydrous mineral of an igneous rock, usually crystallized at high temperature in a magma containing relatively few volatile components.

pyrolusite [MINERAL] MnO_2 An iron-black mineral that crystallizes in the tetragonal system and is the most important ore of manganese; hardness is 1-2 on Mohs scale, and specific gravity is 4.75.

pyromagma [GEOL] A highly mobile lava, oversaturated with gases, that exists at shallower depths than hypomagma.

pyromelane *See* brookite.

pyromeride [PETR] A devitrified rhyolite characterized by spherulitic texture.

pyrometamorphism [PETR] Contact metamorphism at temperatures near the melting points of the component minerals.

pyrometasomatism [PETR] Forming of contact-metamorphic mineral deposits at high temperatures by emanations from the intrusive rock, involving replacement of the enclosing rock with the addition of materials.

pyromorphite [MINERAL] $Pb_5(PO_4)_3Cl$ A green, yellow, brown, gray, or white mineral of the apatite group, crystallizing in the hexagonal system; a minor ore of lead. Also known as green lead ore.

pyrope [MINERAL] $Mg_3Al_2(SiO_4)_3$ A mineral species of the garnet group characterized by a deep fiery-red color and occurring in basic and ultrabasic igneous rocks.

pyrophane *See* fire opal.

pyrophanite [MINERAL] $MnTiO_3$ A blood-red rhombohedral mineral consisting of manganese titanate; it is isomorphous with ilmenite.

pyrophyllite [MINERAL] $AlSi_2O_5(OH)$ A white, greenish, gray, or brown phyllosilicate mineral that resembles talc and occurs in a foliated form or in compact masses in quartz veins, granites, and metamorphic rocks. Also known as pencil stone.

pyroretinite [MINERAL] A type of retinite found in the brown coals of Aussig (Usti and Labem), in Bohemian Czechoslovakia.

pyroschist [PETR] A schist or shale that has a sufficiently high carbon content to burn with a bright flame or to yield volatile hydrocarbons when heated.

pyrosmalite [MINERAL] $(Mn,Fe)_4Si_3O_7(OH,Cl)_6$ A colorless, pale-brown, gray, or gray-green mineral composed mainly of basic iron manganese silicate with chlorine.

pyrosphere [GEOL] The zone of the earth below the lithosphere, consisting of magma. Also known as magmosphere.

pyrostibite *See* kermesite.

pyrostilpnite [MINERAL] Ag_3SbS_3 A hyacinth-red mineral composed of silver antimony sulfide, occurring in monoclinic crystal tufts; it is polymorphous with pyrargerite.

Pyrotheria [PALEON] An extinct monofamilial order of primitive, mastodonlike, herbivorous, hoofed mammals restricted to the Eocene and Oligocene deposits of South America.

Pyrotheriidae [PALEON] The single family of the Pyrotheria.

pyroxene [MINERAL] A family of diverse and important rock-forming minerals having infinite (Si_2O_6) single inosilicate chains as their principal motif; colors range from white through yellow and green to brown and greenish black; hardness is 5.5-6 on Mohs scale, and specific gravity is 3.2–4.0.

pyroxene alkali syenite [GEOL] A quartz-poor (less than 20%) member of the char-nockite series, characterized by the presence of microperthite.

pyroxene monzonite [GEOL] A quartz-poor (less than 20%) member of the char-nockite series, containing approximately equal amounts of microperthite and pla-gioclase.

pyroxene syenite [GEOL] A quartz-poor (less than 20%) member of the charnockite series, containing more microperthite than plagioclase.

pyroxenite [PETR] A heavy, dark-colored, phaneritic igneous rock composed largely of pyroxene with smaller amounts of olivine and hornblende, and formed by crystall-ization of gabbraic magma.

pyroxenoids [MINERAL] A mineral group (including wollastonite and rhodonite) compositionally similar to pyroxene, but SiO_4 tetrahedrons are connected in rings rather than chains.

pyroxferroite [MINERAL] (Fe, Mn, Ca)SiO_3 A yellow mineral found in lunar samples; the iron analog of pyroxmangite.

pyroxmangite [MINERAL] (Mn, Fe, Ca)SiO_3 A brown, triclinic mineral; a variety of rhodonite containing appreciable iron.

pyrrhite *See* pyrochlore.

pyrrhotite [MINERAL] $Fe_{1-x}S$ A common reddish-brown to brownish-bronze min-eral that occurs as rounded grains to large masses, more rarely as tabular pseudohex-agonal crystals and rosettes; hardness is 4 on Mohs scale, and specific gravity is 4.6 (for the composition Fe_7S_8).

pythmic [GEOL] Pertaining to the bottom of a lake.

Q

quadrantal point *See* intercardinal point.

Quadrijugatoridae [PALEON] A monomorphic family of extinct ostracods in the superfamily Hollinacea.

quaking bog [GEOL] A peat bog floating or growing over water-saturated land which shakes or trembles when walked on.

quality of snow [METEOROL] The amount of ice in a snow sample expressed as a percent of the weight of the sample. Also known as thermal quality of snow.

quantitative geomorphology [GEOL] The assignment of dimensions of mass, length, and time to all descriptive parameters of landform geometry and geomorphic processes, followed by the derivation of empirical mathematical relationships and formulation of rational mathematical models relating those parameters.

quaquaversal [GEOL] Of strata and geologic structures, dipping outward in all directions away from a central point.

quarry [ENG] An open or surface working or excavation for the extraction of building stone, ore, coal, gravel, or minerals.

quarry bar [ENG] A horizontal bar with legs at each end, used to carry machine drills.

quarry face [MIN ENG] The freshly split face of ashlar, squared off for the joints only and used for massive work.

quarrying [ENG] The surface exploitation and removal of stone or mineral deposits from the earth's crust. [GEOL] *See* plucking.

quarrying machine [MECH ENG] Any machine used to drill holes or cut tunnels in native rock, such as a gang drill or tunneling machine; most commonly, a small locomotive bearing rock-drilling equipment operating on a track.

quarry powder [MATER] Ammonium nitrate dynamites used in quarrying where blasts of several tons of explosives are needed.

quarry sap *See* quarry water.

quarry water [ENG] Subsurface water retained in freshly quarried rock. Also known as quarry sap.

quartz [MINERAL] SiO_2 A colorless, transparent rock-forming mineral with vitreous luster, crystallizing in the trigonal trapezohedral class of the rhombohedral subsystem; hardness is 7 on Mohs scale, and specific gravity is 2.65; the most abundant and widespread of all minerals.

quartzarenite [PETR] A quartz-rich sandstone with framework grains separated predominantly by cement rather than matrix; essentially an orthoquartzite.

quartz basalt [PETR]   An igneous rock with more than 5% quartz.

quartz-bearing diorite *See* quartz diorite.

quartz claim   [MIN ENG]   In the United States, a mining claim containing ore in veins or lodes, as contrasted with placer claims carrying mineral, usually gold, in alluvium.

quartz crystal *See* rock crystal.

quartz diorite   [PETR]   A group of plutonic rocks having the composition of diorite but with large amounts of quartz (greater than 20%). Also known as quartz-bearing diorite; tonalite.

quartz-flooded limestone [PETR]   A limestone characterized by an abundance of quartz particles that had been imported suddenly from a nearby source by wind or water currents, but that gradually became sparser in an upward direction and completely disappeared within a few centimeters.

quartz graywacke [PETR]   A graywacke containing abundant grains of quartz and chert and less than 10% each of feldspars and rock fragments.

quartz horizontal magnetometer   [ENG]   A type of relative magnetometer used as a geomagnetic field instrument and as an observatory instrument for routine calibration of recording equipment.

quartzine [MINERAL]   Fibrous chalcedony characterized by fibers having a positive crystallographic elongation (fibers parallel to the c axis).

quartzite [PETR]   A granoblastic metamorphic rock consisting largely or entirely of quartz; most quartzites are formed by metamorphism of sandstone.

quartzitic sandstone   [PETR]   Sandstone consisting of 100% quartz grains cemented with silica.

quartz lattice *See* rhyodacite.

quartz mine   [MIN ENG]   A mine in which a valuable constituent, such as gold, is found in veins rather than in placers; so named because quartz is the chief accessory mineral in such deposits.

quartz monzonite [PETR]   Granitic rock in which 10–50% of the felsic constituents are quartz, and in which the ratio of alkali feldspar to total feldspar is between 35% and 65%. Also known as adamellite.

quartz norite [GEOL]   A member of the charckonite series which contains plagioclase but no potassium feldspar.

quartzose [GEOL]   Containing quartz as a principal constituent.

quartzose arkose [PETR]   A sandstone containing 50–85% quartz, chert, and metamorphic quartzite, 15-25% feldspars and feldspathic crystalline rock fragments, and 0-25% micas and micaceous metamorphic rock fragments.

quartzose chert [PETR]   A vitreous, sparkly, shiny chert which under high magnification shows a heterogeneous mixture of pyramids, prisms, and faces of quartz; also includes chert in which the secondary quartz is largely anhedral.

quartzose graywacke [PETR]   **1.** A sandstone containing 50–85% quartz, chert, and metamorphic quartzite, 15-25% micas and micaceous metamorphic rock fragments, and 0-25% feldspars and feldspathic rock fragments. **2.** A graywacke that has lost its micaceous constituents through abrasion and thus tends to approach an orthoquartzite.

quartzose sandstone [PETR]   Sandstone consisting of more than 95% clear quartz grains and less than 5% matrix. Also known as quartz sandstone.

quartzose shale [PETR]   A green or gray shale composed predominantly of rounded quartz grains of silt size, commonly associated with highly mature sandstones (orthoquartzites), representing the reworking of residual clays as transgressive seas encroached on old land areas.

quartzose subgraywacke *See* protoquartzite.

quartz-pebble conglomerate *See* orthoquartzitic conglomerate.

quartz porphyry [PETR]   A porphyritic extrusive or hypabyssal rock containing quartz and alkali feldspar phenocrysts embedded in a microcrystalline or cryptocrystalline matrix. Also known as granite porphyry.

quartz sandstone *See* quartzose sandstone.

quartz schist [PETR]   A schist whose foliation is due mainly to streaks and lenticles of nongranular quartz.

quartz syenite [PETR]   A group of plutonic rocks having the characteristics of syenite but with a greater amount of quartz (5–20%).

quartz topaz *See* citrine.

quasicratonic [GEOL]   Pertaining to a part of the oceanic crust marginal to the continent that is considered to be formerly continental and that stretched and foundered during expansion. Also known as semicratonic.

quasi-hydrostatic approximation [METEOROL]   The use of the hydrostatic equation as the vertical equation of motion, thus implying that the vertical accelerations are small without constraining them to be zero. Also known as quasi-hydrostatic assumption.

quasi-hydrostatic assumption *See* quasi-hydrostatic approximation.

quasi-stationary front [METEOROL]   A front which is stationary or nearly so; conventionally, a front which is moving at a speed less than about 5 knots (0.26 meter per second) is generally considered to be quasi-stationary. Commonly known as stationary front.

Quaternary [GEOL]   The second period of the Cenozoic geologic era, following the Tertiary, and including the last 2–3 million years.

Queenston shale [GEOL]   A red bed series from the Ordovician found in Niagara Gorge; it is composed of deltaic red shale.

queluzite [PETR]   An igneous rock composed chiefly of the mineral spessartite, occasionally with amphiboles, pyroxenes, or micas.

quenite [PETR]   A fine-grained, dark-colored hypabyssal rock composed of anorthite, chrome disopside, with less olivine and a small amount of bronzite.

quenselite [MINERAL] $PbMnO_2(OH)$ A pitch black mineral consisting of an oxide of lead and manganese; occurs in tabular form.

quenstedtite [MINERAL] $Fe_2(SO_4)_3 \cdot 10H_2O$ A pale violet to reddish-violet, triclinic mineral consisting of hydrated ferric

quick clay [GEOL]   Clay that loses its shear strength after being disturbed.

quickflow [HYD]   Direct runoff representing the major runoff during storm periods; it is the primary source of water in most floods.

quicksand [GEOL]   A highly mobile mass of fine sand consisting of smooth, rounded grains with little tendency to mutual adherence, usually thoroughly saturated with upward-flowing water; tends to yield under pressure and to readily swallow heavy objects on the surface. Also known as running sand.

quickstone [PETR]   A consolidated rock that flowed under the influence of gravity before lithification.

quickwater [HYD]   The part of a stream characterized by a strong current.

quilted surface [GEOL]   A land surface characterized by broad, rounded, uniformly convex hills separating valleys that are comparatively narrow.

quitclaim [MIN ENG]   Legal release of a claim, right, title, or interest by one person or estate to another.

rabal [METEOROL] A method of winds-aloft observation, that is, the determination of wind speeds and directions in the atmosphere above a station; it is accomplished by recording the elevation and azimuth angles of the balloon at specified time intervals while visually tracking a radiosonde balloon with the theodolite.

race [OCEANOGR] A rapid current, or a constricted channel in which such a current flows; the term is usually used only in connection with a tidal current, which may be called a tide race.

rack [MIN ENG] An inclined trough or table for washing or separating ore.

radar climatology [CLIMATOL] The statistics in time and space of radar weather echoes.

radar meteorological observation [METEOROL] An evaluation of the echoes which appear on the indicator of a weather radar, in terms of orientation, coverage, intensity, tendency of intensity, height, movement, and unique characteristics of echoes, that may be indicative of certain types of severe storms (such as hurricanes, tornadoes, or thunderstorms) and of anomalous propagation. Also known as radar weather observation.

radar meteorology [METEOROL] The study of the scattering of radar waves by all types of atmospheric phenomena and the use of radar for making weather observations and forecasts.

radar report [METEOROL] The encoded and transmitted report of a radar meteorological observation; these reports usually give the azimuth, distance, altitude, intensity, shape and movement, and other characteristics of precipitation echoes observed by the radar. Also known as rain area report. Abbreviated RAREP.

radarsonde [ENG] **1.** An electronic system for automatically measuring and transmitting high-altitude meteorological data from a balloon, kite, or rocket by pulse-modulated radio waves when triggered by a radar signal. **2.** A system in which radar techniques are used to determine the range, elevation, and azimuth of a radar target carried aloft by a radiosonde.

radar storm detection [METEOROL] The detection of certain storms or stormy conditions by means of radar; liquid or frozen water drops within the storm reflect radar echoes.

radar storm-detection equation [METEOROL] The equation which relates the variables involved in the radar detection of precipitation.

radar surveying [ENG] Surveying in which airborne radar is used to measure accurately the distance between two ground radio beacons positioned along a baseline; this eliminates the need for measuring distance along the baseline in inaccessible or extremely rough terrain.

radar theodolite [ENG]   A theodolite that uses radar to obtain azimuth, elevation, and slant range to a reflecting target, for surveying or other purposes.

radar upper band *See* upper bright band.

radar weather observation *See* radar meteorological observation.

radar wind [METEOROL]   Wind of which the movement, speed, and direction is observed or determined by a radar tracking of a balloon carrying a radiosonde, a radio transmitter, or a radar reflector.

radar wind system [ENG]   Apparatus in which radar techniques are used to determine the range, elevation, and azimuth of a balloon-borne target, and hence to compute upper-air wind data.

radial drainage pattern [GEOL]   A drainage pattern characterized by radiating streams diverging from a high central area. Also known as centrifugal drainage pattern.

radial faults [GEOL]   Faults arranged like the spokes of a wheel, radiating from a central point.

radial-flow [PETRO ENG]   Pertaining to spokelike flow of reservoir fluids radially inward toward a wellbore focal area.

radial percussive coal cutter [MIN ENG]   A heavy coal cutter having a percussive drill, with extension rods; used in headings and rooms in pillar methods of working and for drilling shot-firing holes.

radiate mud crack [GEOL]   A mud crack that displays an incomplete radiate pattern and lacks normal polygonal development.

radiational cooling [METEOROL]   The cooling of the earth's surface and adjacent air, accomplished (mainly at night) whenever the earth's surface suffers a net loss of heat due to terrestrial radiation.

radiation budget [GEOPHYS]   A quantitative statement of the amounts of radiation entering and leaving a given region of the earth.

radiation chart [GEOPHYS]   Any chart or diagram which permits graphical solution of the (generally unintegrable) flux integrals arising in problems of atmospheric infrared radiation transfer.

radiation fog [METEOROL]   A major type of fog, produced over a land area when radiational cooling reduces the air temperature to or below its dew point; thus, strictly, a nighttime occurrence, although the fog may begin to form by evening twilight and often does not dissipate until after sunrise.

radiation well logging *See* radioactive well logging.

radiative diffusivity [METEOROL]   A characteristic property of a given layer of the atmosphere which governs the rate at which that layer will warm or cool as a result of the transfer, within it, of infrared radiation; the radiative diffusivity is dependent upon the temperature and water-vapor content of the layer of air and upon the pressure within the layer.

radiatus [METEOROL]   A cloud variety whose elements are arranged in straight parallel bands; owing to the effect of perspective, these bands seem to converge toward a point on the horizon, or, when the bands cross the entire sky, toward two opposite points.

radioactive mineral   [MINERAL]   Any mineral species that contains uranium or thorium as an essential part of the chemical composition; examples are uraninite, pitchblende, carnotite, coffinite, and autunite.

radioactive snow gage   [ENG]   A device which automatically and continuously records the water equivalent of snow on a given surface as a function of time; a small sample of a radioactive salt is placed in the ground in a lead-shielded collimator which directs a beam of radioactive particles vertically upward; a Geiger-Müller counting system (located above the snow level) measures the amount of depletion of radiation caused by the presence of the snow.

radioactive well logging   [ENG]   The recording of the differences in radioactive content (natural or neutron-induced) of the various rock layers found down an oil well borehole; types include γ-ray, neutron, and photon logging. Also known as radiation well logging; radioactivity prospecting.

radioactivity log   [ENG]   Record of radioactive well logging.

radioactivity prospecting *See* radioactive well logging.

radiochronology   [GEOL]   An absolute-age dating method based on the existing ratio between radioactive parent elements (such as uranium-238) and their radiogenic daughter isotopes (such as lead-206).

radio climatology   [CLIMATOL]   The study of regional and seasonal variations in the manner of propagation of radio energy through the atmosphere.

radio duct   [GEOPHYS]   An atmospheric layer, typically shallow and almost horizontal, in which radio waves propagate in an anomalous fashion; ducts occur when, due to sharp inversions of temperature or humidity, the vertical gradient of the radio index of refraction exceeds a critical value.

radioelectric meteorology *See* radio meteorology.

radiogeology   [GEOCHEM]   The study of the distribution patterns of radioactive elements in the earth's crust, and the role of radioactive processes in geologic phenomena.

radio hole   [GEOPHYS]   Strong fading of the radio signal at some position in space along an air-to-air or air-to-ground path; the effect is caused by the abnormal refraction of radio waves.

radiolarian chert   [GEOL]   A homogeneous cryptocrystalline radiolarite with a well-developed matrix.

radiolarian earth   [GEOL]   A porous, unconsolidated siliceous sediment formed from the opaline silica skeletal remains of Radiolaria; formed from radiolarian ooze.

radiolarian ooze   [GEOL]   A siliceous ooze containing the skeletal remains of the Radiolaria.

radiolarite   [GEOL]   **1.** A whitish, hard, consolidated equivalent of radiolarian earth. **2.** Radiolarian ooze that has been indurated.

radiolitic   [PETR]   **1.** Pertaining to the texture of an igneous rock, characterized by radial, fanlike groupings of acicular crystals, resembling sectors of spherulites. **2.** Referring to limestones in which the components radiate from central points, with the cement comprising less than 50% of the total rock.

radio meteorology   [METEOROL]   That branch of the science of meteorology which embraces the propagation of radio energy through the atmosphere, and the use of radio and radar equipment in meteorology; this is the most general term and includes radar meteorology. Also known as radioelectric meteorology.

radiometric age   [GEOL]   Geologic age expressed in years determined by quantitatively measuring radioactive elements and their decay products.

radio prospecting [ENG] Use of radio and electric equipment to locate mineral or oil deposits.

radiosonde [ENG] A balloon-borne instrument for the simultaneous measurement and transmission of meteorological data; the instrument consists of transducers for the measurement of pressure, temperature, and humidity, a modulator for the conversion of the output of the transducers to a quantity which controls a property of the radio-frequency signal, a selector switch which determines the sequence in which the parameters are to be transmitted, and a transmitter which generates the radio-frequency carrier.

radiosonde balloon [AERO ENG] A balloon used to carry a radiosonde aloft; it is considerably larger than a pilot balloon or a ceiling balloon.

radiosonde commutator [ELECTR] A component of a radiosonde consisting of a series of alternate electrically conducting and insulating strips; as these are scanned by a contact, the radiosonde transmits temperature and humidity signals alternately.

radiosonde observation [METEOROL] An evaluation in terms of temperature, relative humidity, and pressure aloft, of radio signals received from a balloon-borne radiosonde; the height of each mandatory and significant pressure level of the observation is computed from these data. Also known as raob.

radiosonde-radio-wind system [ENG] An apparatus consisting of a standard radiosonde and radiosonde ground equipment to obtain upper-air data on pressure, temperature, and humidity, and a self-tracking radio direction finder to provide the elevation and azimuth angles of the radiosonde so that the wind vectors may be obtained.

radiosonde set [ENG] A complete set for automatically measuring and transmitting high-altitude meteorological data by radio from such carriers as a balloon or rocket.

radio window [GEOPHYS] A band of frequencies extending from about 6 to 30,000 megahertz, in which radiation from the outer universe can enter and travel through the atmosphere of the earth.

radiozone [GEOL] A para-time-rock unit representing a zone or succession of strata established on common radioactivity criteria.

Radstockian [GEOL] A European stage of geologic time forming the upper Upper Carboniferous, above Staffordian and below Stephanian, equivalent to uppermost Westphalian.

rafaelite [PETR] A nepheline-free orthoclase-bearing hypabyssal rock that also contains analcime and calcic plagioclase.

raffiche [METEOROL] In the Mediterranean region, gusts from the mountains; violent gusts of the bora.

raft *See* float coal.

rafted ice [OCEANOGR] A form of pressure ice composed of overlying pieces of ice floe.

rafting [GEOL] Transporting of rock by floating ice or floating organic materials (such as logs) to places not reached by water currents.   [OCEANOGR] The process of forming rafted ice.

raft lake [HYD] A relatively short-lived body of water impounded along a stream by a raft.

raft tectonics *See* plate tectonics.

rag [PETR] Any of various hard, coarse, rubbly, or shell rocks that weather with a rough, irregular surface, such as a flaggy sandstone or limestone used as a building stone. Also known as ragstone.

rag bolt *See* barb bolt.

ragged ceiling *See* indefinite ceiling.

raggiatura [METEOROL] Land squalls descending with great force from ravines and valleys in high land in Italy; they extend only a short distance off the west coast.

raglanite [PETR] A nepheline syenite composed of oligoclase, nepheline, and corundum with minor amounts of mica, calcite, magnetite, and apatite.

rain [METEOROL] Precipitation in the form of liquid water drops with diameters greater than 0.5 millimeter, or if widely scattered the drops may be smaller; the only other form of liquid precipitation is drizzle.

rain and snow mixed [METEOROL] Precipitation consisting of a mixture of rain and wet snow; usually occurs when the temperature of the air layer near the ground is slightly above freezing.

rain area report *See* radar report.

rainbow granite [PETR] A type of granite having either a black or dark-green background with pink, yellowish, or reddish mottling, or a pink background with dark mottling.

rain cloud [METEOROL] Any cloud from which rain falls; a popular term having no technical denotation.

rain crust [HYD] A type of snow crust, formed by refreezing after surface snow crystals have been melted and wetted by liquid precipitation; composed of individual ice particles such as firn.

rain desert [ECOL] A desert in which rainfall is sufficient to maintain a sparse general vegetation.

raindrop [METEOROL] A drop of water of diameter greater than 0.5 millimeter falling through the atmosphere.

raindrop impressions *See* rain prints.

raindrop imprints *See* rain prints.

rain factor [HYD] A coefficient designed to measure the combined effect of temperature and moisture on the formation of soil humus; it is obtained by dividing the annual rainfall (in millimeters) by the mean annual temperature (in degrees Celsius).

rainfall [METEOROL] The amount of precipitation of any type; usually taken as that amount which is measured by means of a rain gage (thus a small, varying amount of direct condensation is included).

rainfall frequency [CLIMATOL] The number of times, during a specified period of years, that precipitation of a certain magnitude or greater occurs or will occur at a station; numerically, the reciprocal of the frequency is usually given.

rainfall inversion *See* precipitation inversion.

rainfall penetration [HYD] The depth below the soil surface to which water from a given rainfall has been able to infiltrate.

rainfall regime [CLIMATOL] The character of the seasonal distribution of rainfall at any place; the chief rainfall regimes, as defined by W. G. Kendrew, are equatorial, tropical, monsoonal, oceanic and continental westerlies, and Mediterranean.

rainforest climate *See* wet climate.

rain gage [ENG] An instrument designed to collect and measure the amount of rain that has fallen. Also known as ombrometer; pluviometer. Also known as statement; udometer.

rain-gage shield [ENG] A device which surrounds a rain gage and acts to maintain horizontal flow in the vicinity of the funnel so that the catch will not be influenced by eddies generated near the gage. Also known as wind shield.

rain gush *See* cloudburst.

rain gust *See* cloudburst.

raininess [METEOROL] Generally, the quantitative character of rainfall for a given place.

rain-intensity gage [ENG] An instrument which measures the instantaneous rate at which rain is falling on a given surface. Also known as rate-of-rainfall gage.

rainmaking [METEOROL] Popular term applied to all activities designed to increase, through any artificial means, the amount of precipitation released from a cloud.

rain pillar [GEOL] A minor landform consisting of a column of soil or soft rock capped and protected by pebbles or concretions, produced by the differential erosion from the impact of falling rain.

rain prints [GEOL] Small, shallow depressions formed in soft sediment or mud by the impact of falling raindrops. Also known as raindrop impressions; raindrop imprints.

rain shadow [METEOROL] An area of diminished precipitation on the lee side of mountains or other topographic obstacles.

rainsquall [METEOROL] A squall associated with heavy convective clouds, frequently the cumulonimbus type; usually sets in shortly before the thunderstorm rain, blowing outward from the storm and generally lasting only a short time. Also known as thundersquall.

rain stage [METEOROL] The thermodynamic process of condensation of water from moist air in an idealized saturation-adiabatic or pseudoadiabatic lifting, at temperatures above the freezing point; begins at the condensation level.

rainwash [GEOL] **1.**The washing away of loose surface material by rainwater after it has reached the ground but before it has been concentrated into definite streams. **2.** Material transported and accumulated, or washed away, by rainwater.

rainwater [HYD] Water that has fallen as rain and is quite soft, as it has not yet collected soluble matter from the soil.

rainy climate [CLIMATOL] In W. Koppen's climatic classification, any climate type other than the dry climates; however, it is generally understood that this refers principally to the tree climates and not the polar climates.

rainy season [CLIMATOL] In certain types of climate, an annually recurring period of one or more months during which precipitation is a maximum for that region. Also known as wet season.

raise [MIN ENG] A shaftlike mine opening, driven upward from a level to connect with a level above, or to the surface.

raise boring machine [MIN ENG] A machine that is used to drill pilot holes between levels in a mine, and to ream the pilot hole to the finished dimension of the raise.

raised beach [GEOL] An ancient beach raised to a level above the present shoreline by uplift or by lowering of the sea level; often bounded by inland cliffs.

raised bog [ECOL] An area of acid, peaty soil, especially that developed from moss, in which the center is relatively higher than the margins.

raise drill [MIN ENG] A circular raise driving machine which bores a pilot hole and reams it to finished raise diameter.

raise driller [MIN ENG] A person who works in a raise. Also known as raiseman.

raiseman *See* raise driller.

rake *See* plunge.

ralstonite [MINERAL] $NaMgAl_5F_{12}(OH)_6 \cdot 3H_2O$ A colorless, white, or yellowish mineral composed of hydrous basic sodium magnesium aluminum fluoride, occurring in octahedral crystals.

ram *See* barney.

Ramapithecinae [PALEON] A subfamily of Hominidae including the protohominids of the Miocene and Pliocene.

Ramapithecus [PALEON] The genus name given to a fossilized upper jaw fragment found in the Siwalik hills, India; closely related to the family of man.

ramdohrite [MINERAL] $Pb_3Ag_2Sb_6S_{13}$ A dark-gray mineral composed of a lead silver antimony sulfur compound.

rammelsbergite [MINERAL] $NiAs_2$ A gray mineral composed of nickel diarsenide; it is dimorphous with pararammelsbergite. Also known as white nickel.

ramp [HYD] An accumulation of snow forming an inclined plane between land or land ice and sea ice or shelf ice. Also known as drift ice foot.   [MIN ENG] A slope between levels in open-pit mining.

ramp mining [MIN ENG] The development of moderately inclined accessways from the surface to mining levels for haulage of ore, materials, waste, men, and equipment.

ramp valley [GEOL] A trough between faults, forced downward by lateral pressure.

ramsdellite [MINERAL] MnO_2 An orthorhombic mineral composed of manganese dioxide; it is dimorphous with pyrolusite.

Rancholabrean [GEOL] A stage of geologic time in southern California, in the upper Pleistocene, above the Irvingtonian.

random forecast [METEOROL] A forecast in which one of a set of meteorological contingencies is selected on the basis of chance; it is often used as a standard of comparison in determining the degree of skill of another forecast method.

rang [PETR] A unit of subdivision in the C.I.P.W. (Cross-Iddings-Pirsson-Washington) classification of igneous rocks.

range of tide [OCEANOGR] The difference in height between consecutive high and low tides at a place.

range zone [GEOL] Formal biostratigraphic zone made up of a body of strata comprising the total horizontal (geographic) and vertical (stratigraphic) range of occurrence of a specified taxon of a group of taxa.

rank [GEOL] **1.** A coal classification based on degree of metamorphism. **2.** *See* stack.

rankinite [MINERAL] $Ca_3Si_2O_7$ A monoclinic mineral composed of calcium silicate.

Ranney oil-mining system [PETRO ENG] A method used to get oil from oil sands that involves driving mine galleries from shafts communicating to the surface in impermeable strata above and below the oil strata; holes drilled at short intervals along the galleries into the oil sands drain the oil or gas through pipes sealed in the drill holes into tanks from which the gas or oil is pumped to the surface.

ransomite [MINERAL] $Cu(Fe,Al)_2(SO_4)_4 \cdot 7H_2O$ A sky-blue mineral composed of hydrous copper iron aluminum sulfate.

raob *See* radiosonde observation.

rapakivi [PETR] Granite or quartz monzonite characterized by orthoclase phenocrysts mantled with plagioclase. Also known as wiborgite.

rapakivi texture [PETR]   An igneous and metamorphic rock texture in which spherical potassium feldspar crystals are surrounded by a rim of sodium feldspar, both within a finer-grained matrix.

rapid [HYD]   A portion of a stream in swift, disturbed motion, but without cascade or waterfall; usually used in the plural.

rare-earth mineral [MINERAL]   A mineral having a high concentration of rare-earth elements; examples are monazite, xenotime, and bastnaesite.

RAREP *See* radar report.

rasorite *See* kernite.

raspite [MINERAL] $PbWO_4$ A yellow or brownish-yellow mineral composed of lead tungstate, occurring as monoclinic crystals.

rate-of-rainfall gage *See* rain-intensity gage.

rate of sedimentation [GEOL]   The amount of sediment accumulated in an aquatic environment over a given period of time, usually expressed as thickness of accumulation per unit time. Also known as sedimentation rate.

rathite [MINERAL] $Pb_{13}As_{18}S_{40}$ A dark-gray mineral with metallic luster composed of sulfide of lead and arsenic; occurs as orthorhombic crystals.

rathole [MIN ENG]   A shallow, small-diameter, auxiliary hole alongside the main borehole, drilled at an angle to the main hole; after core drilling is completed, the rathole is reamed out and the larger-size hole is advanced, usually by some noncoring method.

rating curve [HYD]   For a given point on a stream, a graph of discharge versus stage.

ratio of rise [OCEANOGR]   The ratio of the height of tide at two places.

Rauschelback rotor [ENG]   A free-turning S-shaped propeller used to measure ocean currents; the number of rotations per unit time is proportional to the flow.

ravelly ground [GEOL]   Rock that breaks into small pieces when drilled and tends to cave or slough into the hole when the drill string is pulled, or binds the drill string by becoming wedged or locked between the drill rod and the borehole wall.

ravine [GEOGR]   A small and narrow valley with steeply sloping sides.

ravinement [GEOL]   **1.** The formation of a ravine or ravines. **2.** An irregular junction which marks a break in sedimentation, such as an erosion line occurring where shallow-water marine deposits have cut down into slightly eroded underlying beds.

raw [METEOROL]   Colloquially descriptive of uncomfortably cold weather, usually meaning cold and damp, but sometimes cold and windy.

raw humus *See* ectohumus.

rawin [METEOROL]   A method of winds-aloft observation, that is, the determination of wind speeds and directions in the atmosphere above a station; accomplished by tracking a balloon-borne radar target, responder, or radiosonde transmitter with either radar or a radio direction finder.

rawinsonde [METEOROL]   A method of upper-air observation consisting of an evaluation of the wind speed and direction, temperature, pressure, and relative humidity aloft by means of a balloon-borne radiosonde tracked by a radar or radio direction finder.

Rayleigh atmosphere [METEOROL]   An idealized atmosphere consisting of only those particles, such as molecules, that are smaller than about one-tenth of the wavelength of all radiation incident upon that atmosphere; in such an atmosphere, simple Rayleigh scattering would prevail.

Rayleigh wave [GEOPHYS] In seismology, a surface wave with a retrograde, elliptical motion at the free surface. Also known as R wave.

razor stone *See* novaculite.

reach [GEOGR] **1.** A continuous, unbroken surface of land or water. Also known as stretch. **2.** A bay, estuary, or other arm of the sea extending up into the land. [HYD] A straight, continuous, or extended part of a river, stream, or restricted waterway.

reaction border *See* reaction rim.

reaction pair [MINERAL] Any two minerals, one of which is formed at the expense of the other by reaction with liquid.

reaction principle [MINERAL] The concept of a reaction series for the principal rock-forming minerals.

reaction rim [PETR] A surficial rim around one mineral produced by the reaction of the core mineral with the surrounding magma. Also known as reaction border.

reaction series [MINERAL] Any series of minerals in which early formed varieties react with the melt to yield new minerals; two different types of reaction series exist, continuous and discontinuous.

reactive fluid [PETRO ENG] Any fluid that alters the internal geometry of a reservoir's porosity; for example, water is a reactive fluid when it causes swelling of clays and consequent changes in porosity.

realgar [MINERAL] AsS A red to orange mineral crystallizing in the monoclinic system, having a resinous luster and found in short, vertical striated crystals; specific gravity is 3.48, and hardness is 1.5–2 on Mohs scale. Also known as red arsenic; red orpiment; sandarac.

rebat [METEOROL] The lake breeze of Lake Geneva, Switzerland; it blows from about 10 a.m. to 4 p.m.

rebound [GEOL] The isostatic readjustment upward of a landmass depressed by glacial loading.

reboyo [METEOROL] A persistent (day-long) storm from the southwest during the rainy season on the Brazilian coast.

Recent [GEOL] An epoch of geologic time (late Quaternary) following the Pleistocene; referred to as Holocene in several European countries.

recess [GEOL] **1.** An indentation occurring in a surface, bounded by a straight line. **2.** An area having the axial traces of folds concave toward the outer edge of the folded belt.

recession [GEOL] **1.** The backward movement, or retreat, of an eroded escarpment. **2.** A continuing landward movement of a shoreline or beach undergoing erosion. Also known as retrogression. **3.** The withdrawal of a body of water (as a sea or lake), thereby exposing formerly submerged areas. [HYD] The gradual upstream retreat of a waterfall.

recessional moraine [GEOL] **1.** An end moraine formed during a temporary halt in the final retreat of a glacier. **2.** A moraine formed during a minor readvance of the ice front during a period of glacial recession. Also known as stadial moraine.

recession curve [HYD] A hydrograph showing the decrease of the runoff rate after rainfall or the melting of snow.

recharge [HYD] **1.** The processes involved in the replenishment of water to the zone of saturation. **2.** The amount of water added or absorbed. Also known as groundwater increment; groundwater recharge; groundwater replenishment; increment; intake.

recharge well [HYD]   A well used as a source of water in the process of artificial recharge. Also known as injection well.

reciprocal lattice [CRYSTAL]   A lattice array of points formed by drawing perpendiculars to each plane *(hkl)* in a crystal lattice through a common point as origin; the distance from each point to the origin is inversely proportional to spacing of the specific lattice planes; the axes of the reciprocal lattice are perpendicular to those of the crystal lattice.

reciprocal vectors [CRYSTAL]   For a set of three vectors forming the primitive translations of a lattice, the vectors that form the primitive translations of the reciprocal lattice.

reclined fold *See* recumbent fold.

recompletion [PETRO ENG]   Redrilling an oil well to a new producing zone (new depth) when the current zone is depleted.

recomposed granite [PETR]   An arkose composed of consolidated feldspathic residue that has been reworked and decomposed so slightly that upon cementation the rock resembles granite except that its grain is less even and it contains a greater percentage of quartz. Also known as reconstructed granite.

reconnaissance [ENG]   A mission to secure data concerning the meteorological, hydrographic, or geographic characteristics of a particular area.

reconstructed granite *See* recomposed granite.

recording rain gage [ENG]   A rain gage which automatically records the amount of precipitation collected, as a function of time. Also known as pluviograph.

recording thermometer *See* thermograph.

record observation [METEOROL]   A type of aviation weather observation; the most complete of all such observations and usually taken at regularly specified and equal intervals (hourly, usually on the hour). Also known as hourly observation.

recovery [MIN ENG]   The proportion or percentage of coal or ore mined from the original seam or deposit.   [PETRO ENG]   The removal (recovery) of oil or gas from reservoir formations.

recovery factor [PETRO ENG]   The ratio of recoverable oil reserves to the oil in place in a reservoir.

recrystallization [PETR]   The formation of new mineral grains in crystalline form in a rock under the influence of metamorphic processes.

rectangular chart [MAP]   **1.** A chart in a rectangular shape. **2.** A chart on the rectangular map projection.

rectangular cross ripple mark [GEOL]   An oscillation cross ripple mark consisting of two sets of ripples which intersect at right angles, enclosing a rectangular pit.

rectangular drainage pattern [GEOL]   A drainage pattern characterized by many right-angle bends in both the main streams and their tributaries. Also known as lattice drainage pattern.

rectangular projection [MAP]   A cylindrical map projection with uniform spacing of the parallels; used for the star chart in the Air Almanac.

rectification [GEOL]   The simplification and straightening of the outline of an initially irregular and crenulate shoreline by marine erosion cutting back headlands and offshore islands, and by deposition of waste resulting from erosion or of sediment brought down by neighboring rivers.

recumbent fold [GEOL]   An overturned fold with a nearly horizontal axial surface. Also known as reclined fold.

recurrence interval [HYD] The average time interval between occurrences of a hydrologic event, such as between a given flood and one of the same or greater magnitude.

recurvature [METEOROL] With respect to the motion of severe tropical cyclones (hurricanes and typhoons), the change in direction from westward and poleward to eastward and poleward; such recurvature of the path frequently occurs as the storm moves into middle latitudes.

recurved spit *See* hook.

red antimony *See* kermesite.

red arsenic *See* realgar.

redbed [GEOL] Continentally deposited sediment composed principally of sandstone, siltsone, and shale; red in color due to the presence of ferric oxide (hematite). Also known as red rock.

red clay [GEOL] A fine-grained, reddish-brown pelagic deposit consisting of relatively large proportions of windblown particles, meteoric and volcanic dust, pumice, shark teeth, manganese nodules, and debris transported by ice. Also known as brown clay.

red cobalt *See* erythrite.

red copper ore *See* cuprite.

red earth [GEOL] Leached, red, deep, clayey soil that is characteristic of a tropical climate. Also known as red loam.

red hematite *See* hematite.

redingtonite [MINERAL] $(Fe,Mg,Ni)(Cr,Al)_2(SO_4)_4 \cdot 22H_2O$ A pale-purple mineral composed of a hydrous sulfate of iron, magnesium, nickel, chromium, and aluminum.

red iron ore *See* hematite.

red lead *See* lead tetroxide.

red lead ore *See* crocoite.

red loam *See* red earth.

red magnetism [GEOPHYS] The magnetism of the north-seeking end of a freely suspended magnet; this is the magnetism of the earth's south magnetic pole.

red mud [GEOL] A reddish terrigenous mud composed of up to 25% calcium carbonate and deriving its color from the presence of ferric oxide; found on the sea floor near deserts and near the mouths of large rivers. [MET] An iron oxide–rich residue obtained in purifying bauxite in the Bayer process.

red ocher *See* ferric oxide; hematite.

red orpiment *See* realgar.

red oxide of zinc *See* zincite.

red rock *See* redbed.

redruthite *See* chalcocite.

Red Sea [GEOGR] A body of water that lies between Arabia and northeastern Africa, about 2000 kilometers long, 300 kilometers wide, and a maximum depth of about 2300 meters.

red snow [HYD] A snow surface of reddish color caused by the presence within it of certain microscopic algae or particles of red dust.

redstone [PETR] **1.** Any reddish sedimentary rock, such as red-colored sandstone. **2.** A deep-red, clayey sandstone of siltstone representing a floodplain micaceous arkose.

reduced pressure [METEOROL]   The calculated value of atmospheric pressure at mean sea level or some other specified level, as derived (reduced) from station pressure or actual pressure; thus, sea level pressure is nearly always a reduced pressure.

reduction of tidal current [OCEANOGR]   The processing of observed tidal current data to obtain mean values of tidal current constants.

reduction of tides [OCEANOGR]   The processing of observed tidal data to obtain mean values of tidal constants.

red zinc ore *See* zincite.

reedmergnerite [MINERAL] $NaBSi_3O_8$ A colorless, triclinic borate mineral that represents the boron analog of albite.

reef [GEOL]   **1.** A ridge- or moundlike layered sedimentary rock structure built almost exclusively by organisms.   **2.** An offshore chain or range of rock or sand at or near the surface of the water.   [MIN ENG] A major ore trend or ore body.

reef breccia [PETR]   A rock formed by the consolidation of limestone fragments broken off from a reef by the action of waves and tides.

reef cap [GEOL]   A deposit of fossil-reef material overlying or covering an island or mountain.

reef cluster [GEOL]   A group of reefs of wholly or partly contemporaneous growth, found within a circumscribed area of geologic province.

reef complex [GEOL]   The solid reef core and the heterogeneous and contiguous fragmentary material derived from it by abrasion.

reef conglomerate *See* reef talus.

reef core [GEOL]   The rock mass constructed in place, and within the rigid growth lattice formed by reef-building organisms.

reef detritus [GEOL]   Fragmental material derived from the erosion of an organic reef. Also known as reef debris.

reef edge [GEOL]   The seaward margin of the reef flat, commonly marked by surge channels.

reef flank [GEOL]   The part of the reef that surrounds, interfingers with, and locally overlies the reef core, often indicated by massive or medium beds of reef talus dipping steeply away from the reef core.

reef flat [GEOL]   A flat expanse of dead reef rock which is partly or entirely dry at low tide; shallow pools, potholes, gullies, and patches of coral debris and sand are features of the reef flat.

reef front [GEOL]   The upper part of the outer or seaward slope of a reef, extending to the reef edge from above the dwindle point of abundant living coral and coralline algae.

reef-front terrace [GEOL]   A shelflike or benchlike eroded surface, sometimes veneered with organic growth, sloping seaward to a depth of 8–15 fathoms (15–27 meters).

reef knoll [GEOL]   **1.** A bioherm or fossil coral reef represented by a small, prominent, rounded hill, up to 100 meters high, consisting of resistant reef material, being either a local exhumation of an original reef feature or a feature produced by later erosion.   **2.** A present-day reef in the form of a knoll, that is, a small reef patch developed locally and built upward rather than outward.

reef limestone [PETR]   Limestone composed of the remains of sedentary organisms such as sponges, and of sediment-binding organic constituents such as calcareous algae. Also known as coral rock.

reef milk [GEOL] A very-fine-grained matrix material of the back-reef facies, consisting of white, opaque microcrystalline calcite derived from abrasion of the reef core and reef flank.

reef patch [GEOL] A single large colony of coral formed independently on a shelf at depths less than 70 meters in the lagoon of a barrier reef or of an atoll. Also known as patch reef.

reef pinnacle [GEOL] A small, isolated spire of rock or coral, especially a small reef patch.

reef rock [PETR] A hard, unstratified rock composed of sand, shale, and the calcareous remains of sedentary organisms, cemented by calcium carbonate.

reef segment [GEOL] A part of an organic reef lying between passes, gaps, or channels.

reef slope [GEOL] The face of a reef rising from the sea floor.

reef talus [GEOL] Massive inclined strata composed of reef detritus deposited along the seaward margin of an organic reef. Also known as reef conglomerate.

reef tufa [GEOL] Drusy, prismatic, fibrous calcite deposited directly from supersaturated water upon the void-filling internal sediment of the calcite mudstone of a reef knoll.

reef wall [GEOL] A wall-like upgrowth composed of living coral and the skeletal remains of dead coral and other reef-building organisms, which reaches an intertidal level and acts as a partial barrier between adjacent environments.

reel locomotive [MIN ENG] A trolley locomotive with a wire-rope reel for drawing mining cars out of rooms.

reentrant [GEOL] A prominent, generally angular indentation into a coastline.

reevesite [MINERAL] $Na_6Fe_2(OH)_{16}(CO_3)\cdot4H_2O$ Hydrous oxide mineral known only in meteorites.

reference ellipsoid [MAP] A reference surface used to represent the size and shape of the earth for cartography.

reference level [ENG] *See* datum plane. [OCEANOGR] **1.** Level of no motion. **2.** A level for which current is known; allows determination of absolute current from relative current.

reference plane [ENG] *See* datum plane.

reference signal level *See* reference level.

reference spheroid [GEOD] An ellipsoid of revolution, chosen to approximate the geoid, on which geodetic triangulation measurements are computed.

reference station [OCEANOGR] **1.** A place for which independent daily predictions are given in the tide or current tables, from which corresponding predictions are obtained for other stations by means of differences or factors. **2.** A place for which tidal or tidal current constants have been determined and which is used as a standard for the comparison of simultaneous observations at a second station. Also known as standard station.

refine [ENG] To free from impurities, as the separation of petroleum, ores, or chemical mixtures into their component parts.

reflecting nephoscope *See* mirror nephoscope.

reflection plane *See* plane of mirror symmetry; plane of reflection.

reflection seismology *See* reflection shooting.

reflection shooting [ENG]   A procedure in seismic prospecting based on the measurement of the travel times of waves which, originating from an artificially produced disturbance, have been reflected to detectors from subsurface boundaries separating media of different elastic-wave velocities; used primarily for oil and gas exploration. Also known as reflection seismology.

reflection survey [ENG]   Study of the presence, depth, and configuration of underground formations; a ground-level explosive charge (shot) generates vibratory energy (seismic rays) that strike formation interfaces and are reflected back to ground-level sensors. Also known as seismic survey.

reflection twin [CRYSTAL]   A crystal twin whose symmetry is formed by an apparent mirror image across a plane.

refolding [GEOL]   A process by which folds of one generation are subjected to and stressed by a force of different orientation.

refoliation [GEOL]   A foliation that is subsequent to and oriented differently from an earlier foliation.

refraction coefficient [OCEANOGR]   The square root of the ratio of the spacing between orthogonals in deep water and in shallow water; it is a measure of the effect of refraction in diminishing wave height by increasing the length of the wave crest.

refraction diagram [OCEANOGR]   A chart showing the position of the wave crests at a particular time, or the successive positions of a particular wave crest as it moves shoreward.

refraction process [ENG]   Seismic (reflection) survey in which the distance between the explosive shot and the receivers (sensors) is large with respect to the depths to be mapped.

refractive modulus *See* modified index of refraction.

regelation [HYD]   Phenomenon in which ice melts at the bottom of droplets of highly concentrated saline solution that are trapped in ice which has frozen over polar waters, and freezes at the top of these droplets, so that the droplets move downward through the ice, leaving it hard and clear.

regenerated glacier [HYD]   A glacier that becomes active after a period of stagnation.

regime [GEOL]   The existence in a stream channel of a balance between erosion and deposition over a period of years.

regimen [HYD]   **1.** The behavior characteristic of the total amount of water involved in a drainage basin. **2.** Analysis of the total volume of water involved with a lake, including water losses and gains, over a period of a year. **3.** The flow characteristics of a stream with respect to velocity, volume, form of and alterations in the channel, capacity to transport sediment, and the amount of material supplied for transportation.

regional dip [GEOL]   The nearly uniform and generally low-angle inclination of strata over a wide area. Also known as normal dip.

regional forecast *See* area forecast.

regional geology [GEOL]   The geology of a large region, treated from the viewpoint of the spatial distribution and position of stratigraphic units, structural features, and surface forms.

regional metamorphism [GEOL]   Geological metamorphism affecting an extensive area.

regional metasomatism [GEOL]   Metasomatic processes affecting extensive areas whereby the introduced material may be derived from partial fusion of the rocks involved from deep-seated magmatic sources.

regional migration [PETRO ENG] Horizontal movement of gas or oil through a reservoir formation as a result of artificial pressure differences created by withdrawal of gas or oil at well sites.

regional slope [GEOL] The generally uniform dip of rock strata or land surface over a wide area.

regional slope deposit [GEOL] A sedimentary deposit widely distributed as a thin sheet over a regional slope.

regional snowline [HYD] The level above which, averaged over a large area, snow accumulation exceeds ablation year after year.

regional unconformity [GEOL] A continuous unconformity extending throughout a wide region that may be nearly continentwide, and usually represents a long period of time.

region of escape *See* exosphere.

regmagenesis [GEOL] Diastrophic production of regional strike-slip displacements.

regmaglypt [GEOL] Any of various small, well-defined, characteristic indentations or pits on the surface of meteorites, frequently resembling the imprints of fingertips in soft clay. Also known as pezograph; piezoglypt.

regolith [GEOL] The layer rock or blanket of unconsolidated rocky debris of any thickness that overlies bedrock and forms the surface of the land. Also known as mantle rock.

Regosol [GEOL] In early United States soil classification systems, one of an azonal group of soils that form from deep, unconsolidated deposits and have no definite genetic horizons.

regradation [GEOL] The formation by a stream of a new profile of equilibrium, as when the former profile, after gradation, became deformed by crustal movements.

regression [GEOL] The theory that some rivers have sources on the rainier sides of mountain ranges and gradually erode backward until the ranges are cut through. [OCEANOGR] Retreat of the sea from land areas, and the consequent evidence of such withdrawal.

regression conglomerate [GEOL] A coarse sedimentary deposit formed during a retreat (recession) of the sea.

regressive overlap *See* offlap.

regular dodecahedron *See* pyritohedron.

regular sampling [MIN ENG] The continuous or intermittent sampling of the same coal or coke received regularly at a given point.

regular ventilating circuit [MIN ENG] All places in the mine through which there is a positive natural flow of air.

regulated split [MIN ENG] In mine ventilation, a split where it is necessary to control the volumes in certain low-resistance splits to cause air to flow into the splits of high resistance.

regulator [MIN ENG] An opening in a wall or door in the return airway of a district to increase its resistance and reduce the volume of air flowing.

Reid equation [PETRO ENG] Relation of gas-well flow rate to pitot-tube readings for various impact pressures.

rejuvenate [GEOL] The act of stimulating a stream to renewed erosive activity either by tectonic uplift or a drop in sea level.

rejuvenated fault scarp [GEOL]   A fault scarp revived by renewed movement along an old fault line after partial dissection or erosion of the initial scarp. Also known as revived fault scarp.

rejuvenated stream [HYD]   A mature stream that has reverted to the behavior and forms of a more youthful stage due to rejuvenation, usually as a result of uplift. Also known as revived stream.

rejuvenated water [HYD]   Water returned to the terrestrial water supply as a result of compaction and metamorphism.

rejuvenation [GEOL]   The restoration of youthful features to fluvial landscapes; the renewal of youthful vigor to low-gradient streams is usually caused by regional upwarping of broad areas formerly at or near base level.   [HYD]   **1.** The stimulation of a stream to renew erosive activity. **2.** The renewal of youthful vigor in a mature stream.

rejuvenation head See knickpoint.

relative compaction [ENG]   The percentage ratio of the field density of soil to the maximum density as determined by standard compaction.

relative contour See thickness line.

relative current [OCEANOGR]   The current which is a function of the dynamic slope of an isobaric surface and which is determined from an assumed layer of no motion.

relative dating [GEOL]   The proper chronological placement of a feature, object, or happening in the geologic time scale without reference to its absolute age.

relative deflection See astrogeodetic deflection.

relative divergence See development index.

relative geologic time [GEOL]   Nonabsoluter geological time in which events may be placed relatively to one another.

relative humidity [METEOROL]   The (dimensionless) ratio of the actual vapor pressure of the air to the saturation vapor pressure. Abbreviated RH.

relative hypsography See thickness pattern.

relative isohypse See thickness line.

relative magnetometer [ENG]   Any magnetometer which must be calibrated by measuring the intensity of a field whose strength is accurately determined by other means; opposed to absolute magnetometer.

relative permeability [GEOL]   Specific permeability of a porous rock formation to a particular phase (oil, water, gas) at a particular saturation and a particular saturation distribution; for example, ratio of effective permeability to a specified phase to the rock's absolute permeability.

relative pressure field [OCEANOGR]   The pressure field as it would occur if pressure distribution was based only upon distribution of mass in the sea.

relative relief See local relief.

relative topography See thickness pattern.

relaxation [GEOL]   In experimental structural geology, the diminution of applied stress with time, as the result of any of various creep processes.

relay haulage [MIN ENG]   Single-track, high-speed mine haulage from one relay station to another. Also known as intermediate haulage.

released mineral [MINERAL]   A mineral formed during the crystallization of a magma due to failure of an earlier phase to react with the liquid portion of the magma.

release fracture [GEOL] A fracture formed as a result of a decrease in the maximum principal stress.

release joint *See* sheeting structure.

relic [GEOL] **1.** A landform that remains intact after decay or disintegration or that remains after the disappearance of the major portion of its substance. **2.** A vestige of a particle in a sedimentary rock, such as a trace of a fossil fragment.

relict [GEOL] **1.** Referring to a topographic feature that remains after other parts of the feature have been removed or have disappeared. **2.** Pertaining to a mineral, structure, or feature of rock which represents that of an earlier rock and which persists in spite of processes tending to destroy it, such as metamorphism.

relict dike [GEOL] In a granitized mass, a tabular, crystalloblastic body that represents a dike which was emplaced prior to, and which was relatively resistant to, the granitization process.

relict glacier [HYD] A remnant of an older and larger glacier.

reliction [HYD] The slow and gradual withdrawal or recession of the water in a sea, a lake, or a stream, leaving the former bottom as permanently exposed and uncovered dry land.

relict lake [HYD] A lake that survives in an area formerly covered by the sea or a larger lake, or a lake that represents a remnant resulting from a partial extinction of the original body of water.

relict mineral [MINERAL] A mineral of a rock that persists from an earlier rock.

relict permafrost [GEOL] Permafrost formed in the past which persists in areas where it would not form today.

relict sediment [GEOL] A sediment which was in equilibrium with its environment when first deposited but which is unrelated to its present environment even though it is not buried by later sediments, such as a shallow-marine sediment on the deep ocean floor.

relict soil [GEOL] A soil formed on a preexisting landscape but not subsequently buried under younger sediments.

relict texture [GEOL] In mineral deposits, an original texture that persists after partial replacement.

relief [CRYSTAL] The apparent topography exhibited by minerals in thin section as a consequence of refractive index. [GEOD] The configuration of a part of the earth's surface, with reference to altitude and slope variations and to irregularities of the land surface.

relief hole [ENG] Any of the holes fired after the cut holes and before the lifter holes in breaking ground for tunneling or shaft sinking.

relief limonite [MINERAL] Indigenous limonite that is porous and cavernous in texture.

relief map [MAP] A map of an area showing the topographic relief.

relief model [MAP] A three-dimensional relief map.

Relizean stage [GEOL] A subdivision of the Miocene in the California-Oregon-Washington area.

remanent magnetization [GEOPHYS] That component of a rock's magnetization whose direction is fixed relative to the rock and which is independent of moderate, applied magnetic fields.

renardite [MINERAL] $Pb(UO_2)_4(PO_4)_2(OH)_4 \cdot 7H_2O$ A yellow mineral composed of hydrous basic lead uranyl phosphate.

Rendoll [GEOL] A suborder of the soil order Mollisol, formed in highly calcareous parent materials, mostly restricted to humid, temperate regions; the soil profile consists of a dark upper horizon grading to a pale lower horizon.

Rendzina [GEOL] One of an intrazonal, calcimorphic group of soils characterized by a brown to black, friable surface horizon and a light-gray or yellow, soft underlying horizon; found under grasses or forests in humid to semiarid climates.

repeated reflection *See* multiple reflection.

reperforation [PETRO ENG] Creation of new perforations (holes) in oil well tubing opposite to oil-bearing reservoir zones; creates more opportunity for fluid to drain from the formation into the wellbore.

repi [HYD] A lake, pond, or other standing water body associated with a sink or subsidence of land surface.

replacement [GEOL] Growth of a new or chemically different mineral in the body of an old mineral by simultaneous capillary solution and deposition.   [PALEON] Substitution of inorganic matter for the original organic constituents of an organism during fossilization.

replacement deposit [MINERAL] A mineral deposit formed by the in-position replacement of one mineral for another.

replacement dike [GEOL] A dike which is made by gradual transformation of wall rock by solutions along fractures or permeable zones.

replacement texture [GEOL] The texture exhibited where one mineral has replaced another.

replacement transfusion *See* exchange transfusion.

replacement vein [GEOL] A mineral vein formed by the gradual transformation of an original vein by secondary fluids.

representative sample [MIN ENG] In testing or valuation of a mineral deposit, a sample so large and average in composition as to be considered representative of a specified volume of the surrounding ore body.

reseau [METEOROL] The term adopted by the World Meteorological Organization for the worldwide network of meteorological stations which have been chosen to represent the meteorology of the globe (*réseau mondial*).

resequent fault-line scarp [GEOL] A fault-line scarp which faces in the same direction as the original fault scarp or in which the downthrown block is topographically lower than the upthrown block.

resequent stream [HYD] A stream whose direction follows an original consequent stream but is generally lower; resequent streams are generally tributary to a subsequent stream.

reserved minerals [MIN ENG] Economic minerals that belong to the state, which confers the right to prospect for and to mine them on any applicant.

reserves [MIN ENG] The quantity of workable mineral or of gas or oil which is calculated to lie within given boundaries.

reserves-decline relationship [PETRO ENG] Relationship between production-rate decline over a period of time to the total remaining hydrocarbon reserves in a reservoir.

reservoir cycling [PETRO ENG] Repressuring of an oil reservoir by reinjection of dry gas (gas with liquids stripped out) into the formation.

reservoir drive mechanism [PETRO ENG]   The physical action by which hydrocarbons (gas or liquid) are moved through the porous reservoir structure; for example, gas drive or water drive. Also known as oil-well drive.

reservoir dynamics [PETRO ENG]   Fluid-flow performance within an oil or gas reservoir.

reservoir fluid [GEOL]   The subterranean fluid trapped by a reservoir formation; can include natural gas, liquid and vapor petroleum hydrocarbons, and interstitial water.

reservoir pressure [GEOL]   **1.** The pressure of fluids (water, oil, gas) in a subsurface formation. **2.** The pressure under which fluids are confined in rocks.

reservoir rock [GEOL]   Friable, porous sandstone containing deposits of oil or gas.

reshabar [METEOROL]   A strong, very turbulent, dry northeast wind of bora type which blows down mountain ranges in southern Kurdistan in Persia; it is dry and hot in summer and cold in winter.

residual [GEOL]   **1.** Of a mineral deposit, formed by either mechanical or chemical concentration. **2.** Pertaining to a residue left in place after weathering of rock. **3.** Of a topographic feature, representing the remains of a formerly great mass or area and rising above the surrounding surface.

residual anticline [GEOL]   In salt tectonics, a relative structural high resulting from the depression of two adjacent rim synclines. Also known as residual dome.

residual clay [GEOL]   Very finely divided clay material formed in place by weathering of rock. Also known as primary clay.

residual compaction [GEOL]   The difference between the amount of compaction that will ultimately occur for a given increase in applied stress, and that which has occurred at a specified time.

residual dome *See* residual anticline.

residual free gas [PETRO ENG]   Free gas-cap gas in equilibrium with residual liquid hydrocarbons in a depleted reservoir, such as a reservoir at the end of its primary or economic producing life.

residual liquid [GEOL]   The volatile components of a magma that remain in the magma chamber after much crystallization has taken place.

residual ochre [GEOL]   An earthy, red, yellow, or brownish iron oxide powder of iron oxide (usually the mineral limonite) produced during chemical weathering.

residual sediment *See* resistate.

residual stress field *See* ambient stress field.

residual swelling [GEOL]   The difference between the original prefreezing level of the ground and the level reached by the settling after the ground is completely thawed.

residual tack *See* aftertack.

residual valley [GEOL]   An intervening trough between uplifted mountains.

residue [GEOL]   The in-place accumulation of rock debris which remains after weathering has removed all but the least soluble constituent.

resinite [GEOL]   A variety of exinite composed of resinous compounds, often in elliptical or spindle-shaped bodies.

resinous coal [GEOL]   Coal in which large proportions of resinous material are contained in the attritus.

resinous luster [GEOL]   The luster on the fractured surfaces of certain minerals (such as opal, sulfur, amber, and sphalerite) and rocks (such a pitchstone) that resemble resin.

resin roof bolting [MIN ENG]   The fixation of metal roof bolts in rock holes with a bonding resin.

resin tin *See* rosin tin.

resistance magnetometer [ENG]   A magnetometer that depends for its operation on variations in the electrical resistance of a material immersed in the magnetic field to be measured.

resistate [GEOL]   A sediment consisting of minerals that are chemically resistant and are enriched in the residues of weathering processes. Also known as residual sediment.

resistivity factor *See* formation factor.

resistivity index [PETRO ENG]   Ratio of the true electrical resistivity of a rock system at a specified water saturation, to the resistivity of the rock itself; used for calculation of electrical well-logging data.

resistivity well logging [PETRO ENG]   The measurement of subsurface electrical resistivities (normal and lateral to the borehole) during electrical logging of oil wells.

resonance trough [METEOROL]   A large-scale pressure trough which forms at an appropriate wavelength away from a dominant trough; for example, the mean trough over the Mediterranean in winter is often considered a resonance trough between the two more dynamically active troughs along the east coasts of North America and Asia.

resorption [PETR]   The process by which a magma redissolves previously crystallized minerals.

restite [GEOL]   Any immobile or less mobile part of a migmatite during migmatization.

restricted basin [GEOL]   A depression in the floor of the ocean in which the water circulation is topographically restricted and therefore generally is oxygen-depleted. Also known as barred basin; silled basin.

resultant wind [CLIMATOL]   The vectorial average of all wind directions and speeds for a given level at a given place for a certain period, such as a month.

resurrected [GEOL]   Of or pertaining to a surface, landscape, or feature (such as a mountain, peneplain, or fault scarp) that has been restored by exhumation to its previous status in the existing relief. Also known as exhumed.   [HYD]   Pertaining to a stream that follows an earlier drainage system after a period of brief submergence has slightly masked the old course by a thin film of sediments. Also known as palingenetic.

resurrected-peneplain shoreline [GEOL]   A shoreline of submergence formed where the sea rests against an inclined resurrected peneplain.

retained water [HYD]   The water remaining in rock or soil after gravity groundwater has been drained out.

retardation [OCEANOGR]   The amount of time by which corresponding tidal phases grow later day by day, averaging approximately 50 minutes.

retardation sheet *See* wave plate.

retarded acid [PETRO ENG]   Oil well acidizing solution whose reactivity is slowed by addition of artificial gums and thickening agents, so that the acid penetrates deeper into the formation before being spent.

retgersite [MINERAL] $NiSO_4 \cdot 6H_2O$ A deep emerald green, tetragonal mineral consisting of a hydrated nickel sulfate.

reticular *See* reticulate.

reticulate [GEOL] **1.** Referring to a vein or lode with netlike texture. **2.** Referring to rock texture in which crystals are partly altered to a secondary material, forming a network that encloses the remnants of the original mineral. Also known as mesh-texture; reticular; reticulated.

reticulated *See* reticulate.

Reticulosa [PALEON] An order of Paleozoic hexactinellid sponges with a branching form in the subclass Hexasterophora.

retinite [MINERAL] A fossil resin, such as glessite, krantzite, muckite, and ambrite, composed of 6–15% oxygen, lacking succinic acid, and found in brown coals and peat.

retreat [MIN ENG] Workings in the opposite direction of advance work which, when completed, will permit the area to be abandoned as finished.

retrograde gas-condensate reservoir *See* dew-point reservoir.

retrograde metamorphism [PETR] Formation of metamorphic minerals of a lower grade of metamorphism at the expense of minerals which are characteristic of a higher grade. Also known as diaphthoresis; retrogressive metamorphism.

retrograde reservoir [GEOL] Hydrocarbon reservoir in which hydrocarbons are initially in the vapor phase; as pressure is reduced, the bubble-point line is passed and liquids are formed; upon further pressure reduction, a vapor phase is again formed.

retrograde wave [METEOROL] An atmospheric wave which moves in a direction opposite to that of the flow in which the wave is embedded; retrogression of a particular wave on daily charts is rarely seen, but is frequently observed on 4-day or monthly mean charts.

retrograding shoreline [GEOL] A shoreline that is being moved landward by wave erosion.

retrogression [METEOROL] The motion of an atmospheric wave or pressure system in a direction opposite to that of the basic flow in which it is embedded.

retrogressive metamorphism *See* retrograde metamorphism.

return streamer [GEOPHYS] The intensely luminous streamer which propagates upward from earth to cloud base in the last phase of each lightning stroke of a cloud-to-ground discharge. Also known as main stroke; return stroke.

return stroke *See* return streamer.

return water [HYD] *See* return flow. [PETRO ENG] In a water-injection operation (waterflood) for an oil reservoir, the reinjection of salt water that is produced along with the oil.

retzian [MINERAL] $Mn_2Y(AsO_4)(OH)_4$ A chocolate brown to chestnut brown, orthorhombic mineral consisting of a basic arsenate of calcium, rare earths, and manganese.

reversal of dip [GEOL] Change in the dip direction of bedding near a fault such that the beds curve toward the fault surface in a direction exactly opposite that of the drag folds. Also known as dip reversal.

reverse cell [METEOROL] A circulating fluid system in which the circulation in a vertical plane is thermally indirect; that is, cooler air rises relative to warmer air.

reverse circulation drilling [MIN ENG] **1.** A variation of the rotary drilling method in which the cuttings are pumped up and out of the drill pipe, an advantage in certain large diameter holes. **2.** Diamond core drilling in which the water is injected through a stuffing box into the annular space around the drill rods and thus forced up special large drill rods.

reversed *See* overturned.

reversed arc [GEOL]   A curved belt of islands which is concave toward the open ocean, the opposite of most island arcs.

reversed consequent stream [HYD]   A consequent stream whose direction of flow is contrary to that normally consistent with the geologic structure.

reversed polarity [GEOPHYS]   Natural remanent magnetism opposite that of the present geomagnetic field.

reversed stream [HYD]   A stream whose direction of flow has been reversed, as by glacial action, landsliding, gradual tilting of a region, or capture.

reversed tide [OCEANOGR]   An oceanic tide that is out of phase with the apparent motions of the tide-producing body, so that low tide is directly under the tide-producing body and is accompanied by a low tide on the opposite side of the earth. Also known as inverted tide.

reverse fault *See* thrust fault.

reverse scarplet [GEOL]   An earthquake scarplet facing in toward the mountain slope and enclosing a trench, produced by reversal of earlier movement along a fault. Also known as earthquake rent.

reverse slip fault *See* thrust fault.

reverse slope [GEOL]   A hill descending away from a ridge.

reversing current [OCEANOGR]   Any current that changes direction, with a period of slack water at each reversal of direction.

reversing dune [GEOL]   A dune that tends to develop unusual height but migrates only a limited distance because seasonal shifts in dominant wind direction cause it to move alternately in nearly opposite directions.

reversing water bottle *See* Nansen bottle.

revet-crag [GEOL]   One of a series of narrow, pointed outliers or ridges of eroded strata inclined like a revetment against a mountain spur.

revived fault scarp *See* rejuvenated fault scarp.

revived stream *See* rejuvenated stream.

revolution [GEOL]   A little-used term to describe a time of profound crustal movements, on a continentwide or worldwide scale, which led to abrupt geographic, climatic, and environmental changes that were related to changes in forms of life.

revolving storm [METEOROL]   A cyclonic storm, or one in which the wind revolves about a central low-pressure area.

rework [GEOL]   Any geologic material that has been removed or displaced by natural agents from its origin and incorporated in a younger formation.

Reynolds effect [METEOROL]   A process of drop growth in clouds which involves net evaporation from cloud drops warmer than others and net condensation on the cooler drops.

Reynolds model [OCEANOGR]   A laboratory model of ocean currents in which inertial forces and frictional forces predominate, and in which the Reynolds number is used extensively in calculations.

rezbanyite [MINERAL] $Pb_3Cu_2Bi_{10}S_{19}$ A metallic-gray mineral composed of sulfide of lead, copper, and bismuth.

RH *See* relative humidity.

rhabdite *See* schreibersite.

rhabdoglyph [PALEON] A trace fossil consisting of a presumable worm trail appearing on the undersurface of flysch beds (sandstones) as a nearly straight bulge with little or no branching.

rhabdophane [MINERAL] $(Ce,Y,La,Di)(PO_4)\cdot H_2O$ A brown, pinkish, or yellowish-white mineral consisting of a hydrated phosphate of cerium, yttrium, and rare earths.

Rhachitomi [PALEON] A group of extinct amphibians in the order Temnospondyli in which pleurocentra were retained.

Rhaetian [GEOL] A European stage of geologic time; the uppermost Triassic (above Norian, below Hettangian of Jurassic). Also known as Rhaetic.

Rhaetic *See* Rhaetian.

Rhamphorhynchoidea [PALEON] A Jurassic suborder of the Pterosauria characterized by long, slender tails with an expanded tip.

rhegmagenesis [GEOL] Orogeny characterized by the development of large-scale strike-slip faults.

rheid [GEOL] A substance (below its melting point) which deforms by viscous flow during applied stress at an order of magnitude at least three times that of elastic deformation under similar circumstances.

rheid fold [GEOL] A fold whose strata deform by viscous flow as if they were fluid.

rheidity [GEOL] Relaxation time of a substance, divided by 1000.

Rhenanida [PALEON] An order of extinct marine fishes in the class Placodermi distinguished by mosaics of small bones between the large plates in the head shield.

rheoglyph [GEOL] A hieroglyph produced by syngenetic deformation, such as slumping.

rheoignimbrite [GEOL] An ignimbrite, on the slope of a volcanic crater, that has developed secondary flowage due to high temperatures.

rheomorphic intrusion [PETR] The injection of country rock that has become mobilized into the igneous intrusion that caused the rheomorphism.

rheomorphism [PETR] Mobilization of a rock by at least partial fusion accompanied by, and sometimes promoted by, addition of new material by diffusion.

rhizic water *See* soil water.

Rhizodontidae [PALEON] An extinct family of lobefin fishes in the order Osteolepiformes.

rhizosphere [GEOL] The soil region subject to the influence of plant roots and characterized by a zone of increased microbiological activity.

Rhodanian orogeny [GEOL] A short-lived orogeny that occurred at the end of the Miocene period.

rhodite [MINERAL] A mineral consisting of a native alloy of rhodium (about 40%) and gold.

rhodizite [MINERAL] $CsAl_4Be_4B_{11}O_{25}(OH)_4$ A white mineral composed of a basic borate of cesium, aluminum, and beryllium, occurring as isometric crystals.

rhodochrosite [MINERAL] $MnCO_3$ A rose-red to pink or gray mineral form of manganese carbonate with hexagonal symmetry but occurring in massive or columnar form; isomorphous with calcite and siderite, has a hardness of 3.5–4 on Mohs scale, and a specific gravity of 3.7; a minor ore of manganese.

rhodolite [MINERAL] A violet-red garnet species composed of a mixture of almandite and pyrope in about a 3:1 ratio.

rhodonite [MINERAL] $MnSiO_3$ A pink or brown mineral inosilicate crystallizing in the triclinic system and commonly found in cleavable to compact masses or in embedded grains; luster is vitreous, hardness is 5.5–6 on Mohs scale, and specific gravity is 3.4–3.7.

rhomb *See* rhombohedron.

rhombic dodecahedron [CRYSTAL] A crystal form in the cubic system that is a dodecahedron whose faces are equal rhombuses.

rhombic lattice *See* orthorhombic lattice.

rhombic system *See* orthorhombic system.

Rhombifera [PALEON] An extinct order of Cystoidea in which the thecal canals crossed the sutures at the edges of the plates, so that one-half of any canal lay in one plate and the other half on an adjoining plate.

rhombochasm [GEOL] A parallel-sided gap in the sialic crust occupied by simatic crust, probably caused by spreading and separation.

rhomboclase [MINERAL] $HFe^{3+}(SO_4)_2 \cdot 4H_2O$ A colorless mineral composed of hydrous acid ferric sulfate, occurring in rhombic plates.

rhombohedral [CRYSTAL] **1.** Of or pertaining to the rhombohedral system. **2.** Of or pertaining to crystal cleavage in or a centered lattice of the hexagonal system.

rhombohedral close packing [GEOL] A tight arrangement of uniform solid spheres in a cluster sediment or crystal lattice in which the unit cell has six planes passing through the centers of light spheres located at the corners of a regular rhombohedron; an aggregate so packed has minimum porosity.

rhombohedral iron ore *See* hematite; siderite.

rhombohedral lattice [CRYSTAL] A crystal lattice in which the three axes of a unit cell are of equal length, and the three angles between axes are the same, and are not right angles. Also known as trigonal lattice.

rhombohedral packing [CRYSTAL] The tightest manner of systematic arrangement of uniform solid spheres in a clastic sediment or crystal lattice, characterized by a unit cell of six planes passed through eight sphere centers situated at the corners of a regular rhombohedron.

rhombohedral system [CRYSTAL] A division of the trigonal crystal system in which the rhombohedron is the basic unit cell.

rhombohedron [CRYSTAL] A trigonal crystal form that is a parallelepiped, the six identical faces being rhombs. Also known as rhomb.

rhomboid ripple mark [GEOL] An aqueous current ripple mark characterized by a reticular arrangement of diamond-shaped tongues of sand, with each tongue having two acute angles, one pointing upcurrent and the other pointing downcurrent.

rhomboporoid cryptostome [PALEON] Any of a group of extinct bryozoans in the order Cryptostomata that built twiglike colonies with zooecia opening out in all directions from the central axis of each branch.

rhomb-porphyry [PETR] A porphyritic alkaline syenite composed of an alkali feldspar groundmass with augites having rhombohedral cross sections as the principal phenocryst minerals.

rhumb line [MAP] A line on the surface of the earth making the same oblique angle with all meridians. Also known as loxodrome.

rhyacolite *See* sanidine.

Rhynchosauridae [PALEON] An extinct family of generally large, stout, herbivorous lepidosaurian reptiles in the order Rhynchocephalea.

Rhynchotheriinae [PALEON] A subfamily of extinct elaphantoid mammals in the family Gomphotheriidae comprising the beak-jawed mastodonts.

Rhyniophyta [PALEOBOT] A subkingdom of the Embryobionta including the relatively simple, uppermost Silurian-Devonian vascular plants.

Rhyniopsida [PALEOBOT] A class of extinct plants in the subkingdom Rhyniophyta characterized by leafless, usually dichotomously branched stems that bore terminal sporangia.

rhyodacite [PETR] A group of extrusive porphyritic igneous rocks containing quartz, plagioclase, and biotite phenocrysts in a fine-grained to glassy groundmass composed of alkali feldspar and silica minerals. Also known as dellenite; quartz lattice.

rhyolite [PETR] A light-colored, aphanitic volcanic rock composed largely of alkali feldspar and free silica with minor amounts of mafic minerals; the extrusive equivalent of granite.

rhyolitic glass [GEOL] Volcanic glass that is chemically equivalent to rhyolite.

rhyolitic lava [GEOL] A highly viscous, silica-rich lava.

rhyolitic magma [PETR] A type of magma formed by differentiation from basaltic magma in combination with assimilation of siliceous material, or by melting of portions of the earth's sialic layer.

rhyolitic tuff [GEOL] A tuff composed of fragments of rhyolitic lava.

rhythmic accumulations [GEOL] Regular patterns of ripples and cusps in sediment on the beach or the sea floor, formed by currents and waves.

rhythmic crystallization [PETR] In igneous rocks, a phenomenon in which different minerals crystallize in concentric layers, giving rise to orbicular texture.

rhythmic driving [MIN ENG] Driving carried out between two shifts; that is, the drilling, loading, and blasting are carried out in one shift and the mucking and transportation in the following one.

rhythmic layering [GEOL] A type of layering in an igneous intrusion which is easily observable and in which there is repetition of zones of varying composition.

rhythmic sedimentation [GEOL] A repetitious, regular sequence of rock units formed by sedimentary succession and indicating a frequent, predictable recurrence of the same sequence of conditions.

rhythmic stratification [GEOL] The occurrence of sediment layers in repetitive patterns, such as a regular alternation of layers of lime and clay.

rhythmic succession [GEOL] A succession of rock units showing continual and repeated changes of lithology.

rhythmite [GEOL] An independent unit of a rhythmic succession or of beds that were developed by rhythmic sedimentation.

ria [GEOGR] **1.** Any broad, estuarine river mouth. **2.** A long, narrow coastal inlet, except a fjord, whose depth and width gradually and uniformly diminish inland.

ria coast [GEOGR] A coast with several parallel rias extending far inland and alternating with ridgelike promontories.

ria shoreline [GEOGR] A type of coastline developed along a drowning landmass in which numerous long and narrow arms of the sea extend inland parallel with one another and perpendicular to the coastline.

rib [GEOL]   A layer or dike of rock forming a small ridge on a steep mountainside. [MIN ENG]   **1.** A solid pillar of coal or ore left for support. **2.** A thin stratum in a seam of coal.

rib-and-furrow [GEOL]   The bedding-plane expression for micro-cross-bedding, consisting of sets of small, transverse arcuate markings confined to long, narrow, parallel grooves oriented parallel to the current flow and separated by narrow ridges.

ribbon [PETR]   One of a set of parallel bands in a rock or mineral.

ribbon band [GEOL]   An elongate and flattened volcanic bomb derived from ropes of lava.

ribbon banding [PETR]   A banding produced in the bedding of a sedimentary rock by thin strata of contrasting colors, giving the rock an appearance which suggests bands of ribbons.

ribbon diagram [GEOL]   A continuous geologic cross section that is drawn in perspective along a curved or sinuous line.

ribbon jasper [GEOL]   Banded jasper with parallel, ribbonlike stripes of alternating colors or shades of color. Also known as riband jasper.

ribbon lightning [GEOPHYS]   Ordinary streak lightning that appears to be spread horizontally into a ribbon of parallel luminous streaks when a very strong wind is blowing at right angles to the observer's line of sight; successive strokes of the lightning flash are then displaced by small angular amounts and may appear to the eye or camera as distinct paths. Also known as band lightning; fillet lightning.

ribbon rock [PETR]   A rock showing a succession of thin layers of differing composition or appearance.

ribbon slate [PETR]   Slate produced by incomplete metamorphism of clearly visible residual bedding planes that cut across the cleavage surface.

ribbon structure [GEOL]   A succession of thin layers of different mineralogy and texture often contorted and deformed.

rib hole [MIN ENG]   One of the final holes fired in blasting ground at the sides of a shaft or tunnel. Also known as trimmer.

rib pillar [MIN ENG]   A pillar whose length is large compared with its width.

ribut [METEOROL]   Sharp, short squalls during comparatively calm winds from May to November in Malaya.

rice coal [GEOL]   Anthracite that will pass through circular holes 5/16 inch (7.9 millimeters) in diameter, but not 3/16 inch (4.8 millimeters) in diameter, in a screen.

richellite [MINERAL] $Ca_3Fe_{10}(PO_4)_8(OH,F)_{12} \cdot nH_2O$ A yellow mineral composed of hydrous basic iron calcium fluophosphate; occurs in masses.

Richmondian [GEOL]   A North American stage of geologic time: Upper Ordovician (above Maysvillian, below Lower Silurian).

rich ore [MIN ENG]   Relatively high grade ore.

richterite [MINERAL] $(Na,K)_2(Mg,Mn,Ca)_6Si_8O_{22}(OH)_2$ A brown, yellow, or rose-red monoclinic mineral composed of basic silicate of sodium, potassium, magnesium, manganese, and calcium; a member of the amphibole group.

Richter scale [GEOPHYS]   A scale of numerical values of earthquake magnitude ranging from 1 to 9.

rickardite [MINERAL] Cu_4Te_3 A deep-purple mineral composed of copper telluride, occurring in masses.

rider [MIN ENG] A steel or iron crossbeam which slides between the guides in a sinking shaft; it is carried by the hoppit and serves to guide and steady the hoppit during its movement up and down the shaft.

ridge [GEOL] An elongate, narrow, steep-sided elevation of the earth's surface or the ocean floor. [METEOROL] An elongated area of relatively high atmospheric pressure, almost always associated with, and most clearly identified as, an area of maximum anticyclonic curvature of wind flow. Also known as wedge.

ridged ice [OCEANOGR] Pressure ice having a readily observed surface roughness in the form of a ridge or many ridges.

ridge-top trench [GEOL] A trench, occasionally found at or near the crest of high, steep-sided mountain ridges, formed by the creep displacement of a large slab of rock along shear surfaces more or less parallel with the side slope of the ridge.

riebeckite [MINERAL] $Na_2(Fe,Mg)_5Si_8O_{22}(OH)_2$ A blue or black monoclinic amphibole occurring as a primary constituent in some acid- or sodium-rich igneous rocks.

Riecke's principle [MINERAL] The principle that solution of a mineral occurs most readily at points of greatest external pressure, and crystallization occurs most readily at points of least external pressure; applied to recrystallization in metamorphic rock.

riedenite [PETR] An igneous rock composed of large tabular biotite crystals in a granular groundmass of nosean, biotite, pyroxene, and small amounts of sphene and apatite.

riegel [GEOL] A low, traverse ridge of bedrock on the floor of a glacial valley. Also known as rock bar; threshold; verrou.

riffle [HYD] **1.** A shallows across a stream bed over which water flows swiftly and is broken into waves by submerged obstructions. **2.** Shallow water flowing over a riffle.

rift [GEOL] **1.** A narrow opening in a rock caused by cracking or splitting. **2.** A high, narrow passage in a cave.

rift-block mountain [GEOL] A mountain range which is a horst block bounded by normal faults.

rift-block valley [GEOL] A valley which occupies a graben.

rift valley [GEOL] A deep, central cleft with a mountainous floor in the crest of a midoceanic ridge. Also known as central valley; midocean rift.

right-lateral fault *See* dextral fault.

right-lateral slip fault *See* dextral fault.

right-slip fault *See* dextral fault.

rill [GEOL] A small, transient runnel. [HYD] A small brook or stream.

rillenstein [GEOL] A pattern of tiny solution grooves of about 1 millimeter or less in width, formed on the limestone surface of a karstic region.

rill erosion [GEOL] The formation of numerous, closely spaced rills due to the uneven removal of surface soil by streamlets of running water. Also known as rilling; rill wash; rillwork.

rill flow [HYD] Surface runoff flowing in small irregular channels too small to be considered rivulets.

rilling *See* rill erosion.

rill mark [GEOL] A small, dendritic channel formed on beach mud or sand by a rill, especially if on the lee side of a partially buried obstruction.

rillstone *See* ventifact.

rill wash *See* rill erosion.

rillwork *See* rill erosion.

rim cement [GEOL]   A thin layer of calcium carbonate, hematite, or silica developed on the surface of detrital grains during diagenesis.

rime [HYD]   A white or milky and opaque granular deposit of ice formed by the rapid freezing of supercooled water drops as they impinge upon an exposed object; composed essentially of discrete ice granules, and has densities as low as 0.2–0.3 gram per cubic centimeter.

rime fog *See* ice fog.

rimmed kettle [GEOL]   A morainal depression with raised edges.

rimmed solution pool [GEOL]   A pool in rock with a hardened rim resulting from deposition of lime during evaporation at low tide.

rimming wall [GEOL]   A steep, ridgelike erosional remnant of continuous layers of porous, permeable, poorly cemented, detrital limestones, believed to form under tropical or subtropical conditions by surface-controlled secondary cementation of an original steep slope and followed by differential erosion that brings the cemented zone into relief.

rimpylite [MINERAL]   A group name for several green and brown hornblendes with high contents of $(Al, Fe)_2O_3$.

rim ridge [GEOL]   A minor ridge of till defining the edge of a moraine plateau.

rimrock [GEOL]   A top layer of resistant rock on a plateau outcropping with vertical or near vertical walls.

rimstone [GEOL]   A calcium-containing deposit ringing an overflowing basin such as a hot spring.

rim syncline [GEOL]   In salt tectonics, a local depression that develops as a border around a salt dome, as the salt in the underlying strata is displaced toward the dome. Also known as peripheral sink.

rincon [GEOL]   **1.** A small, secluded valley. **2.** A bend in a stream.

ring current [GEOPHYS]   A westward electric current which is believed to circle the earth at an altitude of several earth radii during the main phase of geomagnetic storms, resulting in a large worldwide decrease in the geomagnetic field horizontal component at low latitudes.

ring depression [GEOL]   The annular, structurally depressed area surrounding the central uplift of a cryptoexplosion structure; faulting and folding may be involved in its formation. Also known as peripheral depression; ring syncline.

ring dike [GEOL]   A roughly circular dike that is vertical or inclined away from the center of the arc. Also known as ring-fracture intrusion.

ring fault [GEOL]   **1.** A fault that bounds a rift valley. **2.** A steep-sided fault pattern that is cylindrical in outline and associated with cauldron subsidence. Also known as ring fracture.

ring fissure [GEOL]   A roughly circular dessication crack formed on a playa around a point source (generally a phreatophyte).

ring fracture *See* ring fault.

ring-fracture intrusion *See* ring dike.

ring-fracture stoping [GEOL] Large-scale magmatic stoping that is associated with cauldron subsidence.

ring holes [MIN ENG] The group of boreholes radially drilled from a common-center setup.

ringite [GEOL] An igneous rock formed by the mixing of silicate and carbonatite magmas.

ring-lifter case *See* lifter case.

ring silicate *See* cyclosilicate.

ring stress [MIN ENG] The zone of stress in rock surrounding all development excavations.

rinkite *See* mosandrite.

rinkolite *See* mosandrite.

rinneite [MINERAL] NaK_3FeCl_6 A colorless, pink, violet, or yellow mineral composed of sodium potassium iron chloride, occurring in granular masses.

Rio Tinto process [MIN ENG] Heap leaching of curiferous sulfides that have been oxidized to sulfates by prolonged atmospheric weathering.

rip [MIN ENG] To break down the roof in mine roadways to increase the headroom for haulage, traffic, and ventilation. [OCEANOGR] A turbulent agitation of water generally caused by the interaction of currents and wind.

riparian water loss [HYD] Discharge of water through evapotranspiration along a watercourse, especially water transpired by vegetation growing along the watercourse.

ripbit *See* detachable bit; jackbit.

rip channel [GEOL] A channel, often more than 2 meters (6.6 feet) deep, carved on the shore by a rip current.

rip current [OCEANOGR] The return flow of water piled up on shore by incoming waves and wind.

ripe [GEOL] Referring to peat, in an advanced state of decay. [HYD] Descriptive of snow that is in a condition to discharge meltwater; ripe snow usually has a coarse crystalline structure, a snow density near 0.5, and a temperature near 32°F (0°C).

ripidolite [MINERAL] $(Mg,Fe^{2+})_9Al_6Si_5O_{20}(OH)_{16}$ A mineral of the chlorite group; consists of basic magnesium iron aluminum silicate. Also known as aphrosiderite.

ripping face support [MIN ENG] A timber or steel support at the ripping lip.

ripping lip [MIN ENG] The end of the enlarged roadway section where work is proceeding.

ripple [GEOL] A very small ridge of sand resembling or suggesting a ripple of water and formed on the bedding surface of a sediment. [OCEANOGR] A small curling or undulating wave controlled to a significant degree by both surface tension and gravity.

ripple bedding [GEOL] A bedding surface characterized by ripple marks.

ripple biscuit [GEOL] A bedding structure produced by lenticular lamination of sand in a bay or lagoon.

ripple drift [GEOL] A pattern of cross-lamination formed by sedimentary deposits on both sides of a migrating ripple.

ripple index [GEOL] On a rippled surface, the ratio of the crest-to-crest distance to the crest-to-trough distance.

ripple lamina [GEOL] An internal sedimentary structure formed in sand or silt by currents or waves, as opposed to a ripple mark formed externally on a surface.

ripple load cast [GEOL] A load cast of a ripple mark showing evidence of penecontemporaneous deformation in the accumulation of its trough and crest and in the oversteepening of the component laminae.

ripple mark [GEOL] **1.** A surface pattern on incoherent sedimentary material, especially loose sand, consisting of alternating ridges and hollows formed by wind or water action. **2.** One of the ridges on a ripple-marked surface.

ripple symmetry index [GEOL] A measure of the degree of symmetry of a ripple mark, equal to the ratio of the length of the gentle (upcurrent) side to the steep (downcurrent) side.

rips [OCEANOGR] A turbulent agitation of water, generally caused by the interaction of currents and wind; in nearshore regions they may be currents flowing swiftly over an irregular bottom; sometimes referred to erroneously as tide rips.

rise [GEOL] A long, broad elevation which rises gently from its surroundings, such as the sea floor.

rise of tide [OCEANOGR] Vertical distance from the chart datum to a higher water datum.

rise pit [GEOL] A pit through which an underground stream rises to the surface with a calm and steady flow.

riser [GEOL] A steplike topographic feature, such as a steep slope between terraces.

rising limb [HYD] The rising portion of the hydrograph resulting from runoff of rainfall or snowmelt.

rising tide [OCEANOGR] The portion of the tide cycle between low water and the following high water.

Riss [GEOL] **1.** A European stage of geologic time: Pleistocene (above Mindel, below Würm). **2.** The third stage of glaciation of the Pleistocene in the Alps.

Rissoacea [PALEON] An extinct superfamily of gastropod mollusks.

Riss-Würm [GEOL] The third interglacial stage of the Pleistocene in the Alps, following the Riss glaciation and preceding the Würm glaciation.

river [HYD] A large, natural freshwater surface stream having a permanent or seasonal flow and moving toward a sea, lake, or another river in a definite channel. [LAP] A pure-white diamond of very high grade.

river bar [GEOL] A ridgelike accumulation of alluvium in the channel, along the banks, or at the mouth of a river.

river basin [GEOL] The area drained by a river and all of its tributaries.

riverbed [GEOL] The channel which contains, or formerly contained, a river.

river bottom [GEOL] The low-lying alluvial land along a river. Also known as river flat.

river breathing [HYD] Fluctuation of the water level of a river.

river capture *See* capture.

river-deposition coast [GEOL] A deltaic coast characterized by lobate seaward bulges crossed by river distributaries and bordered by lowlands.

river drift [GEOL] Rock material deposited by a river in one place after having been moved from another.

river end   [HYD]   The lowest point of a river with no outlet to the sea, situated where its water disappears by percolation or evaporation.

river flat   *See* river bottom.

river forecast   [HYD]   A forecast of the expected stage or discharge at a specified time, or of the total volume of flow within a specified time interval, at one or more points along a stream.

river gage   [ENG]   A device for measuring the river stage; types in common use include the staff gage, the water-stage recorder, and wire-weight gage. Also known as stream gage.

river ice   [HYD]   Any ice formed in or carried by a river.

river mining   [MIN ENG]   Mining or excavating beds of existing rivers after deflecting their course, or by dredging without changing the flow of water.

river morphology   [GEOL]   The study of the channel pattern and the channel geometry at several points along a river channel, including the network of tributaries within the drainage basin. Also known as channel morphology; fluviomorphology; stream morphology.

river-pebble phosphate   [GEOL]   A transported, dark variety of pebble phosphate obtained from bars and floodplains of rivers. Also known as river rock.

river piracy   *See* capture.

river plain   *See* alluvial plain.

river rock   *See* river-pebble phosphate.

river run gravel   [GEOL]   Natural gravel as found in deposits that have been subjected to the action of running water.

river system   [HYD]   The aggregate of stream channels draining a river basin.

river terrace   *See* stream terrace.

river tide   [HYD]   A tide that occurs in rivers emptying directly into the sea, showing three characteristic modifications of ocean tides: the speed at which the tide travels upstream depends on the depth of the channel, the further upstream the longer the duration of the falling tide and shorter the duration of the rising tide, and the range of the tide decreases with distance upstream.

riverwash   [GEOL]   **1.** Soil material that has been transported and deposited by rivers. **2.** An alluvial deposit in a river bed or flood channel, subject to erosion and deposition during recurring flood periods.

river water   [HYD]   Water having carbonate, sulfate, and calcium as its main dissolved constituents; distinguished from seawater by its chloride and sodium content.

riving   [GEOL]   The splitting off, cracking, or fracturing of rock, especially by frost action.

rivulet   [HYD]   A small stream; a brook.

road   [GEOL]   One of a series of erosional terraces in a glacial valley, formed as the water level dropped in an ice-dammed lake.   [MIN ENG]   Any mine passage or tunnel.

roadstead   [GEOGR]   An area near the shore, where vessels can anchor in safety; usually a shallow indentation in the coast.

roaring forties   [METEOROL]   A popular nautical term for the stormy ocean regions between 40° and 50° latitude; it usually refers to the Southern Hemisphere, where there is an almost completely uninterrupted belt of ocean with strong prevailing westerly winds.

roaring sand [GEOL]   A sounding sand, found on a desert dune, that sets up a low roaring sound that sometimes can be heard for a distance of 400 meters (120 feet).

roast sintering *See* blast roasting.

rob [MIN ENG]   To take out ore or coal from a mine with a view to immediate product, and not to subsequent working.

robbery *See* capture.

Robin Hood's wind [METEOROL]   In Great Britain, saturated air with temperatures near freezing; it is raw and penetrating.

robinsonite [MINERAL] $Pb_7Sb_{12}S_{25}$ A mineral composed of lead antimony sulfide.

rocdrumlin *See* rock drumlin.

roche moutonnée [GEOL]   A small, elongate hillock of bedrock sculptured by a large glacier so that its long axis is oriented in the direction of ice movement; the upstream side is gently inclined, smoothly rounded, but striated, and the downstream side is steep, rough, and hackly.

rock [PETR]   **1.** A consolidated or unconsolidated aggregate of mineral grains consisting of one or more mineral species and having some degree of chemical and mineralogic constancy. **2.** In the popular sense, a hard, compact material with some coherence, derived from the earth.

rock asphalt *See* asphalt rock.

rock association [PETR]   A group of igneous rocks within a petrographic province that are related chemically and petrographically, generally in a systematic manner such that chemical data for the rocks plot as smooth curves on variation diagrams. Also known as rock kindred.

rock awash [OCEANOGR]   In U.S. Coast and Geodetic Survey terminology, a rock exposed at any stage of the tide between the datum of mean high water and the sounding datum, or one just bare at these data.

rock-a-well [PETRO ENG]   The procedure of bleeding pressure alternately from the casing of a well and from the tubing until the well starts flowing.

rock bar *See* riegel.

rock-basin lake *See* paternoster lake.

rock bench *See* structural bench.

rock bit [ENG]   Any one of many different types of roller bits used on rotary-type drills for drilling large-size holes in soft to medium-hard rocks.

rockbolt [ENG]   A bar, usually constructed of steel, which is inserted into predrilled holes in rock and secured for the purpose of ground control.

rock bolting [ENG]   A method of securing or strengthening closely jointed or highly fissured rocks in mine workings, tunnels, or rock abutments by inserting and firmly anchoring rock bolts oriented perpendicular to the rock face or mine opening.

rockbridgeite [MINERAL] $Fe^{2+}Fe_6^{3+}(PO_4)_4(OH)_8$ A basic phosphate mineral containing iron; isomorphous with frondelite.

rock-bulk compressibility [GEOL]   One of three types of rock compressibility (matrix, bulk, and pore); the fractional change in volume of the bulk volume of the rock with a unit change in pressure.

rock bump [MIN ENG]   The sudden release of the weight of the rocks over a coal seam or of enormous lateral stresses.

rockburst [MIN ENG] A sudden and violent rock failure around a mining excavation on a sufficiently large scale to be considered a hazard.

rock cleavage [PETR] The capacity of a rock to split along certain parallel surfaces more easily than along others.

rock control [GEOL] The influences of differences in earth materials on development of landforms.

rock cork *See* mountain cork.

rock creep [GEOL] A form of slow flowage in rock materials evident in the downhill bending of layers of bedded or foliated rock and in the slow downslope migration of large blocks of rock away from their parent outcrop.

rock crystal [MINERAL] A transparent, colorless form of quartz with low brilliance; used for lenses, wedges, and prisms in optical instruments. Also known as berg crystal; crystal; mountain crystal; pebble; quartz crystal.

rock cycle [GEOL] The interrelated sequence of events by which rocks are initially formed, altered, destroyed, and reformed as a result of magmatism, erosion, sedimentation, and metamorphism.

rock-defended terrace [GEOL] **1.** A river terrace having a ledge or outcrop of resistant rock at its base which serves as protection against undermining. **2.** A marine terrace having a mass of resistant rock at the base of the cliff which protects against wave erosion.

rock desert [GEOL] An upland desert in which bedrock is either exposed or is covered with a thin veneer of coarse rock fragments.

rock drill [MECH ENG] A machine for boring relatively short holes in rock for blasting purposes; motive power may be compressed air, steam, or electricity.

rock drum *See* rock drumlin.

rock drumlin [GEOL] A smooth, streamlined hill modeled by glacial erosion, which has a core of bedrock usually veneered with a layer of glacial till, and which resembles a true drumlin in outline and form but is generally less symmetrical and less regularly shaped. Also known as drumlinoid; false drumlin; rocdrumlin; rock drum.

rock dust distributor *See* rock duster.

rock duster [MIN ENG] A machine that distributes rock dust over the interior surfaces of a coal mine by means of air to prevent coal dust explosions. Also known as rock dust distributor.

rock element [PETR] The coherent, intact piece of rock that is the basic constituent of the rock system and which has physical, mechanical, and petrographic properties that can be described or measured by laboratory tests.

rocker [MIN ENG] A small digging bucket mounted on two rocker arms in which auriferous alluvial sands are agitated by oscillation, in water, to collect gold.

rocker dump car [MIN ENG] A small-capacity mining car; the most popular and most widely used are the gravity dump types, designed so that the weight of the load tips the body when a locking latch is released by hand.

rocker shovel [MIN ENG] A digging and loading machine consisting of a bucket attached to a pair of semicircular runners; lifts and dumps the bucket load into a car or another materials-transport unit behind the machine.

rocket lightning [GEOPHYS] A rare form of lightning whose luminous channel seems to advance through the air with only the speed of a skyrocket.

rocketsonde *See* meteorological rocket.

rock fabric *See* fabric.

rock failure [GEOL]   Fracture of a rock that has been stressed beyond its ultimate strength.

rockfall [GEOL]   **1.** The fastest-moving landslide; free fall of newly detached bedrock segments from a cliff or other steep slope; usually occurs during spring thaw. **2.** The rock material moving in or moved by a rockfall.

rock fan [GEOL]   A fan-shaped bedrock surface whose apex is where a mountain stream debouches upon a piedmont slope, and which occupies an area where a pediment meets the mountain slope.

rock-floor robbing [GEOL]   A form of sheetflood erosion in which sheetfloods remove crumbling debris from rock surfaces in desert mountains.

rock flour [GEOL]   A fine, chemically unweathered powder of rock-forming minerals produced by pulverization of rock fragments during natural transport or crushing. Also known as glacial flour.

rock flowage *See* flow.

rockforming [GEOL]   Referring to any minerals which commonly occur in important proportions in common rocks.

rock glacier [GEOL]   Boulders and fine material cemented by ice about a meter below the surface. Also known as talus glacier.

rock-glacier creep [GEOL]   A rapid talus creep of tongues of debris in a cold region, caused by the expansive force of the alternate freeze and thaw of ice in the interstices of the debris.

rock gypsum [MINERAL]   Massive, coarsely crystalline to earthy, finely granular type of gypsum found in gyp rock.

rock island *See* meander core.

rock loader [MIN ENG]   Any device or machine used for loading slate or rock inside a mine.

rock magnetism [GEOPHYS]   The natural remanent magnetization of igneous, metamorphic, and sedimentary rocks resulting from the presence of iron oxide minerals.

rock matrix compressibility [GEOL]   One of three types of rock compressibility (matrix, bulk, and pore); the fractional change in volume of the solid rock material (grains) with a unit change in pressure.

rock meal *See* rock milk.

rock mechanics [GEOPHYS]   Application of the principles of mechanics and geology to quantify the response of rock when it is acted upon by environmental forces, particularly when human-induced factors alter the original ambient forces.

rock milk [MINERAL]   A soft, white, earthy or powdery variety of calcite. Also known as agaric mineral; bergmehl; forril farina; rock meal.

rock pediment [GEOL]   A pediment formed on the surface of bedrock.

rock permeability [GEOL]   The ability of a rock to receive, hold, or pass fluid materials (oil, water, and gas) by nature of the interconnections of its internal porosity.

rock phosphate *See* phosphorite.

rock pillar [GEOL]   **1.** A column of rock produced by differential weathering or erosion, as along a joint plane. **2.** In a cave, a pillar-type structure that is residual bedrock rather than a stalacto-stalagmite.

rock pool [GEOL]   A tidal pool formed along a rocky shoreline.

rock pressure [GEOPHYS] **1.** Stress in underground geologic material due to weight of overlying material, residual stresses, and pressures resulting from swelling clays. **2.** *See* ground pressure.

rock river [GEOL] A very long and narrow rock stream.

rock saw [MIN ENG] A type of mechanical miner that is used to remove large blocks of material; cuts narrow slots or channels by the action of a moving steel band or blade and a slurry of abrasive particles (sometimes diamonds) rather than teeth; small flame jets are also used.

rockshaft [MIN ENG] A shaft through which rock can be brought into a mine for filling, stopes, or other excavations.

rock shelter [GEOL] A cave that is formed by a ledge of overhanging rock.

rock silk [MINERAL] A silky variety of asbestos.

Rocksite program [MIN ENG] A U.S. Navy program concerned with undersea mining or consolidated mineral deposits; studied direct sea-floor access at remote sites through shafts drilled in the sea floor.

rockslide [GEOL] The sudden, rapid downward movement of newly detached bedrock segments over a surface of weakness, such as of bedding, jointing, or faulting. Also known as rock slip.

rock slip *See* rockslide.

rock stack [GEOL] A rocky crag that has been uplifted from an old sea floor.

rock step *See* knickpoint.

rock-stratigraphic unit [GEOL] A lithologically homogeneous body of strata characterized by certain observable physical features, or by the dominance of a certain rock type or combination of rock types; rock-stratigraphic units include groups, formations, members, and beds. Also known as geolith; lithologic unit; lithostratic unit; lithostratigraphic unit; rock unit.

rock stream [GEOL] Rocks moving (or already moved) in a mass down a slope under the influence of their own weight.

rock system [GEOPHYS] In rock mechanics, all natural environmental factors that can influence the behavior of that portion of the earth's crust that will become part of an engineering structure.

rock terrace [GEOL] A stream terrace on the side of a valley composed of resistant bedrock which remains during erosion of weaker overlying and underlying beds.

rock type [PETR] **1.** One of the three major rock groups: igneous, sedimentary, metamorphic. **2.** A rock having a unique, identifiable set of characters, such as basalt.

rock unit *See* rock-stratigraphic unit.

rock wood *See* mountain wood.

rod [GEOL] A rodlike sedimentary particle characterized by a width-length ratio less than 2/3 and a thickness-width ratio more than 2/3. Also known as roller.

rodingite [PETR] A medium-to coarse-grained, commonly calcium-enriched gabbroic rock containing grossular and diallage as essential minerals.

rodite *See* diogenite.

rod pump [PETRO ENG] Type of oil well sucker-rod pump that can be inserted into or removed from oil well tubing without moving or disturbing the tubing itself. Also known as insert pump.

rod slide *See* slide.

rod stuffing box [ENG] An annular packing gland fitting between the drill rod and the casing at the borehole collar; allows the rod to rotate freely but prevents the escape of gas or liquid under pressure.

roedderite [MINERAL] $(Na,K)_2(Mg,Fe)_5Si_{12}O_{30}$ A silicate meteorite mineral.

roemerite [MINERAL] $FeFe_2(SO_4)_4 \cdot 14H_2O$ A rust-brown to yellow mineral composed of hydrous ferric and ferrous iron sulfate.

roesslerite [MINERAL] $MgH(AsO_4) \cdot 7H_2O$ A monoclinic mineral composed of hydrous acid magnesium arsenate; it is isomorphous with phosphorroesslerite.

roestone *See* oolite.

rofla [GEOL] An extremely narrow, tortuous gorge, frequently formed by meltwater streams flowing from a glacier.

ROFOR [METEOROL] An international code word used to indicate a route forecast (along an air route).

ROFOT [METEOROL] An international code word used to indicate a route forecast, with units in the English system.

rognon [GEOL] **1.** A small rocky peak or ridge surrounded by glacier ice in a mountainous region. **2.** A similar peak projecting above the bed of a former glacier. **3.** *See* nunakol.

roily water [HYD] **1.** Muddy or sediment-filled water. **2.** Turbulent, agitated, or swirling water.

roll [GEOL] A primary sedimentary structure produced by deformation involving subaqueous slump or vertical foundering.   [MIN ENG]   *See* horseback.

roll cloud *See* rotor cloud.

roller [GEOL] *See* rod.   [OCEANOGR] A long, massive wave which usually retains its form without breaking until it reaches the beach or a shoal.

rollers [OCEANOGR] Swells coming from a great distance and forming large breakers on exposed coasts.

roll flattening *See* flattening.

Romanche trench [GEOL] A 7370-meter-deep trench in the Mid-Atlantic ridge near the equator.

romeite [MINERAL] $(Ca,Fe,Mn,Na)_2(Sb,Ti)_2O_6(O,OH,F)$ A honey-yellow to yellowish-brown mineral composed of oxide of calcium, iron, manganese, sodium, antimony, and titanium, occurring in minute octahedrons.

ROMET [METEOROL] An international code word denoting route forecast, with units in the metric system.

rondada [METEOROL] In Spain, a wind that shifts diurnally from northwest through north, east, south, and west.

rongstockite [GEOL] A medium- to fine-grained plutonic rock composed of zoned plagioclase, orthoclase, some cancrinite, augite, mica, hornblende, magnetite, sphene, and apatite.

roof bolt [MIN ENG] One of the long steel bolts driven into walls or roofs of underground excavations to strengthen the pinning of rock strata.

roof control [MIN ENG] The study of rock behavior when undermined by mining operations, and the most effective measures to control movements.

roof cut [MIN ENG] A machine cut made with a turret coal cutter in the roof immediately above the seam.

roofed dike [GEOL] A dike that has an upward termination.

roof foundering [GEOL] Collapse of overlying rock into a magma chamber following excavation of a large quantity of magma.

roof jack [MIN ENG] A screw- or pump-type extension post used as a temporary roof support.

roof pendant [GEOL] Downward projection or sag into an igneous intrusion of the country rock of the roof. Also known as pendant.

roof stringer [MIN ENG] A lagging bar running parallel with the working place above the header in a weak or scaly top in narrow rooms or entries which have short life.

room [GEOL] An open area in a cave. [MIN ENG] **1.** Space driven off an entry in which coal is produced. **2.** Working place in a flat mine.

room-and-pillar [MIN ENG] A system of mining in which the coal or ore is mined in rooms separated by narrow ribs or pillars; pillars are subsequently worked.

room conveyor [MIN ENG] Any conveyor which carries coal from the face of a room toward the mouth.

room crosscut See breakthrough.

rooseveltite [MINERAL] $BiAsO_4$ A gray mineral consisting of bismuth arsenate; occurs as thin botryoidal crusts.

rooster tail [HYD] A plumelike form of water and sometimes spray that occurs at the intersection of two crossing waves.

root [GEOL] **1.** The lower limit of an ore body. Also known as bottom. **2.** The part of a fold nappe that was originally linked to its root zone.

root clay See underclay.

root zone [GEOL] **1.** The area where a low-angle thrust fault steepens and descends into the crust. **2.** The source of the root of a fold nappe.

ropak [OCEANOGR] An ice cake standing on edge as a result of excessive pressure. Also known as turret ice.

rope boring [ENG] A method similar to rod drilling except that rigid rods are replaced by a steel rope to which the boring tools are attached and allowed to fall by their own weight.

rope cutter See hook tender.

rope rider See trip rider.

ropy lava See pahoehoe.

rosasite [MINERAL] $(Cu,Zn)_2(OH)_2(CO_3)$ A green to bluish-green and sky blue mineral consisting of a carbonate-hydroxide of copper and zinc.

roscherite [MINERAL] $(Ca,Mn,Fe)_2Al(PO_4)(OH)\cdot 2H_2O$ A dark-brown mineral composed of hydrous basic phosphate of aluminum, calcium, manganese, and iron, occurring as monoclinic crystals.

roscoelite [MINERAL] $K(V,Al,Mg)_3Si_3O_{10}(OH)_2$ Tan, grayish-brown, or greenish-brown vanadium-bearing mica mineral occurring in minute scales or flakes.

rose diagram [GEOL] A circular graph indicating values in several classes of vector properties of rocks such as cross-bedding direction.

roselite [MINERAL] $(Ca,Co)_2(Co,Mg)(AsO_4)_2\cdot 2H_2O$ A pink or rose-colored, monoclinic mineral consisting of a hydrated arsenate of calcium, cobalt, and magnesium.

rose quartz [MINERAL] A pink variety of crystalline quartz; commonly massive and used as a gemstone. Also known as Bohemian ruby.

rosette [MINERAL] Rose-shaped, crystalline aggregates of barite, marcasite, or pyrite formed in sedimentary rock.

rosieresite [MINERAL] A yellow to brown mineral composed of hydrous aluminum phosphate containing lead and copper, occurring in stalactitic masses.

rosin tin [MINERAL] A red or yellow variety of cassiterite. Also known as resin tin.

Rosiwal analysis [PETR] A quantitive method of estimating the volume percentages of the minerals in a rock, in which thin sections of a rock are examined under a microscope which has a micrometer to measure the linear intercepts of each mineral along a particular set of lines.

Ross Barrier [OCEANOGR] A wall of shelf ice bordering on the Ross Sea.

Rossby wave [METEOROL] A wave on a uniform current in a two-dimensional non-divergent fluid system, rotating with varying angular speed about the local vertical (beta plane); this is a special case of a barotropic disturbance, conserving absolute vorticity; applied to atmospheric flow, it takes into account the variability of the Coriolis parameter while assuming the motion to be two-dimensional. Also known as planetary wave.

Rossel Current [OCEANOGR] A seasonal Pacific Ocean current flowing westward and north-westward along both the southern and northeastern coasts of New Guinea, the southern part flowing through Torres Strait and losing its identity in the Arafura Sea, and the northern part curving northeastward to join the equatorial countercurrent of the Pacific Ocean.

rossite [MINERAL] $CaV_2O_6 \cdot 4H_2O$ A yellow, triclinic mineral consisting of a hydrated calcium vanadate.

Ross Sea [GEOGR] Arm of the South Pacific Ocean off Antarctica.

rosterite *See* vorobyevite.

rosthornite [MINERAL] A brown to garnet-red variety of retinite with a low (4.5%) oxygen content, found in lenticular masses in coal.

rotary blasthole drilling [MIN ENG] A term applied to two types of drilling: in quarrying and open pit mining it implies rotary drilling with roller-type bits, using compressed air for cuttings removal, either conventional rotary table drive or hydraulic motor to produce rotation, with hydraulic or wire-line mechanisms to add part of the weight of the drill to the weight of the tools to increase bit pressure; and in underground mining and sometimes aboveground, it implies the drilling of small-diameter blastholes with a diamond drill, using either coring or noncoring diamond bits.

rotary breaker [MIN ENG] A breaking machine for coal or ore; consists of a trommel screen with a heavy steel shell fitted with lifts which raise and convey the coal and stone forward and break it; as the material is broken, the undersize passes through the apertures.

rotary current [OCEANOGR] A current with the direction of flow rotating through all points of the compass.

rotary dump car [MIN ENG] A small mine car in which the car body is mounted on a rotary dumper.

rotary dumper [MIN ENG] A steel structure on which a mine car revolves and discharges the contents, usually sideways.

rotary fault [GEOL] A fault in which displacement is downward at one point and upward at another point. Also known as pivotal fault; rotational fault.

rotary rig   [PETRO ENG]   The collective equipment used with a rotary drill; includes prime movers or engines, derrick or mast, hoisting and rotating equipment, drill pipe, drill collars and bit, and the mud system used to circulate drilling fluid.

rotary table   [PETRO ENG]   A circular unit on the floor of a derrick which rotates the drill pipe and bit.

rotary vibrating tippler   [MIN ENG]   A tippler designed to overcome the tendency for coal or dirt to stick to the bottom of the tubs so that when the tippler is inverted, the car rests upon a vibrating frame which frees any material tending to stick to the bottom.

rotating-beam ceilometer   [ENG]   An electronic, automatic-recording meteorological device which determines cloud height by means of triangulation.

rotating models   [OCEANOGR]   Laboratory models for studying ocean currents, the models being rotated to simulate in part the earth's rotation.

rotational fault *See* rotary fault.

rotational wave *See* shear wave; S wave.

rotation anemometer   [ENG]   A type of anemometer in which the rotation of an element serves to measure the wind speed; rotation anemometers are divided into two classes: those in which the axis of rotation is horizontal, as exemplified by the windmill anemometer; and those in which the axis is vertical, such as the cup anemometer.

rotation axis   [CRYSTAL]   A symmetry element of certain crystals in which the crystal can be brought into a position physically indistinguishable from its original position by a rotation through an angle of $360°/n$ about the axis, where n is the multiplicity of the axis, equal to 2, 3, 4, or 6. Also known as symmetry axis.

rotation-inversion axis   [CRYSTAL]   A symmetry element of certain crystals in which a crystal can be brought into a position physically indistinguishable from its original position by a rotation through an angle of $360°/n$ about the axis followed by an inversion, where n is the multiplicity of the axis, equal to 1, 2, 3, 4, or 6. Also known as inversion axis.

rotation-reflection axis   [CRYSTAL]   A symmetry element of certain crystals in which a crystal can be brought into a position physically indistinguishable from its original position by a rotation through an angle of $360°/n$ about the axis followed by a reflection in the plane perpendicular to the axis, where n is the multiplicity of the axis, equal to 1, 2, 3, 4, or 6.

rotation twin   [CRYSTAL]   A twin crystal in which the parts will coincide if one part is rotated 180° (sometimes 30, 60, or 120°).

rotenturm wind   [METEOROL]   A warm south wind blowing through Rotenturm Pass in the Transylvanian Alps.

Rotliegende   [GEOL]   A European series of geologic time: Lower and Middle Permian.

rotor cloud   [METEOROL]   Turbulent, altocumulus-type cloud formation found in the lee of some large mountain barriers, particularly in the Sierra Nevadas near Bishop, California; the air in the cloud rotates around an axis parallel to the range. Also known as roll cloud.

rotoreflection axis   [CRYSTAL]   A type of symmetry element that combines a rotation of 60, 90, 120, or 180° with reflection across the plane perpendicular to the axis. Also known as rotary reflection axis.

rotten ice   [HYD]   Any piece, body, or area of ice which is in the process of melting or disintegrating; it is characterized by honeycomb structure, weak bonding between crystals, or the presence of meltwater or sea water between grains. Also known as spring sludge.

rotten spot *See* pothole.

rougemontite [GEOL]   A coarse-grained igneous rock composed of anorthite, titanaugite, and small amounts of olivine and iron ore.

rougher cell [MIN ENG]   Flotation cells in which the bulk of the gangue is removed from the ore.

roughness elements [OCEANOGR]   Structures attached to laboratory models to simulate the roughness of the ocean floor.

roundness [GEOL]   The degree of abrasion of sedimentary particles; expressed as the radius of the average radius of curvature of the edges or corners to the radius of curvature of the maximum inscribed sphere.

round wind [METEOROL]   A wind that gradually changes direction through approximately 180° during the daylight hours.

route component [METEOROL]   The average forecast wind component parallel to the flight path at flight level for an entire route; it is positive if helping (tailwind), and negative if retarding (headwind).

route forecast [METEOROL]   An aviation weather forecast for one or more specified air routes.

routivarite [GEOL]   A fine-grained igneous rock containing orthoclase, plagioclase, quartz, and garnet.

rouvillite [GEOL]   A light-colored theralite composed predominantly of labradorite and nephiline, with small amounts of titanaugite, hornblende, pyrite, and apatite.

rouvite [MINERAL] $CaU_2V_{12}O_{36}\cdot20H_2O$ A purplish- to bluish-black mineral consisting of a hydrated vanadate of calcium and uranium; occurs as dense masses, crusts, and coatings.

roweite [MINERAL] $(Mn,Mg,Zn)Ca(BO_2)_2(OH)_2$ A light-brown mineral composed of basic borate of calcium, manganese, magnesium, and zinc.

row shooting [MIN ENG]   Setting off a row of holes nearest the face first, and then other rows in succession behind it.

R tectonite [PETR]   A tectonite in which the fabric is believed to have resulted from rotation.

rubber ice [OCEANOGR]   Newly formed sea ice which is weak and elastic.

rubble [GEOL]   **1.** A loose mass of rough, angular rock fragments, coarser than sand. **2.** *See* talus. [HYD]   Fragments of floating or grounded sea ice in hard, roughly spherical blocks measuring 0.5–1.5 meters (1.5–4.5 feet) in diameter, and resulting from the breakup of larger ice formations. Also known as rubble ice.

rubble drift [GEOL]   **1.** A rubbly deposit (or congeliturbate) formed by solifluction under periglacial conditions. **2.** A coarse mass of angular debris and large blocks set in an earthy matrix of glacial origin.

rubble ice *See* rubble.

rubble tract [GEOL]   The part of the reef flat immediately behind and on the lagoon side of the reef front, paved with cobbles, pebbles, blocks, and other coarse reef fragments.

rubellite [MINERAL]   The red to red-violet variety of the gem mineral tourmaline; hardness is 7–7.5 on Mohs scale, and specific gravity is near 3.04.

rubicelle [MINERAL]   A yellow or orange-red gem variety of spinel.

rubidium magnetometer *See* rubidium-vapor magnetometer.

rubidium-strontium dating [GEOL] A method for determining the age of a mineral or rock based on the decay rate of rubidium-87 to strontium-87.

rubidium-vapor magnetometer [ENG] A highly sensitive magnetometer in which the spin precession principle is combined with optical pumping and monitoring for detecting and recording variations as small as 0.01 gamma (0.1 microoersted) in the total magnetic field intensity of the earth. Also known as rubidium magnetometer.

ruby [MINERAL] The red variety of the mineral corundum; in its finest quality, the most valuable of gemstones.

ruby copper ore *See* cuprite.

ruby mica [MINERAL] The finest grade of Indian mica; used for electrical capacitors.

ruby silver [MINERAL] Either of two red silver sulfide minerals: pyrorgyrite (dark-ruby silver) and proustite (light-ruby silver).

ruby spinel [MINERAL] A clear-red gem variety of spinel, containing small amounts of chromium and having the color but none of the other attributes of true ruby.

ruby zinc *See* zincite.

rudaceous [PETR] Of or pertaining to a sedimentary rock composed of a large quantity of fragments that are larger than sand grains (diameter greater than 2 millimeters).

rudistids [PALEON] Fossil sessile bivalves that formed reefs during the Cretaceous in the southern Mediterranean or the Tethyan belt.

rudite [GEOL] A sedimentary rock composed of fragments coarser than sand grains.

Rudzki anomaly [GEOPHYS] A gravity anomaly calculated by replacing the surface topography by its mirror image within the geoid.

ruggedness number [GEOL] A dimensionless number that expresses the geometric characteristics of a drainage system; derived from the product of maximum basin relief and drainage density within the drainage basin.

Rugosa [PALEON] An order of extinct corals having either simple or compound skeletons with internal skeletal structures consisting mainly of three elements, the septa, tabulae, and dissepiments.

ruin agate [MINERAL] A brown variety of agate displaying, on a polished surface, markings that resemble or suggest the outlines of ruins or ruined buildings.

ruin marble [PETR] A brecciated limestone that, when cut and polished, gives a mosaic effect suggesting the appearance of ruins or ruined buildings.

rule of approximation [MIN ENG] A rule applicable to placer mining locations and entries upon surveyed lands, to be applied on the basis of 10-acre (40,469-square-meter) legal subdivisions.

run [GEOL] **1.** A ribbonlike, flat-lying, irregular orebody following the stratification of the host rock. **2.** A branching or fingerlike extension of the feeder of an igneous intrusion. [MIN ENG] *See* slant.

runaround [MIN ENG] A bypass driven in the shaft pillar to permit safe passage from one side of the shaft to the other.

run in [ENG] To lower the assembled drill rods and auxiliary equipment into a borehole.

runite *See* graphic granite.

runnel [GEOL] A troughlike hollow on a tidal sand beach which carries water drainage off the beach as the tide retreats.

runner [MIN ENG] A vertical timber sheet pile used to prevent collapse of an excavation.

running sand *See* quicksand.

runoff [HYD] **1.** Surface streams that appear after precipitation. **2.** The flow of water in a stream, usually expressed in cubic feet per second; the net effect of storms, accumulation, transpiration, meltage, seepage, evaporation, and percolation. [MIN ENG] Collapse of a coal pillar in a mine.

runoff coefficient [HYD] The percentage of precipitation that appears as runoff.

runoff cycle [HYD] The part of the hydrologic cycle involving water between the time it reaches the land as precipitation and its subsequent evapotranspiration or runoff.

runoff desert [ECOL] An arid region in which local rain is insufficient to support any perennial vegetation except in drainage or runoff channels.

runoff intensity [HYD] The excess of rainfall intensity over infiltration capacity, usually expressed in inches of rainfall per hour. Also known as runoff rate.

runoff pit [MIN ENG] Catchment area to which spillage from classifiers, thickeners, and slurry pumps can gravitate if it becomes necessary to dump their contents. Also known as spill pit.

runoff rate *See* runoff intensity.

run-of-mill [MIN ENG] Ore accepted for treatment, after waste and dense media rejection. Also known as mill-head ore.

run-of-mine *See* mine run.

run of the coast [GEOGR] The trend of the coast.

run-up *See* swash.

runway observation [METEOROL] An evaluation of certain meteorological elements observed at a specified point on or near an airport runway; temperature, wind speed and direction, ceiling, and visibility are among the elements frequently observed at such locations, because of the importance of these data to aircraft landing and takeoff operations.

runway temperature [METEOROL] The temperature of the air just above the runway at an airport (usually at about 4 feet but ideally at engine or wing height), used in the determination of density altitude; therefore, runway temperature observations are made and reported at airports when critical values of density altitude prevail.

runway visibility [METEOROL] The visibility along an identified runway, determined from a specified point on the runway with the observer facing in the same direction as a pilot using the runway.

runway visual range [METEOROL] The maximum distance along the runway at which the runway lights are visible to a pilot after touchdown.

Rupelian [GEOL] A European stage of middle Oligocene geologic time, above the Tongrian and below the Chattian. Also known as Stampian.

rupture *See* fracture.

Russell flask [PETRO ENG] Device for volumetric determination of the true volume of sand grains within a unit bulk volume of grains plus voids.

russellite [MINERAL] Bi_2WO_6 A pale yellow to greenish, tetragonal mineral consisting of an oxide of bismuth and tungsten; occurs as fine-grained compact masses.

rutherfordine [MINERAL] $(UO_2)(CO_3)$ A yellow mineral composed of uranyl carbonate, occurring as masses of fibers.

rutilated quartz   [MINERAL]   Sagenitic quartz charactrized by the presence of enclosed needlelike crystals of rutile. Also known as Venus hairstone.

rutile   [MINERAL] TiO_2 A reddish-brown tetragonal mineral common in acid igneous rocks, in metamorphic rocks, and as residual grain in beach sand.

R wave *See* Rayleigh wave.

S

Saalic orogeny [GEOL] A short-lived orogeny that occurred early in the Permian period, between the Autunian and Saxonian stages.

sabach *See* caliche.

Sabinas [GEOL] A North American (Gulf Coast) provincial series in Upper Jurassic gelogic time, below the Coahuilan.

sabkha *See* sebkha.

sabulous *See* arenaceous.

saccharoidal [PETR] The texture of a rock that is crystalline or granular. Also known as sucrosic; sugary.

saccus *See* vesicle.

saddle [GEOL] **1.** A gap that is broad and gently sloping on both sides. **2.** A relatively flat ridge that connects the peaks of two higher elevations. **3.** That part along the surface axis or axial trend of an anticline that is a low point or depression.

saddleback [GEOL] A hill or ridge characterized by a concave outline along its crest. [METEOROL] The cloudless air between the "towers" of two cumulus congestus or cumulonimbus clouds and above a lower cloud mass.

saddle fold [GEOL] A flexural fold perpendicular to the parent fold and having an additional flexure at its crest.

saddle point *See* col.

safety board [PETRO ENG] A board placed in a derrick for a man to stand on when handling drill rods at single, double, triple, or quadruple levels; the boards are placed at suitable heights to handle a stand of drill rods for that number of joints.

safety cable [MIN ENG] A mining machine cable designed to cut off power when the positive conductor insulation is damaged.

safety cage [MIN ENG] A cage, box, or platform used for lowering and hoisting miners, tools, and equipment into and out of mines.

safety car [MIN ENG] Any mine car or hoisting cage provided with safety stops, catches, or other precautionary devices.

safety chain [MIN ENG] A chain connecting the first and last cars of a trip to prevent separation, if a coupling breaks.

safety door [MIN ENG] An extra door ready for use in the event of damage to the existing ventilation door or in any emergency, for example, explosion or fire.

safety gate [MIN ENG] An automatically operated gate at the top of a mine shaft or at landings both to guard the entrance and to prevent falling into the shaft.

safety lamp [MIN ENG] In coal mining, a lamp that is relatively safe to use in atmospheres which may contain flammable gas.

safety post [MIN ENG] A timber placed near the face of workings to protect the workmen. Also known as safety prop.

safety prop *See* safety post.

safflorite [MINERAL] $CoAs_2$ A cobalt arsenide mineral that occurs in tin-white masses, and is dimorphous with smaltite; found in Canada, Morocco, and the United States.

sag [GEOL] **1.** A pass or gap in a ridge or mountain range shaped like a saddle. **2.** A shallow depression in a relatively flat land surface. **3.** A regional basin with gently sloping sides.

sag bolt [MIN ENG] A device to measure roof sag; a 12-foot (3.7-meter) unit installed without a bearing plate and securely anchored with the aid of a heavy nut, extending about 2 inches (5 centimeters) from the hole; three ½-inch (1.3 centimeter) strips of colored tape wrapped around the extending section of the bolt, green at the roof line followed by yellow and then red, help detect roof sag at a glance.

sagenite [MINERAL] A variety of rutile that is acicular and occurs in reticulated twin groups of crystals crossing at 60°.

sagenitic [GEOL] Containing acicular minerals.

Sagenocrinida [PALEON] A large order of extinct, flexible crinoids that occurred from the Silurian to the Permian.

Saghathiinae [PALEON] An extinct subfamily of hyracoids in the family Procaviidae.

sahel [METEOROL] A strong dust-bearing desert wind in Morocco.

sahlinite [MINERAL] $Pb_{14}(AsO_4)_2O_9Cl_4$ A pale sulfur yellow, monoclinic mineral consisting of a basic chloride-arsenate of lead; occurs in aggregates of small scales.

sahlite *See* salite.

Saint Peter sandstone [GEOL] An artesian aquifer of early Lower Paleozoic age which underlies part of Minnesota, Wisconsin, Iowa, Illinois, and Indiana.

Sakmarian [GEOL] A European stage of geologic time; the lowermost Permian, above Stephanian of Carboniferous and below Artinskian.

sal *See* sial.

salable coal [MIN ENG] Total output of a coal mine, less the tonnage rejected or consumed during preparation for market.

Salado formation [GEOL] A red-bed formation from the Permian found in southeast New Mexico; contains rock salt and potash salts.

salammoniac [MINERAL] NH_4Cl A white, isometric, crystalline mineral composed of native ammonium chloride.

salband *See* selvage.

salcrete [GEOL] A thin, hard crust of salt-cemented sand grains, occurring on a marine beach that is occasionally or periodically saturated by saline water.

saléeite [MINERAL] $Mg(UO_2)_2(PO_4)_2 \cdot 10H_2O$ A lemon-yellow mineral composed of hydrous phosphate of magnesium and uranium.

salesite [MINERAL] $Cu(IO_3)(OH)$ A bluish-green mineral composed of basic iodate of copper.

salfemic rock [GEOL] An igneous rock in which the ratio of salic to femic minerals is greater than 3:5 and less than 5:3.

salic [GEOL] A soil horizon enriched with secondary salts, at least 2 percent, and measuring at least 15 centimeters in thickness. [MINERAL] Pertaining to certain light-colored minerals, such as quartz and feldspars, that are rich in silica or magnesium and commonly occur in igneous rock.

salient [GEOL] **1.** A landform that projects or extends outward or upward from its surroundings. **2.** An area in which the axial traces of folds are convex toward the outer edge of the folded belt.

saliferous stratum [GEOL] A stratum that contains, produces, or is impregnated with salt. Also known as saliniferous stratum.

salina [GEOL] An area, such as a salt flat, in which deposits of crystalline salts are formed or found. [HYD] A body of water containing high concentrations of salt.

salinastone [GEOL] A sedimentary rock composed mostly of saline minerals which are usually precipitated but may be fragmental.

saline-alkali soil [GEOL] A salt-affected soil with a content of exchangeable sodium greater than 15%, with much soluble salts, and with a pH value usually less than 9.5.

salinelle [GEOL] A mud volcano erupting saline mud.

saline soil [GEOL] A nonalkali, salt-affected soil with a high content of soluble salts, with exchangeable sodium of less than 15%, and with a pH value less than 8.5.

saliniferous stratum *See* saliferous stratum.

salinity [OCEANOGR] The total quantity of dissolved salts in sea water, measured by weight in parts per thousand.

salinity current [OCEANOGR] A density current in the ocean whose flow is caused, controlled, or maintained by its relatively greater density due to excessive salinity.

salinity logging [PETRO ENG] Technique for measurement and recording of saltwater-bearing zones in an oil or gas reservoir; uses a combination of neutron logging with a chlorine curve.

salinity-temperature-depth recorder [ENG] An instrument consisting of sensing elements usually lowered from a stationary ship, and a recorder on board which simultaneously records measurements of temperature, salinity, and depth. Also known as CTD recorder; STD recorder.

salinization [GEOL] In a soil of an arid, poorly drained region, the accumulation of soluble salts by the evaporation of the waters that bore them to the soil zone.

salinometer [ENG] An instrument that measures water salinity by means of electrical conductivity or by a hydrometer calibrated to give percentage of salt directly.

salite [MINERAL] $(Mg,Fe)_2Si_2O_6$ A grayish-green to black mineral variety of diopside containing more magnesium than iron; member of the clinopyroxene group. Also spelled sahlite.

salitrite [PETR] A lamprophyre composed chiefly of titanite and diopside with acmite, accessory apatite, microcline, and occasionally anorthoclase and baddeleyite.

salmonsite [MINERAL] A buff-colored mineral composed of hydrous phosphate of manganese and iron occurring in cleavable masses.

salt [MIN ENG] **1.** To introduce extra amounts of a valuable or waste mineral into a sample to be assayed. **2.** To artificially enrich, as a mine, usually with fraudulent intent.

salt-affected soil [GEOL] A general term for a soil that is not suitable for the growth of crops because of an excess of salts, exchangeable sodium, or both.

salt-and-pepper sand [GEOL]   A sand composed of a mixture of light- and dark-colored grains.

salt anticline [GEOL]   A structure like a salt dome but with a linear salt core. Also known as salt wall.

saltation [GEOL]   Transport of a sediment in which the particles are moved forward in a series of short intermittent bounces from a bottom surface.

saltation load [GEOL]   The part of the bed load that is bouncing along the stream bed or is moved, directly or indirectly, by the impact of bouncing particles.

salt burst [GEOL]   Rock destruction caused by crystallization of soluble salts that enter the pores.

salt crust [HYD]   A salt deposit formed on an ice surface by crystal growth forcing salt out of young sea ice and pushing it upward.

salt desert [GEOL]   A desert with a salt-bearing soil.

salt dome [GEOL]   A diapiric or piercement structure in which there is a central, equidimensional salt plug.

salt-dome breccia [GEOL]   A breccia found in deep shale sequences and occurring as a dome-shaped mass in a broad zone surrounding a salt plug.

saltern *See* saltworks.

salt field [GEOL]   An area overlying a usually workable salt deposit of economic value.

salt flat [GEOL]   The level, salt-encrusted bottom of a lake or pond that is temporarily or permanently dried up.

salt flowers *See* ice flowers.

salt glacier [GEOL]   A gravitational flow of salt down the slopes of a salt plug, following the preexisting structure.

salt haze [METEOROL]   A haze created by the presence of finely divided particles of sea salt in the air, usually derived from the evaporation of sea spray.

salt hill [GEOL]   An abrupt hill of salt, with sinkholes and pinnacles at its summit.

saltierra [GEOL]   A deposit of salt left by evaporation of a shallow salt lake.

salt lake [HYD]   A confined inland body of water having a high concentration of salts, principally sodium chloride.

salt mine [MIN ENG]   A mine containing deposits of rock salt.

salt-mud combination log [PETRO ENG]   Record of electrical logging of the mud in oil well boreholes in the presence of sodium chloride incursions from adjacent formations.

saltpeter cave [GEOL]   A cave in which there are deposits of saltpeter earth.

saltpeter earth [GEOL]   A deposit containing calcium nitrate and found in caves.

salt pillow [GEOL]   An embryonic salt dome rising from its source bed, still at depth.

salt pit [GEOL]   A pit in which sea water is received and evaporated and from which salt is obtained.

salt plug [GEOL]   The salt core of a salt dome.

salt polygon [GEOL]   A surface of salt on a playa, having three to eight sides marked by ridges of material formed as a result of the expansive forces of crystallizing salt, and ranging in width from several centimeters to 30 meters.

salt tectonics [GEOL]   The study of the structure and mechanism of emplacement of salt domes.

salt wall *See* salt anticline.

salt water *See* seawater.

salt-water front [OCEANOGR] The interface between fresh and salt water in a coastal aquifer or in an estuary.

salt-water intrusion [HYD] Displacement of fresh surface water or groundwater by salt water due to its greater density.

salt-water underrun [OCEANOGR] A type of density current occurring in a tidal estuary, due to the greater salinity of the bottom water.

salt-water wedge [OCEANOGR] A wedge-shaped intrusion of salty ocean water into a fresh-water estuary or tidal river; it slopes downward in the upstream direction, and salinity increases with depth.

salt-water well [PETRO ENG] A well from which salt water flows after the petroleum contents are depleted.

salt weathering [GEOL] The granular disintegration or fragmentation of rock material produced by saline solutions or by salt-crystal growth.

salt wedge [OCEANOGR] A wedge-shaped mass of salt water from an ocean or sea which intrudes the mouth and lower course of a river.

salt well [ENG] A bored or driven well from which brine is obtained.

saltworks [ENG] A building or group of buildings where salt is produced commercially, as by extraction from sea water or from the brine of salt springs. Also known as salina; saltern.

samarskite [MINERAL] $(Y,Ce,U,Ca,Fe,Pb,Th)(Nb,Ta,Ti,Sn)_2O_6$ A velvet-black to brown metamict orthorhombic mineral with splendent vitreous to resinous luster occurring in granite pegmatites. Also known as ampangabeite; uranotantalite.

sampled grade [MIN ENG] The amount of valuable metal in the ore in place as determined by underground, surface, or drill-hole sampling.

sampleite [MINERAL] $NaCaCu_5(PO_4)_4Cl\cdot5H_2O$ A blue mineral composed of hydrous phosphate and chloride of sodium, calcium, and copper.

sampling area ratio [MIN ENG] The volume of the soil displaced in proportion to the volume of the sample; a well-designed tool has an area ratio of about 20 percent.

sampling pipe [MIN ENG] A small pipe built into a seal to take air samples in a sealed area.

sampling spoon [MIN ENG] A cylinder with spoonlike cutting edge for taking soil sam ples.

samsonite [MINERAL] $Ag_4MnSb_2S_6$ A black mineral composed of sulfide of silver, manganese, and antimony occurring in monoclinic prismatic crystals.

sanakite [PETR] A glassy andesite composed of bronzite, augite, magnetite, and a few large plagioclase and garnet crystals.

sanbornite [MINERAL] $BaSi_2O_5$ A white triclinic mineral composed of barium silicate.

sand [GEOL] A loose material consisting of small mineral particles, or rock and mineral particles, distinguishable by the naked eye; grains vary from almost spherical to angular, with a diameter range from $\frac{1}{16}$ to 2 millimeters.

sand apron [GEOL] A deposit of sand along the shore of a lagoon of a reef.

sandarac *See* realgar.

sand auger *See* dust whirl.

sand avalanche [GEOL] Movement of large masses of sand down a dune face when the angle of repose is exceeded or when the dune is disturbed.

sandbank [GEOL] A deposit of sand forming a mound, hillside, bar, or shoal.

sandbar [GEOL] A bar or low ridge of sand bordering the shore and built up, or near, to the surface of the water by currents or wave action. Also known as sand reef.

sandblasting [GEOL] Abrasion affected by the action of hard, windblown mineral grains.

sand boil *See* blowout.

sand cay *See* sandkey.

sand cone [GEOL] **1.** A cone-shaped deposit of sand, produced especially in an alluvial cone. **2.** A low debris cone whose protective veneer consists of sand.

sand count [PETRO ENG] Determination of the total thickness of an oil or gas reservoir's permeable section (excluding shale streaks and other impermeable zones); can be derived from electrical logs.

sand crystal [GEOL] A large crystal loaded up to 60% with detrital sand inclusions formed in a sandstone during or as a result of cementation.

sand devil *See* dust whirl.

sand dike [GEOL] A sedimentary dike consisting of sand that has been squeezed or injected upward into a fissure.

sand drift [GEOL] **1.** Movement of windblown sand along the surface of a desert or shore. **2.** An accumulation of sand against the leeward side of a fixed obstruction.

sand drip [GEOL] A rounded or crescentic surface form on a beach sand, resulting from the sudden absorption of overwash.

sand dune [GEOL] A mound of loose windblown sand commonly found along low-lying seashores above high-tide level.

sand fall [GEOL] An accumulation of sand swept over a cliff or escarpment.

sand flat [GEOL] A sandy tidal flat barren of vegetation.

sand flood [GEOL] A vast body of sand moving or borne along a desert, as in the Arabian deserts.

sand gall *See* sand pipe.

sand glacier [GEOL] **1.** An accumulation of sand that is blown up the side of a hill or mountain and through a pass or saddle, and then spread out on the opposite side to form a wide, fan-shaped plain. **2.** A horizontal plateau of sand terminated by a steep talus slope.

sand-grain volume [PETRO ENG] In oil reservoir porosity calculations, the actual volume filled by sand grains, without allowance for spaces (voids) between the grains.

sand hill [GEOL] A ridge of sand, especially a sand dune in a desert region.

sand hole [GEOL] A small pit (7–8 millimeters in depth and a little less wide than deep) with a raised margin, formed on a beach by waves expelling air from a formerly saturated mass of sand.

sand horn [GEOL] A pointed sand deposit extending from the shore into shallow water.

sandkey [GEOL] A small sandy island parallel with the shore. Also known as sand cay.

sand levee *See* whaleback dune.

sand lobe   [GEOL]   A rounded sand deposit extending from the shore into shallow water.

sand pavement   [GEOL]   A sandy surface derived from coarse-grained sand ripples, developed on the lower, windward slope of a dune or rolling sand area during a period of intermittent light, variable winds.

sand pipe   [GEOL]   A pipe formed in sedimentary rocks, filled with considerable sand and some gravel. Also known as sand gall.

sand plain   [GEOL]   A small outwash plain formed by deposition of sand transported by meltwater streams flowing from a glacier.

sand return   [PETRO ENG]   The return of injected sand to the wellbore following formation fracturing; it constitutes a problem.

sand ridge   [GEOL]   **1.** Any low ridge of sand formed at some distance from the shore, and either submerged or emergent, such as a longshore bar or a barrier beach. **2.** One of a series of long, wide, extremely low, parallel ridges believed to represent the eroded stumps of former longitudinal sand dunes. **3.** A crescent-shaped landform found on a sandy beach, such as a beach cusp. **4.** *See* sand wave.

sand river   [GEOL]   A river that deposits much of its sand load along its middle course, to be subsequently removed by the wind.

sandrock   [GEOL]   A field term for a sandstone that is not firmly cemented.

sand roll *See* pseudonodule.

sand run   [GEOL]   **1.** A fluidlike motion of dry sand. **2.** A mass of dry sand in motion.

sands   [MIN ENG]   The coarser and heavier particles of crushed ore, of such size that they settle readily in water and may be leached by allowing the solution to percolate.

sand sea   [GEOL]   **1.** An extensive assemblage of sand dunes of several types in an area where a great supply of sand is present; characterized by an absence of travel lines, or directional indicators, and by a wavelike appearance of dunes separated by troughs. **2.** The flat, rain-smoothed plain of volcanic ash and other pyroclastics on the floor of a caldera.

sand shadow   [GEOL]   A lee-side accumulation of sand, as a small turret-shaped dune, formed in the shelter of, and immediately behind, a fixed obstruction, such as clumps of vegetation.

sandshale   [GEOL]   A sedimentary deposit consisting of thin alternating beds of sandstone and shale.

sand-shale ratio   [GEOL]   The ratio between the thickness or percentage of sandstone and that of shale in a geologic section.

sand sheet   [GEOL]   A thin accumulation of coarse sand or fine gravel having a flat sur face.

sand snow   [HYD]   Snow that has fallen at very cold temperatures (of the order of $-25°C$); as a surface cover, it has the consistency of dust or light dry sand.

sandspit   [GEOL]   A spit consisting principally of sand.

sand splay   [GEOL]   A floodplain splay consisting of coarse sand particles.

sandstone   [PETR]   A detrital sedimentary rock consisting of individual grains of sand-size particles 0.06 to 2 millimeters in diameter either set in a fine-grained matrix (silt or clay) or bonded by chemical cement.

sandstone-arenite   [GEOL]   A sedarenite composed chiefly of sandstone fragments.

sandstone cylinder *See* sandstone pipe.

sandstone dike [GEOL]   A dike made of sandstone or lithified sand.

sandstone pipe [ECOL]   A clastic pipe consisting of sandstone. Also known as sandstone cylinder.

sandstone sill [GEOL]   A tabular mass of sandstone that has been emplaced by sedimentary injection parallel to the structure or by bedding of preexisting rock in the manner of an igneous sill.

sandstorm [METEOROL]   A strong wind carrying sand through the air, the diameter of most particles ranging from 0.08 to 1 millimeter; in contrast to a duststorm, the sand particles are mostly confined to the lowest 2 meters above ground, rarely rising more than 11 meters.

sand streak [GEOL]   A low, linear ridge formed at the interface of sand and air or water, oriented parallel to the direction of flow, and having a symmetric cross section.

sand stream [GEOL]   A small sand delta spread out at the mouth of a gully, or a deposit of sand along the bed of a small creek, formed by a torrential rain.

sand strip [GEOL]   A long, narrow ridge of sand extending for a long distance downwind from each horn of a dune.

sandur *See* outwash plain.

sandwash [GEOL]   A sandy or gravel stream bed, devoid of vegetation, containing water only during a sudden and heavy rainstorm.

sand wave [GEOL]   A large, ridgelike primary structure resembling a water wave on the upper surface of a sedimentary bed that is formed by high-velocity air or water currents. Also known as sand ridge.

sand wedge [GEOL]   A wedge-shaped accumulation of sand with the apex downward formed by the filling in of winter contraction cracks.

sandy bentonite *See* arkosic bentonite.

sandy chert [PETR]   Chert formed in sandy beds by replacement of cement, or the filling of pore spaces, with silica.

Sangamon [GEOL]   The third interglacial stage of the Pleistocene epoch in North America, following the Illinoian glacial and preceding the Wisconsin.

sanidal [GEOL]   Pertaining to the continental shelf.

sanidine [MINERAL] $KAlSi_3O_8$ An alkali feldspar mineral occurring in clear, glassy crystals embedded in unaltered acid volcanic rocks; a high-temperature, disordered form. Also known as glassy feldspar; ice spar; rhyacolite.

sanidinite [PETR]   A type of igneous rock composed chiefly of sanidine.

sanitary nipper *See* latrine cleaner.

sanmartinite [MINERAL] $ZnWO_4$ A mineral composed of zinc tungstate.

sannaite [PETR]   An extrusive rock containing phenocrysts of barkevikite, pyroxene, and biotite (in order of decreasing abundance) in a fine-grained to dense groundmass of alkali feldspar, acmite, chlorite, calcite, and pseudomorphs of mica after nepheline.

SA node *See* sinoauricular node.

sansar [METEOROL]   A northwest wind of Persia.

sansicl [GEOL]   An unconsolidated sediment, consisting of a mixture of sand, silt, and clay, in which no component forms 50% or more of the whole aggregate.

Santa Ana [METEOROL] A hot, dry, foehnlike desert wind, generally from the northeast or east, especially in the pass and river valley of Santa Ana, California, where it is further modified as a mountain-gap wind.

Santa Rosa storm [METEOROL] In Argentina, an annual storm near the end of August.

Santonian [GEOL] A european stage of geologic time in the Upper Cretaceous, above the Coniacian and below the Campanian.

santorinite [PETR] **1.** A light-colored extrusive rock containing approximately 60–65% silica and calcic plagioclase (labradorite to anorthite) as the only feldspar. **2.** A hypersthene andesite containing plagioclase crystals that have labradorite cores and sodic rims and a groundmass with microlites of sodic oligoclase.

sanukite [PETR] An andesite characterized by orthopyroxene as the mafic mineral, andesine as the plagioclase, and a glassy groundmass.

sanukitoid *See* orthoandesite.

saponite [MINERAL] A soft, soapy, white or light-buff to bluish or reddish trioctahedral montmorillonitic clay mineral consisting of hydrous magnesium aluminosilicate and occurring in masses in serpentine and basaltic rocks. Also known as bowlingite; mountain soap; piotine; soapstone.

sappare *See* kyanite.

sapphire [MINERAL] Any of the gem varieties of the mineral corundum, especially the blue variety, except those that have medium to dark tones of red that characterize ruby; hardness is 9 on Mohs scale, and specific gravity is near 4.00.

sapphire quartz [MINERAL] An indigo-blue opaque variety of quartz.

sapphirine [MINERAL] $(MgFe)_{15}(Al,Fe)_{34}Si_7O_{80}$ A green or pale-blue mineral composed of silicate and oxide of magnesium, iron, and aluminum; usually occurs in granular form.

sapping [GEOL] Erosion along the base of a cliff by the wearing away of softer layers, thus removing the support for the upper mass which breaks off into large blocks and falls from the cliff face. Also known as undermining.

Saprist [GEOL] A suborder of the soil order Histosol consisting of residues in which plant structures have been largely obliterated by decay; saturated with water most of the time.

saprogenous ooze [GEOL] Ooze formed of putrefying organic matter.

saprolite [GEOL] A soft, earthy red or brown, decomposed igneous or metamorphic rock that is rich in clay and formed in place by chemical weathering. Also known as saprolith; sathrolith.

saprolith *See* saprolite.

sapropel [GEOL] A mud, slime, or ooze deposited in more or less open water.

sapropel-clay [GEOL] A sedimentary deposit in which the amount of clay is greater than that of sapropel.

sapropelic coal [GEOL] Coal formed by putrefaction of organic matter under anaerobic conditions in stagnant or standing bodies of water. Also known as sapropelite.

sapropel-peat *See* peat-sapropel.

Saracen stone *See* sarsen.

sàrca [METEOROL] A violent north wind of Lake Garda in Italy.

sarcopside [MINERAL] $(Fe,Mn,Mg)_3(PO_4)_2$ A mineral composed of a phosphate of manganese, magnesium, and iron.

sard [MINERAL] A translucent brown, reddish-brown, or deep orange-red variety of chalcedony. Also known as sardine; sardius.

Sardic orogeny [GEOL] A short-lived orogeny that occurred near the end of the Cambrian period.

sardine *See* sard.

sardius *See* sard.

sardonyx [MINERAL] An onyx characterized by parallel layers of sard, a deep orange-red variety of chalcedony, and a mineral of different color.

Sargasso Sea [GEOGR] A region of the North Atlantic Ocean; boundaries are defined in the west and north by the Gulf Stream, in the east by longitude 40°W, and in the south by latitude 20°N.

sarkinite [MINERAL] $Mn_2(AsO_4)(OH)$ A flesh-red monoclinic mineral composed of hydrous manganese arsenate, occurring in crystals.

Sarmatian [GEOL] A European stage of geologic time: the upper Miocene, above Tortonian, below Pontian.

sarmientite [MINERAL] $Fe_2(AsO_4)(SO_4)(OH)\cdot 5H_2O$ A yellow mineral composed of basic hydrous arsenate and sulfate of iron; it is isomorphous with diadochite.

sarnaite [GEOL] A feldspathoid-bearing syenite composed of cancrinite and acmite.

sarospatakite [GEOL] A micaceous clay mineral composed of mixed layers of illite and montmorillonite.

sarsden stone *See* sarsen.

sarsen [GEOL] A large residual mass of stone left after the erosion of a once continuous bed of which it formed a part. Also known as Saracen stone; sarsden stone.

sartorite [MINERAL] $PbAs_2S_4$ A dark-gray monoclinic mineral, occurring in crystalline form.

sassoline *See* sassolite.

sassolite [MINERAL] H_3BO_3 A white or gray mineral consisting of native boric acid usually occurring in small pearly scales as an incrustation or as tabular triclinic crystals. Also known as sassoline.

sastruga [HYD] A ridge of snow up to 2 inches (5 centimeters) high formed by wind erosion and aligned parallel to the wind. Also known as skavl; zastruga.

satellitic crater *See* secondary crater.

sathrolith *See* saprolite.

satin ice *See* acicular ice.

satin spar [MINERAL] A white, translucent, fine fibrous variety of gypsum having a silky luster. Also known as satin stone.

satin stone *See* satin spar.

saturated air [METEOROL] Moist air in a state of equilibrium with a plane surface of pure water or ice at the same temperature and pressure; that is, air whose vapor pressure is the saturation vapor pressure and whose relative humidity is 100%.

saturated mineral [MINERAL] A mineral that forms in the presence of free silica.

saturated permafrost [GEOL] Permafrost that contains no more ice than the ground could hold if the water were in the liquid state.

saturated rock [PETR]   An igneous rock composed principally of saturated minerals.

saturated surface *See* water table.

saturated zone *See* zone of saturation.

saturation adiabat [METEOROL]   On a thermodynamic diagram, a line of constant wet-bulb potential temperatures; in practice, approximate computations are usually employed, and the resulting lines represent, ambiguously, saturation adiabats and pseudoadiabats. Also known as moist adiabat; wet adiabat.

saturation-adiabatic lapse rate [METEOROL]   A special case of process lapse rate, defined as the rate of decrease of temperature with height of an air parcel lifted in a saturation-adiabatic process through an atmosphere in hydrostatic equilibrium. Also known as moist-adiabatic lapse rate.

saturation-adiabatic process [METEOROL]   An adiabatic process in which the air is maintained at saturation by the evaporation or condensation of water substance, the latent heat being supplied by or to the air respectively; the ascent of cloudy air, for example, is often assumed to be such a process.

saturation curve [GEOL]   A curve showing the weight of solids per unit volume of a saturated soil mass as a function of water content.

saturation deficit [METEOROL]   **1.** The difference between the actual vapor pressure and the saturation vapor pressure at the existing temperature. **2.** The additional amount of water vapor needed to produce saturation at the current temperature and pressure, expressed in grams per cubic meter. Also known as vapor-pressure deficit.

saturation line [HYD]   The boundary on a glacier between the soaked zone and the percolation zone.

saturation mixing ratio [METEOROL]   A thermodynamic function of state; the value of the mixing ratio of saturated air at the given temperature and pressure; this value may be read directly from a thermodynamic diagram.

saturation ratio [METEOROL]   The ratio of the actual specific humidity to the specific humidity of saturated air at the same temperature.

sauconite [MINERAL]   The zinc-bearing end member of the montmorillonite group; a trioctahedral clay mineral.

saucyite [PETR]   A glassy rhyolitic rock composed of large sanidine phenocrysts in a groundmass of orthoclase microlites and minute crystals of biotite, augite, sphene, zircon, and magnetite.

sault [HYD]   A waterfall or rapids in a stream.

Saurichthyidae [PALEON]   A family of extinct chondrostean fishes bearing a superficial resemblance to the Aspidorhynchiformes.

Saurischia [PALEON]   The lizard-hipped dinosaurs, an order of extinct reptiles in the subclass Archosauria characterized by an unspecialized, three-pronged pelvis.

Sauropoda [PALEON]   A group of fully quadrupedal, seemingly herbivorous dinosaurs from the Jurassic and Cretaceous periods in the suborder Sauropodomorpha; members had small heads, spoon-shaped teeth, long necks and tails, and columnar legs.

Sauropodomorpha [PALEON]   A suborder of extinct reptiles in the order Saurischia, including large, solid-limbed forms.

Sauropterygia [PALEON]   An order of Mesozoic marine reptiles in the subclass Eur yapsida.

saussurite [MINERAL]   A white or grayish, tough, compact mineral aggregate composed chiefly of a mixture of albite or oligoclase and zoisite or epidote.

saussuritization [GEOL]   A metamorphic process involving replacement of plagioclase in basalts and gabbros by a fine-grained aggregate of zoisite, epidote, albite, calcite, sericite, and zeolites.

savanna climate *See* tropical savanna climate.

Savic orogeny [GEOL]   A short-lived orogeny that occurred in late Oligocene geologic time, between the Chattian and Aquitanian stages.

saw-cut [GEOL]   A large canyon that cuts abruptly across a terrace, so that it is visible only from places almost at its edge.

sawtooth blasting [MIN ENG]   The cutting of a series of slabs which, in plan, resemble sawteeth by blasting oblique, horizontal holes along a face.

sawtooth floor channeling [MIN ENG]   A method of channeling inclined beds of marble by removing right-angle blocks in succession from the various beds, thus giving the floor a zigzag or sawtooth appearance.

sawtooth stoping [MIN ENG]   In the United States, overhand stoping in which the line of advance is up the dip, and benches are advanced in a line parallel with the drift.

Saxonian [GEOL]   A European stage of geologic time in the Middle Permian, above the Autunian and below the Thuringian.

scabland [GEOL]   Elevated land that is essentially flat-lying and covered with basalt and has only a thin soil cover, sparse vegetation, and usually deep, dry channels.

scabrock [GEOL]   **1.** An outcropping of scabland. **2.** Weathered material of a scabland surface.

scacchite [MINERAL] $MnCl_2$ A mineral composed of native manganese chloride, found in volcanic regions.

scaglia [GEOL]   A dark, very-fine-grained, somewhat calcareous shale usually developed in the Upper Cretaceous and Lower Tertiary periods of the northern Apennines.

scale height [GEOPHYS]   A measure of the decrease of atmospheric pressure with height; when the atmospheric temperature is constant with height, the pressure varies exponentially with height, and the scale height is the height interval over which the pressure changes by a factor of $1/e$.

scales of motion [OCEANOGR]   A series of increasing characteristic magnitudes of motion, ranging from tiny eddies of turbulence to oceanwide currents, each member of the series interacting with the adjacent members.

scaling [MIN ENG]   Removing loose rocks and coal from the roof, walls, or face after blasting.

scalped anticline *See* breached anticline.

scapolite [MINERAL]   A white, gray, or pale-green complex aluminosilicate of sodium and calcium belonging to the tectosilicate group of silicate minerals; crystallizes in the tetragonal system and is vitreous; hardness is 5–6 on Mohs scale, and specific gravity is 2.65–2.74. Also known as wernerite.

scapolitization [GEOL]   Introduction of or replacement by scapolite.

scar [GEOL]   **1.** A steep, rocky eminence, such as a cliff or precipice, where bare rock is well exposed. Also known as scaur; scaw. **2.** *See* shore platform.

scarp *See* escarpment.

scarped plain [GEOL] A terrain characterized by a succession of faintly inclined or gently folded strata.

scarp face *See* scarp slope.

scarp-foot spring [HYD] A spring that flows onto the land surface at or near the foot of an escarpment.

scarpland [GEOGR] A region marked by a succession of nearly parallel cuestas separated by lowlands.

scarplet *See* piedmont scarp.

scarpline [GEOL] A relatively straight line of cliffs of considerable extent, produced by faulting or erosion along a fault.

scarp slope [GEOL] The steep face of a cuesta, or asymmetric ridge, facing in an opposite direction to the dip of the strata. Also known as front slope; inface; scarp face.

scarp stream [HYD] An obsequent stream flowing down a scarp, such as down the scarp slope of a cuesta.

scatter diagram *See* point diagram.

scattered [METEOROL] Descriptive of a sky cover of 0.1 to 0.5 (5 to 54%), applied only when clouds or obscuring phenomena aloft are present, not applied for surface-based obscuring phenomena.

scattering layer [OCEANOGR] A layer of organisms in the sea which causes sound to scatter and to return echoes.

scaur *See* scar.

scavenger well [HYD] A well located between a source of potential contamination and a well (or group of wells) yielding usable water, so that the former well can be pumped (or allowed to flow) as waste to prevent the contaminated water from reaching the good wells.

scaw *See* scar.

schafarzikite [MINERAL] $Fe_5Sb_4O_{11}$ A red to brown mineral composed of iron antimony oxide.

schairerite [MINERAL] $Na_3(SO_4)(F,Cl)$ A colorless rhombohedral mineral composed of sodium sulfate with fluorine and chlorine, occurring in crystals.

schalstein [PETR] A slaty rock formed by shearing basaltic or andesitic tuff or lava.

scharnitzer [METEOROL] A cold, northerly wind of long duration in Tyrol, Austria.

scheelite [MINERAL] $CaWO_4$ A yellowish-white mineral crystallizing in the tetragonal system and occurring in tabular or massive form in pneumatolytic veins associated with quartz; an ore of tungsten.

schefflerite [MINERAL] $(Ca,Mn)(Mg,Fe,Mn)-Si_2O_6$ Brown to black variety of pyroxene that crystallizes in the monoclinic system and contains manganese and frequently iron.

scheteligite [MINERAL] $(Ca,Y,Sb,Mn)_2(Ti,Ta,Nb,W)_2O_6(O,OH)$ A mineral composed of oxide of calcium, rare-earth metals, antimony, manganese, titanium, columbium, and tantalum.

schirmerite [MINERAL] $PbAg_4Bi_4S_9$ A mineral composed of lead, silver, and bismuth sulfide.

schist [GEOL] A large group of coarse-grained metamorphic rocks which readily split into thin plates or slabs as a result of the alignment of lamellar or prismatic minerals.

schist-arenite [PETR]   A light-colored sandstone containing more than 20% rock fragments derived from an area of regionally metamorphosed rocks.

schistose [GEOL]   Pertaining to rocks exhibiting schistosity.

schistosity [GEOL]   A type of cleavage characteristic of metamorphic rocks, notably schists and phyllites, in which the rocks tend to split along parallel planes defined by the distribution and parallel arrangement of platy mineral crystals.

Schlernwind [METEOROL]   East wind blowing down from the Schlern near Bozen in Tyrol, Austria.

schlieren [PETR]   Irregular streaks with shaded borders in some igneous rocks, representing the segregation of light and dark minerals or altered inclusions, elongated by flow.

schlieren arch [GEOL]   An intrusive igneous body with flow layers which occur along its borders but which are poorly developed or absent in its interior.

schlieren dome [GEOL]   An intrusive body more or less completely outlined by flow layers which culminate in one central area.

Schlumberger dipmeter [ENG]   An instrument that measures both the amount and direction of dip by readings taken in the borehole; it consists of a long, cylindrical body with two telescoping parts and three long, springy metal strips, arranged symmetrically round the body, which press outward and make contact with the walls of the hole.

Schlumberger photoclinometer [ENG]   An instrument that measures simultaneously the amount and direction of the deviation of a borehole; the sonde, designed to lie exactly parallel to the axis of the borehole, is fitted with a small camera on the axis of a graduated glass bowl, in which a steel ball rolls freely and a compass is mounted in gimbals; the camera is electrically operated from the surface and takes a photograph of the bowl, the steel ball marks the amount of deviation, the position in relation to the image of the compass needle gives the direction of deviation.

Schmidt net [GEOL]   A coordinate or reference system used to plot a Schmidt projection.

Schmidt projection [GEOL]   A Lambert azimuthal equal-area projection of the lower hemisphere of a sphere onto the plane of a meridian; used in structural geology.

schoepite [MINERAL] $UO_3 \cdot 2H_2O$ A yellow secondary mineral that is composed of hydrous uranium oxide.

schönfelsite [PETR]   A form of basalt containing embedded crystals of olivine and augite in a complex, dense fine-grained groundmass.

Schönflies crystal symbols [CRYSTAL]   Symbols denoting the 32 crystal point groups or symmetry classes; capital letters indicate the general type of class, and subscripts the multiplicity of rotation axes and the existence of additional symmetries.

schorl *See* schorlite.

schorlite [MINERAL]   The black, iron-rich, opaque variety of tourmaline. Also known as schorl.

schorlomite [MINERAL] $Ca_3(Fe,Ti)_2(Si,Ti)_3O_{12}$ Black mineral of the garnet group that has a vitreous luster and usually occurs in masses; hardness is 7–7.5 on Mohs scale, and specific gravity is 3.81–3.88.

schott [GEOGR]   A shallow saline lake in southern Tunisia or on the plateaus of northern Algeria, which is usually dry during the summer.

schreibersite [MINERAL] $(Fe,Ni)_3P$ A silver-white to tin-white magnetic meteorite mineral crystallizing in the tetragonal system and occurring in tables or plates as oriented inclusions in iron meteorites. Also known as rhabdite.

schriesheimite [PETR]   An amphibole peridotite that contains diopside.

schroeckingerite [MINERAL] $NaCa_3(UO_2)(CO_3)(SO_4)F\cdot10H_2O$ A yellowish secondary mineral composed of hydrous sodium calcium uranyl carbonate, sulfate, and fluoride.

schrötterite [MINERAL]   An opaline variety of allophane that is rich in aluminum.

schrund line [GEOL]   The base of the bergschrund, or deep crevasse, at a late stage in the excavation of a cirque; the schrund line separates the steep slope of the cirque wall from the gentler slope below.

Schubertellidae [PALEON]   An extinct family of marine protozoans in the superfamily Fusulinacea.

schultenite [MINERAL] $PbHAsO_4$ A colorless mineral composed of lead hydrogen arsenate occurring in tabular orthorhombic crystals.

schungite [GEOL]   Amorphous carbon-rich material in Precambrian schists.

schuppen structure See imbricate structure.

Schwagerinidae [PALEON]   A family of fusulinacean protozoans that flourished during the Early and Middle Pennsylvanian and became extinct during the Late Permian.

schwartzembergite [MINERAL] $Pb_5(IO_3)Cl_3O_3$ A mineral composed of lead iodate, chloride, and oxide.

scissors fault [GEOL]   A fault on which the offset or separation along the strike increases in one direction from an initial point and decreases in the other direction. Also known as differential fault.

sclerotinite [GEOL]   A variety of inertinite composed of fungal sclerotia.

scolecite [MINERAL] $CaAl_2Si_3O_{10}$ A zeolite mineral that occurs in delicate, radiating groups of white fibrous or acicular crystals; sometimes shows wormlike motion upon heating.

scolecodont [PALEON]   Any of the paired, pincerlike jaws occurring as fossils of annelid worms.

scolite [GEOL]   Any of the small tubes in rock believed to be the fossilized burrows of worms.

scoopfish See underway sampler.

scoria [GEOL]   Vesicular, cindery, dark lava formed by the escape and expansion of gases in basaltic or andesitic magma; generally denser and darker than pumice. [MATER]   Refuse after melting metals or reducing ore.

scoria cone [GEOL]   A volcanic cone composed of a vesicular, cindery crust on the surface of lava that is basaltic or andesitic in nature.

scoria mound [GEOL]   A volcanic knoll composed of vesicular, cindery crust on the surface of lava that is basaltic or andesitic in nature.

scoria tuff [GEOL]   A deposit of fragmented scoria in a fine-grained tuff matrix.

scorodite [MINERAL] $FeAsO_4\cdot2H_2O$ A pale leek-green or liver-brown orthorhombic mineral consisting of ferric arsenate; isomorphous with mansfieldite and represents a minor ore of arsenic.

scorzalite [MINERAL] $FeAl_2(PO_4)_2(OH)_2$ A blue mineral composed of basic iron aluminum phosphate; it is isomorphous with lazulite.

Scotch mist [METEOROL] A combination of thick mist (or fog) and heavy drizzle occurring frequently in Scotland and in parts of England.

Scotch pebble [GEOL] A rounded fragment of agate, carnelian, cairngorm, or other varieties of quartz, found in the gravel of parts of Scotland, and used as a semiprecious stone.

Scotch-type volcano [GEOL] A volcanic form characterized by concentric cuestas and produced by cauldron subsidence.

scour *See* tidal scour.

scour and fill [GEOL] The process of first digging out and then refilling a channel instigated by the action of a stream or tide; refers particularly to the process that occurs during a period of flood.

scour channel [GEOL] A large, groovelike erosional feature produced in sediments by scour.

scour depression [GEOL] A crescent-shaped hollow in the stream bed near the outside of the stream's bend, caused by water that scours below the grade of the stream.

scouring [GEOL] **1.** An erosion process resulting from the action of the flow of air, ice, or water. **2.** *See* glacial scour.

scouring velocity [GEOL] The velocity of water which is necessary to dislodge stranded solids from the stream bed.

scour lineation [GEOL] A smooth, low, narrow (2–5 centimeters or 0.8–2.0 inches wide) ridge formed on a sedimentary surface and believed to result from the scouring action of a current of water.

scour mark [GEOL] A current mark produced by the cutting or scouring action of a current flowing over the bottom of a river or body of water.

scourway [GEOL] A channel created by a powerful water current, particularly the temporary channels formed by streams on the edge of a Pleistocene ice sheet.

scout boring [MIN ENG] A bore made to test a geologic formation being prospected.

scout hole [MIN ENG] **1.** A borehole penetrating only the uppermost part of an ore body in order to delineate the surface configuration. **2.** A shallow borehole used to ascertain the presence of ore or to explore an area in a preliminary manner.

scram drive [MIN ENG] Underground drive above the tramming level, along which scrapers move ore to a discharge chute.

scraper ripper [MIN ENG] A piece of strip-mine equipment with teeth on the lip to rip or break the coal and with a flight conveyor to remove the broken coal.

scratch hardness test [MINERAL] A determination of the resistance of a mineral to scratching by testing it with minerals on the Mohs scale.

scree [GEOL] **1.** A mound of loose, angular material, less than 10 centimeters. **2.** *See* talus.

screened pan [METEOROL] An evaporation pan the top of which is covered by wire-mesh screening (¼-inch or 6-millimeter mesh); the screening reduces air circulation and insolation, and results in a pan coefficient nearer to unity than that for unscreened pans.

screen size [MIN ENG] A standard for determining the size of diamond particles; the size of the screened particle is determined by the size of the opening through which the diamond particle will not pass.

screw axis [CRYSTAL] A symmetry element of some crystal lattices, in which the lattice is unaltered by a rotation about the axis combined with a translation parallel to the axis and equal to a fraction of the unit lattice distance in this direction.

screw dislocation [CRYSTAL] A dislocation in which atomic planes form a spiral ramp winding around the line of the dislocation.

scroll [GEOL] One of a series of crescent-shaped sediments on the inner bank of a moving channel, deposited there by the stream.

scroll meander [GEOL] A type of forced-cut meander, in which the scrolls built on the inner bank cause erosion of the outer bank.

scrubber [MIN ENG] A device, such as a wash screen, wash trommel, log washer, and hydraulic jet or monitor, in which a coarse and sticky material, for example, ore or clay, is either washed free of adherents or mildly disintegrated.

scud [METEOROL] Ragged low clouds, usually stratus fractus; most often applied when such clouds are moving rapidly beneath a layer of nimbostratus.

Scythian stage [GEOL] A stage in the lesser Triassic series of the alpine facies. Also known as Werfenian stage.

sea [GEOGR] A usually salty lake lacking an outlet to the ocean.   [OCEANOGR] **1.** A major subdivision of the ocean. **2.** A heavy swell or ocean wave still under the influence of the wind that produced it. **3.** *See* ocean.

sea arch [GEOL] An opening through a headland, formed by wave erosion or solution (as by the enlargement of a sea cave, or by the meeting of two sea caves from opposite sides), which leaves a bridge of rock over the water. Also known as sea bridge.

sea ball [OCEANOGR] A spherical mass of somewhat fibrous material of living or fossil vegetation (especially algae), produced mechanically in shallow waters along a seashore by the compacting effect of wave movement.

sea bank *See* seawall.

seabeach [GEOL] A beach along the margin of the sea.

seabed *See* sea floor.

sea bottom *See* sea floor.

sea breeze [METEOROL] A coastal, local wind that blows from sea to land, caused by the temperature difference when the sea surface is colder than the adjacent land; it usually blows on relatively calm, sunny summer days, and alternates with the oppositely directed, usually weaker, nighttime land breeze.

sea breeze of the second kind *See* cold-front-like sea breeze.

sea bridge *See* sea arch.

sea-captured stream [HYD] A stream, flowing parallel to the seashore, that is cut in two as a result of marine erosion and that may enter the sea by way of a waterfall.

sea cave [GEOL] A split or hollow opening, usually at sea level, in the base of a sea cliff, formed by waves acting on weak parts of the weathered rock. Also known as marine cave; sea chasm.

sea channel [GEOL] A long, narrow, U-shaped or V-shaped shallow depression of the sea floor, usually occurring on a gently sloping plain or fan.

sea chasm *See* sea cave.

sea cliff [GEOL] An erosional landform, produced by wave action, which is either at the seaward edge of the coast or at the landward side of a wave-cut platform and which denotes the inner limit of the beach erosion.

seacoast [GEOGR] The land adjacent to the sea.

sea fan *See* submarine fan.

sea floor [GEOL] The bottom of the ocean. Also known as seabed; sea bottom.

sea floor spreading [GEOL]   The hypothesis that the ocean floor is spreading away from the midoceanic ridges and is being conveyed landward by convective cells in the earth's mantle, carrying the continental blocks as passive passengers; the ocean floor moves away from the midoceanic ridge at the rate of 1 to 10 centimeters per year and provides the source of power in the hypothesis of plate tectonics. Also known as ocean floor spreading; spreading concept; spreading floor hypothesis.

sea-foam *See* sepiolite.

sea fog [METEOROL]   A type of advection fog formed over the ocean as a result of any of a variety of processes, as when air that has been lying over a warm water surface is transported over a colder water surface, resulting in a cooling of the lower layer of air below its dew point.

sea front [GEOGR]   An area partly bounded by the sea.

sea gate [GEOGR]   A way giving access to the sea such as a gate, channel, or beach.

sea glow [OCEANOGR]   The luminous, cobalt-blue appearance of very clear water in the open ocean, caused by upward-scattered light from which much of the red has been absorbed.

sea gully *See* slope gully.

sea ice [OCEANOGR]   **1.** Ice formed from seawater. **2.** Any ice floating in the sea.

sea-ice shelf [OCEANOGR]   Sea ice floating in the vicinity of its formation and separated from fast ice, of which it may have been a part, by a tide crack or a family of such cracks.

sea level [GEOL]   The level of the surface of the ocean; especially, the mean level halfway between high and low tide, used as a standard in reckoning land elevation or sea depths.

sea-level chart *See* surface chart.

sea-level datum [ENG]   A determination of mean sea level that has been adopted as a standard datum for heights or elevations, based on tidal observations over many years at various tide stations along the coasts.

sea-level pressure [METEOROL]   The atmospheric pressure at mean sea level, either directly measured or, most commonly, empirically determined from the observed station pressure.

sea-level pressure chart *See* surface chart.

seam [GEOL]   **1.** A stratum or bed of coal or other mineral. **2.** A thin layer or stratum of rock. **3.** A very narrow coal vein.

seamanite [MINERAL] $Mn_3(PO_4)(BO_3)\cdot 3H_2O$ A pale- to wine-yellow orthorhombic mineral that is a phosphate and borate of manganese; occurs in crystals.

sea marsh *See* sea meadow.

seam blast [MIN ENG]   A blast made by placing powdered or other explosive along and in a seam or crack between the solid wall and the stone or coal to be removed.

sea meadow [ECOL]   *See* sea marsh.   [OCEANOGR]   Any of the upper layers of the open ocean that have such an abundance of phytoplankton that they provide food for marine organisms.

sea mist *See* steam fog.

seamount [GEOL]   An elevation of the sea floor that is either flat-topped or peaked, rising to about 3000–1000 feet (900–300 meters) or more, with the summit approximately 1000–6000 feet (300–1800 meters) below sea level.

seamount chain [GEOL]   Several seamounts in a line with bases separated by a relatively flat sea floor.

seamount group [GEOL] Several closely spaced seamounts not in a line.

seamount range [GEOL] Three or more seamounts having connected bases and aligned along a ridge or rise.

sea mud [GEOL] A rich, slimy deposit in a salt marsh or along a seashore, sometimes used as a manure. Also known as sea ooze.

sea ooze *See* sea mud.

sea peak [GEOL] A peaked elevation of the sea floor, rising 1000 meters or more from the floor.

sea puss [GEOL] The submerged channel or inlet that develops across a bar due to the action of a rip current.

seaquake [GEOPHYS] An earth tremor whose epicenter is beneath the ocean and can be felt only by ships in the vicinity of the epicenter. Also known as submarine earthquake.

searlesite [MINERAL] $NaB(SiO_3)_2 \cdot H_2O$ A white mineral composed of hydrous sodium borosilicate occurring as spherulites.

sea salt [OCEANOGR] The salt remaining after the evaporation of seawater, containing sodium and magnesium chlorides and magnesium and calcium sulfates.

sea-salt nucleus [OCEANOGR] A condensation nucleus of a highly hygroscopic nature produced by partial or complete desiccation of particles of sea spray or of seawater droplets derived from breaking bubbles.

seascape [OCEANOGR] The surrounding sea as it appears to an observer.

seascarp [GEOL] A submarine cliff that is relatively long, high, and straight.

seashore [GEOL] **1.** The strip of land that borders a sea or ocean. Also known as seaside; shore. **2.** The ground between the usual tide levels. Also known as seastrand.

seashore lake [GEOGR] A lake, containing either fresh or salt water, which lies along a seashore; it is separated from the sea by a river, a delta, or a wall of sediment.

sea slope [GEOL] The slope of land toward the sea.

sea smoke *See* steam fog.

season [CLIMATOL] A division of the year according to some regularly recurrent phenomena, usually astronomical or climatic.

seasonal current [OCEANOGR] An ocean current which has large changes in speed or direction due to seasonal winds.

seasonally frozen ground [GEOL] Ground that is frozen during low temperatures and remains so only during the winter season. Also known as frost zone.

seasonal recovery [HYD] Recharge of groundwater during and after a wet season, with a rise in the level of the water table.

seasonal stream [HYD] A stream whose flow is not constant because it has water in its course only during certain seasons.

seasonal thermocline [OCEANOGR] A thermocline which develops in the oceans in summer at relatively shallow depths due to surface heating and downward transport of heat caused by mixing of water generated by summer winds.

seasonal variation [GEOPHYS] The variation of any parameter of the upper atmosphere with season; for example, the variation of ion densities of different parts of the ionosphere, and the resulting variation in transmission of radio signals over large distances.

sea state [OCEANOGR] The numerical or written description of the ocean-surface roughness.

seastrand *See* seashore.

sea-surface slope [OCEANOGR] A gradual change in the level of the sea surface with distance, caused by Coriolis and wind forces.

seat clay *See* underclay.

seat earth *See* underclay.

sea terrace *See* marine terrace.

sea turn [METEOROL] A wind coming from the sea, often bringing mist; the term is limited mainly to New England, United States.

sea valley [GEOL] A relatively shallow, wide depression with gentle slopes in the sea floor, the bottom of which grades continuously downward.

seawall [GEOL] A steep-faced, long embankment situated by powerful storm waves along a seacoast at high-water mark.

seawater [OCEANOGR] Water of the seas, distinguished by high salinity. Also known as salt water.

seawater thermometer [ENG] A specially designed thermometer to measure the temperature of a sample of seawater; an instrument consisting of a mercury-in-glass thermometer protected by a perforated metal case.

sebcha *See* sebkha.

sebka *See* sebkha.

sebkha [GEOL] A geologic feature, in North Africa, which is a smooth, flat, plain usually high in salt; after a rain the plain may become a marsh or a shallow lake until the water evaporates. Also known as sabkha; sebcha; sebka; sibjet.

seca [METEOROL] A drought, or dry wind, in Brazil.

secant conic chart *See* conic chart with two standard parallels.

secant conic projection *See* conic projection with two standard parallels.

sechard [METEOROL] A dry, warm foehn wind over Lake Geneva in Switzerland.

seclusion [METEOROL] A special case of the process of occlusion, where the point at which the cold front first overtakes the warm front (or quasi-stationary front) is at some distance from the apex of the wave cyclone.

secondary [GEOL] A term with meanings that changed from early to late in the 19th century, when the term was confined to the entire Mesozoic era; it was finally replaced by Mesozoic era.

secondary circle *See* secondary great circle.

secondary clay [GEOL] A clay that has been transported from its place of formation and redeposited elsewhere.

secondary coast [GEOL] A relatively stable seacoast or shoreline whose features are the result of present-day marine processes.

secondary cold front [METEOROL] A front which forms behind a frontal cyclone and within a cold air mass, characterized by an appreciable horizontal temperature gradient.

secondary consequent stream [HYD] A tributary of a subsequent stream, flowng parallel to or down the same slope as the original consequent stream; it is usually developed after the formation of a subsequent stream, but in a direction consistent with that of the original consequent stream. Also known as subconsequent stream.

secondary consolidation [GEOL] Consolidation of sedimentary material, at essentially constant pressure, resulting from internal processes such as recrystallization.

secondary cosmic rays [GEOPHYS] Radiation produced when primary cosmic rays enter the atmosphere and collide with atomic nuclei and electrons.

secondary crater [GEOL] An impact crater produced by the relatively low-velocity impact of fragments ejected from a large primary crater. Also known as satellitic crater.

secondary cyclone [METEOROL] A cyclone which forms near or in association with a primary cyclone. Also known as secondary low.

secondary dispersion [GEOCHEM] The process whereby secondary geochemical halos are produced.

secondary drilling [MIN ENG] The process of drilling the so-called popholes for the purpose of breaking the larger masses of rock thrown down by the primary blast.

secondary enlargement [MINERAL] Overgrowth by chemical deposition on a mineral grain of additional material of identical composition in optical and crystallographic continuity with the original grain; crystal faces characteristic of the original mineral often result. Also known as secondary growth.

secondary enrichment [GEOL] The addition to a vein or ore body of material that originated later in time from the oxidation of decomposed ore masses that overlie the vein.

secondary front [METEOROL] A front which may form within a baroclinic cold air mass which itself is separated from a warm air mass by a primary frontal system; the most common type is the secondary cold front.

secondary geosyncline [GEOL] A geosyncline appearing at the culmination of or after geosynclinal orogeny.

secondary glacier [HYD] A small valley glacier that joins a larger trunk glacier as a tributary glacier.

secondary great circle [GEOD] A great circle perpendicular to a primary great circle, as a meridian. Also known as secondary circle.

secondary growth See secondary enlargement.

secondary haulage [MIN ENG] That portion of the haulage system which collects coal from gathering-haulage delivery points and delivers it to the main portion of the system.

secondary interstices [GEOL] Openings in a rock that formed after the enclosing rock was formed.

secondary limestone [PETR] Limestone deposited from solution in cracks and cavities of other rocks.

secondary low See secondary cyclone.

secondary mineral [MINERAL] A mineral produced in an enclosing rock after the rock was formed as a result of weathering or metamorphic or solution activity, and usually at the expense of a primary material that came into existence earlier.

secondary oil recovery [PETRO ENG] Procedures used to increase the flow of oil from depleted or nearly depleted wells; includes fracturing, acidizing, waterflood, and gas injection.

secondary porosity [GEOL] The interstices that appear in a rock formation after it has formed, because of dissolution or stress distortion taking place naturally or artificially as a result of the effect of acid treatment or the injection of coarse sand.

secondary reflection See multiple reflection; shoot.

secondary reserves [PETRO ENG] Reserves recoverable commercially at current prices and costs as a result of artificial supplementation of the reserve's natural (gas-drive) energy.

secondary shaft [MIN ENG] The shaft which extends a mine downward from the bottom of, but not in line with, the primary shaft.

secondary splits [MIN ENG] Splits formed by separation of the main air splits.

secondary stratification [GEOL] The layering that occurs when sediments that were at one time deposited are resuspended and redeposited. Also known as indirect stratification.

secondary stratigraphic trap *See* stratigraphic trap.

secondary structure [GEOL] A structure such as a fault, fold, or joint resulting from tectonic movement that started after the rock in which it is found was emplaced. [PALEON] A coarse structure usually between the thin sheets in the protective wall of a tintinnid.

secondary tectonite [GEOL] A tectonite having a deformation fabric.

secondary tide station [ENG] A place at which tide observations are made over a short period to obtain data for a specific purpose.

secondary twinning [CRYSTAL] Twinning of a crystal caused by an external influence, such as pressure in rock.

secondary wave *See* S wave.

second bottom [GEOL] The first terrace rising over a floodplain.

second-foot [HYD] A contraction of cubic foot per second (cfs), the unit of stream discharge commonly used in the United States.

second-foot day [HYD] The volume of water represented by a flow of 1 cubic foot per second for 24 hours; equal to 86,400 cubic feet (approximately 2446.58 cubic meters); used extensively as a unit of runoff volume or reservoir capacity, particularly in the eastern United States.

second-order climatological station [CLIMATOL] A station at which observations of atmospheric pressure, temperature, humidity, winds, clouds, and weather are made at least twice daily at fixed hours, and at which the daily maximum and minimum of temperature, the daily amount of precipitation, and the duration of bright sunshine are observed.

second-order station [METEOROL] After U.S. Weather Bureau practice, a station manned by personnel certified to make aviation weather observations or synoptic weather observations.

section [GEOL] **1.** An inclined or vertical surface that is uncovered either naturally (as a sea cliff or stream bank) or artificially (as a strip mine or road cut) through a part of the earth's crust. **2.** A description or scale drawing of the successive rock units or geologic structures shown by the exposed surface, or their appearance if cut through by any intersecting plane. **3.** *See* columnar section. **4.** *See* geologic section. **5.** *See* type section. **6.** *See* thin section.

sector [METEOROL] Something resembling the sector of a circle, as a warm sector between the warm and cold fronts of a cyclone.

sector wind [METEOROL] The average observed or computed wind (direction and speed) at flight level for a given sector of an air route; sectors for over-ocean flights usually consist of 10° of longitude.

secular variation [GEOPHYS] The changes, measured in hundreds of years, in the magnetic field of the earth. Also known as geomagnetic secular variation.

sedarenite [PETR] A litharenite composed chiefly of sedimentary rock fragments.

sedentary soil [GEOL] Soil that still lies on the rock from which it was formed.

sedifluction [GEOL] The subaquatic or subaerial movement of material in unconsolidated sediments, occurring in the primary stages of diagenesis.

sediment [GEOL] **1.** A mass of organic or inorganic solid fragmented material, or the solid fragment itself, that comes from weathering of rock and is carried by, suspended in, or dropped by air, water, or ice; or a mass that is accumulated by any other natural agent and that forms in layers on the earth's surface such as sand, gravel, silt, mud, fill, or loess. **2.** A solid material that is not in solution and either is distributed through the liquid or has settled out of the liquid.

sedimentary breccia [PETR] A rock composed of fragments that are larger than 2 millimeters in diameter and are the result of sedimentary processes; characterized by imperfect mechanical sorting of its materials and by a higher concentration of fragments from one local source or by a wide variety of materials mixed together in no particular pattern. Also known as sharpstone conglomerate.

sedimentary cycle *See* cycle of sedimentation.

sedimentary differentiation [GEOL] The progressive separation (by erosion and transportation) of a well-defined rock mass into physically and chemically unlike products that are resorted and deposited as sediments in more or less separate areas.

sedimentary dike [GEOL] A tabular mass of sedimentary material that cuts across the structure or bedding of preexisting rock in the manner of an igneous dike and that is formed by the filling of a crack or fissure by forcible injection or intrusion of sediments under abnormal pressure, or by simple infilling of sediments.

sedimentary facies [GEOL] A stratigraphic facies differing from another part or parts of the same unit in both lithologic and paleontologic characters.

sedimentary insertion [GEOL] The emplacement of sedimentary material among deposits or rocks already formed, such as by infilling, injection, or intrusion, or through localized subsidence due to solution of underlying rock.

sedimentary intrusion *See* intrusion.

sedimentary laccolith [GEOL] An intrusion of plastic sedimentary material (such as clayey salt breccia) forced up under high pressure and penetrating parallel or nearly parallel to the bedding planes of the invaded formation; characterized by a very irregular thickness.

sedimentary lag [GEOL] Delay between the formation of potential sediment by weathering and its removal and deposition.

sedimentary petrography [PETR] The description and classification of sedimentary rocks. Also known as sedimentography.

sedimentary petrology [PETR] The study of the composition, characteristics, and origin of sediments and sedimentary rocks.

sedimentary quartzite *See* orthoquartzite.

sedimentary rock [PETR] A rock formed by consolidated sediment deposited in layers. Also known as derivative rock; neptunic rock, stratified rock.

sedimentary structure [GEOL] A structure in sedimentary rocks, such as crossbedding, ripple marks, and sandstone dikes, produced either contemporaneously with deposition (primary sedimentary structures) or shortly after deposition (secondary sedimentary structures).

sedimentary tectonics [GEOL] Folding and deformation in geosynclinal basins caused by subsidence and buckling of strata.

sedimentary trap [GEOL] An area in which sedimentary material accumulates instead of being transported farther, as in an area between high-energy and low-energy environments.

sedimentary tuff [GEOL] A tuff containing a small amount of nonvolcanic detrital material.

sedimentary volcanism [GEOL] The expelling, extruding, or breaking through of overlying formations by a mixture of sediment, water, and gas, driven by the gas under pressure.

sedimentation [GEOL] **1.** The act or process of accumulating sediment in layers. **2.** The process of deposition of sediment.   [MET] Classification of metal powders by the rate of settling in a fluid.

sedimentation basin [GEOL] A depression in the ocean floor with a wide, flat bottom in which sediment accumulates.

sedimentation curve [GEOL] A curve showing cumulatively, and in successive units of time, the amount of sediment accumulated or removed from an originally uniform suspension.

sedimentation diameter [GEOL] The diameter of a sedimentary particle, determined from the measurement of a hypothetical sphere of the same gravity and settling velocity as those of a given sedimentary particle in the same fluid.

sedimentation radius [GEOL] One-half of the sedimentation diameter.

sedimentation trend [GEOL] The direction in which sediments were laid down.

sedimentation trough [GEOL] A depression in the ocean floor with a narrow U- or V-shaped bottom in which sediment accumulates.

sedimentation unit [GEOL] A sedimentary deposit formed during one distinct act of sedimentation.

sediment charge [HYD] In a stream, the ratio of the weight or volume of sediment to the weight or volume of water passing a given cross section per unit of time.

sediment concentration [HYD] The ratio of the dry weight of the sediment in a water-sediment mixture (obtained from a stream or other body of water) to the total weight of the mixture.

sediment corer [ENG] A heavy coring tube which punches out a cylindrical sediment section from the ocean bottom.

sediment-delivery ratio [GEOL] The ratio of sediment yield of a drainage basin to the total amount of sediment moved by sheet erosion and channel erosion.

sediment discharge [HYD] The amount of sediment moved by a stream in a given time, measured by dry weight or by volume. Also known as sediment-transport rate.

sediment discharge rating [HYD] A relationship between the discharge of sediment and the total discharge of the stream. Also known as silt discharge rating.

sediment load [HYD] The solid material that is transported by a natural agent, especially by a stream.

sedimentography *See* sedimentary petrography.

sedimentology [GEOL] The science concerned with the description, classification, origin, and interpretation of sediments and sedimentary rock.

sediment-production rate [GEOL] Sediment yield per unit of drainage area, derived by dividing the annual sediment yield by the area of the drainage basin.

sediment station [HYD] A vertical cross-sectional plane of a stream, usually normal to the mean direction of flow, where samples of suspended load are collected on a systematic basis for determining concentration, particle-size distribution, and other characteristics.

sediment-transport rate *See* sediment discharge.

seed fern [PALEOBOT] The common name for the extinct plants classified as Pteridospermae, characterized by naked seeds borne on large, fernlike fronds.

Seelandian [GEOL] A European stage of geologic time in the lowermost Paleocene.

seep [GEOL] An area, generally small, where water, or another liquid such as oil, percolates slowly to the land surface. [PETRO ENG] An oil spring whose daily yield ranges from a few drops to several barrels of oil; usually located at low elevations where water has accumulated.

seepage [HYD] The slow movement of water through small openings and spaces in the surface of unsaturated soil into or out of a body of surface or subsurface water.

seepage face [GEOL] A belt on a slope, such as the bank of a stream, along which water emerges at atmospheric pressure and flows down the slope.

seepage lake [HYD] **1.** A closed lake that loses water mainly by seepage through the walls and floor of its basin. **2.** A lake that receives its water mainly from seepage.

segregated ice [HYD] Ice films, seams, lenses, rods, or layers generally 1 to 150 millimeters thick that grow in permafrost by drawing in water as the ground freezes. Also known as Taber ice.

segregated vein [GEOL] A fissure filled with mineral matter derived from country rock by the action of percolating water. Also known as exudation vein.

segregation [GEOL] The formation of a secondary feature within a sediment after deposition due to chemical rearrangement of minor constituents.

segregation banding [PETR] A compositional band in gneisses that is the result of segregation of material from an originally homogeneous rock.

seiche [OCEANOGR] A standing-wave oscillation of an enclosed or semienclosed water body, continuing pendulum-fashion after cessation of the originating force, which is usually considered to be strong winds or barometric pressure changes.

seif dune [GEOL] A large, tapering, longitudinal dune or chain of sand dunes with a sharp crest that in profile consists of a succession of peaks and cols.

seismic activity *See* seismicity.

seismic anisotropy [GEOPHYS] The dependence of seismic velocity on the direction of propagation.

seismic area *See* earthquake zone.

seismic discontinuity [GEOPHYS] **1.** A surface at which velocities of seismic waves change abruptly. **2.** A boundary between seismic layers of the earth. Also known as interface; velocity discontinuity.

seismic efficiency [GEOPHYS] The proportion of the total available strain energy which is radiated as seismic waves.

seismic-electric effect [GEOPHYS] The variation of resistivity with elastic deformation of rocks.

seismic event [GEOPHYS] An earthquake or a somewhat similar transient earth motion caused by an explosion.

seismic exploration [ENG] The exploration for economic deposits by using seismic techniques, usually involving explosions, to map subsurface structures.

seismic gradient [GEOPHYS] The variation of seismic velocity with distance in a specificed direction. Also known as velocity gradient.

seismic hazard [GEOPHYS] Any physical phenomenon, such as ground shaking or ground failure, that is associated with an earthquake and that may produce adverse effects on human activities.

seismic intensity [GEOPHYS] The average rate of flow of seismic-wave energy through a unit section perpendicular to the direction of propagation.

seismicity [GEOPHYS] The phenomena of earth movements. Also known as seismic activity.

seismic map [GEOPHYS] A contour map constructed from seismic data, the z coordinate of which could be either time or depth.

seismic profiler [ENG] A continuous seismic reflection system used to study the structure beneath the sea floor to depths of 10,000 feet (3000 meters) or more, using a rotating drum to record reflections.

seismic prospecting [GEOPHYS] Geophysical prospecting based on the analysis of elastic waves generated in the earth by artificial means.

seismic ray [GEOL] The path along which seismic energy travels.

seismic risk [GEOPHYS] **1.** An assortment of earthquake effects that range from ground shaking, surface faulting, and landsliding to economic loss and casualties. **2.** The probability that social or economic consequences of earthquakes will equal or exceed specified values at a site, at several sites, or in an area, during a specified exposure time.

seismic sea wave *See* tsunami.

seismic shooting [ENG] A method of geophysical prospecting in which elastic waves are produced in the earth by the firing of explosives.

seismic survey *See* reflection survey.

seismic velocity [GEOPHYS] The rate of propagation of an elastic wave, usually measured in kilometers per second.

seismic vertical [GEOL] **1.** The point on the earth's surface directly over the point within the earth from which an earthquake impulse originates. **2.** The vertical line between the surface point and the point of origin.

seismochronograph [ENG] A chronograph for determining the time at which an earthquake shock appears.

seismogram [ENG] The record made by a seismograph.

seismograph [ENG] An instrument that records vibrations in the earth, especially earthquakes.

seismology [GEOPHYS] **1.** The study of earthquakes. **2.** The science of strain-wave propagation in the earth.

seismometer [ENG] An instrument that detects movements in the earth.

seismoscope [ENG] An instrument for recording only the occurrence or time of occurrence (not the magnitude) of an earthquake.

seistan [METEOROL] A strong wind of monsoon origin which blows from between the northwest and north-northwest and sets in about the end of May or early June in the historic Seistan district of eastern Iran and Afghanistan; it continues almost without cessation until about the end of September; because of its duration it is known as the wind of 120 days (bad-i-sad-o-bistroz).

selagite [PETR]   A mica trachyte characterized by abundant tabular biotite crystals in a holocrystalline groundmass of orthoclase and diopside, and possibly quartz and olivine.

selatan [METEOROL]   Strong, dry, southerly winds of the southeast monsoon in the Netherlands East Indies and the Celebes.

selective acidizing [PETRO ENG]   Oil-reservoir acid treatment (acidizing) in which the acid is injected into specific reservoir zones; contrasted with uncontrolled acidizing in which the acid solution is simply pumped down the casing and is forced into adjacent rock.

selective filling [MIN ENG]   Filling by hand so that stone or dirt is rejected and only clean coal or ore is loaded.

selective flotation [MIN ENG]   The surface or froth selecting of the valuable minerals rather than the gangue.

selective fracturing [PETRO ENG]   Procedures for obtaining multiple formation fractures in a specific reservoir zone by plugging casing perforations or by isolating (with packers) the desired zone prior to fracturing operations.

selective fusion [GEOL]   The fusion of only a portion of a mixture, such as a rock.

selective mining [MIN ENG]   A method of mining whereby ore of unwarranted high value is mined in such manner as to make the low-grade ore left in the mine incapable of future profitable extraction; in other words, the best ore is selected in order to make good mill returns, leaving the low-grade ore in the mine.

selective replacement [GEOL]   The replacement of one mineral by another, preferentially within an altered rock mass.

selenite [MINERAL]   The clear, colorless variety of gypsum crystallizing in the monoclinic system and occurring in crystals or in crystal mass. Also known as spectacle stone.

selenite butte [GEOL]   A small tabular mound, rising 1–3 meters above a playa, composed of lake sediments capped with a veneer of selenite formed by deflation of the playa or by the effects of rising groundwater.

self-acting door [MIN ENG]   A ventilation door that is constructed of two halves which move on small pulleys and which are forced apart centrally as the trams come in contact with the converging beams that operate the door.

self-acting incline *See* gravity haulage.

self-advancing supports [MIN ENG]   An assembly of hydraulically operated steel hydraulic supports, on a long-wall face, which are moved forward as a unit. Also known as walking props.

self-contained portable electric lamp [MIN ENG]   An electric lamp which is operated by an electric battery and is specifically designed to be carried about by its user.

self-dumping cage [MIN ENG]   A cage in which the deck is pivoted so that as the cage is lifted, toward the end of the lift, the deck tilts and the end door is lifted, discharging the coal.

self-dumping car [MIN ENG]   A mine car which can be side-tipped while in motion on the rail track; it is fitted with a spherically contoured wheel which engages a ramp structure and gradually tilts the car.

self-rescuer [MIN ENG]   A small filtering device carried by a miner underground to provide immediate protection against carbon monoxide and smoke in case of a mine fire or explosion; used for escape purposes only, because it does not sustain life in atmospheres containing deficient oxygen.

seligmannite [MINERAL] $PbCuAsS_3$ A metallic gray orthorhombic mineral, occurring in crystals.

sellaite [MINERAL] MgF_2 A colorless mineral composed of magnesium fluoride occurring in tetragonal prismatic crystals.

selvage [PETR] The marginal zone of an igneous mass, generally characterized by a fine-grain, or sometimes glassy, texture. Also known as salband.

semianthracite [GEOL] Coal which is between bituminous coal and anthracite in metamorphic rank, and which has a fixed-carbon content of 86–92%.

semiarid climate See steppe climate.

semibituminous coal [GEOL] Coal that is harder and more brittle than bituminous coal, has a high fuel ratio, contains 10–20% volatile matter, and burns without smoke; ranks between bituminous and semianthracite coals.

semibolson [GEOL] A wide desert basin or valley whose central playa is absent or poorly developed, and which is drained by an intermittent stream that flows through canyons at each end and reaches a surface outlet.

semibright coal [GEOL] A type of banded coal defined microscopically as consisting of between 80 and 61% bright ingredients such as vitrain, clarain, and fusain, with clarodurain and durain composing the remainder.

semicontrolled mosaic [MAP] A mosaic which is composed of photographs of approximately the same scale laid so that major ground features match their geographical coordinates.

semicratonic See quasicratonic.

semicrystalline See hypocrystalline.

semidiurnal current [OCEANOGR] A tidal current in which the tidal-day current cycle consists of two flood currents and two ebb currents, separated by slack water, or of two changes in direction of 360° of a rotary current; this is the most common type of tidal current throughout the world.

semidiurnal tide [OCEANOGR] A tide having two high waters and two low waters during a tidal day.

semidull coal [GEOL] A type of banded coal consisting mainly of clarodurain and durain, with from 40 to 21% bright ingredients such as vitrain, clarain, and fusain.

semifusinite [GEOL] A coal maceral with a well-defined woody structure and optical properties intermediate between those of nitrinite and those of fusinite.

semischist [PETR] A partly metamorphosed sedimentary rock, exhibiting some foliation.

semisplint coal [GEOL] Banded coal that is intermediate between bright-banded and splint coal, and has 20–30% opaque attritus and more than 5% anthraxylon.

semseyite [MINERAL] $Pb_9Sb_8S_{21}$ A gray to black mineral composed of lead antimony sulfide.

senaite [MINERAL] $(Fe,Mn,Pb)TiO_3$ A black mineral consisting of a lead- and manganese-bearing ilmenite; occurs as rough crystals and rounded fragments.

senarmontite [MINERAL] Sb_2O_3 A colorless or grayish mineral composed of native antimony trioxide occurring in masses or as octahedral crystals.

Senecan [GEOL] A North American provincial series of geologic time, forming the lower part of the Upper Devonian, above the Erian and below the Chautauquan.

senescent lake [HYD] A lake that is approaching extinction, as from filling by remains of aquatic vegetation.

senesland [GEOL] A land surface intermediate between a matureland and a peneplain.

sengierite [MINERAL] $Cu(UO_2)_2(VO_4)_2\cdot8-10H_2O$ A yellowish-green mineral composed of hydrous copper uranyl vanadate.

senile [GEOL] Pertaining to the stage of senility of the cycle or erosion.

Senonian [GEOL] A European stage of geologic time, forming the Upper Cretaceous, above the Turonian and below the Danian.

sensible heat flow [METEOROL] In the atmosphere, the poleward transport of sensible heat (enthalpy) across a given latitude belt by fluid flow.

sensible temperature [METEOROL] The temperature at which air with some standard humidity, motion, and radiation would provide the same sensation of human comfort as existing atmospheric conditions.

sensitive clay [GEOL] A clay whose shear strength is reduced to a very small fraction of its former value on remolding at constant moisture content.

separation [GEOL] The apparent relative displacement on a fault, measured in any given direction. [MIN ENG] The removal of gangue from raw ores, as in frothing.

separator *See* gas-oil separator.

sepiolite [MINERAL] $Mg_4(Si_2O_5)_3(OH)_2\cdot6H_2O$ A soft, lightweight, absorbent, white to light-gray or light-yellow clay mineral, found principally in Asia Minor; used for tobacco pipe bowls and ornamental carvings. Also known as meerschaum; seafoam.

septarian [GEOL] Pertaining to the irregular polygonal pattern of internal cracks developed in septaria.

septarian boulder *See* septarium.

septarian nodule *See* septarium.

septarium [GEOL] A large (80–90 centimeters in diameter), spheroidal concretion, usually composed of argillaceous carbonate, characterized by internal cracking into irregular polygonal blocks that become cemented together by crystalline minerals. Also known as beetle stone; septarian boulder; septarian nodule; turtle stone.

Sequanian [GEOL] Upper Lower Jurassic (Upper Lusitanian) geologic time. Also known as Astartian.

sequence [GEOL] **1.** A sequence of geologic events, processes, or rocks, arranged in chronological order. **2.** A geographically discrete, major informal rock-stratigraphic unit of greater than group or supergroup rank. Also known as stratigraphic sequence. [METEOROL] *See* collective.

sequence of current [OCEANOGR] The order of occurrence of the tidal current strengths of a day, with special reference to whether the greater flood immediately precedes or follows the greater ebb.

sequence of tide [OCEANOGR] The order in which the tides of a day occur, with special reference to whether the higher high water immediately precedes or follows the lower low water.

sequent geosyncline [GEOL] The constituent geosynclines of a polygeosyncline, separated from one another by the development of geanticlines.

sequential landform [GEOL] One of an orderly succession of smaller landforms that are developed by the erosion, weathering, and mass wasting of larger initial landforms.

serac [HYD]   A sharp ridge or pinnacle of ice among the crevasses of a glacier.

serandite [MINERAL] $Na(Mn,Ca)_2Si_3O_8(OH)$ A rose-red mineral composed of a basic silicate of manganese, lime, potash, and soda occurring in monoclinic crystals.

serein [METEOROL]   The doubtful phenomenon of fine rain falling from an apparently clear sky, the clouds, if any, being too thin to be visible; frequently, fine rain is observed with a clear sky overhead, but clouds to windward clearly indicate the source of the drops.

serial observation [OCEANOGR]   The procurement of water samples and temperature readings at a number of levels between the surface and the bottom of an ocean.

serial station [OCEANOGR]   An oceanographic station consisting of one or more Nansen casts.

seriate [GEOL]   Having crystals that vary gradually in size.

sericite [MINERAL]   A white, fine-grained potassium mica, usually muscovite in composition, having a silky luster and found as small flakes in various metamorphic rocks.

sericitic sandstone [PETR]   A sandstone in which sericite (derived by decomposition of feldspar) intermingles with finely divided quartz and fills the voids between quartz grains.

sericitization [GEOL]   A hydrothermal or metamorphic process involving the introduction of or replacement by sericite.

series [GEOL]   **1.** A number of rocks, minerals, or fossils that can be arranged in a natural sequence due to certain characteristics, such as succession, composition, or occurrence.   **2.** A time-stratigraphic unit, below system and above stage, composed of rocks formed during an epoch of geologic time.

serpentine [MINERAL] $(Mg,Fe)_3Si_2O_5(OH)_4$ A group of green, greenish-yellow, or greenish-gray ferromagnesian hydrous silicate rock-forming minerals having greasy or silky luster and a slightly soapy feel; translucent varieties are used for gemstones as substitutes for jade.

serpentine jade [MINERAL]   A variety of the mineral serpentine resembling jade in appearance and used as an ornamental stone.

serpentine rock *See* serpentinite.

serpentine spit [GEOGR]   A spit that is extended in more than one direction due to variable or periodically shifting currents.

serpentinite [PETR]   A rock composed almost entirely of serpentine minerals. Also known as serpentine rock.

serpentinization [GEOL]   A hydrothermal process by which magnesium-rich silicate minerals are converted into or replaced by serpentine minerals.

serpent kame *See* esker.

serpierite [MINERAL] $(Cu,Zn,Ca)_5(SO_4)_2(OH)_6 \cdot 3H_2O$ A bluish-green mineral composed of hydrous basic sulfate of copper, zinc, and calcium; occurs in tabular crystals and tufts.

serrate ridge *See* arête.

Serridentinae [PALEON]   An extinct subfamily of elephantoids in the family Gomphotheriidae.

service shaft [MIN ENG]   A shaft used only for hoisting men and materials to and from underground.

seston [OCEANOGR]   Minute living organisms and particles of nonliving matter which float in water and contribute to turbidity.

set [GEOL] A group of essentially conformable strata or cross-strata, separated from other sedimentary units by surfaces of erosion, nondeposition, or abrupt change in character. [MIN ENG] *See* frame set. [OCEANOGR] The direction toward which an oceanic current flows.

settled [METEOROL] Pertaining to weather, devoid of storms for a considerable period.

settled ground [MIN ENG] Ground which has ceased to subside over the waste area of a mine, having reached a state of full subsidence.

settled snow [HYD] An old snow that has been strongly metamorphosed and compacted.

settlement [GEOL] The subsidence of surficial material (such as coastal sediments) due to compaction. [MIN ENG] The gradual lowering of the overlying strata in a mine, due to extraction of the mined material.

settling pond [MIN ENG] A natural or artificial pond for recovering the solids from an effluent.

severe storm [METEOROL] In general, any destructive storm, but usually applied to a severe local storm, that is, an intense thunderstorm, hail storm, or tornado.

severe-storm observation [METEOROL] An observation (and report) of the occurrence, location, time, and direction of movement of severe local storms.

severe weather [METEOROL] A more general term for severe storm.

seybertite *See* clintonite.

Seymouriamorpha [PALEON] An extinct group of labyrinthodont Amphibia of the Upper Carboniferous and Permian in which the intercentra were reduced.

sferics *See* atmospheric interference.

sferics fix [METEOROL] The estimated location of a source of atmospherics, presumably a lightning discharge.

sferics observation [METEOROL] An evaluation, from one or more sferics receivers, of the location of weather conditions with which lightning is associated; such observations are more commonly obtained from networks of two or three widely spaced stations; simultaneous observations of the azimuth of the discharge are made at all stations, and the location of the storm is determined by triangulation.

shaft [MIN ENG] An excavation of limited area compared with its depth, made for finding or mining ore or coal, raising water, ore, rock, or coal, hoisting and lowering men and material, or ventilating underground workings; the term is often specifically applied to approximately vertical shafts as distinguished from an incline or an inclined shaft.

shaft allowance [MIN ENG] The extra space between the excavation diameter and the finished diameter to accommodate the permanent shaft lining.

shaft cable [MIN ENG] A specially armored cable of great mechanical strength running down a mine shaft.

shaft capacity [MIN ENG] The output of ore or coal that can be expected to be raised regularly and in normal circumstances.

shaft column [MIN ENG] A length of pipes installed in a mine shaft for pumping, for hydraulic stowing, or for compressed air.

shaft crusher [MIN ENG] A hard-rock crusher in a shaft, set to reduce large lumps of ore to a convenient size for delivery to the skip.

shaft deformation bar [MIN ENG]   A length of 1½-inch (3.8-centimeter) pipe fitted at one end with a micrometer and at the other end with a hard-steel cone for measuring the deformation in the cross section of a shaft.

shaft drilling [MIN ENG]   The drilling of small shafts up to about 5 feet (1.5 meters) in diameter with the shot drill.

shaft house [MIN ENG]   A building at the mouth of a shaft, where ore or rock is received from the mine.

shaft lining [MIN ENG]   The timber, steel, brick, or concrete structure fixed around a shaft to support the walls.

shaft pillar [MIN ENG]   A large area of a coal or ore seam which is left unworked around the shaft bottom to protect the shaft from damage by subsidence.

shaft plumbing [MIN ENG]   The operation of orienting two plumb bobs, both at surface and at depth in order to transfer the bearing underground.

shaft pocket [MIN ENG]   Ore storage pocket, of one or more compartments, cut into the wall on one or both sides of a vertical shaft or in the hanging wall of an inclined shaft.

shaft siding [MIN ENG]   The station or landing place arranged for buckets or tubs at the bottom of the winding shaft.

shaft signaling [MIN ENG]   The transmission of visible and audible signals between the onsetter or hitcher at the pit bottom and the banksman or hoistman at the pit top.

shaft sinking [MIN ENG]   Excavating a shaft downward, usually from the surface, to the workable coal or ore.

shaft sinking drill [MIN ENG]   A large-diameter drill with multiple rotary cones or cutting bits, used for shaft sinking.

shaft station [MIN ENG]   An enlargement of a level near a shaft from which ore, coal, or rock may be hoisted and supplies unloaded.

shaker conveyor [MIN ENG]   A conveyor consisting of a length of metal troughs, with suitable supports, to which a reciprocating motion is imparted by drives.

shake wave *See* S wave.

shaking table *See* Wilfley table.

shale [PETR]   A fine-grained laminated or fissile sedimentary rock made up of silt- or clay-size particles; generally consists of about one-third quartz, one-third clay materials, and one-third miscellaneous minerals, including carbonates, iron oxides, feldspars, and organic matter.

shale-arenite [PETR]   A sedarenite composed chiefly of shale fragments.

shale reservoir [GEOL]   Underground hydrocarbon reservoir in which the reservoir rock is a brittle, siliceous, fractured shale.

shale shaker [PETRO ENG]   A vibrating screen over which drilling fluid is passed to trap the drill cuttings as the fluid passes through.

shallow-focus earthquake [GEOPHYS]   An earthquake whose focus is located within 70 kilometers of the earth's surface.

shallow fog [METEOROL]   In weather-observing terminology, low-lying fog that does not obstruct horizontal visibility at a level 6 feet (1.8 meters) or more above the surface of the earth; this is, almost invariably, a form of radiation fog.

shallow inland seas [GEOL]   Epeiric seas which periodically cover cratonic areas as a result of continental subsidence or eustatic rises in sea level.

shallow marginal seas   [GEOL]   Epeiric seas along the cratonic margins.

shallows   [HYD]   A shallow place or area in a body of water, or an expanse of shallow water.

shallow water   [HYD]   Water of such a depth that bottom topography affects surface waves.

shallow-water wave   [HYD]   A progressive gravity wave in water whose depth is much less than the wavelength.

shallow well   [HYD]   **1.** A water well, generally dug up by hand or by excavating machinery, or put down by driving or boring, that taps the shallowest aquifer in the vicinity.   **2.** A well whose water level is shallow enough to permit use of a suction pump, the practical lift of which is taken as 22 feet (6.7 meters).

shaluk   [METEOROL]   Any hot desert wind other than simoom.

shaly   [GEOL]   Pertaining to, composed of, containing, or having the properties of shale, especially readily split along close-spaced bedding planes.

shaly bedding   [GEOL]   Laminated bedding varying between 2 and 10 millimeters in thickness.

shamal   [METEOROL]   The northwest wind in the lower valley of the Tigris and Euphrates and the Persian Gulf; it may set in suddenly at any time, and generally lasts from 1 to 5 days, dying down at night and freshening again by day; but in June and early July it continues almost without cessation (the great or 40-day shamal).

shandite   [MINERAL] $Ni_3Pb_2S_2$ A rhombohedral mineral composed of nickel lead sulfide, occurring in crystals.

shantung   [GEOL]   A monadnock in the process of burial by huangho deposits.

Shantung soil   *See* Noncalcic Brown soil.

sharki   *See* kaus.

sharkskin pahoehoe   [GEOL]   A type of pahoehoe displaying numerous tiny spines or spicules on the surface.

shark-tooth projection   [GEOL]   Sharp pointed projections several centimeters in length, formed by the pulling apart of plastic lava.

sharp-edged gust   [METEOROL]   A gust that represents an instantaneous change in wind direction or speed.

sharpite   [MINERAL] $(UO_2)(CO_3)\cdot H_2O$ A greenish-yellow mineral composed of hydrous basic uranyl carbonate.

sharp sand   [GEOL]   An angular-grain sand free of clay, loam, and other foreign particles.

sharpstone   [GEOL]   Any rock fragment having angular edges and corners and being more than 2 millimeters in diameter.

sharpstone conglomerate   *See* sedimentary breccia.

shatter breccia   [PETR]   A tectonic breccia composed of angular fragments that show little rotation.

shatter cone   [GEOL]   A striated conical rock fragment along which fracturing has occurred.

shatter zone   [GEOL]   An area of randomly fissured or cracked rock that may be filled by mineral deposits, forming a network pattern of veins.

shattuckite [MINERAL] $Cu_5(SiO_3)_4H_2O$ A blue mineral composed of basic copper silicate, occurring in fibrous masses.

sheaf structure [GEOL]   A bundled arrangement of crystals that is characteristic of certain fibrous minerals, such as stibnite.

shear [MIN ENG]   To make vertical cuts in a coal seam that has been undercut.

shear burst [MIN ENG]   The explosive breaking of wall rock in a deep mining field by the occurrence of a single shear crack parallel to the face in one of the walls, causing rock behind the shear plane to expand freely into the stope and then to disrupt and fill the place with debris.

shear cleavage *See* slip cleavage.

shear fold [GEOL]   A similar fold whose mechanism is shearing or slipping along closely spaced planes that are parallel to the fold's axial surface. Also known as glide fold; slip fold.

shear-gravity wave [GEOPHYS]   A combination of gravity waves and a Helmholtz wave on a surface of discontinuity of density and velocity.

shear joint [GEOL]   A joint that is a shear fracture; it is a potential plane of shear. Also known as slip joint.

shear line [METEOROL]   A line or narrow zone across which there is an abrupt change in the horizontal wind component parallel to this line; a line of maximum horizontal wind shear.

shear moraine [GEOL]   A debris-laden surface or zone found along the margin of any ice sheet or ice cap, dipping in toward the center of the ice sheet but becoming parallel to the bed at the base.

shear plane [GEOL]   *See* shear surface. [HYD]   A planar surface in a glacier, usually laden with rock debris, attributed to discontinuous shearing or overthrusting.

shear slide [GEOL]   A landslide, especially a slump, produced by shear failure usually along a plane of weakness such as a bedding or cleavage plane.

shear sorting [GEOL]   Sorting of sediments in which the smaller grains tend to move toward the zone of greatest shear strain, and the larger grains toward the zone of least shear.

shear structure [GEOL]   A local structure in which earth stresses have been relieved by many small, closely spaced fractures.

shear surface [GEOL]   A surface along which differential movement has taken place parallel to the surface. Also known as shear plane.

shear wave *See* S wave.

shear zone [GEOL]   A tabular area of rock that has been crushed and brecciated by many parallel fractures resulting from shear strain; often becomes a channel for underground solutions and the seat of ore deposition.

sheet [GEOL]   **1.** A thin flowstone coating of calcite in a cave. **2.** A tabular igneous intrusion, especially when concordant or only slightly discordant. [HYD]   *See* sheet-flood.

sheet crack [GEOL]   A planar crack attributed to shrinkage of sediment due to dewatering.

sheet deposit [GEOL]   A stratiform mineral deposit that is more or less horizontal and extensive relative to its thickness.

sheet drift [GEOL]   An evenly spread deposit of glacial drift that did not significantly alter the form of the underlying rock surface.

sheeted fissure [GEOL] A closely spaced fissure.

sheeted vein [GEOL] A vein filling a shear zone.

sheeted zone [GEOL] An area of mineral deposits consisting of sheeted veins.

sheet erosion [GEOL] Erosion of thin layers of surface materials by continuous sheets of running water. Also known as sheetflood erosion; sheetwash; surface wash; unconcentrated wash.

sheetflood [HYD] A broad expanse of moving, storm-borne water that spreads as a thin, continuous, relatively uniform film over a large area for a short distance and duration. Also known as sheet; sheetwash.

sheetflood erosion *See* sheet erosion.

sheet flow [HYD] An overland flow or downslope movement of water taking the form of a thin, continuous film over relatively smooth soil or rock surfaces and not concentrated into channels larger than rills.

sheet frost [HYD] A thick coating of rime formed on windows and other surfaces.

sheet ice [HYD] A smooth, thin layer of ice formed by rapid freezing of the surface layer of a body of water.

sheeting [GEOL] The process by which thin sheets, slabs, scales, plates, or flakes of rock are successively broken loose or stripped from the outer surface of a large rock mass in response to release of load. Also known as exfoliation.

sheeting caps [MIN ENG] A row of caps put on blocks about 14 inches (36 centimeters) high which are placed on top of the drift sets when constructing the permanent floor in the stope.

sheeting plane [PETR] In igneous rocks, the primary cleavage plane or parting.

sheeting structure [GEOL] A fracture or joint formed by pressure-release jointing or exfoliation. Also known as exfoliation joint; expansion joint; pseudostratification; release joint; sheet joint; sheet structure.

sheet joint *See* sheeting structure.

sheet lightning [GEOPHYS] A diffuse, but sometimes fairly bright, illumination of those parts of a thundercloud that surround the path of a lightning flash, particularly a cloud discharge or cloud-to-cloud discharge. Also known as luminous cloud.

sheet line [MAP] The outermost border line of a map or chart.

sheet mineral *See* phyllosilicate.

sheet sand *See* blanket sand.

sheet sandstone [GEOL] A thin, blanket-shaped deposit of sandstone of regional extent.

sheet silicate *See* phyllosilicate.

sheet spar [GEOL] A sheet crack filled with spar.

sheet structure *See* sheeting structure.

sheetwash [GEOL] **1.** The detritus deposited by a sheetflood. **2.** *See* sheet erosion. [HYD] **1.** A wide, moving expanse of water on an arid plain; the combined result of many streams issuing from the mountains. **2.** *See* sheetflood.

Shelby tube [ENG] A thin-shelled tube used to take deep-soil samples; the tube is pushed into the undisturbed soil at the bottom of the casting of the borehole driven into the ground.

shelf [GEOL] **1.** Solid rock beneath alluvial deposits. **2.** A flat, projecting ledge of rock. **3.** *See* continental shelf.

shelf break [GEOL] An obvious steepening of the gradient between the continental shelf and the continental slope.

shelf channel [GEOL] A valley formed in a shelf by erosion.

shelf edge [GEOL] The demarcation, without dramatic change in gradient, between continental shelf and continental slope.

shelf facies [GEOL] A sedimentary facies characterized by carbonate rocks and fossil shells and produced in the neritic environments of marginal shelf seas. Also known as foreland facies; platform facies.

shelf ice [HYD] The ice of an ice shelf. Also known as barrier ice.

shelf sea [OCEANOGR] A shallow marginal sea located on the continental shelf, usually less than 150 fathoms (275 meters) in depth; an example is the North Sea.

shelfstone [GEOL] A speleothem formed at the water's edge as a horizontally projecting ledge.

shell [GEOL] **1.** The crust of the earth. **2.** A thin hard layer of rock.

shell ice [HYD] Ice, on a body of water, that remains as an unbroken surface when the water level drops so that a cavity is formed between the water surface and the ice.

shell sand [GEOL] A loose aggregate that is largely composed of shell fragments of sand size.

shelly [GEOL] **1.** Pertaining to a sediment or sedimentary rock containing the shells of animals. **2.** Pertaining to land abounding in or covered with shells.

shelly facies [GEOL] A nongeosynclinal sedimentary facies that is commonly characterized by abundant calcareous fossil shells, dominant carbonate rocks (limestones and dolomites), mature orthoquartzitic sandstones, and a paucity of shales.

shelly pahoehoe [GEOL] A type of pahoehoe characterized by open tubes and blisters on the surface.

shelter cave [GEOL] A cave which extends only a short way underground, and whose roof of overlying rock usually extends beyond its sides.

shelter porosity [GEOL] A type of primary interparticle porosity created by the sheltering effect of relatively large sedimentary particles which prevent the infilling of pore space by finer clastic particles.

Shelton loader [MIN ENG] A modified coal-cutting machine in which the picks of the cutter chain are replaced by loading flights, which push the prepared coal up a ramp on to the face conveyor.

shergottite [GEOL] An achondritic stony meteorite that is composed chiefly of pigeonite and maskelynite.

sheridanite [MINERAL] $(Mg,Al)_6(Al,Si)_4O_{10}(OH)_8$ A pale-green to colorless talclike mineral composed of basic magnesium aluminum silicate.

sherry topaz [MINERAL] A brownish-yellow to yellow-brown variety of topaz resembling the color of sherry wine.

shield *See* palette.

shield basalt [GEOL] A basaltic lava flow from a group of small, close-spaced shield-volcano vents that coalesced to form a single unit.

shield cone [GEOL] A cone or dome-shaped volcano built up by successive outpourings of lava.

shielding factor [GEOPHYS] The ratio of the strength of the magnetic field at a directional compass to its strength if there were no disturbing material; usually expressed as a decimal.

shielding layer [METEOROL] The layer of air nearest the earth, with reference to the manner in which this layer shields the earth from activity in the free atmosphere above, or vice versa.

shield volcano [GEOL] A broad, low volcano shaped like a flattened dome and built of basaltic lava. Also known as basaltic dome; lava dome.

shift [GEOL] The relative displacement of the units affected by a fault but outside the fault zone itself.

shimmer [METEOROL] To appear tremulous or wavering, due to varying atmospheric refraction in the line of sight.

shingle [GEOL] Pebbles, cobble, and other beach material, coarser than ordinary gravel but roughly the same size and occurring typically on the higher parts of a beach.

shingle barchan [GEOL] A dunelike ridge formed of shingle perpendicular to the beach in shallow water.

shingle beach [GEOL] A narrow beach composed of shingle and commonly having a steep slope on both its landward and seaward sides. Also known as cobble beach.

shingle-block structure See imbricate structure.

shingle rampart [GEOL] A rampart of shingle built along a reef on the seaward edge.

shingle ridge [GEOL] A steeply sloping bank of shingle heaped upon and parallel with the shore.

shingle structure See imbricate structure.

ship drift [OCEANOGR] A method of measuring ocean currents; the ship itself is used as a current tracer, its motions being measured by navigating equipment on board.

ship report [METEOROL] The encoded and transmitted report of a marine weather observation.

ship synoptic code [METEOROL] A synoptic code for communicating marine weather observations; it is a modification of the international synoptic code.

shoal [GEOL] A submerged elevation rising from the bed of a shallow body of water and consisting of, or covered by, unconsolidated material, and may be exposed at low water.

shoal breccia [PETR] A breccia formed by the action of waves and tides on a shoal, and resulting from diastrophism or aggradation.

shoaling [OCEANOGR] The bottom effect which influences the height of waves moving from deep to shallow water.

shoal patches [OCEANOGR] Individual and scattered elevations of the bottom, with depths of 10 fathoms (18 meters) or less, but composed of any material except rock or coral.

shoal reef [GEOL] A reef formed in irregular masses amid submerged shoals of calcareous reef detritus.

shoal water [OCEANOGR] Shallow water; over a shoal.

shock breccia [PETR] A fragmental rock formed by the action of shock waves, such as suevite formed by meteorite impact.

shock bump [MIN ENG] A rock bump resulting from the sudden collapse of a strong deposit.

shock lithification [GEOL] The conversion of originally loose fragmental materials into coherent aggregates by the action of shock waves, such as those generated by explosions or meteorite impacts.

shock loading [GEOPHYS] The process of subjecting material to the action of high-pressure shock waves generated by artificial explosions or by meteorite impact.

shock melting [GEOPHYS] Fusion of material as a result of the high temperatures produced by the action of high-pressure shock waves.

shock metamorphism [PETR] The complete permanent changes (physical, chemical, mineralogic, morphologic) in rocks caused by transient high-pressure shock waves that act over short-time intervals, ranging from a few microseconds to a fraction of a minute.

shock zone [GEOL] A volume of rock in or around an impact or explosion crater in which a distinctive shock-metamorphic deformation or transformation effect is present.

shoe [MIN ENG] **1.** Pieces of steel fastened to a mine cage and formed to fit over the guides to guide it when it is in motion. **2.** The bottom wedge-shaped piece attached to tubbing when sinking through quicksand. **3.** A trough to convey ore to a crusher. **4.** A coupling of rolled, cast, or forged steel to protect the lower end of the casting or drivepipe in overburden, or the bottom end of a sampler when pressed into a formation being sampled.

shoestring [GEOL] A long, relatively straight and narrow sedimentary body having a width/thickness ratio of less than 5:1, usually 1:1.

shoestring rill [GEOL] One of several long, narrow, uniform channels, closely spaced and roughly parallel with one another, that merely score the homogeneous surface of a relatively steep slope of bare soil or weak, clay-rich bedrock, and that develop wherever overland flow is intense.

shoestring sand [GEOL] A shoestring composed of sand and usually buried in mud or shale, usually a sandbar or channel fill.

shonkinite [PETR] A dark-colored syenite composed principally of augite and or tho clase with some olivine, hornblende, biotite, and nepheline.

shoofly *See* slant.

shoot [GEOL] *See* ore shoot. [GEOPHYS] The energy that goes up through the strata from a seismic profiling shot and is reflected downward at the surface or at the base of the weathering; appears either as a single wave or unites with a wave train that is traveling downward. Also known as secondary reflection. [HYD] **1.** A place where a stream flows or descends swiftly. **2.** A natural or artificial channel, passage, or trough through which water is moved to a lower level. **3.** A rush of water down a steep place or a rapids.

shore [GEOL] **1.** The narrow strip of land immediately bordering a body of water. **2.** *See* seashore.

shore current [HYD] A water current near a shoreline, often flowing parallel to the shore.

shoreface [GEOL] The narrow, steeply sloping zone between the seaward limit of the shore at low water and the nearly horizontal offshore zone.

shoreface terrace [GEOL] A wave-built terrace in the shoreface region, composed of gravel and coarse sand swept from the wave-cut bench into deeper water.

shore ice [OCEANOGR] Sea ice that has been beached by wind, tides, currents, or ice pressure; it is a type of fast ice, and may sometimes be rafted ice.

shore lead [OCEANOGR] A lead between pack ice and fast ice or between floating ice and the shore; it may be closed by wind or currents so that only a tide crack remains.

shoreline [GEOL] The intersection of a specified plane of water, especially mean high water, with the shore; a limit which changes with the tide or water level. Also known as strandline; waterline.

shoreline cycle [GEOL] The cycle of changes through which sequential forms of coastal features pass during shoreline development, from the establishment of a water level to the time when the water can do no more work.

shoreline-development ratio [GEOL] A ratio indicating the degree of irregularity of a lake shoreline, given as the length of the shoreline to the circumference of a circle whose area is equal to that of the lake.

shoreline of depression [GEOL] A shoreline of submergence that implies an absolute subsidence of the land.

shoreline of elevation [GEOL] A shoreline of emergence that implies an absolute rise of the land.

shoreline of emergence [GEOL] A straight or gently curving shoreline formed by the dominant relative emergence of the floor of an ocean or a lake. Also known as emerged shoreline; negative shoreline.

shoreline of submergence [GEOL] A shoreline, characterized by bays, promontories, and other minor features, formed by the dominant relative submergence of a landmass. Also known as positive shoreline; submerged shoreline.

shore platform [GEOL] The horizontal or gently sloping surface produced along a shore by wave erosion. Also known as scar.

shore polyn'ya [OCEANOGR] A polyn'ya between pack ice and the coast, or between pack ice and an ice front, formed by a current or by wind.

shoreside *See* onshore.

shore tank [PETRO ENG] A shoreside storage tank for liquid petroleum products discharged by tankers.

shore terrace [GEOL] **1.** A terrace produced along the shore by wave and current action. **2.** *See* marine terrace.

short-crested wave [OCEANOGR] An ocean wave whose crest is of finite length; that is, the type actually found in nature.

short-delay blasting [ENG] A method of blasting by which explosive charges are detonated in a given sequence with short time intervals.

short fuse [ENG] **1.** Any fuse that is cut too short. **2.** The practice of firing a blast, the fuse on the primer of which is not sufficiently long to reach from the top of the charge to the collar of the borehole; the primer, with fuse attached, is dropped into the charge while burning.

shortite [MINERAL] $Na_2Ca_2(CO_3)_3$ A mineral composed of sodium and calcium carbonate.

short leg [ENG] One of the wires on an electric blasting cap, which has been shortened so that when placed in the borehole, the two splices or connections will not come opposite each other and make a short circuit.

short-range forecast [METEOROL] A weather forecast made for a time period generally not greater than 48 hours in advance.

short sea [OCEANOGR] A sea whose waves are of a short, irregular, broken character.

shortwall [MIN ENG]   **1.** A method of mining in which comparatively small areas are worked separately. **2.** A length of coal face between about 5 and 30 yards (4.6 and 27 meters), generally employed in pillar methods of working.

shortwall coal cutter [MIN ENG]   A machine for undercutting coal which has a long, rigid chain jib fixed in relation to the main body of the machine and which cuts across a heading from right to left, being drawn across by means of a steel-wire rope.

shoshonite [PETR]   A basaltic rock composed of olivine and augite phenocrysts in a groundmass of labradorite with orthoclase rims, olivine, augite, a small amount of leucite, and some dark-colored glass.

shot [MIN ENG]   Coal broken by blasting or other methods.

shot boring [ENG]   The act or process of producing a borehole with a shot drill.

shot copper [GEOL]   Small, rounded particles of native copper, molded by the shape of vesicles in basaltic host rock, and resembling shot in size and shape.

shot drill *See* calyx drill.

shot-firing cable [ELEC]   A two-conductor cable which leads from the exploder to the detonator wires. Also known as firing cable.

shot-firing circuit [ELEC]   The path taken by the electric current from the exploder along the shot-firing cable, the detonator wires, and finally the detonator when a shot is detonated.

shot-firing curtain [MIN ENG]   A steel frame with chains about 6 inches (15 centimeters) apart suspended from the roof about 9 to 12 feet (2.7 to 3.7 meters) from the face of an advancing tunnel to intercept flying debris when shot-firing at the face.

shothole [ENG]   The borehole in which an explosive is placed for blasting.

shothole casing [ENG]   A lightweight pipe, usually about 4 inches (10 centimeters) in diameter and 10 feet (3 meters) long, with threaded connections on both ends, used to prevent the shothole from caving and bridging.

shothole drill [MECH ENG]   A rotary or churn drill for drilling shotholes.

shot point [ENG]   The point at which an explosion (such as in seismic prospecting) originates, generating vibrations in the ground.

shot rock [ENG]   Blasted rock.

shoulder [GEOL]   **1.** A short, rounded spur protruding laterally from the slope of a mountain or hill. **2.** The sloping segment below the summit of a mountain or hill. **3.** A bench on the flanks of a glaciated valley, located at the sharp change of slope where the steep sides of the inner glaciated valley meet the more gradual slope above the level of glaciation. **4.** A joint structure on a joint face produced by the intersection of plume-structure ridges with fringe joints.

shoved moraine *See* push moraine.

Showalter stability index [METEOROL]   A measure of the local static stability of the atmosphere, expressed as a numerical index.

shower [METEOROL]   Precipitation from a convective cloud; characterized by the suddenness with which it starts and stops, by the rapid changes of intensity, and usually by rapid changes in the appearance of the sky.

shrinkage crack [GEOL]   A small crack produced in fine-grained sediment or rock by the loss of contained water during drying or dehydration.

shrinkage index [GEOL]   The numerical difference between the plastic limit of a material and its shrinkage limit.

shrinkage limit [GEOL] That moisture content of a soil below which a decrease in moisture content will not cause a decrease in volume, but above which an increase in moisture will cause an increase in volume.

shrinkage pore [GEOL] An irregular pore formed in muddy sediment by shrinkage.

shrinkage ratio [GEOL] The ratio of a volume change to the moisture-content change above the shrinkage limit.

shrinkage stoping [MIN ENG] A modification of overhead stoping, involving the use of a part of the ore for the purpose of support and as a working platform. Also known as back stoping.

shrub-coppice dune [GEOL] A small dune formed on the leeward side of bush-and-clump vegetation.

shuga [OCEANOGR] A spongy, rather opaque, whitish chunk of ice which forms instead of pancake ice if the freezing takes place in sea water which is considerably agitated.

shungite [GEOL] A hard, black, amorphous, coallike material composed of more than 98% carbon.

shunt [MIN ENG] To shove or turn off to one side, as a car or train from one track to another.

shut-in pressure [PETRO ENG] The equilibrated reservoir pressure measured when all the gas or oil outflow has been shut off.

shut-in well [PETRO ENG] An oil or gas well that is closed off; the well is shut so that it does not produce a fluid product of any kind.

shutterridge [GEOL] A ridge formed by vertical, lateral, or oblique displacement of a fault traversing a ridge-and-valley topography with the displaced part of a ridge shutting in the adjacent ravine or canyon.

shuttle car [MIN ENG] An electrically propelled vehicle on rubber tires or caterpillar treads used to transfer raw materials, such as coal and ore, from loading machines in trackless areas of a mine to the main transportation system.

sial [PETR] A petrologic term for the silica- and alumina-rich upper rock layers of the earth's crust; gives rise to granite magma; the bulk of the continental blocks is sialic. Also known as granitic layer; sal.

Siberian anticyclone *See* Siberian high.

Siberian high [METEOROL] An area of high pressure which forms over Siberia in winter, and which is particularly apparent on mean charts of sea-level pressure; centered near lake Baikal. Also known as Siberian anticyclone.

siberite [MINERAL] A violet-red or purplish lithian variety of tourmaline.

sibjet *See* sebkha.

sicklerite [MINERAL] (Li,Mn)(PO$_4$) A dark-brown mineral composed of hydrous lithium manganese phosphate occurring in cleavable masses.

SID *See* sudden ionospheric disturbance.

side canyon [GEOL] A ravine or other valley smaller than a canyon, through which a tributary flows into the main stream.

side-centered lattice [CRYSTAL] A type of centered lattice that is centered on the side faces only.

side-discharge shovel [MIN ENG] A shovel loader having a 21-cubic-foot (0.59-cubic-meter) bucket, hinged to the chassis to dig, lift, and discharge the material sideways onto a scraper or a belt conveyor.

side drift *See* adit.

side dumper [MIN ENG] An ore, rock, or coal car that can be tilted sidewise and thus emptied.

side-end lines [MIN ENG] Limits of a mining claim looked on as boundary lines, especially in the case of veins originating within but extending outside the claim.

side pinacoid [CRYSTAL] A pinacoid with Miller indices (010) in an orthorhombic, monoclinic, or triclinic crystal.

side plates [MIN ENG] In timbering, where both a cap and a sill are used, and the posts act as spreaders, the cap and the sill are termed side plates.

sideraerolite *See* stony-iron meteorite.

siderite [MINERAL] $FeCO_3$ A brownish, gray, or greenish rhombohedral mineral composed of ferrous carbonate; hardness is 4 on Mohs scale, and specific gravity is 3.9. Also known as chalybite; iron spar; rhombohedral iron ore; siderose; sparry iron; spathic iron; white iron ore.

sideroferrite [GEOL] A variety of native iron occurring as grains in petrified wood.

siderogel [MINERAL] A mineral consisting of truly amorphous $FeO(OH)$ and occurring in some bog iron ores.

siderograph [ENG] An instrument that keeps the time of the Greenwich longitude; consists of a clock and a navigation instrument.

siderolite *See* stony-iron meteorite.

sideromelane [MINERAL] Any iron-rich mafic mineral.

sideronatrite [MINERAL] $Na_2Fe(SO_4)(OH)\cdot3H_2O$ A yellow mineral composed of basic hydrous sodium iron sulfate occurring in fibrous masses.

sideronitic texture [GEOL] In mineral deposits, a mesh of silicate minerals so shattered and pressed as to force out solutions and other volatiles.

siderophyllite [MINERAL] An iron-rich variety of biotite.

siderophyre [GEOL] A stony-iron meteorite containing bronzite and tridymite crystals in a nickel-iron network. Also known as siderophyry.

siderophyry *See* siderophyre.

siderotil [MINERAL] $(Cu,Fe)SO_4\cdot5H_2O$ A white to yellowish or pale greenish-white mineral consisting of ferrous sulfate pentahydrate; occurs as fibrous crusts and groups of needlelike crystals.

side stream *See* tributary.

sideswipe [GEOPHYS] **1.** A phenomenon wherein two cross reflections come from a single seismograph, due to the almost simultaneous arrival of reflection energy from both limbs of a syncline or from two nearby, steeply dipping fault scarps. **2.** In refraction shooting, the lateral deflection of a minimun-time path to include a nearby, steeply dipping, high-velocity boundary such as a flank of a salt dome.

side-tipping loader [MIN ENG] A front-end loading machine which discharges the bucket load by tipping it sideways.

siegenite [MINERAL] $(Co,Ni)_3S_4$ A mineral composed of nickel cobalt sulfide.

sierozem [GEOL] A soil found in cool to temperate arid regions, characterized by a brownish-gray surface on a lighter layer based on a carbonate or hardpan layer.

sierra [GEOGR] A high range of hills or mountains with irregular peaks that give a sawtooth profile.

sieve deposition [GEOL] The formation of coarse-grained lobate masses on an alluvial fan whose material is sufficiently coarse and permeable to permit complete infiltration of water before it reaches the toe of the fan.

sieve lob [GEOL] A coarse-grained lobate mass produced by sieve deposition on an alluvial fan.

sieve texture *See* poikiloblastic.

siffanto [METEOROL] A southwest wind of the Adriatic Sea; it is often violent.

sigma-t [OCEANOGR] An abbreviated value of the density of a sea-water sample of temperature T and salinity S: $\sigma T = [\rho(S, T) - 1] \times 10^3$, where $\rho(S, T)$ is the value of the sea-water density in centimeter-gram-second units at standard atmospheric pressure.

sigmoidal dune [GEOL] A dune with an S-shaped ridge crest formed by the merger of crescentic dunes.

sigmoidal fold [GEOL] A recumbent fold having an axial surface which resembles the Greek letter sigma.

significant wave [OCEANOGR] Statistically, a wave with the average height of the highest third of the waves of a given wave group.

sigua [METEOROL] A straight-blowing monsoon gale of the Philippines.

sikussak [OCEANOGR] Very old sea ice trapped in fjords; it resembles glacier ice because snowfall and snow drifts contribute to its formation.

silcrete [GEOL] A conglomerate of sand and gravel cemented by silica.

silex [MINERAL] A pure or finely ground quartz.

silexite [GEOL] Chert occurring in calcareous beds. [PETR] Igneous rock composed mainly of primary quartz.

silica [MINERAL] SiO_2 Naturally occurring silicon dioxide; occurs in five crystalline polymorphs (quartz, tridymite, cristobalite, coesite, and stishovite), in cryptocrystalline form (as chalcedony), in amorphous and hydrated forms (as opal), and combined in silicates.

silica sand [GEOL] Sand having a very high percentage of silicon dioxide; a source of silicon.

silica stone [PETR] A sedimentary rock composed of siliceous minerals.

silicate [MINERAL] Any of a large group of minerals whose crystal lattice contains SiO_4 tetrahedra, either isolated or joined through one or more of the oxygen atoms.

silication [GEOL] The conversion to or the replacement by silicates.

silicatization [MIN ENG] The sealing off of water by the injection of calcium silicate under pressure; sometimes used to reduce the leakage of water through defective lengths of tubing in a shaft.

siliceous [PETR] Describing a rock containing abundant silica, especially free silica.

siliceous dust [MIN ENG] The dust arising from the dry-working of sand, sandstone, trap, granite, and other igneous rocks; the dust is not soluble in the body fluids, and often results in a form of pneumoconiosis, known as silicosis.

siliceous earth [GEOL] A loose, friable, soft, porous, lightweight, fine-grained, and usually white siliceous sediment, usually derived from the remains of organisms.

siliceous limestone [PETR] **1.** A dense, dark, commonly thin-bedded limestone representing an intimate admixture of calcium carbonate and chemically precipitated silica that are believed to have accumulated simultaneously. **2.** A silicified limestone, bearing evidence of replacement of calcite by silica.

siliceous ooze [GEOL]   An ooze composed of siliceous skeletal remains of organisms, such as radiolarians.

siliceous sediment [GEOL]   A sediment composed of fragmental, concretionary, or precipitated siliceous materials.

siliceous shale [PETR]   A hard, fine-grained rock with the texture of shale and with as much as 85% silica.

siliceous sinter [MINERAL]   A white, lightweight, porous, opaline variety of silica, deposited by a geyser or hot spring. Also known as fiorite; geyserite; pearl sinter; sinter.

silicic [PETR]   Sard of magma or igneous rock rich in silica (usually at least 65%); granite is a silicic rock. Also known as oversaturated; persilicic.

silicification [GEOL]   Introduction of or replacement by silica. Also known as silification.

silicified wood [GEOL]   A material formed by the silicification of wood, generally in the form of opal or chalcedony, in such a manner as to preserve the original form and structure of the wood. Also known as agatized wood; opalized wood; petrified wood; woodstone.

silicinate [GEOL]   Pertaining to the silica cement of a sedimentary rock.

siliclastic [PETR]   Pertaining to clastic noncarbonate rocks which are almost exclusively silicon-bearing, either as forms of quartz or as silicates.

silicomagnesiofluorite [MINERAL] $Ca_4Mg_3Si_2O_5(OH)_2F_{10}$ A mineral composed of basic calcium magnesium fluoride and silicate.

silk [GEOL]   Microscopic needle-shaped crystalline inclusions of rutile in a natural gem from which subsurface reflections produce a whitish sheen resembling that of a silk fabric.

sill [GEOL]   **1.** Submarine ridge in relatively shallow water that separates a partly closed basin from another basin or from an adjacent sea. **2.** A tabular igneous intrusion that is oriented parallel to the planar structure of surrounding rock.   [MIN ENG]   **1.** A piece of wood laid across a drift to constitute a frame to support uprights of timber sets and to carry the track of the tramway. **2.** The floor of a gallery or passage in a mine.

sill depth [OCEANOGR]   The maximum depth at which there is horizontal communication between an ocean basin and the open ocean. Also known as threshold depth.

silled basin *See* restricted basin.

sillenite [MINERAL] Bi_2O_3 A mineral composed of native bismuth oxide, is polymorphous with bismite, and occurs as earthy masses.

sillimanite [MINERAL] Al_2SiO_5 A brown, pale-green, or white neosilicate mineral with vitreous luster crystallizing in the orthorhombic system; commonly occurs in slender crystals, often in fibrous aggregates; hardness is 6–7 on Mohs scale, and specific gravity is 3.23. Also known as fibrolite.

silt [GEOL]   **1.** A rock fragment or a mineral or detrital particle in the soil having a diameter of 0.002–0.05 millimeter that is, smaller than fine sand and larger than coarse clay. **2.** Sediment carried or deposited by water. **3.** Soil containing at least 80% silt and less than 12% clay.

silt discharge rating *See* sediment discharge rating.

silting [GEOL]   The deposition or accumulation of stream-deposited silt that is suspended in a body of standing water.

siltite *See* siltstone.

silt loam [GEOL] A soil containing 50–88% silt, 0–27% clay, and 0–50% sand.

silt shale [PETR] A consolidated sediment consisting of no more than 10% sand and having a silt/clay ratio greater than 2:1.

silt soil [GEOL] A soil containing 80% or more of silt, and not more than 12% of clay and 20% of sand.

siltstone [GEOL] Indurated silt having a shalelike texture and composition. Also known as siltite.

silttil [GEOL] A chemically decomposed and eluviated till consisting of a friable, brownish, open-textured silt that contains a few small siliceous pebbles.

Silurian [GEOL] **1.** A period of geologic time of the Paleozoic era, covering a time span of between 430–440 and 395 million years ago. **2.** The rock system of this period.

silver frost [METEOROL] A deposit of glaze built up on trees, shrubs, and other exposed objects during a fall of freezing precipitation; the product of an ice storm. Also known as silver thaw.

silver glance *See* argentite.

silver jamesonite *See* owyheeite.

silver thaw *See* silver frost.

sima [PETR] A petrologic term for the lower layer of the earth's crust, composed of silica- and magnesia-rich rocks; source of basaltic magma; sima is equivalent to the lower part of the continental crust and the bulk of the oceanic crust. Also known as intermediate layer.

similar fold [GEOL] A fold in deformed beds in which the successive folds resemble each other.

simoom [METEOROL] A strong, dry, dust-laden desert wind which blows in the Sahara, Israel, Syria, and the desert of Arabia; its temperature may exceed 130°F (54°C), and the humidity may fall below 10%.

simple conic chart [MAP] A chart on a simple conic projection.

simple conic projection [MAP] A conic map projection in which the surface of a sphere or spheroid, such as the earth, is developed on a tangent cone which is then spread out to form a plane.

simple crater [GEOL] A meteorite impact crater of relatively small diameter, characterized by a uniformly concave-upward shape and a maximum depth in the center, and lacking a central uplift.

simple cross-bedding [GEOL] Cross-bedding in which the lower bounding surfaces are nonerosional surfaces.

simple cubic lattice [CRYSTAL] A crystal lattice whose unit cell is a cube, and whose lattice points are located at the vertices of the cube.

simple dike [PETR] An igneous dike emplaced in a single episode.

simple shear [GEOPHYS] Strain caused by differential movements on one set of parallel planes which results in internal rotation of fabric elements.

simple twin [CRYSTAL] A twinned crystal composed of only two individuals in twin relation.

simple valley [GEOL] A valley that maintains a constant relation to the general structure of the underlying strata.

simpsonite [MINERAL] $AlTaO_4$ A hexagonal mineral composed of aluminum tantalum oxide and occurring in short crystals.

Simpson's rule [PETRO ENG]   A mathematical relationship for calculating the oil- or gas-bearing net-pay volume of a reservoir; uses the contour lines from a subsurface geological map of the reservoir, including gas-oil and gas-water contacts.

sincosite [MINERAL] $Ca(VO)_2(PO_4)_2 \cdot 5H_2O$ A leek-green mineral composed of hydrous calcium vanadyl phosphate and occurring in tetragonal scales or plates.

sine galvanometer [ENG]   A type of magnetometer in which a small magnet is suspended in the center of a pair of Helmholtz coils, and the rest position of the magnet is measured when various known currents are sent through the coils.

Sinemurian [GEOL]   A European stage of geologic time; Lower Jurassic, above Hattangian and below Pliensbachian.

singing sand *See* sounding sand.

single-base powder [MATER]   An explosive or propellant powder in which nitrocellulose is the only active ingredient.

single completion [PETRO ENG]   An oil or gas well drilled to produce fluids from a single reservoir level or zone, and using a single tubing string.

single-cycle mountain [GEOL]   A fold mountain that has been destroyed without reelevation of any of its important parts.

single packing [MIN ENG]   Strip packing on a longwall face in which the widest pack is along the roadside.

single-shot survey [PETRO ENG]   An oil well directional log or record with a single-reading device that is either run down into the drill pipe or positioned in a nonmagnetic drill collar.

single-station analysis [METEOROL]   The analysis or reconstruction of the weather pattern from more or less continuous meteorological observations made at a single geographic location, or the body of techniques employed in such an analysis.

single-theodolite observation [METEOROL]   The usual type of pilot-balloon observation, that is, using one theodolite.

singular corresponding point [METEOROL]   A center of elevation or depression on a constant-pressure chart (or a center of high or low pressure on a constant-height chart) considered as a reappearing characteristic of successive charts.

singularity [METEOROL]   A characteristic meteorological condition which tends to occur on or near a specific calendar date more frequently than chance would indicate; an example is the January thaw.

sinhalite [MINERAL] $MgAl(BO_4)$ A mineral composed of magnesium aluminum borate; sometimes used as a gem.

sinistral fault *See* left lateral fault.

sinistral fold [GEOL]   An asymmetric fold whose long limb, when viewed along its dip, appears to have a leftward offset.

sink [GEOL]   **1.** A circular or ellipsoidal depression formed by collapse on the flank of or near to a volcano. **2.** A slight, low-lying desert depression containing a central playa or saline lake with no outlet, as where a desert stream comes to an end or disappears by evaporation.   [MIN ENG]   **1.** To excavate strata downward in a vertical line for the purpose of winning and working minerals. **2.** To drill or put down a shaft or borehole.

sinker [MIN ENG]   **1.** A person who sinks mine shafts and puts in framing. **2.** A special movable pump used in shaft sinking. **3.** *See* sinker drill.

sinker bar [MIN ENG]   A short, heavy rod placed above the drill jars to increase the effect of the upward sliding jars in well-drilling with cable tools.

sinker drill   [MIN ENG]   A jackhammer type of rock drill used in shaft sinkings. Also known as sinker.

sinkhole   [GEOL]   Closed surface depressions in regions of karst topography produced by solution of surface limestone or the collapse of cavern roofs.

sinkhole plain   [GEOL]   A regionally extensive plain or plateau characterized by well-developed karst features.

sinking   [OCEANOGR]   The downward movement of surface water generally caused by converging currents or when a water mass becomes denser than the surrounding water. Also known as downwelling.

sinking and walling scaffold   [MIN ENG]   A platform designed for use in shaft sinking to enable sinking and walling to be performed simultaneously. Also known as Galloway sinking and walling stage.

sinking bucket   *See* hoppit.

sinking center   [OCEANOGR]   A subarctic region where very saline surface water from tropical regions sinks through the colder and less saline layer beneath.

sinking pump   [MIN ENG]   A long, narrow, electrically driven centrifugal-type pump designed for keeping a shaft dry during sinking operations.

sinoite   [MINERAL] Si_2N_2O A nitride mineral known only in meteorites.

sinople   [MINERAL]   A blood-red or brownish-red (with a tinge of yellow) variety of quartz containing inclusions of hematite.

sinter   [MINERAL]   *See* siliceous sinter.   [PETR]   A chemical sedimentary rock deposited by precipitation from mineral waters, especially siliceous sinter and calcareous sinter.

siphon barograph   [ENG]   A recording siphon barometer.

siphon barometer   [ENG]   A J-shaped mercury barometer in which the stem of the J is capped and the cusp is open to the atmosphere.

Siphonotretacea   [PALEON]   A superfamily of extinct, inarticulate brachiopods in the suborder Acrotretidina of the order Acrotretida having an enlarged, tear-shaped, apical pedicle valve.

sirocco   [METEOROL]   A warm south or southeast wind in advance of a depression moving eastward across the southern Mediterranean Sea or North Africa.

siserskite   [MINERAL]   A light steel gray mineral consisting of an alloy of osmium and iridium; occurs in tabular form.

sitaparite   *See* bixbyite.

size analysis   *See* particle-size analysis.

size-frequency analysis   *See* particle-size analysis.

sjogrenite   [MINERAL] $Mg_6Fe_2(OH)_{16}(CO_3) \cdot 4H_2O$ A hexagonal mineral composed of hydrous basic magnesium iron carbonate.

skarn   [GEOL]   A lime-bearing silicate derived from nearly pure limestone and dolomite with the introduction of large amounts of silicon, aluminum, iron, and magnesium.

skauk   [HYD]   An extensive field of crevasses in a glacier.

skavl   *See* sastruga.

skeleton grain   [GEOL]   A relatively stable and not readily translocated grain of soil material, concentrated or reorganized by soil-forming processes.

skeleton layer [OCEANOGR]   The structure that is formed at the bottom of sea ice while freezing, and consists of vertically oriented platelets of ice separated by layers of brine.

skeleton texture [PETR]   Descriptive of the texture of limestone that consists of an in-place accumulation of skeletal material, that is, the hard parts secreted by organisms.

skerry [GEOL]   A low, small, rugged and rocky island or reef.

skialith [PETR]   A vague remnant of country rock assimilated in granite.

skid [MIN ENG]   An arrangement upon which certain coal-cutting machines travel along the working faces.

skid boulder [GEOL]   An isolated angular block of stone resting on the floor of a playa, derived from an outcrop near the playa margin, and associated with a trail or mark indicating that the boulder has recently slid across the mud surface.

Skiddavian *See* Arenigian.

skill score [METEOROL]   In synoptic meteorology, an index of the degree of skill of a set of forecasts, expressed with reference to some standard such as forecasts based upon chance, persistence, or climatology.

skim ice [HYD]   First formation of a thin layer of ice on the water surface.

skimming [HYD]   **1.** Diversion of water from a stream or conduit by shallow overflow in order to avoid diverting sand, silt, or other debris carried as bottom load. **2.** Withdrawal of fresh groundwater from a thin body or lens floating on salt water by means of shallow wells or infiltration galleries.

skin effect [PETRO ENG]   The restriction to fluid flow through a reservoir adjacent to the borehole; calculated as a factor of reservoir pressure, product rate, formation volume and thickness, porosity, and other related parameters.

skin-friction coefficient [METEOROL]   A dimensionless drag coefficient expressing the proportionality between the frictional force per unit area, or the shearing stress exerted by the wind at the earth's surface, and the square of the surface wind speed.

skiou *See* morvan.

skip cast [GEOL]   The cast of a skip mark.

skip mark [GEOL]   A crescent-shaped mark that is one of a linear pattern of regularly spaced marks made by an object that skipped along the bottom of a stream.

skip shaft [MIN ENG]   A mine shaft prepared for hauling a skip.

skiron [METEOROL]   The Greek name for the northwest wind, which is cold in winter but hot and dry in summer.

skleropelite [PETR]   An argillaceous or allied rock which has been indurated by low-grade metamorphism, is more massive and dense than shale, and differs from slate by the absence of cleavage.

skolite [MINERAL]   A scaly, dark-green variety of glauconite rich in aluminum and calcium and deficient in ferric iron.

skomerite [PETR]   A fine-grained, compact extrusive rock containing microscopic grains and crystals of augite, olivine, and phenocrysts of decomposed plagioclase (probably albite) in a groundmass of plagioclase, thought to be more calcic than the phenocrysts.

skutterudite [MINERAL] $(Co,Ni)As_3$ A tin-white mineral with metallic luster composed of cobalt and nickel arsenides; crystallizes in the isometric system but commonly is massive; hardness is 5.5–6 on Mohs scale, and specific gravity is 6.6; it is a minor ore of cobalt and nickel.

sky cover [METEOROL] In surface weather observations, the amount of sky covered but not necessarily concealed by clouds or by obscuring phenomena aloft, the amount of sky concealed by obscuring phenomena that reach the ground, or the amount of sky covered or concealed by a combination of the two phenomena.

skyhook [MIN ENG] To drive bolts into the overhead rock of a mine in order to reinforce the ceiling.

sky map [METEOROL] A pattern of variable brightness observable on the underside of a cloud layer, and caused by the different reflectivities of material on the earth's surface immediately beneath the clouds; this term is used mainly in polar regions.

slab [GEOL] A cleaved or finely parallel jointed rock, which splits into tabular plates from 1 to 4 inches (2.5 to 10 centimeters) thick. Also known as slabstone. [HYD] A layer in, or the whole-thickness of, a snowpack that is very hard and has the ability to sustain elastic deformation under stress. [MIN ENG] A slice taken off the rib of an entry or room in a mine.

slabbing machine [MIN ENG] A coal-cutting machine designed to make cuts in the side of a room or entry pillar preparatory to slabbing.

slabbing method [MIN ENG] A method of mining pillars in which successive slabs are cut from one side or rib of the pillar after a room is finished, until as much of the pillar is removed as can safely be recovered.

slab entry [MIN ENG] An entry which is widened or slabbed to provide a working place for a second miner.

slab jointing [GEOL] Jointing produced in rock by the formation of numerous cleaved or closely spaced parallel fissures dividing the rock into thin slabs.

slab pahoehoe [GEOL] A pahoehoe whose surface consists of a jumbled arrangement of slabs of flow crust.

slack barrel [PETRO ENG] A petroleum-industry container used for shipment of petroleum paraffin; generally contains 235 to 245 pounds (107 to 111 kilograms) net, and is of lighter construction than the ordinary oil barrel but of the same general shape.

slack ice [HYD] Ice fragments on still or slow-moving water.

slack water [OCEANOGR] The interval when the speed of the tidal current is very weak or zero; usually refers to the period of reversal between ebb and flood currents.

slaking [GEOL] **1.** Crumbling and disintegration of earth materials when exposed to air or moisture. **2.** The breaking up of dried clay when saturated with water.

slant [MIN ENG] **1.** Any short, inclined crosscut connecting the entry with its air course to facilitate the hauling of coal. Also known as shoofly. **2.** A heading driven diagonally between the dip and the strike of a coal seam. Also known as run.

slate [PETR] A group name for various very-fine-grained rocks derived from mudstone, siltstone, and other clayey sediment as a result of low-degree regional metamorphism; characterized by perfect fissility or slaty cleavage which is a regular or perfect planar schistosity.

slate ribbon [GEOL] A relict ribbon structure on the cleavage surface of slate, in which varicolored and straight, wavy, or crumpled stripes cross the cleavage surface.

slaty cleavage *See* flow cleavage.

slavikite [MINERAL] $MgFe_3^{3+}(SO_4)_4(OH)_3 \cdot 18H_2O$ A greenish-yellow mineral composed of hydrous basic magnesium ferric sulfate and occurring as rhombohedral crystals.

sleet [METEOROL] Colloquially in some parts of the United States, precipitation in the form of a mixture of snow and rain.

slice [GEOL]   An arbitrary section of some uniform standard, such as thickness of a stratigraphic unit that is otherwise indivisible for purposes of analytic study. [MIN ENG]   **1.** A thin broad piece cut off, as a portion of ore cut from a pillar or face.   **2.** To remove ore by successive slices.

slice drift [MIN ENG]   In sublevel caving, the crosscuts driven between every other slice from 18 to 36 feet (5.5 to 11.0 meters) apart.

slice method [METEOROL]   A method of evaluating the static stability over a limited area at any reference level in the atmosphere; unlike the parcel method, the slice method takes into account continuity of mass by considering both upward and downward motion.

slicing method [MIN ENG]   Removal of a horizontal layer from a massive ore body.

slick [OCEANOGR]   Area in which capillary waves are absent or suppressed.

slickens [GEOL]   A layer of fine silt deposited by a flooding stream.   [MIN ENG] The light earth removed by sluicing in hydraulic mining.

slickenside [GEOL]   A surface that is polished and smoothly striated and results from slippage along a fault plane.

slickolite [GEOL]   A vertically discontinuous slip-scratch surface made by slippage and shearing and developed on sharply dipping bedding planes of limestone that shapes the wall of a solution cavity.

slide [GEOL]   **1.** A vein of clay intersecting and dislocating a vein vertically, or the vertical dislocation itself.   **2.** A rotational or planar mass movement of earth, snow, or rock resulting from failure under shear stress along one or more surfaces.   [MIN ENG]   **1.** An upright rail fixed in a shaft with corresponding grooves for steadying the cages.   **2.** A trough used to guide and to support rods in a tripod when drilling an angle hole. Also known as rod slide.

sliding *See* gravitational sliding.

sliding scale [METEOROL]   A set of combinations of ceilings and visibilities which constitute the operational weather limits at an airport; as the observed value of one element increases, the limiting value of the other element decreases, and vice versa.

sliming [OCEANOGR]   The formation of films of algae on submerged structures.

slip [GEOL]   The actual relative displacement along a fault plane of two points which were formerly adjacent on either side of the fault. Also known as actual relative movement; total displacement.

slip bedding [GEOL]   Convolute bedding formed as the result of subaqueous sliding.

slip block [GEOL]   A separate rock mass that has slid away from its original position and come to rest down the without undergoing much deformation.

slip cleavage [GEOL]   Cleavage that is superposed on slaty cleavage or schistosity, characterized by spaced cleavage with thin tabular bodies of rock between the cleavage planes. Also known as close-joints cleavage; crenulation cleavage; shear cleavage; strain-slip cleavage.

slip face [GEOL]   The steeply sloping leeward surface of a sand dune. Also known as sandfall.

slip joint *See* shear joint.

slip-off slope [GEOL]   The long, low, gentle slope on the inside of the downstream face of a stream meander.

slippage [PETRO ENG]   The movement of gas past or through a liquid-phase reservoir front; this movement occurs instead of driving the liquid forward; it can exist in gas-drive reservoirs or in gas-lift oil-well bores.

slipping [MIN ENG]   Enlarging an excavation by breaking one or more walls.

slip plane [CRYSTAL]   *See* glide plane.   [GEOL]   A planar slip surface.

slip sheet [GEOL]   A stratum or rock on the limb of an anticline that has slid down and away from the anticline; a gravity collapse structure.

slip surface [GEOL]   The displacement surface of a landslide.

slip-weld hanger [PETRO ENG]   A type of hanger used to suspend the lower strings of an oil-well casing pipe.

slope [GEOL]   The inclined surface of any part of the earth's surface.

slope correction [GEOL]   A tape correction applied to a distance measured on a slope in order to reduce it to a horizontal distance, between the vertical lines through its end points.

slope current *See* gradient current.

slope engineer [MIN ENG]   In anthracite and bituminous coal mining, one who operates a hoisting engine to haul loaded and empty mine cars along a haulage road in a mine.

slope failure [GEOL]   The downward and outward movement of a mass of soil beneath a natural slope or other inclined surface; four types of slope failure are rockfall, rock flow, plane shear, and rotational shear.

slope gully [GEOL]   A small, discontinuous submarine valley, usually formed by slumping along a fault scarp or the slope of a river delta. Also known as sea gully.

slope mine [MIN ENG]   A mine opened by a slope or an incline.

slope stability [GEOL]   The resistance of an inclined surface to failure by sliding or collapsing.

slope stake [MIN ENG]   Stake set at the point where the finished side slope of an excavation or embankment meets the original grade.

slope wash [GEOL]   **1.** The mass-wasting process, assisted by nonchanneled running water, by which rock and soil is transported down a slope, specifically, sheet erosion. **2.** The material that is or has been transported.

slot [MIN ENG]   To hole; to undercut or channel.

slough [ENG]   The fragments of rocky material from the wall of a borehole.   [HYD]   A minor marshland or tidal waterway which usually connects other tidal areas; often more or less equivalent to a bayou.

slow igniter cord [ENG]   An igniter cord made with a central copper wire around which is extruded a plastic incendiary material with an iron wire embedded to give greater strength; the whole is enclosed in a thin extruded plastic coating.

slow ion *See* large ion.

slud [GEOL]   **1.** Muddy material which has moved downslope by solifluction.   **2.** Ground that behaves as a viscous fluid, including material moved by solifluction and by mechanisms not limited to gravitational flow.

sludge [GEOL]   A soft or muddy bottom deposit as on tideland or in a stream bed.   [OCEANOGR]   A dense, soupy accumulation of new sea ice consisting of incoherent floating frazil crystals. Also known as cream ice; sludge ice; slush.

sludge assay [MIN ENG]   The chemical assaying of drill cuttings for a specific metal or group of metals.

sludge box [MIN ENG]   A wooden box in which sludge settles from the mud flush.

sludge cake [OCEANOGR]   An accumulation of sludge hardened into a cake strong enough to bear the weight of a man.

sludge floe [OCEANOGR] Sludge that is hardened into a floe strong enough to bear the weight of a person.

sludge ice *See* sludge.

sludge lump [OCEANOGR] An irregular mass of sludge formed as a result of strong winds.

sludge pit *See* slushpit.

sludge pond *See* slushpit.

sluff [ENG] The mud cake detached from the wall of a borehole. [MIN ENG] The falling of decomposed, soft rocks from the roof or walls of mine openings.

slug [MIN ENG] To inject a borehole with cement, slurry, or various liquids containing shredded materials in an attempt to restore lost circulation by sealing off the openings in the borehole-wall rocks.

sluice box [MIN ENG] A long, inclined trough or launder with riffles in the bottom that provide a lodging place for heavy minerals in ore concentration.

sluice tender [MIN ENG] In metal mining, a laborer who tends sluice boxes.

sluicing [MIN ENG] **1.** Washing auriferous earth through sluices provided with riffles and other gold-saving appliances. **2.** Separation of minerals in a flowing stream of water. **3.** Moving earth, sand, gravel, or other rock or mineral materials by flowing water.

slump [GEOL] A type of landslide characterized by the downward slipping of a mass of rock or unconsolidated debris, moving as a unit or several subsidiary units, characteristically with backward rotation on a horizontal axis parallel to the slope; common on natural cliffs and banks and on the sides of artificial cuts and fills.

slump ball [GEOL] A relatively flattened mass of sandstone resembling a large concretion, measuring from 2 centimeters to 3 meters across, commonly thinly laminated with internal contortions and a smooth or lumpy external form, and formed by subaqueous slumping.

slump basin [GEOL] A shallow basin near the base of a canyon wall and on a shale hill or ridge, formed by small, irregular slumps.

slump bedding [GEOL] Also known as slurry bedding. **1.** Any disturbed bedding. **2.** Convolute bedding produced by subaqueous slumping or lateral movement of newly deposited sediment.

slump fault *See* normal fault.

slump fold [GEOL] An intraformational fold produced by slumping of soft sediments, as at the edge of the continental shelf.

slump overfold [GEOL] A fold consisting of hook-shaped masses of sandstone produced during slumping.

slump scarp [GEOL] A low cliff or rim of thin solidified lava occurring along the margins of a lava flow and against the valley walls or around steptoes after the central part of the lava crust collapsed due to outflow of still-molten underlying layers.

slump sheet [GEOL] A well-defined bed of limited thickness and wide horizontal extent, containing slump structures.

slump structure [GEOL] Any sedimentary structure produced by subaqueous slumping.

slurry bedding *See* slump bedding.

slurry mining [MIN ENG] The hydraulic breakdown of a subsurface ore matrix with drill-hole equipment, and the eduction of the resulting slurry to the surface for processing.

slurry slump [GEOL] A slump in which the incoherent sliding mass is mixed with water and disintegrates into a quasiliquid slurry.

slush [HYD] Snow or ice on the ground that has been reduced to a soft, watery mixture by rain, warm temperature, or chemical treatment. [OCEANOGR] *See* sludge.

slush avalanche [GEOL] A rapid and far-reaching downslope transport of rock debris released by snow supersaturated with meltwater and marking the catastrophic opening of ice- and snow-dammed brooks to the spring flood.

slush ball [HYD] An extremely compact accretion of snow, frazil, and ice particles.

slush field [HYD] An area of water-saturated snow having a soupy consistency. Also known as snow swamp.

slushflow [HYD] **1.** A mudflow-like outburst of water-saturated snow along a stream course, commonly occurring in the Arctic Zone after intense thawing has produced more meltwater than can drain through the snow, and having a width generally several times greater than that of the stream channel. **2.** A flow of clear slush on a glacier, as in Greenland.

slush icing [METEOROL] The accumulation of ice and water on exposed surfaces of aircraft when the craft is flown through wet snow or snow and liquid drops at temperatures near 0°C.

slushpit [ENG] An excavation or diked area to hold water, mud, sludge, and other discharged matter from an oil well. Also known as mud pit; sludge pit; sludge pond.

slush pond [HYD] A pool or lake containing slush, on the ablation surface of a glacier.

slush pump [PETRO ENG] A pump used to circulate the drilling fluid during rotary drilling.

small circle [GEOD] A circle on the surface of the earth, the plane of which does not pass through the earth's center.

small-craft warning [METEOROL] A warning, for marine interests, of impending winds up to 28 knots (32 miles per hour or 52 kilometers per hour).

small diurnal range [OCEANOGR] The difference in height between mean lower high water and mean higher low water.

small hail [METEOROL] Frozen precipitation consisting of small, semitransparent, roundish grains, each grain consisting of a snow pellet surrounded by a very thin ice covering, giving it a glazed appearance.

small ice floe [OCEANOGR] An ice floe of sea ice 30 to 600 feet (9 to 180 meters) across.

small ion [METEOROL] An atmospheric ion of the type that has the greatest mobility; and hence, collectively, it is the principal agent of atmospheric conduction; evidence indicates that each ion is a singly-charged atmospheric molecule (or, rarely, an atom) about which a few other neutral molecules are held by the electrical attraction of the central ionized molecule; estimates of the number of satellite molecules range as high as 12. Also known as fast ion; light ion.

small-ion combination [METEOROL] Either of two processes by which small ions disappear: the union of a small ion and a neutral Aitken nucleus to form a new large ion, or the neutralization of a large ion by the small ion.

small tropic range [OCEANOGR] The difference in height between tropic lower high water and tropic higher low water.

smaltite [MINERAL] $(Co,Ni)As_{3-x}$ A metallic-gray isometric mineral composed of nickel cobalt arsenide.

smaragd *See* emerald.

smaragdite [MINERAL]   A green amphibole mineral that is pseudomorphous after pyroxene in rocks such as eclogite.

smectite [MINERAL]   Dioctahedral (montmorillonite) and trioctahedral (saponite) clay minerals, and their chemical varieties characterized by swelling properties and high cation-exchange capacities.

smithite [MINERAL] $AgAsS_2$ A red monoclinic mineral composed of silver arsenic sulfide and occurring as small crystals.

smithsonite [MINERAL] $ZnCO_3$ White, yellow, gray, brown, or green secondary carbonate mineral associated with sphalerite and commonly reniform, botryoidal, stalactitic, or granular; hardness is 5 on Mohs scale, and specific gravity is 4.30–4.45; it is an ore of zinc. Also known as calamine; dry-bone ore; szaskaite; zinc spar.

smog [METEOROL]   Air pollution consisting of smoke and fog.

smoke horizon [METEOROL]   The top of a smoke layer which is confined by a low-level temperature inversion in such a way as to give the appearance of the horizon when viewed from above against the sky; in such instances the true horizon is usually obscured by the smoke layer.

smokes [METEOROL]   Dense white haze and dust clouds common in the dry season on the Guinea coast of Africa, particularly at the approach of the harmattan.

smokestone *See* smoky quartz.

smoky quartz [MINERAL]   A smoky-yellow, smoky-brown, or brownish-gray, often transparent variety of crystalline quartz containing inclusions of carbon dioxide; may be used as a semiprecious stone. Also known as cairngorm; smokestone.

smooth chert [GEOL]   A hard, dense, homogeneous chert (insoluble residue) characterized by a conchoidal-to-even fracture surface that is devoid of roughness and by a lack of crystallinity, granularity, or other distinctive structure.

smooth drilling [ENG]   Drilling in a rock formation in which a fast rotation of the drill stem, a fast rate of penetration, and a high recovery of core can be achieved with vibration-free rotation of the drill stem.

smooth phase [GEOL]   The part of stream traction whereby a mass of sediment travels as a sheet with gradually increasing density from the surface downward.

smooth sea [OCEANOGR]   Sea with waves no higher than ripples or small wavelets.

smothered bottom [GEOL]   A sedimentary surface on which complete, well-preserved, and commonly very fragile and delicate fossils were saved by an influx of mud that buried them instantly.

snaking stream *See* meandering stream.

snap [METEOROL]   A brief period of extreme (generally cold) weather setting in suddenly, as in a "cold snap."

snezhura *See* snow slush.

sno *See* elvegust.

snout [GEOGR]   A promontory or protruding mass of rock.   [HYD]   The protruding lower extremity of a glacier.

snow [METEOROL]   The most common form of frozen precipitation, usually flakes of starlike crystals, matted ice needles, or combinations, and often rime-coated.

snow accumulation [METEOROL]   The actual depth of snow on the ground at any instant during a storm, or after any single snowstorm or series of snowstorms.

snow avalanche [HYD]   An avalanche of relatively pure snow; some rock and earth material may also be carried downward. Also known as snowslide.

snow banner   [METEOROL]   Snow being blown from a mountain crest. Also known as snow plume; snow smoke.

snow barchan   [HYD]   A crescentic or horseshoe-shaped snow dune of windblown snow with the ends pointing downwind. Also known as snow medano.

snow bin   [ENG]   A box for measuring the amount of snowfall; a type of snow gage.

snow blink   [METEOROL]   A bright, white glare on the underside of clouds, produced by the reflection of light from a snow-covered surface; this term is used in polar regions with reference to the sky map. Also known as snow sky.

snowbridge   [HYD]   Snow bridging a crevasse in a glacier.

snow cap   [HYD]   **1.** Snow covering a mountain peak when no snow exists at lower elevations. **2.** Snow on the surface of a frozen lake.

snow climate   *See* polar climate.

snow cloud   [METEOROL]   A popular term for any cloud from which snow falls.

snow concrete   [HYD]   Snow that is compacted at low temperatures by heavy objects (as by a vehicle) and that sets into a tough substance of considerably greater strength than uncompressed snow. Also known as snowcrete.

snow course   [HYD]   An established line, usually from several hundred feet to as much as a mile long, traversing representative terrain in a mountainous region of appreciable snow accumulation; along this course, measurements of snow cover are made to determine its water equivalent.

snow cover   [HYD]   **1.** All accumulated snow on the ground, including that derived from snowfall, snowslides, and drifting snow. Also known as snow mantle. **2.** The extent, expressed as a percentage, of snow cover in a particular area.

snow-cover chart   [METEOROL]   A synoptic chart showing areas covered by snow and contour lines of snow depth.

snowcreep   [HYD]   The slow internal deformation of a snowpack resulting from the stress of its own weight and metamorphism of snow crystals.

snowcrete   *See* snow concrete.

snow crust   [HYD]   A crisp, firm, outer surface upon snow.

snow crystal   [METEOROL]   Any of several types of ice crystal found in snow; a snow crystal is a single crystal, in contrast to a snowflake which is usually an aggregate of many single snow crystals.

snow cushion   [HYD]   An accumulation of snow, commonly deep, soft, and unstable, deposited in the lee of a cornice on a steep mountain slope.

snow density   [HYD]   The ratio of the volume of meltwater that can be derived from a sample of snow to the original volume of the sample; strictly speaking, this is the specific gravity of the snow sample.

snowdrift   [HYD]   Snow deposited on the lee of obstacles, lodged in irregularities of a surface, or collected in heaps by eddies in the wind.

snowdrift glacier   [HYD]   A semipermanent mass of firn, formed by drifted snow in depressions in the ground or behind obstructions. Also known as catchment glacier; drift glacier.

snowdrift ice   [HYD]   Permanent or semipermanent masses of ice, formed by the accumulation of drifted snow in the lee of projections, or in depressions of the ground. Also known as glacieret.

snow dune   [HYD]   An accumulation of wind-transported snow resembling the forms of sand dunes.

snow dust [METEOROL]   Fine snow crystals fragmented or driven by the wind.

snow eater [METEOROL]   **1.** Any warm wind blowing over a snow surface; usually applied to a foehn wind. **2.** A fog over a snow surface; so called because of the frequently observed rapidity with which a snow cover disappears after a fog sets in.

snowfall [METEOROL]   **1.** The rate at which snow falls; in surface weather observations, this is usually expressed as inches of snow depth per 6-hour period. **2.** A snow storm.

snowfield [HYD]   **1.** A broad, level, relatively smooth and uniform snow cover on ground or ice at high altitudes or in mountainous regions above the snow line. **2.** The accumulation area of a glacier. **3.** A small glacier or accumulation of perennial ice and snow too small to be designated a glacier.

snowflake [METEOROL]   An ice crystal or, much more commonly, an aggregation of many crystals which falls from a cloud; simple snowflakes (single crystal) exhibit beautiful variety of form, but the symmetrical shapes reproduced so often in photomicrographs are not actually found frequently in snowfalls; broken single crystals, fragments, or clusters of such elements are much more typical of actual snows.

snowflake obsidian [PETR]   An obsidian that contains white, gray, or reddish spherulites ranging in size from microscopic to a meter or more in diameter.

snow flurry [METEOROL]   Popular term for snow shower, particularly of a very light and brief nature.

snowflush [GEOL]   An accumulation of drifted snow, windblown soil, and wind-transported seeds on a lee slope, characteristically marked during the winter by a dark patch of soil.

snow forest climate [CLIMATOL]   A major category in W. Koppen's climatic classification, defined by a coldest-month mean temperature of less than 26.6°F (3°C) and a warmest-month mean temperature of greater than 50°F (10°C).

snow gage [HYD]   An instrument for measuring the amount of water equivalent in a snowpack. Also known as snow sampler.

snow garland [HYD]   A rare phenomenon in which snow is festooned from trees, fences, and so on, in the form of a rope of snow, several feet long and several inches in diameter; produced by surface tension acting in thin films of water bonding individual crystals; such garlands form only when the surface temperature is close to the melting point, for only then will the requisite films of slightly supercooled water exist.

snow geyser [METEOROL]   Fine, powdery snow blown upward by a snow tremor.

snow glide [HYD]   The slow slip of a snowpack over the ground surface caused by the stress of its own weight.

snow grains [METEOROL]   Precipitation in the form of very small, white opaque particles of ice; the solid equivalent of drizzle; the grains resemble snow pellets in external appearance, but are more flattened and elongated, and generally have diameters of less than 1 millimeter; they neither shatter nor bounce when they hit a hard surface. Also known as granular snow.

snow ice [HYD]   Ice crust formed from snow, either by compaction or by the refreezing of partially thawed snow.

snow line [GEOGR]   **1.** A transient line delineating a snow-covered area or altitude. **2.** An area with more than 50% snow cover. **3.** The altitude or geographic line separating areas in which snow melts in summer from areas having perennial ice and snow.

snow mantle *See* snow cover.

snow medano *See* snow barchan.

snowmelt [HYD]   The water resulting from the melting of snow; it may evaporate, seep into the ground, or become a part of runoff.

snow niche *See* nivation hollow.

snowpack [HYD]   The amount of annual accumulation of snow at higher elevations in the western United States, usually expressed in terms of average water equivalent.

snow patch erosion *See* nivation.

snow pellets [METEOROL]   Precipitation consisting of white, opaque, approximately round (sometimes conical) ice particles which have a snowlike structure and are about 2 to 5 millimeters in diameter; snow pellets are crisp and easily crushed, differing in this respect from snow grains, and they rebound when they fall on a hard surface and often break up. Also known as graupel; soft hail; tapioca snow.

snow plume *See* snow banner.

snowquake *See* snow tremor.

snow resistograph [ENG]   An instrument for recording a hardness profile of a snow cover by recording the force required to move a blade up through the snow.

snow ripple *See* wind ripple.

snow roller [HYD]   A mass of snow, shaped somewhat like a lady's muff, rather common in mountainous or hilly regions; it occurs when snow, moist enough to be cohesive, is picked up by wind blowing down a slope and rolled onward and downward until either it becomes too large or the ground levels off too much for the wind to propel it further; snow rollers vary in size from very small cylinders to some as large as 4 feet (1.2 meters) long and 7 feet (2.1 meters) in circumference.

snow sampler [ENG]   A hollow tube for collecting a sample of snow in place. Also known as snow tube.   [HYD]   *See* snow gage.

snow sky *See* snow blink.

snowslide *See* snow avalanche.

snow sludge [OCEANOGR]   Sludge formed mainly from snow.

snow slush [HYD]   Slush formed from snow that has fallen into water that is at a temperature below that of the snow. Also known as snezhura.

snow smoke *See* snow banner.

snow stage [METEOROL]   The thermodynamic process of sublimation of water vapor into snow in an idealized saturation-adiabatic or pseudoadiabatic expansion (lifting) of moist air; the snow stage begins at the condensation level when it is higher than the freezing level.

snowstorm [METEOROL]   A storm in which snow falls.

snow survey [HYD]   The process of determining depth and water content of snow at representative points, for example, along a snow course.

snow swamp *See* slush field.

snow tremor [HYD]   A disturbance in a snowfield, caused by the simultaneous settling of a large area of thick snow crust or surface layer. Also known as snowquake.

snub [MIN ENG]   **1.** To increase the height of an undercut by means of explosives or otherwise. **2.** To check the descent of a car by the turn of a rope around a post.

soaked zone [HYD]   The area on a glacier where considerable surface melting occurs in summer; meltwater percolates through the whole mass of the snow layer, bringing it all to the melting temperature.

soaprock *See* soapstone.

soapstone [MINERAL] **1.** A mineral name applied to steatite or to massive talc. Also known as soaprock. **2.** *See* saponite. [PETR] A metamorphic rock characterized by massive, schistose, or interlaced fibrous texture and a soft unctuous feel.

socked in [METEOROL] In the early days of aviation, pertaining to weather at an airport when ceiling or visibility were of such low values that the airport was effectively closed to aircraft operations.

sodaclase *See* albite.

soda-granite [PETR] **1.** A granite in which soda is more abundant than potash. **2.** A granite that contains soda-plagioclase instead of the orthoclase found in normal granite.

soda lake [HYD] An alkali lake rich in dissolved sodium salts, especially sodium carbonate, sodium chloride, and sodium sulfate. Also known as natron lake.

sodalite [MINERAL] $Na_2Al_3Si_3O_{12}Cl$ A blue or sometimes white, gray, or green mineral tectosilicate of the feldspathoid group, crystallizing in the isometric system, with vitreous luster, hardness of 5 on Mohs scale, and specific gravity of 2.2–2.4; used as an ornamental stone.

soda mica *See* paragonite.

soda microcline *See* anorthoclase.

soda niter [MINERAL] $NaNO_3$ A colorless to white mineral composed of sodium nitrate, crystallizing in the rhombohedral division of the hexagonal system; hardness is 1½ to 2 on Mohs scale and specific gravity is 2.266. Also known as nitratine; Peru saltpeter.

soddyite [MINERAL] $(UO_2)_{12}Si_5O_{22}\cdot14H_2O$ A pale-yellow orthorhombic mineral composed of hydrous uranium silicate and occurring in fine-grained aggregates or crystals.

sodium-calcium feldspar *See* plagioclase.

sodium feldspar *See* albite.

sodium illite *See* brammalite.

soffione [GEOL] A jet of steam and other vapors issuing from the ground in a volcanic area.

soffosian knob *See* frost mound.

soft coal *See* bituminous coal.

soft ground [MIN ENG] **1.** A mineral deposit which can be mined without drilling and shooting hard rock. **2.** The rock about underground openings that does not stand well and requires heavy timbering.

soft hail *See* snow pellets.

soft rime [HYD] A white, opaque coating of fine rime deposited chiefly on vertical surfaces, especially on points and edges of objects, generally in supercooled fog.

soft rock [MIN ENG] Rock that can be removed by air-operated hammers, but cannot be handled economically by a pick. [PETR] **1.** A broad designation for sedimentary rock. **2.** A rock that is relatively nonresistant to erosion.

Sohm Abyssal Plain [GEOL] A basin in the North Atlantic, about 2400 fathoms deep, between Newfoundland and the Mid-Atlantic Ridge.

soil [GEOL] **1.** Unconsolidated rock material over bedrock. **2.** Freely divided rock-derived material containing an admixture of organic matter and capable of supporting vegetation.

soil air [GEOL]   The air and other gases in spaces in the soil; specifically, that which is found within the zone of aeration. Also known as soil atmosphere.

soil atmosphere *See* soil air.

soil blister *See* frost mound.

soil chemistry [GEOCHEM]   The study and analysis of the inorganic and organic components and the life cycles within soils.

soil colloid [GEOL]   Colloidal complex of soils composed principally of clay and humus.

soil complex [GEOL]   A mapping unit used in detailed soil surveys; consists of two or more recognized classifications.

soil creep [GEOL]   The slow, steady downhill movement of soil and loose rock on a slope. Also known as surficial creep.

soil erosion [GEOL]   The detachment and movement of topsoil by the action of wind and flowing water.

soil formation *See* soil genesis.

soil genesis [GEOL]   The mode by which soil originates, with particular reference to processes of soil-forming factors responsible for the development of true soil from unconsolidated parent material. Also known as pedogenesis; soil formation.

soil mechanics [ENG]   The application of the laws of solid and fluid mechanics to soils and similar granular materials as a basis for design, construction, and maintenance of stable foundations and earth structures.

soil moisture *See* soil water.

soil physics [GEOPHYS]   The study of the physical characteristics of soils; concerned also with the methods and instruments used to determine these characteristics.

soil profile [GEOL]   A vertical section of a soil, showing horizons and parent material.

soil science [GEOL]   The study of the formation, properties, and classification of soil; includes mapping. Also known as pedology.

soil series [GEOL]   A family of soils having similar profiles, and developing from similar original materials under the influence of similar climate and vegetation.

soil shear strength [GEOL]   The maximum resistance of a soil to shearing stresses.

soil stripes [GEOL]   Alternating bands of fine and coarse material in a soil structure.

soil structure [GEOL]   Arrangement of soil into various aggregates, each differing in the characteristics of its particles.

soil survey [GEOL]   The systematic examination of soils, their description and classification, mapping of soil types, and the assessment of soils for various agricultural and engineering uses.

soil thermograph [ENG]   A remote-recording thermograph whose sensing element may be buried at various depths in the earth.

soil thermometer [ENG]   A thermometer used to measure the temperature of the soil, usually the mercury-in-glass thermometer. Also known as earth thermometer.

soil water [HYD]   Water in the belt of soil water. Also known as rhizic water; soil moisture.

soil-water belt *See* belt of soil water.

soil-water zone *See* belt of soil water.

solaire [METEOROL]   A name generally applied to winds from an easterly direction (that is, from the rising sun) in central and southern France.

sol-air temperature [METEOROL]   The temperature which, under conditions of no direct solar radiation and no air motion, would cause the same heat transfer into a house as that caused by the interplay of all existing atmospheric conditions.

solano [METEOROL]   A southeasterly or easterly wind on the southeast coast of Spain in summer; usually an extension of the sirocco; it is hot and humid and sometimes brings rain; when dry, it is dusty.

solar absorption index [GEOPHYS]   A relation of the sun's angle at various latitudes and local times with the ionospheric absorption.

solar air mass [METEOROL]   The optical air mass penetrated by light from the sun for any given position of the sun.

solar attachment [ENG]   A device for determining the true meridian directly from the sun; used an an attachment on a surveyor's transit or compass.

solar climate [CLIMATOL]   The hypothetical climate which would prevail on a uniform solid earth with no atmosphere; thus, it is a climate of temperature alone and is determined only by the amount of solar radiation received.

solar constant [METEOROL]   The rate at which energy from the sun is received just outside the earth's atmosphere on a surface normal to the incident radiation and at the earth's mean distance from the sun; it is 0.140 watt per square centimeter.

solar evaporation [HYD]   The evaporation of water due to the sun's heat.

solarimeter [ENG]   **1.** A type of pyranometer consisting of a Moll thermopile shielded from the wind by a bell glass. **2.** *See* pyranometer.

solar-radiation observation [GEOPHYS]   An evaluation of the radiation from the sun that reaches an observation point; the observing instrument is usually a pyrheliometer or pyranometer.

solar-terrestrial phenomena [GEOPHYS]   All observed physical effects that are caused by solar activity; the phenomena may be in the atmosphere or on the earth's surface; an example is the aurora borealis.

solar tide [OCEANOGR]   The tide caused solely by the tide-producing forces of the sun.

solar-topographic theory [CLIMATOL]   The theory that the changes of climate through geologic time (the paleoclimates) have been due to changes of land and sea distribution and orography, combined with fluctuations of solar radiation of the order of 10–20% on either side of the mean.

solar wind [GEOPHYS]   The supersonic flow of gas, composed of ionized hydrogen and helium, which continuously flows from the sun out through the solar system with velocities of 300 to 1000 kilometers per second; it carries magnetic fields from the sun.

sole [GEOGR]   The lowest part of a valley.   [GEOL]   **1.** The bottom of a sedimentary stratum. **2.** The middle and lower portion of the shear surface of a landslide. **3.** The underlying fault plane of a thrust nappe. Also known as sole plane.   [HYD]   The basal ice of a glacier, often dirty in appearance due to contained rock fragments.

sole injection [GEOL]   An igneous intrusion that was put in place along a thrust plane.

sole mark [GEOL]   An irregularity or penetration on the undersurface of a sedimentary stratum.

solenoid *See* meteorological solenoid.

solenoidal index [METEOROL] The difference between the mean virtual temperature from the surface to some specified upper level averaged around the earth at 55° latitude, and the mean virtual temperature for the corresponding layer averaged at 35° latitude.

Solenopora [PALEOBOT] A genus of extinct calcareous red algae in the family Solenoporaceae that appeared in the Late Cambrian and lasted until the Early Tertiary.

Solenoporaceae [PALEOBOT] A family of extinct red algae having compact tissue and the ability to deposit calcium carbonate within and between the cell walls.

sole plane *See* sole.

solfatara [GEOL] A fumarole from which sulfurous gases are emitted.

solid car [MIN ENG] A mine car equipped with a swivel coupling and generally used with a rotary dump.

solid crib timbering [MIN ENG] Shaft timbering with cribs laid solidly upon one another.

solid drilling [ENG] In diamond drilling, using a bit that grinds the whole face, without preserving a core for sampling.

solifluction [GEOL] A rapid soil creep, especially referring to downslope soil movement in periglacial areas. Also known as sludging; soil flow; soil fluction.

solifluction lobe [GEOL] An isolated, tongue-shaped feature of the land surface with a steep front and a smooth upper surface formed by more rapid solifluction on certain sections of the slope. Also known as solifluction tongue.

solifluction mantle [GEOL] The locally derived, unsorted material moved downslope by solifluction. Also known as flow earth.

solifluction sheet [GEOL] A broad deposit of a solifluction mantle.

solifluction stream [GEOL] A narrow, streamlike deposit of a solifluction mantle.

solifluction tongue *See* solifluction lobe.

solodize [GEOL] To improve a soil by removing alkalies from it.

Solod soil *See* Soloth soil.

Solonchak soil [GEOL] One of an intrazonal, balamorphic group of light-colored soils rich in soluble salts.

Solonetz soil [GEOL] One of an intrazonal group of black alkali soils having a columnar structure.

solore [METEOROL] A cold, night wind of the mountains following the course of the Drome River in southeastern France.

Soloth soil [GEOL] One of an intrazonal halomorphic group of soils formed from saline material; the surface layer is soft and friable, and overlies a light-colored leached horizon which, in turn, overlies a dark horizon. Also known as Solod soil.

solstitial tidal currents [OCEANOGR] Tidal currents of especially large tropic diurnal inequality occurring at the time of solstitial tides.

solstitial tides [OCEANOGR] Tides occurring near the times of the solstices, when the tropic range is especially large.

solum [GEOL] The upper part of a soil profile, composed of A and B horizons in mature soil. Also known as true soil.

solution gas [PETRO ENG] Gaseous reservoir hydrocarbons dissolved in liquid reservoir hydrocarbons because of the prevailing pressures in the reservoir. Also known as dissolved gas.

solution gas drive *See* internal gas drive.

solution-gas reservoir [PETRO ENG] Oil reservoir initially at or above the bubble-point pressure of the gas-oil mixture, and produced primarily by the expansion of the oil and its dissolved gas. Also known as dissolved-gas reservoir.

solution groove [GEOL] One of a series of continuous, subparallel furrows developed on an inclined or vertical surface of a soluble and homogeneous rock (such as the limestone walls of a cave) by the slow corroding action of trickling water.

solution mining [MIN ENG] The extraction of soluble minerals from subsurface strata by injection of fluids, and the controlled removal of mineral-laden solutions.

solution pool [GEOL] A pool in a rock that is formed by the dissolution of the rock in ocean water.

solution porosity [PETRO ENG] A generic designation for reservoir-rock porosity created by solution action; some examples are crystalline limestone and dolomite, porous cap rock, and honeycombed anhydrite.

solution potholes [GEOL] Potholes produced in carbonate rocks by dissolution.

solution transfer [GEOL] A process whereby pressure solution of detrital mineral grains at contact areas is followed by recrystallization on the less strained parts of the grain surfaces.

Somali Current *See* East Africa Coast Current.

somma [GEOL] The rim of a volcano.

sonar [ENG] **1.** A system that uses underwater sound, at sonic or ultrasonic frequencies, to detect and locate objects in the sea, or for communication; the commonest type is echo-ranging sonar, other versions are passive sonar, scanning sonar, and searchlight sonar. Derived from sound navigation and ranging. **2.** *See* sonar set.

sonar set [ENG] A complete assembly of sonar equipment for detecting and ranging or for communication. Also known as sonar.

sonde [ENG] An instrument used to obtain weather data during ascent and descent through the atmosphere, in a form suitable for telemetering to a ground station by radio, as in a radiosonde.

sonic anemometer [ENG] An anemometer which measures wind speed by means of the properties of wind-borne sound waves; it operates on the principle that the propagation velocity of a sound wave in a moving medium is equal to the velocity of sound with respect to the medium plus the velocity of the medium.

sonic depth finder [ENG] A sonar-type instrument used to measure ocean depth and to locate underwater objects; a sound pulse is transmitted vertically downward by a piezoelectric or magnetostriction transducer mounted on the hull of the ship; the time required for the pulse to return after reflection is measured electronically. Also known as echo sounder.

sonic pump [PETRO ENG] A type of lifting pump used in a shallow oil well to pump out the crude; consists of a string of tubing equipped with a check valve at each point, and mechanical means on the surface to vibrate the tubing string vertically; creates a harmonic condition that results in several hundred strokes per minute, with the strokes being a small fraction of an inch.

sonic sounding [ENG] Determining the depth of the ocean bottom by measuring the time for an echo to return to a shipboard sound source.

sonic well logging [ENG] A well logging technique that uses a pulse-echo system to measure the distance between the instrument and a sound-reflecting surface; used to measure the size of cavities around brine wells, and capacities of underground liquefied petroleum gas storage chambers.

Sonolog [PETRO ENG]   An acoustical device used for sound-reflection logging of oil well boreholes to determine the fluid level in a pumping well.

sonometer [ENG]   An instrument for measuring rock stress; a piano wire is stretched between two bolts in the rock, and any change of pitch after destressing is observed and used to indicate stress.

sonora [METEOROL]   A summer thunderstorm in the mountains and deserts of southern California and Baja California.

sordawalite *See* tachylite.

sorosilicate [MINERAL]   A structural type of silicate whose crystal lattice has two SiO_4 tetrahedra sharing one oxygen atom.

sorotiite [GEOL]   A type of meteorite similar to the pallasites, with troilite substituting for olivine.

sorted [GEOL]   **1.** Pertaining to a nongenetic group of patterned-ground features displaying a border of stones, including boulders, commonly alternating with very small particles, including silt, sand, and clay. **2.** Pertaining to an unconsolidated sediment or a cemented detrital rock consisting of particles of essentially uniform size or of particles lying within the limits of a single grade.

sorted polygon [GEOL]   A patterned ground having a sorted appearance due to a border of stones and characterized by a polygonal mesh. Also known as stone polygon.

sorting [GEOL]   The process by which similar in size, shape, or specific gravity sedimentary particles are selected and separated from associated but dissimilar particles by the agent of transportation.

sorting coefficient [GEOL]   A sorting index equal to the square root of the ratio of the larger quartile (the diameter having 25% of the cumulative size-frequency distribution larger than itself) to the smaller quartile (the diameter having 75% of the cumulative size-frequency distribution larger than itself).

sorting index [GEOL]   A measure of the degree of sorting in a sediment based on the statistical spread of the frequency curve of particle sizes.

sou'easter *See* southeaster.

sound exclusion [PETRO ENG]   Several techniques used to prevent a borehole from sloughing in drilling a well through a reservoir rock that has an unconsolidated nature, similar to beach sand; the borehole can be lined with a screen, the sand consolidated with a binding material, or prepack gravel liner can be used.

sounding [ENG]   **1.** Determining the depth of a body of water by an echo sounder or sounding line. **2.** Measuring the depth of bedrock by driving a steel rod into the soil. **3.** Any penetration of the natural environment for scientific observation.   [MIN ENG]   **1.** Knocking on a mine roof to see whether it is sound or safe to work under. **2.** Subsurface investigation by observing the penetration resistance of the subsurface material without drilling holes, by driving a rod into the ground or by using a penetrometer.

sounding balloon [ENG]   A small free balloon used for carrying radiosonde equipment aloft.

sounding device [PETRO ENG]   An acoustical device used to measure the liquid level in a wellbore; for example, a Sonolog or an Echometer.

sounding sand [GEOL]   Sand that emits musical, humming, or crunching sounds when disturbed. Also known as singing sand.

sounding sextant *See* hydrographic sextant.

source area *See* provenance.

source bed [GEOL]   The original stratigraphic horizon from which secondary sulfide minerals were derived.

sourceland *See* provenance.

source region [METEOROL]   An extensive area of the earth's surface characterized by essentially uniform surface conditions and so situated with respect to the general atmospheric circulation that an air mass may remain over it long enough to acquire its characteristic properties.

source rock [GEOL]   **1.** Rock from which fragments have been derived which form a later, usually sedimentary rock. Also known as mother rock; parent rock. **2.** Sedimentary rock, usually shale and limestone, deposited together with organic matter which was subsequently transformed to liquid or gaseous hydrocarbons.

sour corrosion [PETRO ENG]   Corrosion occurring in oil or gas wells where there is an iron sulfide corrosion product, and hydrogen sulfide is present in the produced reservoir fluid.

sour dirt [PETRO ENG]   Sulfate-impregnated soil or soil characterized by escaping sulfur dioxide or hydrogen sulfide; considered an indicator of oil in the area.

south [GEOD]   The direction 180° from north.

South African jade *See* Transvaal jade.

South America [GEOGR]   The southernmost of the Western Hemisphere continents, three-fourths of which lies within the tropics.

South Atlantic Current [OCEANOGR]   An eastward-flowing current of the South Atlantic Ocean that is continuous with the northern edge of the West Wind Drift.

Southeast Drift Current [OCEANOGR]   A North Atlantic Ocean current flowing southeastward and southward from a point west of the Bay of Biscay toward southwestern Europe and the Canary Islands, where it continues as the Canary Current.

southeaster [METEOROL]   A southeasterly wind, particularly a strong wind or gale; for example, the winter southeast storms of the Bay of San Francisco. Also spelled sou'easter.

South Equatorial Current [OCEANOGR]   Any of several ocean currents, flowing westward, driven by the southeast trade winds blowing over the tropical oceans of the Southern Hemisphere and extending slightly north of the equator. Also known as Equatorial Current.

souther [METEOROL]   A south wind, especially a strong wind or gale.

southerly burster [METEOROL]   A cold wind from the south in Australia.

Southern Polar Front *See* Antarctic Convergence.

south foehn [METEOROL]   A foehn condition sustained by a strong south-to-north airflow across a transverse mountain barrier; the south foehn of the Alps may well be the most striking foehn in the world.

south frigid zone [GEOGR]   That part of the earth south of the Antarctic Circle.

south geographical pole [GEOGR]   The geographical pole in the Southern Hemisphere, at latitude 90°S. Also known as South Pole.

south geomagnetic pole [GEOPHYS] The geomagnetic pole in the Southern Hemisphere at approximately 78.5°S, longitude 111°E, 180° from the north geomagnetic pole. Also known as south pole.

South Indian Current [OCEANOGR] An eastward-flowing current of the southern Indian Ocean that is continuous with the northern edge of the West Wind Drift.

South Pacific Current [OCEANOGR] An eastward-flowing current of the South Pacific Ocean that is continuous with the northern edge of the West Wind Drift.

south pole *See* south geomagnetic pole.

south temperate zone [GEOGR] That part of the earth between the Tropic of Capricorn and the Antarctic Circle.

southwester [METEOROL] A southwest wind, particularly a strong wind or gale. Also spelled sou'wester.

sou'wester *See* southwester.

souzalite [MINERAL] $(Mg,Fe)_3(Al,Fe)_4(PO_4)_4(OH)_6 \cdot 2H_2O$ A green mineral composed of hydrous basic phosphate of magnesium, iron, and aluminum.

space charge [GEOPHYS] In atmospheric electricity, space charge refers to a preponderance of either negative or positive ions within any given portion of the atmosphere.

spaced loading [ENG] Loading shot holes so that cartridges are separated by open spacers which do not prevent the concussion from one charge from reaching the next.

space lattice *See* crystal lattice; lattice.

spall [GEOL] **1.** A fragment removed from the surface of a rock by weathering. **2.** A relatively thin, sharp-edged fragment produced by exfoliation. **3.** A rock fragment produced by chipping with a hammer. [MIN ENG] To break ore.

spalling [GEOL] The chipping or fracturing with an upward heaving, of rock caused by a compressional wave at a free surface.

spangolite [MINERAL] $Cu_6Al(SO_4)(OH)_{12}Cl \cdot 3H_2O$ A dark-green hexagonal mineral composed of hydrous basic sulfate and chloride of aluminum and copper and occurring as crystals.

spar [MIN ENG] A small clay vein in a coal seam. [MINERAL] Any transparent or translucent, nonmetallic, light-colored, readily cleavable, crystalline mineral; examples are calcspar and fluorspar.

sparagmite [GEOL] Late Precambrian fragmental rocks of Scandinavia, characterized by high proportions of microcline.

sparite *See* sparry calcite.

Sparnacean [GEOL] A European stage of geologic time; upper upper Paleocene, above Thanetian, below Ypresian of Eocene.

sparry calcite [MINERAL] A clean, coarse-grained calcite crystal. Also known as calcsparite; sparite.

sparry cement [GEOL] Clear, relatively coarse-grained calcite in the interstices of any sedimentary rock.

spartalite *See* zincite.

spasmodic turbidity current [GEOPHYS] A single, rapidly developed turbidity current.

spatial dendrite [METEOROL] A complex ice crystal with fernlike arms that extend in many directions (spatially) from a central nucleus; its form is roughly spherical. Also known as spatial dendritic crystal.

spatial dendritic crystal *See* spatial dendrite.

spatter cone [GEOL] A low, steep-sided cone of small pyroclastic fragments built up on a fissure or vent. Also known as agglutinate cone; volcanello.

spatter rampart [GEOL] A low, circular ridge of pyroclastics built up around the margins of small volcanoes.

special observation [METEOROL] A category of aviation weather observation taken to report significant changes in one or more of the observed elements since the last previous record observation.

special weather report [METEOROL] The encoded and transmitted weather report of a special observation.

species number [OCEANOGR] The first number in the argument number in a Doodson tide schedule; indicates approximately the period of a component of tidal potential.

specific energy [HYD] The energy at any cross section of an open channel, measured above the channel bottom as datum; numerically the specific energy is the sum of the water depth plus the velocity head, $v^2/2g$, where v is the velocity of flow and g the acceleration of gravity.

specific humidity [METEOROL] In a system of moist air, the (dimensionless) ratio of the mass of water vapor to the total mass of the system.

specific productivity index [PETRO ENG] Barrels per day of oil produced per pound decline in bottom-hole pressure per foot of effective reservoir thickness.

specific-volume anomaly [OCEANOGR] The excess of the actual specific volume of the sea water at any point in the ocean over the specific volume of sea water of salinity 35 parts per thousand (‰) and temperature 0°C at the same pressure. Also known as steric anomaly.

specific yield [HYD] The quantity of water which a unit volume of aquifer, after being saturated, will yield by gravity; it is expressed either as a ratio or as a percentage of the volume of the aquifer; specific yield is a measure of the water available to wells.

spectacle stone *See* selenite.

spectral hygrometer [ENG] A hygrometer which determines the amount of precipitable moisture in a given region of the atmosphere by measuring the attenuation of radiant energy caused by the absorption bands of water vapor; the instrument consists of a collimated energy source, separated by the region under investigation and a detector which is sensitive to those frequencies that correspond to the absorption bands of water vapor.

specular hematite [MINERAL] A variety of hematite with a blue-gray color and bright metallic luster.

specular iron *See* specularite.

specularite [MINERAL] A black or gray variety of hematite with brilliant metallic luster, occurring in micaceous or foliated masses, or in tabular or disklike crystals. Also known as gray hematite; iron glance; specular iron.

speed-in [PETRO ENG] To start drilling by making a hole.

spelean [GEOL] Of or pertaining to a feature in a cave.

speleology [GEOL] The study and exploration of caves.

speleothem [GEOL] A secondary mineral deposited in a cave by the action of water. Also known as cave formation.

spencerite [MINERAL] $Zn_4(PO_4)_2(OH)_2 \cdot 3H_2O$ A pearly white monoclinic mineral composed of hydrous basic zinc phosphate and occurring in scaly masses and small crystals.

spending beach [GEOL] In a wave basin, the beach on which the entering waves spend themselves, except for the small remainder entering the inner harbor.

spergenite [GEOL] A biocalcarenite containing ooliths and fossil debris and having a maximum quartz content of 10%. Also known as Bedford limestone; Indiana limestone.

sperrylite [MINERAL] $PtAs_2$ A tin-white isometric mineral composed of platinum arsenide; the only platinum compound known to occur in nature; hardness is 6–7 on Mohs scale, and specific gravity is 10.60.

spessartite [MINERAL] $Mn_3Al_2(SiO_4)_3$ A mineral composed of manganese aluminum silicate with small amounts of iron, magnesium, or other elements. [PETR] A lamprophyre composed of a sodic plagioclase groundmass in which green hornblende phenocrysts are embedded; also contains accessory olivine, biotite, apatite, and opaque oxides.

Sphaeractinoidea [PALEON] An extinct group of fossil marine hydrozoans distinguished in part by the relative prominence of either vertical or horizontal trabeculae and by the presence of long, tabulate tubes called autotubes.

sphaerite [MINERAL] Light-gray or bluish mineral composed of hydrous aluminum phosphate and occurring in global concretions.

sphaerolitic *See* spherulitic.

sphalerite [MINERAL] $(Zn,Fe)S$ The low-temperature form and common polymorph of zinc sulfide; a usually brown or black mineral that crystallizes in the hextetrahedral class of the isometric system, occurs most commonly in coarse to fine, granular, cleanable masses, has resinous luster, hardness of 3.5 on Mohs scale, and specific gravity of 4.1. Also known as blende; false galena; jack; lead marcasite; mock lead; mock ore; pseudogalena; steel jack.

Sphenacodontia [PALEON] A suborder of extinct reptiles in the order Pelycosauria which were advanced, active carnivores.

sphene [MINERAL] $CaTiSiO_5$ A brown, green, yellow, gray, or black neosilicate mineral common as an accessory mineral in igneous rocks; it is monoclinic and has resinous luster; hardness is 5–5.5 on Mohs scale; specific gravity is 3.4–3.5. Also known as grothite; titanite.

sphenochasm [GEOL] A triangular gap of oceanic crust separating two continental blocks and converging to a point.

sphenoid [CRYSTAL] An open crystal, occurring in monoclinic crystals of the sphenoidal class, and characterized by two nonparallel faces symmetrical with an axis of twofold symmetry.

sphenolith [GEOL] A wedgelike igneous intrusion that is partly concordant and partly discordant.

Sphenyllopsida [PALEOBOT] An extinct class of embryophytes in the division Equisetophyta.

spherical separator [PETRO ENG] A gas-oil separator in the form of a spherical vessel.

spherical weathering *See* spheroidal weathering.

spheroidal group [CRYSTAL] A group in the tetragonal symmetry system; the sphenoid is the typical form.

spheroidal recovery [GEOPHYS] The hypothetical return of the earth to spheroid form after it has been distorted.

spheroidal weathering [GEOL] Chemical weathering in which concentric or spherical shells of decayed rock are successively separated from a block of rock; commonly results in the formation of a rounded boulder of decomposition. Also known as concentric weathering; spherical weathering.

spherulite [GEOL] A spherical body or coarsely crystalline aggregate having a radial internal structure arranged about one or more centers.

spherulitic [PETR] Relating to the texture of a rock composed of numerous spherulites. Also known as globular; sphaerolitic.

Sphinctozoa [PALEON] A group of fossil sponges in the class Calcarea which have a skeleton of massive calcium carbonate organized in the form of hollow chambers.

spiculite [PETR] A spindle-shaped belonite thought to have formed by the coalescence of globulites.

spile [MIN ENG] **1.** A temporary lagging driven ahead on levels in loose ground. **2.** A short piece of plank sharpened flatwise and used for driving into watery stratums as sheet piling to assist in checking the flow of water.

spiling *See* forepoling.

spilite [PETR] An altered basalt containing albitized feldspar accompanied by low-temperature, hydrous crystallization products such as chlorite, calcite, and epidote.

spilling [OCEANOGR] The process by which steep waves break on approaching the shore; white water appears on the crest and the wave top gradually rolls over, without a crash.

spilling breaker *See* plunging breaker.

spillover [METEOROL] That part of orographic precipitation which is carried along by the wind so that it reaches the ground in the nominal rain shadow on the lee side of the barrier.

spill pit *See* runoff pit.

spill stream *See* overflow stream.

spinel [MINERAL] **1.** $MgAl_2O_4$ A colorless, purplish-red, greenish, yellow, or black mineral, usually forming octahedral crystals, and characterized by great hardness; used as a gemstone. **2.** A group of minerals of general formula AB_2O_4, where A is magnesium, ferrous iron, zinc, or manganese, or a combination of them, and B is aluminum, ferric iron, or chromium.

spiral band [METEOROL] Spiral-shaped radar echoes received from precipitation areas within intense tropical cyclones (hurricanes or typhoons); they curve cyclonically in toward the center of the storm and appear to merge to form the wall around the eye of the storm. Also known as hurricane band; hurricane radar band.

spiral cutterhead [MIN ENG] A rotary digging device which dislodges and feeds alluvial sand or gravel to the intake of a suction dredge.

spiral layer *See* Ekman layer.

Spiriferida [PALEON] An order of fossil articulate brachiopods distinguished by the spiralium, a pair of spirally coiled ribbons of calcite supported by the crura.

Spiriferidina [PALEON] A suborder of the extinct brachiopod order Spiriferida including mainly ribbed forms having laterally or ventrally directed spires, well-developed interareas, and a straight hinge line.

spit [GEOGR] A small point of land commonly consisting of sand or gravel and which terminates in open water.

Spitsbergen Current [OCEANOGR] An ocean current flowing northward and westward from a point south of Spitsbergen, and gradually merging with the East Greenland Current in the Greenland Sea; the Spitsbergen Current is the continuation of the northwestern branch of the Norwegian Current.

spitted fuse [ENG] A slow-burning fuse which has been cut open at the lighting end for ease of ignition.

splash erosion [GEOL] Erosion resulting from the impact of falling raindrops.

splent coal *See* splint coal.

spliced [GEOL] Relating to veins that pinch out and are overlapped at that point by another parallel vein.

splint *See* splint coal.

splint coal [GEOL] A hard, dull, blocky, grayish-black, banded bituminous coal characterized by an uneven fracture and a granular texture; burns with intense heat. Also known as splent coal; splint.

split [GEOL] A coal seam that cannot be mined as a single unit because it is separated by a parting of other sedimentary rock. Also known as coal split; split coal. [MIN ENG] **1.** To divide the air current into separate circuits to ventilate more than one section of the mine. **2.** Any division or branch of the ventilating current.

split coal *See* split.

splitting [MIN ENG] **1.** Lamina of mica with a maximum thickness of 0.0012 inch (30 micrometers), split from blocks and thins. **2.** One of a pair of horizontal level headings driven through a pillar, in pillar workings, in order to mine the pillar coal.

SP logging *See* spontaneous-potential well logging.

spodic horizon [GEOL] A soil horizon characterized by illuviation of amorphous substances.

Spodosol [GEOL] A soil order characterized by accumulations of amorphous materials in subsurface horizons.

spodumene [MINERAL] $LiAlSi_2O_6$ A white to yellowish-, purplish-, or emerald-green clinopyroxene mineral occurring in prismatic crystals; hardness is 6.5–7 on Mohs scale, and specific gravity 3.13–3.20; an ore of lithium. Also known as triphane.

spoil [MIN ENG] **1.** The overburden or nonore material from a coal mine. **2.** A stratum of coal and dirt mixed.

spoil bank [MIN ENG] **1.** In surface mining, the accumulation of overburden. **2.** The place where spoil is deposited. Also known as spoil heap.

spoil dam [MIN ENG] An earthen dike forming a depression, in which returns from a borehole can be collected and retained.

spongework [GEOL] A pattern of small irregular interconnecting cavities on walls of limestone caves.

Spongiomorphida [PALEON] A small, extinct Mesozoic order of fossil colonial Hydrozoa in which the skeleton is a reticulum composed of perforate lamellae parallel to the upper surface and of regularly spaced vertical elements in the form of pillars.

Spongiomorphidae [PALEON] The single family of extinct hydrozoans comprising the order Spongiomorphida.

spongolite [GEOL] A rock or sediment composed chiefly of the remains of sponges. Also known as spongolith.

spongolith *See* spongolite.

spontaneous nucleation [METEOROL] The nucleation of a phase change of a substance without the benefit of any seeding nuclei within or otherwise in contact with that substance; examples of such systems are a pure vapor condensing to its pure liquid state, a pure liquid freezing to its pure solid state, and a pure solution crystallizing to yield pure solute crystals.

spontaneous polarization [GEOPHYS] A phenomenon resulting in differences in static electrical potential between points in the earth due to chemical reactions, differences in solution concentration, or flow of fluids through porous media.

spontaneous-potential well logging [ENG] The recording of the natural electrochemical and electrokinetic potential between two electrodes, one above the other, lowered into a drill hole; used to detect permeable beds and their boundaries. Also known as SP logging.

spoon [MIN ENG] An instrument in which earth or pulp may be delicately tested by washing to detect gold or amalgam.

sporadic E layer [GEOPHYS] A layer of intense ionization that occurs sporadically within the E layer; it is variable in time of occurrence, height, geographical distribution, penetration frequency, and ionization density.

sporinite [GEOL] A variety of exinite composed of spore exines which have been compressed parallel to the stratification.

spot elevation [MAP] Elevation of a point on a map or chart, usually indicated by a dot accompanied by a number indicating the vertical distance of the point from the reference datum; spot elevations are used principally to indicate points higher than their surroundings.

spotting [MIN ENG] Bringing mine cars or surface wagons to the correct spot for loading, discharging, or any other purpose.

spotting hoist [MIN ENG] A small haulage engine used for bringing mine cars into the correct position under a loading chute or feeder or some other point.

spotty ore [MIN ENG] Ore in which the valuable material is concentrated irregularly as small particles.

spot wind [METEOROL] In air navigation, wind direction and speed, either observed or forecast if so specified, at a designated altitude over a fixed location.

spouting horn [GEOL] A sea cave with a rearward or upward opening through which water spurts or sprays after waves enter the cave. Also known as chimney; oven.

sprag [MIN ENG] A prop supporting the roof or ore in a mine.

spragger [MIN ENG] In coal mining, a laborer who rides trains of cars and controls their free movement down gently sloping inclines by throwing switches and by poking sprags between the wheel spokes to stop them.

sprag road [MIN ENG] A road so steep that sprags must be used on the wheels of ore cars during descent.

spray region *See* fringe region.

spreader [MIN ENG] **1.** A horizontal timber below the cap of a set, used to stiffen the legs, and to support the brattice when there are two air courses in the same gangway. **2.** A piece of timber stretched across a shaft as a temporary support of the walls.

spreading concept *See* sea floor spreading.

spreading floor hypothesis *See* sea floor spreading.

spring [HYD] A general name for any discharge of deep-seated, hot or cold, pure or mineralized water.

spring crust [HYD] A type of snow crust, formed when loose firn is recemented by a decrease in temperature; it is most common in late winter and spring.

spring gravimeter [ENG] An instrument used for making relative measurements of gravity; the elongation s of the spring may be considered proportional to gravity g, $s = (1/k)g$, and the basic formula for relative measurements is $g_2 - g_1 = k(s_2 - s_1)$.

spring high water *See* mean high-water springs.

spring-lifter case *See* lifter case.

spring low water *See* mean low-water springs.

spring range [OCEANOGR] The mean semidiurnal range of tide when spring tides are occurring; the mean difference in height between spring high water and spring low water. Also known as mean spring range.

spring rise [OCEANOGR] The height of mean high-water springs above the chart datum.

spring seepage [HYD] A spring of small discharge. Also known as weeping spring.

spring sludge *See* rotten ice.

spring snow [HYD] A coarse, granular snow formed during spring by alternate freezing and thawing. Also known as corn snow.

spring tidal currents [OCEANOGR] Tidal currents of increased speed occurring at the time of spring tides.

spring tide [OCEANOGR] Tide of increased range which occurs about every 2 weeks when the moon is new or full.

spring velocity [OCEANOGR] The average speed of the maximum flood and maximum ebb of a tidal current at the time of spring tides.

sprinkle [METEOROL] A very light shower of rain.

spud [MIN ENG] A nail, resembling a horseshoe nail, with a hole in the head, driven into mine timbering or into a wooden plug inserted in the rock to mark a surveying station.

spudded-in [MIN ENG] A borehole that has been started and has reached bedrock or in which the standpipe has been set.

spurrite [MINERAL] $Ca_5(SiO_4)_2(CO_3)$ A light-gray mineral occurring in granular masses.

squall [METEOROL] A strong wind with sudden onset and more gradual decline, lasting for several minutes; in the United States observational practice, a squall is reported only if a wind speed of 16 knots or higher (8.23 meters per second) is sustained for at least 2 minutes.

squall cloud [METEOROL] A small eddy cloud sometimes formed below the leading edge of a thunderstorm cloud, between the upward and downward currents.

squall line [METEOROL] A line of thunderstorms near whose advancing edge squalls occur along an extensive front; the region of thunderstorms is typically 20 to 50 kilometers wide and a few hundred to 2000 kilometers long.

square set [MIN ENG] A set of timbers composed of a cap, girt, and post which meet so as to form a solid 90° angle; they are so framed at the intersection as to form a compression joint, and join with three other similar sets.

square-set block caving [MIN ENG] Block caving in which the ore is extracted through drifts supported by square sets.

square-set stopes [MIN ENG] The use of square-set timbering to support the ground as ore is extracted.

squeeze [MIN ENG] **1.** The settling, without breaking, of the roof over a considerable area of working. Also known as creep; nip; pinch. **2.** The gradual upheaval of the floor of a mine due to the weight of the overlying strata. **3.** The sections in coal seams that have become constricted by the squeezing in of the overlying or underlying rock.

SSP *See* static spontaneous potential.

stability [GEOL]   **1.** The resistance of a structure, spoil heap, or clay bank to sliding, overturning, or collapsing. **2.** Chemical durability, resistance to weathering.

stability chart [METEOROL]   A synoptic chart that shows the distribution of a stability index.

stability index [METEOROL]   An indication of the local static stability of a layer of air.

stack [ENG]   **1.** To stand and rack drill rods in a drill tripod or derrick.   [GEOL]   An erosional, coastal landform that is a steep-sided, pillarlike rocky island or mass that has been detached by wave action from a shore made up of cliffs; applies particularly to a stack that is columnar in structure and has horizontal stratifications. Also known as marine stack; rank.

stade [GEOL]   A substage of a glacial stage marked by a secondary advance of glaciers.

stadia [ENG]   A surveying instrument consisting of a telescope with special horizontal parallel lines or wires, used in connection with a vertical graduated rod.

stadia hairs [ENG]   Two horizontal lines in the reticle of a theodolite arranged symmetrically above and below the line of sight. Also known as stadia wires.

stadial moraine *See* recessional moraine.

stadia rod [ENG]   A graduated rod used with a stadia to measure the distance from the observation point to the rod by observation of the length of rod subtended by the distance between the stadia hairs.

stadia tables [ENG]   Mathematical tables from which may be found, without computation, the horizontal and vertical components of a reading made with a transit and stadia rod.

stadia wires *See* stadia hairs.

stadimeter [ENG]   An instrument for determining the distance to an object, but its height must be known; the angle, subtended by the object's bottom and top as measured at the observer's position, is proportional to the object's height; the instrument is graduated directly in distance.

Staffellidae [PALEON]   An extinct family of marine protozoans (superfamily Fusulinacea) that persisted during the Pennsylvanian and Early Permian.

Staffordian [GEOL]   A European stage of geologic time forming the middle Upper Carboniferous, above Yorkian and below Radstockian, equivalent to part of the upper Westphalian.

stage [GEOL]   **1.** A developmental phase of an erosion cycle in which landscape features have distinctive characteristic forms. **2.** A phase in the historical development of a geologic feature. **3.** A major subdivision of a glacial epoch. **4.** A time-stratigraphic unit ranking below series and above substage.   [HYD]   The elevation of the water surface in a stream as measured by a river gage with reference to some arbitrarily selected zero datum.   [MIN ENG]   **1.** A certain length of underground roadway worked by one horse. **2.** A narrow thin dike, especially one where the material of which the dike is composed is soft. **3.** A platform on which mine cars stand.

stage acidizing [PETRO ENG]   An oil-reservoir acid treatment (acidizing) in which the formation is treated with two or more separate stages of acid, instead of a single large treatment.

stage separation [PETRO ENG]   A system for gas-oil separation of well fluids by a series of stages instead of a single operation.

staggered blastholes [MIN ENG]   Two rows of holes staggered to a triangular pattern to distribute the burden when shot-firing in thick coal seams.

staggered-line drive [PETRO ENG] In the placement and drilling of water-injection wells for water flood recovery of oil, the staggered (versus in-line) areal arrangement of the injection wells.

stagnant glacier [HYD] A glacier which has ceased to move.

stagnant water [HYD] Motionless water, not flowing in a stream or current.

stagnation [HYD] **1.** The condition of a body of water unstirred by a current or wave. **2.** The condition of a glacier that has stopped flowing.

stagnum [HYD] A pool of water with no outlet.

stainierite *See* heterogenite.

stalactite [GEOL] A conical or roughly cylindrical speleothem formed by dripping water and hanging from the roof of a cave; usually composed of calcium carbonate.

stalacto-stalagmite [GEOL] A columnar deposit formed by the union of a stalactite with its complementary stalagmite.

stalagmite [GEOL] A conical speleothem formed upward from the floor of a cave by the action of dripping water; usually composed of calcium carbonate.

stamp battery [MIN ENG] A machine for crushing very strong ores or rocks; consists essentially of a crushing member (gravity stamp) which is dropped on a die, the ore being crushed in water between the shoe and the die.

stamp copper [MIN ENG] A copper-bearing rock that is stamped and washed before it is smelted.

stamp head [MIN ENG] A heavy and nearly cylindrical cast-iron head fixed on the lower end of the stamp rod, shank, or lifter to give weight in stamping the ore.

stamping [MIN ENG] Reducing to the desired fineness in a stamp mill; the grain is usually not so fine as that produced by grinding in pans.

stamping mill [MIN ENG] A machine in which ore is finely crushed by descending pestles (stamps), usually operated by hydraulic power. Also known as crushing mill.

stamukha [OCEANOGR] An individual piece of stranded ice.

stand [OCEANOGR] The interval at high or low water when there is no appreciable change in the height of the tide. Also known as tidal stand.

standard artillery atmosphere [METEOROL] A set of values describing atmospheric conditions on which ballistic computations are based, namely: no wind, a surface temperature of 15°C, a surface pressure of 1000 millibars, a surface relative humidity of 78%, and a lapse rate which yields a prescribed density-altitude relation.

standard artillery zone [METEOROL] A vertical subdivision of the standard artillery atmosphere; it may be considered a layer of air of prescribed thickness and altitude.

standard atmosphere [METEOROL] A hypothetical vertical distribution of atmospheric temperature, pressure, and density which is taken to be representative of the atmosphere for purposes of pressure altimeter calibrations, aircraft performance calculations, aircraft and missile design, and ballistic tables; the air is assumed to obey the perfect gas law and hydrostatic equation, which, taken together, relate temperature, pressure, and density variations in the vertical; it is further assumed that the air contains no water vapor, and that the acceleration of gravity does not change with height.

standard depth-pressure recorder [PETRO ENG] A device for the measurement of pressures at the bottom of a well bore (that is, bottom-hole pressure); a spring-restrained piston moves a recording stylus on a pressure-sealed chart.

standard meridian [GEOD] The meridian used for reckoning standard time; throughout most of the world the standard meridians are those whose longitudes are exactly divisible by 15°.   [MAP] A meridian of a map projection along which the scale is as stated.

standard mineral [MINERAL] A mineral that, on the basis of chemical analyses, is theoretically capable of being present in a rock. Also known as normative mineral.

standard parallel [MAP] A parallel on a map or chart along which the scale is as stated for that map or chart.   [GEOD] The parallel or parallels of latitude used as control lines in the computation of a map projection.

standard plane [CRYSTAL] The crystal plane whose Miller indices are (111), that is, whose intercepts on the crystal axes are proportional to the corresponding sides of a unit cell.

standard pressure [METEOROL] The arbitrarily selected atmospheric pressure of 1000 millibars to which adiabatic processes are referred for definitions of potential temperature, equivalent potential temperature, and so on.

standard project flood [HYD] The volume of streamflow expected to result from the most severe combination of meteorological and hydrologic conditions which are reasonably characteristic of the geographic region involved, excluding extremely rare combinations.

standard seawater *See* normal water.

standard station *See* reference station.

standard visibility *See* meteorological range.

standard visual range *See* meteorological range.

standing cloud [METEOROL] Any stationary cloud maintaining its position with respect to a mountain peak or ridge.

standing ground [MIN ENG] Ground that will stand firm without timbering.

standing valve [PETRO ENG] A sucker-rod-pump (oil well) discharge valve that remains stationary during the pumping cycle, in contrast to a traveling valve.

stanfieldite [MINERAL] $Ca_4(Mg,Fe,Mn)_5(PO_4)_6$ A phosphate mineral found only in meteorites.

stannite [MINERAL] Cu_2FeSnS_4 A steel-gray or iron-black mineral crystallizing in the tetragonal system and occurring in granular masses; luster is metallic, hardness is 4 on Mohs scale, and specific gravity is 4.3–4.53. Also known as bell-metal ore; tin pyrites.

star ruby [MINERAL] An asteriated variety of ruby with normally six chatoyant rays.

star sapphire [MINERAL] A variety of sapphire exhibiting a six-pointed star resulting from the presence of microscopic crystals in various orientations within the gemstone.

starting barrel [ENG] A short (12 to 24 inches or 30 to 60 centimeters) core barrel used to begin coring operations when the distance between the drill chuck and the bottom of the hole or to the rock surface in which a borehole is to be collared is too short to permit use of a full 5- or 10-foot-long (1.5- or 3.0-meter-long) core barrel.

starved basin [GEOL] A sedimentary basin in which the rate of subsidence exceeds the rate of sedimentation.

state of the sea [OCEANOGR] A description of the properties of the wind-generated waves on the surface of the sea.

state of the sky [METEOROL] The aspect of the sky in reference to the cloud cover; the state of the sky is fully described when the amounts, kinds, directions of movement, and heights of all clouds are given.

static granitization [PETR] The formation of a granitic rock by a metasomatic process in the absence of compressive forces or strains.

static grizzly [MIN ENG] A grizzly in the form of a stationary bar screen which either allows suitable pieces of rock or ore to pass over and unwanted small sizes to drop through, or rejects oversize pieces while allowing suitable material to drop through.

static level [HYD] The height to which water will rise in an artesian well; the static level of a flowing well is above the ground surface.

static metamorphism [GEOL] Regional metamorphism caused by heat and solvents at high lithostatic pressures. Also known as load metamorphism.

static oceanography [OCEANOGR] Branch of oceanography that deals with the physical and chemical nature of water in the ocean and with the shape and composition of the ocean bottom.

static SP *See* static spontaneous potential.

static spontaneous potential [PETRO ENG] Theoretical maximum spontaneous potential current that can be measured in a down-hole, mud log in clean sand. Abbreviated SSP; static SP.

static stability [METEOROL] The stability of an atmosphere in hydrostatic equilibrium with respect to vertical displacements, usually considered by the parcel method. Also known as convectional stability; hydrostatic stability; vertical stability.

station [MIN ENG] **1.** An enlargement in a mining shaft or gallery on any level used for a landing at any desired place and also for receiving loaded mine cars that are to be sent to the surface. **2.** An opening into a level which heads out of the side of an inclined plane; the point at which a surveying instrument is planted or observations are made.

stationary front *See* quasi-stationary front.

station continuity chart [METEOROL] A chart or graph on which time is one coordinate, and one or more of the observed meteorological elements at that station is the other coordinate.

station elevation [METEOROL] The vertical distance above mean sea level that is adopted as the reference datum level for all current measurements of atmospheric pressure at the station.

station model [METEOROL] A specified pattern for entering, on a weather map, the meteorological symbols that represent the state of the weather at a particular observation station.

station pressure [METEOROL] The atmospheric pressure computed for the level of the station elevation.

statistical forecast [METEOROL] A weather forecast based upon a systematic statistical examination of the past behavior of the atmosphere, as distinguished from a forecast based upon thermodynamic and hydrodynamic considerations.

staurolite [MINERAL] $FeAl_4(SiO_4)_2(OH)_2$ A reddish-brown to black neosilicate mineral that crystallizes in the orthorhombic system, has resinous to vitreous luster, hardness is 7–7.5 on Mohs scale, and specific gravity is 3.7. Also known as crossstone; fairy stone; grenatite; staurotide.

staurotide *See* staurolite.

STD recorder *See* salinity-temperature-depth recorder.

steadiness *See* persistence.

steady-state model [PETRO ENG] Electric or electrolytic analogs of a reservoir formation used to study the steady-state flow of fluids through porous media; includes gel, blotter, liquid, potentiometric, and similar models.

steam fog [METEOROL]   Fog formed when water vapor is added to air which is much colder than the vapor's source; most commonly, when very cold air drifts across relatively warm water. Also known as frost smoke; sea mist; sea smoke; steam mist; water smoke.

steam jet [MIN ENG]   A system of ventilating a mine by means of a number of jets of steam at high pressure kept constantly blowing off from a series of pipes in the bottom of the upcast shaft.

steam mist *See* steam fog.

steatite [PETR]   A compact, massive, fine-ground rock composed principally of talc, but with much other material.

steatization [GEOL]   Introduction of or replacement by talc or steatite.

S tectonite [PETR]   A tectonite whose fabric is dominated by planar surfaces of formation or deformation, such as slate.

steel jack [MIN ENG]   A screw jack suitable in mechanical mining; used for legs or upright timbers.   [MINERAL]   *See* sphalerite.

steel sets [MIN ENG]   Steel beam used in main entries of coal mines and in shafts of metal mines; I beams for caps and H beams for posts or wall plates.

steel tunnel support [MIN ENG]   One of the tunnel support systems made of steel; five types are continuous rib; rib and post; rib and wall plate; rib, wall plate, and post; and full circle rib.

steering [METEOROL]   Loosely used for any influence upon the direction of movement of an atmospheric disturbance exerted by another aspect of the state of the atmosphere; for example, a surface pressure system tends to be steered by isotherms, contour lines, or streamlines aloft, or by warm-sector isobars or the orientation of a warm front.

steering level [METEOROL]   A hypothetical level, in the atmosphere, where the velocity of the basic flow bears a direct relationship to the velocity of movement of an atmospheric disturbance embedded in the flow.

Stefan's formula [OCEANOGR]   A formula for the growth of thickness h of an ice cover on the ocean at various freezing temperatures, expressed as

$$h \approx \sqrt{\left(\frac{2l}{\lambda_i \rho_i}\right)\psi}$$

where l is the coefficient of thermal conductivity, λ_i is the latent heat of fusion, ρ_i is the density of ice, and ψ is the cold sum (in degree days below 0°C).

Stegodontinae [PALEON]   An extinct subfamily of elephantoid proboscideans in the family Elephantidae.

Stegosauria [PALEON]   A suborder of extinct reptiles of the order Ornithischia comprising the plated dinosaurs of the Jurassic which had tiny heads, great triangular plates arranged on the back in two alternating rows, and long spikes near the end of the tail.

steigerite [MINERAL] $4AlVO_4 \cdot 13H_2O$ A canary-yellow mineral composed of hydrous aluminum vanadate and occurring in masses.

steinkern [GEOL]   **1.** Rock material formed from consolidated mud or sediment that filled a hollow organic structure, such as a fossil shell. **2.** The fossil formed after dissolution of the mold. Also known as endocast; internal cast.

stellar crystal *See* plane-dendritic crystal.

stellar lightning [GEOPHYS]   Lightning consisting of several flashes seeming to radiate from a single point.

stem [ENG]   **1.** The heavy iron rod acting as the connecting link between the bit and the balance of the string of tools on a churn rod. **2.** To insert packing or tamping material in a shothole.

stem bag [MIN ENG]   A fire-resisting paper bag filled with dry sand for stemming shotholes.

stemming rod [ENG]   A nonmetallic rod used to push explosive cartridges into position in a shothole and to ram tight the stemming.

Stenomasteridae [PALEON]   An extinct family of Euchinoidea in the order Holasteroida comprising oval and heart-shaped forms with fully developed pore pairs.

Stensioellidae [PALEON]   A family of Lower Devonian placoderms of the order Petalichthyida having large pectoral fins and a broad subterminal mouth.

Stenurida [PALEON]   An order of Ophiuroidea, comprising the most primitive brittlestars, known only from Paleozoic sediments.

step [ENG]   A small offset on a piece of core or in a drill hole resulting from a sudden sidewise deviation of the bit as it enters a hard, tilted stratum or rock underlying a softer rock.   [GEOL]   A hitch or dislocation of the strata.   [MIN ENG]   The portion of a longwall face at right angles to the line of the face formed when a place is worked in front of or behind an adjoining place.

step fault [GEOL]   One of a set of closely spaced, parallel faults. Also known as distributive fault; multiple fault.

Stephanian [GEOL]   A European stage of Upper Carboniferous geologic time, forming the Upper Pennsylvanian, above the Westphalian and below the Sakmarian of the Permian.

stephanite [MINERAL] Ag_5SbS_4 An iron-black mineral crystallizing in the orthorhombic system and having a metallic luster; an ore of silver. Also known as black silver; brittle silver ore; goldschmidtine.

step lake *See* paternoster lake.

stepout time [GEOPHYS]   In seismic prospecting, the time differentials in arrivals of a given peak or trough of a reflected or refracted event for successive detector positions on the earth's surface.

step-out well [PETRO ENG]   A well drilled at a later time over remote, undeveloped portions of a partially developed continuous reservoir rock. Also known as delayed development well.

steppe [GEOGR]   An extensive grassland in the semiarid climates of southeastern Europe and Asia; it is similar to but more arid than the prairie of the United States.

steppe climate [CLIMATOL]   The type of climate in which precipitation though very slight, is sufficient for growth of short, sparse grass; typical of the steppe regions of south-central Eurasia. Also known as semiarid climate.

stepped leader [GEOPHYS]   The initial streamer of a lightning discharge; an intermittently advancing column of high ion density which established the channel for subsequent return streamers and dart leaders.

steptoe [GEOL]   An isolated protrusion of bedrock, such as the summit of a hill or mountain, in a lava flow.

stercorite [MINERAL] $Na(NH_4)H(PO_4)\cdot 4H_2O$ A white to yellowish and brown, triclinic mineral consisting of a hydrated acid phosphate of sodium and ammonium.

stereographic chart [MAP] A chart on the stereographic projection.

stereographic net *See* net.

stereographic projection [MAP] A perspective conformal, azimuthal map projection in which points on the surface of a sphere or spheroid, such as the earth, are conceived as projected by radial lines from any point on the surface to a plane tangent to the antipode of the point of projection; circles project as circles through the point of tangency, except for great circles which project as straight lines; the principal navigational use of the projection is for charts of the polar regions. Also known as azimuthal orthomorphic projection.

Stereospondyli [PALEON] A group of labyrinthodont amphibians from the Triassic characterized by a flat body without pleurocentra and with highly developed intercentra.

sternbergite [MINERAL] $AgFe_2S_3$ A dark-brown or black mineral composed of silver iron sulfide and occurring as tabular crystals or flexible laminae.

sterrettite *See* kolbeckite.

stewartite [GEOL] A steel-gray, iron-containing variety of bort that has magnetic properties.   [MINERAL] $Mn_3(DO)_2\cdot 4H_2O$ A brownish-yellow mineral composed of hydrous manganese phosphate occurring in minute crystals or fibrous tufts in pegmatites.

Sthenurinae [PALEON] An extinct subfamily of marsupials of the family Diprotodontidae, including the giant kangaroos.

stibiconite [MINERAL] $Sb_3O_6(OH)$ A pale yellow to yellowish- or reddish-white mineral consisting of a basic or hydrated oxide of antimony; occurs in massive form, as a powder, and in crusts.

stibiocolumbite [MINERAL] $Sb(Nb,Ta,Cb)O_4$ A dark brown to light yellowish- or reddish-brown, orthorhombic mineral consisting of an oxide of antimony and tantalum-columbium.

stibium *See* antimonite.

stibnite *See* antimonite.

stichtite [MINERAL] $Mg_6Cr_2(CO_3)(OH)_{16}\cdot 4H_2O$ A lilac-colored rhombohedral mineral which is composed of hydrous basic carbonate of magnesium and chromium.

stilbite [MINERAL] $Ca(Al_2Si_7O_{18})\cdot 7H_2O$ A white, brown, or yellow mineral belonging to the zeolite family of silicates; crystallizes in the monoclinic system, occurs in sheaflike aggregates of tabular crystals, and has pearly luster; hardness is 3.5–4 on Mohs scale, and specific gravity is 2.1–2.2. Also known as desmine.

Stiles method [PETRO ENG] A technique for computing oil recovery by waterflood methods, taking into account the distribution of varying permeability stratums throughout the reservoir.

stillstand [GEOL] Referring to a land area, a continent, or an island, to remain stationary with respect to the interior of the earth or to sea level.

still water [HYD] A portion of a stream having a very slight gradient and no visible current.

still-water level [OCEANOGR] The level that the sea surface would assume in the absence of wind waves.

still well [METEOROL] A device, used in evaporation pan measurements, which provides an undisturbed water surface and support for the hook gage; the U.S. Weather Bureau model consists of a brass cylinder, 8 inches (20.32 centimeters) high and 3.5 inches (8.89 centimeters) in diameter, mounted over a hole in a triangular galvanized iron base which is provided with leveling screws.

stilpnomelane [MINERAL] $K(Fe,Mg,Al)_3Si_4O_{10}(OH)_2 \cdot H_2O$ A black or greenish-black mineral composed of basic hydrous potassium iron magnesium aluminum silicate; occurs as fibers, iron magnesium aluminum silicate; occurs as fibers, incrustations, and foliated plates.

stimulation treatment [PETRO ENG] One of the techniques to increase (stimulate) oil- or gas-reservoir production, such as acidizing, fracturing, controlled underground explosions, or various cleaning techniques. Also known as well stimulation.

stinger [PETRO ENG] A support which is attached to the stern of a pipe-laying barge and which controls the bending of the pipe as it leaves the barge to enter the water.

stink *See* gob stink.

stinkdamp [MIN ENG] The hydrogen sulfide that occurs in mines.

stinkstone [GEOL] A stone containing decomposing organic matter that gives off an offensive odor when rubbed or struck.

stipoverite *See* stishovite.

stirrup [MIN ENG] **1.** A piece of steel hung from a gallows frame to engage the endgate hooks when a mine car is tilted over; used at dumps. **2.** A screw joint suspended from the brakestaff of a spring pole, by which the boring rods are adjusted to the depth of the borehole. Also known as temper screw.

stishovite [MINERAL] SiO_2 A polymorph of quartz, a dense, fine-grained mineral formed under very high pressure (about 1,000,000 pounds per square inch or 7×10^9 newtons per square meter); it is the only mineral in which the silicon atom has a coordination number of six; specific gravity is 4.28. Also known as stipoverite.

stock [PETR] A usually discordant, batholithlike body of intrusive igneous rock not exceeding 40 square miles (103.6 square kilometers) in surface exposure and usually discordant.

stokesite [MINERAL] $CaSnSi_3O_9 \cdot 2H_2O$ A colorless orthorhombic mineral composed of hydrous calcium tin silicate occurring in crystals.

stolzite [MINERAL] $PbWO_4$ A tetragonal mineral composed of native lead tungstate; it is isomorphous with wulfenite and dimorphous with raspite.

stone [GEOL] **1.** A small fragment of rock or mineral. **2.** *See* stony meteorite.

stoneboat [MIN ENG] A flat runnerless sled for transporting heavy material.

stone bubble *See* lithophysa.

stone coal *See* anthracite.

stone dust [MIN ENG] Inert dust spread on roadways in coal mines as a defense against the danger of coal-dust explosions; effective because the stone dust absorbs heat.

stone-dust barrier [MIN ENG] A device erected in mine roadways to arrest explosions; consists of trays or vee troughs loaded with stone dust, which are upset or overturned by the pressure wave in front of an explosion and the flame, producing a dense cloud of inert dust which blankets the flame and stops further propagation of the explosion.

stone gobber [MIN ENG]   In bituminous coal mining, one who removes stone and other refuse from coal mine floors and dumps the refuse into mine cars for disposal.

stone ice *See* ground ice.

stone polygon *See* sorted polygon.

stone ring [GEOL]   A ring of stones surrounding a central area of finer material; characteristic of sorted circle and sorted polygon.

stony-iron meteorite [GEOL]   Any of the rare meteorites containing at least 25% of both nickel-iron and heavy basic silicates. Also known as iron-stony meteorite; lithosiderite; sideraerolite; siderolite; syssiderite.

stony meteorite [GEOL]   Any meteorite composed principally of silicate minerals, especially olivine, pyroxene, and plagioclase. Also known as aerolite; asiderite; meteoric stone; meteorolite; stone.

stooping [METEOROL]   An atmospheric refraction phenomenon; a special case of sinking in which the curvature of light rays due to atmospheric refraction decreases with elevation so that the visual image of a distant object is foreshortened in the vertical.

stope [MIN ENG]   **1.** To excavate ore in a vein by driving horizontally upon it a series of workings, one immediately over the other, or vice versa; each horizontal working is called a stope because when a number of them are in progress, each working face under attack assumes the shape of a flight of stairs. **2.** Any subterranean extraction of ore except that which is incidentally performed in sinking shafts or driving levels for the purpose of opening the mine.

stope assay plan [MIN ENG]   A plan that details assay value of ore exposures in a stope.

stope board [MIN ENG]   A timber staging on the floor of a stope for setting a rock drill; the stage is tilted so that the bottom holes can be drilled in the same inclined direction.

stope fillings [MIN ENG]   Broken waste material or low-grade matter from a lode or vein used to fill stopes on abandonment.

Stopehammer [MIN ENG]   A trademark for an air-feed hammer drill.

stope hoist [MIN ENG]   A small, portable, compressed-air hoist for operating a scraper-loader or for pulling heavy timbers into position, often used in narrow stopes.

stope pillar [MIN ENG]   An ore column left in place to support the stope.

stoper *See* stoping drill.

stoping drill [MIN ENG]   A small air or electric drill, usually mounted on an extensible column, for working stopes, raises, and narrow workings. Also known as stoper.

stoping ground [MIN ENG]   Part of an ore body opened by drifts and raises, and ready for breaking down.

stopping [MIN ENG]   A brattice, or more commonly, a masonry or brick wall built across old headings, chutes, or airways to confine the ventilating current to certain passages, or to lock up the gas in old workings, or to smother a mine fire.

storage battery locomotive [MIN ENG]   An underground locomotive powered by storage batteries.

storage equation [HYD]   The equation of continuity applied to unsteady flow; it states that the fluid inflow to a given space during an interval of time minus the outflow during the same interval is equal to the change in storage; it is applied in hydrology to the routing of floods through a reservoir or a reach of a stream; the moisture continuity equation applied to the atmosphere is a modification of this.

storage routing *See* flood routing.

storm [METEOROL] An atmospheric disturbance involving perturbations of the prevailing pressure and wind fields on scales ranging from tornadoes (1 kilometer across) to extratropical cyclones (2–3000 kilometers across); also the associated weather (rain storm or blizzard) and the like.

storm beach [GEOL] A ridge composed of gravel or shingle built up by storm waves at the inner margin of a beach.

storm center [METEOROL] The area of lowest atmospheric pressure of a cyclone; this is a more general expression than eye of the storm, which refers only to the center of a well-developed tropical cyclone, in which there is a tendency of the skies to clear.

storm choke [PETRO ENG] A device installed in an oil-well tubing string below the surface to shut in the well when the flow reaches a predetermined rate; provides an automatic shutoff in case Christmas-tree or control valves are damaged. Also known as tubing safety valve.

storm delta *See* washover.

storm detection [METEOROL] Any of the methods and techniques used to detect the formation of severe storms, including procedures for locating, tracking, and forecasting; special tools adapted to this purpose are radar and satellites to supplement meteorological charts and visual observations.

storm ice foot [OCEANOGR] An ice foot produced by the breaking of a heavy sea or the freezing of wind-driven spray.

storm microseism [GEOPHYS] A microseism lasting 25 or more seconds, caused by ocean waves.

storm model [METEOROL] A physical, three-dimensional representation of the inflow, outflow, and vertical motion of air and water vapor in a storm.

storm surge [OCEANOGR] A rise above normal water level on the open coast due only to the action of wind stress on the water surface; includes the rise in level due to atmospheric pressure reduction as well as that due to wind stress. Also known as storm wave; surge.

storm tide [OCEANOGR] Height of a storm surge or hurricane wave above the astronomically predicted sea level.

storm track [METEOROL] The path followed by a center of low atmospheric pressure.

storm transposition [METEOROL] The transfer of precipitation patterns or DDA (depth-duration-area) values from the areas where they actually occurred to areas where they could occur; if necessary, the precipitation values are modified to account for differences in elevation or intervening barriers, and restrictions on change in shape or orientation of the storm may be imposed.

storm warning [METEOROL] A specially worded forecast of severe weather conditions, designed to alert the public to impending dangers; usually, this refers to a warning of potentially dangerous wind conditions for marine interests.

storm-warning signal [METEOROL] An arrangement of flags or pennants (by day) and lanterns (by night) displayed on a coastal storm-warning tower.

storm-warning tower [METEOROL] A tower, generally constructed of steel, for displaying coastal storm-warning signals.

storm wave *See* storm surge.

storm wind [METEOROL] In the Beaufort wind scale, a wind whose speed is from 56 to 63 knots (64 to 72 miles per hour or 104 to 117 kilometers per hour).

Storrow whirling hygrometer [ENG]   A hygrometer in which the two thermometers are mounted side by side on a brass frame and fitted with a loose handle so that it can be whirled in the atmosphere to be tested; the instrument is whirled at some 200 revolutions per minute for about 1 minute and the readings on the wet- and dry-bulb thermometers are recorded; used in conjunction with Glaisher's or Marvin's hygrometrical tables.

stoss [GEOL]   Of the side of a hill, knob, or prominent rock, facing the upstream side of a glacier.

stowboard [MIN ENG]   A mine heading used for storing waste.

strain bursts [MIN ENG]   Rock bursts in which there is spitting, flaking, and sudden fracturing at the face, indicating increased pressure at the site.

strain relief method [MIN ENG]   A method for determining absolute strain and stress within rock in place by boring a smooth hole in the rock and inserting a gage capable of measuring diametral deformation, and overcoring the hole with a large coring bit; the change in the diameter of the hole when the rock cylinder is free to expand is a function of the original stress in the rock and its elastic modulus.

strain restoration method [MIN ENG]   A method for determining absolute strain and stress within rock in place by the installation of strain gages on the rock surface, cutting of a slot in the rock between the strain gages so that the surface rock is free to expand, installation of a flat jack in the slot, and application of hydraulic pressure to the flat jack until the rock is restored to its original state of strain; the original stress in the rock is presumed to be equal to the final pressure in the flat jack.

strain seismograph [ENG]   A seismograph that detects secular strains related to tectonic processes and tidal yielding of the solid earth; also detects strains associated with propagating seismic waves.

strain seismometer [ENG]   A seismometer that measures relative displacement of two points in order to detect deformation of the ground.

strain shadow *See* pressure shadow.

strain-slip [GEOL]   A rock fracture resulting in a slight displacement.

strain-slip cleavage *See* slip cleavage.

strait [GEOGR]   **1.** A neck of land. **2.** A narrow waterway connecting two larger bodies of water.

strake [MIN ENG]   A relatively wide trough set at a slope and covered with a blanket or corduroy for catching comparatively coarse gold and any valuable mineral.

strand [GEOL]   A beach bordering a sea or an arm of an ocean.

stranded floe ice foot *See* stranded ice foot.

stranded ice [OCEANOGR]   Ice held in place by virtue of being grounded. Also known as grounded ice.

stranded ice foot [OCEANOGR]   An ice foot formed by the stranding of floes or small icebergs along a shore; it may be built up by freezing spray or breaking seas. Also known as stranded floe ice foot.

strandflat [GEOL]   **1.** A low, flat, very wide wave-cut platform extending off the coast of western Norway. **2.** A discontinuous shelf of land inside a fjord. **3.** *See* wave-cut platform.

strandline [GEOL]   **1.** A beach raised above the present sea level. **2.** *See* shoreline.

strandline *See* shoreline.

strapping [PETRO ENG]   A petroleum industry procedure in which storage tanks are strapped (measured) on their outside with steel measuring tapes to calculate the volumetric capacity of the tank for increments of height.

strapping table [PETRO ENG]   A tabular record of tank volume versus height so that taped (strapped) measurements of liquid depth can be converted into liquid volumes.

strath [GEOL]   **1.** A broad, elongate depression with steep sides on the continental shelf. **2.** An extensive remnant of a broad, flat valley floor that has undergone degradation following uplift.

strath terrace [GEOL]   An extensive remnant of a strath from a former erosion cycle.

stratification [GEOL]   An arrangement or deposition of sedimentary material in layers, or of sedimentary rock in strata.   [HYD]   **1.** The arrangement of a body of water, as a lake, into two or more horizontal layers of differing characteristics, especially densities. **2.** The formation of layers in a mass of snow, ice, or firn.

stratification index [GEOL]   A measure of the beddedness of a stratigraphic unit, expressed as the number of beds in the unit per 100 feet (30 meters) of section.

stratification plane [GEOL]   A demarcation between two layers of sedimentary rock, often signifying that the layers were deposited under different conditions.

stratified drift [GEOL]   Fluvioglacial drift composed of material deposited by a meltwater stream or settled from suspension.

stratified ocean [OCEANOGR]   An ocean where there is a vertical gradient of density.

stratified rock *See* sedimentary rock.

stratiform [GEOL]   **1.** Descriptive of a layered mineral deposit of either igneous or sedimentary origin. **2.** Consisting of parallel bands, layers, or sheets.   [METEOROL]   Description of clouds of extensive horizontal development, as contrasted to the vertically developed cumuliform types.

stratiformis [METEOROL]   A cloud species consisting of a very extensive horizontal layer or layers which need not be continuous; this species is the most common form of the genera altocumulus and stratocumulus and is occasionally found in cirrocumulus.

stratigrapher [GEOL]   A geologist who deals with stratified rocks, for example, the classification, nomenclature, correlation, and interpretation of rocks.

stratigraphic geology *See* stratigraphy.

stratigraphic map [GEOL]   A map showing the areal distribution, configuration, or aspect of a stratigraphic unit or surface, such as an isopach map or a lithofacies map.

stratigraphic oil fields [GEOL]   Hydrocarbon reserves in stratigraphic (sedimentary) traps formed by the positioning of clastic materials through chemical deposition.

stratigraphic separation *See* stratigraphic throw.

stratigraphic sequence *See* sequence.

stratigraphic throw [GEOL]   The thickness of the strata which originally separated two beds brought into contact at a fault. Also known as stratigraphic separation.

stratigraphic trap [GEOL]   Sealing of a reservoir bed due to lithologic changes rather than geologic structure. Also known as porosity trap; secondary stratigraphic trap.

stratigraphic unit [GEOL]   A stratum of rock or a body of strata classified as a unit on the basis of character, property, or attribute.

stratigraphy [GEOL]   A branch of geology concerned with the form, arrangement, geographic distribution, chronologic succession, classification, correlation, and mutual relationships of rock strata, especially sedimentary. Also known as stratigraphic geology.

stratocumulus [METEOROL]   A principal cloud type predominantly stratiform, in the form of a gray or whitish layer of patch, which nearly always has dark parts.

stratopause [METEOROL]   The boundary or zone of transition separating the stratosphere and the mesosphere; it marks a reversal of temperature change with altitude.

stratosphere [METEOROL]   The atmospheric shell above the troposphere and below the mesosphere; it extends, therefore, from the tropopause to about 55 kilometers, where the temperature begins again to increase with altitude.

stratosphere radiation [GEOPHYS]   Any infrared radiation involved in the complex infrared exchange continually proceeding within the stratosphere.

stratospheric coupling [METEOROL]   The interaction between disturbances in the stratosphere and those in the troposphere.

stratospheric steering [METEOROL]   The steering of lower-level atmospheric disturbances along the contour lines of the tropopause, which lines are presumably roughly parallel to the direction of the wind at the tropopause level.

stratotype [GEOL]   A specifically bounded type section of rock strata to which a time-stratigraphic unit is ascribed, ideally consisting of a complete and continuously exposed and deposited sequence of correlatable strata, and extending from a readily identifiable basal boundary to a readily identifiable top boundary.

stratovolcano [GEOL]   A volcano constructed of lava and pyroclastics, deposited in alternating layers. Also known as composite volcano.

stratus [METEOROL]   A principal cloud type in the form of a gray layer with a rather uniform base; a stratus does not usually produce precipitation, but when it does occur it is in the form of minute particles, such as drizzle, ice crystals, or snow grains.

stray [GEOL]   A lenticular rock formation encountered unexpectedly in drilling an oil or a gas well; it differs from an adjacent persistent formation in lithology and hardness.

strays *See* atmospheric interference.

stray sand [GEOL]   A stray composed of sandstone.

streak [MINERAL]   The color of a powdered mineral, obtained by rubbing the mineral on a streak plate.

streak lightning [GEOPHYS]   Ordinary lightning, of a cloud-to-ground discharge, that appears to be entirely concentrated in a single, relatively straight lightning channel.

stream [HYD]   A body of running water moving under the influence of gravity to lower levels in a narrow, clearly defined natural channel.

stream-built terrace *See* alluvial terrace.

stream capacity [GEOL]   The ability of a stream to carry detritus, measured at a given point per unit of time.

stream capture *See* capture.

stream channel [GEOL]   A long, narrow, sloping troughlike depression where a natural stream flows or may flow. Also known as streamway.

stream channel form ratio [GEOL]   The mathematical relationship between a stream channel width, depth, and channel perimeter.

stream current [HYD]   A steady current in a stream or river.   [OCEANOGR]   A deep, narrow, well-defined fast-moving ocean current.

streamer [GEOPHYS] A sinuous channel of very high ion-density which propagates itself through a gas by continual establishment of an electron avalanche just ahead of its advancing tip; in lightning discharges, the stepped leader, and return streamer all constitute special types of streamers.

stream erosion [GEOL] The progressive removal of exposed matter from the surface of a stream channel by a stream.

streamflow [HYD] A type of channel flow, applied to surface runoff moving in a stream.

streamflow routing *See* flood routing.

stream frequency [GEOL] A measure of topographic texture expressed as the ratio of the number of streams in a drainage basin to the area of the basin. Also known as channel frequency.

stream gage *See* river gage.

stream gradient [GEOL] The angle, measured in the direction of flow, between the water surface (for large streams) or the channel flow (for small streams) and the horizontal. Also known as stream slope.

stream gradient ratio [GEOL] Ratio of the stream gradient of a stream channel of one order to the stream gradient of the next higher order channel in the same drainage basin. Also known as channel gradient ratio.

stream-length ratio [HYD] Ratio of the mean length of a stream of a given order to the mean length of the next lower order stream in the same basin.

stream order [HYD] The designation by a dimensionless integer series (1, 2, 3, . . .) of the relative position of stream segments in the network of a drainage basin. Also known as channel order.

stream piracy *See* capture.

stream profile [HYD] The longitudinal profile of a stream.

stream robbery *See* capture.

stream segment [HYD] The part of a stream extending between designated tributary junctions. Also known as channel segment.

stream slope *See* stream gradient.

stream terrace [GEOL] One of a series of level surfaces on a stream valley flanking and parallel to a stream channel and above the stream level, representing the uneroded remnant of an abandoned floodplain or stream bed. Also known as river terrace.

stream tin [GEOL] The mineral cassiterite occurring as pebbles in alluvial deposits.

stream transport [GEOL] Movement of rock material in and by a stream.

streamway *See* stream channel.

strengite [MINERAL] $FePO_4 \cdot 2H_2O$ A pale-red mineral crystallizing in the orthorhombic system, isomorphous with variscite and dimorphous with phosphosiderite, and specific gravity 2.87.

strength of current [OCEANOGR] **1.** The phase of a tidal current at which the speed is a maximum. **2.** The velocity of the current at this time.

strength of ebb [OCEANOGR] **1.** The ebb current at the time of maximum speed. **2.** The speed of the current at this time.

strength of ebb interval [OCEANOGR] The time interval between the transit (upper or lower) of the moon and the next maximum ebb current at a place.

strength of flood [OCEANOGR] **1.** The flood current at the time of maximum speed. **2.** The speed of the current at this time.

strength of flood interval [OCEANOGR] The time interval between the transit (upper or lower) of the moon and the next maximum flood current at a place.

stress mineral [MINERAL] Any mineral whose formation in metamorphosed rock is favored by shearing stress.

stretch [GEOGR] *See* reach.   [PETRO ENG] The increase in length of oil-well casing or tubing when freely suspended in fluid mediums.

stretched pebbles [GEOL] Pebbles in a sedimentary rock which have been elongated from their original shape by deformation.

stretcher [MIN ENG] A bar used for roof support on roadways and which is either wedged against or pocketed into the sides of the roadway without support of legs or struts.

stretcher bar [MIN ENG] A single screw column, capable of holding one machine drill; used in small drifts.

stretch fault *See* stretch thrust.

stretch thrust [GEOL] A reverse fault developed as a result of shear in the middle limb of an overturned fold. Also known as stretch fault.

striated ground *See* striped ground.

striation [GEOL] One of a series of parallel or subparallel scratches, small furrows, or lines on the surface of a rock or rock fragment; usually inscribed by rock fragments embedded at the base of a moving glacier.   [MINERAL] One of a series of parallel, shallow depressions or narrow bands on the cleavage face of a mineral caused either by growth twinning or oscillatory growth of different crystal faces.

strigovite [MINERAL] $Fe_3(Al,Fe)_3Si_3O_{11}(OH)_7$ A dark-green mineral of the chlorite group, composed of basic aluminum iron silicate; occurs as crystalline incrustations.

strike [GEOL] The direction taken by a structural surface, such as a fault plane, as it intersects the horizontal. Also known as line of strike.

strike board [MIN ENG] A board at the top of a shaft from which the bucket is tipped; used in shaft sinking; formerly, the beam or plank at the shaft top on which the baskets were landed. Also known as strike tree.

strike fault [GEOL] A fault whose strike is parallel with that of the strata involved.

strike joint [GEOL] A joint that strikes parallel to the bedding or cleavage of the constituent rock.

strike separation [GEOL] The distance of separation on either side of a fault surface of two formerly adjacent beds.

strike-shift fault *See* strike-slip fault.

strike slip [GEOL] The component of the slip of a fault that is parallel to the strike of the fault. Also known as horizontal displacement; horizontal separation.

strike-slip fault [GEOL] A fault whose direction of movement is parallel to the strike of the fault. Also known as strike-shift fault.

strike stream *See* subsequent stream.

strike tree *See* strike board.

string [ENG] A piece of pipe, casing, or other down-hole drilling equipment coupled together and lowered into a borehole.   [GEOL] A very small vein, either independent or occurring as a branch of a larger vein. Also known as stringer.

stringer *See* string.

stringer lode [GEOL] A lode that consists of many narrow veins in a mass of country rock.

stringing [PETRO ENG] The connecting of lengths of pipe end to end (tubing or casing) to make a string long enough to reach to the desired depth in a well bore.

string shot [PETRO ENG] An oil-well stimulation technique in which a string of explosive (for example, Prima Cord) is hung opposite to the producing zone down a wellbore and detonated; used to remove deposits (gypsum, mud, or paraffin) from the formation face.

strip [MIN ENG] To remove coal, stone, or other material from a quarry or from a working that is near the surface of the earth.

striped ground [GEOL] A pattern of alternating stripes formed by frost action on a sloping surface. Also known as striated ground; striped soil.

striped soil *See* striped ground.

strip mine [MIN ENG] An opencut mine in which the overburden is removed from a coal bed before the coal is taken out.

strip mining [MIN ENG] The mining of coal by surface mining methods.

stripped illite *See* degraded illite.

stripped plane [GEOL] The upper, exposed surface of a resistant stratum that forms a stripped structural surface when extended over a considerable area.

stripped structural surface [GEOL] An erosion surface formed in an area underlain by horizontal or gently sloping strata of unequal resistance where the overlying softer beds have been removed by erosion. Also known as stripped surface.

stripped surface *See* stripped structural surface.

stripper rubber [PETRO ENG] A pressure-actuated seal used to control gas pressure in the casing-tubing annulus of low-pressure wells while inserting (running) or withdrawing (pulling) tubing.

stripping area [MIN ENG] In stripping operations, an area encompassing the pay material, its bottom depth, the thickness of the layer of waste, the slope of the natural ground surface, and the steepness of the safe slope of cuts.

stripping a shaft [MIN ENG] **1.** Removing the timber from an abandoned shaft. **2.** Trimming or squaring the sides of a shaft.

stripping ratio [MIN ENG] The unit amount of spoil or waste that must be removed to gain access to a similar unit amount of ore or mineral material.

stripping shovel [MIN ENG] A shovel with an especially long boom and stick, enabling it to reach further and pile higher.

strip pit [MIN ENG] **1.** A coal or other mine worked by stripping. **2.** An open-pit mine.

stroke density [GEOPHYS] The areal density of lightning discharges over a given region during some specified period of time, as number per square mile per year.

stromatite [GEOL] Chorismite having flat or folded parallel layers of two or more textural elements. Also known as stromatolith.

stromatolite [GEOL]   A structure in calcareous rocks consisting of concentrically laminated masses of calcium carbonate and calcium-magnesium carbonate which are believed to be of calcareous algal origin; these structures are irregular to columnar and hemispheroidal in shape, and range from 1 millimeter to many meters in thickness. Also known as callenia.

stromatolith [GEOL]   **1.** A complex sill-like igneous intrusion interfingered with sedimentary strata. **2.** *See* stromatite.

Stromatoporoidea [PALEON]   An extinct order of fossil colonial organisms thought to belong to the class Hydrozoa; the skeleton is a coenosteum.

Strombacea [PALEON]   An extinct superfamily of gastropod mollusks in the order Prosobranchia.

strombolian [GEOL]   A type of volcanic eruption characterized by fire fountains of lava from a central crater.

stromeyerite [MIN]   CuAgS   A metallic-gray orthorhombic mineral with a blue tarnish composed of silver copper sulfide occurring in compact masses.

strong breeze [METEOROL]   In the Beaufort wind scale, a wind whose speed is from 22 to 27 knots (25 to 31 miles per hour or 41 to 50 kilometers per hour).

strong gale [METEOROL]   In the Beaufort wind scale, a wind whose speed is from 41 to 47 knots (47 to 54 miles per hour or 76 to 87 kilometers per hour).

strontianite [MINERAL] $SrCO_3$ A pale-green, white, gray, or yellowish mineral of the aragonite group having orthorhombic symmetry and occurring in veins or as masses; hardness is 3.5 on Mohs scale, and specific gravity is 3.76.

Strophomenida [PALEON]   A large diverse order of articulate brachiopods which first appeared in Lower Ordovician times and became extinct in the Late Triassic.

Strophomenidina [PALEON]   A suborder of extinct, articulate brachiopods in the order Strophomenida characterized by a concavo-convex shell, the pseudodeltidium and socket plates disposed subparallel to the hinge.

struck capacity [MIN ENG]   The volume of water a mine car, tram, hoppet, or wagon would hold if the conveyance were of watertight construction.

structural analysis *See* structural petrology.

structural bench [GEOL]   A bench typifying the resistant edge of a terrace that is being reduced by erosion. Also known as rock bench.

structural contour map [GEOL]   A map representation of a subsurface stratigraphic unit; depicts the configuration of a rock surface by means of elevation contour lines.

structural drill [MECH ENG]   A highly mobile diamond- or rotary-drill rig complete with hydraulically controlled derrick mounted on a truck, designed primarily for rapidly drilling holes to determine the structure in subsurface strata or for use as a shallow, slim-hole producer or seismograph drill.

structural drilling [ENG]   Drilling done specifically to obtain detailed information delineating the location of folds, domes, faults, and other subsurface structural features indiscernible by studying strata exposed at the surface.

structural fabric *See* fabric.

structural geology [GEOL]   A branch of geology concerned with the form, arrangement, and internal structure of the rocks.

structural high [GEOL]   Any of various structural features such as a crest, culmination, anticline, or dome.

structural low   [GEOL]   Any of various structural features such as a basin, a syncline, a saddle, or a sag.

structural petrology   [PETR]   The study of the internal structure of a rock to determine its deformational history. Also known as fabric analysis; microtectonics; petrofabric analysis; petrofabrics; petrogeometry; petromorphology; structural analysis.

structural terrace   [GEOL]   A terracelike landform developed where generally steeply inclined and otherwise uniformly dipping strata locally flatten.

structural trap   [GEOL]   Containment in a reservoir bed of oil or gas due to flexure or fracture of the bed.

structural valley   [GEOL]   A valley whose form and origin is attributable to the underlying geologic structure.

structure   [GEOL]   **1.** An assemblage of rocks upon which erosive agents have been or are acting. **2.** The sum total of the structural features of an area.   [MINERAL]   The form taken by a mineral, such as tabular or fibrous.   [PETR]   A macroscopic feature of a rock mass or rock unit, best seen in an outcrop.

structure cell   *See* unit cell.

structure contour   [GEOL]   A contour that portrays a structural surface, such as a fault. Also known as subsurface contour.

structure-contour map   [GEOL]   A map that uses structure contour lines to portray subsurface configuration. Also known as structure map.

structure map   *See* structure-contour map.

structure section   [GEOL]   A vertical section showing the observed or inferred geologic structure on a vertical surface or plane.

strut   [MIN ENG]   A vertical-compression member in a structure or in an underground timber set.

struvite   [MINERAL] $Mg(NH_4)PO_4 \cdot 6H_2O$ A colorless to yellow or pale-brown mineral consisting of a hydrous ammonium magnesium phosphate, and occurring in orthorhombic crystals; hardness is 2 on Mohs scale, and specific gravity is 1.7.

stub entry   [MIN ENG]   A short, narrow entry turned from another entry and driven into the solid coal, but not connected with other mine workings.

stuffed mineral   [MINERAL]   A mineral having extra ions of a foreign element within its larger interstices.

stull   [MIN ENG]   A platform laid on timbers, braced across a working from side to side, to support workers or to carry ore or waste.

stull piece   [MIN ENG]   **1.** A piece of timber placed slanting over the back of a level to prevent rock falling into the level from the stopes above. **2.** Timbers bracing the platform of a stull.

stull stoping   [MIN ENG]   Stull timbers placed between the foot and hanging walls, which constitute the only artificial support provided during the excavation of a stope.

stump   [MIN ENG]   A small pillar of coal left between the gangway or airway and the breasts to protect these passages; any small pillar.

stupp   [MIN ENG]   A black residue from distilled mercury ore, consisting of soot, hydrocarbons, mercury and mercury compounds, and ore dust.

sturtite   [MINERAL]   A black mineral composed of hydrous silicate of iron, manganese, calcium, and magnesium; occurs in compact masses.

Stuve chart [METEOROL]   A thermodynamic diagram with atmospheric temperature as the x axis and atmospheric pressure to the power 0.286 as the y ordinate, increasing downward; named after G. Stuve. Also known as adiabatic chart; pseudoadiabatic chart.

stylolite [GEOL]   An irregular surface, generally parallel to a bedding plane, in which small toothlike projections on one side of the surface fit into cavities of complementary shape on the other surface; interpreted to result diagenetically by pressure solution.

stylotypite *See* tetrahedrite.

subaerial [GEOL]   Pertaining to conditions and processes occurring beneath the atmosphere or in the open air, that is, on or adjacent to the land surface.

subage [GEOL]   A subdivision of a geologic age.

subalkaline [GEOCHEM]   Pertaining to a soil in which the pH is 8.0 to 8.5, usually in a limestone or salt-marsh region.

Subantarctic Intermediate Water [OCEANOGR]   A layer of water above the deep-water layer in the South Atlantic.

subaqueous [HYD]   Pertaining to conditions and processes occurring in, under, or beneath the surface of water, especially fresh water.

subaqueous dune [GEOL]   A dune resulting from entrainment of grains by the flow of moving water.

subaqueous mining [MIN ENG]   Surface mining in which the mined material is removed from the bed of a natural body of water.

subarctic [GEOGR]   Pertaining to regions adjacent to the Arctic Circle or having characteristics somewhat similar to those of these regions.

subarctic climate *See* taiga climate.

subarid [CLIMATOL]   Pertaining to regions that are moderately or slightly arid.

subarkose [GEOL]   Sandstone that is intermediate in composition between arkose and pure quartz sandstone; it contains less feldspar than arkose.

subbituminous coal [GEOL]   Black coal intermediate in rank between lignite and bituminous coal; has more carbon and less moisture than lignite.

subbottom reflection [GEOPHYS]   The return of sound energy from a discontinuity in material below the surface of the sea bottom.

subcapillary interstice [GEOL]   An interstice in which the molecular attraction of its walls extends across the entire opening; it is smaller than a capillary interstice.

subconchoidal [GEOL]   Pertaining to a fracture that is partly or vaguely conchoidal in shape.

subconsequent stream *See* secondary consequent stream.

subcontinent [GEOGR]   **1.** A landmass such as Greenland that is large but not as large as the generally recognized continents. **2.** A large subdivision of a continent (for example, the Indian subcontinent) distinguished geologically or geomorphically from the rest of the continent.

subcrop [GEOL]   An occurrence of strata beneath the subsurface of an inclusive stratigraphic unit that succeeds an unconformity on which there is marked overstep.

subduction [GEOL]   The process by which one crustal block descends beneath another, such as the descent of the Pacific plate beneath the Andean plate along the Andean Trench.

suberinite [GEOL]   A variety of provitrinite composed of corky tissue.

subfeldspathic [GEOL]   Referring to mature lithic wacke or arenite containing an abundance of quartz grains with less than 10% feldspar grains.

subgelisol [GEOL]   Unfrozen ground beneath permafrost.

subgeostrophic wind [METEOROL]   Any wind of lower speed than the geostrophic wind required by the existing pressure gradient.

subglacial [GEOL]   Pertaining to the area in or at the bottom of, or immediately beneath, a glacier.

subglacial moraine *See* ground moraine.

subgradient wind [METEOROL]   A wind of lower speed than the gradient wind required by the existing pressure gradient and centrifugal force.

subgraywacke [PETR]   An argillaceous sandstone with a composition intermediate between graywacke and orthoquartzite; a clay matrix is usually present but it amounts to less than 15%.

subhedral [MINERAL]   **1.** Pertaining to an individual mineral crystal that is partly bounded by its own crystal faces and partly bounded by surfaces formed against preexisting crystals. **2.** Descriptive of a crystal having partially developed crystal faces.

subhumid climate [CLIMATOL]   A humidity province based on its typical vegetation. Also known as grassland climate; prairie climate.

subidiomorphic *See* hypidiomorphic.

subjacent [GEOL]   Being lower than but not directly underneath.

subjacent igneous body [GEOL]   An igneous intrusion without a known floor, and which presumably enlarges downward.

sublacustrine [GEOL]   Existing or formed on the bottom of a lake.

sublacustrine channel [GEOL]   A channel eroded in a lake bed either before the lake existed or by a strong current in the lake.

sublevel [MIN ENG]   An intermediate level opened a short distance below the main level; or in the caving system of mining, a 15–20-foot (4.6–6.1-meter) level below the top of the ore body, preliminary to caving the ore between it and the level above.

sublevel caving [MIN ENG]   A stoping method in which relatively thin blocks of ore are caused to cave by successively undermining small panels.

sublevel drive [MIN ENG]   A drive often made in a section which divides the deposit into narrower panels and zones.

sublevel stoping [MIN ENG]   A mining method involving overhand, underhand, and shrinkage stoping; the characteristic feature is the use of sublevels which are worked simultaneously, the lowest on a given block being farthest advanced and the sublevels above following one another at short intervals.

sublimation nucleus [METEOROL]   Any particle upon which an ice crystal may grow by the process of sublimation.

sublimation vein [GEOL]   A vein of mineral that has condensed from a vapor.

sublittoral zone [OCEANOGR]   The benthic region extending from mean low water (or 40–60 meters, according to some authorities) to a depth of about 100 fathoms (200 meters), or the edge of a continental shelf, beyond which most abundant attached plants do not grow.

submarine [OCEANOGR]   Being or functioning in the sea.

submarine bulge *See* fan.

submarine canyon [GEOL] Steep-sided valleys winding across the continental shelf or continental slope, probably originally produced by Pleistocene stream erosion, but presently the site of turbidity flows.

submarine cave *See* submarine fan.

submarine delta *See* submarine fan.

submarine earthquake *See* seaquake.

submarine fan [GEOL] A shallow marine sediment that is fan- or cone-shaped and lies off the seaward opening of large rivers and submarine canyons. Also known as abyssal cave; abyssal fan; sea fan; submarine cave; submarine delta; subsea apron.

submarine geology *See* geological oceanography.

submarine isthmus [GEOL] A submarine elevation joining two land areas and separating two basins or depressions by a depth less than that of the basins.

submarine mine [MIN ENG] A mine for the extraction of minerals or ores under the sea.

submarine peninsula [GEOL] An elevated portion of the submarine relief resembling a peninsula.

submarine pit [GEOL] A cavity on the bottom of the sea. Also known as submarine well.

submarine plain *See* plain.

submarine relief [GEOL] Relative elevations of the ocean bed, or the representation of them on a chart.

submarine spring [HYD] A spring of water issuing from the bottom of the sea.

submarine station [OCEANOGR] **1.** One of the places for which tide or tidal current predictions are determined by applying a correction to the predictions of a reference station. **2.** A tide or tidal current station at which a short series of observations have been made; these observations are reduced by comparison with simultaneous observations at a reference station.

submarine topography [GEOL] Configuration of a surface such as the sea bottom or of a surface of given characteristics within the water mass.

submarine trench *See* trench.

submarine trough *See* trough.

submarine valley *See* valley.

submarine weathering [GEOL] A slow alteration of the form, texture, and composition of the sea floor from chemical, thermal, and biological causes.

submarine well *See* submarine pit.

submerged breakwater [OCEANOGR] A breakwater with its top below the still water level; when struck by a wave, part of the wave energy is reflected seaward and the remaining energy is largely dissipated in a breaker, transmitted shoreward as a multiple crest system, or transmitted shoreward as a simple wave system.

submerged coastal plain [GEOL] The continental shelf as the seaward extension of a coastal plain on the land. Also known as coast shelf.

submerged lands [GEOL] Lands covered by water at any stage of the tide, as distinguished from tidelands which are attached to the mainland or an island and are covered or uncovered with the tide; tidelands presuppose a high-water line as the upper boundary, submerged lands do not.

submerged shoreline *See* shoreline of submergence.

submergence [GEOL] A change in the relative levels of water and land either from a sinking of the land or a rise of the water level.

subpolar anticyclone *See* subpolar high.

subpolar glacier [HYD] A polar glacier with 10 to 20 meters of firn in the accumulation area where some melting occurs.

subpolar high [METEOROL] A high that forms over the cold continental surfaces of subpolar latitudes, principally in Northern Hemisphere winters; these highs typically migrate eastward and southward. Also known as polar anticyclone; polar high; subpolar anticyclone.

subpolar low-pressure belt [METEOROL] A belt of low pressure located, in the mean, between 50 and 70° latitude; in the Northern Hemisphere, this belt consists of the Aleutian low and the Icelandic low; in the Southern Hemisphere, it is supposed to exist around the periphery of the Antarctic continent.

subpolar westerlies *See* westerlies.

subsea apron *See* submarine fan.

subsequent [GEOL] Referring to a geologic feature that followed in time the development of a consequent feature of which it is a part.

subsequent drainage [HYD] Drainage by a stream developed subsequent to the system of which it is a part; drainage follows belts of weak rocks.

subsequent fold *See* cross fold.

subsequent stream [HYD] A stream that flows in the general direction of the strike of the underlying strata and is subsequent to the formation of the consequent stream of which it is a tributary. Also known as longitudinal stream; strike stream.

subsequent valley [GEOL] A valley eroded by a stream developed subsequent to the system of which it is a part.

subsidence [METEOROL] A descending motion of air in the atmosphere, usually with the implication that the condition extends over a rather broad area. [MIN ENG] A sinking down of a part of the earth's crust due to underground excavations.

subsidence break [MIN ENG] A fracture in the rocks overlying a coal seam or mineral deposit resulting from mining operations.

subsidence inversion [METEOROL] A temperature inversion produced by the adiabatic warming of a layer of subsiding air; this inversion is enhanced by vertical mixing in the air layer below the inversion.

subsidiary fracture *See* tension fracture.

subsidiary transport [MIN ENG] The conveying or haulage of coal or mineral from the working faces to a junction or loading point.

subsoil [GEOL] **1.** Soil underlying surface soil. **2.** *See* B horizon.

subsoil ice *See* ground ice.

substation [MIN ENG] A subsidiary station for the conversion of power to the type, usually direct current, and voltage needed for mining equipment and fed into the mine power system.

substratosphere [METEOROL] A region of indefinite lower limit just below the stratosphere.

substratum [GEOL] Any layer underlying the true soil.

subsurface contour *See* structure contour.

subsurface current [OCEANOGR]   An underwater current which is not present at the surface or whose core (region of maximum velocity) is below the surface.

subsurface flow [HYD]   Interflow plus groundwater flow.

subsurface geology [GEOL]   The study of geologic features beneath the land or sea-floor surface. Also known as underground geology.

subterranean ice *See* ground ice.

subterranean stream [HYD]   A subsurface stream that flows through a cave or a group of communicating caves.

subtropic [METEOROL]   An indefinite belt in each hemisphere between the tropic and temperate regions; the polar boundaries are considered to be roughly 35–40° northern and southern latitudes, but vary greatly according to continental influence, being farther poleward on the western coasts of continents and farther equatorward on the eastern coasts.

subtropical anticyclone *See* subtropical high.

Subtropical Convergence [OCEANOGR]   The zone of converging currents, generally located in midlatitudes.

subtropical cyclone [METEOROL]   The low-level (surface chart) manifestation of a cutoff low.

subtropical easterlies *See* tropical easterlies.

subtropical easterlies index [METEOROL]   A measure of the strength of the easterly wind between the latitudes of 20° and 35°N; the index is computed from the average sea-level pressure difference between these latitudes and is expressed as the east to west component of the corresponding geostrophic wind in meters and tenths of meters per second.

subtropical high [METEOROL]   One of the semipermanent highs of the subtropical high-pressure belt; these highs appear as centers of action on mean charts of surface pressure; they lie over oceans and are best developed in the summer season. Also known as oceanic anticyclone; oceanic high; subtropical anticyclone.

subtropical high-pressure belt [METEOROL]   One of the two belts of high atmospheric pressure that are centered, in the mean, near 30°N and 30°S latitudes; these belts are formed by the subtropical highs.

subtropical westerlies *See* westerlies.

Subulitacea [PALEON]   An extinct superfamily of gastropod mollusks in the order Prosobranchia which possessed a basal fold but lacked an apertural sinus.

successive fracture treatment [PETRO ENG]   A second or third fracturing operation of an oil well in an oil reservoir to fracture a new part or zone.

succinite [MINERAL]   An amber-colored variety of grossularite.

sucker rod [PETRO ENG]   A connecting rod between a down-hole oil-well pump and the lifting or pumping device on the surface.

sucker-rod pump [PETRO ENG]   A cylinder-piston-type pump used to displace oil into the oil-well tubing string, and to the surface.

sucrosic *See* saccharoidal.

suction anemometer [ENG]   An anemometer consisting of an inverted tube which is half-filled with water that measures the change in water level caused by the wind's force.

suctive *See* permissive.

sudburite [GEOL] A basic basalt composed of hypersthene, augite, and magnetite, among other minerals.

sudden commencement [GEOPHYS] Magnetic storms which start suddenly (within a few seconds) and simultaneously all over the earth.

sudden ionospheric disturbance [GEOPHYS] A complex combination of sudden changes in the condition of the ionosphere following the appearance of solar flares, and the effects of these changes. Abbreviated SID.

suestada [METEOROL] Strong southeast winds occurring in winter along the coast of Argentina, Uruguay, and southern Brazil; they cause heavy seas and are accompanied by fog and rain; the counterpart of the northeast storm in North America.

suevite [GEOL] A grayish or yellowish fragmental rock associated with meteorite impact craters; resembles tuff breccia or pumiceous tuff but is of nonvolcanic origin.

sugar berg [OCEANOGR] An iceberg of porous glacier ice.

sugarloaf sea [OCEANOGR] A sea characterized by waves that rise into sugarloaf shapes, with little wind, possibly resulting from intersecting waves.

sugar snow *See* depth hoar.

sugary *See* saccharoidal.

sukhovei [METEOROL] Literally dry wind; a dry, hot, dusty wind in the south Russian steppes, which blows principally from the east and frequently brings a prolonged drought and crop damage.

sulfate mineral [MINERAL] A mineral compound characterized by the sulfate radical SO_4.

sulfide mineral [MINERAL] A mineral compound characterized by the linkage of sulfur with a metal or semimetal.

sulfoborite [MINERAL] $Mg_6H_4(BO_3)_4(SO_4)_2 \cdot 7H_2O$ A mineral composed of hydrous acid sulfate and borate of magnesium.

sulfofication [GEOCHEM] Oxidation of sulfur and sulfur compounds into sulfates, occurring in soils by the agency of bacteria.

sulfohalite [MINERAL] $Na_6(SO_4)_2FCl$ A mineral composed of sulfate, chloride, and fluoride of sodium.

sulfophile element [GEOCHEM] An element occurring preferentially in an oxygen-free mineral. Also known as thiophile element.

sulfur [MINERAL] A yellow orthorhombic mineral occurring in crystals, masses, or layers, and existing in several allotropic forms; the native form of the element.

sulfur ball [GEOL] A bubble of hot volcanic gas encased in a sulfurous mud skin that solidified on contact with air.

sulfur-mud pool *See* mud pot.

sulfur spring [HYD] A spring containing sulfur compounds such as hydrogen sulfide.

sultriness [METEOROL] An oppressively uncomfortable state of the weather which results from the simultaneous occurrence of high temperature and high humidity, and often enhanced by calm air and cloudiness.

sulvanite [MINERAL] Cu_3VS_4 A bronze-yellow mineral composed of copper vanadium sulfide occurring in masses.

sumatra [METEOROL] A squall, with wind speeds occasionally exceeding 30 miles (48 kilometers) per hour, in the Malacca Strait between Malay and Sumatra during the southwest monsoon (April through November).

summation principle [METEOROL] In United States weather observing practice, the rule which governs the assignment of sky cover amount to any layer of cloud or obscuring phenomenon, and to the total sky cover; in essence, this principle states that the sky cover at any level is equal to the summation of the sky cover of the lowest layer plus the additional sky cover provided at all successively higher layers up to and including the layer in question; thus, no layer can be assigned a sky cover less than a lower layer, and no sky cover can be greater than 1.0 (10/10).

summit plain *See* peak plain.

sun crack *See* mud crack.

sun cross [METEOROL] A rare halo phenomenon in which bands of white light intersect over the sun at right angles; it appears probable that most of such observed crosses appear merely as a result of the superposition of a parhelic circle and a sun pillar.

sun crust [HYD] A type of snow crust, formed by refreezing of surface snow crystals after having been melted by the sun.

sun drawing water [METEOROL] Popular designation for a phenomenon of the sun showing through scattered openings in a layer of clouds into a layer of turbid air that is hazy or dusty; bright bands are seen where the several beams of sunlight pass down through the subcloud layer; sailors called the phenomenon the backstays of the sun.

sundtite *See* andorite.

sunk [MIN ENG] Drilled downward, as a shaft.

sunlit aurora [GEOPHYS] An aurora which occurs in the part of the upper atmosphere which is in the sunlight, above the earth's shadow.

sun opal *See* fire opal.

sun pillar [METEOROL] A luminous streak of light, white or slightly reddened, extending above and below the sun, most frequently observed near sunrise or sunset; it may extend to about 20° above the sun, and generally ends in a point. Also known as light pillar.

sunstone [MINERAL] An aventurine feldspar containing minute flakes of hematite; usually brilliant and translucent, it emits reddish or golden billowy reflection. Also known as heliolite.

superadiabatic lapse rate [METEOROL] An environmental lapse rate greater than the dry-adiabatic lapse rate, such that potential temperature decreases with height.

supercapillary interstice [GEOL] An interstice that is too large to hold water above the free water surface by surface tension; it is larger than a capillary interstice.

supercooled cloud [METEOROL] A cloud composed of supercooled liquid water-drops.

superficial deposit *See* surficial deposit.

supergene [MINERAL] Referring to mineral deposits or enrichments formed by descending solutions. Also known as hypergene.

supergeostrophic wind [METEOROL] Any wind of greater speed than the geostrophic wind required by the pressure gradient.

superglacial [HYD] Of or pertaining to the upper surface of a glacier or ice sheet.

supergradient wind [METEOROL] A wind of greater speed than the gradient wind required by the existing pressure gradient and centrifugal force.

superimposed [GEOL] Pertaining to layered or stratified rocks.

superimposed drainage [HYD] Drainage by superimposed streams.

superimposed fan [GEOL] An alluvial fan developed on, and having a steeper gradient than, an older fan.

superimposed fold *See* cross fold.

superimposed glacier [GEOL] A glacier whose course is maintained despite different preexisting structures and lithologies as the glacier erodes downward.

superimposed stream [HYD] A stream, started on a new surface, that kept its course through the different preexisting lithologies and structures encountered as it eroded downward into the underlying rock. Also known as superinduced stream.

superimposed valley [GEOL] A valley eroded by or containing a superimposed stream.

superincumbent [GEOL] Pertaining to a superjacent layer, especialy one that is situated so as to exert pressure.

superinduced stream *See* superimposed stream.

superior air [METEOROL] An exceptionally dry mass of air formed by subsidence and usually found aloft but occasionally reaching the earth's surface during extreme subsidence processes.

superior tide [OCEANOGR] The tide in the hemisphere in which the moon is above the horizon.

superjacent [GEOL] Pertaining to a stratum situated immediately upon or over a particular lower stratum or above an unconformity.

superjacent waters [OCEANOGR] The waters above the continental shelf.

supermature [GEOL] Pertaining to a texturally mature clastic sediment whose grains have become rounded.

superposed stream *See* consequent stream.

superposition [GEOL] **1.** The order in which sedimentary layers are deposited, the highest being the youngest. **2.** The process by which the layering occurs.

superprint *See* overprint.

superrefraction *See* ducting.

superresolution [OCEANOGR] Separation of tides into components of different frequencies, without taking measurements for the full extent of the longest-period component.

supersaturation [METEOROL] The condition existing in a given portion of the atmosphere when the relative humidity is greater than 100%, in respect to a plane surface of pure water or pure ice.

supply current [GEOPHYS] The electrical current in the atmosphere which is required to balance the observed air-earth current of fair-weather regions by transporting positive charge upward or negative charge downward.

supracrustal rocks [GEOL] Rocks that overlie basement rock.

supralateral tangent arcs [METEOROL] Two oblique luminous arcs, concave to the sun and tangent to the halo of 46° at points above the altitude of the sun.

supratidal sediment [GEOL]   The sediment deposited immediately above the high-tide level.

supratidal zone [GEOL]   Pertaining to the shore area immediately marginal to and above the high-tide level.

surf [OCEANOGR]   Wave activity in the area between the shoreline and the outermost limit of breakers, that is, in the surf zone.

surface air leakage [MIN ENG]   The amount of surface air entering the fan through the casing at the top of the upcast shaft, the air-lock doors, and fan-drift walls.

surface boundary layer [METEOROL]   That thin layer of air adjacent to the earth's surface, extending up to the so-called anemometer level (the base of the Ekman layer); within this layer the wind distribution is determined largely by the vertical temperature gradient and the nature and contours of the underlying surface, and shearing stresses are approximately constant. Also known as atmospheric boundary layer; friction layer; ground layer; surface layer.

surface chart [METEOROL]   An analyzed synoptic chart of surface weather observations; essentially, a surface chart shows the distribution of sea-level pressure (therefore, the positions of highs, lows, ridges, and troughs) and the location and nature of fronts and air masses, plus the symbols of occurring weather phenomena, analysis of pressure tendency (isallobars), and indications of the movement of pressure systems and fronts. Also known as sea-level chart; sea-level-pressure chart; surface map.

surface creep [GEOL]   A stage of the wind erosion process in which grains of sand move each other along the surface.

surface current [OCEANOGR]   **1.** Water movement which extends to depths of 3–10 feet (1–3 meters) below the surface in nearshore areas, and to about 33 feet (10 meters) in deep-ocean areas. **2.** Any current whose maximum velocity core is at or near the surface.

surface deposit *See* surficial deposit.

surface detention [HYD]   Water in temporary storage as a thin sheet over the soil surface during the occurrence of overland flow.

surface drainage [HYD]   Natural or artificial removal of excess groundwater.

surface drilling [MIN ENG]   Boreholes collared at the surface of the earth, as opposed to boreholes collared in mine workings or underwater.

surface duct [GEOPHYS]   Atmospheric duct for which the lower boundary is the surface of the earth.

surface flow *See* overland flow.

surface friction [GEOPHYS]   The drag or skin friction of the earth on the atmosphere, usually expressed in terms of the shearing stress of the wind on the earth's surface.

surface geology [GEOL]   The scientific study of the features at the surface of the earth.

surface hoar [HYD]   **1.** Fernlike ice crystals formed directly on a snow surface by sublimation; a type of hoarfrost. **2.** Hoarfrost that has grown primarily in two dimensions, as on a window or other smooth surface.

surface inversion [METEOROL]   A temperature inversion based at the earth's surface; that is, an increase of temperature with height beginning at ground level. Also known as ground inversion.

surface layer *See* surface boundary layer.

surface lift [MIN ENG]   In the freezing method of shaft sinking, freezing and heaving of the surface around the shaft due to the formation of ice and the variation of temperature.

surface map *See* surface chart.

surface mining [MIN ENG]   Mining at or near the surface; includes placer mining, mining in open glory-hole or milling pits, mining and removing ore from opencuts by hand or with mechanical excavating and transportation equipment, and the removal of capping or overburden to uncover the ores.

surface of discontinuity [METEOROL]   An interface, applied to the atmosphere; for example, an atmospheric front is represented ideally by a surface of discontinuity of velocity, density, temperature, and pressure gradient.

surface phase [GEOCHEM]   A thin rock layer differing in geochemical properties from those of the volume phases on either side. Also known as volume phase.

surface pressure [METEOROL]   The atmospheric pressure at a given location on the earth's surface; the expression is applied loosely and about equally to the more specific terms: station pressure and sea-level pressure.

surface retention *See* surface storage.

surface rights [MIN ENG]   **1.** Ownership of the surface land only, mineral rights being reserved. **2.** Ownership of the surface land plus mineral rights. **3.** The right of a mineral owner or an oil and gas lessee to use as much surface land as may be reasonably necessary for the conduct of operations under the lease.

surface soil [GEOL]   The soil extending 5 to 8 inches (13 to 20 centimeters) below the surface.

surface storage [HYD]   The part of precipitation retained temporarily at the ground surface as interception or depression storage so that it does not appear as infiltration or surface runoff either during the rainfall period or shortly thereafter. Also known as initial detention; surface retention.

surface temperature [METEOROL]   Temperature of the air near the surface of the earth. [OCEANOGR]   Temperature of the layer of seawater nearest the atmosphere.

surface visibility [METEOROL]   The visibility determined from a point on the ground, as opposed to control-tower visibility.

surface wash *See* sheet erosion.

surface water [HYD]   All bodies of water occurring on the surface of the earth. [OCEANOGR]   *See* mixed layer.

surface wave [OCEANOGR]   A progressive gravity wave in which the disturbance is of greatest amplitude at the air-water interface.

surface weather observation [METEOROL]   An evaluation of the state of the atmosphere as observed from a point at the surface of the earth, as opposed to an upper-air observation, and applied mainly to observations which are taken for the primary purpose of preparing surface synoptic charts.

surface wind [METEOROL]   The wind measured at a surface observing station; customarily, it is measured at some distance above the ground itself to minimize the distorting effects of local obstacles and terrain.

surfactant flooding *See* micellar flooding.

surf beat [OCEANOGR]   Oscillations of water level near shore, associated with groups of high breakers.

surficial creep *See* soil creep.

surficial deposit   [GEOL]   Unconsolidated alluvial, residual, or glacial deposits overlying bedrock or occurring on or near the surface of the earth. Also known as superficial deposit; surface deposit.

surficial geology   [GEOL]   The scientific study of surficial deposits, including soils.

surf ripple   [GEOL]   A ripple mark formed on a sandy beach by wave-generated currents.

surf zone   [OCEANOGR]   The area between the landward limit of wave uprush and the farthest seaward breaker.

surge   [OCEANOGR]   **1.** Wave motion of low height and short period, from about ½ to 60 minutes. **2.** *See* storm surge.

surge bunker   [MIN ENG]   A large-capacity storage hopper, installed near the pit bottom or at the input end of a processing plant to provide uniform bulk deliveries.

surge column   [PETRO ENG]   A large-sized pipe of sufficient height to provide a static head able to absorb the surging liquid discharge of the process tank to which it is connected. Also known as boot.

surge line   [METEOROL]   A line along which a discontinuity in the wind speed occurs.

surge tank   [MIN ENG]   In pumping of ore pulps, a relatively small tank which maintains a steady loading of the pump.

surging breaker   *See* plunging breaker.

surging glacier   [HYD]   A glacier that alternates periodically between surges (brief periods of rapid flow) and stagnation.

suroet   [METEOROL]   A persistent, rain-bearing southwest wind on the west coast of France.

sursassite   [MINERAL] $Mn_5Al_4Si_5O_{21} \cdot 3H_2O$ A mineral composed of hydrous manganese aluminum silicate.

surveying sextant   *See* hydrographic sextant.

surveyor's compass   [ENG]   An instrument used to measure horizontal angles in surveying.

surveyor's cross   [ENG]   An instrument for setting out right angles in surveying; consists of two bars at right angles with sights at each end.

surveyor's level   [ENG]   A telescope and spirit level mounted on a tripod, rotating vertically and having leveling screws for adjustment.

surveyor's measure   [ENG]   A system of measurement used in surveying having the engineer's, or Gunter's chain, as a unit.

susannite   [MINERAL] $Pb_4(SO_4)(CO_3)_2(OH)_2$ A greenish or yellowish, rhombohedral mineral that is dimorphous with leadhillite.

suspended load   [GEOL]   The part of the stream load that is carried for a long time in suspension. Also known as suspension load.

suspended tubbing   [MIN ENG]   A permanent method of lining a circular shaft, in which the tubbing (German type) is temporarily suspended from the next wedging curb above and for which no temporary supports are required; slurry is run in behind the tubbing by means of a funnel passing through the holes provided in the segments.

suspended water   *See* vadose water.

suspension   [MIN ENG]   The bolting of rock to secure fragments or sections, such as small slabs barred down after blasting blocks of rock broken by fracture or joint patterns, which may subsequently loosen and fall.

suspension current *See* turbidity current.

suspension load *See* suspended load.

suspension roast *See* flash roast.

sussexite [MINERAL] $MnBO_2OH$ A white mineral composed of basic manganese borate occurring in fibrous veins.

svabite [MINERAL] $Ca_5(AsO_4)_3F$ A colorless, yellow, rose, or reddish-brown mineral composed of fluoride-arsenate of calcium.

svanbergite [MINERAL] $SrAl_3(PO_4)(SO_4)(OH)_6$ A colorless to yellow mineral composed of basic phosphate and sulfate of strontium and aluminum; it is isomorphous with corkite, hinsdalite, and woodhouseite.

swab [MIN ENG]   **1.** A pistonlike device provided with a rubber cap ring that is used to clean out debris inside a borehole or casing. **2.** The act of cleaning the inside of a tubular object with a swab.   [PETRO ENG]   In petroleum drilling, to pull the drill string so rapidly that the drill mud is sucked up and overflows the collar of the borehole, thus leaving an undesirably empty borehole.

swamper [MIN ENG]   **1.** A rear brakeman in a metal mine. **2.** A laborer who assists in hauling ore and rock, coupling and uncoupling cars, throwing switches, and loading and unloading carriers.

swartzite [MINERAL] $CaMg(UO_2)(CO_3)_3 \cdot 12H_2O$ A green monoclinic mineral composed of hydrous carbonate of calcium, magnesium, and uranium.

swash [GEOL]   **1.** A narrow channel or ground within a sand bank, or between a sand bank and the shore. **2.** A bar over which the sea washes.   [OCEANOGR]   The rush of water up onto the beach following the breaking of a wave. Also known as run-up; uprush.

swash mark [GEOL]   A fine, wavy or arcuate line or minute ridge consisting of fine sand, seaweed, and other debris on a beach; marks the farthest advance of wave uprush. Also known as debris line; wave line; wavemark.

S wave [GEOPHYS]   A seismic body wave propagated in the crust or mantle of the earth by a shearing motion of material; speed is 3.0–4.0 kilometers per second in the crust and 4.4–4.6 in the mantle. Also known as distortional wave; equivoluminal wave; rotational wave; secondary wave; shake wave; shear wave; tangential wave; transverse wave.

swedenborgite [MINERAL] $NaBe_4SbO_7$ A colorless to wine-yellow mineral composed of sodium beryllium antimony oxide.

sweep-out pattern [PETRO ENG]   The area pattern of water advance in a petroleum reservoir, as for waterflood operation.

sweet corrosion [PETRO ENG]   Corrosion occurring in oil or gas wells where there is no iron-sulfide corrosion product, and there is no odor of hydrogen sulfide in the produced reservoir fluid.

sweet roast *See* dead roast.

swell [GEOL]   **1.** The volumetric increase of soils on being removed from their compacted beds due to an increase in void ratio. **2.** A local enlargement or thickening in a vein or ore deposit. **3.** A low dome or quaquaversal anticline of considerable areal extent; long and generally symmetrical waves contribute to the mixing processes in the surface layer and thus to its sound transmission properties. **4.** Gently rising ground, or a rounded hill above the surrounding ground or ocean floor.   [MIN ENG] *See* horseback.   [OCEANOGR]   Ocean waves which have traveled away from their generating area; these waves are of relatively long length and period, and regular in character.

swell-and-swale topography [GEOGR] A low-relief, undulating landscape characterized by gentle slopes and rounded hills interspersed with shallow depressions.

swell direction [OCEANOGR] The direction from which swell is moving.

swelled ground [GEOL] A soil or rock that expands when wetted.

swell forecast [OCEANOGR] Prediction of the frequency and height of swell waves in a remote area from the characteristics of the waves at their origin.

swelling clay [GEOL] Clay that can absorb large amounts of water, such as bentonite.

swinestone [PETR] Limestone containing black bituminous matter, which gives off an objectionable odor when rubbed.

swing-hammer crusher [MIN ENG] A rotary crusher with rotating hammers that break up ore by impelling it against breaker plates.

switchback [MIN ENG] A zigzag arrangement of railroad tracks by means of which a train can reach a higher or lower level by a succession of easy grades.

switchman [MIN ENG] A laborer who throws switches of mine tracks in a coal mine.

switch plate [MIN ENG] An iron plate on tramroads in mines to change the direction of movement.

swivel [PETRO ENG] A short piece of casing having one end belled over a heavy ring, and having a large hole through both walls, the other end being threaded.

swivel head [MECH ENG] The assembly of a spindle, chuck, feed nut, and feed gears on a diamond-drill machine that surrounds, rotates, and advances the drill rods and drilling stem; on a hydraulic-feed drill the feed gears are replaced by a hydraulically actuated piston assembly.

Sycidales [PALEON] A group of fossil aquatic plants assigned to the Charophyta, characterized by vertically ribbed gyrogonites.

syenite [PETR] A visibly crystalline plutonic rock with granular texture composed largely of alkali feldspar, with subordinate plagioclose and mafic minerals; the intrusive equivalent of trachyte.

syenodiorite [PETR] Plutonic rock consisting of acid plagioclase, orthoclase, and a ferromagnesian mineral.

syenogabbro [PETR] Plutonic rock consisting of basic plagioclase, orthoclase, and a dark mineral such as augite.

sylvanite [MINERAL] $(Au,Ag)Te_2$ A steel-gray, silver-white, or brass-yellow mineral that crystallizes in the monoclinic system and often occurs in implanted crystals. Also known as goldschmidtite; graphic tellurium; white tellurium; yellow tellurium.

sylvester [MIN ENG] A hand-operated device for withdrawing supports from the waste or old workings in a mine by means of a long chain which allows the device to be positioned at a safe distance from the support to be extracted.

sylvine *See* sylvite.

sylvite [MINERAL] KCl A salty-tasting, white or colorless isometric mineral, occurring in cubes or crystalline masses or as a saline residue; the chief ore of potassium. Also known as leopoldite; sylvine.

symmetrical fold [GEOL] A fold whose limbs have approximately the same angle of dip relative to the axial surface. Also known as normal fold.

symmetric ripple mark [GEOL] A ripple mark whose cross-section profile is symmetric.

Symmetrodonta [PALEON]   An order of the extinct mammalian infraclass Pantotheria distinguished by the central high cusp, flanked by two smaller cusps and several low minor cusps, on the upper and lower molars.

symmetry class *See* crystal class.

symmetry plane *See* plane of mirror symmetry.

symmict [GEOL]   Referring to a sedimentation unit that is structureless and in which coarse- and fine-grained particles are mixed more extensively in the lower part.

symmictite [PETR]   An eruptive breccia that is homogenized and is made up of a mixture of country rock and intrusive rock.

symmicton *See* diamicton.

symon fault *See* horseback.

Symon's cone crusher [MIN ENG]   A modified gyratory crusher used in secondary ore crushing that consists of a downward-flaring bowl within which is gyrated a conical crushing head; the main shaft is gyrated by means of a long eccentric which is driven by bevel gears.

Symon's disk crusher [MIN ENG]   A mill in which the crushing is done between two cup-shaped plates that revolve on shafts set at a small angle to each other.

sympathetic detonation [ENG]   Explosion caused by the transmission of a detonation wave through any medium from another explosion.

symplectite *See* symplektite.

symplektite [MINERAL]   An intimate intergrowth of two different minerals. Also spelled symplectite.

symplesite [MINERAL] $Fe_2(AsO_4)_3 \cdot 8H_2O$ A blue to bluish-green triclinic mineral composed of hydrous iron arsenate.

synadelphite [MINERAL] $(Mn,Mg,Ca,Pb)(AsO_4)(OH)_5$ A black mineral composed of basic arsenate of manganese, often with magnesium, calcium, lead, or other metals.

synantectic [MINERAL]   Refers to a mineral that was formed by the reaction of two other minerals.

synantexis [GEOL]   Deuteric alteration.

synchisite *See* synchysite.

synchronous [GEOL]   Geological rock units or features formed at the same time.

synchronous pluton [GEOL]   Any pluton whose time of emplacement coincides with a major orogeny.

synchysite [MINERAL] $(Ce,La)Ca(Co_3)_2F$ A mineral composed of fluoride and carbonate of calcium, cerium, and lanthanum. Also spelled synchisite.

synclinal axis *See* trough surface.

synclinal valley [GEOL]   Pertaining to a topographic valley whose sides coincide with a synclinal fold.

syncline [GEOL]   A fold having stratigraphically younger rock material in its core; it is concave upward.

synclinorium [GEOL]   A composite synclinal structure in a region of lesser folds.

syngenesis [GEOL]   In-place formation of unconsolidated sediments.

syngenetic [GEOL] **1.** Pertaining to a primary sedimentary structure formed contemporaneously with sediment deposition. **2.** Pertaining to a mineral deposit formed contemporaneously with the enclosing rock. Also known as ideogenous.

syngenite [MINERAL] $K_2Ca(SO_4)_2 \cdot H_2O$ A colorless or white mineral composed of hydrous potassium calcium sulfate occurring in tabular crystals.

synkinematic *See* syntectonic.

synoptic [METEOROL] Refers to the use of meteorological data obtained simultaneously over a wide area for the purpose of presenting a comprehensive and nearly instantaneous picture of the state of the atmosphere.

synoptic chart [METEOROL] Any chart or map on which data and analyses are presented that describe the state of the atmosphere over a large area at a given moment in time.

synoptic climatology [CLIMATOL] The study and analysis of climate in terms of synoptic weather information, principally in the form of synoptic charts; the information thus obtained gives the climate (that is, average weather) of a given locality in a given synoptic situation rather than the usual climatic parameters which represent averages over all synoptic conditions.

synoptic code [METEOROL] In general, any code by which synoptic weather observations are communicated; among the synoptic codes in use are the international synoptic code, ship synoptic code, U.S. Airways code, and RECCO code.

synoptic meteorology [METEOROL] The study and analysis of synoptic weather information.

synoptic model [METEOROL] Any model specifying a space distribution of some meteorological elements; the distribution of clouds, precipitation, wind, temperature, and pressure in the vicinity of a front is an example of a synoptic model.

synoptic oceanography [OCEANOGR] The study of the physical spatial parameters of the ocean through analysis of simultaneous observations from many stations.

synoptic report [METEOROL] An encoded and transmitted synoptic weather observation.

synoptic scale *See* cyclonic scale.

synoptic wave chart [OCEANOGR] A chart of an ocean area on which is plotted synoptic wave reports from vessels, along with computed wave heights for areas where reports are lacking; atmospheric fronts, highs, and lows are also shown; isolines of wave height and the boundaries of areas having the same dominant wave direction are drawn.

synoptic weather observation [METEOROL] A surface weather observation, made at periodic times (usually at 3- and 6-hourly intervals specified by the World Meteorological Organization), of sky cover, state of the sky, cloud height, atmospheric pressure reduced to sea level, temperature, dew point, wind speed and direction, amount of precipitation, hydrometeors and lithometeors, and special phenomena that prevail at the time of the observation or have been observed since the previous specified observation.

synorogenic [GEOL] Referring to a geologic process occurring at the same time as orogenic activity.

syntaxial overgrowth [MINERAL] A crystallographically oriented overgrowth of two alternating, chemically identical substances.

syntaxis [MAP] On a map, a sheaflike pattern of mountains converging on a common center.

syntectic [GEOL] *See* syntexis.   [MET] Isothermal, reversible conversion of a solid phase into two conjugate liquid phases on applying heat.

syntectonic [GEOL] Refers to a geologic process or event occurring during tectonic activity. Also known as synkinematic.

syntexis [GEOL] Magma made by the melting of two or more rock types and the assimilation of country rock. Also known as syntectic.

Syntrophiidina [PALEON] A suborder of extinct articulate brachiopods of the order Pentamerida characterized by a strong dorsal median fold.

Synxiphosura [PALEON] An extinct heterogeneous order of arthropods in the subclass Merostomata possibly representing an explosive proliferation of aberrant, terminal, and apparently blind forms.

Syringophyllidae [PALEON] A family of extinct corals in the order Tabulata.

syrinx [PALEON] A tube surrounding the pedicle in certain fossil brachiopods.

syserskite [MINERAL] Mineral composed of an alloy of osmium (50–80%) and iridium (20–50%).

syssiderite *See* stony-iron meteorite.

system [GEOL] **1.** A major time-stratigraphic unit of worldwide significance, representing the basic unit of Phanerozic rocks. **2.** A group of related structures, such as joints.

systematic joints [GEOL] Joints occurring in patterns or sets and oriented perpendicular to the boundaries of the constituent rock unit.

systematic sampling [MIN ENG] Extracting samples at evenly spaced periods or in fixed quantities from a unit of coal.

systematic support [MIN ENG] The regular setting of timber or steel supports at fixed intervals irrespective of the condition of the roof and sides.

szaibelyite [MINERAL] $(Mn,Mg)(BO_2)(OH)$ A white to buff or straw yellow, orthorhombic mineral consisting of a basic borate of manganese and magnesium; occurs as veinlets, masses, or embedded nodules.

szaskaite *See* smithsonite.

szmikite [MINERAL] $MnSO_4 \cdot H_2O$ A monoclinic mineral composed of hydrous manganese sulfate.

szomolnokite [MINERAL] $FeSO_4 \cdot H_2O$ A yellow or brown monoclinic mineral composed of hydrous ferrous sulfate.

T

tabbyite [MINERAL]   A variety of solid asphalt found in the western United States; used as rubber filler and with roofing materials.

taber ice *See* segregated ice.

tabetisol *See* talik.

table [MIN ENG]   **1.** In placer mining, a wide, shallow sluice box designed to recover gold or other valuable material from screened gravel. **2.** A platform or plate on which coal is screened and picked.

table flotation [MIN ENG]   A flotation process in which a slurry of ore is fed to a shaking table where flotable particles become glomerules, held together by minute air bubbles and edge adhesion; the glomerules roll across the table and are discharged nearly opposite the feed end; the process is helped by jets of low-pressure air.

table iceberg *See* tabular iceberg.

table knoll [GEOGR]   A knoll with a comparatively smooth, flat top.

tableland [GEOGR]   A broad, elevated, nearly level, and extensive region of land that has been deeply cut at intervals by valleys or broken by escarpments. Also known as continental plateau.

tablemount *See* guyot.

table mountain [GEOGR]   A flat-topped mountain.

table reef [GEOL]   A small, isolated organic reef which has a flat top and does not enclose a lagoon.

tabling [MIN ENG]   Separation of two materials of different densities by passing a dilute suspension over a slightly inclined table having a reciprocal horizontal motion or shake with a slow forward motion and a fast return.

tabula [PALEON]   A transverse septum that closes off the lower part of the polyp cavity in certain extinct corals and hydroids.

tabular [GEOL]   Referring to a sedimentary particle whose length is two to three times its thickness.

tabular berg *See* tabular iceberg.

tabular crystal [CRYSTAL]   A crystal that appears broad and flat due to two prominent parallel faces.

tabular iceberg [OCEANOGR]   An iceberg with clifflike sides and a flat top; usually arises by detachment from an ice shelf. Also known as table iceberg; tabular berg.

tabular spar *See* wollastonite.

Tabulata [PALEON]   An extinct Paleozoic order of corals of the subclass Zoantharia characterized by an exclusively colonial mode of growth and by secretion of a calcareous exoskeleton of slender tubes.

tachyhydrite [MINERAL] $CaMg_2Cl_6 \cdot 12H_2O$ A honey yellow, hexagonal mineral consisting of a hydrated chloride of calcium and magnesium; occurs in massive form.

tachylite [GEOL]   A black, green, or brown volcanic glass formed from basaltic magma. Also known as basalt glass; basalt obsidian; hyalobasalt; jaspoid; sordawalite; wichtisite.

Taconian orogeny [GEOL]   A process of formation of mountains in the latter part of the Ordovician period, particularly in the northern Appalachians. Also known as Taconic orogeny.

Taconic orogeny *See* Taconian orogeny.

taconite [GEOL]   The siliceous iron formation from which high-grade iron ores of the Lake Superior district have been derived; consists chiefly of fine-grained silica mixed with magnetite and hematite.

tactite [PETR]   A rock with a complex mineralogical composition, formed by contact metamorphism and metasomatism of carbonate rocks.

taele *See* frozen ground.

Taeniodonta [PALEON]   An order of extinct quadrupedal land mammals, known from early Cenozoic deposits in North America.

Taeniolabidoidea [PALEON]   An advanced suborder of the extinct mammalian order Multituberculata having incisors that were self-sharpening in a limited way.

taeniolite [MINERAL] $KLiMg_2Si_4O_{10}F_2$ A white or colorless mica mineral.

taenite [MINERAL]   A meteoritic mineral consisting of a nickel-iron alloy, with a nickel content varying from about 27 to 65%.

tagilite *See* pseudomalachite.

Tahuian [GEOL]   A local Eocene time subdivision in Australia whose identification is based on foraminiferans. Duces taiga vegetation, that is, too cold for prolific tree growth but milder than the tundra climate and moist enough to promote appreciable vegetation. Also known as subarctic climate.

taiga [ECOL]   A zone of forest vegetation encircling the Northern Hemisphere between the arctic-subarctic tundras in the north and the steppes, hardwood forests, and prairies in the south.

tailings [MIN ENG]   **1.** The parts, or a part, of any incoherent or fluid material separated as refuse, or separately treated as inferior in quality or value. **2.** The decomposed outcrop of a vein or bed. **3.** The refuse material resulting from processing ground ore.

tailings settling tank [MIN ENG]   A vessel in which solids are removed from the tailings effluent in mineral processing plants.

tailrace [ENG]   A channel for carrying water away from a turbine, waterwheel, or other industrial application.   [MIN ENG]   A channel for conveying mine tailings.

tail rope [MIN ENG]   **1.** The rope which passes around the return sheave in main-and-tail haulage or a scraper loader layout. **2.** The rope that is used to draw the empty cars back into a mine in a tail-rope system. **3.** A counterbalance rope attached beneath the cage when the cages are hoisted in balance. **4.** A hemp rope used for moving pumps in shafts.

tail-rope system [MIN ENG] Haulage by a hoisting engine and two separate drums in which the main rope is attached to the front end of a trip of cars, and the tail rope is attached to the rear end of the trip.

tail sheave [MIN ENG] The return sheave for an endless rope or the tail rope of the main-and-tail-rope system, placed at the far end of a haulageway.

tail track system [MIN ENG] A form of track layout for car or trip loading in which the track can be extended down the heading, turned right or left, or turned back, U-fashion, in an adjacent heading.

tailwind [METEOROL] A wind which assists the intended progress of an exposed, moving object, for example, rendering an airborne object's ground speed greater than its airspeed; the opposite of a headwind. Also known as following wind.

taino [METEOROL] A tropical cyclone (hurricane) in parts of the Greater Antilles.

taku wind [METEOROL] A strong, gusty, east-northeast wind, occurring in the vicinity of Juneau, Alaska, between October and March; it sometimes attains hurricane force at the mouth of the Taku River, after which it is named.

talc [MINERAL] $Mg_3Si_4O_{10}(OH)_2$ A whitish, greenish, or grayish hydrated magnesium silicate mineral crystallizing in the monoclinic system; it is extremely soft (hardness is 1 on Mohs scale) and has a characteristic soapy or greasy feel.

talcose rock [PETR] A rock having a soft and soapy feel, that is, resembling talc.

talc schist [PETR] A schist in which talc is the dominant schistose material.

talik [GEOL] A Russian term applied to permanently unfrozen ground in regions of permafrost; usually applies to a layer which lies above the permafrost but below the active layer, that is, when the permafrost table is deeper than the depth reached by winter freezing from the surface. Also known as tabetisol.

talking [MIN ENG] A series of small bumps or cracking noises within the mine walls, indicating that the rock is beginning to yield to stresses.

tally [MIN ENG] A mark, number, or tin ticket placed by the miner on each car of coal or ore that is sent from the work place, thus facilitating a count or tally of all filled cars.

talus [GEOL] Also known as rubble; scree. **1.** Coarse and angular rock fragments derived from and accumulated at the base of a cliff or steep, rocky slope. **2.** The accumulated heap of such fragments.

talus creep [GEOL] The slow, downslope movement of talus.

talus glacier *See* rock glacier.

talus slope [GEOL] A steep, concave slope consisting of an accumulation of talus. Also known as debris slope.

tamarugite [MINERAL] $NaAl(SO_4)_2 \cdot 6H_2O$ A colorless, monoclinic mineral consisting of a hydrated sulfate of sodium and aluminum; occurs as crystals and masses.

tamping bag [ENG] A bag filled with stemming material such as sand for use in horizontal and upward sloping shotholes.

tamping bar [ENG] A piece of wood for pushing explosive cartridges or forcing the stemming into shotholes.

tamping plug [ENG] A plug of iron or wood used instead of tamping material to close up a loaded blasthole.

tandem hoisting [MIN ENG] Hoisting in a deep shaft with two skips running in one shaft; the lower skip is suspended from the tail rope of the upper skip.

tangeite *See* calciovolborthite.

tangent arc [METEOROL] Generic name for several types of halo arcs that form as loci tangent to other halos; the halo of 22° occasionally exhibits the horizontal and vertical tangent arcs, and the halo of 46° exhibits the infralateral tangent arcs and the supralateral tangent arcs.

tangent galvanometer [ENG] A galvanometer in which a small compass is mounted horizontally in the center of a large vertical coil of wire; the current through the coil is proportional to the tangent of the angle of deflection of the compass needle from its normal position parallel to the magnetic field of the earth.

tangential wave *See* S wave.

tangent offset [ENG] In surveying, a method of plotting traverse lines; angles are laid out by linear measurement, using a constant times the natural tangent of the angle.

tank battery [PETRO ENG] A grouping of interconnected storage tanks situated to receive the output of one or more oil wells.

tank farm [PETRO ENG] An area in which a number of large-capacity storage tanks are located, generally used for crude oil or petroleum products.

tank switch [PETRO ENG] An automatic control of lease tanks in oil fields, including controls of lines to fill tanks and for pipeline runs; can be electrically or pneumatically actuated.

tantalite [MINERAL] $(Fe,Mn)Ta_2O_6$ An iron-black mineral that crystallizes in the orthorhombic system and commonly occurs in short prismatic crystals; luster is submetallic, hardness is 6 on Mohs scale, and specific gravity is 7.95; principal ore of tantalum.

tanteuxenite [MINERAL] $(Y,Ce,Ca)(Ta,Nb,O)_2(O,OH)_6$ A brown or black variety of euxenite with tantalum substituting for niobium. Also known as delorenzite; eschwegeite.

tap [MIN ENG] To intersect with a borehole and withdraw or drain the contained liquid, as water from a water-bearing formation or from underground workings.

tapered pipeline [PETRO ENG] A changing of the pressure grade, either by change of wall thickness or material, of pipeline sections as working pressure is lessened.

taphrogenesis *See* taphrogeny.

taphrogeny [GEOL] The formation of rift or trench phenomena, characterized by block faulting and associated subsidence. Also known as taphrogenesis.

taphrogeosyncline [GEOL] A geosyncline formed as a rift basin between faults.

taping [ENG] The process of measuring distances with a surveyor's tape.

tapioca snow *See* snow pellets.

tapiolite [MINERAL] $Fe(Ta,Nb)_2O_6$ A mineral that is isomorphous with mossite; occurs in pegmatites or detrital deposits; an ore of tantalum.

taranakite [MINERAL] $KAl_3(PO_4)_3(OH)\cdot9H_2O$ A white, gray, or yellowish-white mineral consisting of a hydrated basic phosphate of potassium and aluminum.

tarantata [METEOROL] A strong breeze from the northwest in the Mediterranean region.

tarapacaite [MINERAL] K_2CrO_4 A bright canary yellow, orthorhombic mineral consisting of potassium chromate; occurs in tabular form.

tarbuttite [MINERAL] $Zn_2(PO_4)(OH)$ A triclinic mineral of varying color, consisting of basic zinc phosphate.

target [ENG] **1.** The sliding weight on a leveling rod used in surveying to enable the staffman to read the line of collimation. **2.** The point that a borehole or an exploratory work is intended to reach.

tarn   [GEOGR]   A landlocked pool or small lake that may occur in a marsh or swamp, or that may occupy a basin amid mountain ranges.

tarnish   [MINERAL]   The altered color and luster of a mineral surface; characteristic of copper-bearing minerals.

tar sand   [GEOL]   A type of oil sand; a sand whose interstices are filled with asphalt that remained after the escape of the lighter fractions of crude oil.

tar seep   [GEOL]   Natural tar that, because of its close proximity to the ground surface, seeps from cracks in the earth or from between rocks, often forming pits or pools.

tasmanite   [GEOL]   An impure coal, transitional between cannel coal and oil shale. Also known as combustible shale; Mersey yellow coal; white coal; yellow coal.

tautirite   [GEOL]   An igneous rock composed of potassium feldspar, andesine, nepheline, and amphibole, with abundant accessory sphene.

tau-value   [METEOROL]   The time rate of change of D value at a fixed point defined by the relation $\tau = (\Delta_t D)/(\Delta t)$, where Δt is the change in time and $\Delta_t D$ is the change in D value during this time interval; tau-values are expressed in terms of feet per hour; tau-value lines are drawn on 4-D charts and constitute the time dimension of these charts.

tavistockite   [MINERAL] $Ca_3Al_2(PO_4)_2(OH)_6$ A white, orthorhombic mineral consisting of a basic phosphate of calcium and aluminum.

Taxocrinida   [PALEON]   An order of flexible crinoids distributed from Ordovician to Mississippian.

taylorite   *See* bentonite.

Taylor-Orowan dislocation   *See* edge dislocation.

Tchernozem   *See* Chernozem.

teallite   [MINERAL] $PbSnS_2$ A grayish-black, orthorhombic mineral consisting of lead tin sulfide.

tear fault   [GEOL]   A very steep to vertical fault associated with and perpendicular to the strike of an overthrust fault.

tectite   *See* tektite.

tectofacies   [GEOL]   A lithofacies that is interpreted tectonically.

tectogene   [GEOL]   A long, relatively narrow downward fold of sialic crust considered to be an early phase in mountain-building processes. Also known as geotectogene.

tectogenesis   *See* orogeny.

tectonic analysis   *See* petrotectonics.

tectonic breccia   [PETR]   A breccia developed from brittle rocks, formed as a result of crustal movements and produced by lateral or vertical pressure. Also known as dynamic breccia; pressure breccia.

tectonic conglomerate   *See* crush conglomerate.

tectonic cycle   [GEOL]   The orogenic cycle which relates larger crustal features, such as mountain belts, to a series of stages of development. Also known as geosynclinal cycle.

tectonic framework   [GEOL]   The relationship in space and time of subsiding, stable, and rising tectonic elements in a sedimentary source area.

tectonic land   [GEOL]   Linear fold ridges and volcanic islands which existed for a short time in the interior sections of an orogenic belt during the geosynclinal phase.

tectonic lens   [GEOL]   An elongate, sausage-shaped body of rock formed by distortion of a continuous incompetent layer enclosed between competent layers, similar to a boudin, but genetically distinct.

tectonic map   [GEOL]   A map which shows the architecture of the upper portion of the earth's crust.

tectonic moraine   [GEOL]   An aggregation of boulders incorporated in the base of an overthrust mass.

tectonic patterns   [GEOL]   The arrangement of the large structural units of the earth's crust, such as mountain systems, shields or stable areas, basins, arches, and volcanic archipelagoes.

tectonic plate   [GEOL]   Any one of the internally rigid crustal blocks of the lithosphere which move horizontally across the earth's surface relative to one another. Also known as crustal plate.

tectonic rotation   [GEOL]   Internal rotation of a tectonite in the direction of transport.

tectonics   [GEOL]   A branch of geology that deals with regional structural and deformational features of the earth's crust, including the mutual relations, origin, and historical evolution of the features. Also known as geotectonics.

tectonite   [PETR]   A rock in which the history of its deformation is reflected in its fabric.

tectonoeustatism   [OCEANOGR]   Fluctuations of sea level due to changes in the capacities of the ocean basins resulting from earth movements.

tectonomagnetism   [GEOPHYS]   Study of magnetic anomalies due to tectonic stress.

tectonometer   [ENG]   An apparatus, including a microammeter, used on the surface to obtain knowledge of the structure of the underlying rocks.

tectonophysicist   [GEOPHYS]   One who studies elastic deformation of flow and rupture of constituent materials of the earth's crust and makes deductions concerning the forces that cause these deformations.

tectonophysics   [GEOPHYS]   A branch of geophysics dealing with the physical processes involved in forming geological structures.

tectosilicate   [MINERAL]   A structural type of silicate in which all four oxygen atoms of the silicate tetrahedra are shared with neighboring tetrahedra; tectosilicates include quartz, the feldspars, the feldspathoids, and zeolites. Also known as framework silicate.

tectosome   [GEOL]   A body of strata representing a tectotope.

tectosphere   [GEOL]   The region of the earth's crust occupied by the tectonic plates.

teepleite   [MINERAL] $Na_2BO_2Cl \cdot 2H_2O$ A mineral composed of hydrous chloride and borate of sodium.

teeth of the gale   [METEOROL]   An old nautical term for the direction from which the wind is blowing (upwind, windward); to sail into the teeth of the gale is to sail to windward.

tehuantepecer   [METEOROL]   A violent squally wind from north or north-northeast in the Gulf of Tehuantepec in winter; it originates in the Gulf of Mexico, as a norther which crosses the isthmus and blows through the gap between the Mexican and Guatemalan mountains.

teineite   [MINERAL] $CuTeO_3 \cdot 2H_2O$ A greenish to yellowish, probably triclinic mineral consisting of a hydrated sulfate-tellurate of copper; occurs as crystals.

tektite   [GEOL]   A collective term applied to certain objects of natural glass of debatable origin that are widely strewn over the land and in sediments under the oceans; composition and size vary, and overall shapes resemble splash forms; most tektites are believed to be of extraterrestrial origin. Also known as obsidianite; tectite.

telain *See* provitrain.

teleconnection [GEOL] Method for constructing a uniform time scale for a part of the Pleistocene, using identification and correlation of a series of varves, usually over great distances or even worldwide.

telemagmatic [GEOL] Pertaining to a hydrothermal mineral deposit that is distant from its magmatic source.

telemeteorograph [ENG] Any meteorological instrument, such as a radiosonde, in which the recording instrument is located at some distance from the measuring apparatus; for example, a meteorological telemeter.

telemeteorography [ENG] The science of the design, construction, and operation of various types of telemeteorographs.

telemeteorometry [METEOROL] The study of making meteorological observations at a distance.

Teleosauridae [PALEON] A family of Jurassic reptiles in the order Crocodilia characterized by a long snout and heavy armor.

telescope structure [GEOL] An alluvial-fan structure characterized by younger fans with flatter gradients spreading out from between fan mesas of older fans with steeper gradients.

telescopic derrick [ENG] A drill derrick divided into two or more sections, with the uppermost sections nesting successively into the lower sections.

telescopic loading trough [MIN ENG] A shaker conveyor trough of two sections, one nested in the other, used near the face for advancing the trough line without the necessity of adding either a standard or a short length of pan after each cut.

telescopic tripod [ENG] A drill or surveyor's tripod each leg of which is a series of two or more closely fitted nesting tubes, which can be locked rigidly together in an extended position to form a long leg or nested one within the other for easy transport.

teleseism [GEOPHYS] An earthquake that is far from the recording station.

teleseismology [GEOPHYS] The aspect of seismology dealing with records made at a distance from the source of the impulse.

telethermal [GEOL] Pertaining to a hydrothermal mineral deposit precipitated at a shallow depth and at a mild temperature.

telinite [GEOL] A variety of provitrinite composed of plant cell-wall material.

tellurite [MINERAL] TeO_2 A white or yellowish orthorhombic mineral consisting of tellurium dioxide, and occurring in crystals; it is dimorphous with paratellurite.

tellurium glance *See* nagyagite.

tellurobismuthite [MINERAL] Bi_2Te_3 A pale lead gray, hexagonal mineral consisting of a bismuth and tellurium compound; occurs as irregular plates or foliated masses.

tellurometer [ENG] A microwave instrument used in surveying to measure distance; the time for a radio wave to travel from one observation point to the other and return is measured and converted into distance by phase comparison, much as in radar.

Temnospondyli [PALEON] An order of extinct amphibians in the subclass Labyrinthodontia having vertebrae with reduced pleurocentra and large intercentra.

temperate belt [CLIMATOL] A belt around the earth within which the annual mean temperature is less than 20°C (68°F) and the mean temperature of the warmest month is higher than 10°C (50°F).

temperate climate [CLIMATOL] The climate of the middle latitudes; the climate between the extremes of tropical climate and polar climate.

temperate glacier [HYD]   A glacier which, at the end of the melting season, is composed of firn and ice at the melting point.

temperate rainy climate [CLIMATOL]   One of the major categories in W. Kippen's climatic classification; the coldest-month mean temperature is less than 64.4°F (18°C) and greater than 26.6°F (−3°C), and the warmest-month mean temperature is more than 50°F (10°C).

temperate westerlies *See* westerlies.

temperate-westerlies index [METEOROL]   A measure of the strength of the westerly wind between latitudes 35°N and 55°N; the index is computed from the average sea-level pressure difference between these latitudes and is expressed as the west to east component of geostrophic wind in meters and tenths of meters per second.

Temperate Zone [CLIMATOL]   Either of the two latitudinal zones on the earth's surface which lie between 23°27' and 66°32' N and S (the North Temperate Zone and South Temperate Zone, respectively).

temperature belt [METEOROL]   The belt which may be drawn on a thermograph trace or other temperature graph by connecting the daily maxima with one line and the daily minima with another.

temperature-humidity index [METEOROL]   An index which gives a numerical value, in the general range of 70–80, reflecting outdoor atmospheric conditions of temperature and humidity as a measure of comfort (or discomfort) during the warm season of the year; equal to 15 plus 0.4 times the sum of the dry-bulb and wet-bulb temperatures in degrees Fahrenheit. Also known as comfort index; discomfort index. Abbreviated CI; DI; THI.

temperature inversion [METEOROL]   A layer in the atmosphere in which temperature increases with altitude; the principal characteristic of an inversion layer is its marked static stability, so that very little turbulent exchange can occur within it; strong wind shears often occur across inversion layers, and abrupt changes in concentrations of atmospheric particulates and atmospheric water vapor may be encountered on ascending through the inversion layer.   [OCEANOGR]   A layer of a large body of water in which temperature increases with depth.

temperature log [PETRO ENG]   A continuous record of temperature versus depth in an oil-well borehole.

temperature profile recorder [ENG]   A portable instrument for measuring temperature as a function of depth in shallow water, particularly in lakes, in which a thermistor element transmits data over an electrical cable to a recording drum and depth is measured by the amount of wire paid out.

temperature province [CLIMATOL]   A major division of C. W. Thornthwaite's schemes of climatic classification, determined as a function of the temperature-efficiency index or the potential evapotranspiration.

temperature-salinity diagram [OCEANOGR]   The plot of temperature versus salinity data of a water column; the resulting diagram identifies the water masses within the column, the column's stability, indicates the σ_T value via lines of constant σ_T printed on paper, and allows an estimate of the accuracy of the temperature and salinity measurements. Also known as T-S curve; T-S diagram; T-S relation.

temperature survey [PETRO ENG]   An analysis of temperature changes or differences down an oil-well borehole, based on a temperature log; used to locate the top of casing cement, lost circulation zones, or gas entry zones.

temperature zone [CLIMATOL]   A portion of the earth's surface defined by relatively uniform temperature characteristics, and usually bounded by selected values of some measure of temperature or temperature effect.

temper screw *See* stirrup.

temporale [METEOROL] A rainy wind from the southwest to west resulting from a deflection of the southeast trades of the eastern South Pacific onto the Pacific coast of Central America.

temporary base level [GEOL] Any base level, other than sea level, below which a land area temporarily cannot be reduced by erosion. Also known as local base level.

tendency [METEOROL] The local rate of change of a vector or scalar quality with time at a given point in space.

tendency chart *See* change chart.

tendency equation [METEOROL] An equation for the local change of pressure at any point in the atmosphere, derived by combining the equation of continuity with an integrated form of the hydrostatic equation.

tendency interval [METEOROL] The finite increment of time over which a change of the value of a meteorological element is measured in order to estimate its tendency; the most familiar example is the three-hour time interval over which local pressure differences are measured in determining pressure tendency.

tenggara [METEOROL] A strong, dry, hazy, east or southeast wind during the east monsoon in the Spermunde Archipelago.

tennantite [MINERAL] $(Cu,Fe)_{12}As_4S_{13}$ A lead-gray mineral crystallizing in the isometric system; it is isomorphous with tetrahedrite; an important ore of copper.

tenorite [MINERAL] CuO A triclinic mineral that occurs in small, shining, steel-gray scales, in black powder, or in black earthy masses; an ore of copper.

tension crack [GEOL] An extension fracture caused by tensile stress.

tension fault [GEOL] A fault in which crustal tension is a factor, such as a normal fault. Also known as extensional fault.

tension fracture [GEOL] A minor rock fracture developed at right angles to the direction of maximum tension. Also known as subsidiary fracture.

tension jack [MIN ENG] A type of jack with a jackscrew for wedging against the mine roof and a ratchet device for applying tension on a chain that is attached to the tail or foot section of a belt conveyor, and used to restore the proper tension to the belt.

tension joint [GEOL] A joint that is a tension fracture.

tension packer [PETRO ENG] A device to pressure-seal the annular space between an oil-well casing and tubing, held in place by tension against an upward push; a type of production packer.

tension-type hanger [PETRO ENG] A type of tubing hanger for multiple-completion oil wells to allow for the varying lengths of tubing strings.

tented ice [OCEANOGR] Pressure-type ice that results as ice is displaced upward and a flat-sided arch with a cavity is formed between the raised ice and the underlying water.

tenting [OCEANOGR] The vertical displacement upward of ice under pressure to form a flat-sided arch with a cavity beneath.

tepee butte [GEOGR] A tepeelike hill or knoll, especially one of soft material capped by more resistant rock.

tepee structure [GEOL] A disharmonic sedimentary structure consisting of a fold that resembles an inverted depressed V in cross section.

tepetate *See* caliche.

tephigram [METEOROL] A thermodynamic diagram designed by Napier Shaw with temperature and logarithm of potential temperature as coordinates; isobars are gently curved lines and the chart is rotated so that pressure increases downward; vapor lines and saturation adiabats are curved; on this chart, energy is proportional to the area enclosed by the curve representing the process.

tephra [GEOL] Denotes all pyroclastics of a volcano.

tephrite [PETR] A group of basaltic extrusive rocks composed chiefly of calcic plagioclase, augite, and nepheline or leucite, with some sodic sanidine.

tephrochronology [GEOL] The dating of different layers of volcanic ash for the establishment of a sequence of geologic and archeologic occurrences.

tephroite [MINERAL] Mn_2SiO_4 An olivine mineral that occurs with zinc and manganese minerals.

Teratornithidae [PALEON] An extinct family of vulturelike birds of the Pleistocene of western North America included in the order Falconiformes.

Terebratellidina [PALEON] An extinct suborder of articulate brachiopods in the order Terebratulida in which the loop is long and offers substantial support to the side arms of the lophophore.

terlinguaite [MINERAL] Hg_2OCl A sulfur yellow to greenish-yellow, monoclinic mineral consisting of an oxychloride of mercury.

terminal forecast [METEOROL] An aviation weather forecast for one or more specified air terminals.

terminal moraine [GEOL] An end moraine that extends as an arcuate or crescentic ridge across a glacial valley; marks the farthest advance of a glacier. Also known as marginal moraine.

ternary diagram [PETR] A triangular diagram that graphically depicts the composition of a three-component mixture or ternary system.

terrace [GEOL] **1.** A horizontal or gently sloping embankment of earth along the contours of a slope to reduce erosion, control runoff, or conserve moisture. **2.** A narrow coastal strip sloping gently toward the water. **3.** A long, narrow, nearly level surface bounded by a steeper descending slope on one side and by a steeper ascending slope on the other side. **4.** A benchlike structure bordering an undersea feature.

terraced pool [GEOGR] A shallow, rimmed pool on the surface of a reef.

terracette [GEOL] A small steplike form developed on the surface of a slumped soil mass along a steep grassy incline.

terrain profile recorder See airborne profile recorder.

terrain sensing [ENG] The gathering and recording of information about terrain surfaces without actual contact with the object or area being investigated; in particular, the use of photography, radar, and infrared sensing in airplanes and artificial satellites.

terral levante [METEOROL] A land breeze of Spain and Brazil, sometimes a northwest squall of foehn character.

terra miraculosa See bole.

terrane [GEOL] A rock formation, a cluster of rock formations, or the general area of outcrops.

terra rossa [GEOL] A reddish-brown soil overlying limestone bedrock.

terrestrial coordinates See geographical coordinates.

terrestrial electricity [GEOPHYS] Electric phenomena and properties of the earth; used in a broad sense to include atmospheric electricity. Also known as geoelectricity.

terrestrial environment   [GEOGR]   The earth's land area, including its man-made and natural surface and subsurface features, and its interfaces and interactions with the atmosphere and the oceans.

terrestrial equator *See* astronomical equator.

terrestrial frozen water   [HYD]   Seasonally or perennially frozen waters of the earth, exclusive of the atmosphere.

terrestrial gravitation   [GEOPHYS]   The effect of gravitational attraction of the earth.

terrestrial meridian *See* astronomical meridian.

terrestrial radiation   [GEOPHYS]   Electromagnetic radiation originating from the earth and its atmosphere at wavelengths determined by their temperature. Also known as earth radiation; eradiation.

terrestrial sediment   [GEOL]   A sedimentary deposit on land above tidal reach.

terrigenous sediment   [GEOL]   Shallow marine sedimentary deposits composed of eroded terrestrial material.

Tertiary   [GEOL]   The older major subdivision (period) of the Cenozoic era, extending from the end of the Cretaceous to the beginning of the Quaternary, from 70,000,000 to 2,000,000 years ago.

tertiary circulation   [METEOROL]   The generally small, localized atmospheric circulations, represented by such phenomena as local winds, thunderstorms, and tornadoes.

tertiary crushing   [MIN ENG]   **1.** The preliminary breaking down of run-of-mine ore and sometimes coal. **2.** The third stage in crushing, following primary and secondary crushing.

tertiary grinding   [MIN ENG]   The third-stage grinding in a ball mill when a particularly fine grinding of ore is needed.

teschemacherite   [MINERAL] $(NH_4)HCO_3$ A colorless to white or yellowish, orthorhombic mineral consisting of ammonium bicarbonate; occurs as compact, crystalline masses.

teschenite   [PETR]   A granular hypabyssal rock composed principally of calcic plagioclase, augite, and sometimes hornblende, with some brotite and analcime.

test   [PETRO ENG]   A procedure for the analysis of current, potential and ultimate product flow, and pressure-decline properties of various types of petroleum reservoirs.

test hole   [MIN ENG]   A drill hole or shallow excavation to assess an ore body or to obtain rock samples to determine their structural and physical characteristics.

test well   [PETRO ENG]   A well to determine the presence of petroleum oil and its potential commercial value in terms of abundance and accessibility.

Tetracorallia   [PALEON]   The equivalent name for Rugosa.

tetradymite   [MINERAL] Bi_2Te_2S A pale steel-gray mineral that usually occurs in foliated masses in auriferous veins; has metallic luster, hardness of 1.5–2 on Mohs scale, and specific gravity of 7.2–7.6.

tetragonal lattice   [CRYSTAL]   A crystal lattice in which the axes of a unit cell are perpendicular, and two of them are equal in length to each other, but not to the third axis.

tetragonal trisoctahedron *See* trapezohedron.

tetragonal tristetrahedron *See* deltohedron.

tetrahedrite   [MINERAL] $(Cu,Fe,Zn,Ag)_{12}Sb_4S_{13}$ A grayish-black mineral crystallizing in the isometric system as tetrahedrons and occurring in massive or granular form;

luster is metallic, hardness is 3.5–4 on Mohs scale, and specific gravity is 4.6–5.1; an important ore of copper. Also known as fahlore; gray copper ore; panabase; stylotypite.

tetrahedron [CRYSTAL]   An isometric crystal form in cubic crystals, in the shape of a four-faced polyhedron, each face of which is a triangle.

tetrahexahedron [CRYSTAL]   A form of regular crystal system with four triangular isosceles faces on each side of a cube; there are altogether 24 congruent faces.

Tetralophodontinae [PALEON]   An extinct subfamily of proboscidean mammals in the family Gomphotheridae.

tetratohedral crystal [CRYSTAL]   A crystal which has one quarter of the maximum number of faces allowed by the crystal system to which the crystal belongs.

texture [GEOL]   The physical nature of the soil according to composition and particle size.   [PETR]   The physical appearance or character of a rock; applied to the megascopic or microscopic surface features of a homogeneous rock or mineral aggregate, such as grain size, shape, and arrangement.

thalassic [OCEANOGR]   Of or pertaining to the smaller seas.

thalassocratic [GEOL]   **1.** Pertaining to a thalassocraton. **2.** Referring to a period of high sea level in the geologic past.

thalassocraton [GEOL]   A craton that is part of the oceanic crust.

thalassophile element [GEOCHEM]   An element that is relatively more abundant in sea water than in normal continental waters, such as sodium and chlorine.

Thalattosauria [PALEON]   A suborder of extinct reptiles in the order Eosuchia from the Middle Triassic.

thalweg [GEOGR]   The middle of the principal navigable waterway which serves as a boundary between two states.   [GEOL]   **1.** A line connecting the lowest points along a stream bed or a valley. Also known as valley line. **2.** A line crossing all contour lines on a land surface perpendicularly.   [HYD]   Water seeping through the ground below the surface in the same direction as a surface stream course.

thanatocoenosis [PALEON]   The assemblage of dead organisms of fossils that occurred together in a given area at a given moment of geologic time. Also known as death assemblage; taphocoenosis.

Thanetian [GEOL]   A European stage of geologic time; uppermost Paleocene, above Montian, below Ypresian of Eocene.

thaw [CLIMATOL]   A warm spell during which ice and snow melt, as a January thaw.

thawing [MIN ENG]   Working permanently frozen ground by pumping water at a temperature of from 50 to 60°F (10 to 15.5°C) through pipes down into the frozen gravel.

thawing index [CLIMATOL]   The number of degree days, above and below 32°F, between the lowest and highest points on the cumulative degree-days time curve for one thawing season.

thawing season [CLIMATOL]   The period of time between the lowest point and the succeeding highest point on the time curve of cumulative degree days above and below 32°F.

thaw pipe [MIN ENG]   A string of pipe drilled into a string of drill rods that is frozen in a borehole in permafrost, through which water is circulated to thaw the ice and free the drill rods.

Thecideidina [PALEON]   An extinct suborder of articulate brachiopods doubtfully included in the order Terebratulida.

Thecodontia [PALEON] An order of archosaurian reptiles, confined to the Triassic and distinguished by the absence of a supratemporal bone, parietal foramen, and palatal teeth, and by the presence of an antorbital fenestra.

thenardite [MINERAL] Na_2SO_4 A colorless, grayish-white, yellowish, yellow-brown, or reddish, orthorhombic mineral consisting of sodium sulfate.

theralite [PETR] A dark-colored, visibly crystalline rock composed chiefly of pyroxene with smaller amounts of calcic plagioclase and nepheline.

Therapsida [PALEON] An order of mammallike reptiles of the subclass Synapsida which first appeared in mid-Permian times and persisted until the end of the Triassic.

thermal [METEOROL] A relatively small-scale, rising current of air produced when the atmosphere is heated enough locally by the earth's surface to produce absolute instability in its lower layers.

thermal aureole *See* aureole.

thermal climate [CLIMATOL] Climate as defined by temperature, and divided regionally into temperature zones.

thermal convection [METEOROL] Atmospheric currents, predominantly vertical, arising from the release of gravitational visibility; commonly produced by solar heating of the ground; the cause of convective (cumulus) clouds. Also known as free convection; gravitational convection.

thermal gas [GEOCHEM] Gas formed from hydrocarbons deep in a sedimentary basin, where a temperature increase has caused thermal cracking of the crude oil.

thermal gradient [GEOPHYS] The rate of temperature change with distance; for example, its increase with depth below the surface of the earth.

thermal high [METEOROL] A high resulting from the cooling of air by a cold underlying surface, and remaining relatively stationary over the cold surface.

thermal jet [METEOROL] A region in the atmosphere where isotherms or thickness lines are closely packed; therefore, a region of very strong thermal wind.

thermal low [METEOROL] An area of low atmospheric pressure due to high temperatures caused by intensive heating at the earth's surface; common to the continental subtropics in summer, thermal lows remain stationary over the area that produces them, their cyclonic circulation is generally weak and diffuse, and they are nonfrontal. Also known as heat low.

thermal metamorphism [PETR] Metamorphism that results from temperature-controlled and induced chemical reconstitution of preexisting rocks, with little influence of pressure. Also known as thermometamorphism.

thermal spring [HYD] A spring whose water temperature is higher than the local mean annual temperature of the atmosphere.

thermal steering [METEOROL] The steering of an atmospheric disturbance in the direction of the thermal wind in its vicinity; equivalent to steering along thickness lines; for this purpose the thermal wind is usually taken from the earth's surface to a level in the middle troposphere.

thermal stratification [HYD] Horizontal layers of differing densities produced in a lake by temperature changes at different depths.

thermal structure [PETR] A distinct structural pattern, such as a dome or anticline, defined by the arrangement of metamorphic zones of increasing grade.

thermal tide [METEOROL] A variation in atmospheric pressure due to the diurnal differential heating of the atmosphere by the sun; so-called in analogy to the conventional gravitational tide.

thermal vorticity [METEOROL]   The vorticity of a thermal wind.

thermal vorticity advection [METEOROL]   The advection or transport of the thermal vorticity by the thermal wind, in analogy to the advection of the vorticity by the wind.

thermal wind [METEOROL]   The mean wind-shear vector in geostrophic balance with the gradient of mean temperature of a layer bounded by two isobaric surfaces.

thermal wind equation [METEOROL]   An equation for the vertical variation of the geostrophic wind in hydrostatic equilibrium which may be written $- (\partial \mathbf{V}/\partial p) = (R/pf) \, \mathbf{k} \times \nabla_p T$, where $\mathbf{V}$ is the vector geostrophic wind, p the pressure (used here as the vertical coordinate), R the gas constant for air, f the Coriolis parameter, $\mathbf{k}$ a vertically directed unit vector, and ∇_p the isobaric del operator.

thermocline [GEOPHYS]   **1.** A temperature gradient as in a layer of sea water, in which the temperature decrease with depth is greater than that of the overlying and underlying water. Also known as metalimnion. **2.** A layer in a thermally stratified body of water in which such a gradient occurs.

thermocyclogenesis [METEOROL]   A theory of cyclogenesis by G. Stüve, in which the disturbance is initiated in the stratosphere and is reflected in the development of a disturbance in the lower troposphere.

thermogram [ENG]   The recording made by a thermgraph.

thermograph [ENG]   An instrument that senses, measures, and records the temperature of the atmosphere. Also known as recording thermometer.   [OPTICS]   A far-infrared image-forming device that provides a thermal photograph by scanning a far-infrared image of an object or scene.

thermograph correction card [ENG]   A table for quick and accurate correction of the reading of a thermograph to that of the more accurate dry-bulb thermometer at the same time and place.

thermohaline [OCEANOGR]   Pertaining to the joint activity of salinity and temperature in the oceans.

thermohaline convection [OCEANOGR]   Vertical water movement observed when sea water, due to conditions of decreasing temperature or increasing salinity, becomes heavier than the water beneath it.

thermointegrator [ENG]   An apparatus, used in studying soil temperatures, for measuring the total supply of heat during a given period; it consists of a long nickel coil (inserted into the soil by an attached rod) forming a 100-ohm resistance thermometer and a 6-volt battery, the current used being recorded on a galvanometer; a mercury thermometer can be used.

thermoisopleth [CLIMATOL]   An isopleth of temperature; specifically, a line on a climatic graph showing the variation of temperature in relation to two coordinates.

thermokarst topography [GEOL]   An irregular land surface formed in a permafrost region by melting ground ice.

thermometamorphism *See* thermal metamorphism.

thermometer [ENG]   An instrument that measures temperature.

thermometer anemometer [ENG]   An anemometer consisting of two thermometers, one with an electric heating element connected to the bulb; the heated bulb cools in an airstream, and the difference in temperature as registered by the heated and unheated thermometers can be translated into air velocity by a conversion chart.

thermometric depth [OCEANOGR]   The ocean depth, in meters, deduced from the difference between the paired protected and unprotected reversing thermometer readings; the unprotected reversing thermometer indicates higher temperature due to pressure effects on the instrument.

thermonatrite [MINERAL] $Na_2CO_3 \cdot H_2O$ A colorless to white, grayish, or yellowish, orthorhombic mineral consisting of sodium carbonate monohydrate; occurs as a crust or efflorescence.

thermoremanent magnetization [GEOPHYS] The permanent magnetization of igneous rocks, acquired at the time of cooling from the molten state.

thermosphere [METEOROL] The atmospheric shell extending from the top of the mesosphere to outer space; it is a region of more or less steadily increasing temperature with height, starting at 70 or 80 kilometers; the thermosphere includes, therefore, the exosphere and most or all of the ionosphere.

thermosteric anomaly [OCEANOGR] Component of the specific volume anomaly for a parcel of sea water at 1-atmosphere pressure due to its temperature being other than the standard temperature of 0°C.

thermotropic model [METEOROL] A model atmosphere used in numerical forecasting, in which the parameters are the height of one constant-pressure surface (usually 500 millibars) and one temperature (usually the mean temperature between 100 and 500 millibars).

Theropoda [PALEON] A suborder of carnivorous bipedal saurischian reptiles which first appeared in the Upper Triassic and culminated in the uppermost Cretaceous.

Theropsida [PALEON] An order of extinct mammallike reptiles in the subclass Synapsida.

THI *See* temperature-humidity index.

thick-bedded [GEOL] Pertaining to a sedimentary bed that ranges in thickness from 60 to 120 centimeters (2–4 feet).

thickness [METEOROL] The vertical depth, measured in geometric or geopotential units, of a layer in the atmosphere bounded by surfaces of two different values of the same physical quantity, usually constant-pressure surfaces.

thickness chart [METEOROL] A type of synoptic chart showing the thickness of a certain physically defined layer in the atmosphere; it almost always refers to an isobaric thickness chart, that is, a chart of vertical distance between two constant-pressure surfaces.

thickness line [METEOROL] A line drawn through all geographic points at which the thickness of a given atmospheric layer is the same. Also known as relative contour; relative isohypse.

thickness pattern [METEOROL] The general geometric distribution of thickness lines on a thickness chart. Also known as relative hypsography; relative topography.

thick-skinned structure [GEOL] Any large-scale structure, such as a fold or fault, believed to have originated as a result of basement movement beneath overlying rocks.

thick-thin chart *See* isentropic thickness chart.

thief [PETRO ENG] In the petroleum industry, a device that permits the taking of samples from a predetermined location in the liquid body to be sampled.

Thiessen polygon method [METEOROL] A method of assigning areal significance to point rainfall values: perpendicular bisectors are constructed to the lines joining each measuring station with those immediately surrounding it; the bisectors form a series of polygons, each polygon containing one station; the value of precipitation measured at a station is assigned to the whole area covered by the enclosing polygon.

thill *See* underclay.

thin [METEOROL] In aviation weather observations, the description of a sky cover that is predominantly transparent.

thin-bedded [GEOL] Pertaining to a sedimentary bed that ranges in thickness from 5 to 60 centimeters (2 inches to 2 feet).

thin-out [GEOL] Gradual thinning of a stratum, vein, or other body of rock until the upper and lower surfaces meet and the rock disappears.

thin section [GEOL] A piece of rock or mineral specifically prepared to study its optical properties; the sample is ground to 0.03-millimeter thickness then polished and placed between two microscope slides. Also known as section.

thin-skinned structure [GEOL] Any large-scale structure, such as a fold or fault, confined to and originating within a thin layer of rocks above a surface of décollement.

thiophile element See sulfophile element.

thiospinel [MINERAL] Any mineral with the spinel structure having the general formula AR_2S_4.

third-order climatological station [CLIMATOL] As defined by the World Meteorological Organization, a station, other than a precipitation station, at which the observations are of the same kind as those at a second-order climatological station, but are not so comprehensive, are made once a day only, and are made at other than the specified hours.

thirling See holing.

thirty-day forecast [METEOROL] A weather forecast for a period of 30 days; as issued by the U.S. Weather Bureau, the forecast concerns expected departures of temperature and precipitation from normal.

thirty-two nucleus [METEOROL] An unidentified type of freezing nucleus which first becomes active when a supercooled cloud is cooled to about $-32°C$.

thixotropic clay [GEOL] A clay that weakens when disturbed and increases in strength upon standing.

Thlipsuridae [PALEON] A Paleozoic family of ostracod crustaceans in the suborder Platycopa.

tholeiite [PETR] **1.** A group of basalts composed principally of plagioclase, pyroxene, and iron oxide minerals as phenocrysts in a glassy groundmass. **2.** Any rock in the group.

thomsenolite [MINERAL] $NaCaAlF_6·H_2O$ A colorless to white, monoclinic mineral consisting of a hydrated aluminofluoride of sodium and calcium; it is dimorphous with pachnolite.

thomsonite [MINERAL] $NaCa_2Al_5Si_5O_{20}·6H_2O$ Snow-white zeolite mineral forming orthorhombic crystals and occurring in masses of radiating crystals; hardness is 5–5.5 on Mohs scale.

thoreaulite [MINERAL] $SnTa_2O_7$ A brown, monoclinic mineral consisting of an oxide of tin and tantalum; occurs as rough, prismatic crystals.

thorianite [MINERAL] ThO_2 A radioactive mineral that crystallizes in the isometric system, occurs in worn cubic crystals, is brownish black to reddish brown in color, and has resinous luster; hardness is 7 on Mohs scale, and specific gravity is 9.7–9.8.

thorite [MINERAL] $ThSiO_4$ A brownish-yellow to brownish-black and black radioactive mineral that is tetragonal in crystallization; hardness is about 4.5 on Mohs scale, and specific gravity is 4.3–5.4.

thorogummite [MINERAL] A silicate mineral and chemical variant of thorium silicate, with similar properties; isostructural with thorite and zircon; it is deficient in silica and contains small amounts of OH in substitution for oxygen.

thortveitite [MINERAL] $(Sc,Y)_2Si_2O_7$ A grayish-green mineral occurring in ortho-rhombic crystals; a source of scandium.

thread [GEOL] An extremely small vein, even thinner than a stringer. [MIN ENG] A more or less straight line of stall faces, having no cuttings, loose ends, fast ends, or steps.

thread-lace scoria [GEOL] Scoria whose vesicle walls have collapsed and are repre-sented only by a network of threads.

threeling *See* trilling.

three-piece set [MIN ENG] A set of timber consisting of a cap and its two supportive posts.

three-point method [GEOL] A method used to determine the dip and strike of a structural surface from three points of varying elevation along the surface.

three-shift cyclic mining [MIN ENG] A system of cyclic mining on a longwall con-veyor face, with coal cutting on one shift, hand filling and conveying on the next, and ripping, packing, and advancement of the face conveyor on the third shift.

threshold *See* riegel.

threshold depth *See* sill depth.

threshold velocity [GEOPHYS] The minimum velocity at which wind or water begins to move particles of soil, sand, or other material at a given place under specified conditions.

through glacier [HYD] A two-ended glacier, consisting of two valley glaciers in a depression, flowing in opposite directions.

throughput [MIN ENG] The quantity of ore or other material passed through a mill or a section of a mill in a given time or at a given rate.

through valley [GEOL] **1.** A depression eroded across a divide by glacier ice or melt-water streams. **2.** A valley excavated by a through glacier.

throw [GEOL] The vertical component of dip separation on a fault, or generally the amount of vertical displacement on any fault.

thrust [GEOL] Overriding movement of one crystal unit over another. [MIN ENG] **1.** A crushing of coal pillars caused by excess weight of the superincumbent rocks, the floor being harder than the roof. **2.** The ruins of the fallen roof, after pillars and stalls have been removed.

thrust fault [GEOL] A low-angle (less than a 45° dip) fault along which the hanging wall has moved up relative to the footwall. Also known as reverse fault; reverse slip fault; thrust slip fault.

thrust moraine *See* push moraine.

thrust nappe [GEOL] The body of rock that makes up the hanging wall of a thrust fault. Also known as thrust block; thrust plate; thrust sheet; thrust slice.

thrust slip fault *See* thrust fault.

thucolite [GEOL] Concentrations of carbonaceous matter in ancient sedimentary rocks.

thulite [MINERAL] A pink, rose-red, or purplish-red variety of epidote that contains manganese; used as an ornamental stone.

thunder [GEOPHYS] The sound emitted by rapidly expanding gases along the channel of a lightning discharge.

thunderbolt [GEOPHYS]   In mythology, a lightning flash accompanied by a material bolt or dart and which causes great damage; it is still used as a popular term for a single lightning discharge accompanied by thunder.

thundercloud [METEOROL]   A convenient and often used term for the cloud mass of a thunderstorm, that is, a cumulonimbus.

thunderhead *See* incus.

thundersquall *See* rainsquall.

thunderstorm [METEOROL]   A convective storm accompanied by lightning and thunder and rain, rarely snow showers but often hail, and gusty squall winds at the onset of precipitation; the characteristic cloud is the cumulonimbus.

thunderstorm cell [METEOROL]   The convection cell of a cumulonimbus cloud.

thunderstorm charge separation [GEOPHYS]   **1.** The process by which the large electric field found within thunderclouds is generated. **2.** The processes by which particles bearing opposite electrical charges are given the charges and are transported to different regions of the active cloud.

thunderstorm day [METEOROL]   An observational day during which thunder is heard at the station; precipitation need not occur.

Thuringian [GEOL]   A European stage of Upper Permian geologic time, above the Saxonian and below the Triassic.

Thylacoleonidae [PALEON]   An extinct family of carnivorous marsupials in the superfamily Phalangeroidea.

tidal channel [OCEANOGR]   A major channel followed by tidal currents, extending from the ocean into a tidal marsh or tidal flat.

tidal component *See* partial tide.

tidal constants [OCEANOGR]   Tidal relations that remain essentially constant for any particular locality.

tidal constituent *See* partial tide.

tidal correction [GEOPHYS]   A correction made in gravity observations to remove the effect of the earth's tides.

tidal current [OCEANOGR]   The alternating horizontal movement of water associated with the rise and fall of the tide caused by the astronomical tide-producing forces.

tidal current chart [OCEANOGR]   A chart showing by arrows and numbers the average direction and speed of tidal currents at a particular part of the current cycle.

tidal current tables [OCEANOGR]   Tables issued annually which give daily predictions of the times of slack water and the times and velocities of the strength of flood and ebb currents for a number of reference stations, together with differences and constants for obtaining predictions at subordinate stations.

tidal cycle *See* tide cycle.

tidal datum [OCEANOGR]   A level of the sea, defined by some phase of the tide, from which water depths and heights of tide are reckoned. Also known as tidal datum plane.

tidal datum plane *See* tidal datum.

tidal day [OCEANOGR]   The interval between two consecutive high waters of the tide at a given place, averaging 24 hours and 51 minutes.

tidal delta [GEOL]   A sand bar or shoal formed in the entrance of an inlet by the action of reversing tidal currents.

tidal difference [OCEANOGR]   The difference in time or height of a high or low water at a subordinate station and at a reference station for which predictions are given in the tide tables; the difference applied to the prediction at the reference station gives the corresponding time or height for the subordinate station.

tidal energy [OCEANOGR]   The energy in a tide flowing from a basin into an open sea.

tidal epoch *See* phase lag.

tidal excursion [OCEANOGR]   The net horizontal distance over which a water particle moves during one tidal cycle of flood and ebb; the distances traversed during ebb and flood are rarely equal in nature, since there is usually a layered circulation in an estuary, with a net surface flow in one direction compensated by an opposite flow at depth.

tidal flat [GEOL]   A marshy, sandy, or muddy nearly horizontal coastal flatland which is alternately covered and exposed as the tide rises and falls.

tidal frequency [OCEANOGR]   The rate of travel, in degrees per day, of a component of a tide, the component being created by a particular juxtaposition of forces in the sun-earth-moon system.

tidal friction [OCEANOGR]   The frictional effect of the tidal wave particularly in shallow waters that lengthens the tidal epoch and tends to slow the rotational velocity of the earth, thus increasing very slowly the length of the day.

tidal glacier *See* tidewater glacier.

tidal harbor [OCEANOGR]   A harbor affected by the tides, in distinction to a harbor in which the water level is maintained by caissons or gates.

tidal inlet [GEOL]   A natural inlet maintained by tidal currents.

tidalite [GEOL]   Any sediment transported and deposited by tidal currents.

tidal marsh [GEOGR]   Any marsh whose surface is covered and uncovered by tidal flow.

tidal platform ice foot [OCEANOGR]   An ice foot between high and low water levels, produced by the the rise and fall of the tide.

tidal pool [OCEANOGR]   An accumulation of sea water remaining in a depression on a beach or reef after the tide recedes.

tidal potential [OCEANOGR]   Tidal forces expressed as components of a vector field.

tidal prism [OCEANOGR]   The difference between the mean high-water volume and the mean low-water volume of an estuary.

tidal range *See* tide range.

tidal scour [GEOL]   Sea-floor erosion caused by strong tidal currents, resulting in removal of inshore sediments and formation of deep holes and channels. Also known as scour.

tidal stand *See* stand.

tidal water [OCEANOGR]   Any water whose level changes periodically due to tidal action.

tidal wave [OCEANOGR]   **1.** Any unusually high and generally destructive sea wave or water level along a shore. **2.** *See* tide wave.

tidal wind [METEOROL]   A very light breeze which occurs in calm weather in inlets where the tide sets strongly; it blows onshore with rising tide and offshore with ebbing tide.

tide [OCEANOGR]   The periodic rising and falling of the oceans resulting from lunar and solar tide-producing forces acting upon the rotating earth.

tide amplitude [OCEANOGR] One-half of the difference in height between consecutive high water and low water; half the tide range.

tide bulge *See* tide wave.

tide crack [OCEANOGR] A crack in sea ice, parallel to the shore, caused by the vertical movement of the water due to tides; several such cracks often appear as a family.

tide curve [OCEANOGR] Any graphic representation of the rise and fall of the tide; time is generally represented by the abscissas, and the height of the tide by the ordinates; for normal tides the curve so produced approximates a sine curve.

tide cycle [OCEANOGR] A period which includes a complete set of tide conditions or characteristics, such as a tidal day or a lunar month. Also known as tidal cycle.

tide gage [ENG] A device for measuring the height of a tide; may be observed visually or may consist of an elaborate recording instrument.

tidehead [OCEANOGR] The inland limit of water affected by a tide.

tide hole [OCEANOGR] A hole made in ice to observe the height of the tide.

tide indicator [ENG] That part of a tide gage which indicates the height of tide at any time; the indicator may be in the immediate vicinity of the tidal water or at some distance from it.

tideland [GEOGR] Land which is under water at high tide and uncovered at low tide.

tidemark [OCEANOGR] **1.** A high-water mark left by tidal water. **2.** The highest point reached by a high tide.

tide notes [OCEANOGR] Notes included on nautical charts which give information on the mean range or the diurnal range of the tide, mean tide level, and extreme low water at key places on the chart.

tide pole [ENG] A graduated spar used for measuring the rise and fall of the tide. Also known as tide staff.

tide prediction [OCEANOGR] The mathematical process by which the times and heights of the tide are determined in advance from the harmonic constituents at a place.

tide-producing force [GEOPHYS] The slight local difference between the gravitational attraction of two astronomical bodies and the centrifugal force that holds them apart.

tide race [OCEANOGR] A strong tidal current or a channel in which such a current flows.

tide range [OCEANOGR] The difference in height between consecutive high and low waters. Also known as tidal range.

tide staff *See* tide pole.

tide station [OCEANOGR] A place where observations of the tides are obtained.

tide table [OCEANOGR] A table giving daily predictions, usually a year in advance, of the times and heights of the tide for a number of reference stations.

tidewater [OCEANOGR] **1.** A body of water, such as a river, affected by tides. **2.** Water inundating land at flood tide.

tidewater glacier [HYD] A glacier that descends into the sea and usually has a terminal ice cliff. Also known as tidal glacier.

tide wave [OCEANOGR] A long-period wave associated with the tide-producing forces of the moon and sun, and identified with the rising and falling of the tide. Also known as tidal wave; tide bulge.

tideway [OCEANOGR] A channel through which a tidal current runs.

tie [MIN ENG] A support for the roof in coal mines.

tieback [MIN ENG] **1.** A beam serving a purpose similar to that of a fend-off beam, but fixed at the opposite side of the shaft or inclined road. **2.** The wire ropes or stay rods that are sometimes used on the side of the tower opposite the hoisting engine, either in place of or to reinforce the engine braces.

tiemannite [MINERAL] HgSe A steel gray to blackish–lead gray mineral consisting of mercuric selenide; commonly occurs in massive form.

tiger's-eye [MINERAL] A yellowish-brown crystalline variety of quartz; a translucent, fibrous, broadly chatoyant gemstone that may be dyed other colors.

tight [ENG] **1.** Unbroken, crack-free, and solid rock in which a naked hole will stand without caving. **2.** A borehole made impermeable to water by cementation or casing.

tight fold *See* closed fold.

tight sand [GEOL] A sand whose interstices are filled with finer grains of the matrix material, thus effectively destroying porosity and permeability. Also known as close sand.

tilasite [MINERAL] CaMg(AsO$_4$)F A gray, gray-violet, olive green, or apple green, monoclinic mineral consisting of a fluorarsenate of calcium and magnesium.

till [GEOL] Unsorted and unstratified drift consisting of a heterogeneous mixture of clay, sand, gravel, and boulders which is deposited by and underneath a glacier. Also known as boulder clay; glacial till; ice-laid drift.

till billow [GEOL] An undulating mass of glacial drift that is disposed in an irregular pattern with regard to the direction of movement of the ice.

tilleyite [MINERAL] Ca$_5$(Si$_2$O$_7$)(CO$_3$)$_2$ A white mineral consisting of a carbonate and silicate of calcium.

tillite [PETR] A sedimentary rock formed by lithification of till, especially pre-Pleistocene till.

Tillodontia [PALEON] An order of extinct quadrupedal land mammals known from early Cenozoic deposits in the Northern Hemisphere and distinguished by large, rodentlike incisors, blunt-cuspid cheek teeth, and five clawed toes.

tilloid [GEOL] A nonglacial till-like deposit.   [PETR] A rock of uncertain origin which resembles tillite.

till plain [GEOL] An extensive, relatively flat area overlying a till.

till sheet [GEOL] A sheet, layer, or bed of till.

tilt [METEOROL] The inclination to the vertical of a significant feature of the circulation (or pressure) pattern or of the field of temperature or moisture; for example, troughs in the westerlies usually display a westward tilt with altitude in the lower and middle troposphere.

tilt block [GEOL] A tilted fault block.

tilted iceberg [OCEANOGR] A tabular iceberg that has become unbalanced, so that the flat, level top is inclined.

tilted interface [GEOL] Oil-water interface in which water moves in a generally linear direction under an oil accumulation which is, for instance, in an anticline.

tiltmeter [ENG] An instrument used to measure small changes in the tilt of the earth's surface, usually in relation to a liquid-level surface or to the rest position of a pendulum.

timbered stope [MIN ENG] A stope made of square-set timbering or any of its variations.

timbering [MIN ENG]   The timber structure used for supporting the faces of an excavation during the progress of construction.

timbering machine [MIN ENG]   An electrically driven machine to raise and hold timber in place while the supporting posts are being set, the posts having been cut to desired length previously by the machine's power-driven saw.

timber mat [MIN ENG]   Broken timber forming the roof of an ore deposit that is being extracted by a caving method, such as top slicing.

timber packer *See* pack builder.

timber puller [MIN ENG]   A machine used to remove the timber supports in a mine.

timber trolley [MIN ENG]   A carriage consisting of a timber or steel base, mounted on wheels, with U-shaped arms.

timber truck [MIN ENG]   Any truck or car used for hauling timber inside of a mine.

time-correlation [GEOL]   A correlation of age or mutual time relations between stratigraphic units in separated areas.

time-interval radiosonde *See* pulse-time-modulated radiosonde.

time line [GEOL]   **1.** A line that indicates equal geologic age in a correlation diagram. **2.** A rock unit represented by a time line.

time-rock unit *See* time-stratigraphic unit.

time-stratigraphic facies [GEOL]   A stratigraphic facies based on the amount of geologic time during which deposition and nondeposition of sediment occurred.

time-stratigraphic unit [GEOL]   A stratigraphic unit based on geologic age or time of origin. Also known as chronolith; chronolithologic unit; chronostratic unit; chronostratigraphic unit; time-rock unit.

time-transgressive *See* diachronous.

time-transgressive unit [GEOL]   A laterally continuous, lithologically homogeneous rock unit whose age differs regionally.

tincal *See* borax.

tincalconite [MINERAL] $Na_2B_4O_7 \cdot 5H_2O$ A colorless to dull-white mineral, crystallizing in the rhombohedral system; one of the principal ores of borax and boron compounds. Also known as mohavite; octahedral borax.

tin pyrites *See* stannite.

tin stone *See* cassiterite.

tinticite [MINERAL] $Fe_3(PO_4)_2(OH)_3 \cdot 3\frac{1}{2}H_2O$ A creamy white mineral with a yellowish-green tint, consisting of a hydrated basic iron phosphate.

tipper [MIN ENG]   An apparatus for emptying coal or ore cars by turning them upside down and then righting them, with a minimum of manual labor.

tipping-bucket rain gage [ENG]   A type of recording rain gage; the precipitation collected by the receiver empties into one side of a chamber which is partitioned transversely at its center and is balanced bistably upon a horizontal axis; when a predetermined amount of water has been collected, the chamber tips, spilling out the water and placing the other half of the chamber under the receiver; each tip of the bucket is recorded on a chronograph, and the record obtained indicates the amount and rate of rainfall.

tipple [MIN ENG]   **1.** The place where the mine cars are tipped and emptied of their coal. **2.** The tracks, trestles, and screens at the entrance to a colliery, where coal is screened and loaded.

titanaugite [MINERAL] $Ca(Mg,Fe,Ti)(Si,Al)_2O_6$ A variety of augite rich in titanium and occurring in basaltic rocks.

titanic iron ore *See* ilmenite.

titanite *See* sphene.

Titanoideidae [PALEON] A family of extinct land mammals in the order Pantodonta.

titanothere [PALEON] Any member of the family Brontotheriidae.

Tithonian [GEOL] Southern European equivalent of the Portlandian stage (uppermost Jurassic) of geologic time.

tjaele *See* frozen ground.

Toarcian [GEOL] A European stage of geologic time; Lower Jurassic (above Pliensbachian, below Bajocian).

tobacco jack *See* wolframite.

toe [GEOL] The leading edge of a thrust nappe. [MIN ENG] **1.** The burden of material between the bottom of the borehole and the free face. **2.** The bottom of the borehole. **3.** A spurn, or small pillar of coal. **4.** The base of a bank in an open-pit mine.

toe hole [ENG] A blasting hole, usually drilled horizontally or at a slight inclination into the base of a bank, bench, or slope of a quarry or open-pit mine.

tofan [METEOROL] A violent spring storm common in the mountains of Indonesia.

toise [GEOD] A unit of length equal to about 6.4 feet (1.95 meters); used in early geodetic surveys.

tombolo [GEOL] A sand or gravel bar or spit that connects an island with another island or an island with the mainland. Also known as connecting bar; tie bar; tying bar.

tombolo cluster *See* complex tombolo.

tombolo series *See* complex tombolo.

tonalite *See* quartz diorite.

tongara [METEOROL] A hazy, southeast wind in the Macassar Strait.

Tongrian [GEOL] A European stage of geologic time; lower Oligocene (above Ludian of Eocene, below Rupelian). Also known as Lattorfian.

tongue [GEOL] **1.** A minor rock-stratigraphic unit of limited geographic extent; it disappears laterally in one direction. **2.** A lava flow branching from a larger flow. [OCEANOGR] **1.** A protrusion of water into a region of different temperature, or salinity, or dissolved oxygen concentrating. **2.** A protrusion of one water mass into a region occupied by a different water mass.

ton-kilometer [MIN ENG] A unit of measurement equal to the weight in tons of material transported in a mine multiplied by the number of kilometers driven.

tonstein [GEOL] Kaolinitic bands in certain coalfields which have characteristic fossil fauna from short-lived but widespread marine invasions.

tool mark [GEOL] Any of the wide variety of current marks, such as groove marks, prod marks, and skip marks, produced by the continuous contact or intermittent impact of solid, current-borne objects against a muddy bottom.

tool nipper [MIN ENG] A person whose duty it is to carry powder, drills, and tools to the various levels of the mine and to bring dull tools and drills to the surface.

top *See* overburden.

topaz [MINERAL] $Al_2SiO_4(F,OH)$　A red, yellow, green, blue, or brown neosilicate mineral that crystallizes in the orthorhombic system and commonly occurs in prismatic crystals with pyramidal terminations; hardness is 8 on Mohs scale, and specific gravity is 3.4–3.6; used as a gemstone.

topaz quartz *See* citrine.

top-benching [MIN ENG]　The method by which the bench is removed from above, as with a dragline.

top cager [MIN ENG]　A person at the top of a mine shaft who superintends the lowering and raising of the cage, and, at most mines, the removing of loaded cars from the placing of empty cars in the cage.

top cut [MIN ENG]　A machine cut made in the coal at or near the top of the working face in a mine.

topographical latitude *See* geodetic latitude.

topographic curl effect [OCEANOGR]　A term in Ekman's differential equation for the effects of variable wind stress, variable depth, variable friction, and variable latitude on the deep current; tends to make the curl G (velocity of deep current) positive when the current flows over increasing depth and negative when the depth decreases in the direction of the current.

topographic infancy *See* infancy.

topographic map [MAP]　A large-scale map showing relief and man-made features of a portion of a land surface distinguished by portrayal of position, relation, size, shape, and elevation of the features.

topographic maturity *See* maturity.

topographic old age *See* old age.

topographic survey [ENG]　A survey that determines ground relief and location of natural and man-made features thereon.

topographic unconformity [GEOGR]　A lack of harmony or conformity between two parts of a landscape or two kinds of topography.

topography [GEOGR]　**1.** The general configuration of a surface, including its relief; may be a land or water-bottom surface. **2.** The natural surface features of a region, treated collectively as to form.

topset bed [GEOL]　One of the nearly horizontal sedimentary layers deposited on the top surface of an advancing delta.

top slicing [MIN ENG]　A method of stoping in which the ore is extracted by excavating a series of horizontal (sometimes inclined) timbered slices alongside each other, beginning at the top of the ore body and working progressively downward.

top slicing and cover caving [MIN ENG]　A mining method that entails the working of the ore body from the top down in successive horizontal slices that may follow one another sequentially or simultaneously; the overburden or cover is caved after mining a unit.

topsoil [GEOL]　**1.** Soil presumed to be fertile and used to cover areas of special planting. **2.** Surface soil, usually corresponding with the A horizon, as distinguished from subsoil.

tor [GEOGR]　An isolated, rough pinnacle or rocky peak.

torbanite [GEOL]　A variety of coal that resembles a carbonaceous shale in outward appearance; it is fine-grained, black to brown, and tough. Also known as bitumenite; kerosine shale.

torbernite [MINERAL] $Cu(UO_2)_2(PO_4)_2 \cdot 8\text{-}12H_2O$ A green radioactive mineral crystallizing in the tetragonal system and occurring in tabular crystals or in foliated form. Also known as chalcolite; copper uranite; cuprouranite; uran-mica.

tornado [METEOROL] An intense rotary storm of small diameter, the most violent of weather phenomena; tornadoes always extend downward from the base of a convective-type cloud, generally in the vicinity of a severe thunderstorm.

tornado belt [METEOROL] The district of the United States in which tornadoes are most frequent; it encompasses the great lowland areas of the central and upper Mississippi, the Ohio, and lower Missouri River valleys.

tornado cloud *See* tuba.

tornado echo [METEOROL] A type of radar precipitation echo which has been observed in connection with a number of tornadoes; it frequently appears, on plan-position-indicator scopes, in the form of the figure 6 in the southwest sector of the storm; this echo has not been noted with all radar-observed tornadoes.

torose load cast [GEOL] One of a group of elongate load casts with alternate contractions and swellings, which may terminate down current in bulbous, teardrop, or spiral forms.

torpedo [ENG] An encased explosive charge slid, lowered, or dropped into a borehole and exploded to clear the hole of obstructions or to open communications with an oil or water supply. Also known as bullet.

torque-coil magnetometer [ENG] A magnetometer that depends for its operation on the torque developed by a known current in a coil that can turn in the field to be measured.

Torrert [GEOL] A suborder of the soil order Vertisol; it is the driest soil of the order and forms cracks that tend to remain open; occurs in arid regions.

torreyite [MINERAL] $(Mg,Mn,Zn)_7(SO_4)(OH)_{12} \cdot 4H_2O$ A bluish-white mineral consisting of a hydrated basic sulfate of magnesium, manganese, and zinc; occurs in massive form.

Torricellian barometer *See* mercury barometer.

Torrid Zone [CLIMATOL] The zone of the earth's surface which lies between the Tropics of Cancer and Capricorn.

Torrox [GEOL] A suborder of the soil order Oxisol that is low in organic matter, well drained, and dry most of the year; believed to have been formed under rainier climates of past eras.

torsion fault *See* wrench fault.

torsion hygrometer [ENG] A hygrometer in which the rotation of the hygrometric element is a function of the humidity; such hygrometers are constructed by taking a substance whose length is a function of the humidity and twisting or spiraling it under tension in such a manner that a change in length will cause a further rotation of the element.

torso mountain *See* monadnock.

Tortonian [GEOL] A European stage of geologic time: Miocene (above Helvetian, below Sarmatian).

tosca [METEOROL] A southwest wind on Lake Garda in Italy.

total conductivity [GEOPHYS] In atmospheric electricity, the sum of the electrical conductivities of the positive and negative ions found in a given portion of the atmosphere.

total displacement *See* slip.

total evaporation *See* evapotranspiration.

total porosity [GEOL]   The ratio of total void space in porous oil–reservoir rock to the bulk volume of the rock itself.

total slip *See* net slip.

touriello [METEOROL]   A south wind of foehn type descending from the Pyrenees in the Ariège valley, France; it is especially violent in February and March, when it melts the snow, flooding the rivers and sometimes causing avalanches.

tourmaline [MINERAL] $(Na,Ca)(Al,Fe,Li,Mg)_3Al_6(BO_3)_3Si_6O_{18}(OH)_4$ Any of a group of cyclosilicate minerals with a complex chemical composition, vitreous to resinous luster, and variable color; crystallizes in the ditrigonal-pyramidal class of the hexagonal system, has piezoelectric properties, and is used as a gemstone.

Tournaisian [GEOL]   European stage of lowermost Carboniferous time.

Toussaint's formula [METEOROL]   A rule for the linear decrease of temperature with height in an atmosphere for which the temperature at mean sea level is 15°C, and given by the formula $t = 15 - 0.0065z$, where t is the temperature in degrees Celsius, and z is the geometric height in meters above mean sea level.

tower excavator [MIN ENG]   A cableway excavator designed specifically for levee work but which is used extensively in the stripping of overburden, spoil, or waste in surface mining: basically, it is a Sauerman-type excavator with towers either fixed or movable, and when the head tower is located on the spoil pile and the tail tower on the unexcavated wall, pits of almost unlimited width can be dug.

towering [METEOROL]   A refraction phenomenon; a special case of looming in which the downward curvature of the light rays due to atmospheric refraction increases with elevation so that the visual image of a distant object appears to be stretched in the vertical direction.

towering cumulus [METEOROL]   A descriptive term, used mostly in weather observing, for the cloud type cumulus congestus.

tower loader [MIN ENG]   A front-end loader whose bucket is lifted along tracks on a more or less vertical tower.

Toxasteridae [PALEON]   A family of Cretaceous echinoderms in the order Spatangoida which lacked fascioles and petals.

Toxodontia [PALEON]   An extinct suborder of mammals representing a central stock of the order Notoungulata.

trace [GEOL]   The intersection of two geological surfaces.   [METEOROL]   A precipitation of less than 0.005 inch (0.127 millimeter).

trace element [GEOCHEM]   A nonessential element found in small quantities (usually less than 1.0%) in a mineral.   Also known as accessory element; guest element.

trace fossil [GEOL]   A trail, track, or burrow made by an animal and found in ancient sediments such as sandstone, shale, or limestone.   Also known as ichnofossil.

trace slip [GEOL]   That component of the net slip in a fault which is parallel to the trace of an index plane on a fault plane.

trace-slip fault [GEOL]   A fault whose net slip is trace slip.

trachybasalt [PETR]   An extrusive rock characterized by calcic plagioclase and sanidine, with augite, olivine, and possibly minor analcime or leucite.

trachyte [PETR]   The light-colored, aphanitic rock (the volcanic equivalent of syenite), composed largely of alkali feldspar with minor amounts of mafic minerals.

trachytoid texture [GEOL] The texture of a phaneritic extrusive igneous rock in which the microlites of a mineral, not necessarily feldspar, in the groundmass have a subparallel or randomly divergent alignment.

track cable scraper [MIN ENG] A type of excavator that uses a bottomless scraper bucket which conveys its load over the ground and is operated by a two-drum hoist which controls a track cable that spans the working area and a haulage cable that leads to the front of the bucket.

track haulage [MIN ENG] Movement or transportation of excavated or mined materials in cars or trucks that run on rails.

trackless mine [MIN ENG] A mine in which rubber-tired vehicles are used for haulage and transport.

trackless tunneling [MIN ENG] Tunneling by means of loaders mounted on caterpillars.

traction [GEOL] Transport of sedimentary particles along and parallel to a bottom surface of a stream channel by means of rolling, sliding, dragging, pushing, or saltation.

trade air [METEOROL] The type of air of which the trade winds consist, and whose chief thermodynamic characteristic is the presence of the trade-wind inversion.

trade cumulus *See* trade-wind cumulus.

trade wind [METEOROL] The wind system, occupying most of the tropics, which blows from the subtropical highs toward the equatorial trough; a major component of the general circulation of the atmosphere; the winds are northeasterly in the Northern Hemisphere and southeasterly in the Southern Hemisphere; hence they are known as the northeast trades and southeast trades, respectively.

trade-wind cumulus [METEOROL] The characteristic cumulus cloud of the trade winds over the oceans in average, undisturbed weather conditions; the individual cloud usually exhibits a blocklike appearance since its vertical growth ends abruptly in the lower stratum of the trade-wind inversion; a group of fully grown clouds shows considerable uniformity in size and shape. Also known as trade cumulus.

trade wind desert [CLIMATOL] **1.** An area of very little rainfall and high temperature which occurs where the trade winds or their equivalent (such as the harmattan) blow over land; the best examples are the Sahara and Kalahari deserts. **2.** The arid coldwater coasts that occur on the western shores of North and South America and Africa.

trade-wind inversion [METEOROL] A characteristic temperature inversion usually present in the trade-wind streams over the eastern portions of the tropical oceans: it is formed by broad-scale subsidence of air from high altitudes in the eastern extremities of the subtropical highs; while descending, the current meets the opposition of the low-level maritime air flowing equatorward; the inversion forms at the meeting point of these two strata which flow horizontally in the same direction.

traersu [METEOROL] A violent east wind of Lake Garda in Italy.

trail [GEOL] A line of rock fragments that were picked up by glacial ice at a localized outcropping and left scattered along a fairly well-defined tract during the movement of a glacier.

trajectory [GEOPHYS] The path of a seismic wave.

tramming [MIN ENG] Pushing tubs, mine cars, or trams by hand.

tramontana [METEOROL] A cold wind from the northeast or north, particularly on the west coast of Italy and northern Corsica, but also in the Balearic Islands and the Ebro valley in Catalonia.

tramp metal [MIN ENG]   Unwanted metal which finds its way into the mill ore stream.

tramp metal detector [MIN ENG]   A sensing device which detects presence of unwanted metal in an ore stream, and sounds an alarm or removes the metal.

transcurrent fault [GEOL]   A strike-slip fault characterized by a steeply inclined surface. Also known as transverse thrust.

transfer [MIN ENG]   A vertical or inclined connection between two or more levels, used as an ore pass.

transfer car [MIN ENG]   A quarry car provided with transverse tracks, on which the gang car may be conveyed to or from the saw gang.

transformation *See* inversion.

transformation twin [CRYSTAL]   A crystal twin developed by a growth transformation from a higher to a lower symmetry.

transform fault [GEOL]   A strike-slip fault with offset ridges characteristic of a mid-oceanic ridge.

transgression [GEOL]   Geologic evidence of landward extension of the sea. Also known as invasion; marine transgression.   [OCEANOGR]   Extension of the sea over land areas.

transgressive deposit [GEOL]   Sediment deposited during transgression of the sea or during subsidence of the land.

transgressive overlap *See* onlap.

transit [ENG]   **1.** A surveying instrument with the telescope mounted so that it can measure horizontal and vertical angles. Also known as transit theodolite. **2.** To reverse the direction of the telescope of a transit by rotating 180° about its horizontal axis. Also known as plunge.

transit theodolite *See* transit.

translational fault [GEOL]   A fault in which there has been uniform movement in one direction and no rotational component of movement. Also known as translatory fault.

translational movement [GEOL]   Movement, as of fault blocks, that is uniform, without rotation, so that parallel features maintain their orientation.

translation gliding *See* crystal gliding.

translation group [CRYSTAL]   The collection of all translation operations which carry a crystal lattice into itself.

translatory fault *See* translational fault.

translucent attritus [GEOL]   Attritus composed principally of transparent humic degradation matter. Also known as humodurite.

translucidus [METEOROL]   A cloud variety occurring in a layer, patch, or extensive sheet, the greater part of which is sufficiently translucent to reveal the position of the sun, or through which higher clouds may be discerned; this variety is found in the general altocumulus, altostratus, stratocumulus, and stratus.

transmission function [GEOPHYS]   A mathematical formulation of relationships between infrared transmission in the atmosphere, the path length, and the concentration of absorbing gases.

transobuoy [ENG]   A free-floating or moored automatic weather station developed for the purpose of providing weather reports from the open oceans; it transmits barometric pressure, air temperature, sea-water temperature, and wind speed and direction.

transparent sky cover　[METEOROL]　In United States weather-observing practice, that portion of sky cover through which higher clouds and blue sky may be observed; opposed to opaque sky cover.

transportation　[GEOL]　A phase of sedimentation concerned with movement by natural agents of sediment or any loose or weathered material from one place to another.

Transvaal jade　[MINERAL]　A mineral that is not a true jade but a green grossularite garnet. Also known as South African jade.

transverse bar　[GEOL]　A slightly submerged sand bar extending perpendicular to the shoreline.

transverse basin　*See* exogeosyncline.

transverse cyclindrical orthomorphic chart　*See* transverse Mercator chart.

transverse cylindrical orthomorphic projection　*See* transverse Mercator projection.

transverse dune　[GEOL]　A sand dune with a nearly straight ridge crest formed by the merger of crescentic dunes; elongated at right angles to the direction of prevailing winds, with a gentle windward slope and a steep leeward slope.

transverse equator　[MAP]　A meridian the plane of which is perpendicular to the axis of a transverse projection; it serves as the origin for measurement of transverse latitude.

transverse fault　[GEOL]　A fault whose strike is more or less perpendicular to the general structural trend of the region.

transverse fold　*See* cross fold.

transverse gallery　[MIN ENG]　An auxiliary crosscut made in thick deposits across the ore body in order to divide it into sections along the strike.

transverse graticule　[MAP]　A fictitious graticule based upon a transverse projection.

transverse joint　*See* cross joint.

transverse latitude　[MAP]　Angular distance from a transverse equator. Also known as inverse latitude.

transverse Mercator chart　[MAP]　A chart on the transverse Mercator projection. Also known as inverse cylindrical orthomorphic chart; inverse Mercator chart; transverse cylindrical orthomorphic chart.

transverse Mercator projection　[MAP]　A conformal map projection in which the regular Mercator projection is rotated (transversed) 90° in azimuth, the central meridian corresponding to the line which represents the equator on the regular Mercator; the characteristics as to scale are identical to those of the regular Mercator, except that the scale is dependent on distances east or west of the meridian instead of north or south of the equator. Also known as inverse cylindrical orthomorphic projection; inverse Mercator projection; transverse cylindrical orthomorphic projection.

transverse meridian　[MAP]　A great circle perpendicular to a transverse equator.

transverse parallel　[MAP]　A circle or line parallel to a transverse equator, connecting all points of equal transverse latitude. Also known as inverse parallel.

transverse pole　[MAP]　One of the two points 90° from a transverse equator.

transverse rhumb line　[MAP]　A line making the same oblique angle with all fictitious meridians of a transverse Mercator projection; transverse parallels and meridians may be considered special cases of the transverse rhumb line. Also known as inverse rhumb line.

transverse ripple mark　[GEOL]　A ripple mark formed nearly perpendicular to the direction of the current.

transverse thrust *See* transcurrent fault.

transverse valley [GEOL]   **1.** A valley perpendicular to the general strike of the underlying strata.   **2.** A valley cutting perpendicularly across a ridge, range, or chain of mountains. Also known as cross valley.

transverse wave *See* S wave.

trap [GEOL]   *See* oil trap.   [PETR]   Any dark-colored, fine-grained, nongranitic, hypabyssal or extrusive rock. Also known as trappide; trap rock.

trapdoor fault [GEOL]   A circular fault that is hinged at one end.

trapezohedron [CRYSTAL]   An isometric crystal form of 24 faces, each face of which is an irregular four-sided figure. Also known as icositetrahedron; leucitohedron; tetragonal trisoctahedron.

trapped radiation [GEOPHYS]   Radiation from space that has become trapped in the magnetic field of the earth, as in the Van Allen belt.

trappide *See* trap.

trap rock *See* trap.

Trauzl test [ENG]   A test to determine the relative disruptive power of explosives, in which a standard quantity of explosive (10 grams) is placed in a cavity in a lead block and exploded; the resulting volume of cavity in the block is compared with the volume produced under the same conditions by a standard explosive, usually trinitrotoluene (TNT).

traveling compartment [MIN ENG]   The section of a mine shaft used for raising and lowering the miners.

traveling road [MIN ENG]   A roadway used by miners for walking to and from the face, that is, from the shaft bottom or main entry to the workings.

traveling valve [PETRO ENG]   A sucker-rod-pump (oil well) discharge valve that moves with the plunger of a stationary-barrel-type pump, and with the barrel of a traveling-barrel-type pump; contrasted with a standing valve.

traverse [GEOL]   A line of survey or sampling across a thin section of geological region.   [METEOROL]   A westerly wind in central France; it is moderate to strong, generally squally, humid and thundery in summer, especially on slopes facing west; it is cold in winter and spring and brings snow or hail showers.

traversia [METEOROL]   A South American nautical term (especially Chile) for a west wind from the sea.

traversier [METEOROL]   In the Mediterranean, a dangerous wind blowing directly into port.

travertine [GEOL]   Concretionary limestone deposited at the mouth of a hot spring.

treanorite *See* allanite.

tree climate [CLIMATOL]   Any type of climate which supports the growth of trees, including the tropical rainy climates, temperate rainy climates, and snow-forest climates.

trellis drainage [HYD]   A drainage pattern characterized by parallel main streams and secondary tributaries intersected at right angles by tributaries. Also known as espalier drainage; grapevine drainage.

Trematosauria [PALEON]   A group of Triassic amphibians in the order Temnospondyli.

tremolite [MINERAL] $Ca_2Mg_5Si_8O_{22}(OH)_2$ Magnesium-rich monoclinic calcium amphibole that forms one end member of a group of solid-solution series with iron, sodium, and aluminum; occurs in long blade-shaped or short stout prismatic crystals and also in masses or compound aggregates.

tremor [GEOPHYS]   A minor earthquake. Also known as earthquake tremor; earth tremor.

trench [GEOGR]   **1.** A narrow, straight, elongate, U-shaped valley between two mountain ranges. **2.** A narrow stream-eroded canyon, gulley, or depression with steep sides.   [GEOL]   A long, narrow, deep depression of the sea floor, with relatively steep sides. Also known as submarine trench.

trench sampling [MIN ENG]   A slight refinement of grab sampling in which the ore material to be sampled is spread out flat and channeled in one direction with a shovel, and the material for the sample is taken at regular intervals along the channel.

trend [GEOL]   The direction of an outcrop of a layer, vein, fold, or other geologic feature. Also known as direction.

Trentonian [GEOL]   A North American stage of geologic time; Middle Ordovician (above Wilderness, below Edenian); equivalent to the upper Mohawkian.

Trepostomata [PALEON]   An extinct order of ectoproct bryozoans in the class Stenolaemata characterized by delicate to massive colonies composed of tightly packed zooecia with solid calcareous zooecial walls.

treptomorphism *See* isochemical metamorphism.

trial pit [MIN ENG]   A shallow hole, 2 to 3 feet (60 to 90 centimeters) in diameter, put down to test shallow minerals or to establish the nature and thickness of superficial deposits and depth to bedrock.

triangle cut [MIN ENG]   A zigzag arrangement of drill holes permitting larger openings to be obtained as the drill holes can break out between the preceding row of holes.

triangular facet [GEOL]   A triangular-shaped steep-sloped hill or cliff formed usually by the erosion of a fault-truncated hill.

triangular method [MIN ENG]   A method of ore reserve estimation based on the assumption that a linear relationship exists between the grade difference and the distance between all drill holes.

triangulation [ENG]   A surveying method for measuring a large area of land by establishing a base line from which a network of triangles is built up; in a series, each triangle has at least one side common with each adjacent triangle.

Triassic [GEOL]   The first period of the Mesozoic era, lying above Permian and below Jurassic, 180–225 million years ago.

tributary [HYD]   A stream that feeds or flows into or joins a larger stream or a lake. Also known as contributory; feeder; side stream; tributary stream.

tributary glacier [GEOL]   A glacier that flows into a larger glacier.

tributary stream *See* tributary.

tributary waterway [HYD]   Any body of water that flows into a larger body, that is, a creek in relation to a river, a river in relation to a bay, and a bay in relation to the open sea.

trichalcite [MINERAL] $Cu_5Ca(AsO_4)_2(CO_3)(OH)_4 \cdot 6H_2O$ A verdigris green to blue-green, orthorhombic mineral consisting of hydrated copper arsenate. Also known as tyrolite.

trichite [PETR]   A black, straight or curved, hairlike crystallite.

triclinic crystal [CRYSTAL]   A crystal whose unit cell has axes which are not at right angles, and are unequal. Also known as anorthic crystal.

triclinic system [CRYSTAL] The most general and least symmetric crystal system, referred to by three axes of different length which are not at right angles to one another.

Triconodonta [PALEON] An extinct mammalian order of small flesh-eating creatures of the Mesozoic era having no angle or a pseudoangle on the lower jaw and triconodont molars.

tridymite [MINERAL] SiO_2 A white or colorless crystal occurring in minute, thin, tabular crystals or scales; a high-temperature polymorph of quartz.

trigonal lattice *See* rhombohedral lattice.

trigonal system [CRYSTAL] A crystal system which is characterized by threefold symmetry, and which is usually considered as part of the hexagonal system since the lattice may be either hexagonal or rhombohedral.

trigonite [MINERAL] $MnPb_3H(AsO_3)_3$ A sulfur yellow to yellowish-brown or dark brown, monoclinic mineral consisting of an acid arsenite of lead and manganese; occurs in domatic form.

trigonometric leveling [ENG] A method of determining the difference of elevation between two points, by using the principles of triangulation and trigonometric calculations.

Trigonostylopoidea [PALEON] A suborder of Paleocene-Eocene ungulate mammals in the order Astrapotheria.

trilateration [ENG] The measurement of a series of distances between points on the surface of the earth, for the purpose of establishing relative positions of the points in surveying.

trill *See* trilling.

trilling [CRYSTAL] A cyclic crystal twin consisting of three individual crystals. Also known as threeling; trill.

Trilobita [PALEON] The trilobites, a class of extinct Cambrian-Permian arthropods characterized by an exoskeleton covering the dorsal surface, delicate biramous appendages, body segments divided by furrows on the dorsal surface, and a pygidium composed of fused segments.

Trilobitoidea [PALEON] A class of Cambrian arthropods that are closely related to the Trilobita.

Trimerellacea [PALEON] A superfamily of extinct inarticulate brachiopods in the order Lingulida; they have valves, usually consisting of calcium carbonate.

Trimerophytopsida [PALEOBOT] A group of extinct land vascular plants with leafless, dichotomously branched stems that bear terminal sporangia.

trimmer [MIN ENG] **1.** A piece of bent wire used to regulate the size of the flame of a safety lamp without removing the top of the lamp. **2.** A worker who arranges coal in the hold of a vessel (miner, ship) as the coal is discharged into it from bins. **3.** A person who cleans miners' lamps. **4.** An apparatus for trimming a pile of coal into a regular form (as a cone or prism). **5.** *See* rib hole.

trip [MIN ENG] **1.** The line of cars hauled by mules or by motor, or run on a slope, plane, or sprag road. **2.** An automatic arrangement for dumping cars.

trip change [MIN ENG] The period during which the loaded cars are taken away and empties are brought back.

triphane *See* spodumene.

triphylite [MINERAL] $Li(Fe^{2+},Mn^{2+})PO_4$ A grayish-green or bluish-gray mineral crystallizing in the orthorhombic system; it is isomorphous with lithiophilite.

trip lamp [MIN ENG] A removable self-contained mine lamp, designed for marking the rear end of a train (trip) of mine cars.

triple entry [MIN ENG] A system of opening a mine by driving three parallel entries as main entries.

triplite [MINERAL] $(Mn,Fe,Mg,Ca)_2(PO)_4(F,OH)$ A dark brown, chestnut brown, reddish-brown, or salmon pink, monoclinic mineral consisting of a fluophosphate of iron, manganese, magnesium, and calcium; occurs in massive form.

tripoli [GEOL] A lightweight, porous, friable, siliceous sedimentary rock that may have a white, gray, pink, red, or yellow color; used for polishing metals and stones.

tripolite *See* diatomaceous earth.

tripping [MIN ENG] **1.** The process of pulling or lowering drill-string equipment in a borehole. **2.** To open a latch or locking device, thereby allowing a door or gate to open to empty the contents of a skip or bailer.

trippkeite [MINERAL] $CuAs_2O_4$ A greenish-blue, tetragonal mineral consisting of copper arsenite.

trip rider [MIN ENG] A rider who throws switches, gives signals, and makes couplings. Also known as rope rider.

tripuhyite [MINERAL] $FeSb_2O_6$ A greenish-yellow to dark brown mineral consisting of iron antimonate; occurs as microcrystalline aggregates.

Trochacea [PALEON] A recent subfamily of primitive gastropod mollusks in the order Aspidobranchia.

Trochiliscales [PALEOBOT] A group of extinct plants belonging to the Charophyta in which the gyrogonites are dextrally spiraled.

troctolite [PETR] A gabbro composed principally of calcic plagioclase and olivine. Also known as forellenstein.

troegerite [MINERAL] $(UO_2)_3(AsO_4)_2\cdot12H_2O$ A lemon yellow, tetragonal mineral consisting of a hydrated uranium arsenate.

troilite [MINERAL] FeS A meteorite mineral crystallizing in the hexagonal system; a variety of pyrrhotite.

trolley [GEOL] A basin-shaped depression in strata. Also known as lum.

trommel [MIN ENG] **1.** A revolving cylindrical screen used to grade coarsely crushed ore: the ore is fed into the trommel at one end, the fine material drops through the holes, and the coarse is delivered at the other end. Also known as trommel screen. **2.** To separate coal into various sizes by passing it through a revolving screen.

trommel screen *See* trommel.

trona [MINERAL] $Na_2(CO_3)\cdot Na(HCO_3)\cdot2H_2O$ A gray-white or yellowish-white mineral that crystallizes in the monoclinic system and occurs in fibrous or columnar layers or masses. Also known as urao.

Tropept [GEOL] A suborder of the order Inceptisol, characterized by moderately dark A horizons with modest additions of organic matter, B horizons with brown or reddish colors, and slightly pale C horizons; restricted to tropical regions with moderate or high rainfall.

tropical air [METEOROL] A type of air whose characteristics are developed over low latitudes.

tropical climate [CLIMATOL] A climate which is typical of equatorial and tropical regions, that is, one with continually high temperatures and with considerable precipitation, at least during part of the year.

tropical cyclone [METEOROL]   The general term for a cyclone that originates over tropical oceans; at maturity, the tropical cyclone is one of the most intense storms of the world; winds exceeding 175 knots (324 kilometers per hour) have been measured, and the rain is torrential.

tropical disturbance [METEOROL]   A cyclonic wind system of the tropics, of lesser intensity than a tropical cyclone.

tropical easterlies [METEOROL]   The trade winds when shallow and exhibiting a strong vertical shear; at about 500 feet (152 meters) the easterlies give way to the upper westerlies, which are sufficiently strong and deep to govern the course of cloudiness and weather. Also known as subtropical easterlies.

tropical front *See* intertropical front.

tropical meteorology [METEOROL]   The study of the tropical atmosphere; the dividing lines, in each hemisphere, between the tropical easterlies and the mid-latitude westerlies in the middle troposphere roughly define the poleward boundaries of this region.

tropical monsoon climate [CLIMATOL]   One of the tropical rainy climates; it is sufficiently warm and rainy to produce tropical rainforest vegetation, but it does exhibit the monsoon climate influences in that it has a winter dry season.

tropical rainforest climate [CLIMATOL]   In general, the climate which produces tropical rainforest vegetation, that is, a climate of unbroken warmth, high humidity, and heavy annual precipitation. Also known as tropical wet climate.

tropical rainy climate [CLIMATOL]   A major category in W. Köppen's climatic classification, characterized by a mean temperature of the coldest month of 64.4°F (18°C) or higher, and by a mean annual precipitation, in inches, greater than $0.44(t - a)$, where t is the mean annual temperature in degrees Fahrenheit, and a equals 32 for precipitation chiefly in winter, 19.4 for evenly distributed precipitation, and 6.8 for precipitation chiefly in summer.

tropical savanna climate [CLIMATOL]   In general, the type of climate which produces the vegetation of the tropical and subtropical savanna; thus, a climate with a winter dry season, a relatively short but heavy rainy summer season, and high year-round temperatures. Also known as savanna climate; tropical wet and dry climate.

tropical wet and dry climate *See* tropical savanna climate.

tropical wet climate *See* tropical rainforest climate.

tropic higher-high-water interval [OCEANOGR]   The lunitidal interval pertaining to the higher high waters at the time of tropic tides.

tropic high-water inequality [OCEANOGR]   The average difference between the heights of the two high waters of the tidal day at the time of tropic tides.

tropic lower-low-water interval [OCEANOGR]   The lunitidal interval pertaining to the lower low waters at the time of tropic tides.

tropic low-water inequality [OCEANOGR]   The average difference between the heights of the two low waters of the tidal day at the time of tropic tides.

tropics [CLIMATOL]   Any portion of the earth characterized by a tropical climate.

tropic tidal currents [OCEANOGR]   Tidal currents of increased diurnal inequality occurring at the time of tropic tides.

tropic tide [OCEANOGR]   A tide occurring when the moon is near maximum declination; the diurnal inequality is then at a maximum.

tropic velocity [OCEANOGR]   The speed of the greater flood or greater ebb at the time of tropic currents.

tropopause [METEOROL] The boundary between the troposphere and stratosphere, usually characterized by an abrupt change of lapse rate; the change is in the direction of increased atmospheric stability from regions below to regions above the tropopause; its height varies from 15 to 20 kilometers in the tropics to about 10 kilometers in polar regions.

tropopause chart [METEOROL] A synoptic chart showing the contour lines of the tropopause and tropopause break lines.

tropopause inversion [METEOROL] The decrease in the lapse rate of temperature encountered at the level of the tropopause. Also known as upper inversion.

troposphere [METEOROL] That portion of the atmosphere from the earth's surface to the tropopause, that is, the lowest 10 to 20 kilometers of the atmosphere.

tropospheric ducting *See* ducting.

tropospheric superrefraction [GEOPHYS] Phenomenon occurring in the troposphere whereby radio waves are bent sufficiently to be returned to the earth.

trough [GEOL] **1.** A small, straight depression formed just offshore on the bottom of a sea or lake and on the landward side of a longshore bar. **2.** Any narrow, elongate depression in the surface of the earth. **3.** An elongate depression on the sea floor that is wider and shallower than a trench. Also known as submarine trench. **4.** The line connecting the lowest points of a fold. [METEOROL] An elongated area of relatively low atmospheric pressure; the opposite of a ridge.

trough aloft *See* upper-level trough.

trough crossbedding [GEOL] A variety of crossbedding in which the lower crossbedding surfaces are smoothly curved, rather than planar.

trough fault [GEOL] One of a set of two faults bounding a graben.

trough plane *See* trough surface.

trough surface [GEOL] A surface or plane connecting the troughs of the bed of a syncline. Also known as synclinal axis; trough plane.

trough valley *See* U-shaped valley.

trough washer [MIN ENG] A sloping wooden trough, 1½ to 2 feet wide, 8 to 12 feet long, and 1 foot deep (about 50 by 300 by 30 centimeters), open at the tail end but closed at the head end; it is used to float adhering clay or fine stuff from the coarser portions of an ore or coal.

Trube's correlation [PETRO ENG] An empirical correlation (based on pseudocritical properties) for compressibilities of undersaturated oil-reservoir fluids.

Trucherognathidae [PALEON] A family of conodonts in the order Conodontophorida in which the attachment scar permits the conodont to rest on the jaw ramus.

truck *See* barney.

trudellite [MINERAL] $Al_{10}(SO_4)_3Cl_{12}(OH)_{12}\cdot30H_2O$ An amber yellow, hexagonal mineral consisting of a hydrated basic sulfate-chloride of aluminum; occurs as compact masses.

true [GEOD] Related to true north.

true air temperature [METEOROL] Basic air temperature corrected for heat of compression error due to high-speed motion of the thermometer through the air, as on an aircraft.

true altitude *See* corrected altitude.

true convergence [GEOD] The angle at which one meridian is inclined to another on the surface of the earth.

true dip *See* dip.

true formation resistivity   [GEOPHYS]   Electrical resistivity of a clean (nonshaly) porous reservoir formation containing hydrocarbons and formation water; value is greater than the resistivity when there is added water incursion.

true mean temperature   [METEOROL]   As adopted by the International Meteorological Organization, a monthly or annual mean air temperature based upon hourly observations at a given place, or on some combination of less frequent observations designed to represent this mean as nearly as possible.

true soil *See* solum.

true width   [MIN ENG]   The width of thickness of a vein or stratum as measured perpendicular to or normal to the dip and the strike; the true width is always the least width.

true wind   [METEOROL]   Wind relative to a fixed point on the earth.

true wind direction   [METEOROL]   The direction, with respect to true north, from which the wind is blowing.

truncated landform   [GEOGR]   A landform cut off, especially by erosion, and forming a steep side or cliff.

trunk roadway   [MIN ENG]   The main developing heading from the pit bottom and is usually driven along the strike of the coal seam.

trunk stream *See* main stream.

Tryblidiidae   [PALEON]   An extinct family of Paleozoic mollusks.

tschermakite   [MINERAL] $Ca_2Mg_3(Al,Fe^{3+})_2(Al_2Si_6)O_{22}(OH,F)_2$ An amphibole mineral.

T-S curve *See* temperature-salinity diagram.

T-S diagram *See* temperature-salinity diagram.

T-S relation *See* temperature-salinity diagram.

tsumebite   [MINERAL] $Pb_2Cu(PO_4)(SO_4)(OH)$ An emerald green, monoclinic mineral consisting of a hydrated basic phosphate and sulfate of lead and copper.

tsunami   [OCEANOGR]   A long-period sea wave produced by a seaquake or volcanic eruption; it may travel for thousands of miles. Also known as seismic sea wave.

Tsushima Current   [OCEANOGR]   That part of the Kuroshio Current flowing northeastward through the Korea Strait and along the Japanese coast in the Sea of Japan.

tuba   [METEOROL]   A cloud column or inverted cloud cone, pendant from a cloud base; this supplementary feature occurs mostly with cumulus and cumulonimbus; when it reaches the earth's surface it constitutes the cloudy manifestation of an intense vortex, namely, a tornado or waterspout. Also known as pendant cloud; tornado cloud.

tubbing   [MIN ENG]   The watertight cast-iron lining of a circular shaft built up of segments with the space outside the tubbing grouted to add strength and to improve watertightness.

tube   [GEOL]   A passage in a cave having smooth sides and an elliptical to nearly circular cross section.

tubing hanger *See* hanger.

tubing head   [PETRO ENG]   A spool-type unit or housing attached to the top flange on the uppermost oil-well-casing head to support the tubing string and to seal the annulus between the tubing string and the production casing string.

tubing-head adapter flange [PETRO ENG] An intermediate flange used in oil wells to connect the top tubing-head flange to the master valve (Christmas tree) and to provide support for the tubing.

tubing pump [PETRO ENG] A type of oil-well, sucker-rod pump in which the pump barrel is attached to the tubing string, and lowered into the well bore with the tubing.

tubing safety valve *See* storm choke.

tufa [GEOL] A spongy, porous limestone formed by precipitation from evaporating spring and river waters, often onto leaves and stems of neighboring plants. Also known as calcareous sinter; calcareous tufa.

tufaceous [GEOL] Pertaining to or similar to tufa.

tuff [GEOL] Consolidated volcanic ash, composed largely of fragments (less than 4 millimeters) produced directly by volcanic eruption; much of the fragmented material represents finely comminuted crystals and rocks.

tuffaceous [GEOL] Pertaining to sediments which contain up to 50% tuff.

tuff ball *See* mud ball.

tuff lava *See* welded tuff.

tugger [MIN ENG] A small portable pneumatic or electric hoist mounted on a column and used in a mine.

tumuli lava [GEOL] A type of lava flow forming ovoid mounds, a few feet high and a few tens of feet long, caused by buckling up of the crust.

tundra climate [CLIMATOL] The climate which produces tundra vegetation; it is too cold for the growth of trees but does not have a permanent snow-ice cover.

tungstenite [MINERAL] WS_2 A dark lead gray mineral consisting of tungsten disulfide; occurs in massive form, in scaly or feathery aggregates.

tungstite [MINERAL] $WO_3 \cdot H_2O$ A bright yellow, golden yellow, or yellowish-green mineral thought to consist of hydrated tungsten oxide; occurs in massive form and as platy crystals.

tunnel blasting [ENG] A method of heavy blasting in which a heading is driven into the rock and afterward filled with explosives in large quantities, similar to a borehole, on a large scale, except that the heading is usually divided in two parts on the same level at right angles to the first heading, forming in plan a T, the ends of which are filled with explosives and the intermediate parts filled with inert material like an ordinary borehole.

tunnel set [MIN ENG] Timbers 6 to 8 inches (15 to 20 centimeters) in diameter and of sufficient height to support the roof of the tunnel.

tunnel system [MIN ENG] A method of mining in which tunnels or drifts are extended at regular intervals from the floor of the pit into the ore body.

turanite [MINERAL] $Cu_5(VO_4)_2(OH)_4$ An olive green, orthorhombic mineral consisting of basic copper vanadate; occurs as reniform crusts and spherical concretions.

turbidite [GEOL] Any sediment or rock transported and deposited by a turbidity current, generally characterized by graded bedding, large amounts of matrix, and commonly exhibiting a Bouma sequence.

turbidity [METEOROL] Any condition of the atmosphere which reduces its transparency to radiation, especially to visible radiation.

turbidity current [OCEANOGR] A highly turbid, relatively dense current carrying large quantities of clay, silt, and sand in suspension which flows down a submarine slope through less dense sea water. Also known as density current; suspension current.

turbidity factor   [GEOPHYS]   A measure of the atmospheric transmission of incident solar radiation; if I_0 is the flux density of the solar beam just outside the earth's atmosphere, I the flux density measured at the earth's surface with the sun at a zenith distance which implies an optical air mass m, and $I_{m,w}$ the intensity which would be observed at the earth's surface for a pure atmosphere containing 1 centimeter of precipitable water viewed through the given optical air mass, then turbidity factor θ is given by $\theta = (\ln I_0 - \ln I)/(\ln I_0 - \ln I_{m,w})$.

turbodrill   [PETRO ENG]   A rotary tool used in drilling oil or gas wells in which the bit is rotated by a turbine motor inside the well.

turbonada   [METEOROL]   A short thundersquall on the north Spanish coast, sometimes accompanied by waterspouts.

turbosphere   [METEOROL]   The region of the atmosphere in which turbulence frequently exists.

turbulent heat conduction   [OCEANOGR]   Conduction of heat in water by lateral and vertical eddy diffusion, with currents.

Turkey stone   *See* turquoise.

turn of the tide   *See* change of tide.

Turonian   [GEOL]   A European stage of geologic time: Upper or Middle Cretaceous (above Cenomanian, below Coniacian).

turquoise   [MINERAL] $CuAl_6(PO_4)_4(OH)_8 \cdot 4H_2O$ A semitranslucent sky-blue, bluish-green, apple-green, or greenish-gray mineral that crystallizes in the triclinic system and occurs in veinlets or as crusts of massive, concretionary, and stalactite shapes; an important gem mineral. Also known as calaite; Turkey stone.

turret coal cutter   [MIN ENG]   A coal cutter in which the horizontal jib can be adjusted vertically to cut at different levels in the seam; for example, an overcut.

turret ice   *See* ropak.

turtle stone   *See* septarium.

twilight arch   *See* bright segment.

twin   *See* twin crystal.

twin crystal   [CRYSTAL]   A compound crystal which has one or more parts whose lattice structure is the mirror image of that in the other parts of the crystal. Also known as twin.

twin entry   [MIN ENG]   A pair of parallel entries, one of which is an intake air course and the other the return air course; rooms can be worked from both entries.

twin law   [CRYSTAL]   A statement relating two or more individuals of a twin to one another in terms of their crystallography (twin plane, twin axis, and so on).

twinning   [CRYSTAL]   The development of a twin crystal by growth, translation, or gliding.

twinning plane   *See* twin plane.

twin plane   [CRYSTAL]   The plane common to and across which the individual crystals or components of a crystal twin are symmetrically arranged or reflected. Also known as twinning plane.

twister   [METEOROL]   In the United States, a colloquial term for tornado.

two-layer ocean [OCEANOGR]   An idealized ocean in which a layer of uniform density near the surface overlays a deep layer of uniform but distinctly higher-density water.

two-piece set [MIN ENG]   A set of timbers consisting of a cap and a single post.

two-stage hoisting [MIN ENG]   Deep shaft hoisting with two winders, one at the surface, and the other at mid-depth in the shaft.

tychite [MINERAL] $Na_6Mg_2(SO_4)(CO_3)_4$ A white, isometric mineral consisting of a sulfate-carbonate of sodium and magnesium.

Tyndall flowers [HYDROL]   Small water-filled cavities, often of basically hexagonal shape, which appear in the interior of ice masses upon which light is falling.

type-α leader [GEOPHYS]   A stepped leader of lightning which exhibits very little branching and whose individual steps are short and so weakly luminous as to be difficult to discern.

type-β leader [GEOPHYS]   A stepped leader of lightning in which the upper portion of the channel is characterized by longer and brighter steps than those found in the lower portion of the channel, a consequence of excessive branching in the upper parts under the influence of strong fields set up by heavy space charges near and around the upper end of the channel.

type locality [GEOL]   **1.** The place at which a stratigraphic unit is typically displayed and from which it derives its name. **2.** The place where a geologic feature was first recognized and described.

type section [GEOL]   That sequence of strata identified as the original sequence for a location or area; the standard against which other stratigraphy of parts of the area are compared. Also known as section.

typhoon [METEOROL]   A severe tropical cyclone in the western Pacific.

typhoon wind *See* hurricane wind.

Typotheria [PALEON]   A suborder of extinct rodentlike herbivores in the order Notoungulata.

tyuyamunite [MINERAL] $Ca(UO_2)_2(VO_4)_2 \cdot 5-8H_2O$ A yellow orthorhombic mineral occurring in incrustations as a secondary mineral; an ore of uranium. Also known as calciocarnotite.

ubac [METEOROL] The shady (usually north) side of an Alpine mountain, characterized by a lower timberline and snow line than the sunny side.

Udalf [GEOL] A suborder of the soil order Alfisol; brown soil formed in a udic moisture regime and in a mesic or warmer temperature regime.

Udert [GEOL] A suborder of the soil order Vertisol; formed in a humid region so that surface cracks remain open only for 2–3 months.

Udoll [GEOL] A suborder of the Mollisol soil order; found in humid, temperate, and warm regions where maximum rainfall comes during growing season; has thick, very dark A horizons, brown B horizons, and paler C horizons.

Udult [GEOL] A suborder of the soil order Ultisol; organic-carbon content is low, argillic horizons are reddish or yellowish; formed in a udic moisture regime.

U figure *See* U index.

uhligite [MINERAL] A black, pseudoisometric mineral consisting of an oxide of titanium and calcium, with zirconium and aluminum replacing titanium.

U index [GEOPHYS] The difference between consecutive daily mean values of the horizontal component of the geomagnetic field. Also known as U figure.

Uintatheriidae [PALEON] The single family of the extinct mammalian order Dinocerata.

Uintatheriinae [PALEON] A subfamily of extinct herbivores in the family Uintatheriidae including all horned forms.

Ulatisian [GEOL] A mammalian age in a local stage classification of the Eocene in use on the Pacific Coast based on foraminifers.

ulexite [MINERAL] $NaCaB_5O_9 \cdot 8H_2O$ A white mineral that crystallizes in the triclinic system and forms rounded reniform masses of extremely fine acicular crystals. Also known as boronatrocalcite; cotton ball; natronborocalcite.

ullmannite [MINERAL] $NiSbS$ A steel-gray to black mineral consisting of nickel antimonide and sulfide, usually with a little arsenic, occurring massive, and having a metallic luster. Also known as nickel-antimony glance.

Ulloa's ring *See* Bouguer's halo.

ulmic acid *See* ulmin.

ulmin [GEOL] Alkali-soluble organic substances derived from decaying vegetable matter; occurs as amorphous brown to black gel material. Also known as carbohumin; fundamental jelly; fundamental substance; gelose; humin; humogelite; jelly; ulmic acid; vegetable jelly.

ulrichite *See* uraninite.

ultimate recovery [PETRO ENG]   Estimated total (ultimate) recovery of hydrocarbon fluids expected from a reservoir during its productive lifetime.

Ultisol [GEOL]   A soil order characterized by typically moist soils, with horizons of clay accumulation and a low supply of bases.

ultrabasic [PETR]   Of igneous rock, having a low silica content, as opposed to the higher silica contents of acidic, basic, and intermediate rocks.

ultrabasite *See* diaphorite.

ultramafic [PETR]   Referring to igneous rock composed principally of mafic minerals, such as olivine and pyroxene.

ultravulcanian [GEOL]   A type of volcanic eruption characterized by periodic violent gaseous explosions of lithic dust and solid blocks, with little if any fiery scoria.

umangite [MINERAL] Cu_3Se_2 A dark cherry red mineral consisting of copper selenide; occurs in massive form, in small grains or fine granular aggregates.

Umbrept [GEOL]   A suborder of the Inceptisol soil order; has dark A horizon more than 10 inches (25 centimeters) thick, brown B horizons, and slightly paler C horizons; soil is strongly acid, and clay minerals are crystalline; occurs in cool or temperate climates.

umpire [MIN ENG]   An assay made by a third party to settle the difference in assays made by the purchaser and the seller of ore.

unaka [GEOL]   A large residual mass rising above a peneplain that is less well developed than one having a monodnock.

unakite [PETR]   An altered igneous rock composed principally of epidote, pink orthoclase, and quartz.

unbalanced cutter chain [MIN ENG]   A cutter chain which carries more picks along the bottom line than along the top line.

unbalanced hoisting [MIN ENG]   The method of hoisting in small one-compartment shafts with only one cage in operation, as opposed to balanced winding.

unbalanced shothole [MIN ENG]   A shothole in which the explosive charge breaks down the coal at the back of the machine cut while leaving the front portion standing or in large blocks.

unconcentrated wash *See* sheet erosion.

unconformable [GEOL]   Pertaining to strata that do not conform in position, dip, or strike to the older underlying rocks.

unconformity [GEOL]   The relation between adjacent rock strata whose time of deposition was separated by a period of nondeposition or of erosion; a break in a stratigraphic sequence.

unconformity iceberg [OCEANOGR]   An iceberg consisting of more than one kind of ice, such as blue water-formed ice and névé; such an iceberg often contains many crevasses and silt bands.

unconsolidated material [GEOL]   Loosely arranged or unstratified sediment whose particles are not cemented together.

undercast [METEOROL]   A cloud layer of ten-tenths (1.0) coverage as viewed from an observation point above the layer; the term is used in pilot reporting of in-flight weather conditions.   [MIN ENG]   An air crossing in which one airway is deflected to pass under the other airway.

underchain haulage [MIN ENG] Haulage in which the chains are placed beneath the mine car at certain intervals with suitable hooks that thrust against the car axle.

underclay [GEOL] A layer of clay or other fine-grained detrital material underlying a coal bed or comprising the floor of a coal seam. Also known as coal clay; root clay; seat clay; seat earth; thill; underearth; warrant.

underclay limestone [GEOL] A thin, fresh-water limestone that is relatively free of fossils and is dense and nodular; found in underlying coal deposits.

undercliff [GEOL] A subordinate cliff or terrace formed by material which has fallen or slid from above.

underconsolidation [GEOL] Less than normal consolidation of sedimentary material for the existing overburden.

undercurrent [OCEANOGR] A water current flowing beneath a surface current at a different speed or in a different direction.

undercut [MIN ENG] To cut below or in the lower part of a coal bed by chipping away the coal with a pick or mining machine; cutting is usually done on the level of the floor of the mine, extending laterally the entire face and 5 or 6 feet (1.5 or 1.8 meters) into the material.

undercutting [GEOL] Erosion of material at the base of a steep slope, cliff, or other exposed rock.

underearth *See* underclay.

underfit stream [HYD] A misfit stream that appears to be too small to have eroded the valley in which it flows.

underflow conduit [GEOL] A permeable deposit underlying a surface stream channel.

underground geology *See* subsurface geology.

underground glory-hole method [MIN ENG] A mining method used in large deposits with a very strong roof: the deposit is divided by levels and on every level chutes are raised to the next level; mining starts from the mouth of the chutes in such a way as to develop a funnel-shaped excavation (mill, or glory) with slopes so steep that the broken ore falls into the chutes and thus to the cars on the lower level; a sufficiently strong pillar is left for protection at the higher level. Also known as underground milling.

underground ice *See* ground ice.

underground milling *See* underground glory-hole method.

underground stream [HYD] A subsurface body of water flowing in a definite current in a distinct channel.

underhand stoping [MIN ENG] Mining downward or from upper to lower level; the stope may start below the floor of a level and be extended by successive horizontal slices, either worked sequentially or simultaneously in a series of steps; the stope may be left as an open stope or supported by stulls or pillars.

underhand work [MIN ENG] Picking or drilling downward.

underhole [MIN ENG] To mine out a portion of the bottom of a seam, by pick or powder, thus leaving the top unsupported and ready to be blown down by shots, broken down by wedges, or mined with a pick or bar.

underlay shaft [MIN ENG] A shaft sunk in the footwall and following the dip of a vein. Also known as footwall shaft; underlier.

underlie [GEOL] To lie or be situated under; to occupy a lower position, or to pass beneath.

underlier *See* underlay shaft.

undermelting [HYD] The melting from below of any floating ice.

undermine [MIN ENG] To excavate the earth beneath, especially for the purpose of causing to fall; to form a mine under.

undermining *See* sapping.

undersaturated [PETR] Pertaining to igneous rock composed of unsaturated minerals, that is, without free silica.

undersea mining [MIN ENG] The working of economic deposits (usually coal) situated in strata or rocks below the seabed.

underthrust [GEOL] A thrust fault in which the lower, active rock mass has been moved under the upper, passive rock mass.

undertow [OCEANOGR] A subsurface seaward movement by gravity flow of water carried up on a sloping beach by waves or breakers.

underway bottom sampler *See* underway sampler.

underway sampler [ENG] A device for collecting samples of sediment on the ocean bottom, consisting of a cup in a hollow tube; on striking the bottom, the cup scoops up a small sample which is forced into the tube which is then closed with a lid, and the device is hoisted to the surface. Also known as scoopfish; underway bottom sampler.

undisturbed [ENG] Pertaining to a sample of material, as of soil, subjected to so little disturbance that it is suitable for determinations of strength, consolidation, permeability characteristics, and other properties of the material in place.

undulatus *See* billow cloud.

unfreezing [GEOL] The upward movement of stones to the surface as a result of repeated freezing and thawing of the containing soil.

ungemachite [MINERAL] $K_3Na_9Fe(SO_4)_6(OH)_3 \cdot 9H_2O$ A colorless to pale yellow, hexagonal mineral consisting of a hydrated basic sulfate of potassium, sodium, and iron; occurs in tabular form.

uniformitarianism [GEOL] Classically, the concept that the present is the key to the past; the principle that contemporary geologic processes have occurred in the same regular manner and with essentially the same intensity throughout geologic time, and that events of the geologic past can be explained by phenomena observable today. Also known as principle of uniformity.

unit cell [CRYSTAL] A parallelepiped which will fill all space under the action of translations which leave the crystal lattice unchanged. Also known as structure cell. [MIN ENG] In flotation, a single cell.

United States airways code [METEOROL] A synoptic code for communicating aviation weather observations. Also known as airways code.

unit of coal [MIN ENG] The quantity of coal from which the sample is taken and which the sample represents.

unit train [MIN ENG] A system for delivering coal in which a string of cars, with distinctive markings and loaded to full visible capacity, is operated without service frills or stops along the way for cars to be cut in and out.

universal instrument *See* altazimuth.

universal transmission function   [GEOPHYS]   A mathematical relationship that attempts to describe quantitatively the complex infrared propagation (including absorption and reradiation) in the atmosphere.

universal transverse Mercator grid   [MAP]   A particular grid based upon a transverse Mercator projection, according to specifications laid down by military authorities; it may be superimposed on any map.

unlimited ceiling   [METEOROL]   A ceiling that exists when the total sky cover is less than 0.6%, or when the total transparent sky cover is 0.5% or more, or when surface-based obscuring phenomena are classed as partial obscuration (that is, they obscure 0.9% or less of the sky) and no layer aloft is reported as broken or overcast.

unproductive development   [MIN ENG]   The drifts, tunnels, and crosscuts driven in stone, preparatory to opening out production faces in a coal seam or ore body.

unproven area   [MIN ENG]   An area in which it has not been established by drilling operations whether oil or gas may be found in commercial quantities.

unreserved minerals   [MIN ENG]   Minerals which belong to the owner of the land on which or in which they are located.

unrestricted visibility   [METEOROL]   The visibility when no obstruction to vision exists in sufficient quantity to reduce the visibility to less than 7 miles (11.3 kilometers).

unsaturated   [MINERAL]   Referring to a mineral that will not form in the presence of free silica.

unsaturated zone   *See* zone of aeration.

unsettled   [METEOROL]   Pertaining to fair weather which may at any time become rainy, cloudy, or stormy.

unsurveyed area   [MAP]   An area on a map or chart where both relief and planimetric data are unavailable, and which is usually labeled unsurveyed; or an area on a map or chart which shows little or no charted data because accurate information is limited or not available.

upcast   [MIN ENG]   **1.** The opening through which the return air ascends and is removed from the mine; the opposite of downcast or intake.   **2.** An upward current of air passing through a shaft.   **3.** Material that has been thrown up, as by digging.

updrift   [OCEANOGR]   The opposite direction from that of the primary movement of littoral material.

upgrade   [MIN ENG]   **1.** To increase the commercial value of a coal or mineral product by appropriate treatment.   **2.** To increase the quality rating of diamonds beyond or above the rating implied by their particular classification.

up-hole   [MIN ENG]   A borehole collared in an underground working place and drilled in a direction pointed above the horizontal plane of the drill-machine swivel head.

uphole time   [GEOPHYS]   The time that a seismic pulse requires from an explosion at some depth in a shot hole to the surface of the earth.

upland   [GEOGR]   **1.** An extensive region of high land.   **2.** The higher ground of a region, in contrast to a valley, plain, or other low-lying land.   **3.** The elevated land above the low areas along a stream or between hills.

upper   [GEOL]   Pertaining to rocks or strata that normally overlie those of earlier formations of the same subdivision of rocks.

upper air   [METEOROL]   The region of the atmosphere which is above the lower troposphere; although no distinct lower limit is set, the term is generally applied to levels above that at which the pressure is 850 millibars.

upper-air chart *See* upper-level chart.

upper-air disturbance [METEOROL]   A disturbance of the flow pattern in the upper air, particularly one which is more strongly developed aloft than near the ground.   Also known as upper-level disturbance.

upper-air observation [METEOROL]   A measurement of atmospheric conditions aloft, above the effective range of a surface weather observation.   Also known as sounding; upper-air sounding.

upper-air sounding *See* upper-air observation.

upper anticyclone *See* upper-level anticyclone.

upper atmosphere [METEOROL]   The general term applied to the atmosphere above the troposphere.

upper atmosphere dynamics [METEOROL]   Motion of the atmosphere above 500 kilometers; predominant dynamical phenomena are internal gravity waves, tides, sound waves, turbulence, and large-scale circulation.

upper band *See* upper bright band.

upper branch [GEOD]   That half of a meridian or celestial meridian from pole to pole which passes through a place or its zenith.

upper bright band [METEOROL]   A level of enhanced radar echo occasionally observed at a higher altitude than the bright band of the melting level; it is attributable to the growth of a layer of ice crystals in a supercooled cloud into snow pellets.   Also known as radar upper band; upper band.

Upper Cambrian [GEOL]   The latest epoch of the Cambrian period of geologic time, beginning approximately 510 million years ago.

Upper Carboniferous [GEOL]   The European epoch of geologic time equivalent to the Pennsylvanian of North America.

Upper Cretaceous [GEOL]   The late epoch of the Cretaceous period of geologic time, beginning about 90 million years ago.

upper cyclone *See* upper-level cyclone.

Upper Devonian [GEOL]   The latest epoch of the Devonian period of geologic time, beginning about 365 million years ago.

upper front [METEOROL]   A front which is present in the upper air but does not extend to the ground.

upper high *See* upper-level anticyclone.

Upper Huronian *See* Animikean.

upper inversion *See* tropopause inversion.

Upper Jurassic [GEOL]   The latest epoch of the Jurassic period of geologic time, beginning approximately 155 million years ago.

upper-level anticyclone [METEOROL]   An anticyclonic circulation existing in the upper air; this often refers to such anticyclones only when they are much more pronounced at upper levels than at and near the earth's surface.   Also known as high aloft; high-level anticyclone; upper anticyclone; upper high; upper-level high.

upper-level chart [METEOROL]   A synoptic chart of meteorological conditions in the upper air, almost invariably referring to a standard constant-pressure chart.   Also known as upper-air chart.

upper-level cyclone [METEOROL]   A cyclonic circulation existing in the upper air, and specifically, as seen on an upper-level constant-pressure chart; often restricted to describe cyclones associated with relatively little cyclonic circulation in the lower atmosphere. Also known as high-level cyclone; low aloft; upper cyclone; upper-level low; upper low.

upper-level disturbance *See* upper-air disturbance.

upper-level high *See* upper-level anticyclone.

upper-level low *See* upper-level cyclone.

upper-level ridge [METEOROL] A pressure ridge existing in the upper air, especially one that is stronger aloft than near the earth's surface. Also known as high-level ridge; ridge aloft; upper ridge.

upper-level trough [METEOROL] A pressure trough existing in the upper air, but sometimes restricted to the troughs that are much more pronounced aloft than near the earth's surface. Also known as high-level trough; trough aloft; upper trough.

upper-level winds *See* winds aloft.

upper mantle [GEOL] The portion of the mantle lying above a depth of about 1000 kilometers. Also known as outer mantle; peridotite shell.

Upper Mississippian [GEOL] The latest epoch of the Mississippian period of geologic time.

upper mixing layer [METEOROL] The region of the upper mesophere between about 50 and 80 kilometers (that is, immediately above the mesopeak) through which there is a rapid decrease of temperature with height and where there appears to be considerable turbulence.

Upper Ordovician [GEOL] The latest epoch of the Ordovician period of geologic time, beginning approximately 440 million years ago.

Upper Pennsylvanian [GEOL] The latest epoch of the Pennsylvanian period of geologic time.

Upper Permian [GEOL] The latest epoch of the Permian period of geologic time, beginning about 245 million years ago.

upper ridge *See* upper-level ridge.

Upper Silurian [GEOL] The latest epoch of the Silurian period of geologic time.

Upper Triassic [GEOL] The latest epoch of the Triassic period of geologic time, beginning about 200 million years ago.

upper trough *See* upper-level trough.

upper winds *See* winds aloft.

uprush [METEOROL] The strong upward-flow air current in cumulus clouds during their stage of rapid development, often preceding a thunderstorm. Also known as vertical jet. [OCEANOGR] *See* swash.

upset [MIN ENG] **1.** A narrow heading connecting two levels in inclined coal. **2.** A capsized or broken skip.

upsetted moraine *See* push moraine.

upslope fog [METEOROL] A type of fog formed when air flows upward over rising terrain and, consequently, is adiabatically cooled to or below its dew point.

upstream [HYD] Toward the source of a stream.

upthrow [GEOL] **1.** The fault side that has been thrown upward. **2.** The amount of vertical fault displacement.

upwarp [GEOL]   A broad anticline with gently sloping limbs formed as a result of differential uplift.

upwelling [OCEANOGR]   The process by which water rises from a deeper to a shallower depth, usually as a result of divergence of offshore currents.

upwind [METEOROL]   In the direction from which the wind is flowing.

upwind effect [METEOROL]   The effect of an orographic barrier in producing orographic precipitation windward of the base of the barrier, because the airflow is forced upward before the barrier slope is actually reached.

Uralean [GEOL]   A stage of geologic time in Russia: uppermost Carboniferous (above Gzhelian, below Sakmarian of Permian).

uralite [MINERAL]   A green variety of secondary amphibole; it is usually fibrous or acicular and is formed by alteration of pyroxene.

uralitization [GEOL]   **1.** A process of replacement whereby pyroxene undergoes alteration resulting in uralite. **2.** Development of amphibole from pyroxene.

uraninite [MINERAL] UO_2 A black, brownish-black, or dark-brown radioactive mineral that is isometric in crystallization; often contains impurities such as thorium, radium, cerium, and yttrium metals, and lead; the chief ore of uranium; hardness is 5.5–6 on Mohs scale, and specific gravity of pure UO_2 is 10.9, but that of most natural material is 9.7–7.5. Also known as coracite; ulrichite.

uranium age [GEOL]   The age of a mineral as calculated from the numbers of ionium atoms present originally, now, and when equilibrium is established with uranium.

uranium-lead dating [GEOL]   A method for calculating the geologic age of a material in years based on the radioactive decay rate of uranium-238 to lead-206 and of uranium-235 to lead-207.

uranium ocher *See* gummite.

uran-mica *See* torbernite.

uranocircite [MINERAL] $Ba(UO_2)_2(PO_4)_2 \cdot 8H_2O$ A yellow-green, tetragonal mineral consisting of a hydrated phosphate of barium and uranium; occurs as crystals.

uranophane [MINERAL] $Ca(UO_2)_2Si_2O_7 \cdot 6H_2O$ A yellow or orange-yellow radioactive secondary mineral; it is dimorphous with β-uranophane. Also known as uranotile.

uranopilite [MINERAL] $(UO_2)_6(SO_4)(OH)_{10} \cdot 12H_2O$ A bright yellow, lemon yellow, or golden yellow, monoclinic mineral consisting of a hydrated basic sulfate of uranium; occurs as encrustations and masses.

uranosphaerite [MINERAL] $Bi_2O_3 \cdot 2UO_3 \cdot 3H_2O$ An orange-yellow or brick red, orthorhombic mineral consisting of a hydrated oxide of bismuth and uranium.

uranospinite [MINERAL] $Ca(UO_2)_2(AsO_4)_2 \cdot 8H_2O$ A lemon yellow to siskin green, tetragonal mineral consisting of a hydrated arsenate of calcium and uranium; occurs in tabular form.

uranotantalite *See* samarskite.

uranothorite [MINERAL]   A uranium-bearing variety of thorite.

uranotile *See* uranophane.

urao *See* trona.

urban heat island [METEOROL]   Increased urban temperatures of 1–2°C higher for daily maxima and 1–9°C for daily minima compared to rural environs resulting from changes in moisture balance due to impermeable surfaces, decreased humidity, or alteration in heat balance.

ureilite [GEOL]   An achondritic stony meteorite consisting principally of olivine and clinobronzite, with some nickel-iron, troilite, diamond, and graphite.

ureyite [MINERAL] $NaCrSi_2O_6$ A meteoritic mineral of the pyroxene group.

urstromthal [GEOL]   A large channel cut by a stream of water from melting ice, flowing along the edge of an ice sheet.

U-shaped valley [GEOL]   A type of valley with a broad floor and steep walls produced by glacial erosion. Also known as trough valley; U valley.

U.S. Survey foot [GEOD]   The foot used by the U.S. Coast and Geodetic Survey in which 1 inch is equal to 2.540005 centimeters.

Ustalf [GEOL]   A suborder of the soil order Alfisol; red or brown soil formed in a ustic moisture regime and in a mesic or warmer temperature regime.

Ustert [GEOL]   A suborder of the Vertisol soil order; has a faint horizon and is dry for an appreciable period or more than one period of the year.

Ustoll [GEOL]   A suborder of the soil order Mollisol; formed in a ustic moisture regime and in a mesic or warmer temperature regime; may have a calcic, petrocalcic, or gypsic horizon.

Ustox [GEOL]   A suborder of the soil order Oxisol that is low to moderate in organic matter, well drained, and dry for at least 90 days per year.

Ustult [GEOL]   A suborder of the soil order Ultisol; brownish or reddish, with low to moderate organic-carbon content; a well-drained soil of warm-temperate and tropical climates with moderate or low rainfall.

utahite *See* jarosite.

uvala [GEOGR]   Broad-bottomed lowlands.

U valley *See* U-shaped valley.

uvanite [MINERAL] $U_2V_6O_{21}\cdot15H_2O$ A brownish-yellow, orthorhombic mineral consisting of a hydrated uranium vanadate; occurs as crystalline masses and coatings.

uvarovite [MINERAL] $Ca_3Cr_2(SiO_4)_3$ The emerald-green, calcium-chromium end member of the garnet group. Also known as ouvarovite; uwarowite.

uwarowite *See* uvarovite.

V

vacuole *See* vesicle.

vacuum [PHYS] **1.** Theoretically, a space in which there is no matter. **2.** Practically, a space in which the pressure is far below normal atmospheric pressure so that the remaining gases do not affect processes being carried on in the space.

vadose water [HYD] Water in the zone of aeration. Also known as kremastic water; suspended water; wandering water.

vadose zone *See* zone of aeration.

vaesite [MINERAL] NiS_2 An isometric mineral with pyrite structure composed of sulfide of nickel.

valais wind [METEOROL] The notable valley wind that blows along the Rhone Valley from the upper end of Lake Geneva (Valais Canton); it is sufficiently strong and regular to distort the growth of trees.

valence crystal *See* covalent crystal.

valencianite [MINERAL] A variety of potassium feldspar from Mexico.

valentinite [MINERAL] Sb_2O_3 A colorless to snow white mineral consisting of antimony trioxide.

vallerite [MINERAL] $CuFeS_2$ A sulfide mineral found in meteorites.

valley [GEOGR] A generally broad area of flat, low-lying land bordered by higher ground. [GEOL] A relatively shallow, wide depression of the sea floor with gentle slopes. Also known as submarine valley.

valley bottom *See* valley floor.

valley breeze [METEOROL] A gentle wind blowing up a valley or mountain slope in the absence of cyclonic or anticyclonic winds, caused by the warming of the mountainside and valley floor by the sun.

valley fill [GEOL] Unconsolidated sedimentary deposit which fills or partly fills a valley.

valley flat [GEOL] The small plain at the bottom of a narrow valley with steep sides.

valley floor [GEOL] The broad, flat bottom of a valley. Also known as valley bottom; valley plain.

valley glacier [HYD] A glacier that flows down the walls of a mountain valley.

valley iceberg [OCEANOGR] An iceberg weathered in such a manner that a large U-shaped slot extends through the iceberg. Also known as drydock iceberg.

valley line *See* thalweg.

valley plain *See* valley floor.

valley train [GEOL]   A long, narrow body of outwash, deposited by meltwater far beyond the margin of an active glacier and extending along the floor of a valley. Also known as outwash train.

valley wind [METEOROL]   A wind which ascends a mountain valley (up-valley wind) during the day; the daytime component of a mountain and valley wind system.

van [MIN ENG]   **1.** A test of the value of an ore, made by washing (vanning) a small quantity, after powdering it, on the point of a shovel. **2.** To separate, as ore from veinstone, by washing it on the point of a shovel. **3.** A shovel used in ore dressing.

vanadate [MINERAL]   Any of several mineral compounds characterized by pentavalent vanadium and oxygen in the anion; an example is vanadinite.

vanadinite [MINERAL] $Pb_5(VO_4)_3Cl$ A red, yellow, or brown opatite mineral often occurring as globular masses encrusting other minerals in lead mines; an ore of vanadium and lead hardness is 2.75–3 on Mohs scale, and specific gravity is 6.66–7.10.

Van Allen radiation belt [GEOPHYS]   One of the belts of intense ionizing radiation in space about the earth formed by high-energy charged particles which are trapped by the geomagnetic field.

vandenbrandite [MINERAL] $CuO \cdot UO_3 \cdot 2H_2O$ A dark green to black mineral consisting of a hydrated oxide of copper and uranium; occurs in small crystals and massive form.

vane anemometer [ENG]   A portable instrument used to measure low wind speeds and airspeeds in large ducts; consists of a number of vanes radiating from a common shaft and set to rotate when facing the wind.

vanishing tide [OCEANOGR]   When a high water and low water "melt" together into a period of several hours with a nearly constant water level.

vanner [MIN ENG]   A machine for dressing ore; the name is given to various patented devices in which the peculiar motions of the shovel in the miner's hands in the operation of making a van are, or are supposed to be, successfully imitated. Also known as vanning machine.

vanning machine *See* vanner.

vanoxite [MINERAL] $V_4{}^{+}V_2{}^{5}O_{13} \cdot 8H_2O$ A black mineral consisting of a hydrous oxide of vanadium; occurs as microscopic crystals and in massive form.

vanthoffite [MINERAL] $Na_6Mg(SO_4)_4$ A colorless mineral consisting of a sulfate of sodium and magnesium; occurs in massive form.

vapor-dominated hydrothermal reservoir [GEOL]   Any geothermal system mainly producing dry steam; the Geysers area of northern California and the Larderelle region of Italy are two examples.

vapor pressure [METEOROL]   The partial pressure of water vapor in the atmosphere.

vapor-pressure deficit *See* saturation deficit.

vapor trail *See* condensation trail.

vapor volume equivalent [PETRO ENG]   The volume of vapor to which a specified amount of liquid would be equivalent at designated standard conditions (for example, 14.65 psia and 60°F, or 15.5°C); used in the petroleum industry to calculate the specific gravity of fluids from gas-condensate wells.

vardar [METEOROL]   A cold fall wind blowing from the northwest down the Vardar valley in Greece to the Gulf of Salonica; it occurs when atmospheric pressure over eastern Europe is higher than over the Aegean Sea, as is often the case in winter. Also known as vardarac.

vardarac *See* vardar.

variable ceiling   [METEOROL]   After United States weather-observing practice, a condition in which the ceiling rapidly increases and decreases while the ceiling observation is being made; the average of the observed values is used as the reported ceiling, and it is reported only for ceilings of less than 3000 feet (914 meters).

variable-depth sonar   [ENG]   Sonar in which the projector and receiving transducer are mounted in a watertight pod that can be lowered below a vessel to an optimum depth for minimizing thermal effects when detecting underwater targets.

variable radio-frequency radiosonde   [ENG]   A radiosonde whose carrier frequency is modulated by the magnitude of the meteorological variables being sensed.

variable visibility   [METEOROL]   After United States weather observing practice, a condition in which the prevailing visibility fluctuates rapidly while the observation is being made; the average of the observed values is used as the reported visibility, and it is reported only for visibilities of less than 3 miles (4.8 kilometers).

variation *See* declination.

variation of latitude   [GEOPHYS]   Change of the latitude of a place on earth because of the irregular movement of the north and south poles; the movement is caused by the earth's shifting on its axis.

variation per day   [GEOPHYS]   The change in the value of any geophysical quantity during 1 day.

variation per hour   [GEOPHYS]   The change in the value of any geophysical quantity during 1 hour.

variation per minute   [GEOPHYS]   The change in the value of any geophysical quantity during 1 minute.

variole   [GEOL]   A spherule the size of a pea, usually consisting of radiating plagioclase or pyroxene crystals.

variolitic   [PETR]   Referring to the texture of basic igneous rock composed of varioles in a finer-grained matrix.

variometer   [ENG]   A geomagnetic device for detecting and indicating changes in one of the components of the terrestrial magnetic field vector, usually magnetic declination, the horizontal intensity component, or the vertical intensity component.

Variscan orogeny   [GEOL]   The late Paleozoic orogenic era in Europe, extending through the Carboniferous and Permian. Also known as Hercynian orogeny.

varulite   [MINERAL] $(Na,Ca)(Mn,Fe)_2(PO_4)_2$ An olive green, orthorhombic mineral consisting of a phosphate of sodium, calcium, manganese, and iron; occurs in massive form.

varve   [GEOL]   A sedimentary bed, layer, or sequence of layers deposited in a body of still water within a year's time, and usually during a season. Also known as glacial varve.

varve clay *See* varved clay.

varved clay   [GEOL]   A lacustrine sediment of distinct layers consisting of varves. Also known as varve clay.

vashegyite   [MINERAL] $2Al_4(PO_4)_3(OH)_3 \cdot 27H_2O$ A white or pale green to yellow and brownish mineral consisting of a hydrous basic aluminum phosphate; occurs in massive and microcrystalline forms.

vaterite [MINERAL] $CaCO_3$ A rare hexagonal mineral consisting of unstable calcium carbonate; it is trimorphous with calcite and aragonite.

vaudaire [METEOROL] A violent south wind; a foehn of Lake Geneva in Switzerland. Also known as vauderon.

vauderon *See* vaudaire.

vauquelinite [MINERAL] $Pb_2Cu(CrO_4)PO_4(OH)$ A monoclinic mineral of varying color, consisting of a basic chromate-phosphate of lead and copper.

vauxite [MINERAL] $FeAl_2(PO_4)_2(OH)_2 \cdot 7H_2O$ A sky blue to Venetian blue, triclinic mineral consisting of a hydrated basic phosphate of iron and aluminum.

V cut [ENG] In mining and tunneling, a cut where the material blasted out in plan is like the letter V; usually consists of six or eight holes drilled into the face, half of which form an acute angle with the other half.

veatchite [MINERAL] $Sr_2B_{11}O_{16}(OH)_5 \cdot H_2O$ A white mineral consisting of hydrous strontium borate.

vectopluviometer [ENG] A rain gage or array of rain gages designed to measure the inclination and direction of falling rain; vectopluviometers may be constructed in the fashion of a wind vane so that the receiver always faces the wind, or they may consist of four or more receivers arranged to point in cardinal directions.

vectorial structure *See* directional structure.

veering [METEOROL] **1.** In international usage, a change in wind direction in a clockwise sense (for example, south to southwest to west) in either hemisphere of the earth. **2.** According to widespread usage among United States meteorologists, a change in wind direction in a clockwise sense in the Northern Hemisphere, counterclockwise in the Southern Hemisphere.

vegetable jelly *See* ulmin.

veil [METEOROL] A very thin cloud through which objects are visible.

vein [GEOL] A mineral deposit in tabular or shell-like form filling a fracture in a host rock.

veined gneiss [PETR] A composite gneiss with irregular layering.

veinite [GEOL] A genetic type of veined gneiss in which the vein material was secreted from the rock itself.

vein quartz [PETR] A rock composed chiefly of sutured quartz crystals of pegmatitic or hydrothermal origin of variable size.

velocity discontinuity *See* seismic discontinuity.

velocity gradient *See* seismic gradient.

velocity ratio [OCEANOGR] The ratio of the speed of tidal current at a subordinate station to the speed of the corresponding current at the reference station.

velum [METEOROL] An accessory cloud veil of great horizontal extent draped over or penetrated by cumuliform clouds; velum occurs with cumulus and cumulonimbus.

vendaval [METEOROL] A stormy southwest wind on the southern Mediterranean coast of Spain and in the Straits of Gibraltar; it occurs with a low advancing from the west in late autumn, winter, or early spring, and is often accompanied by thunderstorms and violent squalls.

Vening Meinesz zone [GEOD] A belt exhibiting negative anomalies, generally in relation to island arcs or oceanic deeps. Also known as negative strip.

venite [PETR]   Migmatite having mobile portions which were formed by exudation from the rock itself.

vent [ENG]   **1.** A small passage made with a needle through stemming, for admitting a squib to enable the charge to be lighted.   **2.** A hole, extending up through the bearing at the top of the core-barrel inner tube, which allows the water and air in the upper part of the inner tube to escape into the borehole.   **3.** A small hole in the upper end of a core-barrel inner tube that allows water and air in the inner tube to escape into the annular space between the inner and outer barrels.   [GEOL]   The opening of a volcano on the surface of the earth.

vent da Mùt [METEOROL]   A strong, wet wind of Lake Garda in Italy.

vent des dames [METEOROL]   A daily sea breeze of about 15 miles (24 kilometers) per hour from the southwest in summer on the Mediterranean coast east of the Rhone delta, extending some 20 miles (32 kilometers) inland.

vent du midi [METEOROL]   A south wind in the center of the Massif Central and the southern Cevennes (France); it is warm, moist, and generally followed by a southwest wind with heavy rain.

ventifact [GEOL]   A stone or pebble whose shape, wear, faceting, cut, or polish is the result of sandblasting. Also known as glyptolith; rillstone; wind-cut stone; wind-grooved stone; wind-polished stone; wind-scoured stone; wind-shaped stone.

vento di sotto [METEOROL]   Breezes blowing up-lake on Lake Garda in Italy.

veranillo [CLIMAT]   The lesser dry season, made up of a few weeks of hot dry weather, that breaks up the summer rainy season on the Pacific coast of Mexico and Central America.

verano [CLIMAT]   In Mexico and Central America, the main dry season, generally occurring from November through April.

Verbeekinidae [PALEON]   A family of extinct marine protozoans in the superfamily Fusulinacea.

vergence [GEOL]   The direction of overturning or of inclination of a fold.

verglas *See* glaze.

vermiculite [MINERAL] $(Mg,Fe,Al)_3(Al,Si)_4O_{10}(OH)_2 \cdot 4H_2O$ A clay mineral constituent similar to chlorite and montmorillonite, and consisting of trioctahedral mica sheets separated by double water layers; sometimes used as a textural material in painting, or as an aggregate in certain plaster formulations used in sculpture.

vernadskite *See* antlerite.

vernal [GEOPHYS]   Pertaining to spring.

verrou *See* riegel.

vertebratus [METEOROL]   A cloud variety (applied mainly to the genus cirrus), the elements of which are arranged in a manner suggestive of vertebrae, ribs, or a fish skeleton.

vertical anemometer [METEOROL]   An instrument which records the vertical component of the wind speed.

vertical differential chart [METEOROL]   A synoptic chart showing the difference in value of a meteorological element between two levels in the atmosphere; a common example is the thickness chart.

vertical intensity [GEOPHYS]   The magnetic intensity of the vertical component of the earth's magnetic field, reckoned positive if downward, negative if upward.

vertical jet *See* uprush.

vertical obstacle sonar [ENG] An active sonar used to determine heights of objects in the path of a submersible vehicle; its beam sweeps along a vertical plane, about 30° above and below the direction of the vehicle's motion. Abbreviated VOS.

vertical seismograph [ENG] An instrument for recording the vertical component of the ground motion of an earthquake.

vertical separator [PETRO ENG] A gas-oil separator in the form of a vertical cylindrical tank.

vertical slip [GEOL] The vertical component of the net slip in a fault. Also known as vertical dip slip.

vertical stability *See* static stability.

vertical stretching [METEOROL] A process in which ascending vertical motion of air increases with altitude, or descending motion decreases with (increasing) altitude.

vertical visibility [METEOROL] According to United States weather observing practice, the distance that an observer can see vertically into a surface-based obscuring phenomenon, such as fog, rain, or snow.

Vertisol [GEOL] A soil order formed in regoliths high in clay; subject to marked shrinking and swelling with changes in soil water content; low in organic content and high in bases.

very close pack ice [OCEANOGR] Sea ice so concentrated that there is little if any open water.

very open pack ice [OCEANOGR] Sea ice whose concentration ranges between one-tenth and three-tenths of the sea surface.

vesicle [GEOL] A cavity in lava formed by entrapment of a gas bubble during solidification. Also known as air sac; bladder; saccus; vacuole; wing.

vesicular structure [PETR] A structure that is common in many volcanic rocks and which forms when magma is brought to or near the earth's surface; may form a structure with small cavities, or produce a pumiceous structure or a scoriaceous structure.

vesuvian *See* leucite; vesuvianite.

Vesuvian eruption *See* Vulcanian eruption.

Vesuvian garnet *See* leucite.

vesuvianite [MINERAL] $Ca_{10}Mg_2Al_4(SiO_4)_5(Si_2O_7)_2(OH)_4$ A brown, yellow, or green mineral found in contact-metamorphosed limestones. Also known as idocrase; vesuvian.

veszelyite [MINERAL] $(Cu,Zn)_3(PO_4)(OH)_3 \cdot 2H_2O$ A greenish-blue to dark blue, monoclinic mineral consisting of a hydrated basic phosphate of copper and zinc.

V flume [MIN ENG] A V-shaped flume, supported by trestlework, and used by miners for bringing down timber and wood from the high mountains; the flume water is also used for mining purposes.

VFR weather [METEOROL] In aviation terminology, route or terminal weather conditions which allow operation of aircraft under visual flight rules.

vibrating-reed magnetometer [ENG] An instrument that measures magnetic fields by noting their effect on the vibration of reeds excited by an alternating magnetic field.

villiaumite [MINERAL] NaF A carmine, isometric mineral consisting of sodium fluoride; occurs in masssive form.

Vindobonian [GEOL] A European stage of geologic time, middle Miocene.

violarite   [MINERAL] Ni_2FeS_4 A violet-gray mineral of the linnaeite group consisting of a sulfide of nickel and iron; found in meteorites.

virazon   [METEOROL]   **1.** The very strong southwesterly sea breeze experienced where the coastal chains of the Andes Mountains descend steeply to the sea; it sets in about 10 a.m. and reaches its greatest strength at about 3 p.m.   **2.** A westerly sea breeze of Spain and Portugal.

virga   [METEOROL]   Wisps or streaks of water or ice particles falling out of a cloud but evaporating before reaching the earth's surface as precipitation. Also known as fall streaks; Fallstreifen; precipitation trails.

virtual gravity   [METEOROL]   The force of gravity on a parcel of air, reduced by centrifugal force due to the motion of the parcel relative to the earth.

virtual height   [GEOPHYS]   The apparent height of a layer in the ionosphere, determined from the time required for a radio pulse to travel to the layer and return, assuming that the pulse propagates at the speed of light. Also known as equivalent height.

virtual pressure   [METEOROL]   The pressure of a parcel of moist air when it has the same density as a parcel of dry air at the same temperature.

virtual temperature   [METEOROL]   In a system of moist air, the temperature of dry air having the same density and pressure as the moist air.

viscous magnetization   *See* viscous remanent magnetization.

viscous remanent magnetization   [GEOPHYS]   A process in which grains of magnetic minerals, which are either too small or too finely divided by undergrowths of different chemical composition to retain a permanent magnetization indefinitely, acquire a new direction of magnetization when the direction of the earth's magnetic field changes. Abbreviated VRM. Also known as viscous magnetization.

Viséan   [GEOL]   A European stage of lower Carboniferous geologic time forming the lowermost Upper Mississippian, above Tournalaisian and below lower Namurian.

visibility   [METEOROL]   In weather observing practice, the greatest distance in a given direction at which it is just possible to see and identify with the unaided eye, in the daytime, a prominent dark object against the sky at the horizon and, at nighttime, a known, preferably unfocused, moderately intense light source.

visor tin   [MINERAL]   Twin crystals of cassiterite characterized by a notch.

visual range   [METEOROL]   The distance, under daylight conditions, at which the apparent contrast between a specified type of target and its background becomes just equal to the threshold contrast of an observer.

vitavite   *See* moldavite.

vitrain   [GEOL]   A brilliant black coal lithotype with vitreous luster and cubical cleavage. Also known as pure coal.

vitreous copper   *See* chalcocite.

vitreous sliver   *See* argentite.

vitric tuff   [GEOL]   Tuff composed principally of volcanic glass fragments.

vitrification   [GEOL]   Formation of a glassy or noncrystalline material.

vitrinite   [GEOL]   A maceral group that is rich in oxygen and composed of humic material associated with peat formation; characteristic of vitrain.

vitrinoid   [GEOL]   Vitrinite occurring in bituminous coking coals; characterized by a reflectance of 0.5–2.0%.

vitrophyre [PETR]   Any porphyritic igneous rock whose groundmass is glassy. Also known as glass porphyry.

viuga [METEOROL]   A cold north or northeast storm of the Russian steppes, lasting about 3 days.

vivianite [MINERAL] $Fe_3(PO_4)_2 \cdot 8H_2O$ A colorless, blue, or green mineral in the unaltered state (darkens upon oxidation); crystallizes in the monoclinic system and occurs in earth form and as globular and encrusting fibrous masses. Also known as blue iron earth; blue ocher.

void ratio [GEOL]   The ratio of the volume of void space to the volume of solid substance in any material consisting of void space and solid material, such as a soil sample, a sediment, or a sedimentary rock.

volatile component [GEOL]   A component of magma whose vapor pressures are high enough to allow them to be concentrated in any gaseous phase. Also known as volatile flux.

volatile flux *See* volatile component.

volatile-oil reservoir [PETRO ENG]   A type of bubble-point oil reservoir in which the temperature is high and the liquid density is low (leading to a volatilized oil situation), reducing the amount of producible liquids.

volatile transfer *See* gaseous transfer.

volborthite [MINERAL] $Cu_3(UO_4)_2 \cdot 3H_2O$ An olive green to green and yellowish-green, monoclinic mineral consisting of hydrated copper vanadate.

volcanello *See* spatter cone.

volcanic ash [GEOL]   Fine pyroclastic material; particle diameter is less than 4 millimeters.

volcanic bombs [GEOL]   Pyroclastic ejecta; the lava fragments, liquid or plastic at the time of ejection, acquire rounded forms, markings, or internal structure during flight or upon landing.

volcanic breccia [PETR]   A pyroclastic rock that is composed of angular volcanic fragments having a diameter larger than 2 millimeters and that may or may not have a matrix.

volcanic dome [GEOL]   A steep-sided heap of viscous lava accreted in or near the vent of a volcano.

volcanic foam *See* pumice.

volcanic gases [GEOL]   Volatile matter composed principally of about 90% water vapor, and carbon dioxide, sulfur dioxide, hydrogen, carbon monoxide, and nitrogen, released during an eruption of a volcano.

volcanic glass [GEOL]   Natural glass formed by the cooling of molten lava, or one of its liquid fractions, too rapidly to allow crystallization.

volcanicity *See* volcanism.

volcaniclastic rock [PETR]   Clastic rock containing volcanic material in any proportion.

volcanic mud [GEOL]   Sediment containing large quantities of ash from a volcanic eruption, mixed with water.

volcanic mudflow [GEOL]   The flow of volcanic mud down the slope of a volcano.

volcanic rift zone [GEOL]   A zone comprising volcanic fissures with underlying dike assemblages; occurs in Hawaii.

volcanic rock [GEOL] Finely crystalline or glassy igneous rock resulting from volcanic activity at or near the surface of the earth. Also known as extrusive rock.

volcanics [PETR] Igneous rocks that solidified after reaching or nearing the earth's surface.

volcanic vent [GEOL] The channelway or opening of a volcano through which magma ascends to the surface; two general types are fissure and pipelike vents.

volcanism [GEOL] The movement of magma and its associated gases from the interior into the crust and to the surface of the earth. Also known as volcanicity.

volcano [GEOL] **1.** A mountain or hill, generally with steep sides, formed by the accumulation of magma extruded through openings or volcanic vents. **2.** The vent itself.

volcanology [GEOL] The branch of geology that deals with volcanism.

voltaite [MINERAL] A greenish-black to black, isometric mineral consisting of a hydrated potassium iron sulfate.

voltzite [MINERAL] Zn_5S_4O A rose red, yellowish, or brownish mineral consisting of an oxysulfide of zinc; occurs in implanted spherical globules and as a crust.

volume phase *See* surface phase.

volume transport [OCEANOGR] The volume of moving water measured between two points of reference and expressed in cubic meters per second.

von Arx current meter [ENG] A type of current-measuring device using electromagnetic induction to determine speed and, in some models, direction of deep-sea currents.

vorobievite *See* vorobyevite.

vorobyevite [MINERAL] A rose-red, purplish-red, or pinkish cesium-containing variety of beryl; used as a gem. Also known as morganite; rosterite; vorobievite; worobieffite.

VOS *See* vertical obstacle sonar.

vougesite [PETR] A lamprophyre having an orthoclase and hornblende groundmass in which are embedded hornblende phenocrysts.

vrbaite [MINERAL] $Tl_4Hg_3Sb_2As_8S_{20}$ A dark gray-black, orthorhombic mineral that occurs in small crystals.

vriajem *See* friagem.

VRM *See* viscous remanent magnetization.

V-shaped depression [METEOROL] On a surface chart, a low or trough about which the isobars display a pronounced V shape, with the point of the V usually extending equatorward from the parent low.

V-shaped valley [GEOL] A valley having a cross-sectional profile in the form of the letter V, commonly produced by stream erosion. Also known as V valley.

vug [PETR] A small cavity in a vein or rock usually lined with minerals differing in composition from those of the enclosing rock. Also known as bughole.

Vulcanian eruption [GEOL] A volcanic eruption characterized by periodic explosive events. Also known as paroxysmal eruption; Plinian eruption; Vesuvian eruption.

vulgar establishment *See* high-water full and change.

vuthan [METEOROL] In southern South America, an intense storm.

V valley *See* V-shaped valley.

wacke [PETR]   Sandstone composed of a mixture of angular and unsorted or poorly sorted fragments of minerals and rocks and an abundant matrix of clay and fine silt.

wad [MINERAL]   A massive, generally soft, amorphous, earthy, dark-brown or black mineral composed principally of manganese oxides with some other minerals, and formed by decomposition of manganese minerals. Also known as black ocher; bog manganese; earthy manganese.

wadi [GEOL]   In the desert regions of southwestern Asia and northern Africa, a stream bed or channel, or a steep-sided ravine, gulley, or valley, which carries water only during the rainy season. Also spelled wady.

wady *See* wadi.

wagnerite [MINERAL] $Mg_2(PO_4)F$ A yellow, grayish, flash-red, or greenish, monoclinic mineral consisting of magnesium fluophosphate.

Wahl correlation [PETRO ENG]   A pressure-volume-temperature (PVT) correlation used to estimate the total oil recovery from a solution-gas-drive oil reservoir; it is based on assumed PVT data, and may be in error.

wairakite [MINERAL] $CaAl_2Si_4O_{12} \cdot 2H_2O$ A zeolite mineral that is isostructural with analcime.

wake stream theory [OCEANOGR]   The theory that, in a stratified ocean, a compensation current must develop on the right side of a wake stream, flowing in the same direction, and a countercurrent in the opposite direction must appear to the left.

walking props *See* self-advancing supports.

wall [GEOL]   The side of a cave passage.   [MIN ENG]   **1.** The side of a level or drift. **2.** The country rock bounding a vein laterally. **3.** The face of a longwall working or stall, commonly called coal wall.

wall cloud [METEOROL]   A rotating lowered cloud base from which a tornado develops.

waller *See* pack builder.

wallplate [MIN ENG]   A horizontal timber supported by posts resting on sills and extending lengthwise on each side of the tunnel; roof supports rest on the wallplates.

wall reef [GEOL]   A linear, steep-sided coral reef constructed on a reef wall.

wall rock [GEOL]   Rock that encloses a vein.

wall rock alteration [GEOL]   Alteration of wall rock adjacent to hydrothermal veins by the fluid responsible for formation of the mineral deposit.

wall-sided glacier [HYD]   A glacier unconfined by a marked ravine or valley.

walpurgite [MINERAL] $Bi_4(UO_2)(AsO_4)_2O_4 \cdot 3H_2O$ A wax yellow to straw yellow, triclinic mineral consisting of a hydrated arsenate of bismuth and uranium. Also known as waltherite.

waltherite *See* walpurgite.

wander *See* apparent wander.

wandering dune [GEOL]   A sand dune that has moved as a unit in the leeward direction of the prevailing winds, and that is characterized by the lack of vegetation to anchor it. Also known as migratory dune; traveling dune.

wandering water *See* vadose water.

want *See* nip.

wardite [MINERAL] $Na_4CaAl_{12}(PO_4)_8(OH)_{18} \cdot 6H_2O$ A blue-green to pale green, tetragonal mineral consisting of a hydrated basic phosphate of sodium, calcium, and aluminum.

warm-air drop *See* warm pool.

warm air mass [METEOROL]   An air mass that is warmer than the surrounding air; an implication that the air mass is warmer than the surface over which it is moving.

warm anticyclone *See* warm high.

warm braw [METEOROL]   A warm, dry, foehn wind which persists for up to 8 days during the east monsoon in the Schouten Islands off the north coast of New Guinea.

warm-core anticyclone *See* warm high.

warm-core cyclone *See* warm low.

warm-core high *See* warm high.

warm-core low *See* warm low.

warm cyclone *See* warm low.

warm drop *See* warm pool.

warm front [METEOROL]   Any nonoccluded front, or portion thereof, which moves in such a way that warmer air replaces colder air.

warm high [METEOROL]   At a given level in the atmosphere, any high that is warmer at its center than at its periphery. Also known as warm anticyclone; warm-core anticyclone; warm-core high.

warm low [METEOROL]   At a given level in the atmosphere, any low that is warmer at its center than at its periphery; the opposite of a cold low. Also known as warm-core cyclone; warm-core low; warm cyclone.

warm pool [METEOROL]   A region, or pool, of relatively warm air surrounded by colder air; the opposite of a cold pool; commonly applied to warm air of appreciable vertical extent isolated in high latitudes when a cutoff high is formed. Also known as warm-air drop; warm drop.

warm sector [METEOROL]   The area of warm air, within the circulation of a wave cyclone, which lies between the cold front and warm front of a storm.

warm tongue [METEOROL]   A pronounced poleward extension or protrusion of warm air.

warm-tongue steering [METEOROL]   The steering influence apparently exerted upon a tropical cyclone by an upper-level warm tongue which often extends a considerable distance into regions adjacent to the cyclone.

warning stage [HYD] The stage, on a fixed river gage, at which it is necessary to begin issuing warnings or river forecasts if adequate precautionary measures are to be taken before flood stage is reached.

warp [GEOL] **1.** An upward or downward flexure of the earth's crust. **2.** A layer of sediment deposited by water.

warrant *See* underclay.

warringtonite *See* brochantite.

warwickite [MINERAL] $(Mg,Fe)_3Ti(BO_4)_2$ A hair brown to dull black, orthorhombic mineral consisting of a titanoborate of magnesium and iron; occurs as prismatic crystals.

Wasatch winds [METEOROL] Strong, easterly, jet-effect winds blowing out of the mouths of the canyons of the Wasatch Mountains onto the plains of Utah.

wash [ENG] **1.** To clean cuttings or other fragmental rock materials out of a borehole by the jetting and buoyant action of a copious flow of water or a mud-laden liquid. **2.** The erosion of core or drill string equipment by the action of a rapidly flowing stream of water or mud-laden drill-circulation liquid. [GEOL] **1.** An alluvial placer. **2.** A piece of land washed by a sea or river. **3.** *See* alluvial cone.

washability [MIN ENG] Coal properties determining the amenability of a coal to improvement in quality by cleaning.

wash-and-strain ice foot [OCEANOGR] An ice foot formed from ice casts and slush and attached to a shelving beach, between the high and low waterlines; high waves and spray may cause it to build up above the high waterline.

wash-built terrace *See* alluvial terrace.

washing plant [MIN ENG] A plant where slimes are removed from relatively coarse ore by washing, tumbling, or scrubbing.

wash load [GEOL] The finer part of the total sediment load of a stream which is supplied from bank erosion or an external upstream source, and which can be carried in large quantities.

Washoe zephyr [METEOROL] The chinook on the Nevada side of the Sierra Nevada Mountains of northern California.

washout *See* horseback.

washover [GEOL] Material deposited by overwash, especially a small delta produced by storm waves and built on the landward side of a bar or barrier. Also known as storm delta; wave delta.

wash plain *See* alluvial plain.

waste [MIN ENG] **1.** The barren rock in a mine. **2.** The refuse from ore dressing and smelting plants. **3.** The fine coal made in mining and preparing coal for market.

waste bank [MIN ENG] A bank made of earth excavated during the digging of a ditch and laid parallel to it.

waste filling [MIN ENG] Material used for support in heavy ground and in large stopes to prevent failure of rock walls and to minimize or control subsidence and to make it possible to extract pillars of ore left in the earlier stages of mining; material used for filling includes waste rock sorted out in the stopes or mined from rock walls, milltailing, sand and gravel, smelter slag, and rock from surface open cuts or quarries.

waste plain *See* alluvial plain.

waste raise [MIN ENG] An excavation in the mine in which barren rock and other material is broken up for use as filling at a stope.

waste rock [MIN ENG]   Valueless rock that must be fractured and removed in order to gain access to or upgrade ore. Also known as muck; mullock.

water atmosphere [METEOROL]   The concept of a separate atmosphere composed only of water vapor.

water-base mud [PETRO ENG]   Oil-well drilling mud in which the liquid component is water, into which are mixed the thickeners and other additives.

water-bearing strata [GEOL]   Ground layers below the standing water level.

water block [PETRO ENG]   The tendency of accumulated water-oil emulsion around the lower (producing) end of an oil well borehole to block the movement of formation fluids through the formation and toward the borehole.

water budget *See* hydrologic accounting.

water cloud [METEOROL]   Any cloud composed entirely of liquid water drops; to be distinguished from an ice-crystal cloud and from a mixed cloud.

water content [HYD]   The liquid water present within a sample of snow (or soil) usually expressed in percent by weight; the water content in percent of water equivalent is 100 minus the quality of snow. Also known as free-water content; liquid-water content.

watercourse [HYD]   **1.** A stream of water. **2.** A natural channel through which water may run or does run.

water curb *See* garland.

water cycle *See* hydrologic cycle.

water-drive reservoir [PETRO ENG]   An oil or gas reservoir in which pressure is maintained to a greater or lesser extent by an influx of water as the oil or gas is removed.

water equivalent [METEOROL]   The depth of water that would result from the melting of the snowpack or of a snow sample; thus, the water equivalent of a new snowfall is the same as the amount of precipitation represented by that snowfall.

water exchange [OCEANOGR]   The volume and rate of water exchange between air and a body of water in a specific location, or between several bodies of water, controlled by such factors as tides, winds, river discharge, and currents.

waterfall [HYD]   A perpendicular or nearly perpendicular descent of water in a stream.

waterfall lake *See* plunge pool.

waterflooding *See* flooding.

water front [GEOGR]   An area partly bounded by water.

water gap [GEOL]   A deep and narrow pass that cuts to the base of a mountain ridge, and through which a stream flows; the Delaware Water Gap is an example.

water garland *See* garland.

water influx [PETRO ENG]   **1.** The incursion of water (natural or injected) into oil- or gas-bearing formations. **2.** One of the mechanisms of oil production in which the water movement (drive) displaces and moves the reservoir fluids toward the well borehole.

water knockout drum [PETRO ENG]   A device for removal of water from oil well fluids (gas, or gas with oil). Also known as water knockout trap; water knockout vessel.

water knockout trap *See* water knockout drum.

water knockout vessel *See* water knockout drum.

water level *See* water table.

waterline *See* shoreline; water table.

water loss *See* evapotranspiration.

water mass   [OCEANOGR]   A body of water identified by its temperature-salinity curve or chemical composition, and normally consisting of a mixture of two or more water types.

water opening   [OCEANOGR]   A break in sea ice, revealing the sea surface.

water requirement   [HYD]   The total quantity of water required to mature a specified crop under field conditions; includes applied irrigation, water precipitation, and groundwater available to the crop.

water ring *See* garland.

water sky   [METEOROL]   The dark appearance of the underside of a cloud layer when it is over a surface of open water.

water smoke *See* steam fog.

water snow   [HYD]   Snow that, when melted, yields a more than average amount of water; thus, any snow with a high water content.

waterspout   [METEOROL]   A tornado occurring over water; rarely, a lesser whirlwind over water, comparable in intensity to a dust devil over land.

water table   [HYD]   The planar surface between the zone of saturation and the zone of aeration. Also known as free-water elevation; free-water surface; groundwater level; groundwater surface; groundwater table; level of saturation; phreatic surface; plane of saturation; saturated surface; water level; waterline.

water type   [OCEANOGR]   Ocean water of a specified temperature and salinity.

water-vapor absorption   [METEOROL]   The absorption of certain wavelengths of infrared radiation by atmospheric water vapor; a process of fundamental importance in the energy budget of the earth's atmosphere.

water year   [HYD]   Any 12-month period, usually selected to begin and end during a relatively dry season, used as a basis for processing streamflow and other hydrologic data; the period from October 1 to September 30 is most widely used in the United States.

wattevilleite   [MINERAL] $Na_2Ca(SO_4)_2 \cdot 4H_2O$ A snow white mineral consisting of a hydrated sulfate of sodium and calcium; occurs as aggregates of acicular or hairlike crystals.

wave base   [HYD]   The depth at which sediments are not stirred by wave action, usually about 10 meters. Also known as wave depth.

wave basin   [GEOGR]   A basin close to the inner entrance of a harbor in which the waves from the outer entrance are absorbed, thus reducing the size of the waves entering the inner harbor.

wave-built platform *See* alluvial terrace.

wave-built terrace   [GEOL]   A gently sloping coastal surface built up by sediment and loose material at the seaward or lakeward edge of a wave-cut platform. Also known as built terrace; wave-built platform.

wave-cut bench   [GEOL]   A level or nearly level narrow platform produced by wave erosion and extending outward from the base of a wave-cut cliff. Also known as beach platform; high-water platform.

wave-cut cliff   [GEOL]   A cliff formed by the erosive action of waves on rock.

wave-cut plain *See* wave-cut platform.

wave-cut platform [GEOL] A gently sloping surface which is produced by wave erosion and which extends into the sea for a considerable distance from the base of the wave-cut cliff. Also known as cut platform; erosion platform; strand flat; wave-cut plain; wave-cut terrace; wave platform.

wave-cut terrace *See* wave-cut platform.

wave cyclone [METEOROL] A cyclone which forms and moves along a front; the circulation about the cyclone center tends to produce a wavelike deformation of the front. Also known as wave depression.

wave delta *See* washover.

wave depression *See* wave cyclone.

wave depth *See* wave base.

wave disturbance [METEOROL] In synoptic meteorology, the same as wave cyclone, but usually denoting an early state in the development of a wave cyclone, or a poorly developed one.

wave erosion *See* marine abrasion.

wave forecasting [OCEANOGR] The theoretical determination of future wave characteristics based on observed or forecasted meteorological phenomena.

wave height [OCEANOGR] The height of a water-surface wave is generally taken as the height difference between the wave crest and the preceding trough.

wave line *See* swash mark.

wavellite [MINERAL] $Al_3(PO_4)_2(OH_3) \cdot 5H_2O$ A white to yellow, green, or black mineral crystallizing in the orthorhombic system and occurring in small hemispherical aggregates.

wavemark *See* swash mark.

wave platform *See* wave-cut platform.

wave ripple mark *See* oscillation ripple mark.

wave setdown [OCEANOGR] A decrease in the mean water level in the region in which breakers form near the seashore, caused by the presence of a pressure field.

wave setup [OCEANOGR] An increase in the mean water level shoreward of the region in which breakers form at the seashore, caused by the onshore flux of momentum against the beach.

wave shaper [ENG] Of explosives, an insert or core of inert material or of explosives having different detonation rates, used for changing the shape of the detonation wave.

wave system [OCEANOGR] In ocean wave studies, a group of waves which have the same height, length, and direction of movement.

wave theory of cyclones [METEOROL] A theory of cyclone development based upon the principle of wave formation on an interface between two fluids; in the atmosphere, a front is taken as such an interface.

waxy [MINERAL] A type of mineral luster that is soft like that of wax.

weak ground [MIN ENG] Roof and walls of underground excavations which would be in danger of collapse unless suitably supported.

weather [METEOROL] **1.** The state of the atmosphere, mainly with respect to its effects upon life and human activities; as distinguished from climate, weather consists of the short-term (minutes to months) variations of the atmosphere. **2.** As used in the making of surface weather observations, a category of individual and combined atmospheric phenomena which must be drawn upon to describe the local atmospheric activity at the time of observation.

weather central [METEOROL] An organization which collects, collates, evaluates, and disseminates meteorological information in such a manner that it becomes a principal source of such information for a given area.

weathered iceberg [OCEANOGR] An iceberg which is irregular in shape, due to an advanced stage of ablation; it may have overturned.

weathered layer [GEOPHYS] The zone of the earth which lies immediately below the surface and is characterized by low wave velocities.

weather forecast [METEOROL] A forecast of the future state of the atmosphere with specific reference to one or more associated weather elements.

weathering [GEOL] Physical disintegration and chemical decomposition of earthy and rocky materials on exposure to atmospheric agents, producing an in-place mantle of waste. Also known as clastation; demorphism.

weathering correction [GEOPHYS] A velocity correction which is applied to seismic data, necessitated by the diminished velocity of seismic wave propagation in weathered rock.

weathering-potential index [GEOL] A measure of the susceptibility of a rock or mineral to weathering.

weathering velocity [GEOPHYS] The velocity of propagation of seismic waves through weathered rock.

weather map [METEOROL] A chart portraying the state of the atmospheric circulation and weather at a particular time over a wide area; it is derived from a careful analysis of simultaneous weather observations made at many observing points in the area.

weather-map type *See* weather type.

weather minimum [METEOROL] The worst weather conditions under which aviation operations may be conducted under either visual or instrument flight rules; usually prescribed by directives and standing operating procedures in terms of minimum ceiling, visibility, or specific hazards to flight.

weather modification [METEOROL] The changing of natural weather phenomena by technical means; so far, only on the microscale of condensation and freezing nuclei has it been possible to exert modifying influences.

weather observation [METEOROL] An evaluation of one or more meteorological elements that describe the state of the atmosphere either at the earth's surface or aloft.

weather observation radar *See* weather radar.

weather pit [GEOL] A shallow depression (depth up to 15 centimeters) on the flat or gently sloping summit of large exposures of granite or granitic rocks, attributed to strongly localized solvent action of impounded water.

weather radar [ENG] Generally, any radar which is suitable or can be used for the detection of precipitation or clouds. Also known as weather observation radar.

weather resistance [ENG] The ability of a material, paint, film, or the like to withstand the effects of wind, rain, or sun and to retain its appearance and integrity.

weather shore [METEOROL] As observed from a vessel, the shore lying in the direction from which the wind is blowing.

weather side [METEOROL] The side of a ship exposed to the wind or weather.

weather signal [METEOROL] A visual signal displayed to indicate a weather forecast.

weather station [METEOROL]   A place and facility for the bservation, measurement, and recording and transmission of data of the variable elements of weather; one of the most effective network facilities is that of the U.S. Weather Bureau.

weather type [METEOROL]   A series of generalized synoptic situations, usually presented in chart form; weather types are selected to represent typical pressure patterns, and were originally devised as a method for lengthening the effective time-range of forecasts. Also known as weather-map type.

weather window [PETRO ENG]   That part of the year when the weather is suitable for operations, such as pipelaying or platform installation, which cannot be undertaken in adverse sea conditions.

weberite [MINERAL] Na_2MgAlF_7 A light gray, orthorhombic mineral consisting of an aluminofluoride of sodium and magnesium; occurs as grains and masses.

websterite *See* aluminite.

Weddell Current [OCEANOGR]   A surface current which flows in an easterly direction from the Weddell Sea outside the limit of the West Wind Drift.

weddellite [MINERAL] $CaC_2O_4 \cdot 2H_2O$ A colorless to white or yellowish-brown to brown, tetragonal mineral consisting of calcium oxalate dihydrate.

Wedener-Bergeron process *See* Bergeron-Findeisen theory.

wedge *See* ridge.

wedging [ENG]   **1.** A method used in quarrying to obtain large, regular blocks of building stones; a row of holes is drilled, either by hand or by pneumatic drills, close to each other so that a longitudinal crevice is formed into which a gently sloping steel wedge is driven, and the block of stone can be detached without shattering. **2.** The act of changing the course of a borehole by using a deflecting wedge. **3.** The lodging of two or more wedge-shaped pieces of core inside a core barrel, and therefore blocking it. **4.** The material, moss, or wood used to render the shaft lining tight.

weeping core [PETRO ENG]   A core cut that is covered with tearlike fluid drops when it is raised to the surface; usually indicates that the formation will be poor in oil yield.

weeping spring *See* spring seepage.

wehrlite [MINERAL] $BiTe$ A mineral that is a native alloy of bismuth and tellurium. Also known as mirror glance.   [PETR]   A peridotite composed principally of olivine and clinopyroxene with accessory opaque oxides.

weibullite [MINERAL] $Pb_4Bi_6S_9Se_4$ A steel gray mineral consisting of lead bismuth sulfide with selenium replacing the sulfide; occurs in indistinct prismatic crystals in massive form.

weighing rain gage [ENG]   A type of recording rain gage, consisting of a receiver in the shape of a funnel which empties into a bucket mounted upon a weighing mechanism; the weight of the catch is recorded, on a clock-driven chart, as inches of precipitation; used at climatological stations.

weight barometer [ENG]   A mercury barometer which measures atmospheric pressure by weighing the mercury in the column or the cistern.

weinschenkite [MINERAL]   **1.** $YPO_4 \cdot 2H_2O$ A white mineral consisting of a hydrous yttrium phosphate. Also known as churchite. **2.** A dark-brown variety of hornblende high in ferric iron, aluminum, and water.

weir tank [PETRO ENG]   A type of oil-field storage tank with high- and low-level weir boxes and liquid-level controls for metering the liquid content of the tank.

weissite [MINERAL] Cu_5Te_3 A dark bluish-black mineral consisting of copper telluride; occurs in massive form.

welded tuff [PETR]   A pyroclastic deposit hardened by the action of heat, pressure from overlying material, and hot gases. Also known as tuff lava.

welding [GEOL]   Consolidation of sediments by pressure; water is squeezed out and cohering particles are brought within the limits of mutual molecular attraction.

Welge method [PETRO ENG]   A method of calculation of the anticipated oil-recovery performance of a gas-cap-drive oil reservoir.

wellbore hydraulics [PETRO ENG]   A branch of oil production engineering that deals with the motion of fluids (oil, gas, or water) in wellbore tubing or casing, or the annulus between tubing and casing.

well completion [PETRO ENG]   The final sealing off of a drilled well (after drilling apparatus is removed from the borehole) with valving, safety, and flow-control devices.

well conditioning [PETRO ENG]   **1.** Preparation of a well for sampling procedures by control of production rate and associated pressure drawdown. **2.** Removal of accumulated scale, wax, mud, and sand from the inner surfaces of a wellbore, or breakage of water blocks to increase production of oil or gas.

well core [ENG]   A sample of rock penetrated in a well or other borehole obtained by use of a hollow bit that cuts a circular channel around a central column or core.

wellhole [MIN ENG]   **1.** A large-diameter vertical hole used in quarries and opencast pits for taking heavy explosive charges in blasting. **2.** The sump, or portion of a shaft below the place where skips are caged at the bottom of the shaft, in which water collects.

well injectivity [PETRO ENG]   The ability of an injection well (water or gas) to receive injected fluid; can be negatively influenced by formation plugging, borehole scale, or liquid blocking around the lower end of the borehole.

well logging [ENG]   The technique of analyzing and recording the character of a formation penetrated by a drill hole in mineral exploration and exploitation work.

well performance [PETRO ENG]   The measurement of a well's production of oil or gas as related to the well's anticipated productive capacity, pressure drop, or flow rate.

well shooting [ENG]   The firing of a charge of nitroglycerin, or other high explosive, in the bottom of a well for the purpose of increasing the flow of water, oil, or gas.

well-sorted [GEOL]   Referring to a sorted sediment that consists of particles of approximately the same size and has a sorting coefficient of less than 2.5.

well spacing [PETRO ENG]   Areal location and interrelationship between producing oil or gas wells in an oil field; calculated for the maximum ultimate production from a given reservoir.

well stimulation *See* stimulation treatment.

well-type manometer [ENG]   A type of double-leg, glass-tube manometer; one leg has a relatively small diameter, and the second leg is a reservoir; the level of the liquid in the reservoir does not change appreciably with change of pressure; a mercury barometer is a common example.

Wenlockian [GEOL]   A European stage of geologic time: Middle Silurian (above Tarannon, below Ludlovian).

Wentworth classification [GEOL]   A logarithmic grade for size classification of sediment particles starting at 1 millimeter and using the ratio of $\frac{1}{2}$ in one direction (and 2 in the other), providing diameter limits to the size classes of 1, $\frac{1}{2}$, $\frac{1}{4}$, etc. and 1, 2, 4, etc.

Wentworth scale [GEOL]   A geometric grade scale for sedimentary particles ranging from clay particles (diameter less than $\frac{1}{250}$ millimeter) to boulders (diameters greater than 256 millimeters), in which the size classes are related to one another by a constant ratio of ½ (4, 2, 1, ½, etc.).

Werfenian stage *See* Scythian stage.

wernerite *See* scapolite.

west [GEOGR]   The direction 90° to the left or 270° to the right of north.

West Australia Current [OCEANOGR]   The complex current flowing northward along the west coast of Australia; it is strongest from November to January, and weakest and variable from May to July; it curves toward the west to join the South Equatorial Current.

westerlies [METEOROL]   The dominant west-to-east motion of the atmosphere, centered over the middle latitudes of both hemispheres; at the earth's surface, the westerly belt (or west-wind belt) extends, on the average, from about 35 to 65° latitude. Also known as circumpolar westerlies; middle-latitude westerlies; mid-latitude westerlies; polar westerlies; subpolar westerlies; subtropical westerlies; temperate westerlies; zonal westerlies; zonal winds.

westerly wave [METEOROL]   An atmospheric wave disturbance embedded in the mid-latitude westerlies.

Western Equatorial Countercurrent [OCEANOGR]   Weak, arrow bands of eastward-flowing water observed in some winter months in the western Atlantic near the equator.

West Greenland Current [OCEANOGR]   The current flowing northward along the west coast of Greenland into the Davis Strait; part of this current joins the Labrador Current, while the other part continues into Baffin Bay.

Westphalian [GEOL]   A European stage of Upper Carboniferous geologic time, forming the Middle Pennsylvanian, above upper Namurian and below Stephanian.

westward intensification [OCEANOGR]   The intensification of ocean currents to the west, derived from a mathematical model that includes the effects of zonal wind stress at the sea surface and internal friction.

West Wind Drift *See* Antarctic Circumpolar Current.

wet adiabat *See* saturation adiabat.

wet assay [MIN ENG]   The determination of the quantity of a desired constituent in ores, metallurgical residues, and alloys by the use of the processes of solution, flotation, or other liquid means.

wet blasting [ENG]   Shot firing in wet holes.

wet-bulb depression [METEOROL]   The difference in degrees between the dry-bulb temperature and the wet-bulb temperature.

wet-bulb temperature [METEOROL]   **1.** Isobaric wet-bulb temperature, that is, the temperature an air parcel would have if cooled adiabatically to saturation at constant pressure by evaporation of water into it, all latent heat being supplied by the parcel. **2.** The temperature read from the wet-bulb thermometer; for practical purposes, the temperature so obtained is identified with the isobaric wet-bulb temperature.

wet-bulb thermometer [ENG]   A thermometer having the bulb covered with a cloth, usually muslin or cambric, saturated with water.

wet-cell caplight [MIN ENG]   A rechargeable head lamp; the batteries are worn on the belt.

wet climate [CLIMATOL]   A climate whose vegetation is of the rainforest type. Also known as rainforest climate.

wet hole [ENG] A borehole that traverses a water-bearing formation from which the flow of water is great enough to keep the hole almost full of water.

wet season *See* rainy season.

wet snow [METEOROL] Deposited snow that contains a great deal of liquid water.

wet tabling [MIN ENG] A tabling process in which a pulp of two or more minerals flows across an inclined, riffled plane surface, is shaken longwise, and is water-washed crosswise.

wetted perimeter [GEOL] The portion of the perimeter of a steam channel cross section which is in contact with the water.

wetting phase [PETRO ENG] In a two-phase oil reservoir system (oil and water), one phase (water) will wet the pore surfaces of the reservoir formation, the other (oil) will not.

whaleback dune [GEOL] A smooth, elongated mound or hill of desert sand shaped generally like a whale's back; formed by passage of a succession of longitudinal dunes along the same path. Also known as sand levee.

wherryite [MINERAL] a light green mineral consisting of a basic carbonate-sulfate of lead and copper; occurs in massive form.

whewellite [MINERAL] $Ca(C_2O_4) \cdot H_2O$ A colorless or yellowish or brownish, monoclinic mineral consisting of calcium oxalate monohydrate; occurs as crystals.

whipstock [PETRO ENG] A long wedge dropped or placed in a petroleum well in order to deflect the drill from some obstruction.

whirlpool [OCEANOGR] Water in rapid rotary motion.

whirly [METEOROL] A small violent storm, a few yards (or meters) to 100 yards (91 meters) or more in diameter, frequent in Antarctica near the time of the equinoxes.

whistler [GEOPHYS] An effect that occurs when a plasma disturbance, caused by a lightning discharge, travels out along lines of magnetic force of the earth's field and is reflected back to its origin from a magnetically conjugate point on the earth's surface; the disturbance may be picked up electromagnetically and converted directly to sound; the characteristic drawn-out descending pitch of the whistler is a dispersion effect due to the greater velocity of the higher-frequency components of the disturbance.

whitecap [OCEANOGR] A cloud of bubbles at the sea surface caused by a breaking wave.

white clay *See* kaolin.

white cobalt *See* cobaltite.

white damp [MIN ENG] In mining, carbon monoxide (CO); a gas that may be present in the afterdamp of a gas or coal-dust explosion, or in the gases given off by a mine fire; it is an important constituent of illuminating gas, supports combustion, and is very poisonous.

white feldspar *See* albite.

white frost *See* hoarfrost.

white garnet *See* leucite.

white mica *See* muscovite.

white nickel *See* rammelsbergite.

white olivine *See* forsterite.

whiteout [METEOROL] An atmospheric optical phenomenon of the polar regions in which the observer appears to be engulfed in a uniformly white glow: shadows, horizon, and clouds are not discernible; sense of depth and orientation are lost; dark objects in the field of view appear to float at an indeterminable distance. Also known as milky weather.

Whiterock [GEOL] A North American stage of lowermost Middle Ordovician time, above lower Ordovician and below Marmor.

white schorl *See* albite.

white squall [METEOROL] A sudden squall in tropical or subtropical waters, which lacks the usual squall cloud and whose approach is signaled only by the whiteness of a line of broken water or whitecaps.

white tellurium *See* krannerite; sylvanite.

white water [OCEANOGR] Frothy water, as in whitecaps or breakers.

whitleyite [GEOL] An achondritic stony meteorite consisting essentially of enstatite with fragments of black chondrite.

whitlockite [MINERAL] $Ca_9(Mg,Fe)H(PO_4)_7$ A rare mineral that forms hexagonal crystals.

whizzer mill *See* Jeffrey crusher.

whole gale [METEOROL] **1.** In storm-warning terminology, a wind of 48 to 63 knots (55 to 72 miles, or 89 to 133 kilometers, per hour). **2.** In the Beaufort wind scale, a wind whose speed is from 48 to 55 knots (55 to 63 miles, or 89 to 102 kilometers, per hour).

wiborgite *See* rapakivi.

wichtisite *See* tachylite.

Widmanstatten patterns [GEOL] Characteristic figures that appear on the surface of an iron meteorite when the meteorite is cut, polished, and etched with acid.

wiggle stick *See* divining rod.

Wilderness [GEOL] A North American stage of Middle Ordovician geologic time, above Porterfield and below Trentonian.

wildflysch [GEOL] A type of flysch facies that represents a stratigraphic unit with irregularly sorted boulders resulting from fragmentation, and twisted, confused beds resulting from slumping or sliding due to the influence of gravity.

wild snow [METEOROL] Newly deposited snow which is very fluffy and unstable; in general, it falls only during a dead calm at very low air temperatures.

Wilfley table [MIN ENG] A flat, rectangular surface that can be tilted and shaken about the long axis and has horizontal riffles for imposing restraint in removing minerals from classified sand. Also known as shaking table.

wilkeite [MINERAL] $Ca_5(SiO_4,PO_4,SO_4)_3(O,OH,F)$ A rose red or yellow, hexagonal mineral consisting of a basic sulfate-silicate-phosphate of calcium.

willemite [MINERAL] Zn_2SiO_4 A white, greenish-yellow, green, reddish, or brown mineral that forms rhombohedral crystals and exhibits intense bright-yellow fluorescence in ultraviolet light; a minor ore of zinc.

Williamsoniaceae [PALEOBOT] A family of extinct plants in the order Cycadeoidales distinguished by profuse branching.

williwaw [METEOROL] A very violent squall in the Straits of Magellan; it may occur in any month but occurs most frequently in winter.

willy-willy [METEOROL] In Australia, a severe tropical cyclone.

winch operator *See* hoistman.

wind [METEOROL]   The motion of air relative to the earth's surface; usually means horizontal air motion, as distinguished from vertical motion, and air motion averaged over the response period of the particular anemometer.

wind chill [METEOROL]   That part of the total cooling of a body caused by air motion.

wind-chill index [METEOROL]   The cooling effect of any combination of temperature and wind, expressed as the loss of body heat in kilogram calories per hour per square meter of skin surface; it is only an approximation because of individual body variations in shape, size, and metabolic rate.

wind crust [HYD]   A type of snow crust, formed by the packing action of wind on previously deposited snow; wind crust may break locally but, unlike wind slab, does not constitute an avalanche hazard.

wind current [METEOROL]   Generally, any of the quasi-permanent, large-scale wind systems of the atmosphere, for example, the westerlies, trade winds, equatorial easterlies, or polar easterlies.

wind-cut stone *See* ventifact.

wind direction [METEOROL]   The direction from which wind blows.

wind direction indicator [ENG]   A device to indicate the direction from which the wind blows; an example is a weather vane.

wind-direction shaft [METEOROL]   A representational mark for wind direction on a synoptic chart, it is a straight line drawn directly upwind from the station circle; the wind arrow is completed by adding the wind-speed barbs and pennants to the outer end of the shaft.

wind divide [METEOROL]   A semipermanent feature of the atmospheric circulation (usually a high-pressure ridge) on opposite sides of which the prevailing wind directions differ greatly.

wind erosion [GEOL]   Detachment, transportation, and deposition of loose topsoil or sand by the action of wind.

wind gap [GEOL]   A shallow, relatively high-level notch in the upper part of a mountain ridge, usually an abandoned water gap. Also known as air gap; wind valley.

wind-grooved stone *See* ventifact.

wind measurement [METEOROL]   The determination of three parameters: the size of an air sample, its speed, and its direction of motion.

windmill anemometer [ENG]   A rotation anemometer in which the axis of rotation is horizontal; the instrument has either flat vanes (as in the air meter) or helicoidal vanes (as in the propeller anemometer); the relation between wind speed and angular rotation is almost linear.

window [GEOL]   A break caused by erosion of a thrust sheet or a large recumbent anticline that exposes the rocks that lie beneath the thrust sheet. Also known as fenster.   [GEOPHYS]   Any range of wavelengths in the electromagnetic spectrum to which the atmosphere is transparent.   [HYD]   The unfrozen part of a river surrounded by river ice during the winter.

window frost [HYD]   A thin deposit of hoarfrost often found on interior surfaces of windows in winter, and frequently exhibiting beautiful fernlike patterns.

window ice [HYD]   A thin deposit of ice which forms by the freezing of many tiny drops of water that have condensed on the indoors side of a cold window surface.

wind-polished stone *See* ventifact.

wind ripple [METEOROL]   One of a series of wavelike formations on a snow surface, an inch or so in height, at right angles to the direction of wind. Also known as snow ripple.

wind rose [METEOROL]   A diagram in which statistical information concerning direction and speed of the wind at a location may be summarized; a line segment is drawn in each of perhaps eight compass directions from a common origin; the length of a particular segment is proportional to the frequency with which winds blow from that direction; thicknesses of a segment indicate frequencies of occurrence of various classes of wind speed.

windrow [GEOL]   Any accumulation of material formed by wind or tide action.

winds aloft [METEOROL]   Generally, the wind speeds and directions at various levels in the atmosphere above the domain of surface weather observations, as determined by any method of winds-aloft observation. Also known as upper-level winds; upper winds.

winds-aloft observation [METEOROL]   The measurement and computation of wind speeds and directions at various levels above the surface of the earth.

wind scoop [METEOROL]   A saucerlike depression in the snow near obstructions such as trees, houses, and rocks, caused by the eddying action of the deflected wind.

wind-scoured stone *See* ventifact.

wind-shaped stone *See* ventifact.

wind shear [METEOROL]   The local variation of the wind vector or any of its components in a given direction.

wind shield *See* rain gage shield.

wind-shift line [METEOROL]   A line or narrow zone along which there is an abrupt change of wind direction.

wind slab [HYD]   A type of snow crust; a patch of hard-packed snow, which is packed as it is deposited in favored spots by the wind, in contrast to wind crust.

wind speed [METEOROL]   The rate of motion of air.

windstorm [METEOROL]   A storm in which strong wind is the most prominent characteristic.

wind stress [METEOROL]   The drag or tangential force per unit area exerted on the surface of the earth by the adjacent layer of moving air.

wind valley *See* wind gap.

wind vane [ENG]   An instrument used to indicate wind direction, consisting basically of an asymmetrically shaped object mounted at its center of gravity about a vertical axis; the end which offers the greater resistance to the motion of air moves to the downwind position; the direction of the wind is determined by reference to an attached oriented compass rose.

wind velocity [METEOROL]   The speed and direction of wind.

windward [METEOROL]   In the general direction from which the wind blows.

wind wave [OCEANOGR]   A wave resulting from the action of wind on a water surface.

wing *See* vesicle.

winged headland [GEOGR]   A seacliff with two bays or spits, one on either side.

winning [MIN ENG]   **1.** A new mine opening. **2.** The portion of a coal field laid out for working. **3.** Mining.

winnowing gold [MIN ENG] Tossing up dry powdered auriferous material in air, and catching the heavier particles not blown away.

winter ice [OCEANOGR] Level sea ice more than 8 inches (20 centimeters) thick, and less than 1 year old; the stage which follows young ice.

winter-talus ridge [GEOL] A wall-like arcuate ridge on the floor of a cirque formed by freezing activity that dislodged boulders from a cirque wall covered with a snowbank. Also known as nivation ridge.

winze [MIN ENG] A vertical or inclined opening or excavation connecting two levels in a mine, differing from a raise only in construction; a winze is sunk underhand, and a raise is put up overhand.

wire drag [ENG] An apparatus for surveying rocky underwater areas where normal sounding methods are insufficient to ensure the discovery of all existing submerged obstructions, small shoals, or rocks above a given depth or for determining the least depth of an area; it consists essentially of a buoyed wire towed at the desired depth by two launches.

wire line [PETRO ENG] A line or cable used to lower and raise devices and gages in oil well boreholes; used for logging instruments and bottom-hole pressure gages.

wire-line coring [PETRO ENG] A method for obtaining samples of reservoir rocks during the drilling phase of oil wells.

Wisconsin [GEOL] Pertaining to the fourth, and last, glacial stage of the Pleistocene epoch in North America; followed the Sangamon interglacial, beginning about 85,000 ± 15,000 years ago and ending 7000 years ago.

wisper wind [METEOROL] A cold night wind, blowing out of the valley of the Wisper River in Germany during clear weather.

witherite [MINERAL] $BaCO_3$ A yellowish- or grayish-white mineral of the aragonite group that has orthorhombic symmetry, hardness of $3\frac{1}{4}$ on Mohs scale, and specific gravity 4.3.

Witte-Margules equation [OCEANOGR] A formula expressing the slope of the boundary layer between two water masses of different densities and velocities, taking into account the rotation of the earth. Also known as Margules equation.

wittichenite [MINERAL] Cu_3BiS_3 A steel gray to tin white, orthorhombic mineral consisting of copper bismuth sulfide; occurs in tabular and massive form.

wittite [MINERAL] $Pb_5Bi_6(S,Se)_{14}$ A light lead gray, orthorhombic or monoclinic mineral consisting of a sulfide of lead and bismuth.

wolfachite [MINERAL] $Ni(As,Sb)S$ A silver white to tin white mineral consisting of nickel, arsenic, and antimony sulfide; occurs in small crystals and in aggregates.

Wolfcampian [GEOL] A North American provincial series of geologic time; lowermost Permian (below Leonardian, above Virgilian of Pennsylvania).

wolfeite [MINERAL] $(Fe,Mn)_2(PO_4)(OH)$ A pinkish, wine yellow to yellowish-brown or reddish-brown, monoclinic mineral consisting of a basic phosphate of iron and manganese.

wolfram See tungsten; wolframite.

wolframine See wolframite.

wolframite [MINERAL] $(Fe,Mn)WO_4$ A brownish- or grayish-black mineral occurring in short monoclinic, prismatic, bladed crystals; the most important ore of tungsten. Also known as tobacco jack; wolfram; wolframine.

wollastonite [MINERAL] $CaSiO_3$ A white to gray inosilicate mineral (a pyroxenoid) that crystallizes in the triclinic system in tabular crystals and has a pearly or silky luster on the cleavages; hardness is 5–5.5 on Mohs scale, and specific gravity is 2.85. Also known as tabular spar.

wood copper *See* olivenite.

woodhouseite [MINERAL] $CaAl_3(PO_4)(SO_4)(OH)_6$ A colorless to flesh-colored or white, hexagonal mineral consisting of a basic sulfate-phosphate of calcium and aluminum; occurs in small crystals and tabular form.

woodstone *See* silicified wood.

wood tin [MINERAL] A riniform, brownish variety of cassiterite with fibers radiating concentrically and resembling dry wood. Also known as dneprovskite.

woodwardite [MINERAL] $Cu_4Al_2(SO_4)(OH)_{12} \cdot 2–4H_2O$ A greenish-blue to turquoise blue mineral consisting of a hydrated basic sulfate of copper and aluminum; occurs as botryoidal concretions and in spherulitic form.

woody lignite *See* bituminous wood.

worked-out [MIN ENG] Exhausted, referring to a coal seam or ore deposit.

working [MIN ENG] **1.** The whole strata excavated in working a seam. **2.** Ground or rocks shifting under pressure and producing noise.

working place [MIN ENG] The place in a mine at which coal or ore is being actually mined.

Workman-Reynolds effect [GEOPHYS] A mechanism for electric charge separation during freezing of slightly impure water; when a very dilute solution of certain salts freezes rapidly, a strong potential difference is established between the solid and liquid phases; for some salts, the ice attains negative charge, for others, positive; this mechanism has been suggested as one possible mode of thunderstorm charge separation in those portions of a thunderstorm downdraft where snow-pellet or hail particles sweep out supercooled waterdrops.

world rift system [GEOL] The system of interconnected midocean ridges which is the locus of tensional splitting and magma upwelling believed responsible for sea-floor spreading.

worobieffite *See* vorobyevite.

wrap-around hanger [PETRO ENG] An oil-well tubing hanger made up of two hinged halves with a resilient sealing element between two steel mandrels.

wrench fault [GEOL] A lateral fault with a more or less vertical fault surface. Also known as basculating fault; torsion fault.

Wright system [PETRO ENG] A method for mining oil from partially drained sands that involves drilling a shaft through the productive strata, followed by long, slanting holes drilled radially in all directions from the shaft bottom into the oil sands.

wulfenite [MINERAL] $PbMoO_4$ A yellow, orange, orange-yellow, or orange-red tetragonal mineral occurring in tabular crystals or granular masses; an ore of molybdenum. Also known as yellow lead ore.

Würm [GEOL] **1.** A European stage of geologic time: uppermost Pleistocene (above Riss, below Holocene). **2.** Pertaining to the fourth glaciation of the Pleistocene epoch in the Alps, equivalent to the Wisconsin glaciation in North America, following the Riss-Würm interglacial.

wurtzilite [GEOL] A black, massive, sectile, infusible, asphaltic pyrobitumen derived from the metamorphosis of petroleum.

wurtzite [MINERAL] (Zn,Fe)S   A brownish-black hexagonal mineral consisting of zinc sulfide and occurring in hemimorphic pyramidal crystals, or in radiating needles and bundles.

wustite [MINERAL] FeO   An artificial mineral that consists of ferric oxide.

wye level *See* Y level.

Wynyardiidae [PALEON]   An extinct family of herbivorous marsupial mammals in the order Diprotodonta.

xanthochroite *See* greenockite.

xanthoconite [MINERAL] Ag_3AsS_3 A dark red to dull orange to clove brown mineral consisting of silver arsenic sulfide.

xanthophyllite *See* clintonite.

xanthosiderite *See* goethite.

xanthoxenite [MINERAL] $Ca_2Fe(PO_4)_2(OH)\cdot1\frac{1}{2}H_2O$ A pale yellow to brownish-yellow, monoclinic or triclinic mineral consisting of a hydrated basic phosphate of calcium and iron; occurs as masses and crusts.

x axis [CRYSTAL] A reference axis in a quartz crystal.

xenoblast [MINERAL] A mineral which has grown during metamorphism without development of its characteristic crystal faces. Also known as allotrioblast.

xenocryst [CRYSTAL] A crystal in igneous rock that resembles a phenocryst and is foreign to the enclosing body of rock. Also known as chadacryst.

xenolith [PETR] An inclusion in an igneous rock which is not genetically related, such as an unmelted fragment of country rock. Also known as accidental inclusion; exogenous inclusion.

xenomorphic *See* allotriomorphic.

xenomorphic-granular rock [PETR] An igneous rock having a granular texture characterized by a xenomorphic fabric. Also known as allotriomorphic-granular rock.

xenothermal [MINERAL] Pertaining to a mineral deposit formed at high temperature but at shallow to moderate depth.

xenotime [MINERAL] $Y(PO_4)$ A tetragonal mineral of varying color, consisting of yttrium phosphate.

Xenungulata [PALEON] An order of large, digitigrade, extinct, tapirlike mammals with relatively short, slender limbs and five-toed feet with broad, flat phalanges; restricted to the Paleocene deposits of Brazil and Argentina.

Xerert [GEOL] A suborder of the soil order Vertisol, formed in a Mediterranean climate; wide surface cracks open and close once a year.

Xeroll [GEOL] A suborder of the soil order Mollisol, formed in a xeric moisture regime; may have a calcic, petrocalcic, or gypsic horizon, or a duripan.

xerothermal period *See* xerothermic period.

xerothermic period [GEOL] A postglacial interval of a warmer, drier climate. Also known as xerothermal period.

Xerult [GEOL] A suborder of the soil order Ultisol, formed in a xeric moisture regime; brownish or reddish soil with a low to moderate organic-carbon content.

Xiphodontidae [PALEON] A family of primitive tylopod ruminants in the super-family Anaplotherioidea from the late Eocene to the middle Oligocene of Europe.

x-ray powder diffractometer *See* powder diffraction camera.

X wave *See* extraordinary wave.

xylinite [GEOL] A variety of provitrinite consisting of xylem or lignified tissue.

xyloid coal *See* bituminous wood.

xyloid lignite *See* bituminous wood.

Y

yalca [METEOROL] A local name for a severe snowstorm with a strong squally wind which occurs in the Andes Mountain passes of northern Peru.

yamase [METEOROL] A cool, onshore, easterly wind in the Senriku district of Japan in summer.

Yarmouth interglacial [GEOL] The second interglacial stage of the Pleistocene epoch in North America, following the Kansan glacial stage and before the Illinoian.

y axis [CRYSTAL] A line perpendicular to two opposite parallel faces of a quartz crystal.

yellow arsenic *See* orpiment.

yellow cake [MIN ENG] The final precipitate formed in the milling of uranium ores.

yellow lead ore *See* wulfenite.

yellow mud [GEOL] Mud containing sediment having a characteristic yellow color, resulting from certain iron compounds.

yellow pyrite *See* chalcopyrite.

yellow quartz *See* citrine.

Yellow Sea [GEOGR] An inlet of the Pacific Ocean between northeastern China and Korea.

yellow snow [HYD] Snow with a golden or yellow appearance because of the presence of pine or cypress pollen.

yellow tellurium *See* sylvanite.

Y factor [PETRO ENG] An empirical relationship of bubble-point data (pressure and formation volume) used to smooth oil reservoir solution-gas/oil-ratio data for graphical presentation.

yielding arches [MIN ENG] Steel arches installed in underground openings as the ground is removed to support loads caused by changing ground movement or by faulted and fractured rock; when the ground load exceeds the design load of the arch as installed, yielding takes places in the joint of the arch, permitting the overburden to settle into a natural arch of its own and thus tending to bring all forces into equilibrium.

yielding floor [MIN ENG] A soft floor which heaves and flows into open spaces when subjected to heavy pressure from packs or pillars.

yielding prop [MIN ENG] A steel prop which is adjustable in length and incorporates a sliding or flexible joint which comes into operation when the roof pressure exceeds a set load or value.

yield-pillar system [MIN ENG] A method of roof control whereby the natural strength of the roof strata is maintained by the relief of pressure in working areas and the controlled transference of load to abutments which are clear of the workings and roadways.

Y level [ENG] A surveyor's level with Y-shaped rests to support the telescope. Also known as wye level.

yoked basin *See* zeugogeosyncline.

Yorkian [GEOL] A European stage of geologic time forming part of the lower Upper Carboniferous, above Lanarkian and below Staffordian, equivalent to part of the lower Westphalian.

youg [METEOROL] A hot wind during unsettled summer weather in the Mediterranean.

young ice [HYD] Newly formed ice in the transitional stage of development from ice crust to winter ice.

youth [GEOL] The first stage of the cycle of erosion in which the original surface or structure is the dominant topographic feature; characterized by broad, flat-topped interstream divides, numerous swamps and shallow lakes, and progressive increase of local relief. Also known as topographic youth.

yttrocrasite [MINERAL] $(Y,Th,U,Ca)_2Ti_4O_{11}$ A black, orthorhombic mineral consisting of an oxide of rare earths and titanium.

yttrotantalite [MINERAL] $(Y,U,Fe)(Ta,Nb)O_4$ A black or brown, orthorhombic mineral consisting of an oxide of iron, yttrium, uranium, columbium, and tantalum; occurs in prismatic and tabular form.

Yucatán Current [OCEANOGR] A rapid northward flowing current along the western side of the Yucatán Strait; generally loops to the north and exits as the Florida Current.

yugawaralite [MINERAL] $CaAl_2Si_6O_{16}\cdot4H_2O$ A zeolite mineral consisting of hydrous calcium aluminum silicate.

Z

Zalambdalestidae [PALEON] A family of extinct insectivorous mammals belonging to the group Proteutherea; they occur in the Late Cretaceous of Mongolia.

zaratite [MINERAL] $Ni_3(CO_3)(OH)_4 \cdot 4H_2O$ An emerald-green mineral consisting of a hydrous basic nickel carbonate and occurring in incrustations or compact masses.

zastruga *See* sastruga.

z axis [CRYSTAL] The optical axis of a quartz crystal, perpendicular to both the x and y axes.

Zechstein [GEOL] A European series of geologic time, especially in Germany: Upper Permian (above Rothliegende).

Zemorrian [GEOL] A North American stage of Oligocene and Miocene geologic time, above Refugian and below Saucesian.

zenithal rain [METEOROL] In the tropics or subtropics, the rainy season which recurs annually or semiannually at about the time that the sun is most nearly overhead (at zenith).

zeolite [MINERAL] **1.** A group of white or colorless, sometimes red or yellow, hydrous tectosilicate minerals characterized by an aluminosilicate tetrahedral framework, ion-exchangeable large cations, and loosely held water molecules permitting reversible dehydration. **2.** Any mineral of the zeolite group, such as analcime, chabazite, natrolite, and stilbite.

zeolite facies [PETR] Metamorphic rocks formed in the transitional period from diagenesis to metamorphism, at pressures of about 2000–3000 bars and temperatures of 200–300°C.

zero curtain [GEOL] The layer of ground between the active layer and permafrost where the temperature remains nearly constant at 0°C.

zero layer [OCEANOGR] A reference level in the ocean, at which horizontal motion is at a minimum.

zeugogeosyncline [GEOL] A geosyncline in a craton or stable area, within which is also an uplifted area, receiving clastic sediments. Also known as yoked basin.

zeunerite [MINERAL] $Cu(UO_2)_2(AsO_4)_2 \cdot 10\text{--}16H_2O$ A green secondary mineral of the autunite group consisting of a hydrous copper uranium arsenate; it is isomorphous with uranospinite.

zeylanite *See* ceylonite.

zigzag lightning [GEOPHYS] Ordinary lightning of a cloud-to-ground discharge that appears to have a single, but very irregular, lightning channel; viewed from the right angle, this may be observed as beaded lightning.

zincaluminite [MINERAL] $Zn_6Al_6(SO_4)_2(OH)_{26} \cdot 5H_2O$ A white to bluish-white and pale blue mineral consisting of a basic hydrated sulfate of zinc and aluminum; occurs in tufts and crusts.

zincite [MINERAL] $(Zn,Mn)O$ A deep-red to orange-yellow brittle mineral; an ore of zinc. Also known as red oxide of zinc; red zinc ore; ruby zinc; spartalite.

zinckenite *See* zinkenite.

zinc spar *See* smithsonite.

zinc spinel *See* gahnite.

zinkenite [MINERAL] $Pb_6Sb_{14}S_{27}$ A steel-gray orthorhombic mineral consisting of a lead antimony sulfide and occurring in crystals and in masses; has metallic luster, hardness of 3–3.5 on Mohs scale, and specific gravity of 5.30–5.35. Also spelled zinckenite.

zinnwaldite [MINERAL] $K_2(Li,Fe,Al)_6(Si,Al)_8O_{20}(OH,F)_4$ A pale-violet, yellowish, brown, or dark-gray mica mineral; an iron-bearing variety of lepidolite; the characteristic mica of greisens.

zippeite [MINERAL] $(UO_2)_2(SO_4)(OH)_2 \cdot nH_2O$ An orange-yellow, orthorhombic mineral consisting of a hydrated basic sulfate of uranium.

zircon [MINERAL] $ZrSiO_4$ A brown, green, pale-blue, red, orange, golden-yellow, grayish, or colorless neosilicate mineral occurring in tetragonal prisms; it is the chief source of zirconium; the colorless varieties provide brilliant gemstones. Also known as hyacinth; jacinth; zirconite.

zirconite *See* zircon.

zirkelite [MINERAL]   A black mineral consisting of an oxide of zirconium, titanium, calcium, ferrous iron, thorium, uranium, and rare earths.

zobaa [METEOROL]   In Egypt, a lofty whirlwind of sand resembling a pillar, moving with great velocity.

zodiacal light [GEOPHYS]   A diffuse band of luminosity occasionally visible on the ecliptic; it is sunlight diffracted and reflected by dust particles in the solar system within and beyond the orbit of the earth.

zodiacal pyramid [GEOPHYS]   The pattern formed by the zodiacal light. Also known as zodiacal cone.

zoisite [MINERAL] $Ca_2Al_3Si_3O_{12}(OH)$ A white, gray, brown, green, or rose-red orthorhombic mineral of the epidote group consisting of a basic calcium aluminum silicate and occurring massive or in prismatic crystals.

zonal [METEOROL]   Latitudinal, easterly or westerly, opposed to meridional.

zonal circulation *See* zonal flow.

zonal flow [METEOROL]   The flow of air along a latitude circle; more specifically, the latitudinal (east or west) component of existing flow. Also known as zonal circulation.

zonal index [METEOROL]   A measure of strength of the middle-latitude westerlies, expressed as the horizontal pressure difference between 35° and 55° latitude, or as the corresponding geostrophic wind.

zonal soil [GEOL]   In early classification systems in the United States, a soil order including soils with well-developed characteristics that reflect the influence of agents of soil genesis. Also known as mature soil.

zonal theory [GEOL]   A theory of the formation of mineral deposition and sequence patterns, based on the changes in a mineral-bearing fluid as it passes upward from a magmatic source.

zonal westerlies *See* westerlies.

zonal wind [METEOROL]   The wind, or wind component, along the local parallel of latitude, as distinguished from the meridional wind.

zonal winds *See* westerlies.

zonal wind-speed profile [METEOROL]   A diagram in which the speed of the zonal flow is one coordinate and latitude the other.

zonation [GEOL]   The condition of being arranged in zones.

zonda [METEOROL]   A hot wind in Argentina.

zone [CRYSTAL]   A set of crystal faces which intersect (or would intersect, if extended) along edges which are all parallel.   [GEOGR]   An area or region of latitudinal character.   [GEOL]   A belt, layer, band, or strip of earth material such as rock or soil.

zone axis [CRYSTAL]   A line through the center of a crystal which is parallel to all the faces of a zone.

zone indices [CRYSTAL]   Three integers identifying a zone of a crystal; they are the crystallographic coordinates of a point joined to the origin by a line parallel to the zone axis.

zone law [CRYSTAL]   A law which states that the Miller indices (h, k, l) of any crystal plane lying in a zone with zone indices (u, v, w) satisfy the equation $hu + lv + kw = 0$.

zone of aeration [GEOL]   A subsurface zone containing water below atmospheric pressure and air or gases at atmospheric pressure. Also known as unsaturated zone; vadose zone; zone of suspended water.

zone of cementation [GEOL]   The layer of the earth's crust in which unconsolidated deposits are cemented by percolating water containing dissolved minerals from the overlying zone of weathering. Also known as belt of cementation.

zone of illuviation *See* B horizon.

zone of maximum precipitation [METEOROL]   In a mountain region, the belt of elevation at which the annual precipitation is greatest.

zone of saturation [HYD]   A subsurface zone in which water fills the interstices and is under pressure greater than atmospheric pressure. Also known as phreatic zone; saturated zone.

zone of soil water *See* belt of soil water.

zone of suspended water *See* zone of aeration.

zonochlorite *See* pumpellyite.

zorsite [MINERAL] $Ca_2Al_3Si_3O_{12}(OH)$ White, gray, brown, green, or rose-red orthorhombic mineral of the epidote group; an essential constituent of saussurite.

Zosterophyllatae [PALEOBOT]   *See* Zosterophyllopsida.

Zosterophyllopsida [PALEOBOT]   A group of early land vascular plants ranging from the Lower to the Upper Devonian; individuals were leafless and rootless.

Zwischengebirge *See* median mass.